2 CDS

CIRCUITS

CIRCUITS

Fawwaz T. Ulaby
The University of Michigan

Michel M. Maharbiz
The University of California, Berkeley

ISBN: 978-1-934891-10-0

10 9 8 7 6 5 4 3 2

Publisher: Tom Robbins
General Manager: Erik Luther
Marketing Manager: Brad Armstrong
Technology Manager : Mark Walters
Compositor: Paul Mailhot, PreTeX Inc.
Series Advisors: James H. McClellan, Georgia Institute of Technology
Charles G. Sodini, M.I.T.

Cover illustration: Courtesy of Canon U.S.A. All rights reserved.

Library of Congress Control Number: 2010936655

To an academic, writing a book is an endeavor of love.

We dedicate this book to Jean and Anissa.

About NTS Press

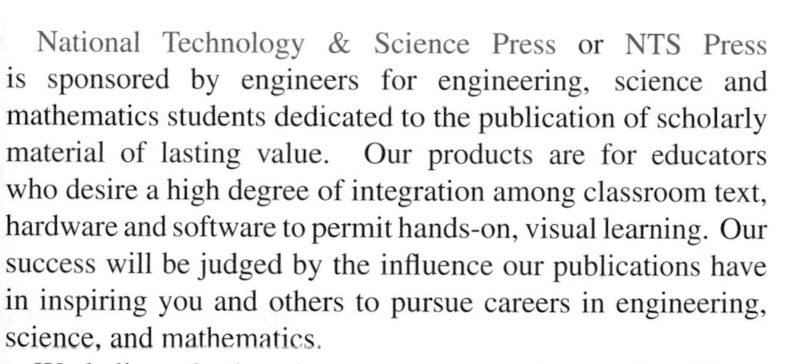

National Technology & Science Press or NTS Press is sponsored by engineers for engineering, science and mathematics students dedicated to the publication of scholarly material of lasting value. Our products are for educators who desire a high degree of integration among classroom text, hardware and software to permit hands-on, visual learning. Our success will be judged by the influence our publications have in inspiring you and others to pursue careers in engineering, science, and mathematics.

We believe the learning process is most rewarding if you build up an intuitive understanding of concepts by pausing to tinker, explore and reflect. Thus, our publishing philosophy follows the belief that learning takes place when you are actively involved and we attempt to encourage this process in several ways. We build our textbooks with a plethora of worked-out examples, with reinforcing and advanced problems, and with interactive computer visuals. We address the computational aspects of problem solving by integrating computer-based learning tools into these presentations to enhance the discussion and analysis of engineering and science applications. In several cases our educational materials are developed with hardware experimentation platforms in mind, such as Universal Serial Bus (USB) devices for acquiring, generating and analyzing information. Owning your own portable laboratory equipment permits you to perform experimentation and take measurements anywhere and at anytime, thereby reinforcing your understanding of theoretical concepts. Having a little fun is okay too.

Clearly, rising textbook costs have an enormous effect on the way you view educational material. And there is little doubt that the Internet has been the most influential agent of change in education in recent time. The prevalence of search engine technology combined with a variety of open-source, open-content, and Wiki sites designed to deliver content has created a proliferation of freely available information and has circulated it more widely. Authored, edited, printed material is being overtaken by community libraries of digitized content from which new content is constructed, remixed, reordered and reassembled, often absent of continuity, flow, and accountability. This free content seems to address concerns about rising textbook prices, but it still has a cost. By contrast we try carefully to design products that come together in one piece with the author's judgment, passion, and imagination intact. These books are available with a reasonable price, and may be counted on as "reputable islands of knowledge in the vast ocean of unscrutinized information." www.ntspress.com

Brief Contents

Contents

List of Technology Briefs

Preface

As the foundational course in the majority of electrical and computer engineering curricula, an *Electric Circuits* course should serve three vital objectives:

(1) It should introduce the fundamental principles of circuit analysis and equip the student with the skills necessary to analyze any planar, linear circuit, including those driven by dc or ac sources, or by more complicated waveforms such as pulses and exponentials.

(2) It should guide the student into the seemingly magical world of *domain transformations*—such as the Laplace and Fourier transforms, not only as circuit analysis tools, but also as mathematical languages that are "spoken" by many fields of science and engineering.

(3) It should expand the student's technical horizon by introducing him/her to some of the many allied fields of science and technology.

This book aims to accomplish exactly those objectives. Among its distinctive features are:

Technology Briefs The book contains 23 Technology Briefs, each providing an overview of a topic that every electrical and computer engineering professional should become familiar with. Electronic displays, data storage media, sensors and actuators, supercapacitors, and 3-D imaging are typical of the topics shared with the reader. The Briefs are presented at a technical level intended to challenge the reader to pursue the subject further on his/her own.

Application Notes Each chapter (except for Chapter 1) includes a section focused on how certain devices or circuits might be used in practical applications. Examples include A/D and D/A converters, three-phase power networks, CMOS inverters in computer processors, signal modulators, and several others.

Multisim SPICE circuit simulators have been part of teaching and learning how circuits respond to electrical stimuli for at least the past two decades. Multisim, a relatively recent SPICE-based software simulator, has the distinct advantage over its predecessors that it offers a friendlier computer-use interface, thereby making it easier to use and manipulate. In addition to introducing its functionality through examples throughout the book, Multisim is highlighted through 43 modules contained on the DVD-ROM accompanying the book. The student is strongly encouraged to take advantage of this rich resource.

Digital Camera Through an independent segment called "Putting It All Together," nestled between Chapters 9 and 10, a generic digital camera is used as an example of a system that incorporates many of the circuits and technologies covered in the book. Presented in block-diagram form, the various sensing, imaging, and computational operations of the digital camera provide a bridge between the fundamentals covered in the book and real-world systems and circuits.

DVD-ROM The two DVD-ROMs accompanying the book contain:

(1) All Figures and Tables, and many of the major equations.

(2) Solutions to all of the Exercises contained in the book. The icon on the text pages indicates that related material can be found on the enclosed DVD.

(3) 43 Multisim Modules (see Appendix C for details).

(4) NI Multisim and LabVIEW Student Edition software.

(5) MathScript software, which can perform matrix inversion and many other calculations, much like The MathWorks, Inc. MATLAB® software.

Acknowledgments

A science or engineering textbook is the product of an integrated effort by many professionals. Invariably, the authors receive far more of the credit than they deserve, for if it were not for the creative talents of so many others, the book would never have been possible, much less a success. We are indebted to many students and colleagues, most notably the following trio.

Richard Carnes: For his meticulous typing of the manuscript, careful drafting of its figures, and overall stewardship of the project. Richard imparted the same combination of precision and passion to the manuscript as he always does when playing Chopin on the piano.

Adib Nashashibi: For his superb attention to detail as the "Quality Control Officer" of the project. He checked many of the derivations in the text, as well as the solutions of numerous end-of-chapter problems.

Joe Steinmeyer: For testing the Multisim problems contained in the text and single-handedly developing all of the Multisim modules on the DVD-ROMs. Shortly thereafter, Joe went to MIT to pursue a Ph.D. in electrical engineering.

For their reviews of the overall manuscript and for offering many constructive criticisms, we are grateful to Professors Fred Terry and Jamie Phillips of the University of Michigan, Keith Holbert of Arizona State University, Ahmad Safaai-Jazi of Virginia Polytechnic Institute and State University, Robin Strickland of the University of Arizona, and Frank Merat of Case Western Reserve University. The manuscript was also scrutinized by a highly discerning group of University of Michigan graduate students: Mike Benson, Fikadu Dagefu, Scott Rudolph, and Jane Whitcomb. Multisim sections were reviewed by Peter Ledochowitsch.

Editing and compositioning the manuscript to generate an appealing look in a functional format is an art unto itself. Our thanks go to Paul Mailhot of PreTeX, Inc.

NTS Press offers an innovative approach to publishing science and engineering textbooks. With today's computer-savvy student in mind, NTS's goal is to publish textbooks that help the student understand how the fundamentals connect to real-world applications, and to market its books at affordable prices. NTS Press is the brainchild of Tom Robbins, an old hand in the textbook publishing business, who recently decided that the time is ripe for a different publishing paradigm. We support Tom's endeavor and we are grateful for the opportunity to publish this book under NTS Press, which provides a dedicated web site for the book (www.ntspress.com).

We enjoyed writing this book, and we hope you enjoy learning from it.

FAWWAZ ULABY AND MICHEL MAHARBIZ

Photo Credits

Page 2 (Figure 1-1): Fawwaz Ulaby

Page 4 (bottom left) Dorling Kindersley/Getty Images; (top right) © Bettmann/CORBIS

Page 5 (top to bottom left) Chuck Eby; John Jenkins, sparkmuseum.com

Page 5 (top to bottom, right) IEEE History Center; History San José; LC-USZ62-39702, Library of Congress

Page 6 (left, top to bottom) History San José; © Bettmann/CORBIS

Page 6 (right, top to bottom) MIT Museum; © Bettmann/CORBIS; Emilio Segre Visual Archives/American Institute of Physics/Science Photo Library

Page 7 (left, top to bottom) Courtesy of Dr. Steve Reyer; Courtesy of Texas Instruments Incorporated (right, top to bottom) NASA; Digital Equipment Corporation

Page 8 (left, top to bottom) Used with permission of SRI International; Courtesy of Texas Instruments Incorporated (right, top to bottom) Courtesy National Instruments

Page 10 Courtesy Office of Basic Energy Sciences, Office of Science, U.S. Department of Energy

Page 12 From "When will computer hardware match the human brain?" by Hans Moravec, *Journal of Transhumanism*, Vol. 1, 1998 Used with permission

Page 35 Fawwaz Ulaby

Page 38 Pacific Northwest National Laboratory

Page 39 Courtesy of Central Japan Railway Company; Courtesy General Electric Healthcare

Page 57 Courtesy of Khalil Najafi, University of Michigan

Page 135 Courtesy of Veljko Milanovic

Page 138 Courtesy of International Business Machines Corporation

Page 163 Courtesy of Intel Corporation

Page 166 Courtesy of Prof. Ken Wise, University of Michigan

Page 190 Michel M. Maharbiz

Page 199 (photo) AP Photo/Rob Widdis; (figure) from Science August, 18 2006, Vol. 313 (#5789) Reprinted with permission of AAAS

Page 203 Fawwaz Ulaby

Page 271 (left to right) Analog Devices; Courtesy of Prof. Khalil Najafi, University of Michigan

Page 272 Analog Devices

Page 325 (left to right) © Reuters/CORBIS; © Colin Anderson/Brand X/Corbis

Page 343 Altzone

Page 442 Courtesy of Prof. Kristopher J. Pister of the University of California at Berkeley

Page 481 Courtesy of Argonne National Laboratory

Page 482 Courtesy Alex Zettl Research Group, Lawrence Berkeley National Laboratory and University of California at Berkeley

Page 510 © Steve Allen/Brand X/Corbis

Page 536 Courtesy of Wendell Lim, Associate Director of National Science Foundation's Synthetic Biology Engineering Research Center, University of California, Berkeley.

Page 537 Aaron Chevalier and Nature (Nov. 24, 2005)

Page 551 NASA

Page 552 NASA

Page 553 ROBYN BECK/AFP/Getty Images

Page 560 After A. V. Oppenheim and J. S. Lim, "The importance of phase in signals," *Proceedings of the IEEE*, v. 69, no. 5, May 1981, pp. 529–541

Page 562 Courtesy of Renaldi Winoto

CHAPTER 1

Circuit Terminology

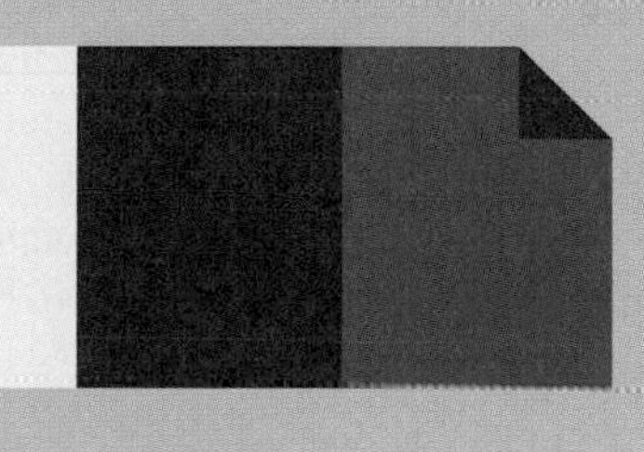

Chapter Contents

Objectives

Upon learning the material presented in this chapter, you should be able to:

1. Differentiate between active and passive devices; analysis and synthesis; device, circuit, and system; and dc and ac.
2. Point to important milestones in the history of electrical and computer engineering.
3. Use multiple and submultiple prefixes.
4. Relate electric charge to current; voltage to energy; power to current and voltage; and apply the passive sign convention.
5. Describe the properties of dependent and independent sources.
6. Define the i-v relationship for: a voltage source; a current source; a resistor; a capacitor; and an inductor.
7. Describe the operation of SPST and SPDT switches.

Cell-Phone Circuit Architecture

Electronic circuits are contained in just about every gadget we use in daily living. In fact, electronic sensors, computers, and displays are at the operational heart of most major industries, from agricultural production and transportation to healthcare and entertainment. The ubiquitous cell phone (Fig. 1-1), which has become practically indispensable, is a perfect example of an integrated electronic architecture made up of a large number of interconnected circuits. It includes amplifier circuits, oscillators, frequency up- and down-converters, and circuits with many other types of functions (Fig. 1-2). Factors such as compatibility among the various circuits and proper electrical connections between them are critically important to the overall operation and integrity of the cell phone.

Usually, we approach electronic analysis and design through a hierarchical arrangement where we refer to the overall entity as a ***system***, its subsystems as ***circuits***, and the individual circuit elements as ***devices*** or components. Thus, we may regard the cell phone as a system (which is part of a much larger communication system); its audio-frequency amplifier, for example, as a circuit, and the resistors, integrated circuits (ICs), and other constituents of the amplifier as devices. In actuality, an IC is a fairly complex circuit in its own right, but its input/output functionality is such that usually it can be represented by a relatively simple equivalent circuit, thereby allowing us to treat it like a device. Generally, we refer to

Figure 1-1: Cell phone.

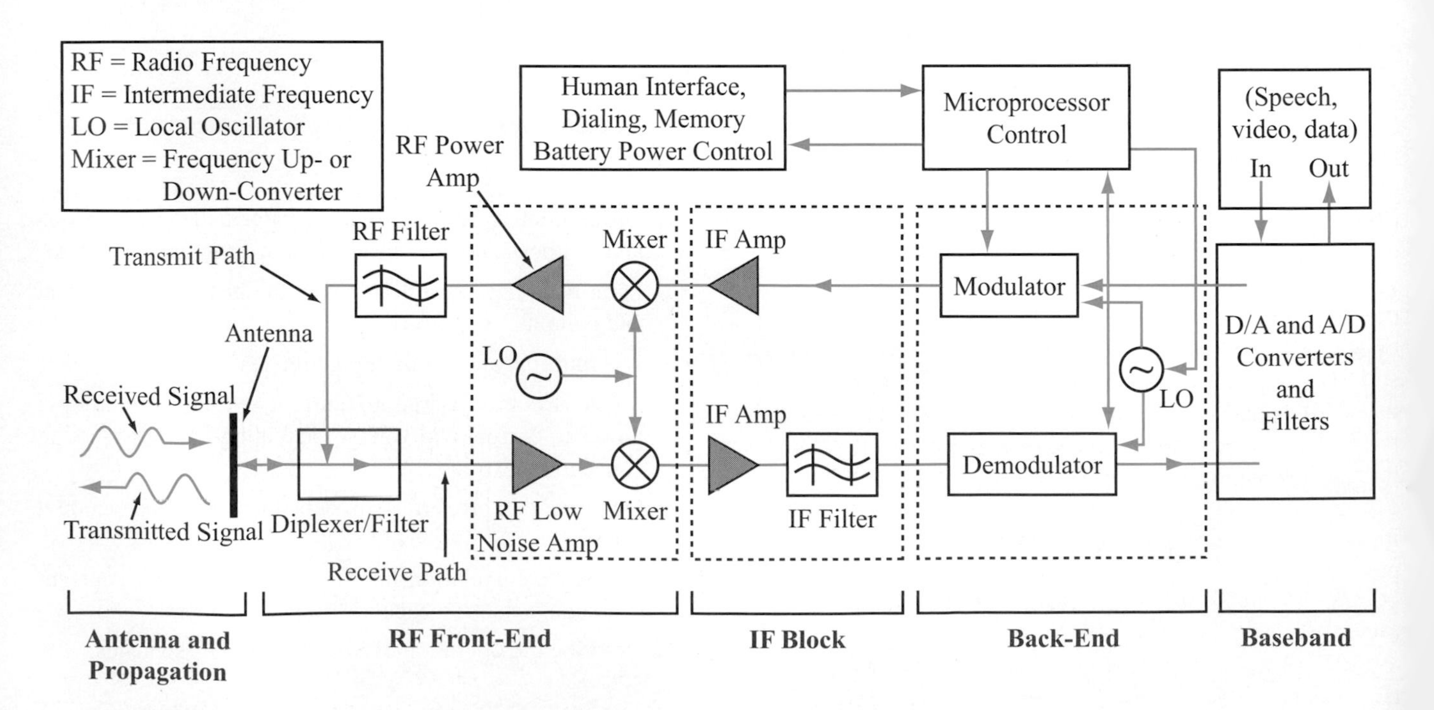

Figure 1-2: Cell-phone block diagram.

devices that do not require an external power source in order to operate as ***passive devices***; these include resistors, capacitors, and inductors. In contrast, an ***active device*** (such as a transistor or an IC) cannot function without a power source.

This book is about ***electric circuits***. A student once asked: "What is the difference between an ***electric*** circuit and an ***electronic*** circuit? Are they the same or different?" Strictly speaking, both refer to the flow of electric charge carried by electrons, but historically, the term "electric" preceded "electronic," and over time the two terms have come to signify different things:

> An electric circuit is one composed of passive devices, in addition to voltage and current sources, and possibly some types of switches. In contrast, the term "electronic" has become synonymous with transistors and other active devices.

The study of electric circuits usually precedes and sets the stage for the study of electronic circuits, and even though a course on electric circuits usually does not deal with the internal operation of an active device, it does incorporate active devices in the circuit examples considered for analysis, but it does so by representing the active devices in terms of equivalent circuits.

An ***electric circuit***, as defined by *Webster's English Dictionary*, is a "complete or partial path over which current may flow." The path may be confined to a physical structure (such as a metal wire connecting two components), or it may be an unbounded channel carrying electrons through it. An example of the latter is when a lightning bolt strikes the ground, creating an electric current between a highly charged atmospheric cloud and the earth's surface.

The study of electric circuits consists of two complementary parts: ***analysis*** and ***synthesis*** (Fig. 1-3). Through analysis, we develop an understanding of "how" a given circuit works. If we think of a circuit as having an input—a stimulus—and an output—a response, the tools we use in circuit analysis allow us to relate mathematically the output response to the input stimulus, enabling us to analytically and graphically "observe" the behavior of the output as we vary the relevant parameters of the input. An example might be a specific amplifier circuit, in which case the objective of circuit analysis might be to establish how the output voltage varies as a function of the input voltage over the full operational range of the amplifier parameters. By analyzing the operation of each circuit in a system containing multiple circuits, we can characterize the operation of the overall system.

As a process, synthesis is the reverse of analysis. In engineering, we tend to use the term ***design*** as a synonym for synthesis. The design process usually starts by defining the operational specifications that a gadget or system should meet, and then we work backwards (relative to the analysis process) to develop circuits that will satisfy those specifications. In analysis, we are dealing with a single circuit with a specific set of operational characteristics. We may employ different analysis tools and techniques, but the circuit is unique, and so are its operational characteristics. That is not necessarily the case for synthesis; the design process may lead to multiple circuit realizations—each one of which exhibits or satisfies the desired specifications.

Given the complementary natures of analysis and synthesis, it stands to reason that developing proficiency with the tools of circuit analysis is a necessary prerequisite to becoming a successful design engineer. This textbook is intended to provide the student with a solid foundation of the primary set of tools and mathematical techniques commonly used to analyze both ***direct current (dc)*** and ***alternating current (ac)*** circuits, as well as circuits driven by pulses and other types of waveforms. A dc circuit is one in which voltage and current sources are constant as a function of time, whereas in ac circuits, sources vary sinusoidally with time. Even though this is not a book on circuit design, design problems occasionally are introduced into the discussion as a way to illustrate how the analysis and synthesis processes complement each other.

> **Review Question 1-1:** What are the differences between a device, a circuit, and a system?
>
> **Review Question 1-2:** What is the difference between analysis and synthesis?

1-1 Historical Timeline

We live today in the age of electronics. No field of science or technology has had as profound an influence in shaping the

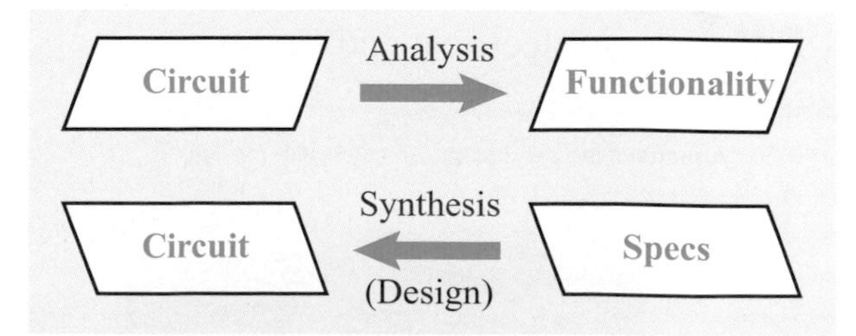

Figure 1-3: The functionality of a circuit is discerned by applying the tools of circuit analysis. The reverse process, namely the realization of a circuit whose functionality meets a set of specifications, is called circuit synthesis or design.

operational infrastructure of modern society as has the field of electronics. Our computers and communication systems are at the nexus of every major industry, from food production and transportation to health care and entertainment. Even though no single event marks the beginning of a discipline, electrical engineering became a recognized profession sometime in the late 1800s (see chronology). ***Alexander Graham Bell*** invented the telephone (1876); ***Thomas Edison*** perfected his incandescent light bulb (1880) and built an electrical distribution system in a small area in New York City; ***Heinrich Hertz*** generated radio waves (1887); and ***Guglielmo Marconi*** demonstrated radio telegraphy (1901). The next 50 years witnessed numerous developments, including radio communication, TV broadcasting, and radar for civilian and military applications—all supported by electronic circuitry that relied entirely on vacuum tubes. The invention of the ***transistor*** in 1947 and the development of the ***integrated circuit*** (IC) shortly thereafter (1958) transformed the field of electronics by setting it on an exponentially changing course towards "smaller, faster, and cheaper."

Computer engineering is a relatively young discipline. The first ***all-electronic computer***, the ENIAC, was built and demonstrated in 1945, but computers did not become available for business applications until the late 1960s and for personal use until the introduction of Apple I in 1976. Over the past 20 years, not only have computer and communication technologies expanded at a truly impressive rate (see Technology Brief 2 on page 20), but more importantly, it is the seamless integration of the two technologies that has made so many business and personal applications possible.

Generating a comprehensive chronology of the events and discoveries that have led to today's technologies is beyond the scope of this book, but ignoring the subject altogether would be a disservice to both the reader and the subject of electric circuits. The abbreviated chronology presented on the next few pages represents our compromise solution.

Chronology: Major Discoveries, Inventions, and Developments in Electrical and Computer Engineering

ca. 1100 BC **Abacus** is the earliest known calculating device.

ca. 900 BC According to legend, a shepherd in northern Greece, **Magnus**, experiences a pull on the iron nails in his sandals by the black rock he was standing on. The rock later became known as magnetite [a form of iron with permanent magnetism].

ca. 600 BC Greek philosopher **Thales** describes how amber, after being rubbed with cat fur, can pick up feathers [static electricity].

1600 **William Gilbert** (English) coins the term electric after the Greek word for amber (*elektron*) and observes that a compass needle points north to south because the Earth acts as a bar magnet.

1614 **John Napier** (Scottish) develops the logarithm system.

1642 **Blaise Pascal** (French) builds the first adding machine using multiple dials.

1733 **Charles François du Fay** (French) discovers that electric charges are of two forms and that like charges repel and unlike charges attract.

1745 **Pieter van Musschenbroek** (Dutch) invents the Leyden jar, the first electrical capacitor.

1800 **Alessandro Volta** (Italian) develops the first electric battery.

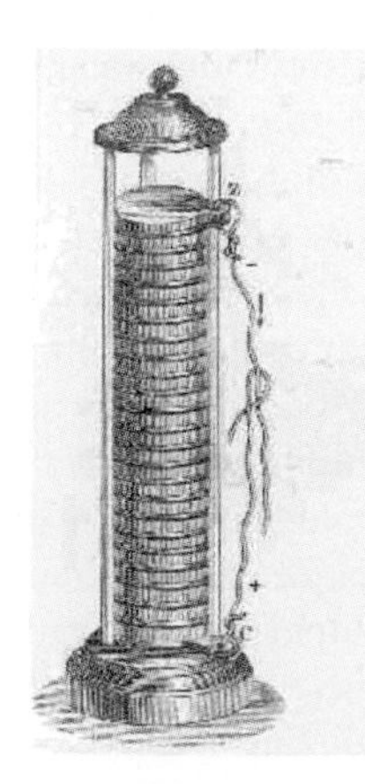

1827 **Georg Simon Ohm** (German) formulates Ohm's law relating electric potential to current and resistance.

1827 **Joseph Henry** (American) introduces the concept of inductance and builds one of the earliest electric motors. He also assisted Samuel Morse in the development of the telegraph.

1837 **Samuel Morse** (American) patents the electromagnetic telegraph using a code of dots and dashes to represent letters and numbers.

1876 **Alexander Graham Bell** (Scottish-American) invents the telephone: the rotary dial becomes available in 1890, and by 1900, telephone systems are installed in many communities.

1879 **Thomas Edison** (American) demonstrates the operation of the incandescent light bulb, and in 1880, his power distribution system provided dc power to 59 customers in New York City.

1887 **Heinrich Hertz** (German) builds a system that can generate electromagnetic waves (at radio frequencies) and detect them.

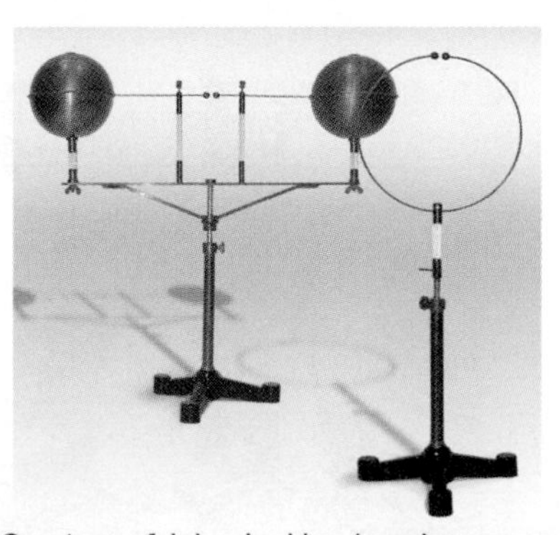

Courtesy of John Jenkins (sparkmuseum.com)

1888 **Nikola Tesla** (Croatian-American) invents the ac motor.

1893 **Valdemar Poulsen** (Danish) invents the first magnetic sound recorder using steel wire as recording medium.

1895 **Wilhelm Röntgen** (German) discovers X-rays. One of his first X-ray images was of the bones in his wife's hands. [1901 Nobel prize in physics.]

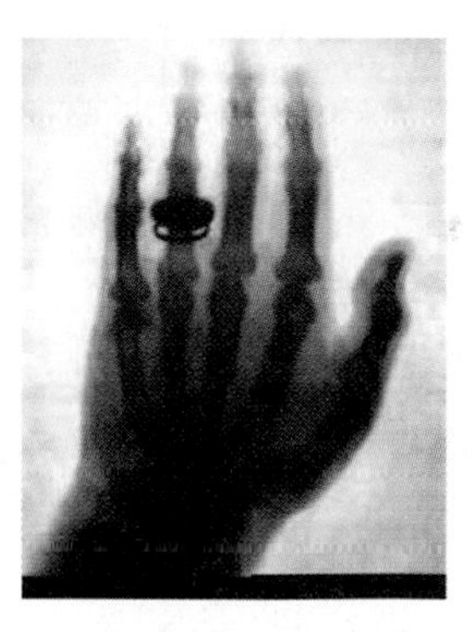

1896 **Guglielmo Marconi** (Italian) files his first of many patents on wireless transmission by radio. In 1901, he demonstrates radio telegraphy across the Atlantic Ocean. [1909 Nobel prize in physics, shared with Karl Braun (German).]

1897 **Karl Braun** (German) invents the cathode ray tube (CRT). [1909 Nobel prize, shared with Marconi.]

1897 **Joseph John Thomson** (English) discovers the electron and measures its charge-to-mass ratio. [1906 Nobel prize in physics.]

1902 **Reginald Fessenden** (American) invents amplitude modulation for telephone transmission. In 1906, he introduces AM radio broadcasting of speech and music on Christmas Eve.

1904 **John Fleming** (British) patents the diode vacuum tube.

1907 **Lee De Forest** (American) develops the triode tube amplifier for wireless telegraphy, setting the stage for long-distance phone service, radio, and television.

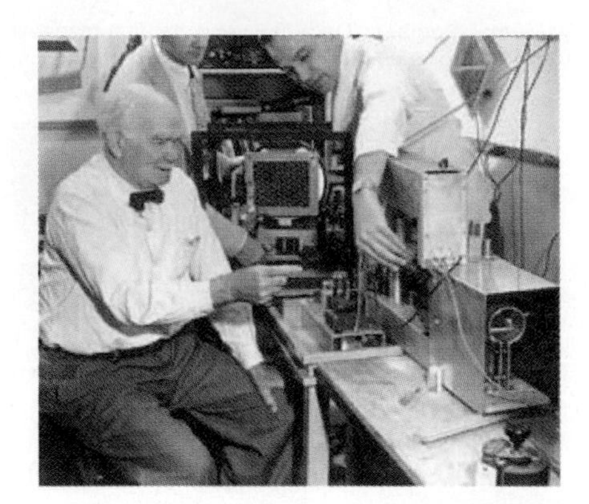

1917 **Edwin Howard Armstrong** (American) invents the superheterodyne radio receiver, dramatically improving signal reception. In 1933, he develops frequency modulation (FM), providing superior sound quality of radio transmissions over AM radio.

1920 Birth of commercial radio broadcasting; **Westinghouse Corporation** establishes radio station KDKA in Pittsburgh, Pennsylvania.

1923 **Vladimir Zworykin** (Russian-American) invents television. In 1926, **John Baird** (Scottish) transmits TV images over telephone wires from London to Glasgow. Regular TV broadcasting began in Germany (1935), England (1936), and the United States (1939).

1926 Transatlantic telephone service established between London and New York.

1930 **Vannevar Bush** (American) develops the differential analyzer, an analog computer for solving differential equations.

1935 **Robert Watson-Watt** (Scottish) invents radar.

1945 **John Mauchly** and **J. Presper Eckert** (both American) develop the ENIAC, the first all-electronic computer.

1947 **William Shockley**, **Walter Brattain**, and **John Bardeen** (all Americans) invent the junction transistor at Bell Labs. [1956 Nobel prize in physics.]

1948 **Claude Shannon** (American) publishes his *Mathematical Theory of Communication*, which formed the foundation of information theory, coding, cryptography, and other related fields.

1950 **Yoshiro Nakama** (Japanese) patents the floppy disk as a magnetic medium for storing data.

1954 **Texas Instruments** introduces the first AM transistor radio.

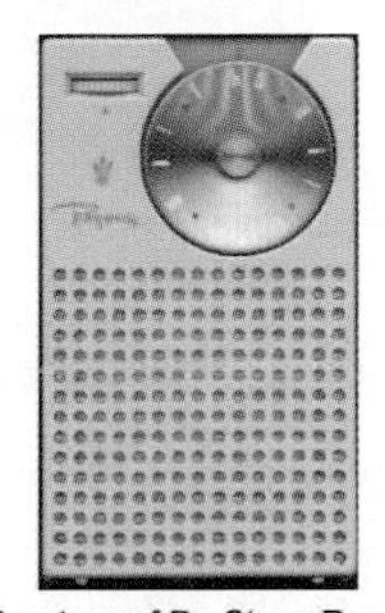

Courtesy of Dr. Steve Reyer

1955 The pager is introduced as a radio communication product in hospitals and factories.

1955 **Narinder Kapany** (Indian-American) demonstrates optical fiber as a low-loss, light-transmission medium.

1956 **John Backus** (American) develops FORTRAN, the first major programming language.

```
C      FORTRAN PROGRAM FOR
PRINTING A TABLE OF CUBES
       DO 5 I = 1, 64
       ICUBE = I * I * I
       PRINT 2, I, ICUBE
    2  FORMAT (1H, I3,I7)
    5  CONTINUE
       STOP
```

1958 **Charles Townes** and **Arthur Schawlow** (both Americans) develop the conceptual framework for the laser. [Townes shared 1964 Nobel prize in physics with Aleksandr Prokhorov and Nicolay Bazov (both Soviets).] In 1960 **Theodore Maiman** (American) builds the first working model of a laser.

1958 **Bell Labs** develops the modem.

1958 **Jack Kilby** (American) builds the first integrated circuit (IC) on germanium, and independently, **Robert Noyce** (American) builds the first IC on silicon.

1959 **Ian Donald** (Scottish) develops an ultrasound diagnostic system.

1960 **Echo**, the first passive communication satellite is launched and successfully reflects radio signals back to Earth. In 1962, the first communication satellite, Telstar, is placed in geosynchronous orbit.

1960 **Digital Equipment Corporation** introduces the first minicomputer, the PDP-1, which was followed with the PDP-8 in 1965.

1962 **Steven Hofstein** and **Frederic Heiman** (both American) invent the MOSFET, which became the workhorse of computer microprocessors.

1964 **IBM**'s 360 mainframe becomes the standard computer for major businesses.

1965 **John Kemeny** and **Thomas Kurtz** (both American) develop the BASIC computer language.

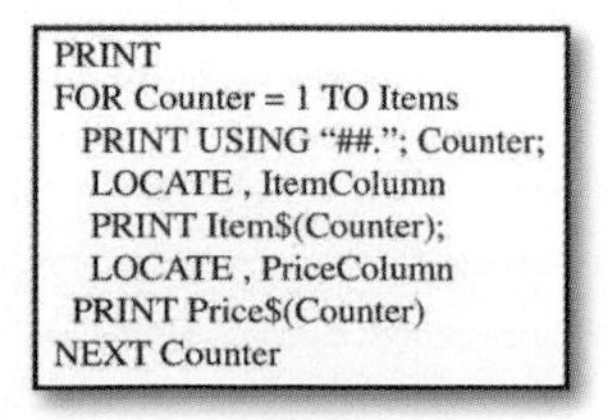

```
PRINT
FOR Counter = 1 TO Items
  PRINT USING "##."; Counter;
   LOCATE , ItemColumn
   PRINT Item$(Counter);
   LOCATE , PriceColumn
  PRINT Price$(Counter)
NEXT Counter
```

1965 **Konrad Zuse** (German) develops the first programmable digital computer using binary arithmetic and electric relays.

1968 **Douglas Engelbart** (American) demonstrates a word-processor system, the mouse pointing device, and the use of a Windows-like operating system.

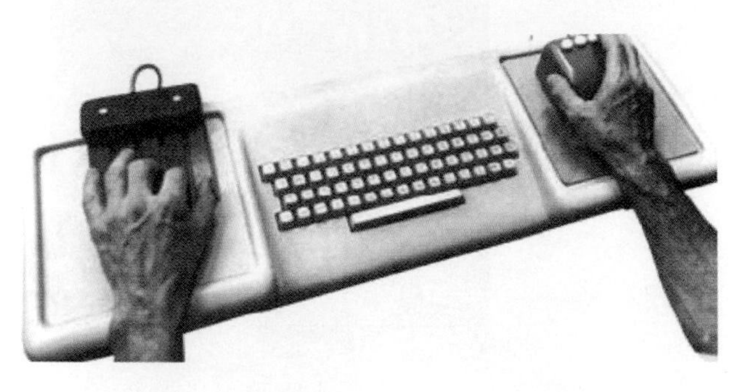

1969 ARPANET is established by the U.S. Department of Defense, which is to evolve later into the Internet.

1970 **James Russell** (American) patents the CD-ROM, as the first system capable of digital-to-optical recording and playback.

1971 **Texas Instruments** introduces the pocket calculator.

Courtesy of Texas Instruments

1971 **Intel** introduces the 4004 four-bit microprocessor, which is capable of executing 60,000 operations per second.

1972 **Godfrey Hounsfield** (British) and **Alan Cormack** (South African–American) develop the computerized axial tomography scanner (CAT scan) as a diagnostic tool. [1979 Nobel Prize in physiology or medicine.]

1976 **IBM** introduces the laser printer.

1976 **Apple Computer** sells Apple I in kit form, followed by the fully assembled Apple II in 1977, and the Macintosh in 1984.

1979 Japan builds the first cellular telephone network:

- 1983 cellular phone networks start in the United States.
- 1990 electronic beepers become common.
- 1995 cell phones become widely available.

1980 **Microsoft** introduces the MS-DOS computer disk operating system. Microsoft Windows is marketed in 1985.

1981 **IBM** introduces the PC.

1984 Worldwide Internet becomes operational.

1988 First transatlantic optical fiber cable between the U.S. and Europe is operational.

1989 **Tim Berners-Lee** (British) invents the World Wide Web by introducing a networking hypertext system.

1996 **Sabeer Bhatia** (Indian-American) and **Jack Smith** (American) launch Hotmail as the first webmail service.

1997 Palm Pilot becomes widely available.

1997 The 17,500-mile fiber-optic cable extending from England to Japan is operational.

2002 Cell phones support video and the Internet.

2007 The power-efficient White LED invented by **Shuji Nakamura** (Japanese) in the 1990s promises to replace Edison's lightbulb in most lighting applications.

1-2 Units, Dimensions, and Notation

The standard system used in today's scientific literature to express the units of physical quantities is the ***International System of Units (SI)***, abbreviated after its French name *Système Internationale*. Time is a fundamental ***dimension***, and the second is the ***unit*** by which it is expressed relative to a specific reference standard. The SI configuration is based on the six ***fundamental dimensions*** listed in Table 1-1, and their units are called ***Fundamental SI units***. All other dimensions, such as velocity, force, and energy, are regarded as ***secondary*** because their units are based on and can be expressed in terms of the six fundamental units. Appendix A provides a list of the quantities used in this book, together with their symbols and units.

In science and engineering, a set of ***prefixes*** commonly are used to denote multiples and submultiples of units. These prefixes, ranging in value between 10^{-18} and 10^{18}, are listed in Table 1-2. An electric current of 3×10^{-6} A, for example, may be written as 3 μA.

Table 1-1: Fundamental SI units.

Dimension	Unit	Symbol
Length	meter	m
Mass	kilogram	kg
Time	second	s
Electric Current	ampere	A
Temperature	kelvin	K
Amount of substance	mole	mol

Table 1-2: Multiple and submultiple prefixes.

Prefix	Symbol	Magnitude
exa	E	10^{18}
peta	P	10^{15}
tera	T	10^{12}
giga	G	10^{9}
mega	M	10^{6}
kilo	k	10^{3}
milli	m	10^{-3}
micro	μ	10^{-6}
nano	n	10^{-9}
pico	p	10^{-12}
femto	f	10^{-15}
atto	a	10^{-18}

The physical quantities we will discuss in this book (such as voltage and current) may be constant in time or may vary with time.

As a general rule, we shall use:

- A lowercase letter, such as i for current, to represent the general case:

  i *may or may not be time varying*

- A lowercase letter followed with (t) to emphasize time:

  $i(t)$ *is a time-varying quantity*

- An uppercase letter if the quantity is not time varying; thus:

  I *is of constant value (dc quantity)*

- A letter printed in boldface to denote that:

  $\mathbf{I}$ *has a specific meaning, such as a vector, a matrix, the phasor counterpart of* $i(t)$*, or the Laplace or Fourier transform of* $i(t)$

Exercise 1-1: Convert the following quantities to scientific notation: (a) 52 mV, (b) 0.3 MV, (c) 136 nA, and (d) 0.05 Gbits/s.

Answer: (a) 5.2×10^{-2} V, (b) 3×10^{5} V, (c) 1.36×10^{-7} A, and (d) 5×10^{7} bits/s. (See)

Exercise 1-2: Convert the following quantities to a prefix format such that the number preceding the prefix is between 1 and 999: (a) 8.32×10^{7} Hz, (b) 1.67×10^{-8} m, (c) 9.79×10^{-16} g, (d) 4.48×10^{13}V, and (e) 762 bits/s.

Answer: (a) 83.2 MHz, (b) 16.7 nm, (c) 979 ag, (d) 44.8 TV, and (e) 762 bits/s. (See)

Exercise 1-3: Simplify the following operations into a single number, expressed in prefix format: (a) $A = 10\ \mu\text{V} + 2.3$ mV, (b) $B = 4\text{THz} - 230$ GHz, (c) $C = 3\ \text{mm}/60\ \mu\text{m}$.

Answer: (a) $A = 2.31$ mV, (b) $B = 3.77$ THz, (c) $C = 50$. (See)

Technology Brief 1: Micro- and Nanotechnology

History and Scale

As humans and our civilizations developed, our ability to control the environment around us improved dramatically. The use and construction of tools was essential to this increased control. A quick glance at the scale (or size) of manmade and natural things is very illustrative (Fig. TF1-1). Early tools (such as flint, stone, and metal hunting gear) were on the order of tens of centimeters. Over time, we began to build ever-smaller and ever-larger tools. The pyramids of Giza (ca. 2600 BCE) are 100-m tall; the largest modern construction crane is the K10,000 Kroll Giant Crane at 100 m long and 82 m tall; and the current (2007) tallest man-made structure is the KVLY-TV antenna mast in Blanchard, North Dakota at 0.63 km! Miniaturization also proceeded apace; for example, the first hydraulic valves may have been Sinhalese valve pits of Sri Lanka (ca. 400 BCE), which were a few meters in length; the first toilet valve (ca. 1596) was tens of centimeters in size; and by comparison, the largest dimension in a modern microfluidic valve used in biomedical analysis chips is less than 100 μm!

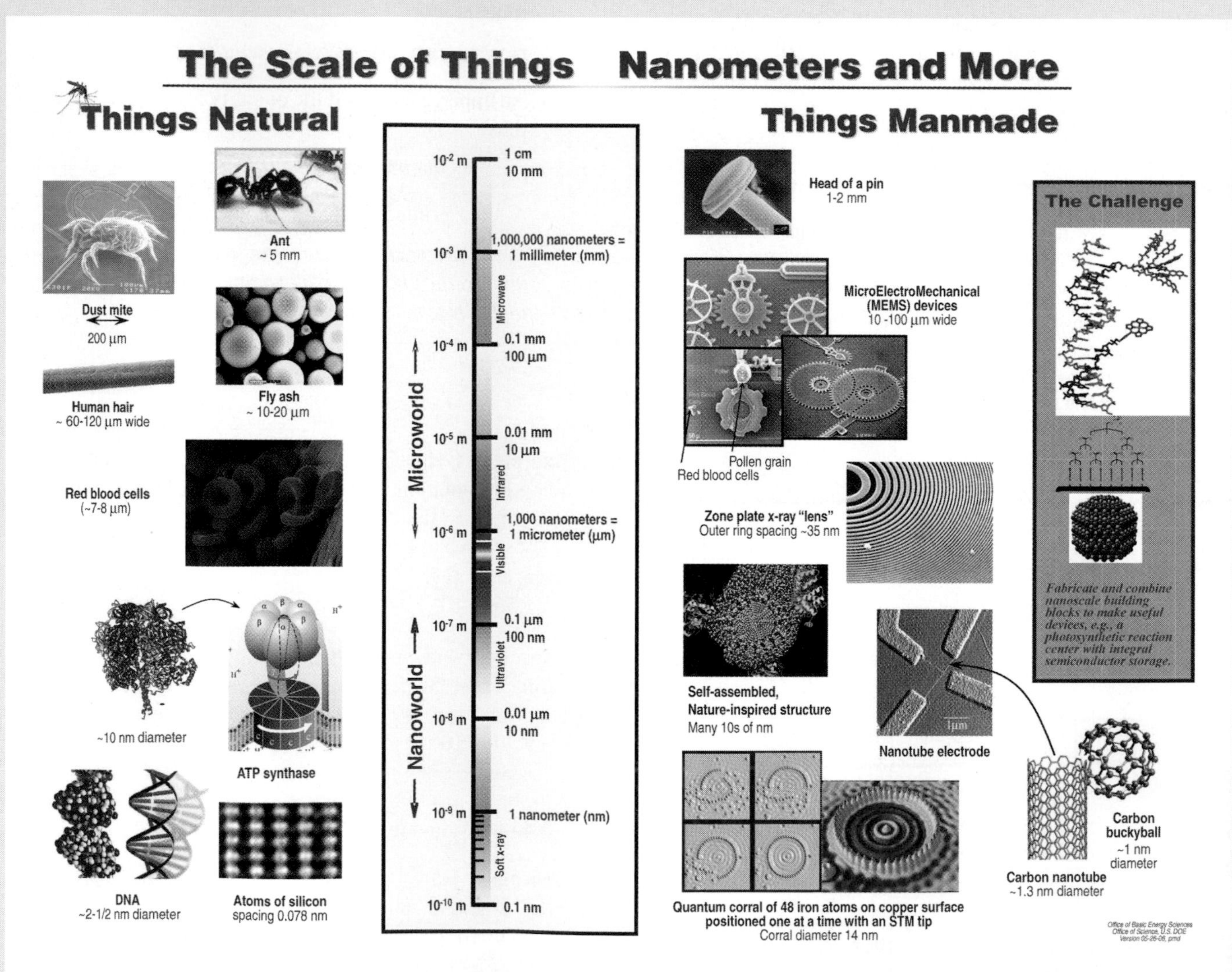

Figure TF1-1: The scale of natural and man-made objects, sized from nanometers to centimeters. (Courtesy of U.S. Department of Energy.)

In electronic devices, miniaturization has been a key enabler in almost all of the technologies that shape the world around us. Consider computation and radio-frequency communications, two foundations of 21st-century civilization. The first true automated computer was arguably the first Babbage Difference Engine, proposed by Charles Babbage to the Royal Astronomical Society (1822). The complete engine would have had 25,000 moving parts and measured approximately 2.4 m $\times$ 2.3 m $\times$ 1 m. Only a segment with 2000 parts was completed and today is considered the first modern calculator. The first general-purpose electronic computer was the Electronic Numerical Integrator and Computer (ENIAC), which was constructed at the University of Pennsylvania between 1943 and 1946. The ENIAC was 10 ft tall, occupied 1,000 square feet, weighed 30 tons, used $\sim$100,000 components, and required 150 kilowatts of power! What could it do? It could perform simple mathematical operations on 10-digit numbers at approximately 2,000 cycles per second (addition took 1 cycle, multiplication 14 cycles, and division and square roots 143 cycles). With the invention of the semiconductor transistor in 1947 and the development of the ***integrated circuit*** in 1959 (see Technology Brief 7 on page 135 on IC Fabrication Process), it became possible to build thousands (now trillions) of electronic components onto a single substrate or ***chip***. The 4004 microprocessor chip (ca. 1971) had 2250 transistors and could execute 60,000 instructions per second; each transistor had a "gate" on the order of 10 μm (10^{-5} m). In comparison, the 2006 Intel Core has 151 million transistors with each transistor gate measuring 65 nm (6.5×10^{-8} m), and it can perform 27 billion instructions per second.

Similar miniaturization trends are obvious in the technology used to manipulate the electromagnetic spectrum. The ability of a circuit component to interact with electromagnetic waves depends on how its size compares with the wavelength (λ) of the signal it is trying to manipulate. For example, to efficiently transmit or receive signals, a wire antenna must be comparable to λ in length. Some of the first electromagnetic waves used for communication were in the 1-MHz range (corresponding to $\lambda = 300$ m) which today is allocated primarily to AM radio broadcasting. [The frequency f (in Hz) is related to the wavelength λ (in meters) by $\lambda f = c$, where $c = 3 \times 10^8$ m/s is the velocity of light in vacuum.] With the advent of portable radio and television, the usable spectrum was extended into the megahertz range (10^2 to 10^3 MHz or $\lambda = 3$ m to 30 cm). Modern cell phones operate in the low gigahertz (GHz) range (1 GHz $= 10^9$ Hz). Each of these shifts has necessitated technological revolutions as components and devices continue to shrink. The future of electronics looks bright (and tiny) as the processing and communication of signals approaches the terahertz (THz) range (10^{12} Hz)!

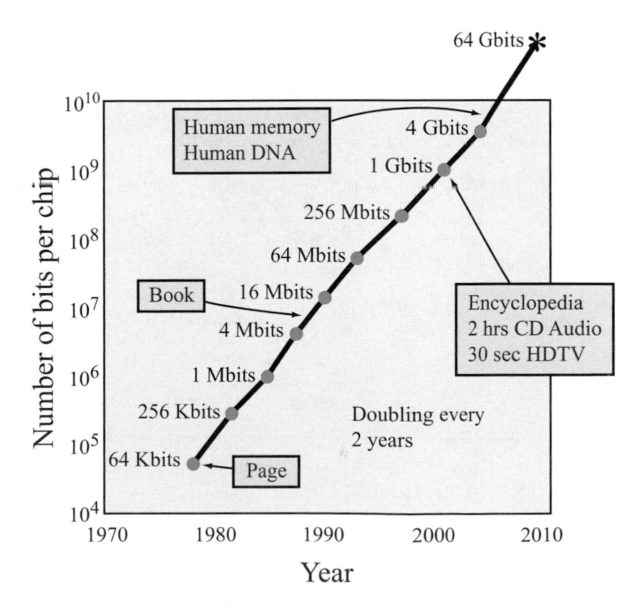

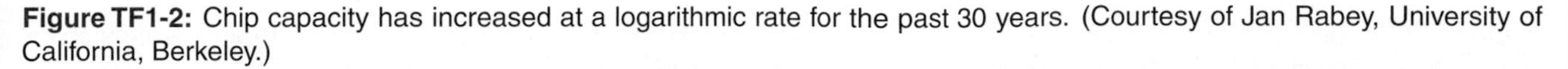
Figure TF1-2: Chip capacity has increased at a logarithmic rate for the past 30 years. (Courtesy of Jan Rabey, University of California, Berkeley.)

Scaling Trends and Nanotechnology

It is an observable fact that each generation of tools enables the construction of a new, smaller, more powerful generation of tools. This is true not just of mechanical devices, but electronic ones as well. Today's high-power processors could not have been designed, much less tested, without the use of previous processors that were employed to draw and simulate the next generation. Two observations can be made in this regard. First, we now have the technology to build tools to manipulate the environment at atomic resolution. At least one generation of micro-scale techniques (ranging from microelectromechanical systems—or MEMS—to micro-chemical methods) has been developed that, while useful in themselves, are also enabling the construction of newer, nano-scale devices. These newer devices range from 5-nm ($1 \text{ nm} = 10^{-9}$ m) transistors to femtoliter (10^{-15}) microfluidic devices that can manipulate single protein molecules. At these scales, the lines between mechanics, electronics, and chemistry begin to blur! It is to these ever-increasing interdisciplinary innovations that the term ***nanotechnology*** rightfully belongs.

Second, the rate at which these innovations are occurring seems to be increasing exponentially! Consider Fig. TF1-2 and TF1-3 and note that the y-axis is logarithmic and the plots are very close to straight lines. This phenomenon, which was observed to hold true for the number of transistors that can be fabricated into a single processor, was noted by Gordon Moore in 1965 and was quickly named "Moore's Law" (see Technology Brief 2: Moore's Law on page 20).

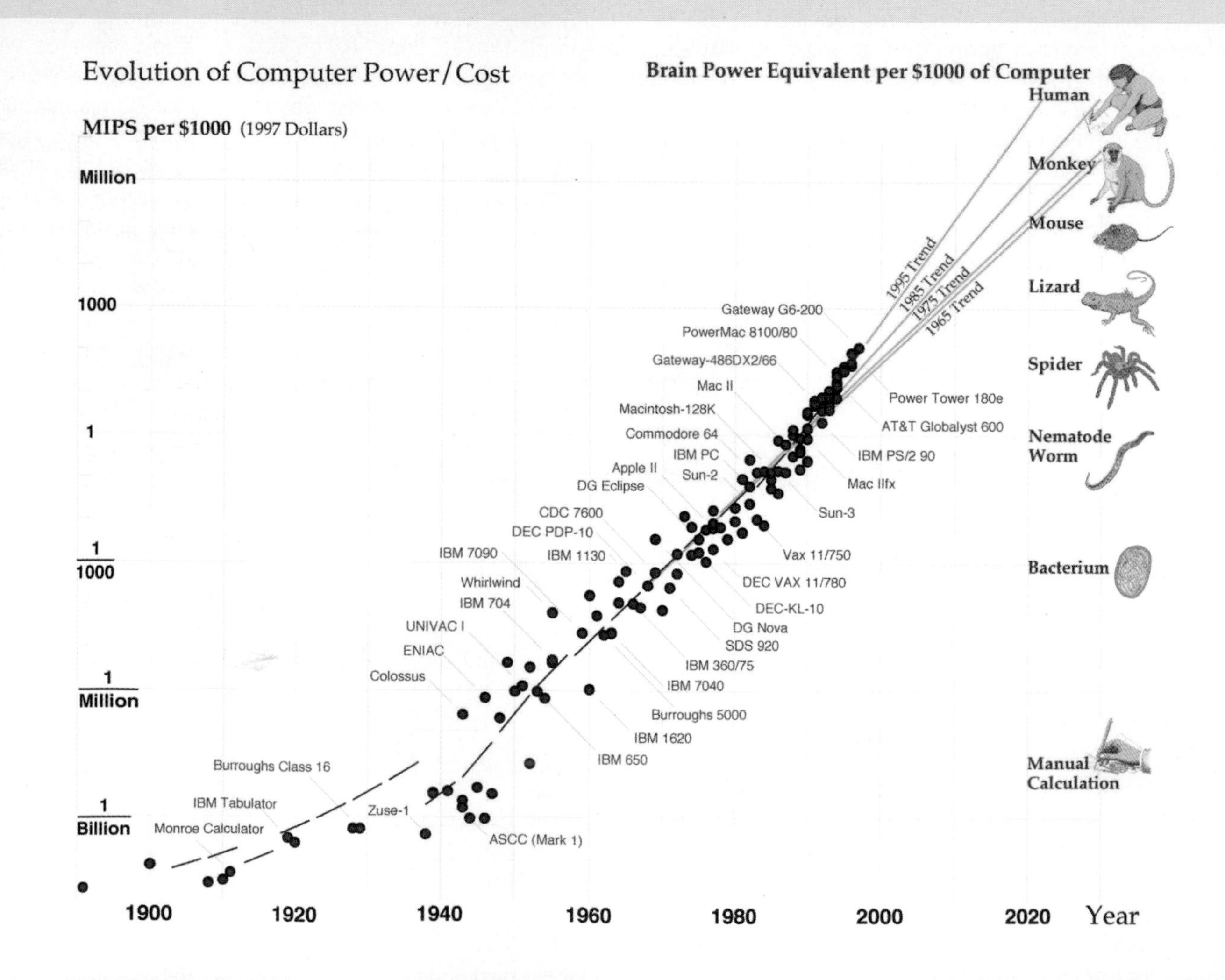

Figure TF1-3: Time plot of computer processing power in MIPS per $1000. (From "When will computer hardware match the human brain?" by Hans Moravec, *Journal of Transhumanism*, Vol. 1, 1998.)

1-3 Electric Charge and Current

1-3.1 Charge

At the atomic scale, all matter contains a mixture of neutrons, positively charged protons, and negatively charged electrons. The nature of the force induced by electric charge was established by the French scientist Charles Augustin de Coulomb (1736–1806) during the latter part of the 18th century. This was followed by a series of experiments on electricity and magnetism over the next 100 years, culminating in J. J. Thompson's discovery of the electron in 1897. Through these and more recent investigations, we can ascribe to electric charge the following fundamental properties:

1. Charge can be either positive or negative.
2. The fundamental (smallest) quantity of charge is that of a single electron or proton. Its magnitude usually is denoted by the letter e.
3. According to the law of conservation of charge, the (net) charge in a closed region can neither be created nor destroyed.
4. Two like charges repel one another, whereas two charges of opposite polarity attract.

The unit for charge is the coulomb (C) and the magnitude of e is

$$e = 1.6 \times 10^{-19} \quad \text{(C)}. \tag{1.1}$$

The symbol commonly used to represent charge is q. The charge of a single proton is $q_p = e$, and that of an electron, which is equal in magnitude but opposite in polarity, is $q_e = -e$. It is important to note that the term *charge* implies "net charge," which is equal to the combined charge of all protons present in any given region of space minus the combined charge of all electrons in that region. Hence, *charge is always an integral multiple of e*.

The last of the preceding properties is responsible for the movement of charge from one location to another, thereby constituting an ***electric current***. Consider the simple circuit in Fig. 1-4 depicting a battery of voltage V connected across a resistor R using metal wires. The arrangement gives rise to an electric current given by ***Ohm's law*** (which will be discussed in some detail in Chapter 2):

$$I = \frac{V}{R}. \tag{1.2}$$

As shown in Fig. 1-4:

The current flows from the positive (+) terminal of the battery to the negative (−) terminal, along the path *external* to the battery.

Through chemical or other means, the battery generates a supply of electrons at its negatively labeled terminal by ionizing some of the molecules of its constituent material. A convenient model for characterizing the functionality of a battery is to regard the internal path between its terminals as unavailable for the flow of charge, forcing the electrons to flow from the (−) terminal, through the external path, and towards the (+) terminal to achieve neutrality. It is important to note that:

The direction of electric current is defined to be the same as the direction of flow that positive charges would follow, which is opposite to the direction of flow of electrons.

Even though we talk about electrons *flowing* through the wires and the resistor, in reality the process is a ***drift*** movement rather than free-flow. The wire material consists of atoms with loosely attached electrons. The positive polarity of the (+) terminal exerts an attractive force on the electrons of the hitherto neutral atoms adjacent to that terminal, causing some of the loosely attached electrons to detach and jump to the (+) terminal. The atoms that have lost those electrons now become positively charged (ionized), thereby attracting electrons from their neighbors and compelling them to detach from their hosts and to attach themselves to the ionized atoms instead. This process continues throughout the wire segment (between the

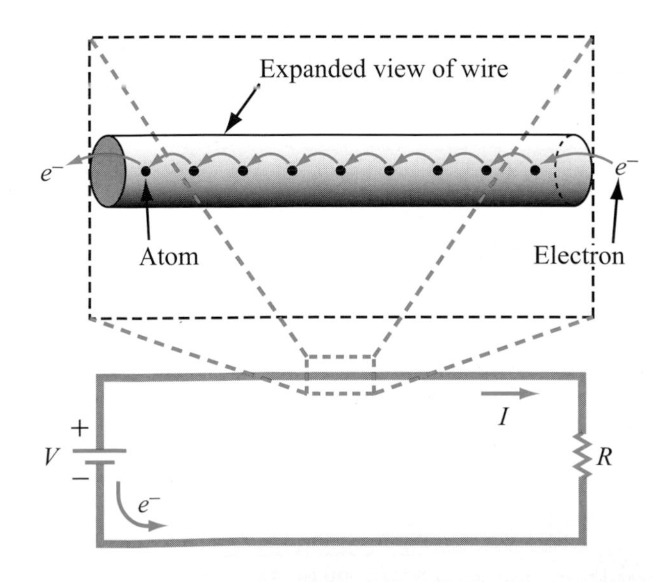

Figure 1-4: The current flowing in the wire is due to electron transport through a drift process, as illustrated by the magnified structure of the wire.

(+) battery terminal and the resistor), into the longitudinal path of the resistor, and finally through the wire segment between the resistor and the (−) terminal. The net result is that the (−) terminal loses an electron and the (+) terminal gains one, making it *appear* as if the very same electron that left the (−) terminal actually flowed through the wires and the resistor and finally appeared at the (+) terminal. It is as if the path itself were not involved in the electron transfer, which is not the case.

The process of sequential migration of electrons from one atom to the next is called ***electron drift***, and it is this process that gives rise to the flow of ***conduction current*** through a circuit. To illustrate how important this process is in terms of the electronic transmission of information, let us examine the elementary transmission experiment represented by the circuit shown in Fig. 1-5. The circuit consists of an 8-volt battery and a switch on one end, a resistor on the other end, and a 60-m-long two-wire transmission line in between. The wires are made of copper, and they have a circular cross section with a 2-mm diameter. After closing the switch, a current will start to flow through the circuit. It is instructive to compare two velocities associated with the consequence of closing the switch, namely the electron drift velocity inside the copper wires and the transmission velocity (of the information announcing that the switch has been closed) between the battery and the resistor. For the specified parameters of the circuit shown in Fig. 1-5, the electron drift velocity—which is the actual physical velocity of the electrons along the wire—can be calculated readily and shown to be on the order of only 10^{-4} m/s. Hence, it would take about 1 million seconds ($\sim$ 10 days) for an electron to physically travel over a distance of 120 m. In contrast, the time delay between closing the switch at the sending end and observing a response at the receiving end (in the form of current flow through the resistor) is extremely short ($\approx 0.2\ \mu$s). This is because the transmission velocity is on the order of the velocity of light $c = 3 \times 10^8$ m/s. Thus:

The rate at which information can be transmitted electronically using conducting wires is about 12 orders of magnitude faster than the actual transport velocity of the electrons flowing through those wires!

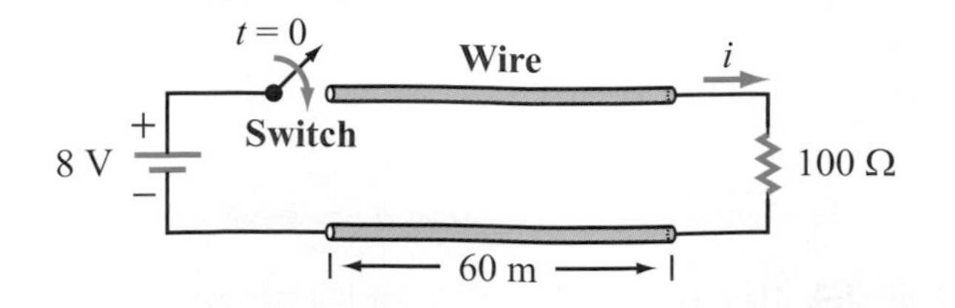

Figure 1-5: After closing the switch, it takes only 0.2 μs to observe a current in the resistor.

This fact is at the heart of what makes electronic communication systems possible.

1-3.2 Current

Moving charge gives rise to current.

Electric current is defined as the time rate of transfer of electric charge across a specified boundary.

For the wire segment depicted in Fig. 1-6, the current i flowing through it is equal to the amount of charge dq that crosses the wire's cross section over an infinitesimal time duration dt, given as

$$i = \frac{dq}{dt} \qquad \text{(A)}, \tag{1.3}$$

and the unit for current is the ampere (A). In general, both positive and negative charges may flow across the hypothetical interface, and the flow may occur in both directions. By convention, *the direction of i is defined to be the direction of the net flow of (net) charge (positive minus negative).* The circuit segment denoted with an arrow in Fig. 1-7(a) signifies that a current of 5 A is flowing through that wire segment in the direction of the arrow. The same information about the current magnitude and direction may be displayed as in Fig. 1-7(b), where the arrow points in the opposite direction and the current is expressed as −5 A.

When a battery is connected to a circuit, the resultant current that flows through it usually is constant in time (Fig. 1-8(a))—at least over the time duration of interest—in which case we refer to it as a ***direct current*** or ***dc*** for short. In contrast, the currents flowing in household systems (as well as in many electrical systems) are called ***alternating currents*** or simply ***ac***, because they vary sinusoidally with time (Fig. 1-8(b)). Other time variations also may occur in circuits, such as exponential

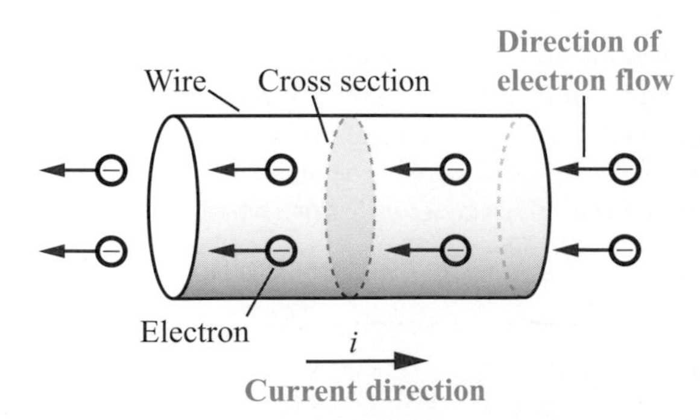

Figure 1-6: Direction of (positive) current flow through a conductor is opposite that of electrons.

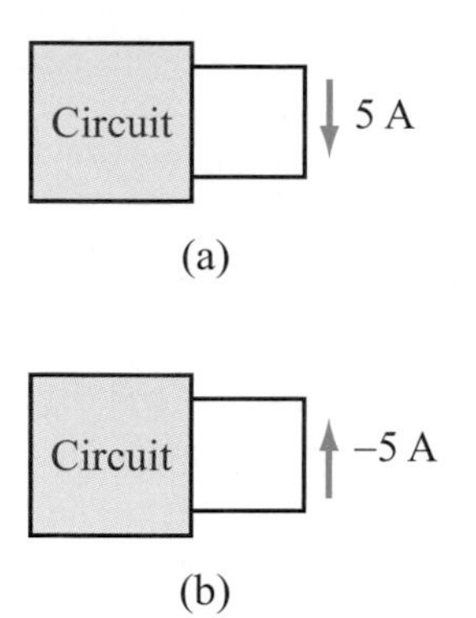

Figure 1-7: A current of 5 A flowing "downward" is the same as −5 A flowing "upward" through the wire.

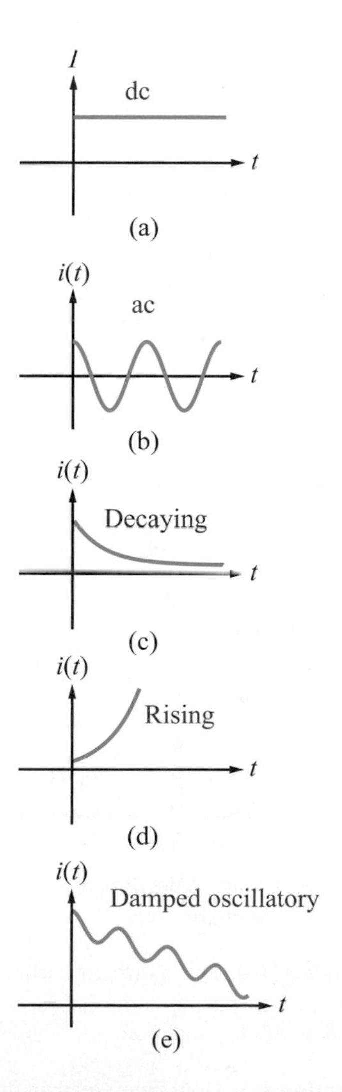

Figure 1-8: Graphical illustrations of various types of current variations with time.

rises and decays (Fig. 1-8(c) and (d)), exponentially damped oscillations (Fig. 1-8(e)), and many others.

Even though in the overwhelming majority of cases the current flowing through a material is dominated by the movement of electrons (as opposed to positively charged ions), it is advisable to *start thinking of the current in terms of positive charge*, primarily to avoid having to keep track of the fact that current direction is defined to be in opposition to the direction of flow of negative charges.

Example 1-1: Charge Transfer

In terms of the current $i(t)$ flowing past a reference cross section in a wire:

(a) Develop an expression for the ***cumulative charge*** $q(t)$ that has been transferred past that cross section up to time t. Apply the result to the exponential current displayed in Fig. 1-9(a), which is given by

$$i(t) = \begin{cases} 0 & \text{for } t < 0, \\ 6e^{-0.2t} \text{ A} & \text{for } t \geq 0. \end{cases} \tag{1.4}$$

(b) Develop an expression for the ***net charge*** $\Delta Q(t_1, t_2)$ that flowed through the cross section between times t_1 and t_2, and then compute ΔQ for $t_1 = 1$ s and $t_2 = 2$ s.

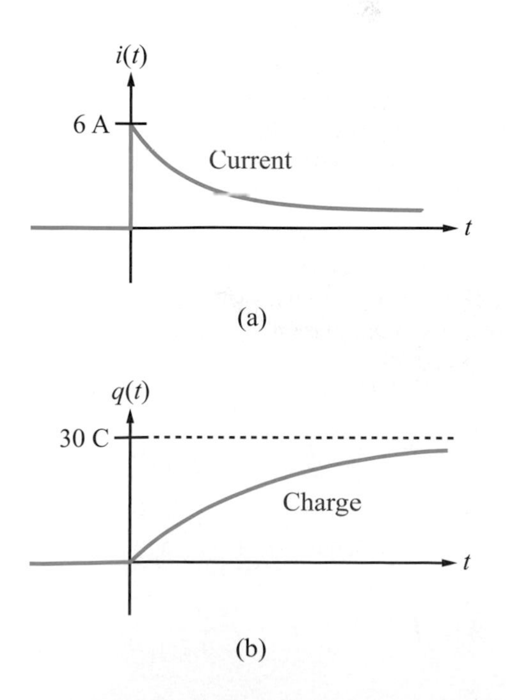

Figure 1-9: The current $i(t)$ displayed in (a) generates the cumulative charge $q(t)$ displayed in (b).

Solution:

(a) We start by rewriting Eq. (1.3) in the form:

$$dq = i\ dt.$$

Then by integrating both sides over the limits $-\infty$ to t, we have

$$\int_{-\infty}^{t} dq = \int_{-\infty}^{t} i\ dt,$$

which yields

$$q(t) - q(-\infty) = \int_{-\infty}^{t} i\ dt, \tag{1.5}$$

where $q(-\infty)$ represents the charge that was transferred through the wire "at the beginning of time." We choose $-\infty$ as a reference limit in the integration, because it allows us to set $q(-\infty) = 0$, implying that no charge had been transferred prior to that point in time. Hence, Eq. (1.5) becomes

$$q(t) = \int_{-\infty}^{t} i\ dt \quad \text{(C)}. \tag{1.6}$$

For $i(t)$ as given by Eq. (1.4), $i(t) = 0$ for $t < 0$. Hence,

$$q(t) = \int_{0}^{t} 6e^{-0.2t}\ dt = \frac{-6}{0.2}\ e^{-0.2t}\Big|_0^t = 30[1 - e^{-0.2t}]\ \text{C}.$$

A plot of $q(t)$ versus t is displayed in Fig. 1-9(b). The cumulative charge that would transfer after a long period of time is obtained by setting $t = +\infty$, which would yield $q(+\infty) = 30$ C.

(b) The cumulative charge that has flowed through the cross section up to time t_1 is $q(t_1)$, and a similar definition applies to $q(t_2)$. Hence, the ***net charge*** that flowed through the cross section over the time interval between t_1 and t_2 is

$$\begin{aligned} \Delta Q(t_1, t_2) &= q(t_2) - q(t_1) \\ &= \int_{-\infty}^{t_2} i\ dt - \int_{-\infty}^{t_1} i\ dt = \int_{t_1}^{t_2} i\ dt. \end{aligned}$$

For $t_1 = 1$ s, $t_2 = 2$ s, and $i(t)$ as given by Eq. (1.4),

$$\begin{aligned} \Delta Q(1, 2) &= \int_{1}^{2} 6e^{-0.2t}\ dt = \frac{6e^{-0.2t}}{-0.2}\Big|_1^2 \\ &= -30(e^{-0.4} - e^{-0.2}) = 4.45\ \text{C}. \end{aligned}$$

Example 1-2: Current

The charge flowing past a certain location in a wire is given by

$$q(t) = \begin{cases} 0 & \text{for } t < 0, \\ 5te^{-0.1t}\ \text{C} & \text{for } t \geq 0. \end{cases}$$

Determine (a) the current at $t = 0$ and (b) the instant at which $q(t)$ is a maximum and the corresponding value of q.

Solution:

(a) Application of Eq. (1.3) yields

$$\begin{aligned} i &= \frac{dq}{dt} \\ &= \frac{d}{dt}(5te^{-0.1t}) \\ &= 5e^{-0.1t} - 0.5te^{-0.1t} \\ &= (5 - 0.5t)e^{-0.1t}\ \text{A}. \end{aligned}$$

Setting $t = 0$ in the expression gives $i(0) = 5$ A.

Note that $i \neq 0$, even though $q(t) = 0$ at $t = 0$.

(b) To determine the value of t at which $q(t)$ is a maximum, we find dq/dt and then set it equal to zero:

$$\begin{aligned} \frac{dq}{dt} &= (5 - 0.5t)e^{-0.1t} \\ &= 0, \end{aligned}$$

which is satisfied when

$$5 - 0.5t = 0 \quad \text{or} \quad t = 10\ \text{s},$$

as well as when

$$e^{-0.1t} = 0 \quad \text{or} \quad t = \infty.$$

The first value ($t = 10$ s) corresponds to a maximum and $t = \infty$ corresponds to a minimum (which can be verified either by graphing $q(t)$ or by taking the second derivative of $q(t)$ and evaluating it at $t = 10$ s and $t = \infty$).

At $t = 10$ s,

$$q(10) = 5 \times 10e^{-0.1 \times 10} = 50e^{-1} = 18.4\ \text{C}.$$

Review Question 1-3: What are the four fundamental properties of electric charge?

Review Question 1-4: Is the direction of electric current in a wire defined to be the same as or opposite to the direction of flow of electrons?

Review Question 1-5: How does electron drift lead to the conduction of electric current?

Exercise 1-4: If the current flowing through a given resistor in a circuit is given by $i(t) = 5[1 - e^{-2t}]$ A for $t \geq 0$, determine the total amount of charge that passed through the resistor between $t = 0$ and $t = 0.2$ s.

Answer: $\Delta Q(0, 0.2) = 0.18$ C. (See CD)

Exercise 1-5: If $q(t)$ has the waveform shown in Fig. E1.5, determine the corresponding current waveform.

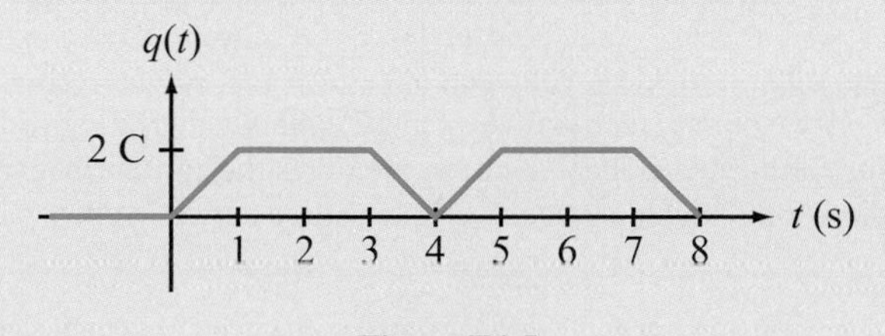

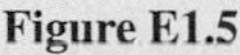

Figure E1.5

Answer:

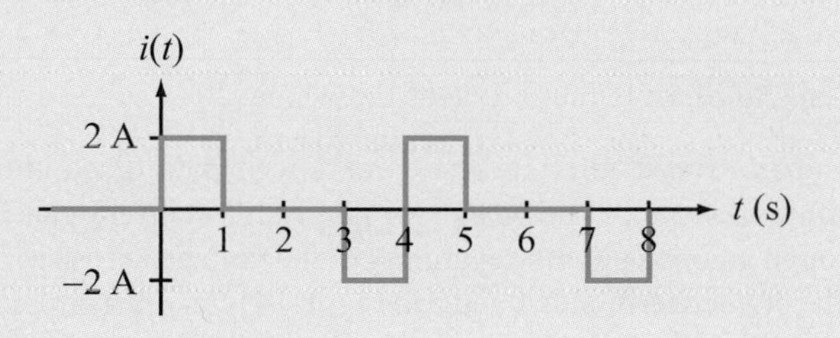

1-4 Voltage and Power

1-4.1 Voltage

The two primary quantities used in circuit analysis are current and voltage. Current is associated with the movement of electric charge and voltage is associated with the ***polarity*** of charge. Before we offer a formal definition for voltage, let us examine the energy implications of polarizing a hitherto neutral material, thereby establishing opposite electrical polarities on its two ends. Suppose we have a piece of material (such as a resistor) to which we connect two short wires and label their end points a and b, as shown in Fig. 1-10. Starting out with an electrically neutral structure, assume that we are able to detach an electron from one of the atoms at point a and move it to point b. Moving a negative charge from the positively charged atom against the attraction force between them requires the expenditure of a certain amount of energy. Voltage is a measure of this expenditure of energy relative to the amount of charge involved, and it always involves two spatial locations:

> Voltage often is denoted v_{ab} to emphasize the fact that it is the voltage difference *between* points $\boldsymbol{a}$ and $\boldsymbol{b}$.

The two points may be two locations in a circuit or any two points in space.

Against this background, we now offer the following formal definition for voltage:

> The voltage between location $\boldsymbol{a}$ and location $\boldsymbol{b}$ is the ratio of $\boldsymbol{dw}$ to $\boldsymbol{dq}$, where $\boldsymbol{dw}$ is the energy in joules (J) required to move (positive) charge $\boldsymbol{dq}$ from $\boldsymbol{b}$ to $\boldsymbol{a}$ (or negative charge from a to b).

That is,

$$v_{ab} = \frac{dw}{dq}, \tag{1.7}$$

and the unit for voltage is the volt (V), named after the inventor of the first battery, Alessandro Volta (1745–1827). Voltage also is called ***potential difference***. In terms of that terminology, if v_{ab} has a positive value, it means that point a is at a potential higher than that of point b. Accordingly, points a and b in Fig. 1-10 are denoted with (+) and (−) signs, respectively. If $v_{ab} = 5$ V, we often use the terminology: "The ***voltage rise*** from b to a is 5 V", or "The ***voltage drop*** from a to b is 5 V".

Just as 5 A of current flowing from a to b in a circuit conveys the same information as −5 A flowing in the opposite direction, a similar analogy applies to voltage. Thus, the two representations in Fig. 1-11 convey the same information with regard to the voltage between terminals a and b. Also, the

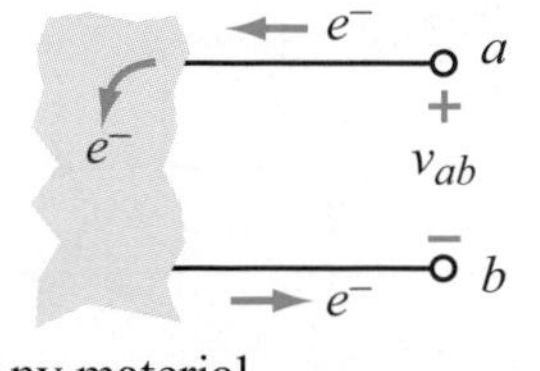

Figure 1-10: The voltage v_{ab} is equal to the amount of energy required to move one unit of negative charge from a to b through the material.

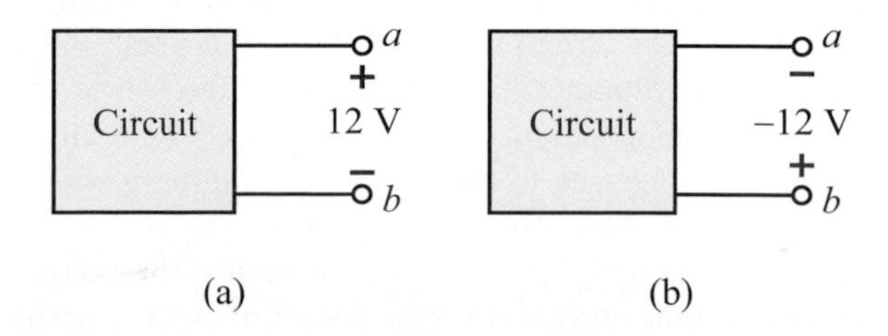

Figure 1-11: In (a), with the (+) designation at node a, $V_{ab} = 12$ V. In (b), with the (+) designation at node b, $V_{ba} = -12$ V, which is equivalent to $V_{ab} = 12$ V. [That is, $V_{ab} = -V_{ba}$.]

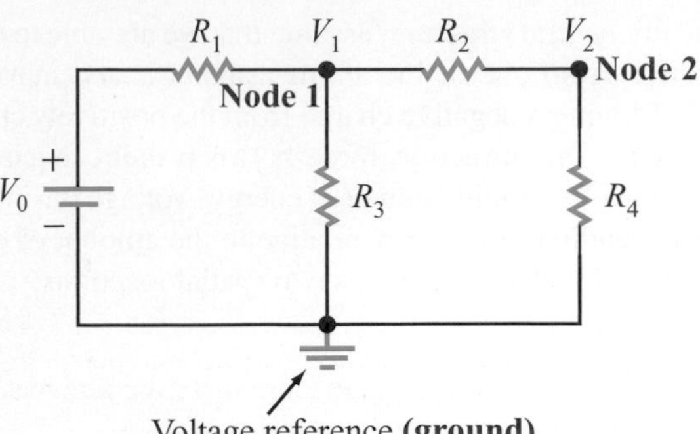

Figure 1-12: Ground is any point in the circuit selected to serve as a reference point for all points in the circuit.

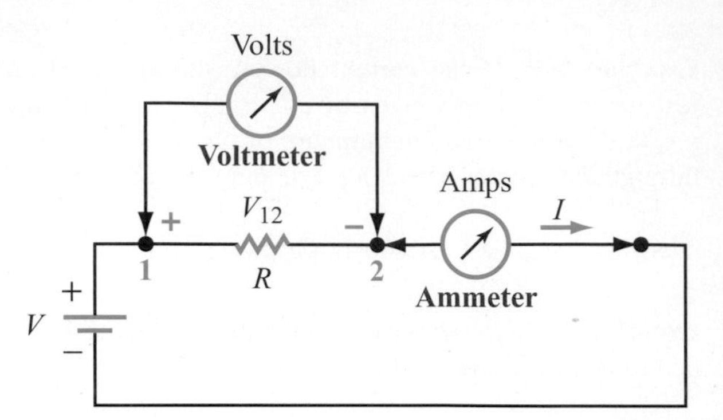

Figure 1-13: An ideal voltmeter measures the voltage difference between two points (such as nodes 1 and 2) without interfering with the circuit. Similarly, an ideal ammeter measures the current magnitude and direction without extracting a voltage drop across itself.

terms *dc* and *ac* defined earlier for current apply to voltage as well; a constant voltage is called a dc voltage and a sinusoidally time-varying voltage is called an ac voltage.

Ground

Since by definition voltage is not an absolute quantity but rather the difference in electric potential between two locations, it is sometimes convenient to select a reference point in the circuit, label it ***ground***, and then define the voltage at any point in the circuit with respect to that ground point. Thus, when we say that the voltage V_1 at node 1 in Fig. 1-12 is 6 V, we mean that the potential difference between node 1 and the ground reference point is 6 V, which is equivalent to having assigned the ground point a voltage of zero.

When a circuit is constructed in a laboratory, the chassis often is used as the common ground point—in which case it is called ***chassis ground***. As discussed later in Section 8-6, in a household electrical network, outlets are connected to three wires—one of which is called ***Earth ground*** because it is connected to the physical ground next to the house.

Voltmeter and Ammeter

The voltmeter is the standard instrument used to measure the voltage difference between two points in a circuit. To measure V_{12} in the circuit of Fig. 1-13, we connect the (+) terminal of the voltmeter to terminal 1 and the (−) terminal to terminal 2. Connecting the voltmeter to the circuit does not require any changes to the circuit, and *in the ideal case, the voltmeter will have no effect on any of the voltages and currents associated with the circuit.* In reality, the voltmeter has to extract some current from the circuit in order to perform the voltage measurement, but the voltmeter is designed such that the amount of extracted current is so small as to have a negligible effect on the circuit.

To measure the current flowing through a wire, it is necessary to *insert* an ammeter in that path, as illustrated by Fig. 1-13. *The voltage drop across an ideal ammeter is zero.*

Open and Short Circuits

An ***open circuit*** refers to the condition of path discontinuity (infinite resistance) between two points. No current can flow through an open circuit, regardless of the voltage across it. The path between terminals 1 and 2 in Fig. 1-14 is an open circuit. In contrast, a ***short circuit*** constitutes the condition of complete path continuity (with zero electrical resistance) between two points, such as between terminals 3 and 4 in Fig. 1-14. No voltage drop occurs across a short circuit, regardless of the magnitude of the current flowing through it.

Switches come in many varieties, depending on the intended function. The simple ON/OFF switch depicted in Fig. 1-15(a) is known as a ***single-pole single-throw (SPST)*** switch. The ON (closed) position acts like a short circuit, allowing current to flow while extracting no voltage drop across the switch's terminals; the OFF (open) position acts like an open circuit. The specific time $t = t_0$ denoted below or above the switch (Fig. 1-15(a)) refers to the time t_0 at which it opens or closes.

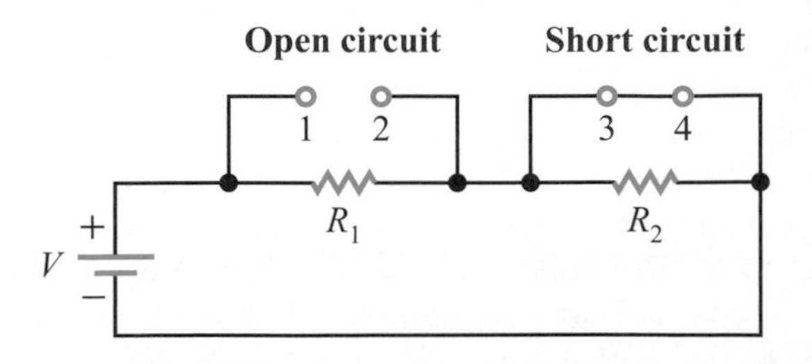

Figure 1-14: Open circuit between terminals 1 and 2, and short circuit between terminals 3 and 4.

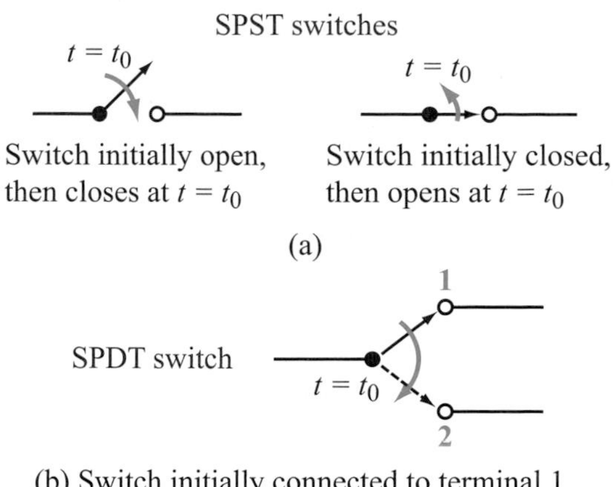

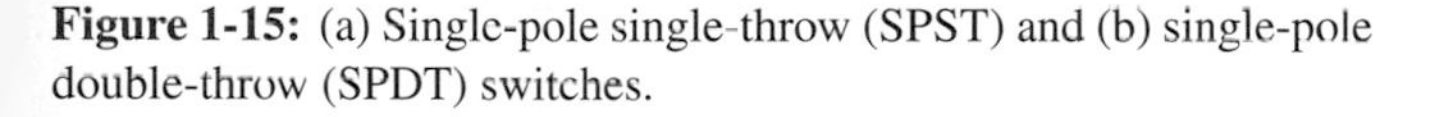

Figure 1-15: (a) Single-pole single-throw (SPST) and (b) single-pole double-throw (SPDT) switches.

If the purpose of the switch is to combine two switching functions so as to connect a common terminal to either of two other terminals, then we need to use the ***single-pole double-throw*** (SPDT) switch illustrated in Fig. 1-15(b). Before $t = t_0$, the common terminal is connected to terminal 1; then at $t = t_0$, that connection ceases (becomes open), and it is replaced with a connection between the common terminal and terminal 2.

1-4.2 Power

The circuit shown in Fig. 1-16(a) consists of a battery and a light bulb connected by an SPST switch in the open position. No current flows through the open circuit, but the battery has a voltage V_{bat} across it, due to the excess positive and negative charges it has at its two terminals. After the switch is closed at $t = 5$ s, as indicated in Fig. 1-16(b), a current I will flow through the circuit along the indicated direction. The battery's excess positive charges will flow from its positive terminal downward through the light bulb towards the battery's negative terminal, and (since current direction is defined to coincide with the direction of flow of positive charge) the current direction will be as indicated in the figure.

The consequences of current flow through the circuit are: (1) The battery acts as a supplier of power and (2) The light bulb acts as a recipient of power, which gets absorbed by its filament, causing it to heat up and glow, resulting in the conversion of electrical power into light and heat. A power supply, such as a battery, offers a ***voltage rise*** across it as we follow the current from the terminal at which it enters (denoted with a (−) sign) to the terminal from which it leaves (denoted with a (+) sign). In contrast, a power recipient (such as a light bulb) exhibits a ***voltage drop*** across its corresponding terminals. This set of assignments of voltage polarities relative to the

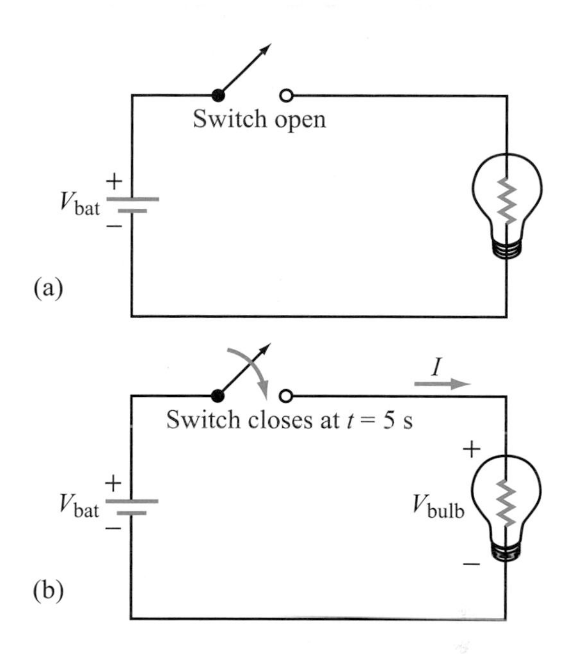

Figure 1-16: Current flow through a resistor (light-bulb filament) after closing the switch.

direction of current flow for devices generating power versus those consuming power is known as the ***passive sign convention*** (Fig. 1-17). We will adhere to it throughout the book.

Our next task is to establish an expression for the power p delivered to or received by an electrical device. By definition, power is the time rate of change of energy,

$$p = \frac{dw}{dt} \qquad \text{(W)}, \tag{1.8}$$

and its unit is the watt (W), named after the Scottish engineer and inventor James Watt (1736–1819) who is credited with the development of the steam engine from an embryonic stage into

Passive Sign Convention

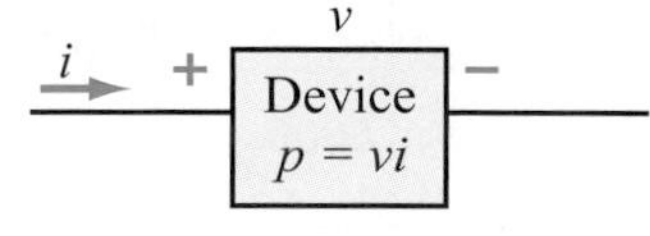

$p > 0$ power delivered to device
$p < 0$ power supplied by device

*Note that i direction is defined as entering (+) side of v.

Figure 1-17: Passive sign convention.

Technology Brief 2: Moore's Law and Scaling

In 1965, Gordon Moore, co-founder of Intel, predicted that the number of transistors in the minimum-cost processor would double every two years (initially, he had guessed they would double every year). Amazingly, this prediction has proven true of semiconductor processors for 40 years, as demonstrated by Fig. TF2-1.

In order to understand ***Moore's Law***, we have to understand the basics about how transistors work. As we will see later in Section 3-7, the basic switching element in semiconductor microprocessors is the ***transistor***. All of the complex components in the microprocessor (including logic gates, arithmetic logic units, and counters) are constructed from combinations of transistors. Within a processor, transistors have different dimensions depending on the component's function; larger transistors can handle more current, so the sub-circuit in the processor that distributes power may be built from larger transistors than, say, the sub-circuit that adds two bits together. In general, the smaller the transistor, the less power it consumes and the faster it can switch between binary states (0 and 1). Hence, an important goal of a circuit designer is to use the smallest transistors possible in a given circuit. We can quantify transistor size according to the smallest drawn dimension of the transistor, sometimes called the ***feature size***. In the Intel 4004, for example, the feature size was approximately 10 μm, which means that it was not possible to make transistors reliably with less than 10-μm features drawn in the CAD program. In modern processors, the feature size is 0.065 μm or 65 nm. (Remember that 1 nm $= 10^{-9}$ m.)

The questions then arise: How small can we go? What is the fundamental limit to shrinking down the size of a transistor? As we ponder this, we immediately observe that we likely cannot make a transistor smaller than the diameter of one silicon or metal atom (i.e., ~ 0.2 to 0.8 nm). But is there a limit prior to this? Well, as we shrink transistors down to the point that they are made of just one or a few atomic layers (~ 1 to 5 nm), we run into issues

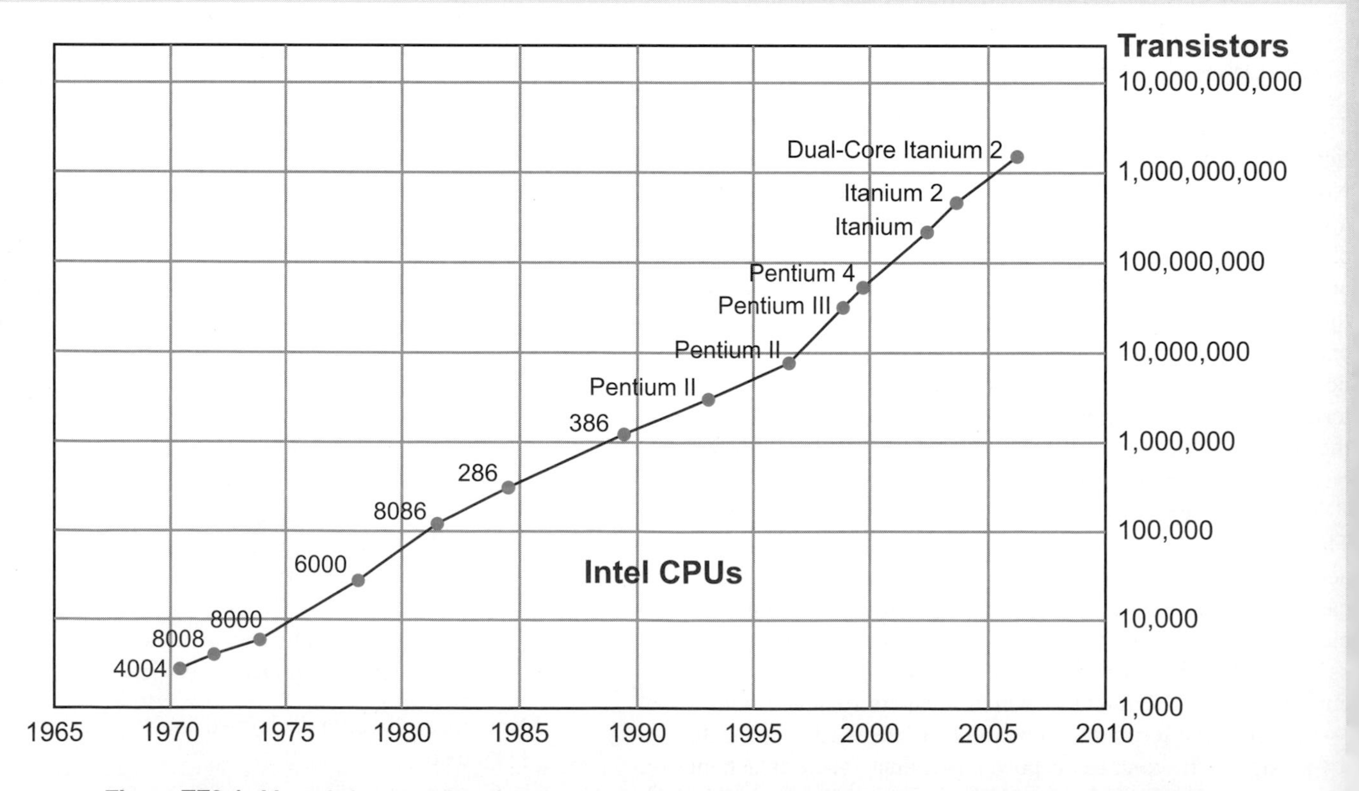

Figure TF2-1: Moore's Law predicts that the number of transistors per processor doubles every two years.

related to the stochastic nature of quantum physics. At these scales, the random motion of electrons between both physical space and energy levels becomes significant with respect to the size of the transistor, and we start to get spurious or random signals in the circuit. There are even more subtle problems related to the statistics of yield. If a certain piece of a transistor contained only 10 atoms, a deviation of just one atom in the device (to a 9-atom or an 11-atom transistor) represents a huge change in the device properties! (Can you imagine your local car dealer telling you your sedan will vary in length by ± 10 percent when it comes from the factory!?) This would make it increasingly difficult to economically fabricate chips with hundreds of millions of transistors. Additionally, there is an interesting issue of heat generation: Like any dissipative device, each transistor gives off a small amount of heat. But when you add up the heat produced by 100 million transistors, you get a very large number! Figure TF2-1 compares the power density (due to heat) produced by different processors with the heat produced by rocket engines and nuclear reactors.

None of these issues are insurmountable. Challenges simply spur driven people to come up with innovative solutions. Many of these problems will be solved, and in the process, provide engineers (like you) with jobs and opportunities. But, more importantly, the minimum feature size of a processor is not the end goal of innovation: It is the means to it. Innovation seeks simply to make *increasingly powerful* processors, not smaller feature sizes. In recent years, processor companies have lessened their attempts at smaller, faster processors and started lumping more of them together to distribute the work among them. This is the idea behind the dual and quad processor cores that power the computers of the last few years. By sharing the workload among various processors (called ***distributed computing***) we increase processor performance while using less energy, generating less heat, and without needing to run at warp speed. So it seems, as we approach ever-smaller features, we simply will transition into new physical technologies and also new computational techniques. As Gordon Moore himself said, "It will not be like we hit a brick wall and stop."

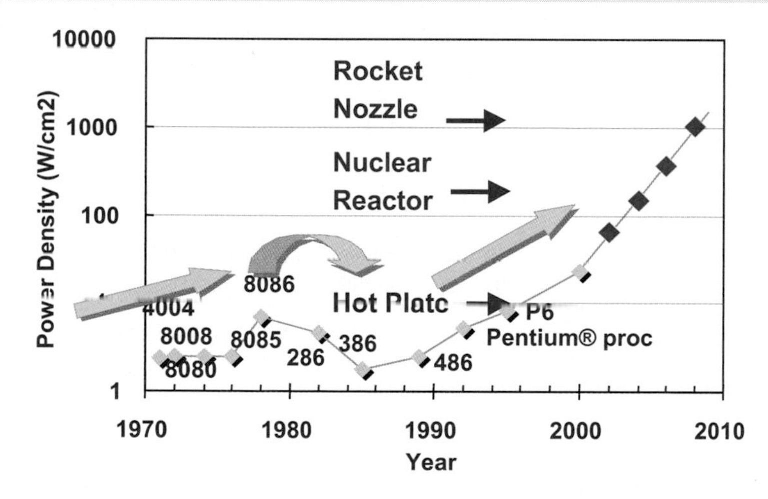

	Light Bulb	Integrated Circuit
Power dissipation	100 W	50 W
Surface area	106 cm^2 (bulb surface area)	1.5 cm^2 (die area)
Heat flux	0.9 W/cm^2	33.3 W/cm^2

Figure TF2-2: The power density generated by an IC in the form of heat is approaching the densities produced by a nuclear reactor. (Courtesy of Jan Rabey, University of California, Berkeley.)

a viable and efficient source of power. Using Eqs. (1.3) and (1.7), we can rewrite Eq. (1.8) as

$$p = \frac{dw}{dt} = \frac{dw}{dq} \cdot \frac{dq}{dt}$$

or simply

$$p = vi \qquad \text{(W)}. \tag{1.9}$$

Consistent with the passive sign convention:

The power delivered to a device is equal to the voltage across it multiplied by the current entering through its (+) voltage terminal.

If the algebraic value of p is negative, then the device is a supplier of energy. For an isolated electric circuit composed of multiple elements, the ***law of conservation of power*** requires that the algebraic sum of power for the entire circuit be always zero. That is, for a circuit with n elements,

$$\sum_{k=1}^{n} p_k = 0, \tag{1.10}$$

which means that the total power supplied by the circuit always must equal the total power absorbed by it.

Power supplies are sometimes assigned ratings to describe their capacities to deliver energy. A battery may be rated as having an output capacity of 200 ***ampere-hours*** (Ah) at 9 volts, which means that it can deliver a current i over a period of time Δt (measured in hours) such that $i\ \Delta t = 200$ Ah, and it can do so while maintaining a voltage of 9 V. Alternatively, its output capacity may be expressed as 1.8 ***kilowatt-hours*** (kWh), which represents the total amount of energy it can supply, namely, $W = vi\ \Delta t$ (with Δt in hours).

Example 1-3: Conservation of Power

For each of the two circuits shown in Fig. 1-18, determine how much power is being delivered to each device and whether it is a power supplier or recipient.

Solution:

(a) For the circuit in Fig. 1-18(a), the current entering the (+) terminal of the device is 0.2 A. Hence, the power P (where we use an uppercase letter because both the current and voltage are dc) is:

$$P = VI = 12 \times 0.2 = 2.4 \text{ W},$$

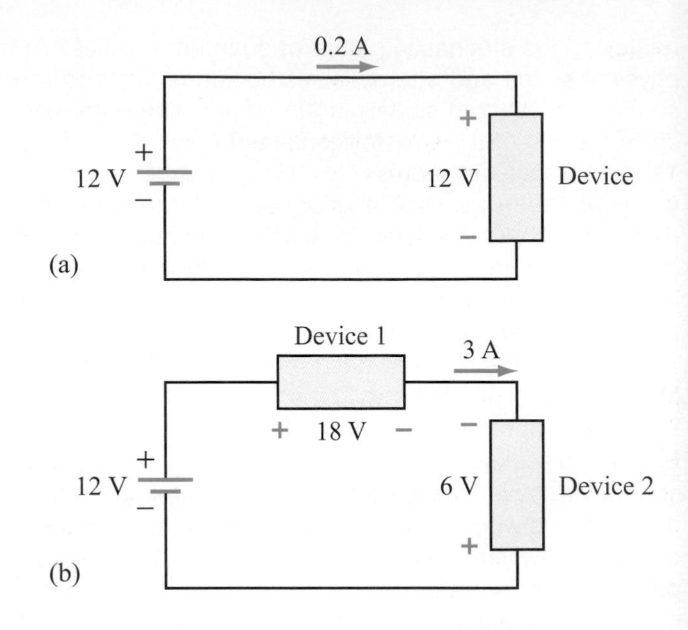

Figure 1-18: Circuits for Example 1-3.

and since $P > 0$, the device is a recipient of power. As we know, the law of conservation of power requires that if the device receives 2.4 W of power then the battery has to deliver exactly that same amount of power. For the battery, the current entering its (+) terminal is −0.2 A (because 0.2 A of current is shown leaving that terminal), so according to the passive sign convention, the power that would be absorbed by the battery (had it been a passive device) is

$$P_{\text{bat}} = 12(-0.2) = -2.4 \text{ W}.$$

The fact that P_{bat} is negative is confirmation that the battery is indeed a supplier of power.

(b) For device 1 in Fig. 1-18(b), the current entering its (+) terminal is 3 A. Hence,

$$P_1 = V_1 I_1 = 18 \times 3 = 54 \text{ W},$$

and the device is a power recipient.

For device 2,

$$P_2 = V_2 I_2 = (-6) \times 3 = -18 \text{ W},$$

and the device is a supplier of power (because P_2 is negative).

By way of confirmation, the power associated with the battery is

$$P_{\text{bat}} = 12(-3) = -36 \text{ W},$$

thereby satisfying the law of conservation of power, which requires the net power of the overall circuit to be exactly zero.

Example 1-4: Energy Consumption

A resistor connected to a 100-V dc power supply was consuming 20 W of power until the switch was turned off, after which the voltage decayed exponentially to zero. If $t = 0$ is defined as the time at which the switch was turned to the *off* position and if the subsequent voltage variation was given by

$$v(t) = 100e^{-2t} \text{ V} \qquad \text{for } t \geq 0$$

(where t is in seconds), determine the total amount of energy consumed by the resistor after the switch was turned off.

Solution:
Before $t = 0$, the current flowing through the resistor was $I = P/V = 20/100 = 0.2$ A. Using this value as the initial amplitude of the current at $t = 0$ and assuming that the current will exhibit the same time variation as the voltage, $i(t)$ can be expressed as

$$i(t) = 0.2e^{-2t} \text{ A} \qquad \text{for } t \geq 0.$$

The instantaneous power is

$$p(t) = v(t) \cdot i(t) = (100e^{-2t})(0.2e^{-2t}) = 20e^{-4t} \text{ W}.$$

We note that the power decays at a rate (e^{-4t}) much faster than the rate for current and voltage (e^{-2t}). The total energy dissipated in the resistor after engaging the switch is obtained by integrating $p(t)$ from $t = 0$ to infinity, namely

$$W = \int_0^\infty p(t)\,dt = \int_0^\infty 20e^{-4t}\,dt = -\frac{20}{4}\,e^{-4t}\Big|_0^\infty = 5 \text{ J}.$$

Exercise 1-6: If a positive current is flowing through a resistor from its terminal a to its terminal b, is v_{ab} positive or negative?

Answer: $v_{ab} > 0$. (See)

Exercise 1-7: A certain device has a voltage difference of 5 V across it. If 2 A of current is flowing through it from its (−) voltage terminal to its (+) terminal, is the device a power supplier or a power recipient, and how much energy does it supply or receive in 1 hour?

Answer: $P = VI = 5(-2) = -10$ W. Hence, the device is a power supplier. $|W| = |P|\,\Delta t = 36$ kJ. (See)

Exercise 1-8: A car radio draws 0.5 A of dc current when connected to a 12-V battery. How long does it take for the radio to consume 1.44 kJ?

Answer: 4 minutes. (See)

1-5 Circuit Elements

Electronic circuits used in functional systems employ a wide range of circuit elements, including transistors and integrated circuits. The operation of most electronic circuits and devices—no matter how complex—can be modeled (represented) in terms of an ***equivalent circuit*** composed of ***basic elements*** with idealized characteristics. The equivalent circuit offers a circuit behavior that closely resembles the behavior of the actual electronic circuit or device over a certain range of specified conditions, such as the range of input signal level or output load resistance. The set of basic elements commonly used in circuit analysis include voltage and current sources; passive elements (which include resistors), capacitors, and inductors; and various types of switches. The basic attributes of switches were covered in Section 1-4.1. The nomenclature and current–voltage relationships associated with the other two groups are the subject of this section.

1-5.1 i–v Relationship

The relationship between the current flowing through a device and the voltage across it defines the fundamental operation of that device. As was stated earlier, Ohm's law states that the current i entering into the (+) terminal of the voltage v across a resistor is given by

$$i = \frac{v}{R}.$$

This is called the ***i–v relationship*** for the resistor. We note that the resistor exhibits a ***linear i–v relationship***, meaning that i and v always vary in a proportional manner, as shown in Fig. 1-19(a), so long as R remains constant. A circuit composed exclusively of elements with linear i–v responses is called a ***linear circuit***. The linearity property of a circuit is an underlying requirement for the various circuit analysis techniques presented in this and future chapters. Diodes and transistors exhibit nonlinear i–v relationships, but we still can apply the analysis techniques specific to linear circuits to circuits containing nonlinear devices by representing those devices in terms of linear subcircuits that contain ***dependent sources***. The concept of a dependent source and how it is used is introduced in Section 1-5.3.

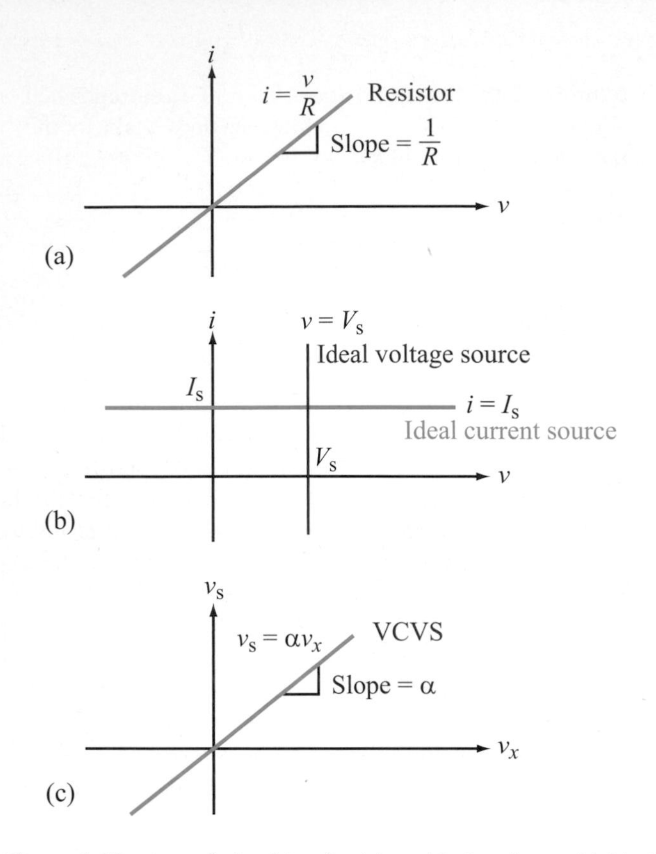

Figure 1-19: i–v relationships for (a) an ideal resistor, (b) ideal, independent current and voltage sources, and (c) a dependent, voltage-controlled voltage source (VCVS).

1-5.2 Independent Sources

An ***ideal, independent voltage source*** provides a specified voltage across its terminals, regardless of the type of load or circuit connected to it. Hence, for a voltage source with a specified voltage V_s, its i–v relationship is given by

$$v = V_s \qquad \text{for any } i,$$

so long as it is not connected to a short circuit. Similarly, an ***ideal, independent current source*** provides a specified current flowing through it, regardless of the voltage across it (but it cannot do so if connected to an open circuit). Its i–v relationship is

$$i = I_s \qquad \text{for any } v.$$

The i–v profile of an ideal voltage source is a vertical line, as illustrated in Fig. 1-19(b), whereas the profile for the ideal current source is a horizontal line.

The circuit symbol used for independent sources is a circle, as shown in Table 1-3, although for dc voltage sources the traditional "battery" symbol is used as well. A household electrical outlet connected through an electrical power-distribution network to a hydroelectric- or nuclear-power generating station provides continuous power at an approximately constant voltage level. Hence, it may be classified appropriately as an independent voltage source. On a shorter time scale, a flashlight's 9-volt battery may be regarded as a voltage source, but only until its stored charge has been used up by the light bulb. Thus, strictly speaking, a battery is a storage device (not a generator), but we tend to treat it as a generator so long as it acts like a constant voltage source.

In reality, no sources can provide the performance specifications ascribed to ideal sources. If a 5-V voltage source is connected across a short circuit, for example, we run into a serious problem of ambiguity. From the standpoint of the source, the voltage is 5 V, but by definition, the voltage is zero across the short circuit. How can it be both zero and 5 V simultaneously? The answer resides in the fact that our description of the ideal voltage source breaks down in this situation. *More realistic models for voltage and current sources include a series resistor in the case of the voltage source, and a* ***shunt*** *(parallel) resistor in the case of the current source*, as shown in Table 1-3. The real voltage source (which may have an elaborate circuit configuration) behaves like a combination of an ***equivalent, ideal voltage source*** v_s in series with an ***equivalent resistance*** R_s. Usually, R_s has a very small value for the voltage source and a very large value for the current source.

1-5.3 Dependent Sources

As alluded to in the opening paragraph of Section 1-5, we often use equivalent circuits to model the behavior of transistors and other electronic devices. The ability to represent complicated devices by equivalent circuits composed of basic elements greatly facilitates not only the circuit analysis process but the design process as well. Such circuit models incorporate the relationships between various parts of the device through the use of a set of *artificial* sources known as ***dependent sources***. The voltage level of a ***dependent voltage source*** is defined in terms of a specific voltage or current elsewhere in the circuit. An example of circuit equivalence is illustrated in Fig. 1-20. In part (a) of the figure, we have a Model 741 ***operational amplifier*** (***op amp***), denoted by the triangular circuit symbol, used in a simple amplifier circuit intended to provide a voltage amplification factor of -2; that is, the output voltage $v_o = -2v_s$, where v_s is the input signal voltage. The op amp, which we will examine later in Chapter 4, is an electronic device with a complex architecture composed of transistors, resistors, capacitors, and diodes, but in practice, its circuit behavior can be represented by a rather simple circuit consisting of two resistors (input resistor R_i and output resistor R_o) and a dependent voltage source, as shown in Fig. 1-20(b). The voltage v_2 on the right-hand side of the circuit in Fig. 1-20(b)

Table 1-3: Voltage and current sources.

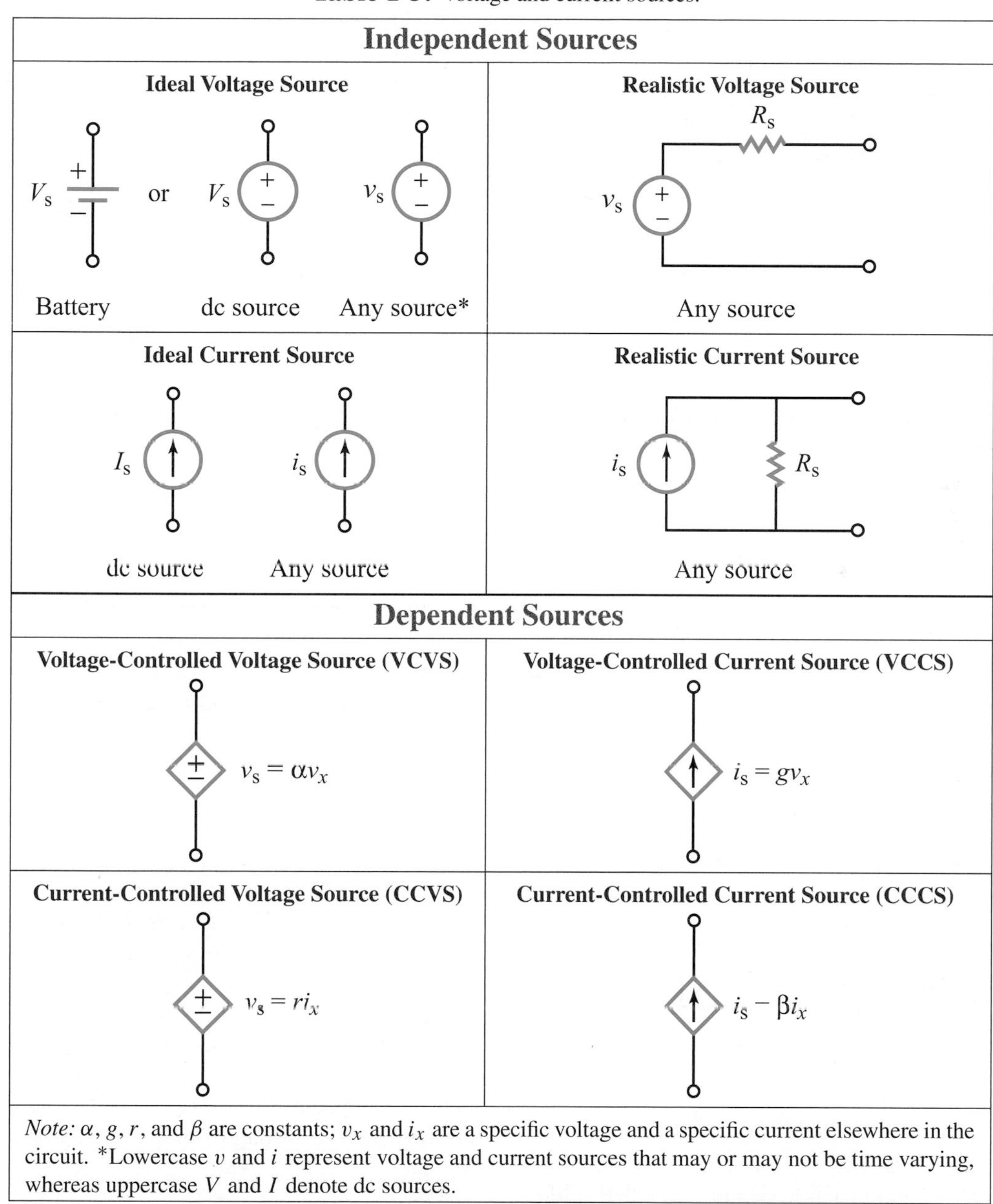

is given by $v_2 = Av_i$, where A is a constant and v_i is the voltage across the resistor R_i located on the left-hand side of the equivalent circuit. In this case, the magnitude of v_2 always depends on the magnitude of v_i, which depends in turn on the input signal voltage v_s and on the values chosen for some of the resistors in the circuit. Since the controlling quantity v_i is a voltage, v_2 is called a ***voltage-controlled voltage source (VCVS)***. Had the controlling quantity been a current source, the dependent source would have been called a ***current-controlled voltage source (CCVS)*** instead. A parallel analogy exists for voltage-controlled and current-controlled current sources. The characteristic symbol for a dependent source is the diamond (Table 1-3). Proportionality constant α in Table 1-3 relates voltage to voltage. Hence, it is dimensionless, as is β, since it relates current to current. Constants g and r have units of (A/V) and (V/A), respectively. Because dependent sources are

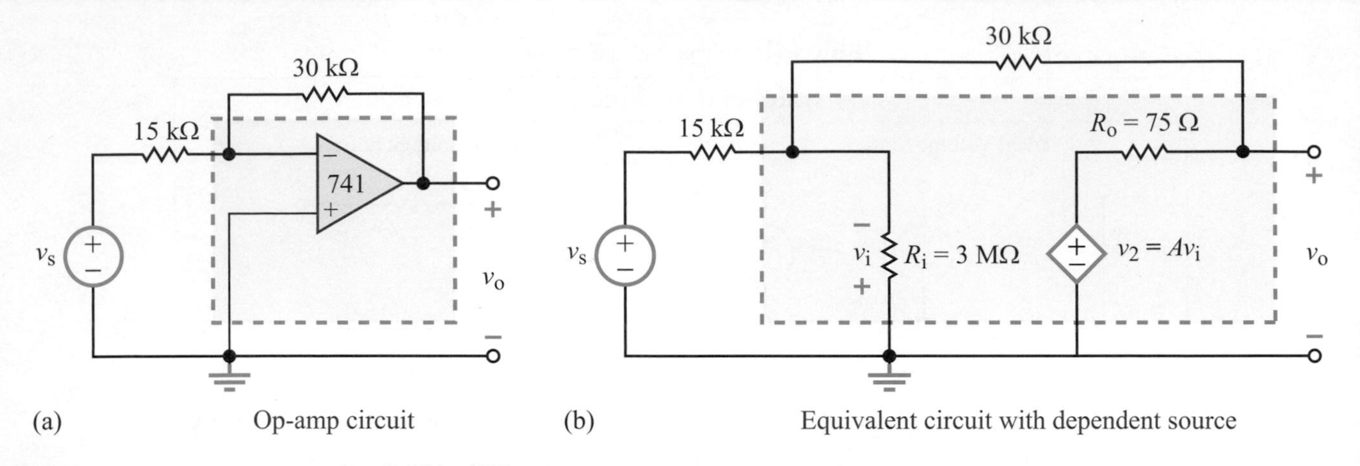

Figure 1-20: An operational amplifier is a complex device, but its circuit behavior can be represented in terms of a simple equivalent circuit that includes a dependent voltage source.

characterized by linear relationships, so are their i–v profiles. An example is shown in Fig. 1-19(c) for the VCVS.

Example 1-5: Dependent Source

Find the magnitude of the voltage V_1 of the dependent source in Fig. 1-21. What type of source is it?

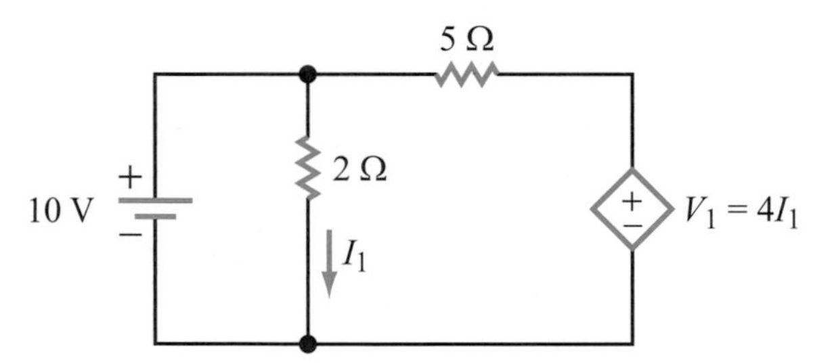

Figure 1-21: Circuit for Example 1-5.

Solution:
Since V_1 depends on current I_1, it is a current-controlled voltage source with a coefficient of 4 V/A.

The 10-V dc voltage is connected across the 2-Ω resistor. Hence, the current I along the designated direction is

$$I_1 = \frac{10}{2} = 5 \text{ A}.$$

Consequently,

$$V_1 = 4I_1 = 4 \times 5 = 20 \text{ V}.$$

1-5.4 Passive Elements

Table 1-4 lists three passive elements. For the resistor, capacitor, and inductor, their ***current–voltage (i–v) relationships*** are given by

$$v_R = Ri_R, \tag{1.11a}$$

$$i_C = C\,\frac{dv_C}{dt}, \tag{1.11b}$$

$$v_L = L\,\frac{di_L}{dt}, \tag{1.11c}$$

where R, C, and L are the resistance in ohms (Ω), capacitance in farads (F), and inductance in henrys (H), respectively; v_R, v_C, and v_L are the voltages across the resistor, capacitor, and inductor, respectively; and i_R, i_C, and i_L are the corresponding currents flowing through them. It should be noted that:

> The i–v relationships are defined such that for any of the three devices, the current direction is into the positive terminal of the voltage across it and out of the negative terminal.

We observe that the resistor exhibits a ***linear i–v relationship*** meaning that i and v always vary in a proportional manner. The capacitor and inductor are characterized by i–v relationships that involve the time derivative d/dt. In fact, if the voltage across the capacitor (v_C) is constant with time (dc), then $dv_C/dt = 0$, and consequently, the current $i_C = 0$ (no matter

Table 1-4: Passive circuit elements and their symbols.

Element	Symbol	i–v Relationship
Resistor	i_R, v_R (+ −), R	$v_R = Ri_R$
Capacitor	i_C, v_C (+ −), C	$i_C = C\dfrac{dv_C}{dt}$
Inductor	i_L, v_L (+ −), L	$v_L = L\dfrac{di_L}{dt}$

how large or small v_C might be). Similarly, if the current i_L flowing through the inductor is dc, then the voltage across it (v_L) is zero. Because the time variations of voltage and current are at the core of what makes capacitors and inductors useful devices, they are used primarily in time-varying circuits. The resistor is used in both dc and time-varying circuits. Our examination of circuits containing capacitors and inductors—both of which are energy storage devices—begins in Chapter 5.

Example 1-6: Switches

The circuit in Fig. 1-22 contains one SPDT switch that changes position at $t = 0$, one SPST switch that opens at $t = 0$, and one SPST switch that closes at $t = 5$ s. Generate circuit diagrams that include only those elements that have current flowing through them for (a) $t < 0$, (b) $0 \le t < 5$ s, and (c) $t \ge 5$ s.

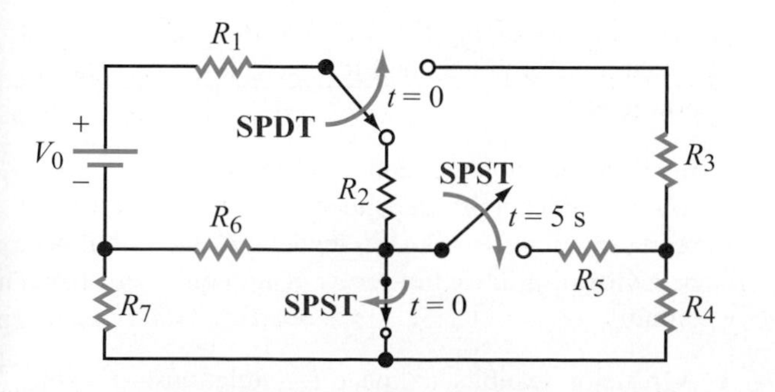

Figure 1-22: Circuit for Example 1-6.

Solution:
See Fig. 1-23.

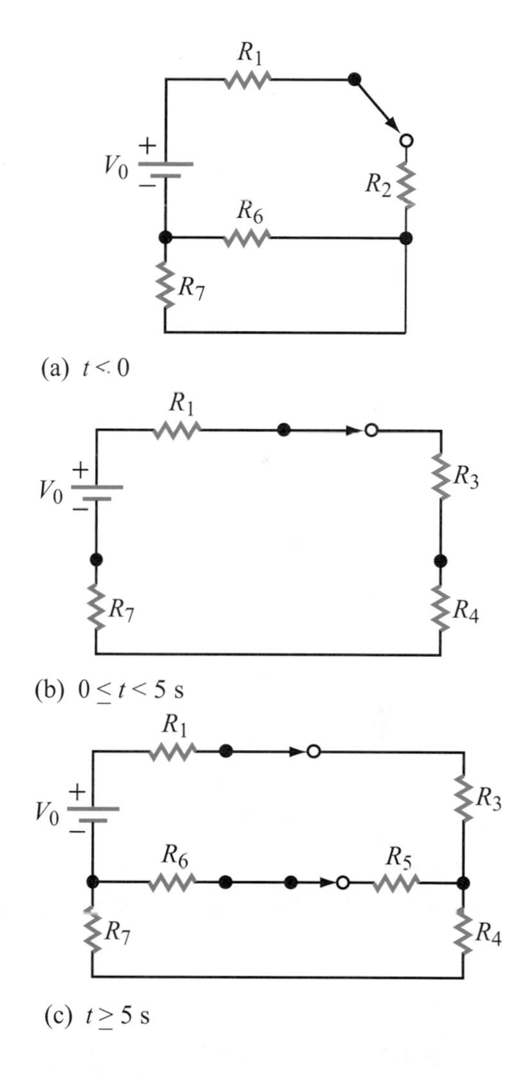

Figure 1-23: Solutions for circuit in Fig. 1-22.

Review Question 1-6: What is the difference between an SPST switch and an SPDT switch?

Review Question 1-7: What is the difference between an independent voltage source and a dependent voltage source? Is a dependent voltage source a real source of power?

Review Question 1-8: What is an "equivalent-circuit" model? How is it used?

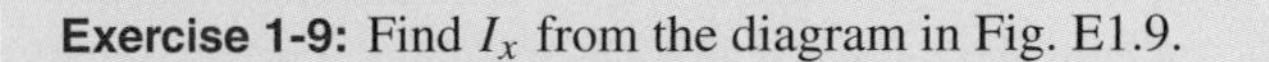

Exercise 1-9: Find I_x from the diagram in Fig. E1.9.

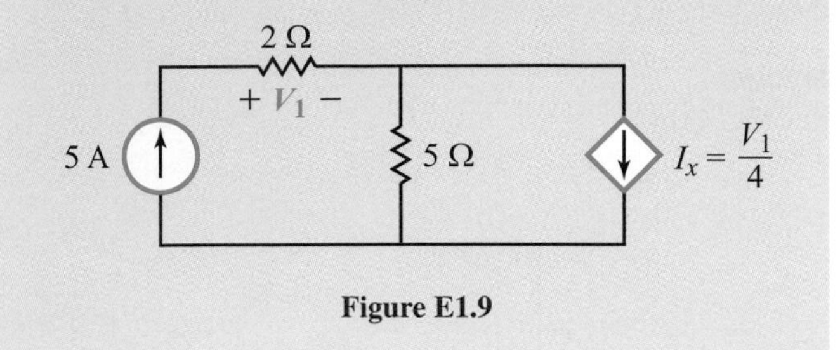

Figure E1.9

Answer: $I_x = 2.5$ A. (See)

Exercise 1-10: In the circuit of Fig. E1.10, find I at (a) $t < 0$ and (b) $t > 0$.

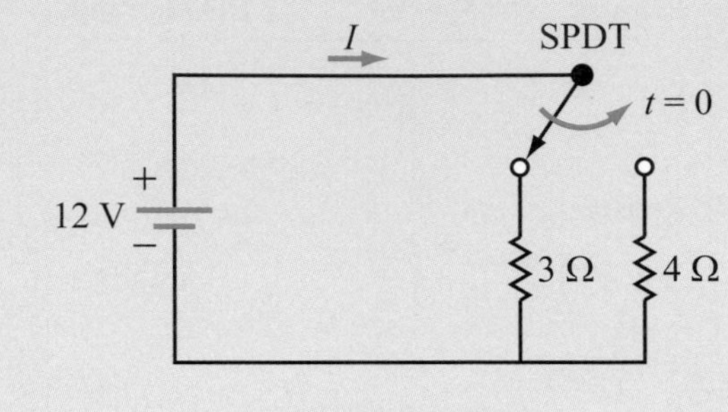

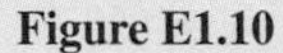

Figure E1.10

Answer: (a) $I = 4$ A, (b) $I = 3$ A. (See)

Chapter 1 Relationships

Ohm's law $i = v/R$

Current $i = dq/dt$
Direction of i = direction of flow of (+) charge

Charge transfer $q(t) = \int_{-\infty}^{t} i\,dt$

$$\Delta Q = q(t_2) - q(t_1) = \int_{t_1}^{t_2} i\,dt$$

Voltage = potential energy difference per unit charge

Passive sign convention Direction of i is into $+v$ terminal of device

Power $p = vi$

If $p > 0$ ➡ device absorbs power
If $p < 0$ ➡ device delivers power

i–v relationships

Resistor	$v_R = Ri_R$
Capacitor	$i_C = C\,dv_C/dt$
Inductor	$v_L = L\,di_L/dt$

CHAPTER HIGHLIGHTS

- Active devices (such as transistors and ICs) require an external power source to operate; in contrast, passive devices (resistors, capacitors, and inductors) do not.
- Analysis and synthesis (design) are complementary processes.
- Current is related to charge by $i = dq/dt$; voltage between locations a and b is $v_{ab} = dw/dq$, where dw is the work (energy) required to move dq from b to a; and power $p = vi$.
- Passive sign convention assigns i direction as entering the (+) side of v; if $p > 0$, the device is recipient (consumer) of power, and if $p < 0$, it is a supplier of power.
- Independent voltage and current sources are real sources of energy; dependent sources are artificial representations used in modeling the nonlinear behavior of a device in terms of an equivalent linear circuit.
- A resistor exhibits a linear i–v relationship, while diodes and transistors do not; the i–v relationships for a capacitor and an inductor involve d/dt.

GLOSSARY OF IMPORTANT TERMS

Provide definitions or explain the meaning of the following terms:

ac
active device
ampere-hours
analysis
conduction current
cumulative charge
dc
dependent source
design
electric charge
electric circuit
electric current
electron drift
equivalent circuit
independent source
i–v characteristic
kilowatt-hours
linear response
open circuit
passive device
passive sign convention
polarization
potential difference
power
prefix
short circuit
SI units
SPST
SPDT
synthesis
voltage

PROBLEMS

Sections 1-2 and 1-3: Dimensions, Charge, and Current

1.1 Use appropriate multiple and submultiple prefixes to express the following quantities:

(a) 3,620 watts (W)

(b) 0.000004 amps (A)

(c) 5.2×10^{-6} ohms (Ω)

(d) 3.9×10^{11} volts (V)

(e) 0.02 meters (m)

(f) 32×10^{5} volts (V)

1.2 Use appropriate multiple and submultiple prefixes to express the following quantities:

(a) 4.71×10^{-8} seconds (s)

(b) 10.3×10^{8} watts (W)

(c) 0.00000000321 amps (A)

(d) 0.1 meters (m)

(e) 8,760,000 volts (V)

(f) 3.16×10^{-16} hertz (Hz)

1.3 Convert:

(a) 16.3 m to mm

(b) 16.3 m to km

(c) 4×10^{-6} μF (microfarad) to pF (picofarad)

(d) 2.3 ns to μs

(e) 3.6×10^{7} V to MV

(f) 0.03 mA (milliamp) to μA

1.4 Convert:

(a) 4.2 m to μm

(b) 3 hours to μseconds

(c) 4.2 m to km

(d) 173 nm to m

(e) 173 nm to μm

(f) 12 pF (picofarad) to F (farad)

1.5 The total charge contained in a certain region of space is -1 C. If that region contains only electrons, how many does it contain?

1.6 A certain cross section lies in the x–y plane. If 3×10^{20} electrons go through the cross section in the z-direction in 4 seconds, and simultaneously 1.5×10^{20} protons go through the same cross section in the negative z-direction, what is the magnitude and direction of the current flowing through the cross section?

1.7 Determine the current $i(t)$ flowing through a resistor if the cumulative charge that has flowed through it up to time t is given by

(a) $q(t) = 3.6t$ mC

(b) $q(t) = 5\sin(377t)$ μC

*__(c)__ $q(t) = 0.3[1 - e^{-0.4t}]$ pC

(d) $q(t) = 0.2t\sin(120\pi t)$ nC

*Answer(s) in Appendix E.

1.8 Determine the current $i(t)$ flowing through a certain device if the cumulative charge that has flowed through it up to time t is given by

(a) $q(t) = -0.45t^3\ \mu\text{C}$

(b) $q(t) = 12\sin^2(800\pi t)$ mC

(c) $q(t) = -3.2\sin(377t)\cos(377t)$ pC

(d) $q(t) = 1.7t[1 - e^{-1.2t}]$ nC

1.9 Determine the net charge ΔQ that flowed through a resistor over the specified time interval for each of the following currents:

(a) $i(t) = 0.36$ A, from $t = 0$ to $t = 3$ s

*__(b)__ $i(t) = [40t + 8]$ mA, from $t = 1$ s to $t = 12$ s

(c) $i(t) = 5\sin(4\pi t)$ nA, from $t = 0$ to $t = 0.05$ s

(d) $i(t) = 12e^{-0.3t}$ mA, from $t = 0$ to $t = \infty$

1.10 Determine the net charge ΔQ that flowed through a certain device over the specified time intervals for each of the following currents:

(a) $i(t) = [3t + 6t^3]$ mA, from $t = 0$ to $t = 4$ s

(b) $i(t) = 4\sin(40\pi t)\cos(40\pi t)\ \mu\text{A}$, from $t = 0$ to $t = 0.05$ s

(c) $i(t) = [4e^{-t} - 3e^{-2t}]$ A, from $t = 0$ to $t = \infty$

(d) $i(t) = 12e^{-3t}\cos(40\pi t)$ nA, from $t = 0$ to $t = 0.05$ s

1.11 If the current flowing through a wire is given by $i(t) = 3e^{-0.1t}$ mA, how many electrons pass through the wire's cross section over the time interval from $t = 0$ to $t = 0.3$ ms?

1.12 The cumulative charge in mC that entered a certain device is given by

$$q(t) = \begin{cases} 0 & \text{for } t < 0, \\ 5t & \text{for } 0 \le t \le 10 \text{ s}, \\ 60 - t & \text{for } 10 \text{ s} \le t \le 60 \text{ s} \end{cases}$$

(a) Plot $q(t)$ versus t from $t = 0$ to $t = 60$ s.

(b) Plot the corresponding current $i(t)$ entering the device.

1.13 A steady flow resulted in 3×10^{15} electrons entering a device in 0.1 ms. What is the current?

1.14 Given that the current in (mA) flowing through a wire is given by:

$$i(t) = \begin{cases} 0 & \text{for } t < 0 \\ 6t & \text{for } 0 \le t \le 5 \text{ s} \\ 30e^{-0.6(t-5)} & \text{for } t \ge 5 \text{ s}, \end{cases}$$

(a) Sketch $i(t)$ versus t.

(b) Sketch $q(t)$ versus t.

1.15 The plot in Fig. P1.15 displays the cumulative amount of charge $q(t)$ that has entered a certain device up to time t. What is the current at

(a) $t = 1$ s

*__(b)__ $t = 3$ s

(c) $t = 6$ s

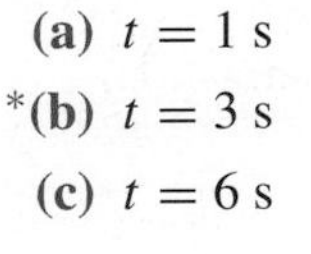

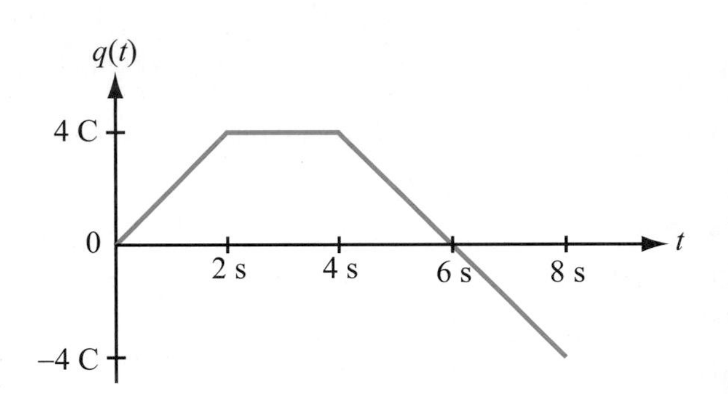

Figure P1.15: $q(t)$ for Problem 1.15.

1.16 The plot in Fig. P1.16 displays the cumulative amount of charge $q(t)$ that has exited a certain device up to time t. What is the current at

(a) $t = 2$ s

(b) $t = 6$ s

(c) $t = 12$ s

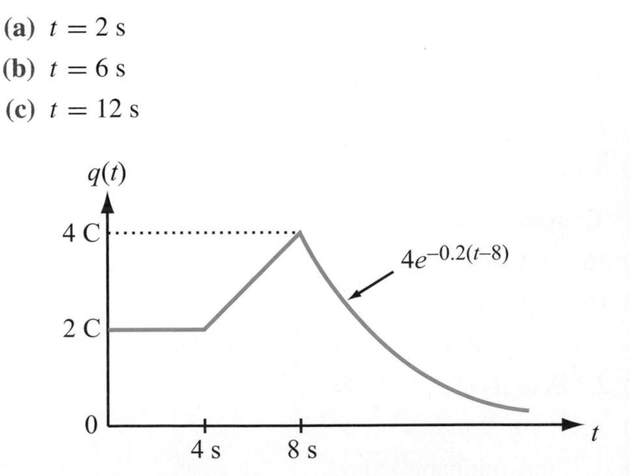

Figure P1.16: $q(t)$ for Problem 1.16.

Sections 1-4 and 1-5: Voltage, Power, and Circuit Elements

1.17 For each of the eight devices in the circuit of Fig. P1.17, determine whether the device is a supplier or a recipient of power and how much power it is supplying or receiving.

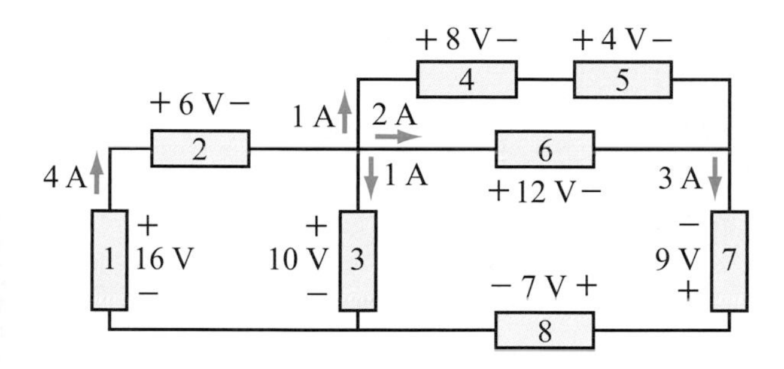

Figure P1.17: Circuit for Problem 1.17.

1.18 For each of the seven devices in the circuit of Fig. P1.18, determine whether the device is a supplier or a recipient of power and how much power it is supplying or receiving.

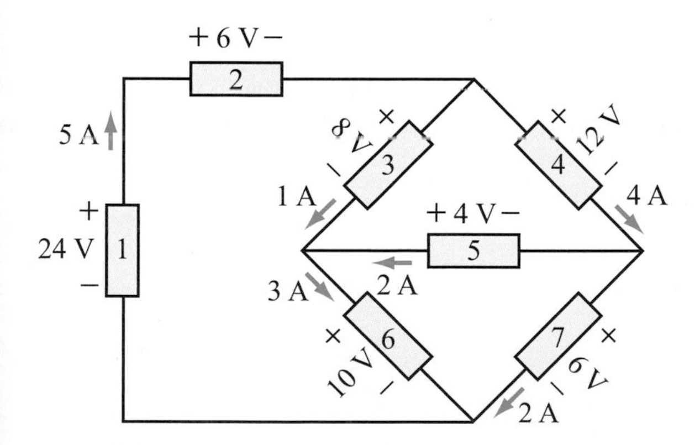

Figure P1.18: Circuit for Problem 1.18.

1.19 An electric oven operates at 120 V. If its power rating is 0.6 kW, what amount of current does it draw, and how much energy does it consume in 12 minutes of operation?

1.20 A 9-V flashlight battery has a rating of 1.8 kWh. If the bulb draws a current of 100 mA when lit; determine the following:

(a) For how long will the flashlight provide illumination?

(b) How much energy in joules is contained in the battery?

(c) What is the battery's rating in ampere-hours?

1.21 The voltage across and current through a certain device are given by

$$v(t) = 5\cos(4\pi t) \text{ V}, \qquad i(t) = 0.1\cos(4\pi t) \text{ A}.$$

Determine:

*__(a)__ The instantaneous power $p(t)$ at $t = 0$ and $t = 0.25$ s.

(b) The average power p_{av}, defined as the average value of $p(t)$ over a full time period of the cosine function (0 to 0.5 s).

1.22 The voltage across and current through a certain device are given by

$$v(t) = 100(1 - e^{-0.2t}) \text{ V}, \qquad i(t) = 30e^{-0.2t} \text{ mA}.$$

Determine:

(a) The instantaneous power $p(t)$ at $t = 0$ and $t = 3$ s.

(b) The cumulative energy delivered to the device from $t = 0$ to $t = \infty$.

1.23 The voltage across a device and the current through it are shown graphically in Fig. P1.23. Sketch the corresponding power delivered to the device and calculate the energy absorbed by it.

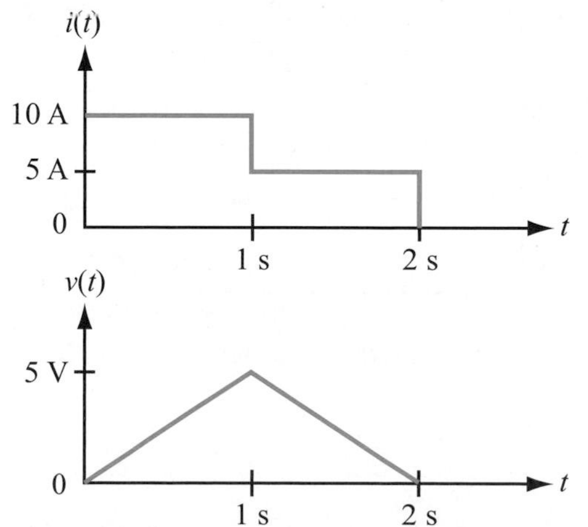

Figure P1.23: $i(t)$ and $v(t)$ of the device in Problem 1.23.

1.24 The voltage across a device and the current through it are shown graphically in Fig. P1.24. Sketch the corresponding power delivered to the device and calculate the energy absorbed by it.

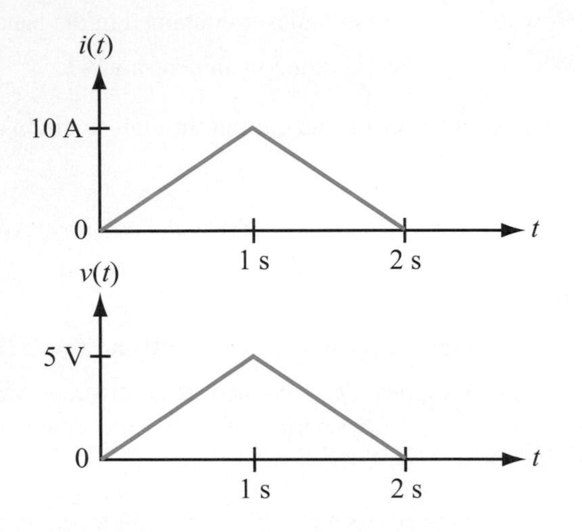

Figure P1.24: $i(t)$ and $v(t)$ of the device in Problem 1.24.

1.25 For the circuit in Fig. P1.25, generate circuit diagrams that include only those elements that have current flowing through them for

(a) $t < 0$

(b) $0 < t < 2$ s

(c) $t > 2$ s

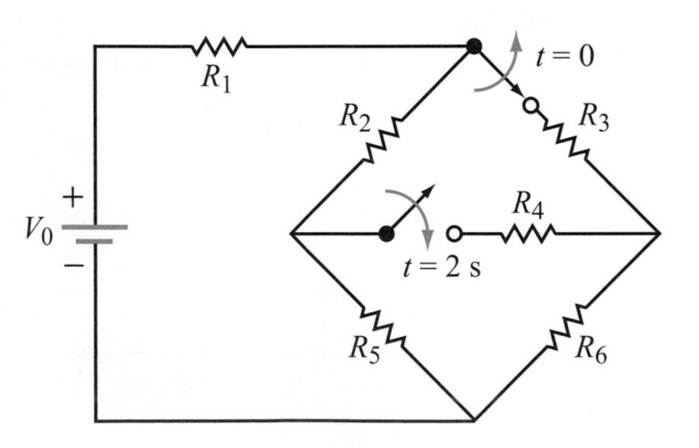

Figure P1.25: Circuit for Problem 1.25.

1.26 For the circuit in Fig. P1.26, generate circuit diagrams that include only those elements that have current flowing through them for

(a) $t < 0$

(b) $0 < t < 2$ s

(c) $t > 2$ s

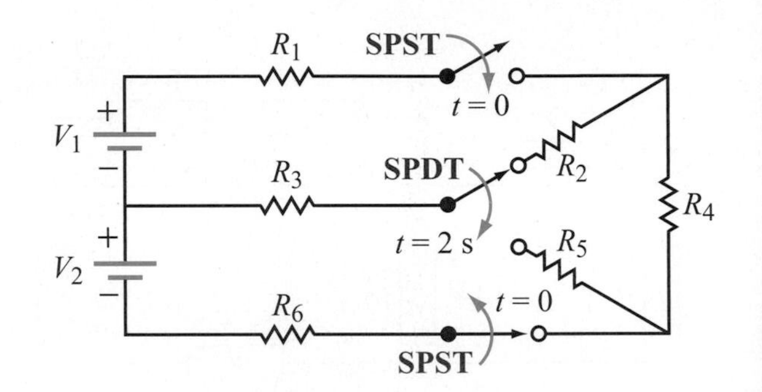

Figure P1.26: Circuit for Problem 1.26.

CHAPTER 2

Resistive Circuits

Chapter Contents

Objectives

Upon learning the material presented in this chapter, you should be able to:

1. Apply Ohm's law and explain the basic properties of piezoresistivity and superconductivity.
2. Define terms used in circuit topology, such as node, extraordinary node, loop, path, and mesh.
3. State Kirchhoff's current and voltage laws; apply them to resistive circuits.
4. Define what is meant when two circuits are said to be "equivalent."
5. Combine resistors in series and in parallel; apply voltage and current division.
6. Apply source transformation between voltage and current sources.
7. Apply Y–Δ transformations.
8. Describe the operation of the Wheatstone-bridge circuit and how it is used to measure small deviations.
9. Use Multisim to analyze simple circuits.

Overview

The study of any field of inquiry starts with nomenclature: defining the terms specific to that field. That is exactly what we did in the preceding chapter. We introduced and defined electric current, voltage, power, open and closed circuits, and dependent and independent voltage and current sources, among others. Now, we are ready to acquire our first set of circuit-analysis tools, which will enable us to analyze a variety of different types of circuits. We will limit our discussion to resistive circuits, namely those circuits containing only sources and resistors. (In future chapters, we will extend those tools to circuits containing capacitors, inductors, and other elements.) Our new toolbox will include three simple, yet powerful laws—Ohm's law and Kirchhoff's voltage and current laws—and several circuit simplification and transformation techniques.

2-1 Ohm's Law

The ***conductivity*** σ of a material is a measure of how easily electrons can drift through the material when an external voltage is applied across it. Materials are classified as ***conductors*** (primarily metals), ***semiconductors***, or ***dielectrics*** (insulators) according to the magnitudes of their conductivities. Tabulated values of σ expressed in units of siemens per meter (S/m) are given in Table 2-1 for a select group of materials. The siemen is the inverse of the ohm, $S = 1/\Omega$, and the inverse of σ is called the ***resistivity*** ρ,

$$\rho = \frac{1}{\sigma} \quad (\Omega\text{-m}), \tag{2.1}$$

which is a measure of how well a material ***impedes*** the flow of current through it. The conductivity of most metals is on the order of 10^7 S/m, which is 20 or more orders of magnitude greater than the conductivity of typical insulators. Common semiconductors, such as silicon and germanium, fall in the in-between range on the conductivity scale.

The values of σ and ρ given in Table 2-1 are specific to room temperature at 20°C. In general, the conductivity of a metal increases with decreasing temperature. At very low temperatures (in the neighborhood of absolute zero), some conductors become ***superconductors***, because their conductivities become practically infinite and their corresponding resistivities approach zero. To learn more about superconductivity, refer to Technology Brief 3 on page 38.

Table 2-1: Conductivity and resistivity of some common materials at 20°C.

Material	Conductivity σ (S/m)	Resistivity ρ (Ω-m)
Conductors		
Silver	6.17×10^7	1.62×10^{-8}
Copper	5.81×10^7	1.72×10^{-8}
Gold	4.10×10^7	2.44×10^{-8}
Aluminum	3.82×10^7	2.62×10^{-8}
Iron	1.03×10^7	9.71×10^{-8}
Mercury (liquid)	1.04×10^6	9.58×10^{-8}
Semiconductors		
Carbon (graphite)	7.14×10^4	1.40×10^{-5}
Pure germanium	2.13	0.47
Pure silicon	4.35×10^{-4}	2.30×10^3
Insulators		
Paper	$\sim 10^{-10}$	$\sim 10^{10}$
Glass	$\sim 10^{-12}$	$\sim 10^{12}$
Teflon	$\sim 3.3 \times 10^{-13}$	$\sim 3 \times 10^{12}$
Porcelain	$\sim 10^{-14}$	$\sim 10^{14}$
Mica	$\sim 10^{-15}$	$\sim 10^{15}$
Polystyrene	$\sim 10^{-16}$	$\sim 10^{16}$
Fused quartz	$\sim 10^{-17}$	$\sim 10^{17}$

2-1.1 Resistance

The ***resistance*** R of a device incorporates two factors: (a) the inherent bulk property of its material to conduct (or impede) current, represented by the conductivity σ (or resistivity ρ), and (b) the shape and size of the device. For a longitudinal resistor (Fig. 2-1), R is given by

$$R = \frac{\ell}{\sigma A} = \rho \frac{\ell}{A} \quad (\Omega), \tag{2.2}$$

where ℓ is the length of the device and A is its cross-sectional area. In addition to its direct dependence on the resistivity ρ, R is directly proportional to ℓ, which is the length of the path that the current has to flow through, and inversely proportional

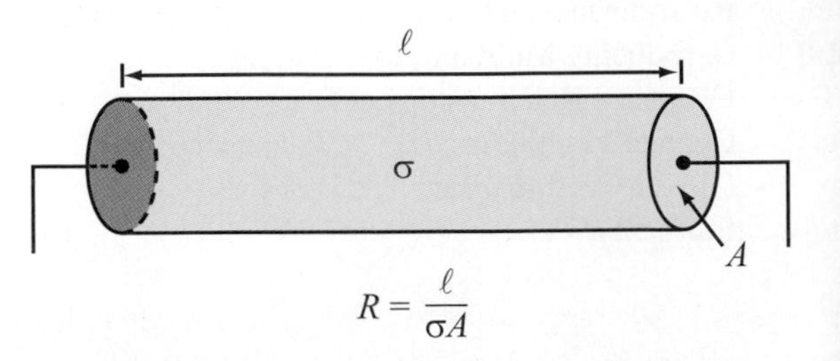

Figure 2-1: Longitudinal resistor of conductivity σ, length ℓ, and cross-sectional area A.

Table 2-2: Diameter d of wires, according to the American Wire Gauge (AWG) system.

AWG Size Designation	Diameter d (mm)
0	8.3
2	6.5
4	5.2
6	4.1
10	2.6
14	1.6
18	1.0
20	0.8

to A, because the larger A is, the easier it is for the electrons to drift through the material.

Every element of an electric circuit has a certain resistance associated with it. This even includes the wires used to connect devices to each other, but we usually treat them like zero-resistance segments because their resistances are so much smaller than the resistances of the other devices in the circuit. To illustrate with an example, let us consider a 10-cm-long segment of one of the wire sizes commonly found in circuit boards, such as the AWG-18 copper wire. According to Table 2-2, which lists the diameter d for various wire sizes as specified by the American Wire Gauge (AWG) system, the AWG-18 wire has a diameter $d = 1$ mm. Using the values specified for ℓ and d and the value for ρ of copper given in Table 2-1, we have

$$\begin{aligned} R &= \rho \frac{\ell}{A} = \rho \frac{\ell}{\pi (d/2)^2} \\ &= 1.72 \times 10^{-8} \times \frac{0.1}{\pi (0.5 \times 10^{-3})^2} \\ &= 2.2 \times 10^{-3}\ \Omega \\ &= 2.2\ \text{m}\Omega. \end{aligned}$$

Thus, R is on the order of milliohms. If the wire segment connects to circuit elements with resistances of ohms or larger, ignoring the resistance of the wire would have no significant impact on the overall behavior of the circuit.

The preceding justification should be treated with some degree of caution. While it is true that a piece of wire may be treated like a short circuit in the majority of circuit configurations, there are certain situations for which such an assumption may not be valid. One obvious example is when the wire is very long, as in the case of a multi-kilometer long electric power-transmission cable. Another is when very thin wires or channels with micron-size diameters are used in microfabricated circuits.

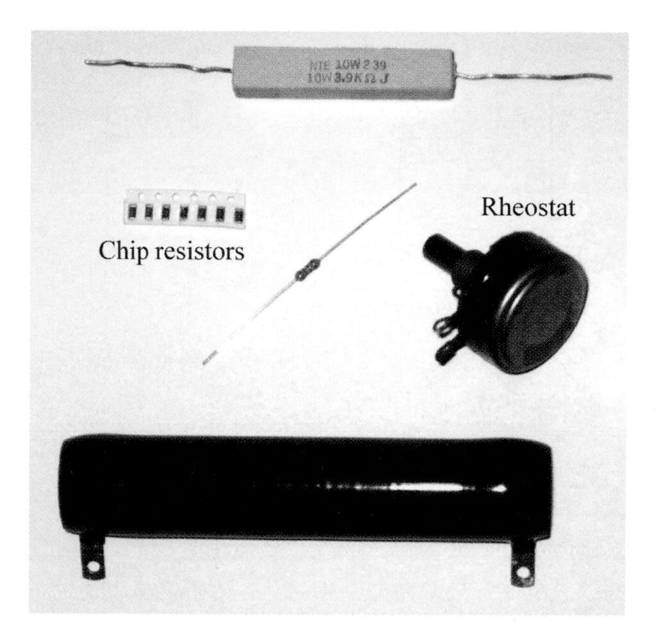

Figure 2-2: Photograph of various types of resistors.

Resistive elements used in electronic circuits are fabricated in many different sizes and shapes to suit the intended application and requisite circuit architecture. Discrete resistors usually are cylindrical in shape and made of a carbon composite, examples of which are shown in Fig. 2-2. Hybrid and miniaturized circuits use film-shaped metal or carbon resistors. In integrated circuits, resistive elements are fabricated through a diffusion process (see Technology Brief 7 on page 135).

For some metal oxides, the resistivity ρ exhibits a strong sensitivity to temperature. A resistor manufactured of such materials is called a ***thermistor*** (Table 2-3), and it is used for temperature measurement, temperature compensation, and related applications. Another interesting type of resistor is the ***piezoresistor***, which is used as a pressure sensor in many household appliances, automotive systems, and biomedical devices. More coverage on piezoresistivity is available in Technology Brief 4 on page 56.

Certain applications, such as volume adjustment on a radio, may call for the use of a resistor with ***variable resistance***. The rheostat and the potentiometer are two standard types of

Table 2-3: Common resistor terminology.

Thermistor	R sensitive to temperature
Piezoresistor	R sensitive to pressure
Rheostat	2-terminal variable resistor
Potentiometer	3-terminal variable resistor

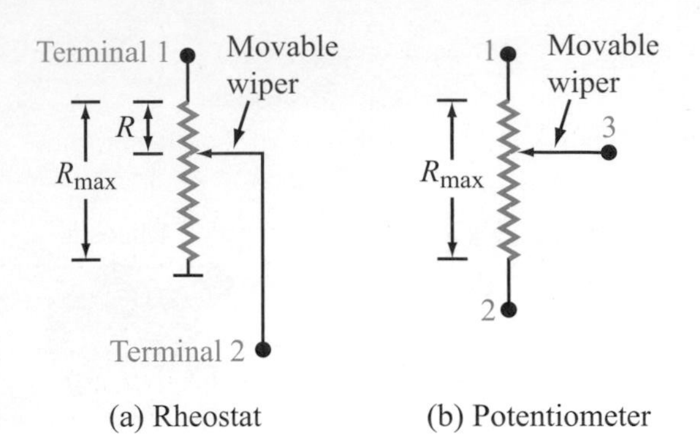

Figure 2-3: (a) A rheostat is used to set the resistance between terminals 1 and 2 at any value between zero and R_{max}; (b) the wiper in a potentiometer divides the resistance R_{max} among R_{13} and R_{23}.

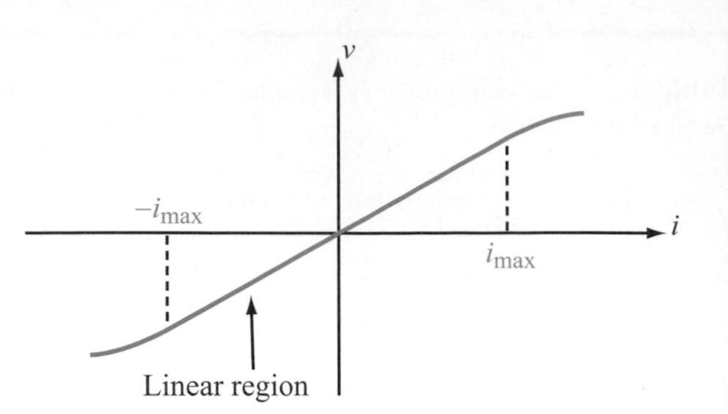

Figure 2-4: i–v response of a typical resistor includes a linear region extending between $-i_{max}$ and i_{max}.

variable resistors in common use. The ***rheostat*** (Fig. 2-3(a)) is a two-terminal device with one of its terminals connected to one end of a resistive track and the other terminal connected to a movable wiper. Movement of the wiper across the resistive track, through rotation of a shaft, can change the resistance between the two terminals from (theoretically) zero resistance to the full resistance value of the track. Thus, if the total resistance of the track is R_{max}, the rheostat can provide any resistance between zero and R_{max}.

The ***potentiometer*** is a three-terminal device. Terminals 1 and 2 in Fig. 2-3(b)) are connected to the two ends of the track (with total resistance R_{max}) and terminal 3 is connected to a movable wiper. When terminal 3 is at the end next to terminal 1, the resistance between terminals 1 and 3 is zero and that between terminals 2 and 3 is R_{max}. Moving terminal 3 away from terminal 1 increases the resistance between terminals 1 and 3 and decreases the resistance between terminals 2 and 3.

2-1.2 i–v Characteristic

Based on the results of his experiments on the nature of conduction in circuits, German physicist Georg Simon Ohm (1787–1854) formulated in 1826 the i–v relationship for a resistor, which has become known as ***Ohm's law***. He discovered that the voltage v across a resistor is directly proportional to the current i flowing through it, namely

$$v = iR, \tag{2.3}$$

with the resistance R being the proportionality factor.

> In compliance with the passive sign convention, the polarity of v is such that the current enters the resistor at the "+" side.

An ideal ***linear resistor*** is one whose resistance R is constant and independent of the magnitude of the current flowing through it, in which case its ***i–v response*** is a straight line. In practice, the i–v response of a real linear resistor is indeed approximately linear, as illustrated in Fig. 2-4, so long as i remains within the ***linear region*** defined by $-i_{max}$ to i_{max}. Outside this range, the response deviates from the straight-line model. When we use Ohm's law as expressed by Eq. (2.3), we tacitly assume that the resistor is being used in its linear range of operation.

Some resistive devices exhibit highly nonlinear i–v characteristics. These include diode elements and light-bulb filaments, among others. Unless noted otherwise, the common use of the term *resistor* in circuit analysis and design usually refers to the linear resistor exclusively.

The flow of current in a resistor leads to power dissipation in the form of heat (or the combination of heat and light in the case of a light bulb's filament). Using Eq. (2.3) in Eq. (1.9) provides the following expression for the power p dissipated in a resistor:

$$p = iv = i^2R \quad \text{(W)}. \tag{2.4}$$

The ***power rating*** of a resistor defines the maximum continuous power level that the resistor can dissipate without getting damaged. Excessive heat can cause melting, smoke, and even fire.

Example 2-1: dc Motor

A 12-V car battery is connected via a 6-m-long, twin-wire cable to a dc motor that drives the wiper blade on the rear window. The cable is copper AWG-10 and the motor exhibits to the rest of the circuit an equivalent resistance $R_m = 2\ \Omega$. Determine (a) the resistance of the cable and (b) the fraction of the power contributed by the battery that gets delivered to the motor.

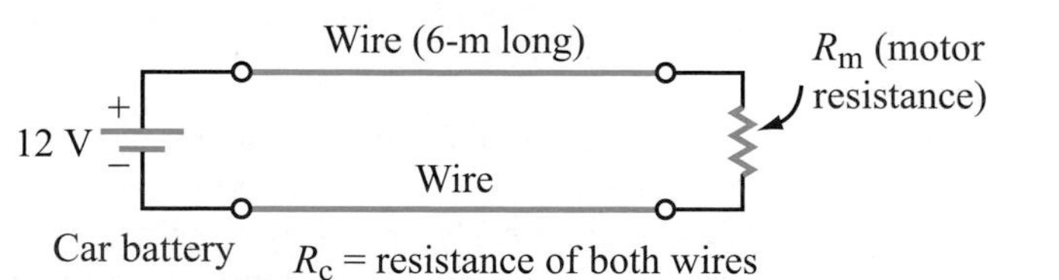

Figure 2-5: Circuit for Example 2-1.

Solution: The circuit described in the problem statement is represented by Fig. 2-5.

(a) With $\ell = 12$ m (total for twin wires), $\rho = 1.72 \times 10^{-8}$ Ω-m for copper, $A = \pi(d/2)^2$, and $d = 2.6$ mm for AWG-10, the cable resistance R_c is

$$\begin{aligned} R_c &= \rho \frac{\ell}{A} \\ &= 1.72 \times 10^{-8} \times \frac{12}{\pi(1.3 \times 10^{-3})^2} \\ &= 0.04\ \Omega. \end{aligned}$$

(b) The total resistance in the circuit is equal to the sum of the cable and motor resistances. [In a later section, we will learn that the resistance of two resistors connected in series is simply equal to the sum of their resistances.] Hence,

$$\begin{aligned} R &= R_c + R_m \\ &= 0.04 + 2 \\ &= 2.04\ \Omega. \end{aligned}$$

Consequently, the current flowing through the circuit is

$$\begin{aligned} I &= \frac{V}{R} \\ &= \frac{12}{2.04} \\ &= 5.88\ \text{A}, \end{aligned}$$

and the power contributed by the battery P and the power delivered to the motor P_m are:

$$\begin{aligned} P &= IV \\ &= 5.88 \times 12 \\ &= 70.56\ \text{W} \end{aligned}$$

and

$$\begin{aligned} P_m &= I^2 R_m \\ &= (5.88)^2 \times 2 \\ &= 69.15\ \text{W}, \end{aligned}$$

and the fraction of P delivered to the load (motor) is

$$\text{Fraction} = \frac{P_m}{P} = \frac{69.15}{70.56} = 0.98 \text{ or } 98 \text{ percent.}$$

Thus, 2 percent of the power is dissipated in the cable.

Review Question 2-1: How does the magnitude of the conductivity of a metal, such as copper, compare with that of a typical insulator, such as mica? What is a superconductor, and why might it be useful?

Review Question 2-2: What is piezoresistivity, and how is it used?

Review Question 2-3: What is meant by the *linear region* of a resistor? Is it related to its power rating?

Exercise 2-1: A cylindrical resistor made of carbon has a power rating of 2 W. If its length is 10 cm and its circular cross section has a diameter of 1 mm, what is the maximum current that can flow through the resistor without damaging it?

Answer: 1.06 A. (See CD)

Exercise 2-2: A rectangular bar made of aluminum has a current of 3 A flowing through it along its length. If its length is 2.5 m and its square cross section has 1-cm sides, how much power is dissipated in the bar at 20°C?

Answer: 5.9 mW. (See CD)

Exercise 2-3: A certain type of diode exhibits a nonlinear relationship between v—the voltage across it—and i—the current entering into its (+) voltage terminal. Over its operational voltage range (0 to 1 V), the current is given by

$$i = 0.5v^2 \qquad \text{for } 0 \le v \le 1\ \text{V}.$$

Determine how the diode's effective resistance varies with v and calculate its value at $v = 0$, 0.01 V, 0.1 V, 0.5 V, and 1 V.

Answer: $R = \dfrac{2}{v}$,

v	R
0	∞
0.01 V	200 Ω
0.1 V	20 Ω
0.5 V	4 Ω
1 V	2 Ω

(See CD)

Technology Brief 3: Superconductivity

When an electric voltage is applied across two points in a conductor, such as copper or silver, current flows between them. The relationship between the voltage difference V and the current I is given by Ohm's law, $V = IR$, where R is the resistance of the conducting material between the two points. It is helpful to visualize the electric current as a fluid of electrons flowing through a dense forest of sturdy metal atoms, called the ***lattice***. Under the influence of the electric field (induced by the applied voltage), the electrons can attain very high instantaneous velocities, but their overall forward progress is impeded by the frequent collisions with the lattice atoms. Every time an electron collides and bounces off an atom, some of that electron's kinetic energy is transferred to the atom, causing the atom to vibrate---which heats up the material---and causing the electron to slow down. The resistance R is a measure of how much of an obstacle the resistor poses to the flow of current, as well as a measure of how much heat it generates for a given current.

Can a conductor ever have zero resistance? The answer is most definitely yes! In 1911, the Dutch physicist Heike Kamerlingh Onnes developed a refrigeration technique so powerful that it could cool helium down low enough to condense into liquid form at 4.2 K (0 kelvin $= -373^\circ$C). Into his new liquid helium container, he immersed (among other things) mercury; he soon discovered that the resistance of a solid piece of mercury at 4.2 K was *zero*! The phenomenon, which was completely unexpected and not predicted by classical physics, was coined ***superconductivity***. According to quantum physics, many materials experience an abrupt change in behavior (called a ***phase transition***) when cooled below a certain ***critical temperature*** T_C.

Superconductors have some amazing properties. The current in a superconductor can persist with no external voltage applied. Even more interesting, currents have been observed to persist in superconductors for many years without decaying; in fact, some theoretical estimates predict that superconductor currents can persist for periods longer than the estimated lifetime of the universe! When a magnet is brought close to the surface of a superconductor, the currents induced by the magnetic field are mirrored exactly by the superconductor (because the superconductor's resistance is zero) and the magnet is repelled (Fig. TF3-1). This property has been used to demonstrate magnetic

Figure TF3-1: The Meissner effect, or strong diamagnetism, seen between a high-temperature superconductor and a rare earth magnet. (Courtesy of Pacific Northwest National Laboratory.)

levitation and is the basis of some super-fast ***maglev trains*** (Fig. TF3-2) being developed around the world. The same phenomenon is used in the ***Magnetic Resonance Imaging (MRI)*** machines that hospitals use to perform 3-D scans of organs and tissues (Fig. TF3-3) and in ***Superconducting Quantum Interference Devices (SQUIDs)*** to examine brain activity at high resolution.

Superconductivity is one of the last frontiers in solid-state physics. Even though the physics of low-temperature superconductors (like mercury, lead, niobium nitride, and others) is now fairly well understood, a different class of ***high-temperature superconductors*** still defies complete theoretical explanation. This class of materials was discovered in 1986 when Alex Müller and Georg Bednorz, at IBM Research Laboratory in Switzerland, created a ceramic compound that superconducted at 30 K. This discovery was followed by the discovery of other ceramics with even higher T_C values; the now-famous YBCO ceramic discovered at the University of Alabama-Huntsville (1987) has a T_C of 92 K, and the world record holder is a group of mercury-cuprate compounds with a T_C of 138 K (1993). New superconducting materials and conditions are still being found; carbon nanotubes, for example, were recently shown to have a T_C of 15 K (Hong Kong University, 2001). Are there higher-temperature superconductors? What theory will explain this higher-temperature phenomenon? Can so-called *room-temperature* superconductors exist? For engineers (like you) the challenges are just beginning: How can these materials be made into useful circuits, devices, and machines? What new designs will emerge? The race is on!

Figure TF3-2: Maglev train. (Courtesy of Central Japan Railway Company.)

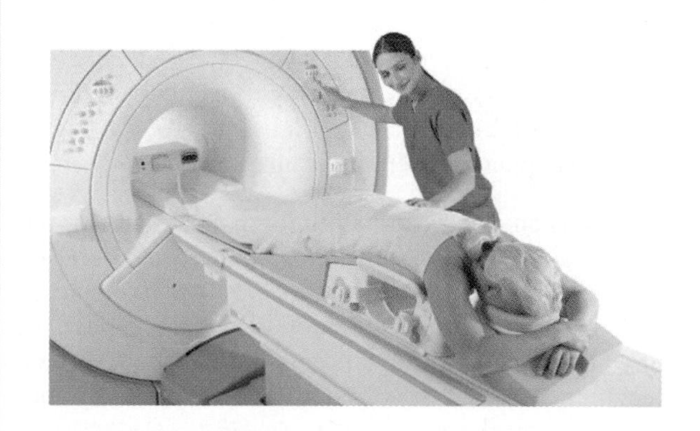

Figure TF3-3: Magnetic Resonance Imaging machine. (Courtesy GE Healthcare.)

2-1.3 Conductance

The reciprocal of resistance is called ***conductance***,

$$G = \frac{1}{R} \quad \text{(S)}, \tag{2.5}$$

and its unit is the siemen (S). In terms of G, Ohm's law can be rewritten in the form

$$i = \frac{v}{R} = Gv, \tag{2.6}$$

and the expression for power becomes

$$p = iv = Gv^2 \quad \text{(W)}. \tag{2.7}$$

2-2 Circuit Topology

The term ***topology*** refers to a branch of mathematics concerned with the properties of geometrical configurations. The study of the geometrical properties of electrical circuits is the subject of circuit topology, also called ***network topology***. The logical approach for introducing any new topic is to start by providing clear and concise definitions for the particular set of terms constituting its vocabulary. The vocabulary of circuit topology includes terms like node, branch, and loop, as well as other derivatives. Before we proceed with describing these terms and the relationships between them, we should note that our treatment will be limited to planar circuits. A ***planar circuit*** is defined as a circuit that can be drawn schematically in two-dimensional space such that no two branches have to cross one another. Greater elaboration of what that means is given in Section 2-2.3.

2-2.1 Nomenclature

A reference summary of key terms used in circuit topology is available in Table 2-4. The planar circuit shown in Fig. 2-6 shall serve to explain the definitions in more detail.

- A ***node*** is an electrical connection point for two or more elements (devices). An ***ordinary node*** connects to only two elements, whereas an ***extraordinary node*** connects to three or more elements. Figure 2-6 contains seven nodes, of which N_1, N_3, N_5, and N_6 are ordinary nodes, and the remaining three (N_2, N_4, and N_7) are extraordinary nodes.
- A ***branch*** is the trace between two consecutive nodes containing only one element between them. That is, a branch contains one and only one element. Figure 2-6 includes nine elements; hence, it includes nine branches.

Table 2-4: Circuit terminology.

Term	Definition
Ordinary node	An electrical connection point that connects to only two elements.
Extraordinary node	An electrical connection point that connects to three or more elements.
Branch	Trace between two consecutive nodes with only one element between them.
Path	Continuous sequence of branches with no node encountered more than once.
Extraordinary path	Path between two adjacent extraordinary nodes.
Loop	Closed path with the same start and end node.
Independent loop	Loop containing one or more branches not contained in any other independent loop.
Mesh	Loop that encloses no other loops.
In-series	Elements that share the same current.
In-parallel	Elements that share the same voltage.

- A ***path*** is any continuous sequence of branches, provided that no one node is encountered more than once. A simple example is path N_1 to N_2 through R_1. Alternative paths between the same two nodes include: v_s–C_1–R_2; v_s–R_5–C_2–L; and v_s–R_5–R_4–R_3.

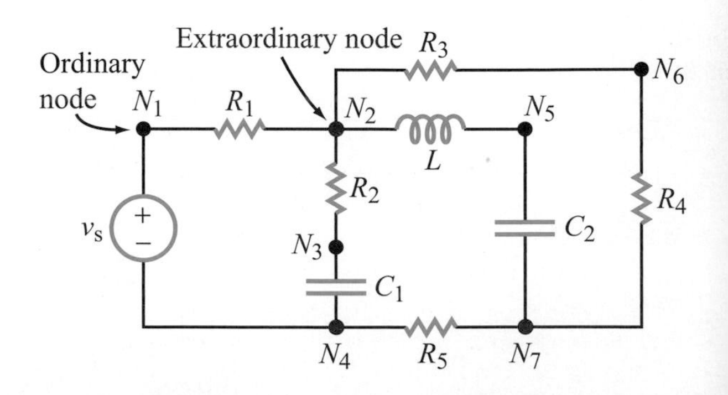

Figure 2-6: Circuit with seven nodes, of which three are extraordinary nodes (N_2, N_4, and N_7).

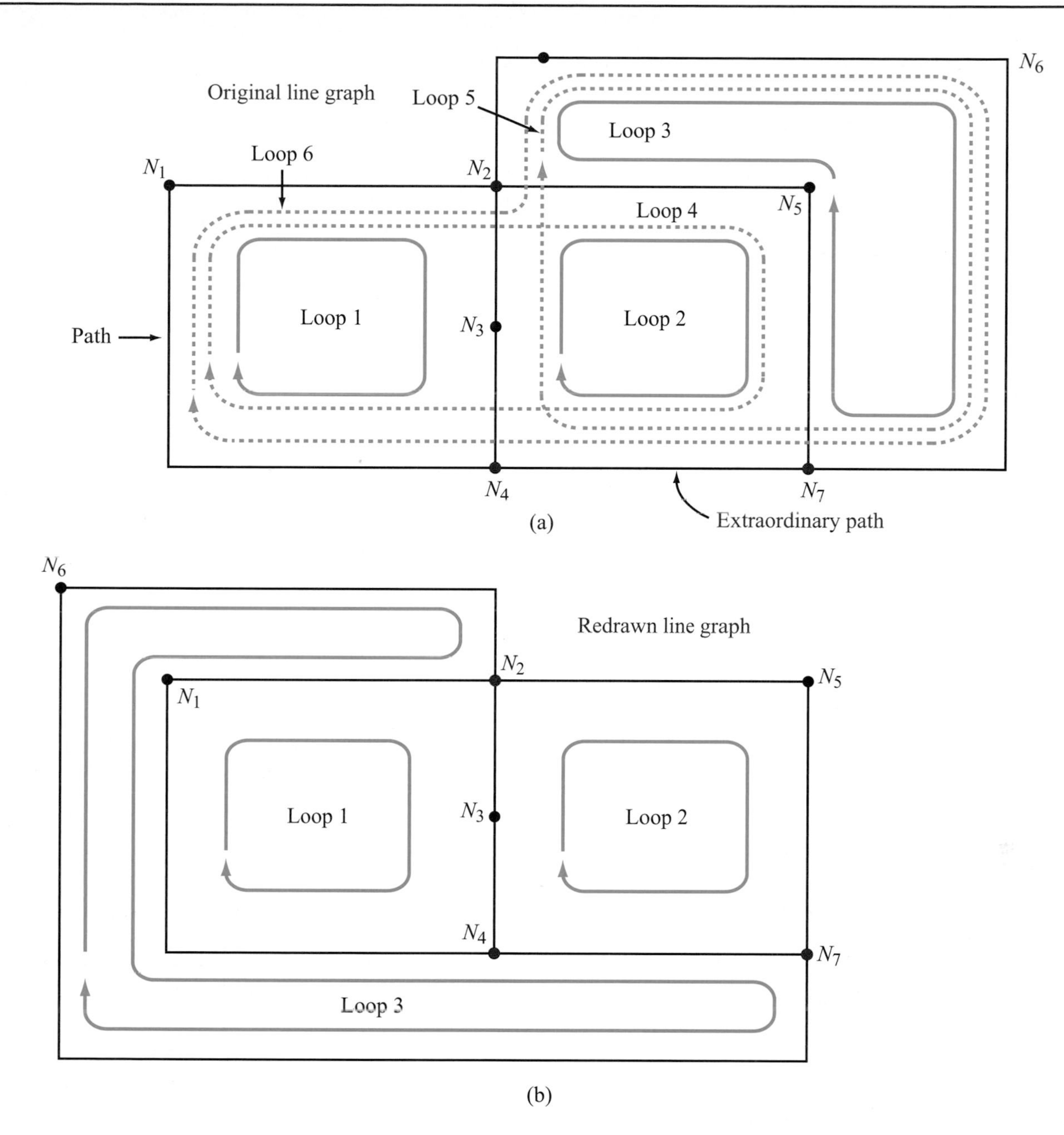

Figure 2-7: (a) Linear graph (circuit with suppressed circuit elements) containing six loops, of which three are independent; (b) redrawing the circuit changes the loops but does not change the circuit.

- An ***extraordinary path*** is a path between two extraordinary nodes, provided that no other extraordinary node exists along that path. The extraordinary nodes in Fig. 2-6 are nodes N_2, N_4, and N_7. Hence, the extraordinary paths in the circuit are: (1) N_2 to N_4 through R_2–C_1, (2) N_2 to N_4 through R_1–υ_s, (3) N_4 to N_7 through R_5, (4) N_2 to N_7 through L–C_2, and (5) N_2 to N_7 through R_3–R_4.

- A ***loop*** is a closed path such that the start and end node is one and the same. The drawing shown in Fig. 2-7(a) is a ***linear graph*** of the circuit in Fig. 2-6, wherein we have suppressed the circuit elements, retained all of the nodes, and replaced the elements with simple lines. The linear graph, which is a useful tool for demonstrating certain concepts without getting distracted with the specifics of the elements in the circuit, shows that the circuit contains six loops. In clockwise direction, these are: (1) Loop 1: N_1–N_2–N_3–N_4–N_1, (2) Loop 2: N_2–N_5–N_7–N_4–N_3–N_2, (3) Loop 3: N_2–N_6–N_7–N_5–N_2, (4) Loop 4: N_1–N_2–N_5–N_7–N_4–N_1, (5) Loop 5: N_2–N_6–N_7–N_4–N_3–N_2, and (6) Loop 6: N_1–N_2–N_6–N_7–N_4–N_1.

- An ***independent loop*** is a loop that contains (at least) one or more branches that are not part of any other independent loop. A circuit may contain several combinations of independent loops. Loops 1, 2, and 3 in Fig. 2-7(a) are indeed all independent loops, because each contains at least one branch that is not contained in the other two. Moreover, because those three loops contain all of the branches in the circuit, none of loops 4, 5, and 6 is an independent loop. Alternative group choices of independent loops include: (1) loops 1, 3, and 4, (2) loops 1, 2, and 5, (3) loops 1, 2, and 6, and so on, but in all cases, the total number of independent loops characterizing the circuit is exactly three.

- A ***mesh*** is a loop that encloses no other loop. Hence, for the specific way in which the circuit was drawn in Fig. 2-7(a), the only loops that are also meshes are loops 1, 2, and 3.

A planar circuit may be drawn in many different ways, as a result of which the loops and meshes may be different for the different representations. To illustrate with an example, the linear graph shown in Fig. 2-7(b) is a different (but equally acceptable) representation of the circuit in Fig. 2-6. Comparison of the linear graph in part (a) with that in part (b) leads to the conclusion that loops 1 and 2 remain unchanged (in terms of the elements they contain) but loops 3 of the two representations do not contain the same elements. Nevertheless, in both representations, the circuit contains exactly three meshes.

For any planar circuit or network composed of discrete elements, the ***number of branches b*** is related to the ***number of nodes n*** and the ***number of independent loops*** ℓ_{ind} by the following theorem:

$$b = n + \ell_{\text{ind}} - 1. \tag{2.8}$$

In the case of the circuit in Fig. 2-6, $n = 7$, $\ell_{\text{ind}} = 3$, and $b = 9$.

A similar expression applies to the relationship between the numbers of ***extraordinary nodes*** n_{ex}, ***extraordinary paths*** p_{ex}, and ***independent loops*** ℓ_{ind}. Namely,

$$p_{\text{ex}} = n_{\text{ex}} + \ell_{\text{ind}} - 1. \tag{2.9}$$

As a check, Fig. 2-6 has three extraordinary nodes, so with $\ell_{\text{ind}} = 3$ we have $p_{\text{ex}} = 5$. The significance and use of Eqs. (2.8) and (2.9) will become evident later when we apply node and loop analysis to solve electric circuits.

2-2.2 In-Series and In-Parallel Connections

The terms ***in-series*** and ***in-parallel*** connections find extensive use in circuit analysis, deserving of specific note. *Two or more devices are said to be connected in series if the same current flows through all of them*, requiring that all nodes along the path containing the in-series elements be ordinary nodes. *Multiple elements connected in parallel share the same pair of nodes, thereby having the same voltage across them.* In Fig. 2-6, the circuit contains four combinations of in-series connections, namely: υ_s–R_1, R_2–C_1, L–C_2, and R_3–R_4. The in-series combination υ_s–R_1 is in parallel with the combination C_1–R_2. Similarly, R_3–R_4 is in parallel with L–C_2.

Exercise 2-4: Which elements in the circuit of Fig. E2.4 are connected (a) in-series or (b) in-parallel?

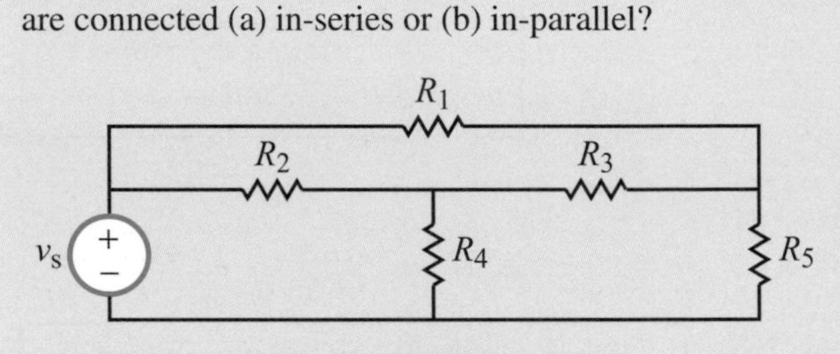

Figure E2.4

Answer: (a) none, (b) none. (See)

Exercise 2-5: The switch in the circuit of Fig. E2.5 closes at $t = 0$. Which elements are in-series and which are in-parallel at (a) $t < 0$ and (b) $t > 0$?

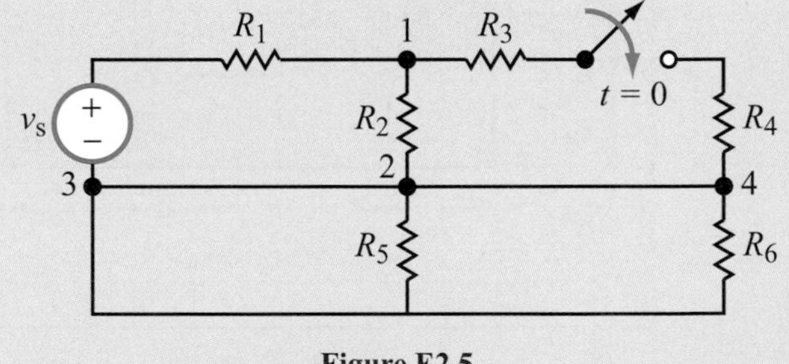

Figure E2.5

Answer: (a) υ_s, R_1, and R_2 are all in series; R_5 and R_6 are in parallel with each other and with the short circuit between nodes 2 and 3; (b) υ_s and R_1 are in series; R_3 and R_4 are in series and their combination is in parallel with R_2; R_5 and R_6 are in parallel with each other and with the short circuit between nodes 2 and 3. (See)

2-2.3 Planar Circuits

The relationships between branches, nodes, and loops expressed by Eqs. (2.8) and (2.9) are specific to planar circuits. In fact, the planar-circuit condition shall be presumed to be true throughout the material covered in this book. Hence, it is important to be clear on what a planar circuit is and what it is not:

A circuit is planar if it is possible to draw it on a two-dimensional plane without having any two of its branches cross over or under one another.

If such a crossing is unavoidable, then the circuit is ***nonplanar***. To clarify what we mean, we start by examining the circuit in Fig. 2-8(a). An initial examination of the circuit topology might suggest that the circuit is nonplanar because the branches containing resistors R_3 and R_4 appear to cross one another without having physical contact between them (absence of a solid dot at crossover point). However, if we redraw the branch containing R_4 on the outside, as shown in configuration (b) of Fig. 2-8, we would then conclude that the circuit is planar after all, and that is so because *it is possible* to draw it in a single plane without crossovers. In contrast, the circuit in Fig. 2-9 is indeed nonplanar because no matter how we might try to redraw it, it will always include at least one crossover of branches.

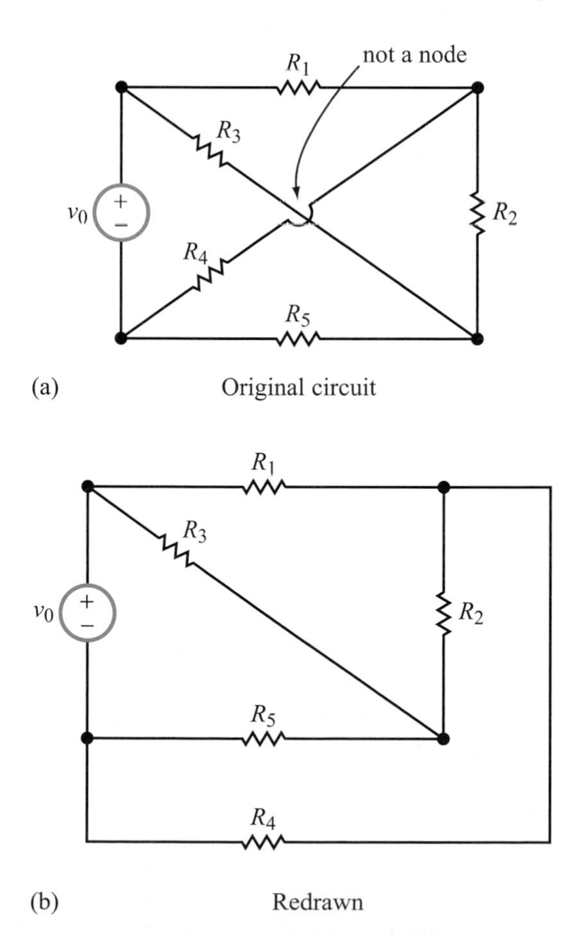

Figure 2-8: The branches containing R_3 and R_4 in (a) *appear* to cross over one another, but redrawing the circuit as in (b) avoids the crossover, thereby demonstrating that the circuit is planar.

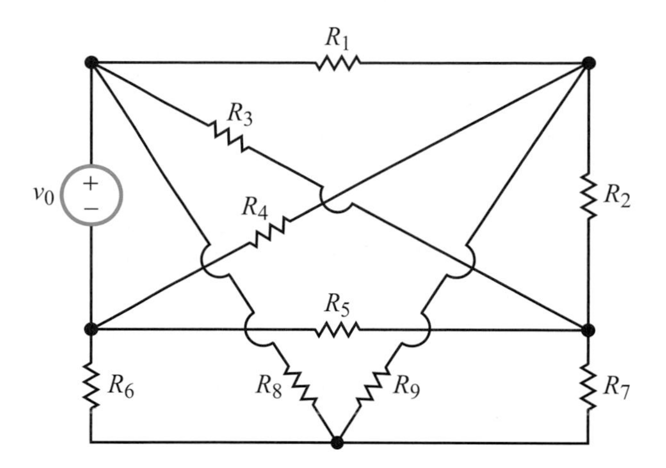

Figure 2-9: Nonplanar circuit.

Review Question 2-4: What is the definition of a planar circuit?

Review Question 2-5: "An extraordinary path is any path between two extraordinary nodes." This statement is not quite correct. Why not?

Review Question 2-6: What is the difference between a loop and a mesh?

Exercise 2-6: A network has six extraordinary paths and eight nodes, of which three are extraordinary. How many branches does it have?

Answer: $b = 11$. (See)

Exercise 2-7: Nodes 1, 2, and 3 are all extraordinary nodes. If resistor R_1 is connected between nodes 1 and 2, R_2 between nodes 1 and 3, and R_3 between nodes 2 and 3, which of the three resistors are connected in series and which are connected in parallel?

Answer: None, and none. (See)

2-3 Kirchhoff's Laws

Circuit theory—encompassing both analysis and synthesis—is built upon a foundation comprised of a small number of fundamental laws. Among the cornerstones are ***Kirchhoff's***

current and ***voltage laws***. Kirchhoff's laws, which constitute the subject of this section, were introduced by the German physicist Gustav Robert Kirchhoff (1824–1887) in 1847, some 21 years after a fellow German, Georg Simon Ohm, developed his famous law.

2-3.1 Kirchhoff's Current Law (KCL)

As defined earlier, a node is a connection point for two or more branches. As such, it is not a real circuit element, and therefore it cannot generate, store, or consume electric charge. This assertion, which follows from the ***law of conservation of charge***, forms the basis of ***Kirchhoff's current law (KCL)***, which states that:

> The algebraic sum of the currents entering a node must always be zero.

Mathematically, KCL can be expressed by the compact form:

$$\sum_{n=1}^{N} i_n = 0 \qquad \text{(KCL)}, \tag{2.10}$$

where N is the total number of branches connected to the node, and i_n is the nth current.

> Common convention is to assign a positive "+" sign to a current if it is entering the node and a negative "−" sign if it is leaving it.

For the node in Fig. 2-10,

$$i_1 - i_2 - i_3 + i_4 = 0, \tag{2.11}$$

where currents i_1 and i_4 were assigned positive signs because they are labeled in the figure as entering the node, and i_2 and i_3 were assigned negative signs because they are leaving the node.

Figure 2-10: Currents at a node.

> Alternatively, we can adopt the opposite convention, namely to assign a "+" to a current leaving the node and a "−" to a current entering it.

Either convention is equally valid so long as it is applied consistently to all currents entering and leaving the node.

By moving i_2 and i_3 to the right-hand side of Eq. (2.11), we obtain the alternative form of KCL, namely

$$i_1 + i_4 = i_2 + i_3, \tag{2.12}$$

which states that:

> The total current entering a node must be equal to the total current leaving it.

Example 2-2: KCL Equations

Write the KCL equations at nodes 1 through 5 in the circuit of Fig. 2-11.

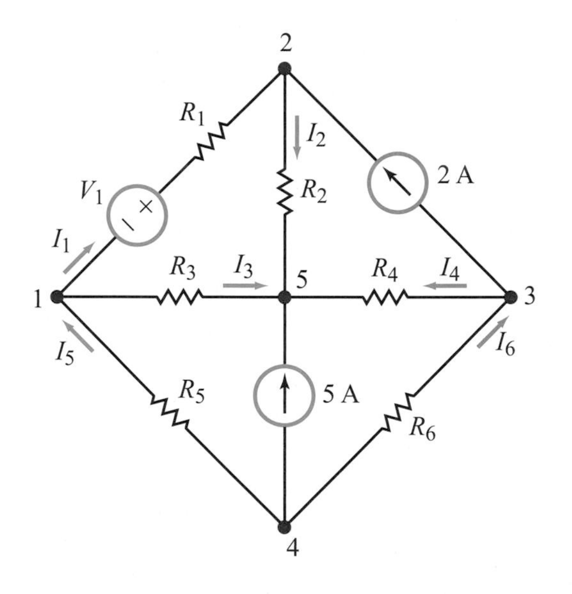

Figure 2-11: Circuit for Example 2-2.

Solution:

At node 1:	$-I_1 - I_3 + I_5 = 0$
At node 2:	$I_1 - I_2 + 2 = 0$
At node 3:	$-2 - I_4 + I_6 = 0$
At node 4:	$-5 - I_5 - I_6 = 0$
At node 5:	$I_3 + I_4 + I_2 + 5 = 0$

Example 2-3: Applying KCL

If V_4, the voltage across the 4–Ω resistor in Fig. 2-12, is 8 V, determine I_1 and I_2.

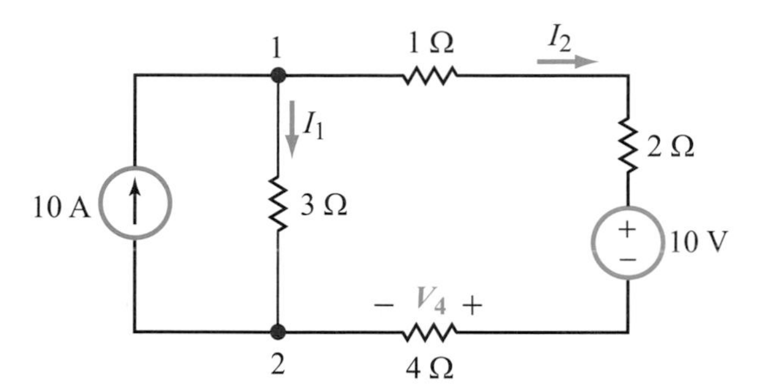

Figure 2-12: Circuit for Example 2-3.

Solution: Applying Ohm's law,

$$I_2 = \frac{V_4}{4} = \frac{8}{4} = 2 \text{ A}.$$

At node 1: $10 - I_1 - I_2 = 0$.

Hence,

$$I_1 = 10 - I_2 = 10 - 2 = 8 \text{ A}.$$

2-3.2 Kirchhoff's Voltage Law (KVL)

The voltage across an element represents the amount of energy expended in moving positive charge from the negative terminal to the positive terminal, thereby establishing a potential energy difference between those terminals. The ***law of conservation of energy*** mandates that if we move electric charge around a closed loop, starting and ending at exactly the same location, the net gain or loss of energy must be zero. Since voltage is a surrogate for potential energy:

The algebraic sum of the voltages around a closed loop must always be zero.

This statement defines ***Kirchhoff's voltage law (KVL)***. In equation form, KVL is given by

$$\sum_{n=1}^{N} v_n = 0 \qquad \text{(KVL)}, \tag{2.13}$$

where N is the total number of branches in the loop and v_n is the nth voltage across the nth branch. Application of Eq. (2.13) requires the specification of a sign convention to use with it. Of those used in circuit analysis, the sign convention we chose to use in this book consists of two steps.

Sign Convention

- Add up the voltages in a systematic clockwise movement around the loop.
- Assign a positive sign to the voltage across an element if the (+) side of that voltage is encountered first, and assign a negative sign if the (−) side is encountered first.

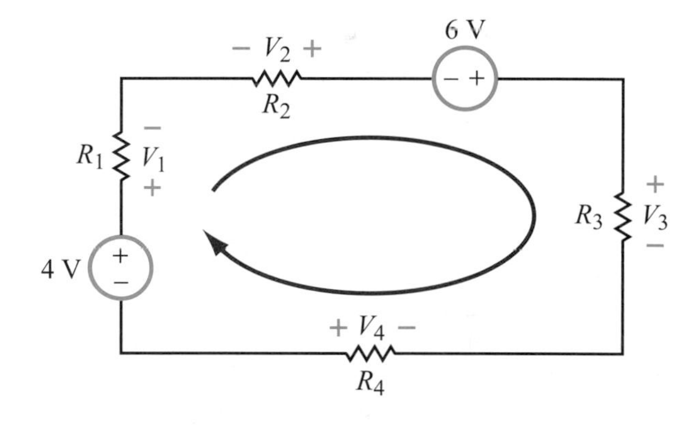

Figure 2-13: One-loop circuit.

Hence, for the loop in Fig. 2-13, starting at the negative terminal of the 4-V voltage source, application of Eq. (2.13) yields

$$-4 + V_1 - V_2 - 6 + V_3 - V_4 = 0. \tag{2.14}$$

An alternative statement of KVL is that *the total voltage rise around a closed loop must equal the total voltage drop around the loop*. Recalling that a voltage rise is realized by moving from the (−) voltage terminal to the (+) terminal across the element, and voltage drop is the converse of that, the clockwise movement around the loop in Fig. 2-13 gives

$$4 + V_2 + 6 + V_4 = V_1 + V_3, \tag{2.15}$$

which mathematically conveys the same information contained in Eq. (2.14).

Table 2-5 provides a summary of KCL and KVL statements.

Table 2-5: Equally valid, multiple statements of Kirchhoff's Current Law (KCL) and Kirchhoff's Voltage Law (KVL).

KCL	• Sum of all currents entering a node = 0 [i = "+" if entering; i = "−" if leaving] • Sum of all currents leaving a node = 0 [i = "+" if leaving; i = "−" if entering] • Total of currents entering = Total of currents leaving
KVL	• Sum of voltages around closed loop = 0 [v = "+" if + side encountered first in clockwise direction] • Total voltage rise = Total voltage drop

Example 2-4: Applying KCL and KVL Equations

For the circuit in Fig. 2-14(a), (a) identify all loops and write their KVL equations, and then (b) solve for the voltage across the 2-A current source.

Solution:

(a) The circuit contains three loops, as shown in Fig. 2-14(b). Before we proceed with writing KVL equations, we should label the voltages across all elements. It is also helpful to label the currents in the circuit and to make sure that their directions are consistent with the voltage polarities across passive elements (in a resistor, the current flows from the positive voltage terminal to the negative terminal). In terms of the voltages so labeled, the KVL equations are

$$\text{Loop 1:} \quad -50 + V_1 + V_2 + V_3 = 0, \tag{2.16}$$

$$\text{Loop 2:} \quad -V_2 + V_c - V_4 = 0, \tag{2.17}$$

and

$$\text{Loop 3:} \quad -50 + V_1 + V_c - V_4 + V_3 = 0. \tag{2.18}$$

We note that these three equations are not entirely independent; in fact, the equation for loop 3 is equal to the sum of the equations for loops 1 and 2.

(b) Using Ohm's law for the 1-Ω, 4-Ω, and 5-Ω resistors and recognizing that the current through the 10-Ω resistor is 2 A, Eqs. (2.16) and (2.17) can be written as

$$-50 + I_1 + 5I_2 + 4I_1 = 0, \tag{2.19}$$

and

$$-5I_2 + V_c - (10 \times 2) = 0. \tag{2.20}$$

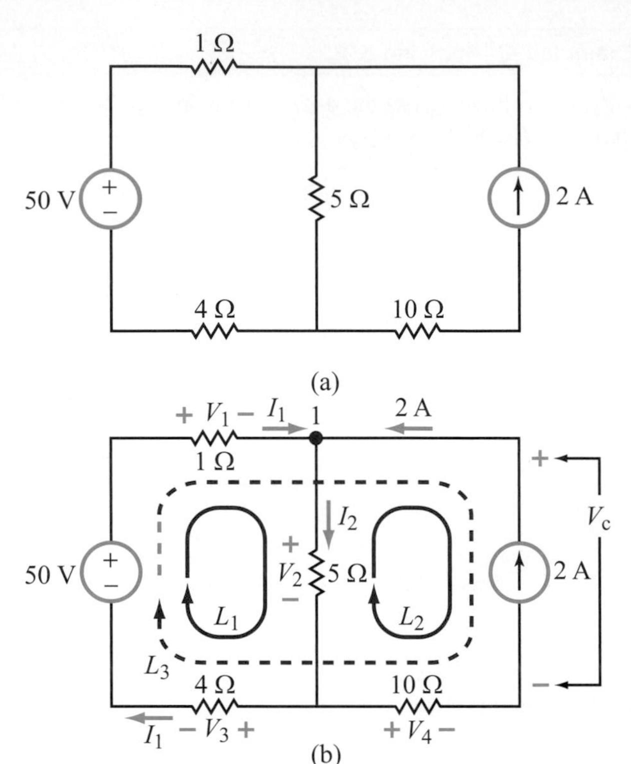

Figure 2-14: Circuit for Example 2-4 (a) before and (b) after labeling currents and voltages.

Next, we apply KCL at node 1 which gives

$$I_1 - I_2 + 2 = 0. \tag{2.21}$$

We now have three equations with three unknowns, namely I_1, I_2, and V_c. Through a simple substitution process, we obtain the following solutions:

$$I_1 = 4 \text{ A},$$

$$I_2 = 6 \text{ A},$$

and

$$V_c = 50 \text{ V}.$$

Review Question 2-7: Explain why KCL is (in essence) a statement of the law of conservation of charge.

Review Question 2-8: Explain why KVL is a statement of conservation of energy. What sign convention is used with KVL?

Example 2-5: Operational Amplifier Circuit

For the operational-amplifier equivalent circuit shown in Fig. 2-15, (a) obtain an expression for υ_0/υ_s—the ratio of the output voltage to the signal voltage—in terms of the indicated circuit elements, and then (b) evaluate the ratio for $R_1 = 15$ kΩ, $R_2 = 30$ kΩ, $R_3 = 75$ Ω, $R_i = 3$ MΩ, and $A = 10^6$.

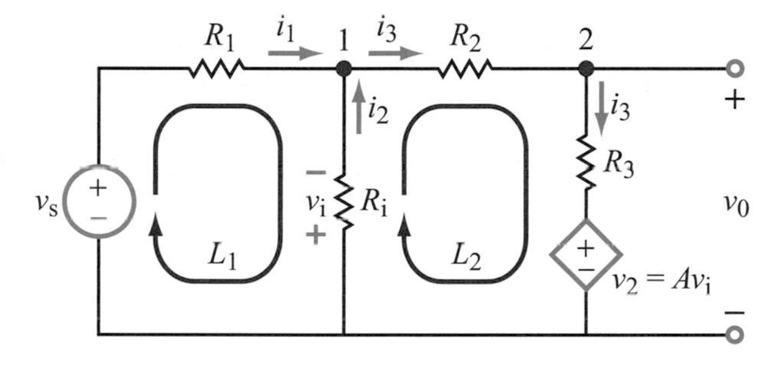

Figure 2-15: Operational amplifier equivalent circuit for Example 2-5.

Solution:

(a) We begin by labeling currents at node 1 and writing the KCL relationship among them; namely,

$$i_1 = i_3 - i_2. \qquad (2.22)$$

We also note that the dependent voltage source υ_2 is given by

$$\upsilon_2 = A\upsilon_i = AR_i i_2. \qquad (2.23)$$

Next, we write the KVL equations for the two loops as

$$-\upsilon_s + R_1 i_1 - R_i i_2 = 0 \qquad (2.24)$$

and

$$R_i i_2 + R_2 i_3 + R_3 i_3 + \upsilon_2 = 0. \qquad (2.25)$$

Our goal is to obtain an expression for υ_0 in terms of υ_s and the indicated circuit elements. From the right-hand side of the circuit,

$$\upsilon_0 = R_3 i_3 + \upsilon_2 = R_3 i_3 + AR_i i_2, \qquad (2.26)$$

which requires knowledge of the unknowns i_2 and i_3.

Upon using Eq. (2.22) in Eq. (2.24) to eliminate i_1, replacing υ_2 in Eq. (2.25) with the expression given by Eq. (2.23), and then grouping terms together to form simultaneous linear equations for i_2 and i_3, we have

$$-i_2(R_1 + R_i) + i_3 R_1 = \upsilon_s \qquad (2.27)$$

and

$$i_2(1 + A)R_i + i_3(R_2 + R_3) = 0. \qquad (2.28)$$

Simultaneous solution of these two equations yields

$$i_2 = -\left[\frac{(R_2 + R_3)}{(R_1 + R_i)(R_2 + R_3) + (1 + A)R_1 R_i}\right]\upsilon_s \qquad (2.29)$$

and

$$i_3 = \left[\frac{(1 + A)R_i}{(R_1 + R_i)(R_2 + R_3) + (1 + A)R_1 R_i}\right]\upsilon_s, \qquad (2.30)$$

Upon inserting the expressions for i_2 and i_3 into Eq. (2.26) and simplifying, we get

$$\frac{\upsilon_0}{\upsilon_s} = \frac{R_i(R_3 - AR_2)}{(R_1 + R_i)(R_2 + R_3) + (1 + A)R_1 R_i}. \qquad (2.31)$$

(b) Recognizing that the magnitudes of R_i and A are on the order of 10^6, those of R_1 and R_2 are on the order of 10^4, and R_3 is on the order of 10^2, it is easy to show that Eq. (2.31) can be reduced to the approximate expression

$$\frac{\upsilon_0}{\upsilon_s} \simeq -\frac{R_2}{R_1}. \qquad (2.32)$$

Given that $R_2 = 30$ kΩ and $R_1 = 15$ kΩ, it follows that

$$\frac{\upsilon_0}{\upsilon_s} \approx -2.$$

Exercise 2-8: If $I_1 = 3$ A in Fig. E2.8, what is I_2?

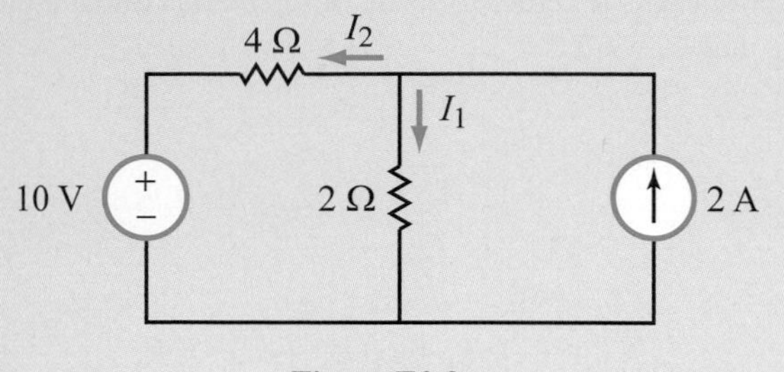

Figure E2.8

Answer: $I_2 = -1$ A. (See)

Exercise 2-9: Apply KCL and KVL to find I_1 and I_2 in Fig. E2.9.

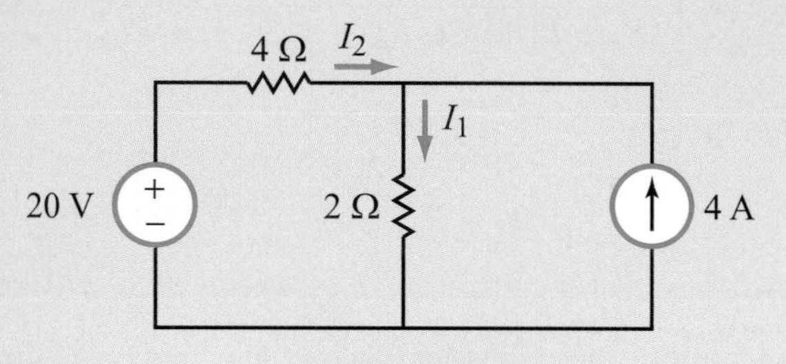

Figure E2.9

Answer: $I_1 = 6$ A, $I_2 = 2$ A. (See)

Exercise 2-10: Determine I_x in the circuit of Fig. E2.10.

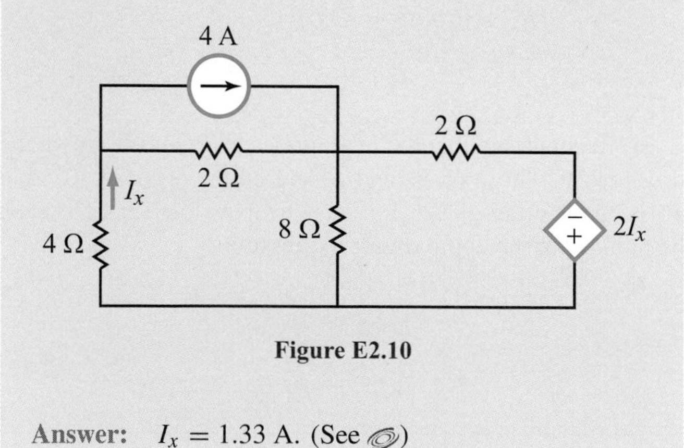

Figure E2.10

Answer: $I_x = 1.33$ A. (See)

2-4 Equivalent Circuits

Even though Kirchhoff's current and voltage laws can be used to write down the requisite number of node and loop equations that are necessary to solve for all of the voltages and currents in a circuit, it is often easier to determine a certain unknown voltage or current by first simplifying the other parts of the circuit. The simplification process involves the use of ***circuit equivalence***, wherein a circuit segment connected between two nodes (such as nodes 1 and 2 in Fig. 2-16) is replaced with another, simpler, circuit whose behavior is such that the voltage difference ($v_1 - v_2$) between the two nodes—as well as the currents entering into them (or exiting from them)—remain unchanged. That is:

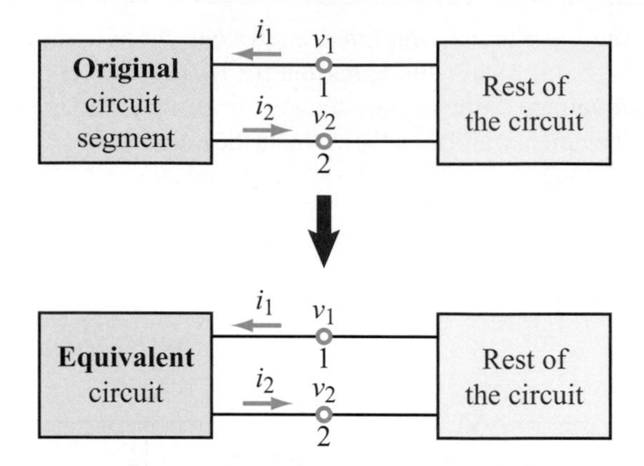

Figure 2-16: Circuit equivalence requires that the equivalent circuit exhibit the same i–v characteristic as the original circuit.

> Two circuits connected between a pair of nodes are considered to be equivalent if they exhibit identical i–v characteristics at those nodes.

To the rest of the circuit, the original and equivalent circuit segments *appear* identical.

We now will examine several types of equivalent circuits and then provide an overall summary at the conclusion of this section.

2-4.1 Resistors in Series

Consider the single-loop circuit of Fig. 2-17(a) in which a voltage source v_s is connected ***in series*** with five resistors. The KVL equation is given by

$$-v_s + R_1 i_s + R_2 i_s + R_3 i_s + R_4 i_s + R_5 i_s = 0, \qquad (2.33)$$

which can be rewritten as

$$\begin{aligned} v_s &= R_1 i_s + R_2 i_s + R_3 i_s + R_4 i_s + R_5 i_s \\ &= (R_1 + R_2 + R_3 + R_4 + R_5) i_s \\ &= R_{eq} i_s, \end{aligned} \qquad (2.34)$$

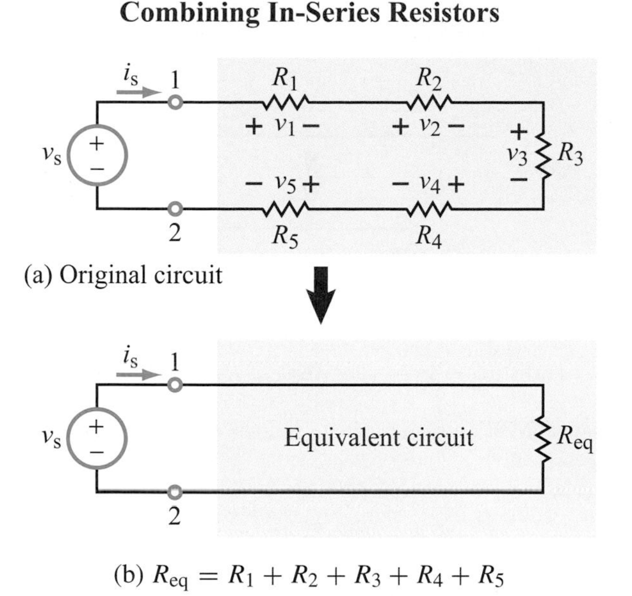

Figure 2-17: In a single-loop circuit, R_{eq} is equal to the sum of the resistors.

where R_{eq} is an ***equivalent resistor*** whose resistance is equal to the sum of the five in-series resistances,

$$R_{eq} = R_1 + R_2 + R_3 + R_4 + R_5. \tag{2.35}$$

From the standpoint of the source voltage v_s and the current i_s it supplies, the circuit in Fig. 2-17(a) is equivalent to that in Fig. 2-17(b). That is,

$$i_s = \frac{v_s}{R_{eq}}. \tag{2.36}$$

For any of the individual resistors, such as R_2, the voltage across it is given by

$$\begin{aligned} v_2 &= R_2 i_s \\ &= \left(\frac{R_2}{R_{eq}}\right) v_s. \end{aligned} \tag{2.37}$$

Similar expressions apply to the other resistors, wherein the voltage across a resistor is equal to v_s multiplied by the ratio of its own resistance to the sum total R_{eq}. Thus, *the single-loop circuit, in effect, divides the source voltage among the series resistors.*

Two basic conclusions can be drawn from the preceding discussion.

1. Multiple resistors connected in series (experiencing the same current) can be combined into a single equivalent resistor R_{eq} whose resistance is equal to the sum of all of their individual resistances.

Mathematically,

$$R_{eq} = \sum_{i=1}^{N} R_i \qquad \text{(resistors in series)}, \tag{2.38a}$$

where N is the total number of resistors in the group.

The second conclusion is known as ***voltage division***:

2. The voltage across any individual resistor R_i in a series circuit is a proportionate fraction (R_i/R_{eq}) of the voltage across the entire group

$$v_i = \left(\frac{R_i}{R_{eq}}\right) v_s. \tag{2.38b}$$

Example 2-6: The Voltage Divider

The term ***voltage divider*** is used commonly in reference to a circuit of the type shown in Fig. 2-18, whose purpose is to supply a secondary load circuit a specific voltage v_2 that is smaller than the available source voltage v_s. In other words, the goal is to scale v_s down to v_2. If $v_s = 100$ V, choose appropriate values for R_1 and R_2 such that $v_2 = 60$ V.

Solution: In view of Eq. (2.37), application of the voltage-division property gives

$$v_2 = \left(\frac{R_2}{R_1 + R_2}\right) v_s.$$

To obtain the desired division, we require

$$\frac{R_2}{R_1 + R_2} = \frac{v_2}{v_s} = \frac{60}{100} = 0.6,$$

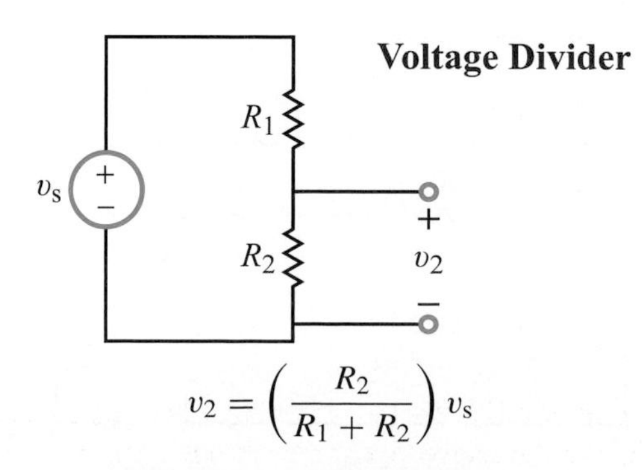

Figure 2-18: Voltage divider.

which can be satisfied through an infinite combination of choices of R_1 and R_2. Hence, we arbitrarily choose

$$R_1 = 2\ \text{k}\Omega \qquad \textit{and} \qquad R_2 = 3\ \text{k}\Omega.$$

2-4.2 Sources in Series

Figure 2-19 contains a single-loop circuit composed of a voltage source, a resistor, and two current sources, all connected in series. One of the current sources indicates that the current flowing through it is 4 A in magnitude and clockwise in direction, while the other current source indicates that the current is 6 A in magnitude and counterclockwise in direction. Continuity of current flow mandates that the current flowing through the loop be exactly the same in both magnitude and direction at every location over the full extent of the loop. So our dilemma is: Is the current 4 A, 6 A, or the difference between the two? It is none of those guesses. The true answer is that the circuit is *unrealizable*, meaning that it is not possible to construct a circuit with two current sources of different magnitudes or different directions that are connected in series. The problem with the circuit of Fig. 2-19 has to do with our representation of ideal current sources. As was stated in Section 1-5.2 and described in Table 1-3, a real current source can be modeled as the parallel combination of an ideal current source and a shunt resistor R_s. Usually, R_s is very large, so very little current flows through it in comparison with the current flowing through the other part of the circuit, in which case it can be deleted without much consequence. In the present case, however, had such shunt resistors been included in the circuit of Fig. 2-19, the dilemma would not have arisen. The lesson we should learn from this discussion is that when we idealize current sources by deleting their parallel resistors, we should never connect them in series in circuit diagrams.

Whereas current sources cannot be connected in series, voltage sources can. In fact, it follows from KVL that the circuit in Fig. 2-20(a) can be simplified into the equivalent circuit of Fig. 2-20(b) with

$$v_{eq} = v_1 - v_2 + v_3 \tag{2.39}$$

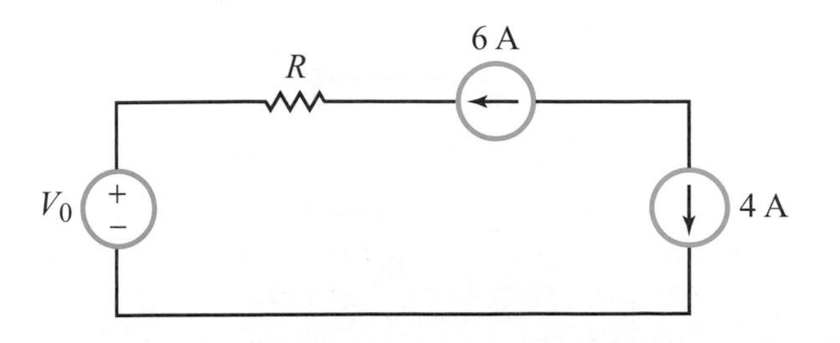

Figure 2-19: Unrealizable circuit; two current sources with different magnitudes or directions cannot be connected in series.

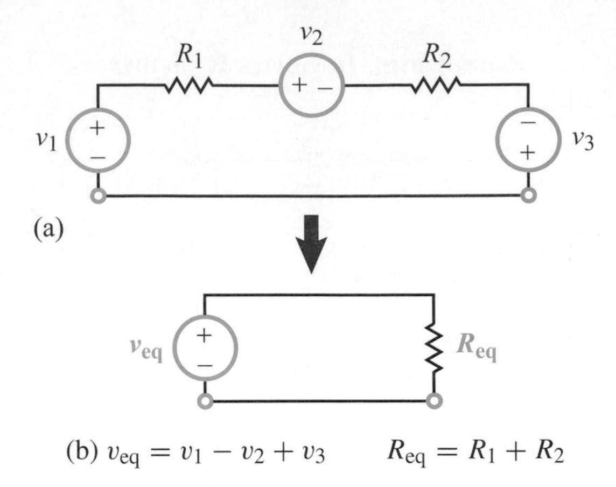

Figure 2-20: In-series voltage sources can be added together algebraically.

and

$$R_{eq} = R_1 + R_2. \tag{2.40}$$

Thus:

> Multiple voltage sources connected in series can be combined into an equivalent voltage source whose voltage is equal to the algebraic sum of the voltages of the individual sources.

2-4.3 Resistors and Sources in Parallel

When multiple resistors are connected in series, they all share the same current, but each has its own individual voltage across it. The converse is true for multiple resistors connected in parallel: The three resistors in Fig. 2-21(a) experience the same voltage across all of them, namely v_s, but each carries its own individual current. The current supplied by the source is *divided* among the branches containing the three resistors. Thus,

$$i_s = i_1 + i_2 + i_3. \tag{2.41}$$

Application of Ohm's law provides

$$i_1 = \frac{v_s}{R_1}, \qquad i_2 = \frac{v_s}{R_2}, \qquad \textit{and} \qquad i_3 = \frac{v_s}{R_3}, \tag{2.42}$$

which when used in Eq. (2.41) leads to

$$i_s = \frac{v_s}{R_1} + \frac{v_s}{R_2} + \frac{v_s}{R_3}. \tag{2.43}$$

We wish to replace the parallel combination of the three resistors with a single equivalent resistor R_{eq}, as depicted in Fig. 2-21(b), such that the current i_s remains unchanged. For the equivalent circuit,

$$i_s = \frac{v_s}{R_{eq}}. \tag{2.44}$$

Combining In-Parallel Resistors

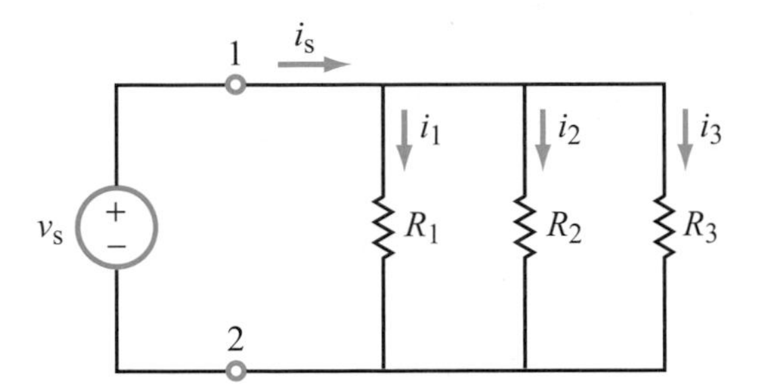

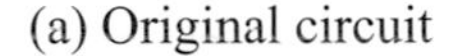
(a) Original circuit

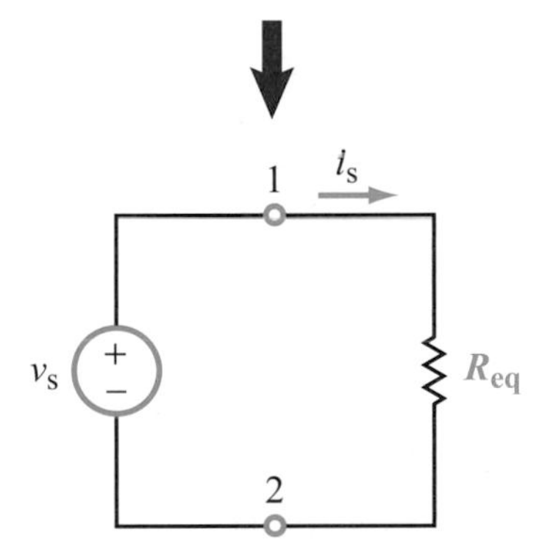

(b) Equivalent circuit

$$R_{eq} = \left(\frac{1}{R_1} + \frac{1}{R_2} + \frac{1}{R_3}\right)^{-1} \qquad i_2 = \left(\frac{R_{eq}}{R_2}\right) i_s$$

Figure 2-21: Voltage source connected to a parallel combination of three resistors.

If the two circuits in Fig. 2-21 are to function the same, as regards the source, then i_s as given by Eq. (2.43) for the original circuit should be equal to the expression for i_s given by Eq. (2.44) for the equivalent circuit. Thus,

$$\frac{v_s}{R_{eq}} = \frac{v_s}{R_1} + \frac{v_s}{R_2} + \frac{v_s}{R_3}, \tag{2.45}$$

from which we conclude that

$$\frac{1}{R_{eq}} = \frac{1}{R_1} + \frac{1}{R_2} + \frac{1}{R_3}. \tag{2.46}$$

This result can be generalized to any N resistors connected in parallel

$$\frac{1}{R_{eq}} = \sum_{i=1}^{N} \frac{1}{R_i} \quad \text{(resistors in parallel)}. \tag{2.47}$$

Multiple resistors connected in parallel divide the input current among them.

For R_2 in Fig. 2-21(a),

$$i_2 = \frac{v_s}{R_2} = \left(\frac{R_{eq}}{R_2}\right) i_s. \tag{2.48}$$

By extension, for a ***current divider*** composed of N in-parallel resistors, the current flowing through R_i is a proportionate fraction (R_{eq}/R_i) of the input current.

It is useful to note that the equivalent resistance for a parallel combination of two resistors R_1 and R_2 (Fig. 2-22) is given by

$$R_{eq} = \frac{R_1 R_2}{R_1 + R_2}. \tag{2.49}$$

As a short-hand notation, we will sometimes denote such a parallel combination $R_1 \parallel R_2$.

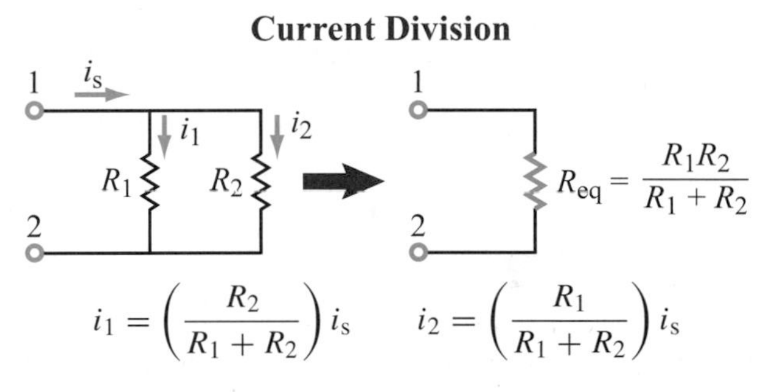

Figure 2-22: Equivalent circuit for two resistors in parallel.

As was noted earlier in Section 2-1.3, *the inverse of the resistance R is the conductance G; $G = 1/R$.* For N conductances connected in parallel, Eq. (2.47) assumes the linear sum

$$G_{eq} = \sum_{i=1}^{N} G_i \quad \text{(conductances in parallel)}. \tag{2.50}$$

Two resistors always can be combined together, whether they are connected in series (sharing the same current) or in parallel (sharing the same voltage). Two voltage sources can be combined when connected in series, but they cannot be connected in parallel. Two current sources can be combined when connected in parallel (as illustrated by Fig. 2-23), but they cannot be connected in series.

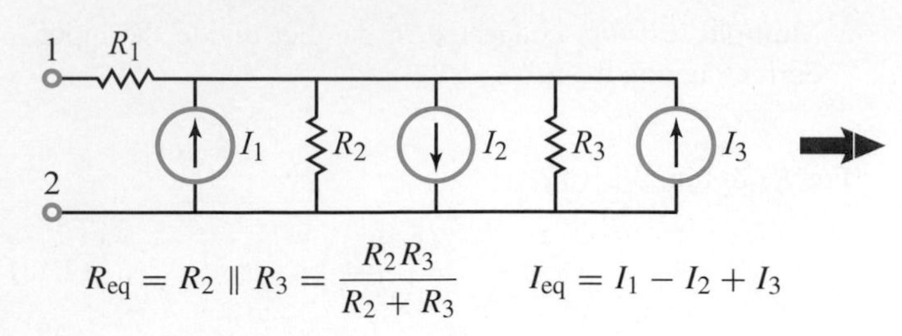

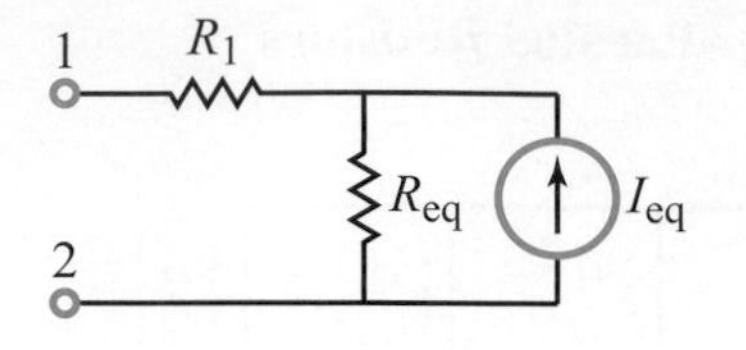

Figure 2-23: Adding current sources connected in parallel.

Example 2-7: Equivalent-Circuit Solution

Use the equivalent-resistance approach to determine V_2, I_1, I_2, and I_3 in the circuit of Fig. 2-24(a).

Solution: Our first step is to combine the 2-Ω and 4-Ω in-series resistances into a 6-Ω resistance and to combine the two 6-Ω in-parallel resistances into a 3-Ω resistance (by applying Eq. (2.49)). The simplifications lead to the circuit in Fig. 2-24(b). Next, we calculate the parallel combination of the 3-Ω and 6-Ω resistors, (3 ‖ 6), again using Eq. (2.49), to get $(3 \times 6)/(3 + 6) = 18/9 = 2\ \Omega$. The new equivalent circuit is displayed in Fig. 2-24(c), from which we deduce that

$$I_1 = \frac{24}{10 + 2} = 2\text{ A}$$

and

$$V_2 = 2I_1 = 2 \times 2 = 4\text{ V}.$$

Returning to Fig. 2-24(b), we apply Ohm's law to find I_2 and I_3.

$$I_2 = \frac{V_2}{3} = \frac{4}{3} = 1.33\text{ A},$$

and

$$I_3 = \frac{V_2}{6} = \frac{4}{6} = 0.67\text{ A}.$$

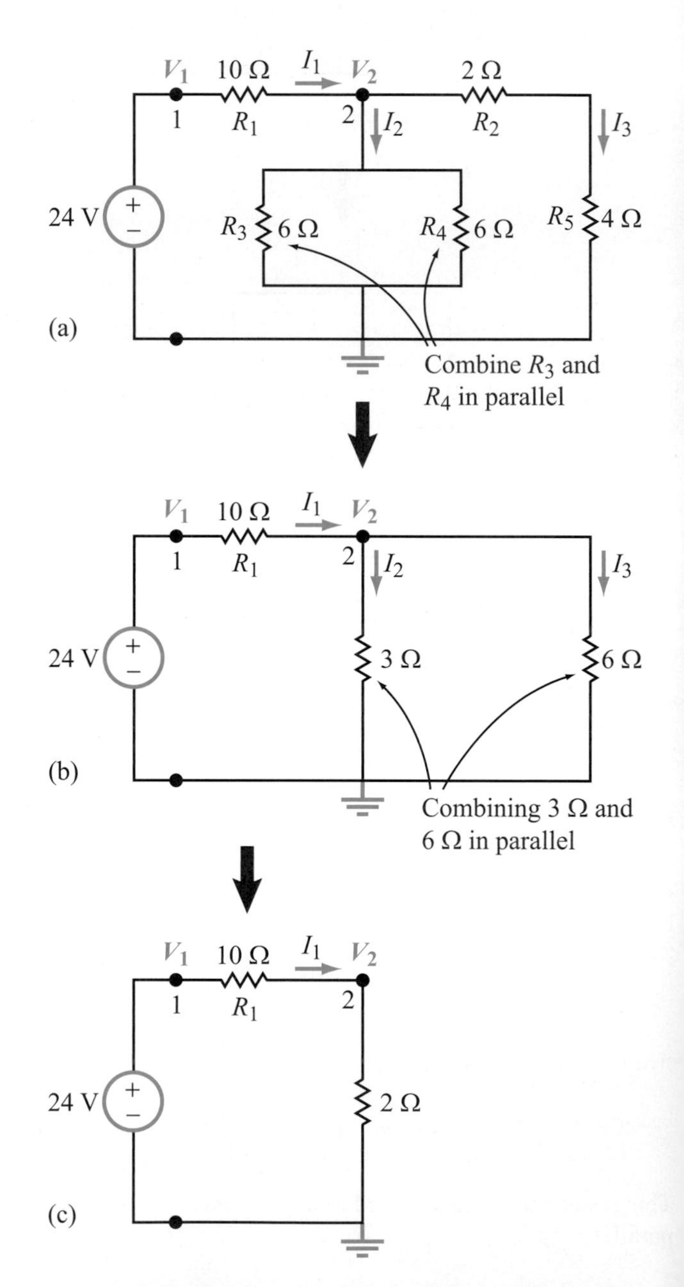

Figure 2-24: Example 2-7. (a) Original circuit, (b) after combining R_3 and R_4 in parallel and combining R_2 and R_5 in series, and (c) after combining the 3-Ω and 6-Ω resistances in parallel.

Review Question 2-9: What conditions must be satisfied in order for two circuits to be considered *equivalent*?

Review Question 2-10: What is a voltage divider and what is a current divider?

Review Question 2-11: What is the i–v relationship for a conductance G?

Exercise 2-11: Apply resistance combining to simplify the circuit of Fig. E2.11 in order to find I. All resistor values are in ohms.

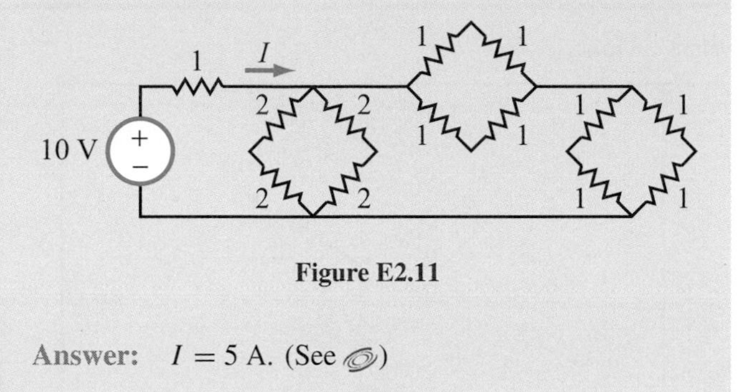

Figure E2.11

Answer: $I = 5$ A. (See)

2-4.4 Source Transformation

We now will demonstrate how a realistic voltage source composed of an ideal voltage source in series with a resistor can be exchanged for a realistic current source composed of an ideal current source in parallel with a shunt resistor, or vice versa. The two circuits are shown in parts (a) and (b) of Fig. 2-25.

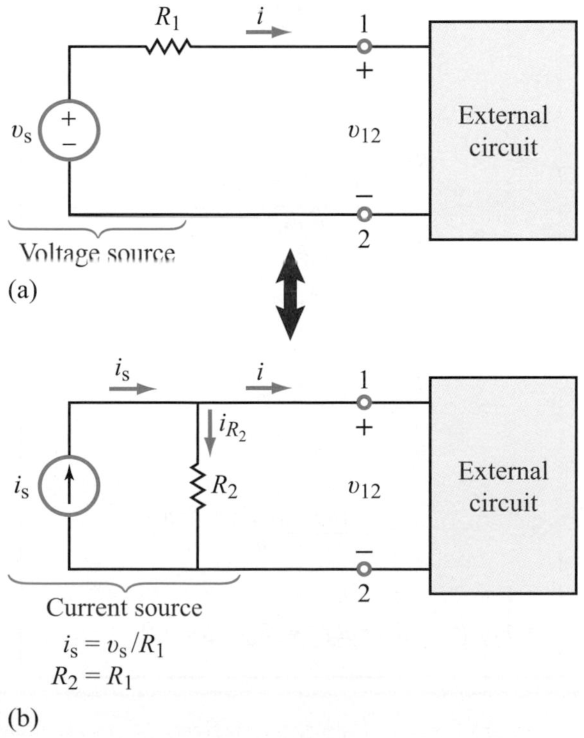

Source Transformation

(a) Voltage source

(b) Current source

$i_s = v_s/R_1$
$R_2 = R_1$

Figure 2-25: Realistic voltage and current sources connected to an external circuit. Equivalence requires that $i_s = v_s/R_1$ and $R_2 = R_1$.

Exchanging the one source for the other requires that they be equivalent—from the vantage point of the external circuit.

> A voltage-source circuit and a current-source circuit are considered equivalent and interchangeable if they deliver the same input current i and voltage v_{12} to the external circuit.

For the voltage-source circuit, application of KVL gives

$$-v_s + iR_1 + v_{12} = 0, \tag{2.51}$$

from which we obtain the following expression for i:

$$i = \frac{v_s}{R_1} - \frac{v_{12}}{R_1}. \tag{2.52}$$

Application of KCL to the current-source circuit gives

$$\begin{aligned} i &= i_s - i_{R_2} \\ &= i_s - \frac{v_{12}}{R_2}, \end{aligned} \tag{2.53}$$

where we used Ohm's law to relate i_{R_2} to v_{12}. Equivalence of Eqs. (2.52) and (2.53) is satisfied for all values of i and v_{12} if and only if:

$$R_1 = R_2 \tag{2.54a}$$

and

$$i_s = \frac{v_s}{R_1}. \tag{2.54b}$$

In summary:

> A voltage source v_s in series with a source resistance R_s is equivalent to the combination of a current source $i_s = v_s/R_s$, in parallel with a shunt resistance R_s.

This equivalence is called ***source transformation*** because it allows us to replace a realistic voltage source with a realistic current source, or vice versa.

A summary of in-series and in-parallel equivalent circuits involving sources and resistors is available in Table 2-6.

Table 2-6: Equivalent circuits.

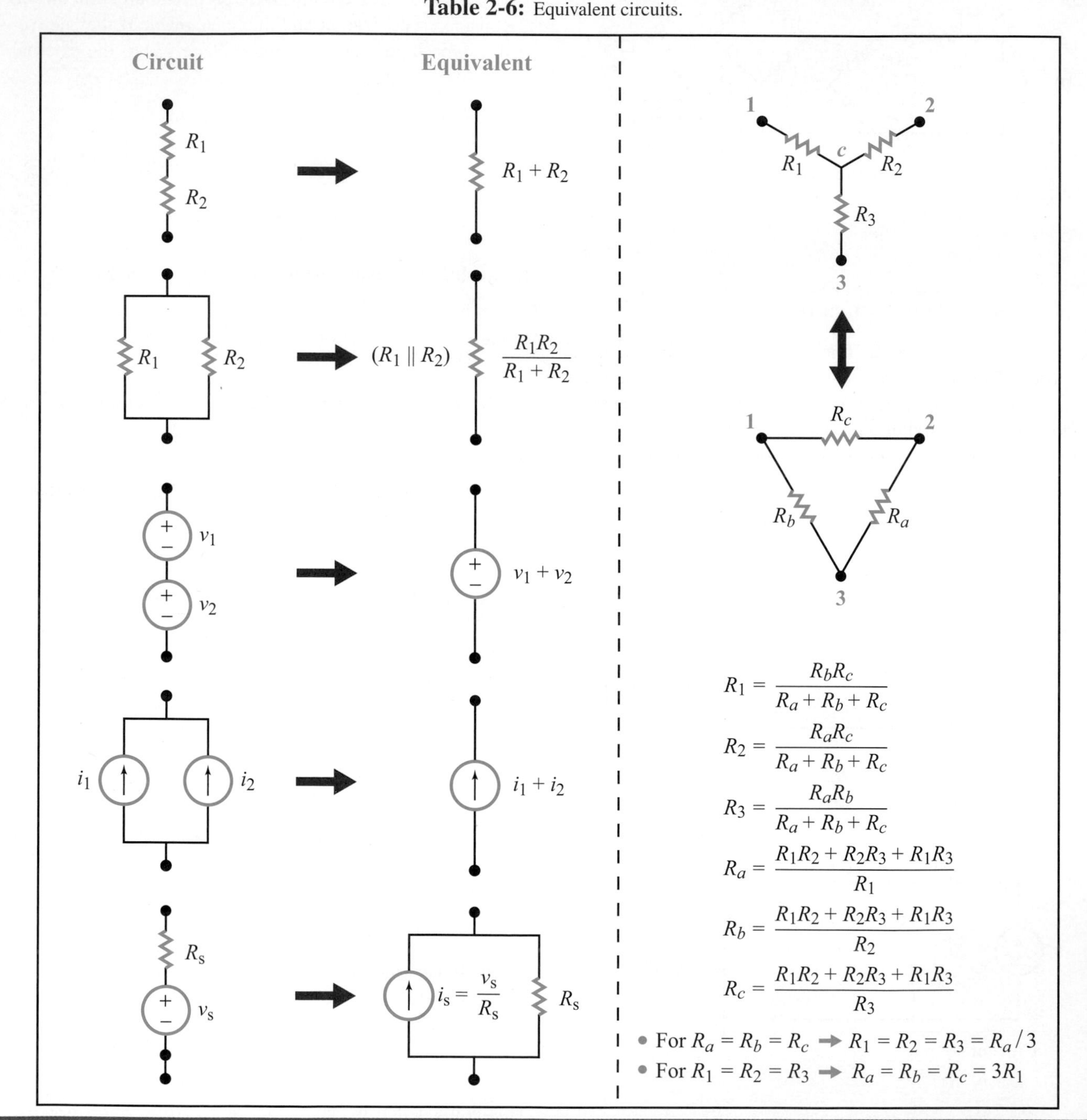

Example 2-8: Source Transformation

Determine the current I in the circuit of Fig. 2-26(a).

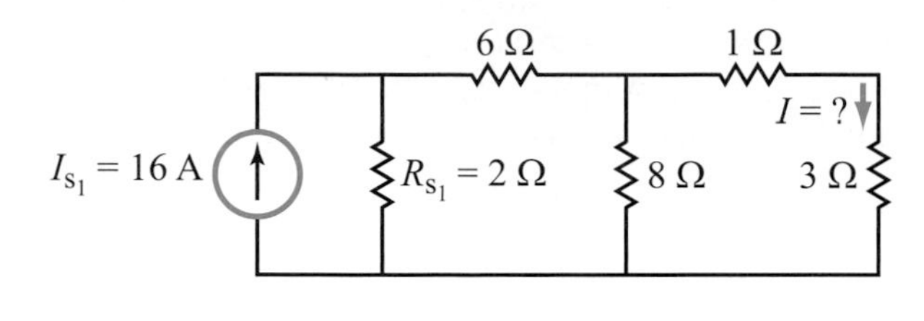

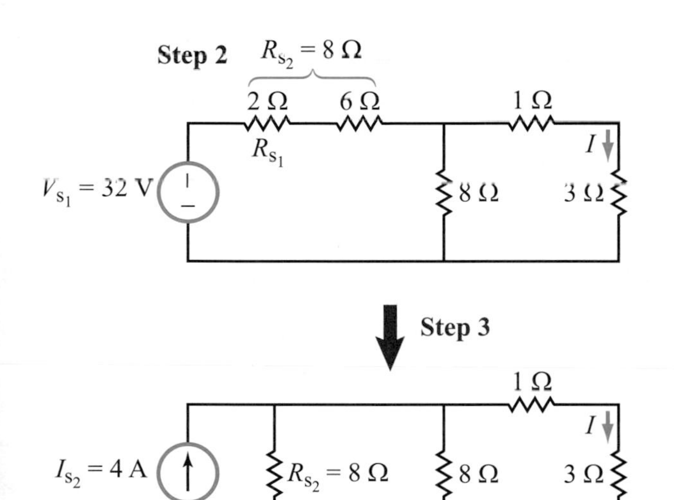

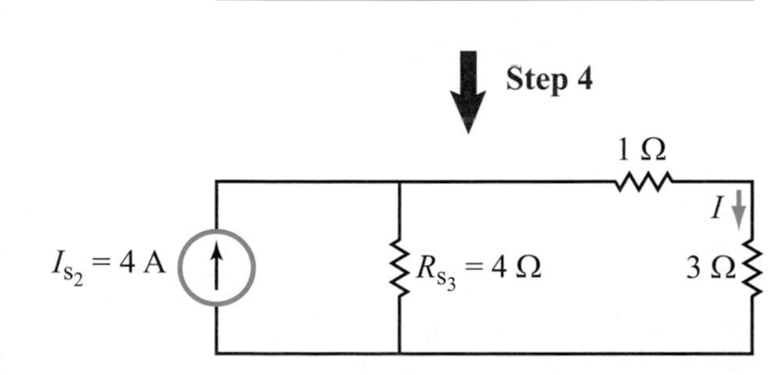

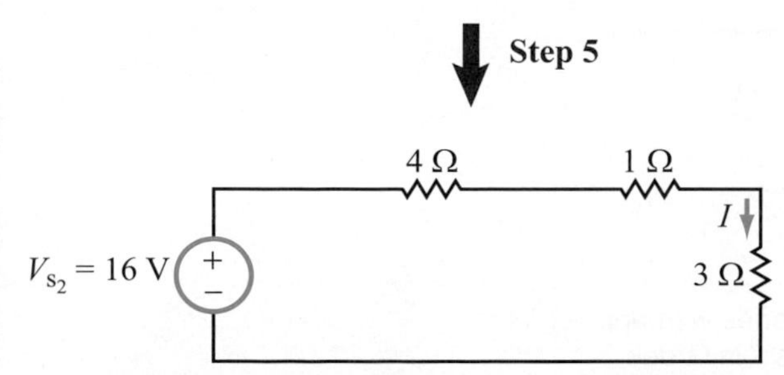

Figure 2-26: Example 2-8 circuit evolution.

Solution: It is best to avoid transformations that would involve the 3-Ω resistor with the unknown current I. Hence, we will apply multiple source-transformation steps, moving from the left end of the circuit towards the 3-Ω resistor.

Step 1: Current to voltage transformation allows us to convert the combination (I_{s_1}, R_{s_1}) to a voltage source

$$V_{s_1} = I_{s_1} R_{s_1} = 16 \times 2 = 32 \text{ V},$$

in series with R_{s_1}.

Step 2: Combining R_{s_1} in series with the 6-Ω resistor results in

$$R_{s_2} = 2 + 6 = 8 \ \Omega.$$

Hence, the new input source becomes (V_{s_1}, R_{s_2}).

Step 3: Convert (V_{s_1}, R_{s_2}) back into a current source

$$I_{s_2} = V_{s_1}/R_{s_2} = 32/8 = 4 \text{ A},$$

in parallel with R_{s_2}.

Step 4: Combine $R_{s_2} = 8\ \Omega$ in parallel with the other 8-Ω resistor $(8 \parallel 8)$ to obtain an equivalent resistance $R_{s_3} = 4\ \Omega$.

Step 5: Convert again to a voltage source

$$V_{s_2} = I_{s_2} R_{s_3} = 4 \times 4 = 16 \text{ V},$$

in series with R_{s_3}.

For the single loop realized in the final step,

$$I = \frac{V_{s_2}}{4 + 1 + 3} = \frac{16}{8} = 2 \text{ A}.$$

Exercise 2-12: Apply source transformation to the circuit in Fig. E2.12 to find I.

Answer: $I = 4$ A. (See ◎)

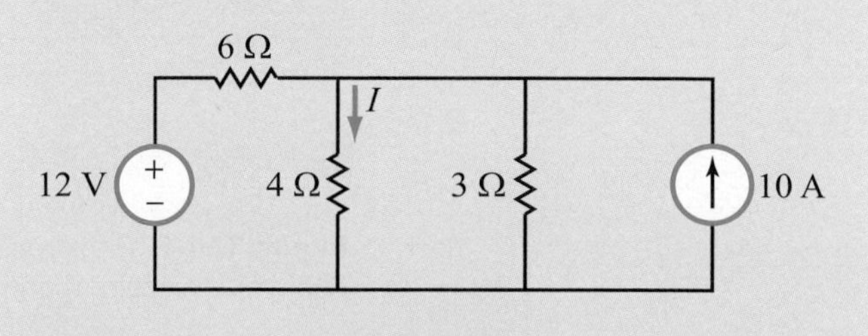

Figure E2.12

Technology Brief 4: Resistors as Sensors

The relationship between the voltage across a conductor and the current through it is given by Ohm's law, $V = IR$. The resistance R of the conductor accounts for the reduction in the electrons' velocities due to collisions with the much larger atoms of the conducting material (see Technology Brief 3 on page 38). The question is: What happens to R if we disturb the atoms of the conductor by applying an external, non-electrical ***stimulus***, such as heating or cooling it, stretching or compressing it, or shining light on it? Through proper choice of materials, we actually can ***modulate*** (change) the magnitude of R by applying such external stimuli, and this forms the basis of many common sensors.

Piezoresistive Sensors

In 1856, Lord Kelvin discovered that applying a mechanical load on a bar of metal changed its resistance. Over the next 150 years, both theoretical and practical advances made it possible to describe the physics behind this effect in both conductors and semiconductors. The phenomenon is referred to as the ***piezoresistive effect*** (Fig. TF4-1) and is used in many practical devices to convert a mechanical signal into an electrical one. Such sensors (Fig. TF4-2) are called ***strain gauges***. Piezoresistive sensors are used in a wide variety of consumer applications, including robot toy "skins" that sense force, microscale gas-pressure sensors, and micromachined accelerometers that sense acceleration. They all use piezoresistors in electrical circuits to generate a signal from a mechanical stimulus.

In its simplest form, a resistance change ΔR occurs when a mechanical pressure P (N/m^2) is applied along the axis of the resistor (Fig. TF4-1)

$$\Delta R = R_0 \alpha P,$$

where R_0 is the unstressed resistance and α is known as the ***piezoresistive coefficient*** (m^2/N). The piezoresistive coefficient is a material property, and for crystalline materials (such as silicon), the piezoresistive coefficient also varies depending on the direction of the applied pressure (relative to the crystal planes of the material). The total resistance of a piezoresistor under stress is therefore given by

$$R = R_0 + \Delta R = R_0(1 + \alpha P).$$

The pressure P, which usually is called the ***mechanical stress*** or ***mechanical load***, is equal to F/A, where F is the force acting on the piezoresistor and A is the cross-sectional area it is acting on. The sign of P is defined as positive for a compressional force and negative for a stretching force. The piezoresistive coefficient α usually has a negative value, so the product αP leads to a decrease in R for compression and an increase for stretching.

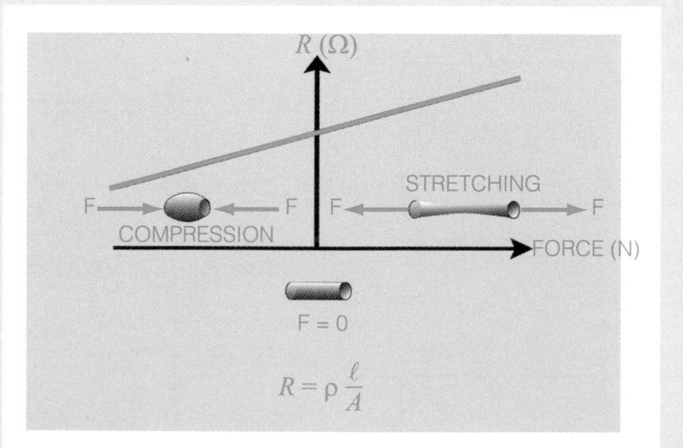

Figure TF4-1: Piezoresistance varies with applied force. The word "piezein" means "to press" in Greek.

Thermistor Sensors

Changes in temperature also can lead to changes in the resistance of a piece of conductor or semiconductor; when used as a sensor, such an element is called a ***thermistor***. As a simple approximation, the change in resistance can be modeled as

$$\Delta R = k\,\Delta T,$$

where ΔT is the temperature change (in degrees C) and k is the first-order temperature coefficient of resistance (Ω/°C). Thermistors are classified according to whether k is negative or positive (i.e., if an increase in temperature decreases or increases the resistance). This approximation works only for small temperature changes; for larger swings, higher-order terms must be included in the equation. Resistors used in electrical circuits that are not intended to be used as sensors are manufactured from materials with the lowest k possible, since circuit designers do not want their resistors changing during operation. In contrast, materials with high values of k are desirable for sensing temperature variations. Care must be taken, however, to incorporate into the sensor response the self-heating effect that occurs due to having a current passing through the resistor itself.

Thermistors are used routinely in modern thermostats and in battery-pack chargers (to prevent batteries from overheating). Thermistors also have found niche applications (Fig. TF4-3) in low-temperature sensing and as fuse replacements (for thermistors with large, positive k values). In the case of current-limiting fuse replacements, a large enough current self-heats the thermistor, and the resistance increases. There is a threshold current above which the thermistor cannot be cooled off by its environment; as it continues to get hotter, the resistance continues to increase, which in turn, causes even more self-heating. This "runaway" effect rapidly shuts current off almost entirely.

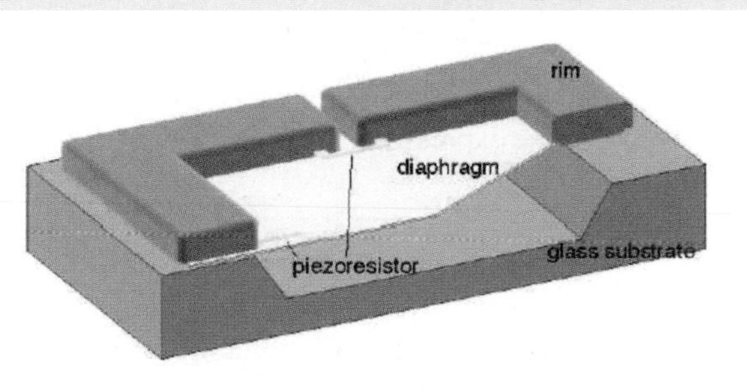

(a) Schematic

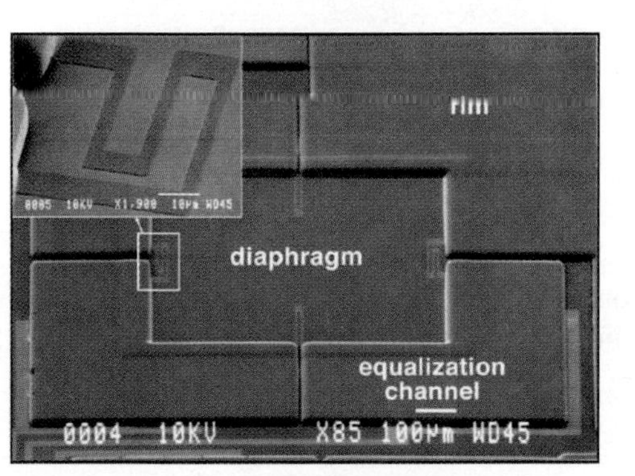

(b) Scanning electron micrograph (SEM) of the sensor

Figure TF4-2: A microfabricated pressure sensor developed at the University of Michigan. It uses piezoresistors to detect deformation of a membrane; when the membrane (white) deflects, it stretches the piezoresistor and the resistance changes. (Courtesy of Khalil Najafi, University of Michigan.)

Figure TF4-3: This micromachined anemometer is a thermistor that measures fluid velocity; as fluid flows by, it cools the thermistor at different rates, depending on the fluid velocity. (Courtesy of Khalil Najafi, University of Michigan.)

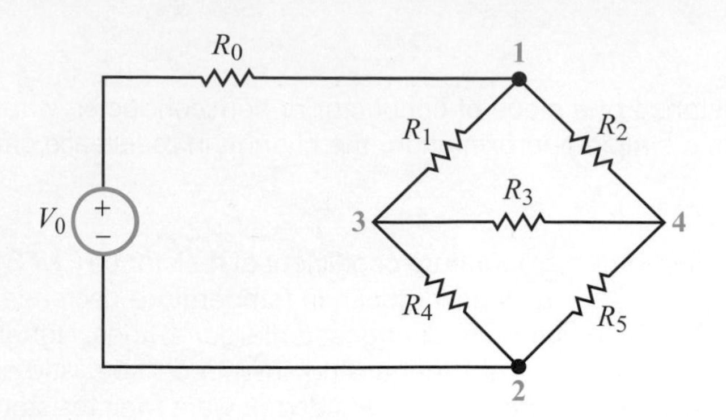

Figure 2-27: No two resistors of this circuit share the same current (connected in series) or voltage (connected in parallel).

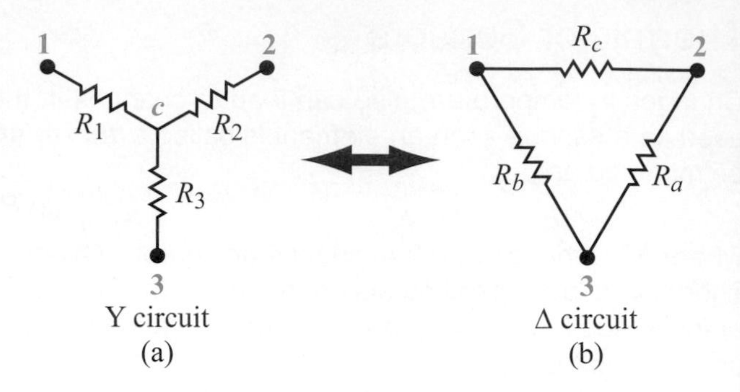

Figure 2-28: Y–Δ equivalent circuits.

2-5 Wye–Delta (Y–Δ) Transformation

In principle, it always is possible to simplify the behavior of a resistive circuit when measured across any two nodes—no matter how complex its topology—down to a simple equivalent circuit composed of an equivalent voltage source in series with an equivalent resistor. The preceding sections offered us tools for combining resistors together whenever they are connected in series or in parallel, as well as for combining in-series voltage sources and in-parallel current sources. Sometimes, however, we may encounter circuit topologies that cannot be simplified using those tools because their resistors are connected neither in series nor in parallel. A case in point is the circuit in Fig. 2-27, in which no two resistors share the same current or voltage. This section introduces a new circuit-simplification tool—known as the ***Wye-Delta (Y–Δ) transformation***—for dealing specifically with such a circuit arrangement.

To that end, let us start by considering the Y and Δ circuit segments shown in Fig. 2-28(a) and (b), respectively. Let us assume that the same external circuit is connected to the Y and Δ circuits at nodes 1, 2, and 3. Our task is to develop a set of transformation relations between the resistor set (R_1, R_2, R_3) of the Y circuit and the resistor set (R_a, R_b, R_c) of the Δ circuit that will allow us to replace the Y circuit with the Δ circuit (or vice versa) without affecting the terminal characteristics (currents and voltages) at nodes 1, 2, and 3. That is, from the standpoint of the external circuit, the Y and Δ circuits should behave equivalently.

The standard procedure employed in deriving the transformation relations is to (a) set one node as an open circuit (i.e., not connected to an external circuit), (b) derive an expression for the resistance between the other two nodes (as if a voltage source were connected between them) of the Y circuit, (c) follow the same procedure for the Δ circuit, and then (d) equate the expressions obtained in steps (b) and (c). For example, with node 3 open-circuited, the Y circuit reduces to just two in-series resistors R_1 and R_2, in which case the resistance between nodes 1 and 2 is simply

$$R_{12} = R_1 + R_2 \qquad \text{(Y-circuit).} \tag{2.55}$$

Repeating the procedure for the Δ circuit (again with node 3 not connected to the external circuit) leads to a configuration between nodes 1 and 2 consisting of R_c in parallel with the series combination of R_a and R_b. Hence,

$$R_{12} = \frac{R_c(R_a + R_b)}{R_a + R_b + R_c} \qquad (\Delta\text{-circuit}). \tag{2.56}$$

Upon equating the expressions for R_{12} given by Eqs. (2.55) and (2.56), we have

$$R_1 + R_2 = \frac{R_c(R_a + R_b)}{R_a + R_b + R_c}. \tag{2.57a}$$

When applied to the other two combinations of nodes, the foregoing procedure leads to:

$$R_2 + R_3 = \frac{R_a(R_b + R_c)}{R_a + R_b + R_c} \tag{2.57b}$$

and

$$R_1 + R_3 = \frac{R_b(R_a + R_c)}{R_a + R_b + R_c}. \tag{2.57c}$$

2-5.1 Δ →Y Transformation

Solution of the preceding set of equations provides the following expressions for R_1, R_2, and R_3:

$$R_1 = \frac{R_b R_c}{R_a + R_b + R_c} \tag{2.58a}$$

$$R_2 = \frac{R_a R_c}{R_a + R_b + R_c} \tag{2.58b}$$

$$R_3 = \frac{R_a R_b}{R_a + R_b + R_c} \tag{2.58c}$$

Note the symmetry associated with the form of these expressions: R_1 *of the Y circuit, which is connected to node 1, is given by an expression (Eq. (2.58a)) whose numerator is the product of the two resistors connected to node 1 in the* Δ *circuit*, namely R_b and R_c. The same form of symmetry applies to R_2 and R_3.

The transformation represented by the three parts of Eq. (2.58) enables us to replace the Δ circuit with a Y circuit without having any impact on the external circuit.

2-5.2 Y→ Δ Transformation

When applied in the reverse direction, from Y to Δ, the associated transformation relations are given by the following expressions.

$$R_a = \frac{R_1R_2 + R_2R_3 + R_1R_3}{R_1} \tag{2.59a}$$

$$R_b = \frac{R_1R_2 + R_2R_3 + R_1R_3}{R_2} \tag{2.59b}$$

$$R_c = \frac{R_1R_2 + R_2R_3 + R_1R_3}{R_3} \tag{2.59c}$$

For this transformation, the symmetry is as follows: R_a *of the* Δ *circuit, which is connected between nodes 2 and 3, is given by an expression (Eq. (2.59a)) whose denominator is* R_1*, the resistor connected to node 1 of the Y circuit*. This form of symmetry also applies to R_b and R_c.

When we started our examination of the Y–Δ transformation, we referred to Fig. 2-27. Returning to that figure, we note that the circuit contains two obvious Δ circuits, namely R_1–R_2–R_3 and R_3–R_5–R_4, as well as two not-so-obvious Y circuits: R_1–R_3–R_4 and R_2–R_3–R_5. To demonstrate that those two combinations are indeed Y circuits, we have redrawn the circuit in the form shown in Fig. 2-29(a) where we stretched nodes 1 and 2 from single points into two horizontal lines. Electrically, we did not change the circuit whatsoever. Figure 2-29(b) depicts another rendition of the same circuit. In this case, the Y circuit given by R_1–R_3–R_4 resembles a sideways T rather than a Y, and the Δ circuit given by R_1–R_3–R_2 resembles a Π. Hence, it is not surprising that the Y–Δ transformation is oftentimes called the T–Π transformation. It is instructive to note that the shape in which a circuit is drawn is irrelevant electrically; what does matter is how the branches are connected to the nodes.

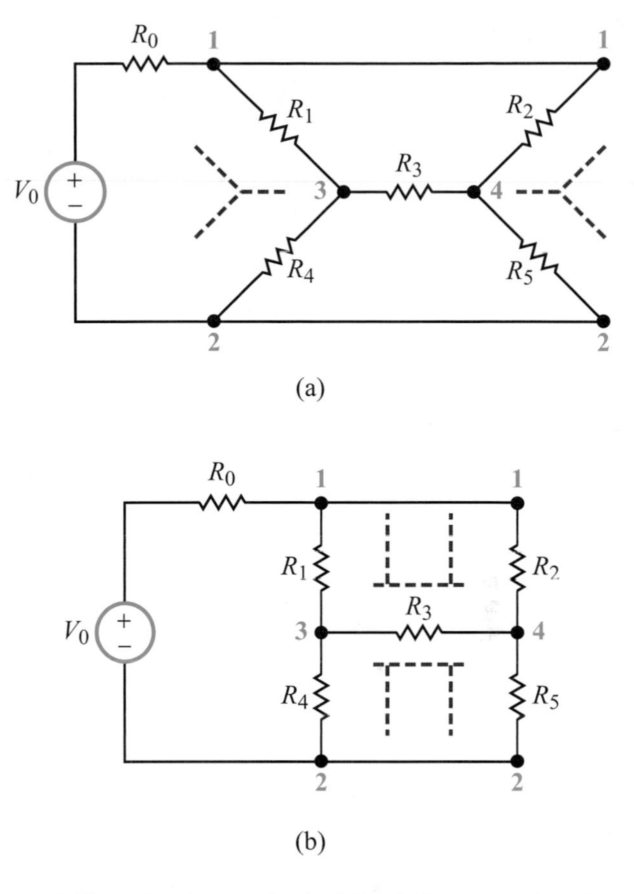

Figure 2-29: Redrawing the circuit of Fig. 2-27 to resemble (a) Y and (b) T and Π subcircuits.

2-5.3 Balanced Circuits

If the resistors of the Δ circuit are all equal, the circuit is said to be ***balanced***, as a result of which the Y circuit will also be balanced and will have equal resistors given by

$$R_1 = R_2 = R_3 = \frac{R_a}{3} \quad (\text{if } R_a = R_b = R_c), \tag{2.60a}$$

and conversely

$$R_a = R_b = R_c = 3R_1 \quad (\text{if } R_1 = R_2 = R_3). \tag{2.60b}$$

Example 2-9: Applying Y–Δ Transformation

Simplify the circuit in Fig. 2-30(a) by applying the Y–Δ transformation so as to determine the current I.

Solution: Noting the symmetry rules associated with the transformation, the Δ circuit connected to nodes 1, 3, and 4 can be replaced with a Y circuit, as shown in Fig. 2-30(b), with resistances

$$R_1 = \frac{24 \times 36}{24 + 36 + 12} = 12\ \Omega,$$

$$R_2 = \frac{24 \times 12}{24 + 36 + 12} = 4\ \Omega,$$

and

$$R_3 = \frac{36 \times 12}{24 + 36 + 12} = 6\ \Omega.$$

Next, we add the 4-Ω and 20-Ω resistors in series, obtaining 24 Ω for the right branch of the trapezoid. Similarly, the left branch combines into 12 Ω and the two in-parallel branches reduce to a resistance equal to $(24 \times 12)/(24 + 12) = 8\ \Omega$. When added to the 5-Ω and 12-Ω in-series resistances, this leads to the final circuit in Fig. 2-30(c). Hence,

$$I = \frac{100}{25} = 4\ \text{A}.$$

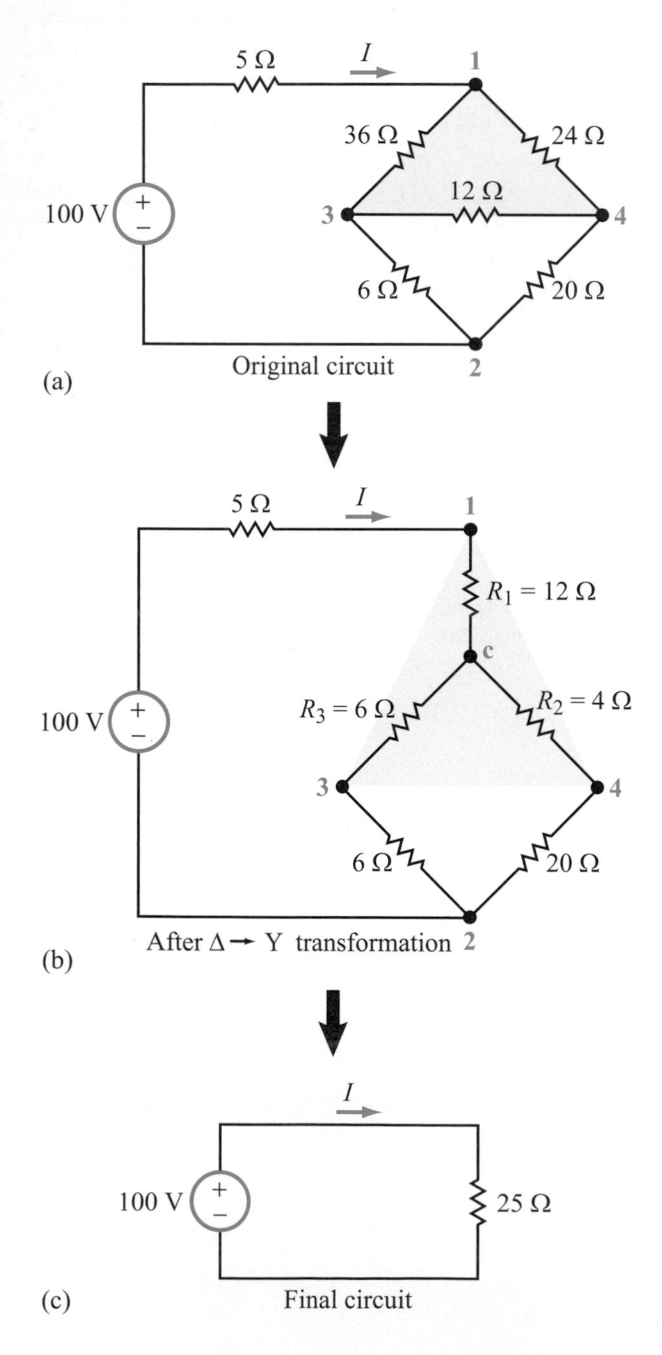

Figure 2-30: Example 2-9 circuit evolution.

Review Question 2-12: When is the Y–Δ transformation used? Describe the inherent symmetry between the resistance values of the Y circuit and those of the Δ circuit.

Review Question 2-13: How are the elements of a balanced Y circuit related to those of its equivalent Δ circuit?

Exercise 2-13: For each of the circuits shown in Fig. E2.13, determine the equivalent resistance between terminals (a, b).

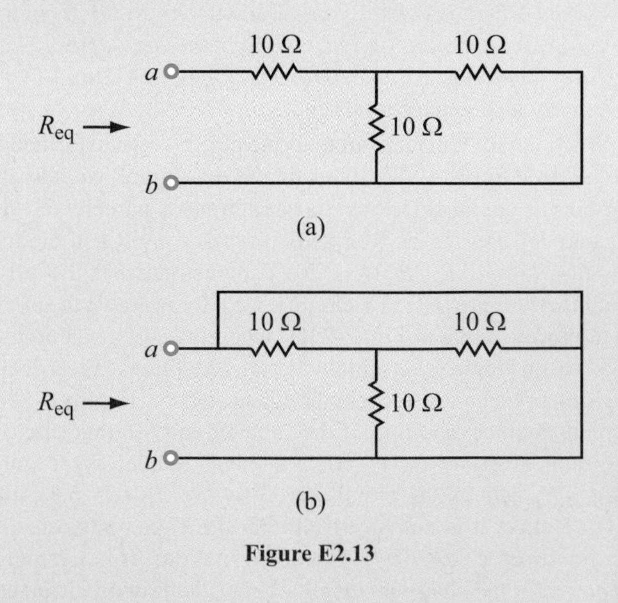

Figure E2.13

Answer: (a) $R_{eq} = 15\ \Omega$, (b) $R_{eq} = 0$. (See)

2-6 The Wheatstone Bridge

Developed initially by Samuel Christie (1784–1865) in 1833 as an accurate ohmmeter for measuring resistance, the Wheatstone bridge subsequently was popularized by Sir Charles Wheatstone (1802–1875), who used it in a variety of practical applications. Today, the Wheatstone-bridge circuit is integral to numerous sensing devices, including strain gauges, force and torque sensors, and inertial gyros. The reader is referred to Technology Brief 4 on page 56 for an illustrative example.

The Wheatstone-bridge circuit shown in Fig. 2-31 consists of four resistors: two fixed resistors (R_1 and R_2) of known values, an adjustable resistor R_3 whose value also is known, and a resistor R_x of unknown resistance. A dc voltage source V_0 is connected between the top node and ground, and an ammeter is connected between nodes 1 and 2. The standard procedure for determining R_x starts by adjusting R_3 so as to make $I_a = 0$. The absence of current flow between nodes 1 and 2, called the ***balanced condition***, implies that $V_1 = V_2$. From voltage division, $V_1 = R_3 V_0/(R_1 + R_3)$, and $V_2 = R_x V_0/(R_2 + R_x)$. Hence,

$$\frac{R_3 V_0}{R_1 + R_3} = \frac{R_x V_0}{R_2 + R_x}. \tag{2.61}$$

A balanced bridge also implies that the voltages across R_1 and R_2 are equal,

$$\frac{R_1 V_0}{R_1 + R_3} = \frac{R_2 V_0}{R_2 + R_x}. \tag{2.62}$$

Dividing Eq. (2.61) by Eq. (2.62) leads to

$$\frac{R_3}{R_1} = \frac{R_x}{R_2},$$

from which we have

$$R_x = \left(\frac{R_2}{R_1}\right) R_3 \quad \text{(Balanced condition).} \tag{2.63}$$

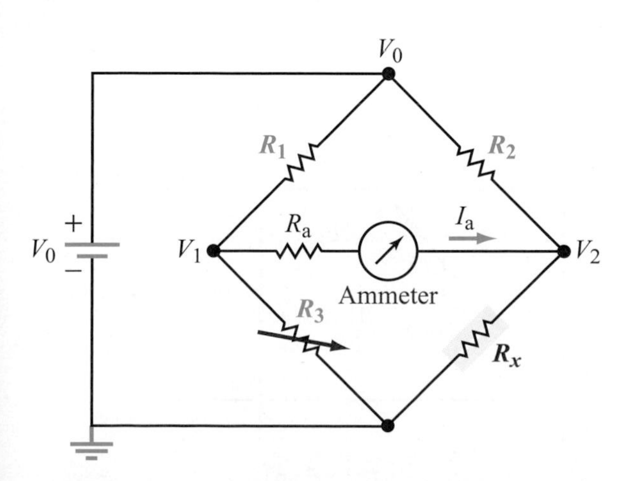

Figure 2-31: Wheatstone-bridge circuit containing an adjustable variable resistor R_3 and an unknown resistor R_x. When R_3 is adjusted to make $I_a = 0$, R_x is determined from $R_x = (R_2/R_1)R_3$.

Example 2-10: Wheatstone-Bridge Sensor

A special version of the Wheatstone bridge (Fig. 2-32) is configured specifically for *measuring small deviations from a reference condition*. An example of a reference condition might be a highway bridge with no load on it. A strain gauge employing a high-sensitivity flexible resistor can measure the small deflection in the bridge surface caused by the weight (force) of a car or truck when present on it. As the force deflects the surface of the bridge to which the resistor is attached, the resistor stretches in length, causing its resistance to increase from a nominal value R (under no stress) to $R + \Delta R$. The other three resistors in the Wheatstone-bridge circuit are all identical and equal to R. Thus, when no vehicles are present on the bridge, the circuit is in the balanced condition.

Develop an approximate expression for V_{out} (the output voltage between nodes 1 and 2) for $\Delta R/R \ll 1$.

Solution: Voltage division gives

$$V_1 = \frac{V_0 R}{R + R} = \frac{V_0}{2}$$

and

$$V_2 = \frac{V_0(R + \Delta R)}{R + (R + \Delta R)} = \frac{V_0(R + \Delta R)}{2R + \Delta R}.$$

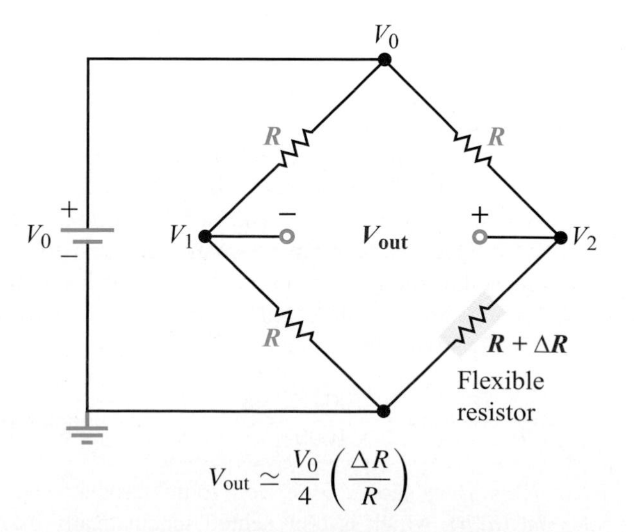

Figure 2-32: Circuit for Wheatstone-bridge sensor.

Hence,

$$\begin{aligned} V_{\text{out}} &= V_2 - V_1 \\ &= \frac{V_0(R+\Delta R)}{2R+\Delta R} - \frac{V_0}{2} \\ &= \frac{2V_0(R+\Delta R) - V_0(2R+\Delta R)}{2(2R+\Delta R)} \\ &= \frac{V_0\,\Delta R}{4R+2\,\Delta R} = \frac{V_0\,\Delta R}{4R(1+\Delta R/2R)}. \end{aligned}$$

Since $\Delta R/R \ll 1$, ignoring the second term in the denominator would incur negligible error. Such an approximation leads to

$$V_{\text{out}} \simeq \frac{V_0}{4}\left(\frac{\Delta R}{R}\right), \tag{2.64}$$

providing a simple linear relationship between the change in resistance ΔR and the output voltage V_{out}.

Review Question 2-14: What is a Wheatstone bridge used for?

Review Question 2-15: What is the *balanced condition* in a Wheatstone bridge?

Exercise 2-14: If in the sensor circuit of Fig. 2-32, $V_0 = 4$ V and the smallest value of V_{out} that can be measured reliably is 1 μV, what is the corresponding accuracy with which $(\Delta R/R)$ can be measured?

Answer: 10^{-6} or 1 part in a million. (See ◎)

2-7 Application Note: Linear versus Nonlinear *i–v* Relationships

Ideal resistors and voltage and current sources are all considered linear elements; the relationship between the current and the voltage across any one of them is described by a straight line. The i–v relationships plotted in Fig. 2-33 for the current source, the voltage source, and the resistor have slopes of 0, ∞, and $1/R$, respectively.

2-7.1 The Fuse: A Simple Nonlinear Element

Many very useful circuit elements do not have linear i–v relationships. Consider Fig. 2-34(a). A realistic voltage source is connected to a load R_{L} at terminals (a,b). Note that the resistance value of the source resistor R_{s} is much smaller than that of the load (1 Ω versus 1 kΩ). It is typical of a well-designed voltage source to have a small source resistor so as to minimize the voltage drop across it. The switch simulates an accidental short circuit. Application of KVL to the loop in Fig. 2-34(a) (with the switch in the open position) leads to

$$I_{\text{s}} = \frac{V_{\text{s}}}{R_{\text{s}}+R_{\text{L}}} = \frac{100}{1+1000} \approx 0.1 \text{ A} \qquad \text{(switch open)}.$$

If, *accidentally*, a short circuit were to be introduced across terminal (a,b), which is represented schematically by the closing of the SPST switch, the current I_{s} will flow entirely through the short circuit, resulting in

$$I_{\text{s}} = \frac{V_{\text{s}}}{R_{\text{s}}} = 100 \text{ A!} \qquad \text{(switch closed)}.$$

This is a very large current. Many household wires would begin to overheat and melt off their insulation at such high currents.

It is precisely for this reason that the ***fuse*** (and later, the ***breaker***) came into heavy use in power-distribution circuits (Fig. 2-34(b)). The i–v curve for a fuse, shown in (Fig. 2-34(c)), is decidedly nonlinear: Above a certain

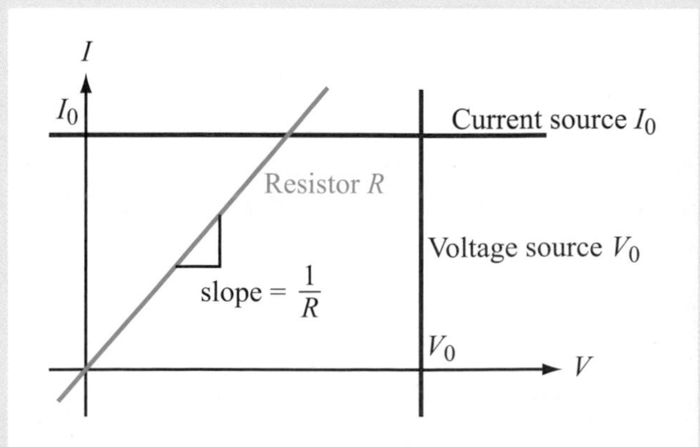

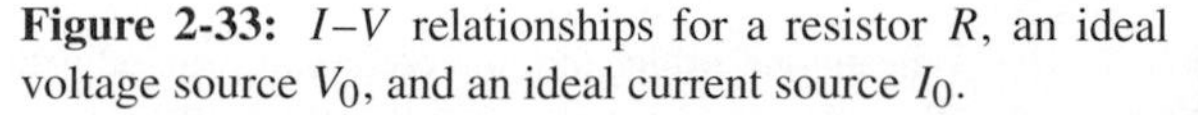
Figure 2-33: I–V relationships for a resistor R, an ideal voltage source V_0, and an ideal current source I_0.

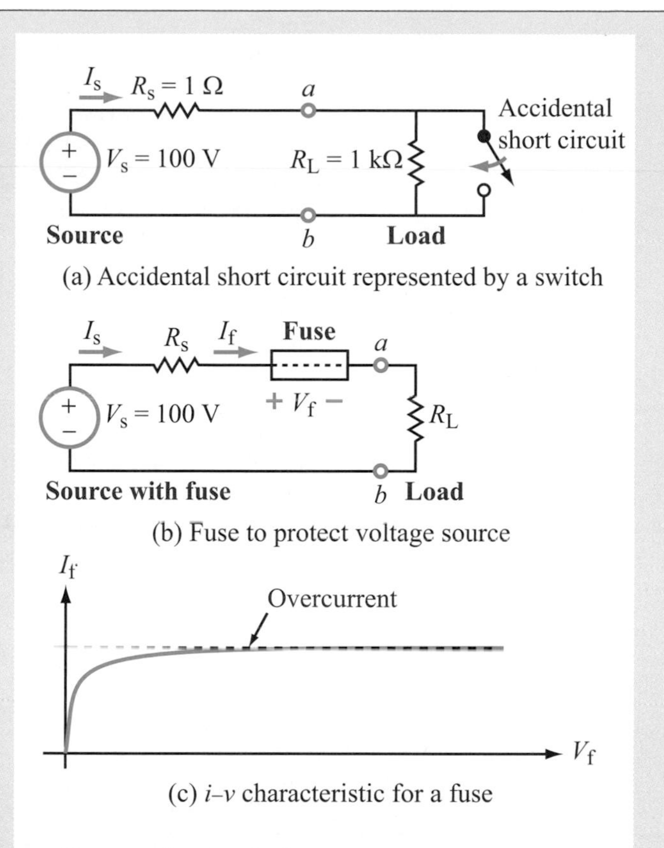

(a) Accidental short circuit represented by a switch

(b) Fuse to protect voltage source

(c) i–v characteristic for a fuse

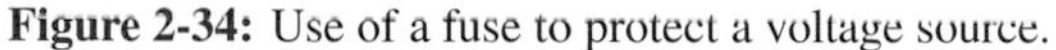

Figure 2-34: Use of a fuse to protect a voltage source.

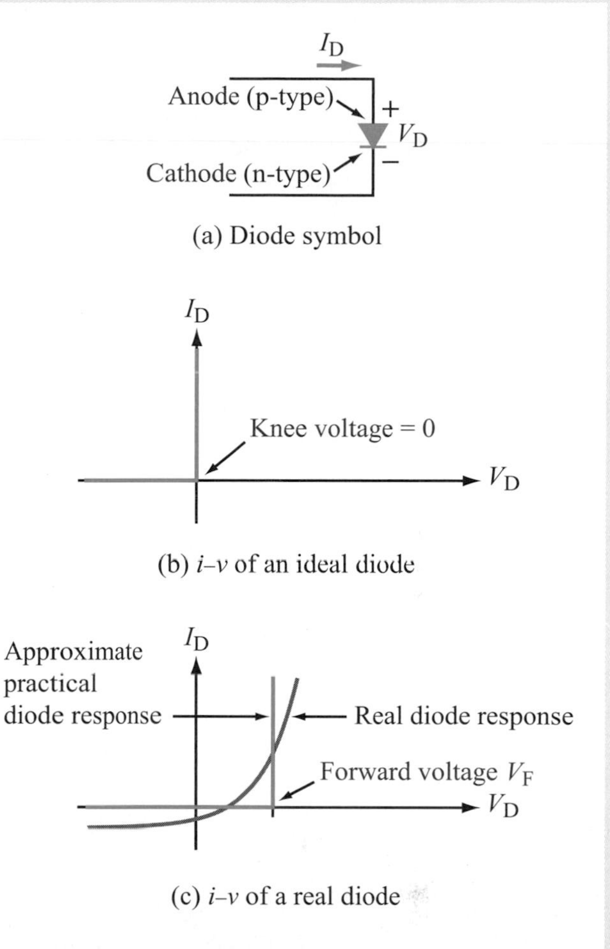

(a) Diode symbol

(b) i–v of an ideal diode

(c) i–v of a real diode

Figure 2-35: pn-junction diode schematic symbol and i–v characteristics.

current level, the fuse will cease to allow more current to pass through it, acting like a current limiter. The physical device contains a small metal wire that is designed to melt away at a specific current level (called its ***overcurrent***), thereby becoming an open circuit and preventing large currents from flowing through the circuit. Note that Fig. 2-34(c) does not explain the fuse's time-dependent behavior; it describes the fuse's behavior only until the moment at which the current exceeds the overcurrent. After that, the fuse just looks like an open circuit.

Fuses also are rated for several other important characteristics such as how fast they can respond. Ultra-fast fuses can trip in milli- to micro-seconds. Another important attribute is the maximum voltage it can sustain across its terminals. Note that in Fig. 2-34(b), once the fuse assumes the role of an open circuit, the voltage across it becomes V_s. If this voltage is too high, arcing and sparks might develop between the terminals (we know from physics that a large-enough voltage in air will break down the air molecules, causing them to conduct and generate a bright spark). Clearly, that is an important rating factor to keep in mind when selecting a fuse.

2-7.2 The Diode: A Solid-State Nonlinear Element

The ***diode*** is a mainstay of solid-state circuits. Its circuit schematic symbol is shown in Fig. 2-35(a) with V_D as the voltage across the diode, defined such that the (+) side is at the anode terminal of the diode and the (−) side at its cathode terminal. There are many types of diodes, including the basic ***pn-junction diode***, the Zener and Schottky diodes, and the ubiquitous ***light-emitting diode*** (LED) used in consumer electronics. An overview of the operation and uses of the LED is available in Technology Brief 5 on page 96. For the present, we will limit our discussion to the pn-junction diode, commonly referred to simply as *the diode*. The *pn* diode consists of a ***p-type*** semiconductor placed in contact with an ***n-type*** semiconductor, thereby forming a *junction*. The p-type material is so named because the impurities that have been added to its bulk material result in a crystalline structure in which the available charged carriers are predominantly ***positive*** charges. The opposite is true for the n-type material; different types of impurities are added to the bulk material, as a result of which the predominant carriers are ***negative*** charges (electrons). In the absence of a voltage across the diode, the two sets of carriers diffuse away from each other at the edge of the junction, generating an associated built-in potential barrier (voltage), called the ***forward-bias voltage*** or ***offset voltage*** V_F.

The main use of the diode is as a *one-way valve* for current. Figure 2-35(b) displays the i–v relationship for an ***ideal diode***, which conveys the following behavior:

> Current can flow through the diode from the (+) terminal to the (−) terminal unimpeded, regardless of its magnitude, but it cannot flow in the opposite direction.

In other words, an ideal diode looks like a short circuit for positive values of V_D and like an open circuit for negative values of V_D. These two states are called ***forward bias*** and ***reverse bias***, respectively. When a positive-bias voltage exceeding V_F is applied to the diode, the potential barrier is counteracted, allowing the flow of current from p to n (which includes positive charges flowing in that direction as well as negative charges flowing in the opposite direction). On the other hand, if a negative-bias voltage is applied to the diode, it adds to the potential barrier, further restricting the flow of charges across the barrier and resulting in no current flow from n to p.

The voltage level at which the diode switches from reverse bias to forward bias is called the ***knee voltage*** or ***forward-bias voltage***. For the ideal diode, $V_F = 0$ and the knee is at $V_D = 0$, which means that the forward-bias segment of its i–v characteristic is aligned perfectly along the I_D-axis, as shown in Fig. 2-35(b).

Real diodes differ from the ideal diode model in two important respects: (1) the knee in the curve is not at $V_D = 0$, and (2) the diode does not behave exactly like a perfect short circuit when in forward bias nor like a perfect open circuit when in reverse bias. Figure 2-35(c) shows a real diode i–v curve, and an approximate, equivalent, diode model for use in practical applications. Note how nonlinear a real diode really is! For many electrical engineering applications, however, the nonlinearities are not so important, and the approximate ideal-like diode model is quite sufficient. The only difference between the ideal diode model of Fig. 2-35(b) and the approximate diode model of Fig. 2-35(c) is that in the latter the transition from reverse to forward bias occurs at a non-zero, positive value of V_D, namely the ***forward-bias voltage*** V_F. For a silicon pn-junction diode, a typical value of V_F is 0.7 V. We always should remember that V_F is a property of the diode itself, not of the circuit it is a part of.

Example 2-11: Diode Circuit

The circuit in Fig. 2-36 contains a diode with $V_F = 0.7$ V. Determine I_D.

Solution: Initially, we do not know whether the diode is forward biased or reverse biased. We will first assume it is forward biased in order to compute I_D. Then, if it turns out that I_D is positive, our assumption will have been validated, but if I_D is negative, we will conclude that the diode is reverse biased and no current flows through the circuit.

Application of KVL around the loop gives

$$-V_s + I_D R + V_D = 0.$$

If the diode is forward biased, $V_D = 0.7$ V, which leads to

$$I_D = \frac{V_s - V_D}{R} = \frac{5 - 0.7}{100} = 43 \text{ mA}.$$

The positive sign of I_D confirms our assumption that the diode is indeed forward biased.

As an interesting aside, one could use this circuit to control the current through a light-emitting diode (LED). As explained in Technology Brief 5 on page 96, the amount of light emitted by an LED (i.e., how bright it appears) is proportional directly to the current I_D passing through it when it is forward biased. By using the circuit in Fig. 2-36 and choosing an appropriate value for R, we can build a circuit that forward biases an LED and controls its brightness.

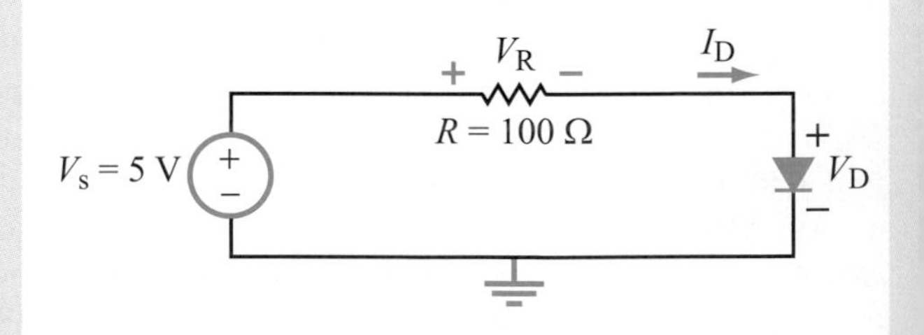

Figure 2-36: Diode circuit of Example 2-11.

Review Question 2-16: What is the *overcurrent* of a fuse?

Review Question 2-17: Why does a pn-junction diode have a non-zero forward-bias voltage V_F?

Exercise 2-15: Determine I in the two circuits of Fig. E2.15. Assume $V_F = 0.7$ V for all diodes.

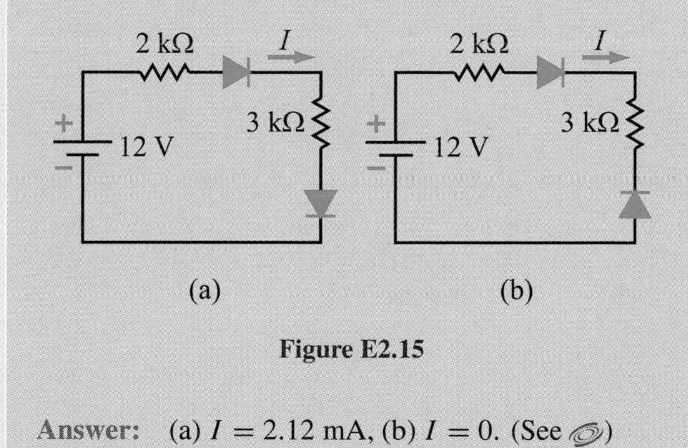

Figure E2.15

Answer: (a) $I = 2.12$ mA, (b) $I = 0$. (See)

2-7.3 Piezoresistor Circuit

According to Technology Brief 4 on page 56, if we apply a force on a resistor along its axis (Fig. 2-37), the resistance changes from R_0, which is the resistance with no stress (pressure) applied, to R as

$$R = R_0 + \Delta R, \tag{2.65}$$

and the deviation ΔR is given by

$$\Delta R = R_0 \alpha P, \tag{2.66}$$

where α is a property of the material that the resistor is made of and is called its ***piezoresistive coefficient***, and P is the ***mechanical stress*** applied to the resistor. The unit for P is newtons/m^2 (N/m^2) and the unit for α is the inverse of that. Compression decreases the length of the resistor and increases its cross section, so in view of Eq. (2.2), which states that the resistance of a longitudinal resistor is given by $R = \rho\ell/A$, the consequence of a compressive force—namely reduction in ℓ and increase in A—leads to a reduction in the magnitude of R. Hence, for compression, ΔR is negative, requiring that α in Eq. (2.66) be defined as a negative quantity.

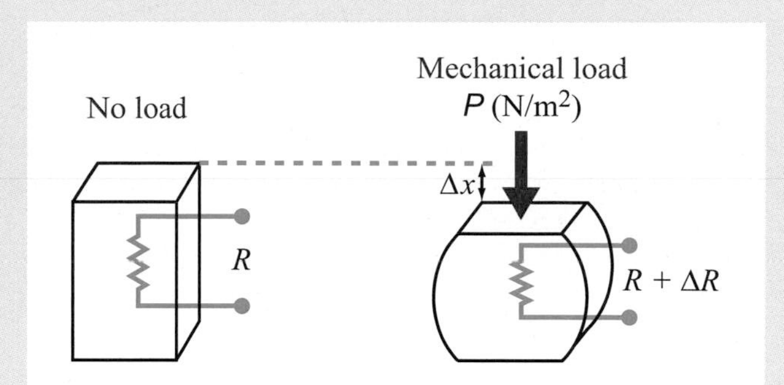

Figure 2-37: The resistance of a piezoresistor changes when mechanical stress is applied.

If a piezoresistor is integrated into a Wheatstone-bridge circuit (as in Fig. 2-32), such that all three other resistors are given by R_0, the expression for the voltage output given by Eq. (2.64) becomes

$$V_{out} = \frac{V_0}{4}\left(\frac{\Delta R}{R_0}\right) = \frac{V_0}{4}\,\alpha P. \tag{2.67}$$

Since V_0 and α are both constants, the linear relationship between the applied stress P and the output voltage V_{out} makes the piezoresistor a natural sensor for detecting or measuring mechanical stress. However, we should examine the sensitivity of such a sensor. As a reference, a finger can apply about 50 N of force across an area of 1 cm^2 (10^{-4} m^2), which is equivalent to a pressure $P = 5 \times 10^5$ N/m^2. If the piezoresistor is made of silicon with $\alpha = -1 \times 10^{-9}$ m^2/N and if the dc source in the Wheatstone bridge is $V_0 = 1$ V, Eq. (2.67) yields the result that $V_{out} = 125\ \mu$V, which is not impossible to measure but quite small nevertheless. How then are such pressure sensors used?

The answer is simple: We need a mechanism to amplify the signal. We can do so electronically by feeding V_{out} into a high-gain amplifier, or we can amplify the mechanical pressure itself before applying it to the piezoresistor. The latter approach can be realized by constructing the piezoresistor into a cantilever structure, as shown in Fig. 2-38 (a cantilever is a fancy name for a "diving board" with one end fixed and the other free). Deflection of the cantilever tip induces stress at the base of the cantilever near the attachment point. If properly designed, the cantilever—which usually is made of silicon or metal—can amplify the applied stress by several orders of magnitude (see Example 2-12).

Example 2-12: A Realistic Piezoresistor Sensor

When a force F is applied on the tip of a cantilever of width W, thickness H, and length L (as shown in Fig. 2-38) the corresponding stress exerted on the piezoresistor attached to the cantilever base is given by

$$P = \frac{FL}{WH^2}. \tag{2.68}$$

Determine the output voltage of a Wheatstone-bridge circuit if $F = 50$ N, $V_0 = 1$ V, the piezoresistor is made of silicon, and the cantilever dimensions are $W = 0.5$ cm, $H = 0.5$ mm, and $L = 1$ cm.

Solution: Combining Eqs. (2.67) and (2.68) gives

$$\begin{aligned} V_{\text{out}} &= \frac{V_0}{4}\,\alpha \cdot \frac{FL}{WH^2} \\ &= \frac{1}{4} \times (-1 \times 10^{-9}) \times \frac{50 \times 10^{-2}}{(5 \times 10^{-3}) \times (5 \times 10^{-4})^2} \\ &= -0.1 \text{ V}. \end{aligned}$$

The integrated piezoresistor–cantilever arrangement generates an output voltage whose magnitude is on the order of 400 times greater than that generated by pressing on the resistor directly!

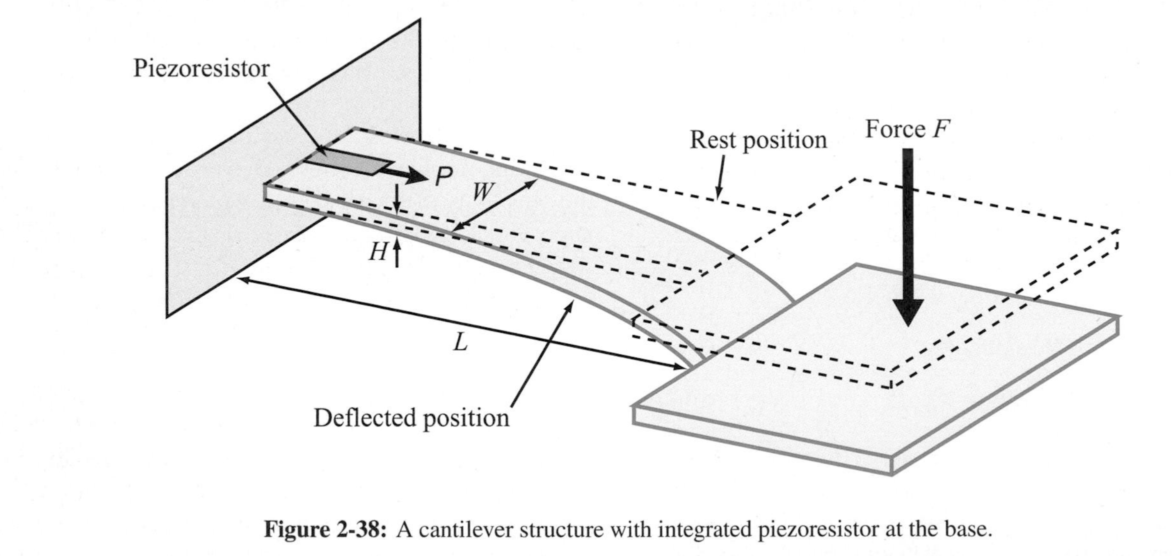

Figure 2-38: A cantilever structure with integrated piezoresistor at the base.

Review Question 2-18: Does compression along the current direction increase or decrease the resistance? Why?

Review Question 2-19: Why are piezoresistors placed at the base of cantilevers and other deflecting structures?

Exercise 2-16: What would the output voltage associated with the circuit of Example 2-12 change to, if the cantilever thickness is reduced by a factor of 2?

Answer: $V_{\text{out}} = -0.4$ V. (See ◎)

2-8 Introducing Multisim

Multisim 11 is the latest edition of National Instruments ***SPICE*** simulator software. SPICE, originally short for Simulation Program with Integrated Circuit Emphasis, was developed by Larry Nagel at the University of California, Berkeley, in the early 1970s. It since has inspired and been used in many academic and commercial software packages to simulate analog, digital, and mixed-signal circuits. Modern SPICE simulators like Multisim are indispensable in integrated circuit design; ICs are so complex that they cannot be built and tested on a breadboard

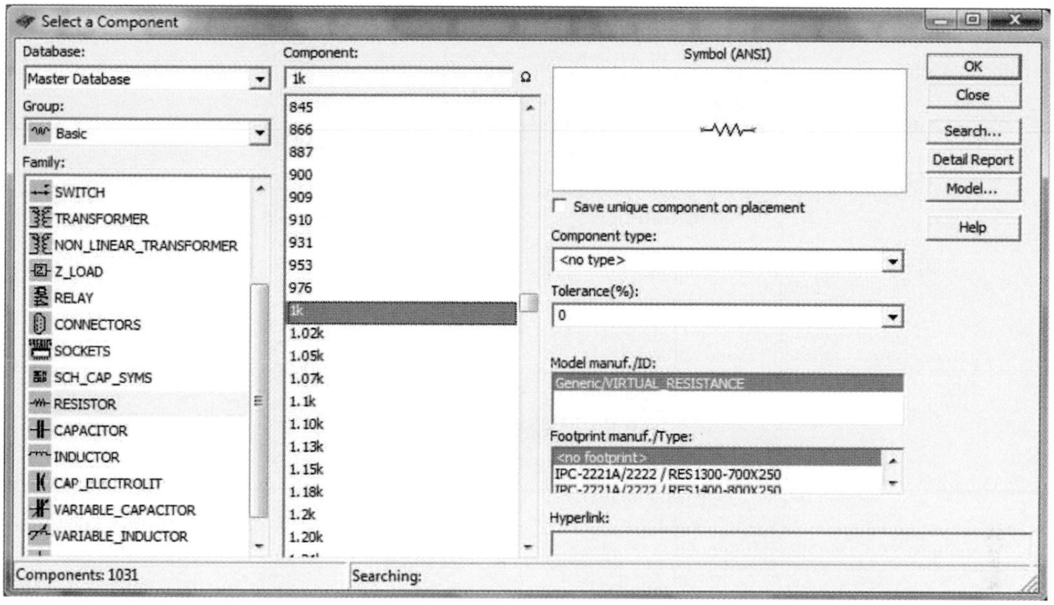

Figure 2-39: Multisim screen for selecting and placing a resistor.

ahead of production (see Technology Brief 7 on page 135). With SPICE, you can draw a circuit from a library of components, specify how the components are connected, and ask the program to solve for all voltages and currents at any point in time. Modern SPICE packages like Multisim include very intuitive graphic user interface (GUI) tools that make both circuit design and analysis very easy. Multisim allows the user to simulate a laboratory experience on his/her computer ahead of actually working with real components.

In this section, you will learn how to:

- Set up and analyze a simple dc circuit in Multisim.
- Use the Measurement Probe tool to quickly solve for voltages and currents.
- Use the Analysis tools for more comprehensive solutions.

We will return to these concepts and learn to apply many other analysis tools throughout the book. Appendix C provides an introduction to the Multisim Tutorial available on the CD that accompanies the book. The Tutorial is a useful reference if you have never used Multisim before. When defining menu selections starting from the main window, the format Menu → Sub-Menu1 → Sub-Menu2 will be used.

2-8.1 Drawing the Circuit

After installing and running Multisim, you will be presented with the ***basic user interface*** window, also referred to as the ***circuit window*** or the ***schematic capture window*** (see Multisim Tutorial on accompanying CD). Here, we will draw our circuits much as if we were drawing them on paper.

Step 1: Placing Resistors in the Circuit

Components in Multisim are organized into a hierarchy going in a descending general order from Database → Group → Family → Component. Every component that you use in Multisim will fit into this hierarchy somewhere.

Place → Component opens the Select a Component window. (Ctrl-W is the shortcut key for the place-component command. Multisim has many shortcut keys, and it will be worthwhile for you to learn some of the basic ones to improve your efficiency in creating and testing circuits.)

Choose Database: Master Database and Group: Basic in the pulldown menus.

Now select Family: RESISTOR.

You should see a long list of resistor values under Component and the schematic symbol for a resistor (Fig. 2-39). Note that the Family menu contains other components like inductors, capacitors, potentiometers, and many more. We will use these in later chapters.

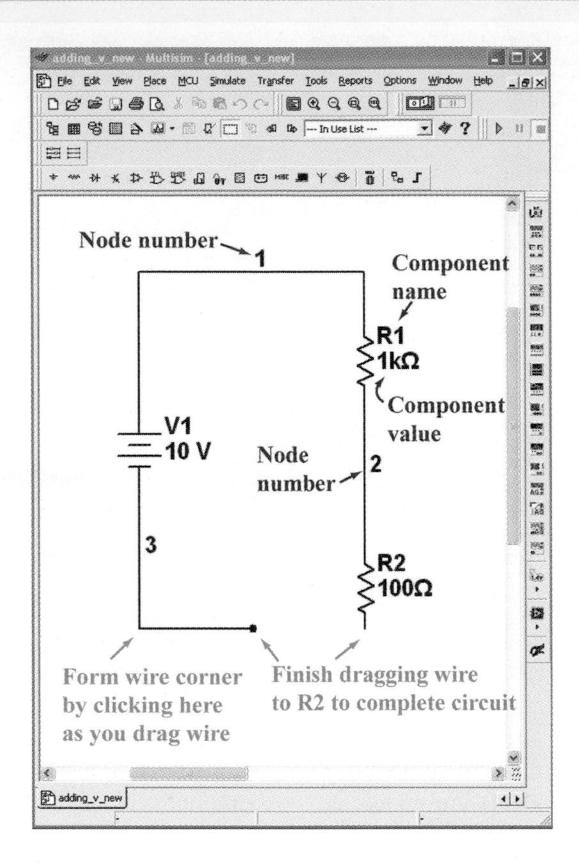

Figure 2-40: Adding a voltage source and completing the circuit.

Scroll down and select a 1k value (the units are in ohms) and then click OK. You should see a resistor in the capture window. Before clicking in the window, Ctrl-R allows you to rotate the resistor in the window. Rotate the resistor such that it is vertical and then click anywhere on the window to place it. Repeat this operation; this time place a vertical 100-ohm resistor directly below the first one (as in Fig. 2-40). How to connect them together will be described shortly. Once you are finished placing components, click Close to return to the schematic capture window.

Note that the components have symbolic names (R1 and R2) and values displayed next to them (1k and 100). Also, by double-clicking on a specific component, you can access many details of the component model and its values. For now, it is sufficient to know that the Resistance value can be altered at any time through the Value menu.

Step 2: Placing an Independent Voltage Source

Just as you did with the resistors, open up the Select a Component window.

Choose Database: Master Database and Group: Sources in the pulldown menus.

Select Family: POWER_SOURCES.

Under Component select DC_POWER and click OK.

Place the part somewhere to the left of the two resistors (Fig. 2-40).

Once placed, close the component window, then double-click on component V1. Under the Value tab, change the Voltage to 10 V. Click OK.

Step 3: Wiring Components Together

Place → Wire allows you to use your mouse to wire components together with click-and-drag motions (Ctrl-Q is the shortcut key for the wire command). You can also enable the wire tool automatically by moving the cursor very close to a component node; you should see the mouse pointer change into a black circle with a cross-hair.

Click on one of the nodes of the dc source with the wire tool activated (you should see the mouse pointer change from a black cross to a black circle with a cross hair when you hover it over a node). Additional clicks anywhere in the schematic window will make corners in the wire. Double-clicking will terminate the wire. Additionally, when not already dragging a wire, double-clicking on any blank spot of the schematic will generate a wire based at the origin of clicking.

Wire the components as shown in Fig. 2-40. Add a GROUND reference point as shown in Fig. 2-41. The Ground can be found in the Component list of POWER_SOURCES. We now have a resistive divider.

2-8.2 Solving the Circuit

In Multisim, there are two broad ways in which to solve a circuit. The first, called ***Interactive Simulation***, allows you to utilize virtual instruments (such as ohmmeters, oscilloscopes, and function generators) to measure aspects of a circuit in a time-based environment. It is best to think of the Interactive Simulation as a simulated "in-lab" experience. Just as in real life, time proceeds in the Interactive Simulation as you analyze the circuit (although the rate at which time proceeds is heavily dependent on your computer's processor speed and the resolution of the simulation). The Interactive Simulation is started using the F5 key, the button, or the toggle switch. The simulation is paused using the F6 key, the button, or the button. The simulation is terminated using either the button or the toggle switch.

The other main way in which to solve a circuit in Multisim is through ***Analyses***. These simulations display their outputs not in instruments, but rather in the ***Grapher*** window (which may produce tables in some instances). These simulations are run for controlled amounts of time or over controlled sweeps of specific variables or other aspects of the circuit. For example, a dc sweep simulates the values of a specified voltage or current in the circuit over a defined range of dc input values.

Each of the methods described has its own advantages and disadvantages, and in fact, both varieties can perform many of the same simulations, albeit with different advantages. The choice of method to be used for a given circuit really comes down to your preferences, which will be formed as you gain more experience with Multisim.

For the circuit in Fig. 2-41, we wish to solve for the voltages at every node and the currents running through every branch. As you will often see in Multisim, the solution can be obtained using either the Interactive Simulation or through one of the Analyses. We will demonstrate both approaches.

Interactive Simulation

Selecting Simulate → Instruments → Measurement Probe allows you to drag and place a measurement probe onto any node in the circuit. (Note that the Instruments menu contains many common types of equipment used in an electronics laboratory.) The Measurement Probe constantly reports both the current running through the branch to which it is assigned and the voltage at that node. Place two probes into the circuit as shown in Fig. 2-41. When placed, by default, the probes should be pointing in the direction shown in Fig. 2-41. If they are not, you can reverse a probe's direction by right-clicking on it and pressing Reverse Probe Direction. Once the probes are in place, you must run the simulation using the commands for Interactive Simulations.

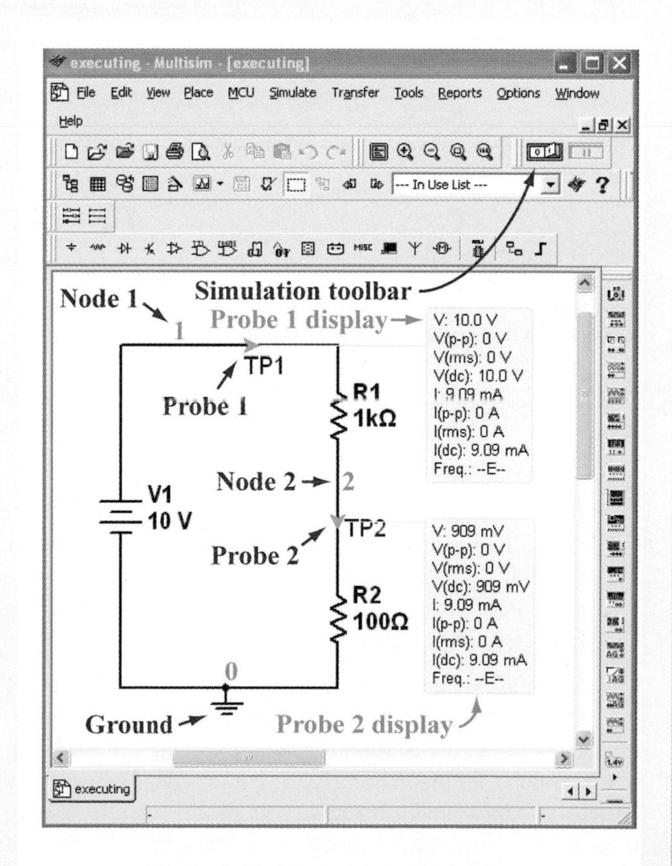

Figure 2-41: Executing a simulation.

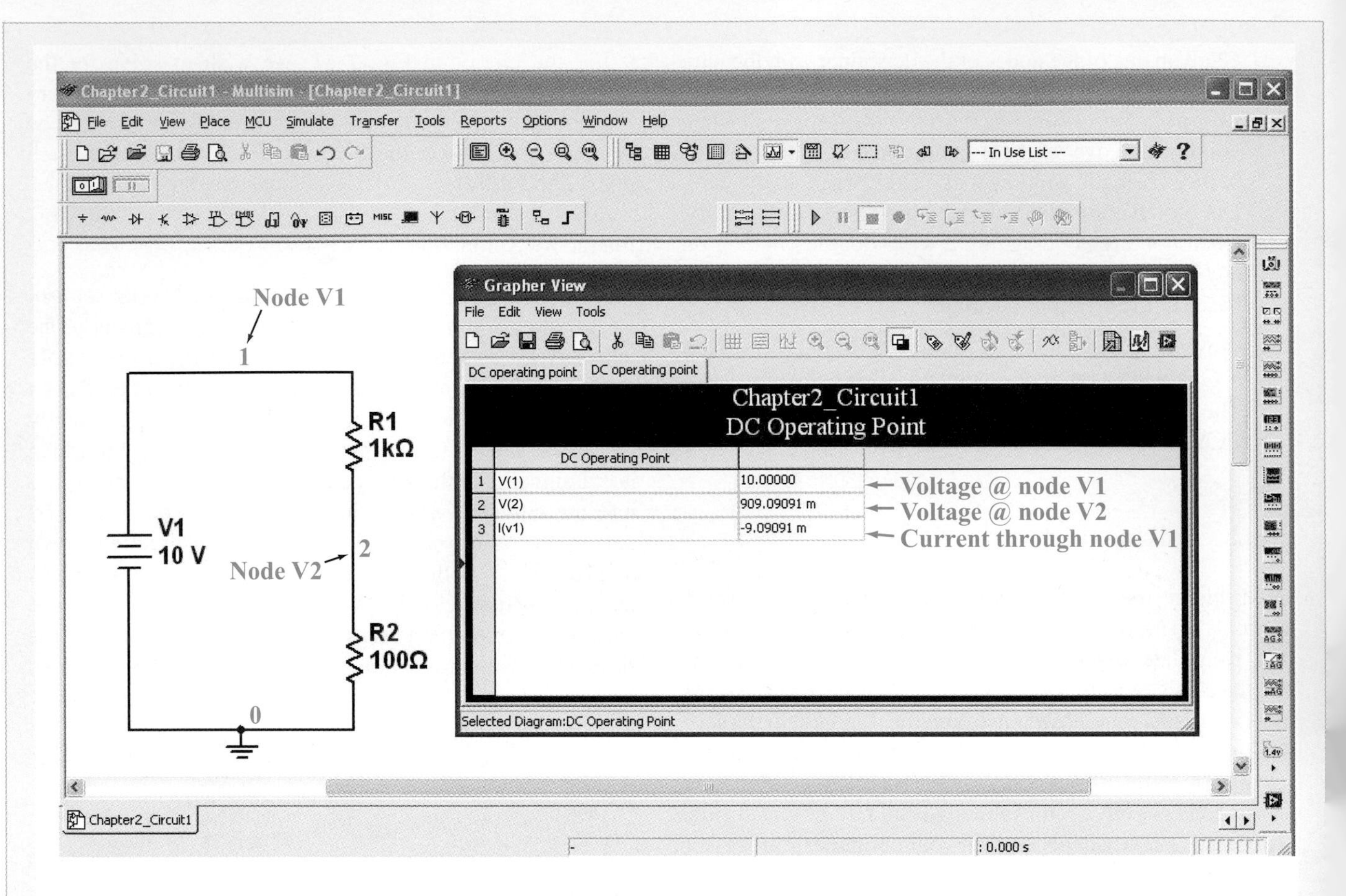

Figure 2-42: Solution window.

As expected, the current running through both wires is the same since the circuit has only one loop.

$$I = \frac{V_1}{R_1 + R_2} = \frac{10}{1000 + 100} = 9.09 \text{ mA}.$$

The voltage at node 1 is 10 V, as defined by the source. Application of voltage division (Fig. 2-18) gives

$$V_2 = \left(\frac{R_2}{R_1 + R_2}\right) V_1 = \left(\frac{100}{1100}\right) 10 = 0.909 \text{ V}.$$

DC Operating Point Analysis

The circuit also can be solved using Simulate → Analyses → DC Operating Point. This method is more convenient than the Interactive Simulation when solving circuits with many nodes. After opening this window, you can specify which voltages and currents you want solved. [The Interactive Simulation mode must be stopped, not just paused, in order for the DC Operating Point Analysis mode to work.] Under the Output tab, select the two node voltages and the branch current in the Variables in Circuit window. Make sure the Variables in Circuit pull-down menu is set to All Variables. Once selected, click Add and they will appear in the Selected variables for analysis window. Once you have selected all of the variables for which you want solutions, simply click Simulate. Multisim then solves the entire circuit and opens a window showing the values of the selected voltages and currents (Fig. 2-42).

2-8.3 Dependent Sources

Multisim provides both defined dependent sources (voltage-controlled current, current-controlled current, etc.) and a generic dependent source whose definition can be entered as a mathematical equation. We will use this second type in the following example.

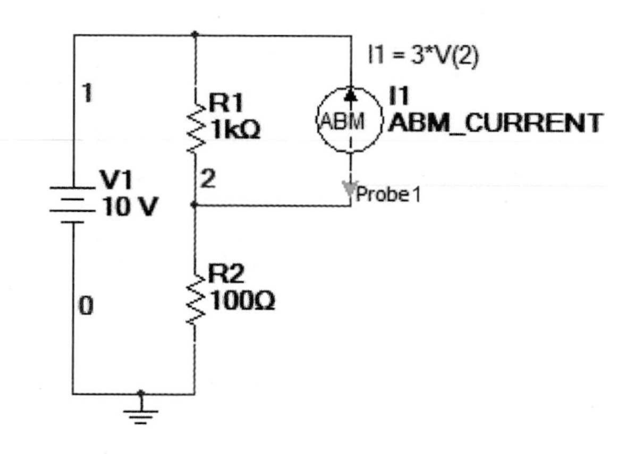

Figure 2-43: Creating a dependent source.

Step 1: The dependent sources are established as follows: Place → Component opens the Select a Component window.

Choose Database: Master Database and Group: Sources in the pulldown menus.

Select Family: CONTROLLED_VOLTAGE or CONTROLLED_CURRENT.

Under Component, select ABM_VOLTAGE or ABM_CURRENT and click OK.

The value of ABM sources (which stands for Analog Behavioral Modeling) can be set directly with mathematical expressions using any variables in the circuit. For information on the variable nomenclature, which may be somewhat confusing, see the Multisim Tutorial on the CD-ROM.

Step 2: Using what you learned in Section 2-8.1, draw the circuit shown in Fig. 2-43 (including the probe at node 2).

Step 3: Double-click the ABM_CURRENT source. Under the value tab, enter: 3*V(2). The expression V(2) refers to the voltage at node 2. This effectively defines this source as a voltage-controlled current source. Note that when making the circuit, if the node numbering in your circuit differs from that in the example (e.g., if nodes 1 and 2 are switched), then take care to keep track of the differences so that you will use the proper node voltage when writing the equation. To edit or change node labels, double-click any wire to open the Net Window. Under Net name enter the label you like for that node.

To write the expression for I1 next to the current source, go to Place → Text, and then type in the expression at a location near the current source. [Ctrl-T is the shortcut key for the place-text command.]

Referencing Currents in Arbitrary Branches

Now let us analyze the circuit using the DC Operating Point Analysis. Our goal is to solve for the voltages at every node and the current running through each branch. Remove the probe from the circuit if you still have it in there by clicking on it so it is highlighted and pressing the Delete key.

To perform a DC operating point analysis, just as we did earlier in Section 2-8.2, go to Simulate → Analyses → DC Operating Point and transfer all available variables into the Selected variables for analysis window. You should notice that the only variables available are V(1), V(2), and I(v1); if Probe 1 is still connected to your circuit, you should also see I(Probe 1) and V(Probe 1). Where are the other currents, such as the current flowing through R1, the current through R2, or even the current coming out of the dependent source? In Multisim and most SPICE software in general, you can only measure/manipulate currents through a Voltage Source (there are some exceptions, but we will ignore them for now). This is why the current through V1, denoted I(v1), is available but the currents through the other components are not. A simple trick, however, to obtain these currents is to add a 0 V dc source into the branches where you want to measure current. Do this to your circuit, so that it ends up looking like that shown in Fig. 2-44.

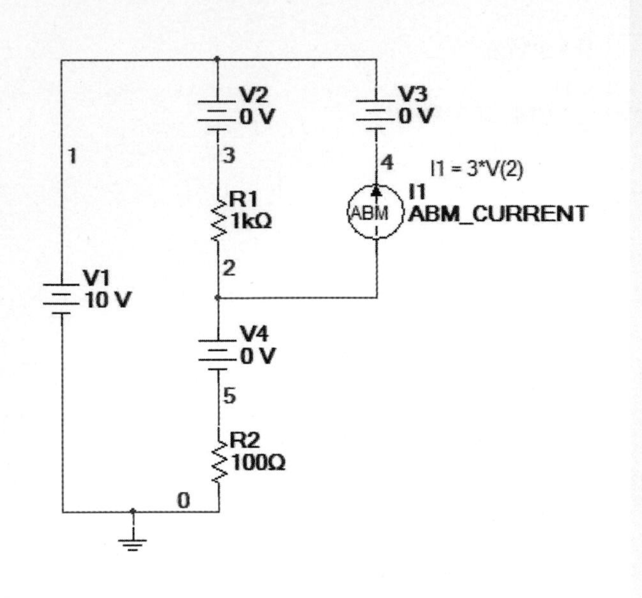

Figure 2-44: Circuit from Fig. 2-43 adapted to read out the currents through R1, R2, and the dependent source.

You will notice that there are new nodes in the circuit now, but since V2, V3, and V4 are 0 V sources, V(3) = V(4) = V(1) and V(5) = V(2).

Go back to the DC Operating Point Analysis window and under the Variables in Circuit window there should now be four currents (I(v1), I(v2), I(v3), and I(v4)) and the five voltages. Highlight all four currents as well as V(1) and V(2) and click Add and then click OK. This will bring up the Grapher window with the results of the analysis.

Note that when we analyze the currents through the branches, the current through a voltage source is defined as going *into* the positive terminal. For example, in source V1, this corresponds to the current flowing *from* Node 1 into V1 and then *out* of V1 to Node 0.

Review Question 2-20: In Multisim, how are components placed and wired into circuits?

Review Question 2-21: How do you obtain and visualize the circuit solution?

Exercise 2-17: The circuit in Fig. E2.17 is called a resistive bridge. How does $V_x = (V_3 - V_2)$ vary with the value of potentiometer R_1?

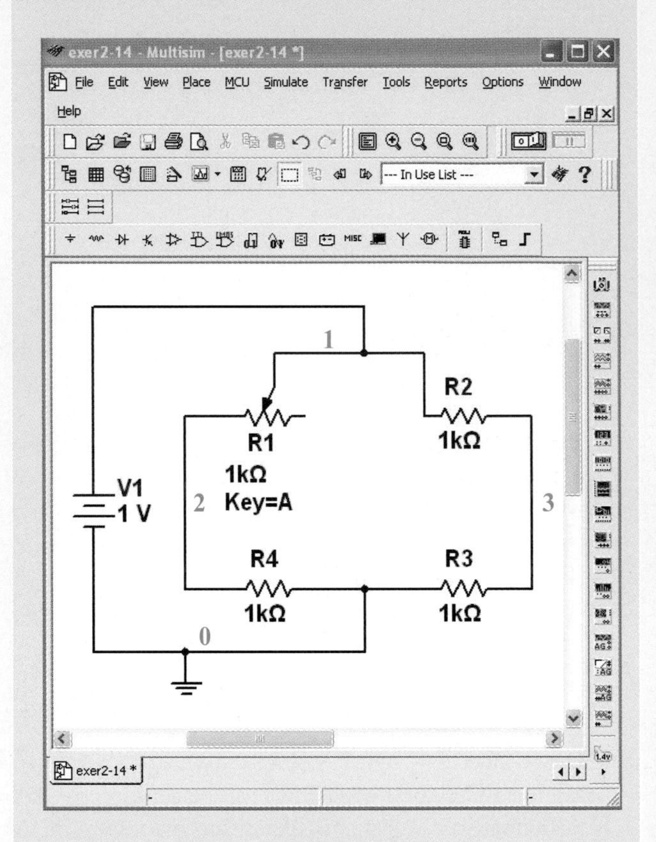

Figure E2.17

Answer: (See ◎)

Exercise 2-18: Simulate the circuit shown in Fig. E2.18 and solve it for the voltage across R_3. The magnitude of the dependent current source is $V_1/100$.

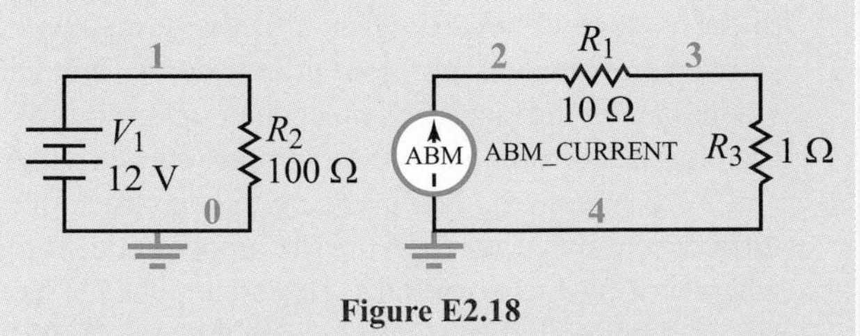

Figure E2.18

Answer: (See ◎)

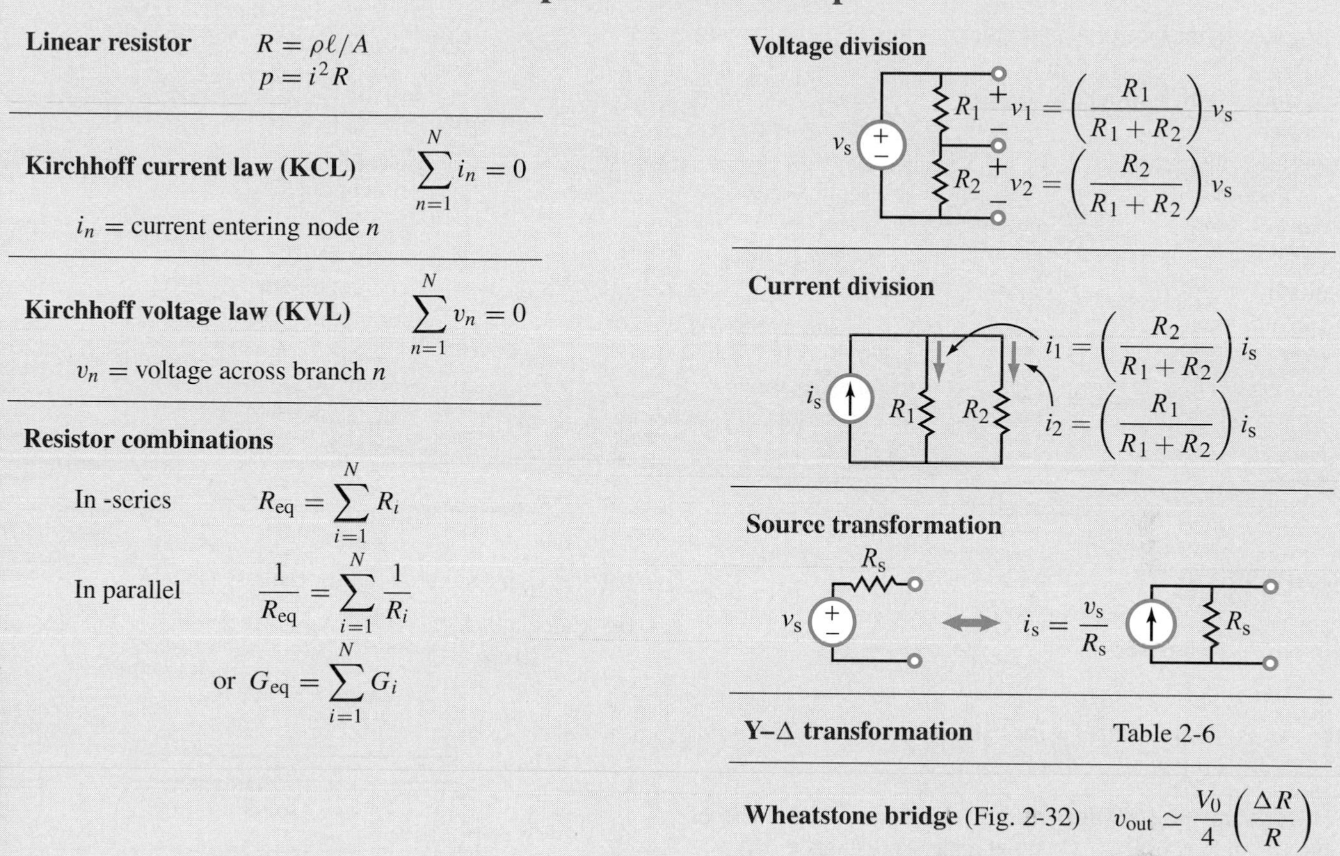

Chapter 2 Relationships

Linear resistor $R = \rho\ell/A$

$p = i^2R$

Kirchhoff current law (KCL) $\sum_{n=1}^{N} i_n = 0$

i_n = current entering node n

Kirchhoff voltage law (KVL) $\sum_{n=1}^{N} v_n = 0$

v_n = voltage across branch n

Resistor combinations

In series $R_{eq} = \sum_{i=1}^{N} R_i$

In parallel $\frac{1}{R_{eq}} = \sum_{i=1}^{N} \frac{1}{R_i}$

or $G_{eq} = \sum_{i=1}^{N} G_i$

Voltage division

$v_1 = \left(\frac{R_1}{R_1 + R_2}\right) v_s$

$v_2 = \left(\frac{R_2}{R_1 + R_2}\right) v_s$

Current division

$i_1 = \left(\frac{R_2}{R_1 + R_2}\right) i_s$

$i_2 = \left(\frac{R_1}{R_1 + R_2}\right) i_s$

Source transformation

$i_s = \frac{v_s}{R_s}$

Y–Δ transformation Table 2-6

Wheatstone bridge (Fig. 2-32) $v_{out} \simeq \frac{V_0}{4}\left(\frac{\Delta R}{R}\right)$

CHAPTER HIGHLIGHTS

- As described by Ohm's law, the i–v relationship of a resistor is linear over a specific range ($-i_{max}$ to $+i_{max}$); however, R may vary with temperature (thermistors) and pressure (piezoresistors).
- Circuit topology defines the relationships between nodes, loops, and branches.
- Kirchhoff's current and voltage laws form the foundation of circuit analysis and synthesis.
- Two circuits are considered equivalent if they exhibit identical i–v characteristics relative to an external circuit.
- Source transformation allows us to represent a real voltage source by an equivalent real current source, and vice versa.
- A Y circuit configuration can be transformed into a Δ configuration, and vice versa.
- The Wheatstone bridge is a circuit used to measure resistance, as well as to detect small deviations (from a reference condition), as in strain gauges and other types of sensors.
- Nonlinear resistive elements include the light bulb, the fuse, the diode, and the light-emitting diode (LED).
- Multisim is a software simulation program capable of simulating electric circuits and analyzing their behavior.

GLOSSARY OF IMPORTANT TERMS

Provide definitions or explain the meaning of the following terms:

balanced bridge circuit
branch
circuit equivalence
conductivity
current divider
dielectric
diode
forward bias
fuse
insulator
KCL
KVL
LED
loop
mechanical stress
mesh
Multisim
node
overcurrent
path
piezoresistive coefficient
piezoresistor
planar circuit
potentiometer
power rating
resistance
resistivity
reverse bias
rheostat
semiconductor
sensitivity
source transformation
superconductor
thermistor
topology
voltage divider
Wheatstone bridge
Y–Δ transformation

PROBLEMS

Section 2-1: Ohm's Law

*2.1 An AWG-14 copper wire has a resistance of 17.1 Ω at 20°C. How long is it?

2.2 A 3-km long AWG-6 metallic wire has a resistance of approximately 6 Ω at 20°C. What material is it made of?

2.3 A thin-film resistor made of germanium is 2 mm in length and its rectangular cross section is 0.2 mm × 1 mm, as shown in Fig. P2.3. Determine the resistance that an ohmmeter would measure if connected across its:

(a) Top and bottom surfaces

*(b) Front and back surfaces

(c) Right and left surfaces

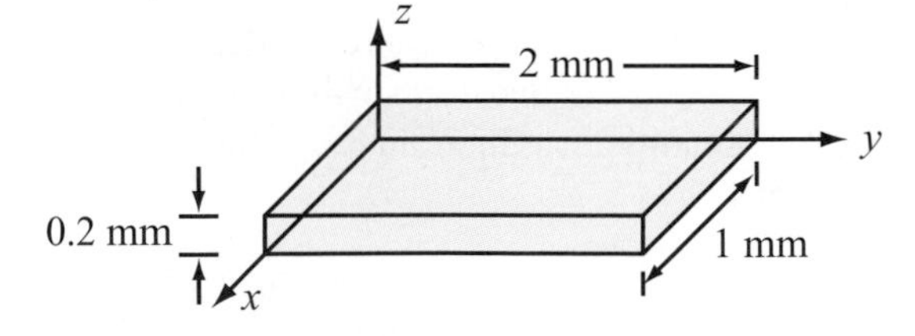

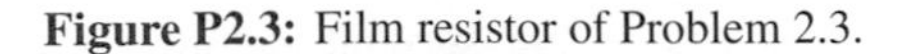

Figure P2.3: Film resistor of Problem 2.3.

2.4 A resistor of length ℓ consists of a hollow cylinder of radius a surrounded by a layer of carbon that extends from $r = a$ to $r = b$, as shown in Fig. P2.4.

(a) Develop an expression for the resistance R.

(b) Calculate R at 20°C for $a = 2$ cm, $b = 3$ cm and $\ell = 10$ cm.

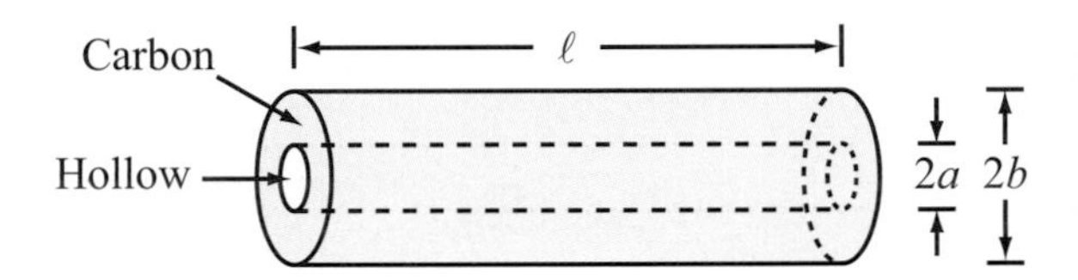

Figure P2.4: Carbon resistor of Problem 2.4.

2.5 A standard model used to describe the variation o resistance with temperature T is given by

$$R = R_0(1 + \alpha T),$$

where R is the resistance at temperature T (measured i °C), R_0 is the resistance at $T = 0$°C, and α is a temperatur coefficient. For copper, $\alpha = 4 \times 10^{-3}$°C^{-1}. At wha temperature is the resistance greater than R_0 by 1 percent?

2.6 A light bulb has a filament whose resistance i characterized by a temperature coefficient $\alpha = 6 \times 10^{-3}$°C^{-} (see resistance model given in Problem 2.5). The bulb i connected to a 100-V household voltage source via a switc After turning on the switch, the temperature of the filame increases rapidly from the initial room temperature of 20° to an operating temperature of 1800°C. When it reaches i operating temperature, it consumes 80 W of power.

*Answer(s) in Appendix E.

(a) Determine the filament resistance at 1800°C.

(b) Determine the filament resistance at room temperature.

(c) Determine the current that the filament draws at room temperature and also at 1800°C.

(d) If the filament deteriorates when the current through it approaches 10 A, is the damage done to the filament greater when it is first turned on or later when it arrives at its operating temperature?

***2.7** A 110-V heating element in a stove can boil a standard-size pot of water in 1.2 minutes, consuming a total of 136 kJ of energy. Determine the resistance of the heating element and the current flowing through it.

2.8 A certain copper wire has a resistance R characterized by the model given in Problem 2.5 with $\alpha = 4 \times 10^{-3}$°C^{-1}. If $R = 60\ \Omega$ at 20°C and the wire is used in a circuit that cannot tolerate an increase in the magnitude of R by more than 10 percent over its value at 20°C, what would be the highest temperature at which the circuit can be operated within its tolerance limits?

Sections 2-3 and 2-4: Topology and Kirchhoff's Laws

2.9 Verify Eq. (2.8) for the circuit in Fig. P2.9.

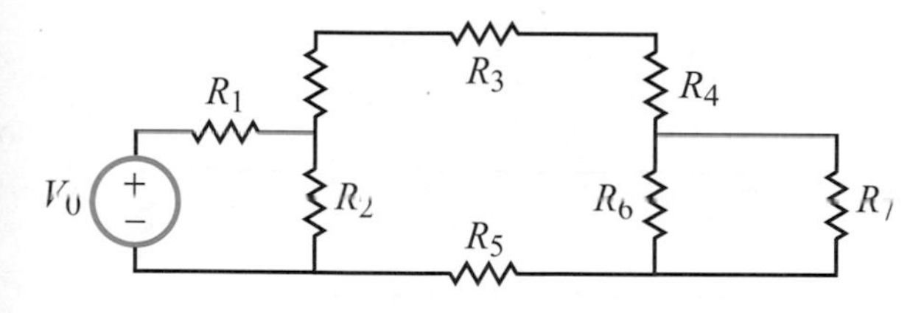

Figure P2.9: Circuit for Problems 2.9 and 2.10.

2.10 Verify Eq. (2.9) for the circuit in Fig. P2.9.

2.11 Verify Eq. (2.8) for the circuit in Fig. P2.11.

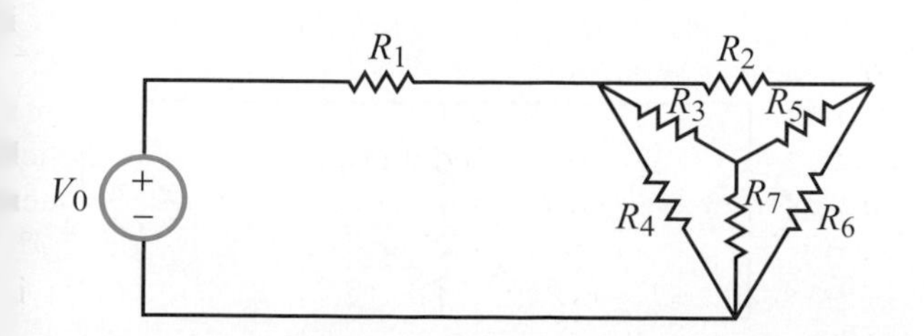

Figure P2.11: Circuit for Problems 2.11 and 2.12.

2.12 Verify Eq. (2.9) for the circuit in Fig. P2.11.

2.13 Determine the current I in the circuit of Fig. P2.13 given that $I_0 = 0$.

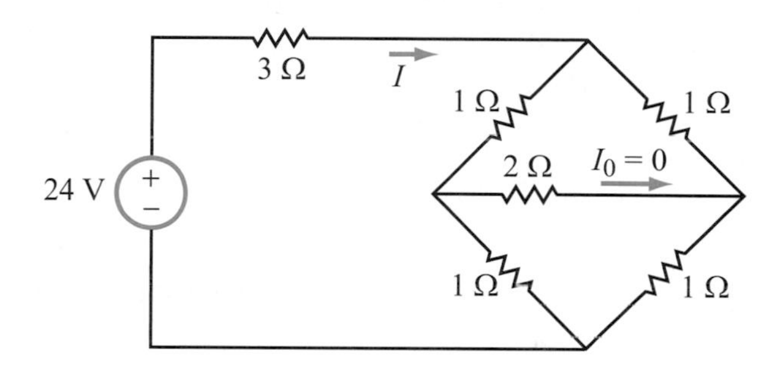

Figure P2.13: Circuit for Problem 2.13.

2.14 Determine currents I_1 to I_3 in the circuit of Fig. P2.14.

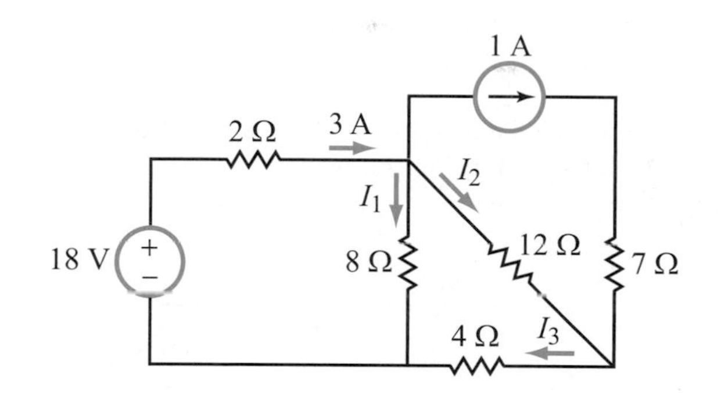

Figure P2.14: Circuit for Problem 2.14.

***2.15** Determine I_x in the circuit of Fig. P2.15.

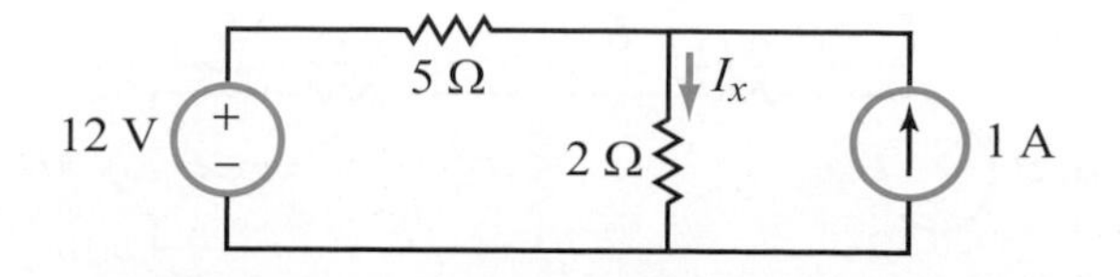

Figure P2.15: Circuit for Problem 2.15.

2.16 Determine currents I_1 to I_4 in the circuit of Fig. P2.16.

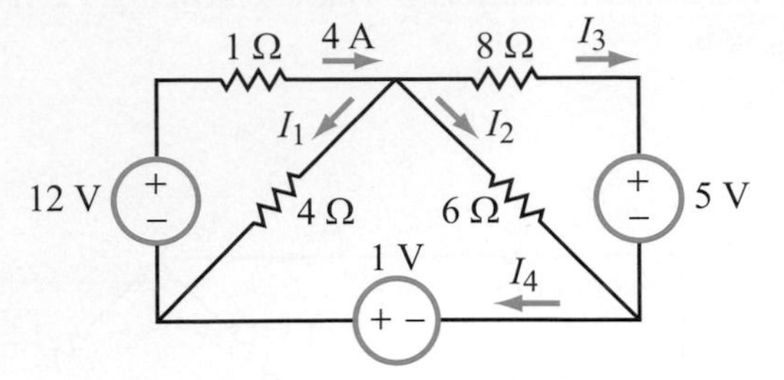

Figure P2.16: Circuit for Problem 2.16.

2.17 Determine currents I_1 to I_4 in the circuit of Fig. P2.17.

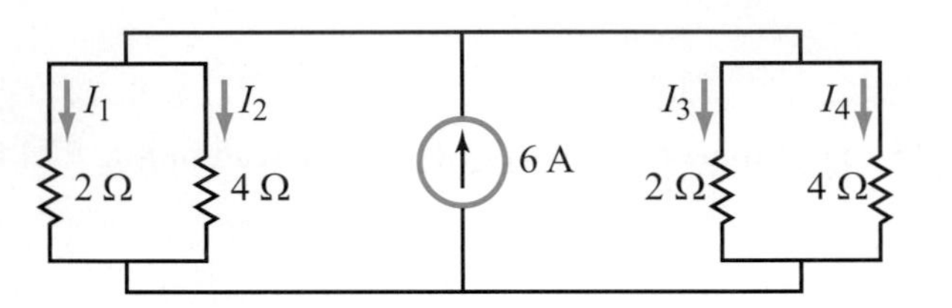

Figure P2.17: Circuit for Problem 2.17.

2.18 Determine the amount of power dissipated in the 3-kΩ resistor in the circuit of Fig. P2.18.

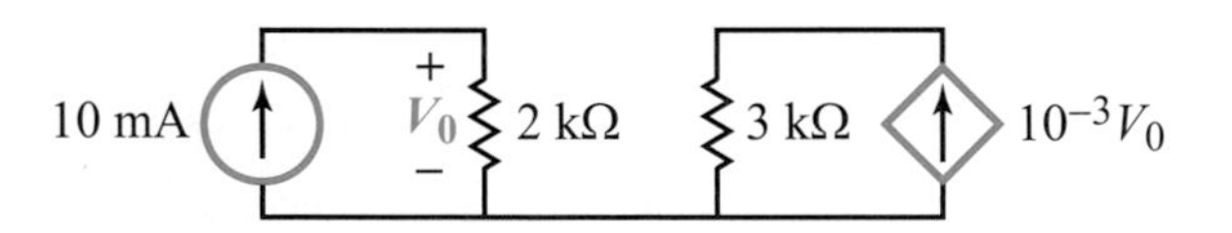

Figure P2.18: Circuit for Problem 2.18.

*__2.19__ Determine I_x and I_y in the circuit of Fig. P2.19.

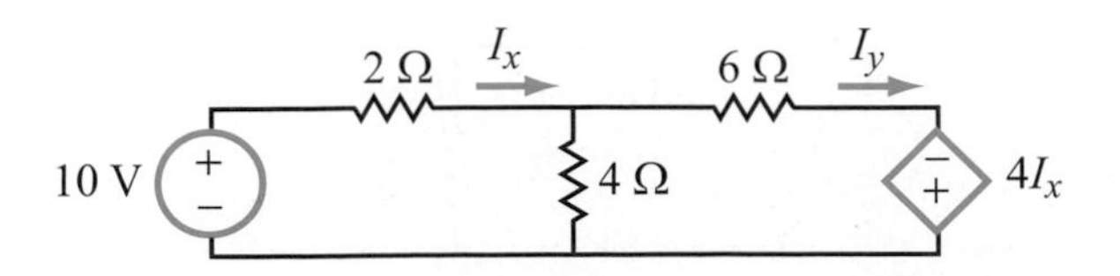

Figure P2.19: Circuit for Problem 2.19.

2.20 Find V_{ab} in the circuit of Fig. P2.20.

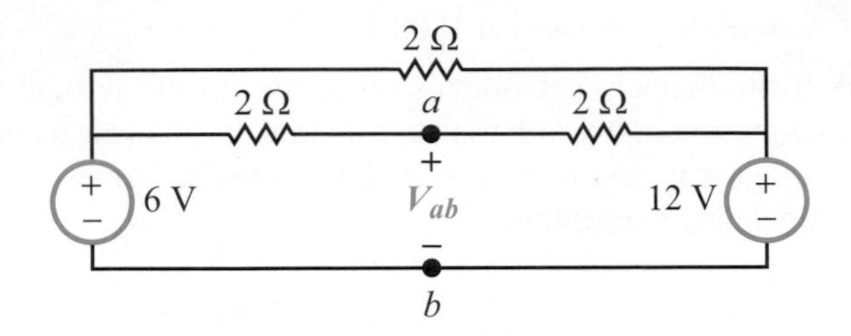

Figure P2.20: Circuit for Problem 2.20.

2.21 Find I_1 to I_3 in the circuit of Fig. P2.21.

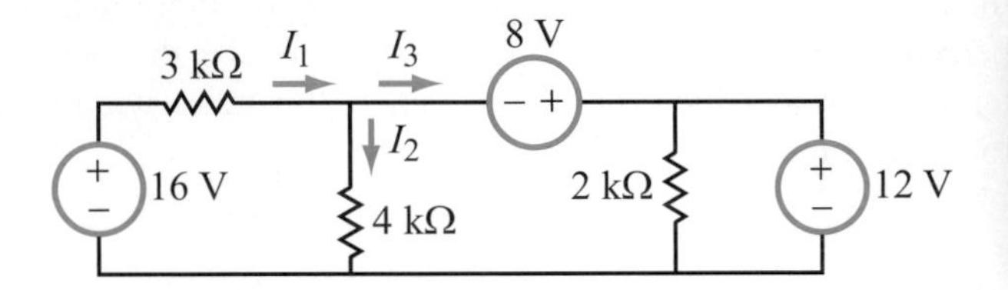

Figure P2.21: Circuit for Problem 2.21.

2.22 Find I in the circuit of Fig. P2.22.

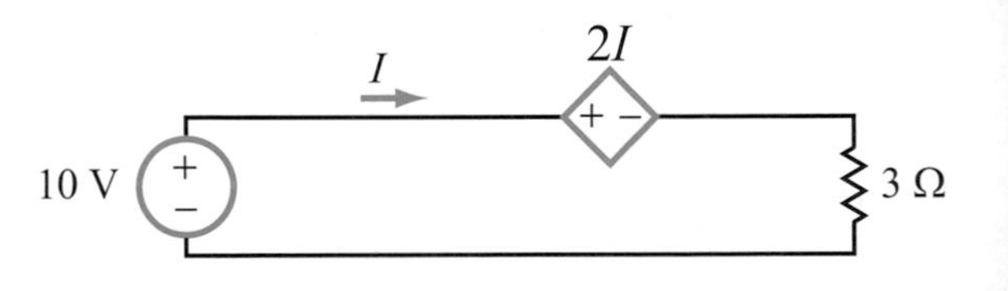

Figure P2.22: Circuit for Problem 2.22.

*__2.23__ Determine the amount of power supplied by the independent current source in the circuit of Fig. P2.23.

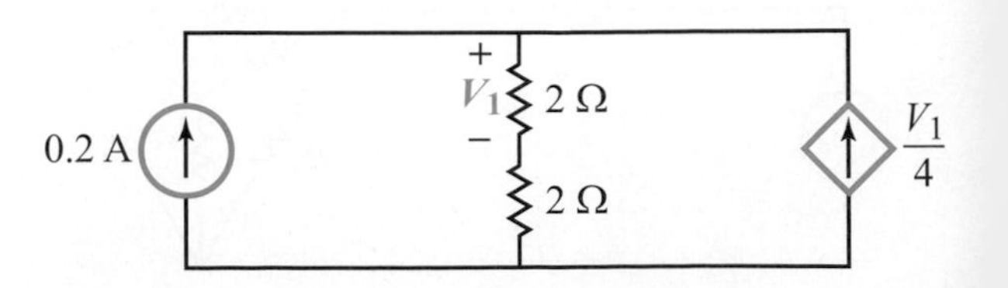

Figure P2.23: Circuit for Problem 2.23.

Section 2-4: Equivalent Circuits

2.24 Given that $I_1 = 1$ A in the circuit of Fig. P2.24, determine I_0.

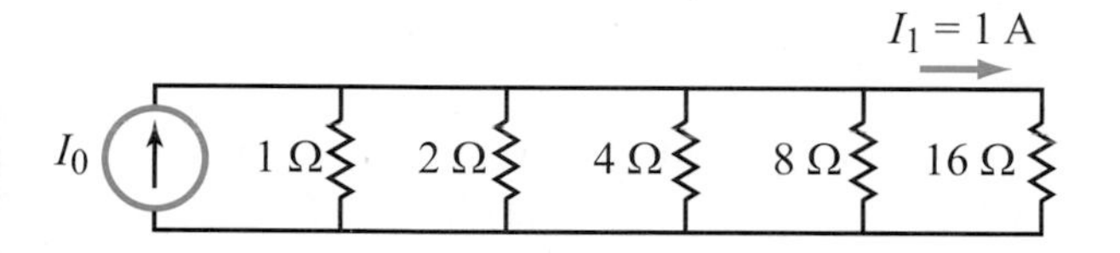

Figure P2.24: Circuit for Problem 2.24.

2.25 What should R be in the circuit of Fig. P2.25 so that $R_{eq} = 4\ \Omega$?

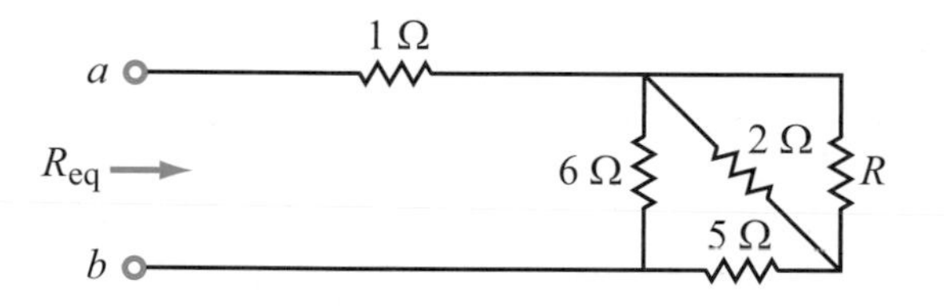

Figure P2.25: Circuit for Problem 2.25.

2.26 Find I_0 in the circuit of Fig. P2.26.

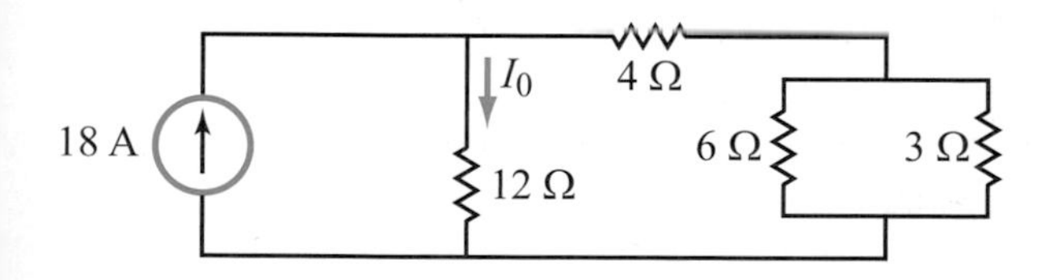

Figure P2.26: Circuit for Problem 2.26.

.27 For the circuit in Fig. P2.27, find I_x for $t < 0$ and $t > 0$.

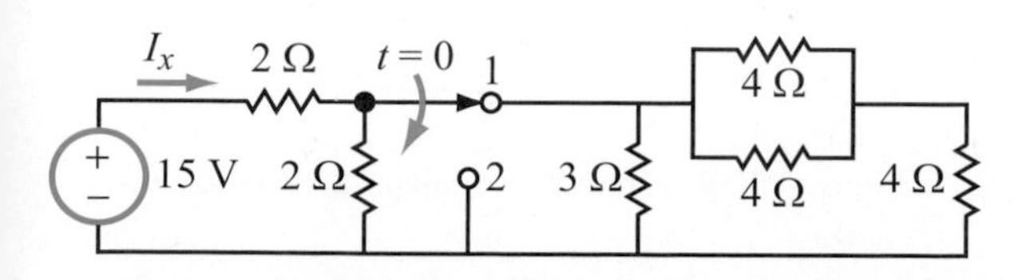

Figure P2.27: Circuit with SPDT switch for Problem 2.27.

2.28 Determine R_{eq} at terminals (a, b) in the circuit of Fig. P2.28.

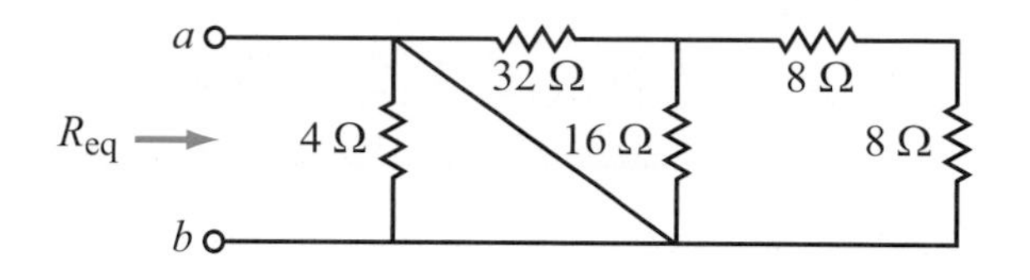

Figure P2.28: Circuit for Problem 2.28.

2.29 Select R in the circuit of Fig. P2.29 so that $V_L = 5$ V.

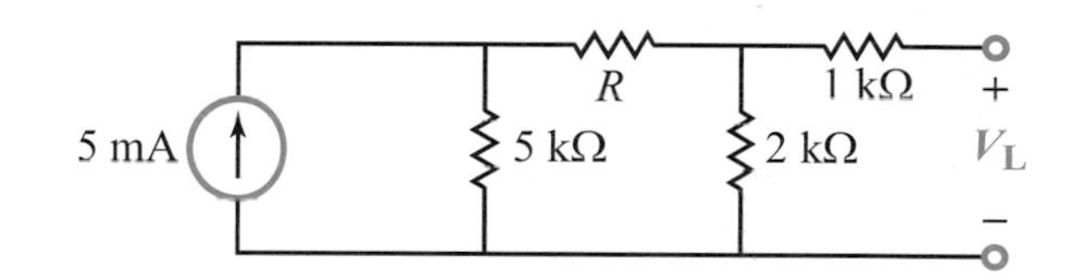

Figure P2.29: Circuit for Problem 2.29.

2.30 If $R = 12\ \Omega$ in the circuit of Fig. P2.30, find I.

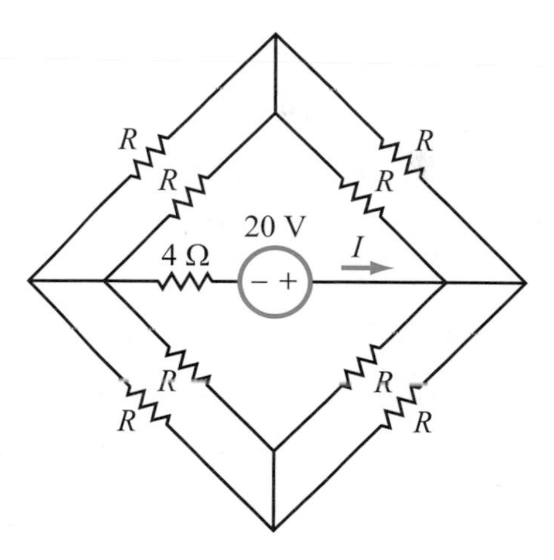

Figure P2.30: Circuit for Problem 2.30.

2.31 Use resistance reduction and source transformation to find V_x in the circuit of Fig. P2.31. All resistance values are in ohms.

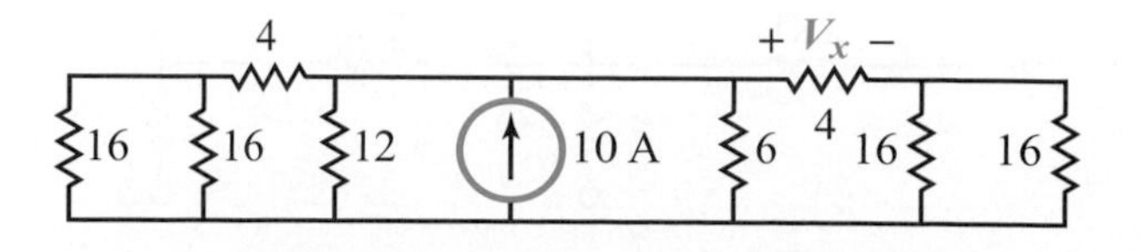

Figure P2.31: Circuit for Problem 2.31.

2.32 Determine A if $V_{\text{out}}/V_s = 9$ in the circuit of Fig. P2.32.

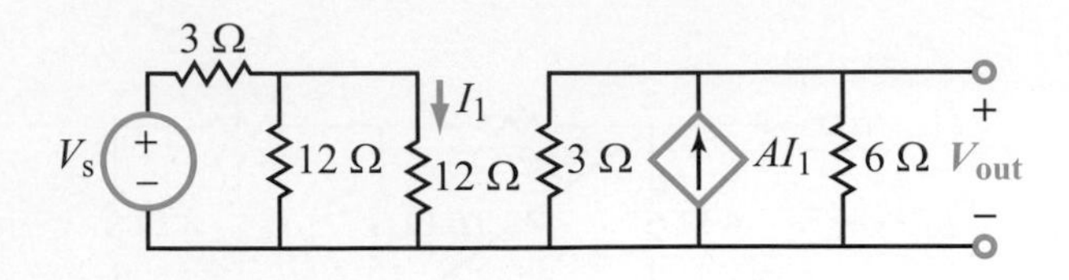

Figure P2.32: Circuit for Problem 2.32.

*__2.33__ For the circuit in Fig. P2.33, find R_{eq} at terminals (a, b).

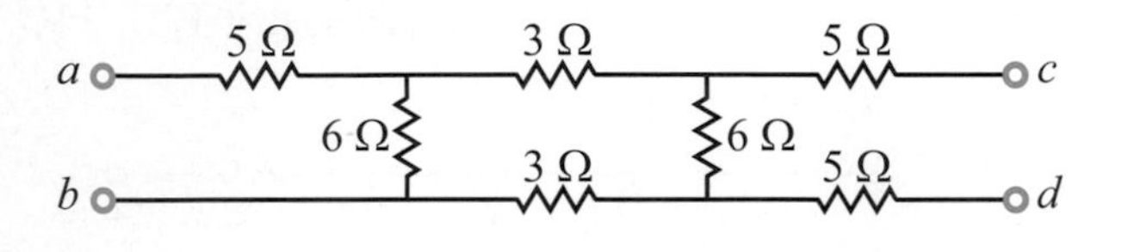

Figure P2.33: Circuit for Problems 2.33 and 2.34.

2.34 Find R_{eq} at terminals (c, d) in the circuit of Fig. P2.33.

2.35 Simplify the circuit to the right of terminals (a, b) in Fig. P2.35 to find R_{eq}, and then determine the amount of power supplied by the voltage source. All resistances are in ohms.

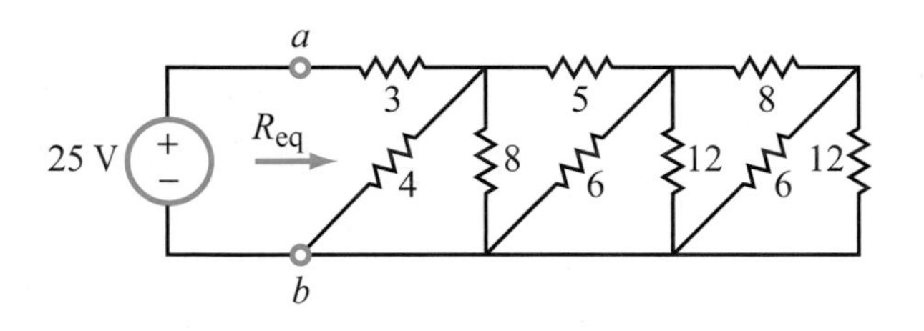

Figure P2.35: Circuit for Problem 2.35.

2.36 For the circuit in Fig. P2.36, determine R_{eq} at

*__(a)__ Terminals (a, b)

(b) Terminals (a, c)

(c) Terminals (a, d)

(d) Terminals (a, f)

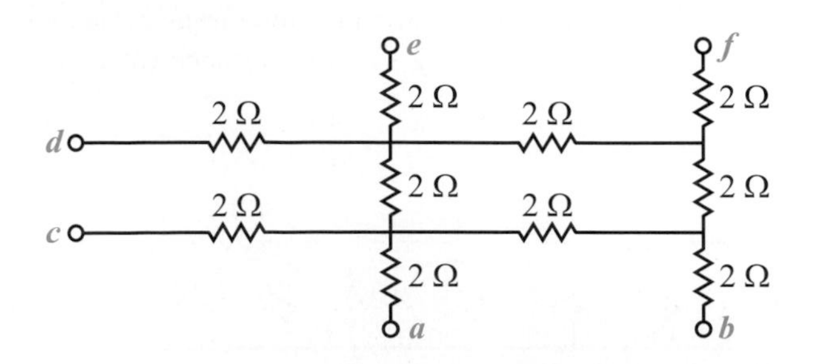

Figure P2.36: Circuit for Problem 2.36.

2.37 Find R_{eq} for the circuit in Fig. P2.37. All resistances are in ohms.

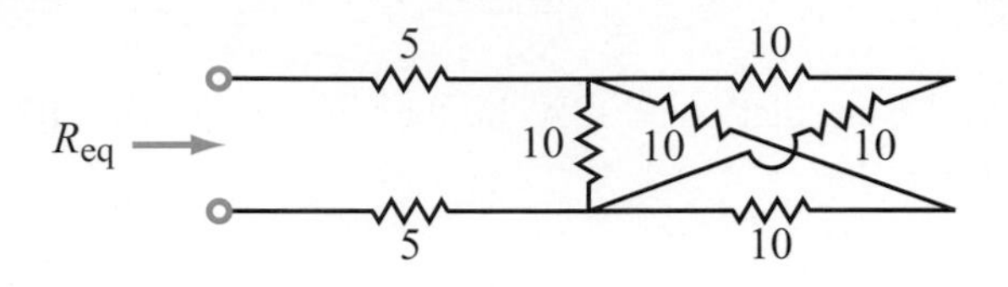

Figure P2.37: Circuit for Problem 2.37.

2.38 Apply voltage and current division to determine V_0 in the circuit of Fig. P2.38 given that $V_{\text{out}} = 0.2$ V.

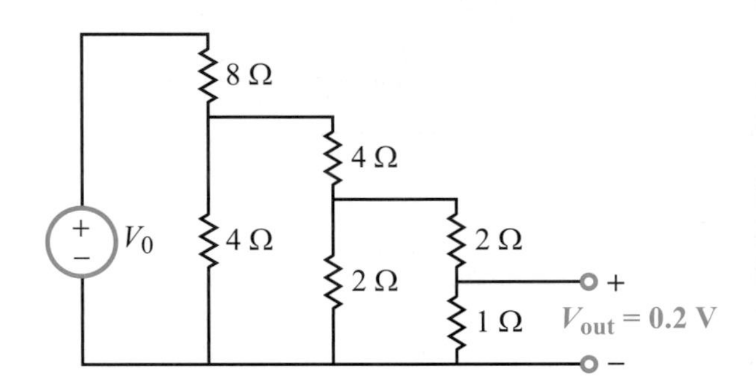

Figure P2.38: Circuit for Problem 2.38.

Sections 2-5 and 2-6: Y–**Δ** and Wheatstone Bridge

2.39 Convert the circuit in Fig. P2.39(a) from a Δ to a Y configuration.

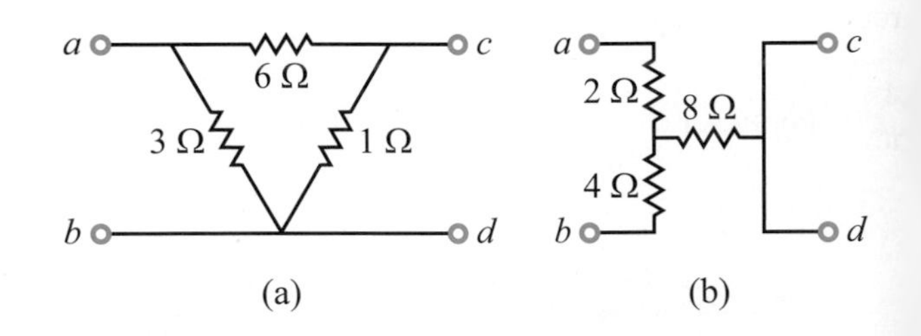

Figure P2.39: Circuit for Problems 2.39 and 2.40.

2.40 Convert the circuit in Fig. P2.39(b) from a T to a Π configuration.

*__2.41__ Find the power supplied by the generator in Fig. P2.41.

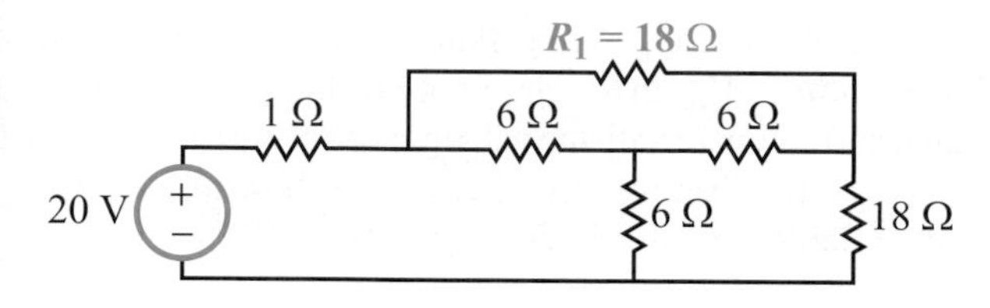

Figure P2.41: Circuit for Problems 2.41 and 2.42.

2.42 Repeat Problem 2.41 after replacing R_1 with a short circuit.

2.43 Find I in the circuit of Fig. P2.43.

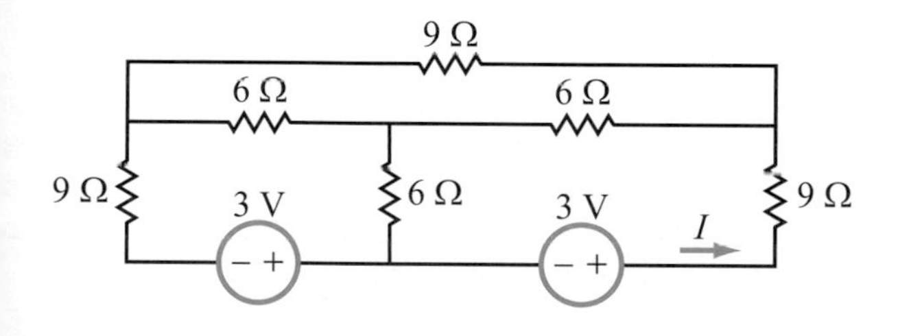

Figure P2.43: Circuit for Problem 2.43.

2.44 Find the power supplied by the voltage source in Fig. P2.44.

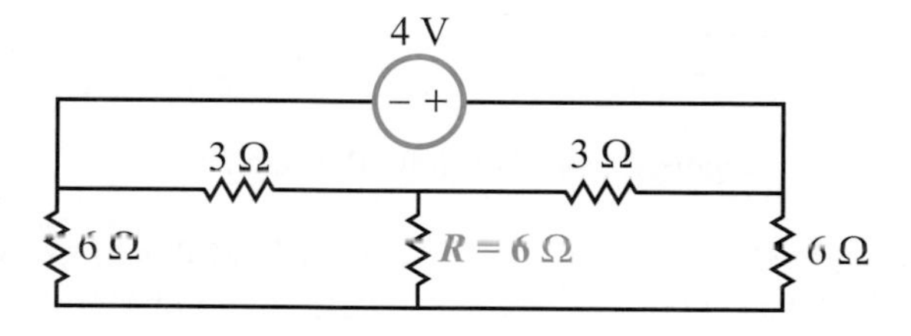

Figure P2.44: Circuit for Problems 2.44 and 2.45.

2.45 Repeat Problem 2.44 after replacing R with a short circuit.

2.46 Find I in the circuit of Fig. P2.46. All resistances are in ohms.

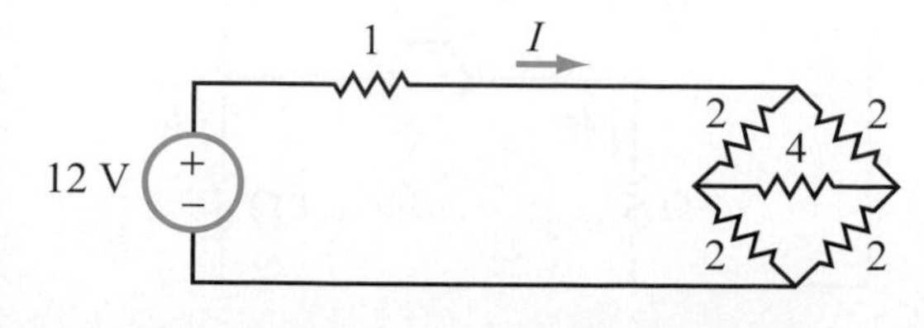

Figure P2.46: Circuit for Problem 2.46.

2.47 Find R_{eq} for the circuit in Fig. P2.47.

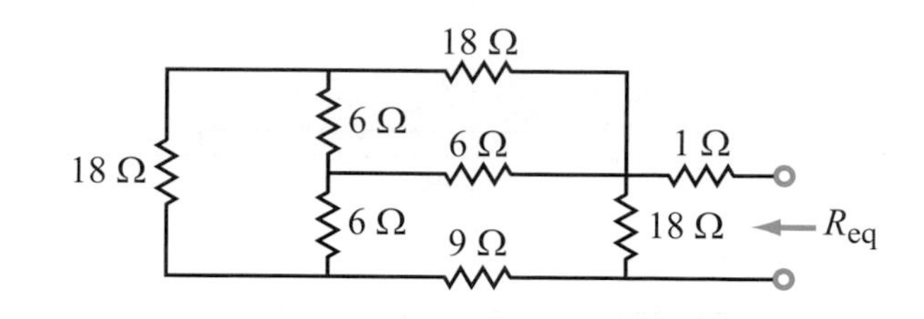

Figure P2.47: Circuit for Problem 2.47.

2.48 Find R_{eq} at terminals (a, b) in Fig. P2.48 if

(a) Terminal c is connected to terminal d by a short circuit

(b) Terminal e is connected to terminal f by a short circuit

(c) Terminal c is connected to terminal e by a short circuit

All resistance values are in ohms.

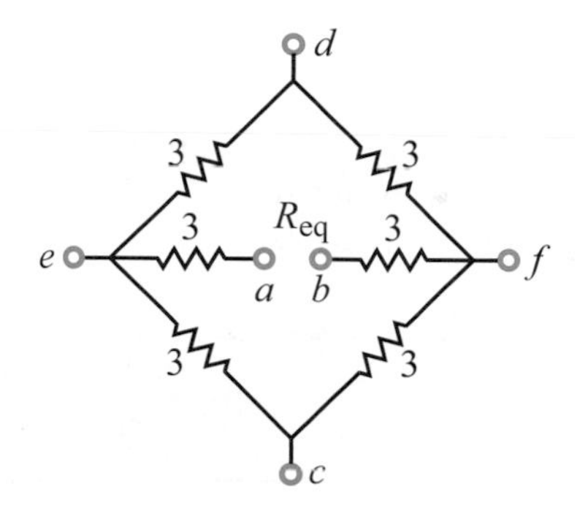

Figure P2.48: Circuit for Problem 2.48.

2.49 For the Wheatstone-bridge circuit of Fig. 2-31, solve the following problems.

*__(a)__ If $R_1 = 1\ \Omega$, $R_2 = 2\ \Omega$, and $R_x = 3\ \Omega$, to what value should R_3 be adjusted so as to achieve a balanced condition?

(b) If $V_0 = 6$ V, $R_a = 0.1\ \Omega$, and R_x were then to deviate by a small amount to $R_x = 3.01\ \Omega$, what would be the reading on the ammeter?

2.50 If $V_0 = 10$ V in the Wheatstone-bridge circuit of Fig. 2-32 and the minimum voltage V_{out} that a voltmeter can read is 1 mV, what is the smallest resistance fraction $(\Delta R/R)$ that can be measured by the circuit?

Section 2-7: i–v Relationships

2.51 Determine I_1 and I_2 in the circuit of Fig. P2.51. Assume $V_F = 0.7$ V for both diodes.

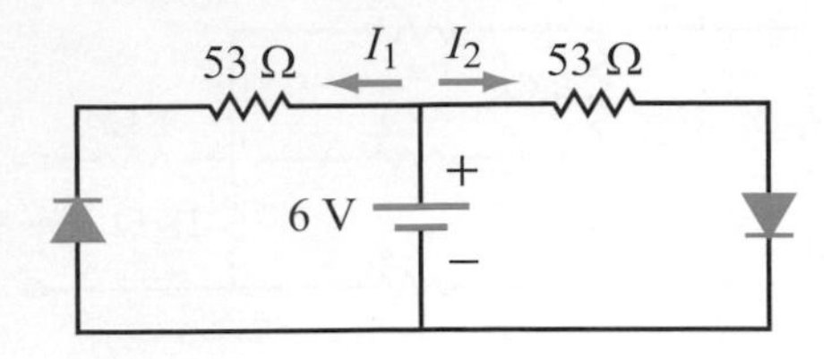

Figure P2.51: Circuit for Problem 2.51.

2.52 Determine V_1 in the circuit of Fig. P2.52. Assume $V_F = 0.7$ V for all diodes.

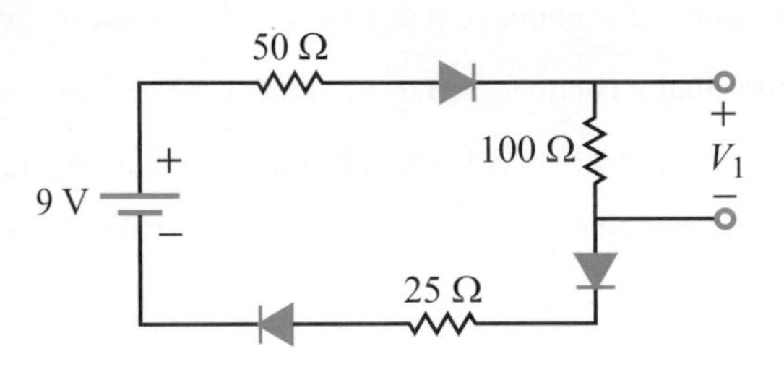

Figure P2.52: Circuit for Problem 2.52.

2.53 If the voltage source in the circuit of Fig. P2.53 generates a single square wave with an amplitude of 2 V, generate a plot for v_{out} for the same time period.

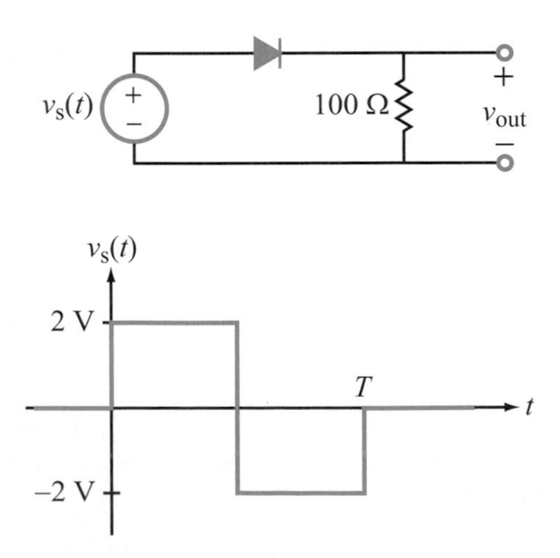

Figure P2.53: Circuit and voltage waveform for Problem 2.53.

2.54 A touch sensor based on a piezoresistor built into a micromechanical cantilever made of silicon is connected in a Wheatstone-bridge configuration with a $V_0 = 1$ V. If $L = 1.44$ cm and $W = 1$ cm, what should the thickness H be so that the touch sensor registers a voltage magnitude of 10 mV when the touch pressure is 10 N?

Section 2-8: Multisim

2.55 Use the DC Operating Point Analysis in Multisim to solve for voltage V_{out} in the circuit of Fig. P2.55. Solve for V_{out} by hand and compare with the value generated by Multisim. See the solution for Exercise 2.17 (on CD) for how to incorporate circuit variables into algebraic expressions.

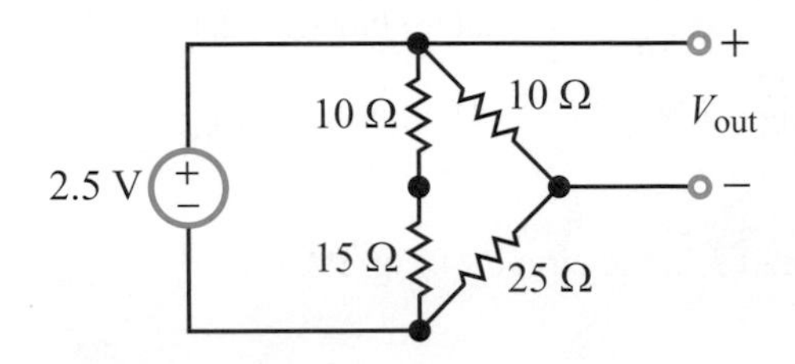

Figure P2.55: Circuit for Problem 2.55.

2.56 Find the ratio V_{out}/V_{in} for the circuit in Fig. P2.56 using DC Operating Point Analysis in Multisim. See the Multisim Tutorial included on the CD on how to reference currents in ABM sources (you should not just type in I(V1)).

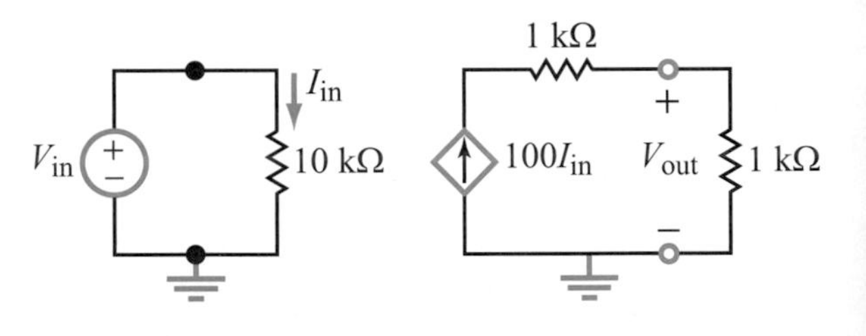

Figure P2.56: Circuit for Problem 2.56.

2.57 Use DC Operating Point Analysis in Multisim to solve for all six labeled resistor currents in the circuit of Fig. P2.57

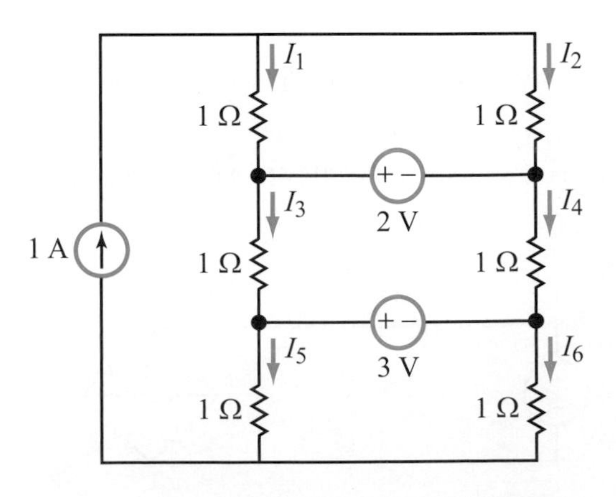

Figure P2.57: Circuit for Problem 2.57.

2.58 Find the voltages across R_1, R_2, and R_3 in the circuit of Fig. P2.58 using the DC Operating Point Analysis tool in Multisim.

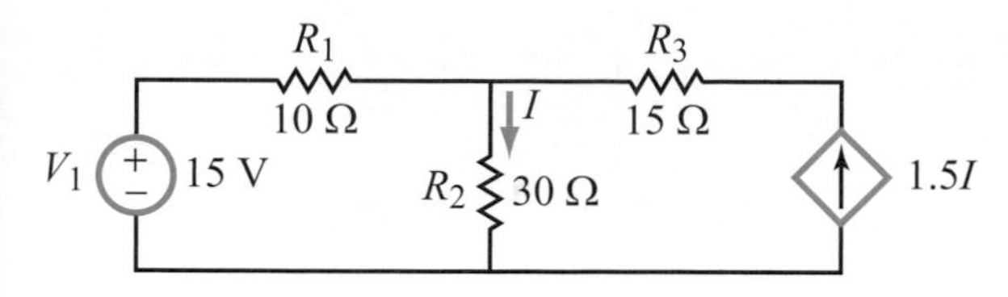

Figure P2.58: Circuit for Problem 2.58.

2.59 Find the equivalent resistance looking into the terminals of the circuit in Fig. P2.59 using a test voltage source and current probes in the Interactive Simulation in Multisim. Compare the answer you get to what you obtain from series and parallel combining of resistors carried out by hand.

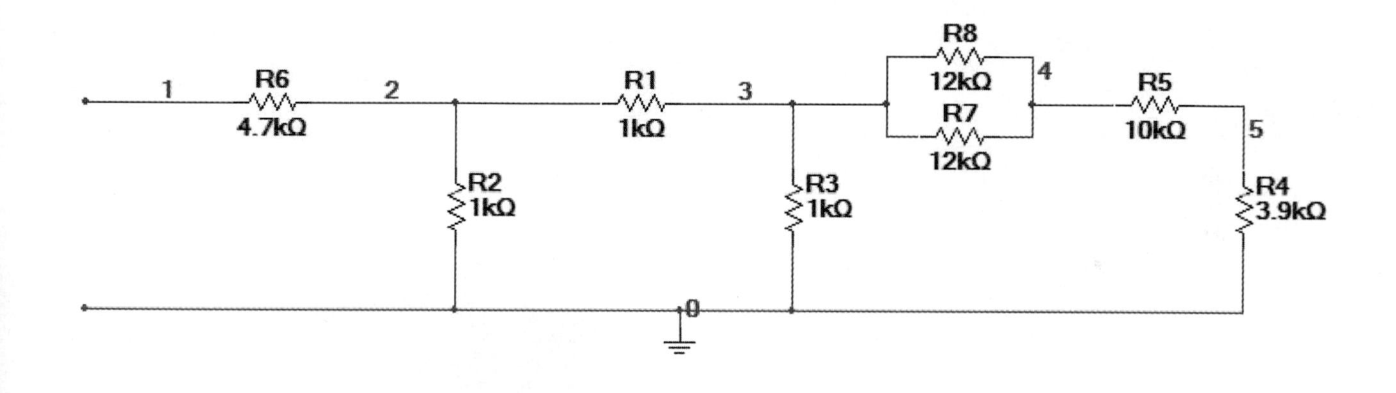

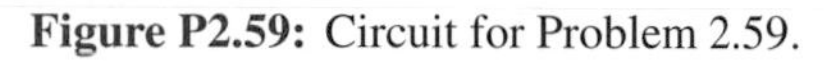

Figure P2.59: Circuit for Problem 2.59.

CHAPTER 3

Analysis Techniques

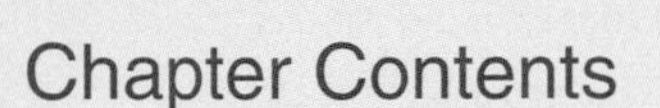

Chapter Contents

Objectives

Upon learning the material presented in this chapter, you should be able to:

1. Apply the node-voltage method to analyze an electric circuit of any configuration, so long as it is linear and planar.
2. Apply the mesh-current method.
3. Apply the by-inspection methods to circuits that satisfy certain conditions.
4. Use the source-superposition method as a tool for evaluating the sensitivity of a circuit to the various sources in the circuit.
5. Determine the Thévenin and Norton-equivalent circuits of any input circuit and use them to evaluate the response of an external load (or an output circuit) to the input circuit.
6. Establish the conditions for maximum transfer of current, voltage, and power from an input circuit to an external load.
7. Learn the basic properties of the bipolar junction transistor.
8. Use Multisim to analyze electric circuits.

Overview

By applying the circuit-analysis skills we developed in the preceding chapter, we now will extend our capability further so we may tackle any linear, planar circuit—no matter how complex. Nodal-voltage and mesh-current equations will be cast into a systematic structure in Sections 3-1 through 3-3, so we may take advantage of standard methods for solving linear, simultaneous equations, either by the use of determinants and matrices (Appendix B) or the execution of computer simulation packages such as MATLAB® or MathScript (Appendix B-3). The nodal and mesh analysis techniques are followed with treatments of two special tools: the source superposition method and the Thévenin/Norton equivalent-circuit method. This chapter concludes with a solution of a fundamental problem aimed at answering the question: When an input circuit is connected to a load (resistor), under what condition(s) is the amount of power transferred from the circuit to the load a maximum?

3-1 Node-Voltage Method

3-1.1 General Procedure

According to Kirchhoff's current law (KCL), the algebraic sum of all currents entering any node in an electric circuit is equal to zero. Built on that principle, the node-voltage analysis method provides a systematic and efficient procedure for determining all of the currents and voltages in a circuit. This determination is realized through the solution of a system of linear, simultaneous equations in which the unknown variables are the voltages at the extraordinary nodes in the circuit. As a reminder, in Section 2-2 we defined *an extraordinary node as a node connected to three or more elements.* For a circuit containing n_{ex} extraordinary nodes, implementation of the node-voltage method consists of three basic steps:

Solution Procedure: Node Voltage

Step 1: Identify all extraordinary nodes, select one of them as a reference node (ground), and then assign node voltages to the remaining $(n_{ex} - 1)$ extraordinary nodes.

Step 2: At each of the $(n_{ex} - 1)$ extraordinary nodes, apply the form of KCL requiring the sum of all currents *leaving* a node to be zero.

Step 3: Solve the $(n_{ex} - 1)$ independent simultaneous equations to determine the unknown node voltages.

Once the node voltages have been determined, all currents through branches and voltages across elements can be calculated readily.

Example 3-1: Circuit with Two Sources

For the circuit in Fig. 3-1, (a) use the node-voltage method to establish a system of node-voltage equations, (b) determine the node voltages if $V_0 = 6$ V, $I_0 = 3$ A, $R_1 = 3$ Ω, $R_2 = R_3 = 2$ Ω, $R_4 = R_5 = 12$ Ω, and $R_6 = 6$ Ω, and (c) calculate the power dissipated in R_5.

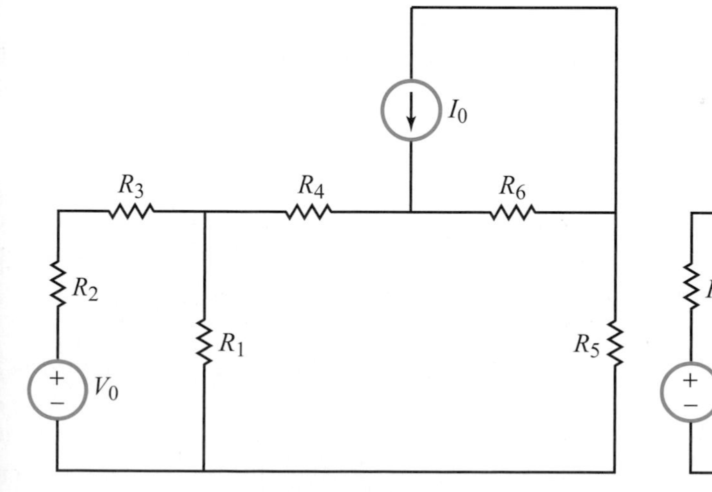

(a) Original circuit

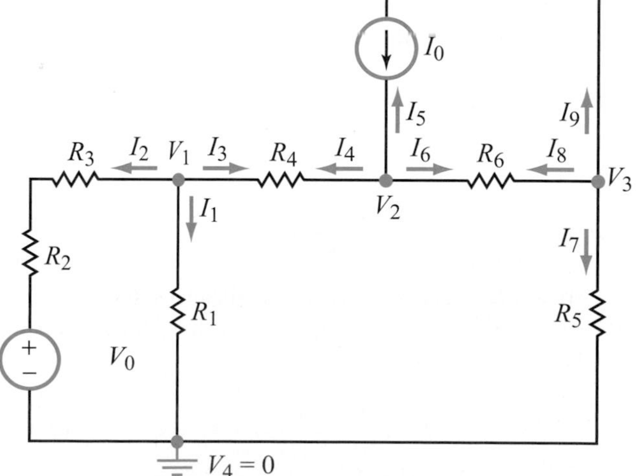

(b) Circuit with designated node voltages

Figure 3-1: Circuit for Example 3-1.

Solution:

(a) Step 1: Identify extraordinary nodes and assign node voltages

The circuit has four extraordinary nodes, one of which (node 4) is selected as the ground node, as shown in Fig. 3-1(b). At nodes 1, 2, and 3, we assign correspondingly node voltages V_1, V_2, and V_3, which are defined with respect to $V_4 = 0$ at the ground node.

Step 2: Apply KCL at nodes 1 through 3

At each node, we designate currents and we choose their directions as *leaving* the node. We realize that $I_3 = -I_4$, for example, but for the sake of consistency we will treat each node the same by designating a current leaving it through every branch connected to it.

Node 1:

$$I_1 + I_2 + I_3 = 0. \quad (3.1)$$

Unless we already know the value of a current, we should express it in terms of the node voltages connected to the branch through which it is flowing. We do so by applying Ohm's law, while reminding ourselves that the convention we adopted for the current direction is that it flows through a resistor from the (+) voltage terminal to the (−) terminal. Hence:

> The current leaving a node is equal to the voltage at that node, minus the voltage at the node to which the current is going, and divided by the resistance.

Consequently, I_1 flowing through R_1 is given by

$$I_1 = \frac{V_1 - 0}{R_1} = \frac{V_1}{R_1}. \quad (3.2)$$

Similarly,

$$I_3 = \frac{V_1 - V_2}{R_4}. \quad (3.3)$$

For I_2, the voltage across the series resistances, $(R_2 + R_3)$, is equal to V_1, reduced by the voltage V_0 of the voltage source in that branch. That is,

$$I_2 = \frac{V_1 - V_0}{R_2 + R_3}. \quad (3.4)$$

Inserting Eqs. (3.2) through (3.4) into Eq. (3.1) gives

$$\frac{V_1}{R_1} + \frac{V_1 - V_0}{R_2 + R_3} + \frac{V_1 - V_2}{R_4} = 0 \quad \text{(node 1)}. \quad (3.5)$$

Node 2:

$$I_4 + I_5 + I_6 = 0,$$

or equivalently,

$$\frac{V_2 - V_1}{R_4} - I_0 + \frac{V_2 - V_3}{R_6} = 0 \quad \text{(node 2)}, \quad (3.6)$$

where we incorporated the fact that $I_5 = -I_0$, as required by the current source.

Node 3:

$$I_7 + I_8 + I_9 = 0,$$

or equivalently,

$$\frac{V_3}{R_5} + \frac{V_3 - V_2}{R_6} + I_0 = 0 \quad \text{(node 3)}. \quad (3.7)$$

We note that by designating all current directions at a node as leaving that node:

> The node-voltage expression for any node always has V of that node preceded with a plus (+) sign. Also, the node voltages of the other nodes are preceded with negative (−) signs.

Thus, V_1 in Eq. (3.5)—which is specific to node 1—has a positive sign wherever it appears in that equation, whereas V_2 and V_3 always will have negative signs if they appear in that equation. A similar pattern applies to the other nodes.

Step 3: Solve simultaneous equations

As a prelude to solving Eqs. (3.5) through (3.7) to determine the unknown voltages V_1 to V_3, we need to reorganize them into standard system of equations as

$$\left(\frac{1}{R_1} + \frac{1}{R_2 + R_3} + \frac{1}{R_4}\right) V_1 - \left(\frac{1}{R_4}\right) V_2 = \frac{V_0}{R_2 + R_3}, \quad (3.8a)$$

$$-\left(\frac{1}{R_4}\right) V_1 + \left(\frac{1}{R_4} + \frac{1}{R_6}\right) V_2 - \frac{V_3}{R_6} = I_0, \quad (3.8b)$$

and

$$-\left(\frac{1}{R_6}\right) V_2 + \left(\frac{1}{R_5} + \frac{1}{R_6}\right) V_3 = -I_0. \quad (3.8c)$$

These are equivalent to

$$a_{11}V_1 + a_{12}V_2 + a_{13}V_3 = b_1, \quad (3.9a)$$

$$a_{21}V_1 + a_{22}V_2 + a_{23}V_3 = b_2, \quad (3.9b)$$

and

$$a_{31}V_1 + a_{32}V_2 + a_{33}V_3 = b_3, \quad (3.9c)$$

with

$$a_{11} = \left(\frac{1}{R_1} + \frac{1}{R_2 + R_3} + \frac{1}{R_4}\right),$$

$$a_{12} = -\frac{1}{R_4},$$

$$a_{13} = 0,$$

$$a_{21} = -\frac{1}{R_4},$$

$$a_{22} = \left(\frac{1}{R_4} + \frac{1}{R_6}\right),$$

$$a_{23} = -\frac{1}{R_6},$$

$$a_{31} = 0,$$

$$a_{32} = -\frac{1}{R_6},$$

$$a_{33} = \left(\frac{1}{R_5} + \frac{1}{R_6}\right),$$

$$b_1 = \frac{V_0}{R_2 + R_3},$$

$$b_2 = I_0,$$

$$b_3 = -I_0.$$

The system of equations given by Eq. (3.9) is now amenable for solution through the application of Cramer's rule or matrix inversion (as illustrated in Appendix B) either manually or by using MATLAB® software.

b) Application of Cramer's rule to Eq. (3.9) while recognizing that $a_{13} = a_{31} = 0$, $a_{23} = a_{32}$, and $a_{12} = a_{21}$, leads to

$$V_1 = \frac{b_1(a_{22}a_{33} - a_{23}^2) - a_{12}(b_2a_{33} - a_{23}b_3)}{a_{11}(a_{22}a_{33} - a_{23}^2) - a_{12}^2a_{33}}, \quad (3.10a)$$

$$V_2 = \frac{a_{11}(b_2a_{33} - a_{23}b_3) - b_1a_{21}a_{33}}{a_{11}(a_{22}a_{33} - a_{23}^2) - a_{12}^2a_{33}}, \quad (3.10b)$$

and

$$V_3 = \frac{a_{11}(a_{22}b_3 - b_2a_{32}) - a_{12}^2b_3 + b_1a_{21}a_{32}}{a_{11}(a_{22}a_{33} - a_{23}^2) - a_{12}^2a_{33}}. \quad (3.10c)$$

(c) Upon using the problem-specified values for V_0, I_0, and the six resistors to evaluate the a- and b-coefficients and then using those in Eq. (3.10), we obtain $V_1 = 126/37$ V, $V_2 = 342/37$ V, and $V_3 = -216/37$ V.

The current flowing through R_5 in Fig. 3-1(b) is

$$I_7 = \frac{V_3}{R_5} = -\frac{216}{37 \times 12} = -0.486 \text{ A},$$

and the power dissipated in R_5 is

$$P = I_7^2 R_5 = (-0.486)^2 \times 12 = 2.84 \text{ W}.$$

Review Question 3-1: The node-*voltage* method relies on the application of Kirchhoff's *current* law. Explain.

Review Question 3-2: Why does a circuit with n_{ex} extraordinary nodes require only $(n_{ex} - 1)$ node-voltage equations to analyze it?

Exercise 3-1: Apply nodal analysis to determine the current I.

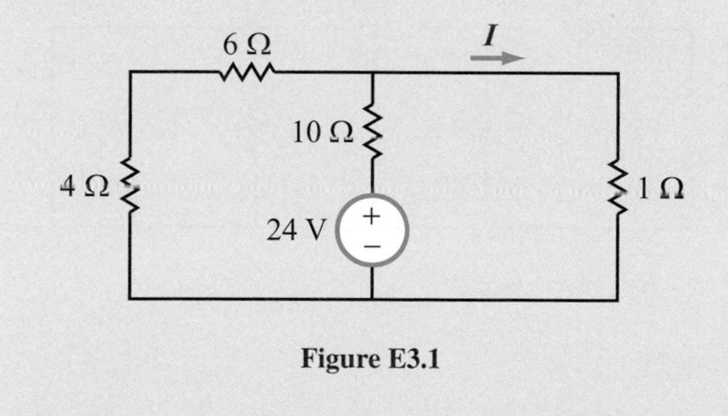

Figure E3.1

Answer: $I = 2$ A. (See)

3-1.2 Circuits Containing Dependent Sources

When a circuit contains dependent sources, the node-voltage analysis method remains applicable, as does the solution procedure outlined in the preceding subsection. However, each dependent source defines a relationship between its own magnitude and some current or voltage elsewhere in the circuit, and that relationship needs to be incorporated into the solution.

Example 3-2: Dependent Current Source

The circuit of Fig. 3-2 contains a current-controlled current source (CCCS) whose magnitude I_x is governed by the current flowing through the 6-Ω resistor in the direction shown. Determine I_x.

Solution:
Following the standard procedure outlined earlier, we start by selecting a ground node and assigning node voltages to the other extraordinary nodes in the circuit, as shown in Fig. 3-2(b). We also designate currents with their directions out of the nodes for all branches connected to nodes 1 and 2.

Next, we write down the node-voltage equations for nodes 1 and 2 as

$$\frac{V_1 - 5.3}{4} + \frac{V_1}{3} + \frac{V_1 - V_2}{6} = 0 \qquad \text{(node 1)},$$

and

$$\frac{V_2 - V_1}{6} + \frac{V_2}{12} - I_x = 0 \qquad \text{(node 2)}.$$

In order to solve these equations, we first need to express I_x in terms of the unknown variables, V_1 and V_2. The dependent source I_x is given in terms of I, which in turn is dependent on the voltage difference between V_1 and V_2. That is,

$$I_x = 2I = 2\frac{(V_1 - V_2)}{6} = \frac{V_1 - V_2}{3}.$$

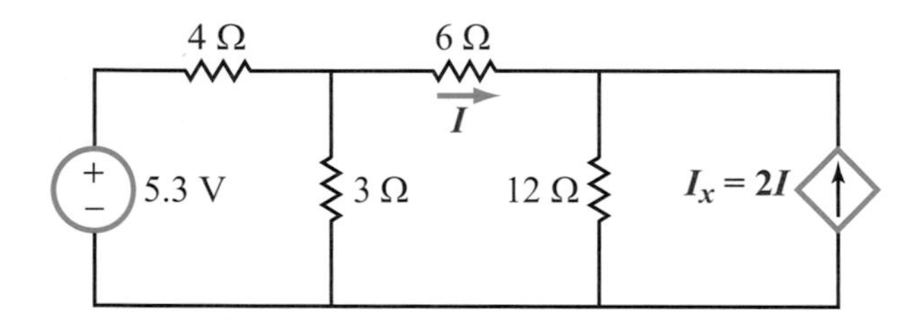

(a) Original circuit

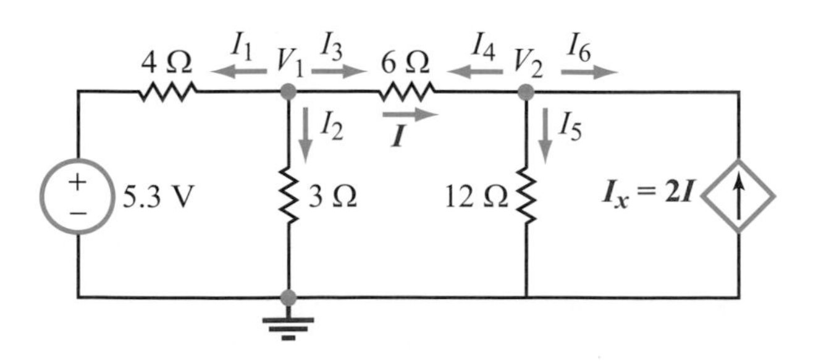

(b) Circuit with designated node voltages

Figure 3-2: Example 3-2.

Upon substituting this expression for I_x into the second of the node-voltage equations and rearranging the node-voltage equations, we end up with

$$9V_1 - 2V_2 = 15.9,$$
$$-6V_1 + 7V_2 = 0.$$

Simultaneous solution of the preceding pair of equations gives $V_1 = 2.18$ V and $V_2 = 1.87$ V. Hence,

$$I_x = \frac{V_1 - V_2}{3} = \frac{2.18 - 1.87}{3} = 0.1 \text{ A}.$$

Exercise 3-2: Apply nodal analysis to find V_a.

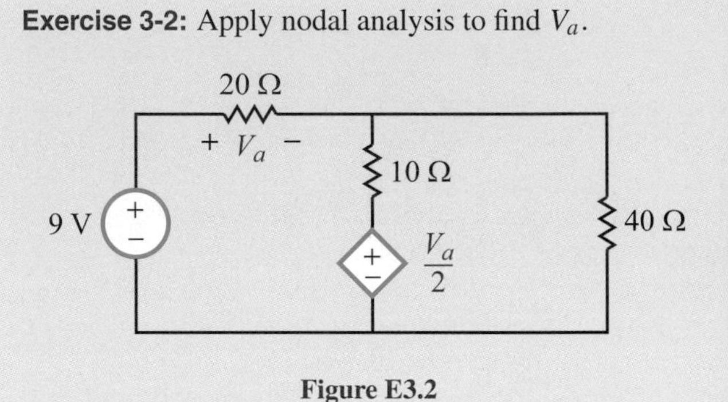

Figure E3.2

Answer: $V_a = 5$ V. (See ◎)

3-1.3 Supernodes

Occasionally, a circuit may contain a solitary voltage source nestled between two extraordinary nodes, with no other elements in series with it between those nodes. Such an arrangement is called a ***supernode***. Examples of supernodes are shown in Fig. 3-3. Formally:

> A supernode is the combination of two extraordinary nodes (excluding the reference node) between which a voltage source exists.

The voltage source may be of the independent or dependent type, and the voltage source may include elements in parallel with it (such as R_4 in parallel with the 16-V source of supernode B) but not in series with it. If one of the two nodes of a supernode is a reference (ground) node, it is called a ***quasi-supernode***.

In addition to the reference node, the circuit of Fig. 3-3 contains five extraordinary nodes designated with node voltages V_1 to V_5. Analyzing the circuit by following the standard procedure outlined in Section 3-1.1 leads to five independent

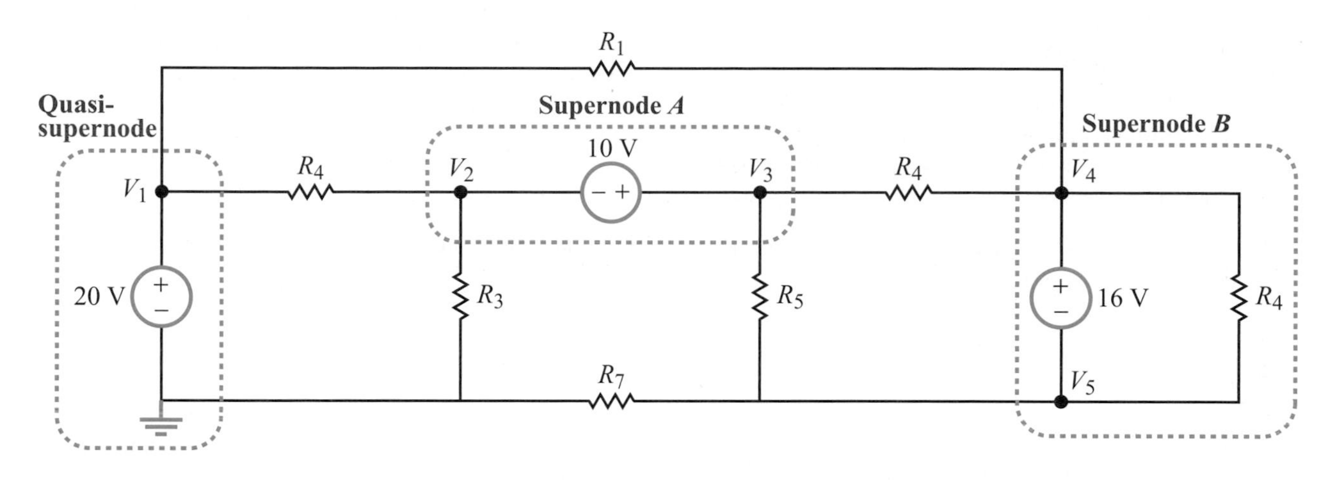

Figure 3-3: Circuit containing two supernodes and one quasi-supernode.

equations in five unknowns. A supernode allows us to collapse its two extraordinary nodes into one, thereby reducing the associated number of node-voltage equations from two to one. Hence, for the circuit of Fig. 3-3, supernodes A and B would lead to two node-voltage equations instead of 4. *For a quasi-supernode, the only relevant information is that the voltage of the non-reference node is equal to the voltage magnitude of the voltage source.* Thus, $V_1 = 20$ V in Fig. 3-3.

To explain the properties of a supernode and how we use it, let us analyze supernode A, all on its own. In Fig. 3-4(a), we show currents I_1 to I_3 leaving node 2 and currents I_4 to I_6 leaving node 3. KCL requires that

$$I_1 + I_2 + I_3 = 0, \qquad \text{(node } V_2\text{)}, \qquad (3.11a)$$

and

$$I_4 + I_5 + I_6 = 0. \qquad \text{(node } V_3\text{)}. \qquad (3.11b)$$

Adding the two equations together and recognizing that $I_3 = -I_4$ leads to

$$I_1 + I_2 + I_5 + I_6 = 0, \qquad \text{(supernode } A\text{)}, \qquad (3.12)$$

which constitutes the four currents leaving supernode A. The implication of Eq. (3.12) is that we can treat nodes 2 and 3 as a combined single node, connected by a dashed line (Fig. 3-4(b)), but we also should acknowledge the fact that

$$V_3 - V_2 = 10 \text{ V}, \qquad \text{(Auxiliary Eq.)}$$

which is a much simpler equation than the typical node-voltage equation.

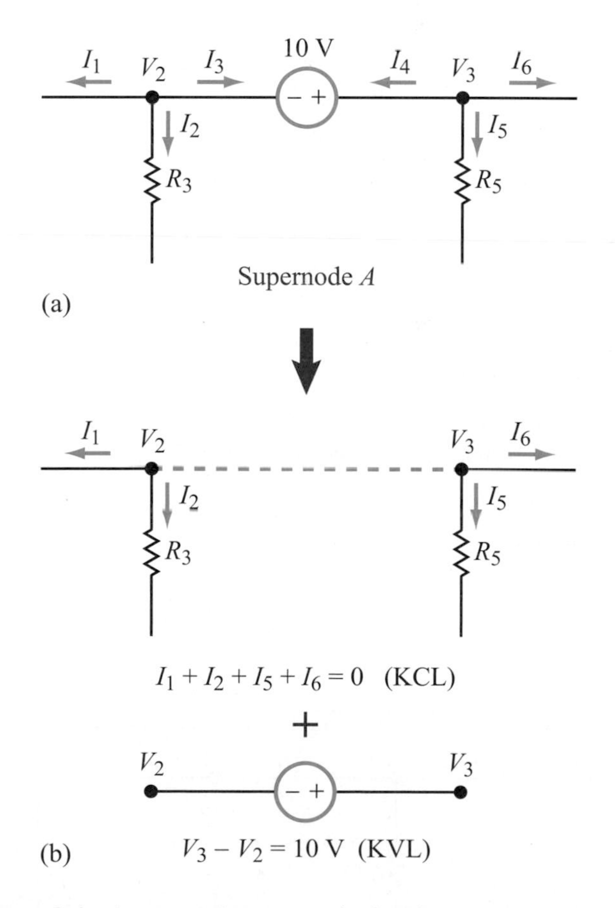

Figure 3-4: A supernode composed of nodes V_2 and V_3 can be represented as a single node, in terms of summing currents flowing out of them, plus an auxiliary equation that defines the voltage difference between V_3 and V_2.

Supernode Attributes

(1) At a supernode, Kirchhoff's current law (KCL) can be applied to the combination of the two nodes as if they are a single node, but the two nodes retain their own identities.

(2) Kirchhoff's voltage law (KVL) is used to express the voltage difference between the two nodes in terms of the voltage of the source between them.

(3) If a supernode contains a resistor in parallel with the voltage source, the resistor will exercise no influence on the currents and voltages in the other parts of the circuit, and therefore, it may be ignored altogether.

(4) For a quasi-supernode, the node-voltage of the non-reference node is equal to the voltage magnitude of the source.

In the circuit of Fig. 3-3, the voltage difference between nodes 4 and 5 is specified by the 16-V source, regardless of the value of R_4 (so long as R_4 is not a short circuit). Conversely, the current flowing through R_4 is equal to 16 V$/R_4$, and it is independent of the voltages and currents elsewhere in the circuit.

Example 3-3: Circuit with a Supernode

Use the supernode concept to solve for the node voltages in Fig. 3-5.

Solution:
The combination of nodes 1 and 2 constitutes a supernode, with an associated node-voltage equation given by

$$I_1 + I_2 + I_3 + I_4 = 0$$

or

$$\frac{V_1 - 4}{2} + \frac{V_1}{4} + \frac{V_2}{8} - 2 = 0,$$

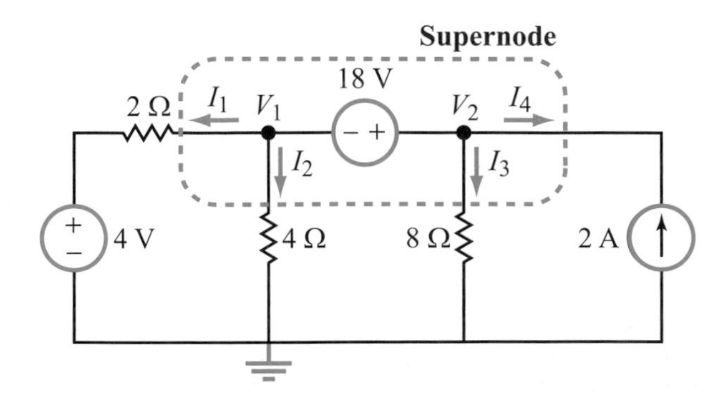

Figure 3-5: Circuit for Example 3-3.

which may be simplified to

$$6V_1 + V_2 = 32.$$

Additionally, the supernode KVL equation is

$$V_2 - V_1 = 18.$$

Simultaneous solution of the two equations yields

$$V_1 = 2 \text{ V}, \qquad V_2 = 20 \text{ V}.$$

Review Question 3-3: What impact does the presence of a dependent source have on the implementation of the node-voltage method?

Review Question 3-4: What is a supernode? How is it treated in nodal analysis?

Exercise 3-3: Apply the supernode concept to determine I in the circuit of Fig. E3.3.

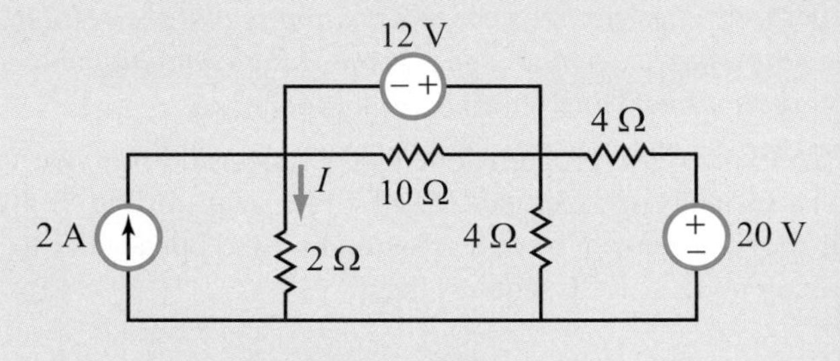

Figure E3.3

Answer: $I = 0.5$ A. (See ◎)

3-2 Mesh-Current Method

3-2.1 General Procedure

A ***mesh*** was defined in Section 2-2 as a loop that encloses no other loop. The current associated with a mesh is called its ***mesh current.*** The circuit in Fig. 3-6 contains two meshes with

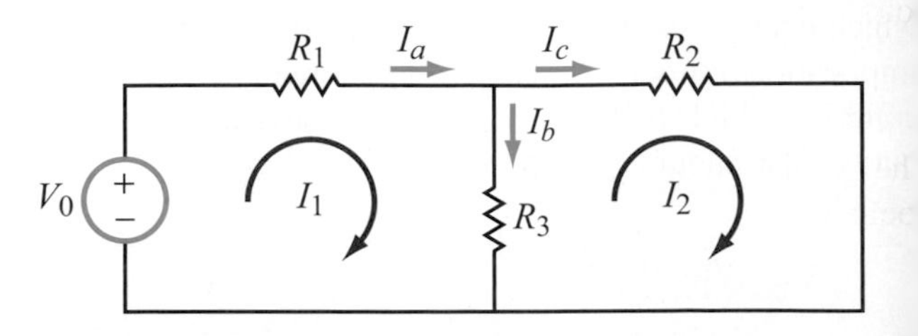

Figure 3-6: Circuit containing two meshes with mesh currents I and I_2.

mesh currents I_1 and I_2. A mesh current may be thought of as the current flowing through the branches of that mesh, with no regard for the currents in neighboring meshes. That does not mean, however, that the mesh current is the same as the actual currents flowing through the elements of that mesh. For an element that belongs to only one mesh, such as R_1 in Fig. 3-6, the current through it is indeed identical with the current in mesh 1. That is,

$$I_a = I_1.$$

On the other hand, if an element is shared by two meshes, as is the case for R_3, the current through it is

$$I_b = I_1 - I_2.$$

Current I_1 is assigned a positive sign because its direction through R_3 is the same as that of I_b, but I_2 is assigned a negative sign because it flows "upward" through I_b. The mesh-current analysis method is based on the application of KVL to all of the meshes in the circuit. The solution procedure, which is analogous with that discussed earlier in Section 3-1 for the node-voltage method, consists of the following steps.

Solution Procedure: Mesh Current

Step 1: Identify all meshes and assign each of them an unknown mesh current. For convenience, define the mesh currents to be clockwise in direction.

Step 2: Apply Kirchhoff's voltage law (KVL) to each mesh.

Step 3: Solve the resultant simultaneous equations to determine the mesh currents.

For the circuit in Fig. 3-6, application of KVL to mesh 1, starting at the bottom left-hand corner and moving clockwise around the loop, gives

$$-V_0 + I_1 R_1 + (I_1 - I_2) R_3 = 0 \qquad \text{(mesh 1)}, \qquad (3.13)$$

where for each term we assigned a (+) or (−) sign to it depending on which of its voltage terminals is encountered first. Also, for a resistor, current flows into the (+) terminal of the voltage across it. For mesh 2,

$$(I_2 - I_1) R_3 + I_2 R_2 = 0 \qquad \text{(mesh 2)}. \qquad (3.14)$$

The two simultaneous equations can be rearranged by collecting coefficients of I_1 and I_2 as

$$(R_1 + R_3) I_1 - I_2 R_3 = V_0 \qquad \text{(mesh 1)}, \qquad (3.15a)$$

and

$$-R_3 I_1 + (R_2 + R_3) I_2 = 0 \qquad \text{(mesh 2)}. \qquad (3.15b)$$

Note the built-in symmetry reflected by the structure of Eqs. (3.15a and b). For mesh 1, the coefficient of I_1 in Eq. (3.15a) is the sum of all of the resistors contained in mesh 1, and the coefficient of I_2 contains the resistor that mesh 1 shares with mesh 2. Furthermore, the coefficients of I_1 and I_2 have opposite signs. The same pattern applies for mesh 2 in Eq. (3.15b); the coefficient of I_2 contains all of the resistors of mesh 2, and the coefficient of I_1 contains the resistor shared by the two meshes. This structural pattern allows us to write the mesh-current equations directly, as discussed in more detail later in Section 3-3.

Example 3-4: Circuit with Three Meshes

Use mesh analysis to (a) obtain mesh-current equations for the circuit in Fig. 3-7 and then (b) determine the current in R_4, given that $V_0 = 18$ V, $R_1 = 6\ \Omega$, $R_2 = R_3 = 2\ \Omega$, $R_4 = 4\ \Omega$ and $R_5 = R_6 = 4\ \Omega$.

Solution:

(a) Applying the symmetry pattern inherent in the structure of the mesh-current equations, we have

$$(R_1 + R_2 + R_5) I_1 - R_2 I_2 - R_5 I_3 = V_0, \qquad (3.16a)$$

$$-R_2 I_1 + (R_2 + R_3 + R_4) I_2 - R_4 I_3 = 0, \qquad (3.16b)$$

and

$$-R_5 I_1 - R_4 I_2 + (R_4 + R_5 + R_6) I_3 = 0. \qquad (3.16c)$$

We note that in Eq. (3.16a) the coefficient of I_1 is positive and is composed of the sum of all resistors in mesh 1 and the coefficients of I_2 and I_3 are negative and include the resistors that meshes 2 and 3 share with mesh 1, respectively. An equivalent pattern pertains to Eqs. (3.16b and c).

(b) For the specified values of V_0 and the six resistors, solution of the simultaneous equations leads to

$$I_1 = 2\text{ A}, \qquad I_2 = 1\text{ A}, \qquad I_3 = 1\text{ A}.$$

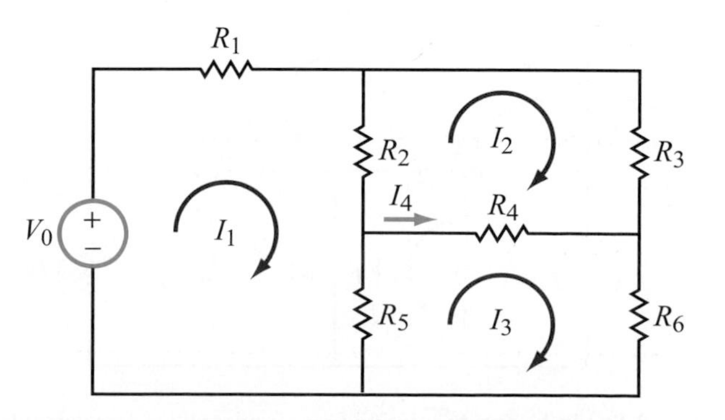

Figure 3-7: Circuit for Example 3-4.

The current through R_4 is

$$I_4 = I_3 - I_2 = 1 - 1 = 0.$$

Given that the circuit is a Wheatstone bridge (Section 2-6) operated under the balanced condition ($R_2 R_6 = R_3 R_5$), the result $I_4 = 0$ is exactly what we should have expected.

Exercise 3-4: Apply mesh analysis to determine I.

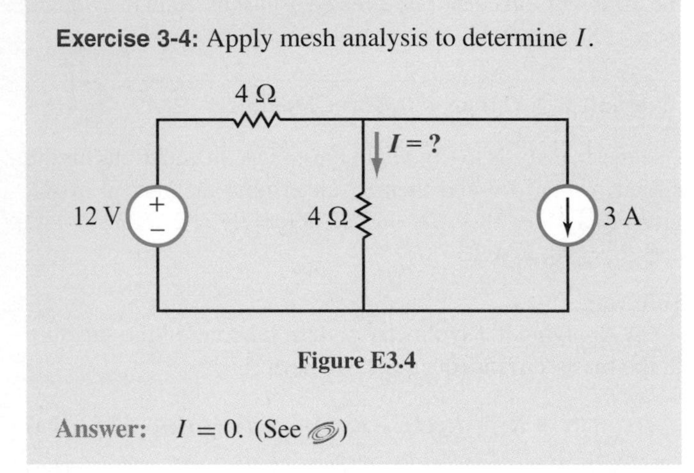

Figure E3.4

Answer: $I = 0$. (See ◎)

3-2.2 Circuit with Dependent Sources

The presence of a dependent source in a circuit does not alter the basic procedure of the mesh-current method, but it requires the addition of a supplemental equation expressing the relationship between the dependent source and the other parts of the circuit.

Example 3-5: Dependent Current Source

Use mesh-current analysis to determine the magnitude of the dependent source I_x in Fig. 3-8.

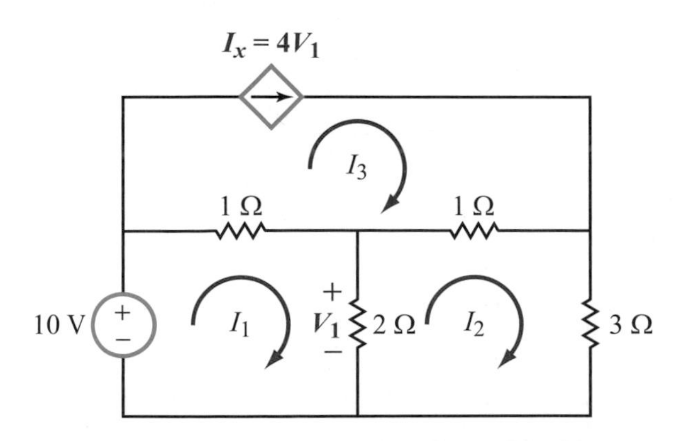

Figure 3-8: Mesh-current solution for a circuit containing a dependent source (Example 3-5).

Solution:
For the meshes with mesh currents I_1 and I_2,

$$(1+2)I_1 - 2I_2 - I_3 = 10, \quad (3.17a)$$

and

$$-2I_1 + (2+1+3)I_2 - I_3 = 0. \quad (3.17b)$$

For mesh 3, we do not need to write a mesh-current equation, because I_3 is specified by the current source as

$$I_3 = I_x = 4V_1.$$

The voltage V_1 across the 2-Ω resistor is given by

$$V_1 = 2(I_1 - I_2).$$

Hence,

$$I_3 = 4V_1 = 8(I_1 - I_2). \quad (3.18)$$

After inserting Eq. (3.18) into Eqs. (3.17a and b) and collecting terms in I_1 and I_2, we end up with

$$-5I_1 + 6I_2 = 10,$$
$$-10I_1 + 14I_2 = 0.$$

Solution of this pair of simultaneous equations gives

$$I_1 = -14 \text{ A}, \qquad I_2 = -10 \text{ A}.$$

Hence,

$$I_x = 8(I_1 - I_2) = 8(-14 + 10) = -32 \text{ A}.$$

Exercise 3-5: Determine the current I in the circuit of Fig. E3.5.

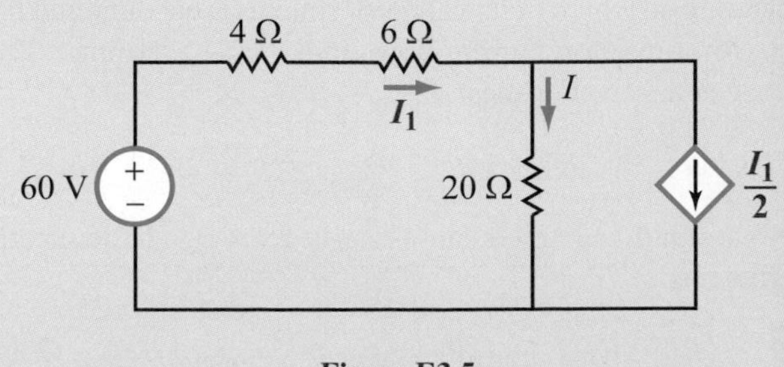

Figure E3.5

Answer: $I = 1.5$ A. (See ◎)

3-2.3 Supermeshes

Two adjoining meshes that share a current source constitute a supermesh.

The current source may be of the independent or dependent type, and it may include a resistor in series with it. The presence of a supermesh in a circuit, such as the one shown in Fig. 3-9(a), simplifies the solution by (a) combining the two mesh-current equations into one and (b) adding a simpler supplemental equation that relates the current of the source to the mesh currents of the two meshes.

In Fig. 3-9(b), the current source of the supermesh has been removed (as has the series resistor R_4) and replaced with a dashed line. The dashed line is a reminder to relate I_0 to the mesh currents, namely

$$I_0 = I_2 - I_3. \tag{3.19}$$

The mesh-current equations for mesh 1 and the joint combination of meshes 2 and 3 are

$$(R_1 + R_2 + R_5)I_1 - R_2 I_2 - R_5 I_3 = V_0, \tag{3.20}$$

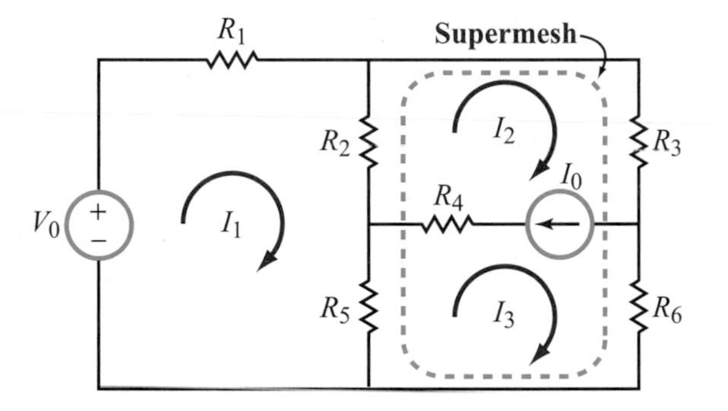

(a) Two adjoining meshes sharing a current source constitute a supermesh.

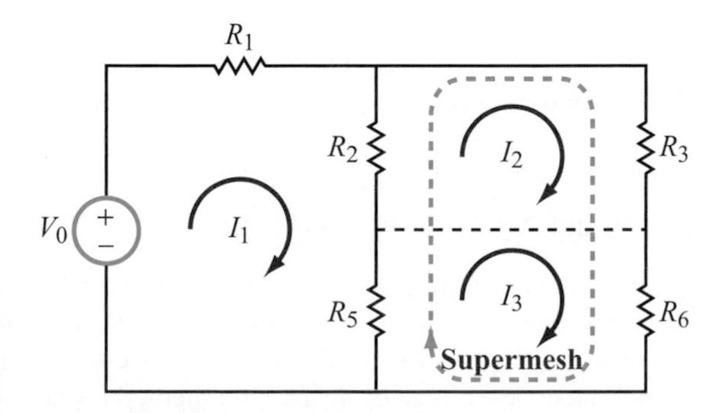

(b) Meshes 2 and 3 can be combined into a single supermesh equation, plus an auxiliary equation $I_0 = I_2 - I_3$.

Figure 3-9: Concept of a supermesh.

and

$$-(R_2 + R_5)I_1 + (R_2 + R_3)I_2 + (R_5 + R_6)I_3 = 0. \tag{3.21}$$

The two mesh-current equations, together with the supplemental equation given by Eq. (3.19) are sufficient to solve for the three mesh currents.

It is instructive to note that the series resistor R_4 played no role in the solution. This is because the current through it is specified by I_0, regardless of the magnitude of R_4 (so long as it is not an open circuit).

Example 3-6: Circuit with a Supermesh

For the circuit in Fig. 3-10(a), determine (a) the mesh currents and (b) the power supplied by each of the two sources.

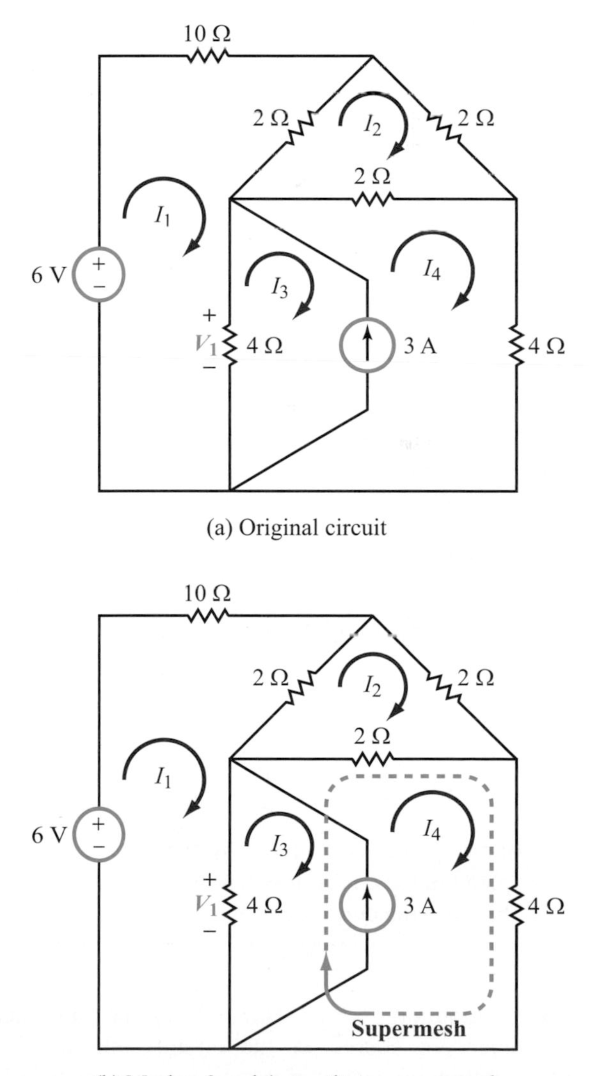

Figure 3-10: Using the supermesh concept to simplify solution of the circuit in Example 3-6.

Solution:

(a) Meshes 3 and 4 share a current source, thereby forming a supermesh. Figure 3-10(b) shows the circuit redrawn such that meshes 3 and 4 can be combined into a single supermesh equation. Consequently, the mesh-current equations for mesh 1, mesh 2, and supermesh 3 and 4 respectively, are

$$(10+2+4)I_1 - 2I_2 - 4I_3 = 6, \tag{3.22a}$$

$$-2I_1 + (2+2+2)I_2 - 2I_4 = 0, \tag{3.22b}$$

and

$$-4I_1 - 2I_2 + 4I_3 + (2+4)I_4 = 0. \tag{3.22c}$$

The supplemental equation associated with the current source is given by

$$I_4 - I_3 = 3. \tag{3.23}$$

Inserting Eq. (3.23) to eliminate I_4 in Eqs. (3.22b and c) followed with a solution of the three simultaneous equations leads to

$$I_1 = 0,$$

$$I_2 = \frac{3}{7} \text{ A},$$

$$I_3 = -\frac{12}{7} \text{ A},$$

$$I_4 = \frac{9}{7} \text{ A}.$$

(b) Since $I_1 = 0$, the power supplied by the 6-V source is

$$P_1 = 6I_1 = 0.$$

To calculate the power supplied by the 3-A current source, we need to know the voltage V_1 across it, which is also the voltage across the 4-Ω resistor given as

$$V_1 = 4(I_1 - I_3) = 4\left(0 - \left(-\frac{12}{7}\right)\right) = \frac{48}{7} \text{ V}.$$

Hence,

$$P_2 = 3V_1 = 3 \times \frac{48}{7} = 20.6 \text{ W}.$$

Thus, all of the power is supplied by the 3-A source alone and is dissipated in the circuit resistances, except for the 10-Ω resistance (because the current through it is $I_1 = 0$).

Review Question 3-5: How does the presence of a dependent source in the circuit influence the implementation procedure of the mesh-current method?

Review Question 3-6: What is a supermesh, and how is it used in mesh analysis?

Exercise 3-6: Apply mesh analysis to determine I in the circuit of Fig. E3.6.

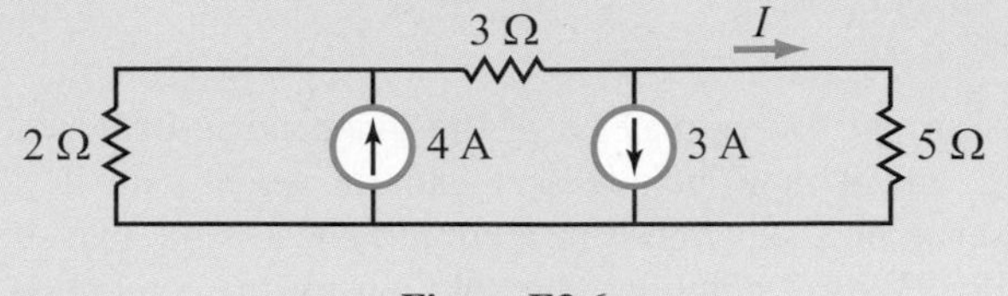

Figure E3.6

Answer: $I = -0.7$ A. (See)

3-3 By-Inspection Methods

The node-voltage and mesh-current methods can be used to analyze any planar circuit, including those containing dependent sources. The solution process relies on the application of KCL and KVL to generate the requisite number of equations necessary to solve for the unknown currents and voltages.

For circuits that contain only independent sources, their KCL and KVL equations exhibit standard patterns, allowing us to write them down by direct *inspection* of the circuit. The method of ***nodal analysis by inspection*** is easy to implement, but it requires that all sources in the circuit be independent current sources. Similarly, ***mesh analysis by inspection*** requires that all sources be independent voltage sources.

If a circuit contains a mixture of independent current and voltage sources, implementation of the by-inspection methods will require a prerequisite step in which some of the current sources are converted to voltage sources, or vice versa, so as to secure the requirement that all sources exclusively are current sources or voltage sources. The conversion process can be realized with the help of the source-transformation technique of Section 2-4.4.

3-3.1 Nodal Analysis by Inspection

Even though it is common practice to characterize the i–v relationship of a resistor in terms of its resistance R, it is more convenient in some cases to work in terms of its conductance $G = 1/R$. The node-voltage by-inspection method is one such case.

We shortly will demonstrate the method for the general case of a circuit composed of n (nonreference) extraordinary nodes. As noted earlier, applicability of the method is limited to circuits with independent current sources. By way of introducing the method, let us consider the simple circuit of Fig. 3-11(a), whose resistances have been relabeled in terms of conductances in Fig. 3-11(b). The circuit has two extraordinary nodes. According to the node-voltage by-inspection method, the circuit is characterized by two node-voltage equations given by

$$G_{11}v_1 + G_{12}v_2 = i_{t_1}, \tag{3.24a}$$

and

$$G_{21}v_1 + G_{22}v_2 = i_{t_2}, \tag{3.24b}$$

where

G_{11} and G_{22} = sum of all conductances connected to nodes 1 and 2, respectively

$G_{12} = G_{21}$ = *negative* of the sum of all conductances connected between nodes 1 and 2

i_{t_1} and i_{t_2} = total of all independent current sources *entering* nodes 1 and 2, respectively (a negative sign applies to a current source leaving a node).

Application of these definitions to Fig. 3-11(b) gives

$$G_{11} = G_1 + G_2,$$
$$G_{22} = G_2 + G_3,$$
$$G_{12} = G_{21} = -G_2,$$
$$i_{t_1} = -i_1,$$

and

$$i_{t_2} = i_1 + i_2.$$

Hence,

$$(G_1 + G_2)v_1 - G_2v_2 = -i_1 \tag{3.25a}$$

and

$$-G_2v_1 + (G_2 + G_3)v_2 = i_1 + i_2. \tag{3.25b}$$

It is a straightforward task to ascertain that Eqs. (3.25a and b) are indeed the correct node-voltage equations for the circuit in Fig. 3-11(b).

Generalizing to the n-node case, the node-voltage equations can be cast in matrix form as

$$\begin{bmatrix} G_{11} & G_{12} & \cdots & G_{1n} \\ G_{21} & G_{22} & \cdots & G_{2n} \\ \vdots & & & \\ G_{n1} & G_{n2} & \cdots & G_{nn} \end{bmatrix} \begin{bmatrix} v_1 \\ v_2 \\ \vdots \\ v_n \end{bmatrix} = \begin{bmatrix} i_{t_1} \\ i_{t_2} \\ \vdots \\ i_{t_n} \end{bmatrix}, \tag{3.26}$$

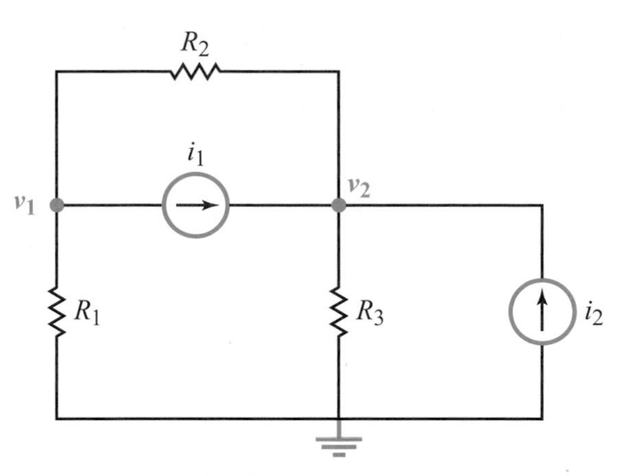

(a) Original circuit

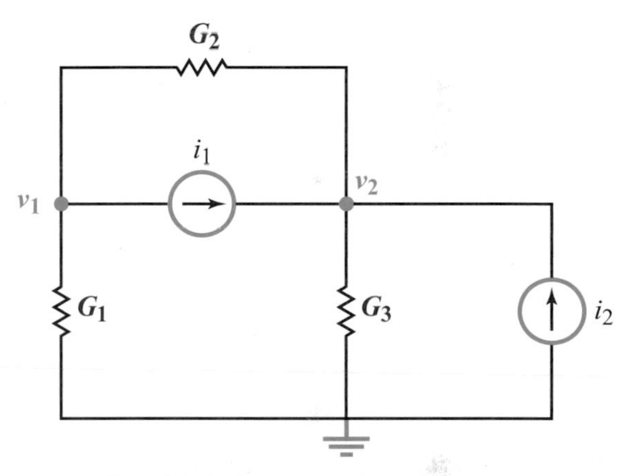

(b) Circuit in terms of conductances

Figure 3-11: Application of the nodal-analysis by-inspection method is facilitated by replacing resistors with conductances.

and abbreviated as

$$\mathbf{GV} = \mathbf{I}_t, \tag{3.27}$$

where $\mathbf{G}$ is the ***conductance matrix*** of the circuit, $\mathbf{V}$ is an unknown ***voltage vector*** representing the node voltages, and $\mathbf{I}_t$ is the ***source vector***. The elements of these matrices are defined as

G_{kk} = sum of all conductances connected to node k

$G_{k\ell} = G_{\ell k}$ = *negative* of conductance(s) connecting nodes k and ℓ, with $k \neq \ell$

V_k = voltage at node k

I_{t_k} = total of current sources *entering* node k (a negative sign applies to a current source leaving the node).

Solution of Eq. (3.27) for the elements of vector $\mathbf{V}$ can be obtained through matrix inversion (Appendix B) or the application of MATLAB® or MathScript (Appendix B).

Example 3-7: Four-Node Circuit

Obtain the node-voltage matrix equation for the circuit in Fig. 3-12 by inspection.

Solution:
At node 1,

$$G_{11} = \frac{1}{1} + \frac{1}{5} + \frac{1}{10} = 1.3.$$

Similarly, at nodes 2, 3, and 4,

$$G_{22} = \frac{1}{5} + \frac{1}{2} + \frac{1}{10} = 0.8,$$

$$G_{33} = \frac{1}{10} + \frac{1}{20} = 0.15,$$

and

$$G_{44} = \frac{1}{10} + \frac{1}{20} = 0.15.$$

The off-diagonal elements of the matrix are

$$G_{12} = G_{21} = -\frac{1}{5} = -0.2,$$

$$G_{13} = G_{31} = -\frac{1}{10} = -0.1,$$

$$G_{14} = G_{41} = 0,$$

$$G_{23} = G_{32} = 0,$$

$$G_{24} = G_{42} = -\frac{1}{10} = -0.1,$$

and

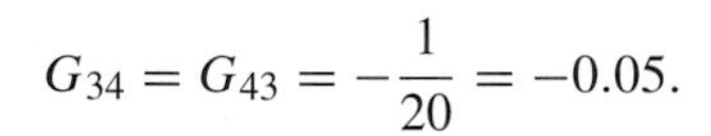

$$G_{34} = G_{43} = -\frac{1}{20} = -0.05.$$

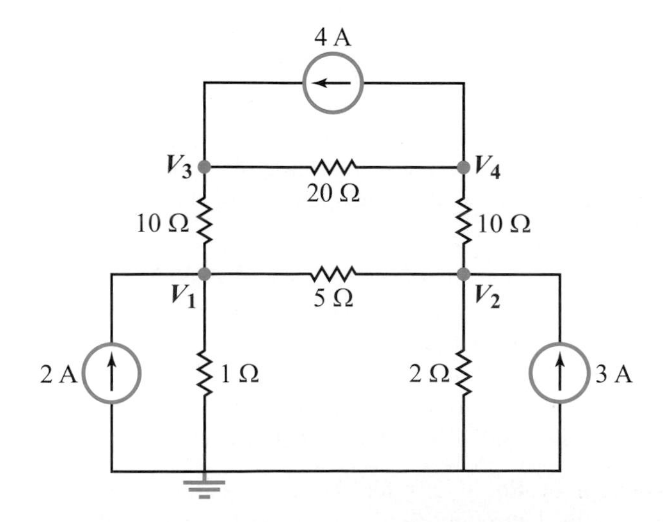

Figure 3-12: Circuit for Example 3-7.

The total currents entering nodes 1 to 4 are

$$I_{t_1} = 2\text{ A},$$

$$I_{t_2} = 3\text{ A},$$

$$I_{t_3} = 4\text{ A},$$

and

$$I_{t_4} = -4\text{ A}.$$

Hence, the node-voltage matrix equation is given by

$$\begin{bmatrix} 1.3 & -0.2 & -0.1 & 0 \\ -0.2 & 0.8 & 0 & -0.1 \\ -0.1 & 0 & 0.15 & -0.05 \\ 0 & -0.1 & -0.05 & 0.15 \end{bmatrix} \begin{bmatrix} V_1 \\ V_2 \\ V_3 \\ V_4 \end{bmatrix} = \begin{bmatrix} 2 \\ 3 \\ 4 \\ -4 \end{bmatrix},$$

Solution by matrix inversion or MATLAB software gives

$$V_1 = 3.73\text{ V},$$

$$V_2 = 2.54\text{ V},$$

$$V_3 = 23.43\text{ V},$$

$$V_4 = -17.16\text{ V}.$$

Exercise 3-7: Apply the node-analysis by-inspection method to generate the node-voltage matrix for the circuit in Fig. E3.7.

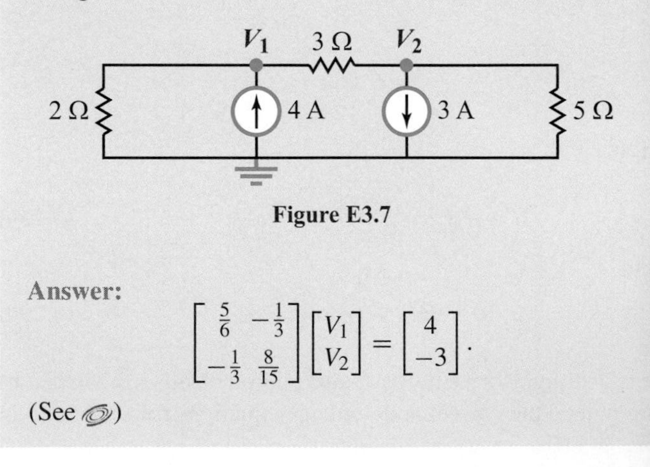

Figure E3.7

Answer:

$$\begin{bmatrix} \frac{5}{6} & -\frac{1}{3} \\ -\frac{1}{3} & \frac{8}{15} \end{bmatrix} \begin{bmatrix} V_1 \\ V_2 \end{bmatrix} = \begin{bmatrix} 4 \\ -3 \end{bmatrix}.$$

(See)

3-3.2 Mesh Analysis by Inspection

By analogy with the node-voltage by-inspection method, for a circuit containing independent voltage sources its n mesh-current equations can be cast in matrix form as

$$\mathbf{RI} = \mathbf{V}_t, \tag{3.28}$$

where **R** is the ***resistance matrix*** of the circuit, **I** is a vector representing the unknown mesh currents, and **V** is the ***source vector***. Equation (3.28) is an abbreviation for

$$\begin{bmatrix} R_{11} & R_{12} & \cdots & R_{1n} \\ R_{21} & R_{22} & \cdots & R_{2n} \\ \vdots & & & \\ R_{n1} & R_{n2} & \cdots & R_{nn} \end{bmatrix} \begin{bmatrix} i_1 \\ i_2 \\ \vdots \\ i_n \end{bmatrix} = \begin{bmatrix} \upsilon_{t_1} \\ \upsilon_{t_2} \\ \vdots \\ \upsilon_{t_n} \end{bmatrix}, \qquad (3.29)$$

where

R_{kk} = sum of all resistances in mesh k,
$R_{k\ell} = R_{\ell k}$ = *negative* of the sum of all resistances shared between meshes k and ℓ (with $k \neq \ell$)
i_k = current of mesh k
υ_{t_k} = total of all independent voltage sources in mesh k, with positive assigned to a voltage rise when moving around the mesh in a clockwise direction.

Example 3-8: Three-Mesh Circuit

Obtain the mesh-current matrix equation for the circuit in Fig. 3-13, by inspection.

Solution:
Application of the definitions for the elements of the matrix **R** and vector $\mathbf{V}_t$ leads to

$$\begin{bmatrix} (2+3+6) & -3 & -6 \\ -3 & (3+4+5) & -5 \\ -6 & -5 & (5+6+7) \end{bmatrix} \begin{bmatrix} I_1 \\ I_2 \\ I_3 \end{bmatrix} = \begin{bmatrix} 6-4 \\ 0 \\ 4 \end{bmatrix},$$

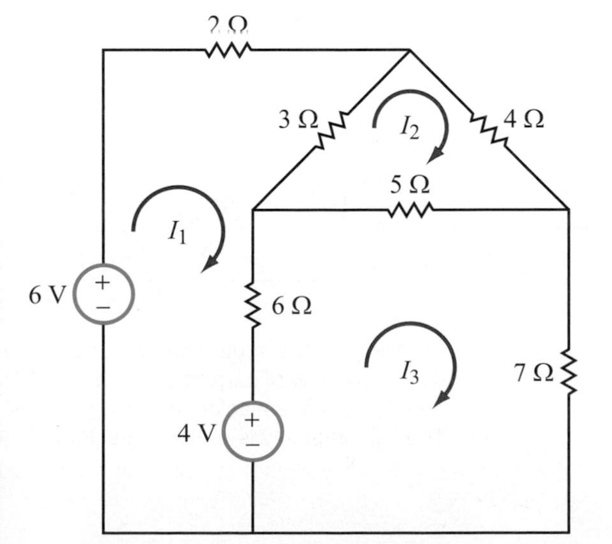

Figure 3-13: Three-mesh circuit of Example 3-8.

which simplifies to

$$\begin{bmatrix} 11 & -3 & -6 \\ -3 & 12 & -5 \\ -6 & -5 & 18 \end{bmatrix} \begin{bmatrix} I_1 \\ I_2 \\ I_3 \end{bmatrix} = \begin{bmatrix} 2 \\ 0 \\ 4 \end{bmatrix}.$$

Solution of the matrix equation gives $I_1 = 0.55$ A, $I_2 = 0.35$ A, and $I_3 = 0.50$ A.

Review Question 3-7: Are the by-inspection methods applicable to (a) circuits containing a mixture of independent voltage and current sources or (b) circuits containing a mixture of independent and dependent voltage sources?

Review Question 3-8: If the circuit contains a mixture of real voltage and current sources, what step should be taken to prepare the circuit for application of one of the two by-inspection methods?

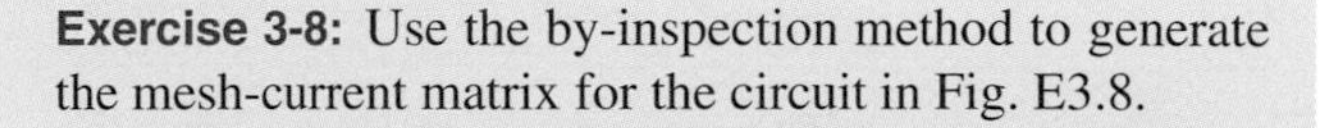

Exercise 3-8: Use the by-inspection method to generate the mesh-current matrix for the circuit in Fig. E3.8.

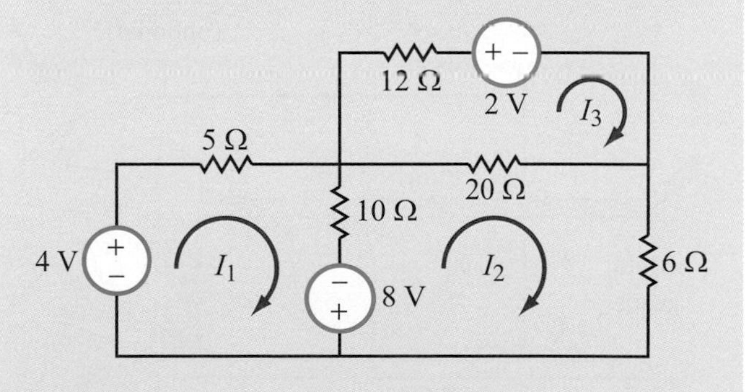

Figure E3.8

Answer:

$$\begin{bmatrix} 15 & -10 & 0 \\ -10 & 36 & -20 \\ 0 & -20 & 32 \end{bmatrix} \begin{bmatrix} I_1 \\ I_2 \\ I_3 \end{bmatrix} = \begin{bmatrix} 12 \\ -8 \\ -2 \end{bmatrix}$$

(See ◎)

Technology Brief 5: Light-Emitting Diodes (LEDs)

Light-emitting diodes (LEDs) are a mainstay of lighting in many manufactured products, from consumer electronics to home appliances to high-efficiency tail-lights for cars. These solid-state semiconductor devices can be fabricated to emit light (Fig. TF5-1) in very narrow bands, centered around any desired wavelength in the spectral range encompassing the infrared, visible, and ultraviolet segments of the electromagnetic spectrum. Modern LEDs are manufactured in a staggering variety of shapes, sizes, and colors. Compared with conventional phosphorescent light bulbs, LEDs have many advantages. Light-emitting diodes respond much faster (microseconds or less), can be made to emit light in a very narrow wavelength band (appear to be a single color), emit more light per watt of electrical energy input, have very long lifetimes (>100,000 hours), and can be integrated directly into semiconductor circuits, printed circuit boards, and in light-focusing packages. LEDs are now inexpensive enough that they are being integrated routinely into street lights, automobile lights, high-efficiency flashlights, and even woven into clothes (e.g., Philips Research Lumalive textiles)!

LEDs are a specific type of the much-larger diode family, whose basic behavior we discussed earlier in Section 2-7. When a voltage is applied in the forward-biased direction across an LED, current flows and photons are emitted (Fig. TF5-2). This occurs because as electrons surge through the diode material they recombine with charge carriers in the material and release energy. The energy is released in the form of photons (quanta of light). The energy of the emitted photon (and hence, its wavelength) depends on the type of material used to make the diode. For example, a diode made of aluminum gallium arsenide (AlGaAs) emits red light, while a diode made from indium gallium nitride (InGaN) emits bluish light. Extensive research over many decades has yielded materials that can emit photons at practically any wavelength across a broad spectrum, extending from the infrared to the ultraviolet. Various "tricks" also have been employed to modify the emitted light after emission. To make white light diodes, for example, certain blue light diodes can be coated with crystal powders which convert the blue light to broad-spectrum "white" light. Other coatings that modify the emitted light (such as quantum dots) are still the subject of research.

In addition to semiconductor LEDs, a newer class of devices called ***Organic Light Emitting Diodes (OLEDs)*** are the subject of intense research efforts. OLEDs operate in an manner analogous to conventional LEDs, but the material used is composed of organic molecules (often polymers). OLEDs are lighter, often flexible, and have the potential to revolutionize handheld and lightweight displays, such as those used in phones, PDAs and flexible screens (see Technology Brief 6 on page 106).

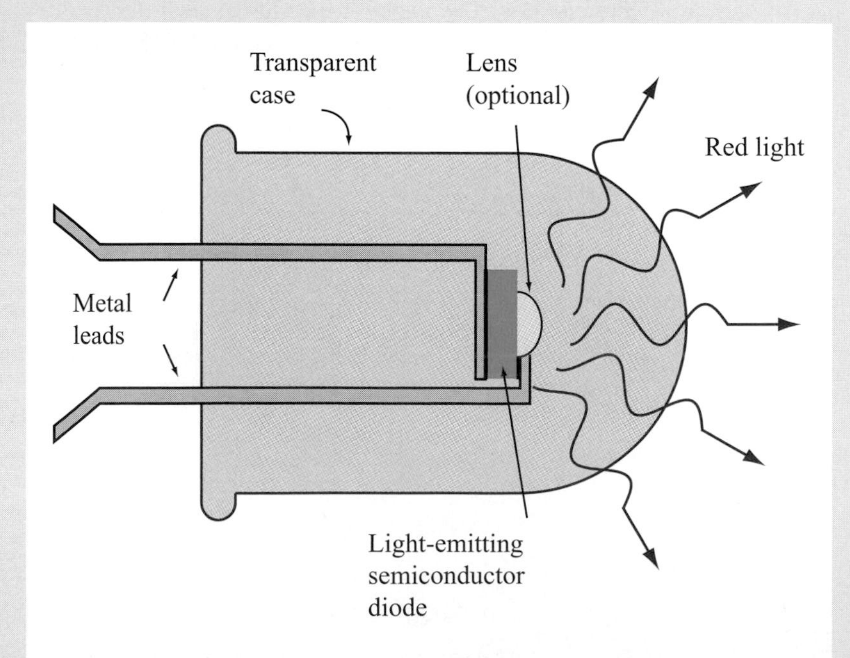

Figure TF5-1: Basic configuration of an LED.

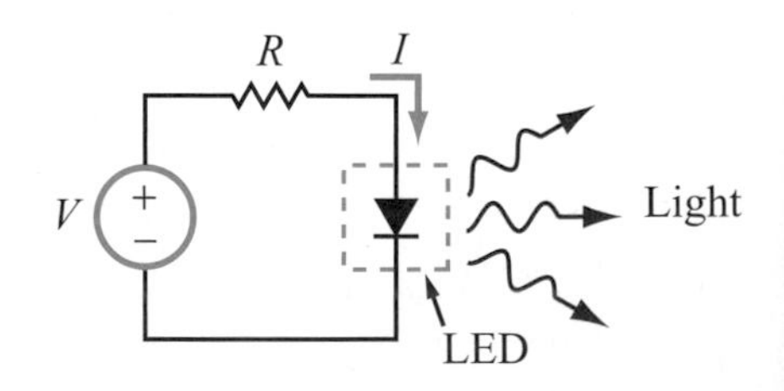

Figure TF5-2: Circuit diagram for an LED circuit; the flow of current through the LED results in the emission of light photons. The light intensity is proportional to I.

3-4 Source Superposition

A system is said to be linear if its output response is directly proportional to the excitation at its input.

In the case of a resistive circuit, the input excitation consists of the combination of all independent voltage and current sources in the circuit, and the output response consists of the set of all voltages across all passive elements in the circuit (namely, the resistors), or all currents through them. Circuits with ideal elements (including those containing capacitors and inductors) satisfy the linearity property, and therefore qualify as linear systems. A linear system obeys the superposition principle, which for a linear circuit translates into:

If a circuit contains more than one independent source, the voltage (or current) response of any element in the circuit is equal to the algebraic sum of the individual responses associated with the individual independent sources, as if each had been acting alone.

Thus, for a circuit with n independent voltage or current sources labeled as sources 1 to n, the voltage υ across a given passive circuit element is given by

$$\upsilon = \upsilon_1 + \upsilon_2 + \cdots + \upsilon_n, \tag{3.30}$$

where υ_k is the response when all sources have been set to zero, except for source k. A similar expression applies to the current i through the circuit,

$$i = i_1 + i_2 + \cdots + i_n. \tag{3.31}$$

The superposition principle can be used to find υ (or i) by executing the following steps:

Solution Procedure: Source Superposition

Step 1: Set all independent sources equal to zero (by replacing voltage sources with short circuits and current sources with open circuits), except for source 1.

Step 2: Apply node-voltage, mesh-current, or any other convenient analysis technique to solve for the response υ_1 due to source 1.

Step 3: Repeat the process for sources 2 through n, calculating in each case the response due to that one source acting alone.

Step 4: Use Eq. (3.30) to determine the total response υ.

Alternatively, the procedure can be used to find currents i_1 to i_n and then to add them up algebraically to find the total current i using Eq. (3.31).

Because it entails solving a circuit multiple times, the source-superposition method may not be attractive, particularly for analyzing circuits with many sources. However, it is a useful tool in both analysis and design for evaluating the sensitivity of a response (such as the current in a load resistor) to specific sources in the circuit.

We should note that whereas the source-superposition method is applicable for voltage and current, it is not applicable for power (see Example 3-9).

Example 3-9: Circuit Analysis by Source Superposition

(a) Use source superposition to determine the current I in the circuit of Fig. 3-14. (b) Determine the amount of power dissipated in the 10-Ω resistor due to each source acting alone and due to both sources acting simultaneously.

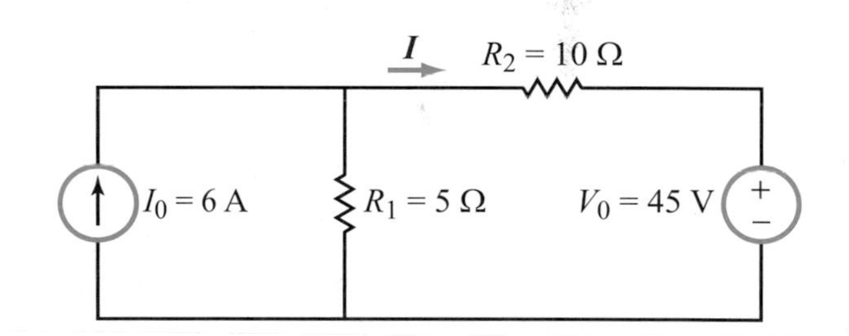

(a) Original circuit

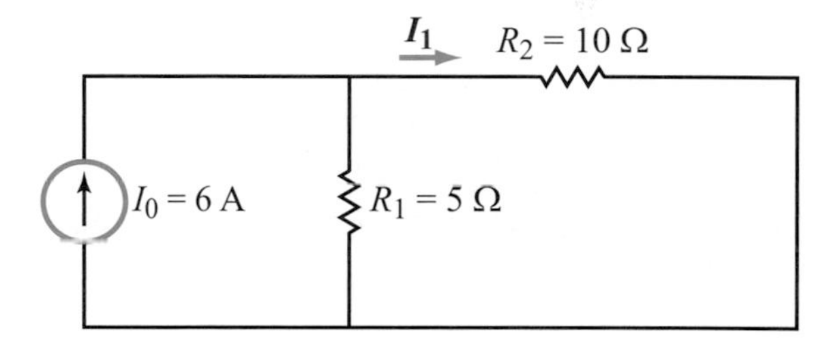

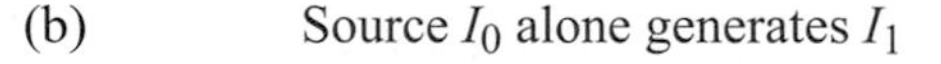
(b) Source I_0 alone generates I_1

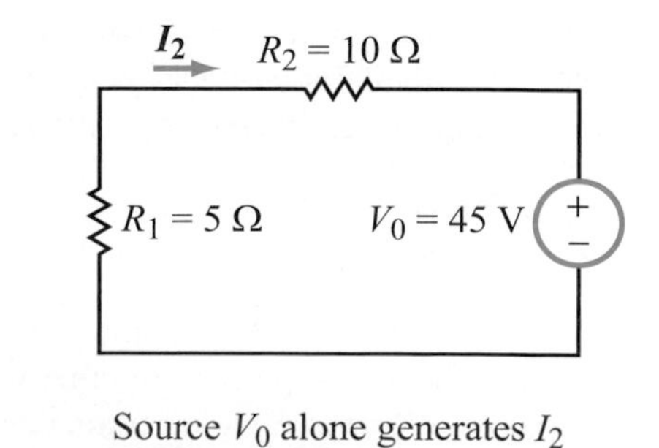

(c) Source V_0 alone generates I_2

Figure 3-14: Application of the source-superposition method to the circuit of Example 3-9.

Solution:

(a) The circuit contains two sources, I_0 and V_0. We start by transforming the circuit into the sum of two new circuits (one with I_0 alone and another with V_0 alone), as shown in parts (b) and (c) of Fig. 3-14, respectively. The current through R_2 due to I_0 alone is labeled I_1, and that due to V_0 alone is labeled I_2. Solution of the circuit in Fig. 3-14(b) [by any of the standard analysis methods] leads to

$$I_1 = 2 \text{ A}.$$

Similarly, solution of the circuit in Fig. 3-14(c) leads to

$$I_2 = -3 \text{ A}.$$

Hence,

$$I = I_1 + I_2 = 2 - 3 = -1 \text{ A}.$$

(b) The amounts of power dissipated in the 10-Ω resistor due to I_1 alone, I_2 alone, and the total current I are, respectively;

$$P_1 = I_1^2 R = 2^2 \times 10 = 40 \text{ W},$$

$$P_2 = I_2^2 R = (-3)^2 \times 10 = 90 \text{ W},$$

and

$$P = I^2 R = 1^2 \times 10 = 10 \text{ W}.$$

Note that $P \neq P_1 + P_2$, because the linearity property does not apply to power.

Review Question 3-9: Explain why the linearity property of electric circuits is an underlying requirement for the application of the source-superposition method.

Review Question 3-10: How is the superposition method used as a sensitivity tool in circuit analysis and design?

Review Question 3-11: Is the source-superposition method applicable to power? In other words, if source 1 alone supplies power P_1 to a certain device and source 2 alone supplies power P_2 to the same device, will the two sources acting simultaneously supply power $P_1 + P_2$ to the device?

Exercise 3-9: Apply the source-superposition method to determine the current I in the circuit of Fig. E3.9.

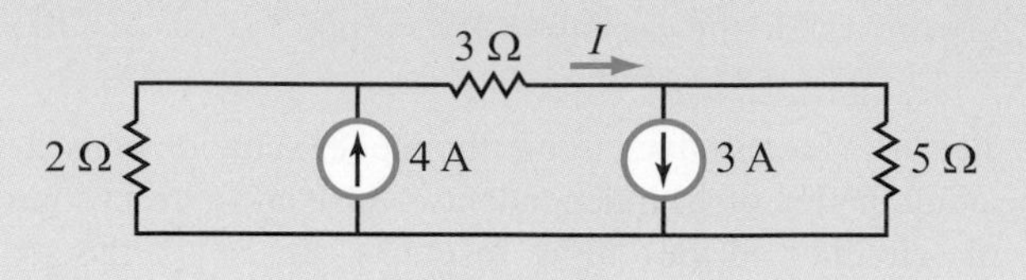

Figure E3.9

Answer: $I = 2.3$ A. (See)

Exercise 3-10: Apply source superposition to determine V_{out} in the circuit of Fig. E3.10.

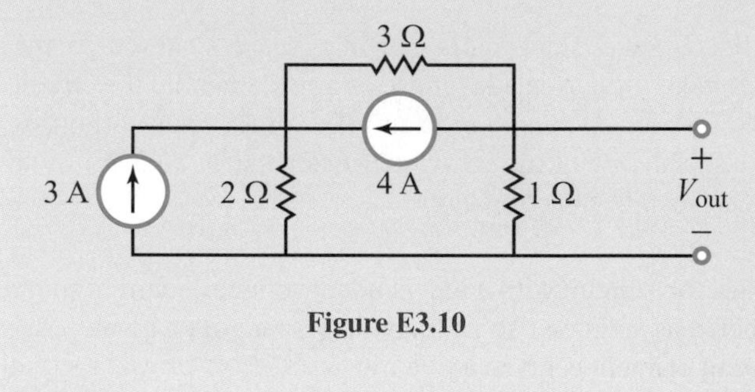

Figure E3.10

Answer: $V_{out} = -1$ V. (See)

3-5 Thévenin and Norton Equivalent Circuits

As depicted by the block diagram shown in Fig. 3-15, a generic ***cell-phone circuit*** consists of several individual circuits, including amplifiers, oscillators, analog-to-digital (A/D) and digital-to-analog (D/A) converters, an antenna, a diplexer that allows the antenna to be used for both transmission and reception, a microprocessor, and other auxiliary circuits. Many of these circuits are quite complex and may contain a large number of active and passive elements, in both discrete and integrated form. So the question one might ask is: *How does an engineer approach an analysis or design task involving such a complex architecture?*

Dealing with the entire circuit all at once would be next to impossible, not only because of its daunting complexity, but also because the individual circuits call for engineers with different specializations.

Fortunately, we have a straightforward answer to the question, namely that each circuit gets modeled as a "black box" with specified input and output terminal characteristics allowing the engineer working with a particular circuit to treat the other circuits connected to it in terms of only those characteristics without much regard to the details of their internal architectures

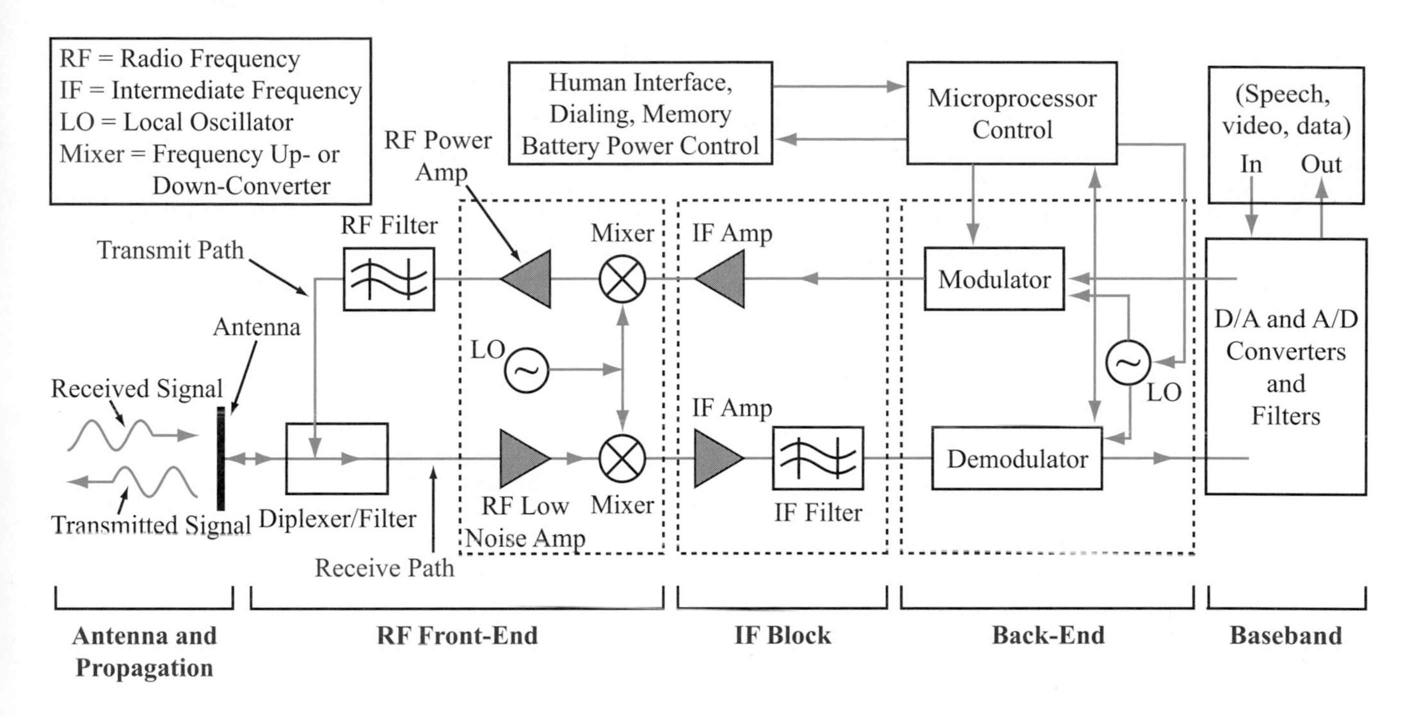

Figure 3-15: Cell-phone block diagram.

For an amplifier, for example, its overall specifications might include voltage gain and frequency bandwidth, among other attributes, but its terminal characteristics refer to how it would "appear" from the perspective of other circuits.

Conversely, from the amplifier's perspective, other circuits are specified in terms of how they appear to the amplifier. Figure 3-16 illustrates the concept from the perspective of the radio-frequency (RF) low-noise amplifier in the receive channel of the cell-phone circuit. The combination of the antenna and diplexer (including the input signal picked up by the antenna) is represented at the input side of the amplifier by an equivalent circuit composed of a voltage source v_s in series with an ***impedance*** $\mathbf{Z}_s$. Impedance (which we shall introduce in a later chapter) is the ac-equivalent of resistance in dc circuits. At the output side of the amplifier, the mixer (whose function is to shift the center frequency of the input signal from 834 MHz down to 70 MHz) is represented by a ***load impedance*** $\mathbf{Z}_L$. Thus, the output terminal characteristics of the antenna/diplexer combination become the input source to the amplifier, and the input impedance of the mixer becomes the load to which the amplifier is connected. *Isolating the amplifier, while keeping it in the context of its input and output neighbors, facilitates both the analysis and design processes.*

Another example of the blackbox approach is the common household outlet; we treat it as a constant-voltage source (in series with a small impedance) regardless of the load impedance

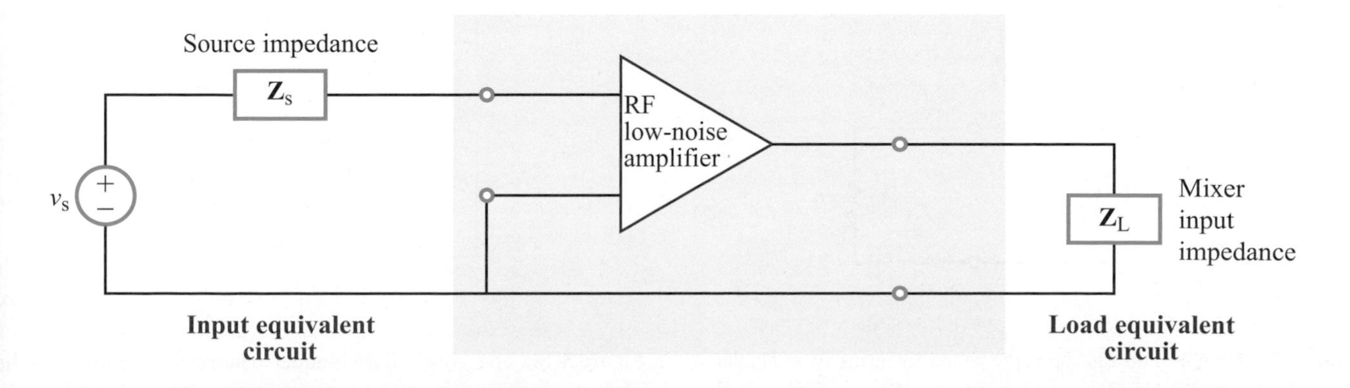

Figure 3-16: Input and output circuits as seen from the perspective of a Radio-Frequency amplifier circuit.

of the appliance we connect to it. By so doing, we avoid having to re-analyze the entire circuit (including the power lines and the power-generating station) every time we change loads (appliances). Our ability to develop equivalent-circuit representations is made possible (in part) by a pair of theorems of fundamental significance known as Thévenin's and Norton's theorems. These constitute the subject of this section.

3-5.1 Thévenin's Theorem

In the 1880s, a French engineer named Leon Thévenin introduced the concept known today as ***Thévenin's theorem***, which asserts:

> A linear circuit can be represented at its output terminals by an equivalent circuit consisting of a series combination of a voltage source v_{Th} and a resistor R_{Th}, where v_{Th} is the open-circuit voltage at those terminals (no load) and R_{Th} is the equivalent resistance between the same terminals when all independent sources in the circuit have been deactivated.

A pictorial representation of Thévenin's theorem is shown in Fig. 3-17, where the actual circuit in part (a) has been replaced with the Thévenin equivalent circuit in part (b). The implication of this model is that when the circuit is connected to a load resistor R_{L}, the current i_{L} running through it will be identical for both the actual circuit and the equivalent circuit. This equivalence holds true for any value of R_{L}, from zero (short circuit) to ∞ (open circuit). Thus, from the standpoint of the load, the two circuits are indistinguishable.

Even though the present discussion pertains to dc currents, the Thévenin concept extends to ac circuits as well. We will revisit the concept in a future chapter for circuits containing capacitors and inductors.

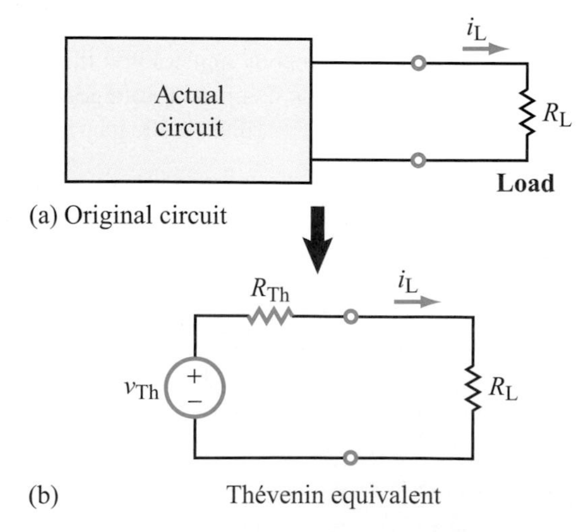

Figure 3-17: A circuit can be represented in terms of a Thévenin equivalent comprised of a voltage source v_{Th} in series with a resistance R_{Th}.

3-5.2 Finding v_{Th}

Given the equivalence of the two circuits shown in Fig. 3-17, the Thévenin voltage v_{Th} can be determined by analyzing the given circuit to calculate the voltage v_{oc} at its output terminals when no load is connected between them (i.e., $R_{\text{L}} = \infty$). Thus,

$$v_{\text{Th}} = v_{\text{oc}}, \tag{3.32}$$

as indicated in Fig. 3-18(a). *The procedure is equally valid for circuits with or without dependent sources. For a circuit with no independent sources, $V_{\text{Th}} = 0$.*

3-5.3 Finding R_{Th}—Short-Circuit Method

Multiple methods are available for finding the Thévenin resistance R_{Th}. We start with the short-circuit method. From Fig. 3-17(b),

$$i_{\text{L}} = \frac{v_{\text{Th}}}{R_{\text{Th}} + R_{\text{L}}}. \tag{3.33}$$

If $R_{\text{L}} = 0$ (short-circuit load), we call i_{L} the short-circuit current i_{sc}, which would be given by

$$i_{\text{sc}} = \frac{v_{\text{Th}}}{R_{\text{Th}}}. \tag{3.34}$$

By analyzing the circuit configuration in Fig. 3-18(b) to find i_{sc}, we can apply Eq. (3.34) to find R_{Th},

$$R_{\text{Th}} = \frac{v_{\text{Th}}}{i_{\text{sc}}}. \tag{3.35}$$

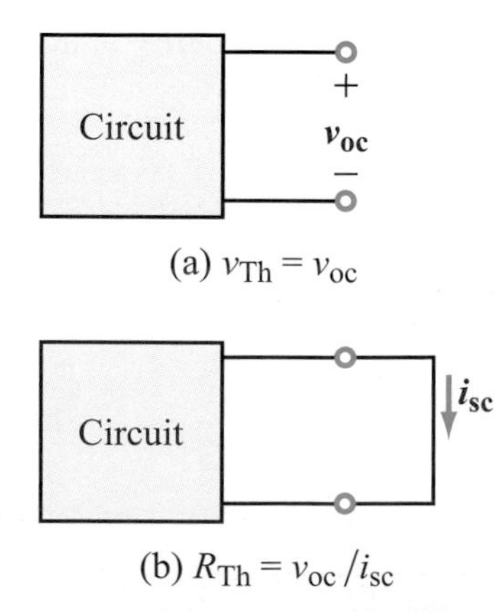

Figure 3-18: Thévenin voltage is equal to the open-circuit voltage and Thévenin resistance is equal to the ratio of v_{oc} to i_{sc}, where i_{sc} is the short-circuit current between the output terminals.

This method is applicable to any circuit with at least one independent source, regardless of whether or not it contains dependent sources.

Example 3-10: Open Circuit/Short Circuit Method

The input circuit to the left of terminals (a, b) in Fig. 3-19(a) is connected to a variable load resistor R_L. Determine (a) the Thévenin equivalent circuit at terminals (a, b) and (b) use it to find the value of R_L that will cause the magnitude of the current through it to be 0.5 A.

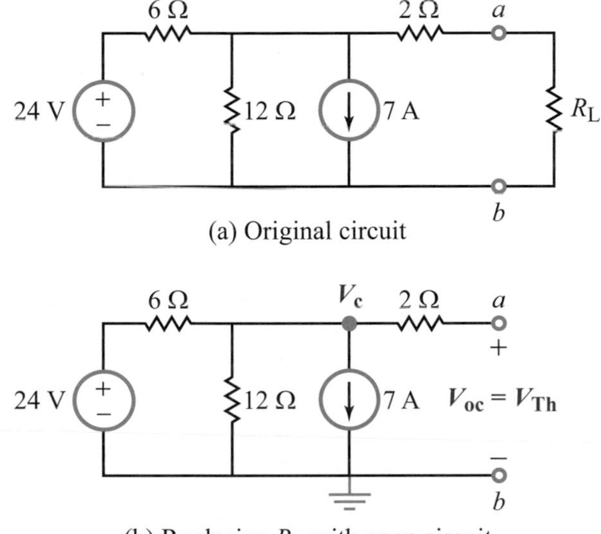

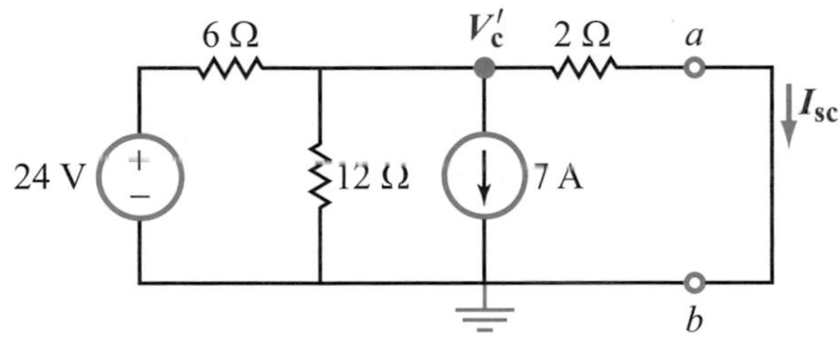

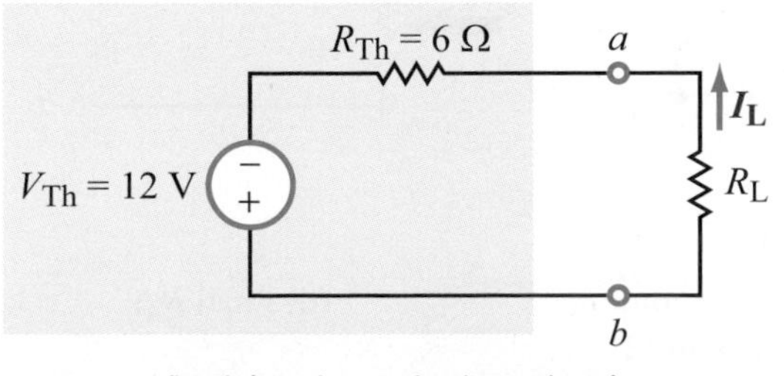

Figure 3-19: Applying open circuit/short circuit method to find the Thévenin equivalent for the circuit of Example 3-10.

Solution: **(a)** With R_L replaced with an open circuit in Fig. 3-19(b), V_{Th} is the open-circuit voltage between terminals (a, b). Since no current flows through the 2-Ω resistor, $V_{Th} = V_c$ at node c. The node-voltage equation at node c is

$$\frac{V_c - 24}{6} + \frac{V_c}{12} + 7 = 0,$$

which leads to $V_c = -12$ V. Hence,

$$V_{Th} = -12 \text{ V}.$$

Next, we replace R_L with a short circuit (Fig. 3-19(c)) and repeat the process to find V'_c:

$$\frac{V'_c - 24}{6} + \frac{V'_c}{12} + 7 + \frac{V'_c}{2} = 0,$$

whose solution gives $V'_c = -4$ V, and

$$I_{sc} = \frac{V'_c}{2} = -\frac{4}{2} = -2 \text{ A}.$$

Hence,

$$R_{Th} = \frac{V_{Th}}{I_{sc}} = \frac{-12}{-2} = 6 \ \Omega,$$

and the Thévenin equivalent circuit is shown in Fig. 3-19(d).

(b) In view of Fig. 3-19(d), for I_L to be 0.5 A, it is necessary that

$$I_L = \frac{12}{6 + R_L} = 0.5 \text{ A}$$

or

$$R_L = 18 \ \Omega.$$

Exercise 3-11: Determine the Thévenin-equivalent circuit at terminals (a, b) in Fig. E3.11.

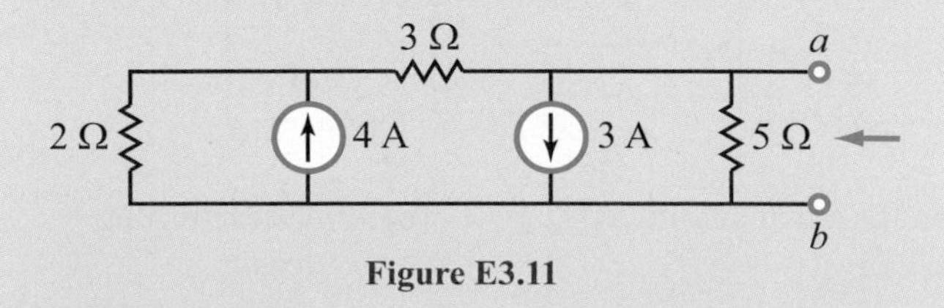

Figure E3.11

Answer: $V_{Th} = -3.5$ V, $I_{sc} = -1.4$ A, $R_{Th} = 2.5$ Ω. (See)

Equivalent-Resistance Method

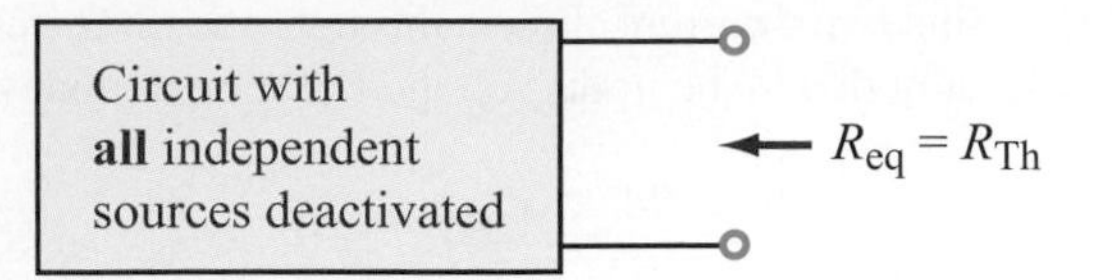

Figure 3-20: For a circuit that does not contain dependent sources, R_{Th} can be determined by deactivating all sources and then simplifying the circuit down to a single resistance R_{eq}.

3-5.4 Finding R_{Th}—Equivalent Resistance Method

If the circuit does not contain dependent sources, R_{Th} can be determined by deactivating all sources (replacing voltage sources with short circuits and current sources with open circuits) and then simplifying the circuit down to a single equivalent resistance between its output terminals, as portrayed by Fig. 3-20. In that case,

$$R_{Th} = R_{eq}. \tag{3.36}$$

This method does not apply to circuits that contain dependent sources.

Example 3-11: Thévenin Resistance

Find R_{Th} at terminals (a, b) for the circuit in Fig. 3-21(a).

Solution:
Since the circuit has no dependent sources, we can apply the equivalent-resistance method. We start by deactivating all of the sources (as shown in Fig. 3-21(b)) where we replaced the voltage source with a short circuit and the current source with an open circuit. After (a) combining the two 50-Ω resistors in parallel, (b) combining their 25-Ω combination in series with the 35-Ω resistance, and (c) finally combining the resultant 60-Ω with the 30-Ω resistance in parallel, we obtain

$$R_{Th} = 20\ \Omega.$$

Exercise 3-12: Find the Thévenin equivalent of the circuit to the left of terminals (a, b) in Fig. E3.12, and then determine the current I.

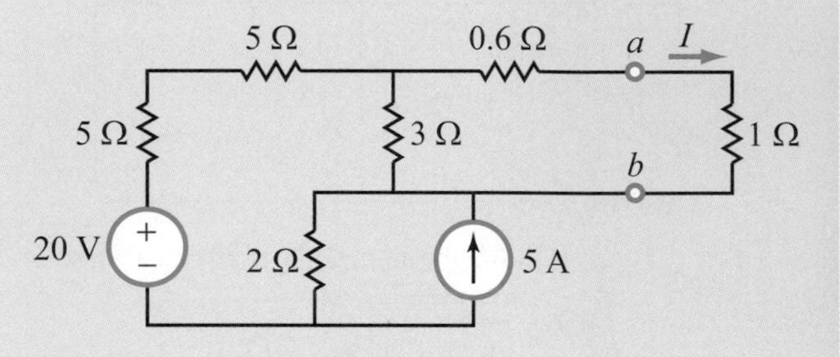

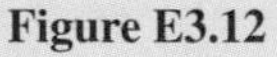
Figure E3.12

Answer:

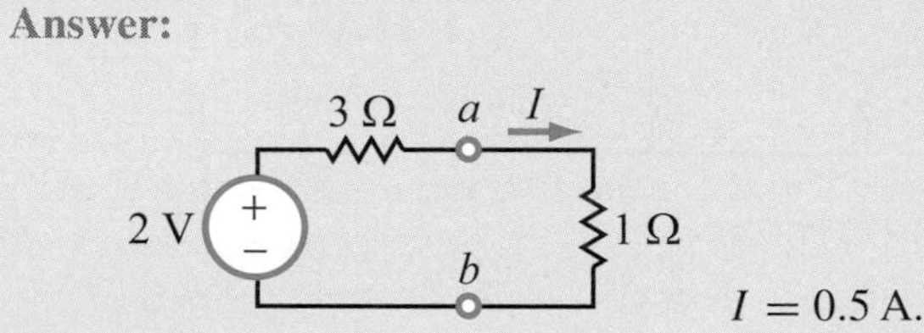

$I = 0.5$ A.

(See ◎)

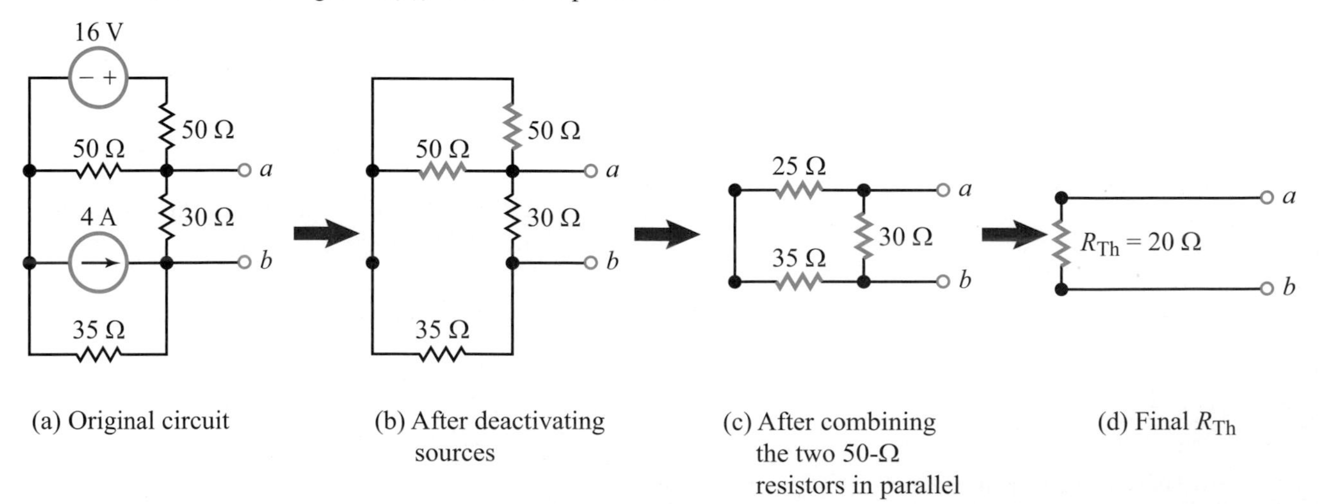

(a) Original circuit

(b) After deactivating sources

(c) After combining the two 50-Ω resistors in parallel

(d) Final R_{Th}

Figure 3-21: After deactivation of sources, systematic simplification leads to R_{Th} (Example 3-11).

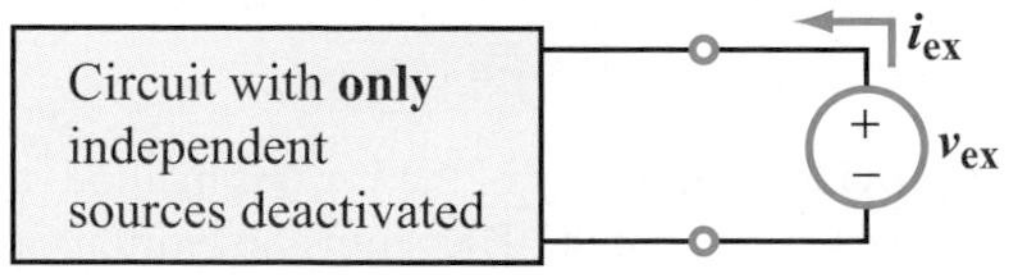

Figure 3-22: If a circuit contains both dependent and independent sources, R_{Th} can be determined by (a) deactivating independent sources (only), (b) adding an external source v_{ex}, and then (c) solving the circuit to determine i_{ex}. The solution is $R_{Th} = v_{ex}/i_{ex}$.

3-5.5 Finding R_{Th}—External-Source Method

The equivalent-resistance method described previously does not apply to circuits containing dependent sources. Hence, an alternative variation is called for. Independent sources again are deactivated (but dependent sources are left alone) and an external voltage source v_{ex} is introduced to excite the circuit, as shown in Fig. 3-22. After analyzing the circuit to determine the current i_{ex}, R_{Th} is found by applying

$$R_{Th} = \frac{v_{ex}}{i_{ex}}. \tag{3.37}$$

Example 3-12: Circuit with Dependent Source

Find the Thévenin equivalent circuit at terminals (a, b) for the circuit in Fig. 3-23(a) by applying the combination of open-circuit-voltage and external-source methods.

Solution:
The equations for mesh currents I_1 and I_2 in Fig. 3-23(a) are given by

$$-68 + 6I_1 + 2(I_1 - I_2) + 4I_x = 0$$

and

$$-4I_x + 2(I_2 - I_1) + 6I_2 + 4I_2 = 0.$$

Recognizing that $I_x = I_2$, solution of these two simultaneous equations leads to

$$I_1 = 8 \text{ A},$$

and

$$I_2 = 2 \text{ A}.$$

The Thévenin voltage is V_{ab}. Hence,

$$\begin{aligned} V_{Th} &= V_{ab} \\ &= 4I_2 \\ &= 8 \text{ V}. \end{aligned}$$

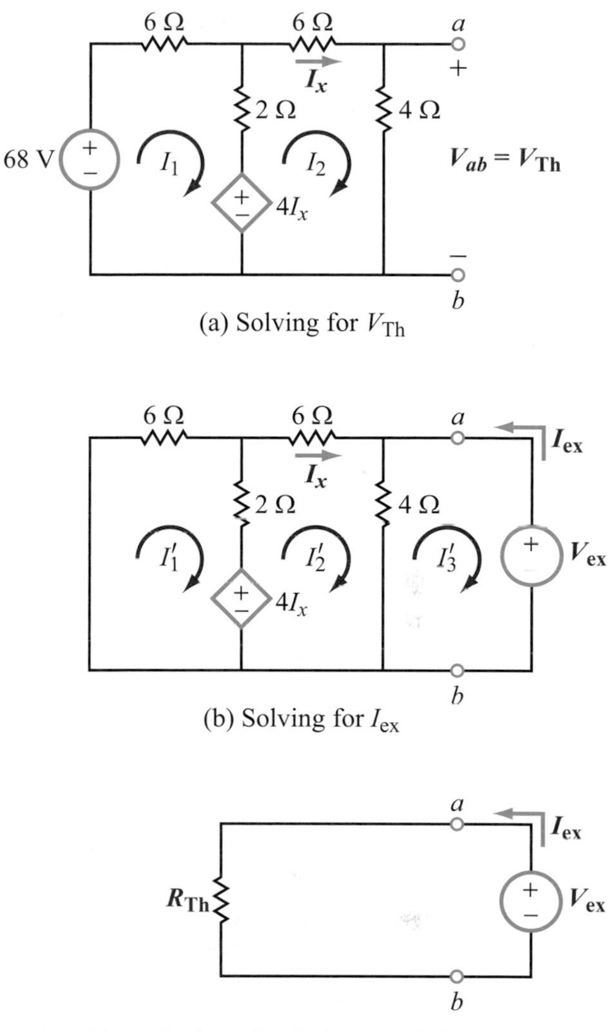

Figure 3-23: Solution of the open-circuit voltage gives $V_{ab} = V_{Th} = 8$ V. Use of the external-voltage method leads to $R_{Th} = 36/17\ \Omega$ (Example 3-12).

To find R_{Th} using the external-source method, we deactivate the 68-V voltage source and we add an external voltage source V_{ex}, as shown in Fig. 3-23(b). Our task is to obtain an expression for I_{ex} in terms of V_{ex}. We now have three mesh currents, which we have labeled I'_1, I'_2, and I'_3. Their equations are given by

$$6I'_1 + 2(I'_1 - I'_2) + 4I_x = 0,$$

$$-4I_x + 2(I'_2 - I'_1) + 6I'_2 + 4(I'_2 - I'_3) = 0,$$

and

$$4(I'_3 - I'_2) + V_{ex} = 0.$$

After replacing I_x with I'_2 and solving the three simultaneous equations, we obtain

$$I'_1 = \frac{1}{18} V_{ex},$$

$$I'_2 = -\frac{2}{9} V_{ex},$$

and

$$I'_3 = -\frac{17}{36} V_{ex}.$$

For the equivalent circuit shown in Fig. 3-23(c),

$$R_{Th} = \frac{V_{ex}}{I_{ex}}.$$

In terms of our solution, $I_{ex} = -I'_3$. Hence,

$$R_{Th} = -\frac{V_{ex}}{I'_3} = \frac{36}{17} \ \Omega.$$

3-5.6 Norton's Theorem

A corollary of Thévenin's theorem, Norton's theorem states that a linear circuit can be represented at its output terminals by an equivalent circuit composed of a parallel combination of a current source i_N and a resistor R_N. Application of source transformation on the Thévenin equivalent circuit shown in Fig. 3-24 leads to the straightforward conclusion that i_N and R_N of the Norton equivalent circuit are given by

$$i_N = \frac{v_{Th}}{R_{Th}} \qquad (3.38a)$$

and

$$R_N = R_{Th}. \qquad (3.38b)$$

Table 3-1 provides a summary of the various methods available for finding the elements of the Thévenin and Norton equivalent circuits.

Question: *How should one decide which analysis method to use?*

Table 3-2 provides a summary of the various analysis methods covered in this chapter. As noted in the table, only the

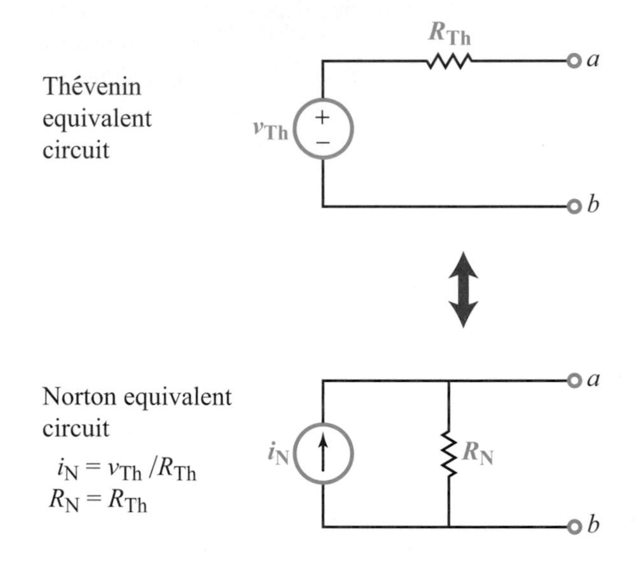

Figure 3-24: Equivalence between Thévenin and Norton equivalent circuits.

Table 3-1: Properties of Thévenin/Norton analysis techniques.

To Determine	Method	Can Circuit Contain Dependent Sources?	Relationship
v_{Th}	Open-circuit v	Yes	$v_{Th} = v_{oc}$
v_{Th}	Short-circuit i (if R_{Th} is known)	Yes	$v_{Th} = R_{Th} i_{sc}$
R_{Th}	Open/short	Yes	$R_{Th} = v_{oc}/i_{sc}$
R_{Th}	Equivalent R	No	$R_{Th} = R_{eq}$
R_{Th}	External source	Yes	$R_{Th} = v_{ex}/i_{ex}$
$i_N = v_{Th}/R_{Th}$; $R_N = R_{Th}$			

Table 3-2: Summary of the applicability conditions of several circuit analysis methods.

		Basic Analysis Methods			
		Applicable to Circuits Containing			
Section	**Method**	**Dependent Source?**	**Voltage and Current Sources?**	**Supernode?**	**Supermesh?**
3-1.1	**Node-voltage/standard**	Yes	Both	No	Yes
3-1.3	**Node-voltage/supernode**	Yes	Both	Yes	Yes
3-2.1	**Mesh-current/standard**	Yes	Both	Yes	No
3-2.3	**Mesh-current/supermesh**	Yes	Both	Yes	Yes
3-3.1	**Node analysis/by-inspection**	No	Current only	No	Yes
3-3.2	**Mesh analysis/by-inspection**	No	Voltage only	Yes	No
		Other Analysis Methods			
3-4	**Source superposition**	For a circuit with n independent sources, method uses applicable basic analysis methods (top section) to analyze the circuit for each source separately.			
3-5	**Thévenin/Norton**	Uses applicable basic analysis methods to generate a simple equivalent circuit (see Table 3-1 for details).			

node-voltage/supernode and mesh-current/supermesh methods can accommodate all possible combinations of dependent and independent sources, with or without supernodes or supermeshes. The by-inspection methods are easier to apply and to construct in matrix form, but they are constrained to circuits containing either exclusive independent current sources or exclusive independent voltage sources.

No simple answer exists to the question as to how one should decide which method to use, but generally speaking, it is advantageous to use a node-voltage method if the circuit is dominated by voltage sources and to use a mesh-current method if most of the sources are current sources. Beyond that simple guideline, consult Table 3-2 before proceeding with your solution.

Review Question 3-12: Why is the Thévenin-equivalent circuit method such a powerful tool when analyzing a complex circuit, such as that of a cell phone?

Review Question 3-13: Section 3-5 offers three different approaches for finding R_{Th}. Which ones apply to circuits containing dependent sources?

Exercise 3-13: Find the Norton equivalent at terminals (a, b) of the circuit in Fig. E3.13.

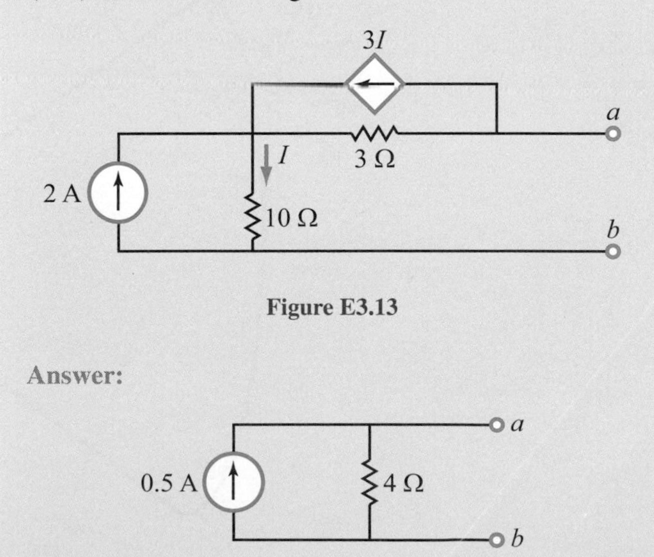

Figure E3.13

(See CD)

Technology Brief 6: Display Technologies

From cuneiform-marked clay balls to the abacus to today's digital projection technology, advances in visual displays have accompanied almost every major leap in information technology. While the earliest "modern" computers relied on cathode ray tubes (CRT) to project interactive images, today's computers can access a wide variety of displays ranging from plasma screens and LED arrays to digital micromirror projectors, electronic ink, and virtual reality interfaces. In this Technology Brief, we will review the major technologies currently available for two-dimensional visual displays.

Cathode Ray Tube (CRT)

The earliest computers relied on the same technology that made the television possible. In a CRT television or monitor (Fig. TF6-1), an ***electron gun*** is placed behind a positively charged glass screen, and a negatively charged electrode (the ***cathode***) is mounted at the input of the electron gun. During operation, the cathode emits streams of electrons into the electron gun. The emitted electron stream is steered onto different parts of the positively charged screen by the electron gun; the direction of the electron stream is controlled by the electric field of the deflecting coils through which the beam passes. The screen is composed of thousands of tiny dots of phosphorescent material arranged in a two-dimensional array. Every time an electron hits a phosphor dot, it glows a specific color (red, blue, or green). A pixel on the screen is composed of phosphors of these three colors. In order to make an image appear to move on the screen, the electron gun constantly steers the electron stream onto different phosphors, lighting them up faster than the eye can detect the changes, and thus, the images appear to move. In modern color CRT displays, three electron guns shoot different electron streams for the three colors.

Interestingly, the basic concept behind the CRT is still very relevant. A new technology called ***Field Emission Display (FED)*** has emerged, which uses a thin film of atomically sharp electron emitter tips (carbon nanotubes can be used for this purpose) to generate the electrons. The electrons emitted by the film collide with phosphor elements just as in the traditional CRT. The primary advantage of this type of "flat-panel" display is that it can provide a wider viewing angle (i.e., one can look at an FED screen at a sharp angle and still see a good image) than possible with conventional LCD or LED technology (discussed next).

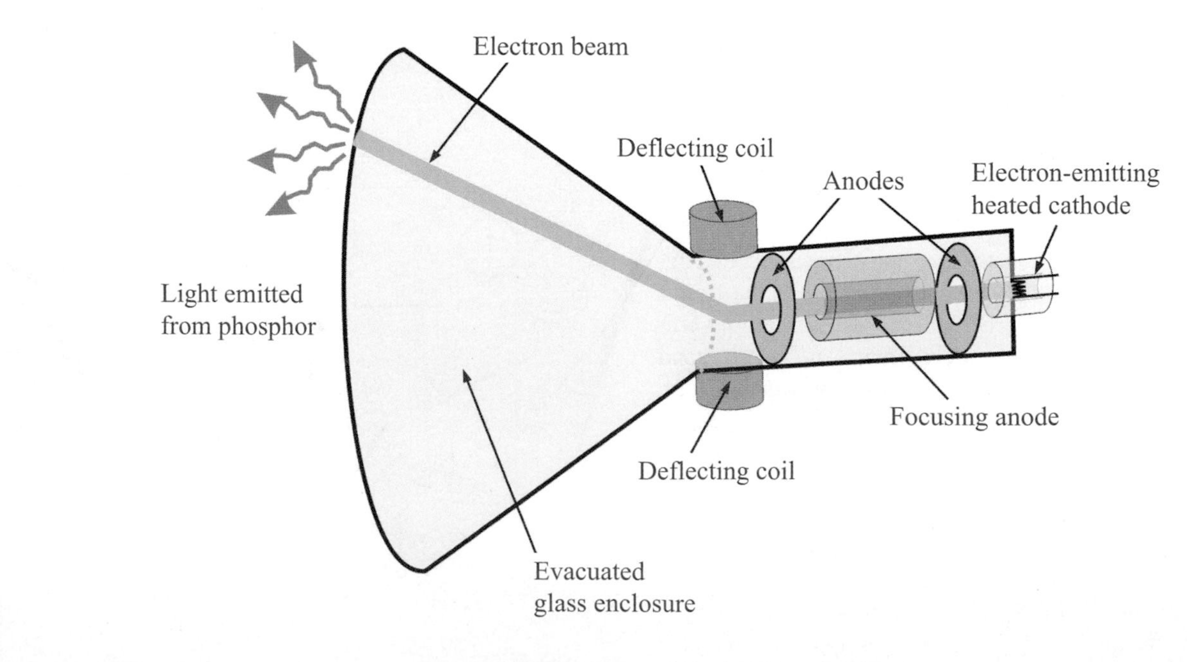

Figure TF6-1: Schematic of CRT operation.

Liquid Crystal Displays (LCD)

LCDs are used in digital clocks, cellular phones, desktop and laptop computers, and some televisions and other electronic systems. They offer a decided advantage over other display technologies (such as cathode ray tubes) in that they are lighter and thinner and consume a lot less power to operate. LCD technology relies on special electrical and optical properties of a class of materials known as ***liquid crystals***, first discovered in the 1880s by botanist Friedrich Reinitzer. In the basic LCD display, light shines through a thin stack of layers as shown in Fig. TF6-2. Each stack consists of layers in the following order (starting from the viewer's eye): color filter, vertical (or horizontal) polarizer filter, glass plate with transparent electrodes, liquid crystal layer, second glass plate with transparent electrodes, horizontal (or vertical) polarizer filter. Light is shone from behind the stack (called the ***backlight***). As light crosses through the layer stack, it is polarized along one direction by the first filter. If no voltage is applied on any of the electrodes, the liquid crystal molecules align the filtered light so that it can pass through the second filter. Once through the second filter, it crosses the color filter (which allows only one color of light through) and the viewer sees light of that color. If a voltage is applied between the electrodes on the glass plates (which are on either side of the liquid crystal), the induced electric field causes the liquid crystal molecules to rotate. Once rotated, the crystals no longer align the light coming through the first filter so that it can pass through the second filter plate. If light cannot cross, the area with the applied voltage looks dark. This is precisely how simple hand-held calculator displays work; usually the bright background is made dark every time a character is displayed.

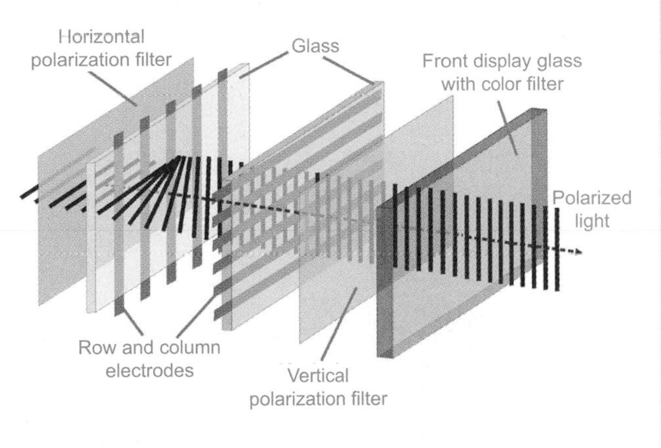

Figure TF6-2: Schematic of LCD operation.

Modern flat-profile computer screens and laptops use a version of the LCD called ***thin-film transistor (TFT)*** LCD; these also are known as ***active matrix*** displays. In TFT LCDs, several thin films are deposited on one of the glass substrates and patterned into transistors. Each color component of a pixel has its own microscale transistor that controls the voltage across the liquid crystal; since the transistors only take up a tiny portion of the pixel area, they effectively are invisible. Thus, each pixel has its own electrode driver built directly into it. This specific feature enabled the construction of the flat high-resolution screens now in common use with desktops (and made the CRT display increasingly obsolete). Since LCD displays also weigh considerably less than a CRT tube, they enabled the emergence of laptop computers in the 1980s. Early laptops used large, heavy monochrome LCDs; most of today's palmtop and laptop systems use active-matrix displays.

Light-Emitting Diode (LED) Displays

A fundamental advance over LCDs is the use of tiny light-emitting diodes (LED) in large pixel arrays on flat screens (see Technology Brief 5: LEDs on page 96). Each pixel in an LED display is composed of three LEDs (one each of red, green, and blue). Whenever a current is made to pass through a particular LED, it emits light at its particular color. In this way, displays can be made flatter (i.e., the LED circuitry takes up less room than an electron gun or LCD) and larger (since making large, flat LED arrays technically is less challenging than giant CRT tubes or LCD displays). Unlike LCDs, LED displays do not need a backlight to function and easily can be made multi-color.

Modern LED research is focused mostly on flexible and ***organic LEDs (OLEDs)***, which are made from polymer light-emitting materials and can be fabricated on flexible substrates (such as an overhead transparency). Flexible displays of this type have been demonstrated by several groups around the world. Organic LEDs emit less light than traditional solid-state LEDs, however, and based on current technology, the lifetime of an OLED usually is shorter than desired for commercial applications.

Plasma Displays

Plasma displays have been around since 1964 when invented at the University of Illinois. While attractive due to their low profile, large viewing angle, brightness, and large screen size, they largely were displaced in the 1980s in the consumer market by LCD displays for manufacturing-cost reasons. In the late 1990s, plasma displays became popular for ***high-definition television (HDTV)*** systems. Each pixel in a plasma display contains one or more microscale pocket(s) of trapped noble gas (usually neon or xenon); electrodes patterned on a glass substrate are placed in front and behind each pocket of gas (Fig. TF6-3). The back of one of the glass plates is coated with light-emitting phosphors. When a sufficient voltage is applied across the electrodes, a large electric field is generated across the noble gas, and a plasma (ionized gas) is ignited. The plasma emits ultraviolet light which impacts the phosphors; when impacted with UV light, the phosphors emit light of a certain color (blue, green, or red). In this way, each pocket can generate one color.

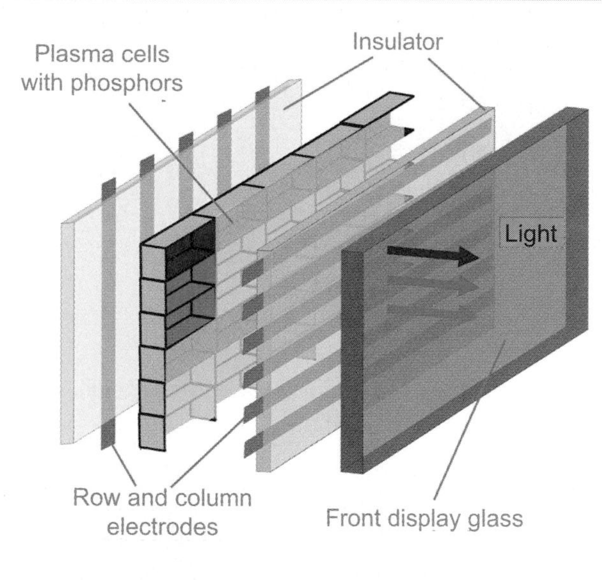

Figure TF6-3: Plasma display.

Electronic Ink

Electronic ink, ***e-paper***, or ***e-ink*** are all names for a set of display technologies made to look like paper with ink on it. In all cases, the display is very thin (almost as thin as real paper), does not use a backlight (ambient light is reflected off the display, just like real paper), and little to no power is consumed when the image is kept constant. The first version of e-paper was invented in the 1970s at Xerox, but it was not until the 1990s that a commercially viable version was developed at MIT. Many companies are now in the process of developing displays for laptops, "e-newspapers," e-books, cell phones, and the like. As of 2006, the largest mass-market display is an 8-inch XGA display from E-Ink, but many newer versions (including color displays) are in the pipeline.

Digital Light Processing (DLP)

Digital light processing (DLP) is the name given to a technology that uses arrays of individual, micro-mechanical mirrors to manipulate light at each pixel position. Invented in 1987 by Dr. Hornbeck at Texas Instruments, this technology has revolutionized projection technology; most of today's small, inexpensive digital projectors are made possible by DLP chips. DLP also is used heavily in large, rear-projection televisions (where it competes with LCD and a variant of LCD).

A basic DLP consists of an array of metal micromirrors, each about 100 micrometers on a side (Fig. TF6-4(inset)). One micromirror corresponds to one pixel on a digital image. Each micromirror is mounted on micromechanical hinges and can be tilted towards or away from a light source several thousand times per second! The mirrors are used to reflect light from a light source (housed within the television or projector case) and through a lens to project it either from behind a screen (as is the case in rear-projection televisions) or onto a flat surface (in the case of projectors), as in (Fig. TF6-4). If a micromirror is tilted away from the light source, that pixel on the projected image becomes dark (since the mirror is not passing the light onto the lens). If it is tilted towards the light source, the pixel lights up. By varying the relative time a given mirror is in each position, grey values can be generated as well. Color can be added by using multiple light sources and either one chip (with a filter wheel) or three chips. The three-chip color DLP used in high-resolution cinema systems can purportedly generate 35 trillion different colors!

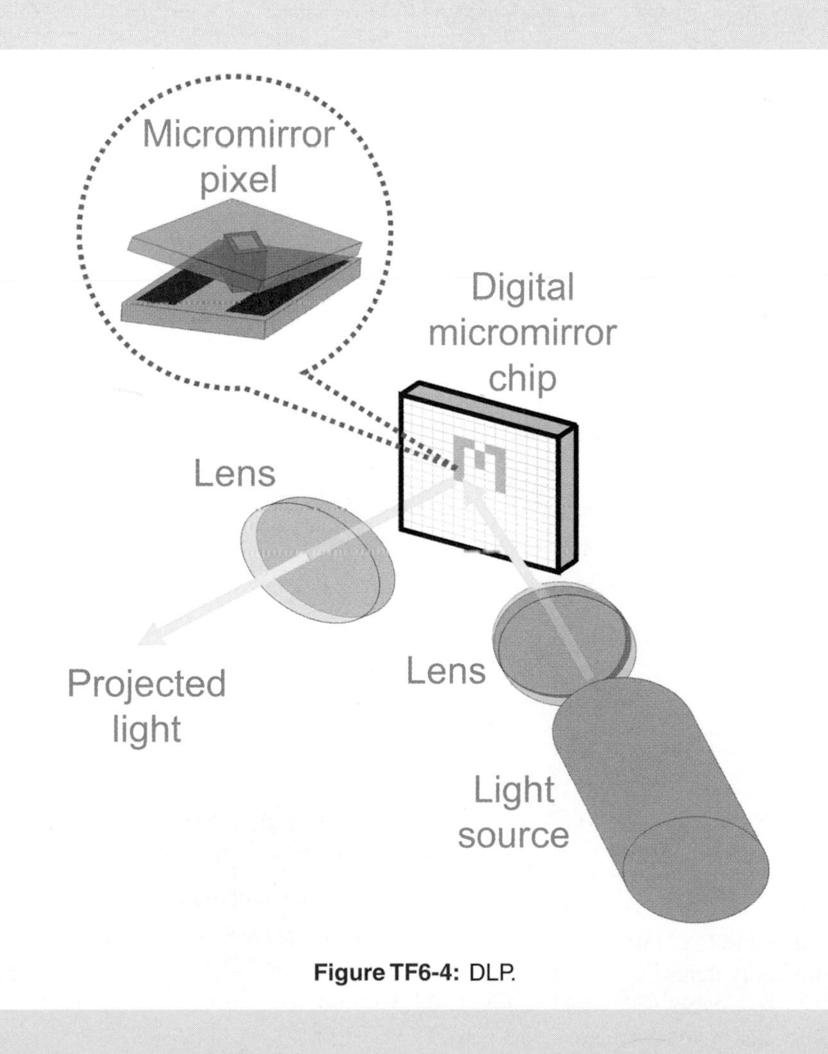

Figure TF6-4: DLP.

3-6 Maximum Power Transfer

Suppose an ***active*** linear circuit is connected to a ***passive*** linear circuit, as depicted by Fig. 3-25(a). An active circuit is one that contains at least one independent source, whereas a passive circuit may contain dependent sources, but no independent sources. For convenience, we shall refer to them as the source and load circuits, respectively. For certain applications, it is desirable to maximize the magnitude of the current i_L that flows from the source circuit to the load circuit, while other applications may call for maximizing the voltage υ_L at the input to the load circuit or maximizing the power P_L that gets transferred from the source to the load. Given a specified source circuit, how, then, does one approach the design of the load circuit so as to achieve these different goals?

The solution to the problem posed by our question is facilitated by the equivalence offered by Thévenin's theorem. We demonstrated in the preceding section that any active, linear circuit always can be represented by an equivalent circuit composed of a Thévenin voltage υ_Th connected in series with a Thévenin resistance R_Th. In the case of the passive load circuit, its equivalent circuit consists of only a Thévenin resistance. To avoid confusion between the two circuits, we denote υ_Th and R_Th of the source circuit as υ_s and R_s, and we denote R_Th of the load circuit as R_L, as shown in Fig. 3-25(b). The current i_L and associated voltage υ_L are given by

$$i_\mathrm{L} = \frac{\upsilon_\mathrm{s}}{R_\mathrm{s} + R_\mathrm{L}} \tag{3.39a}$$

and

$$\upsilon_\mathrm{L} = \frac{\upsilon_\mathrm{s} R_\mathrm{L}}{R_\mathrm{s} + R_\mathrm{L}}. \tag{3.39b}$$

If the source-circuit parameters υ_s and R_s are fixed and the intent is to *transfer* maximum current to the load circuit, then R_L should be zero (short circuit). For a real circuit with a functional purpose, the circuit will need to receive some energy in order to function. Hence, R_L cannot be exactly zero, but it can be made to be very small in comparison with R_s. Thus, to maximize current transfer, the load circuit should be designed such that

$$R_\mathrm{L} \ll R_\mathrm{s} \quad \text{(maximum current transfer)}. \tag{3.40}$$

Based on Eq. (3.39b), the opposite is true for maximum voltage transfer, namely

$$R_\mathrm{L} \gg R_\mathrm{s} \quad \text{(maximum voltage transfer)}. \tag{3.41}$$

The situation for power transfer calls for maximizing the product of i_L and υ_L,

$$p_\mathrm{L} = i_\mathrm{L}\upsilon_\mathrm{L} = \frac{\upsilon_\mathrm{s}^2 R_\mathrm{L}}{(R_\mathrm{s} + R_\mathrm{L})^2}. \tag{3.42}$$

The expression given by Eq. (3.42) is a nonlinear function of R_L. The power p_L goes to zero as R_L approaches either end of its range (0 and ∞), as illustrated by the plot in Fig. 3-26, and it is at a maximum when

$$R_\mathrm{L} = R_\mathrm{s} \qquad \text{(maximum power transfer)}. \tag{3.43}$$

The proof of Eq. (3.43) is given in Example 3-13.

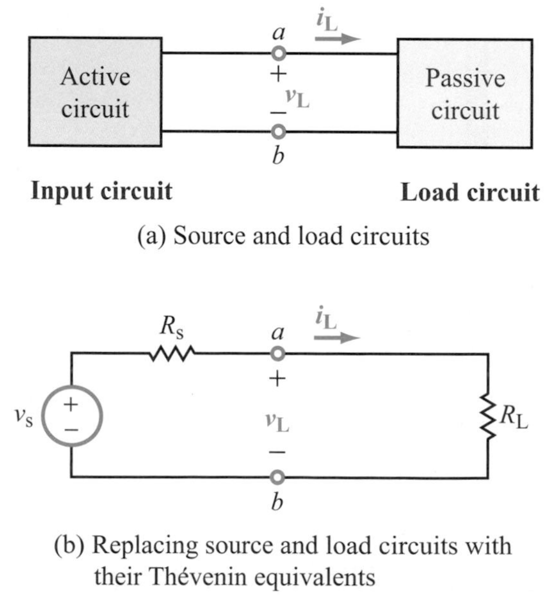

Input circuit: Active circuit — Load circuit: Passive circuit

(a) Source and load circuits

(b) Replacing source and load circuits with their Thévenin equivalents

Figure 3-25: To analyze the transfer of voltage, current, and power from the source circuit to the load circuit, we first replace them with their Thévenin equivalents.

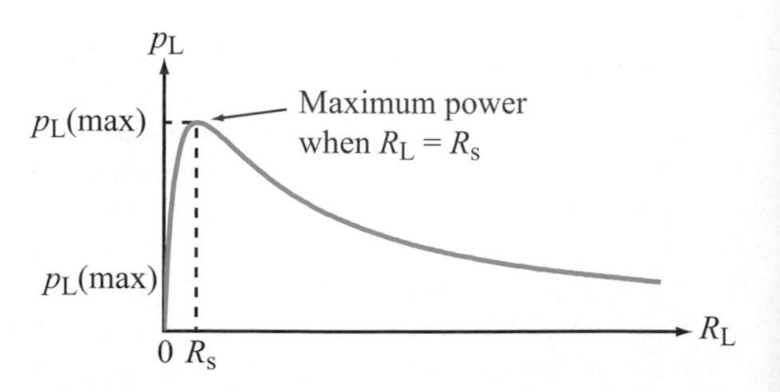

Figure 3-26: Variation of power p_L dissipated in the load R_L, as function of R_L.

Use of $R_L = R_s$ in Eq. (3.42) leads to

$$p_L(\max) = \frac{v_s^2 R_L}{(R_L + R_L)^2} = \frac{v_s^2}{4R_L}, \quad (3.45)$$

which represents 50 percent of the total power generated by the equivalent input source v_s. The other 50 percent is dissipated in R_s.

Example 3-13: Maximum Power Transfer

Prove that p_L, as given by Eq. (3.42), is at a maximum when $R_L = R_s$.

Solution:
To find the value of R_L at which the expression for p_L is at a maximum, we differentiate the expression with respect to R_L and then set the result equal to zero. That is,

$$\frac{dp_L}{dR_L} = \frac{d}{dR_L}\left[\frac{v_s^2 R_L}{(R_s + R_L)^2}\right]$$
$$= v_s^2 \left[\frac{1}{(R_s + R_L)^2} - \frac{2R_L}{(R_s + R_L)^3}\right] = 0.$$

A few simple steps of algebra lead to

$$R_L = R_s.$$

Review Question 3-14: Under what conditions is the power transferred from a power source to a load resistor a maximum?

Review Question 3-15: Of the power generated by an input circuit, what is the maximum fraction that can be transferred to an external load?

Exercise 3-14: The bridge circuit of Fig. E3.14 is connected to a load R_L between terminals (a, b). Choose R_L such that maximum power is delivered to R_L. If $R = 3\ \Omega$, how much power is delivered to R_L?

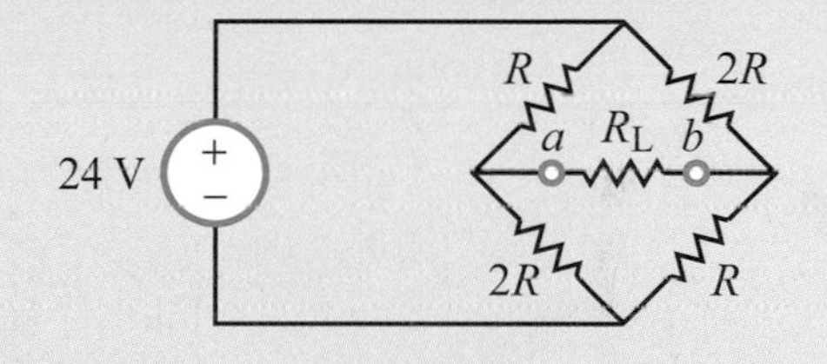

Figure E3.14

Answer: $R_L = 4R/3 = 4\ \Omega$, $P_{max} = 4$ W. (See)

3-7 Application Note: Bipolar Junction Transistor (BJT)

With the exception of the SPDT switch, all of the elements we have discussed thus far have been two-terminal devices, each characterized by a single i–v relationship. These include resistors, voltage and current sources, as well as the pn-junction diode of Section 2-7.2. The potentiometer (Fig. 2-3(b)) may appear to be like a three-terminal device, but in reality it is no more than two resistors—each with its own pair of terminals. This section introduces a true three-terminal device, the ***bipolar junction transistor*** (BJT).

The BJT is a three-layer semiconductor structure commonly made of silicon. Other compounds sometimes are used for special-purpose applications (such as for operation at microwave and optical frequencies), but for the present, we will limit our examination to silicon-based transistors and their uses in dc circuits. The three terminals of a BJT are called the ***emitter***, ***collector***, and ***base***. The emitter and collector are made of the same material—either p-type or n-type—and the base is made of the other material. Thus, the BJT can be constructed to have either a ***pnp configuration*** or an ***npn configuration***, as shown in the diagrams of Fig. 3-27. The geometries and fabrication details of real transistors are far more elaborate than the simple diagrams suggest, but the basic idea that the BJT consists of three alternating layers of p- and n-type material is quite sufficient from the standpoint of its external electrical behavior.

Figure 3-27 also shows schematic symbols used for the pnp and npn transistors. The center terminal is always the base. One of the three leads includes an arrow. *The lead containing the arrow identifies the emitter terminal* and whether the transistor is a pnp or npn. The arrow always points towards an n-type material, so *in the pnp transistor, the arrow points towards the base, whereas in an npn transistor, the arrow points away from the base.*

The directions of the terminal currents shown in Fig. 3-27 are defined such that the base and collector currents I_B and I_C, respectively, flow into the transistor, and the emitter current I_E flows out of it. KCL requires that

$$I_E = I_B + I_C. \quad (3.44)$$

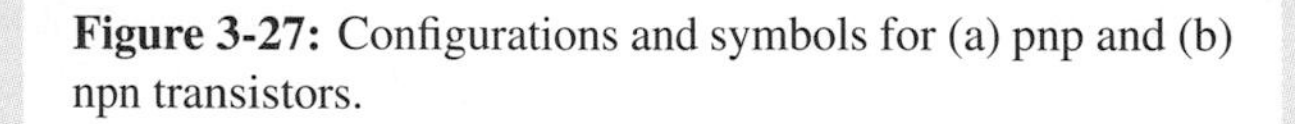

Figure 3-27: Configurations and symbols for (a) pnp and (b) npn transistors.

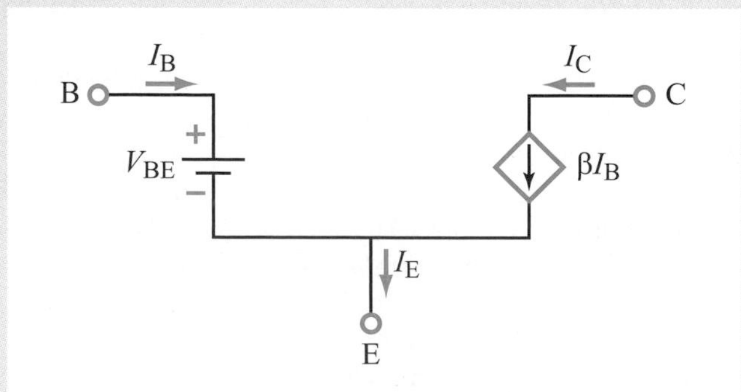

Figure 3-28: dc equivalent model for the npn transistor. The equivalent dc source $V_{BE} \simeq 0.7$ V.

Under normal operating conditions, I_E has the largest magnitude of the three currents, and I_B is much smaller than either I_C or I_E. The transistor can operate under both dc and ac conditions, but we will limit our present discussion to dc circuits. For simplicity, we will consider only the npn common-emitter configuration. Accordingly, we can describe the operation of the npn transistor by the dc equivalent model shown in Fig. 3-28. The circuit contains a constant dc voltage source V_{BE} and a dependent current-controlled current source that relates I_C to I_B by

$$I_C = \beta I_B, \tag{3.46}$$

where β is a transistor parameter called the ***common-emitter current gain***. Under normal operation, $V_{BE} \simeq 0.7$ V, and β may assume values in the range between 30 and 1000, depending on its specific design configuration. To operate in its active mode, the transistor requires that certain dc voltages be applied at its base and collector terminals. We shall refer to these voltages as V_{BB} and V_{CC}, respectively.

Example 3-14: BJT Circuit

Apply the equivalent-circuit model with $V_{BE} \simeq 0.7$ V and $\beta = 200$ to determine I_B, I_C, and V_{CE} in the circuit of Fig. 3-29. Assume that $V_{BB} = 2$ V, $V_{CC} = 10$ V, $R_B = 26$ kΩ, and $R_C = 200$ Ω.

Solution:
Upon replacing the npn transistor with its equivalent circuit, we end up with the circuit shown in Fig. 3-29(b). In the left-hand loop, KVL gives

$$-V_{BB} + R_B I_B + V_{BE} = 0,$$

which leads to

$$I_B = \frac{V_{BB} - V_{BE}}{R_B} = \frac{2 - 0.7}{26 \times 10^3} = 5 \times 10^{-5} \text{ A} = 50\ \mu\text{A}.$$

Given that $\beta = 200$,

$$I_C = \beta I_B = 200 \times 50 \times 10^{-6} = 10 \text{ mA},$$

and

$$V_{CE} = V_{CC} - I_C R_C = 10 - 10^{-2} \times 200 = 8 \text{ V}.$$

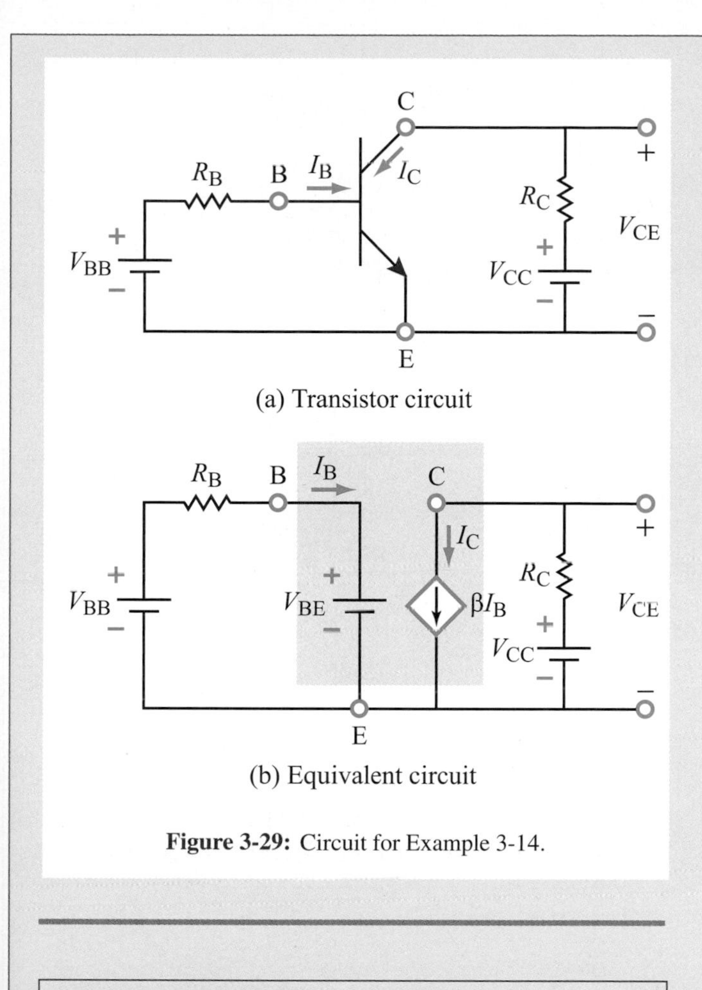

Figure 3-29: Circuit for Example 3-14.

Example 3-15: Digital-Inverter Circuit

Digital logic deals with two states, "0" and "1" (or equivalently "low" and "high"). A digital-inverter circuit provides one of the logic operations performed by a computer processor, namely to invert the state of an input bit from low to high or from high to low. Demonstrate that the transistor circuit shown in Fig. 3-30 functions as a digital inverter by plotting its output voltage V_{out} versus the input voltage V_{in}. A bit is assumed to be in state 0 (low) if its voltage is between 0 and 0.5 V and in state 1 (high) if its voltage is greater than 4 V. Assume that the equivalent model given by Fig. 3-28 is applicable (with $\beta = 20$) with the following qualifications: neither I_B nor V_{out} can have negative values, so if the analysis using the equivalent-circuit model generates a negative value for either one of them, it should be replaced with zero.

Solution:
The equivalent circuit shown in Fig. 3-30(b) provides the following expressions:

$$I_B = \frac{V_{\text{in}} - 0.7}{20\text{k}}, \tag{3.47}$$

$$I_C = \beta I_B = 200 I_B, \tag{3.48}$$

and

$$V_{\text{out}} = V_{CC} - I_C R_C. \tag{3.49}$$

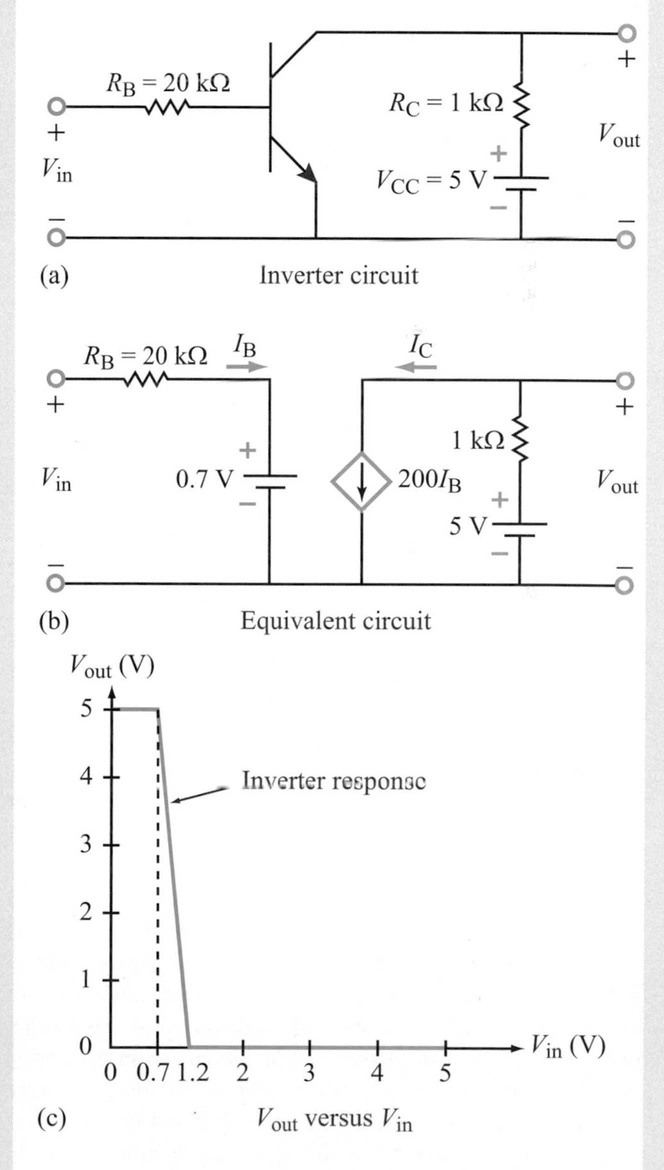

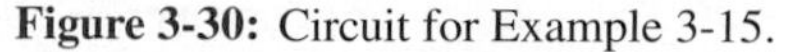

Figure 3-30: Circuit for Example 3-15.

Combining the three equations leads to

$$V_{\text{out}} = V_{\text{CC}} - \frac{\beta R_{\text{C}}}{R_{\text{B}}}(V_{\text{in}} - 0.7)$$
$$= 12 - 10V_{\text{in}} \qquad \text{(V)}. \qquad (3.50)$$

Since V_{out} is linearly related to V_{in}, the plot would be a straight line, as shown in Fig. 3-30(c), but we also have to incorporate the provisions that I_{B} cannot be negative (which occurs when $V_{\text{in}} < 0.7$ V), and V_{out} cannot be negative (which occurs when $V_{\text{in}} = 1.2$ V). The resultant transfer function clearly satisfies the digital inverter requirements:

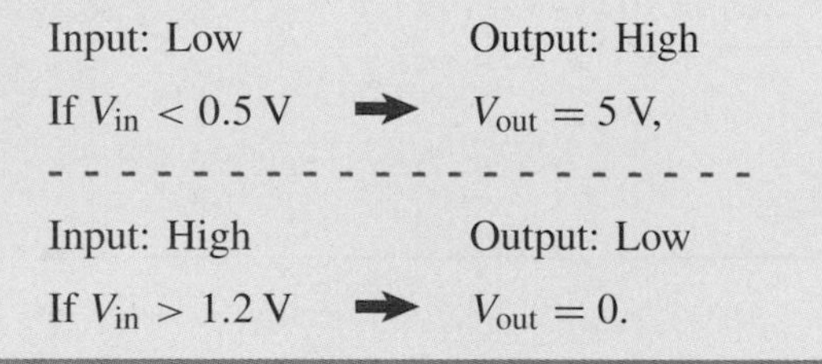

Review Question 3-16: How is the collector current related to the base current in a BJT?

Review Question 3-17: What is a digital inverter? How are its input and output voltages related to one another?

Exercise 3-15: Determine I_{B}, V_{out_1}, and V_{out_2} in the transistor circuit of Fig. E3.15, given that $V_{\text{BE}} = 0.7$ V and $\beta = 200$.

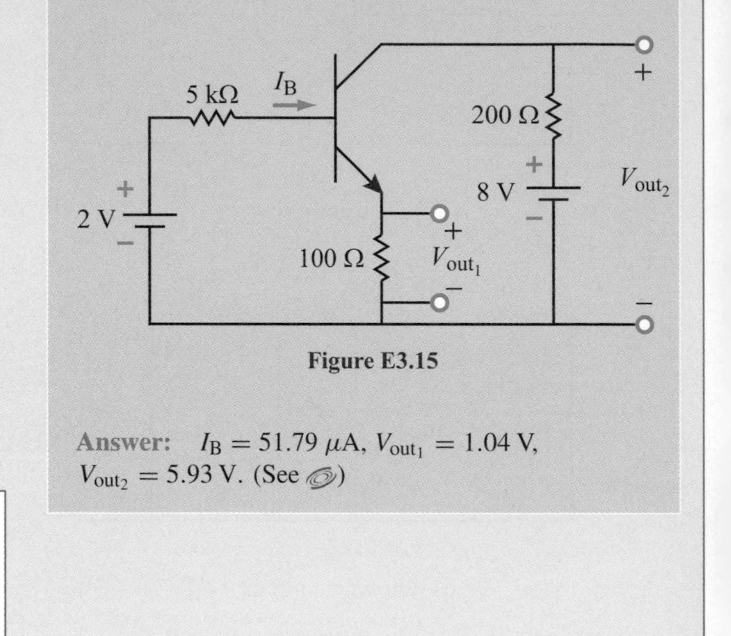

Figure E3.15

Answer: $I_{\text{B}} = 51.79\ \mu$A, $V_{\text{out}_1} = 1.04$ V, $V_{\text{out}_2} = 5.93$ V. (See)

3-8 Nodal Analysis with Multisim

Multisim is a particularly useful tool for analyzing circuits with many nodes. Consider the six-node circuit shown in Fig. 3-31(a) in which the voltages and currents are designated in accordance with the Multisim notation system. In Multisim, V1 refers to the voltage of source 1 and V(1) refers to the voltage at node 1. Application of nodal analysis would generate five equations with five unknowns, V(1) to V(5), whose solution would require the use of matrix algebra or several steps of elimination of variables. [For this simple two-loop circuit, mesh analysis is much easier to apply, as it involves only two mesh equations and one auxiliary equation for the dependent current source, but the objective of the present section is to illustrate how Multisim can be used for circuits involving a large number of nodes.] When drawn in Multisim, the circuit appears in the form shown in Fig. 3-31(b). Application of either Measurement Probes or DC Operating Point Analysis generates the values of V(1) to V(5) listed in the inset of Fig. 3-31(b).

For circuits containing more than four or five nodes, analyzing the circuit *by hand* becomes unwieldy. Moreover, some circuits may contain time-varying sources or elements. Consider, for example, the circuit in Fig. 3-32(a), which is a replica of the circuit in Fig. 3-31 except for the addition of an SPDT switch. [In Multisim, the switch can be *toggled* between positions 1 and 2 using the space bar on your computer.] When connected to position 1, the state of the circuit is identical with that in Fig. 3-31, but when the SPDT switch is moved to position 2, the new circuit configuration includes two additional elements and one extra node.

The circuit drawn in Multisim is shown in Fig. 3-32(b). The SPDT is available in the Select a Component window under the Basic group in the SWITCH family. Measurement Probes were added to nodes 4, 5, and 6. Using the Interactive Simulation feature of Multisim, the circuit can be analyzed in each of its two states by pressing F5 (or the button or toggle switch) to start the simulation, and then toggling the switch by pressing the space bar. This live-action switching capability is why this particular tool is known as Interactive Simulation.

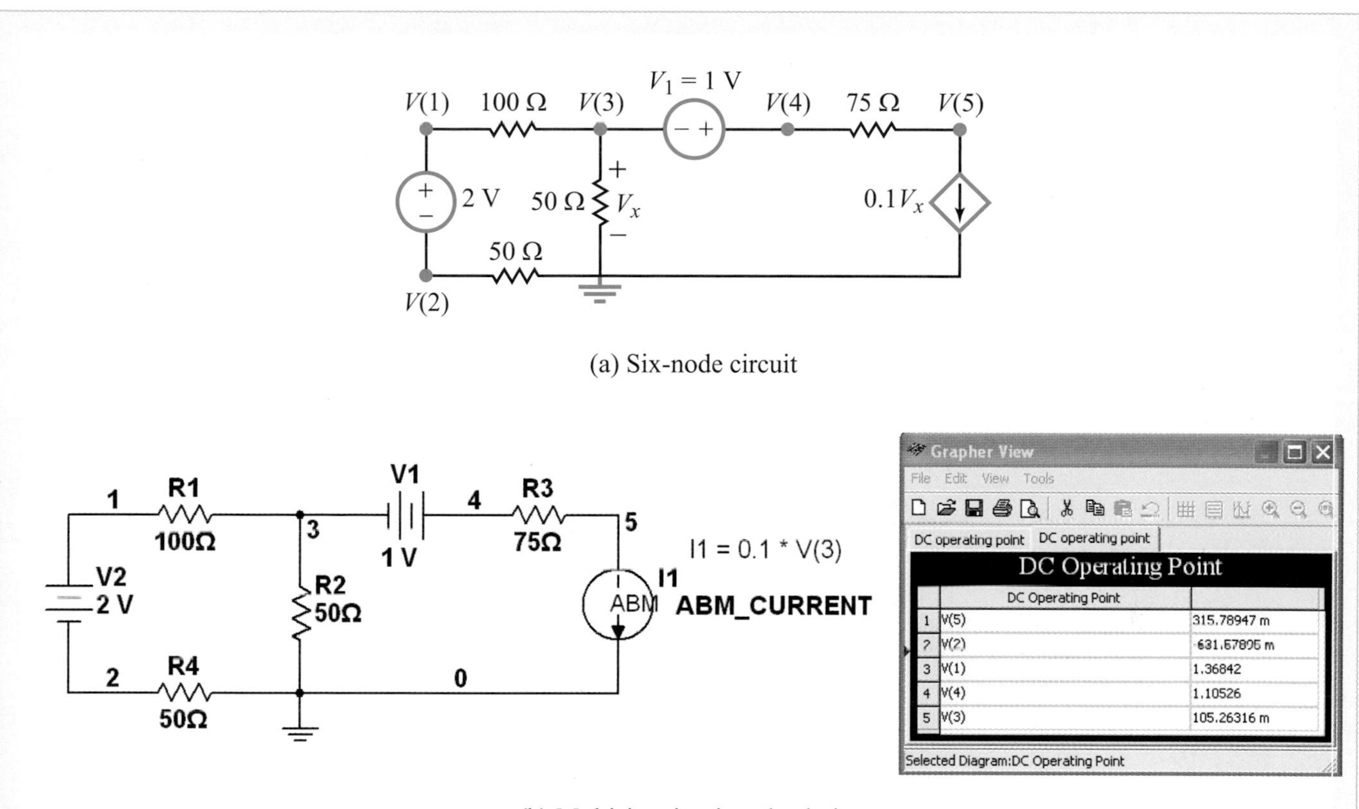

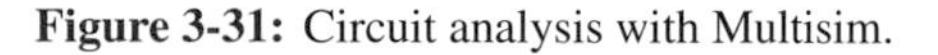
Figure 3-31: Circuit analysis with Multisim.

In the Multisim section of Chapter 2, we examined how the DC Operating Point Analysis tool can be used to determine differences between node voltages. In addition to basic subtraction, there are many operators that you can apply to variables (or combinations of variables) to obtain the desired quantities. [See the Multisim Tutorial on the CD for a list of the basic operators].

We will now use variable manipulation in the DC Operating Point Analysis to calculate the power dissipated or supplied in each component in the circuit in Fig. 3-31(a). To find the power for each component, we need to know both the current through and voltage across each component. While voltage is easy to access, we need a voltage source in series with each branch of the circuit in order to access all of the branch currents. In looking at the circuit, we realize that we need only one additional voltage source, which we set to zero, as shown in Fig. 3-33(a).

Open up the DC Operating Point Analysis window and under the output tab enter equations via the Add Expression... button to express the power in each component. Click OK after entering the expressions. [Remember proper sign notation and current direction.] The equations for power should be

Source V1:	(V(4)-V(3))*I(v1)
Source V2:	(V(1)-V(2))*I(v2)
Source I1:	-V(5)*I(v1)
Resistor R1:	(V(3)-V(1))*I(v2)
Resistor R2:	V(3)*I(v3)
Resistor R3:	(V(5)-V(4))*I(v1)
Resistor R4:	V(2)*I(v2)

Note: Remember that these variable names apply to the circuit shown in Fig. 3-33(a). If your circuit has a different numbering for nodes or voltage sources, your equations will differ in number accordingly.

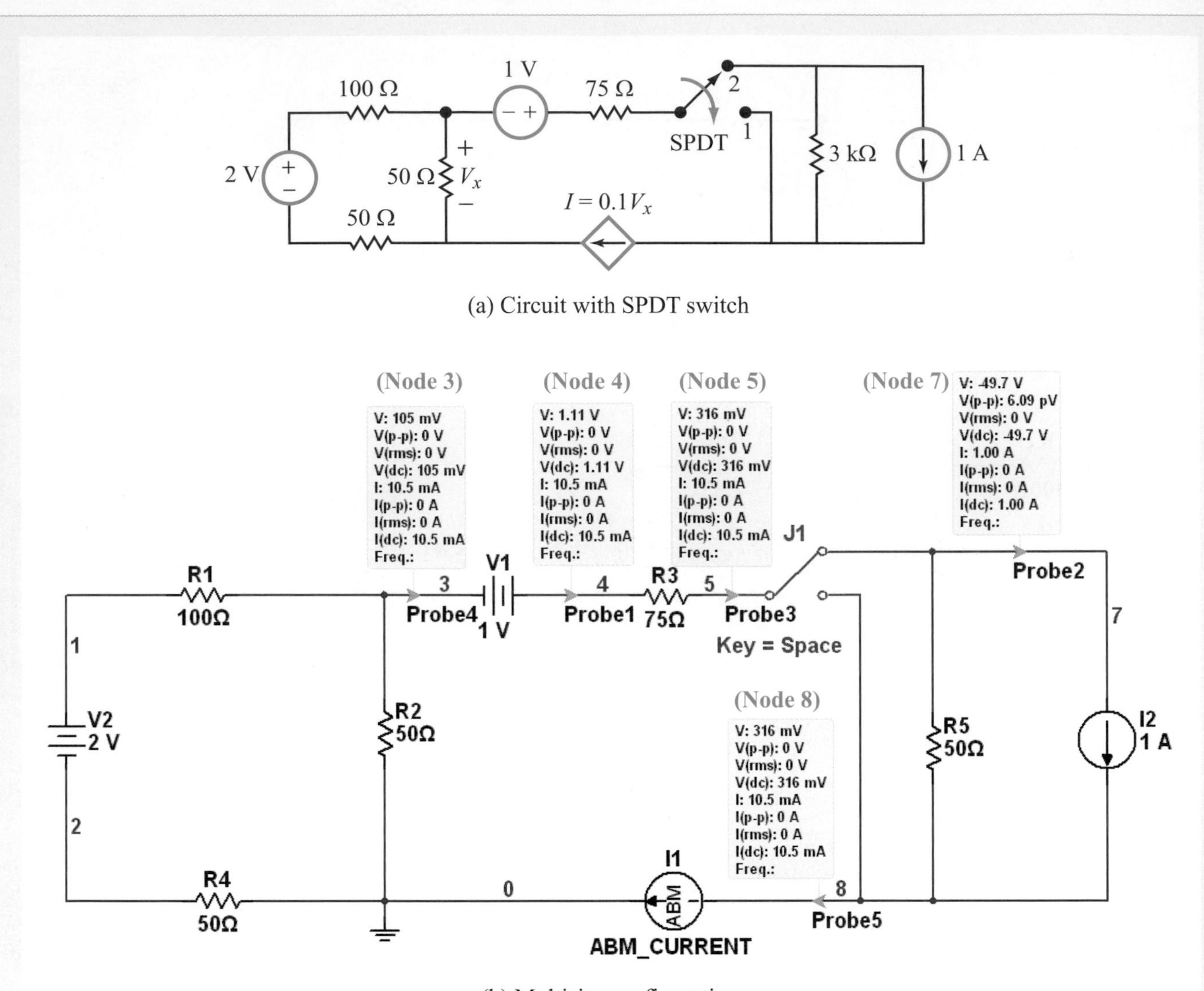

(a) Circuit with SPDT switch

(b) Multisim configuration

Figure 3-32: (a) Circuit with a switch, and (b) its Multisim representation.

Once these equations are entered, the Selected Variables for Analysis field should resemble that in Fig. 3-33(b). To obtain the values, press the Simulate button. The results should agree with those shown in Fig. 3-33(c).

Knowing how to write equations such as these in Multisim is very important, because many other Analyses which you will encounter later in the book utilize identical syntax to that used for the DC Operating Point Analysis.

Review Question 3-18: What is the difference between the Measurement Probe tool and the DC Operating Point Analysis?

Exercise 3-16: Use Multisim to calculate the voltage at node 3 in Fig. 3-32(b) when the SPDT switch is connected to position 2.

Answer: (See 💿)

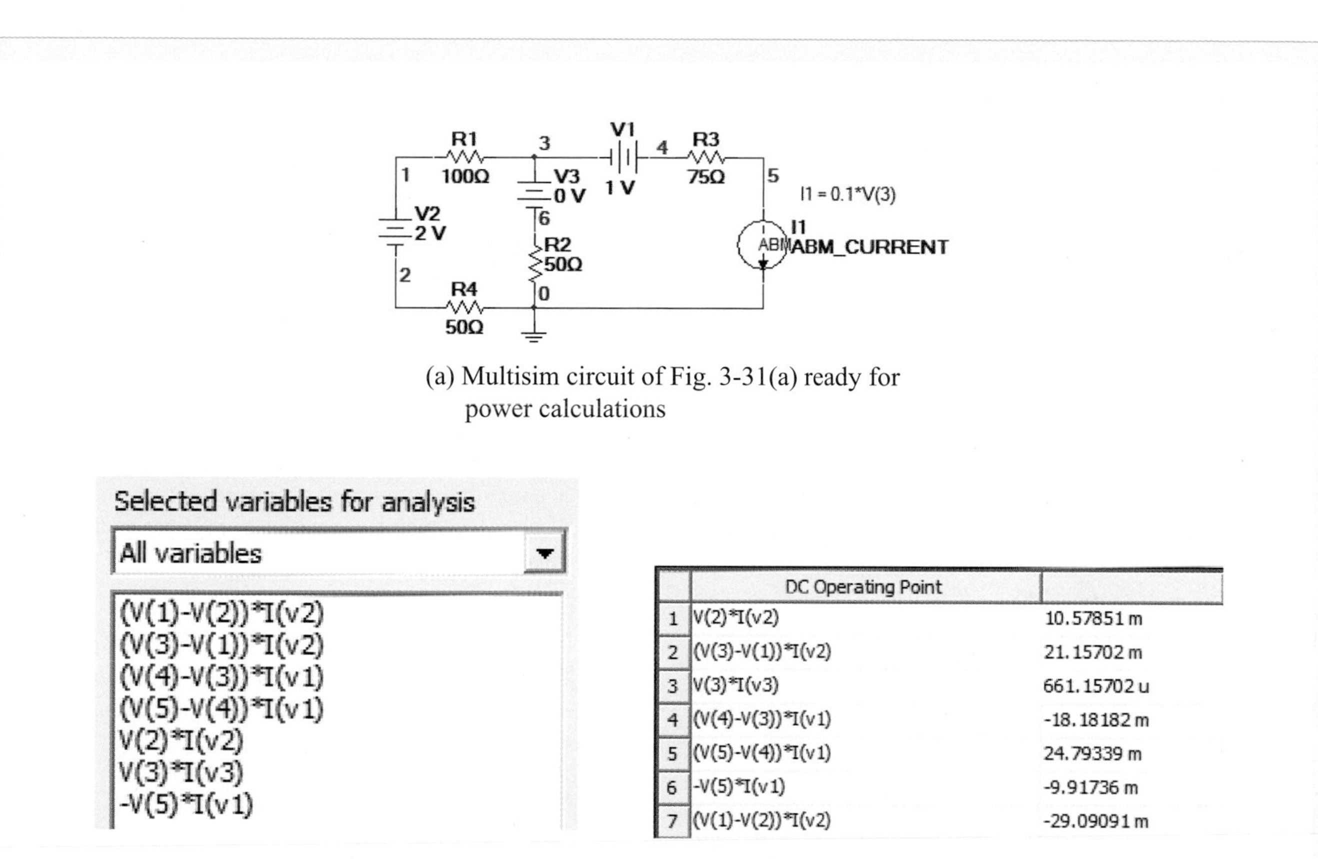

(a) Multisim circuit of Fig. 3-31(a) ready for power calculations

	DC Operating Point	
1	V(2)*I(v2)	10.57851 m
2	(V(3)-V(1))*I(v2)	21.15702 m
3	V(3)*I(v3)	661.15702 u
4	(V(4)-V(3))*I(v1)	-18.18182 m
5	(V(5)-V(4))*I(v1)	24.79339 m
6	-V(5)*I(v1)	-9.91736 m
7	(V(1)-V(2))*I(v2)	-29.09091 m

(b) Selected variables for analysis visible in DC Operating Point Analysis window

(c) Output of simulations (remember that all values are in watts)

Figure 3-33: Multisim procedure for calculating power consumed (or generated) by the seven elements in the circuit of Fig. 3-31(a).

Chapter 3 Relationships

Node-voltage method
$\sum$ of all current leaving a node = 0
[current entering a node is (−)]

Mesh-current method
$\sum$ of all voltages around a loop = 0
[passive sign convention applied to mesh currents in clockwise direction]

Nodal analysis by inspection $\quad \mathbf{GV} = \mathbf{I}_t$

Mesh analysis by inspection $\quad \mathbf{RI} = \mathbf{V}_t$

Thévenin equivalent circuit
$v_{Th} = v_{oc}$
$R_{Th} = v_{oc}/i_{sc}$

Norton equivalent circuit
$i_N = i_{sc}$
$R_N = R_{Th}$

Maximum power transfer
$R_L = R_s$
$$P_L(\max) = \frac{v_s^2}{4R_L}$$

CHAPTER HIGHLIGHTS

- After assigning one of the extraordinary nodes in a circuit the role of voltage reference (ground), application of KCL at the remaining extraordinary nodes provides the requisite number of simultaneous equations for determining the voltages at those nodes.
- Two extraordinary nodes connected by a solitary voltage source constitute a supernode. When applying KCL in the node-voltage method, the two nodes are treated as a single node, which is augmented by an auxiliary relation specifying the voltage difference between the two nodes.
- By assigning a mesh current to each independent loop in a circuit, application of KVL leads to the requisite number of equations for determining the unknown values of the mesh currents.
- Two adjoining loops sharing a branch containing a solitary current source constitute a supermesh. When applying KVL, the two loops are treated as a single loop, which is augmented by an auxiliary relation specifying the current relationship between the mesh currents in the two loops.
- A circuit containing no dependent sources and only current sources can be analyzed by applying the node-voltage by-inspection method, which casts the node-voltage equations in an easy-to-implement matrix form.
- Similarly, a circuit containing no dependent sources and only voltage sources is amenable to application of the mesh-current by-inspection method, which results in an easy-to-implement matrix equation.
- If a circuit contains multiple, real, voltage, or current sources, the linearity property of the circuit implies that the current flowing through any element in the circuit is equal to the sum of multiple currents, bearing a one-to-one correspondence with the multiple sources in the circuit. The same is true for the voltage across the element.
- Thévenin's theorem states that a linear circuit can be represented in terms of a simple equivalent circuit composed of a voltage source in series with a resistor.
- Equivalently, a circuit can be represented by a Norton-equivalent circuit composed of a current source in parallel with a shunt resistor.
- The power transferred by an input circuit to an external load is at a maximum when the load resistance is equal to the Thévenin resistance of the input circuit. The fraction of the power thus transferred is 50 percent of the power supplied by the generator.
- Multisim is a useful tool for simulating the behavior of a circuit and examining its sensitivity to specific variables of interest.

GLOSSARY OF IMPORTANT TERMS

Provide definitions or explain the meaning of the following terms:

bipolar junction transistor (BJT)
by-inspection method
conductance matrix
extraordinary node
load impedance
maximum power transfer
mesh current
node-voltage method
Norton's theorem
resistance matrix
source superposition
source vector
supermesh
supernode
Thévenin's theorem
Thévenin's voltage
Thévenin's resistance

PROBLEMS

Section 3-1: Node-Voltage Method

*__3.1__ Apply nodal analysis to find the node voltage V in the circuit of Fig. P3.1. Use the information to determine the current I.

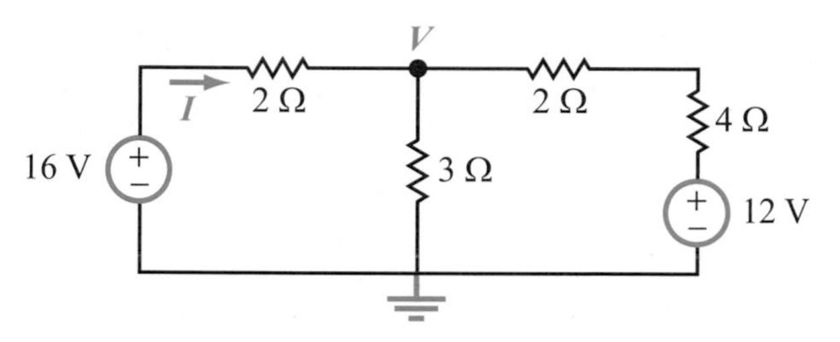

Figure P3.1: Circuit for Problem 3.1.

3.2 Apply nodal analysis to determine V_x in the circuit of Fig. P3.2.

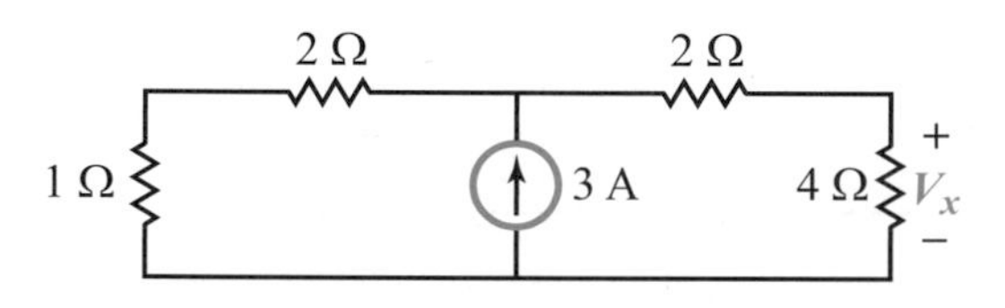

Figure P3.2: Circuit for Problem 3.2.

3.3 Use nodal analysis to determine the current I_x and amount of power supplied by the voltage source in the circuit of Fig. P3.3.

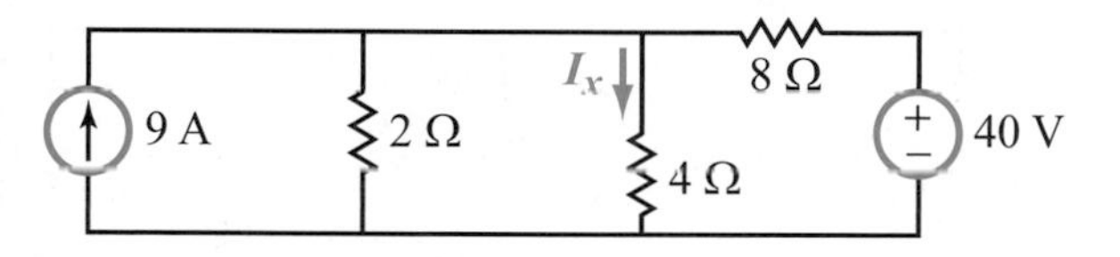

Figure P3.3: Circuit for Problem 3.3.

3.4 For the circuit in Fig. P3.4:

(a) Apply nodal analysis to find node voltages V_1 and V_2.

(b) Determine the voltage V_R and current I.

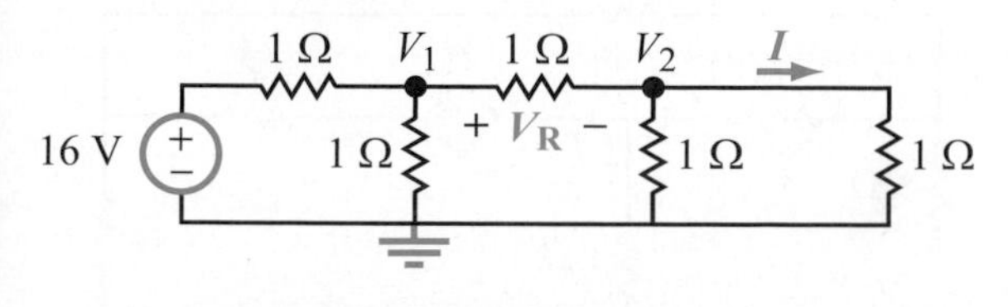

Figure P3.4: Circuit for Problem 3.4.

*Answer(s) in Appendix E.

3.5 Apply nodal analysis to determine the voltage V_R in the circuit of Fig. P3.5.

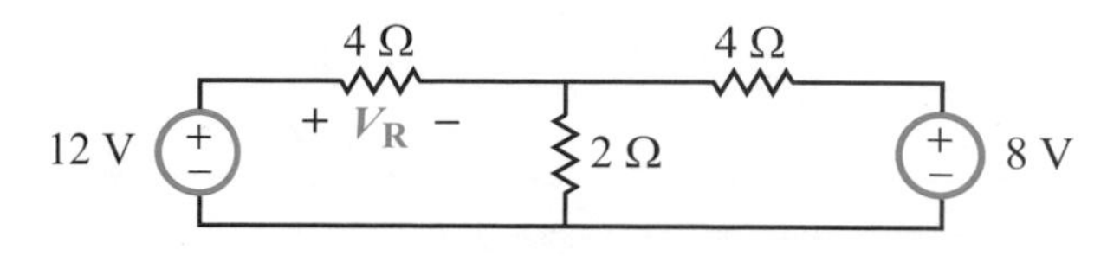

Figure P3.5: Circuit for Problem 3.5.

3.6 Use the nodal-analysis method to find V_1 and V_2 in the circuit of Fig. P3.6, and then apply that to determine I_x.

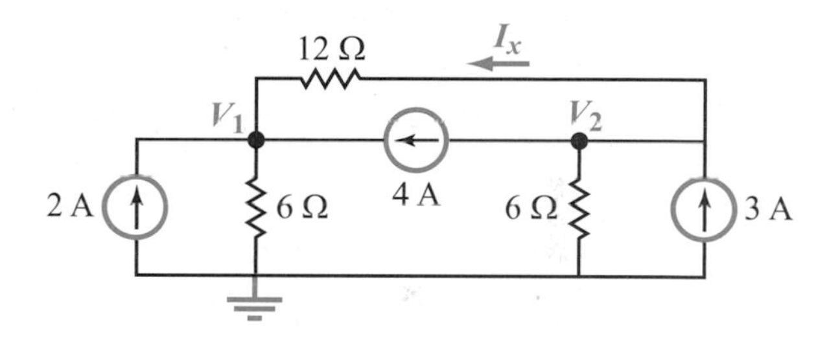

Figure P3.6: Circuit for Problem 3.6.

*__3.7__ Find I_x in the circuit for Fig. P3.7.

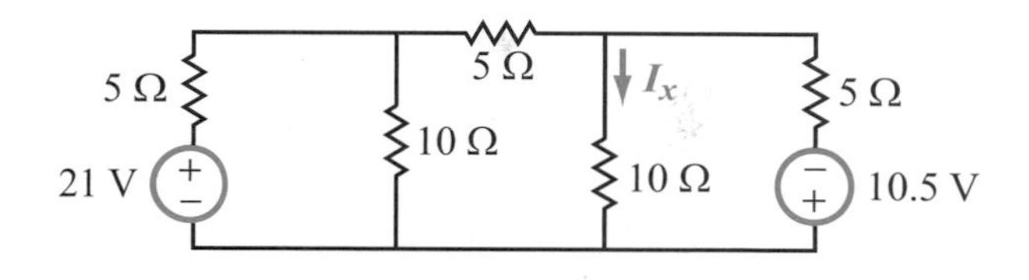

Figure P3.7: Circuit for Problem 3.7.

3.8 For the circuit in Fig. P3.8:

(a) Determine I.

(b) Determine the amount of power supplied by the voltage source.

(c) How much influence does the 4-A source have on the circuit to the left of the 3-A source?

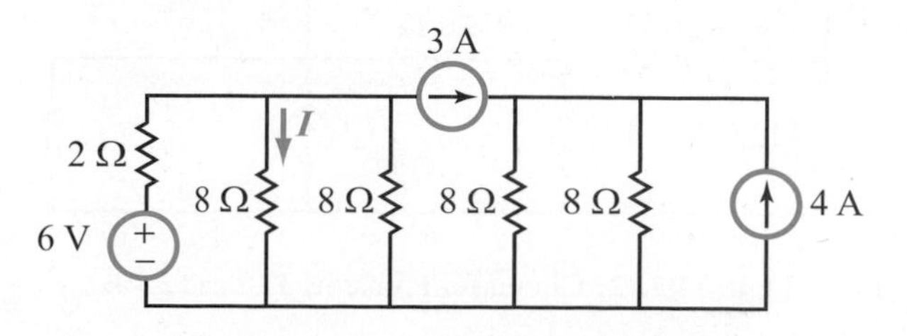

Figure P3.8: Circuit for Problem 3.8.

3.9 Apply nodal analysis to find node voltages V_1 to V_3 in the circuit of Fig. P3.9 and then determine I_x.

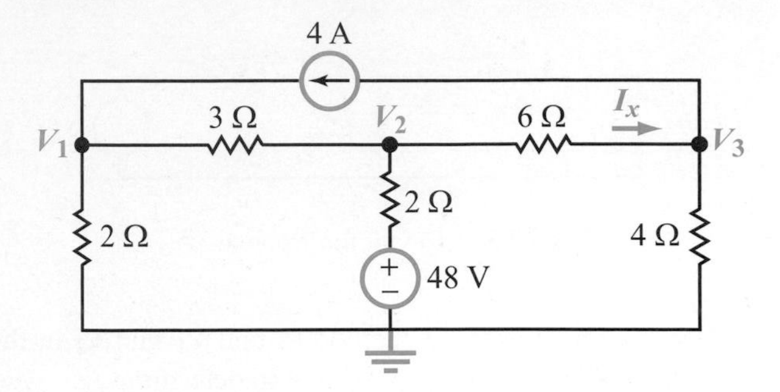

Figure P3.9: Circuit for Problem 3.9.

3.10 The circuit in Fig. P3.10 contains a dependent current source. Determine the voltage V_x.

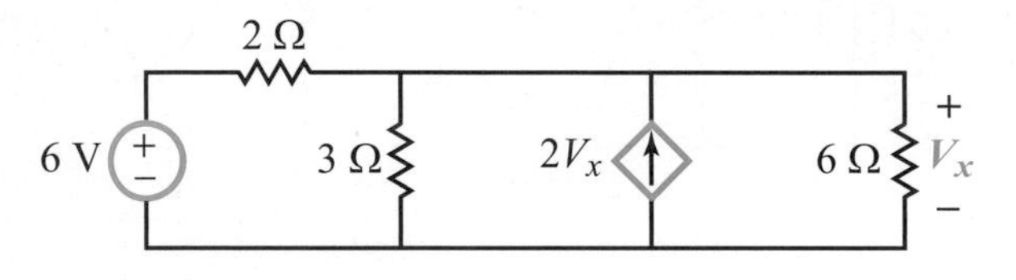

Figure P3.10: Circuit for Problem 3.10.

*__3.11__ Determine the power supplied by the independent voltage source in the circuit of Fig. P3.11.

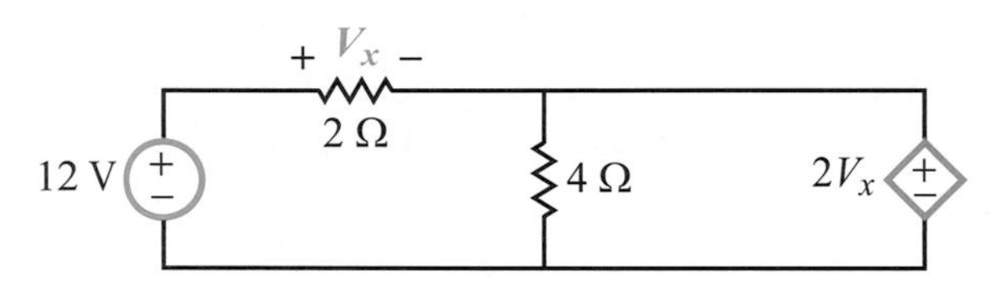

Figure P3.11: Circuit for Problem 3.11.

3.12 The magnitude of the dependent current source in the circuit of Fig. P3.12 depends on the current I_x flowing through the 10-Ω resistor. Determine I_x.

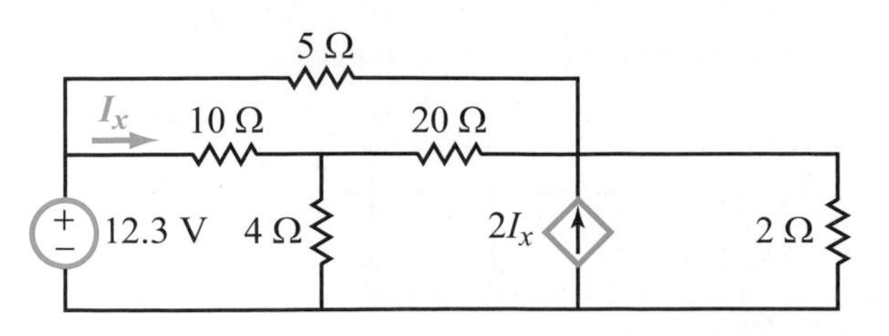

Figure P3.12: Circuit for Problems 3.12 and 3.13.

3.13 Repeat Problem 3-12 after replacing the 5-Ω resistor in Fig. P3.12 with a short circuit.

3.14 Apply nodal analysis to find the current I_x in the circuit of Fig. P3.14.

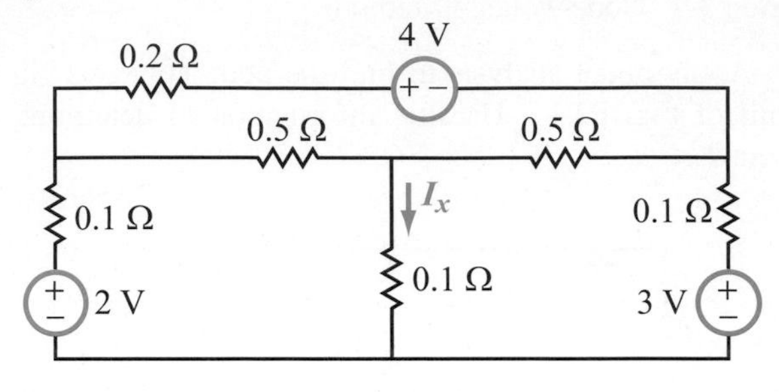

Figure P3.14: Circuit for Problem 3.14.

3.15 Use the supernode concept to find the current I_x in the circuit of Fig. P3.15.

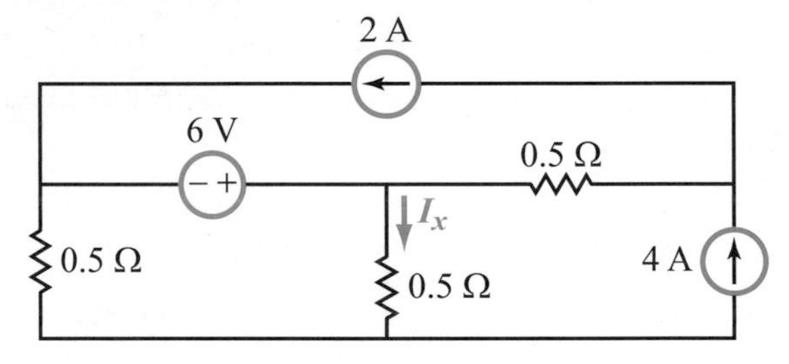

Figure P3.15: Circuit for Problem 3.15.

3.16 Apply the supernode technique to determine V_x in the circuit of Fig. P3.16.

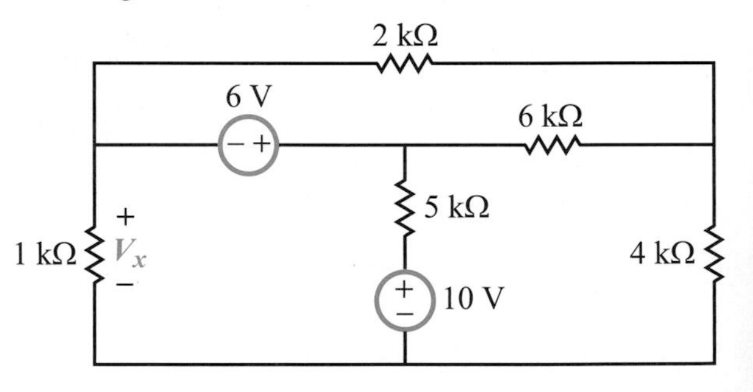

Figure P3.16: Circuit for Problem 3.16.

*__3.17__ Determine V_x in the circuit of Fig. P3.17.

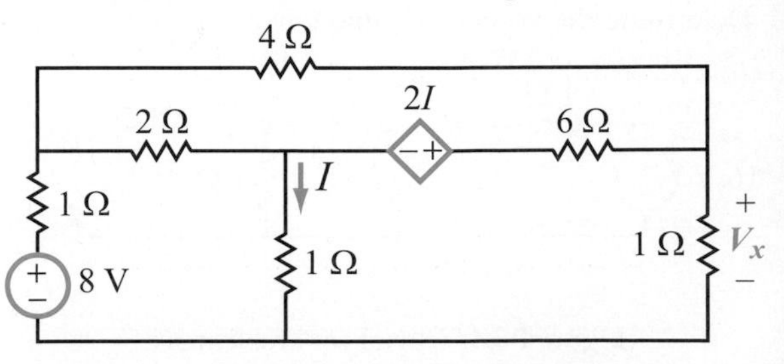

Figure P3.17: Circuit for Problems 3.17 and 3.18.

3.18 Repeat Problem 3-17 after replacing the 2-Ω resistor in Fig. P3.17 with a short circuit.

3.19 For the circuit shown in Fig. P3.19:

(a) Determine R_{eq} between terminals (a, b).

(b) Determine the current I using the result of (a).

(c) Apply nodal analysis to the original circuit to determine the node voltages and then use them to determine I. Compare the result with the answer of part (b).

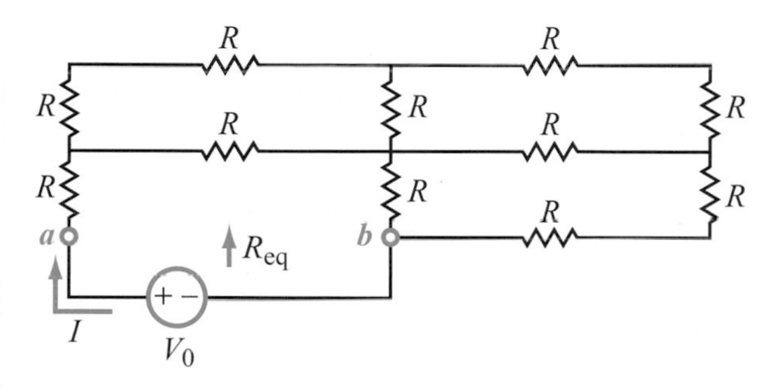

Figure P3.19: Circuit for Problem 3.19.

3.20 For the circuit in Fig. P3.20, determine the current I_x.

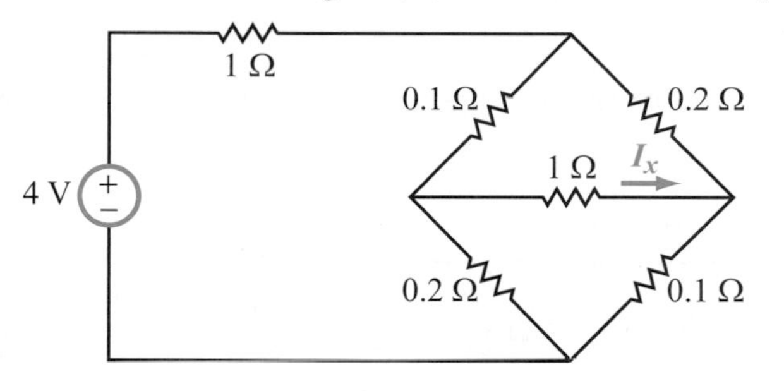

Figure P3.20: Circuit for Problem 3.20.

ection 3-2: Mesh-Current Method

.21 Apply mesh analysis to find the mesh currents in the ircuit of Fig. P3.21. Use the information to determine the oltage V.

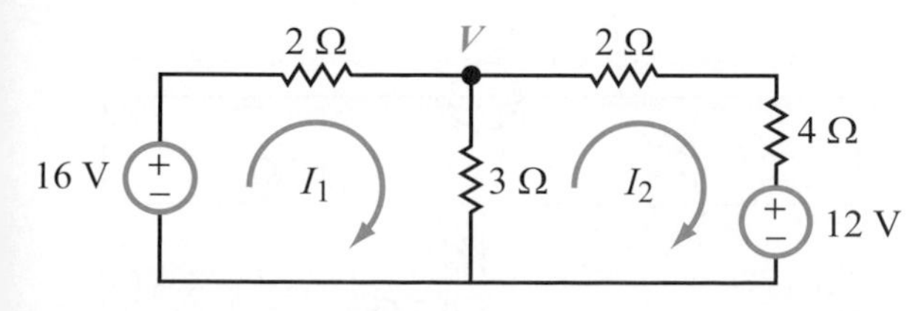

Figure P3.21: Circuit for Problem 3.21.

3.22 Use mesh analysis to determine the amount of power supplied by the voltage source in the circuit of Fig. P3.22.

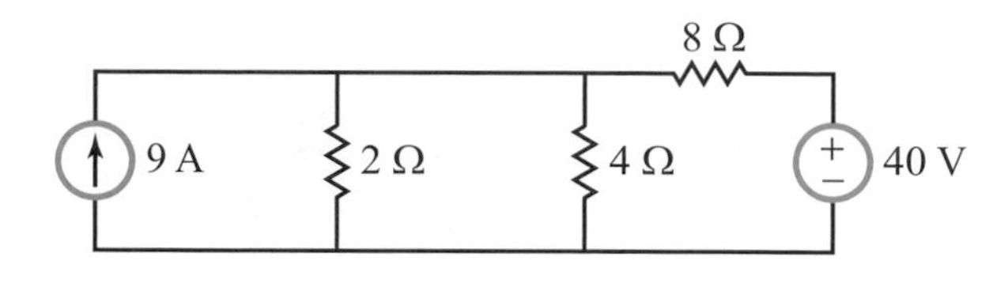

Figure P3.22: Circuit for Problem 3.22.

***3.23** Determine V in the circuit of Fig. P3.23 using mesh analysis.

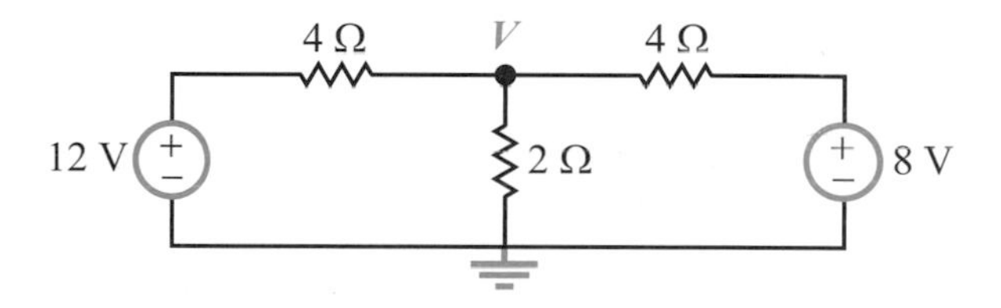

Figure P3.23: Circuit for Problem 3.23.

3.24 Apply mesh analysis to find I in the circuit of Fig. P3.24.

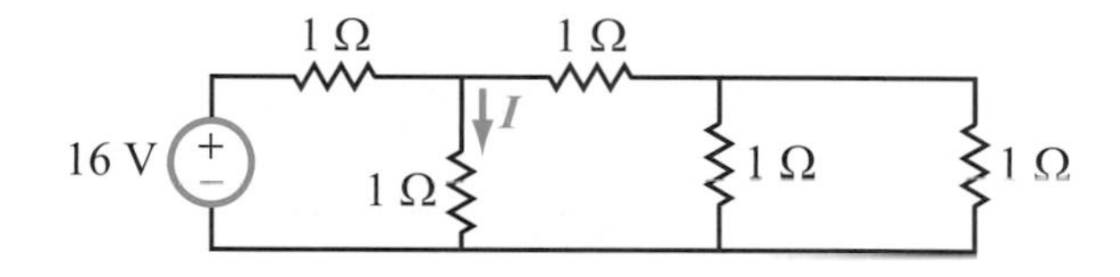

Figure P3.24: Circuit for Problem 3.24.

3.25 Apply mesh analysis to find I_x in the circuit of Fig. P3.25.

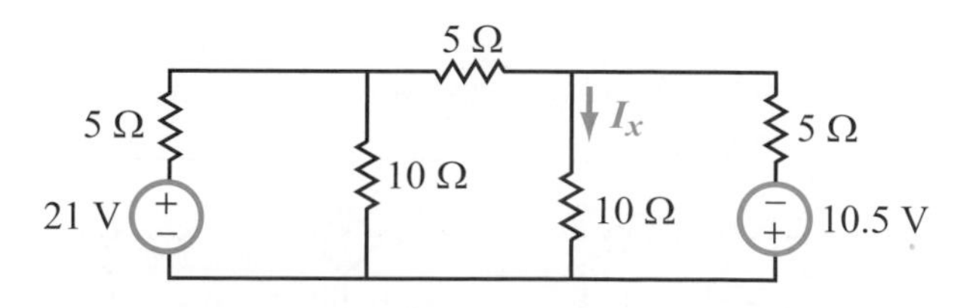

Figure P3.25: Circuit for Problem 3.25.

3.26 Apply mesh analysis to determine the amount of power supplied by the voltage source in Fig. P3.26.

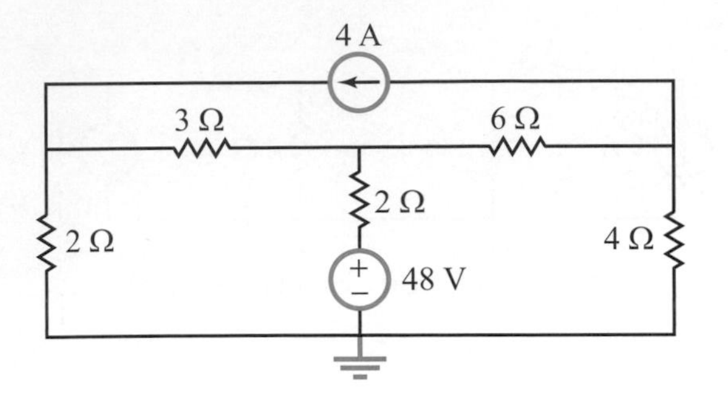

Figure P3.26: Circuit for Problem 3.26.

***3.27** Use the supermesh concept to solve for V_x in the circuit of Fig. P3.27.

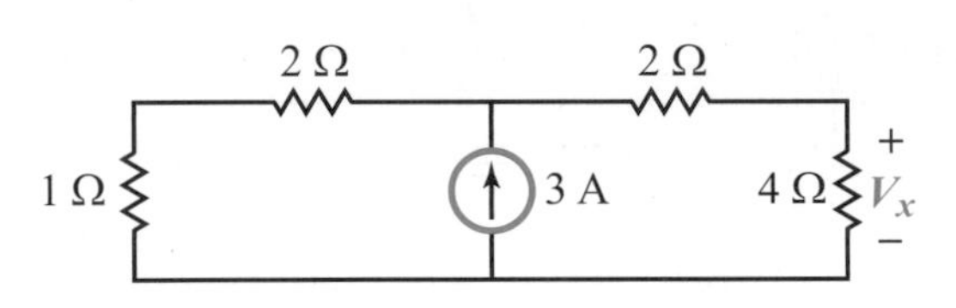

Figure P3.27: Circuit for Problem 3.27.

3.28 Use the supermesh concept to solve for I_x in the circuit of Fig. P3.28.

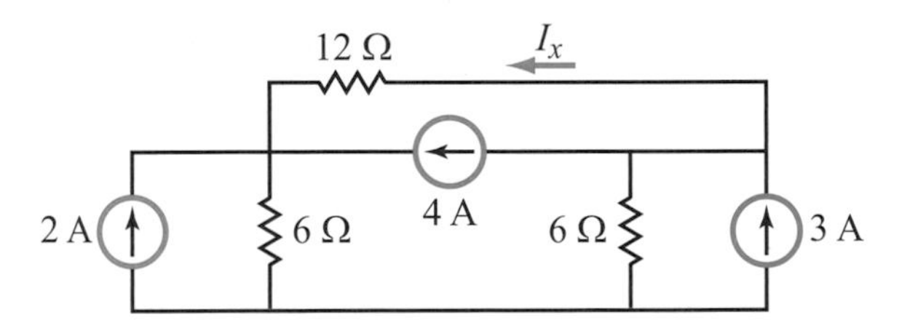

Figure P3.28: Circuit for Problem 3.28.

3.29 Apply mesh analysis to the circuit in Fig. P3.29 to determine V_x.

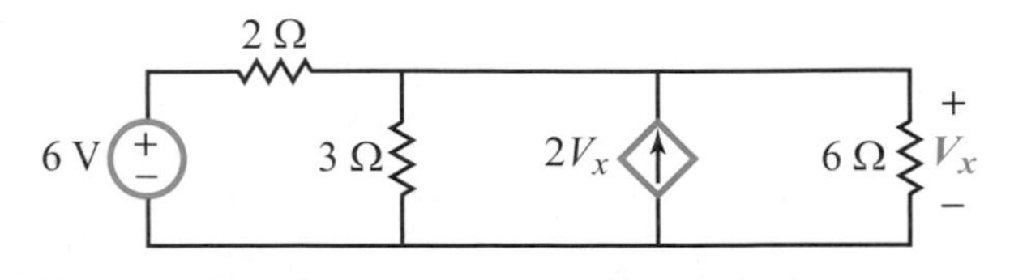

Figure P3.29: Circuit for Problem 3.29.

3.30 Determine the amount of power supplied by the independent voltage source in Fig. P3.30 by applying the mesh-analysis method.

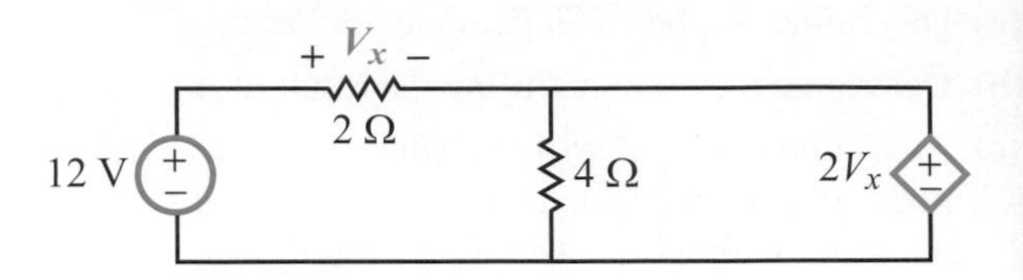

Figure P3.30: Circuit for Problem 3.30.

***3.31** Use mesh analysis to find I_x in the circuit of Fig. P3.31.

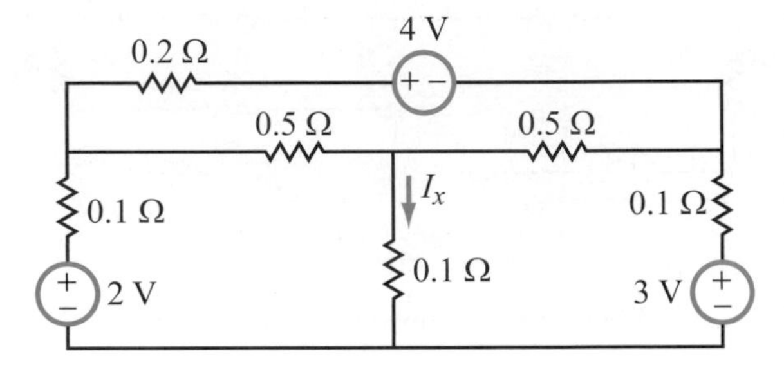

Figure P3.31: Circuit for Problem 3.31.

3.32 The circuit in Fig. P3.32 includes a dependent current source. Apply mesh analysis to determine I_x.

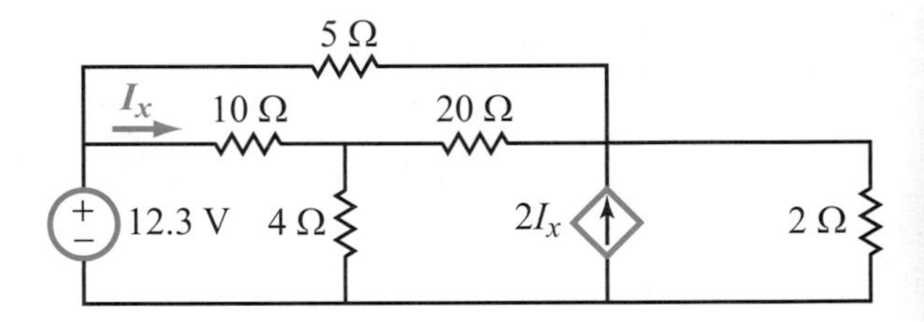

Figure P3.32: Circuit for Problems 3.32 and 3.33.

3.33 Repeat Problem 3.32 after replacing the 5-Ω resistor in Fig. P3.32 with a short circuit.

3.34 Apply mesh analysis to the circuit of Fig. P3.34 to determine I_x.

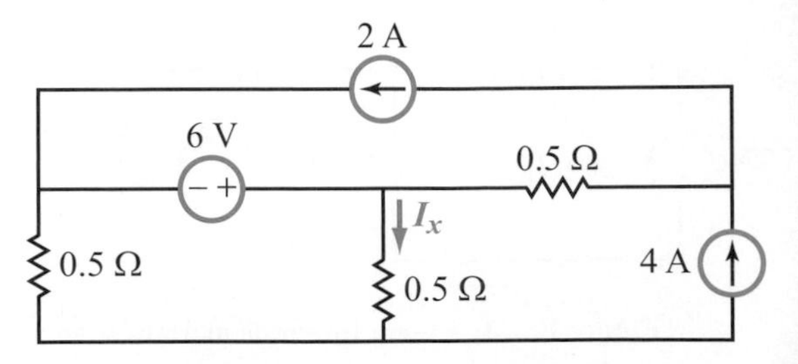

Figure P3.34: Circuit for Problem 3.34.

3.35 Determine V_x in the circuit of Fig. P3.35.

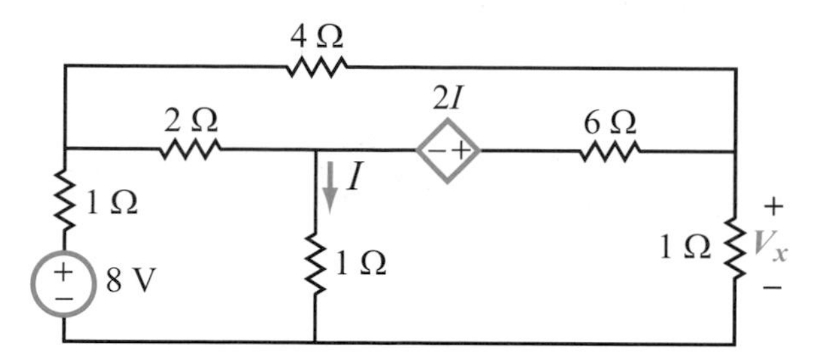

Figure P3.35: Circuit for Problems 3.35 and 3.37.

3.36 Apply the supermesh technique to find V_x in the circuit of Fig. P3.36.

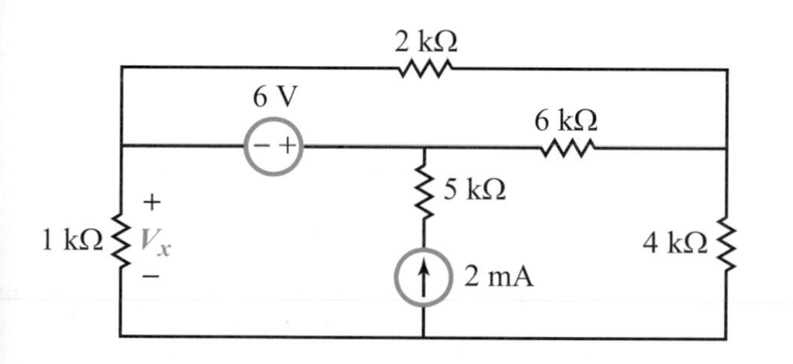

Figure P3.36: Circuit for Problem 3.36.

3.37 Repeat Problem 3.35 after replacing the 2-Ω resistor in Fig. P3.35 with a short circuit.

3.38 Apply mesh analysis to the circuit of Fig. P3.38 to find I_x.

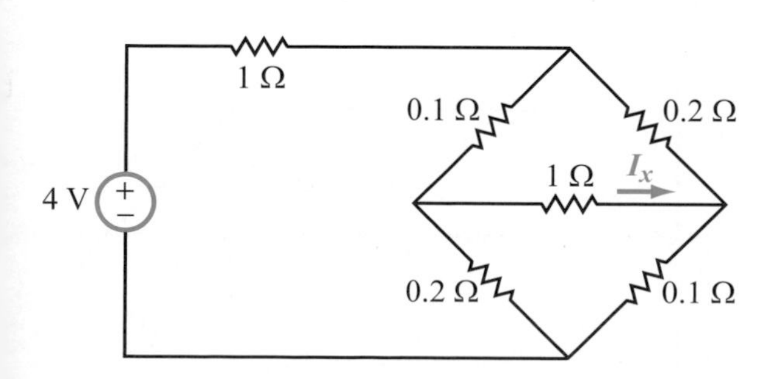

Figure P3.38: Circuit for Problem 3.38.

3.39 Determine I_0 in Fig. P3.39 through mesh analysis.

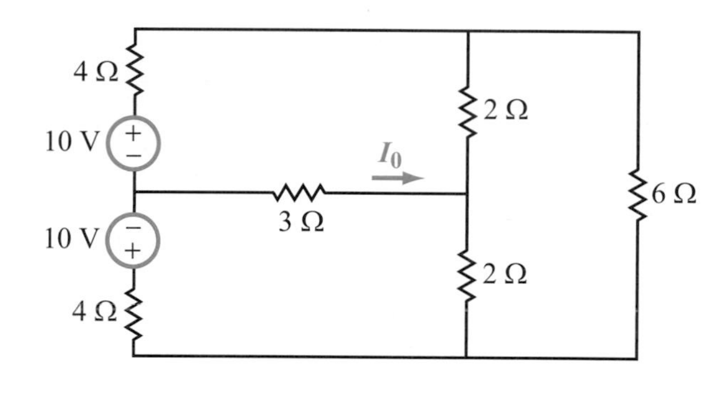

Figure P3.39: Circuit for Problem 3.39.

3.40 Use an analysis method of your choice to determine I_0 in the circuit of Fig. P3.40.

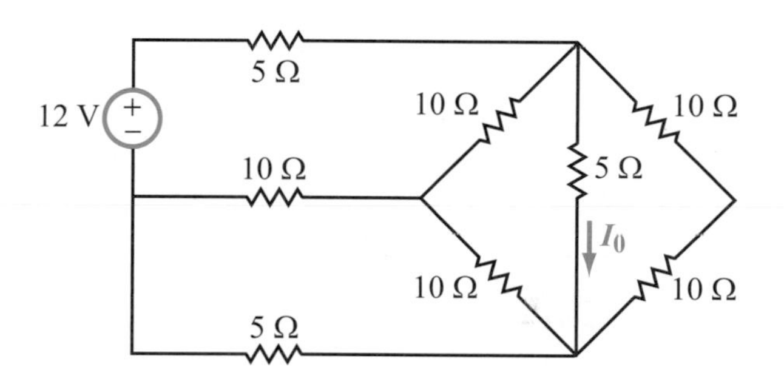

Figure P3.40: Circuit for Problem 3.40.

Sections 3-3 and 3-4: By-Inspection and Superposition Methods

*__3.41__ Apply the by-inspection method to develop a node-voltage matrix equation for the circuit in Fig. P3.41 and then use MATLAB® or MathScript software to solve for V_1 and V_2.

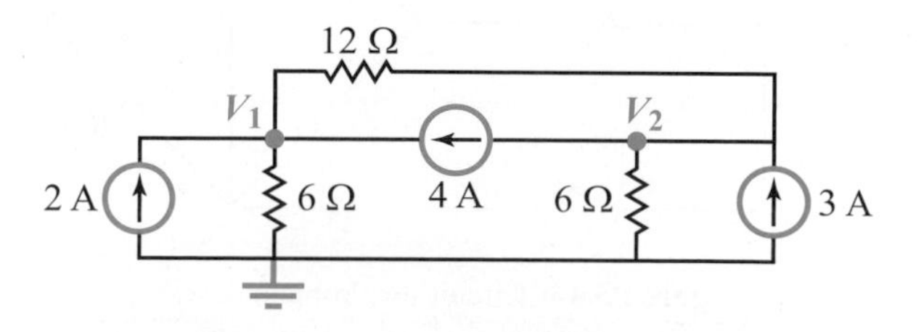

Figure P3.41: Circuit for Problem 3.41.

*3.42 Use the by-inspection method to establish a node-voltage matrix equation for the circuit in Fig. P3.42. Solve the matrix equation by MATLAB® or MathScript software to find V_1 to V_4.

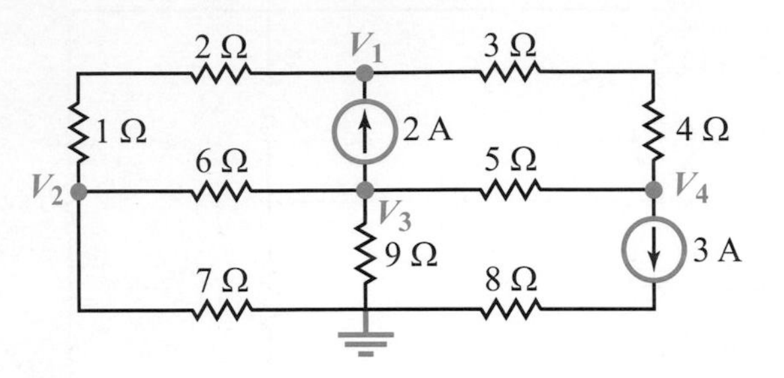

Figure P3.42: Circuit for Problem 3.42.

3.43 Develop a mesh-current matrix equation for the circuit in Fig. P3.43 by applying the by-inspection method. Solve for I_1 to I_3.

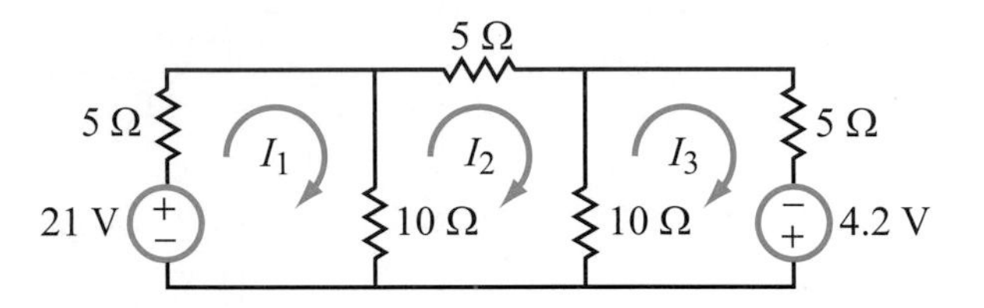

Figure P3.43: Circuit for Problem 3.43.

3.44 Find I_0 in the circuit of Fig. P3.44 by developing a mesh-current matrix equation and then solving it using MATLAB® or MathScript software.

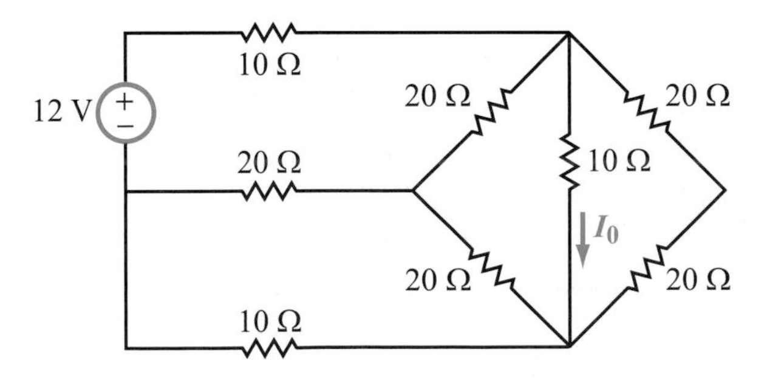

Figure P3.44: Circuit for Problem 3.44.

3.45 Apply the by-inspection method to derive a node-voltage matrix equation for the circuit in Fig. P3.45 and then solve it using MATLAB® or MathScript software to find V_x.

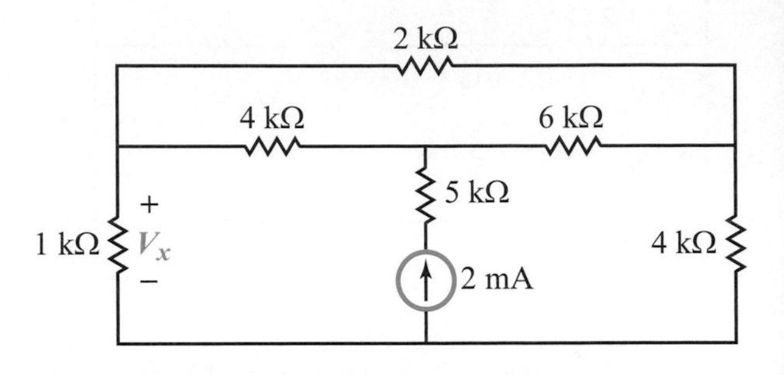

Figure P3.45: Circuit for Problem 3.45.

3.46 Use the by-inspection method to establish the mesh-current matrix equation for the circuit in Fig. P3.46 and then solve the equation to determine V_{out}.

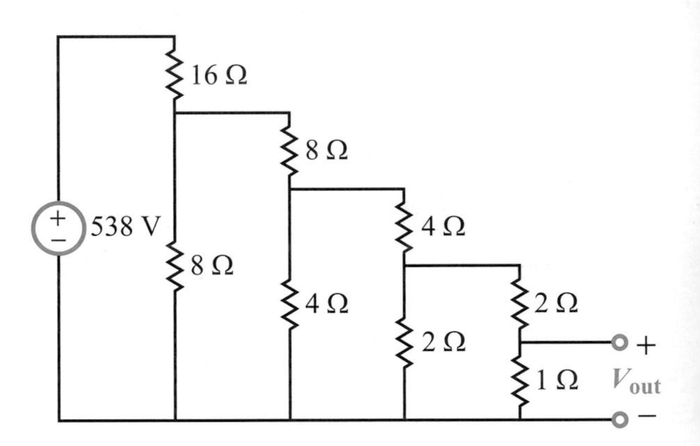

Figure P3.46: Circuit for Problem 3.46.

*3.47 Develop a node-voltage matrix equation for the circui in Fig. P3.47. Solve it to determine I.

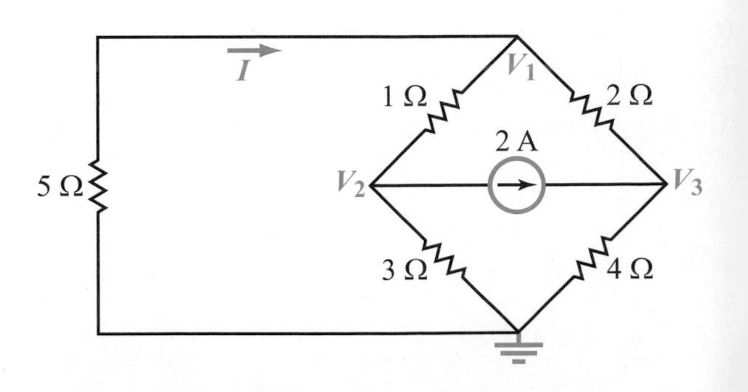

Figure P3.47: Circuit for Problem 3.47.

3.48 Determine the amount of power supplied by the voltage source in Fig. P3.48 by establishing and then solving the mesh-current matrix equation of the circuit.

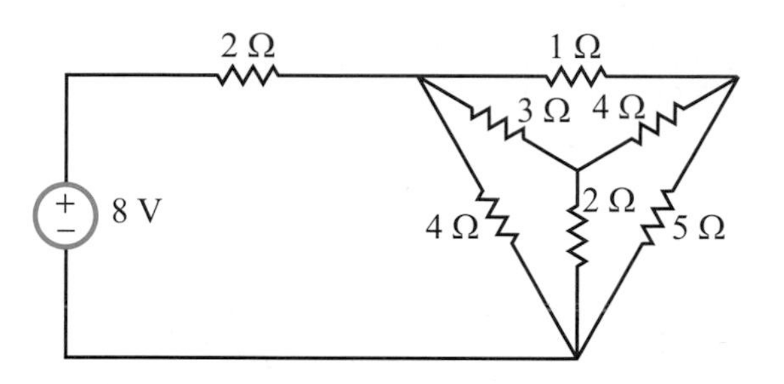

Figure P3.48: Circuit for Problem 3.48.

3.49 Determine the current I_x in the circuit of Fig. P3.49 by applying the source-superposition method. Call I'_x the component of I_x due to the voltage source alone, and I''_x the component due to the current source alone. Show that $I_x = I'_x + I''_x$ is the same as the answer to Problem 3.9.

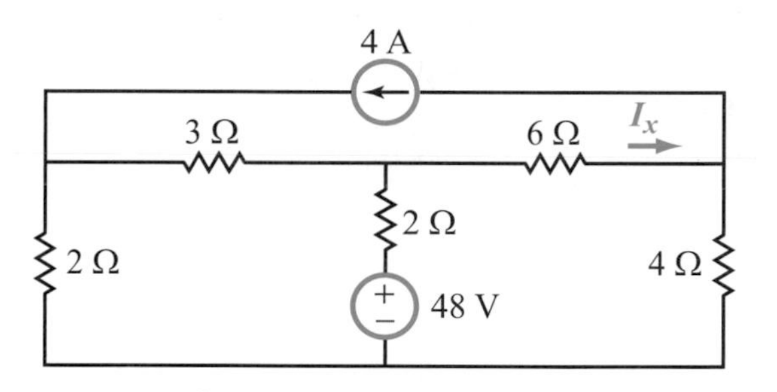

Figure P3.49: Circuit for Problem 3.49.

3.50 Apply the source-superposition method to the circuit in Fig. P3.50 to determine:

(a) I'_x, the component of I_x due to the voltage source alone

(b) I''_x, the component of I_x due to the current source alone

(c) The total current $I_x = I'_x + I''_x$

(d) P', the power dissipated in the 4-Ω resistor due to I'_x

(e) P'', the power dissipated in the 4-Ω resistor due to I''_x

(f) P, the power dissipated in the 4-Ω resistor due to the total current I. Is $P = P' + P''$? If not, why not?

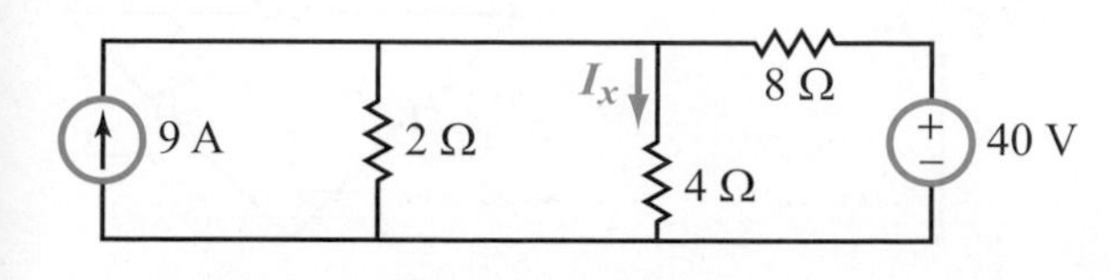

Figure P3.50: Circuit for Problem 3.50.

Section 3-5: Thévenin and Norton Equivalents

*__3.51__ Find the Thévenin equivalent circuit at terminals (a, b) for the circuit in Fig. P3.51.

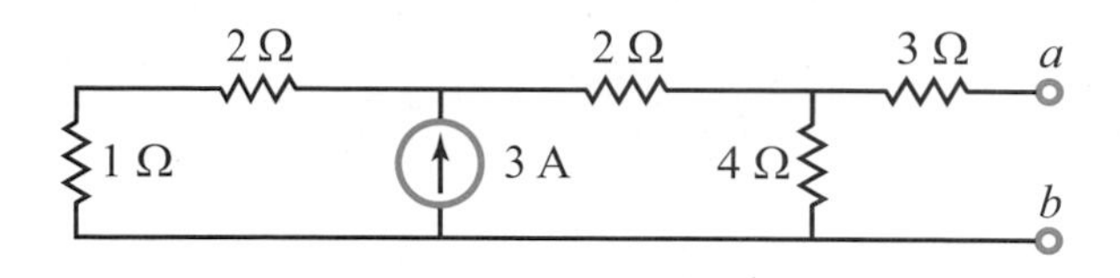

Figure P3.51: Circuit for Problem 3.51.

3.52 Find the Thévenin equivalent circuit at terminals (a, b) for the circuit in Fig. P3.52.

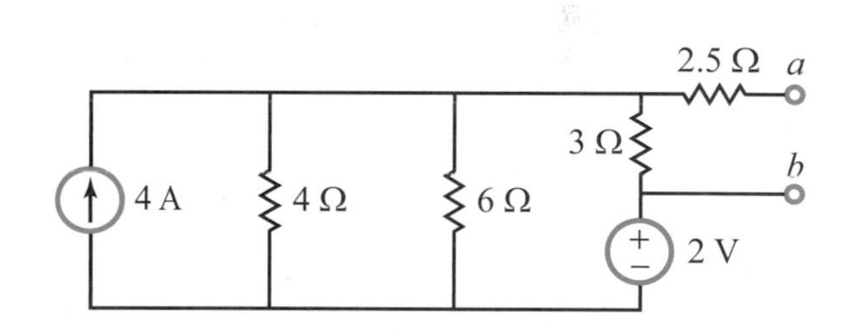

Figure P3.52: Circuit for Problem 3.52.

3.53 The circuit in Fig. P3.53 is to be connected to a load resistor R_L between terminals (a, b).

(a) Find the Thévenin equivalent circuit at terminals (a, b).

(b) Choose R_L so that the current flowing through it is 0.5 A.

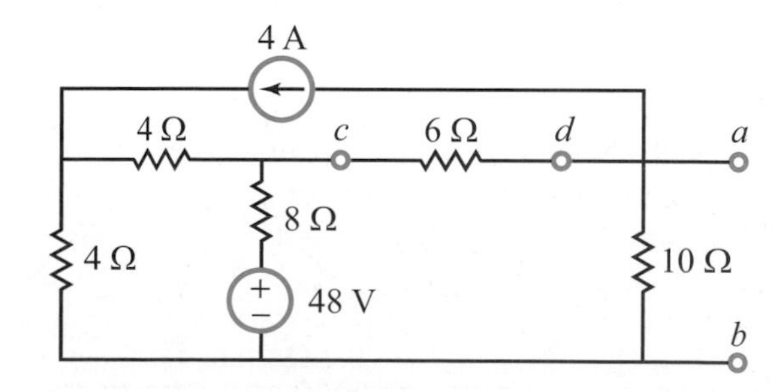

Figure P3.53: Circuit for Problems 3.53 and 3.54.

3.54 For the circuit in Fig. P3.53, find the Thévenin equivalent circuit as seen by the 6-Ω resistor connected between terminals (c, d) as if the 6-Ω resistor is a load resistor connected to (but external to) the circuit. Determine the current flowing through that resistor.

***3.55** Find the Thévenin equivalent circuit at terminals (a, b) for the circuit in Fig. P3.55.

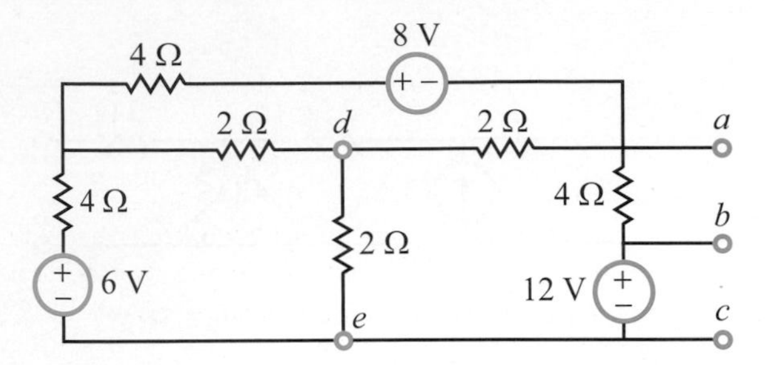

Figure P3.55: Circuit for Problems 3.55 through 3.57.

3.56 Repeat Problem 3-55 for terminals (a, c).

3.57 Repeat Problem 3-55 for terminals (d, e) as seen by the 2-Ω resistor between them (as if it were a load resistor external to the circuit).

3.58 Find the Thévenin equivalent circuit at terminals (a, b) of the circuit in Fig. P3.58.

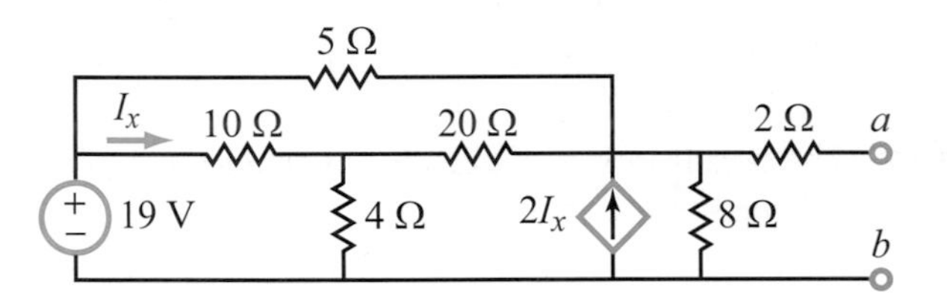

Figure P3.58: Circuit for Problems 3.58 and 3.59.

3.59 Find the Norton equivalent circuit of the circuit in Fig. P3.58 after increasing the magnitude of the voltage source to 38 V.

3.60 Find the Norton equivalent circuit at terminals (a, b) for the circuit in Fig. P3.60.

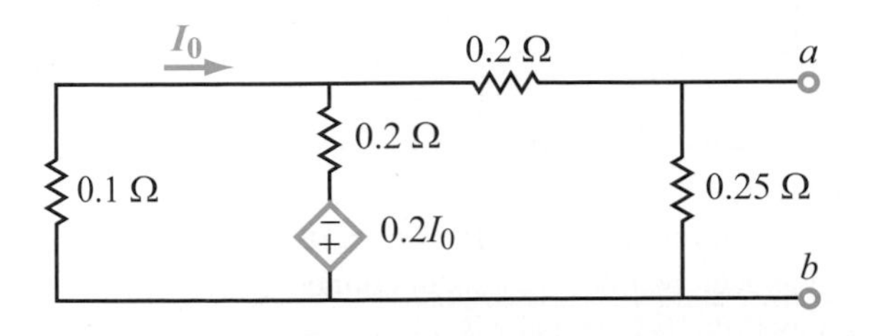

Figure P3.60: Circuit for Problem 3.60.

***3.61** Find the Norton equivalent circuit at terminals (a, b) of the circuit in Fig. P3.61.

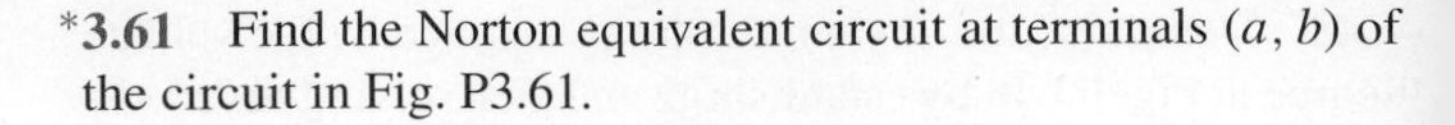

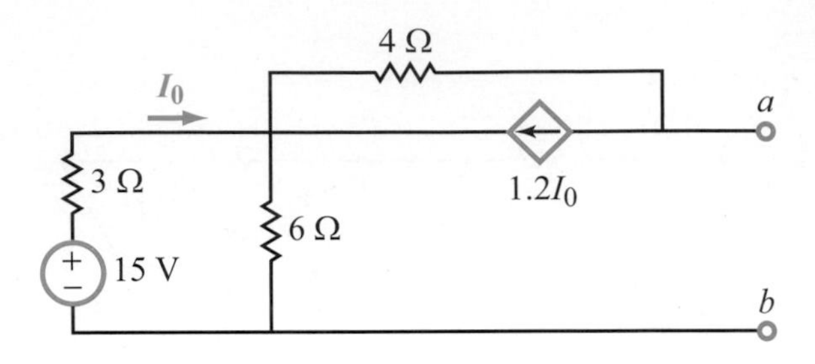

Figure P3.61: Circuit for Problems 3.61 and 3.62.

3.62 Repeat Problem 3.61 after replacing the 6-Ω resistor with an open circuit.

3.63 Find the Norton equivalent circuit at terminals (a, b) of the circuit in Fig. P3.63.

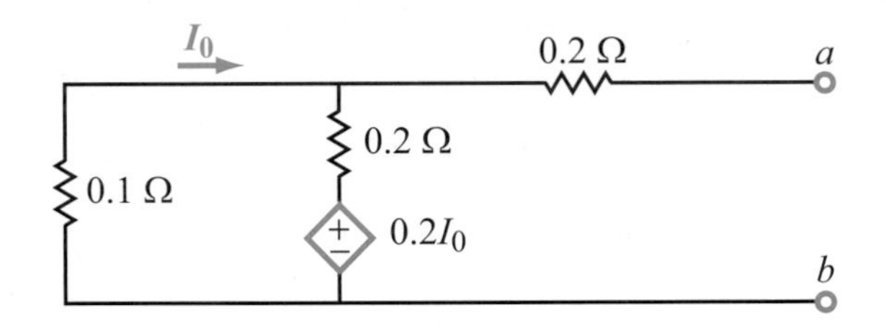

Figure P3.63: Circuit for Problems 3.63.

Section 3-6: Maximum Power Transfer

3.64 For the bridge circuit in Fig. P3.64:

(a) Find the Thévenin equivalent circuit at terminals (a, b) as seen by a load resistor R_L.

(b) Choose R_L so that the current flowing through it is 0.4 A.

(c) Choose R_L so that the power delivered to it is a maximum. How much power will that be?

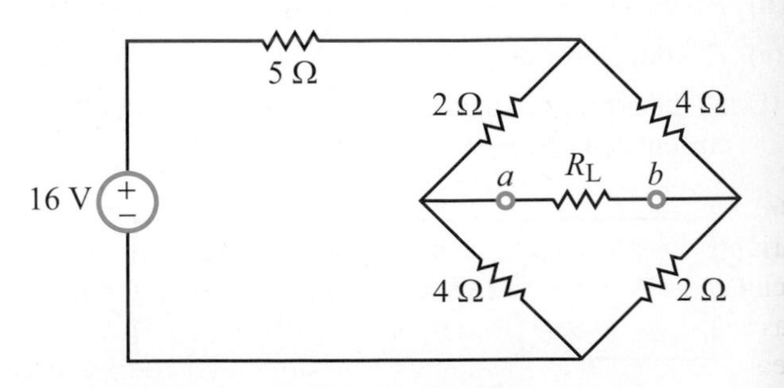

Figure P3.64: Circuit for Problem 3.64.

3.65 What value of the load resistor R_L will extract the maximum amount of power from the circuit in Fig. P3.65, and how much power will that be?

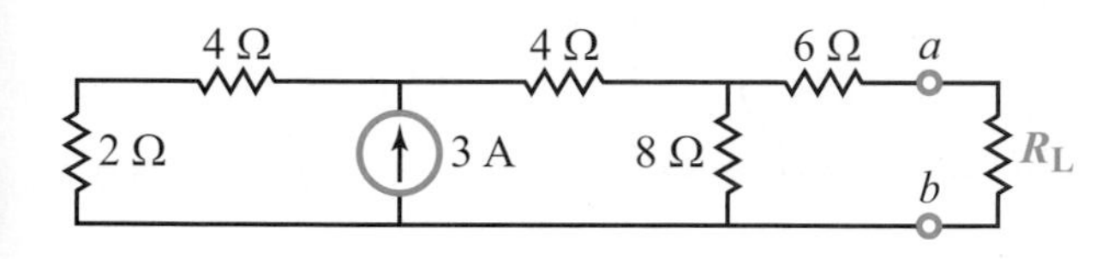

Figure P3.65: Circuit for Problem 3.65.

3.66 For the circuit in Fig. P3.66, choose the value of R_L so that the power dissipated in it is a maximum.

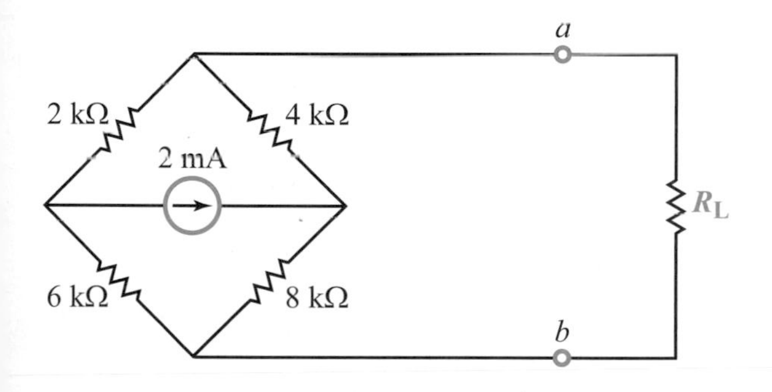

Figure P3.66: Circuit for Problem 3.66.

.67 Determine the maximum power that can be extracted by he load resistor from the circuit in Fig. P3.67.

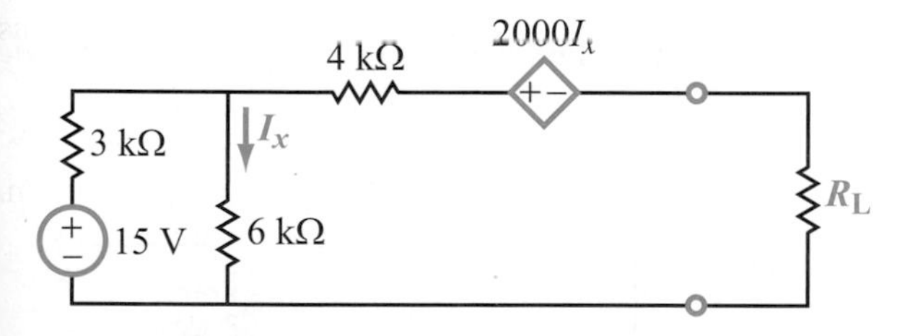

Figure P3.67: Circuit for Problem 3.67.

68 Figure P3.68 depicts a 0-to-10-kΩ potentiometer as a riable load resistor R_L connected to a circuit of an unknown chitecture. When the wiper position on the potentiometer as adjusted such that $R_L = 1.2$ kΩ, the current through it was easured to be 3 mA, and when the wiper was lowered so that $_L = 2$ kΩ, the current decreased to 2.5 mA. Determine the lue of R_L that would extract maximum power from the circuit.

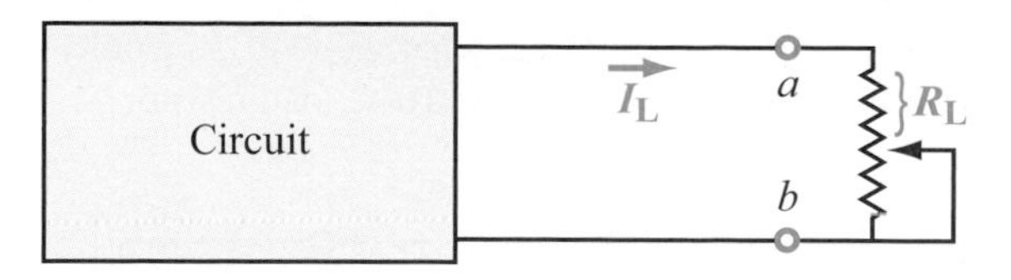

Figure P3.68: Circuit for Problem 3.68.

3.69 The circuit shown in Fig. P3.69 is connected to a variable load R_L through a resistor R_s. Choose R_s so that I_L never exceeds 4 mA, regardless of the value of R_L. Given that choice, what is the maximum power that R_L can extract from the circuit?

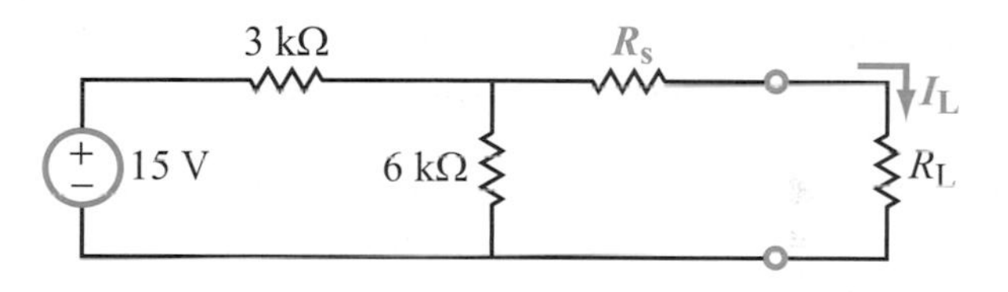

Figure P3.69: Circuit for Problem 3.69.

3.70 In the circuit shown in Fig. P3.70, a potentiometer is connected across the load resistor R_L. The total resistance of the potentiometer is $R = R_1 + R_2 = 5$ kΩ.

(a) Obtain an expression for the power P_L dissipated in R_L for any value of R_1.

(b) Plot P_L versus R_1 over the full range made possible by the potentiometer's wiper.

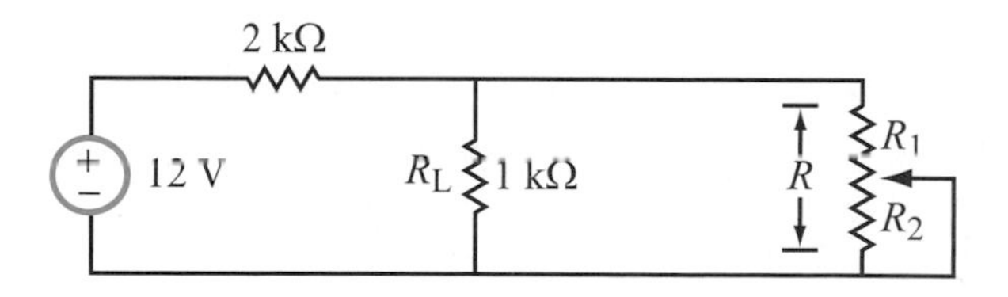

Figure P3.70: Circuit for Problem 3.70.

3.71 Determine the maximum power extractable from the circuit in Fig. P3.71 by the load resistor R_L.

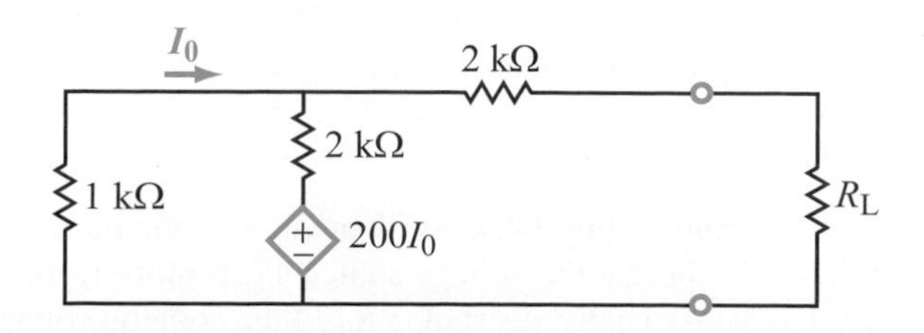

Figure P3.71: Circuit for Problem 3.71.

3.72 In the circuit Fig. P3.72, what value of R_s would result in maximum power transfer to the 10-Ω load resistor?

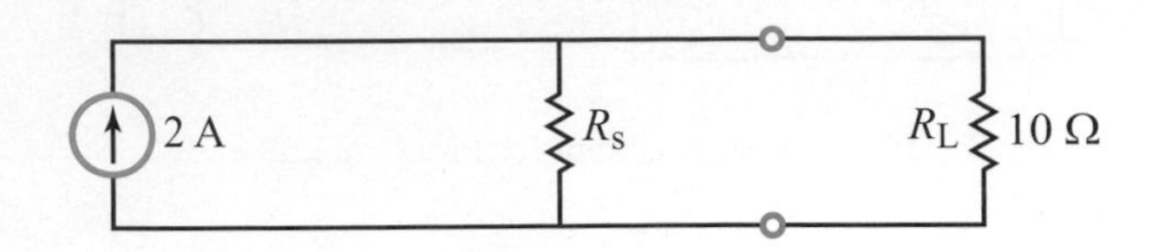

Figure P3.72: Circuit for Problem 3.72.

Section 3-7: Bipolar Junction Transistor

*__3.73__ The two-transistor circuit in Fig. P3.73 is known as a ***current mirror***. It is useful because the current I_0 controls the current I_{REF} regardless of external connections to the circuit. In other words, this circuit behaves like a current-controlled current source. Assume both transistors are the same size such that $I_{B_1} = I_{B_2}$. Find the relationship between I_0 and I_{REF}. (*Hint:* You do not need to know what is connected above or below the transistors. Nodal analysis will suffice.)

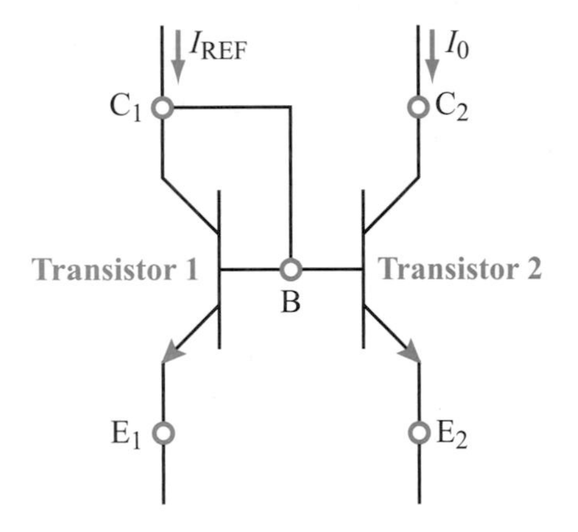

Figure P3.73: A simple current mirror (Problem 3.73).

3.74 The circuit in Fig. P3.74 is a BJT ***common collector amplifier***. Obtain expressions for both the voltage gain ($A_V = V_{out}/V_{in}$) and the current gain ($A_I = I_{out}/I_{in}$). Assume $V_{in} \gg V_{BE}$.

3.75 The circuit in Fig. P3.75 is identical with the circuit in Fig. P3.74, except that the voltage source V_{in} is more realistic in that it has an associated resistance R_{in}. Find both the voltage gain ($A_V = V_{out}/V_{in}$) and the current gain ($A_I = I_{out}/I_{in}$). Assume $V_{in} \gg V_{BE}$.

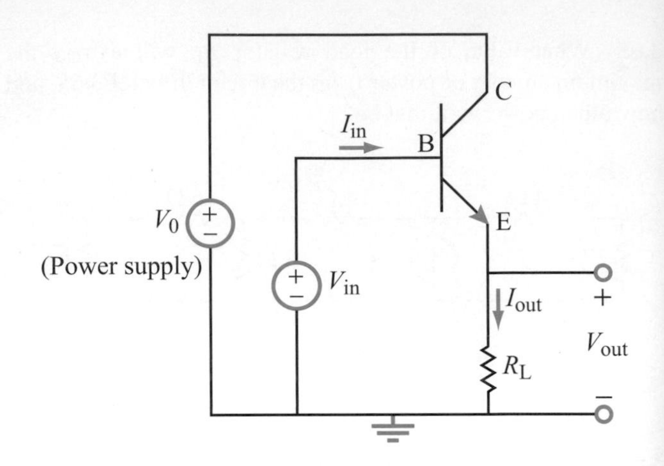

Figure P3.74: Circuit for Problem 3.74.

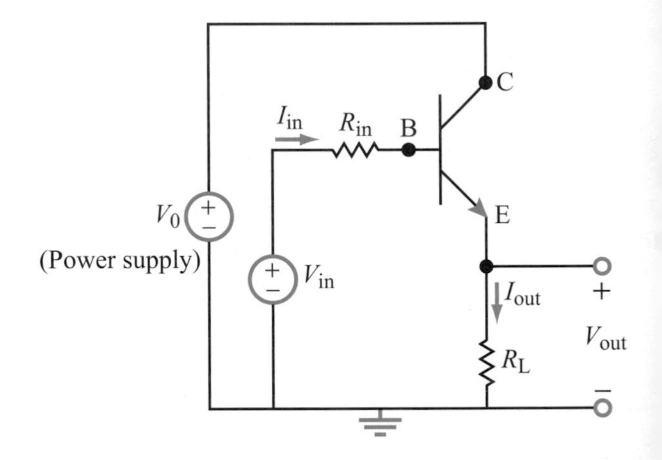

Figure P3.75: Circuit for Problem 3.75.

3.76 The circuit in Fig. P3.76 is a BJT ***common emitter amplifier***. Find V_{out} as a function of V_{in}.

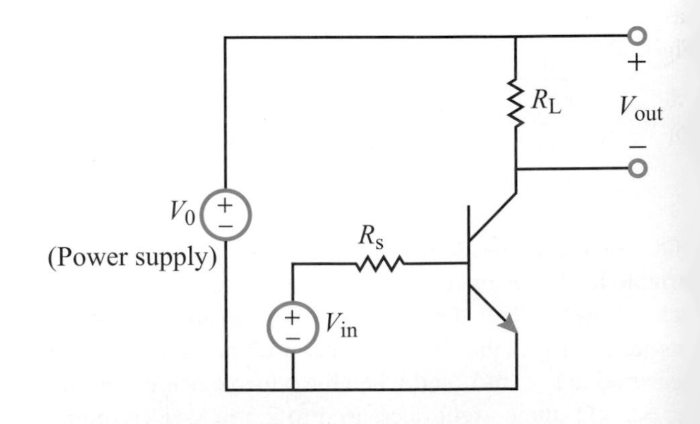

Figure P3.76: Circuit for Problem 3.76.

.77 Obtain an expression for V_{out} in terms of V_{in} for the ommon emitter-amplifier circuit in Fig. P3.77. Assume $'_{\text{in}} \gg V_{\text{BE}}$.

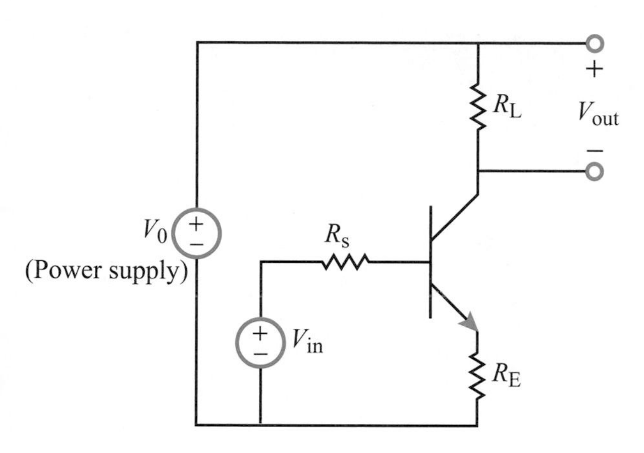

Figure P3.77: Circuit for Problem 3.77.

ection 3-8: Multisim Analysis

.78 Using Multisim, draw the circuit in Fig. P3.78 and solve or voltages V_1 and V_2.

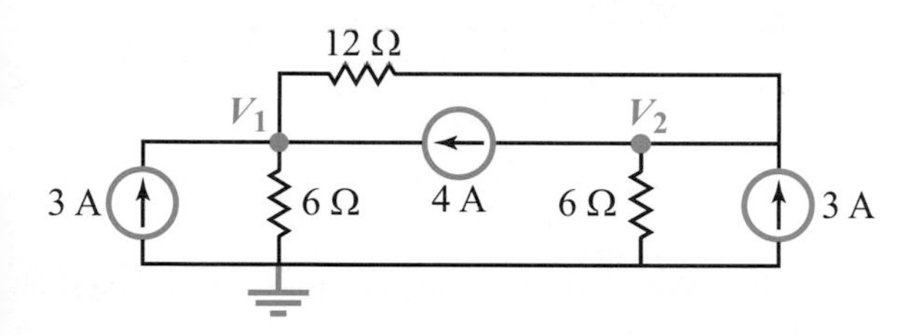

Figure P3.78: Circuit for Problem 3.78.

79 The circuit in Problem 3.44 was solved using ATLAB® or MathScript software. It can be solved just as sily using Multisim. Using Multisim, draw the circuit in g. P3.44 and solve for all node voltages and the current I_0.

80 Using Multisim, draw the circuit in Fig. P3.80 and solve r V_x.

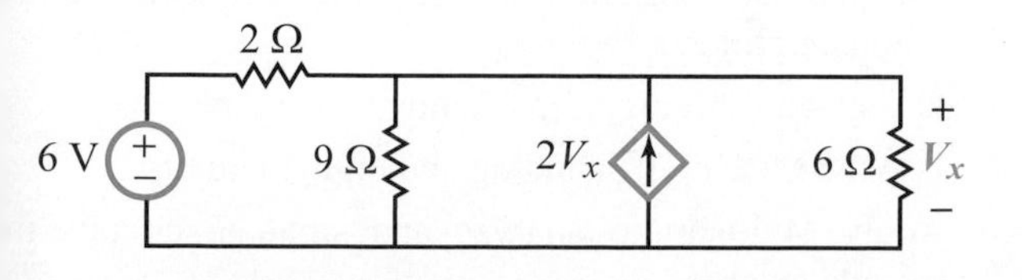

Figure P3.80: Circuit for Problem 3.80.

3.81 Use Multisim to draw the circuit in Fig. P3.81 and solve for V_x.

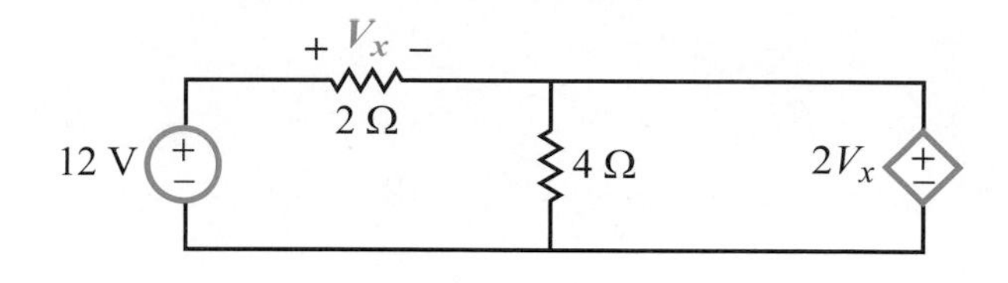

Figure P3.81: Circuit for Problem 3.81.

3.82 Use the DC Operating Point Analysis in Multisim to find the power dissipated or supplied by each component in the circuit in Fig. P3.82 and show that the sum of all powers is zero.

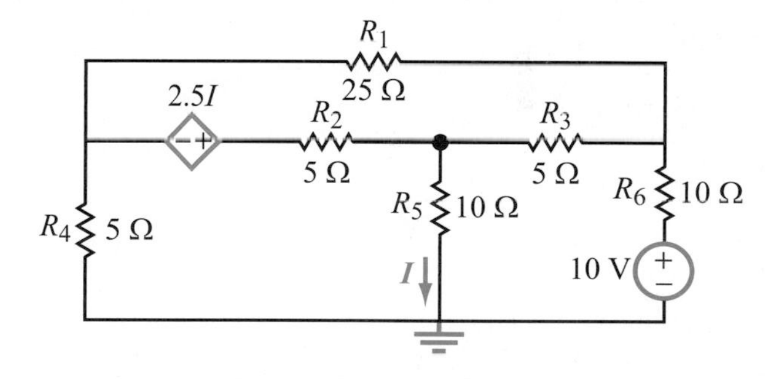

Figure P3.82: Circuit for Problem 3.82.

3.83 Simulate the circuit found in Fig. P3.83 with a 10-Ω resistor placed across the terminals (a, b). Then either by hand or by using tools in Multisim (see Multsim Demo 3.3), find the Thévenin and Norton equivalent circuits and simulate both of those circuits in Multisim with 10-Ω resistors across their output terminals. Show that the voltage drop across and current through the 10-Ω load resistor is the same in all three simulations.

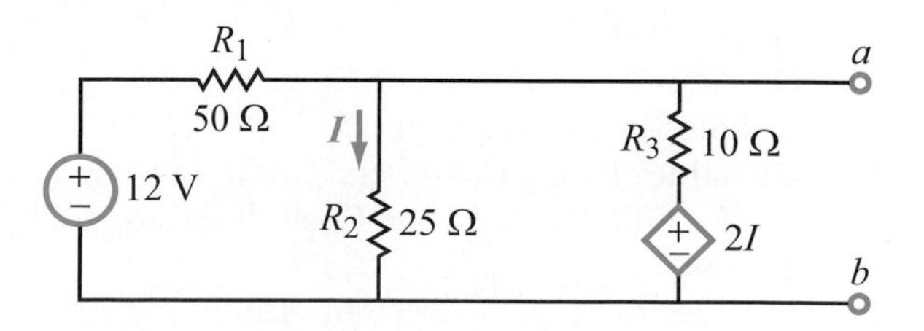

Figure P3.83: Circuit for Problem 3.83.

C H A P T E R

4

Operational Amplifiers

Chapter Contents

Objectives

Upon learning the material presented in this chapter, you shoul
be able to:

1. Describe the basic properties of an op amp and state th constraints of the ideal op-amp model.
2. Explain the role of negative feedback and the tradeo between circuit gain and dynamic range.
3. Analyze and design inverting amplifiers, summin amplifiers, difference amplifiers, and voltage followers.
4. Combine multiple op-amp circuits together to perfor signal processing operations.
5. Analyze and design high-gain, high-sensitivity instrume tation amplifiers.
6. Design an n-bit digital-to-analog converter.
7. Use the MOSFET in analog and digital circuits.
8. Apply Multisim to analyze and simulate circuits th include op amps.

Overview

Since its first realization by Bob Widlar in 1963 and then its introduction by Fairchild Semiconductor in 1968, the operational amplifier, or ***op amp*** for short, has become the workhorse of many signal-processing circuits. It acquired the adjective *operational* because it is a versatile device capable not only of amplifying a signal but also inverting it (reversing its polarity), integrating it, or differentiating it. When multiple signals are connected to its input, the op amp can perform additional mathematical operations—including addition and subtraction. Consequently, op-amp circuits often are cascaded together in various arrangements to support a variety of different applications. In this chapter, we will explore several op-amp circuit configurations, including amplifiers, summers, voltage followers, and digital-to-analog converters.

4-1 Op-Amp Characteristics

The internal architecture of an op-amp circuit consists of many interconnected transistors, diodes, resistors and capacitors—all fabricated on a chip of silicon. Despite its internal complexity, however, *an op amp can be modeled in terms of a relatively simple equivalent circuit that exhibits a **linear** input–output response.* This equivalence allows us to apply the tools we developed in the preceding chapters to analyze (as well as design) a large array of op-amp circuits and to do so with relative ease.

4-1.1 Nomenclature

Commercially available op amps are fabricated in encapsulated packages of various shapes. A typical example is the eight-pin ***DIP configuration*** shown in Fig. 4-1(a) [DIP stands for dual-in-line package]. The pin diagram for the op amp is shown in Fig. 4-1(b), and its circuit symbol (the triangle) is displayed in Fig. 4-1(c). Of the eight pins (terminals) only five need to be connected to an outside circuit, namely:

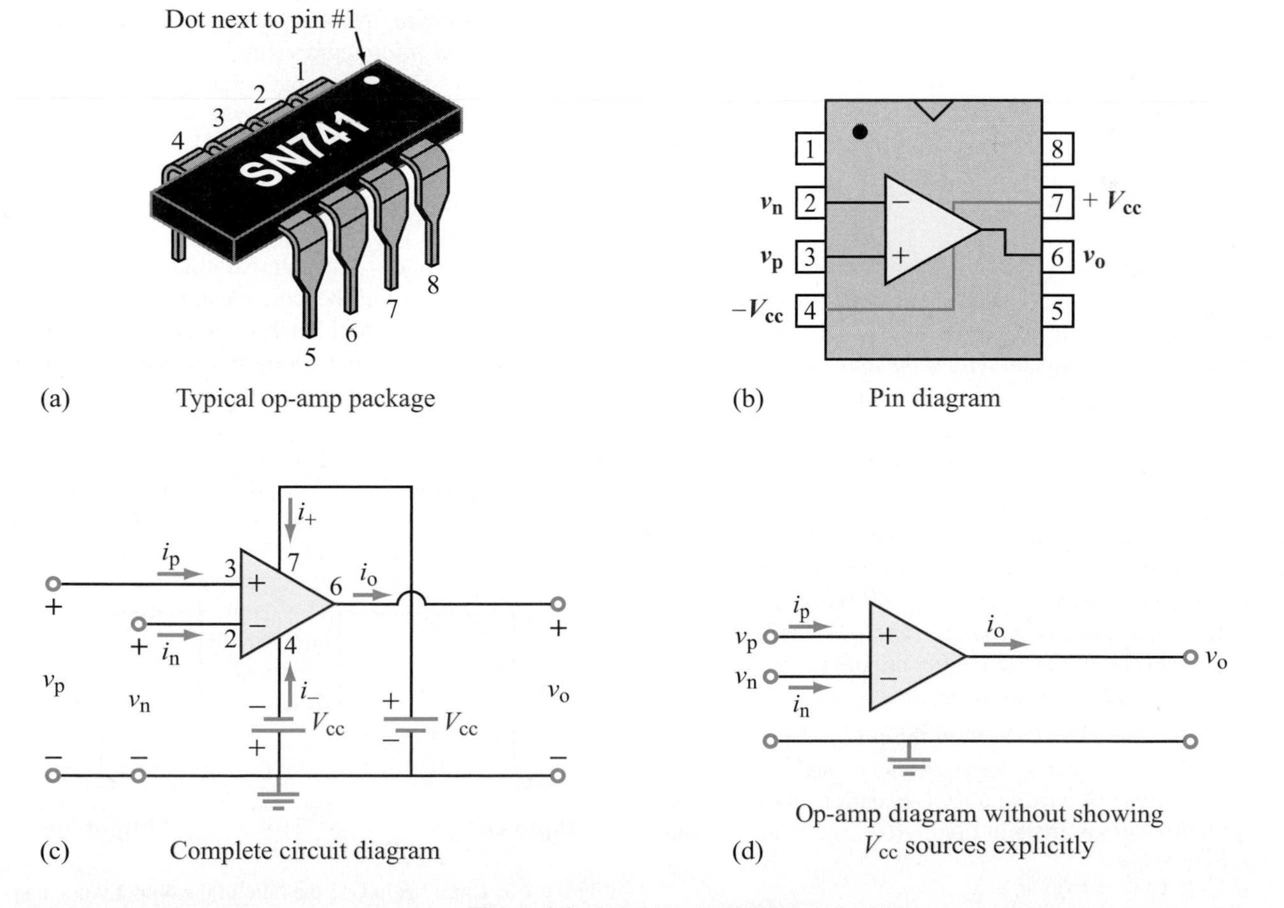

Figure 4-1: Operational amplifier.

Op-Amp Pin Designation

Pin 2	inverting (or *negative*) input voltage, v_n
Pin 3	noninverting (or *positive*) input voltage, v_p
Pin 4	negative (−) terminal of power supply V_{cc}
Pin 7	positive (+) terminal of power supply V_{cc}
Pin 6	output voltage, v_o

The op amp has two input voltage terminals (v_p and v_n) and one output voltage terminal (v_o).

The terms ***noninverting*** and ***inverting*** are associated with the property of the op amp that its output voltage v_o is directly proportional to the noninverting input voltage v_p, whereas for the inverting input voltage v_n, v_o is proportional to $-v_n$.

Kirchhoff's current law applies to any volume of space, including an op amp. Hence, for the five terminals connected to the op amp, KCL mandates that

$$i_o = i_p + i_n + i_+ + i_-, \tag{4.1}$$

where i_p, i_n, and i_o may be constant (dc) or time-varying currents. Currents i_+ and i_- are dc currents generated by the dc power supply V_{cc}. It is important to note at this juncture that from here on forward: *We will ignore the pins connected to V_{cc} when we draw circuit diagrams involving op amps, because so long as the op amp is operated in its linear region* (as will be discussed shortly), V_{cc} *will have no bearing on the operation of the circuit.*

Hence, in the future, the op-amp triangle usually will be drawn with only three terminals, as shown in Fig. 4-1(d). Moreover, the voltages v_p, v_n, and v_o will be defined relative to a common reference or ground. The (+) and (−) labels printed on the op-amp triangle simply denote the noninverting and inverting pins of the op amp not the polarities of v_p or v_n.

Ignoring the pins associated with the power-supply voltage V_{cc} does not mean we can ignore currents i_+ and i_-. To avoid making the mistake of writing a KCL equation on the basis of the simplified diagram given in Fig. 4-1(d), we explicitly state that fact by writing

$$i_o \neq i_p + i_n. \tag{4.2}$$

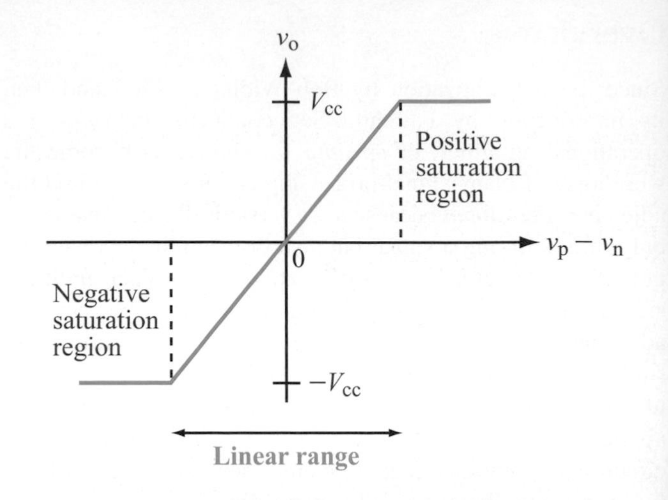

Figure 4-2: Op-amp transfer characteristics. The linear range extends between $v_o = -V_{cc}$ and $+V_{cc}$.

4-1.2 Transfer Characteristics

The plot shown in Fig. 4-2, which depicts the input–output voltage-transfer characteristic of an op amp, is divided into three regions of operation, denoted the ***negative saturation***, ***linear*** and ***positive saturation*** regions. In the linear region, the output voltage v_o is related to the input voltages v_p and v_n by

$$v_o = A(v_p - v_n), \tag{4.3}$$

where A is called the ***op-amp gain***, or the ***open-loop gain***. Strictly speaking, this relationship is valid only when the op amp is not connected to an external circuit on the output side (open loop), but as will become clearer in future sections, it continues to hold (approximately) if the output circuit satisfies certain conditions. The open-loop gain is specific to the op-amp device itself, in contrast with the ***circuit gain*** or ***closed-loop gain***, G, which defines the gain of the entire circuit. Thus, if v_s is the signal voltage of the circuit connected at the input side of the op-amp circuit (Fig. 4-3), and v_L is the voltage across the

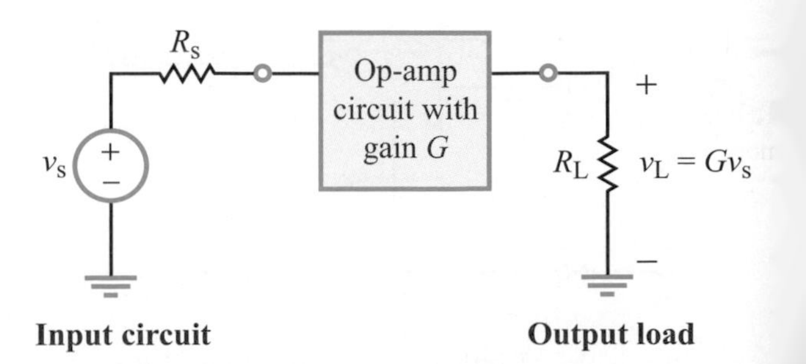

Figure 4-3: Circuit gain G is the ratio of the output voltage v_L to the signal input voltage v_s.

ad connected at its output side, then

$$v_L = G v_s. \qquad (4.4)$$

ccording to Eq. (4.3), v_o is related linearly to the difference etween v_p and v_n or to either one of them if the other is held onstant. Excluding circuits that contain magnetically coupled ansformers, in a regular circuit, no voltage can exceed the net oltage level of the power supply. Hence, the maximum value at v_o can attain is $|V_{cc}|$. The op amp goes into a saturation ode if $|A(v_p - v_n)| > |V_{cc}|$, which can occur on both the egative and positive sides of the linear region.

As we will discuss shortly, the op-amp gain A is typically n the order of 10^5 or greater, and the supply voltage is on e order of volts or tens of volts. In the linear region, v_o is ounded between $-V_{cc}$ and $+V_{cc}$, which means that $(v_p - v_n)$ bounded between $-V_{cc}/A$ and $+V_{cc}/A$. For $V_{cc} = 10$ V nd $A = 10^6$, the operating range of $(v_p - v_n)$ is -10 μV to -10 μV. It is important to keep this in mind as we deal with ircuits containing operational amplifiers.

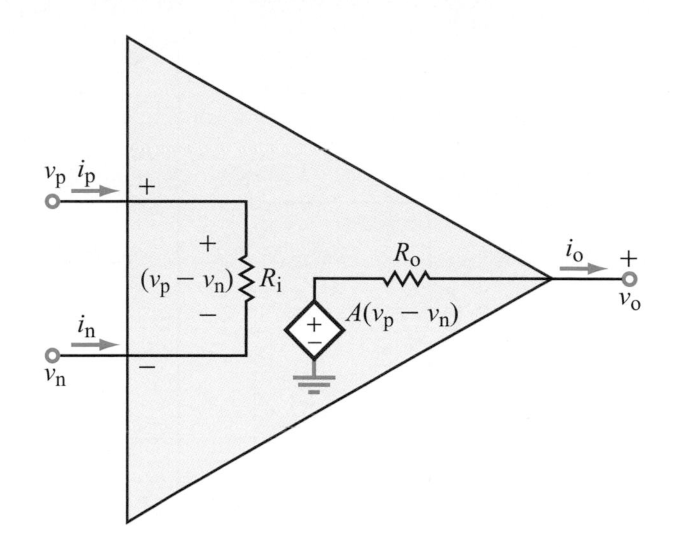

Figure 4-4: Equivalent circuit model for an op amp operating in the linear range.

-1.3 Equivalent-Circuit Model

hen operated in its linear region, the op-amp input–output ehavior can be modeled in terms of the equivalent circuit hown in Fig. 4-4. The equivalent circuit consists of a voltage-ontrolled voltage source of magnitude $A(v_p - v_n)$, an input esistance R_i, and an output resistance R_o. Table 4-1 lists the pical range of values that each of these op-amp parameters ay assume. Based on these values, we note that an op amp is haracterized by:

1) *High input resistance* R_i: at least 1 MΩ, which is highly desirable from the standpoint of voltage transfer from an input circuit (as discussed previously in Section 3-6).

2) *Low output resistance* R_o: which is desirable from the standpoint of transfering the op-amp's output voltage to a load circuit.

(3) *High voltage gain* A: which is the key, as we will see later, to allowing us to further simplify the equivalent circuit into an "ideal" op-amp model with infinite gain.

Example 4-1: Noninverting Amplifier

The circuit shown in Fig. 4-5 uses an op amp to amplify the input signal voltage v_s. Obtain an expression for the circuit gain $G = v_o/v_s$, and then evaluate it for $V_{cc} = 10$ V, $A = 10^6$, $R_i = 10$ MΩ, $R_o = 10$ Ω, $R_1 = 80$ kΩ, and $R_2 = 20$ kΩ.

able 4-1: Characteristics and typical ranges of op-amp parameters. The rightmost column represents the values assumed by the ideal op-amp odel.

Op-Amp Characteristics	Parameter	Typical Range	Ideal Op Amp
• Linear input–output response	Open-loop gain A	10^4 to 10^8 (V/V)	∞
• High input resistance	Input resistance R_i	10^6 to 10^{13} Ω	∞ Ω
• Low output resistance	Output resistance R_o	1 to 100 Ω	0 Ω
• Very high gain	Supply voltage V_{cc}	5 to 24 V	As specified by manufacturer

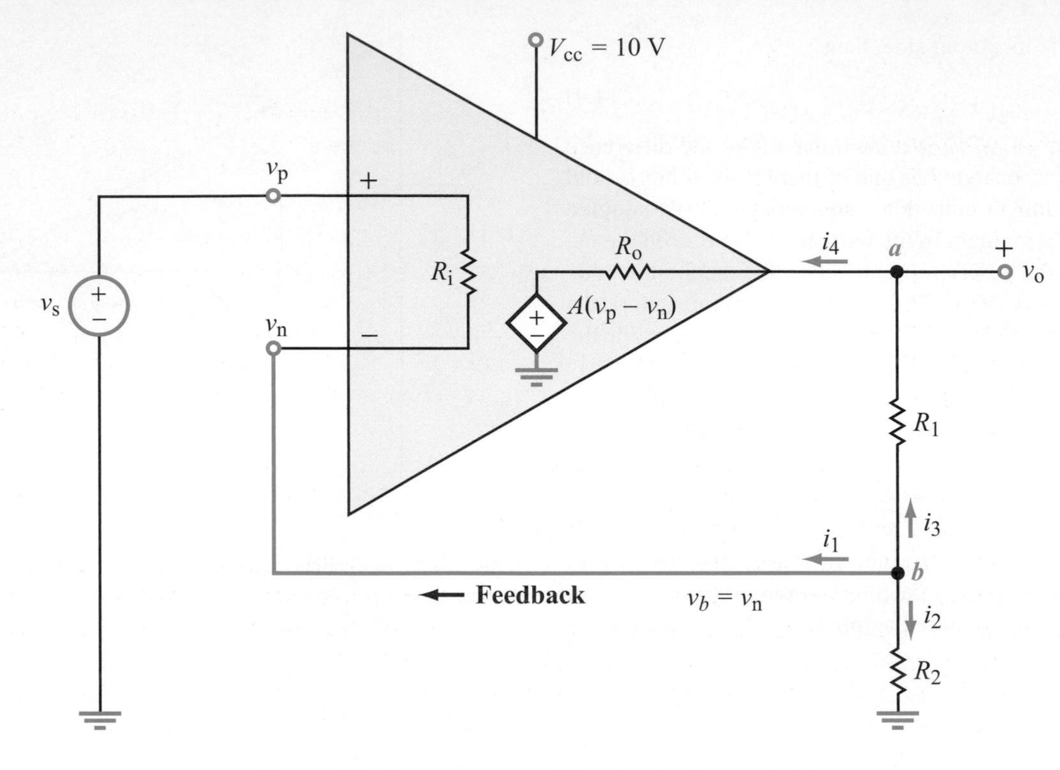

Figure 4-5: Noninverting amplifier circuit of Example 4-1.

Solution: For reference purposes, we label the output as terminal a and the node from which a current is fed back into the op amp as terminal b. The current i_3 flowing from terminal b to terminal a is the same as the current i_4 flowing from terminal a towards R_o. When expressed in terms of node voltages, the equality $i_3 = i_4$ gives

$$\frac{v_n - v_o}{R_1} = \frac{v_o - A(v_p - v_n)}{R_o}. \tag{4.5}$$

At node b, KCL gives $i_1 + i_2 + i_3 = 0$, or

$$\frac{v_n - v_p}{R_i} + \frac{v_n}{R_2} + \frac{v_n - v_o}{R_1} = 0. \tag{4.6}$$

Additionally,

$$v_p = v_s. \tag{4.7}$$

Solution of these simultaneous equations leads to the following expression for G:

$$G = \frac{v_o}{v_s} = \frac{[AR_i(R_1 + R_2) + R_2 R_o]}{AR_2 R_i + R_o(R_2 + R_i) + R_1 R_2 + R_i(R_1 + R_2)}. \tag{4.8}$$

For $V_{cc} = 10$ V, $A = 10^6$, $R_i = 10^7$ Ω, $R_o = 10$ Ω, $R_1 = 80$ kΩ, and $R_2 = 20$ kΩ,

$$G = \frac{v_o}{v_s} = 4.999975 \simeq 5.0. \tag{4.9}$$

In the expression for G, the two parameters A and R_i are several orders of magnitude larger than all of the others. Also, R_o is in series with R_1, which is 8000 times larger. Hence, we would incur minimal error if we let $A \to \infty$, $R_i \to \infty$, and $R_o \to 0$ in which case the expression for G reduces to

$$G = \frac{R_1 + R_2}{R_2} \quad \text{(ideal op-amp model).} \tag{4.10}$$

This approximation, based on the ideal op-amp model that will be introduced in Section 4-3, gives

$$G = \frac{80 \text{ k}\Omega + 20 \text{ k}\Omega}{20 \text{ k}\Omega} = 5.$$

Technology Brief 7: Integrated Circuit Fabrication Process

Do you ever wonder how the processor in your computer was actually fabricated? How is it that engineers can put hundreds of millions of transistors into one device that measures only a few centimeters on a side (and with so few errors) so the devices actually function as expected?

Devices such as modern computer processors and semiconductor memories fall into a class known as ***integrated circuits (IC)***. They are so named because all of the components in the circuit (and their "wires") are fabricated simultaneously onto a circuit during the manufacturing process. This is in contrast to circuits where each component is fabricated separately and then soldered or wired together onto a common board (such as those you probably build in your lab classes). Integrated circuits were first demonstrated independently by Jack Kilby at Texas Instruments and Robert Noyce at Fairchild Semiconductor in the late 1950s. Once developed, the ability to manufacture components and their connections in parallel with good quality control meant that circuits with thousands (then millions, then billions) of components could be designed and built reliably.

Semiconductor Processing Basics

All mainstream semiconductor integrated-circuit processes start with a thin slice of silicon, known as a ***substrate*** or ***wafer***. This wafer is circular and ranges from 4 to 18 inches in diameter and is approximately 1 mm thick (hence its name). Each wafer is cut from a single crystal of the element silicon and polished to its final thickness with atomic smoothness (Fig. TF7-1). Most circuit designs (like your processor) fit into a few square centimeters of silicon area; each self-contained area is known as a ***die***. After fabrication, the wafer is cut to produce independent, rectangular dies often known as ***chips***, which are then packaged to produce the final component you buy at the store.

Figure TF7-1: A single 4-inch silicon wafer. Note the wafer's mirror-like surface. (Courtesy of Veljko Milanovic.)

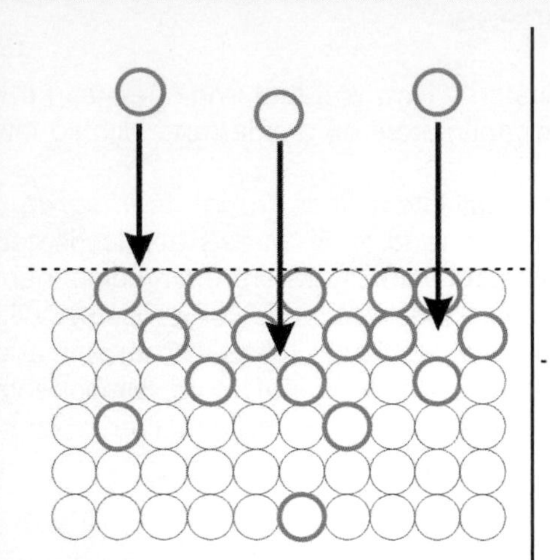

(a) ***Implantation:*** High-energy ions are driven into the silicon. Most become lodged in the first few nanometers, with decreasing concentration away from the surface. In this example, boron (an electron donor) is implanted into a silicon substrate.

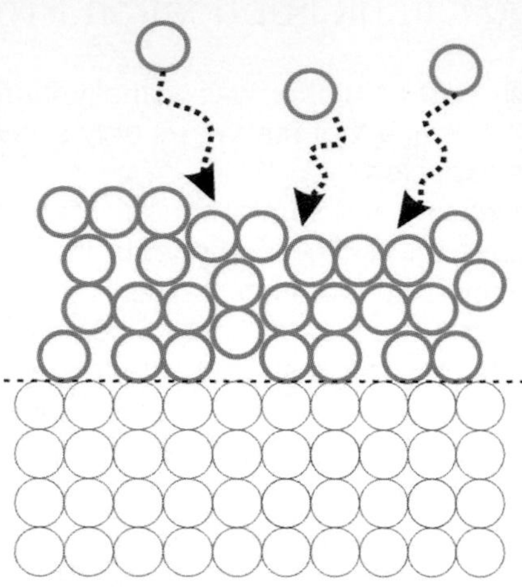

(b) ***Deposition:*** Atoms (or molecules) impact the surface but do not have the energy required to penetrate the surface. They accumulate on the surface in ***thin films***. In this example, aluminum is deposited in a conductive film onto the silicon.

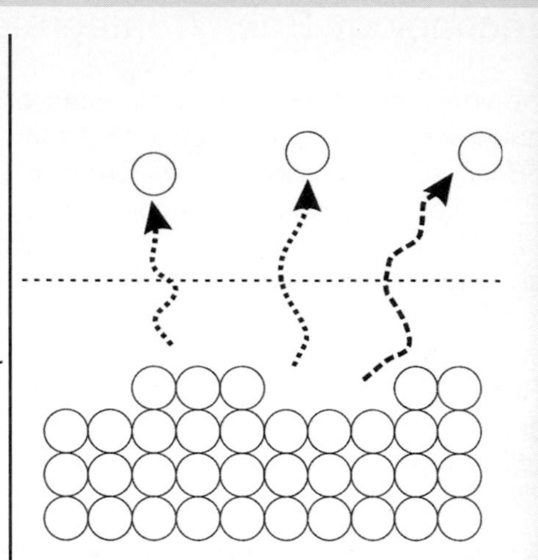

(a) ***Etching:*** Chemical, mechanical, or high-energy plasma methods are used to remove silicon (or other material) from the surface. In this example, silicon is etched away from the substrate.

Figure TF7-2: Cross-section of basic fabrication processes. The dashed line in each drawing indicates the original surface of the wafer.

A specific sequence or ***process*** of chemical and mechanical modifications is performed on certain areas of the wafer. Although complex processes employ a variety of techniques, a basic IC process will employ one of the following three modifications to the wafer:

- ***Implantation***: Atoms or molecules are added to the silicon wafer, changing its electronic properties (Fig. TF7-2(a)).
- ***Deposition***: Materials such as metals, insulators, or semiconductors are added in thin layers (like painting) onto the wafer (Fig. TF7-2(b)).
- ***Etching***: Material is removed from the wafer through chemical reactions or mechanical motion (Fig. TF7-2(c)).

Lithography

When building a multi-component IC, we need to perform different modifications to differents areas of the wafer. We may want to etch some areas and add metal to others, for example. The method by which we define which areas will be modified is known as ***lithography***.

Lithography has evolved much over the last 40 years and will continue to do so. Modern lithography employs all of the basic principles described below, but uses complex computation, specialized materials, and optical devices to achieve the very high resolutions required to reach modern feature sizes.

At its heart, lithography is simply a stencil process. In an old-fashioned stencil process, when a plastic sheet with cut-out letters or numbers is laid on a flat surface and painted, only the cutout areas would be painted. Once the stencil is removed, the design left behind consists of only the painted areas with clean edges and a uniform surface. With that in mind, consider Fig. TF7-3. Given a flat wafer, we first apply a thin coating of liquid polymer known as ***photoresist (PR)***. This layer usually is several hundred nanometers thick and is applied by placing a drop in the center of the wafer and then spinning the wafer very fast (1000 to 5000 rpm) so that the drop spreads out evenly over the surface. Once coated, the PR is heated (usually between 60 to 100°C) in a process known as ***baking***; this allows the PR to solidify

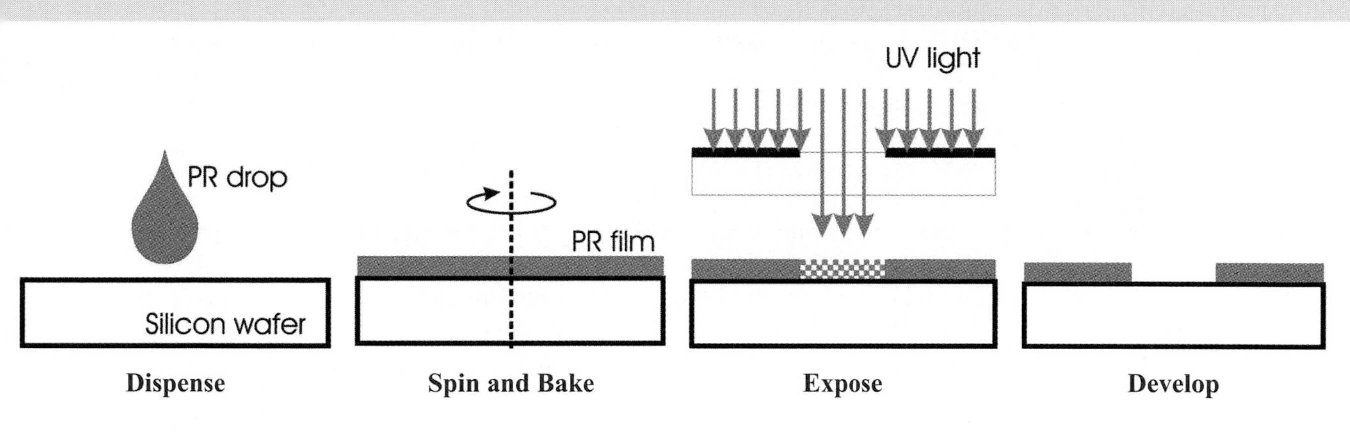

Figure TF7-3: Basic lithography steps.

lightly to a plastic-like consistency. Once baked and when exposed to ultraviolet (UV) light, the bonds that hold the R molecules together are "chopped" up; this makes it easy to wash away the UV-exposed areas (some varieties of R behave in exactly the opposite manner: UV light makes the PR very strong or cross-linked, but we will ignore that echnique here). In lithography, UV light is focused through a glass plate with patterns on it; this is known as ***exposure***. hese patterns act as a "light stencil" for the PR. Wherever UV light hits the PR, that area subsequently can be washed way in a process called ***development***. After development, the PR film remains behind with holes in certain areas.

How is this helpful? Let's look at how the modifications presented earlier can be masked with PR to produce atterned effects (Fig. TF7-4). In each case, we first use lithography to pattern areas onto the wafer (Fig. TF7-4(a)) hen we perform one of our three processes (Fig. TF7-4(b)), and finally, we use a strong solvent such as acetone (nail olish remover) to completely wash away the PR (Fig. TF7-4(c)). The PR allows us to implant, deposit, or etch only in efined areas.

abricating a Diode

n Section 2-7, we discussed the functional performance of the diode as a circuit component. Here, we will examine riefly how a diode is fabricated. Similar but more complex multi-step processes are used to make transistors and ntegrated circuits. Conceptually, the simplest diode is made from two slabs of silicon—each implanted with different toms—pressed together such that they share a boundary (Fig. TF7-5). The n and p areas are pieces of silicon that

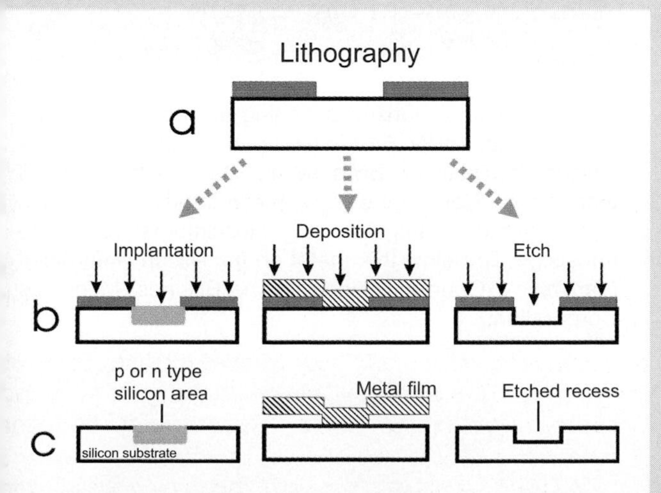

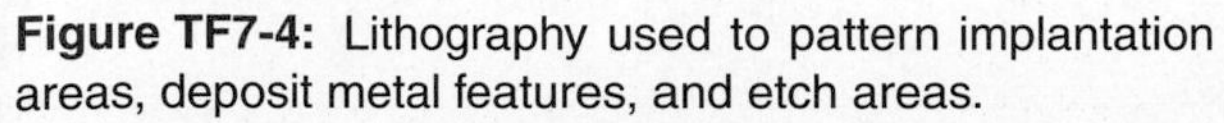

Figure TF7-4: Lithography used to pattern implantation areas, deposit metal features, and etch areas.

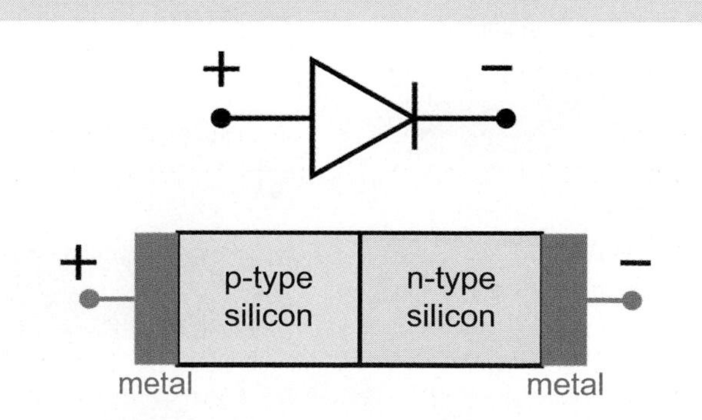

Figure TF7-5: The basic diode (top) circuit symbol and (bottom) conceptual depiction of the physical structure.

have been implanted with atoms (known as impurities) that increase or decrease the number of electrons capable of flowing freely through the silicon. This changes the semiconducting properties of the silicon and creates an electrically active boundary (called a ***junction***) between the n and the p areas of silicon. If both the n and p pieces of silicon are connected to metal wires, this two-terminal device exhibits the diode i–v curve shown in Fig. 2-35(c).

Figure TF7-6 shows the process for making a single diode. Only one step needs further definition: ***oxidation***. During oxidation, the silicon wafer is heated to $> 1000°$C in an oxygen atmosphere. At this temperature, the oxygen atoms and the silicon react and form a layer of SiO_2 on the surface (this layer is often called an ***oxide layer***). SiO_2 is a type of glass and is used as an insulator.

Wires are made by depositing metal layers on top of the device; these are called ***interconnects***. Modern ICs have 6 to 7 such interconnect layers (Fig. TF7-7). These layers are used to make electrical connections between all of the various components in the IC in the same way that macroscopic wires are used to link components on a breadboard.

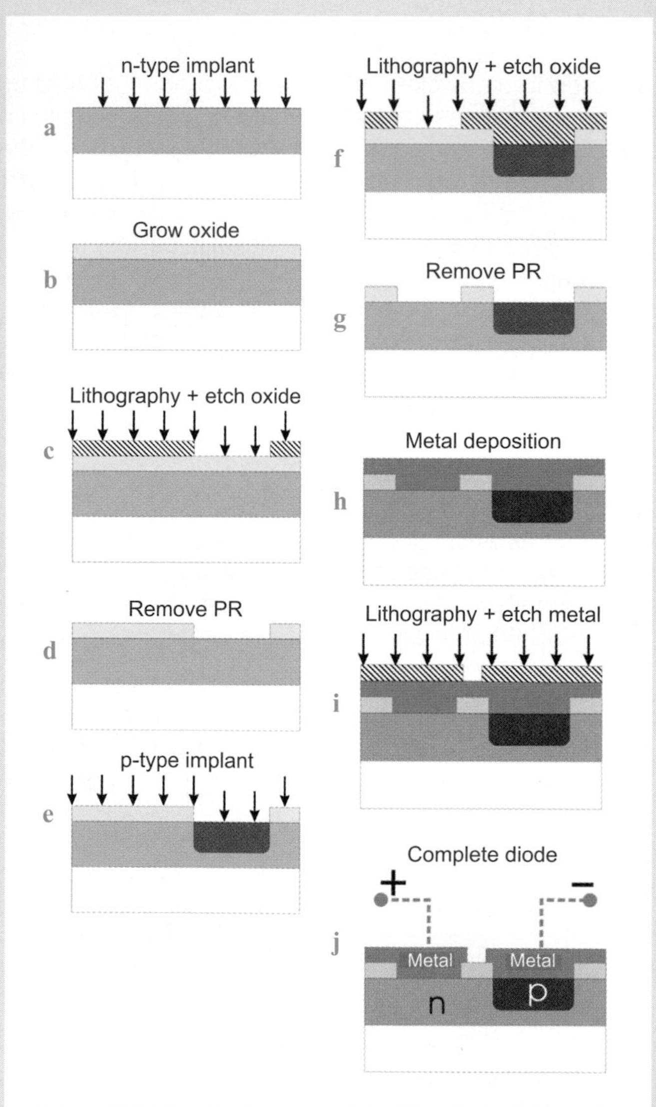

Figure TF7-6: A simple pn-junction diode fabrication process.

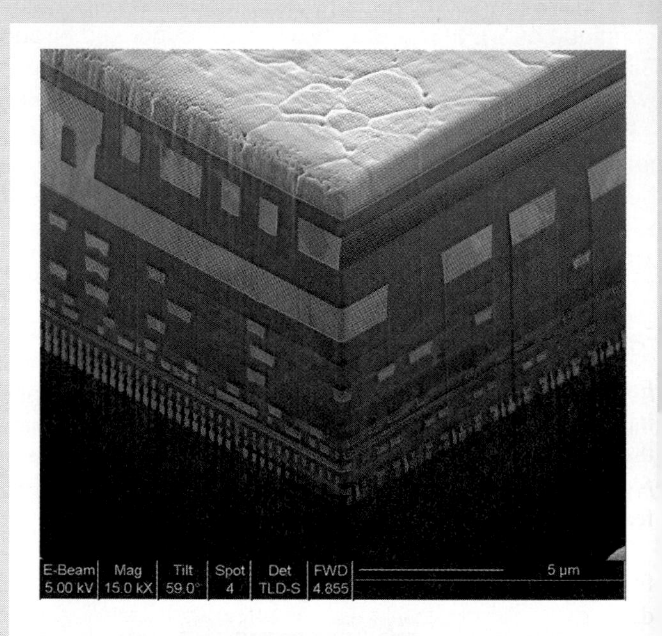

Figure TF7-7: Colorized scanning electron-microscope cross section of a 64-bit high-performance microprocessor chip built in IBM's 90-nm Server-Class CMOS technology. Note that several metal interconnect levels are used (metal lines are orange, insulator is green); the transistors lie below this metal on the silicon wafer itself (dark blue). (Courtesy of International Business Machines Corporation.)

Review Question 4-1: How is the linear range of an op amp defined?

Review Question 4-2: What is the difference between the op-amp gain A and the circuit gain G?

Review Question 4-3: An op amp is characterized by three important input–output attributes. What are they?

Exercise 4-1: In the circuit of Example 4-1 shown in Fig. 4-5, insert a series resistance R_s between v_s and v_p and then repeat the solution to obtain an expression for G. Evaluate G for $R_s = 10\ \Omega$ and use the same values listed in Example 4-1 for the other quantities. What impact does the insertion of R_s have on the magnitude of G?

Answer:

$$G = \frac{[A(R_i + R_s)(R_1 + R_2) + R_2 R_o]}{[AR_2(R_i + R_s) + R_o(R_2 + R_i + R_s) + R_1 R_2 + (R_i + R_s)(R_1 + R_2)]}$$

$$= 4.999977 \qquad \text{(negligible impact).}$$

(See)

4-2 Negative Feedback

Feedback refers to taking a part of the output signal and feeding it back into the input. It is called ***positive feedback*** if it increases the intensity of the input signal, and it is called ***negative feedback*** if it decreases it. Negative feedback is an essential feature of op-amp circuits.

Why do op-amp circuits need feedback and why negative feedback specifically? It seems counter-intuitive to want to decrease the input signal when the intent is to amplify it! We will answer this question by examining the circuit of Example 4-1 in some detail.

Were we to solve Eqs. (4.5) to (4.7) to find an expression for v_n and then evaluate it for the specified numerical values of the circuit parameters, we would have ended up with

$$v_n = 0.999995 v_s. \tag{4.11}$$

With $v_p = v_s$, the net voltage at the input to the op amp is

$$v_i = v_p - v_n = v_s - 0.999995 v_s \simeq 5 \times 10^{-6} v_s. \tag{4.12}$$

Hence, the feedback provided by the wire connecting terminal b in the output circuit to v_n at the input side in effect reduces the input signal from v_s to v_i with the latter being only 5×10^{-6} of the former. What is the payoff?

The op amp uses a supply voltage $V_{cc} = 10$ V. This means that the linear dynamic range for v_o is from -10 V to $+10$ V. Since according to Eq. (4.9) $v_o = 5v_s$, it follows that the corresponding ***linear dynamic range*** for v_s is from -2 V to $+2$ V. Thus, the payoff is that the op-amp circuit in Fig. 4-5 can amplify the signal v_s by a factor of 5 over the full range bounded by ± 2 V.

To appreciate the role of feedback, let us now consider what would have happened had we not had a feedback connection between the output and input sides of the circuit. The circuit in Fig. 4-6 is identical to the one we examined earlier in Example 4-1, except that now terminal v_n is connected to ground instead of to terminal b. Hence,

$$v_p = v_s, \tag{4.13a}$$

$$v_n = 0, \tag{4.13b}$$

and voltage division at the output side gives

$$v_o = A(v_p - v_n)\left(\frac{R_1 + R_2}{R_o + R_1 + R_2}\right) = A\left(\frac{R_1 + R_2}{R_o + R_1 + R_2}\right) v_s. \tag{4.14}$$

For $A = 10^6$, $R_o = 10\ \Omega$, $R_1 = 80$ kΩ, and $R_2 = 20$ kΩ,

$$v_o \simeq 10^6 v_s, \tag{4.15}$$

which means that the circuit gain $G = v_o/v_s = 10^6$. This is a huge gain, but it comes with a severe limitation on the range of values that v_s can assume. Since v_o is bounded by ± 10 V, the linear range of v_s will have to be bounded by $\pm 10\ \mu$V. In the absence of feedback, the circuit gain is very large, but the linear range of the input signal is limited to microvolts.

Application of negative feedback offers a tradeoff between circuit gain and dynamic range.

In the case of the circuit in Example 4-1, $G = 5$, which is 2×10^5 times smaller than the gain of the circuit with no feedback. On the other hand, the dynamic range with feedback is 2×10^5 times greater than without feedback. Other choices of the ratio $(R_1 + R_2)/R_2$ offer other options for tradeoff between circuit gain and dynamic range.

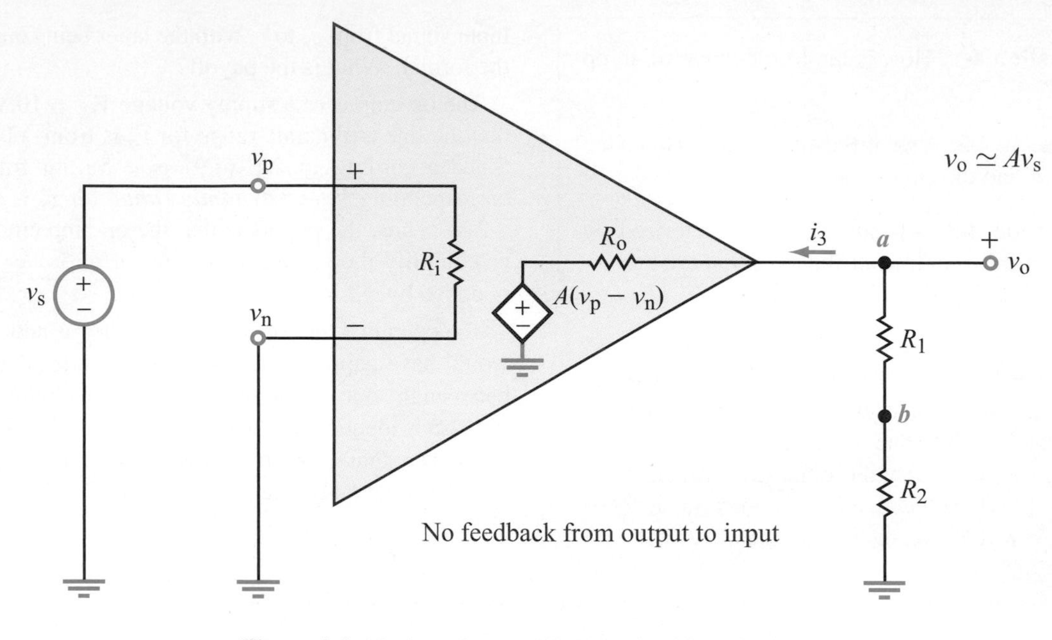

Figure 4-6: Noninverting amplifier circuit with no feedback.

Review Question 4-4: Why is negative feedback used in op-amp circuits?

Review Question 4-5: How large is the circuit gain G in the absence of feedback? How large is it with 100 percent feedback (equivalent to setting $R_1 = 0$ in the circuit of Fig. 4-5)?

Exercise 4-2: To evaluate the tradeoff between the circuit gain G and the linear dynamic range of v_s, apply Eq. (4.8) to find the magnitude of G and then determine the corresponding dynamic range of v_s for each of the following values of R_2: 0 (no feedback), 800 Ω, 8.8 kΩ, 40 kΩ, 80 kΩ, and 1 MΩ. Except for R_2, all other quantities remain unchanged.

Answer:

R_2	G	v_s Range
0	10^6	−10 μV to +10 μV
800 Ω	101	−99 mV to +99 mV
8.8 kΩ	10.1	−0.99 V to +0.99 V
40 kΩ	3	−3.3 V to +3.3 V
80 kΩ	2	−5 V to +5 V
1 MΩ	1.08	−9.26 V to +9.26 V

(See)

4-3 Ideal Op-Amp Model

We noted in Section 4-1 that the op amp has a very large input resistance R_i on the order of 10^7 Ω, a relatively small output resistance R_o on the order of 1–100 Ω, and an open-loop gain $A \simeq 10^6$. Usually, the series resistances of the input circuit connected to terminals v_p and v_n are several orders of magnitude smaller than R_i. Consequently, not only will very little current flow through the input circuit, but also the voltage drop across the input-circuit resistors will be negligibly small in comparison with the voltage drop across R_i. These considerations allow us to simplify the equivalent circuit of the op amp by replacing it with the ideal op-amp circuit model shown in Fig. 4-7, in which R_i has been replaced with on open circuit. An open circuit between terminals v_p and v_n implies the following ***ideal op-amp current constraint***:

$$i_p = i_n = 0 \qquad \text{(Ideal op-amp model).} \tag{4.16}$$

In reality, i_p and i_n are very small but not identically zero; for if they were, there would be no amplification through the op amp. Nevertheless, the current condition given by Eq. (4.16) will prove quite useful.

Similarly, at the output side, if the load resistor connected in series with R_o is several orders of magnitude larger than R_o, then R_o can be ignored by setting it equal to zero. Finally, in the ideal op-amp model, the large open-loop gain A is made infinite—the consequence of which is that

$$v_p - v_n = \frac{v_o}{A} \to 0 \qquad \text{as } A \to \infty.$$

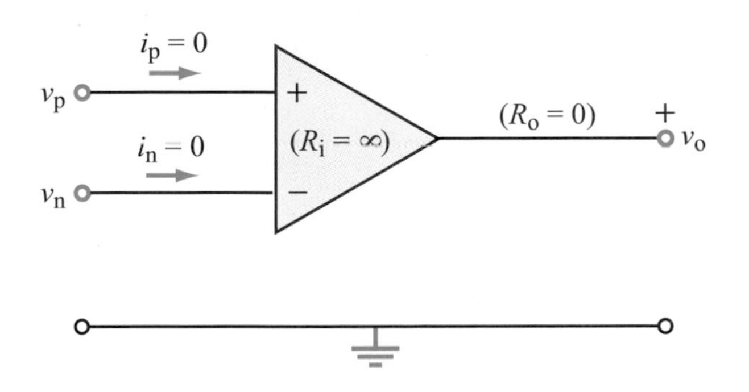

Figure 4-7: Ideal op-amp model.

Hence, we obtain the ***ideal op-amp voltage constraint***

$$v_p = v_n \qquad \text{(Ideal op-amp model).} \tag{4.17}$$

In summary:

The ideal op-amp model characterizes the op amp in terms of an equivalent circuit in which $R_i = \infty$, $R_o = 0$, and $A = \infty$.

The operative consequences are given by Eqs. (4.16) and (4.17) and in Table 4-2.

Table 4-2: Characteristics of the ideal op-amp model.

Ideal Op Amp	
• Current constraint	$i_p = i_n = 0$
• Voltage constraint	$v_p = v_n$
• $A = \infty$ $\quad R_i = \infty$	$R_o = 0$

To illustrate the utility of the ideal op-amp model, let us re-examine the circuit we analyzed earlier in Example 4-1, but we will do so this time using the ideal model. The new circuit, as shown in Fig. 4-8, includes a source resistance R_s, but because the op amp draws no current ($i_p = 0$), there is no voltage drop across R_s. Hence,

$$v_p = v_s, \tag{4.18}$$

and on the output side, v_o and v_n are related through voltage division by

$$v_o = \left(\frac{R_1 + R_2}{R_2}\right) v_n. \tag{4.19}$$

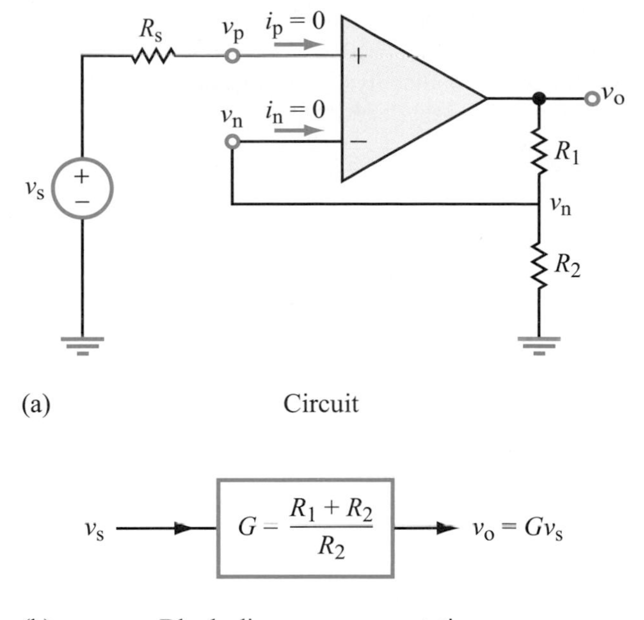

Figure 4-8: Noninverting amplifier circuit: (a) using ideal op-amp model and (b) equivalent block-diagram representation.

Using these two equations, in conjunction with $v_p = v_n$ (from Eq. (4.17)), we end up with the following result for the circuit gain G:

$$G = \frac{v_o}{v_s} = \left(\frac{R_1 + R_2}{R_2}\right), \tag{4.20}$$

which is identical with Eq. (4.10). *Unless stated otherwise, from here on forward, we will use the ideal op-amp model exclusively.*

The ***input resistance*** of the noninverting amplifier circuit is the Thévenin resistance of the op-amp circuit seen by the input source v_s. Because $i_p = 0$, it is easy to show that $R_{\text{input}} = R_{\text{Th}} = R_i = \infty$.

Review Question 4-6: What are the current and voltage constraints of the ideal op amp?

Review Question 4-7: What are the values of the input and output resistances of the ideal op amp?

Review Question 4-8: In the ideal op-amp model, R_o is set equal to zero. To satisfy such an approximation, does the load resistance need to be much larger or much smaller than R_o? Explain.

Exercise 4-3: Consider the noninverting amplifier circuit of Fig. 4-8(a) under the conditions of the ideal op-amp model. Assume $V_{cc} = 10$ V. Determine the value of G and the corresponding dynamic range of v_s for each of the following values of R_1/R_2: 0, 1, 9, 99, 10^3, 10^6.

Answer:

R_1/R_2	G	v_s Range
0	1	-10 V to $+10$ V
1	2	-5 V to $+5$ V
9	10	-1 V to $+1$ V
99	100	-0.1 V to $+0.1$ V
1000	~ 1000	-10 mV to $+10$ mV (Approx.)
10^6	$\sim 10^6$	$-10\ \mu$V to $+10\ \mu$V (Approx.)

4-4 Inverting Amplifier

In an inverting amplifier op-amp circuit, the input source is connected to terminal v_n (instead of to terminal v_p) through an ***input source resistance*** R_s, and terminal v_p is connected to ground. Feedback from the output continues to be applied at v_n (through a ***feedback resistance*** R_f), as shown in Fig. 4-9. It is called an *inverting* amplifier because (as we will see shortly) the circuit gain G is negative.

To relate the output voltage v_o to the input signal voltage v_s, we start by writing down the node-voltage equation at terminal v_n as

$$i_1 + i_2 + i_n = 0 \tag{4.21}$$

or

$$\frac{v_n - v_s}{R_s} + \frac{v_n - v_o}{R_f} + i_n = 0. \tag{4.22}$$

Upon invoking the op-amp current constraint given by Eq. (4.16), namely $i_n = 0$, and the voltage constraint $v_n = v_p$, as well as recognizing that $v_p = 0$ (because terminal v_p is connected to ground), we obtain the relationship

$$v_o = -\left(\frac{R_f}{R_s}\right) v_s. \tag{4.23}$$

The circuit voltage gain of the inverting amplifier therefore is given by

$$G = \frac{v_o}{v_s} = -\left(\frac{R_f}{R_s}\right). \tag{4.24}$$

In addition to amplifying v_s by the ratio (R_f/R_s), the inverting amplifier also reverses the polarity of v_s.

Inverting Amplifier

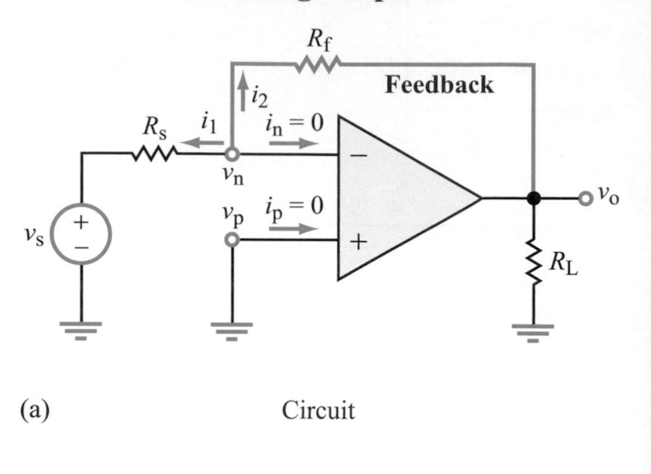

(a) Circuit

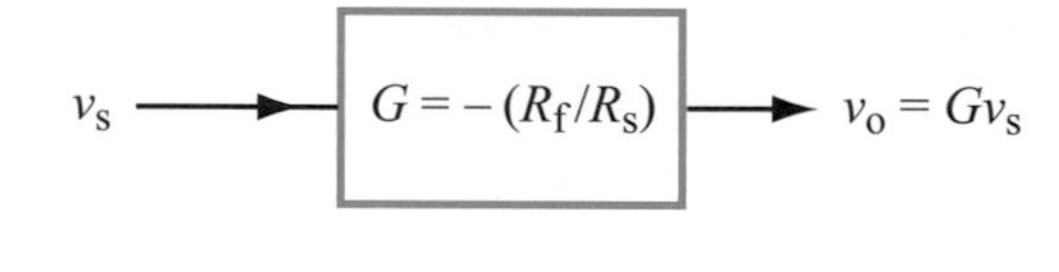

(b) Block diagram

Figure 4-9: Inverting amplifier circuit and its block-diagram equivalent.

It is important to note that v_o is independent of the magnitude of the load resistance R_L, so long as R_L is much larger than the op-amp output resistance R_o (which is an implicit assumption of the ideal op-amp model).

Because $v_n = 0$, a Thévenin analysis of the circuit in Fig. 4-9(a) would reveal that the ***input resistance*** of the inverting amplifier circuit (as seen by source v_s) is $R_{input} = R_{Th} = R_s$.

Example 4-2: Amplifier with Input Current Source

For the circuit shown in Fig. 4-10(a); (a) obtain an expression for the input–output transfer function $K_t = v_o/i_s$ and evaluate it for $R_1 = 1$ kΩ, $R_2 = 2$ kΩ, $R_f = 30$ kΩ, and $R_L = 10$ kΩ and (b) determine the linear dynamic range of i_s if $V_{cc} = 20$ V.

Solution:

(a) Application of the source transformation method converts the combination of i_s and R_2 into a voltage source $v_s = i_s R_2$, in series with a resistance R_2. Upon combining R_2 in series with R_1, we obtain the new circuit shown in Fig. 4-10(b), which is identical in form with the inverting amplifier circuit of Fig. 4-9, except that now the source resistance is

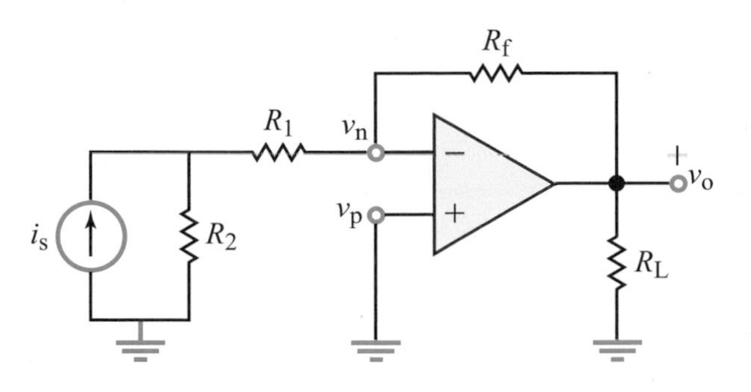

(a) Original circuit

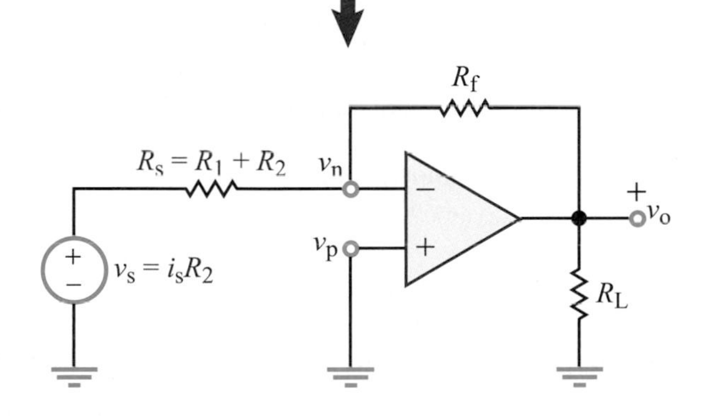

(b) After source transformation

Figure 4-10: Inverting amplifier circuit of Example 4-2.

$R_s = (R_1 + R_2)$. Hence, application of Eq. (4.23) gives

$$v_o = -\left(\frac{R_f}{R_1 + R_2}\right) v_s = -\left(\frac{R_f}{R_1 + R_2}\right) R_2 i_s, \tag{4.25}$$

from which we obtain the transfer function

$$K_t = \frac{v_o}{i_s} = -\frac{R_f R_2}{R_1 + R_2}. \tag{4.26}$$

For $R_1 = 1$ kΩ, $R_2 = 2$ kΩ, and $R_f = 30$ kΩ,

$$K_t = \frac{v_o}{i_s} = -2 \times 10^4 \quad \text{(V/A)}.$$

(b) From the expression for K_t,

$$i_s = -\frac{v_o}{2 \times 10^4},$$

and since $|v_o|$ is bounded by $V_{cc} = 20$ V, the linear range for i_s is bounded by

$$|i_s| = \left|\frac{V_{cc}}{2 \times 10^4}\right| = \left|\frac{20}{2 \times 10^4}\right| = 1 \text{ mA}.$$

Thus, the linear range of i_s extends from -1 mA to $+1$ mA.

Example 4-3: Thévenin Equivalent

For the op-amp circuit shown in Fig. 4-11(a), obtain the Thévenin equivalent of the circuit at (a) its input terminals (a, b) as it would appear to the input source (v_s, R_s) and (b) at its output terminals (c, d) as it would appear to the load resistor R_L.

Solution:

(a) The circuit to the right of terminals (a, b) contains no independent sources (except for the supply voltages V_{cc} and $-V_{cc}$, which supply dc power to the op amp as well as specify its saturation levels but do not control its input–output transfer characteristics). Hence,

$$v_{Th_1} = 0 \qquad @ \text{ terminals } (a, b).$$

To find R_{Th_1} at terminals (a, b), we apply the external-source method of Section 3-5.5, as shown in Fig. 4-11(b). Since $i_p = 0$, i_{ex} flows only through R_1. Consequently,

$$i_{ex} = \frac{v_{ex}}{R_1},$$

and

$$R_{Th_1} = \frac{v_{ex}}{i_{ex}} = R_1 \qquad @ \text{ terminals } (a, b).$$

Hence, R_1 is the *input resistance* of the op-amp circuit at terminals (a, b). Had R_1 not been there (i.e., R_1 = open circuit), R_{Th_1} would have been infinite, which is to be expected since the input resistance (between v_n and v_p) of an ideal op amp is infinite.

(b) The configuration in Fig. 4-11(c) represents the perspective as seen by R_L at terminals (c, d). The Thévenin voltage v_{Th_2} is the open-circuit voltage at terminals (c, d). Since $i = 0$ (an open circuit exists between terminals (c, d)), no voltage drop exists across R_4, and therefore,

$$v_{Th_2} = v_o.$$

At the input side, voltage division gives

$$v_p = \frac{v_s R_1}{R_s + R_1},$$

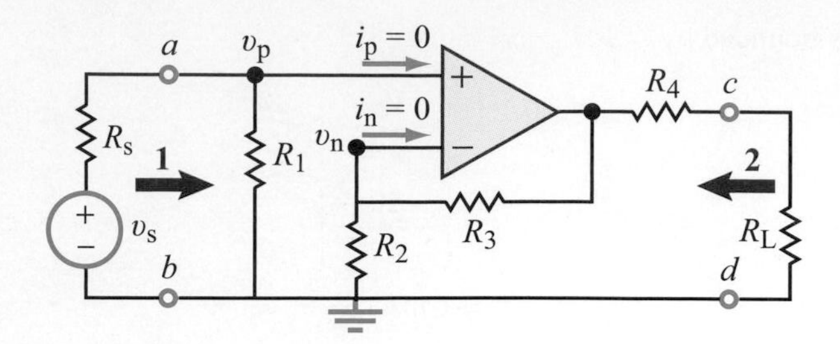

(a) Op-amp circuit

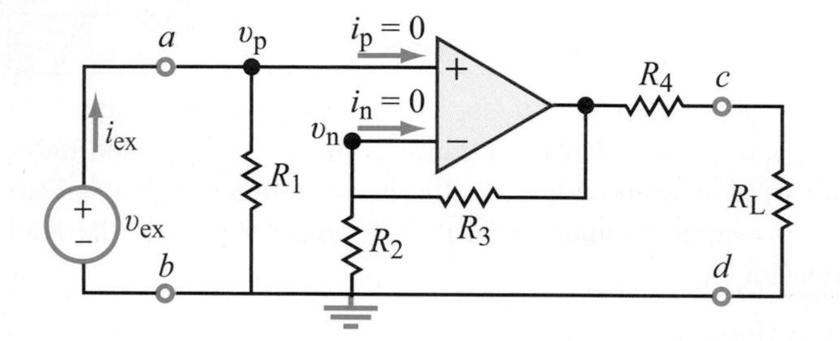

(b) $i_{ex} = v_{ex}/R_1$

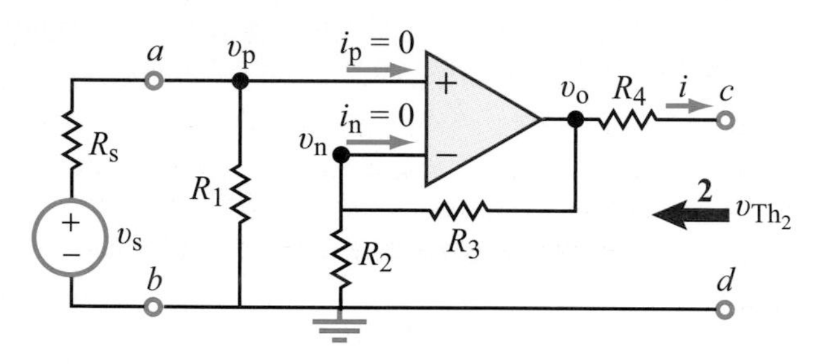

(c) Looking to the left of (c, d)

Figure 4-11: Circuit for Example 4-3.

and similarly, at the output side,

$$v_n = \frac{v_o R_2}{R_2 + R_3}.$$

Applying the ideal op-amp condition $v_p = v_n$ leads to

$$v_{Th_2} = v_o = \left(\frac{R_2 + R_3}{R_2}\right)\left(\frac{R_1}{R_1 + R_s}\right) v_s$$

$$@ \text{ terminals } (c, d).$$

Next, we need to find the Thévenin resistance. If we connect terminal c to terminal d with a short circuit, the current i becomes i_{sc}, and its magnitude would be given by

$$i_{sc} = \frac{v_o}{R_4}.$$

Moreover, the expressions for v_n, v_p, and v_o will remain unchanged. Consequently, application of the open-circuit/short-circuit method of Section 3-5.3 gives

$$R_{Th_2} = \frac{v_{Th_2}}{i_{sc}} = \frac{v_o}{i_{sc}} = R_4 \qquad @ \text{ terminals } (c, d).$$

Review Question 4-9: How does feedback control the gain of the inverting-amplifier circuit?

Review Question 4-10: The expression given by Eq. (4.24) states that the gain of the inverting amplifier is independent of the magnitude of R_L. Would the expression remain valid if $R_L = 0$? Explain.

Exercise 4-4: The input to an inverting-amplifier circuit consists of $v_s = 0.2$ V and $R_s = 10\ \Omega$. If $V_{cc} = 12$ V, what is the maximum value that R_f can assume before saturating the op amp?

Answer: $G_{max} = -60$, $R_f = 600\ \Omega$. (See)

4-5 Summing Amplifier

By connecting multiple sources in parallel at terminal v_n of the inverting amplifier, the circuit becomes an ***adder*** (or more precisely a ***scaled inverting adder***). We will demonstrate how such a circuit (usually called a ***summing amplifier***) works for two input voltages v_1 and v_2, and then we will generalize the concept to multiple sources.

For the circuit shown in Fig. 4-12(a), our goal is to relate the output voltage v_o to v_1 and v_2. To do so, we first will apply the source-transformation technique so as to cast the input circuit in the form of a single voltage source v_s in series with a source resistance R_s. The steps involved in the transformation are illustrated in Fig. 4-12(b) and (c). Voltage to current transformation gives $i_{s_1} = v_1/R_1$ and $i_{s_2} = v_2/R_2$, which can be combined together into a single current source as

$$i_s = i_{s_1} + i_{s_2} = \frac{v_1}{R_1} + \frac{v_2}{R_2} = \frac{v_1 R_2 + v_2 R_1}{R_1 R_2}. \tag{4.27}$$

Similarly, the two parallel resistors add up to

$$R_s = \frac{R_1 R_2}{R_1 + R_2}. \tag{4.28}$$

If we transform (i_s, R_s) into a voltage source (v_s, R_s), we get

$$v_s = i_s R_s = \left(\frac{v_1 R_2 + v_2 R_1}{R_1 R_2}\right) \frac{R_1 R_2}{R_1 + R_2} = \frac{v_1 R_2 + v_2 R_1}{R_1 + R_2}. \tag{4.29}$$

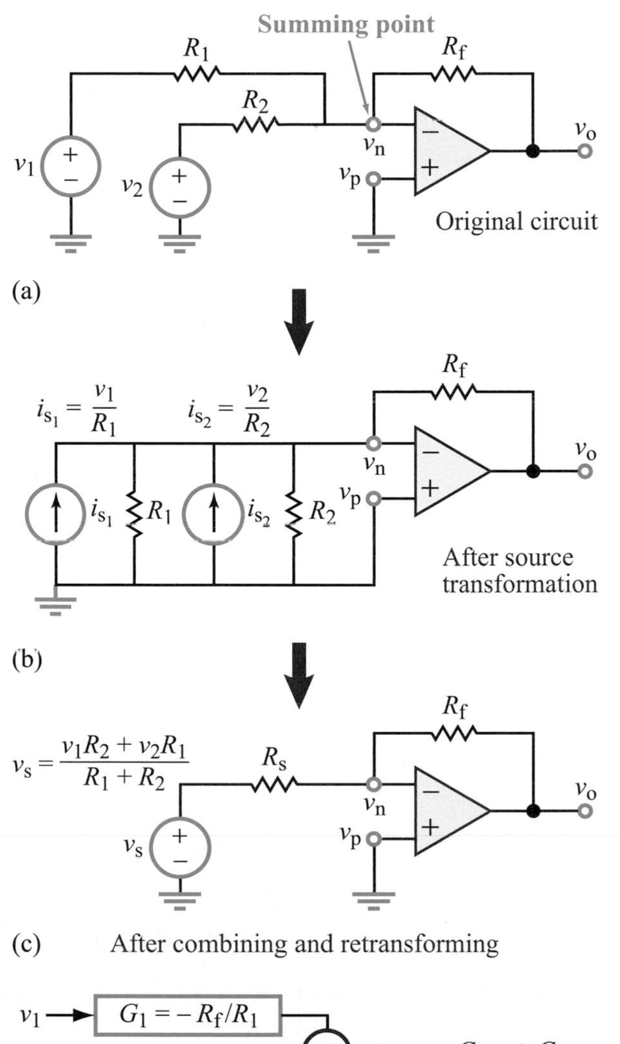

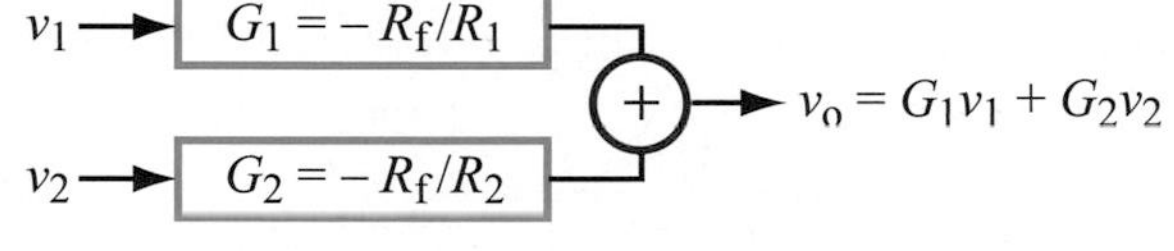

Figure 4-12: Summing amplifier.

he circuit in Fig. 4-12(c) is identical in form with that of the nverting amplifier of Fig. 4-9. Hence, by applying the input–utput voltage relationship given by Eq. (4.23), we have

$$v_o = -\left(\frac{R_f}{R_s}\right) v_s = -\frac{R_f}{\left(\frac{R_1 R_2}{R_1 + R_2}\right)} \left(\frac{v_1 R_2 + v_2 R_1}{R_1 + R_2}\right)$$

$$= -\left(\frac{R_f}{R_1}\right) v_1 - \left(\frac{R_f}{R_2}\right) v_2. \tag{4.30}$$

This expression for v_o can be written in the form

$$v_o = G_1 v_1 + G_2 v_2, \tag{4.31}$$

where $G_1 = -(R_f/R_1)$ is the (negative) gain applied to source voltage v_1, and $G_2 = -(R_f/R_2)$ is the gain applied to v_2. Thus:

The summing amplifier scales v_1 by G_1 and v_2 by G_2 and adds them together.

For the special case where $R_1 = R_2 = R$,

$$v_o = -\left(\frac{R_f}{R}\right)[v_1 + v_2] \qquad \text{(equal gain)}, \tag{4.32}$$

and if additionally, $R_f = R_1 = R_2$, then $G_1 = G_2 = -1$. In this case, the summing amplifier becomes an inverted adder as characterized by

$$v_o = -(v_1 + v_2) \qquad \text{(inverted adder)}. \tag{4.33}$$

Generalizing to the case where the input consists of n input voltage sources v_1 to v_n (associated with source resistances R_1 to R_n, respectively) and all are connected in parallel at the same summing point (terminal v_n), the output voltage becomes

$$v_o = \left(-\frac{R_f}{R_1}\right) v_1 + \left(-\frac{R_f}{R_2}\right) v_2 + \cdots + \left(-\frac{R_f}{R_n}\right) v_n. \tag{4.34}$$

Example 4-4: Summing Circuit

Design a circuit that performs the operation

$$v_o = 4v_1 + 7v_2.$$

Solution: The desired circuit has to amplify v_1 by a factor of 4, amplify v_2 by a factor of 7, and add the two together. A summing amplifier can do that, but it also inverts the sum. Hence, we will need to use a two-stage circuit with the first stage providing the desired operation within a "−" sign and then follow it up with an inverting amplifier with a gain of (−1). The two-stage circuit is shown in Fig. 4-13.

For the first stage, we need to select values for R_1, R_2, and R_{f_1} such that

$$\frac{R_{f_1}}{R_1} = 4 \qquad \text{and} \qquad \frac{R_{f_1}}{R_2} = 7.$$

Since we have only two constraints, we can satisfy the specified ratios with an infinite number of combinations. Arbitrarily,

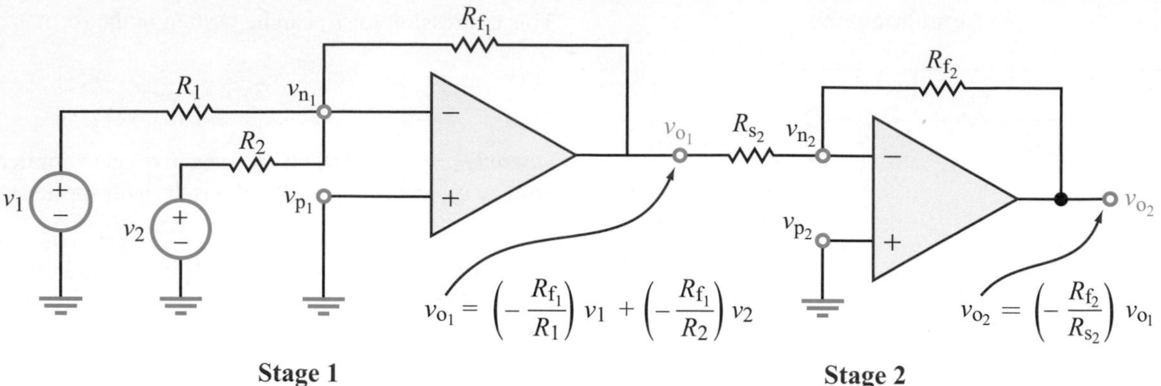

(a) Two-stage circuit

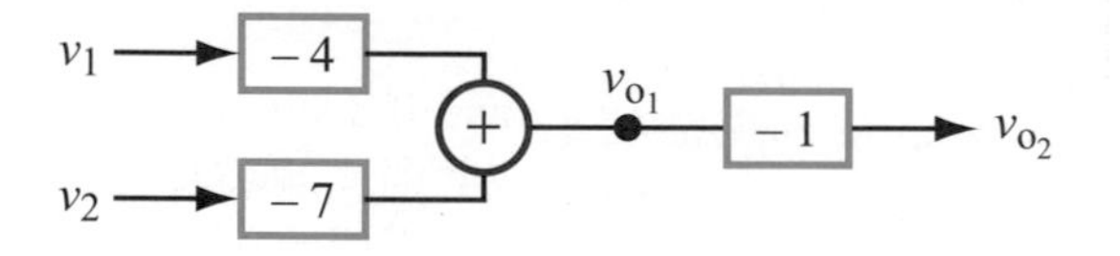

(b) Block diagram

Figure 4-13: Two-stage circuit realization of $v_o = 4v_1 + 7v_2$.

we will choose $R_{f_1} = 56$ kΩ, which then specifies the other resistors as

$$R_1 = 14 \text{ k}\Omega \qquad \text{and} \qquad R_2 = 8 \text{ k}\Omega.$$

For the second stage, a gain of (-1) requires that

$$\frac{R_{f_2}}{R_{s_2}} = 1.$$

Arbitrarily, we choose $R_{f_2} = R_{s_2} = 20$ kΩ.

To perform the summing operation, the solution offered in Example 4-4 employed two inverting amplifier circuits—one to perform an inverted sum, and a second one to provide multiplication by (-1). Alternatively, the same result can be achieved by using a single op amp in a noninverting amplifier circuit, as shown in Fig. 4-14.

From our analysis in Section 4-3, we established that the output voltage v_o of the noninverting amplifier circuit is related to v_p by

$$\frac{v_o}{v_p} = G = \frac{R_1 + R_2}{R_2}. \tag{4.35}$$

For the circuit in Fig. 4-14, in view of the ideal op-amp constraint that the op amp draws no current ($i_p = 0$), it is a straightforward task to show that

$$v_p = \frac{v_1 R_{s_2} + v_2 R_{s_1}}{R_{s_1} + R_{s_2}}. \tag{4.36}$$

Combining Eqs. (4.35) and (4.36) leads to

$$v_o = G\left[\left(\frac{R_{s_2}}{R_{s_2} + R_{s_1}}\right) v_1 + \left(\frac{R_{s_1}}{R_{s_1} + R_{s_2}}\right) v_2\right]. \tag{4.37}$$

To realize a coefficient of 4 for v_1 and a coefficient of 7 for v_2, it is necessary that

$$\frac{G R_{s_2}}{R_{s_1} + R_{s_2}} = 4$$

and

$$\frac{G R_{s_1}}{R_{s_1} + R_{s_2}} = 7.$$

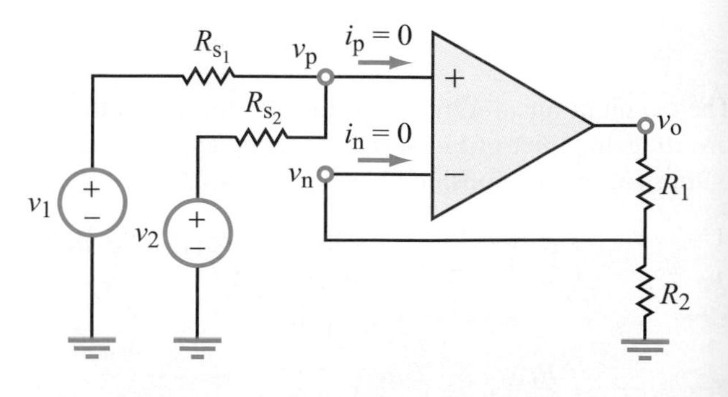

Figure 4-14: Noninverting summer.

A possible solution that satisfies these two constraints is $R_{s_1} = 7\ \text{k}\Omega$, $R_{s_2} = 4\ \text{k}\Omega$, and $G = 11$. Furthermore, the specified value of G can be satisfied by choosing $R_1 = 50\ \text{k}\Omega$ and $R_2 = 5\ \text{k}\Omega$.

> **Review Question 4-11:** What type of op-amp circuits (inverting, noninverting, and others) might one use to perform the operation $v_o = G_1 v_1 + G_2 v_2$ with G_1 and G_2 both positive?
>
> **Review Question 4-12:** What is an inverted adder?

> **Exercise 4-5:** The circuit shown in Fig. 4-13(a) is to be used to perform the operation
>
> $$v_o = 3v_1 + 6v_2.$$
>
> If $R_1 = 1.2\ \text{k}\Omega$, $R_{s_2} = 2\ \text{k}\Omega$, and $R_{f_2} = 4\ \text{k}\Omega$, select values for R_2 and R_{f_1} so as to realize the desired result.
>
> **Answer:** $R_{f_1} = 1.8\ \text{k}\Omega$, $R_2 = 600\ \Omega$. (See ◎)

4-6 Difference Amplifier

When an input signal v_2 is connected to terminal v_p of a noninverting amplifier circuit, the output is a scaled version of v_2. A similar outcome is generated by an inverting amplifier circuit when an input voltage v_1 is connected to the op amp's v_n terminal, except that in addition to scaling v_1 its polarity is reversed as well. The ***difference amplifier*** circuit combines these two functions to perform ***subtraction***.

In the difference amplifier circuit of Fig. 4-15(a), the input signals are v_1 and v_2, R_2 is the feedback resistance, R_1 is the source resistance of v_1, and resistances R_3 and R_4 serve to control the scaling factor (gain) of v_2. To obtain an expression that relates the output voltage v_o to the inputs v_1 and v_2, we apply KCL at nodes v_n and v_p. At v_n, $i_1 + i_2 + i_n = 0$, which is equivalent to

$$\frac{v_n - v_1}{R_1} + \frac{v_n - v_o}{R_2} + i_n = 0. \tag{4.38}$$

At v_p, $i_3 + i_4 + i_p = 0$, or

$$\frac{v_p - v_2}{R_3} + \frac{v_p}{R_4} + i_p = 0. \tag{4.39}$$

Upon imposing the ideal op-amp constraints $i_p = i_n = 0$ and $v_p = v_n$, we end up with

$$v_o = \left[\left(\frac{R_4}{R_3 + R_4}\right)\left(\frac{R_1 + R_2}{R_1}\right)\right] v_2 - \left(\frac{R_2}{R_1}\right) v_1, \tag{4.40}$$

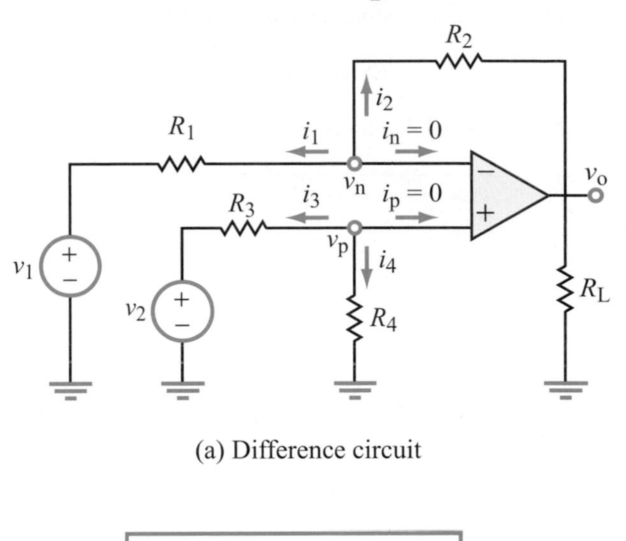

(a) Difference circuit

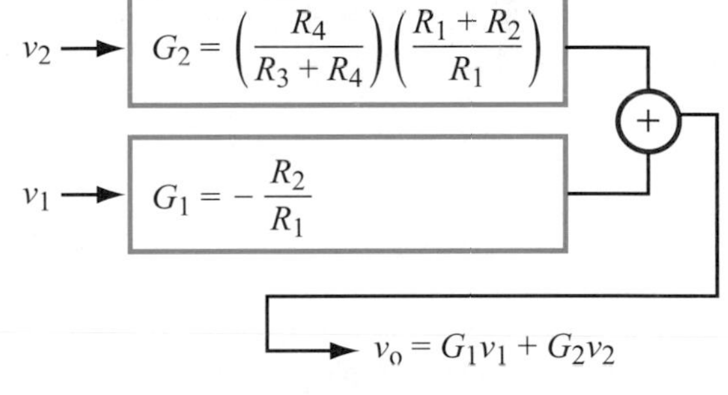

(b) Block diagram

Figure 4-15: Difference-amplifier circuit.

which can be cast in the form

$$v_o = G_2 v_2 + G_1 v_1, \tag{4.41}$$

where the scale factors (gains) are given by

$$G_2 = \left(\frac{R_4}{R_3 + R_4}\right)\left(\frac{R_1 + R_2}{R_1}\right) \tag{4.42a}$$

and

$$G_1 = -\left(\frac{R_2}{R_1}\right). \tag{4.42b}$$

According to Fig. 4-15(b) which is a block-diagram representation of the difference amplifier circuit:

> The difference amplifier scales v_2 by positive gain G_2, v_1 by negative gain G_1 and adds them together.

For the difference amplifier to function as a subtraction circuit with equal gain, its resistors have to be interrelated by

$$R_2 R_3 = R_1 R_4, \tag{4.43}$$

in which case Eq. (4.41) reduces to

$$v_o = \left(\frac{R_2}{R_1}\right)(v_2 - v_1) \qquad \text{(equal gain)}. \tag{4.44}$$

Exact subtraction with no scaling requires that $R_1 = R_2$.

Exercise 4-6: The difference-amplifier circuit of Fig. 4-15 is used to realize the operation

$$v_o = (6v_2 - 2) \text{ V}.$$

Given that $R_3 = 5$ kΩ, $R_4 = 6$ kΩ, and $R_2 = 20$ kΩ, specify values for v_1 and R_1.

Answer: $v_1 = 0.2$ V, $R_1 = 2$ kΩ. (See)

4-7 Voltage Follower

In electronic circuits, we often need to incorporate the functionality of a relatively simple (but important) circuit that serves to insulate the input source from variations in the load resistance R_L. Such a circuit is called a ***voltage follower*** or ***buffer***. To appreciate the utility of the voltage follower, let us first examine the circuit shown in Fig. 4-16(a). An input circuit represented by its Thévenin equivalent (v_s, R_s), is connected to a load R_L. The output voltage is

$$v_o = \frac{v_s R_L}{R_s + R_L} \qquad \text{(without voltage follower)}, \tag{4.45}$$

which obviously is dependent on both R_s and R_L. The voltage-follower circuit is inserted in between the input circuit and the load, as shown in Fig. 4-16(b). Because $i_p = 0$, $v_p = v_s$. Furthermore, in view of the op-amp constraint $v_p = v_n$ and because the output node is connected directly to v_n, it follows that

$$v_o = v_p = v_s \qquad \text{(with voltage follower)}, \tag{4.46}$$

and this is true regardless of the values of R_s and R_L (excluding R_s = open circuit and/or R_L = short circuit, either of which would invalidate the entire circuit). Thus:

The output of the voltage follower ***follows*** the input signal while remaining immune to changes in R_L.

Buffer (Voltage Follower)

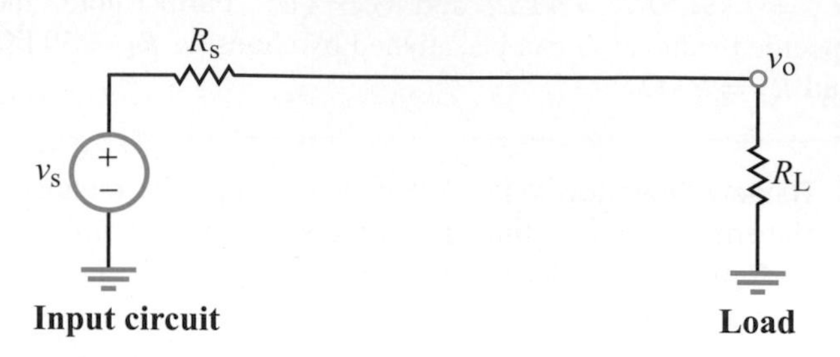

(a) Input circuit connected directly to a load

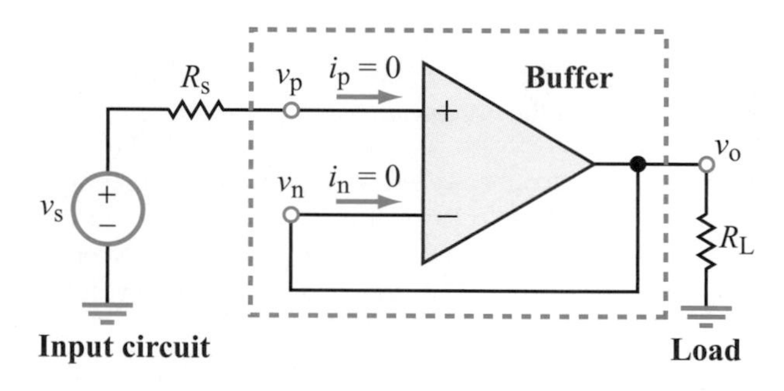

(b) Input circuit separated by a buffer

Figure 4-16: The voltage follower provides no voltage gain ($v_o = v_s$), but it insulates the input circuit from the load.

A circuit that offers this type of protection is often called a ***buffer***.

Review Question 4-13: What is the function of a voltage follower, and why is it called a "buffer"?

Review Question 4-14: How much voltage gain is provided by the voltage follower?

Exercise 4-7: Express v_o in terms of v_1, v_2, and v_3 for the circuit in Fig. E4.7.

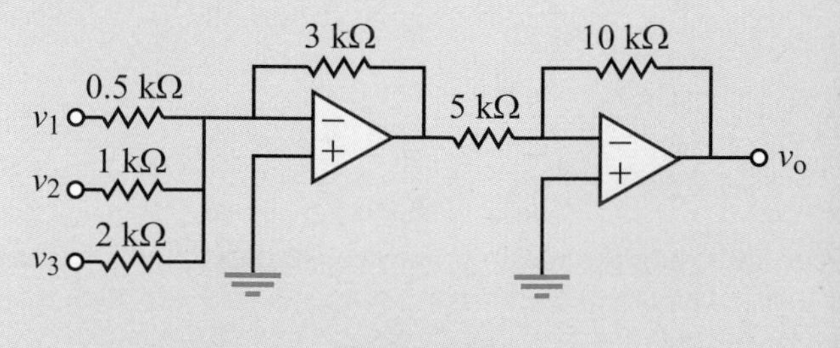

Figure E4.7

Answer: $v_o = 12v_1 + 6v_2 + 3v_3$. (See)

4-8 Op-Amp Signal-Processing Circuits

Table 4-3 provides a summary of the op-amp circuits we have considered thus far, together with their functional characteristics in the form of block-diagram representations. These circuits can be used in various combinations to realize specific signal-processing operations. We note that the input–output transfer functions are independent of the load resistance R_L that may be connected between the output terminal v_o and ground. In the case of the noninverting amplifier, the transfer function is also independent of the source resistance R_s. *When cascading multiple stages of op-amp circuits in series, care must be exercised to ensure that none of the op amps is driven into saturation by the cumulative gain of the multiple stages.*

When analyzing circuits that involve op amps, whether in configurations similar to or different from those we encountered in this chapter, the basic rules to remember are as follows:

Basic Rules of Op-Amp Circuits

(1) KCL and KVL always apply everywhere in the circuit, but KCL fails at the output node when applying the ideal op-amp model.

(2) The op amp will operate in the linear range so long as $|v_o| < |V_{cc}|$.

(3) The ideal op-amp model assumes that the source resistance R_s (connected to terminals v_p or v_n) is much smaller than the op-amp input resistance R_i (which usually is no less than 10 MΩ), and the load resistance R_L is much larger than the op-amp output resistance R_o (which is on the order of tens of ohms).

(4) The ideal op-amp constraints are $i_p = i_n = 0$ and $v_p = v_n$.

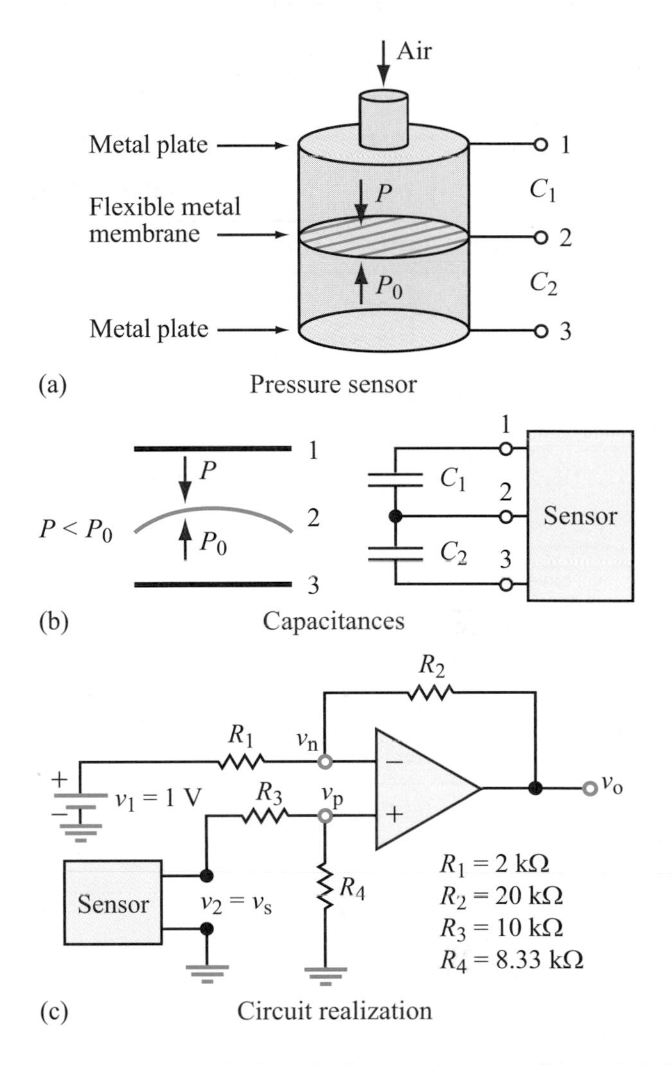

Figure 4-17: Design of a circuit for the pressure sensor of Example 4-5 with P_0 = pressure at sea level and P = pressure at height h.

Example 4-5: Elevation Sensor

A hand-held elevation sensor uses a pair of capacitors separated by a flexible metallic membrane (Fig. 4-17(a)) to measure the height h above sea level. The lower chamber in Fig. 4-17(a) is sealed, and its pressure is P_0, which is the standard atmospheric pressure at sea level. The pressure in the upper chamber, which is open to the outside air, is P. When at sea level, $P = P_0$, so the membrane assumes a flat shape and the two capacitances are equal. Since atmospheric pressure decreases with elevation, a rise in altitude results in a change in the pressure P in the upper chamber, causing the membrane to bend upwards (Fig. 4-17(b)), thereby changing the capacitances of the two capacitors. The sensor measures a voltage v_s that is proportional to the change in capacitance.

Based on measurements of v_s as a function of h, the data was found to exhibit an approximately linear variation given by

$$v_s = 2 + 0.2h \qquad \text{(V)}, \tag{4.47}$$

where h is in km. The sensor is designed to operate over the range $0 \leq h \leq 10$ km. Design a circuit whose output voltage v_o (in volts) is an exact indicator of the height h (in km).

Solution: Based on the given information, the sensor voltage v_s will serve as the input to the circuit we are asked to design, and the output v_o will represent the height elevation h. We therefore need a circuit that can perform the operation

$$v_o = h = \frac{1}{0.2}\, v_s - \frac{2}{0.2} = 5v_s - 10, \tag{4.48}$$

Table 4-3: Summary of op-amp circuits.

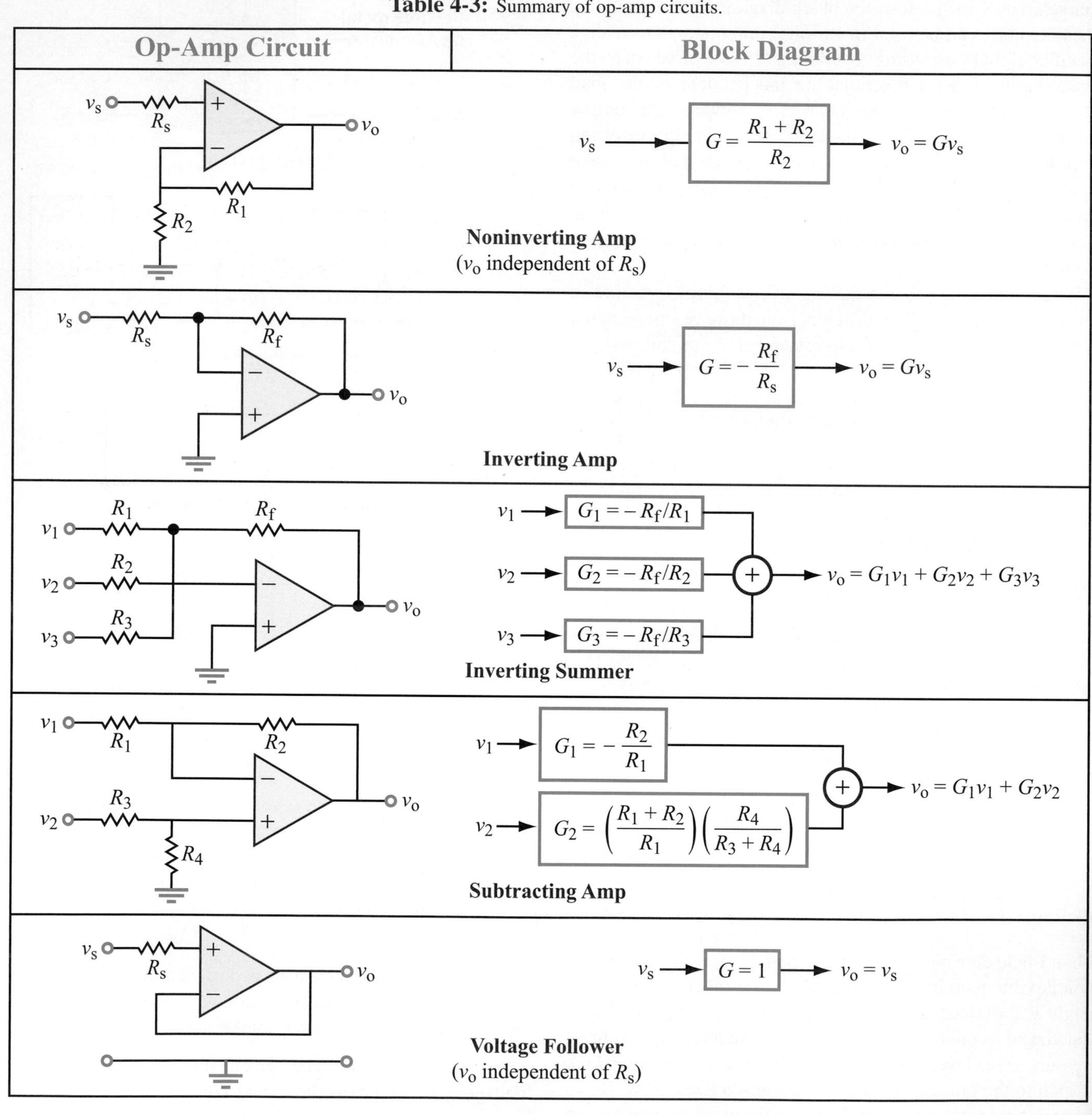

where we have inverted Eq. (4.47) to solve for h in terms of v_s. The functional form of Eq. (4.48) indicates that we have only one active (variable) input, namely v_s, which we need to amplify by a factor of 5 but also need to subtract 10 V from it. There are multiple circuit configurations that can achieve the desired operation, including the subtractor circuit shown in Fig. 4-17(c). According to Eq. (4.40), the output of the difference amplifier is given by

$$v_o = \left[\left(\frac{R_4}{R_3 + R_4}\right)\left(\frac{R_1 + R_2}{R_1}\right)\right] v_2 - \left(\frac{R_2}{R_1}\right) v_1. \qquad (4.49)$$

Equation (4.49) can be made to correspond to Eq. (4.48) if we select the following

(a) $v_s = v_2$

(b) v_1 as a dc voltage source such that $(R_2/R_1)v_1 = 10$ V, which can be satisfied by arbitrarily selecting $v_1 = 1$ V and $(R_2/R_1) = 10$

(c) values for R_1 through R_4 that simultaneously satisfy the conditions

$$\frac{R_2}{R_1} = 10 \quad \text{and} \quad \left(\frac{R_4}{R_3 + R_4}\right)\left(\frac{R_1 + R_2}{R_1}\right) = 5$$

A possible set of values that meets these conditions is

$$R_1 = 2\text{ k}\Omega, \quad R_2 = 20\text{ k}\Omega, \quad R_3 = 10\text{ k}\Omega, \quad R_4 = 8.33\text{ k}\Omega.$$

Before we conclude the design, we should check to make sure that the op amp will operate in its linear range over the full range of operation of the sensor. According to Eq. (4.47), as h varies from zero to 10 km, v_s varies from 2 V to 4 V. The corresponding range of variation of v_o, from Eq. (4.48), is from zero to 10 V. Hence, we should choose an op amp designed to function with a dc supply voltage V_{cc} that exceeds 10 V.

Example 4-6: Circuit with Multiple Op Amps

Relate the output voltage v_o to the input voltages v_1 and v_2 of the circuit in Fig. 4-18.

Solution: By comparing the circuit connections surrounding the four op amps with those given in Table 4-3, we recognize op amps 1 and 2 as noninverting amplifiers, op amp 3 as an inverting amplifier with a gain of −1 (equal input and feedback resistors R_4), and op amp 4 as an inverting summing amplifier with equal gain (same input resistances R_6 at summing point).

We start by examining the pair of input op amps. For op amp 1, $v_{p_1} = v_1$ and $v_{p_1} = v_{n_1}$ (op amp voltage constraint). Hence,

$$v_a = v_{n_1} = v_1.$$

Similarly, for op amp 2,

$$v_b = v_{n_2} = v_2.$$

Since $i_{n_1} = i_{n_2} = 0$ (op amp current constraint),

$$i_2 = \frac{v_b - v_a}{R_2} = \frac{v_2 - v_1}{R_2},$$

and

$$v_{o_2} - v_{o_1} = i_2(R_1 + R_2 + R_3) = \left(\frac{R_1 + R_2 + R_3}{R_2}\right)(v_2 - v_1). \qquad (4.50)$$

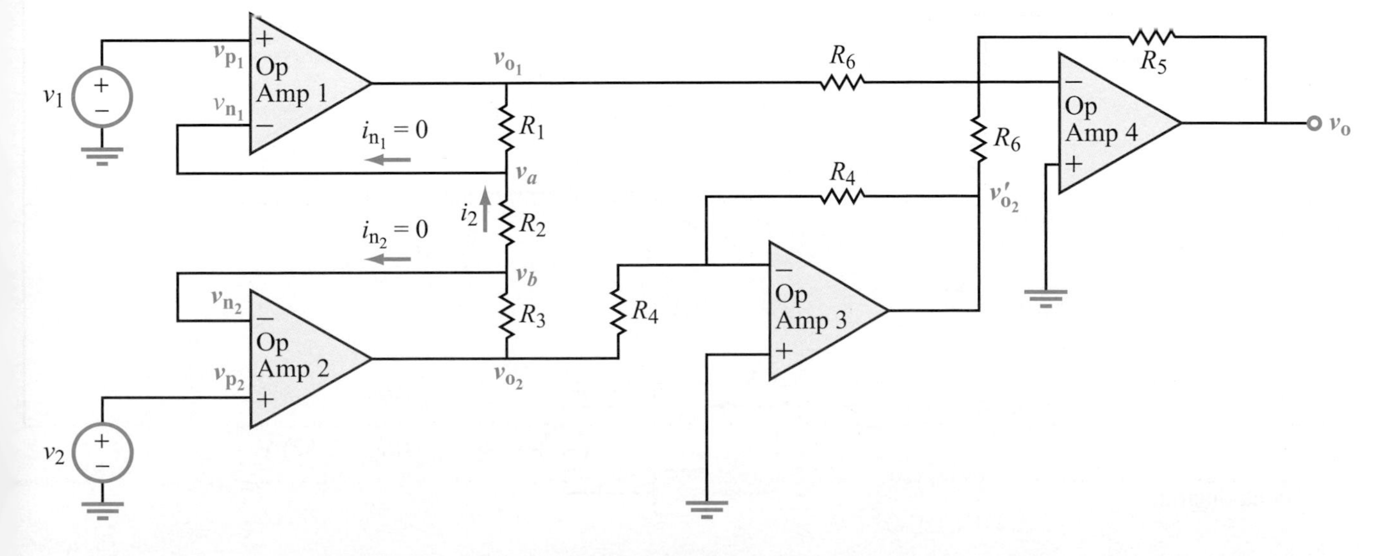

Figure 4-18: Example 4-6.

For op amp 3,

$$v'_{o_2} = -v_{o_2},$$

and for op amp 4,

$$v_o = -\frac{R_5}{R_6}(v_{o_1} + v'_{o_2}) = -\frac{R_5}{R_6}(v_{o_1} - v_{o_2}) = \frac{R_5}{R_6}(v_{o_2} - v_{o_1})$$
$$= R_5\left(\frac{R_1 + R_2 + R_3}{R_6 R_2}\right)(v_2 - v_1). \quad (4.51)$$

Example 4-7: Block-Diagram Representation

Generate a block-diagram representation for the circuit shown in Fig. 4-19(a).

Solution: The first op amp is an inverting amplifier with a dc input voltage $v_1 = 0.42$ V. Its circuit gain G_i (with the subscript added to denote "inverting amp") is

$$G_i = -\frac{30K}{10K} = -3,$$

and its output is

$$v_{o_1} = G_i v_1 = -3(0.42) = -1.26 \text{ V}.$$

The second op amp is a difference amplifier. The gains of its positive and negative channels are given in Table 4-3 as

$$G_2 = \left(\frac{R_4}{R_3 + R_4}\right)\left(\frac{R_1 + R_2}{R_1}\right)$$
$$= \left(\frac{2K}{1K + 2K}\right)\left(\frac{10K + 20K}{10K}\right) = 2$$

and

$$G_1 = -\frac{R_2}{R_1} = -\frac{20K}{10K} = -2.$$

Hence,

$$v_o = G_2 v_2 + G_1 v_{o_1}$$
$$= 2v_2 - 2(-1.26)$$
$$= (2v_2 + 2.52) \text{ V}.$$

4-9 Instrumentation Amplifier

An electric ***sensor*** is a circuit used to measure a physical quantity, such as distance, motion, temperature, pressure, or humidity. *In some applications, the intent is not to measure the magnitude of a certain quantity, but rather to sense small deviations from a nominal value.* For example, if the temperature in a room is to be maintained at 20°C, the functional goal of the temperature sensor is to measure the difference between the room temperature T and the reference temperature $T_0 = 20$°C and then to activate an air conditioning or heating unit if the deviation exceeds a certain prespecified threshold. Let us assume the threshold is 0.1°C. Instead of requiring the sensor to be able to measure T with an absolute accuracy of no less than 0.1°C, an alternative approach would be to design the sensor to measure $\Delta v = v_2 - v_1$, where v_2 is the voltage output of a thermocouple circuit responding to the room temperature T and v_1 is the voltage corresponding to what a calibrated thermocouple would measure when $T_0 = 20$°C. Thus, the sensor is designed to measure the deviation of T from T_0, rather than T itself, with an absolute accuracy of no

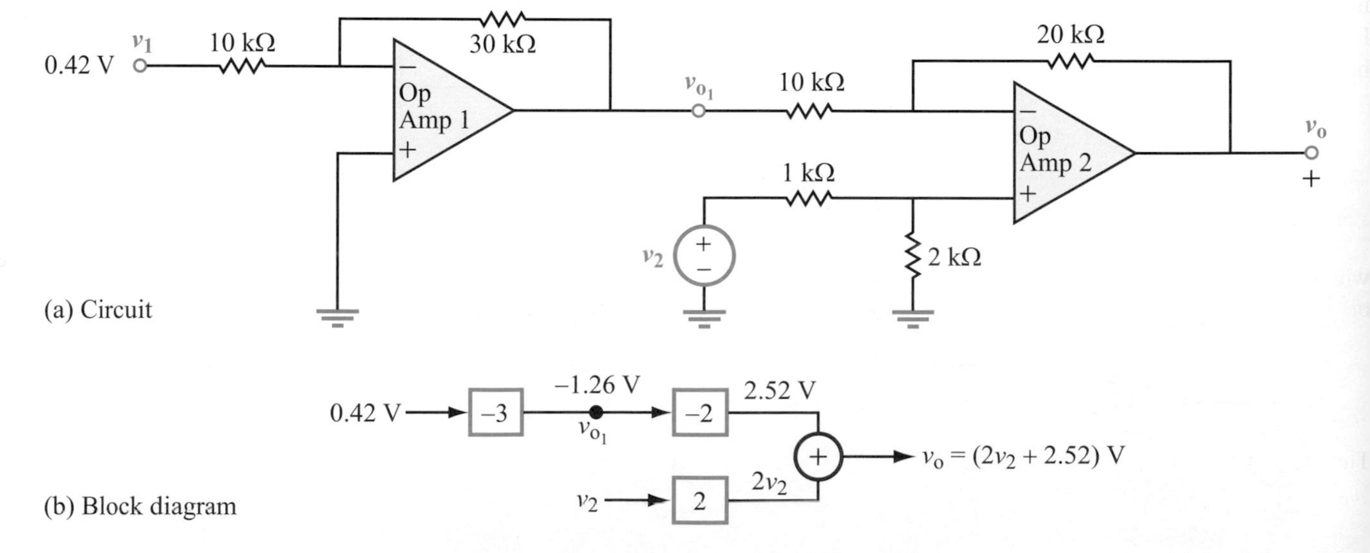

Figure 4-19: Block-diagram representation (Example 4-7).

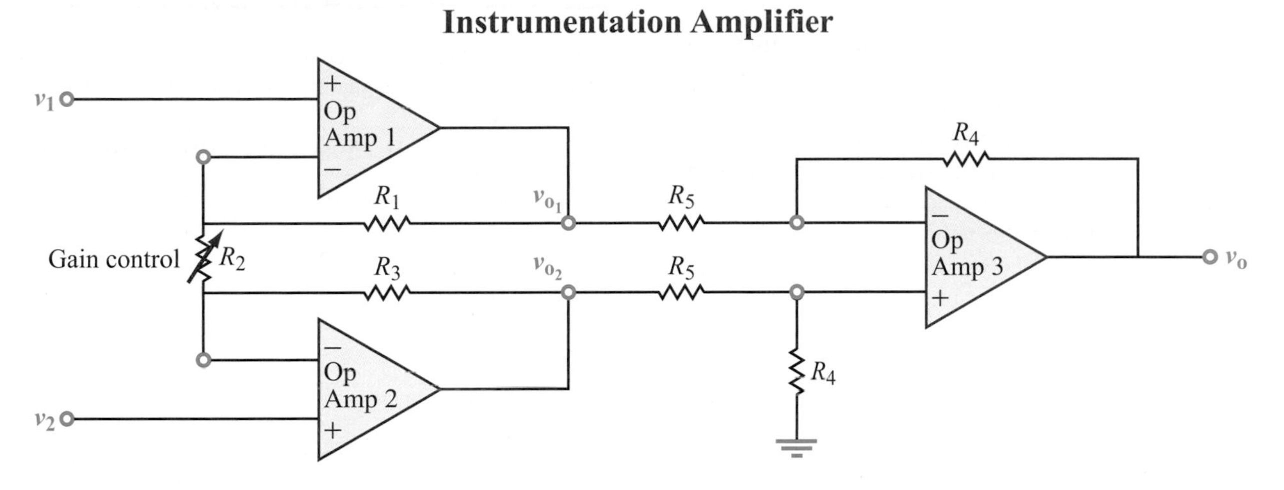

Figure 4-20: Instrumentation-amplifier circuit.

less than 0.1°C. The advantage of such an approach is that the signal is now Δv, which is more than two orders of magnitude smaller than v_2. A circuit with a precision of 10 percent is not good enough for measuring v_2, but it is plenty good for measuring Δv.

The instrumentation amplifier is suited perfectly for detecting and amplifying a small signal deviation when superimposed on one or the other of two much larger (and otherwise identical) signals.

An instrumentation amplifier consists of three op amps, as shown in Fig. 4-20. The circuit configuration for the first two is the same as the one we examined earlier in connection with Example 4-6. According to Eq. (4.50), the voltage difference between the outputs of op amps 1 and 2 is

$$v_{o_2} - v_{o_1} = \left(\frac{R_1 + R_2 + R_3}{R_2}\right)(v_2 - v_1) = G_1(v_2 - v_1), \tag{4.52}$$

where G_1 is the circuit gain of the first stage (which includes op amps 1 and 2) and is given by

$$G_1 = \frac{R_1 + R_2 + R_3}{R_2}. \tag{4.53}$$

The third op amp is a difference amplifier that amplifies $(v_{o_2} - v_{o_1})$ by a gain factor G_2 given by

$$G_2 = \frac{R_4}{R_5}. \tag{4.54}$$

Hence,

$$v_o = G_2 G_1(v_2 - v_1) = \left(\frac{R_4}{R_5}\right)\left(\frac{R_1 + R_2 + R_3}{R_2}\right)(v_2 - v_1). \tag{4.55}$$

To simplify the circuit—and improve precision—all resistors—with the exception of R_2—often are chosen to be identical in design and construction, thereby minimizing deviations between their resistances. If we set $R_1 = R_3 = R_4 = R_5 = R$ in Eq. (4.55), the expression for v_o reduces to

$$v_o = \left(1 + \frac{2R}{R_2}\right)(v_2 - v_1). \tag{4.56}$$

In that case, R_2 becomes the ***gain-control resistance*** of the circuit; its value (relative to R) sets the gain. If the expected signal deviation $(v_2 - v_1)$ is on the order of microvolts to millivolts, the instrumentation amplifier is designed to have an overall gain that would amplify the signal to the order of volts. Thus, *the instrumentation amplifier is a high-sensitivity, high-gain, deviation sensor.* Several semiconductor manufacturers offer instrumentation-amplifier circuits in the form of integrated packages.

Review Question 4-15: When designing a multistage op-amp circuit, what should the design engineer do to insure that none of the op amps is driven into saturation?

Review Question 4-16: If the goal is to measure small deviations between a pair of input signals, what is the advantage of using an instrumentation amplifier over using a difference amplifier?

Exercise 4-8: To monitor brain activity, an instrumentation-amplifier sensor uses a pair of needle-like probes inserted at different locations in the brain to measure the voltage difference between them. If the circuit is of the type shown in Fig. 4-20 with $R_1 = R_3 = R_4 = R_5 = R = 50$ kΩ, $V_{cc} = 12$ V, and the maximum magnitude of the voltage difference that the brain is likely to exhibit is 3 mV, what should R_2 be to maximize the sensitivity of the brain sensor?

Answer: $R_2 = 25$ Ω. (See ◎)

4-10 Digital-to-Analog Converters (DAC)

An n-bit digital signal is described by the sequence $[V_1 V_2 V_3 \ldots V_n]$, where V_1 is called the ***most significant bit (MSB)*** and V_n is the ***least significant bit (LSB)***. Voltages V_1 through V_n can each assume only two possible states—either a 0 or a 1. When a bit is in the 1 state, its decimal value is 2^m, where m depends on the location of that bit in the sequence. For the most significant bit (V_1), its decimal value is $2^{(n-1)}$; for V_2 it is $2^{(n-2)}$; and so on. The decimal value of the least significant bit is $2^{n-n} = 2^0 = 1$, when that bit is in state 1. Any bit in state 0 has a decimal value of 0. Table 4-4 illustrates the correspondence between the binary sequences of a 4-bit digital signal and their decimal values. The binary sequences start at [0000] and end at [1111], representing 16 decimal values extending from 0 to 15 and inclusive of both ends.

A ***digital-to-analog converter (DAC)*** is a circuit that transforms a digital sequence presented to its input into an analog output voltage whose magnitude is proportional to the decimal value of the input signal.

To do so, the DAC in Fig. 4-21 has to sum V_1 to V_n after weighting each by a factor equal to its decimal value. Thus,

Table 4-4: Correspondence between binary sequence and decimal value for a 4-bit digital signal and output of a DAC with $G = -0.5$.

$V_1V_2V_3V_4$	Decimal Value	DAC Output (V)
0000	0	0
0001	1	−0.5
0010	2	−1
0011	3	−1.5
0100	4	−2
0101	5	−2.5
0110	6	−3
0111	7	−3.5
1000	8	−4
1001	9	−4.5
1010	10	−5
1011	11	−5.5
1100	12	−6
1101	13	−6.5
1110	14	−7
1111	15	−7.5

for a 4-bit digital sequence, for example, the output voltage of the DAC has to be related to the input by

$$\begin{aligned} V_{out} &= G(2^{4-1}V_1 + 2^{4-2}V_2 + 2^{4-3}V_3 + 2^{4-4}V_4) \\ &= G(8V_1 + 4V_2 + 2V_3 + V_4), \end{aligned} \quad (4.57)$$

where G is a scale factor that has no influence on the relative weights of the four terms. The magnitude of G is selected to suit the range of the output voltage. If the input is a 3-bit sequence whose range of decimal values extends from 0 to 7, one might design the circuit so that $G = 1$, because in that case, the maximum output voltage is 7 V, which is below V_{cc} for most op amps. For digital signals with longer sequences, G needs to be smaller than 1 in order to avoid saturating the op amp.

The weighted-sum operation of a DAC can be realized by many different signal-processing circuits. A rather straightforward implementation is shown in Fig. 4-22, where an inverting summer uses the ratios of R_f to the individual

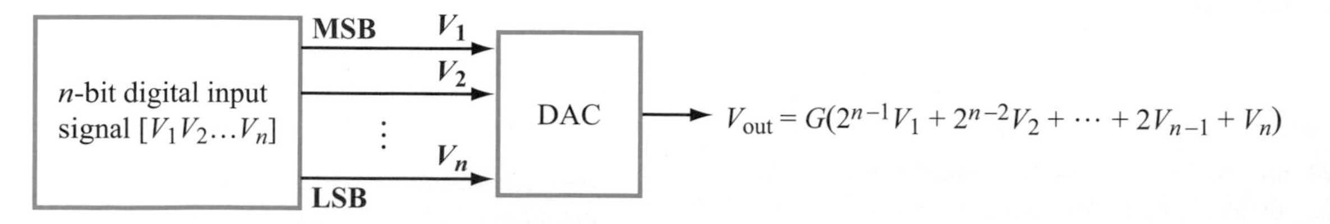

Figure 4-21: A digital-to-analog converter transforms a digital signal into an analog voltage proportional to the decimal value of the digital sequence.

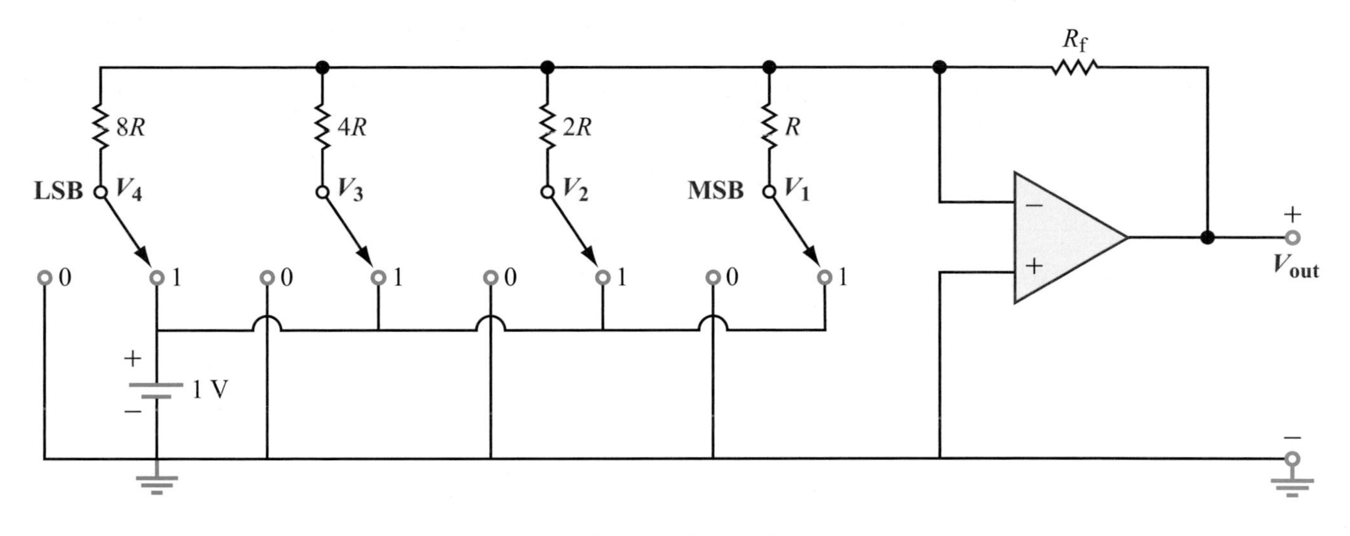

Figure 4-22: Circuit implementation of a DAC.

resistances to realize the necessary weights, and the positions of the switches determine the 0/1 states of the 4 bits. Reference to either Table 4-3 or Eq. (4.34) yields

$$\begin{aligned} V_{out} &= -\frac{R_f}{R} V_1 - \frac{R_f}{2R} V_2 - \frac{R_f}{4R} V_3 - \frac{R_f}{8R} V_4 \\ &= \frac{-R_f}{8R}(8V_1 + 4V_2 + 2V_3 + V_4), \end{aligned} \tag{4.58}$$

which satisfies the relative weights given in Eq. (4.57). Also, in this case,

$$G = -\frac{R_f}{8R}. \tag{4.59}$$

For $[V_1 V_2 V_3 V_4] = [1111]$, $V_{out} = 15G$. By selecting $G_2 = -0.5$ (corresponding to $R_f = 4R$), the output will vary from 0 to -7.5.

Example 4-8: R--2R Ladder

The circuit in Fig. 4-23(a) offers an alternative approach to realizing digital-to-analog conversion of a 4-bit signal. It is called an R–$2R$ ladder, because all of the resistors of its input circuit have values of R or $2R$, thereby limiting the input resistance seen by the dc source to a 2:1 range no matter how many bits are contained in the digital sequence. This is in contrast with the DAC of Fig. 4-22, whose input-resistance range is dependent on the number of bits; 8:1 for a 4-bit converter, and 128:1 for an 8-bit converter. Additionally, circuit performance and precision are superior when fewer groups of resistors are involved in the input circuit. Resistors fabricated in the same production process are likely to exhibit less variability among them than resistors fabricated by different processes.

Show that the R–$2R$ ladder in Fig. 4-23(a) does indeed provide the appropriate weighting for a 4-bit DAC. If $R = 2$ kΩ and $V_{cc} = 10$ V, what is the maximum realistic value that R_f can have?

Solution: Even though we know that (depending on the positions of the switches) V_1 to V_4 can each assume only 2 binary values, namely 0 or 1 V, let us treat V_1 to V_4 as dc power supplies and apply multiple iterations of voltage–current transformations to arrive at the Thévenin equivalent circuit at the input side of the op amp. The result of such a transformation process is shown in Fig. 4-23(b), in which

$$V_{Th} = \frac{V_1}{2} + \frac{V_2}{4} + \frac{V_3}{8} + \frac{V_4}{16} \tag{4.60a}$$

and

$$R_{Th} = R. \tag{4.60b}$$

Consequently,

$$\begin{aligned} V_{out} &= -\frac{R_f}{R_{Th}} V_{Th} \\ &= -\frac{R_f}{R}\left(\frac{V_1}{2} + \frac{V_2}{4} + \frac{V_3}{8} + \frac{V_4}{16}\right) \\ &= -\frac{R_f}{16R}(8V_1 + 4V_2 + 2V_3 + V_4). \end{aligned} \tag{4.61}$$

The voltage $|V_{out}|$ is a maximum when $[V_1 V_2 V_3 V_4] = [1111]$, in which case

$$V_{out} = -\frac{15}{16}\frac{R_f}{R}.$$

To insure that $|V_{out}|$ does not exceed $|V_{cc}| = 10$ V as well as to provide a safety margin of 2 V it is necessary that

$$8 \geq \frac{15}{16}\frac{R_f}{2k},$$

which gives $R_f \leq 17.1$ kΩ.

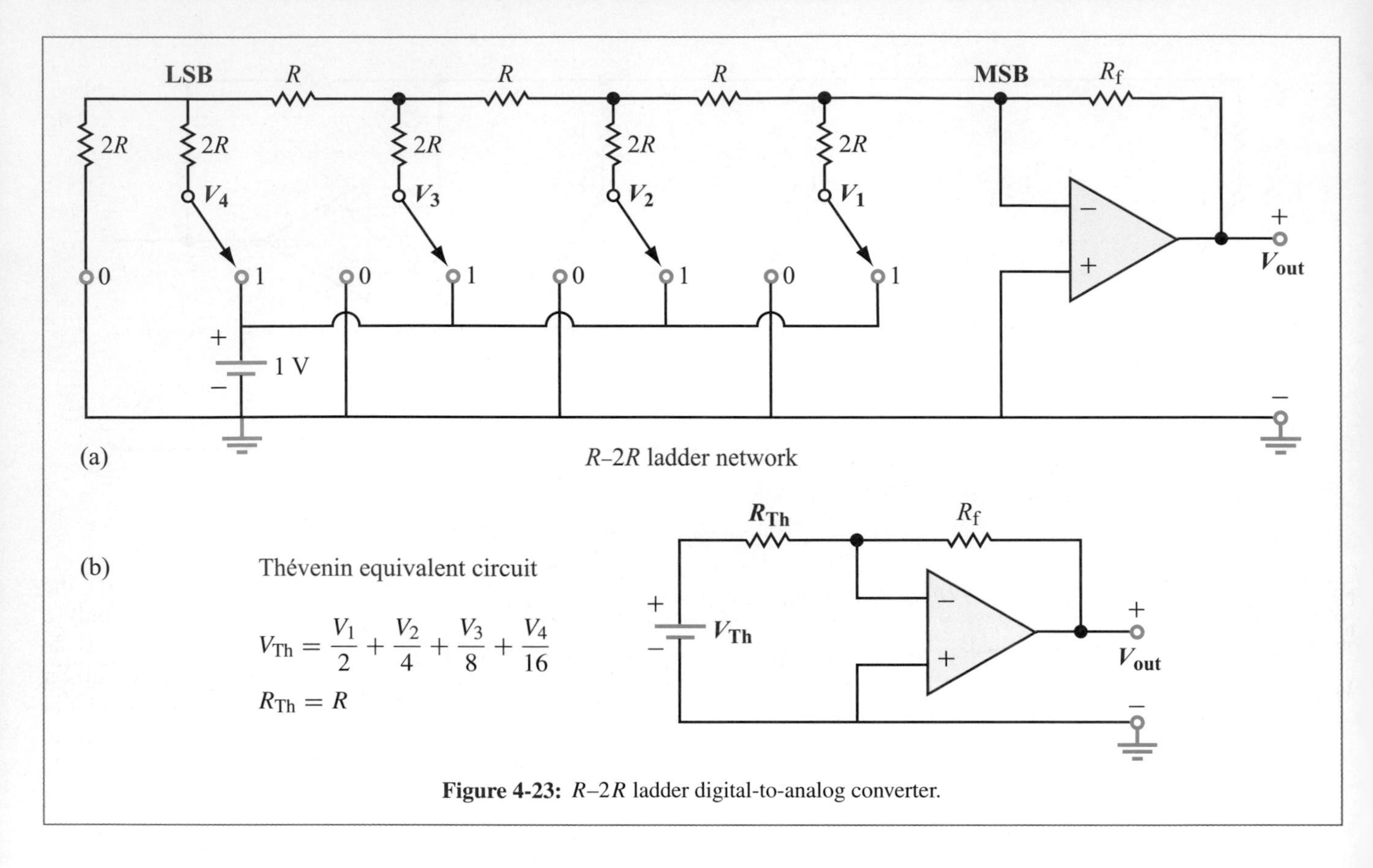

Figure 4-23: R–$2R$ ladder digital-to-analog converter.

Review Question 4-17: In a digital-to-analog converter, what dictates the maximum value that R_f can assume?

Review Question 4-18: What is the advantage of the R–$2R$ ladder (Fig. 4-23) over the traditional DAC (Fig. 4-22)?

Exercise 4-9: A 3-bit DAC uses an R–$2R$ ladder design with $R = 3$ kΩ and $R_f = 24$ kΩ. If $V_{cc} = 10$ V, write an expression for V_{out} and evaluate it for $[V_1 V_2 V_3] = [111]$.

Answer:

$$V_{out} = -\frac{R_f}{8R}(4V_1 + 2V_2 + V_3)$$
$$= -(4V_1 + 2V_2 + V_3).$$

For $[V_1 V_2 V_3] = [111]$, $V_{out} = -7$ V, whose magnitude is smaller than $V_{cc} = 10$ V. (See CD)

4-11 The MOSFET as a Voltage-Controlled Current Source

The simplest model of a ***MOSFET***, which stands for ***metal-oxide semiconductor field-effect transistor***, is shown in Fig. 4-24(a). The vast majority of commercial computer processors are built with MOSFETs; as mentioned in Technology Brief 1 on page 10, a 2006 Intel Core processor contains 151 million independent MOSFETs. A MOSFET has three terminals: the ***gate*** (G), the ***source*** (S), and the ***drain*** (D). Actually, it has a fourth terminal, namely its body (B), but we will ignore it for now because for many applications it is simply connected to the ground terminal. The circuit symbol for the MOSFET may look somewhat unusual, but it is actually a stylized depiction of the physical cross section of a real MOSFET. In a real MOSFET, the gate consists of a very thin layer (< 500 nm thick) of a conducting material adjacent to an even thinner layer (< 100 nm) of insulator. The insulator in turn is placed directly on the surface of a relatively large slab of semiconductor material, usually referred to as "the chip" in everyday conversation (usually silicon 0.5 to 1.5 mm thick). The drain and the source sections are fabricated into this semiconductor chip on either side of the gate.

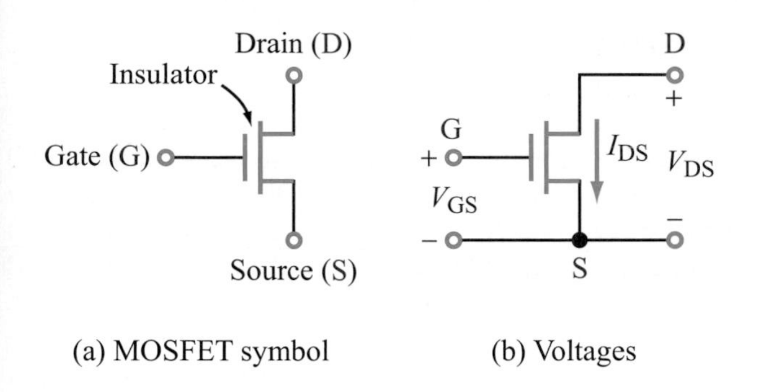

Figure 4-24: MOSFET symbol and voltage designations.

> Because the gate G is separated from the rest of the transistor by the thin insulating layer, no current can flow from G to either D or S.

Nonetheless, it turns out that the voltage difference between terminals G and S is key to the operation of the MOSFET.

Using terminal S as a reference in Fig. 4-24(b), we denote V_{DS} and V_{GS} as the voltages at terminals D and G, respectively. We also denote the current that flows through the MOSFET from D to S as I_{DS}. This simplification is justified by the assumption that no current flows through the gate node to either the drain or source node. The operation of the MOSFET can be analyzed by placing it in the simple circuit shown in Fig. 4-25(a), in which V_{DD} is a dc power supply voltage usually set at a level close to but not greater than, the maximum rated value of V_{DS} for the specific MOSFET model under consideration. The resistance R_D is external to the MOSFET, and its role will be discussed later. The input voltage is synonymous with V_{GS} and the output voltage is synonymous with V_{DS},

$$V_{in} = V_{GS}, \quad \text{and} \quad V_{out} = V_{DS}. \tag{4.62}$$

Moreover V_{out} is related to V_{DD} by

$$V_{out} = V_{DD} - I_{DS} R_D. \tag{4.63}$$

Since current cannot flow from G to either D or S, the only current that can flow through the MOSFET is I_{DS}. The dependence of I_{DS} on V_{GS} and V_{DS} is shown for a typical MOSFET in Fig. 4-25(b) in the form of characteristic curves displaying the response of I_{DS} to V_{DS} at specific

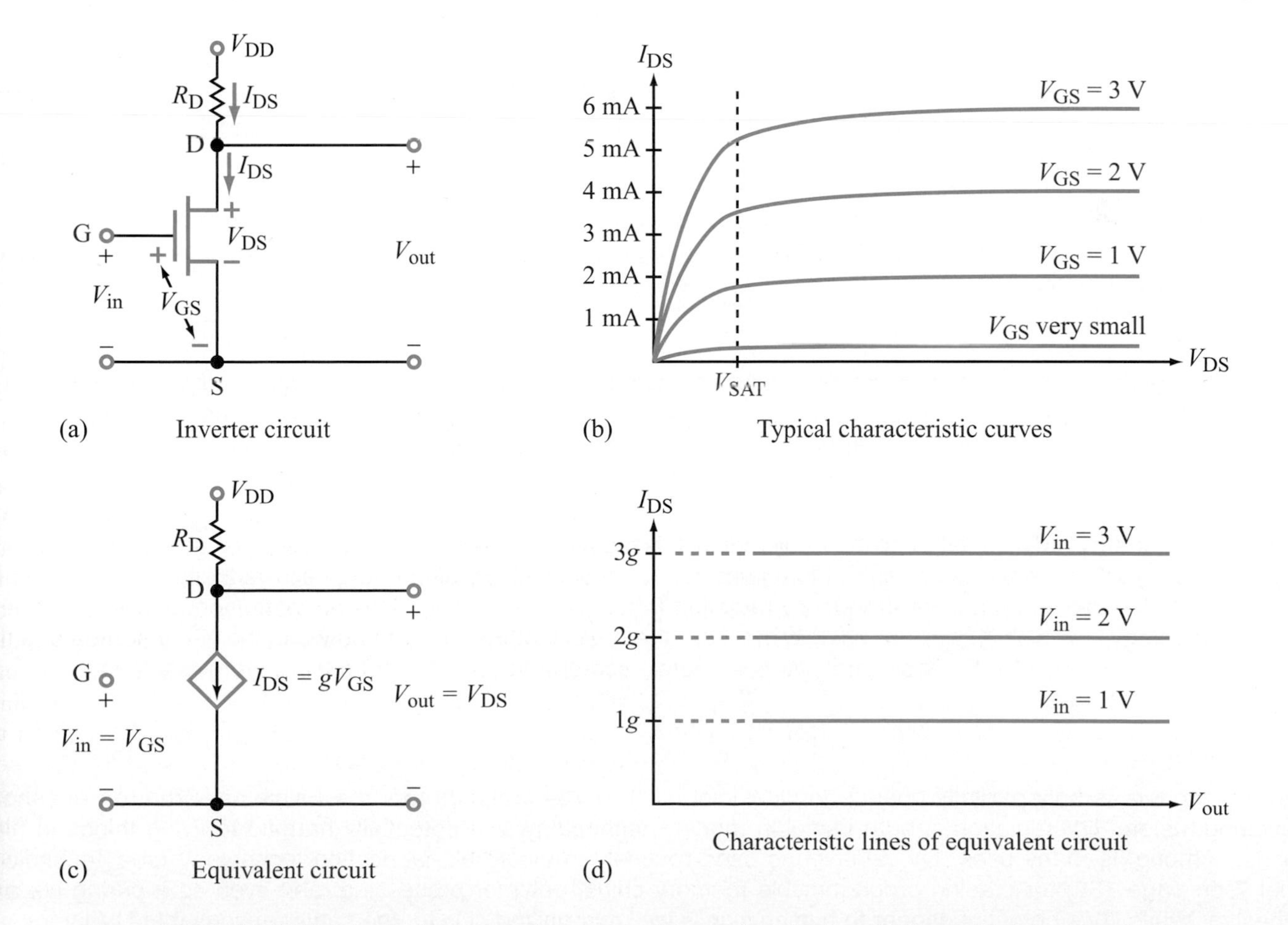

Figure 4-25: MOSFET (a) circuit, (b) characteristic curves, (c) equivalent circuit, and (d) associated characteristic lines.

Technology Brief 8: The Electromagnetic Spectrum

Electromagnetic Energy

The sun's rays, the signal transmitted by a cell phone, and the radiation emitted by plutonium share a fundamental property: they all carry electromagnetic (EM) energy. It is an interesting and fundamental observation that this energy can be described both as a ***wave*** moving through space and as a ***particle***. Neither model alone is sufficient to explain the phenomena we observe in the world around us. This correspondence, called the ***wave-particle duality***, sparked scientific debate as far back as the 1600s, and it was not until the 20th century and the advent of quantum mechanics that this duality was fully incorporated into modern physics.

When we treat EM energy as a wave with alternating electric and magnetic fields, we ascribe to the wave a wavelength λ and an oscillation frequency f, whose product defines the velocity of the wave u as

$$u = f\lambda.$$

If the propagation medium is free space, then u is equal to c, which is the speed of light in vacuum at 3×10^8 m/s. Because of the wave-particle duality, when EM energy is regarded as a particle, each such particle will have the same velocity u as its wave counterpart and will carry energy E whose magnitude is specified by the frequency f through

$$E = hf,$$

where h is Planck's constant (6.6×10^{-34} J·s). In view of the direct link between E and f, we can refer to an EM particle (also called a ***photon***) either by its energy E or by the frequency f of its wave counterpart. The higher the frequency is, the higher is the energy carried by a photon, but also the shorter is its wavelength λ.

The Spectrum

In terms of the wavelength λ, the EM spectrum extends across many orders of magnitude (Fig. TF8-1), from the radio region on one end to the gamma-ray region on the other. The degree to which an EM wave is absorbed or scattered as it travels through a medium depends on the types of constituents present in that medium and their sizes relative to λ of the wave. For Earth's atmosphere, the composition and relative distributions of its gases are responsible for the near total opacity of the atmosphere to EM waves across most of the EM spectrum, except for narrow "windows" in the visible, infrared, and radio spectral regions (Fig. TF8-1). It is precisely because EM waves with these wavelengths can propagate well through the atmosphere that human sight, thermal infrared imaging, and radio communication are possible through the air.

1. Cosmic Rays

Emitted by the decay of the nuclei of unstable elements and by cosmic, high-energy sources in the universe, cosmic rays—which include gamma, beta, and alpha radiation—are highly energetic particles that can be dangerous to organisms and destructive to matter. Earth emits gamma rays of its own, but at very weak levels.

2. X-Rays

Slightly lower energy radiation falls into the X-ray region; this radiation is energetic enough to be dangerous to organisms in large doses, but small doses are safe. More importantly, their relatively high energy allows them to traverse much farther into solid objects than lower frequency radiation (such as visible light). This phenomenon allows for modern medical radiology, in which X-rays are used to measure the opacity of the medium between the X-ray source and the detector or film. Thankfully, Earth's atmosphere efficiently screens the surface from high-energy radiation, such as cosmic rays and X-rays.

3. Ultraviolet Rays

The atmosphere is only partially opaque to ultraviolet (UV) waves, which border the visible spectrum on the short wavelength side. UV radiation is both useful in modern technology and potentially harmful to living things in high doses. Among its many uses, UV radiation is used routinely in electronic fabrication technology (see Technology Brief 7 on page 135) for erasing programmable memory chips, polymer processing, and even as a curing ink and adhesive. While UV's potential danger to human skin is well recognized, it is for the same reasons that UV lamps are used to sterilize hospital and laboratory equipment.

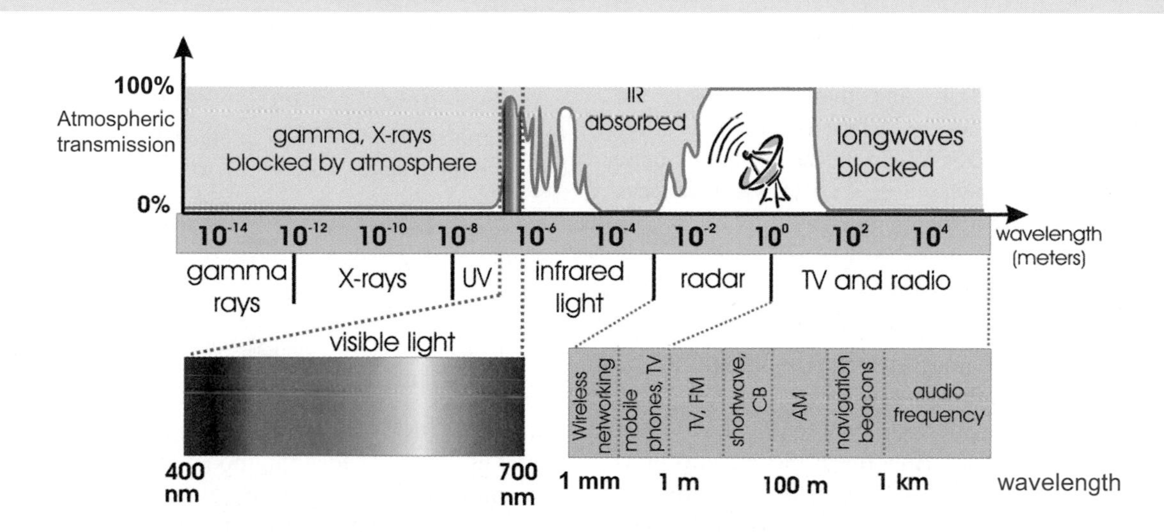

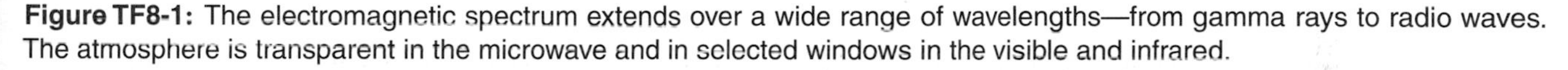

Figure TF8-1: The electromagnetic spectrum extends over a wide range of wavelengths—from gamma rays to radio waves. The atmosphere is transparent in the microwave and in selected windows in the visible and infrared.

4. Visible Light Rays

The wavelength range of visible light extends from about 380 nm (violet color) to 740 nm (red/brown color), although the exact range varies from one human to another. Some species can see well into the infrared (IR) or the UV, so the definition of *visible* is completely anthropocentric. It is no coincidence that evolution led to the development of sight organs that are sensitive to precisely that part of the spectrum where atmospheric absorption is very low. In the visible spectrum, blue light is more susceptible to scattering by atmospheric particles than the longer wavelengths, which is why the sky appears blue to us.

5. Infrared Rays

The infrared (IR) region, straddled in-between the visible spectrum and the radio region, is particularly useful for thermal applications. When an object is heated, the added energy increases the vibrations of its molecules. These molecular vibrations, in turn, release electromagnetic radiation at many frequencies. Within the range of our thermal environment, the peak of the radiated spectrum is in the IR region. This feature has led to the development of IR detectors and cameras for both civilian and military thermal-imaging applications. Nightvision systems use IR detector arrays to image a scene when the intensity of visible-wavelength light is insufficient for standard cameras (see Technology Brief 15 on page 325). This is because material objects emit IR energy even in pitch-black darkness. Conversely, IR energy can be used to heat an object, because a good radiator of IR is also a good absorber. Additionally, IR beams are used extensively in short-distance communication, such as in the remote control of most modern TV sets.

6. Radio Waves

The frequency range of the radio spectrum extends from essentially dc (or zero frequency) to $f = 1$ THz $= 10^{12}$ Hz. It is subdivided into many bands with formal designations (Fig. TF8-1) such as VHF (30 to 300 MHz) and UHF (300 to 3000 MHz), and some of those bands combine together to form bands commonly known by historic designations, such as the microwave band (300 MHz to 30 GHz). All major free-space communication systems operate at frequencies in the radio region, including wireless local area networks (LANs), cell phones, satellite communication, and television and radio transmissions. Because the radio spectrum is used so heavily, spectrum allocation is controlled by various national and international agencies that set standards for what types of devices are permitted to operate, within what frequency bands, and at what maximum-power transmission levels. Cell phones, for example, are allowed to transmit and receive in the 2.11 to 2.2 GHz band and in the 1.885 to 2.025 GHz band.

values of V_{GS}. We observe that if V_{DS} is greater than a certain ***saturation threshold value*** V_{SAT}, the curves assume approximately constant levels, and that these levels are approximately proportional to V_{GS}. These observations allow us to characterize the MOSFET in terms of the simple, equivalent circuit model shown in Fig. 4-25(c), which consists of a single dependent current source given by

$$I_{DS} = gV_{GS}, \tag{4.64}$$

where g is a ***MOSFET gain constant***. The characteristic curves associated with this model, which is valid only if V_{DS} exceeds V_{SAT}, are shown in Fig. 4-25(d).

In real MOSFETs, the relationship between I_{DS} and V_{GS} at saturation is not strictly linear. How linear the relationship is depends (in part) on the size of the transistor. Modern sub-micron transistors used in digital processors exhibit a linear relationship between I_{DS} and V_{GS} at saturation, whereas larger MOSFETs used for power switching may behave nonlinearly. For our purposes, the simplification denoted by Eq. (4.64) will suffice.

4-11.1 Digital Inverter

We now will use the model given by Eq. (4.64) to demonstrate how the MOSFET can function as a ***digital inverter*** by generating an output state of "0" when the input state is "1", and vice versa. Combining Eqs. (4.62) to (4.64) gives

$$V_{out} = V_{DD} - gR_D V_{in}. \tag{4.65}$$

The constant g is a MOSFET parameter, so if we choose R_D such that $gR_D \approx 1$, Eq. (4.65) simplifies to

$$\frac{V_{out}}{V_{DD}} = 1 - \frac{V_{in}}{V_{DD}}. \tag{4.66}$$

In a digital inverter, we are interested in output responses to only two input states. According to Eq. (4.66):

$$\text{If } \frac{V_{in}}{V_{DD}} = 1, \quad \Rightarrow \quad \frac{V_{out}}{V_{DD}} = 0, \tag{4.67a}$$

and

$$\text{if } \frac{V_{in}}{V_{DD}} = 0, \quad \Rightarrow \quad \frac{V_{out}}{V_{DD}} = 1. \tag{4.67b}$$

Hence, the MOSFET circuit in Fig. 4-25(a) behaves like a digital inverter, provided the model given by Eq. (4.64) holds true and requiring that V_{DS} exceeds V_{SAT}. In a real circuit, V_{in} and V_{out} are not given by the simple results indicated by Eq. (4.67), but each can be categorized easily into high and low voltage values to satisfy the functionality of a digital inverter.

4-11.2 NMOS versus PMOS Transistors

The MOSFET circuit of Fig. 4-25(a) actually is called an n-channel MOSFET or ***NMOS*** for short. Its operation is limited to the first quadrant in Fig. 4-25(d), where both I_{DS} and V_{DS} can assume positive values only. A second type of MOSFET called ***PMOS*** (p-channel MOSFET) is designed and fabricated to operate in the third quadrant, corresponding to negative values for I_{DS} and V_{DS}, as illustrated in Fig. 4-26. To distinguish between the two types, the symbol for PMOS includes a small open circle at terminal G.

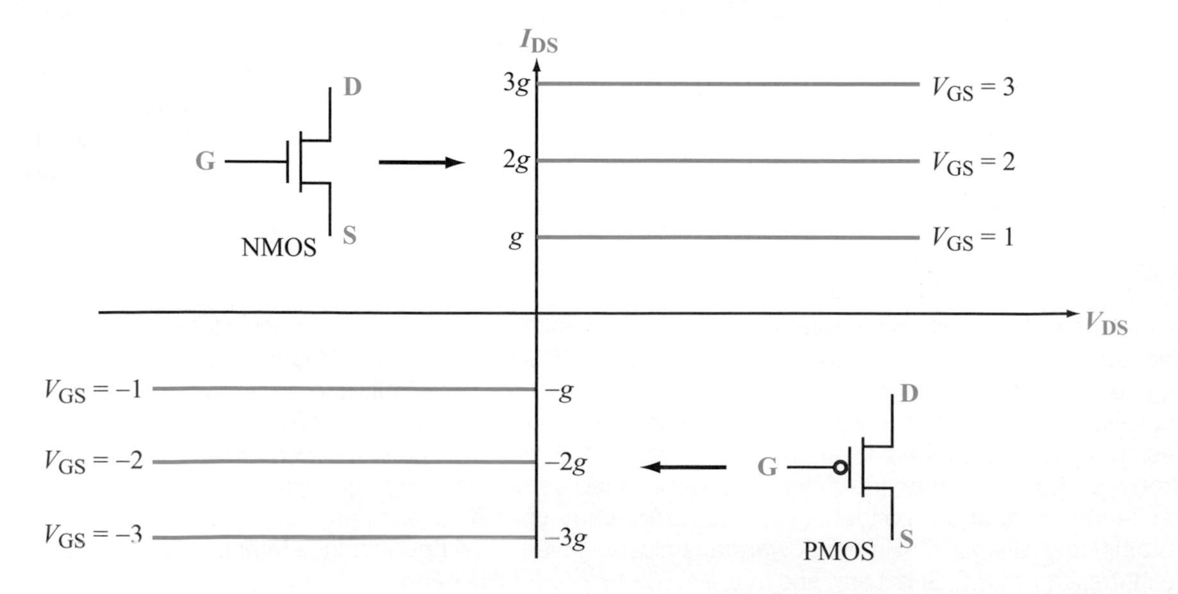

Figure 4-26: Complementary characteristic curves for NMOS and PMOS.

The NMOS inverter circuit of Fig. 4-25(a) provides the correct functionality required from a digital inverter, but it suffers from a serious power-dissipation problem. Let us consider the power consumed by R_D under realistic conditions:

Input State 0:

$$\frac{V_{in}}{V_{DD}} = 0 \Rightarrow I_{DS} \simeq 0 \Rightarrow P_{R_D} = I_{DS}^2 R_D \simeq 0 \qquad (4.68a)$$

Input State 1:

$$\frac{V_{in}}{V_{DD}} = 1 \Rightarrow I_{DS} = \frac{V_{DD}}{R_D} \Rightarrow P_{R_D} = \frac{V_{DD}^2}{R_D}. \qquad (4.68b)$$

Heat dissipation in R_D is practically zero for input state 0, but for input state 1, it is equal to V_{DD}^2/R_D. The value of V_{DD}, which is dictated by the MOSFET specifications, is typically on the order of volts, and R_D can be made very large–on the order of kΩ or tens of kΩ. If R_D is much larger than that, I_{DS} becomes too small for the MOSFET to function as an inverter. For V_{DD} on the order of 10 V and R_D on the order of 10 kΩ, P_{R_D} for an individual NMOS is on the order of 10 mW. This amount of heat generation is trivial for a single transistor, but when we consider that a typical computer processor contains on the order of 10^9 transistors, all confined to a relatively small volume of space, the total amount of heat that would be generated by such an NMOS-based processor would likely *burn a hole through the computer!* To address this heat-dissipation problem, a new technology was introduced in the 1980s called ***CMOS***, which stands for ***complementary MOS***.

CMOS has revolutionized the microprocessor industry and led to the rise of the x86 family of PC processors.

CMOS is a configuration that attaches an NMOS to a PMOS at their drain terminals, as shown in Fig. 4-27. The CMOS inverter provides the same functionality as the simpler NMOS inverter, but it has the distinct advantage in that it dissipates negligible power for *both* input states. The significance of the inverter is in the role it plays as a basic building block for more complicated logic circuits, such as those that perform AND and OR operations.

4-11.3 MOSFETs in Analog Circuits

In addition to their use in digital circuits, MOSFETs also can be used in analog circuits as buffers and amplifiers, as demonstrated by Examples 4-9 and 4-10. As we discussed earlier in Section 4-7, a buffer is a circuit that insulates the input voltage from variations in the load resistance.

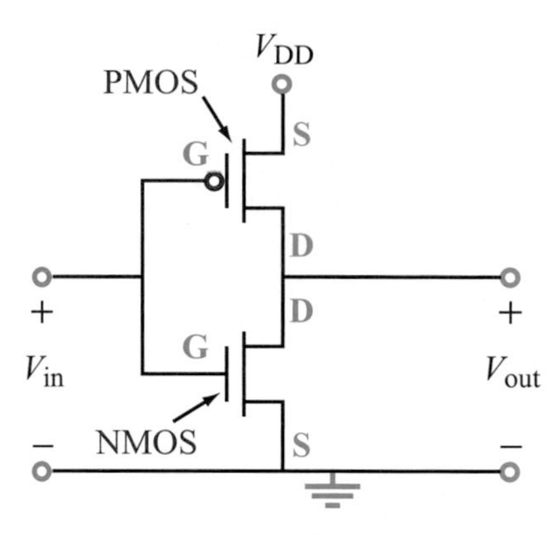

Figure 4-27: CMOS inverter.

Example 4-9: MOSFET Amplifier

The circuit shown in Fig. 4-28(a) is known as a common-source amplifier and uses a MOSFET with a dc drain voltage $V_{DD} = 10$ V and a drain resistance $R_D = 1$ kΩ. The input signal $v_s(t)$ is an ac voltage given by

$$v_s(t) = [500 + 40\cos 300t] \qquad (\mu\text{V}).$$

Note that the amplitude of the input ac signal is several orders of magnitude smaller than that of the dc voltage V_{DD}. Apply

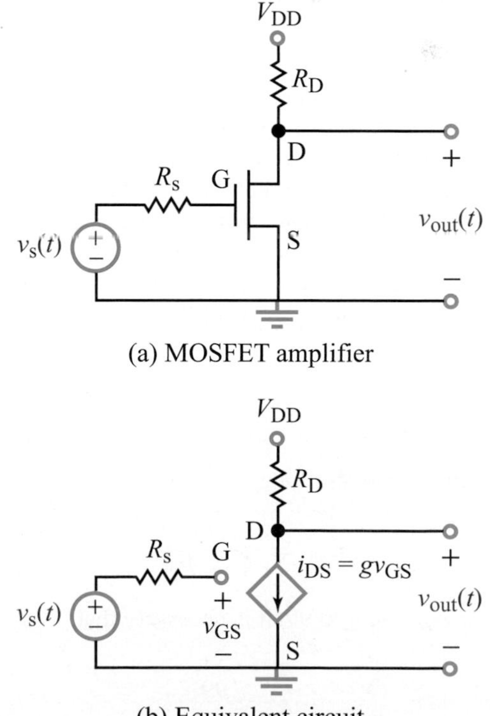

Figure 4-28: MOSFET amplifier circuit for Example 4-9.

the MOSFET equivalent model with $g = 10$ A/V to obtain an expression for $v_{out}(t)$.

Solution: Upon replacing the MOSFET with its equivalent circuit, we end up with the circuit in Fig. 4-28(b). At the input side, because no current flows through R_s, it follows that

$$v_{GS}(t) = v_s(t),$$

and at the output side,

$$v_{out}(t) = V_{DD} - i_{DS}R_D = V_{DD} - gR_D v_{GS}(t)$$
$$= V_{DD} - gR_D v_s(t).$$

We observe that the output voltage consists of a constant dc component (namely V_{DD}) and an ac component that is directly proportional to the input signal $v_s(t)$. For the element values specified in the problem,

$$v_{out}(t) = 10 - 10 \times 10^3 \times (500 + 40\cos 300t) \times 10^{-6}$$
$$= 5 - 0.4\cos 300t \quad \text{V}.$$

The 5-V dc component is simply a level shift superimposed on which is a cosinusoidal signal that is identical with the input signal but is inverted and amplified by an ac gain of 10^4 (from 40 μV to 0.4 V).

Example 4-10: MOSFET Buffer

The circuit in Fig. 4-29(a) consists of a real voltage source (v_s, R_s) connected directly to a load resistor R_L. In contrast, the circuit in Fig. 4-29(b) uses a common-drain MOSFET circuit inbetween the source and the load to *buffer* (insulate) the source from the load. Let us define the source as being buffered from the load if the output voltage across the load is equal to at least 99 percent of v_s. For each circuit, determine the condition on R_L that will satisfy this criterion. Assume $R_s = 100\ \Omega$ and the MOSFET gain factor $g = 10$ A/V.

Solution:

(a) No-Buffer Circuit

For the circuit in Fig. 4-29(a),

$$v_{out_1} = \frac{v_s R_L}{R_s + R_L}.$$

In order for $v_{out_1}/v_s \geq 0.99$, it is necessary that

$$\frac{R_L}{R_s} \geq 99$$

or

$$R_L \geq 9.9\ \text{k}\Omega \qquad (\text{for } R_s = 100\ \Omega).$$

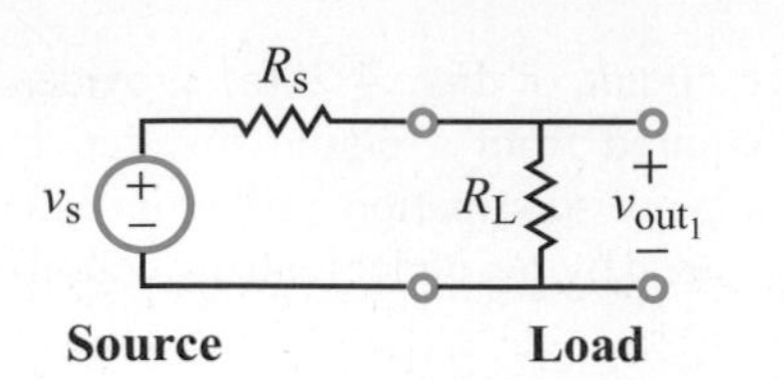

(a) Source connected to load directly

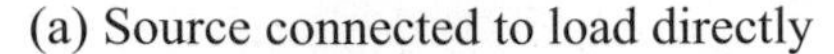

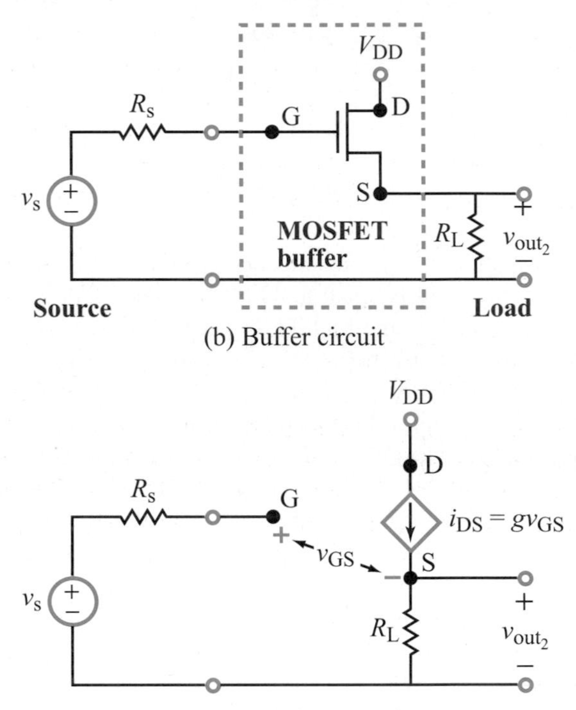

(c) Equivalent circuit

Figure 4-29: Buffer circuit for Example 4-10.

(b) With MOSFET Buffer

For the circuit in Fig. 4-29(c), in which the MOSFET has been replaced with its equivalent circuit, KVL gives

$$-v_s + v_{GS} + v_{out_2} = 0.$$

Also,

$$v_{out_2} = I_{DS}R_L = gR_L v_{GS}.$$

Simultaneous solution of the two equations gives

$$v_{out_2} = \left(\frac{gR_L}{1 + gR_L}\right) v_s.$$

With $g = 10$ A/V and in order for v_{out_2} to be no less than $0.99v_s$ it is necessary that

$$R_L \geq 9.9\ \Omega,$$

which is three orders of magnitude smaller than the requiremen for the un-buffered circuit.

Technology Brief 9: Computer Memory Circuits

The storage of information in electronically addressable devices is one of the hallmarks of all modern computer systems. Among these devices are a class of storage media, collectively called ***solid-state*** or ***semiconductor memories***, which store information by changing the state of an electronic circuit. The state of the circuit usually has two possibilities (0 or 1) and is termed a ***bit*** [see Section 4-10]. Values in memories are represented by a string of ***binary*** bits; a 5-bit sequence $[V_1 V_2 V_3 V_4 V_5]$, for example, can be used to represent any integer decimal value between 0 and 31. How do computers store these bits? Many types of technologies have emerged over the last 40 years, so in this Brief, we will highlight some of the principal technologies in use today or under development. It is worth noting that memory devices usually store these values in arrays. For example, a small memory might store sixteen different 16-bit numbers; this memory usually would be referred to as a 16×16 block or a 256-bit memory. Of course, modern 256-Megabyte computer memories use thousands of much larger blocks to store very large numbers of bits!

Read-Only Memories (ROMs)

One of the oldest, still-employed, memory architectures is the ***Read-Only Memory (ROM)***. The ROM is so termed because it can only be "written" once, and after that it can only be read. ROMs usually are used to store information that will not need to be changed (such as certain startup information on your computer or the short code that runs your digital-camera microcontroller). Each bit in the ROM is held by a single MOSFET transistor.

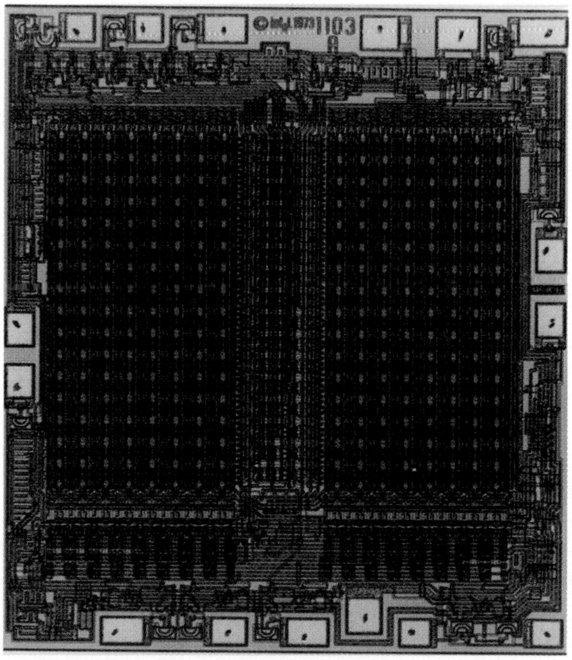

Figure TF9-1: The Intel 1103 DRAM chip, circa 1973. The 1103 could store 1000 bits and used PMOS technology. (Courtesy of Intel Corporation.)

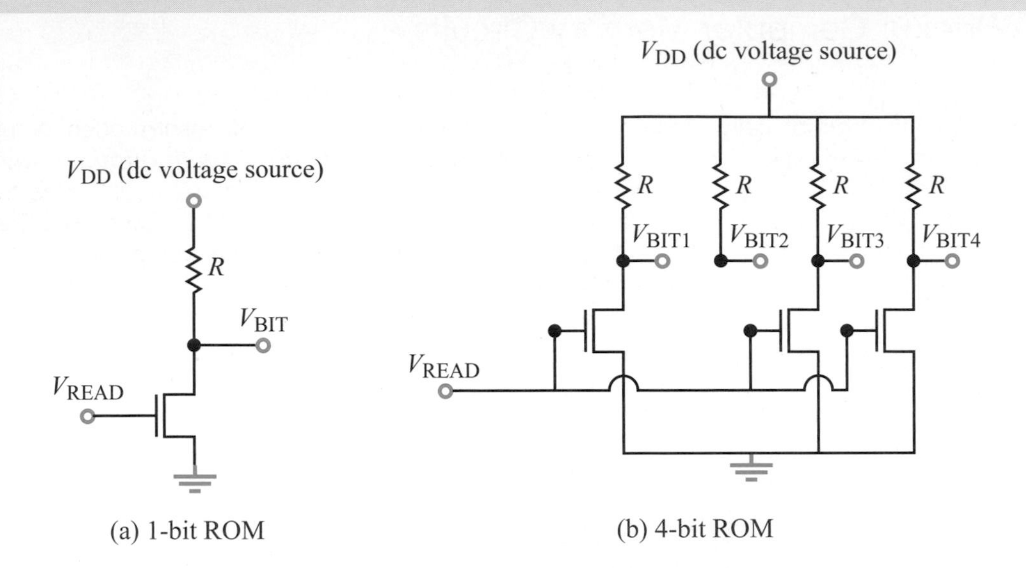

Figure TF9-2: (a) 1-bit ROM that uses a MOSFET transistor, and (b) 4-bit ROM configured to store the sequence [0100], whose decimal value is 4.

Consider the circuit in Fig. TF9-2(a), which operates much like the circuit in Figure 4-25. The MOSFET has three voltages, all referenced to ground. For convenience, the input voltage is labeled V_{READ} and the output voltage is labeled V_{BIT}. The third voltage, V_{DD}, is the voltage of the dc power supply connected to the drain terminal via a resistor R. If $V_{\text{READ}} \ll V_{\text{DD}}$, then the output registers a voltage $V_{\text{BIT}} = V_{\text{DD}}$ denoting the binary state "1," but if $V_{\text{READ}} \geq V_{\text{DD}}$, then the output terminal shorts to ground, generating $V_{\text{BIT}} = 0$ denoting the binary state "0." But how does this translate into a permanent memory on a chip? Let us examine the 4-bit ROM diagrammed in Fig. TF9-2(b). In this case, some bits simply do not have transistors; V_{BIT2}, for example, is permanently connected to V_{DD} via a resistor. This may seem trivial, but this specific 4-bit memory configuration always stores the value [0100]. In this same way, thousands of such components can be strung together in rows and columns in $N \times N$ arrays. As long as a power supply of voltage V_{DD} is connected to the circuit, the memory will report its contents to an external circuit as [0100]. Importantly, even if you remove power altogether, the values are not lost; as soon as you add power back to the chip, the same values appear again (i.e., you would have to break the chip to make it forget what it is storing!). Because of this, these memories also often are called ***non-volatile memories (NVM)***.

Random-Access Memories (RAMs)

RAMs are a class of memories that can be read to and written from constantly. RAMs generally fall into two categories: ***static RAMs*** and ***dynamic RAMs (DRAMs)***. Because RAMs lose the state of their bits if the power is removed, they are termed ***volatile memories***. Static RAMs not only can be read from and written to, but also do not forget their state as long as power is supplied. These circuits also are composed of transistors, but each single bit in a modern static RAM consists of four transistors wired up in a bi-stable circuit (the explanation of which we will leave to your intermediate digital components classes!). Dynamic RAMs, on the other hand, are illustrated more easily. Dynamic RAMs usually hold more bits per area than static RAMs, but they need to be refreshed constantly (even when power is supplied continuously to the chip).

Figure TF9-3 shows a simple one-transistor dynamic RAM. Again, we will treat the transistor as we did in Section 4-11. Note that if we make $V_{\text{ROW}} > V_{\text{DD}}$, then the transistor will conduct and the capacitor C will start charging whatever value we select for V_{COLUMN}. When writing a bit, V_{COLUMN} usually is set at either 0 (GND) or 1 (V_{DD}). We can calculate how long this charging-up process will require, because we know the value of C and the transistor's current gain g_{m} (see Section 5-7). When the capacitor is charged to V_{DD}, a value of 1 is stored in the DRAM. Had we applied instead a value of zero volts to V_{COLUMN}, the transistor would have discharged to ground (instead of charged to V_{DD}) and the bit would have a value of 0. However, note that unlike the ROM, the state of the bit is not "hardwired." That is, if even tiny leakage currents were to flow through the transistor when it is not on (that is, when $V_{\text{ROW}} < V_{\text{DD}}$), then charge will constantly leak away and the voltage of the transistor will drop slowly with time. After a short time (on the order of a few milliseconds in the dynamic RAM in your computer), the capacitor will have irrecoverably lost its value. How is that mitigated? Well, it turns out that a modern memory will read and then re-write every one of its (several billion) bits every 64 milliseconds to keep them refreshed! Because each bit is so simple (one transistor and one capacitor), it is possible to manufacture DRAMs with very high memory densities (which is why 1-Gbit DRAMs are now available in packages of reasonable size). Other variations of DRAMs also exist whose architectures deviate slightly from the previous model—at either the transistor or system level. ***SGRAM (Synchronous Graphics RAM)***, for example, is a DRAM modified for use with graphics adaptors; ***DDR2RAM (Double Data Rate 2 RAM)*** is a second-generation enhancement over DRAM which allows for faster clock speeds and dual-processor operation.

Advanced Memories

Several substantially different technologies are emerging that likely will change the market landscape—just as Flash memories (Technology Brief 8: Flash Memory on page 158) revolutionized portable memory (like your USB memory stick). Apart from the drive to increase storage density and access speed, one of the principal drivers in today's memory research is the development of non-volatile memories that do not degrade over time (unlike Flash).

The ***FeRAM (Ferroelectric RAM)*** is the first of these technologies to enter mainstream production; FeRAM replaces the capacitor in DRAM (Fig. TF9-3) with a ferroelectric capacitor that can hold the binary state even with power removed. While FeRAM can be faster than Flash memories, FeRAM densities are still much smaller than modern Flash (and Flash densities continue to increase rapidly). FeRAM currently is used in niche applications where the increased speed is important. ***MRAM (Magnetoresistive RAM)*** is another emerging technology, currently commercialized by Freescale Semiconductor, which relies on magnetic plates to store bits of data. In MRAM, each cell is composed of two ferromagnetic plates separated by an insulator. The storage and retrieval of bits occurs by manipulation of the magnetic polarization of the plates with associated circuits. Like FeRAM, MRAM currently is overshadowed by Flash memories, but improvements in density, speed, and fabrication methods may make it a viable alternative in the mainstream consumer market in the future. Even more speculative is the idea of using single carbon nanotubes to store binary bits by changing their configuration electronically; this technology is currently known as ***NRAM (Nano RAM)***.

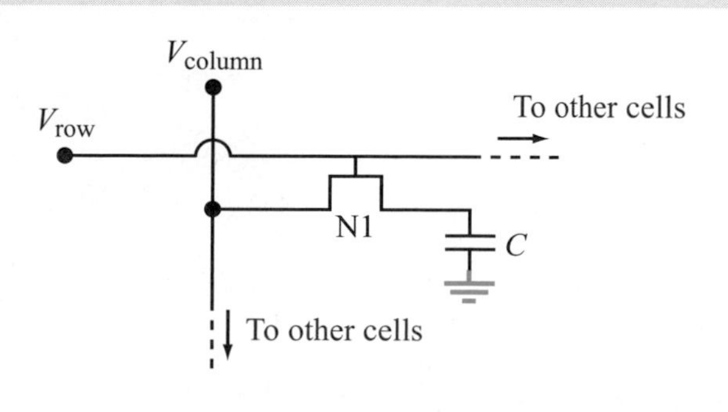

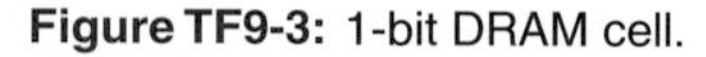
Figure TF9-3: 1-bit DRAM cell.

Review Question 4-19: What is the major advantage of a CMOS over an NMOS as a digital inverter?

Review Question 4-20: When a MOSFET is used in a buffer circuit, $v_{\text{out}} \simeq v_{\text{s}}$, where v_{s} is the input signal voltage. So, why is it used?

Exercise 4-10: In the circuit of Example 4-9, what value of R_{D} will give the highest possible ac gain while keeping $v_{\text{out}}(t)$ always positive?

Answer: $R_{\text{D}} = 1.85$ kΩ. (See ◎)

Exercise 4-11: Repeat Example 4-10, but require that v_{out} be at least 99.9 percent of v_{s}. What should R_{L} be (a) without the buffer and (b) with the buffer?

Answer: (a) $R_{\text{L}} \geq 99.9$ kΩ, (b) $R_{\text{L}} \geq 99.9$ Ω. (See ◎)

4-12 Application Note: Neural Probes

The human brain is composed, in part, of interconnected networks of individual, information-processing cells known as ***neurons***. There are about one trillion (10^{12}) neurons in the human brain with each neuron having on average 7000 connections to other neurons. Although the working of the neural system is well beyond the scope of this book, it is important to note that when a neuron transmits information, it causes a change in the concentrations of various ions in its vicinity. This movement of ions gives rise to an electric current through the neuron's membrane which in turn generates a change in potential (voltage) between various parts of the cell and its surroundings. Thus, when a given neuron fires, a small ($\sim$ 100 mV) but detectable potential drop develops between the cell and its surroundings.

Over the past few decades, various types of devices were built for measuring this electrical phenomenon in neurons. In recent years, however, the field has achieved phenomenal success due in part to the successful development of ***neural probes*** (also known as ***neural interfaces***) with very high sensitivities. An example of a 3-dimensional probe is shown in Fig. 4-30. It consists of a 2-D array of very thin probes—each instrumented with a sensor at each of several locations along its length. With such a probe, it is now possible to measure the ***action potentials*** of firing neurons at a large number of brain locations simultaneously. Modern neural interface systems also have been developed to stimulate or change the electrical state of specific neurons, thereby affecting their operation in the brain. These types of devices not only offer the potential of unraveling aspects of brain development and operation, but they also are beginning to see use in clinical applications for the treatment of chronic neurological disorders, such as Parkinson's disease.

Because these voltage signals are so small, on-board amplification, noise-removal, and analog-to-digital circuitry are needed to process the signal from the brain to the recording device.

Example 4-11: Neural Probe

The neural probe shown in Fig. 4-31 consists of a long shank at the end of which lie two metal electrodes. This shank is inserted a short distance into the brain and the signal coming from these electrodes is recorded. For simplicity, we will model the brain activity between the two probes just like a realistic voltage source V_{s} in series with a resistance R_{s}. The source produces inverted pulses with -100 mV amplitudes. Note that neither V_a nor V_b are grounded relative to the ground level of the circuit. The neural signal needs to be inverted and amplified so that it can be presented to an analog-to-digital converter (ADC) which only operates in the 0 to 5 V range. Design the amplifier circuit.

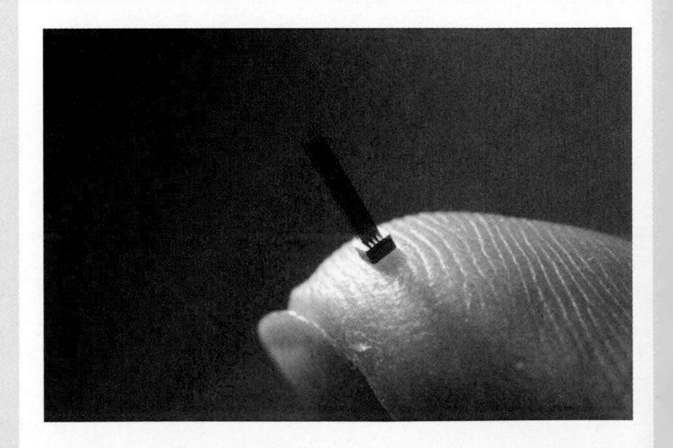

Figure 4-30: Three-dimensional neural probe (5 mm × 5 mm × 3 mm). (Courtesy of Prof. Ken Wise and Gayatri Perlin, University of Michigan.)

Solution: The input signal is represented by the difference between V_a and V_b, and since neither of those terminals is grounded, some sort of differential amplifier is the logical choice for the intended application.

The amplifier should invert the input signal and amplify it into the 0 to 5 V range required by the ADC. Given these constraints, we propose to use the op-amp instrumentation amplifier circuit of Fig. 4-20 with V_a as input v_1 and V_b as input v_2. The amplifier output is proportional to $(v_2 - v_1)$, so the choice of connections we made will realize the inversion requirement automatically. According to Eq. (4.56), if we choose the circuit resistors such that $R_1 = R_3 = R_4 = R_5 = R$, the output voltage is given by

$$v_0 = \left(1 + \frac{2R}{R_2}\right)(v_2 - v_1) = \left(1 + \frac{2R}{R_2}\right)(V_b - V_a)$$

$$= -\left(1 + \frac{2R}{R_2}\right)(V_a - V_b).$$

To amplify $(V_a - V_b)$ from -100 mV to $+5$ V, the ratio (R/R_2) should be chosen such that

$$5 = -\left(1 + \frac{2R}{R_2}\right) \times (-100 \times 10^{-3})$$

or, equivalently,

$$\frac{R}{R_2} = 24.5.$$

If we set $R = 100$ kΩ, then R_2 should be 4.08 kΩ. This will yield a 5-V pulse to the ADC every time a -100-mV pulse is generated by the neuron.

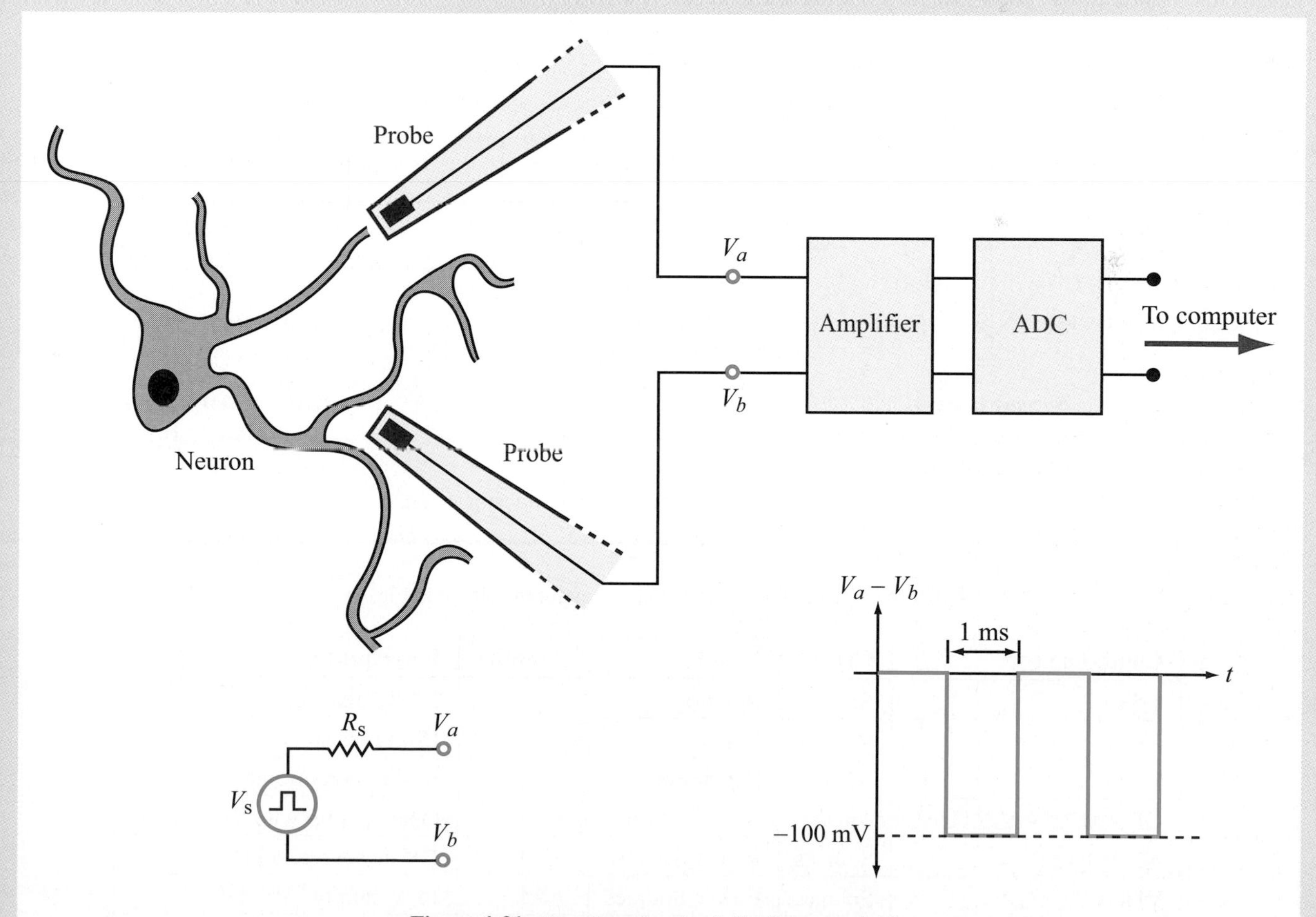

Figure 4-31: Neural-probe circuit for Example 4-11.

4-13 Multisim Analysis

One of the most attractive features of Multisim is its interactive-simulation mode, which we began to utilize in Sections 2-8 and 3-8. The simulation mode allows you to connect virtual test instruments to your circuit and to operate them in real time as Multisim simulates the circuit behavior. In this section, we will explore this feature with an op-amp circuit and two MOSFET circuits.

4-13.1 Op Amps and Virtual Instruments

The circuit shown in Fig. 4-32 uses a resistive Wheatstone bridge (Section 2-6) to detect the change of resistance induced in a sensor modeled as a variable resistor (see Technology Brief 4 on page 56). The output of the bridge is fed into a pair of voltage followers and then into a differential amplifier. The circuit can be constructed and tested in Multisim using the components listed in Table 4-5. The resistance value of the potentiometer component is adjustable with a keystroke (the default is the key "a" to change the resistance in one direction and the default key combination Shift-a to change the resistance in the opposite direction) or by using the mouse slider under the component. In order to observe how changes in the potentiometer cause changes in the output, we need to connect the output to an ***oscilloscope***. Multisim provides several oscilloscopes to choose from, including a generic instrument and virtual versions of commercial oscilloscopes made by Agilent and Tektronix. For starters, it is easiest to use the generic instrument by selecting Simulate → Instruments → Oscilloscope, or by selecting and dragging an oscilloscope from the instrument dock. Figure 4-33 shows the complete circuit drawn in Multisim. The power supplies for the op amps can be found under Components → Sources → POWER SOURCES → VDD (or VSS). Once placed, double-click the VDD (or VSS) component, select the values tab and set the voltage to 15 V for VDD and −15 V for VSS. Once the circuit is complete, you can begin the simulation by pressing F5 (or

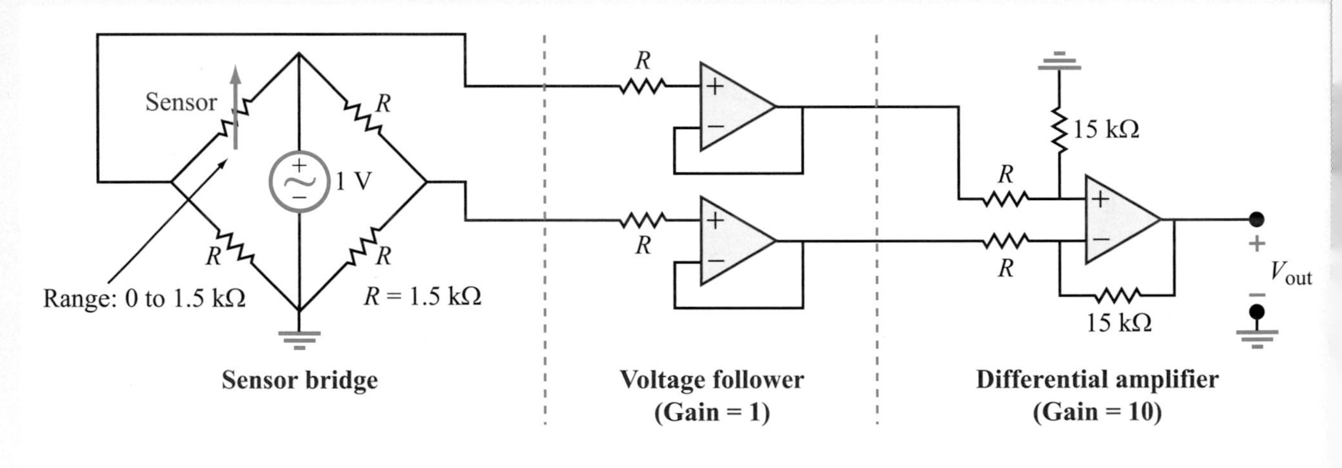

Figure 4-32: Wheatstone-bridge op-amp circuit.

Table 4-5: List of Multisim components for the circuit in Fig. 4-32.

Component	Group	Family	Quantity	Description
1.5 k	Basic	Resistor	7	1.5-kΩ resistor
15 k	Basic	Resistor	2	15-kΩ resistor
1.5 k	Basic	Potentiometer	1	1.5-kΩ potentiometer
OP_AMP_5T_VIRTUAL	Analog	Analog_Virtual	3	Ideal op amp with 5 terminals
AC_POWER	Sources	Power_Sources	1	1-V ac source, 60 Hz
VDD	Sources	Power_Sources	1	15-V supply
VSS	Sources	Power_Sources	1	−15-V supply

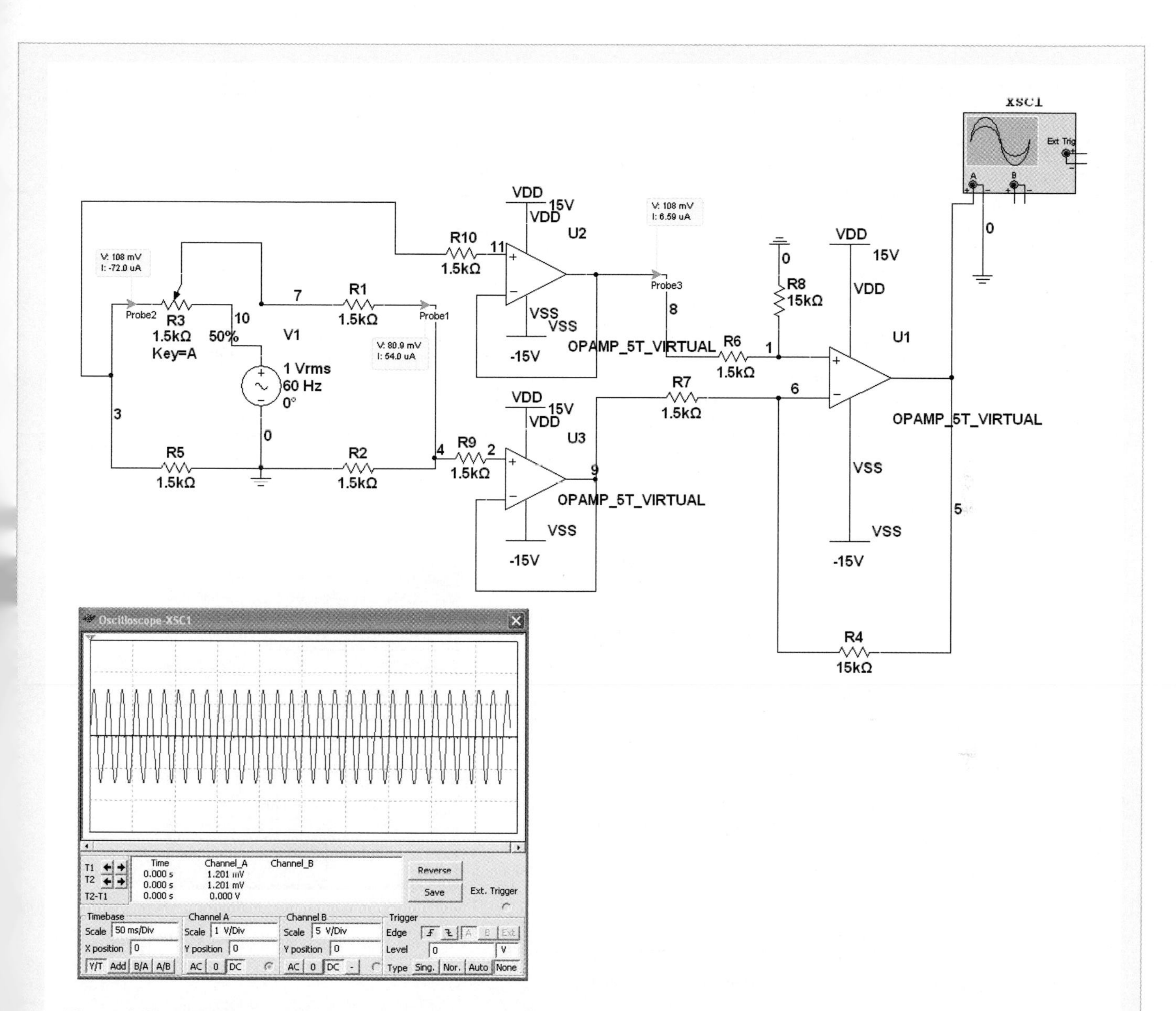

Figure 4-33: Multisim window of the circuit of Fig. 4-32. The oscilloscope trace shows the 60-Hz waveform of the output voltage. Had the voltage source been a dc source, the oscilloscope trace would have been a horizontal line.

Simulate → Run) and pause it by pressing F6. Double-click on the oscilloscope element in the schematic to bring up the oscilloscope's screen. The output voltage should be visible as Channel A in the oscilloscope window. In order to get a good view of the trace, you might need to adjust both its timebase and voltage scale using the controls found at the bottom of the Oscilloscope window. Observe the change in the amplitude of the output by shifting the resistance value of the sensor potentiometer.

With Multisim, you can modify different parts of the circuit and observe the consequent changes in behavior. Make sure to stop your simulation (not just pause it) before changing components or wiring.

Review Question 4-21: What types of Multisim instruments are available for testing a circuit?

Review Question 4-22: Explain what the timebase is on the oscilloscope.

Exercise 4-12: Why are the voltage followers necessary in the circuit of Fig. 4-33? Remove them from the Multisim circuit and connect the resistive bridge directly to the two inputs of the differential amplifier. How does the output vary with the potentiometer setting?

Answer: (See)

4-13.2 The Digital Inverter

The MOSFET inverter introduced in Section 4-11.2 provides a good opportunity to explore the difference between steady-state and time-dependent analysis techniques. Consider again the MOSFET digital inverter of Fig. 4-27. When analyzing this type of logic gate, we usually are interested in both the response of the output voltage to a change in input voltage and in how fast the gate generates the output voltage in response to a change in input voltage. Both types of analyses are possible with Multisim.

Figure 4-34 shows a MOSFET inverter circuit in Multisim. To draw this circuit, you need the components listed in Table 4-6.

Transient Analysis

We can use a function generator (Simulate → Instruments → Function Generator) to observe the inverter output as a

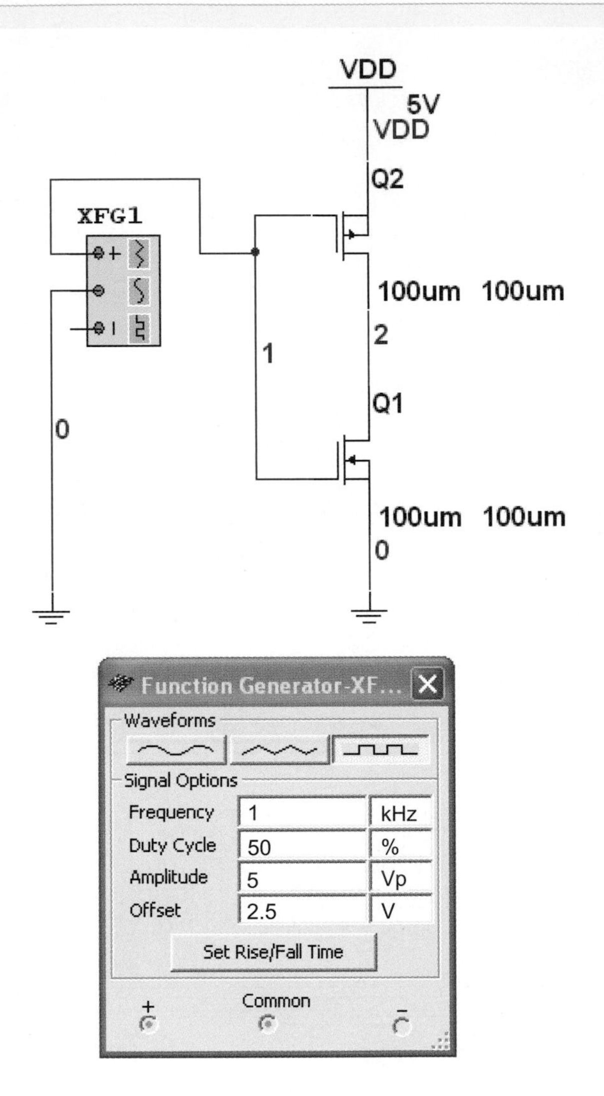

Figure 4-34: Multisim equivalent of the MOSFET circuit of Fig. 4-27.

Table 4-6: Components for the circuit in Fig. 4-34.

Component	Group	Family	Quantity	Description
MOS_3TDN_VIRTUAL	Transistors	Transistors_VIRTUAL	1	3-terminal N-MOSFET
MOS_3TDP_VIRTUAL	Transistors	Transistors_VIRTUAL	1	3-terminal P-MOSFET
VDD	Sources	Power Sources	1	5-V supply
GND	Sources	Power Sources	2	Ground node

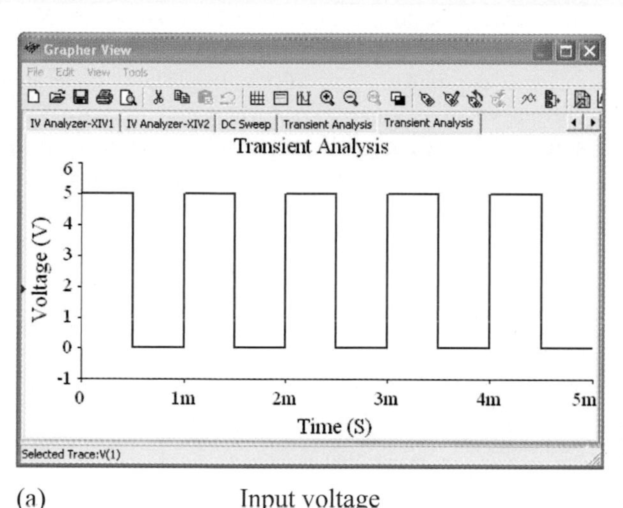

(a) Input voltage

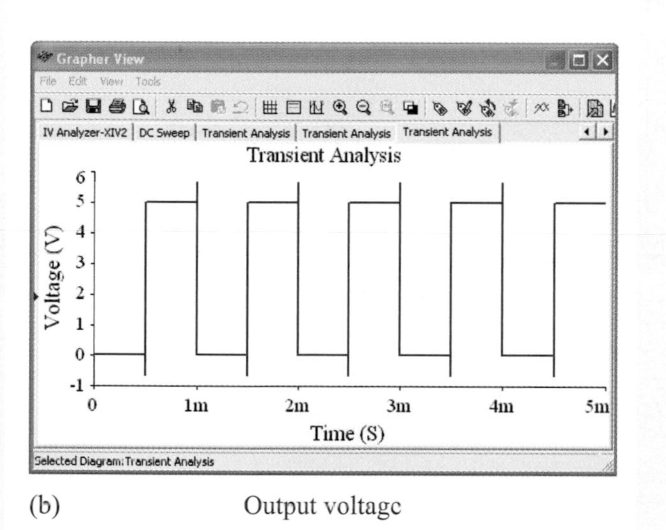

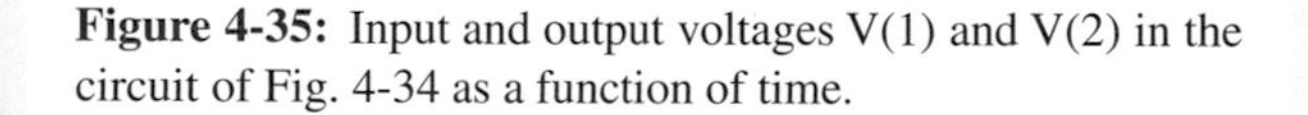
(b) Output voltage

Figure 4-35: Input and output voltages V(1) and V(2) in the circuit of Fig. 4-34 as a function of time.

function of time. Double-click on the function generator to bring up its control window. Set the function generator to Square Wave mode with a frequency of 1 kHz, amplitude of 2.5 V, and an offset of 2.5 V. This will generate a 0–5-V square-wave input. The input and output can be plotted separately as a function of time using Simulate → Analyses → Transient Analysis. Whereas in Interactive Simulation the course of time is open ended (by default it is limited to a duration of 1×10^{30} s), when using Transient Analysis we can define the start and stop times. Maintain the start time at 0 s, set the final time to 0.005 s, and under the Output tab select the input voltage V(1) as the voltage to plot. Click Simulate. The input voltage is plotted as a function of time, as in Fig. 4-35(a). Repeat the simulation after removing V(1) and adding V(2) under the Output tab. Figure 4-35(b) shows the output voltage as a function of time. The input and output plots are essentially mirror images of one another.

Steady-State Analysis

In order to analyze the steady-state output behavior, we first must remove the function generator and replace it with a dc voltage source. The actual voltage value of the source is unimportant. Once wired, select Simulate → Analyses → DC Sweep. This analysis is similar to the DC Operating Point Analysis, but it sweeps through a range of voltages at a node of your choice and solves for the resultant steady-state voltage (or current) at any other node you select. In this way, you can generate and plot input–output relationships for circuits and components.

Choose the source name vv1 as the input and enter 0 V, 5 V, and 0.5 V for the start, stop, and increment values, respectively. Under the Output tab, select the output voltage V(2) as the voltage to plot. Click Simulate. Figure 4-36 shows that the output displays the expected inverter behavior: an input in the 0 to 2 V range generates an output of ~ 5 V; conversely, when the input is in the range between 3 and 5 V, the circuit generates an output voltage of ~ 0 V. In between, we see a gradual transition zone.

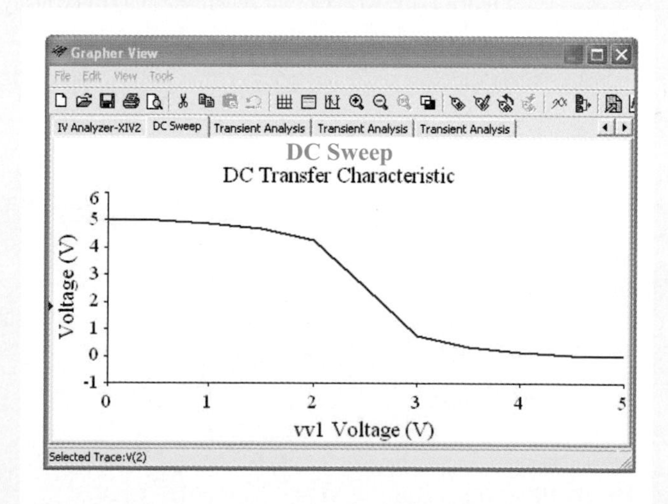

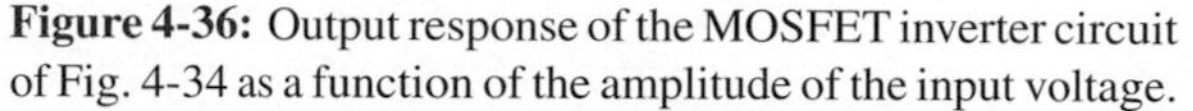
Figure 4-36: Output response of the MOSFET inverter circuit of Fig. 4-34 as a function of the amplitude of the input voltage.

Review Question 4-23: How do the DC Operating Point Analysis, Transient Analysis, and DC Sweep analyses differ?

Review Question 4-24: How many types of waveforms can the generic function-generator instrument provide?

Exercise 4-13: The *IV Analyzer* is another useful Multisim instrument for analyzing circuit performance. To demonstrate its utility, let us use it to generate characteristic curves for an NMOS transistor similar to those in Fig. 4-25(b). Figure E4.13(a) shows an NMOS connected to an IV Analyzer. The instrument sweeps through a range of gate (G) voltages and generates a current-versus-voltage (IV) plot between the drain (D) and source (S) for each gate voltage. Show that the display of the IV analyzer is the same as that shown in Fig. E4.13(b).

Answer: (See)

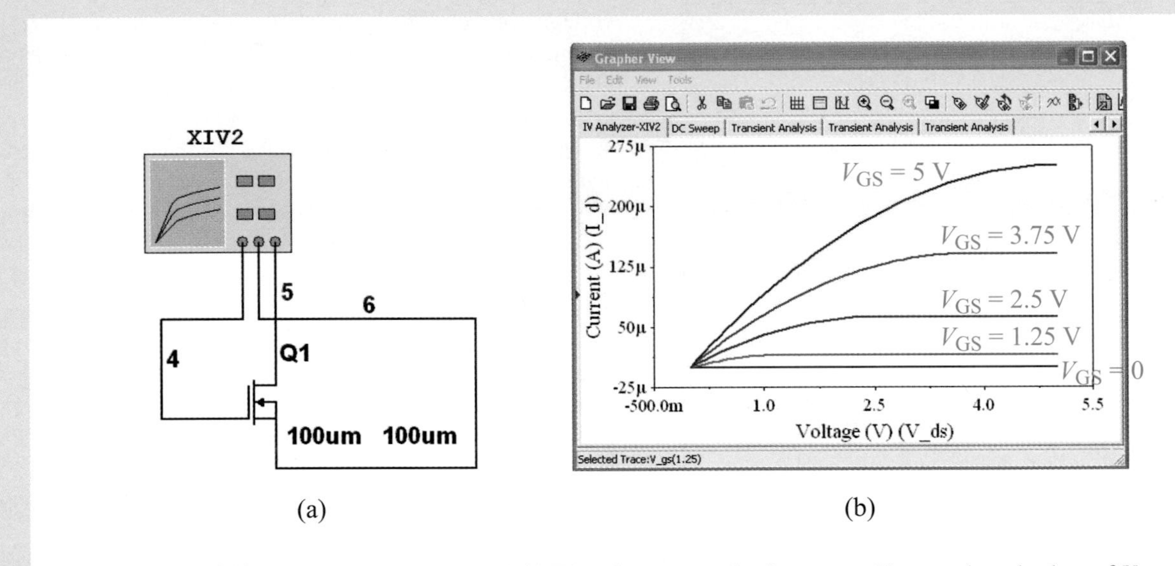

Figure E4.13: (a) Circuit schematic and (b) IV analyzer traces for I_{DS} versus V_{DS} at selected values of V_{GS}.

Chapter 4 Relationships

Ideal op amp	$v_p = v_n$ $i_p = i_n = 0$
Noninverting amp	$G = \dfrac{v_o}{v_s} = \dfrac{R_1 + R_2}{R_2}$
Inverting amp	$G = \dfrac{v_o}{v_s} = -\left(\dfrac{R_f}{R_s}\right)$
Summing amp	$v_o = -R_f\left(\dfrac{v_1}{R_1} + \dfrac{v_2}{R_2}\right)$
Difference amp	$v_o = G_2 v_2 + G_1 v_2$
Voltage follower	$v_o = v_s$
Instrumentation amp (with gain-control resistor R_2)	$v_o = \left(1 + \dfrac{2R}{R_2}\right)(v_2 - v_1)$
MOSFET	$V_{out} = V_{DD} - g R_D V_{in}$

CHAPTER HIGHLIGHTS

- Despite its complex circuit architecture, the op amp can be modeled in terms of a relatively simple, linear equivalent circuit.
- The ideal op amp has infinite gain, infinite input resistance, and zero output resistance.
- Through resistive feedback connections between its output and its two inputs, the op amp can be made to amplify, sum, and subtract multiplc input signals.
- Multistage op-amp circuits can be configured to support a variety of signal-processing functions.
- The instrumentation amplifier is a high-gain, high-sensitivity detector of small signals, making it particularly suitable for sensing deviations from reference conditions.
- Multisim can accommodate op-amp circuits and simulate their input–output responses.

GLOSSARY OF IMPORTANT TERMS

Provide definitions or explain the meaning of the following terms:

buffer
closed-loop gain
current constraint
difference amplifier
digital-to-analog converter (DAC)
dynamic range
feedback
feedback resistance
instrumentation amplifier
inverting adder
inverting amplifier
inverting input
least significant bit
MOSFET
most significant bit
negative feedback
noninverting input
op-amp gain
open-loop gain
R–$2R$ ladder
signal-processing circuit
summing amplifier
voltage constraint
voltage follower

PROBLEMS

Sections 4-1 and 4-2: Op-Amp Characteristics and Negative Feedback

4.1 An op amp with an open-loop gain of 10^6 and $V_{cc} = 12$ V has an inverting-input voltage of 20 μV and a noninverting-input voltage of 10 μV. What is its output voltage?

4.2 An op amp with an open-loop gain of 6×10^5 and $V_{cc} = 10$ V has an output voltage of 3 V. If the voltage at the inverting input is -1 μV, what is the magnitude of the noninverting-input voltage?

4.3 What is the output voltage for an op amp whose noninverting input is connected to ground and its inverting-input voltage is 4 mV? Assume that the op-amp open-loop gain is 2×10^5 and its supply voltage is $V_{cc} = 10$ V.

4.4 With its noninverting-input voltage at 10 μV, the output voltage of an op amp is -15 V. If $A = 5 \times 10^5$ and $V_{cc} = 15$ V, can you determine the magnitude of the inverting-input voltage? If not, can you determine its possible range?

*Answer(s) in Appendix E.

4.5 For the op-amp circuit shown in Fig. P4.5:

(a) Use the model given in Fig. 4-4 to develop an expression for the current gain $G_i = i_L / i_s$.

(b) Simplify the expression by applying the ideal op-amp model (taking $A \to \infty$, $R_i \to \infty$, and $R_o \to 0$).

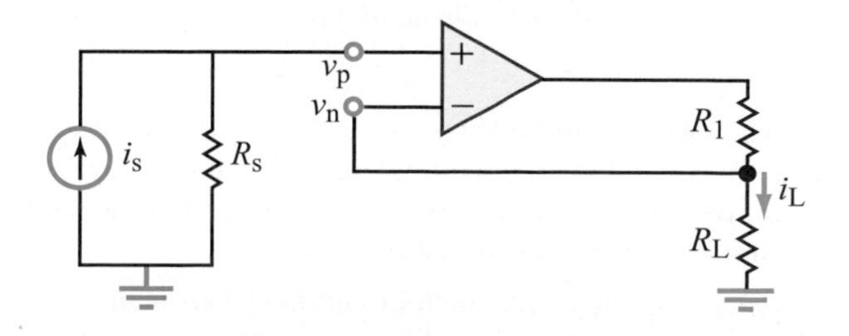

Figure P4.5: Circuit for Problem 4.5.

4.6 The inverting-amplifier circuit shown in Fig. P4.6 uses a resistor R_f to provide feedback from the output terminal to the inverting-input terminal.

(a) Use the equivalent-circuit model of Fig. 4-4 to obtain an expression for the closed-loop gain $G = v_o/v_s$ in terms of R_s, R_i, R_o, R_L, R_f, and A.

(b) Determine the value of G for $R_s = 10\ \Omega$, $R_i = 10\ \text{M}\Omega$, $R_f = 1\ \text{k}\Omega$, $R_o = 50\ \Omega$, $R_L = 1\ \text{k}\Omega$, and $A = 10^6$.

(c) Simplify the expression for G obtained in (a) by letting $A \to \infty$, $R_i \to \infty$, and $R_o \to 0$ (ideal op-amp model).

(d) Evaluate the approximate expression obtained in (c) and compare the result with the value obtained in (b).

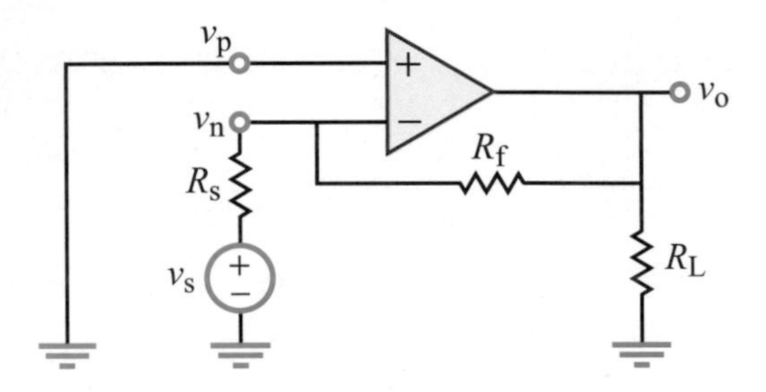

Figure P4.6: Circuit for Problem 4.6.

4.7 For the circuit in Fig. P4.7:

(a) Use the op-amp equivalent-circuit model to develop an expression for $G = v_o/v_s$.

(b) Simplify the expression by applying the ideal op-amp model parameters, namely $A \to \infty$, $R_i \to \infty$, and $R_o \to 0$.

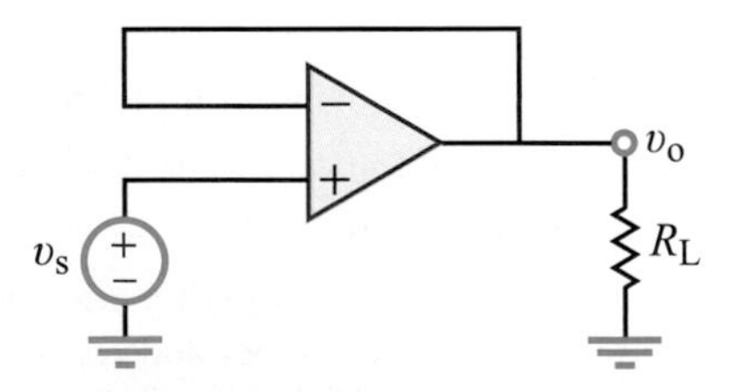

Figure P4.7: Circuit for Problem 4.7.

4.8 The op-amp circuit shown in Fig. P4.8 has a constant dc voltage of 6 V at the noninverting input. The inverting input is the sum of two voltage sources consisting of a 6-V dc source and a small time-varying signal v_s.

(a) Use the op-amp equivalent-circuit model given in Fig. 4-4 to develop an expression for v_o.

(b) Simplify the expression by applying the ideal op-amp model, which lets $A \to \infty$, $R_i \to \infty$, and $R_o \to 0$.

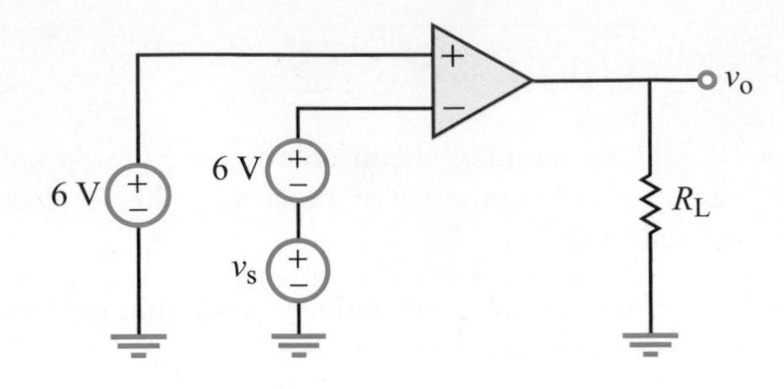

Figure P4.8: Circuit for Problem 4.8.

Sections 4-3 and 4-4: Ideal Op-Amp and Inverting Amp

Assume all op amps to be ideal from here on forward.

*__4.9__ The supply voltage of the op amp in the circuit of Fig. P4.9 is 16 V. If $R_L = 3\ \text{k}\Omega$, assign a resistance value to R_f so that the circuit would deliver 75 mW of power to R_L.

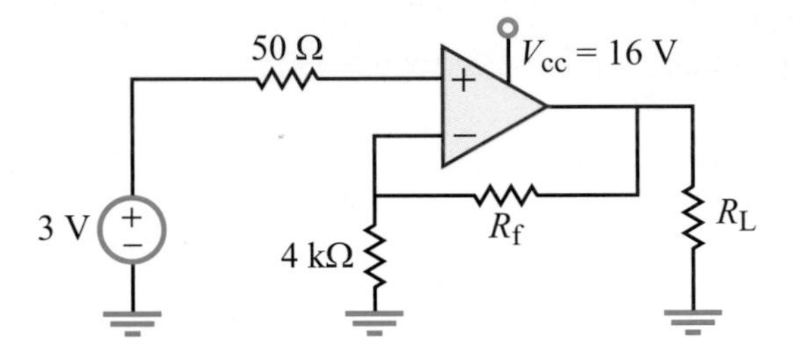

Figure P4.9: Circuit for Problem 4.9.

4.10 In the circuit of Fig. P4.10, a bridge circuit is connected at the input side of an inverting op-amp circuit.

(a) Obtain the Thévenin equivalent at terminals (a, b) for the bridge circuit.

(b) Use the result in (a) to obtain an expression for $G = v_o/v_s$

(c) Evaluate G for $R_1 = R_4 = 100\ \Omega$, $R_2 = R_3 = 101\ \Omega$ and $R_f = 100\ \text{k}\Omega$.

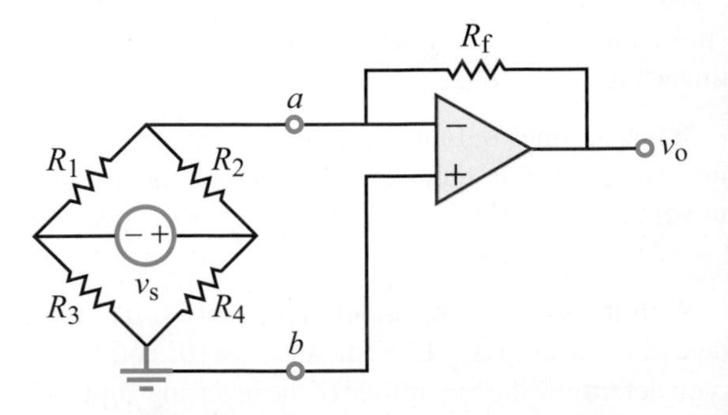

Figure P4.10: Circuit for Problem 4.10.

4.11 Determine the output voltage for the circuit in Fig. P4.11 and specify the linear range for v_s, given that $V_{cc} = 15$ V and $V_0 = 0$.

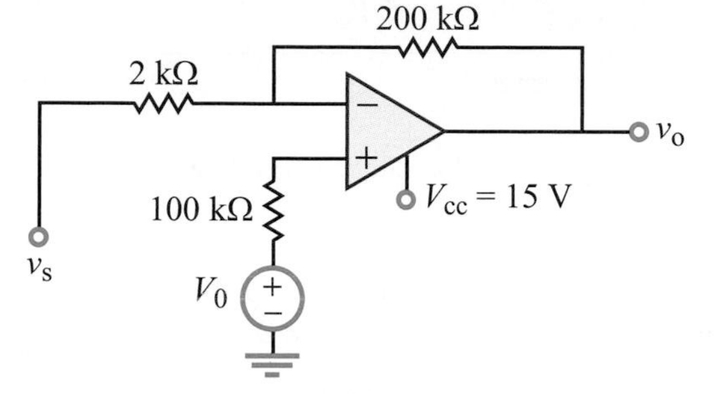

Figure P4.11: Circuit for Problems 4.11 and 4.12.

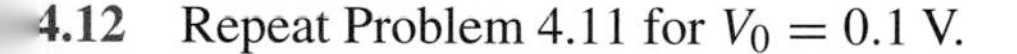

4.12 Repeat Problem 4.11 for $V_0 = 0.1$ V.

4.13 Obtain an expression for the voltage gain $G = v_o/v_s$ for the circuit in Fig. P4.13.

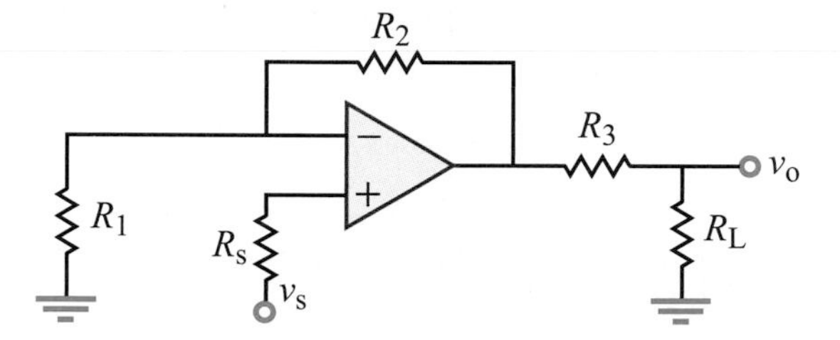

Figure P4.13: Circuit for Problem 4.13.

.14 For the op-amp circuit shown in Fig. P4.14:

(a) Obtain an expression for the current gain $G_i = i_L/i_s$.

b) If $R_L = 12$ kΩ, choose R_f so that $G_i = -15$.

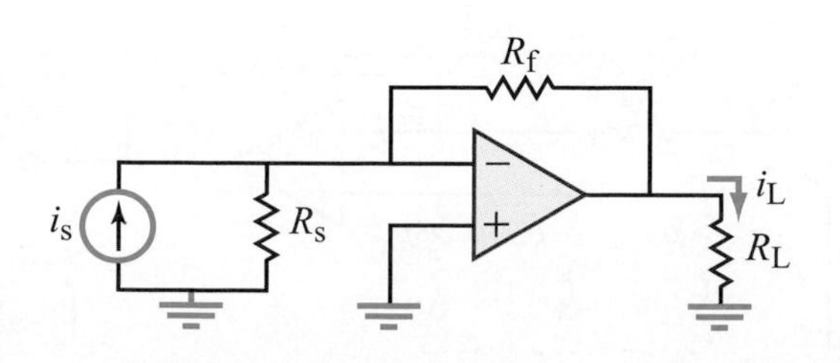

Figure P4.14: Circuit for Problem 4.14.

4.15 Determine the gain $G = v_L/v_s$ for the circuit in Fig. P4.15 and specify the linear range of v_s for $R_L = 4$ kΩ.

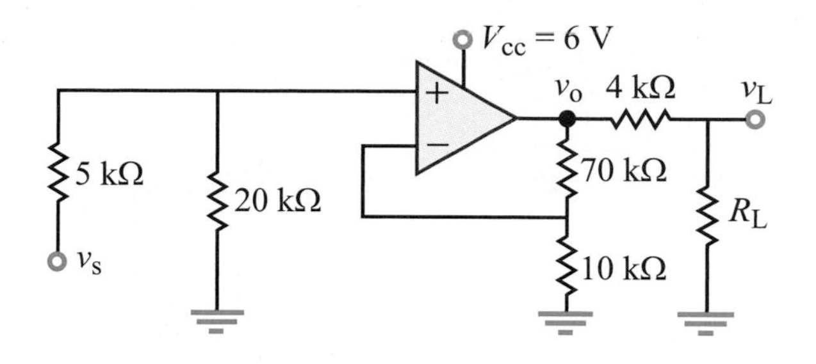

Figure P4.15: Circuit for Problems 4.15 and 4.16.

4.16 For the circuit of Fig. P4.15, what should the resistance value of R_L be so as to have maximum transfer of power into it?

4.17 Determine v_o across the 10-kΩ resistor in the circuit of Fig. P4.17.

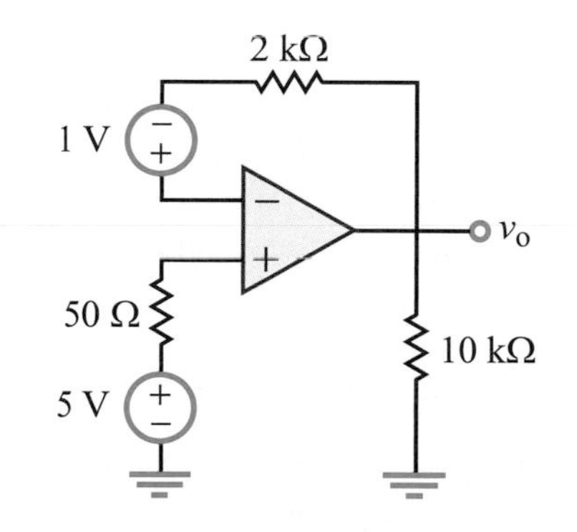

Figure P4.17: Circuit for Problem 4.17.

4.18 Evaluate $G = v_o/v_s$ for the circuit in Fig. P4.18, and specify the linear range of v_s. Assume $R_f = 2400$ Ω.

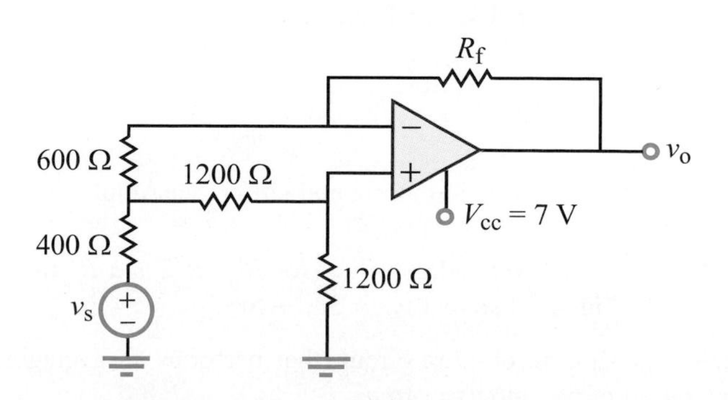

Figure P4.18: Circuit for Problems 4.18 and 4.19.

*__4.19__ Repeat Problem 4.18 for $R_f = 0$.

4.20 Determine the linear range of the source v_s in the circuit of Fig. P4.20.

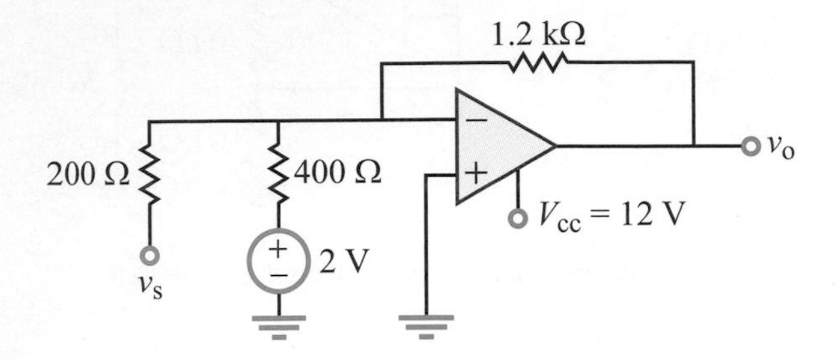

Figure P4.20: Circuit for Problems 4.20 and 4.21.

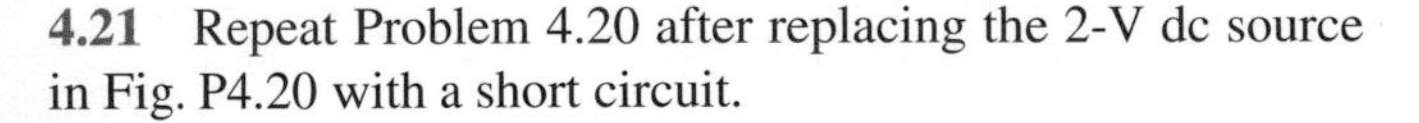

4.21 Repeat Problem 4.20 after replacing the 2-V dc source in Fig. P4.20 with a short circuit.

4.22 The circuit in Fig. P4.22 uses a potentiometer whose total resistance is $R = 10$ kΩ with the upper section being βR and the bottom section $(1-\beta)R$. The stylus can change β from 0 to 0.9. Obtain an expression for $G = v_o/v_s$ in terms of β and evaluate the range of G (as β is varied over its own allowable range).

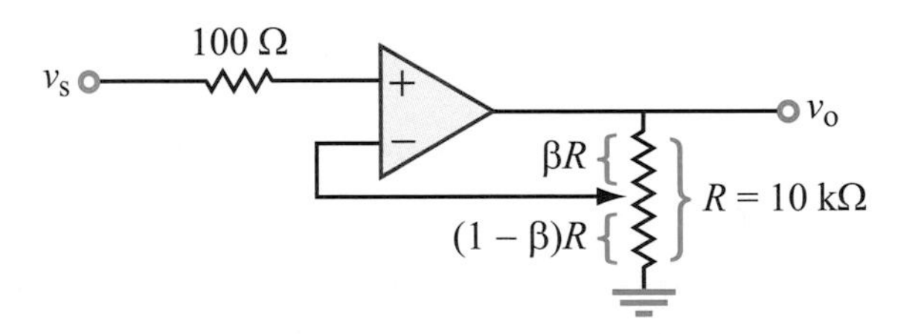

Figure P4.22: Circuit for Problem 4.22.

Sections 4-5 and 4-6: Summing and Difference Amplifiers

4.23 If $R_2 = 4$ kΩ, select values for R_{s_1}, R_{s_2}, and R_1 in the circuit of Fig. 4-14 so that $v_o = 3v_1 + 5v_2$.

4.24 Design an op-amp circuit that performs an averaging operation of five inputs v_1 to v_5.

4.25 For the circuit in Fig. P4.25, generate a plot for v_L as a function of v_s over the full linear range of v_s.

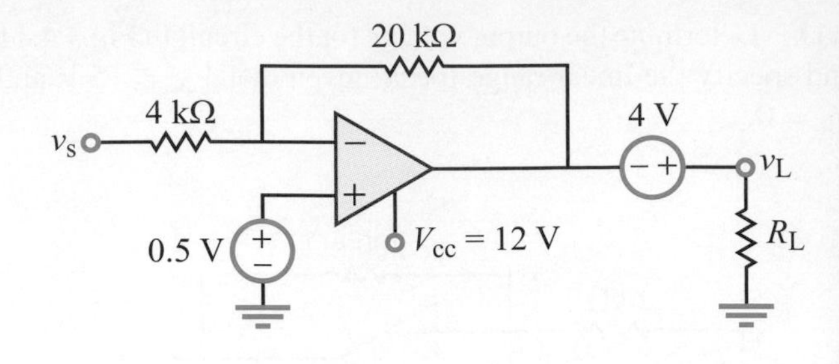

Figure P4.25: Circuit for Problem 4.25.

4.26 Relate v_o in the circuit of Fig. P4.26 to v_s and specify the linear range of v_s. Assume $V_0 = 0$.

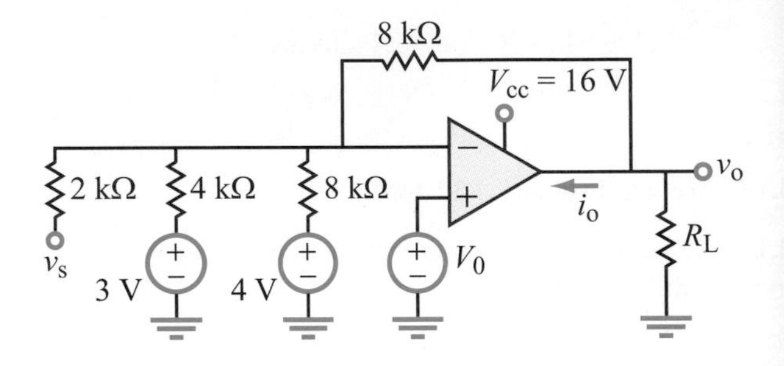

Figure P4.26: Circuit for Problems 4.26 through 4.28.

*__4.27__ Repeat Problem 4.26 for $V_0 = 6$ V.

4.28 Determine the current i_o flowing into the op-amp of the circuit in Fig. P4.26 under the conditions $v_s = 0.5$ V, $V_0 = 0$ and $R_L = 10$ kΩ.

4.29 Design a circuit containing a single op amp that can perform the operation $v_o = 3 \times 10^4(i_2 - i_1)$, where i_2 and i_1 are input current sources.

4.30 Design a circuit that can perform the operation $v_o = 3v_1 + 4v_2 - 5v_3 - 8v_4$, where v_1 to v_4 are input voltage signals.

4.31 Relate v_o in the circuit of Fig. P4.31 to v_1, v_2, and v_3.

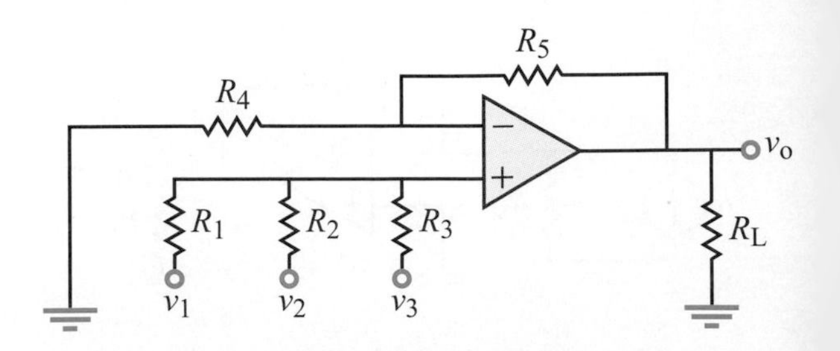

Figure P4.31: Circuit for Problem 4.31.

4.32 For the circuit in Fig. P4.32, obtain an expression for v_o in terms of v_1, v_2, and the four resistors. Evaluate v_o if $v_1 = 0.1$ V, $v_2 = 0.5$ V, $R_1 = 100$ Ω, $R_2 = 200$ Ω, $R_3 = 2.4$ kΩ, and $R_4 = 1.2$ kΩ.

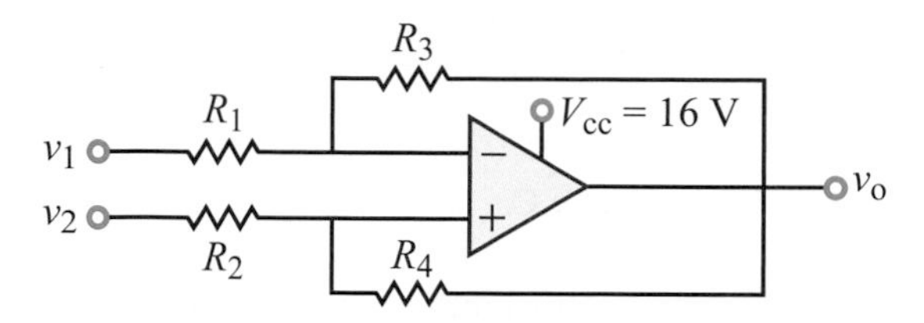

Figure P4.32: Circuit for Problem 4.32.

4.33 Generate a plot for i_L at the output side of the circuit in Fig. P4.33 versus v_s, covering the full linear range of v_s.

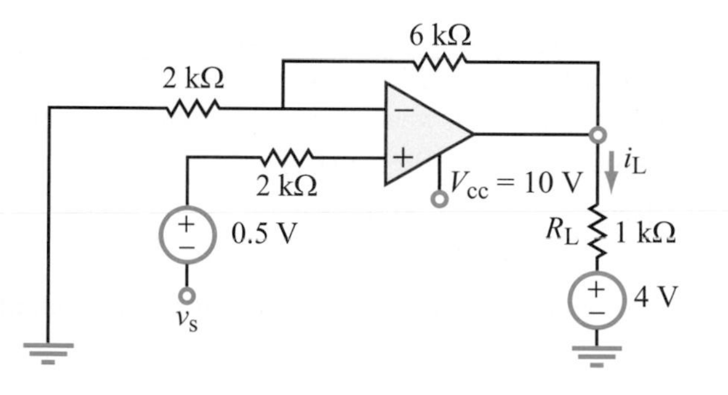

Figure P4.33: Circuit for Problem 4.33.

4.34 The circuit in Fig. P4.34 contains two single-pole single-throw switches, S_1 and S_2. Determine the closed-circuit gain $G = v_o/v_s$ for each of the four possible closed/open switch combinations.

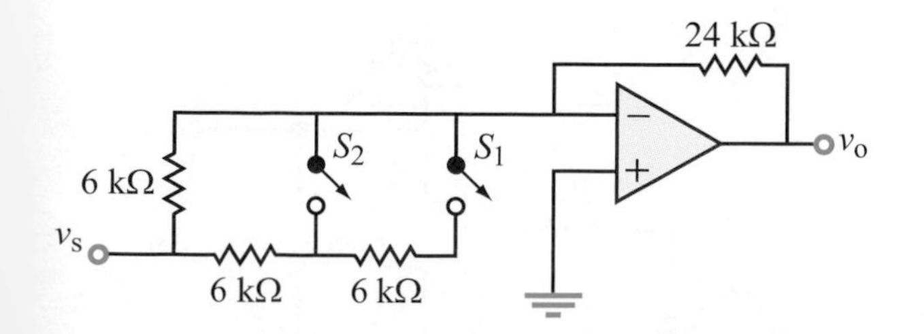

Figure P4.34: Circuit for Problem 4.34.

Section 4-8: Op-Amp Signal-Processing Circuits

4.35 Develop a block-diagram representation for the circuit in Fig. P4.35 for $v_{s_2} = v_{s_3} = 0$ and

(a) R_1 = open circuit

(b) $R_1 = 10$ kΩ.

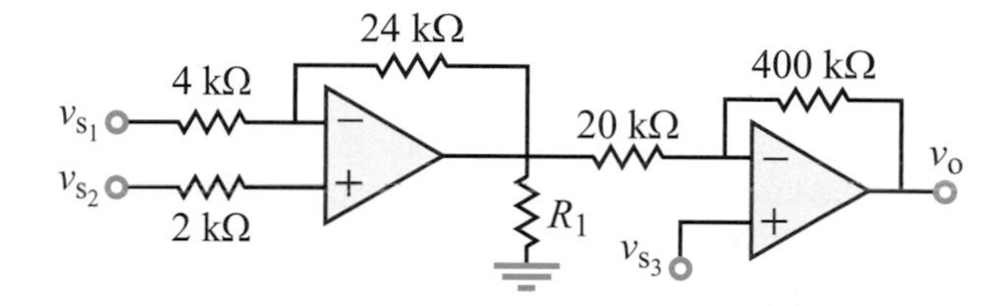

Figure P4.35: Circuit for Problems 4.35 through 4.37.

4.36 Develop a block-diagram representation for the circuit in Fig. P4.35 for $v_{s_3} = 0$ and $R_1 = \infty$.

4.37 Develop a block-diagram representation for the circuit in Fig. P4.35 for $v_{s_2} = 0$ and $R_1 = \infty$.

4.38 For the circuit in Fig. P4.38:

(a) Develop a block-diagram representation with R_L as a variable parameter.

(b) Specify the linear range of v_s.

(c) Determine v_o for $v_s = 0.3$ V and $R_L = 10$ kΩ.

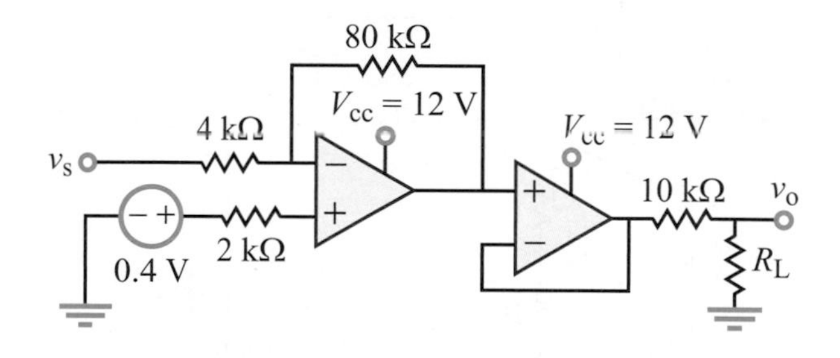

Figure P4.38: Circuit for Problem 4.38.

4.39 Design an op-amp circuit that can perform the operation $v_o = 12v_{s_1} + 3v_{s_2}$, while simultaneously presenting an input resistance of 50 kΩ on the input side for source v_{s_1} and an input resistance of 25 kΩ on the input side for source v_{s_2}.

4.40 Design an op-amp circuit that can perform the operation $v_o = 4v_{s_1} - 3v_{s_2}$, while simultaneously presenting an input resistance of 10 kΩ on the input side for source v_{s_1} and an input resistance of 5 kΩ on the input side for source v_{s_2}.

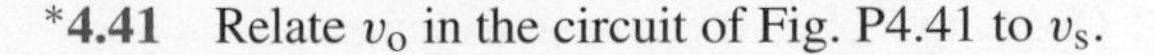

*__4.41__ Relate v_o in the circuit of Fig. P4.41 to v_s.

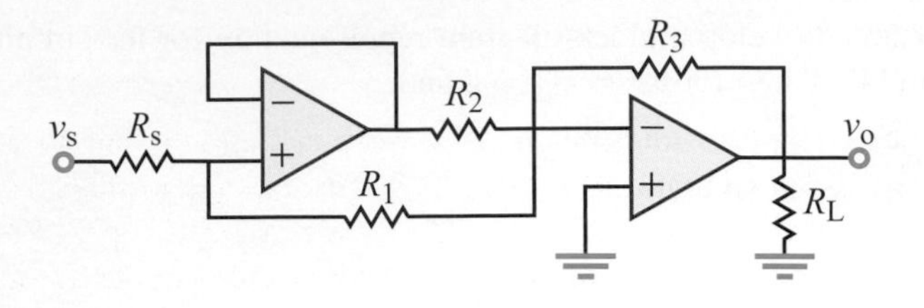

Figure P4.41: Circuit for Problem 4.41.

4.42 In the circuit of Fig. P4.42, op amp 1 receives feedback at its input from its own output as well as from the output of op amp 2. Relate v_o to v_s.

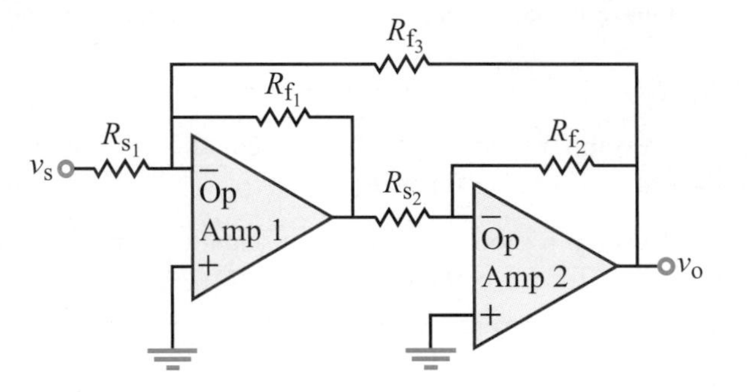

Figure P4.42: Circuit for Problem 4.42.

4.43 Relate v_o in the circuit of Fig. P4.43 to v_1 and v_2.

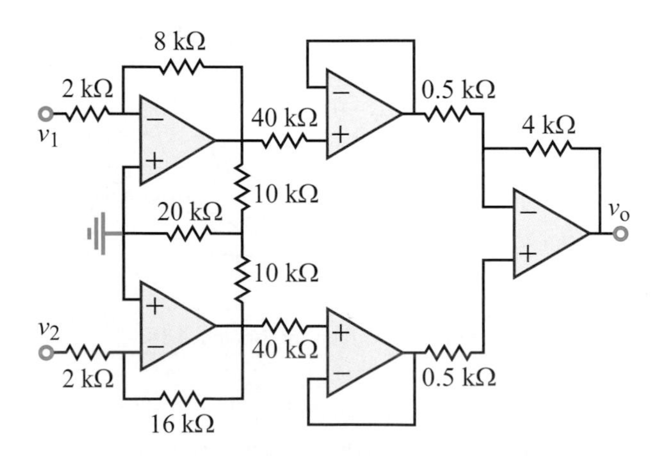

Figure P4.43: Circuit for Problem 4.43.

4.44 Design an op-amp circuit that can perform the operation $i_o = (30i_1 - 8i_2 + 0.6)$ A where i_1 and i_2 are input current sources.

*__4.45__ Relate the output voltage v_o in Fig. P4.45 to v_s.

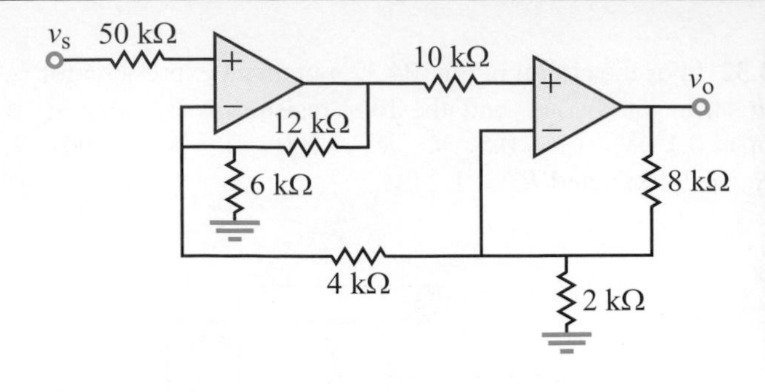

Figure P4.45: Circuit for Problem 4.45.

Sections 4-9 and 4-10: Instrumentation Amp and D/A Converter

4.46 The instrumentation-amplifier circuit shown in Fig. 4-20 is used to measure the voltage differential $\Delta v = v_2 - v_1$. If the range of variation of Δv is from -10 to $+10$ mV and $R_1 = R_3 = R_4 = R_5 = 100$ kΩ, choose R_2 so that the corresponding range of v_o is from -5 to $+5$ V.

4.47 An instrumentation amplifier with $R_1 = R_3 = 10$ kΩ, $R_4 = 1$ MΩ, and $R_5 = 1$ kΩ uses a potentiometer for the gain-control resistor R_2. If the potentiometer resistance can be varied between 10 and 100 Ω, what is the corresponding variation of the circuit gain $G = v_o/(v_2 - v_1)$?

4.48 Design a five-bit DAC using a circuit configuration similar to that in Fig. 4-22.

4.49 Design a six-bit DAC using a R–$2R$ ladder configuration.

Section 4-11: MOSFET

4.50 In Example 4-9, we analyzed a common-source amplifier without a load resistance. Consider the amplifier in Fig. P4.50; it is identical with the circuit in Fig. 4-28, except that we have added a load resistor R_L. Obtain an expression for v_{out} as a function of v_s.

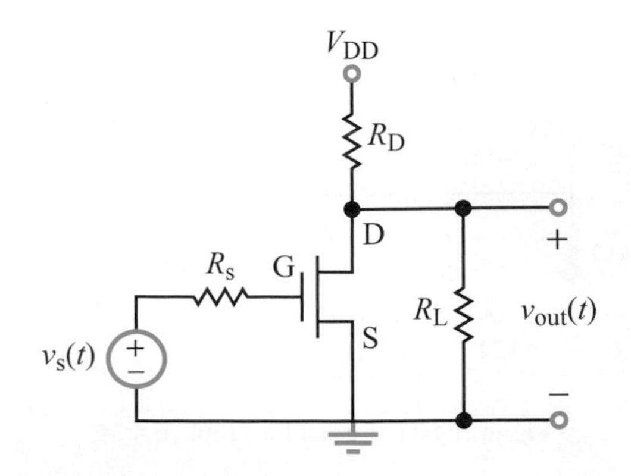

Figure P4.50: MOSFET circuit for Problem 4.50.

*4.51 Determine $v_{\text{out}}(t)$ as a function of $v_{\text{s}}(t)$ for the circuit in Fig. P4.51. Assume $V_{\text{DD}} = 2.5$ V.

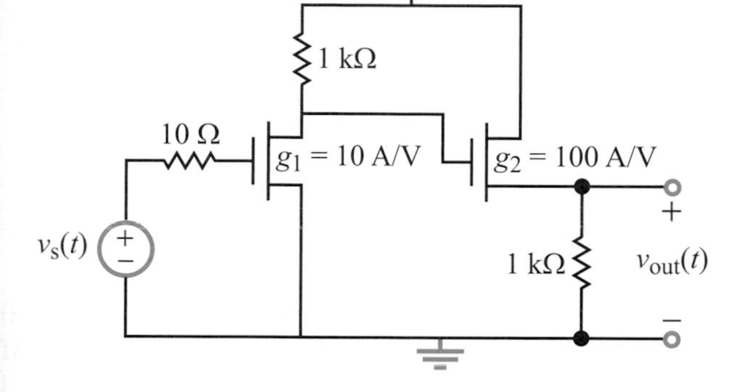

Figure P4.51: Two-MOSFET circuit for Problem 4.51.

4.52 In Problem 3.73 of Chapter 3, we analyzed a current mirror circuit containing BJTs. Current mirror circuits also can be designed using MOSFETs, as shown in Fig. P4.52. Determine the relationship between I_0 and I_{REF}.

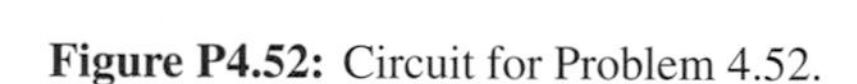

Figure P4.52: Circuit for Problem 4.52.

ection 4-13: Multisim Analysis

.53 Draw a non-inverting amplifier (Fig. 4-5) with a gain f 2 in Multisim. Show that the circuit works as expected y connecting a 1-V pulse source and plotting both the input nd the output voltages using the Grapher Tool and Transient nalysis. Use the 3-terminal virtual op-amp component.

54 Draw an inverting amplifier (Fig. 4-9) with a gain of 3.5 in Multisim. Show that the circuit works as expected by nnecting a 1-V dc voltage source and solving the circuit using e DC Operating Point analysis. Use the 3-terminal virtual op-np component.

4.55 In Multisim, draw a summing amplifier that adds the values of four different dc voltage sources, each with an inverting gain of 4. Use the DC Operating Point analysis tool to verify the circuit performance.

4.56 In Multisim, draw a non-inverting summing amplifier that adds the values of three different dc voltage sources V_1, V_2, and V_3 with gains of 1, 2, and 5, respectively. Apply the DC Operating Point Solution tool to demonstrate that the circuit functions as specified.

4.57 Draw the op-amp circuit shown in Fig. P4.57 in Multisim, provide a DC Operating Point Analysis solution that demonstrates its operation, and state what function the circuit performs.

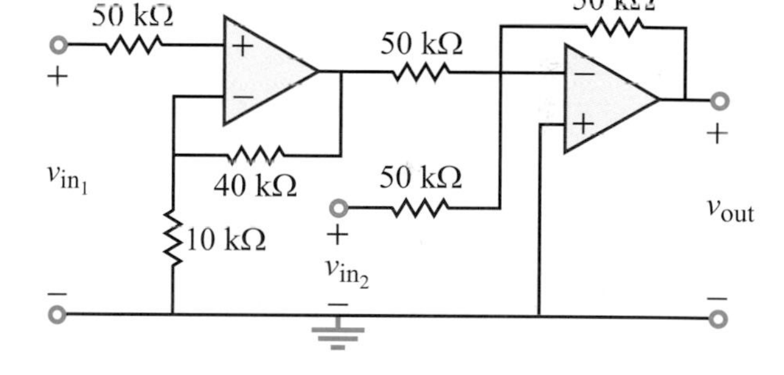

Figure P4.57: Circuit for Problem 4.57.

4.58 Construct the non-inverting amplifier circuit shown in Fig. P4.58 in Multisim. Set the value of R to 50 kΩ and then perform a DC Sweep analysis of the input voltage from −5 to +5 V. Plot the Output. Now change the value of R to 80 kΩ and repeat the DC Sweep analysis. Compare the two plots either side by side or by overlapping them using the Overlay Traces button on the Grapher toolbar. (Use the three-terminal virtual op amp for the simulation.)

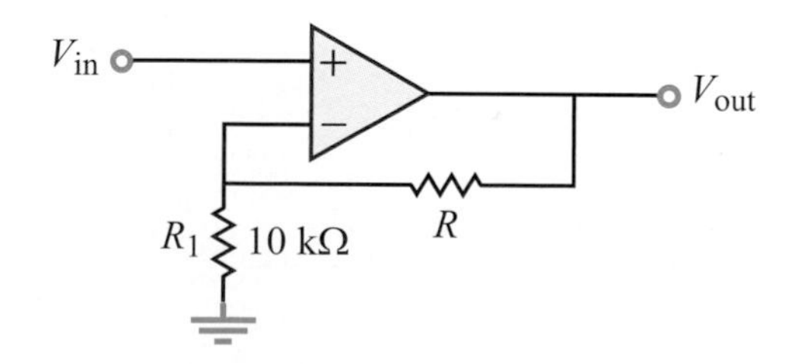

Figure P4.58: Circuit for Problem 4.58.

4.59 Until the 1970s, much research was carried out on analog computers (as distinguished from the digital computers found everywhere today). In fact, analog computers were one of the originally intended users of operational amplifiers. Op amps easily can be incorporated to perform many mathematical operations.

Using the basic op-amp circuits shown in this chapter, construct a circuit that expresses the following algebraic equation in voltage:

$$v = 2x - 3.5y + 0.2z,$$

where v is the output voltage and x, y, and z are three input voltages. Once you have the circuit designed, build it in Multisim and demonstrate that the circuit behaves appropriately by giving it the following inputs: $x = 1.2$, $y = 0.4$, and $z = 0.9$.

C H A P T E R

5

RC and RL First-Order Circuits

Chapter Contents

Objectives

Upon learning the material presented in this chapter, you should be able to:

1. Use mathematical functions to describe several types of nonperiodic waveforms.
2. Define the electrical properties of a capacitor, including its i–v relationship and the relationship between the electrical energy stored in it and the voltage across it.
3. Combine multiple capacitors when connected in series or in parallel.
4. Define the electrical properties of an inductor, including its i–v relationship and the relationship between the magnetic energy stored in it and the current passing through it.
5. Combine multiple inductors when connected in series or in parallel.
6. Analyze the response of a series RC or parallel RL circuit to a sudden change (usually caused by a switch) that occurs in the circuit to which it is connected.
7. Design RC op-amp circuits to perform differentiation and integration and related operations.
8. Apply Multisim to analyze RC and RL circuits.

Overview

Because a resistor is characterized by an i–υ relationship that does not involve time explicitly, namely $\upsilon = iR$, when we apply Kirchhoff's current and voltage laws to resistive circuits, we end up with one or more simultaneous *linear* equations. The process of solving a set of linear equations is relatively straightforward and does not involve time explicitly. If i varies with time, so will υ, in a linearly proportionate manner, and the character of the time variation remains the same for both. Hence, even when a certain voltage or current source in the circuit varies with time—and the resistive elements follow suit—we tend to think of the resistive circuit as *static* rather than dynamic, because the time variation is merely a scale change. Another important feature of resistive elements is that they consume electrical energy by converting it into heat.

Capacitors and inductors represent a contrasting (yet complimentary) class of electrical devices. Not only is time t (or more precisely d/dt) at the heart of how capacitors and inductors function, but they also differ from resistors in that they do not dissipate energy. They can store energy and then release it—but not consume it.

The addition of capacitors and inductors to circuits containing time-varying sources opens the door to ***dynamic circuits*** with a wide range of practical applications. The dynamic response of a circuit to a certain voltage or current source depends on both the architecture of the circuit and the waveform characterizing the time variation of that source. In general, the response consists of a ***transient component*** and a ***steady-state component***.

> The transient response represents the initial reaction immediately after a sudden change, such as closing or opening a switch to connect a source to the circuit.

Most (but not all) electronic circuits are designed such that the transient response usually dies out or reaches an approximately constant level within a fraction of a second after the introduction of the external excitation. Figure 5-1 shows examples of two

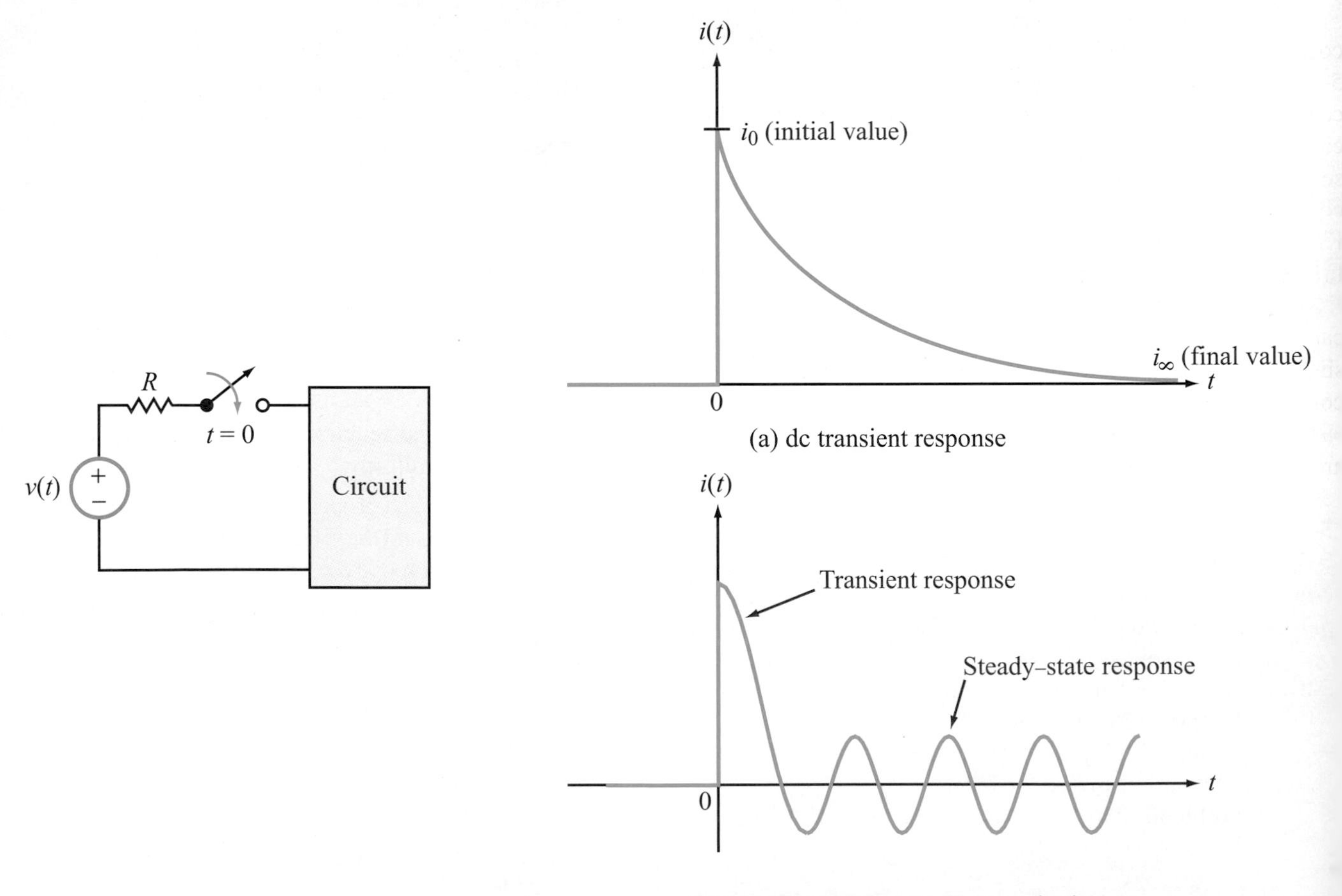

Figure 5-1: Circuit response to (a) dc source $\upsilon(t) = V_0$ and (b) ac source $\upsilon(t) = V_0 \cos \omega t$.

typical circuit responses. In part (a), the external excitation is a dc voltage source, and the displayed response represents the current flowing through a certain capacitor in the circuit, starting at when the switch is closed. The current levels labeled i_0 and i_∞ denote the values exhibited by the transient response at the onset of the change (closing the switch) and a long time afterwards, respectively. They are called the ***initial*** and ***final*** values of $i(t)$.

Our second example displays in Fig. 5-1(b) the response of another circuit to a sinusoidally time-varying source. The combination of the ac source and switch action initially elicit a transient response that quickly transitions into a steady-state response. This ac case belongs to a class of external excitations and circuit responses called ***periodic waveforms***. In contrast, a dc waveform is ***nonperiodic***. As we shall see later, the tools of circuit analysis and design lend themselves to different mathematical approaches when dealing with periodic versus nonperiodic waveforms. Consequently, we will first examine the behavior of circuits excited by nonperiodic external excitations in this and the following chapter before we pursue the treatment of ac circuits starting in Chapter 7.

Section 5-1 introduces some of the nonperiodic waveforms commonly used in electric circuits, followed in Sections 5-2 and 5-3 with presentations of the circuit properties of capacitors and inductors, respectively. Our treatment of the circuit response to nonperiodic excitations is divided into two segments. The first, covered in Sections 5-4 through 5-6 of this chapter, deals with ***first-order circuits***, so named because their Kirchhoff voltage and current equations are characterized by first-order differential equations. First-order circuits include ***RC circuits***—composed of sources, resistors, and a single capacitor (or multiple capacitors that can be combined into a single equivalent capacitor)—and ***RL circuits***, but not circuits containing capacitors and inductors simultaneously. ***RLC circuits***, which give rise to second-order differential equations, are the subject of Chapter 6.

Review Question 5-1: What is the difference between the transient and steady-state components of the circuit response?

Review Question 5-2: Why do we study the circuit response to dc and ac sources separately?

5-1 Nonperiodic Waveforms

Among the multitudes of possible nonperiodic waveforms, the step, ramp, pulse, and exponential waveforms are encountered most frequently in electrical circuits. In this section, we shall review the geometrical properties and corresponding mathematical expressions associated with each of these four waveforms, as well as introduce some of the connections between them.

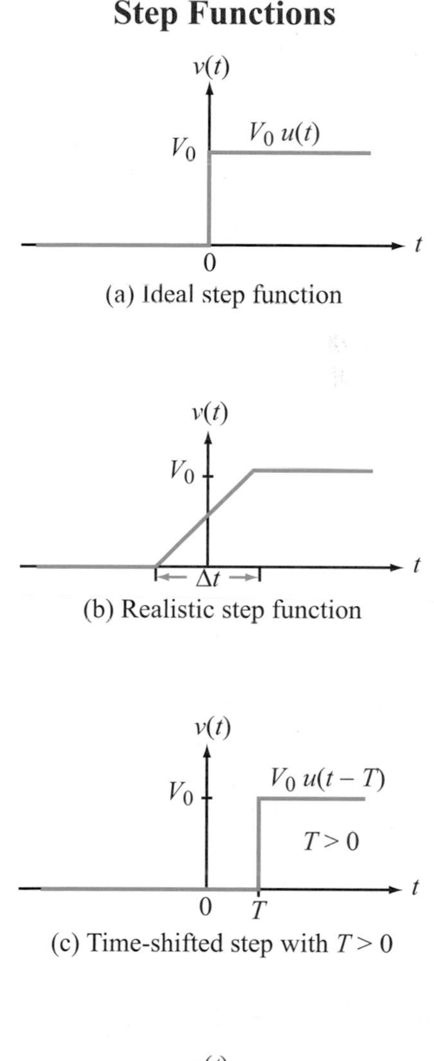

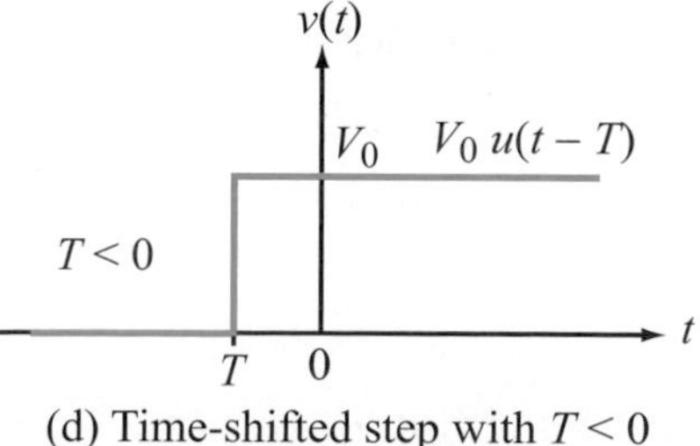

Figure 5-2: Step functions: (a) ideal step function, (b) realistic step function with transition duration Δt, (c) time-shifted step function with $T > 0$, (d) time-shifted step function with $T < 0$.

5-1.1 Step-Function Waveform

The waveform $v(t)$ shown in Fig. 5-2(a) is an (ideal) ***step function***: it is equal to zero for $t < 0$, at $t = 0$ it makes a discontinuous jump to V_0, and from there on forward it remains at V_0. Mathematically, it can be described as

$$v(t) = V_0\, u(t), \tag{5.1}$$

where $u(t)$ is known as the ***unit step function*** and is defined as

$$u(t) = \begin{cases} 0 & \text{for } t < 0, \\ 1 & \text{for } t > 0. \end{cases} \tag{5.2}$$

In reality, it is not possible to construct an (ideal) step function, because that would require changing the value of $v(t)$ from 0 to V_0 in zero time. A more realistic shape of the step function is illustrated in Fig. 5-2(b); the discontinuous jump is replaced with a ramp waveform of duration Δt.

An example of a step function is when a switch is closed so as to connect a voltage source to a circuit. If the time associated with closing the switch is very short in comparison with the time scale of interest, then it may be acceptable to approximate the switch closing by an ideal step function. On the other hand, if we are interested in analyzing the circuit response at a sampling rate whose interval is shorter than or comparable with the transition interval associated with closing the switch, then it may be necessary to use a more realistic, continuous, step function to represent the switch action.

If $v(t)$ transitions between its two levels at a time other than zero, such as at $t = T$, it is written as

$$v(t) = V_0\, u(t - T) = \begin{cases} 0 & \text{for } t < T, \\ V_0 & \text{for } t > T. \end{cases} \tag{5.3}$$

$u(t - T)$ is called the ***time-shifted step function***, which is defined to be zero when its argument $(t - T)$ is less than zero and 1 when its argument is greater than zero.

Figures 5-2(c) and (d) display step-function waveforms with $T > 0$ and $T < 0$, respectively.

5-1.2 Ramp-Function Waveform

A waveform that varies linearly with time, starting at a specific time $t = T$, is called a ***time-shifted ramp function*** and is denoted by $r(t - T)$. If $T = 0$, it simply is called a ***ramp function*** and is denoted by $r(t)$. Formally, $r(t - T)$ is defined as

$$r(t - T) = \begin{cases} 0 & \text{for } t \le T, \\ (t - T) & \text{for } t \ge T. \end{cases} \tag{5.4}$$

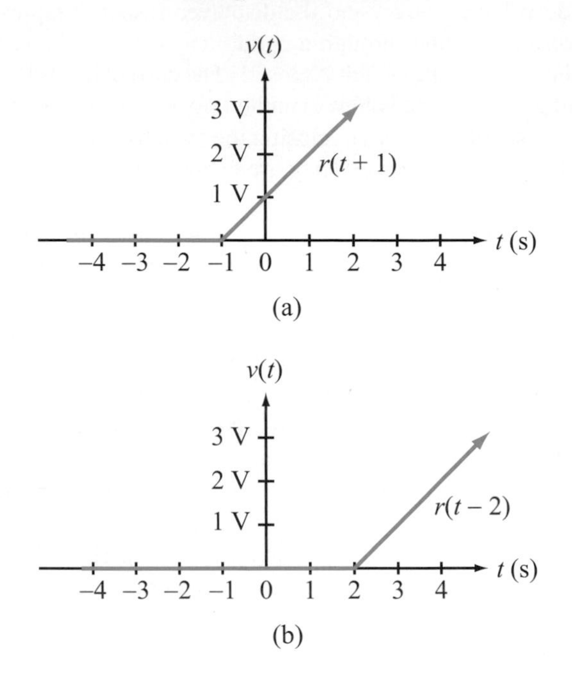

Figure 5-3: Time-shifted ramp functions $r(t - T)$ for (a) $T = -1$ s and (b) $T = 2$ s.

Plots of $v(t) = r(t - T)$ are displayed in Figs. 5-3(a) and (b) for $T = -1$ s and $T = 2$ s, respectively. A voltage $v(t)$ that ramps up at 3 V per second, starting at $t = 1$ s, is shown graphically in Fig. 5-4(a). Mathematically, $v(t)$ can be expressed as

$$v(t) = 3r(t - 1) \qquad \text{V}. \tag{5.5}$$

If the coefficient of $r(t - T)$ is negative, $v(t)$ would exhibit a negative slope, as illustrated by Fig. 5-4(b) for $v(t) = -2r(t + 1)$.

A unit ramp function is related to the unit step function by

$$r(t) = \int_{-\infty}^{t} u(t)\, dt = t\, u(t), \tag{5.6}$$

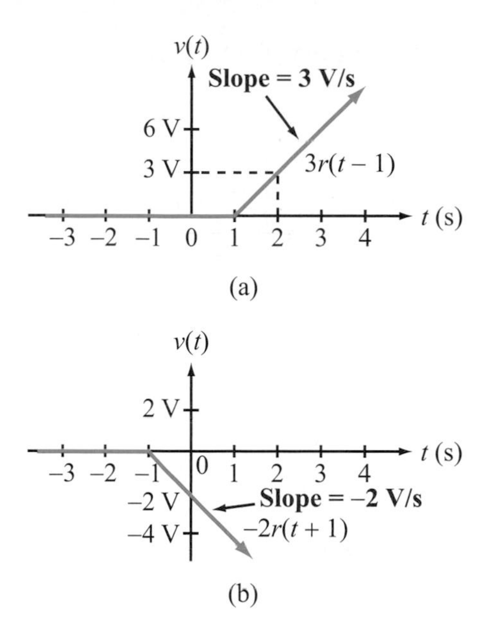

Figure 5-4: Examples of ramp functions.

and for the case where the ramping action starts at $t = T$,

$$\begin{aligned} r(t-T) &= \int_{-\infty}^{t} u(t-T)\, dt \\ &= (t-T)\, u(t-T). \end{aligned} \tag{5.7}$$

Example 5-1: Realistic Step Waveform

Generate an expression to describe the waveform shown in Fig. 5-5(a). Note that the time scale is in ms.

Solution: The voltage $v(t)$ can be synthesized as the sum of two time-shifted ramp functions (Fig. 5-5(b)): one with a positive slope of 3 V/s and a ramp start-up time $T = -2$ ms and a second ramp function that starts at $T = +2$ ms but its slope is -3 V/s. Thus,

$$\begin{aligned} v(t) &= v_1(t) + v_2(t) \\ &= 3r(t+2 \text{ ms}) - 3r(t-2 \text{ ms}) \qquad \text{V.} \end{aligned}$$

In view of Eq. (5.7), $v(t)$ also can be expressed in terms of time-shifted step functions as

$$\begin{aligned} v(t) = 3(t+2 \text{ ms})\, &u(t+2 \text{ ms}) \\ &- 3(t-2 \text{ ms})\, u(t-2 \text{ ms}) \qquad \text{V.} \end{aligned}$$

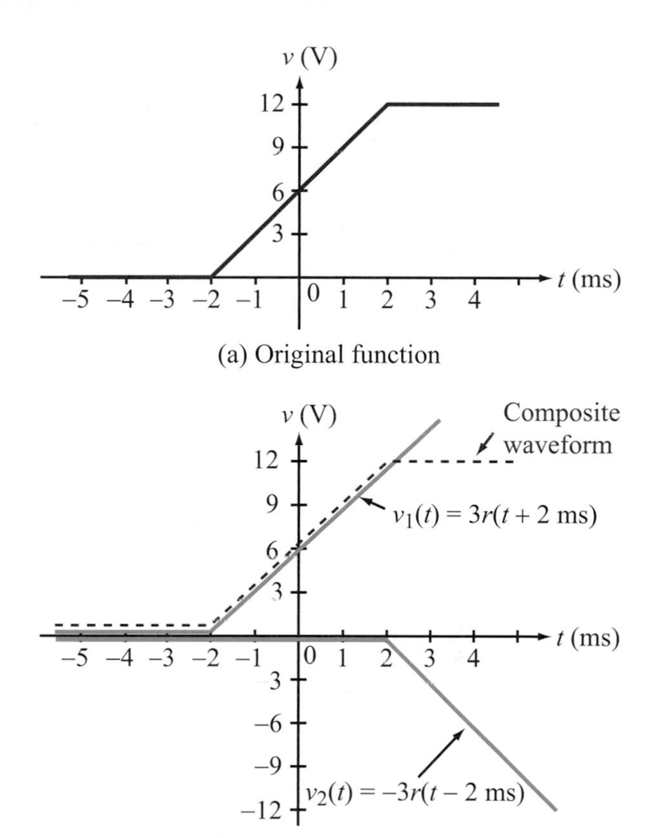

Figure 5-5: Step waveform of Example 5-1.

5-1.3 Pulse Waveform

The diagram in Fig. 5-6(a) depicts a SPDT switch that moves from position 1 to position 2 at $t = 1$ s, connects a dc voltage source to an electric circuit, and then returns to position 1 at $t = 5$ s. From the standpoint of the circuit, the switch actions constitute the introduction of a ***rectangular pulse*** of voltage V_0, as illustrated in Fig. 5-6(b). A pulse also may be triangular or Gaussian in shape or may assume other forms, but in all cases, it usually is assumed that a pulse rises from some specified base level up to a peak value, remains constant for a while, and then declines back to its original base level.

A rectangular pulse can be described in terms of the ***unit rectangular function*** $\text{rect}[(t-T)/\tau]$, which is characterized by two parameters: location of the center of the pulse T and the ***duration of the pulse*** τ, as shown in Fig. 5-7. Its mathematical definition is given by

$$\text{rect}\left(\frac{t-T}{\tau}\right) = \begin{cases} 0 & \text{for } t < (T-\tau/2), \\ 1 & \text{for } (T-\tau/2) \leq t \leq (T+\tau/2), \\ 0 & \text{for } t > (T+\tau/2). \end{cases} \tag{5.8}$$

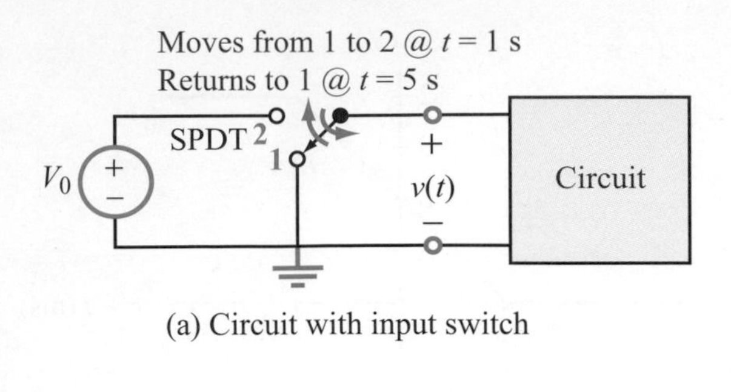

(a) Circuit with input switch

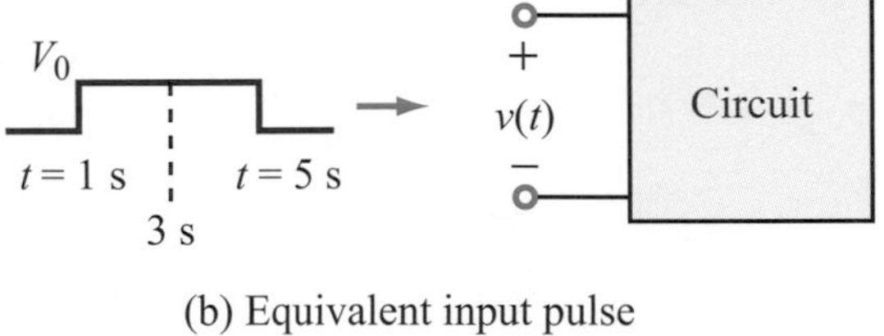

(b) Equivalent input pulse

$\mathrm{rect}\left(\frac{t-3}{4}\right)$

Figure 5-6: Connecting a switch to a dc source at $t = 1$ s and then returning it to ground at $t = 5$ s constitutes a voltage pulse centered at $T = 3$ s and of duration $\tau = 4$ s.

A rectangular pulse can be constructed out of two time-shifted step functions: one that causes the rise in level and another (delayed in time) that cancels the first one. The details are given in Example 5-2.

Example 5-2: Pulses

Construct expressions for (a) the rectangular pulse shown in Fig. 5-8(a) and (b) the trapezoidal pulse shown in Fig. 5-8(b) in terms of step and ramp functions.

Solution:

(a) From Fig. 5-8(a), it is evident that the amplitude of the rectangular pulse is 4 V and its duration is 2 s, extending from $T_1 = 2$ s to $T_2 = 4$ s. Hence, with its center at 3 s and its duration equal to 2 s,

$$v_a(t) = 4\,\mathrm{rect}\left(\frac{t-3}{2}\right) \quad \text{V.} \tag{5.9}$$

The sequential addition of two time-shifted step functions, $v_1(t)$ at $t = 2$ s and $v_2(t)$ at $t = 4$ s, as demonstrated graphically in Fig. 5-8(c), accomplishes the task of synthesizing the rectangle function in terms of two step functions. Specifically,

$$\begin{aligned} v_a(t) &= v_1(t) + v_2(t) \\ &= 4[u(t-2) - u(t-4)] \quad \text{V.} \end{aligned} \tag{5.10}$$

Rectangular Pulses

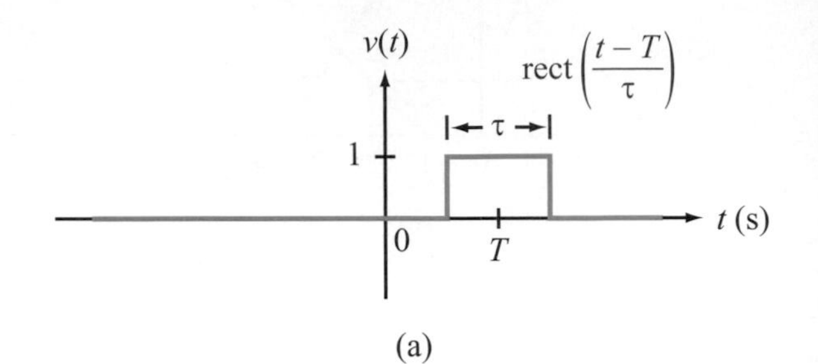

(a)

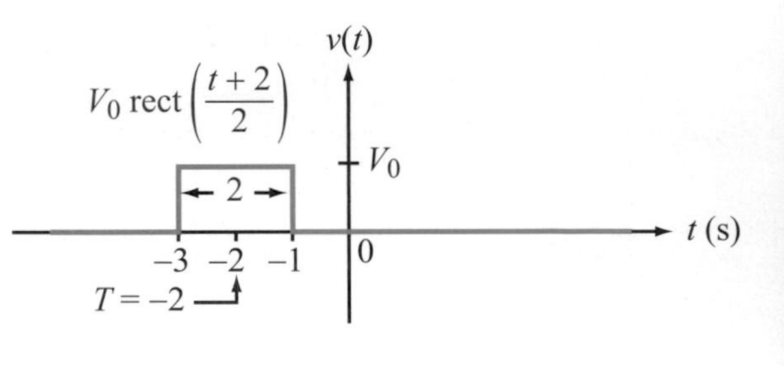

(b)

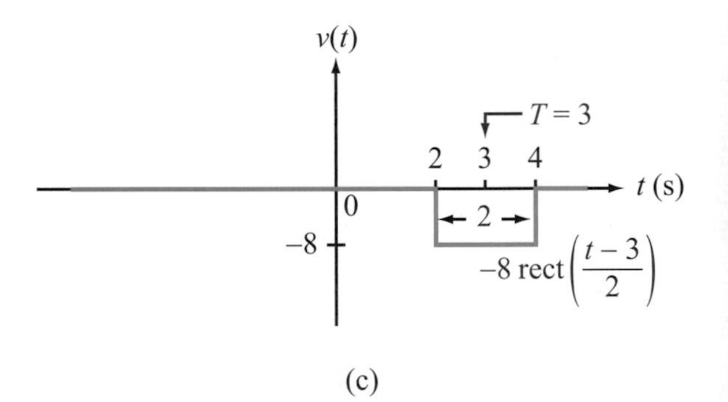
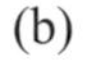

(c)

Figure 5-7: Rectangular pulses.

(b) The trapezoidal pulse consists of three segments, a ramp with a positive slope that starts at $t = 0$ and ends at $t = 1$ s followed by a plateau that extends to $t = 3$ s, and finally, ramp with a negative slope that ends at 4 s. Building on the experience gained from Example 5-1, we can synthesize the trapezoidal pulse in terms of four ramp functions. The process which is illustrated graphically in Fig. 5-8(d), leads to

$$\begin{aligned} v_b(t) &= v_1(t) + v_2(t) + v_3(t) + v_4(t) \\ &= 5[r(t) - r(t-1) - r(t-3) + r(t-4)] \quad \text{V.} \end{aligned} \tag{5.11}$$

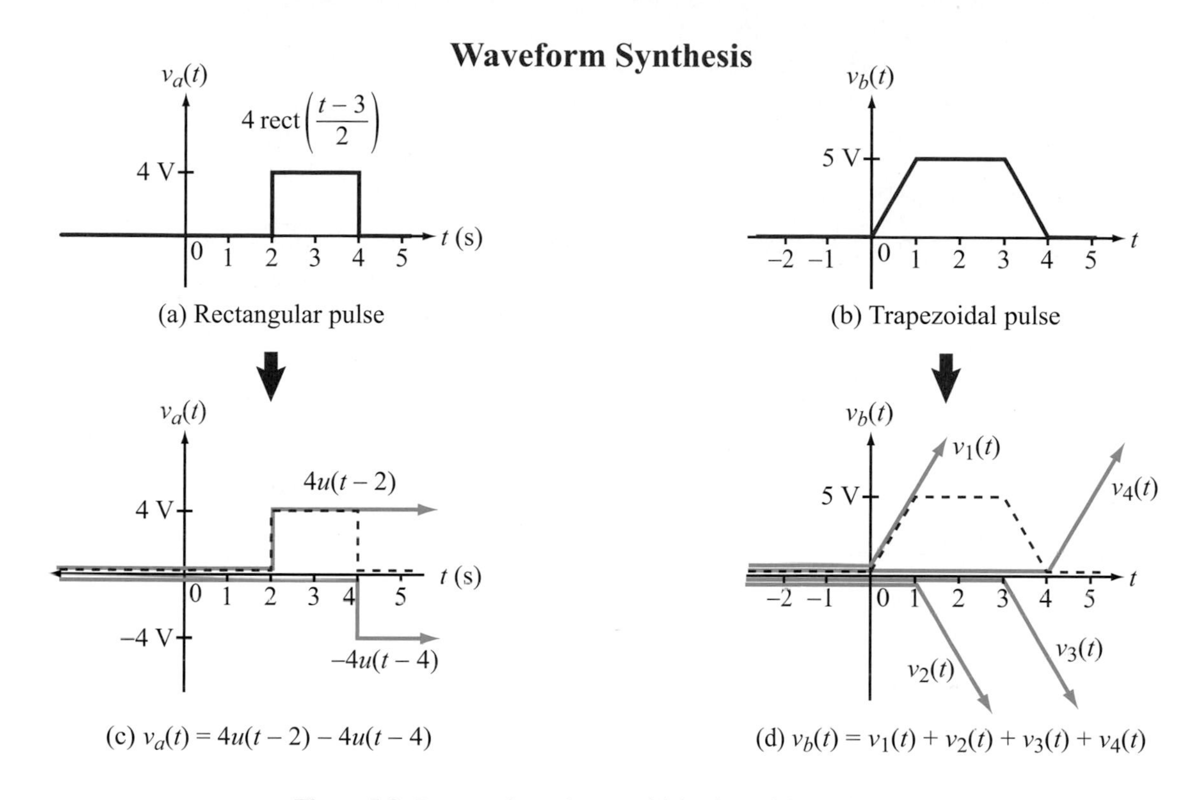

Figure 5-8: Rectangular and trapezoidal pulses of Example 5-2.

Equivalently, using the relationship between the ramp and step functions given by Eq. (5.7), $v_b(t)$ can be expressed as

$$v_b(t) = 5[t\,u(t) - (t-1)\,u(t-1) - (t-3)\,u(t-3) + (t-4)\,u(t-4)] \quad \text{V.} \tag{5.12}$$

Review Question 5-3: What determines the slope of a ramp waveform?

Review Question 5-4: How are the ramp and rectangle functions related to the step function?

Review Question 5-5: A unit step function $u(t)$ is equivalent to closing an SPST switch at $t = 0$. What is $u(-t)$ equivalent to?

Exercise 5-1: Express the waveforms shown in Fig. E5.1 in terms of unit step functions.

Answer: (a) $v(t) = 10\,u(t) - 20\,u(t-2) + 10\,u(t-4)$, (b) $v(t) = 2.5\,r(t) - 10\,u(t-2) - 2.5\,r(t-4)$. (See ◎)

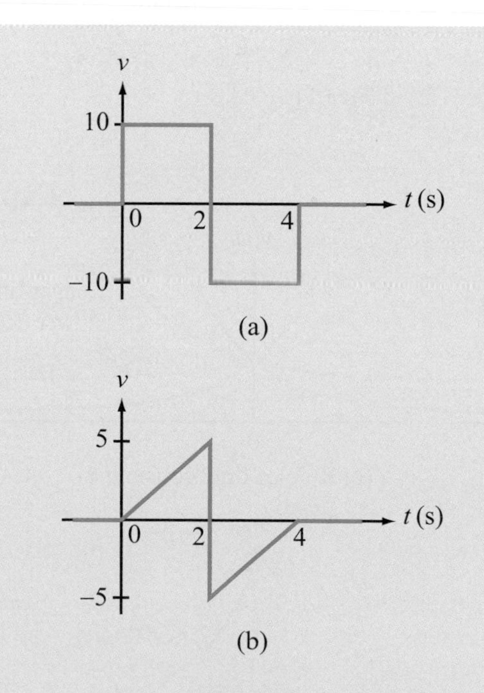

Figure E5.1

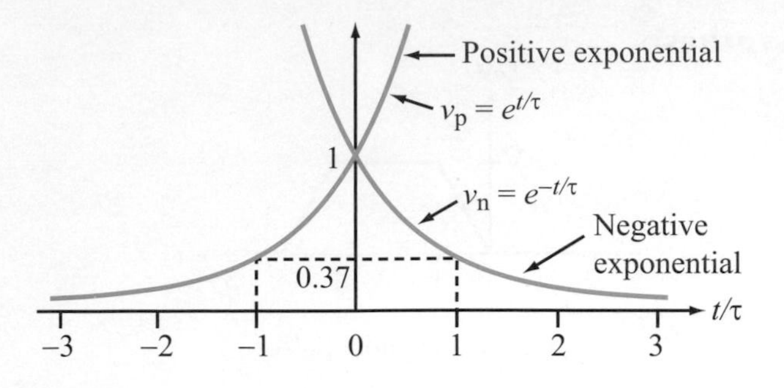

Figure 5-9: By $t = \tau$, the exponential function $e^{-t/\tau}$ has decayed to 37 percent of its original value at $t = 0$.

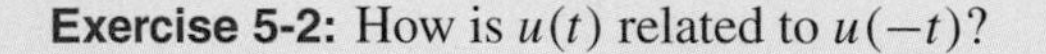

Exercise 5-2: How is $u(t)$ related to $u(-t)$?

Answer: They are mirror images of one another (with respect to the y-axis). (See)

Exercise 5-3: Consider the SPDT switch in Fig. 5-6(a). Assume that it started out at position 2, was moved to position 1 at $t = 1$ s, and then moved back to position 2 at $t = 5$ s. This is the reverse of the sequence shown in Fig. 5-6(a). Express $v(t)$ in terms of (a) units step functions and (b) the rectangle function.

Answer: (a) $v(t) = V_0[u(1-t) + u(t-5)]$,
(b) $v(t) = V_0\left[1 - \text{rect}\left(\frac{t-3}{4}\right)\right]$.

5-1.4 Exponential Waveform

The ***exponential function*** is a particularly useful tool for characterizing fast-rising and fast-decaying waveforms. The (positive) exponential function given by

$$v_p(t) = e^{t/\tau} \tag{5.13}$$

is shown graphically in Fig. 5-9 for a positive value of the ***time constant*** τ. The figure also includes a plot of the ***negative exponential function***, where

$$v_n(t) = e^{-t/\tau}. \tag{5.14}$$

When $t = \tau$, $v_n = e^{-1} = 0.37$. Thus, if a certain quantity (such as a voltage or current) is said to decay exponentially with time, it means that after τ seconds its amplitude decreases to $1/e$ or 37 percent of its initial value. Symmetrically, $v_p = e^{-1} = 0.37$ when $t = -\tau$.

An exponential function with a short time constant rises or decays faster than an exponential function with a longer time constant, as illustrated by the plots in Fig. 5-10(a). Replacing t in the exponential with $(t - T)$ shifts the exponential curve to the right if T has a positive value and to the left if T is negative (Fig. 5-10(b)). In Fig. 5-10(c), the range of the exponential function has been limited to $t > 0$ by multiplying $e^{-t/\tau}$ by $u(t)$, and in Fig. 5-10(d) the function $v(t) = V_0(1 - e^{-t/\tau})\, u(t)$ is used to describe a waveform that builds up as a function of time towards a saturation value V_0.

Table 5-1 provides a summary of common waveform shapes and their equivalent expressions.

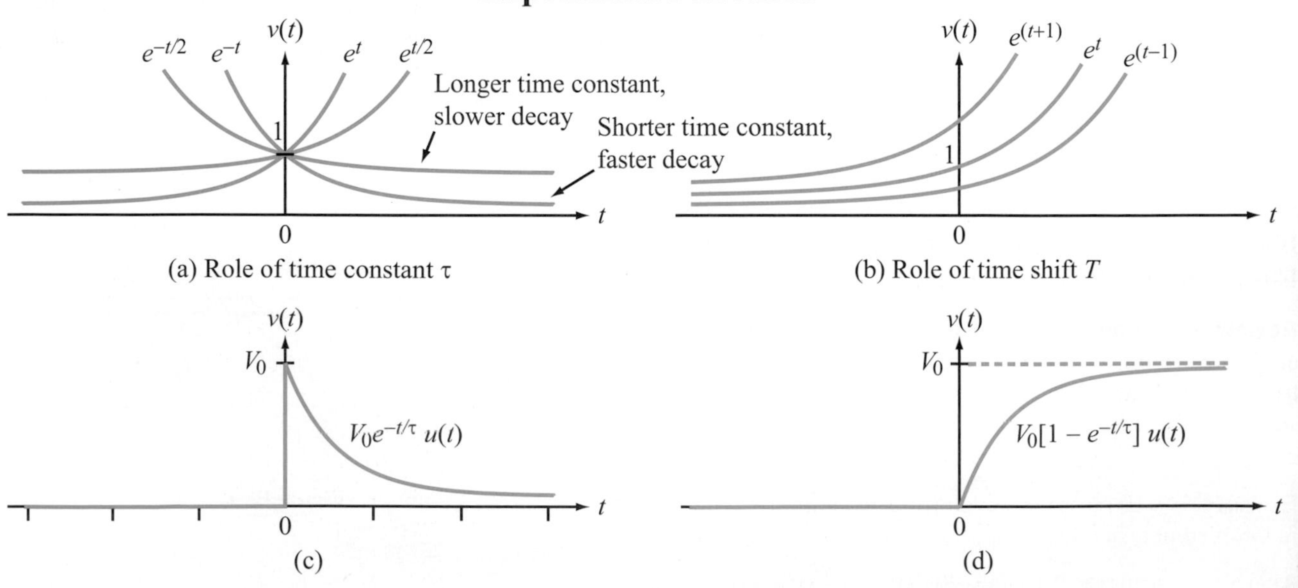

Figure 5-10: Properties of the exponential function.

Table 5-1: Common nonperiodic waveforms.

Waveform	Expression	General Shape
Step	$u(t-T) = \begin{cases} 0 & \text{for } t < T \\ 1 & \text{for } t > T \end{cases}$	
Ramp	$r(t-T) = (t-T)\, u(t-T)$	
Rectangle	$\text{rect}\left(\frac{t-T}{\tau}\right) = u(t-T_1) - u(t-T_2)$ $T_1 = T - \frac{\tau}{2}; \quad T_2 = T + \frac{\tau}{2}$	
Exponential	$\exp[-(t-T)/\tau]\, u(t-T)$	

Review Question 5-6: If the time constant of a negative exponential function is doubled in value, will the corresponding waveform decay faster or slower?

Review Question 5-7: What is the approximate shape of the waveform described by the function $(1 - e^{-|t|})$?

Exercise 5-4: The radioactive decay equation for a certain material is given by $n(t) = n_0 e^{-t/\tau}$, where n_0 is the initial count at $t = 0$. If $\tau = 2 \times 10^8$ s, how long is its half-life? [Half-life $t_{1/2}$ is the time it takes a material to decay to 50 percent of its initial value.]

Answer: $t_{1/2} = 1.386 \times 10^8$ s $= 4$ years, 144 days, 12 hours, 10 minutes, 36 s. (See CD)

Exercise 5-5: If the current $i(t)$ through a resistor R decays exponentially with a time constant τ, what is the value of the power dissipated in the resistor at $t = \tau$, compared with its value at $t = 0$?

Answer: $p(t) = i^2 R = I_0^2 R (e^{-t/\tau})^2 = I_0^2 R e^{-2t/\tau}$, $p(\tau)/p(0) = e^{-2} = 0.135$ or 13.5 percent.

5-2 Capacitors

When separated by an insulating medium, any two conducting bodies (regardless of their shapes and sizes) form a ***capacitor***. The ***parallel-plate capacitor*** shown in Fig. 5-11 represents a simple configuration in which two identical conducting plates (each of area A) are separated by a distance d containing an insulating (dielectric) material of ***electrical permittivity*** ϵ. The permittivity of a material is usually referenced to that of free

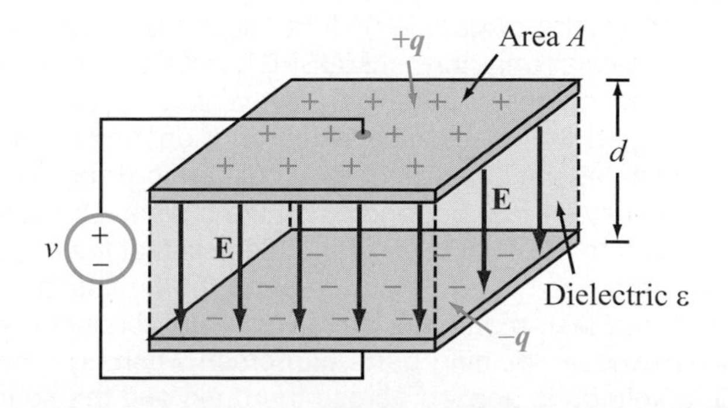

Figure 5-11: Parallel-plate capacitor with plates of area A, separated by a distance d, and filled with an insulating dielectric material of permittivity ϵ.

Technology Brief 10: Flash Memory

Flash memory is a non-volatile type of memory that has gained widespread popularity for portable applications (USB memory sticks, camera memory cards, mobile phones, etc.). Although not faster than DRAM (as of 2006), Flash memory devices may soon replace conventional hard drives (magnetic storage disks) even in personal computers. Figure TF10-1 shows Flash memory chips inside a typical USB memory stick.

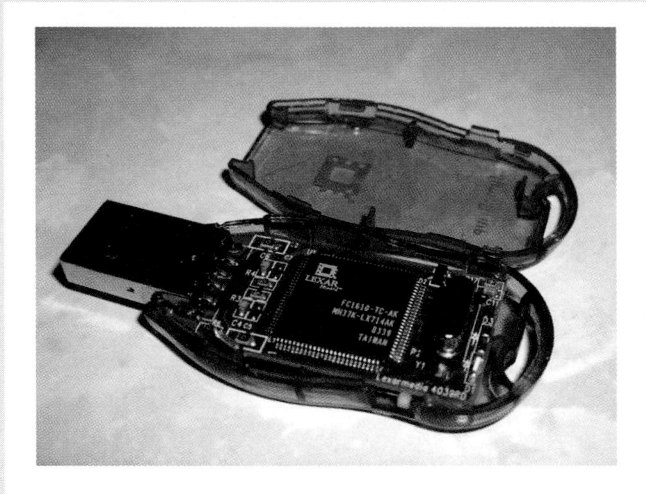

Figure TF10-1: Flash memory chips.

At its heart, a single Flash memory cell is very similar to a MOSFET transistor. Figure TF10-2 shows a Flash cell in cross-section and a typical MOSFET for comparison. The ***Gate (G)***, ***Source (S)***, and ***Drain (D)*** are illustrated on both The principal difference between the MOSFET and the Flash is the addition of a second gate, the ***Floating Gate (FG)*** beneath the normal Gate (G). In a normal MOSFET (Section 4-11), current can flow between the source and the drain when $V_{GS} > V_T$ (where V_T is a threshold voltage of a few millivolts). In a Flash, the floating gate can be charged with electrons (see below); once the electrons enter the FG, they are trapped there (because the floating gate is surrounded by an insulating oxide layer). If no electrons are present in the FG, V_T is relatively low and the device operates in a manner similar to a normal MOSFET, but if electrons are introduced to the FG, V_T is much higher ($V_T > 15$ V). In an array of Flash cells, a V_{GS} of a few volts is applied to all cells simultaneously. Bits whose floating gates have been pre-charged with electrons will not "turn on" and no current will flow between drain and source ($I_{DS} = 0$); bits whose floating gates are not pre-charged will "turn on" and conduct current between drain and source. Bits can thus represent logic states of 1 or 0 depending on the charge loaded into FG.

The key to the popularity of the Flash is that the FG can be charged and discharged on demand (and very quickly). Thus, data can be loaded into and out of the floating gates of each Flash transistor and interrogated when needed. Note that Flash memories are purely semiconductor circuits with no moving parts, making them preferable to hard drives (whose spinning parts can crash when agitated) in portable applications. In order to "write" electrons in the FG, a voltage is applied across the drain and the source ($V_{DS} > 0$). Simultaneously, a very large voltage is applied between the gate and drain ($V_{GS} > 20$ V). This large voltage "sucks" electrons traveling between D and S up through the insulating oxide and into the FG through a process called ***hot electron injection***. The electrons so injected into the FG remain trapped there permanently once V_{GS} is lowered. In order to "erase" electrons from the FG, a large voltag is applied between G and S ($V_{GS} > 20$ V) without a voltage between drain and source ($V_{DS} = 0$). This allows electrons to "bleed" away from FG and into S via quantum tunneling through the oxide. These charging and discharging events to and from the FG eventually degrade the device beyond the point where it can continue to operate reliably; current Flash memories can handle several million re-write cycles before they become unreliable.

As of 2008, commercial Flash memories have reached capacities of 64 gigabytes, with 1 to 4 Gb as the most common for consumer applications. Access rates range from 1 to 12 megabytes per second (MB/s). Further densification is beginning to be hampered by the fact that individual cells are so small (the gate on Samsung's 8-gigabyte chip is 40 nm wide!) that electrons begin hopping from bit to bit (an effect predicted at these small scales by quantum mechanics). Many efforts are underway to ameliorate this through better cell-cell isolation, multi-bit storage in one cell (a cell's FG can be set to more than one state, allowing multiple V_T levels), and system-level circuit optimization.

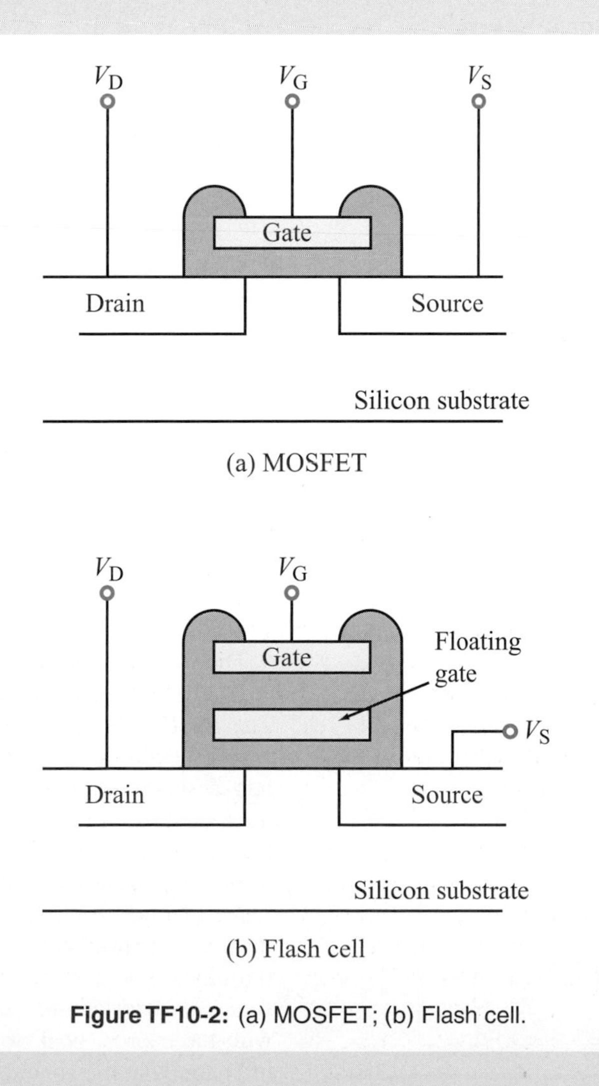

Figure TF10-2: (a) MOSFET; (b) Flash cell.

space, namely $\epsilon_0 = 8.85 \times 10^{-12}$ farads/m (F/m). Hence, the ***relative permittivity*** of a material is defined as

$$\epsilon_r = \frac{\epsilon}{\epsilon_0}. \tag{5.15}$$

When a dielectric material is subjected to an electric field, its atoms become partially polarized. The ***electric field*** E induced in the space between the conducting plates is the result of the voltage v applied across the plates. The ***electrical susceptibility*** χ_e of a material is a measure of how susceptible that material is to electrical polarization. The permittivity ϵ and susceptibility χ_e are related by

$$\epsilon = \epsilon_0(1 + \chi_e). \tag{5.16}$$

In view of Eq. (5.15),

$$\epsilon_r = \frac{\epsilon}{\epsilon_0} = 1 + \chi_e. \tag{5.17}$$

Free space contains no atoms; hence, its $\chi_e = 0$ and $\epsilon_r = 1$. For air at sea level, $\epsilon_r = 1.0006 \simeq 1.0$. Table 5-2 provides typical values of ϵ_r for common types of insulators.

Returning to the parallel-plate capacitor, if a voltage source is connected across the two plates, as shown in Fig. 5-11, charge of equal and opposite polarity is transferred to the conducting surfaces. The plate connected to the (+) terminal of the voltage source will accumulate charge $+q$, and charge $-q$ will accumulate on the other plate. The charges induce an electric field E in the dielectric medium, given by

$$E = \frac{q}{\epsilon A}, \tag{5.18}$$

with the direction of E being from the plate with $+q$ to the plate with $-q$. Moreover, E, whose unit is V/m, is related to the voltage v through

$$E = \frac{v}{d} \quad \text{(V/m)}. \tag{5.19}$$

Table 5-2: Relative electrical permittivity of common insulators.

Material	Relative Permittivity ϵ_r
Air (at sea level)	1.0006
Teflon	2.1
Polystyrene	2.6
Paper	2–4
Glass	4.5–10
Quartz	3.8–5
Bakelite	5
Mica	5.4–6
Porcelain	5.7

For any capacitor, its ***capacitance*** C, measured in farads (F), is defined as the amount of charge q that its positive-polarity plate holds, normalized to the applied voltage responsible for that charge accumulation.

Thus,

$$C = \frac{q}{v} \quad \text{(F)} \quad \text{(any capacitor)}. \tag{5.20}$$

For the parallel-plate capacitor, combining Eqs. (5.18) and (5.19) leads to $q = \epsilon Av/d$. Upon inserting this expression for q in Eq. (5.20), we have

$$C = \frac{\epsilon A}{d} \quad \text{(parallel-plate capacitor)}. \tag{5.21}$$

Even though the expression given by Eq. (5.21) is specific to the parallel-plate capacitor, the general tenor of the expression holds true for other geometrical configurations as well. In general, the capacitance C of any two-conductor system increases with the area of the conducting surfaces, decreases with the separation between them, and is directly proportional to ϵ of the insulating material. For example, the capacitance of a ***coaxial capacitor*** consisting of two concentric conducting cylinders of radii a and b (Fig. 5-12(a)) and separated by a dielectric material of permittivity ϵ is given by

$$C = \frac{2\pi\epsilon\ell}{\ln(b/a)} \quad \text{(coaxial capacitor)}, \tag{5.22}$$

where ℓ is the length of the capacitor. The spacing between the cylinders is $(b - a)$; reducing this spacing, while holding ℓ constant, requires reducing the ratio (b/a), which reduces the value of $\ln(b/a)$, thereby increasing the magnitude of C.

The ***mica capacitor*** shown in Fig. 5-12(b) consists of a stack of conducting plates, interleaved by sheets of mica (dielectric). The ***plastic-foil capacitor*** in Fig. 5-12(c) is constructed by rolling flexible conducting foils (separated by a plastic layer) into a spindle-like configuration. Small capacitors used in microcircuits typically have capacitances in the picofarad (10^{-12} F) to microfarad (10^{-6} F) range. Large capacitors used in power-transmission substations may have capacitors in the range of millifarads (10^{-3} F). Using thin-film polymers for the dielectric insulator and carbon nanotubes for the electrodes (terminals), a new type of capacitor (sometimes called ***supercapacitor*** or ***nanocapacitor***) was developed in the 1990s with the express goal of significantly increasing the amount of charge that the conductors can hold (at a specified voltage

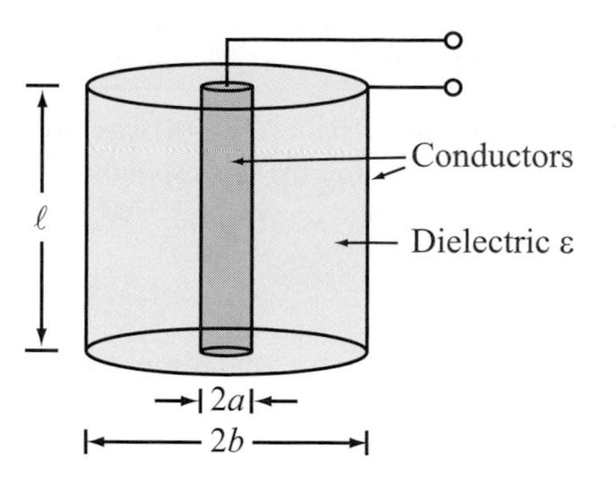

(a) Coaxial capacitor

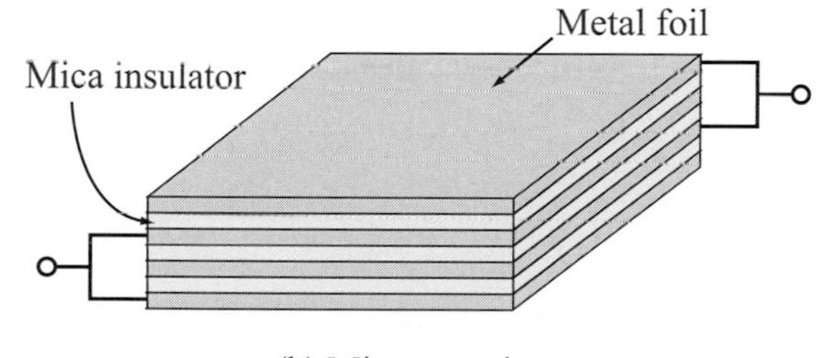

(b) Mica capacitor

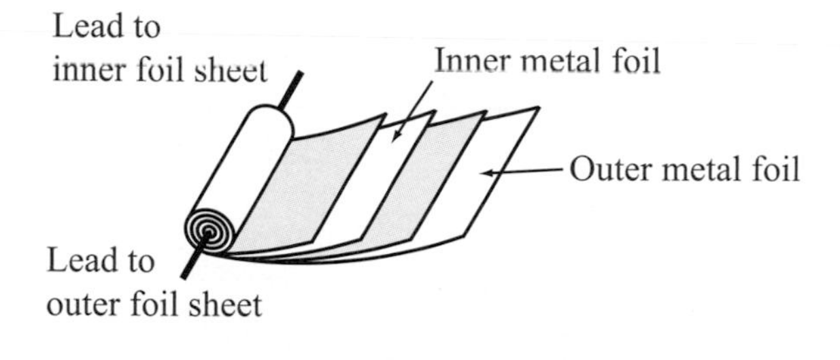

(c) Plastic foil capacitor

Figure 5-12: Various types of capacitors.

evel). Such capacitors have capacitance values that are several rders of magnitude greater than conventional capacitors of omparable size. The new fabrication techniques have not only xpanded the versatility of capacitors in electronic circuits, ut they have also introduced the use of supercapacitors as nergy-storage devices in many electronic applications (see echnology Brief 11: Supercapacitors on page 199).

-2.1 Electrical Properties of Capacitors

ccording to Eq. (5.20), $q = Cv$. Application of the standard efinition for current provides the expression for the current i through a capacitor as

C $\quad + \; v \; - \qquad i = C \dfrac{dv}{dt}$

Figure 5-13: Passive sign convention for capacitor: if current i is entering the (+) voltage terminal across the capacitor, then power is getting transferred into the capacitor. Conversely, if i is leaving the (+) terminal, then power is getting released from the capacitor.

$$\begin{aligned} i &= \frac{dq}{dt} \\ &= C\,\frac{dv}{dt}, \end{aligned} \tag{5.23}$$

where the direction of i and the polarity of v are defined in accordance with the passive sign convention (Fig. 5-13). The i–v relationship expressed by Eq. (5.23) conveys a very important condition, namely:

> The voltage across a capacitor cannot change instantaneously.

This assertion is supported by the observation that if v were to change values in zero time, dv/dt would be infinite, as a result of which the current i would be infinite also. Since i cannot be infinite, v cannot change instantaneously.

Another attribute of Eq. (5.23) relates to the behavior of a capacitor under dc conditions (constant voltage across it). Since $dv/dt = 0$ for a dc voltage, it follows that $i = 0$. Such a behavior is characteristic of an open circuit, through which no current flows even when a non-zero voltage exists across it. Thus:

> Under dc conditions, a capacitor behaves like an open circuit.

To express v in terms of i, we rewrite Eq. (5.23) and integrate both sides from t_0 to t for

$$\int_{t_0}^{t} \left(\frac{dv}{dt}\right) dt = \frac{1}{C} \int_{t_0}^{t} i\; dt, \tag{5.24}$$

where t_0 is a reference point in time at which $v(t_0)$ is known. Integrating the left-hand side and rearranging terms leads to

$$v(t) = v(t_0) + \frac{1}{C} \int_{t_0}^{t} i\; dt. \tag{5.25}$$

In view of $dq = i\,dt$, we recognize that the integral $\int_{t_0}^{t} i\,dt$ represents the amount of charge accumulation on the capacitor. If we are dealing with a capacitor that had no charge on it until a switch was closed or a signal was injected into the circuit and if we conveniently set our time reference such that the signal injection commenced at $t_0 = 0$, then Eq. (5.25) simplifies to

$$v(t) = \frac{1}{C}\int_0^t i\,dt \qquad \text{(capacitor uncharged before } t = 0\text{)}. \tag{5.26}$$

Charging up a capacitor creates an electric field in the dielectric medium between the capacitor's conductors. The electric field becomes the mechanism for storage of electrical energy in that medium. The stored energy can be released by discharging the capacitor. Thus, a capacitor can store energy and release previously stored energy but cannot dissipate energy.

The instantaneous power $p(t)$ transferring into or out of a capacitor is given by

$$p(t) = vi = Cv\,\frac{dv}{dt} \qquad \text{(W)}. \tag{5.27}$$

If the magnitude of $p(t)$ is positive, then by the passive sign convention, the capacitor is receiving power, and if $p(t)$ is negative, it is delivering power.

Energy is the integral of the product of power and time. Hence, the amount of energy stored in the capacitor at any time t is equal to the time integral of $p(t)$ from $-\infty$ (at which time the capacitor was uncharged) to t and is given by

$$w = \int_{-\infty}^{t} p\,dt = C\int_{-\infty}^{t}\left(v\,\frac{dv}{dt}\right)dt = C\int_{-\infty}^{t}\left[\frac{d}{dt}\left(\frac{1}{2}v^2\right)\right]dt, \tag{5.28}$$

which yields

$$w = \frac{1}{2}Cv^2 \qquad \text{(J)}. \tag{5.29}$$

We note that since the capacitor had no charge at $t = -\infty$, then its voltage also was zero at $t = -\infty$.

Equation (5.29) states:

The electrical energy stored in a capacitor at a given instant in time depends on the voltage across the capacitor at that instant, without regard to prior history.

Example 5-3: Capacitor Response to Voltage Waveform

The voltage waveform shown in Fig. 5-14(a) was applied across a 0.6-μF capacitor. Determine the corresponding waveforms for (a) the current $i(t)$, (b) the power $p(t)$, and (c) the energy stored in the capacitor $w(t)$.

Solution:

(a) We start by establishing a suitable expression for the waveform of $v(t)$, shown in Fig. 5-14(a), in terms of ramp functions. Noting that the ramp starts at $t = 0$ and has a slope

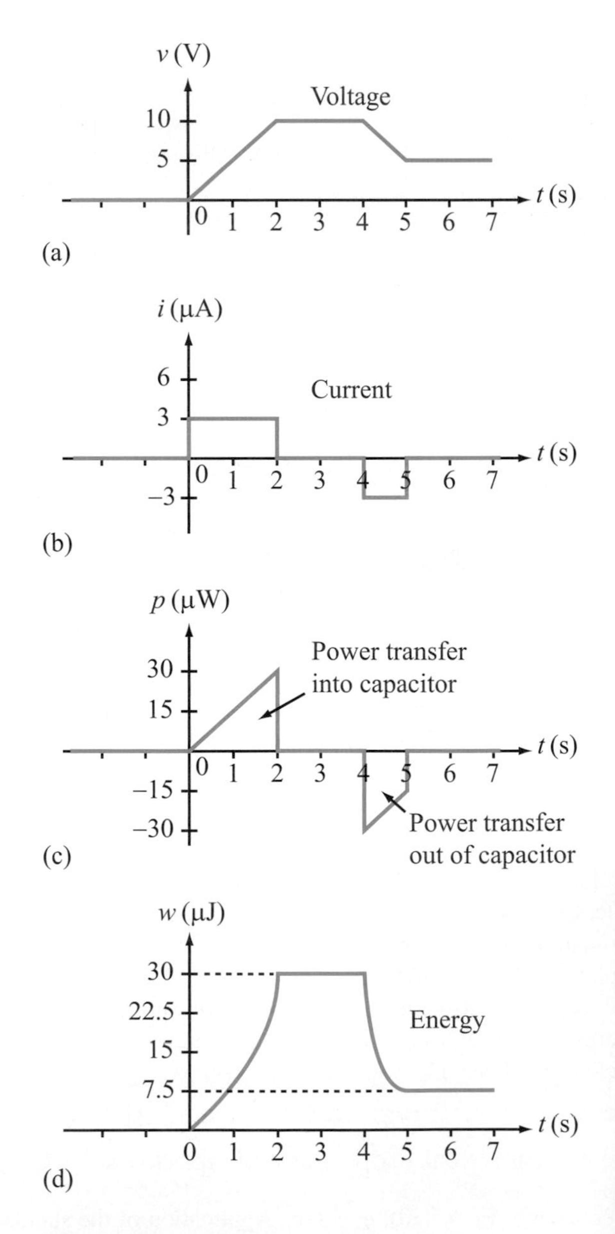

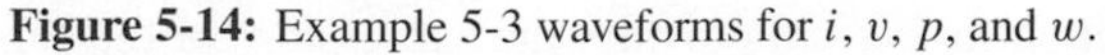

Figure 5-14: Example 5-3 waveforms for i, v, p, and w.

of $10/2 = 5$ V/s, $v(t)$ can be written as

$$v(t) = 5r(t) - 5r(t-2) - 5r(t-4) + 5r(t-5) \quad \text{V}.$$

Recalling that according to Eq. (5.7), $r(t-T) = (t-T)\,u(t-T)$, the expression for $v(t)$ corresponds to

$$v(t) = \begin{cases} 0 & \text{for } t \leq 0, \\ 5t \text{ V} & \text{for } 0 \leq t \leq 2 \text{ s}, \\ 10 \text{ V} & \text{for } 2 \text{ s} \leq t \leq 4 \text{ s}, \\ (-5t+30) \text{ V} & \text{for } 4 \text{ s} \leq t \leq 5 \text{ s}, \\ 5 \text{ V} & \text{for } t \geq 5 \text{ s}. \end{cases} \tag{5.30}$$

Application of Eq. (5.23) gives:

$$i(t) = C\,\frac{dv}{dt} = \begin{cases} 0 & \text{for } t \leq 0, \\ 3\ \mu\text{A} & \text{for } 0 \leq t \leq 2 \text{ s}, \\ 0 & \text{for } 2 \text{ s} \leq t \leq 4 \text{ s}, \\ -3\ \mu\text{A} & \text{for } 4 \text{ s} \leq t \leq 5 \text{ s}, \\ 0 & \text{for } t \geq 5 \text{ s}. \end{cases} \tag{5.31}$$

A plot of the current waveform is displayed in Fig. 5-14(b). We note that $i(t) > 0$ when $v(t)$ has a positive slope, and $i(t) < 0$ when $v(t)$ has a negative slope.

(b) The power $p(t)$, which is equal to the product of Eqs. (5.30) and (5.31), is shown in Fig. 5-14(c).

(c) We can calculate the stored energy $w(t)$ either by integrating $p(t)$ or by applying Eq. (5.29). In either case, we end up with the plot displayed in Fig. 5-14(d).

We note that after $t = 5$ s, the current is zero, the voltage is constant, the power getting transferred into the capacitor is zero (because $i = 0$), and the stored energy remains unchanged at 7.5 μJ.

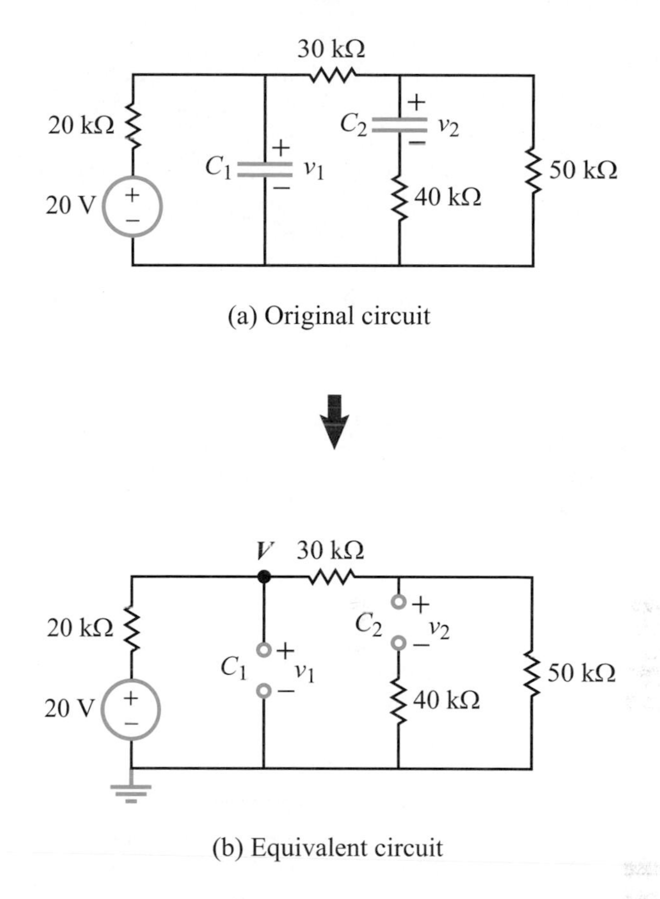

Figure 5-15: Under dc conditions, capacitors behave like open circuits.

Example 5-4: RC Circuit Under dc Conditions

Determine voltages v_1 and v_2 across capacitors C_1 and C_2 in the circuit of Fig. 5-15(a). Assume that the circuit has been in its present condition for a long time.

Solution: Under steady-state dc conditions, no current flows through a capacitor. Replacing capacitors C_1 and C_2 with open circuits, as in Fig. 5-15(b), allows us to apply KCL at node V as

$$\frac{V-20}{20 \times 10^3} + \frac{V}{(30+50) \times 10^3} = 0,$$

which gives $V = 16$ V. Hence,

$$v_1 = V = 16 \text{ V}.$$

Through voltage division, v_2 across the 50-kΩ resistor is given by

$$v_2 = \frac{V \times 50\text{k}}{(30+50)\text{k}} = \frac{16 \times 50}{80} = 10 \text{ V}.$$

Review Question 5-8: Explain why a capacitor behaves like an open circuit under dc conditions.

Review Question 5-9: The voltage across a capacitor cannot change instantaneously. Can the current change instantaneously, and why?

Review Question 5-10: For the capacitor, can $p(t)$ be negative? Can $w(t)$ be negative? Explain.

Exercise 5-6: It is desired to build a parallel-plate capacitor capable of storing 1 mJ of energy when the voltage across it is 1 V. If the capacitor plates are 2 cm × 2 cm each and its insulating material is teflon, what should the separation d be? Is such a capacitor practical?

Answer: $d = 3.72 \times 10^{-12}$ m. No, it is not practical to build a capacitor with such a small d, because it is about two orders of magnitude smaller than the typical spacing between two adjacent atoms in a solid material. (See)

Exercise 5-7: Instead of specifying A and calculating the spacing d needed to meet the 1-mJ requirement in Exercise 5.6, suppose we specify d as 1 μm and then calculate A. How large would A have to be?

Answer: $A = 10.4$ m × 10.4 m, equally impractical! (See)

Exercise 5-8: Determine the current i in the circuit of Fig. E5.8, under dc conditions.

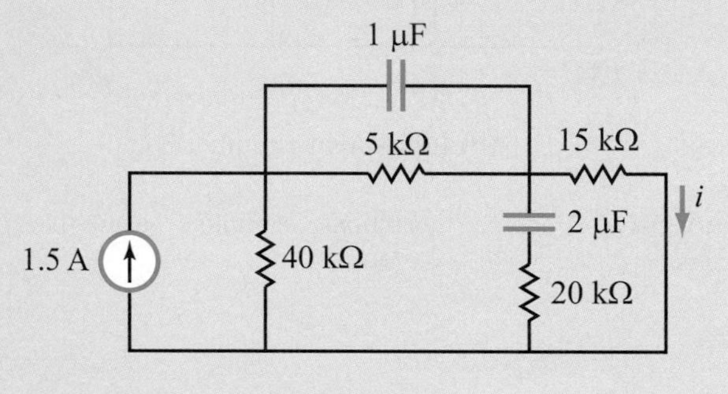

Figure E5.8

Answer: $i = 1$ A. (See)

5-2.2 Series and Parallel Combinations of Capacitors

In Chapter 2, we established that multiple resistors connected in series are equivalent to a single resistor whose resistance is equal to the algebraic sum of the resistances of the individual resistors. This equivalence relationship does not hold true for capacitors. In fact, we will shortly determine that:

The equivalence relationship for capacitors connected in series is similar in form to the relationship for resistors connected in parallel, and vice versa.

Combining In-Series Capacitors

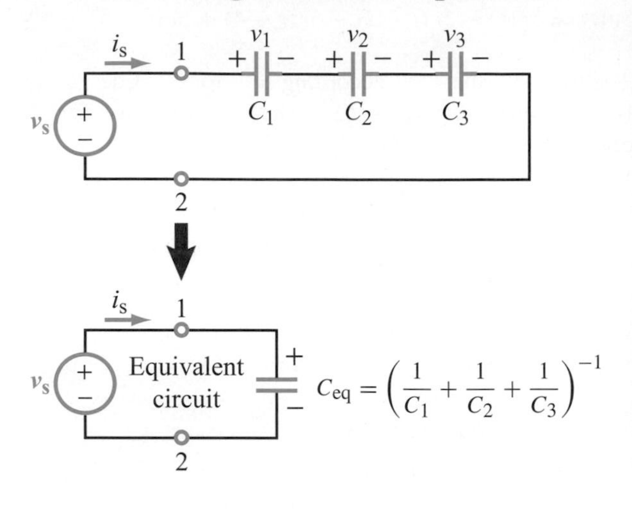

Figure 5-16: Capacitors in series.

Capacitors in Series

Consider the three capacitors shown in Fig. 5-16. They share the same current i_s, which is related to their individual voltages by

$$i_s = C_1 \frac{dv_1}{dt} = C_2 \frac{dv_2}{dt} = C_3 \frac{dv_3}{dt}. \tag{5.32}$$

Also,

$$v_s = v_1 + v_2 + v_3. \tag{5.33}$$

We wish to relate C_{eq} of the equivalent circuit to C_1, C_2, and C_3, subject to the requirement that the actual circuit and its equivalent exhibit identical i–v characteristics at terminals (1, 2). For the equivalent circuit,

$$\begin{aligned} i_s &= C_{eq} \frac{dv_s}{dt} \\ &= C_{eq} \left(\frac{dv_1}{dt} + \frac{dv_2}{dt} + \frac{dv_3}{dt}\right) \\ &= C_{eq} \left(\frac{i_s}{C_1} + \frac{i_s}{C_2} + \frac{i_s}{C_3}\right), \end{aligned} \tag{5.34}$$

which leads to

$$\frac{1}{C_{eq}} = \frac{1}{C_1} + \frac{1}{C_2} + \frac{1}{C_3}. \tag{5.35}$$

Generalizing to the case of N capacitors in series,

$$\frac{1}{C_{eq}} = \sum_{i=1}^{N} \frac{1}{C_i} = \frac{1}{C_1} + \frac{1}{C_2} + \cdots + \frac{1}{C_N} \tag{5.36}$$

(capacitors in series).

Additionally, if at reference time t_0 the capacitors had initial voltages $\upsilon_1(t_0)$ to $\upsilon_N(t_0)$, the initial voltage of the equivalent capacitor is

$$\upsilon_{eq}(t_0) = \sum_{i=1}^{N} \upsilon_i(t_0). \tag{5.37}$$

Capacitors in Parallel

The three capacitors shown in Fig. 5-17 are connected in parallel. Hence, they share the same voltage υ_s, and the source current i_s is equal to the sum of their currents,

$$\begin{aligned} i_s &= i_1 + i_2 + i_3 \\ &= C_1 \frac{d\upsilon_s}{dt} + C_2 \frac{d\upsilon_s}{dt} + C_3 \frac{d\upsilon_s}{dt}. \end{aligned} \tag{5.38}$$

For the equivalent circuit with equivalent capacitor C_{eq},

$$i_s = C_{eq} \frac{d\upsilon_s}{dt}. \tag{5.39}$$

Equating the expressions given by Eqs. (5.38) and (5.39) leads to

$$C_{eq} = C_1 + C_2 + C_3, \tag{5.40}$$

which can be generalized to N capacitors in parallel as

$$C_{eq} = \sum_{i=1}^{N} C_i \quad \text{(capacitors in parallel).} \tag{5.41}$$

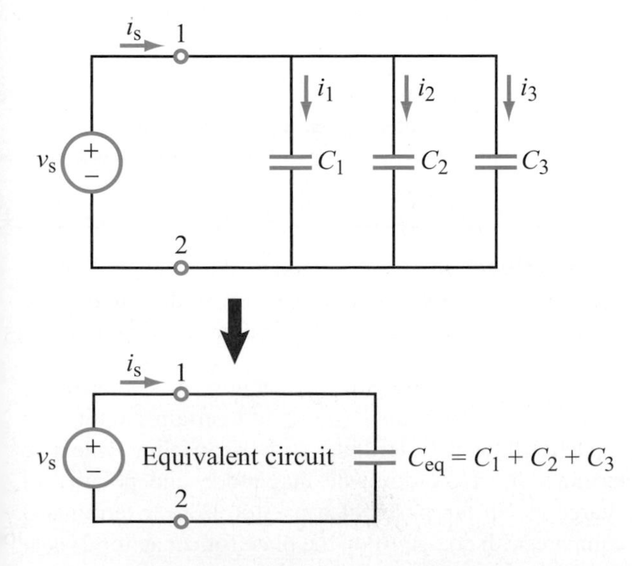

Figure 5-17: Capacitors in parallel.

Since the capacitors are connected in parallel, they shared the same voltage $\upsilon(t_0)$ at reference time t_0. Hence, for the equivalent capacitor

$$\upsilon_{eq}(t_0) = \upsilon(t_0). \tag{5.42}$$

Example 5-5: Equivalent Circuit

Reduce the circuit of Fig. 5-18(a).

Solution: For the resistors, we first can combine R_2 and R_3 in parallel, and then add the result to R_1 in series, noting that interchanging the locations of two elements connected in series is perfectly permissible, as such an action has no influence on either the current flowing through them or the voltages across them. A similar procedure can be followed for the capacitors, but we have to keep in mind that the equivalence relationships for resistors and capacitors are the reciprocal of one another:

$$\begin{aligned} R_2 \parallel R_3 &= \frac{R_2 R_3}{R_2 + R_3} \\ &= \frac{3\text{k} \times 6\text{k}}{3\text{k} + 6\text{k}} \\ &= 2 \text{ k}\Omega. \end{aligned}$$

$$\begin{aligned} R_{eq} &= R_1 + 2 \text{ k}\Omega \\ &= 8 \text{ k}\Omega + 2 \text{ k}\Omega \\ &= 10 \text{ k}\Omega. \end{aligned}$$

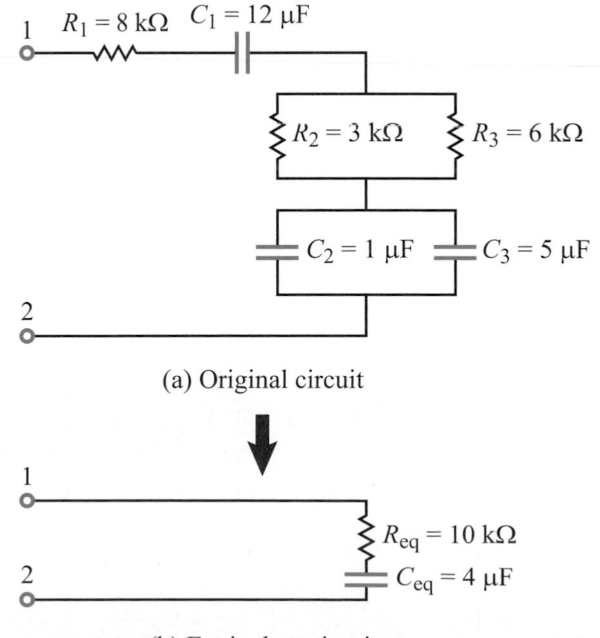

Figure 5-18: Circuit of Example 5-5.

$$C_2 \parallel C_3 = C_2 + C_3$$
$$= 1\ \mu\text{F} + 5\ \mu\text{F}$$
$$= 6\ \mu\text{F}.$$
$$C_{eq} = \frac{C_1 \times 6 \times 10^{-6}}{C_1 + 6 \times 10^{-6}}$$
$$= \left(\frac{12 \times 6}{12 + 6}\right) \times 10^{-6}$$
$$= 4\ \mu\text{F}.$$

The equivalent circuit is shown Fig. 5-18(b).

Example 5-6: Voltage Division

Figure 5-19(a) contains two resistors R_1 and R_2 connected in series to a voltage source υ_s. In Chapter 2, we demonstrated that the voltage υ_s is divided among the two resistors and, for example, υ_1 is given by

$$\upsilon_1 = \left(\frac{R_1}{R_1 + R_2}\right) \upsilon_s. \tag{5.43}$$

Derive the equivalent voltage-division equation for the series capacitors C_1 and C_2 in Fig. 5-19(b). Assume that the capacitors had no charge on them before they were connected to υ_s.

Solution: From the standpoint of the source υ_s, it "sees" an equivalent, single capacitor C given by the series combination of C_1 and C_2, namely

$$C = \frac{C_1 C_2}{C_1 + C_2}. \tag{5.44}$$

The voltage across C is υ_s. The law of conservation of energy requires that the energy that would be stored in the equivalent capacitor C be equal to the sum of the energies stored in C_1 and C_2. Hence, application of Eq. (5.29) gives

$$\frac{1}{2} C \upsilon_s^2 = \frac{1}{2}\ C_1 \upsilon_1^2 + \frac{1}{2}\ C_2 \upsilon_2^2. \tag{5.45}$$

Upon replacing C with the expression given by Eq. (5.44) and replacing the source voltage with $\upsilon_s = \upsilon_1 + \upsilon_2$, we have

$$\frac{1}{2}\left(\frac{C_1 C_2}{C_1 + C_2}\right)(\upsilon_1 + \upsilon_2)^2 = \frac{1}{2} C_1 \upsilon_1^2 + \frac{1}{2} C_2 \upsilon_2^2, \tag{5.46}$$

which reduces to

$$C_1 \upsilon_1 = C_2 \upsilon_2. \tag{5.47}$$

Using $\upsilon_2 = \upsilon_s - \upsilon_1$ in Eq. (5.47) leads to

$$C_1 \upsilon_1 = C_2(\upsilon_s - \upsilon_1)$$

or

$$\upsilon_1 = \left(\frac{C_2}{C_1 + C_2}\right) \upsilon_s. \tag{5.48}$$

We note that in the voltage-division equation for resistors, υ_1 is directly proportional to R_1, whereas in the capacitor case, υ_1 is directly proportional to C_2 (instead of to C_1). Additionally, in view of the relationship given by Eq. (5.47), application of the basic definition for capacitance, namely $C = q/\upsilon$, leads to

$$q_1 = q_2. \tag{5.49}$$

This result is exactly what one would expect when viewing the circuit from the perspective of the voltage source υ_s.

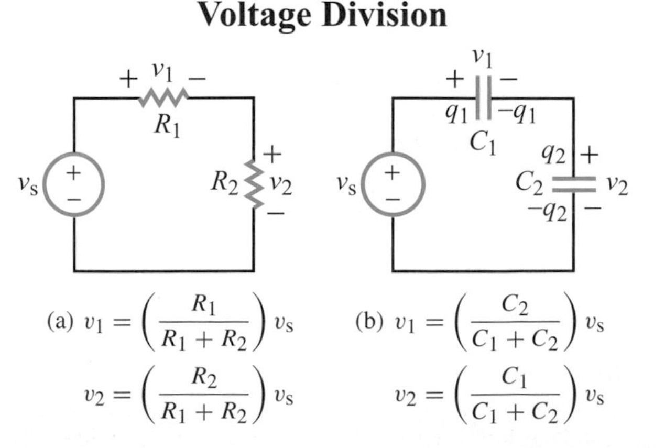

Figure 5-19: Voltage-division rules for (a) in-series resistors and (b) in-series capacitors.

Review Question 5-11: Compare the voltage-division equation for two capacitors in series with that for two resistors in series. Are they identical or different in form?

Review Question 5-12: Two capacitors are connected in series between terminals (a, b) in a certain circuit with capacitor 1 next to terminal a and capacitor 2 next to terminal b. How does the magnitude and polarity of charge q_1 on the plate (of capacitor 1) near terminal a compare with charge q_2 on the plate (of capacitor 2) near terminal b?

Technology Brief 11: Supercapacitors

According to Section 5-2.1, the energy (in joules) stored in a capacitor is given by $w = \frac{1}{2}CV^2$, where C is the capacitance and V is the voltage across it. Why then do we not charge capacitors by applying a voltage across them and then use them instead of batteries in support of everyday gadgets and systems? To help answer this question, we refer the reader to Fig. TF11-1, whose axes represent two critical attributes of storage devices. It is the combination (intersection)

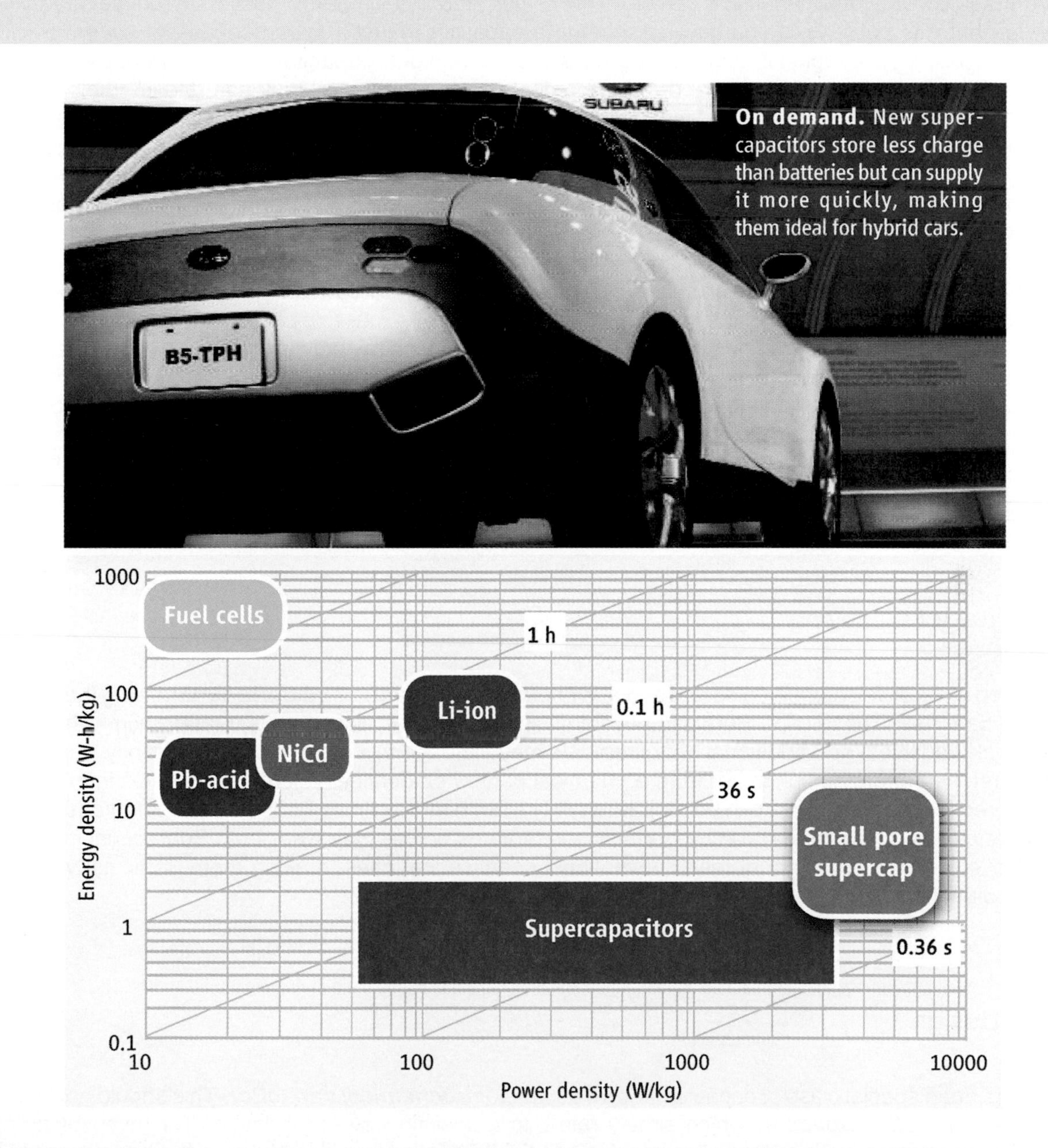

Figure TF11-1: Energy and power densities of modern energy-storage technologies. Even though supercapacitors store less charge than batteries, they can discharge their energy more quickly, making them more suitable for hybrid cars. (*Science,* Vol. 313, p. 902.)

of these attributes that determines the type of applications best suited for each of the various energy devices displayed in the figure. ***Energy density*** W' is a measure of how much energy a device or material can store per unit weight. [Alternatively, energy density can be defined in terms of volume (instead of weight) for applications where minimizing the volume of the energy source is more important than minimizing its weight.] Even though the formal SI unit for energy density is (J/kg), a more common unit is the watt-hour/kg (Wh/kg) with 1 Wh = 3600 J. The second dimension in Fig. TF11-1 is the ***power density*** P' (W/kg), which is a measure of how fast energy can be added to or removed from an energy-storage device (also per unit weight). Power is defined as energy per unit time as $P' = dW'/dt$. According to Fig. TF11-1, fuel cells can store large amounts of energy, but they can deliver that energy only relatively slowly (several hours). In contrast, conventional capacitors can store only small amounts of energy—several orders of magnitude less than fuel cells—but it is possible to charge or discharge a capacitor in just a few seconds—or even a fraction of a second. Batteries occupy the region in-between fuel cells and conventional capacitors; they can store more energy per unit weight than the ordinary capacitor by about three orders of magnitude, and they can release their energy faster than fuel cells by about a factor of 10. Thus, capacitors are partly superior to other energy devices because they can accomodate very fast rates of energy transfer, but the amount of energy that can be "packed into" a capacitor is limited by its size and weight. To appreciate what that means, let us examine the relation $w = \frac{1}{2}CV^2$. To increase w, we need to increase either C or V. For a parallel-plate capacitor, $C = \epsilon A/d$, where ϵ is the permittivity of the material between the plates, A is the area of each of the two plates, and d is the separation between them. The material between the plates should be a good insulator, and for most such insulators, the value of ϵ is in the range between ϵ_0 (permittivity of vacuum) and $6\epsilon_0$ (for mica), so the choice of material can at best increase C by a factor of 6. Making A larger increases both the volume and weight of the capacitor. In fact, since the mass m of the plates is proportional directly to A, the energy density $W' = w/m$ is independent of A. That leaves d as the only remaining variable. Reducing d will indeed increase C, but such a course will run into two serious obstacles: (a) to avoid voltage breakdown (arcing), V has to be reduced along with d such that V/d remains greater than the breakdown value of the insulator; (b) Eventually d approaches subatomic dimensions, making it infeasible to construct such a capacitor. Another serious limitation of the capacitor as an energy storage device is that its voltage does not remain constant as energy is transferred to and from it.

Supercapacitor Technology

A new generation of capacitor technologies, termed ***supercapacitors*** or ***ultracapacitors***, is narrowing the gap between capacitors and batteries. These capacitors can have sufficiently high energy densities to approach within 10 percent of battery storage densities, and additional improvements may increase this even more. Importantly, supercapacitors can absorb or release energy much faster than a chemical battery of identical volume. This helps immensely during recharging. Moreover, most batteries can be recharged only a few hundred times before they are degraded completely; supercapacitors can be charged and discharged millions of times before they wear out. Supercapacitors also have a much smaller environmental footprint than conventional chemical batteries, making them particularly attractive for green energy solutions.

History and Design

Supercapacitors are a special class of capacitor known as an ***electrochemical capacitor***. This should not be confused with the term ***electrolytic capacitor***, which simply refers to a specific way of fabricating a conventional capacitor. Electrochemical capacitors work by making use of a special property of water solutions (and some polymers and gels). When a metal electrode is immersed in water and a potential is applied, the water molecules (and any dissolved ions) immediately align themselves to the charges present at the surface of the metal electrode, as illustrated in Fig. TF11-2(a). This re-arrangement generates a thin layer of organized water molecules (and ions), called a ***double***

layer, that extends over the entire surface of the metal. The very high charge density, separated by a tiny distance on the order of a few nanometers, effectively looks like a capacitor (and a very large one: capacitive densities on the order of ~ 10 μF/cm^2 are common for water solutions). This phenomenon has been known to physicists and chemists since the work of von Helmholtz in 1853, and later Guoy, Chapman, and Stern in the early 20th century. In order to make capacitors useful for commercial applications, several technological innovations were required. Principal among these were various methods for increasing the total surface area that forms the double layer. The first working capacitor based on the electrochemical double layer (patented by General Electric in 1957) used very porous conductive carbon. Modern electrochemical capacitors employ ***carbon aerogels***, and more recently ***carbon nanotubes*** have been shown to effectively increase the total double layer area (Fig. TF11-2(b)).

Supercapacitors are beginning to see commercial use in applications ranging from transportation to low-power consumer electronics. Several bus lines around the world now run with buses powered with supercapacitors; train systems are also in development. Supercapacitors intended for small portable electronics (like your MP3 player) are in the pipeline as well!

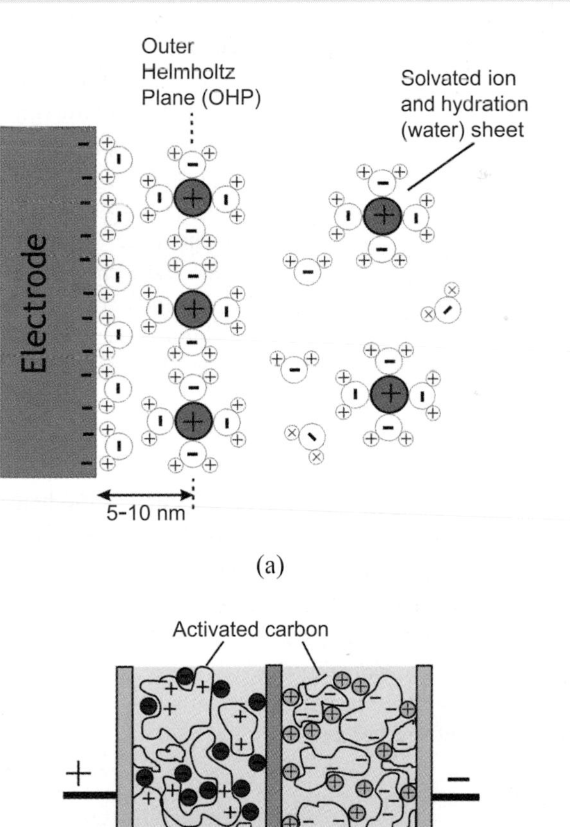

(a)

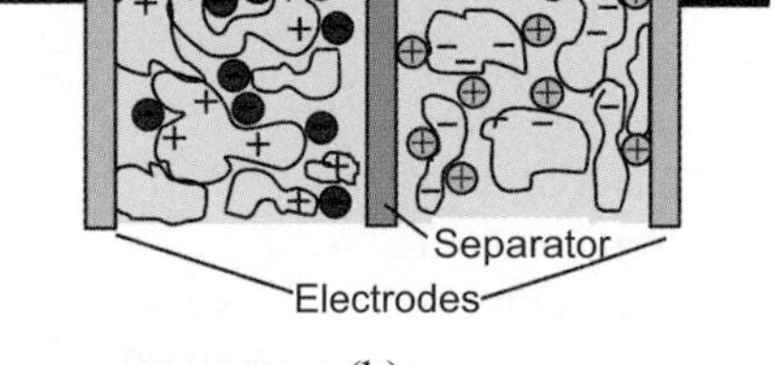

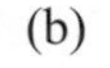

(b)

Figure TF11-2: (a) Conceptual illustration of the water double layer at a charged metal surface; (b) conceptual illustration of an electrochemical capacitor.

Exercise 5-9: Determine C_{eq} and $V_{eq}(0)$ at terminals (a, b) for the circuit in Fig. E5.9 given that $C_1 = 6\ \mu$F, $C_2 = 4\ \mu$F, $C_3 = 8\ \mu$F, and the initial voltages on the three capacitors are $\upsilon_1(0) = 5$ V and $\upsilon_2(0) = \upsilon_3(0) = 10$ V, respectively.

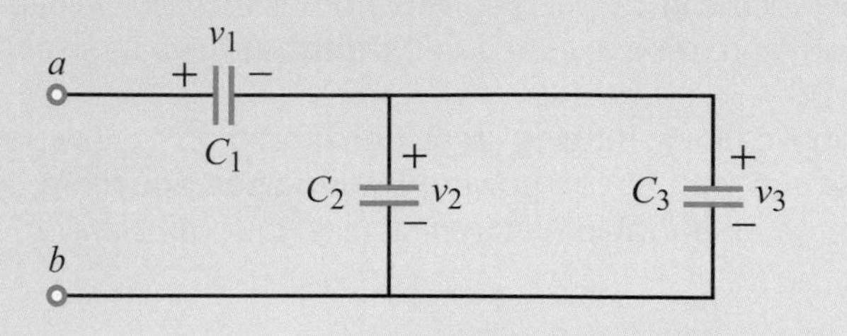

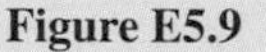

Figure E5.9

Answer: $C_{eq} = 4\ \mu$F, $V_{eq}(0) = 15$ V. (See)

Exercise 5-10: Suppose the circuit of Fig. E5.9 is connected to a dc voltage source $V_0 = 12$ V. Assuming that the capacitors had no charge before they were connected to the voltage source, determine υ_1 and υ_2 given that $C_1 = 6\ \mu$F, $C_2 = 4\ \mu$F, and $C_3 = 8\ \mu$F.

Answer: $\upsilon_1 = 8$ V, $\upsilon_2 = 4$ V. (See)

5-3 Inductors

Capacitors and inductors constitute a canonical pair of devices. Whereas capacitors can store energy through the electric field induced by the voltage imposed across its terminals, inductors can store magnetic energy through the magnetic field induced by the current flowing through its wires. The i–υ relationship is $i = C\ d\upsilon/dt$ for a capacitor; the converse is true for an inductor with $\upsilon = L\ di/dt$. As we will see in Chapter 7, the capacitor acts like an open circuit to low-frequency signals and like a short circuit to high-frequency signals; the exact opposite behavior is exhibited by the inductor.

A typical example of an inductor is the ***solenoid*** configuration shown in Fig. 5-20. The solenoid consists of multiple turns of wire wound in a helical geometry around a cylindrical core. The core may be air filled or may contain a magnetic material with ***magnetic permeability*** μ. If the wire carries a current $i(t)$ and the turns are closely spaced, the solenoid produces a relatively uniform ***magnetic field*** B within its interior region.

Magnetic-flux linkage Λ is defined as the total magnetic flux linking a coil or a given circuit. For a solenoid with N turns carrying a current i,

$$\Lambda = \left(\frac{\mu N^2 S}{\ell}\right) i \qquad \text{(Wb)}, \tag{5.50}$$

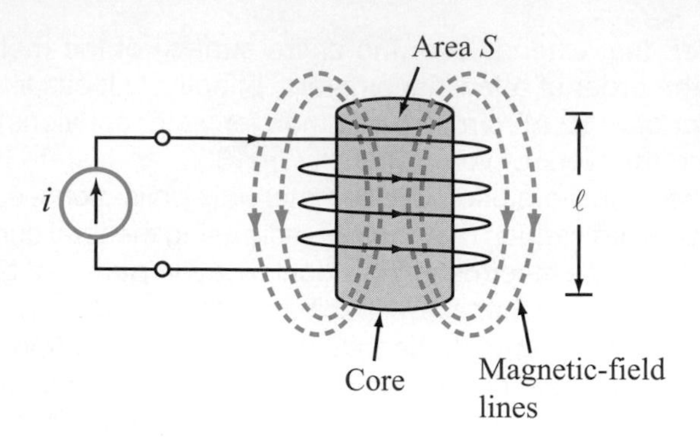

Figure 5-20: The inductance of a solenoid of length ℓ and cross-sectional area S is $L = \mu N^2 S/\ell$, where N is the number of turns and μ is the magnetic permeability of the core material.

where ℓ is the length of the solenoid and S is its cross-sectional area. The unit for Λ is the weber (Wb), named after the German scientist Wilhelm Weber (1804–1891).

Self-inductance refers to the magnetic-flux linkage of a coil (or circuit) with itself, in contrast with ***mutual inductance***, which refers to magnetic-flux linkage in a coil due to the magnetic field generated by another coil (or circuit). Usually, when the term ***inductance*** is used, the intended reference is to self-inductance. The inductance of any conducting system is defined as the ratio of Λ to the current i responsible for generating it, given as

$$L = \frac{\Lambda}{i} \qquad \text{(H)}, \tag{5.51}$$

and its unit is the henry (H), so named to honor the American inventor Joseph Henry (1797–1878). Using the expression for Λ given by Eq. (5.50), we have

$$L = \frac{\mu N^2 S}{\ell} \qquad \text{(solenoid)}. \tag{5.52}$$

The inductance L is directly proportional to μ, the magnetic permeability of the core material. The relative magnetic permeability μ_r is defined as

$$\mu_r = \frac{\mu}{\mu_0}, \tag{5.53}$$

where $\mu_0 \simeq 4\pi \times 10^{-7}$ (H/m) is the magnetic permeability of free space. *Except for ferromagnetic materials, $\mu_r \simeq 1$ for all dielectrics and conductors.* According to Table 5-3, μ_r of ferromagnetic materials (which include iron, nickel, and cobalt) can be as much as five orders of magnitude larger than that of other materials. Consequently, L of an ***iron-core solenoid*** is about 5000 times that of an ***air-core solenoid*** of the same size and shape.

Table 5-3: Relative magnetic permeability of materials.

Material	Relative Permeability μ_r
All Dielectrics and Non-Ferromagnetic Metals	≈ 1.0
Ferromagnetic Metals	
Cobalt	250
Nickel	600
Mild steel	2,000
Iron (pure)	4,000–5,000
Silicon iron	7,000
Mumetal	$\sim$ 100, 000
Purified iron	$\sim$ 200, 000

Air-core inductors have relatively low inductances, on the order of 10 μH or smaller. Consequently, they are used mostly in high-frequency circuits, such as those designed to support AM and FM radio, cell phones, TV, and similar types of transmitters and receivers. ***Ferrite-core inductors*** have the inductance-size advantage over air-core inductors, but they also have the disadvantage that the ferrite material is subject to hysteresis effects. One of the consequences of magnetic hysteresis is that the inductance L becomes a function of the current flowing through it. Magnetic hysteresis is outside the scope of this book; hence, *we always will assume that an inductor is an ideal linear device and its inductance is constant and independent of the current flowing through it.*

In modern circuit design and manufacturing, it is highly desirable to contain circuit size down to the smallest dimensions possible. To that end, it is advantageous to use planar integrated-circuit (IC) devices whenever possible. It is relatively easy to manufacture resistors and capacitors in a planar IC format and to do so for a wide range of resistance and capacitance values, but the same is not true for inductors. Even though inductors can be manufactured in planar form, as illustrated by the coil shown in Fig. 5-21, their inductance values are too small for most circuit applications, necessitating the use of the more bulky, discrete form instead.

Inductors exhibit a number of useful properties, including magnetic coupling and electromagnetic induction. They are employed in microphones and loudspeakers, magnetic relays and sensors, motors and generators, and numerous other applications.

5-3.1 Electrical Properties

According to Faraday's law, if the magnetic-flux linkage in an inductor (or circuit) changes with time, it induces a voltage v across the inductor's terminals given by

$$v = \frac{d\Lambda}{dt}. \tag{5.54}$$

In view of Eq. (5.51),

$$v = \frac{d}{dt}(Li) = L\,\frac{di}{dt}. \tag{5.55}$$

This i–v relationship adheres to the passive sign convention introduced earlier for resistors and capacitors. If the direction of i is into the (+) voltage terminal of the inductor (Fig. 5-22), then the inductor is receiving power. Also, the same logic that led us earlier to the conclusion that the voltage across a capacitor cannot change instantaneously leads us now to the conclusion:

> The current through an inductor cannot change instantaneously.

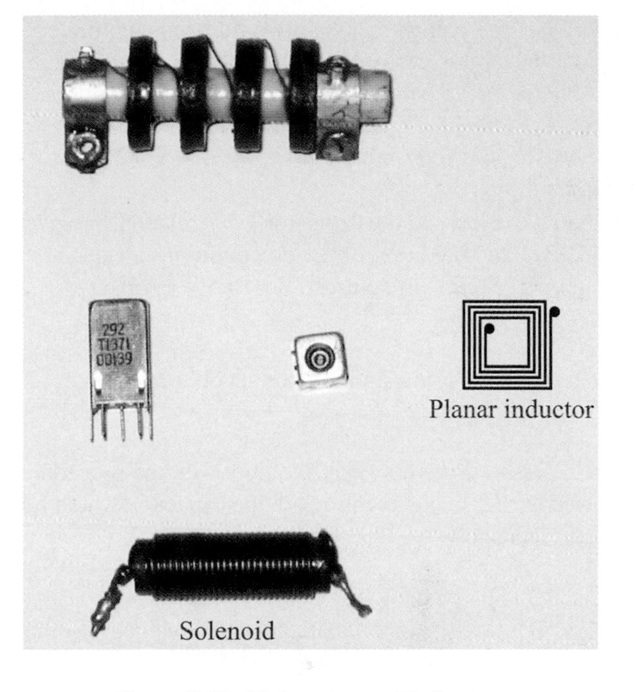

Figure 5-21: Various types of inductors.

L, i, $+$ v $-$, $\quad v = L\,\dfrac{di}{dt}$

Figure 5-22: Passive sign convention for an inductor.

(Otherwise, the voltage across it would become infinite.) The implication of this restriction is that *when a current source connected to an inductor is disconnected by a switch, the current continues to flow for a short amount of time through the air between the switch terminals, manifesting itself in the form of a spark!*

When we discussed the capacitor's i–v relationship given by Eq. (5.23), we noted that under dc conditions a capacitor acts like an open circuit. In contrast, Eq. (5.55) asserts:

> Under dc conditions, an inductor acts like a short circuit.

To express i in terms of v, we duplicate the procedure we followed earlier in connection with the capacitor, which for the inductor leads to

$$i(t) = i(t_0) + \frac{1}{L} \int_{t_0}^{t} v\, dt, \tag{5.56}$$

where t_0 is a reference point in time.

The power delivered to the inductor is given by

$$p(t) = vi = Li\, \frac{di}{dt}, \tag{5.57}$$

and as with the resistor and the capacitor, the sign of p determines whether the inductor is receiving power ($p > 0$) or delivering it ($p < 0$). The accumulation of power over time constitutes the storage of energy. The magnetic energy stored in an inductor is

$$w = \int_{-\infty}^{t} p\, dt = \int_{-\infty}^{t} \left(Li\, \frac{di}{dt} \right) dt, \tag{5.58}$$

which yields

$$w = \frac{1}{2} Li^2 \quad \text{(J)}, \tag{5.59}$$

where it is presumed that at $t = -\infty$ no current was flowing through the inductor. Note the analogy with the capacitor for which $w = \frac{1}{2} Cv^2$.

> The magnetic energy stored in an inductor at a given instant in time depends on the current flowing through the inductor at that instant—without regard to prior history.

Example 5-7: Inductor Response to Current Waveform

Upon closing the switch at $t = 0$ in the circuit of Fig. 5-23(a), the voltage source generates a current waveform through the circuit given by

$$i(t) = 10e^{-0.8t} \sin(\pi t/2) \text{ A}, \qquad (\text{for } t \geq 0).$$

(a) Plot the waveform $i(t)$ versus t and determine the locations of its first maximum, first minimum, and their corresponding amplitudes.

(b) Given that $L = 50$ mH, obtain an expression for $v(t)$ across the inductor and plot its waveform.

(c) Generate a plot of the power $p(t)$ delivered to the inductor.

Solution:

(a) The waveform of $i(t)$ is shown in Fig. 5-23(b). To determine the locations of its maxima and minima, we take the derivative of $i(t)$ and equate it to zero:

$$-0.8 \times 10e^{-0.8t} \sin(\pi t/2) + \left(\frac{\pi}{2}\right) \times 10e^{-0.8t} \cos\left(\frac{\pi t}{2}\right) = 0$$

which simplifies to

$$\tan\left(\frac{\pi t}{2}\right) = \frac{\pi}{1.6}.$$

Its solution is

$$\frac{\pi t}{2} = 1.1 + n\pi \qquad (\text{for } n = 0, 1, 2, \ldots).$$

For $n = 0$, $t = 0.7$ s, which is the location in time of the first maximum of $i(t)$. The next solution, corresponding to $n = 1$, gives the location of the first minimum of $i(t)$ at 2.7 s. The amplitudes of $i(t)$ at these locations are

$$i_{max} = i(t = 0.7 \text{ s}) = 10e^{-0.8\times0.7} \sin(\pi \times 0.7/2) = 5.09 \text{ A}.$$

and

$$i_{min} = i(t = 2.7 \text{ s}) = 10e^{-0.8\times2.7} \sin(\pi \times 2.7/2) = -1.03 \text{ A}$$

(b)

$$\begin{aligned} v(t) &= L\, \frac{di}{dt} \\ &= L\, \frac{d}{dt}[10e^{-0.8t} \sin(\pi t/2)] \\ &= 50 \times 10^{-3} \cdot [-8e^{-0.8t} \sin(\pi t/2) + 5\pi e^{-0.8t} \cos(\pi t/2)] \\ &= [-0.4 \sin(\pi t/2) + 0.25\pi \cos(\pi t/2)]e^{-0.8t} \text{ V}. \end{aligned}$$

The waveform of $v(t)$ is shown in Fig. 5-23(c).

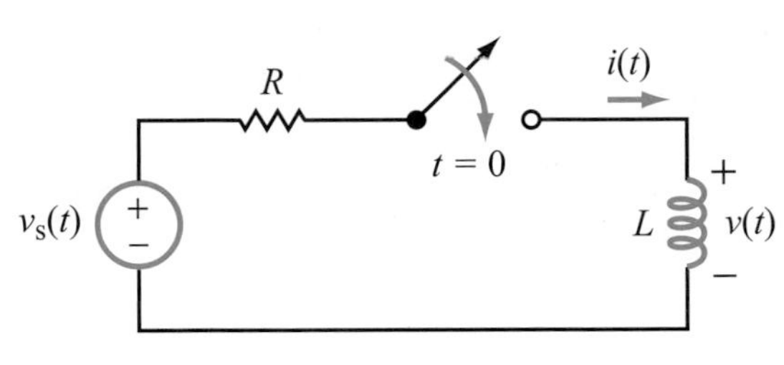

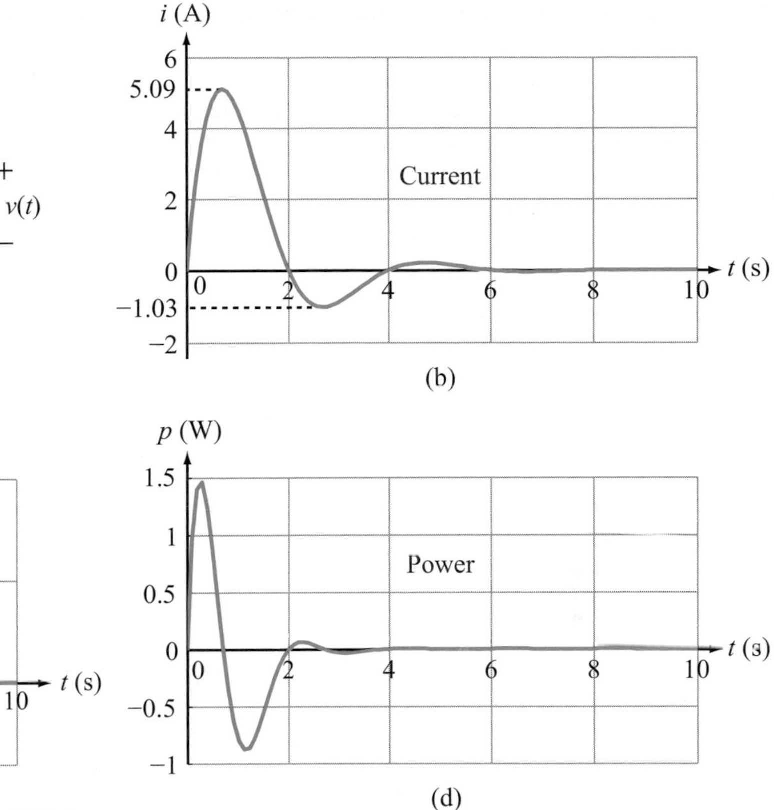

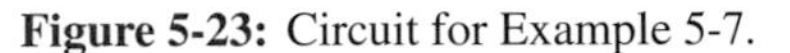

Figure 5-23: Circuit for Example 5-7.

(c)

$$\begin{aligned} p(t) &= v(t)\, i(t) \\ &= [-0.4 \sin(\pi t/2) + 0.25\pi \cos(\pi t/2)]e^{-0.8t} \\ &\quad \times 10e^{-0.8t} \sin(\pi t/2) \\ &= [-4 \sin^2(\pi t/2) + 2.5\pi \cos(\pi t/2) \sin(\pi t/2)] \\ &\quad \times e^{-1.6t} \text{ W}. \end{aligned}$$

The waveform of $p(t)$ shown in Fig. 5-23(d) includes both positive and negative values. During periods when $p(t) > 0$, magnetic energy is getting stored in the inductor. Conversely, when $p(t) < 0$, the inductor is releasing some of its previously stored energy.

Review Question 5-13: What type of material exhibits a magnetic permeability that deviates from μ_0?

Review Question 5-14: Can the voltage across an inductor change instantaneously?

Exercise 5-11: Calculate the inductance of a 20-turn air-core solenoid if its length is 4 cm and the radius of its circular cross section is 0.5 cm.

Answer: $L = 9.87 \times 10^{-7}$ H $= 0.987$ μH. (See)

Exercise 5-12: Determine currents i_1 and i_2 in the circuit of Fig. E5.12, under dc conditions.

Answer: $i_1 = 0$, $i_2 = 6$ A. (See)

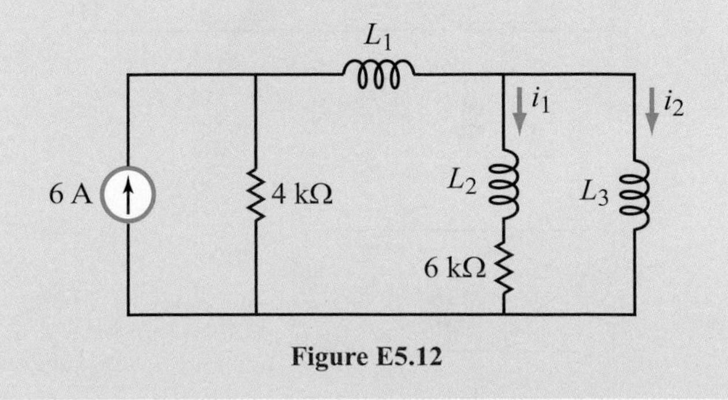

Figure E5.12

5-3.2 Series and Parallel Combinations of Inductors

The rules for combining multiple inductors in series or in parallel are the same as those for resistors.

Inductors in Series

For the three inductors in series in Fig. 5-24,

$$\begin{aligned} v_s &= v_1 + v_2 + v_3 \\ &= L_1 \frac{di_s}{dt} + L_2 \frac{di_s}{dt} + L_3 \frac{di_s}{dt} \\ &= (L_1 + L_2 + L_3) \frac{di_s}{dt}, \end{aligned} \tag{5.60}$$

and for the equivalent circuit,

$$v_s = L_{eq} \frac{di_s}{dt}. \tag{5.61}$$

Hence,

$$L_{eq} = L_1 + L_2 + L_3, \tag{5.62}$$

and for N inductors in series,

$$L_{eq} = \sum_{i=1}^{N} L_i = L_1 + L_2 + \cdots + L_N \tag{5.63}$$

(inductors in series).

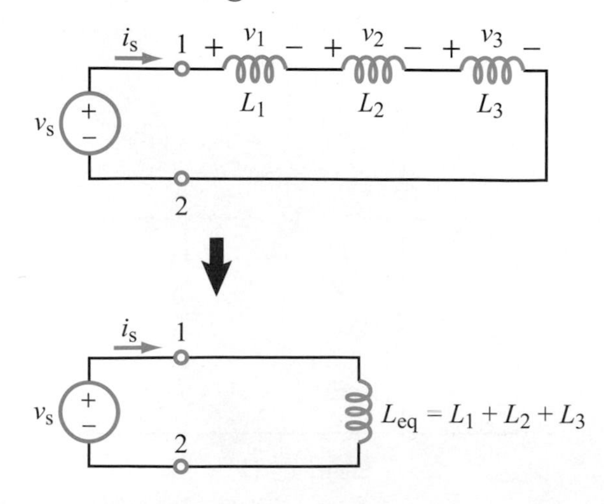

Figure 5-24: Inductors in series.

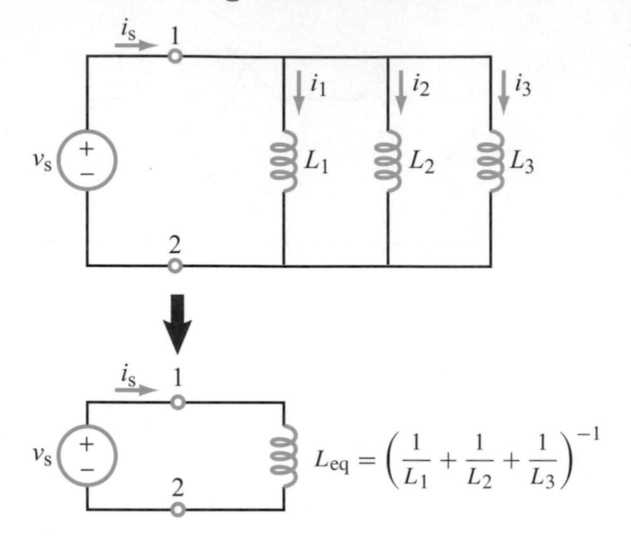

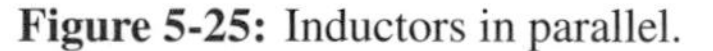

Figure 5-25: Inductors in parallel.

Inductors in Parallel

A similar analysis for the currents in the parallel circuit of Fig. 5-25 leads to

$$\frac{1}{L_{eq}} = \frac{1}{L_1} + \frac{1}{L_2} + \frac{1}{L_3}. \tag{5.64}$$

Generalizing to the case of N inductors,

$$\frac{1}{L_{eq}} = \sum_{i=1}^{N} \frac{1}{L_i} = \frac{1}{L_1} + \frac{1}{L_2} + \cdots + \frac{1}{L_N} \tag{5.65}$$

(inductors in parallel).

If $i_1(t_0)$ through $i_N(t_0)$ are the initial currents flowing through the parallel inductors L_1 to L_N at t_0, then the initial current $i_{eq}(t_0)$ that would be flowing through the equivalent inductor L_{eq} is given by

$$i_{eq}(t_0) = \sum_{j=1}^{N} i_j(t_0). \tag{5.66}$$

A summary of the electrical properties of resistors, inductors and capacitors is available in Table 5-4.

Table 5-4: Basic properties of R, L, and C.

Property	R	L	C
i–v relation	$i = \frac{v}{R}$	$i = \frac{1}{L}\int_{t_0}^{t} v\,dt + i(t_0)$	$i = C\,\frac{dv}{dt}$
v-i relation	$v = iR$	$v = L\,\frac{di}{dt}$	$v = \frac{1}{C}\int_{t_0}^{t} i\,dt + v(t_0)$
p (power transfer in)	$p = i^2 R$	$p = Li\,\frac{di}{dt}$	$p = Cv\,\frac{dv}{dt}$
w (stored energy)	0	$w = \frac{1}{2}Li^2$	$w = \frac{1}{2}Cv^2$
Series combination	$R_{eq} = R_1 + R_2$	$L_{eq} = L_1 + L_2$	$C_{eq} = \frac{C_1 C_2}{C_1 + C_2}$
Parallel combination	$R_{eq} = \frac{R_1 R_2}{R_1 + R_2}$	$L_{eq} = \frac{L_1 L_2}{L_1 + L_2}$	$C_{eq} = C_1 + C_2$
dc behavior	no change	short circuit	open circuit
Can v change instantaneously?	yes	yes	no
Can i change instantaneously?	yes	no	yes

Example 5-8: Energy Storage Under dc Conditions

The circuit in Fig. 5-26(a) has been in its present state for a long time. Determine the amount of energy stored in the capacitors and inductors.

Solution: Our first step is to replace capacitors with open circuits and inductors with short circuits. The process leads to the circuit in Fig. 5-26(b). Current I_1 then is given by

$$I_1 = \frac{24}{(2+4)\text{k}} = 4 \text{ mA},$$

and node voltage V is

$$V = 24 - (4 \times 10^{-3} \times 4 \times 10^3) = 8 \text{ V}.$$

Hence, the amounts of energy stored in C_1, C_2, L_1, L_2, and L_3 are

$$C_1: \quad W = \frac{1}{2}C_1 V^2 = \frac{1}{2} \times 10^{-5} \times 64 = 0.32 \text{ mJ},$$

$$C_2: \quad W = \frac{1}{2}C_2 V^2 = \frac{1}{2} \times 4 \times 10^{-6} \times 64 = 0.128 \text{ mJ},$$

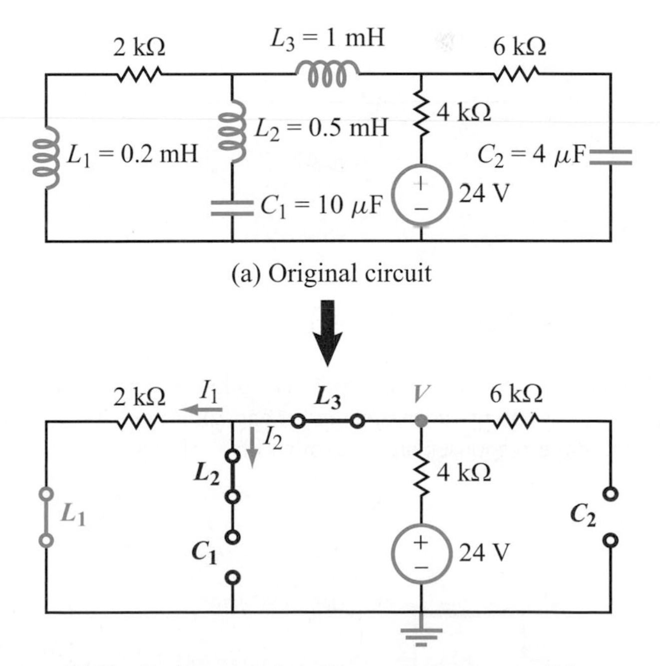

Figure 5-26: Under steady-state dc conditions, capacitors act like open circuits, and inductors act like short circuits.

$$L_1: \quad W = \frac{1}{2}L_1 I_1^2 = \frac{1}{2} \times 0.2 \times 10^{-3} \times (4 \times 10^{-3})^2 = 1.6 \text{ nJ},$$

$$L_2: \quad W = \frac{1}{2}L_2 I_2^2 = \frac{1}{2} \times 0.5 \times 10^{-3} \times (0) = 0,$$

and

$$L_3: \quad W = \frac{1}{2}L_3 I_1^2 = \frac{1}{2} \times 10^{-3} \times (4 \times 10^{-3})^2 = 8 \text{ nJ}.$$

Review Question 5-15: How do the rules for adding inductors in series and in parallel compare with those for resistors and capacitors?

Review Question 5-16: An inductor stores energy through the magnetic field B, but the equation for the energy stored in an inductor is $w = \frac{1}{2}Li^2$. Explain.

Exercise 5-13: Determine L_{eq} at terminals (a, b) in the circuit of Fig. E5.13.

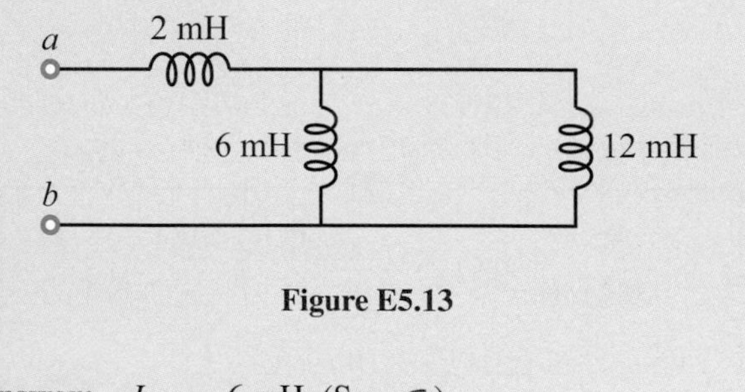

Figure E5.13

Answer: $L_{eq} = 6$ mH. (See)

5-4 Response of the RC Circuit

The circuit shown in Fig. 5-27 is called a ***first-order RC circuit***; it contains a resistor and a capacitor, and its current and voltage responses are determined by solving a first-order differential equation. The name also applies to any other circuit containing sources, resistors, and capacitors—provided it can be reduced to the form of the generic RC circuit of Fig. 5-27 or its Norton equivalent. The voltage source exciting the circuit is a rectangular pulse of amplitude V_s and duration T_0. The objective of the present section is to develop a methodology appropriate for RC circuits, so we may apply it to evaluate the circuit's response to the rectangular-pulse waveform or to other types of nonperiodic waveforms.

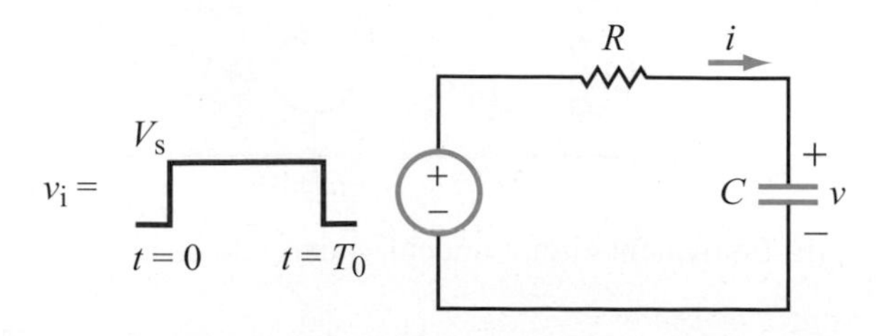

Figure 5-27: Generic first-order RC circuit.

5-4.1 Natural Response of a Charged Capacitor

The series RC circuit shown in Fig. 5-28(a) is connected to a dc voltage source V_s via an SPDT switch. At $t = 0$, the switch disconnects the RC circuit from the source and connects it to terminal 2. We seek to determine the voltage response of the capacitor $v(t)$ for $t \geq 0$.

Before we start our solution, it is important to consider the implication of the information we are given about the state of

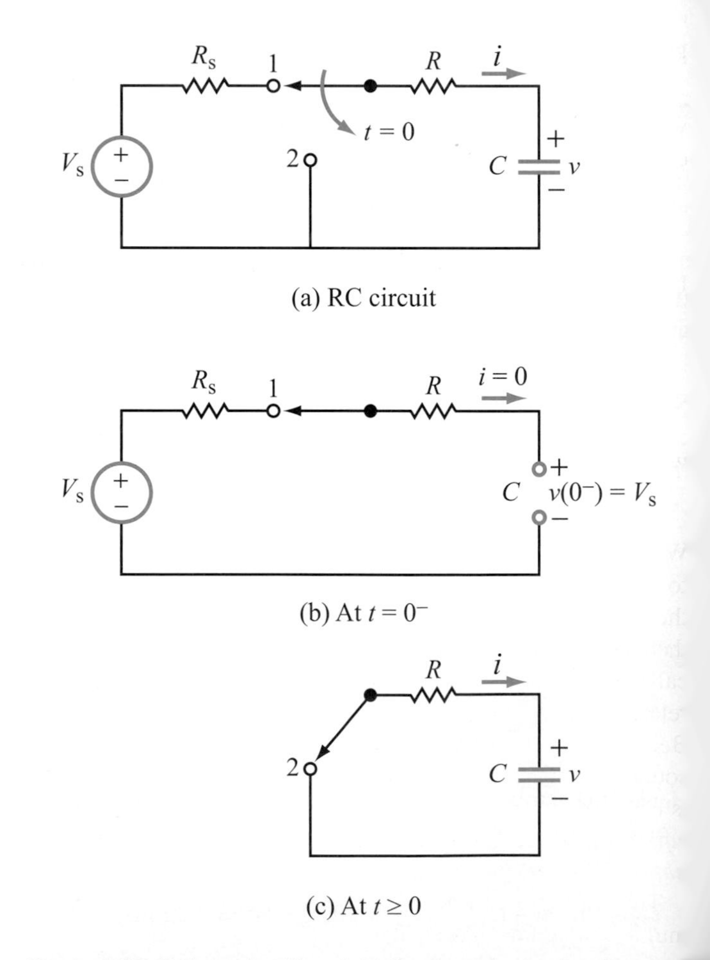

Figure 5-28: RC circuit with an initially charged capacitor that start to discharge its energy after $t = 0$.

the capacitor before and after moving the switch. For purposes of clarity, we define:

(a) $t = 0^-$ as the instant just before the switch is moved from terminal 1 to terminal 2

(b) $t = 0$ as the instant just after it was moved; *$t = 0$ is synonymous with $t = 0^+$*

At $t = 0^-$, the circuit had been in the condition shown in Fig. 5-28(a) for a long time. As we noted earlier in Section 5-2.1, when a dc circuit is in a steady state, its capacitors act like open circuits. Consequently, the open circuit in Fig. 5-28(b), representing the state of the circuit at $t = 0^-$, allows no current to flow through the loop. Hence, $v(0^-) = V_s$, and since the voltage across the capacitor cannot change instantaneously, it follows that $v(0)$, the voltage after moving the switch, is given by

$$v(0) = v(0^-) = V_s. \tag{5.67}$$

As we will see shortly, we will need this piece of information for when we apply initial conditions to the solution of the differential equation of $v(t)$.

For $t \geq 0$, application of KVL to the loop in Fig. 5-28(c) gives

$$Ri + v = 0 \qquad (\text{for } t \geq 0), \tag{5.68}$$

where i is the current through and v is the voltage across the capacitor. Since $i = C\, dv/dt$,

$$RC\,\frac{dv}{dt} + v = 0. \tag{5.69}$$

Upon dividing both terms by RC, Eq. (5.69) takes the form

$$\frac{dv}{dt} + av = 0 \qquad (\text{source-free}), \tag{5.70}$$

where

$$a = \frac{1}{RC}. \tag{5.71}$$

When arranging a differential equation in $v(t)$, it is customary to place all terms that involve $v(t)$ on the left-hand side of the equation and to place terms that do not involve $v(t)$ on the right-hand side. The term(s) on the right-hand side is (are) called the ***forcing function***. For a circuit, the forcing function is related directly to the voltage and current sources in the circuit. Because the RC circuit in Fig. 5-28(c) does not contain any sources, Eq. (5.70) has a zero on its right-hand side and it is called (appropriately) a ***source-free, first-order differential equation***. The solution of the source-free equation is called the ***natural response*** of the circuit.

The standard procedure for solving Eq. (5.70) starts by multiplying both sides by e^{at},

$$\frac{dv}{dt}\,e^{at} + ave^{at} = 0. \tag{5.72}$$

Next, we recognize that the sum of the two terms on the left-hand side is equal to the expansion of the differential of (ve^{at}),

$$\frac{d}{dt}(ve^{at}) = \frac{dv}{dt}\,e^{at} + ave^{at}. \tag{5.73}$$

Hence, Eq. (5.72) becomes

$$\frac{d}{dt}(ve^{at}) = 0. \tag{5.74}$$

Integrating both sides, we have

$$\int_0^t \frac{d}{dt}(ve^{at})\,dt = 0, \tag{5.75}$$

where we have chosen the lower limit to be $t = 0$ (because we are given specific information on the state of the circuit at that point in time). Performing the integration gives

$$ve^{at}\Big|_0^t = 0$$

or

$$v(t)\,e^{at} - v(0) = 0. \tag{5.76}$$

Solving for $v(t)$, we have

$$\begin{aligned} v(t) &= v(0)\,e^{-at}, \\ &= v(0)\,e^{-t/RC} \qquad (\text{for } t \geq 0), \end{aligned} \tag{5.77}$$

where we used Eq. (5.71) for a and appended the inequality $t \geq 0$ to indicate that the expression given by Eq. (5.77) is valid only for $t \geq 0$.

The coefficient of t in the exponent is a critically important parameter, because it determines the temporal rate of $v(t)$. It is customary to rewrite Eq. (5.77) in the form

$$v(t) = v(0)\,e^{-t/\tau} \qquad (\text{natural response}), \tag{5.78}$$

with

$$\tau = RC \qquad (\text{s}), \tag{5.79}$$

where τ is called the ***time constant*** of the circuit, and it is measured in seconds (s).

In view of the initial condition given by Eq. (5.67), namely $v(0) = V_s$, the expression for $v(t)$ becomes

$$v(t) = V_s e^{-t/\tau} \qquad (\text{for } t \geq 0). \tag{5.80}$$

The plot shown in Fig. 5-29(a) indicates that in response to the switch action $v(t)$ decays exponentially with time from V_s at $t = 0$ down to its final value of zero as $t \to \infty$. The decay rate is dictated by the time constant τ. At $t = \tau$,

$$v(t = \tau) = V_s e^{-1} = 0.37 V_s, \tag{5.81}$$

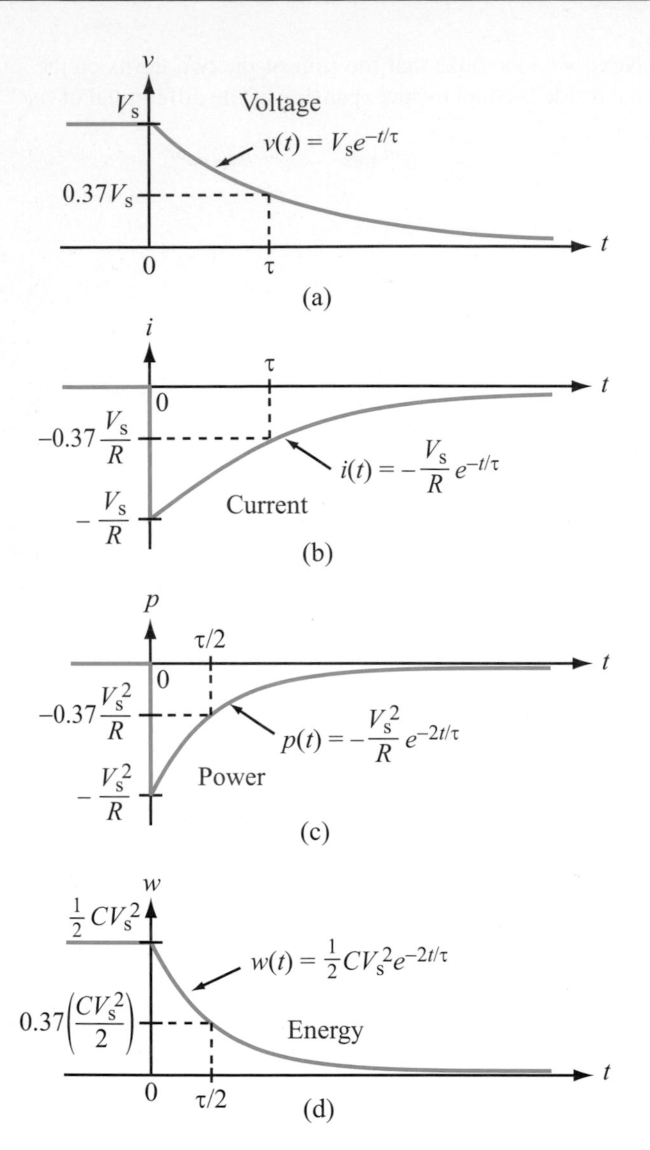

Figure 5-29: Response of the RC circuit in Fig. 5-28(a) to moving the SPDT switch to terminal 2.

which means that at τ seconds after activating the switch, the voltage across the capacitor is down to 37 percent of its initial value. At $t = 2\tau$, it reaches 14 percent, and at $t = 5\tau$, it is less than 1 percent of its initial value. Hence, for all practical purposes, we can treat the circuit as having reached its final state when the switch has been in its new configuration for a time equal to or longer than 5τ.

> The magnitude of the time constant τ is a measure of how fast or how slowly a circuit responds to a sudden change.

As we will see later in Section 5-7, the *clock speed* of a computer processor is to first order proportional to $1/\tau$. Hence, a *slow* circuit with $\tau = 1$ ms would have a clock speed on the order of 1 kHz, whereas a *fast* circuit with $\tau = 1$ ns can support clock speeds as high as 1 GHz.

The current $i(t)$ flowing through the capacitor is given by

$$\begin{aligned} i(t) &= C\,\frac{dv}{dt} = C\,\frac{d}{dt}(V_s e^{-t/\tau}) \\ &= -C\,\frac{V_s}{\tau}\,e^{-t/\tau} \qquad (\text{for } t \geq 0), \end{aligned} \tag{5.82}$$

which simplifies to

$$i(t) = -\frac{V_s}{R}\,e^{-t/\tau}\,u(t) \qquad (\text{for } t \geq 0) \quad (\text{natural response}). \tag{5.83}$$

The plot of $i(t)$ shown in Fig. 5-29(b) indicates that after closing the switch at $t = 0$, the current changes instantly to $(-V_s/R)$—as if the capacitor were a voltage source V_s—and then it decays exponentially down to zero. The negative sign of i signifies that it flows in a counterclockwise direction through the loop, consistent with the behavior of the capacitor as a voltage source.

Given $v(t)$ and $i(t)$, we can provide an expression for $p(t)$, the instantaneous power getting transferred to the capacitor, as

$$\begin{aligned} p(t) = iv &= -\frac{V_s}{R}\,e^{-t/\tau} \times V_s e^{-t/\tau} \\ &= -\frac{V_s^2}{R}\,e^{-2t/\tau} \qquad (\text{for } t \geq 0). \end{aligned} \tag{5.84}$$

In general, power transfer is into a device if $p > 0$ and out of it if $p < 0$. Prior to $t = 0$, the capacitor had been connected to the voltage source for a long time. Hence, power already had flowed into the capacitor and was stored as electrical energy. The minus sign in Eq. (5.84) denotes that after $t = 0$ power flows out of the capacitor and gets dissipated in the resistor. The decay rate for $p(t)$ is $2/\tau$, which is twice as fast as that for $v(t)$ or $i(t)$.

The amount of energy $w(t)$ contained in the medium between the capacitor's oppositely charged conducting plates can be calculated either by integrating $p(t)$ over time from 0 to t or by applying Eq. (5.29). The latter approach gives

$$\begin{aligned} w(t) &= \frac{1}{2}\,Cv^2 \\ &= \frac{CV_s^2}{2}\,e^{-2t/\tau} \qquad (\text{for } t \geq 0). \end{aligned} \tag{5.85}$$

Parts (c) and (d) of Fig. 5-29 display the time waveforms of $p(t)$ and $w(t)$, respectively.

Review Question 5-17: What specific characteristic defines a first-order circuit?

Review Question 5-18: What does the time constant of an RC circuit represent?

Review Question 5-19: For the natural response of an RC circuit, how does the decay rate for voltage compare with that for power?

Exercise 5-14: If in the circuit of Fig. E5.14 $v(0^-) = 24$ V, determine $v(t)$ for $t \geq 0$.

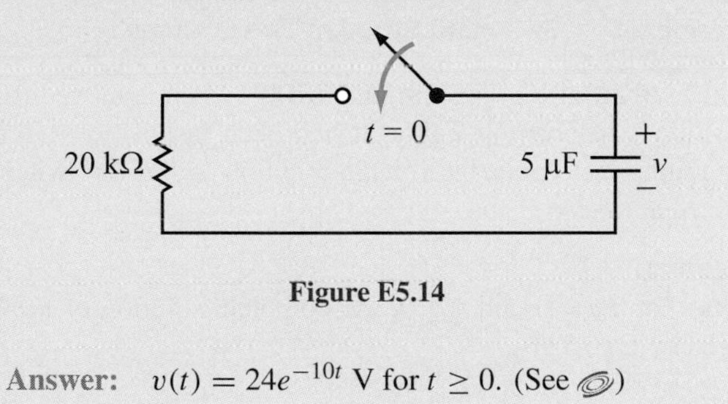

Figure E5.14

Answer: $v(t) = 24e^{-10t}$ V for $t \geq 0$. (See)

5-4.2 General Form of the Step Response of the RC Circuit

When we use the term ***circuit response***, we mean the reaction of a certain voltage or current in the circuit to change, such as the introduction of a new source, the elimination of a source, or some other change in the circuit configuration. Whenever possible, we usually designate $t = 0$ as the instant at which the change occurred and $t \geq 0$ as the time interval over which we seek the circuit response. In the general case, the capacitor may start with a voltage $v(0)$ at $t = 0$ (immediately after the sudden change) and may approach $v(\infty)$ as $t \to \infty$. A circuit configuration that can represent such a scenario is shown in Fig. 5-30(a). Prior to $t = 0$, the RC circuit is connected to a source V_{s_1}, and after $t = 0$, it is connected to a different source V_{s_2}. The circuit can be reduced to the following special cases.

(a) Step response (due to V_{s_2}) of an uncharged capacitor (if $V_{s_1} = 0$)

(b) Step response (due to V_{s_2}) of a charged capacitor (if $V_{s_1} \neq 0$)

(c) Natural response (if $V_{s_2} = 0$) of a charged capacitor ($V_{s_1} \neq 0$)

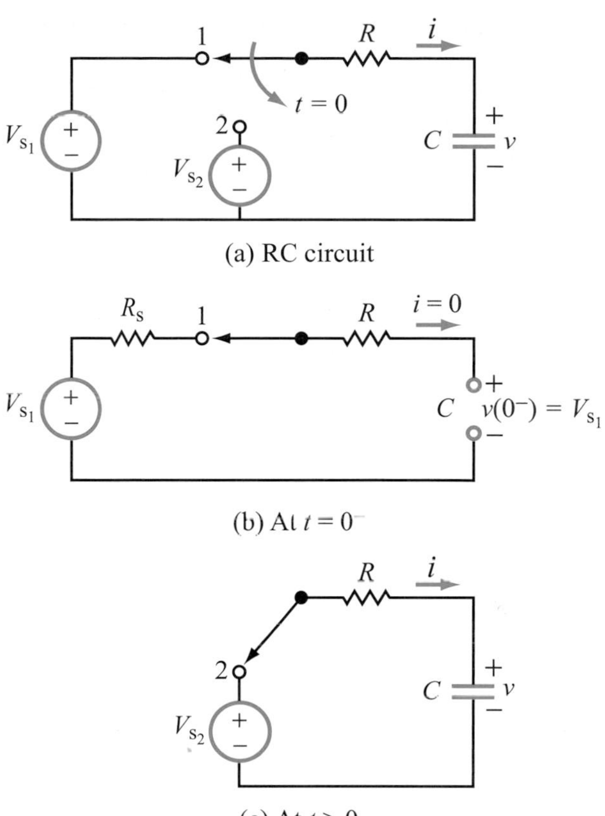

Figure 5-30: RC circuit switched from source V_{s_1} to source V_{s_2} at $t = 0$.

For obvious reasons, we excluded the trivial case where both V_{s_1} and V_{s_2} are zero, and we will now treat the general case where neither V_{s_1} nor V_{s_2} is zero.

At $t = 0^-$ (Fig. 5-30b), the capacitor acts like an open circuit. Consequently, $i(0^-) = 0$, and $v(0^-) = V_{s_1}$. Since v across the capacitor cannot change in zero time, the voltage $v(0)$ after moving the switch to terminal 2 is

$$v(0) = v(0^-) = V_{s_1}. \tag{5.86}$$

For $t \geq 0$, the voltage equation for the loop in Fig. 5-30(c) is

$$-V_{s_2} + iR + v = 0. \tag{5.87}$$

Upon using $i = C\, dv/dt$ and rearranging its terms, Eq. (5.87) can be written in the differential-equation form

$$\frac{dv}{dt} + av = b, \tag{5.88}$$

where

$$a = \frac{1}{RC} \quad \text{and} \quad b = \frac{V_{s_2}}{RC}. \tag{5.89}$$

We note that Eq. (5.88) is similar to Eq. (5.70), except that now we have a non-zero term on the right-hand side of the equation. Nevertheless, the method of solution remains the same. After multiplying both sides of Eq. (5.88) by e^{at} and using the expansion given by Eq. (5.73), we have

$$\frac{d}{dt}(ve^{at}) = be^{at}. \quad (5.90)$$

Integrating both sides,

$$\int_0^t \frac{d}{dt}(ve^{at})\,dt = \int_0^t be^{at} \quad (5.91)$$

gives

$$ve^{at}\big|_0^t = \frac{b}{a}\,e^{at}\Big|_0^t. \quad (5.92)$$

Upon evaluating the functions at the two limits, we have

$$v(t)\,e^{at} - v(0) = \frac{b}{a}\,e^{at} - \frac{b}{a}, \quad (5.93)$$

and then solving for $v(t)$, we have

$$v(t) = v(0)\,e^{-at} + \frac{b}{a}\,(1 - e^{-at}). \quad (5.94)$$

As $t \to \infty$, $v(t)$ reduces to

$$v(\infty) = \frac{b}{a} = V_{s_2}. \quad (5.95)$$

By reintroducing the time constant $\tau = RC = 1/a$ and replacing b/a with $v(\infty)$, we can rewrite Eq. (5.94) in the general form:

$$v(t) = v(\infty) + [v(0) - v(\infty)]e^{-t/\tau} \quad (\text{for } t \geq 0)$$
$$(\text{switch action at } t = 0). \quad (5.96)$$

According to Eq. (5.96):

The voltage response of the RC circuit is determined by three parameters: the initial voltage $v(0)$, the final voltage $v(\infty)$, and the time constant τ.

For the specific circuit in Fig. 5-30(a), Eqs. (5.86) and (5.95) give $v(0) = V_{s_1}$ and $v(\infty) = V_{s_2}$. Hence,

$$v(t) = V_{s_2} + (V_{s_1} - V_{s_2})e^{-t/\tau}. \quad (5.97)$$

If the switch action causing the change in voltage across the capacitor occurs at time T_0 instead of at $t = 0$, Eq. (5.96) assumes the form

$$v(t) = v(\infty) + [v(T_0) - v(\infty)]e^{-(t-T_0)/\tau} \quad (\text{for } t \geq T_0)$$
$$(\text{switch action at } t = T_0), \quad (5.98)$$

where we have replaced t with $(t - T_0)$ on the right-hand side of Eq. (5.96). Now $v(T_0)$ is the initial voltage at $t = T_0$. For easy reference, this expression is made available in Table 5-5, along with expressions for three other types of circuits discussed in future sections.

Example 5-9: Switching Between Two Sources

In the circuit of Fig. 5-31(a), the SPDT switch is moved from position 1 to position 2 after it had been in position 1 for a long time. Determine the voltage $v(t)$ for $t \geq 0$ if the switch is moved at (a) $t = 0$ and (b) $t = 3$ s.

Solution:

(a) For $T_0 = 0$ and $t \geq 0$, the complete solution of $v(t)$ is given by Eq. (5.96),

$$v(t) = v(\infty) + [v(0) - v(\infty)]e^{-t/\tau}. \quad (5.99)$$

We need to determine three quantities: the initial voltage $v(0)$, the final voltage $v(\infty)$, and the time constant τ. The initial voltage is the voltage that existed across the capacitor before moving the switch. Since the switch had been in that position for a long time, we presume that the circuit in Fig. 5-31(b) had reached its steady-state condition long before the switch was moved. Hence, at $t = 0^-$ (just before moving the switch), the capacitor behaves like an open circuit. The voltage $v(0^-)$ across the capacitor is the same as that across the 8-kΩ resistor, and since $i_1 = 0$ at $t = 0^-$, application of voltage division yields

$$v(0^-) = \left(\frac{8\text{k}}{4\text{k} + 8\text{k}}\right) \times 45 = 30 \text{ V}.$$

Incidentally, we could have obtained the same result by transforming the circuit in Fig. 5-31(b) into its Thévenin equivalent.

Incorporating the constraint that the voltage across the capacitor cannot change instantaneously, it follows that

$$v(0) = v(0^-) = 30 \text{ V}.$$

Now we turn our attention to finding $v(\infty)$. After moving the switch to position 2 (Fig. 5-31(c)) and allowing the circuit sufficient time to reach its final state, the capacitor again wil

Table 5-5: Response forms of basic first-order circuits.

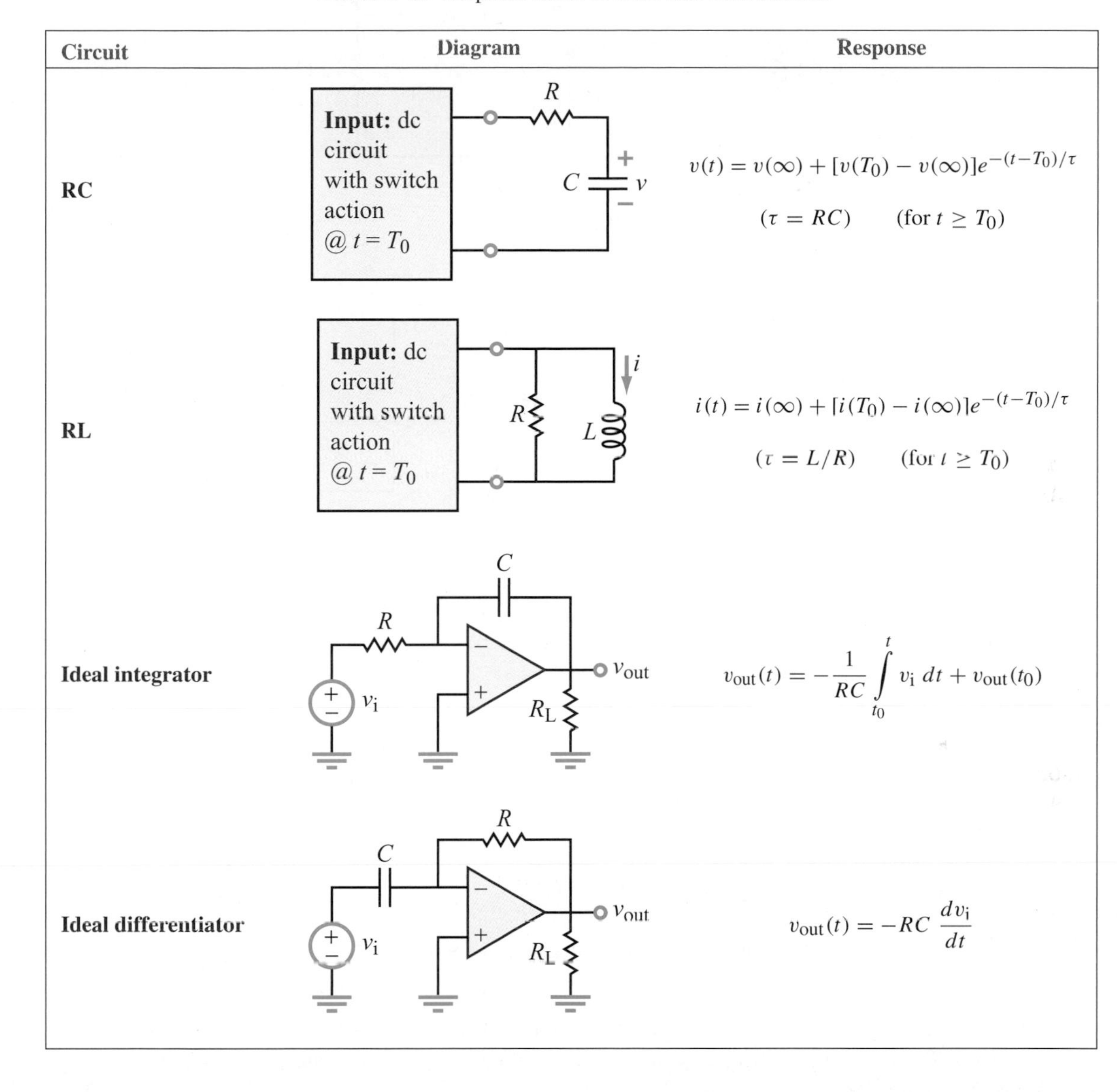

Circuit	Diagram	Response
RC	**Input:** dc circuit with switch action @ $t = T_0$	$v(t) = v(\infty) + [v(T_0) - v(\infty)]e^{-(t-T_0)/\tau}$ $(\tau = RC)$ (for $t \geq T_0$)
RL	**Input:** dc circuit with switch action @ $t = T_0$	$i(t) = i(\infty) + [i(T_0) - i(\infty)]e^{-(t-T_0)/\tau}$ $(\tau = L/R)$ (for $t \geq T_0$)
Ideal integrator		$v_{\text{out}}(t) = -\dfrac{1}{RC}\displaystyle\int_{t_0}^{t} v_{\text{i}}\, dt + v_{\text{out}}(t_0)$
Ideal differentiator		$v_{\text{out}}(t) = -RC\,\dfrac{dv_{\text{i}}}{dt}$

ehave like on open circuit, which means that $i_2 = 0$ at $t = \infty$. oltage division gives

$$v(\infty) = \left(\frac{12\text{k}}{12\text{k} + 24\text{k}}\right) \times 60 = 20 \text{ V}.$$

he time constant of the circuit to the right of terminal 2 is iven by $\tau = RC$, with R being the Thévenin resistance of that circuit. After suppressing the 60-V source, we get

$$R = R_{\text{Th}} = 2 \text{ k}\Omega + 12 \text{ k}\Omega \parallel 24 \text{ k}\Omega$$
$$= 2 \text{ k}\Omega + \frac{12\text{k} \times 24\text{k}}{12\text{k} + 24\text{k}} = 10 \text{ k}\Omega.$$

Hence,

$$\tau = RC = 10 \times 10^3 \times 20 \times 10^{-6} = 0.2 \text{ s}.$$

Substituting the values we obtained for $v(0)$, $v(\infty)$, and τ in Eq. (5.99) leads to

$$v(t) = (20 + 10e^{-5t}) \text{ V} \qquad (\text{for } t \geq 0).$$

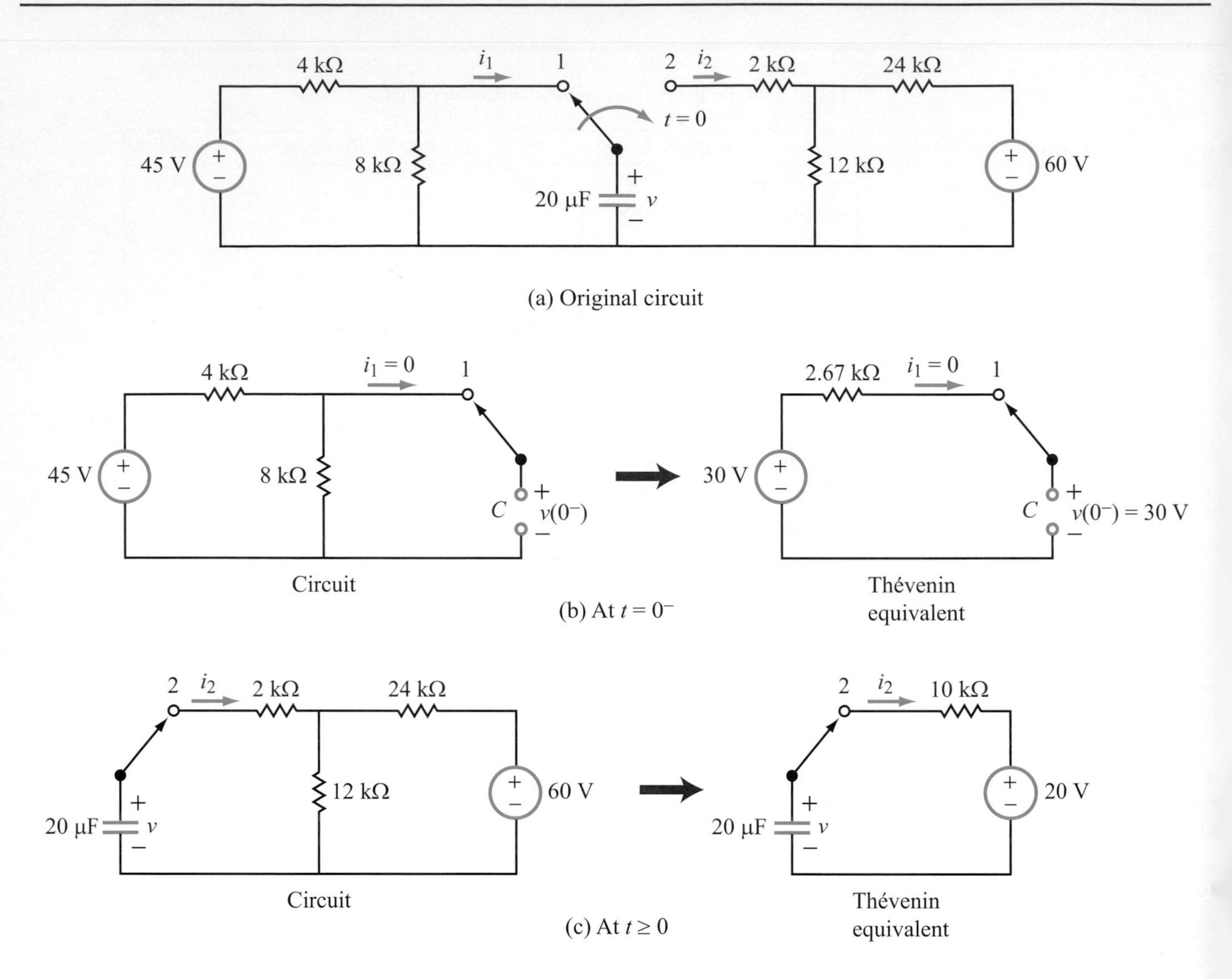

Figure 5-31: Circuit for Example 5-9 (part (a)).

(b) This is a repetition of the previous case except that now the switch action takes place at $T_0 = 3$ s. The applicable expression is given by Eq. (5.98),

$$v(t) = v(\infty) + [v(3) - v(\infty)]e^{-(t-3)/\tau} \quad \text{(for } t \geq 3 \text{ s)}$$

$$= \begin{cases} 30 \text{ V} & \text{for } 0 \leq t \leq 3 \text{ s,} \\ [20 + 10e^{-5(t-3)}] \text{ V} & \text{for } t \geq 3 \text{ s.} \end{cases}$$

Example 5-10: Charge/Discharge Action

Given that the switch in Fig. 5-32 was moved to position 2 at $t = 0$ (after it had been in position 1 for a long time) and then returned to position 1 at $t = 10$ s, determine the voltage response $v(t)$ for $t \geq 0$ and evaluate it for $V_1 = 20$ V, $R_1 = 80$ kΩ, $R_2 = 20$ kΩ, and $C = 0.25$ mF.

Solution: We will divide our solution into two time segments: $v_1(t)$ for $0 \leq t \leq 10$ s and $v_2(t)$ for $t \geq 10$ s.

Time Segment 1: $0 \leq t \leq 10$ s

When the switch is in position 2 (Fig. 5-32(b)), the resistance of the circuit is $R = R_1 + R_2$. Hence, the time constant during this first time segment is

$$\tau_1 = (R_1 + R_2)C$$
$$= (80 + 20) \times 10^3 \times 0.25 \times 10^{-3} = 25 \text{ s.}$$

Application of Eq. (5.96) with $v_1(0) = 0$ (the capacitor had no charge prior to $t = 0$), $v_1(\infty) = V_1 = 20$ V, and $\tau_1 = 25$ s leads

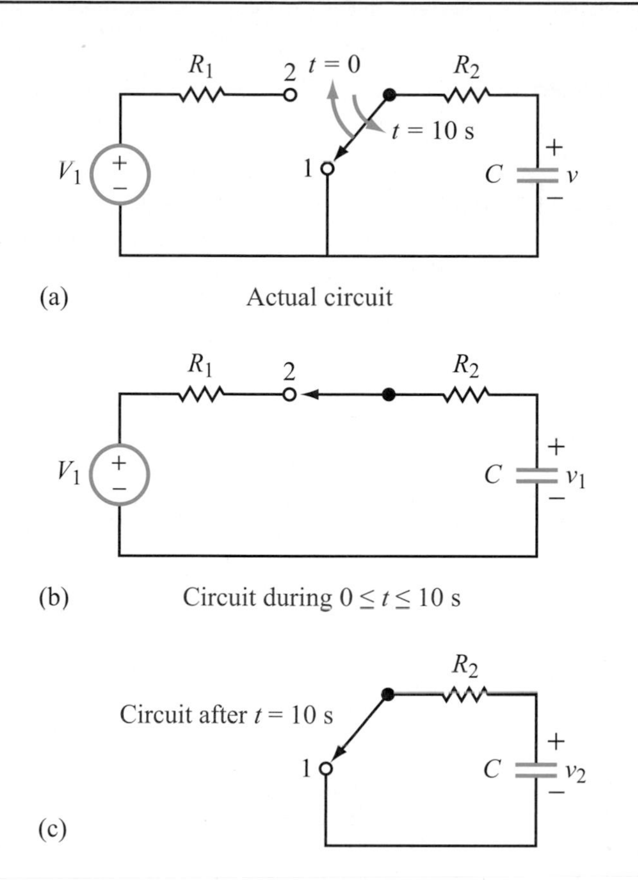

Figure 5-32: After having been in position 1 for a long time, the switch is moved to position 2 at $t = 0$ and then returned to position 1 at $t = 10$ s (Example 5-10).

to

$$v_1(t) = v_1(\infty) + [v_1(0) - v_1(\infty)]e^{-t/\tau_1}$$
$$= 20(1 - e^{-0.04t}) \text{ V} \quad (\text{for } 0 \le t \le 10 \text{ s}).$$

Time Segment 2: $t \ge 10$ s

Voltage $v_2(t)$, corresponding to the second time segment Fig. 5-32(c)), is given by Eq. (5.98) with a new time constant τ_2 as

$$v_2(t) = v_2(\infty) + [v_2(10) - v_2(\infty)]e^{-(t-10)/\tau_2}.$$

The new time constant is associated with the capacitor circuit remaining after returning the switch to position 1,

$$\tau_2 = R_2C$$
$$= 20 \times 10^3 \times 0.25 \times 10^{-3} = 5 \text{ s}.$$

The initial voltage $v_2(10)$ is equal to the capacitor voltage v_1 at the end of time segment 1, namely

$$v_2(10) = v_1(10) = 20(1 - e^{-0.04\times10})$$
$$= 6.59 \text{ V}.$$

With no voltage source present in the R_2C circuit, the charged capacitor will dissipate its energy into R_2, exhibiting a ***natural response*** with a final voltage of $v_2(\infty) = 0$. Consequently,

$$v_2(t) = v_2(10)\, e^{-(t-10)/\tau_2}$$
$$= 6.59e^{-0.2(t-10)} \text{ V} \quad (\text{for } t \ge 10 \text{ s}).$$

Example 5-11: RC-Circuit Response to Rectangular Pulse

Determine the voltage response of a previously uncharged RC circuit to a rectangular pulse $v_i(t)$ of amplitude V_s and duration T_0, as depicted in Fig. 5-33(a). Evaluate and plot the response for $R = 25$ kΩ, $C = 0.2$ mF, $V_s = 10$ V, and $T_0 = 4$ s.

Solution: According to Example 5-2, a rectangular pulse is equivalent to the sum of two step functions. Thus

$$v_i(t) = V_s[u(t - T_1) - u(t - T_2)],$$

where $u(t - T_1)$ accounts for the rise in level from 0 to 1 at $t = T_1$ and the second term (with negative amplitude) serves to counteract (cancel) the first term after $t = T_2$. For the present problem, $T_1 = 0$, and $T_2 = 4$ s. Hence, the input pulse can be written as

$$v_i(t) = V_s\, u(t) - V_s\, u(t - 4).$$

Since the circuit is linear, we can apply the superposition theorem to determine the capacitor response $v(t)$. Thus,

$$v(t) = v_1(t) + v_2(t),$$

where $v_1(t)$ is the response to $V_s\, u(t)$ acting alone and, similarly, $v_2(t)$ is the response to $-V_s\, u(t-4)$ also acting alone.

The response $v_1(t)$ is given by Eq. (5.96) with $v_1(0) = 0$, $v_1(\infty) = V_s$, and $\tau = RC$. Hence,

$$v_1(t) = v_1(\infty) + [v_1(0) - v_1(\infty)]e^{-t/\tau}$$
$$= V_s(1 - e^{-t/\tau}) \qquad (\text{for } t \ge 0).$$

For $V_s = 10$ V and $\tau = RC = 25 \times 10^3 \times 0.2 \times 10^{-3} = 5$ s,

$$v_1(t) = 10(1 - e^{-0.2t}) \text{ V} \qquad (\text{for } t \ge 0).$$

The second step function has an amplitude of $-V_s$ and is delayed in time by 4 s. Upon reversing the polarity of V_s and replacing t with $(t - 4)$, we have

$$v_2(t) = -10[1 - e^{-0.2(t-4)}] \text{ V} \qquad (\text{for } t \ge 4 \text{ s}).$$

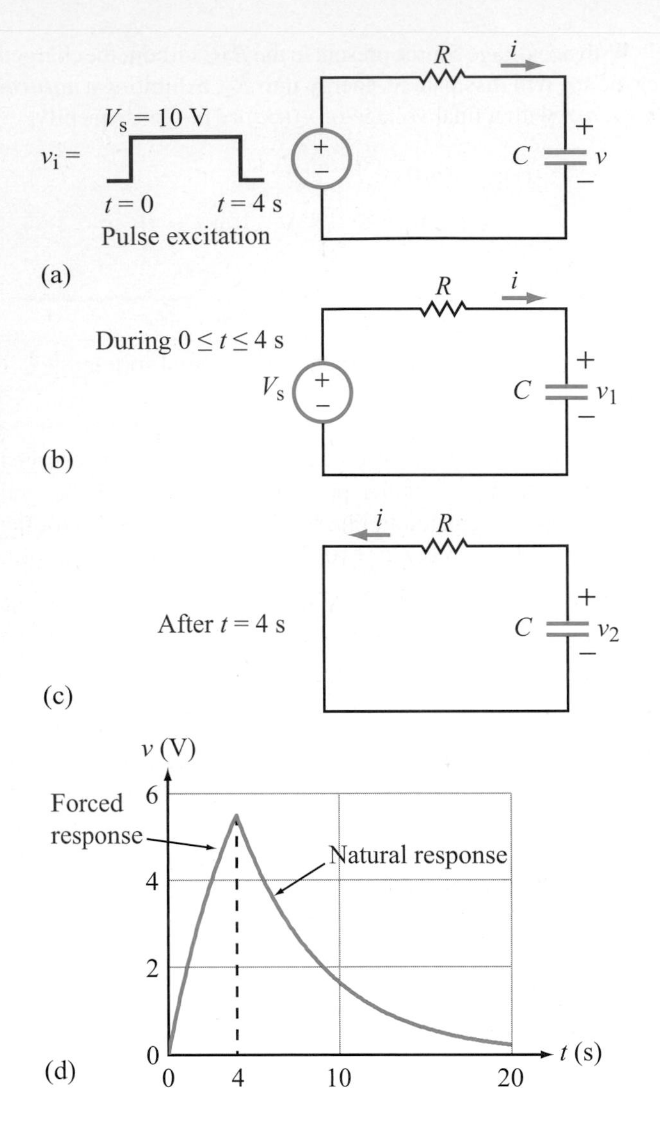

Figure 5-33: RC-circuit response to a 4-s-long rectangular pulse.

The total response for $t \ge 0$ therefore is given by

$$\begin{aligned} v(t) &= v_1(t) + v_2(t) \\ &= 10[1 - e^{-0.2t}] - 10[1 - e^{-0.2(t-4)}]\, u(t-4) \text{ V}, \end{aligned} \tag{5.100}$$

where we introduced the time-shifted step function $u(t-4)$ to assert that the second term is zero for $t \le 4$ s. The plot of $v(t)$ displayed in Fig. 5-33(d) shows that $v(t)$ builds up to a maximum of 5.5 V by the end of the pulse (at $t = 4$ s) and then decays exponentially back to zero thereafter. The build-up part is due to the external excitation and often is called the ***forced response***. In contrast, during the time period after $t = 4$ s, $v(t)$ exhibits a ***natural decay response*** as the capacitor discharges its energy into the resistor. During this latter time segment, $i(t)$ flows in a counterclockwise direction.

Review Question 5-20: What are the three quantities needed to establish $v(t)$ across a capacitor in an RC circuit?

Review Question 5-21: If $V_{s_2} < V_{s_1}$ in the circuit of Fig. 5-30, what would you expect the direction of the current to be after the switch is moved from position 1 to 2? Analyze the process in terms of charge accumulation on the capacitor.

Exercise 5-15: Determine $v_1(t)$ and $v_2(t)$ for $t \ge 0$, given that in the circuit of Fig. E5.15 $C_1 = 6\ \mu$F, $C_2 = 3\ \mu$F, $R = 100$ kΩ, and neither capacitor had any charge prior to $t = 0$.

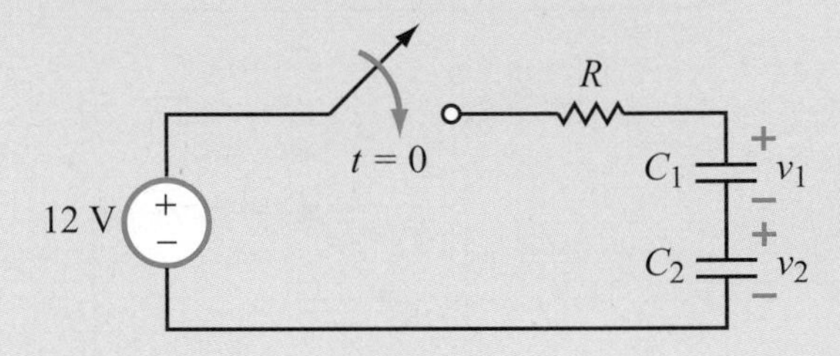

Figure E5.15

Answer: $v_1(t) = 4(1 - e^{-5t})$ V, for $t \ge 0$,
$v_2(t) = 8(1 - e^{-5t})$ V, for $t \ge 0$. (See)

5-5 Response of the RL Circuit

With RC circuits, we developed a first-order differential equation for $v(t)$, the voltage across the capacitor, and then we solved it (subject to initial and final conditions) to obtain a complete expression for $v(t)$. By applying $i = C\, dv/dt$, $p = iv$, and $w = \frac{1}{2}Cv^2$, we were able to determine the corresponding current passing through the capacitor, the power getting transferred to it, and the net energy stored in it. We now will follow an analogous procedure for the RL circuit, but our analysis will focus on the current $i(t)$ through the inductor.

5-5.1 Natural Response of the RL Circuit

After having been in the closed position for a long time, the switch in the RL circuit of Fig. 5-34(a) was opened at $t = 0$, thereby disconnecting the RL circuit from the current source I_s. What happens to the current i flowing through the inductor after the sudden change caused by opening the switch? That is, what is the waveform of $i(t)$ for $t \ge 0$? To answer this question, we first note that at $t = 0^-$ (just before opening the switch), the RL circuit can be represented by the circuit in Fig. 5-34(b), in which the inductor has been replaced with a short circuit. This

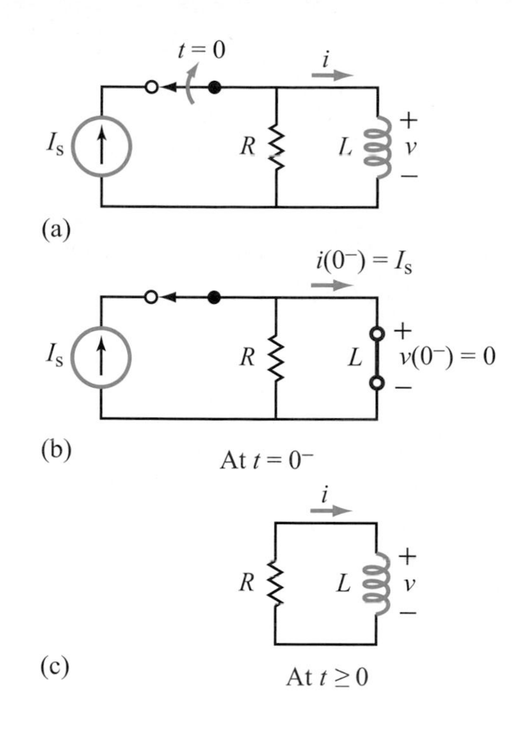

Figure 5-34: RL circuit disconnected from a current source at $t = 0$. *(Since a current source not connected to any load is not physically possible, we always presume that upon disconnecting a current source from a circuit, it gets connected to some external load.)*

is because under steady-state conditions i no longer changes with time, which leads to $v = L\,di/dt = 0$. We also know that a current source entering a node connected to another node via a parallel combination of a resistor R and a short circuit will flow entirely through the short circuit. Hence, $i(0^-) = I_s$. Moreover, since the current through an inductor cannot change instantaneously, the current at $t = 0$ (after opening the switch) has to be

$$i(0) = i(0^-) = I_s.$$

For the time period $t \geq 0$, the loop equation for the RL circuit in Fig. 5-34(c) is given by

$$Ri + L\,\frac{di}{dt} = 0,$$

which can be cast in the form

$$\frac{di}{dt} + ai = 0, \tag{5.101}$$

where a is a temporary constant given by

$$a = \frac{R}{L}. \tag{5.102}$$

The form of Eq. (5.101) is identical with that of Eq. (5.70) for the source-free RC circuit, except that now the variable is $i(t)$, whereas then it was $v(t)$. By analogy with the solution given by Eq. (5.78), our solution for $i(t)$ is given by

$$i(t) = i(0)\,e^{-t/\tau} \qquad (\text{for } t \geq 0) \quad (\text{natural response}), \tag{5.103}$$

where for the RL circuit, the ***time constant*** is given by

$$\tau = \frac{1}{a} = \frac{L}{R}. \tag{5.104}$$

5-5.2 General Form of the Step Response of the RL Circuit

To generalize our solution to the case where the RL circuit may contain sources both before and after the sudden change in the circuit configuration, we adopt the basic circuit shown in Fig. 5-35(a) in which a SPDT switch is moved at $t = 0$ from position 1—where it was connected to current source I_{s_1}—to position 2—which connects the RL circuit to current source I_{s_2}. The state of the circuit at $t = 0^-$ (Fig. 5-35(b)) leads to the conclusion that

$$i(0) = i(0^-) = I_{s_1}.$$

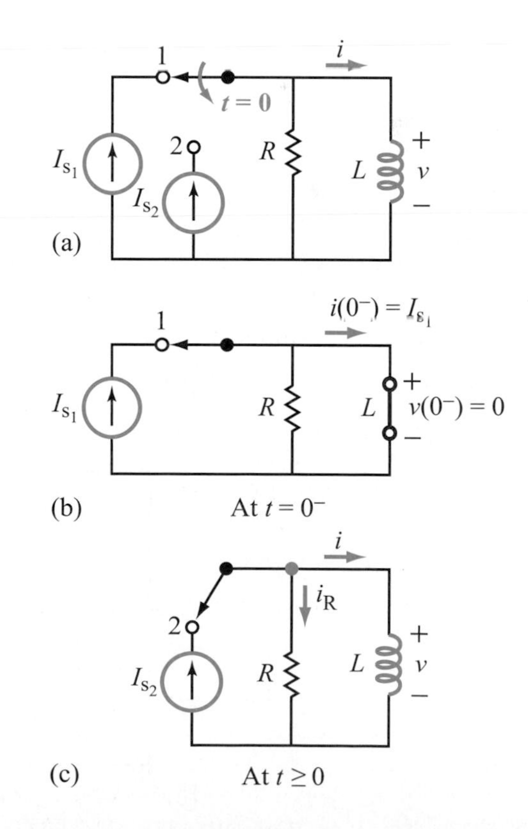

Figure 5-35: RL circuit switched between two current sources at $t = 0$.

The circuit in Fig. 5-35(c) represents the arrangement at $t \geq 0$. Application of KCL at the common node gives

$$-I_{s_2} + i_R + i = 0.$$

Since v is common to R and L, $i_R = v/R$, and by applying $v = L\, di/dt$, the KCL equation becomes

$$\frac{di}{dt} + ai = b, \tag{5.105}$$

where a is as given previously by Eq. (5.102) and

$$b = aI_{s_2} = \frac{R}{L}\, I_{s_2}. \tag{5.106}$$

Not surprisingly, Eq. (5.105) has the same form as Eq. (5.88) for the RC circuit and therefore exhibits a solution analogous to the expression given by Eq. (5.96). Thus, the general form for the current through an inductor in an RL circuit is given by

$$i(t) = i(\infty) + [i(0) - i(\infty)]e^{-t/\tau} \quad (\text{for } t \geq 0)$$
$$(\text{switch action at } t = 0), \tag{5.107}$$

with $\tau = L/R$. For the specific circuit in Fig. 5-35(a), $i(0) = I_{s_1}$ and $i(\infty) = I_{s_2}$.

If the sudden change in the circuit configuration happens at $t = T_0$ instead of at $t = 0$, the general expression for $i(t)$ becomes

$$i(t) = i(\infty) + [i(T_0) - i(\infty)]e^{-(t-T_0)/\tau} \quad (\text{for } t \geq T_0)$$
$$(\text{switch action at } t = T_0), \tag{5.108}$$

where $i(T_0)$ is the current at T_0. This expression is the analogue of Eq. (5.98) for the voltage across the capacitor.

Example 5-12: Circuit with Two RL Branches

After having been in position 1 for a long time, the SPDT switch in Fig. 5-36(a) was moved to position 2 at $t = 0$. Determine i_1, i_2, and i_3 for $t \geq 0$ given that $V_s = 9.6$ V, $R_s = 4$ kΩ, $R_1 = 6$ kΩ, $R_2 = 12$ kΩ, $L_1 = 1.2$ H, and $L_2 = 0.36$ H.

Solution: We start by examining the state of the circuit before moving the switch. At $t = 0^-$, the inductors behave like short

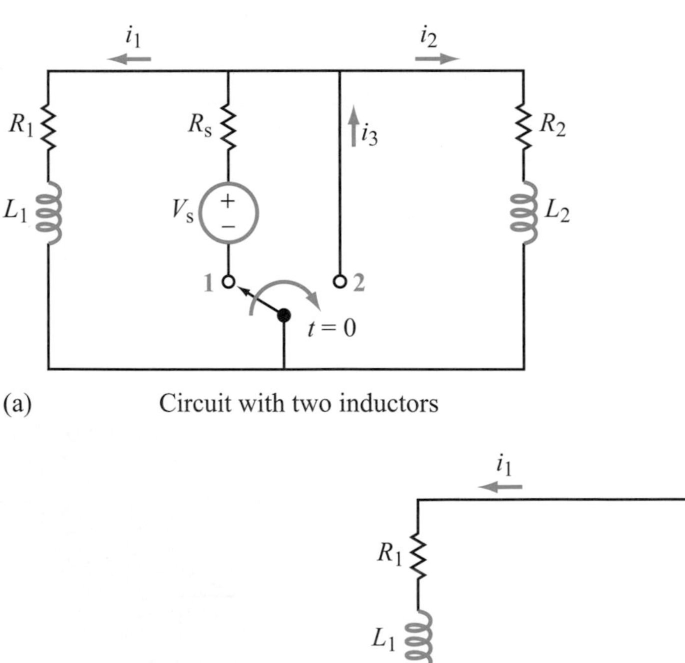

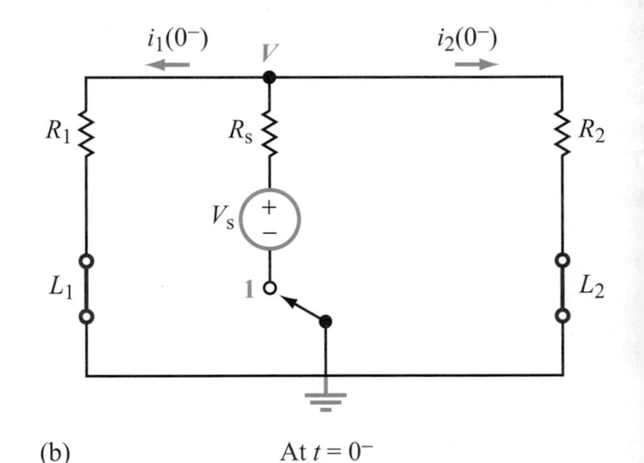

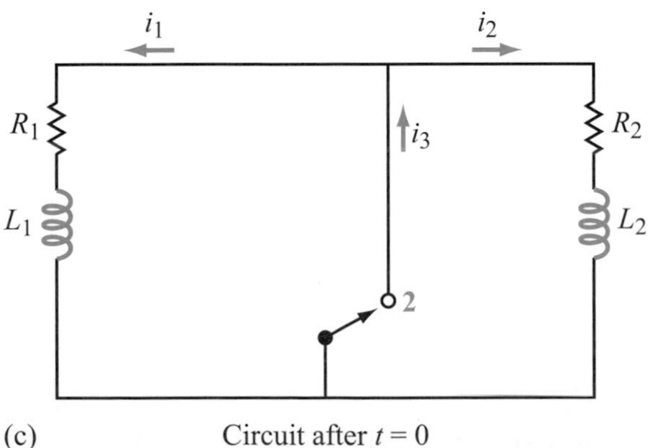

Figure 5-36: Circuit for Example 5-12.

circuits, resulting in the equivalent circuit shown in Fig. 5-36(b). Application of KCL to node V gives

$$\frac{V}{R_1} + \frac{V - V_s}{R_s} + \frac{V}{R_2} = 0,$$

whose solution is

$$V = \frac{R_1 R_2 V_s}{R_1 R_2 + R_1 R_s + R_2 R_s} = \frac{6 \times 12 \times 9.6}{6 \times 12 + 6 \times 4 + 12 \times 4} = 4.8 \text{ V}.$$

Hence, the initial currents $i_1(0)$ and $i_2(0)$ are given by

$$i_1(0) = i_1(0^-) = \frac{V}{R_1} = \frac{4.8}{6 \times 10^3} = 0.8 \text{ mA}$$

and

$$i_2(0) = i_2(0^-) = \frac{V}{R_2} = \frac{4.8}{12 \times 10^3} = 0.4 \text{ mA}.$$

The circuit in Fig. 5-36(c) represents the circuit condition after $t = 0$. Even though we have two resistors and two inductors in the overall circuit, it can be treated as two independent RL circuits because each RL branch is connected across a short circuit. In both cases, the inductors will dissipate their magnetic energy (that they had stored previous to moving the switch) through their respective resistors. Hence, $i_1(\infty) = i_2(\infty) = 0$. The complete expressions for $i_1(t)$ and $i_2(t)$ for $t \geq 0$ then are given by

$$i_1(t) = [i_1(\infty) + [i_1(0) - i_1(\infty)]e^{-t/\tau_1}] = 0.8e^{-t/\tau_1} \text{ mA}$$

and

$$i_2(t) = [i_2(\infty) + [i_2(0) - i_2(\infty)]e^{-t/\tau_2}] = 0.4e^{-t/\tau_2} \text{ mA},$$

where τ_1 and τ_2 are the time constants of the two RL circuits, namely

$$\tau_1 = \frac{L_1}{R_1} = \frac{1.2}{6 \times 10^3} = 2 \times 10^{-4} \text{ s}$$

and

$$\tau_2 = \frac{L_2}{R_2} = \frac{0.36}{12 \times 10^3} = 3 \times 10^{-5} \text{ s}.$$

The current flowing through the short circuit is simply

$$i_3 = i_1 + i_2 = (0.8e^{-t/\tau_1} + 0.4e^{-t/\tau_2}) \text{ mA} \quad (\text{for } t \geq 0).$$

Example 5-13: Response to a Triangle Excitation

The source voltage in the circuit of Fig. 5-37(a) generates a triangular ramp function that starts at $t = 0$, rises linearly to 12 V at $t = 3$ ms, and then drops abruptly down to zero. Additionally, $R = 250\ \Omega$, $L = 0.5$ H, and no current was flowing through L prior to $t = 0$.

(a) Synthesize $v_s(t)$ in terms of unit step functions and plot it.

(b) Develop the differential equation for $i(t)$ for $t \geq 0$.

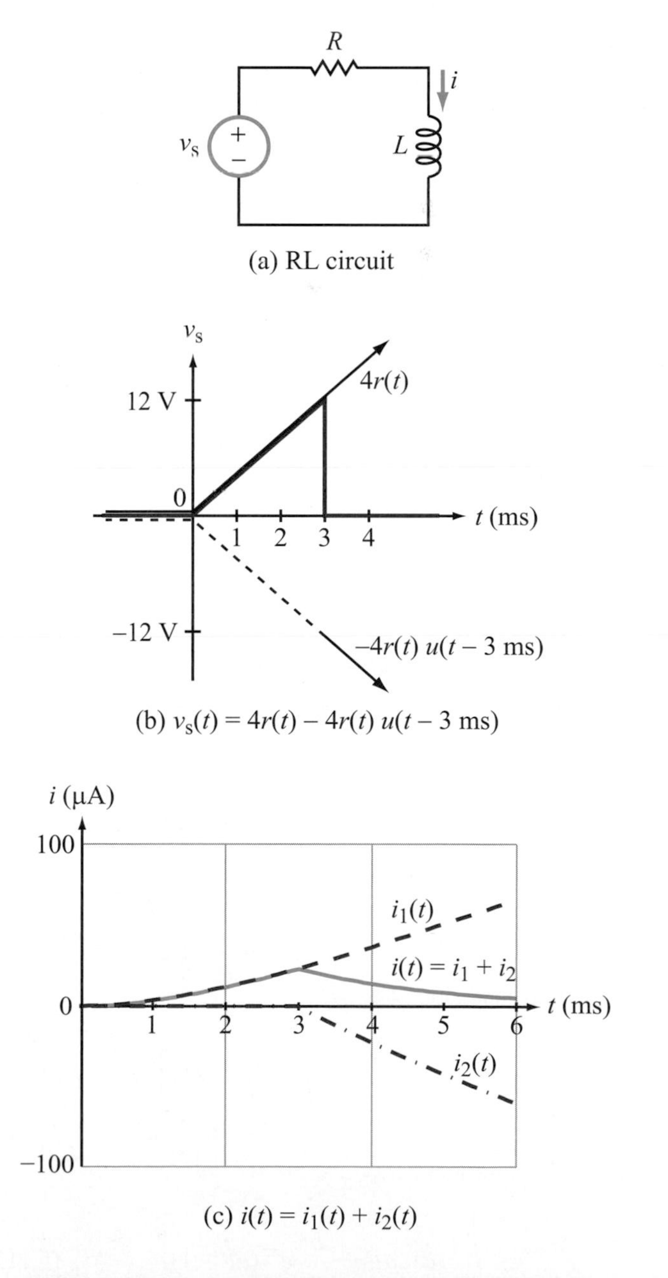

Figure 5-37: Circuit and associated plot for Example 5-13.

(c) Solve the equation and plot $i(t)$ for $t \geq 0$.

Solution:

(a) The waveform of $\upsilon_s(t)$ shown in Fig. 5-37(b) can be synthesized as the sum of two ramp functions:

$$\begin{aligned} \upsilon_s(t) &= 4r(t) - 4r(t)\, u(t - 3 \text{ ms}) \\ &= 4t\, u(t) - 4t\, u(t)\, u(t - 3 \text{ ms}) \\ &= 4t\, u(t) - 4t\, u(t - 3 \text{ ms}) \text{ V}. \end{aligned} \tag{5.109}$$

(b) For $t \geq 0$, the KVL loop equation is given by

$$-\upsilon_s + Ri + L\frac{di}{dt} = 0,$$

which can be rearranged into the form

$$\frac{di}{dt} + ai = \frac{\upsilon_s}{L}, \tag{5.110}$$

where $a = R/L$. Since $\upsilon_s(t)$ is composed of two components, we will write $i(t)$ as the sum of two components,

$$i(t) = i_1(t) + i_2(t), \tag{5.111}$$

where $i_1(t)$ is the solution of Eq. (5.110) with $\upsilon_s = 4t\, u(t)$ and $i_2(t)$ is the solution of Eq. (5.110) with $\upsilon_s = -4t\, u(t - 3 \text{ ms})$. That is,

$$\frac{di_1}{dt} + ai_1 = \frac{4t}{L} = bt \qquad \text{for } t \geq 0 \tag{5.112a}$$

and

$$\frac{di_2}{dt} + ai_2 = \frac{-4t}{L} = -bt \qquad \text{for } t \geq 3 \text{ ms} \tag{5.112b}$$

with $b = 4/L$.

(c) For $i_1(t)$: We start by multiplying both sides of Eq. (5.112a) by e^{at} and then integrating from 0 to t:

$$\int_0^t \left(e^{at}\frac{di_1}{dt} + ai_1e^{at} \right) dt = \int_0^t bte^{at}\, dt. \tag{5.113}$$

For the left-hand side,

$$\int_0^t \left[e^{at}\frac{di_1}{dt} + ai_1e^{at} \right] dt = \int_0^t \left[\frac{d}{dt}(i_1e^{at}) \right] dt = i_1e^{at}\Big|_0^t, \tag{5.114}$$

and for the right-hand side,

$$\int_0^t bte^{at}\, dt = \frac{b}{a^2}\, e^{at}(at - 1)\Big|_0^t. \tag{5.115}$$

In view of Eqs. (5.114) and (5.115), Eq. (5.113) becomes

$$i_1e^{at}\Big|_0^t = \frac{b}{a^2}\, e^{at}(at - 1)\Big|_0^t, \tag{5.116}$$

which leads to

$$i_1(t)\, e^{at} - i_1(0) = \frac{b}{a_2}\, [e^{at}(at - 1) + 1]. \tag{5.117}$$

Given that $i_1(0) = 0$, the expression for $i_1(t)$ becomes

$$i_1(t) = \frac{b}{a^2}\, [(at - 1) + e^{-at}] \qquad (\text{for } t \geq 0). \tag{5.118}$$

For $i_2(t)$: Equations (5.112a) and (5.112b) are identical in form, except for two important differences:

(a) The forcing function for $i_1(t)$ is bt whereas the forcing function for $i_2(t)$ is $-bt$.

(b) The temporal domain of applicability for $i_2(t)$ starts at $t = 3$ ms, instead of at $t = 0$.

Hence, Eq. (5.116) can be adapted to i_2 by replacing b with $-b$ and changing the lower limit of integration to 3 ms, which gives

$$i_2e^{at}\Big|_{3 \text{ ms}}^t = \frac{-b}{a^2}\, e^{at}(at - 1)\Big|_{3 \text{ ms}}^t, \tag{5.119}$$

which leads to

$$\begin{aligned} &i_2(t)\, e^{at} - i_2(3 \text{ ms})\, e^{0.003a} \\ &\quad = -\frac{b}{a^2}[e^{at}(at - 1) - e^{0.003a}(0.003a - 1)]. \end{aligned} \tag{5.120}$$

When we apply superposition, we apply the same initial condition to both RL circuits (corresponding to the two components of $\upsilon_s(t)$). Thus, $i_1(0) = i_2(3 \text{ ms}) = 0$, and Eq. (5.120) simplifies to

$$i_2(t) = -\frac{b}{a^2}[(at - 1) - (0.003a - 1)e^{-a(t-0.003)}] \qquad (\text{for } t \geq 3 \text{ ms}). \tag{5.121}$$

Total solution for $i(t)$: For $R = 250\ \Omega$ and $L = 0.5$ H, $a = R/L = 500$, $b = 4/L = 8$, and

$$i(t) = \begin{cases} i_1(t) & \text{for } 0 \leq t \leq 3 \text{ ms}, \\ i_1(t) + i_2(t) & \text{for } t \geq 3 \text{ ms}, \end{cases}$$

$$= \begin{cases} 32[(500t - 1) + e^{-500t}]\ \mu\text{A} \\ \qquad \text{for } 0 \leq t < 3 \text{ ms}, \\ 103.7e^{-500t}\ \mu\text{A} \\ \qquad \text{for } t \geq 3 \text{ ms}. \end{cases} \tag{5.122}$$

Figure 5-37(c) displays a plot of $i(t)$ versus t.

Review Question 5-22: Compare Eq. (5.96) with Eq. (5.107) to draw an analogy between RC and RL circuits. v, R, and C of the RC circuit correspond to which parameters of the RL circuit?

Review Question 5-23: Suppose the switch in the circuit of Fig. 5-34(a) had been open for a long time, and then it was closed suddenly. Will I_s initially flow through R or L?

Exercise 5-16: Determine $i_1(t)$ and $i_2(t)$ for $t \geq 0$ given that, in the circuit of Fig. E5.17, $L_1 = 6$ mH, $L_2 = 12$ mH, and $R = 2\ \Omega$. Assume $i_1(0^-) = i_2(0^-) = 0$.

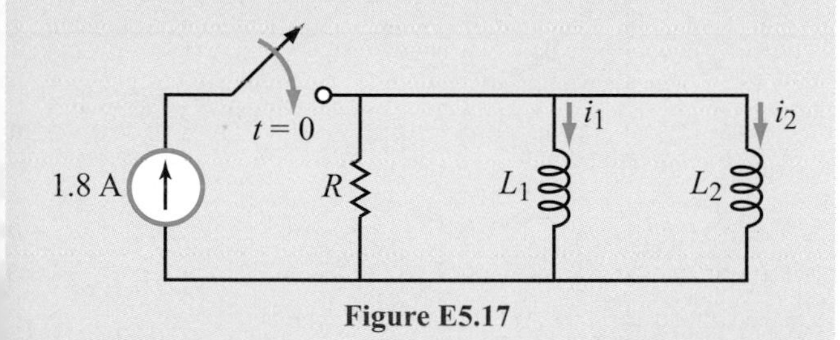

Figure E5.17

Answer: $i_1(t) = 1.2(1 - e^{-500t})\, u(t)$ A, $i_2(t) = 0.6(1 - e^{-500t})\, u(t)$ A. (See CD)

5-6 RC Op-Amp Circuits

Adding capacitors and inductors to resistive circuits vastly expands their utility and versatility. In this section, we will consider a few examples of circuits in which capacitors are used in conjunction with op amps to perform integration, differentiation, and related operations. Even though these specific functions also can be realized through the use of inductors, capacitors are usually the preferred option (whenever such a choice is possible) because of their smaller physical size and availability in planar form.

5-6.1 Ideal Op-Amp Integrator

The circuit shown in Fig. 5-38 resembles the standard inverting-amplifier circuit of Section 4-4, except that its feedback resistor R_f has been replaced with a capacitor C, converting into an op-amp integrator. As we will show shortly:

The output voltage v_{out} of such an integrator circuit is directly proportional to the time integral of the input signal v_i.

The ideal op-amp model has two constraints. The voltage constraint states that $v_p = v_n$, and since $v_p = 0$ in the circuit of Fig. 5-38, it follows that $v_n = 0$. Hence, the current i_R flowing through R is given by

$$i_R = \frac{v_i}{R}. \quad (5.123)$$

Given that $v_n = 0$, the voltage v_C across C is simply v_{out}, and the current flowing through it is

$$i_C = C\,\frac{dv_{out}}{dt}. \quad (5.124)$$

At node v_n,

$$i_R + i_C - i_n = 0. \quad (5.125)$$

In view of the second op-amp constraint, namely $i_n = i_p = 0$, it follows that

$$i_C = -i_R \quad (5.126)$$

or

$$\frac{dv_{out}}{dt} = -\frac{1}{RC}\,v_i. \quad (5.127)$$

Upon integrating both sides of Eq. (5.127) from a reference time t_0 to time t, we have

$$\int_{t_0}^{t} \left(\frac{dv_{out}}{dt}\right) dt = -\frac{1}{RC}\int_{t_0}^{t} v_i\, dt, \quad (5.128)$$

which leads to

$$v_{out}(t) = -\frac{1}{RC}\int_{t_0}^{t} v_i\, dt + v_{out}(t_0). \quad (5.129)$$

Time t_0 is the time at which the integration process begins, and $v_{out}(t_0)$ is the voltage across the capacitor at that instant in time. Thus, according to Eq. (5.129), the output voltage (which is also

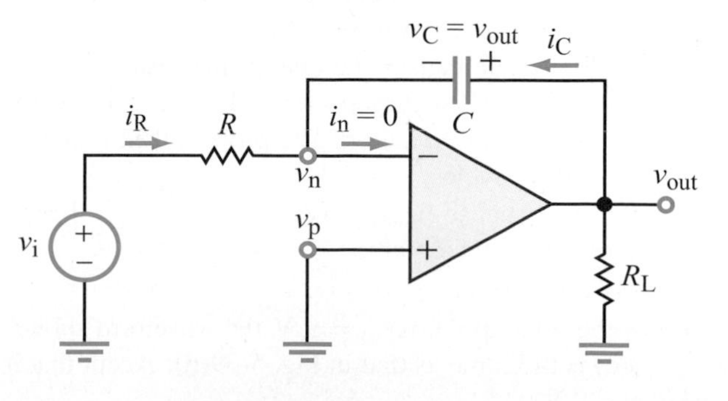

Figure 5-38: Integrator circuit.

the voltage across the capacitor) is equal to whatever voltage existed across the capacitor at the start of the integration process, $v_{out}(t_0)$, incremented by an amount equal to the integrated value of the input voltage (from t_0 to present time t) and multiplied by a (negative) ***scaling factor*** $(-1/RC)$. *Since the magnitude of the output voltage, $|v_{out}|$, cannot exceed the supply voltage V_{cc}, the values of R and C have to be chosen carefully so as to avoid saturating the op amp.*

If the time scale can be conveniently chosen such that the reference time $t_0 = 0$ and the capacitor was uncharged at that point in time (i.e., $v_{out}(0) = 0$), then Eq. (5.129) simplifies to

$$v_{out}(t) = -\frac{1}{RC}\int_0^t v_i\, dt \quad (\text{if } v_{out}(0) = 0). \tag{5.130}$$

Example 5-14: Square-Wave Input Signal

The square-wave signal shown in Fig. 5-39(a) is applied at the input of an ideal integrator circuit with an initial capacitor voltage of zero at $t = 0$. If $R = 200$ kΩ and $C = 2.5$ μF, determine the waveform of the corresponding output voltage for an amp with (a) $V_{cc} = 14$ V and (b) $V_{cc} = 9$ V.

Solution:

(a) The scaling factor is given by

$$-\frac{1}{RC} = -\frac{1}{2\times 10^5 \times 2.5\times 10^{-6}} = -2\ \text{s}^{-1}.$$

For the time period $0 \le t \le 2$ s (first half of the first cycle),

$$v_{out}(t) = -2\int_0^t v_i\, dt = -2\int_0^t 3\, dt = -6t \text{ V} \qquad (0 \le t \le 2 \text{ s}),$$

which is represented by the first ramp function shown in Fig. 5-39(b). The polarity reversal of v_i during the second half of the first cycle causes the energy that had been stored in the capacitor to be discharged, concluding the cycle with no net voltage across the capacitor. The process then is repeated during succeeding cycles.

We note that because $|v_{out}|$ never exceeds $|V_{cc}| = 14$ V, no saturation occurs in the op amp.

(b) For the op amp with $V_{cc} = 9$ V, the waveform shown in Fig. 5-39(c) is the same as that in Fig. 5-39(b), except that it is ***clipped*** at -9 V.

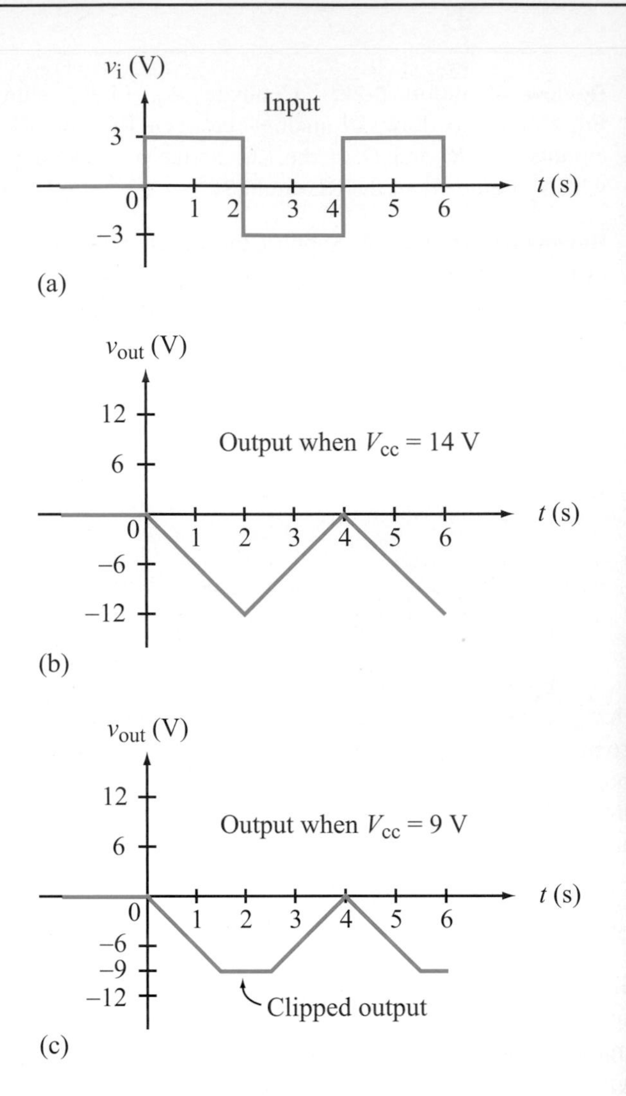

Figure 5-39: Example 5-14 (a) input signal, (b) output signal with no op-amp saturation, and (c) output signal with op-amp saturation at −9 V.

5-6.2 Ideal Op Amp Differentiator

The integrator circuit of Fig. 5-38 can be converted into the differentiator circuit of Fig. 5-40 by simply interchanging the locations of R and C. For the differentiator circuit, application of the voltage and current constraints leads to

$$i_C = C\,\frac{dv_i}{dt},$$

$$i_R = \frac{v_{out}}{R},$$

and

$$i_C = -i_R.$$

RC Differentiator

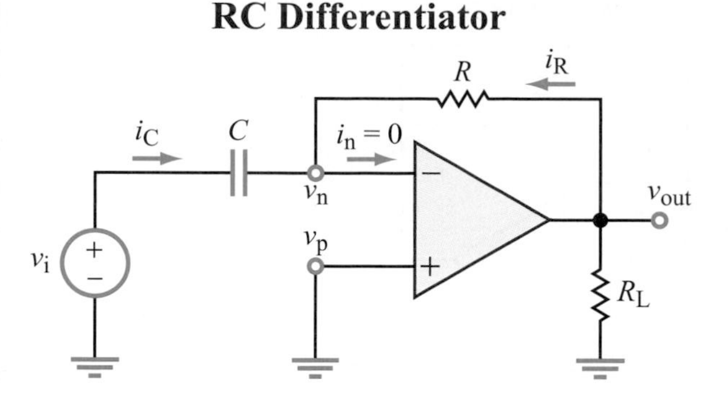

Figure 5-40: Differentiator circuit.

Consequently,

$$v_{out} = -RC\,\frac{dv_i}{dt}, \tag{5.131}$$

which states that the output voltage of the differentiator circuit is proportional directly to the time derivative of its input voltage v_i, and the proportionality factor is $(-RC)$. The differentiator circuit performs the inverse function of that performed by the integrator circuit.

5-6.3 Other Op-Amp Circuits

The relative ease with which we were able to develop input–output relationships for the ideal integrator and differentiator circuits is attributed (at least in part) to the relative simplicity of those circuits. Aside from the load resistor R_L (which exercised no influence on the solutions), the circuits in Figs. 5-38 and 5-40 consisted each of one resistor and one capacitor. Now, through two examples, we will demonstrate how to approach the analysis of RC op-amp circuits with slightly more complicated architecture.

Example 5-15: Pulse Response of an Op-Amp Circuit

The op-amp circuit shown in Fig. 5-41(a) is subjected to an input pulse of amplitude $V_s = 2.4$ V and duration $T_0 = 0.3$ s. Determine and plot the output voltage $v_{out}(t)$ for $t \geq 0$, assuming that the capacitor was uncharged before $t = 0$.

Solution: One possible approach to solving the problem is to analyze the circuit twice—once for the duration of the pulse (0 to 0.3 s) and a second time for $t > 0.3$ s. An alternative approach is to synthesize the rectangular pulse as the sum of two step functions, to seek an independent solution for each step function, and then to add up the solutions. We will illustrate both methods.

Method 1: Two Time Segments

Time Segment 1: $0 \leq t \leq 0.3$ s, and $v_i = V_s = 2.4$ V.

At node v_n,

$$i_1 + i_2 + i_3 = 0,$$

or

$$\frac{v_n - V_s}{R_1} + C\,\frac{d}{dt}(v_n - v_{out_1}) + \frac{v_n - v_{out_1}}{R_2} = 0,$$

where v_{out_1} is the output voltage during time segment 1. Since $v_p = 0$, injection of the ideal op-amp voltage constraint $v_p = v_n$ leads to

$$C\,\frac{dv_{out_1}}{dt} + \frac{v_{out_1}}{R_2} = -\frac{V_s}{R_1},$$

which can be cast in the standard first-order differential-equation form given by

$$\frac{dv_{out_1}}{dt} + av_{out_1} = b, \tag{5.132}$$

where

$$a = \frac{1}{R_2 C}, \quad \text{and} \quad b = -\frac{V_s}{R_1 C}.$$

The solution for Eq. (5.132) is analogous to that given by Eq. (5.94), namely

$$\begin{aligned} v_{out_1}(t) &= v_{out_1}(0)\,e^{-at} + \frac{b}{a}(1 - e^{-at}) \\ &= v_{out_1}(0)\,e^{-t/\tau} - \frac{V_s R_2}{R_1}(1 - e^{-t/\tau}), \end{aligned} \tag{5.133}$$

where

$$\tau = \frac{1}{a} = R_2 C = 0.25 \text{ s}.$$

Given that $v_n = 0$, it is evident from the circuit in Fig. 5-41(a) that

$$v_{out_1} = -v_C,$$

where v_C is the voltage across the capacitor. According to the problem statement, $v_C(0^-) = 0$, and since the voltage across a capacitor cannot change instantaneously, it follows that

$$v_{out_1}(0) = -v_C(0) = -v_C(0^-) = 0.$$

Upon incorporating this piece of information into our solution, we have

$$\begin{aligned} v_{out_1}(t) &= -\frac{V_s R_2}{R_1}(1 - e^{-t/\tau}) \\ &= -12(1 - e^{-4t}) \text{ V} \quad (\text{for } 0 \leq t \leq 0.3 \text{ s}). \end{aligned} \tag{5.134}$$

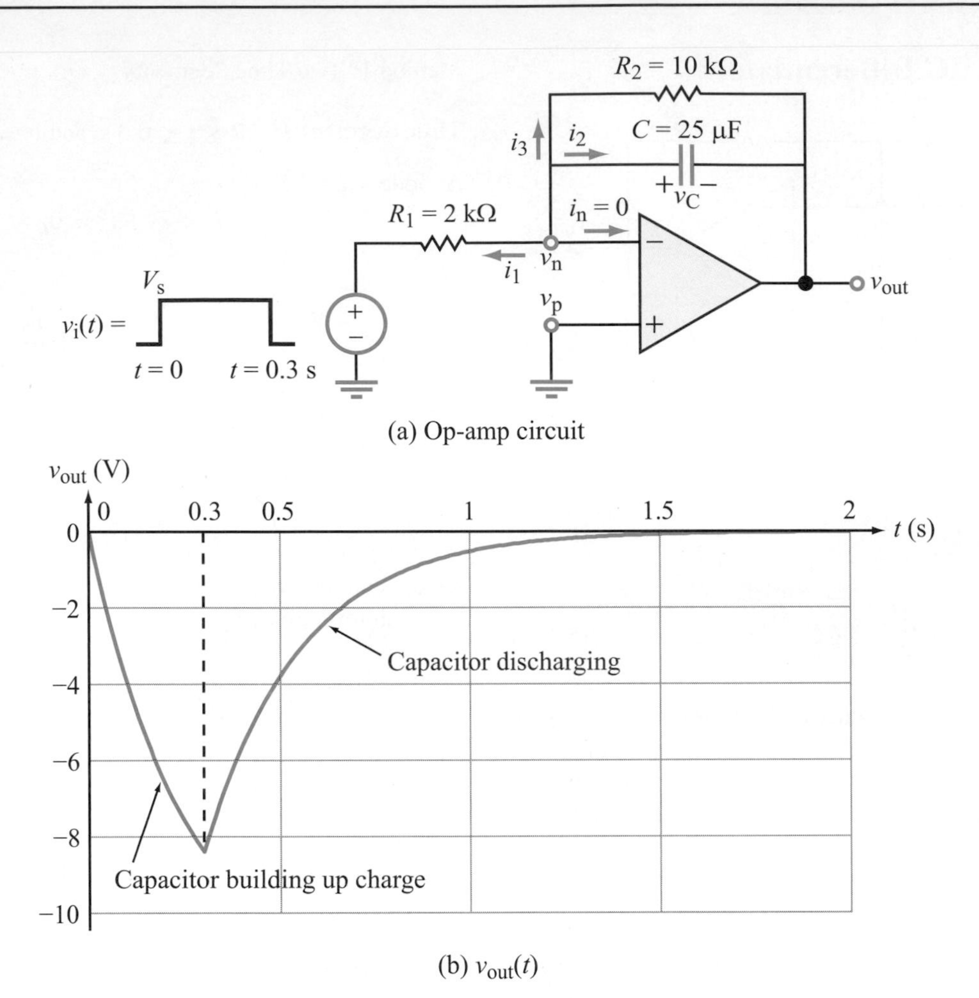

Figure 5-41: Op-amp circuit of Example 5-15.

Time Segment 2: $t > 0.3$ s, and $v_i = 0$.

The form of the solution for this time segment is the same as that given by Eq. (5.133) for the preceding time segment, except for three modifications:

(a) The input voltage is now zero, so we should set $V_s = 0$.

(b) The time variable t should be replaced with $(t - 0.3$ s) to reflect the fact that our starting (reference) time is $t = 0.3$ s, not $t = 0$.

(c) The initial voltage $v_{out_2}(0.3$ s) is not zero (because the capacitor had been building up charge during the previous time segment).

Hence, for time segment 2, v_{out_2} is given by

$$v_{out_2}(t) = v_{out_2}(0.3)\, e^{-4(t-0.3)} \qquad (\text{for } t > 0.3 \text{ s}).$$

The initial voltage $v_{out_2}(0.3)$ is equal to the voltage that existed during the previous time segment at $t = 0.3$ s. Hence,

$$v_{out_2}(0.3) = v_{out_1}(0.3) = -12(1 - e^{-4\times 0.3}) = -8.4 \text{ V}.$$

Hence,

$$v_{out_2}(t) = -8.4e^{-4(t-0.3\text{ s})} \text{ V} \qquad (\text{for } t > 0.3 \text{ s}). \qquad (5.135)$$

The combined output response to the input pulse is displaye[d] in Fig. 5-41(b).

Method 2: Two Step Functions

By modeling the rectangular pulse as

$$v_i(t) = V_s[u(t) - u(t - 0.3 \text{ s})], \qquad (5.136)$$

we can develop a generic solution to a step-function input ar[d] then use it to find

$$v_{out}(t) = v_{out_a}(t) + v_{out_b}(t).$$

We will treat the two step functions as two independent sources, and we will apply the same initial-condition information to both cases; that is, when treating the case of the second step function, we do so as if the first step function had never existed.

To that end, the response of the first step function is given by Eq. (5.134) as

$$v_{\text{out}_a}(t) = -12(1 - e^{-4t}) \text{ V} \qquad (\text{for } t \geq 0). \qquad (5.137)$$

Similarly, after reversing the polarity of V_s and incorporating a time delay of 0.3 s,

$$v_{\text{out}_b}(t) = 12(1 - e^{-4(t-0.3)}) \text{ V} \qquad (\text{for } t \geq 0.3 \text{ s}). \quad (5.138)$$

In view of the definition of the step function, the complete solution is given by

$$v_{\text{out}}(t) = v_{\text{out}_a}(t) + v_{\text{out}_b}(t)$$
$$= \begin{cases} v_{\text{out}_a}(t) & \text{for } 0 \leq t \leq 0.3 \text{ s} \\ v_{\text{out}_a}(t) + v_{\text{out}_b}(t) & \text{for } t > 0.3 \text{ s}. \end{cases} \qquad (5.139)$$

It is a relatively straightforward exercise to demonstrate that the two methods do indeed provide the same solution.

Example 5-16: Op-Amp Circuit with Output Capacitor

Determine $v_C(t)$, the voltage across the capacitor in Fig. 5-42(a), given that $v_i(t) = 3u(t)$ V, the capacitor had no charge on it prior to $t = 0$, $R_1 = 1$ kΩ, $R_2 = 15$ kΩ, $R_3 = 30$ kΩ, $R_4 = 12$ kΩ, $R_5 = 24$ kΩ, and $C = 50$ μF.

Solution: The capacitor is on the output side of the op amp, so one possible approach to solving the problem is to:

(a) Temporarily replace the capacitor with an open circuit

(b) Determine the Thévenin equivalent circuit at terminals (a, b)

(c) Reinsert the capacitor as in Fig. 5-42(c).

To that end, we start by relating v_{out} to v_i. Given that for the ideal op amp $v_n = v_p$ and $i_p = 0$, it follows that

$$v_n = v_p = v_i.$$

Moreover, since $i_n = 0$, v_n and v_{out} are related by

$$v_{\text{out}} = \left(\frac{R_2 + R_3}{R_2}\right) v_n = \left(\frac{R_2 + R_3}{R_2}\right) v_i.$$

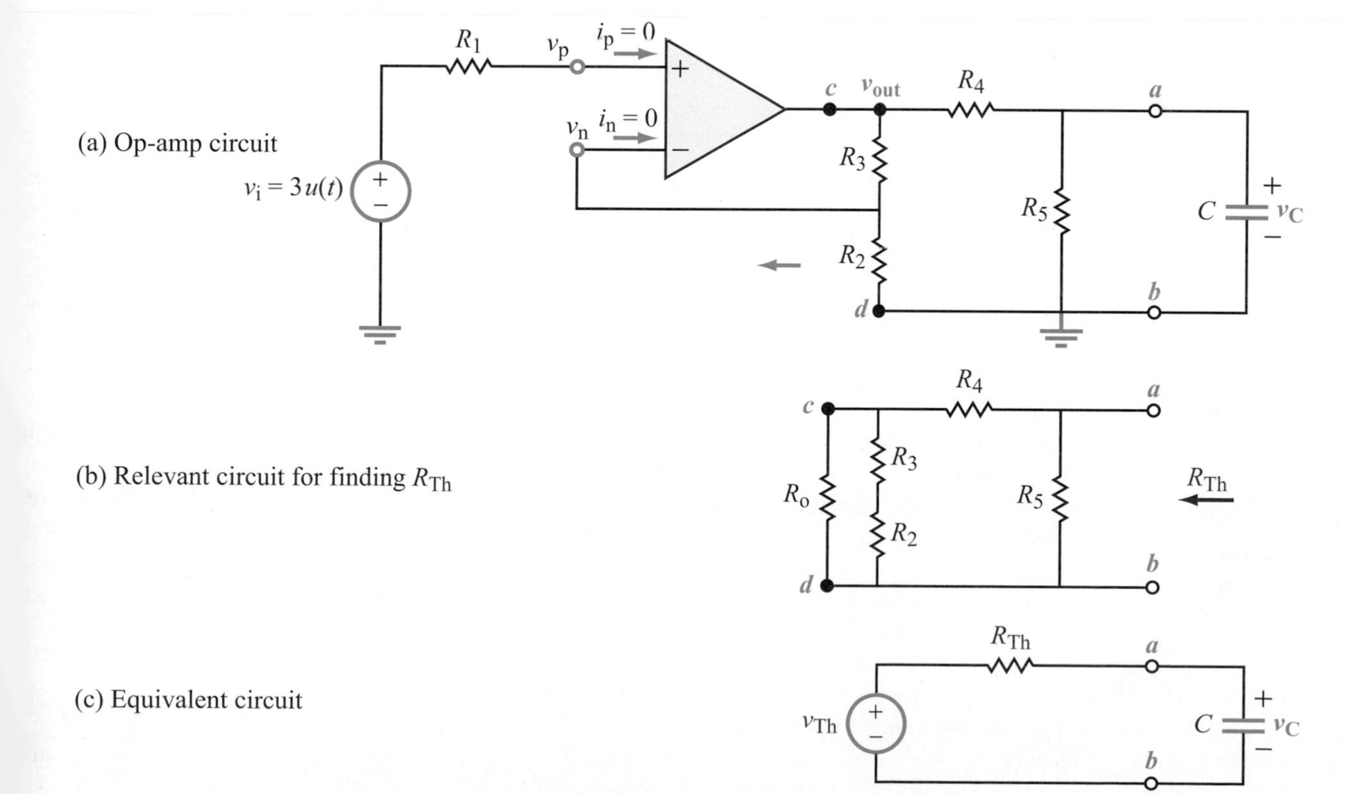

Figure 5-42: Circuit for Example 5-16.

With the capacitor removed, the Thévenin voltage across terminals (a, b) in Fig. 5-42(a) is equal to the voltage across R_5, which is related to v_{out} by the voltage-division rule

$$v_{Th} = \left(\frac{R_5}{R_4 + R_5}\right) v_{out} = \left(\frac{R_5}{R_4 + R_5}\right)\left(\frac{R_2 + R_3}{R_2}\right) v_i$$

$$= \left(\frac{24}{12 + 24}\right)\left(\frac{15 + 30}{15}\right) \times 3 = 6 \text{ V} \qquad (\text{for } t \geq 0).$$

Our next task is to determine the value of R_{Th}. Since $v_p - v_n = 0$, the op-amp's equivalent circuit at terminals (c, d) consists of only the output resistance R_O. Figure 5-42(b) contains the relevant part of the overall circuit seen by terminals (a, b). For the real op amp, R_O is on the order of 10 to 100 Ω, which is at least two orders of magnitude smaller than any of the other resistors in the circuit, lending justification to the ideal op-amp model which sets $R_O = 0$. Consequently,

$$R_{Th} = R_4 \parallel R_5 = \frac{R_4 R_5}{R_4 + R_5} = \frac{12 \times 24}{12 + 24} = 8 \text{ k}\Omega.$$

With v_{Th} and R_{Th} known, we now have a circuit (Fig. 5-42(c)) that resembles the step-function circuit of Fig. 5-30(a). Its solution is given by Eq. (5.97) using $V_{s_1} = 0$ and $V_{s_2} = V_s$, namely

$$v_C(t) = V_s(1 - e^{-t/\tau}).$$

In the present case, $V_s = v_{Th} = 6$ V, and

$$\tau = R_{Th} C = 8 \times 10^3 \times 50 \times 10^{-6} = 0.4 \text{ s}.$$

The capacitor response therefore is given by

$$v_C(t) = 6(1 - e^{-2.5t}) \text{ V} \qquad (\text{for } t \geq 0).$$

Review Question 5-24: What causes clipping of the waveform at the output of an op-amp integrator circuit? Can clipping occur at the output of a differentiator circuit?

Review Question 5-25: If $v_s(t)$ is the input signal to a two-stage op-amp circuit with the first stage being an integrator with $R_1 C_1 = 0.01$ and the second stage being a differentiator with $R_2 C_2 = 0.01$, under what circumstances will the output waveform $v_{out}(t)$ be the same or different from $v_s(t)$?

Exercise 5-17: The input signal to an ideal integrator circuit with $RC = 2 \times 10^{-3}$ s and $V_{cc} = 15$ V is given by $v_s(t) = 2 \sin 100t$ V. What is $v_{out}(t)$?

Answer: $v_{out}(t) = 10[\cos(100t) - 1]$ V. (See ◎)

Exercise 5-18: Repeat Exercise 5.17 for a differentiator instead of an integrator.

Answer: $v_{out}(t) = -0.4 \cos 100t$ V. (See ◎)

5-7 Application Note: Parasitic Capacitance and Computer Processor Speed

As was noted in Section 4-11 and in Technology Brief 9 on page 163, the primary computational element in modern computer processors is the CMOS transistor. How quickly a single logic gate is able to switch its output between logic states 0 and 1 determines how fast the entire processor can perform complex calculations. Figure 5-43(a) displays a sample of a digital sequence, perhaps at the output of a digital inverter. The individual pulses, each denoting a logic state of 0 or 1, are each of duration T. *If it were possible to switch between states instantaneously, the maximum number of pulses that can be sequenced per 1 second is $1/T$. We refer to this rate by several names, including the* ***pulse repetition frequency***, ***switching frequency***, *and* ***clock speed***. *In the present case, we shall call it the switching frequency* and assign it the symbol f_s. That is,

$$f_s = \frac{1}{T} \qquad (\text{Hz}). \qquad (5.140)$$

So if $T = 1$ ns, $f_s = 1/10^{-9} = 1$ GHz, and if we can make the pulse duration narrower, we can increase f_s accordingly. Such a conclusion would be true if we can indeed arrange to have the logic circuit switch between states instantaneously, but it cannot. In Fig. 5-43(b), we show an expanded view of three pulses representing the sequence 101. We observe that the switching process is represented by ramp functions (rather than step functions) and it takes a finite amount of time for the voltage to change between a 0 state and a 1 state, which we shall call the ***rise time*** t_{rise}. Similarly, the fall time between states 1 and 0 is t_{fall}. [The linear rise and fall responses are actually artifacts of certain simplifying assumptions. In general, the responses involve exponentials, in which case it is more appropriate to define t_{rise} and t_{fall} as the durations between the 10 percent level and 90 percent level of the change in voltage.] The total time associated with a pulse is

$$T_{total} = T + t_{rise} + t_{fall}$$
$$= T + 2t_{rise} \qquad (\text{if } t_{rise} = t_{fall}), \qquad (5.141)$$

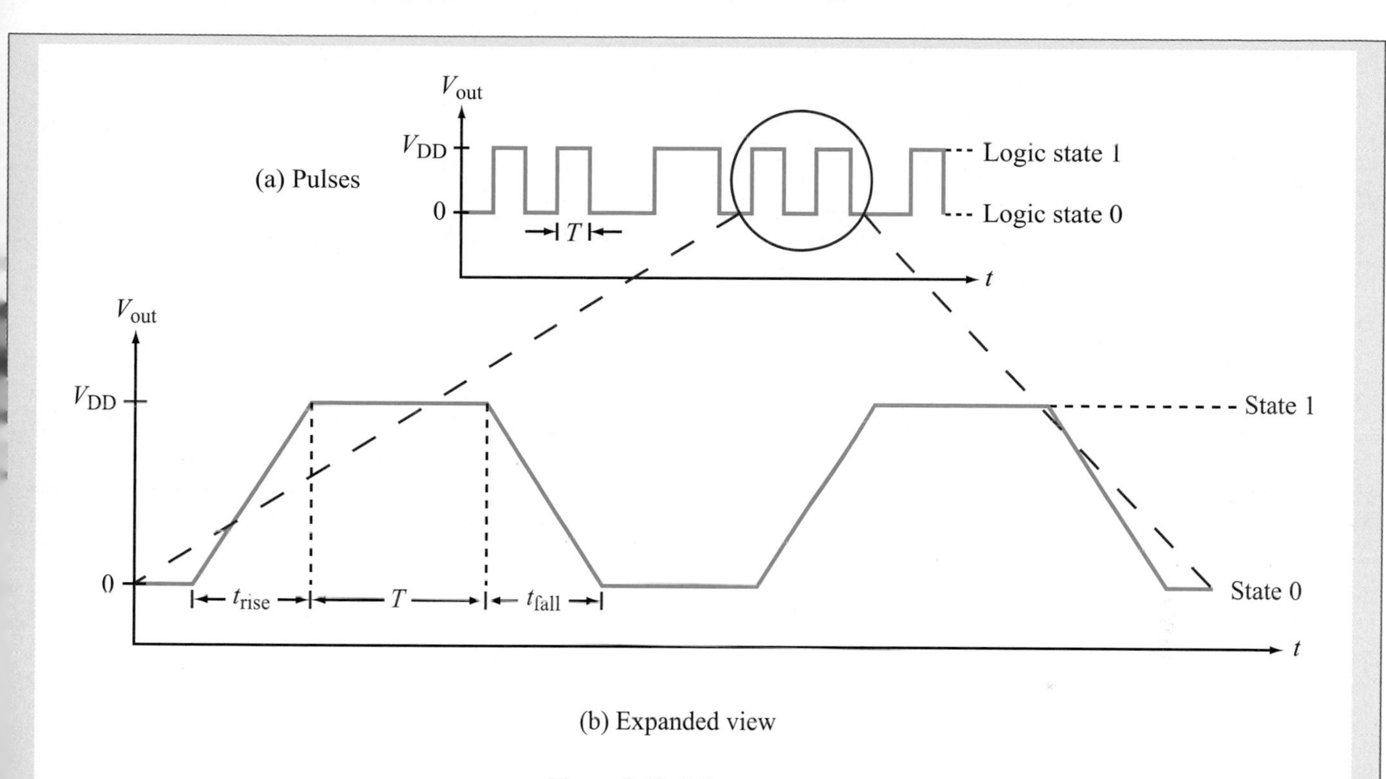

Figure 5-43: Pulse sequence.

and the associated switching frequency is

$$f_s = \frac{1}{T_{total}} = \frac{1}{T + 2t_{rise}}. \qquad (5.142)$$

Even if T can be reduced to zero, the maximum possible switching speed (without overlap between adjacent pulses) would be

$$f_s(\max) = \frac{1}{2t_{rise}}. \qquad (5.143)$$

As we shall see shortly, the switching times (t_{rise} and t_{fall}) are governed in part by the capacitances in the circuit. Consequently, capacitances play a major role in determining the ultimate switching speed of a digital circuit. In fact, capacitances also govern the switching speeds of the *wires*—often referred to as the ***bus***—that connect the processor to the various other devices on a computer ***motherboard***.

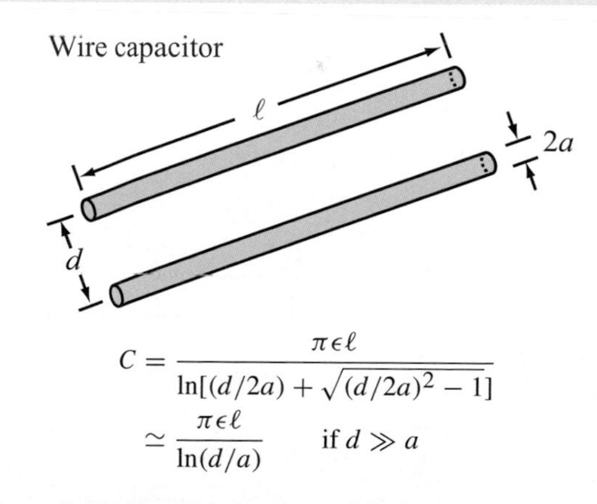

Figure 5-44: Capacitance of a two-wire configuration where ϵ is the permittivity of the material separating the wires.

> Whereas the processor speed of a modern computer is in the GHz range, the ***bus speed*** usually is slower by a factor of 3 to 10.

This is (in part) why a computer appears to slow down when the processor needs to access data through the bus. The following section will examine why this is so.

5-7.1 Parasitic Capacitance

Functionally, any two conducting bodies separated by an insulating material (including air, plastic, and all non-conductors) form a capacitor. The capacitors we have considered thus far are the type designed and fabricated intentionally for use as components in circuits.

In some situations, however, *unintentional* capacitance may exist in the circuit, in which case it usually is called ***parasitic capacitance***. Consider, for example, the capacitance formed by two parallel wires running side by side on a circuit board. The capacitance of such a two-wire ***transmission line*** (Fig. 5-44) is proportional directly to the length of the wires ℓ and inversely proportional to a logarithmic function involving d, the spacing between the wires. Thus, C increases with ℓ and decreases with d. If the wires are sufficiently long, or sufficiently close to one another, or some combination of the two [as to result in a capacitance of significant magnitude relative to the other capacitances in the circuit] such a wire capacitor can slow down the response time of the circuit. In a digital circuit, slower response time means slower switching speed. To explore this subject further, we will now examine the impact of parasitic capacitance on the operation of a MOSFET.

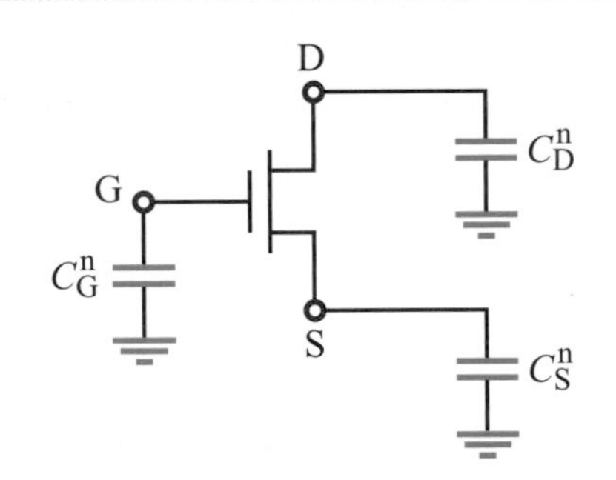

(a) NMOS

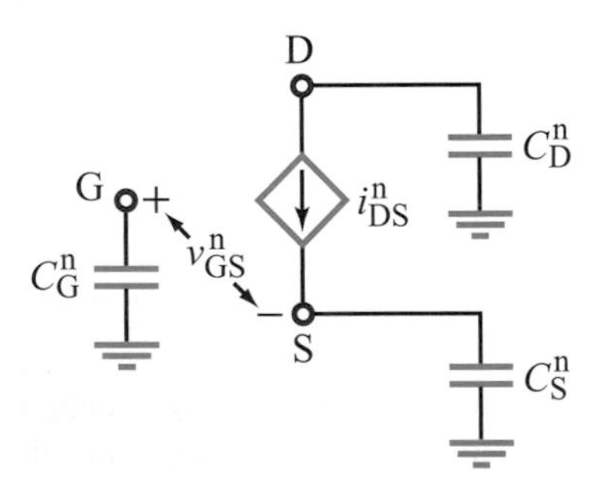

(b) Equivalent circuit

Figure 5-45: n-channel MOSFET (NMOS): (a) circuit symbol with added parasitic capacitances and (b) equivalent circuit. [In a PMOS, parasitic capacitances C^p_D and C^p_S should be shown connected to V_{DD} instead of to ground.]

5-7.2 CMOS Switching Speed

Recall from Section 4-11 that the gate node in a MOSFET is composed of a metal and a semiconductor separated by a thin layer of silicon dioxide that serves as a dielectric insulator. This geometry is somewhat similar to that of the parallel-plate capacitor of Fig. 5-11. Hence, during normal operation, the gate (G) and the source (S) nodes form a capacitor between them, as do the gate and the drain (D) nodes. Other parasitic capacitances also exist in a MOSFET, mainly due to charges separated between the source and the large silicon chip and between the drain and the chip. For simplicity, the various parasitic capacitances can be lumped together into an equivalent model containing three capacitances (all connected to ground) from G, S, and D. As shown in Fig. 5-45, these capacitances are designated C^n_G, C^n_S, and C^n_D, respectively, with the superscript "n" denoting that the circuit configuration applies to the n-channel MOSFET (or NMOS for short) whose body node usually is connected to ground. In a p-channel MOSFET, the body node is connected to V_{DD}. Hence, the model for PMOS would show parasitic capacitances C^p_D and C^p_S connected to V_{DD}, instead of to ground.

Now we are ready to analyze the operation of a CMOS inverter in the presence of parasitic capacitances. The circuit in Fig. 5-46(a) is essentially the same CMOS circuit of Fig. 4-27, except with added parasitic capacitances. The capacitances associated with the n-channel MOSFET are shown connected from terminals G^n, D^n, and S^n to ground. For the p-channel MOSFET, capacitance C^p_G is also connected to ground, but for the other two terminals, the capacitances are shown connected to V_{DD}. The two MOSFETs share a common gate terminal at the input side and a common drain terminal at the output side. Terminal S^n of the NMOS is connected directly to ground, which renders capacitance C^n_S irrelevant. Terminal S^p of the PMOS is connected directly to V_{DD}, which similarly renders C^p_S irrelevant. Capacitances C^n_G and C^p_G both are connected from the common gate terminal to ground and therefore can be combined into an ***equivalent capacitance*** C_{IN}. Incorporating these simplifications leads to the circuit shown in Fig. 5-46(b).

Our next step is to determine the output response $v_{out}(t)$ to a sudden change of state at the input from $v_{in} = 0$ to $v_{in} = V_{DD}$. Let us assume that the change happens at $t = 0$ and that the circuit was already in a steady-state condition by then.

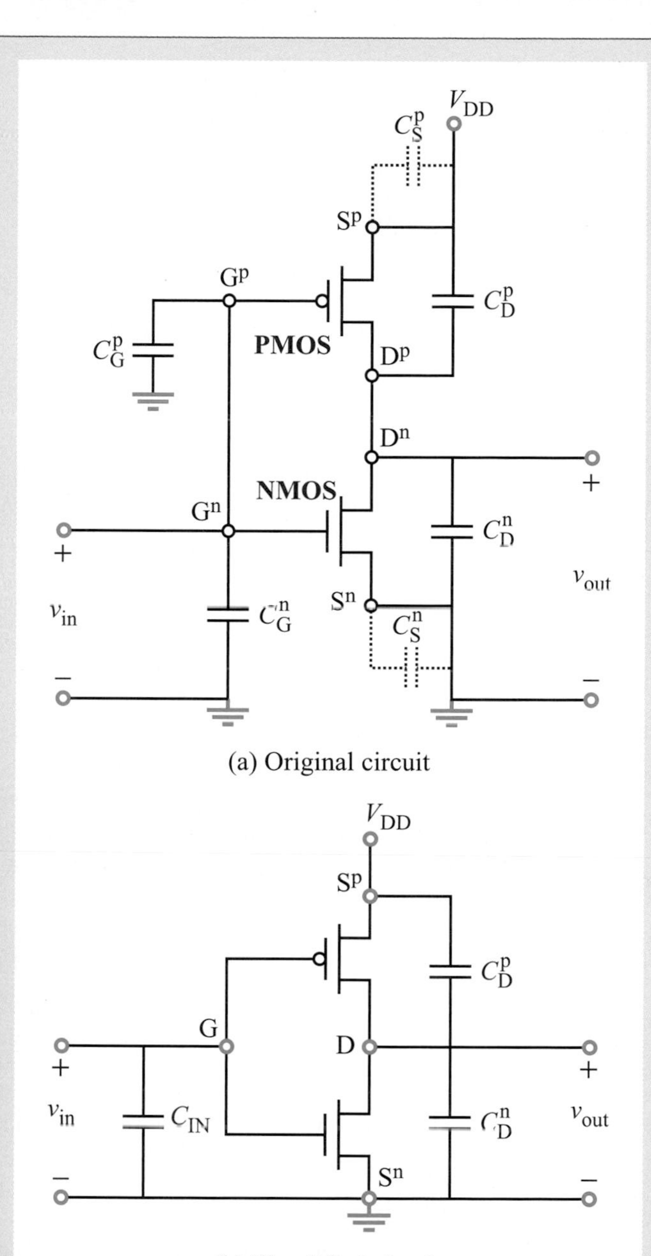

Figure 5-46: Common drain inverter circuit with parasitic capacitances. Superscripts "n" and "p" refer to the NMOS and PMOS transistors, respectively.

(a) At $t = 0^-$:

The capacitances in Fig. 5-47(a) act like open circuits. Also, $v_{in} = 0$, which means that $V_{GS}^n = 0$ for the NMOS and $V_{SG}^p = V_{DD}$ for the PMOS. Under such circumstances,

$$i_{DS}^n - gV_{GS}^n = 0, \quad \text{and} \quad i_{DS}^p = gV_{SG}^p = gV_{DD}, \tag{5.144}$$

where g is the MOSFET gain constant. Furthermore, the PMOS behavior is such that, if V_{SG}^p approaches V_{DD}, the voltage V_{DS}^p across the dependent current source goes to zero. With i_{DS}^n not conducting and i_{DS}^p acting like a short circuit, it follows that the voltage across capacitor C_D^n is

$$v_{out}(0^-) = V_{DD}. \tag{5.145}$$

Since the voltage across a capacitor cannot change instantaneously,

$$v_{out}(0) = V_{DD}. \tag{5.146}$$

(b) At $t \geq 0$:

If v_{in} is a step function that changes from 0 to V_{DD} at $t = 0$, the following pair of responses will take place:

(a) At the input side in the circuit of Fig. 5-47(a), we have an isolated loop comprised of v_{in}, R_s, and C_{IN}. In response to the change in v_{in}, capacitor C_{IN} will charge up to a final voltage V_{DD} at a rate governed by the time constant $\tau = R_s C_{IN}$. Through proper choice of R_s (very small), C_{IN} can charge up to V_{DD} so quickly (in comparison with the response time of the output) that it can be assumed that $V_{GS}^n = V_{DD}$ immediately after $t = 0$.

(b) At the output side, with $V_{GS}^n = V_{DD}$, it follows that $V_{SG}^p = 0$. Hence,

$$i_{DS}^n = gV_{DD}, \quad \text{and} \quad i_{DS}^p = gV_{SG}^p = 0. \tag{5.147}$$

At node D',

$$i_1 + i_2 + i_3 = 0, \tag{5.148}$$

and at node D,

$$i_3 = i_{DS}^n + i_{DS}^p = gV_{DD}. \tag{5.149}$$

Also,

$$i_1 = C_D^p \frac{d}{dt}(v_{out} - V_{DD}) = C_D^p \frac{d}{dt} v_{out}, \tag{5.150}$$

and

$$i_2 = C_D^n \frac{d}{dt} v_{out}. \tag{5.151}$$

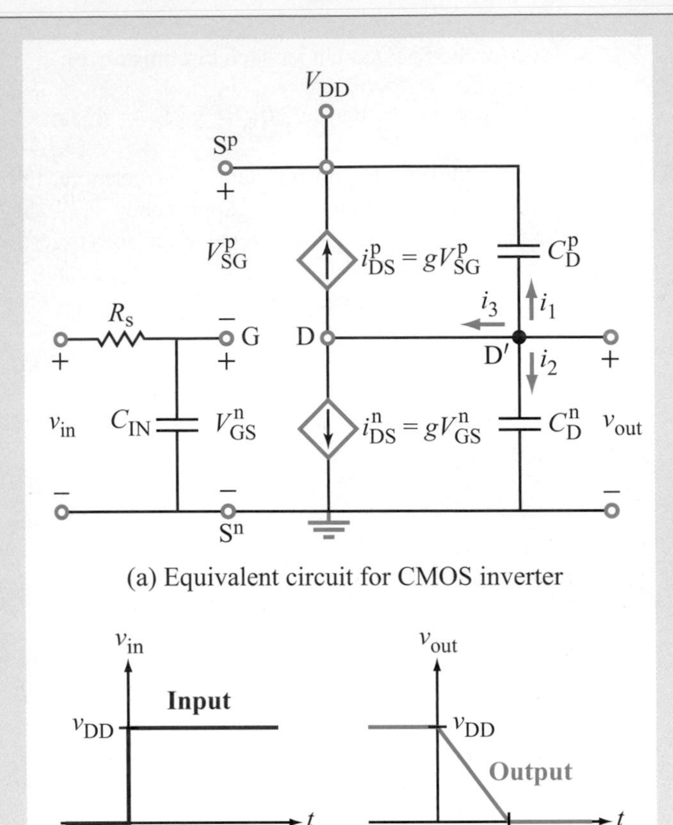

Figure 5-47: (a) Equivalent circuit for the CMOS inverter; (b) the response of $v_{\text{out}}(t)$ to v_{in} changing states from 0 to V_{DD} at $t = 0$.

Upon inserting the expressions given by Eqs. (5.149) through (5.151) into Eq. (5.148) and then rearranging terms, we have

$$\frac{dv_{\text{out}}}{dt} = \frac{-gV_{\text{DD}}}{C_{\text{D}}^{\text{n}} + C_{\text{D}}^{\text{p}}}. \quad (5.152)$$

Integrating both sides from 0 to t gives

$$v_{\text{out}}\Big|_0^t = \frac{-gV_{\text{DD}}}{C_{\text{D}}^{\text{n}} + C_{\text{D}}^{\text{p}}} \int_0^t dt, \quad (5.153)$$

which leads to

$$v_{\text{out}}(t) = v_{\text{out}}(0) - \left(\frac{gV_{\text{DD}}}{C_{\text{D}}^{\text{n}} + C_{\text{D}}^{\text{p}}}\right) t. \quad (5.154)$$

In view of Eq. (5.146), the expression for $v_{\text{out}}(t)$ becomes

$$v_{\text{out}}(t) = V_{\text{DD}}\left[1 - \left(\frac{g}{C_{\text{D}}^{\text{n}} + C_{\text{D}}^{\text{p}}}\right) t\right]. \quad (5.155)$$

Plots of $v_{\text{in}}(t)$ changing states from 0 to V_{DD} at $t = 0$ and of the corresponding response $v_{\text{out}}(t)$ are displayed in Fig. 5-47(b). We observe that t_{fall} is the time it takes for v_{out} to change states from V_{DD} to zero. From Eq. (5.155), we deduce that

$$t_{\text{fall}} = \frac{C_{\text{D}}^{\text{n}} + C_{\text{D}}^{\text{p}}}{g}. \quad (5.156)$$

Example 5-17: Processor Speed

The input to a CMOS inverter consists of a sequence of bits, each 25 picoseconds in duration. Determine the maximum switching frequency at which the CMOS inverter can be operated without causing overlap between adjacent bits (pulses) under each of the following conditions: (a) parasitic capacitances totally ignored and (b) parasitic capacitances included. In both cases, $g = 10^{-5}$ A/V, and $C_{\text{D}}^{\text{n}} = C_{\text{D}}^{\text{p}} = 0.5$ fF.

Solution:

(a) With $T = 25$ ps $= 25 \times 10^{-12}$ s and no capacitances to slow down the switching process, the maximum switching frequency is

$$f_{\text{s}} = \frac{1}{T} = \frac{1}{25 \times 10^{-12}} = 40 \text{ GHz}.$$

(b) From Eq. (5.156),

$$t_{\text{fall}} = \frac{C_{\text{D}}^{\text{n}} + C_{\text{D}}^{\text{p}}}{g} = \frac{(0.5 + 0.5) \times 10^{-15}}{10^{-5}} = 10^{-10} \text{ s}.$$

To determine t_{rise}, we have to repeat the solution that led to Eq. (5.156) but with v_{in} starting in state 1 (i.e., $v_{\text{in}} = V_{\text{DD}}$) and switching to state 0 at $t = 0$. Such a process would lead to

$$v_{\text{out}}(t) = V_{\text{DD}}\left(\frac{g}{C_{\text{D}}^{\text{n}} + C_{\text{D}}^{\text{p}}}\right) t.$$

The time duration that it takes $v_{\text{out}}(t)$ to reach V_{DD} is

$$t_{\text{rise}} = \frac{C_{\text{D}}^{\text{n}} + C_{\text{D}}^{\text{p}}}{g} = t_{\text{fall}}.$$

Hence, in the presence of parasitic capacitances, Eq. (5.142) is applicable. Namely,

$$f_s = \frac{1}{T + 2t_{\text{rise}}} = \frac{1}{25 \times 10^{-12} + 2 \times 10^{-10}} = 4.44 \text{ GHz}.$$

In this example, the parasitic capacitances are responsible for slowing down the switching speed of the CMOS processor by about one order of magnitude.

In the preceding example, we essentially ignored the input capacitances of the CMOS. Since logic gates are strung along in series such that one gate's output is the next gate's input, input capacitances usually are lumped together with the previous gate's output capacitances. To properly incorporate the roles of both input and output parasitic capacitances, a more thorough treatment is needed than the first-order approximation we carried out in this section. Nevertheless, the approximation did succeed in making the point that at high switching rates parasitic capacitances are important and should not be ignored.

Review Question 5-26: What is the rationale for adding parasitic capacitances to nodes G, D, and S in Fig. 5-45?

Review Question 5-27: What determines the maximum switching frequency for a CMOS inverter?

Exercise 5-19: A CMOS inverter with $C_D^n + C_D^p = 20$ fF has a fall time of 1 ps. What is the value of its gain constant?

Answer: $g = 2 \times 10^{-2}$ A/V. (See)

5-8 Analyzing Circuit Response with Multisim

5-8.1 Modeling Switches in Multisim

Determining the time-dependent behavior of large, complex circuits often is difficult to do and extremely time-consuming. Accordingly, designs of commercial circuits rely heavily on SPICE simulators for evaluating the response of a candidate circuit design before constructing the real version. In this section, we will demonstrate how Multisim can be used to analyze the transient response of a circuit driven by a time-dependent source.

Because the first-order RC circuit is straightforward to analyze *by hand*, it makes for a useful example with which we can compare Multisim simulation results to hand calculations. Consider the circuit shown in Fig. 5-48, in which the switch is opened at $t = 0$ after it had been in the closed position for a long time. Hence, prior to $t = 0$, the circuit was in a steady state and the capacitor was fully charged with no current flowing through it (behaving like an open circuit). The voltage across the capacitor is designated V(3) (so as to match the Multisim circuit that we will be constructing soon) and is given by

$$\text{V}(3) = \frac{2.5 \times 10 \text{ k}}{1 \text{ k} + 10 \text{ k}} = 2.27 \text{ V} \quad (@\, t = 0^-).$$

Upon opening the switch, the capacitor will discharge through the 10-kΩ resistor with a time constant given by

$$\tau_{\text{discharge}} = R_1 C_1 = 10^4 \times 5 \times 10^{-15} = 50 \text{ ps}.$$

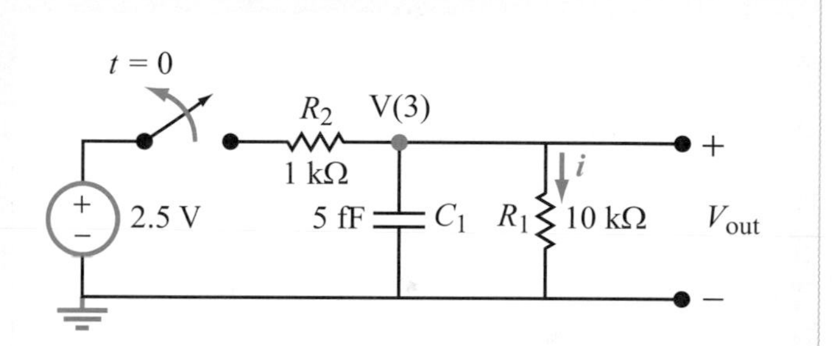

Figure 5-48: RC circuit with an SPST switch.

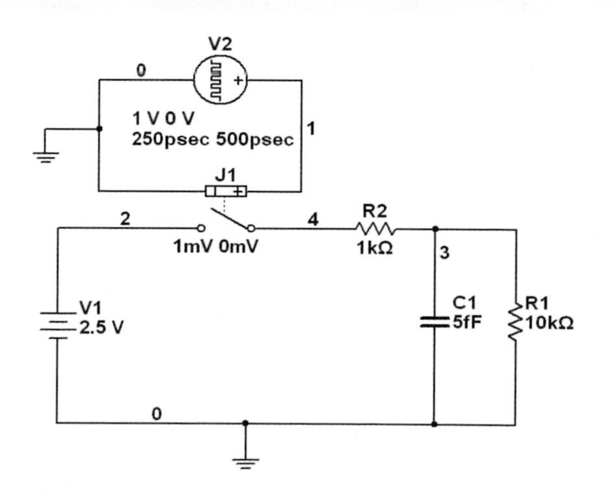

Figure 5-49: Multisim equivalent of the RC circuit in Fig. 5-48.

Table 5-6: Multisim component list for the circuit in Fig. 5-49.

Component	Group	Family	Quantity	Description
1 k	Basic	Resistor	1	1-kΩ resistor
10 k	Basic	Resistor	1	10-kΩ resistor
5 f	Basic	Capacitor	1	5-fF capacitor
VOLTAGE_CONTROLLED_SWITCH	Basic	Switch	1	Switch
DC_POWER	Sources	Power_Sources	1	2.5-V dc source
PULSE_VOLTAGE	Sources	Signal_Voltage_Source	1	Pulse-generating voltage source

Likewise, if the switch were to close at a later time after the circuit had fully discharged, the capacitor would again charge up to 2.27 V, but in this case, the time constant would be

$$\tau_{\text{charge}} = (R_1 \parallel R_2)C_2 = \frac{1\text{ k} \times 10\text{ k}}{11\text{ k}} \times 5 \times 10^{-15} = 4.54\text{ ps}.$$

Thus, the charge-up response of the circuit is much faster (by about one order of magnitude) than its discharge response.

To demonstrate the transient behavior of the circuit with Multisim, we construct the circuit model shown in Fig. 5-49 using the component list given in Table 5-6. The only oddity in the circuit is the use of a Voltage-Controlled Switch and a Pulse Generator source to drive it. Multisim does not provide the user the option to use time-programmable switches, so in order to observe the circuit response to multiple opening and closing events of the switch, we use a voltage-controlled switch in combination with an appropriately configured pulse generator. The exact voltage amplitude of the pulse (V2 in Fig. 5-49 on page 231) is not important (so long as it is larger than the 1-mV threshold of the switch), but the timing of the pulse is critically important, as we want to allow enough time between opening and closing events to observe the complete transient responses of the circuit. Since the longest time constant is 50 ps, double-click on the Pulse Generator and set the Pulse width at 250 ps and the Period at 500 ps so as to provide an adequate time window. Also set the Rise Time and Fall Time to 1 ps.

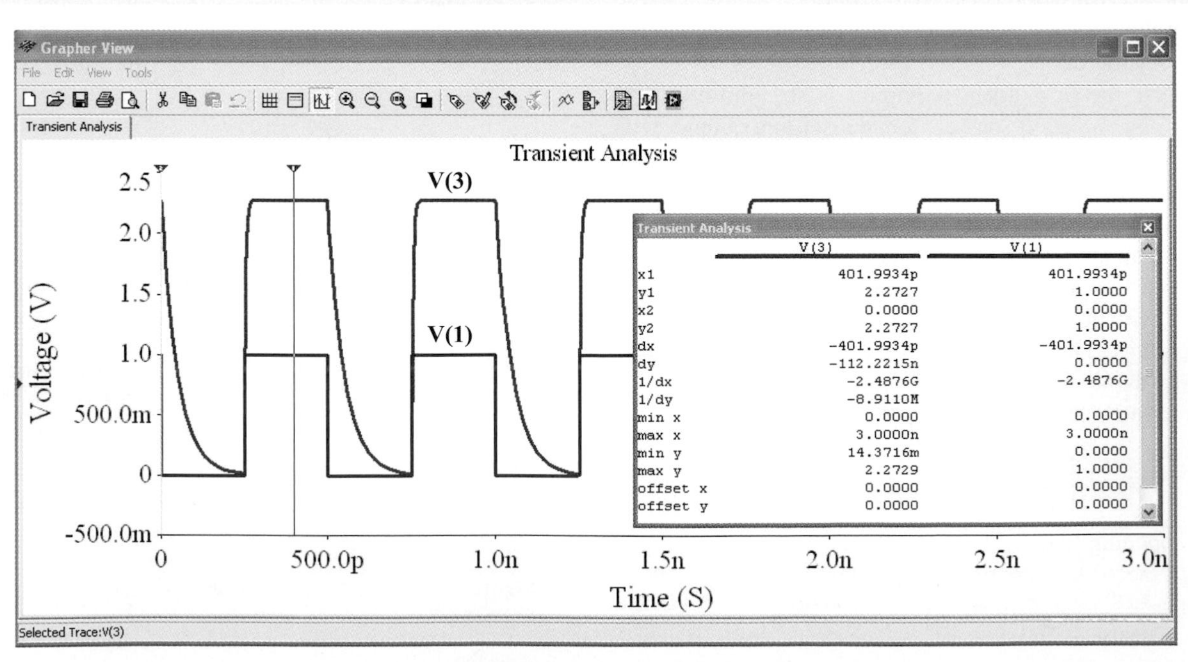

Figure 5-50: Transient response of the circuit in Fig. 5-49.

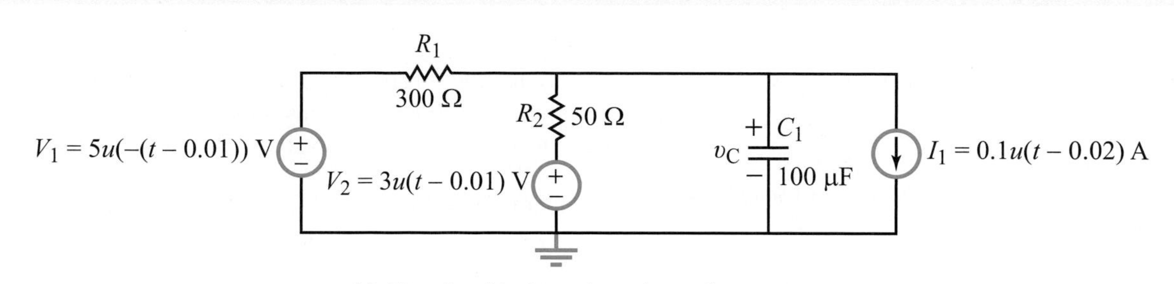

(a) Circuit with three time-dependent sources

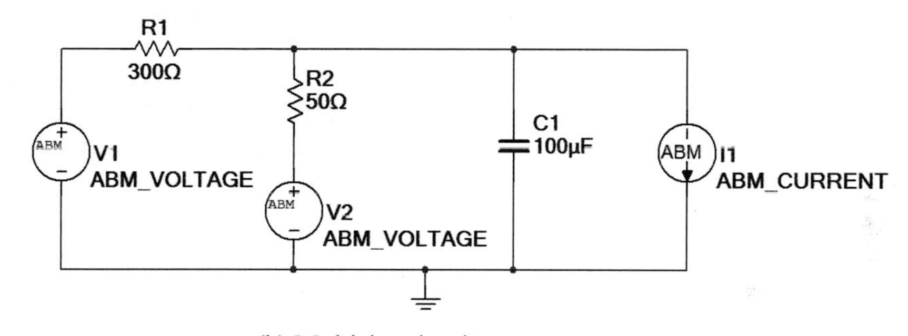

(b) Multisim circuit

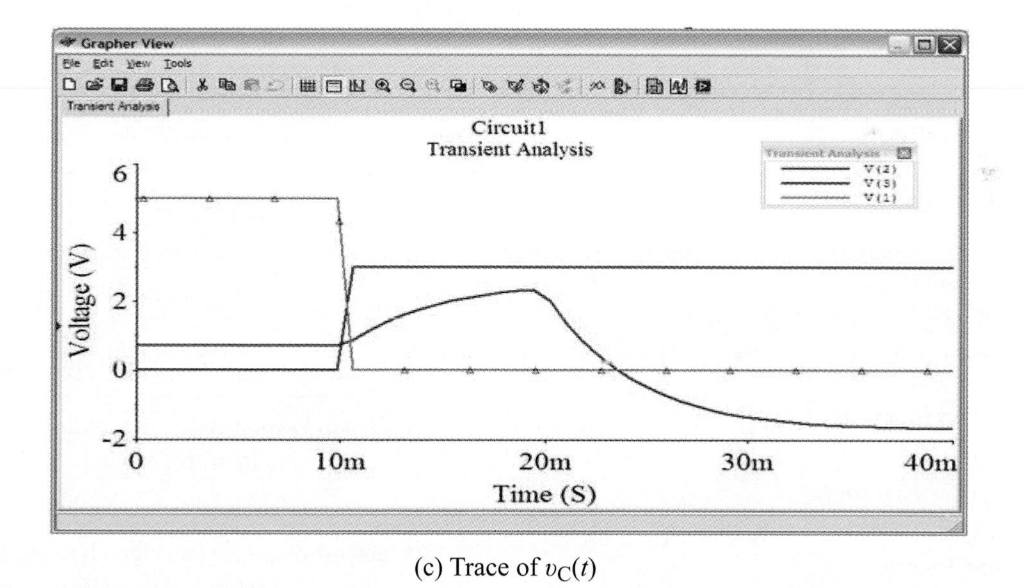

(c) Trace of $v_C(t)$

Figure 5-51: Multisim analysis of a circuit containing time-dependent sources.

To analyze the behavior, we select Simulate → Analyses → Transient Analysis. Make sure to select an End Time equal to a few periods; 3 ns should suffice. (If you forget this, you may need to abort the simulation to prevent it from running for a long time since the default value is 0.001 s! To abort the simulation or any general Analyses which may be taking too long, go to Simulate → Analyses → Stop Analysis.) In the Output tab, select the non-ground node of the capacitor V(3) and the pulse voltage V(1) for time references. Figure 5-50 on page 232 shows the output of the transient analysis. Enabling the Cursor tool in the Grapher window allows the user to read out the exact voltage and time values for any trace.

5-8.2 Modeling Time-Dependent Sources in Multisim

In the previous subsection, we examined how to create switches that toggle with time. What if we wanted to simulate the circuit shown in Fig. 5-51(a) on page 233 and plot v_C over a certain time duration? The circuit has three time-dependent sources, which would make adding switches and pulse generators rather complicated. Multisim allows us to create the time-dependent sources found in this circuit by using the ABM Voltage and Current sources.

In Multisim's ABM syntax, the step function $u(t)$ is represented by the stp(TIME) function. Also, to guard against Multisim calculating incorrect initial conditions prior to the step function, it is advisable to shift the step-function transition to occur 10 ms after the start of the simulation. Hence, we use the following ABM expressions:

For V1 $= 5u(-(t-0.01))$ V: ➡ 5*stp(-TIME+0.01)

For V2 $= 3u(t-0.01)$ V: ➡ 3*stp(TIME-0.01)

For I1 $= 0.1u(t-0.02)$ A: ➡ 0.1*stp(TIME-0.02)

Once these expressions have been entered, go to Simulate → Analyses → Transient Analysis. Leave the Start Time at 0 s, and set the End Time to 0.04 s. Under the Output tab, select the voltages V(1), V(2), and V(3) and press Simulate. This generates the plots shown in Fig. 5-51(c) on page 233.

Chapter 5 Relationships

Unit step function

$$u(t) = \begin{cases} 0 & \text{for } t < 0 \\ 1 & \text{for } t > 0 \end{cases}$$

Time-shifted step function

$$u(t-T) = \begin{cases} 0 & \text{for } t < T \\ 1 & \text{for } t > T \end{cases}$$

Unit ramp function

$$r(t) = \begin{cases} 0 & \text{for } t \le 0 \\ t & \text{for } t \ge 0 \end{cases}$$

Time-shifted ramp function

$$r(t-T) = \begin{cases} 0 & \text{for } t \le T \\ (t-T) & \text{for } t \ge T \end{cases}$$

Unit rectangular function
(pulse center at $t = T$; pulse length $= \tau$)

$$\text{rect}\left[\frac{(t-T)}{\tau}\right] = \begin{cases} 0 & \text{for } t < (T-\tau/2), \\ 1 & \text{for } (T-\tau/2) \le t \le (T+\tau/2), \\ 0 & \text{for } t > (T+\tau/2). \end{cases}$$

Capacitor $i = C\,\dfrac{dv}{dt}$

$$v(t) = v(t_0) + \frac{1}{C}\int_{t_0}^{t} i\,dt$$

$$w = \tfrac{1}{2}\,Cv^2 \qquad \text{(stored energy)}$$

Inductors $v = L\,\dfrac{di}{dt}$

$$i(t) = i(t_0) + \frac{1}{L}\int_{t_0}^{t} v\,dt$$

$$w = \tfrac{1}{2}\,Li^2 \qquad \text{(stored energy)}$$

RC circuit response (sudden change at $t = 0$)

$$v_C(t) = v_C(\infty) + [v(0) - v(\infty)]e^{-t/\tau}$$
$$\tau = RC$$

RL circuit response (sudden change at $t = 0$)

$$i_L(t) = i_L(\infty) + [i_L(0) - i_L(\infty)]e^{-t/\tau}$$
$$\tau = L/R$$

Op-amp integrator

$$v_{out}(t) = -\frac{1}{RC}\int_{t_0}^{t} v_i\,dt + v_{out}(t_0)$$

Op-amp differentiator

$$v_{out}(t) = -RC\,\frac{dv_i}{dt}$$

CHAPTER HIGHLIGHTS

- The step, ramp, rectangle, and exponential functions can be used to characterize a variety of nonperiodic waveforms.
- A capacitor consists of two conducting surfaces. The accumulation of $+q$ on one of the surfaces and $-q$ on the other creates a voltage difference between them. Capacitance is defined as $C = q/v$.
- For a capacitor, $i = C\,dv/dt$, and the electrical energy stored in it is $w = \frac{1}{2}\,Cv^2$.
- The rule for combining capacitors connected in series is the same as the rule for combining resistors connected in parallel, and vice versa.
- An inductor stores magnetic energy when a current passes through it. Its i–v relationship is $v = L\,di/dt$, and its expression for stored energy is $w = \frac{1}{2}\,Li^2$.
- The rules for combining inductors connected in series or in parallel are the same as the rules for resistors.
- A series RC circuit excited by a dc source exhibits a voltage response (across the capacitor) characterized by an exponential function containing a time constant $\tau = RC$.
- A parallel RL circuit exhibits a current response (through the inductor) that has the same form as the voltage response of the series RC circuit, but for the RL circuit, $\tau = L/R$.
- The ideal op-amp integrator circuit—whose output voltage is directly proportional to the time integral of the input signal—is nothing more than an inverting-amplifier circuit with the feedback resistor replaced with a capacitor.
- An integrator circuit becomes a differentiator circuit upon interchanging the locations of R and C.
- Parasitic capacitance is often the factor that ultimately limits the processor speed of a computer.
- Multisim allows us to evaluate the switching response of a circuit.

GLOSSARY OF IMPORTANT TERMS

Provide definitions or explain the meaning of the following terms:

bus speed
clock speed
coaxial capacitor
electric field
electrical permittivity
electrical susceptibility
exponential function
final value
first-order circuit
forced response
forcing function
inductance
initial value
magnetic field
magnetic flux linkage
magnetic permeability
natural response
op-amp differentiator
op-amp integrator
parallel-plate capacitor
parasitic capacitance
periodic waveform
pulse repetition frequency
pulse waveform
ramp function
rectangle function
relative permittivity
solenoid
source free
steady-state response
step function
switching frequency (speed)
time constant
transient response

PROBLEMS

Section 5-1: Nonperiodic Waveforms

5.1 Generate plots for each of the following step-function waveforms over the time span from -5 to $+5$ s.

(a) $v_1(t) = -6u(t+3)$

(b) $v_2(t) = 10u(t-4)$

(c) $v_3(t) = 4u(t+2) - 4u(t-2)$

(d) $v_4(t) = 8u(t-2) + 2u(t-4)$

(e) $v_5(t) = 8u(t-2) - 2u(t-4)$

***5.2** Provide expressions in terms of step functions for the waveforms displayed in Fig. P5.2.

5.3 A 10-V rectangular pulse with a duration of 5 μs starts at $t = 2$ μs. Provide an expression for the pulse in terms of step functions.

5.4 Generate plots for each of the following functions over the time span from -4 to $+4$ s.

(a) $v_1(t) = 5r(t+2) - 5r(t)$

(b) $v_2(t) = 5r(t+2) - 5r(t) - 10u(t)$

(c) $v_3(t) = 10 - 5r(t+2) + 5r(t)$

(d) $v_4(t) = 10 \text{ rect}\left(\dfrac{t+1}{2}\right) - 10 \text{ rect}\left(\dfrac{t-3}{2}\right)$

(e) $v_5(t) = 5 \text{ rect}\left(\dfrac{t-1}{2}\right) - 5 \text{ rect}\left(\dfrac{t-3}{2}\right)$

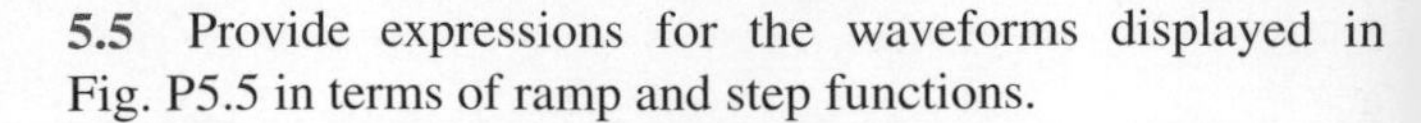

5.5 Provide expressions for the waveforms displayed in Fig. P5.5 in terms of ramp and step functions.

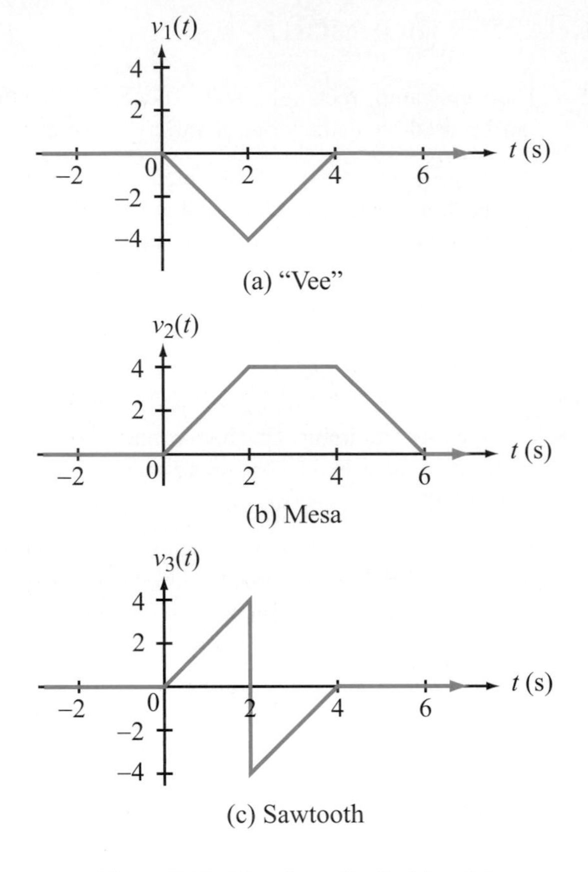

Figure P5.5: Waveforms for Problem 5.5.

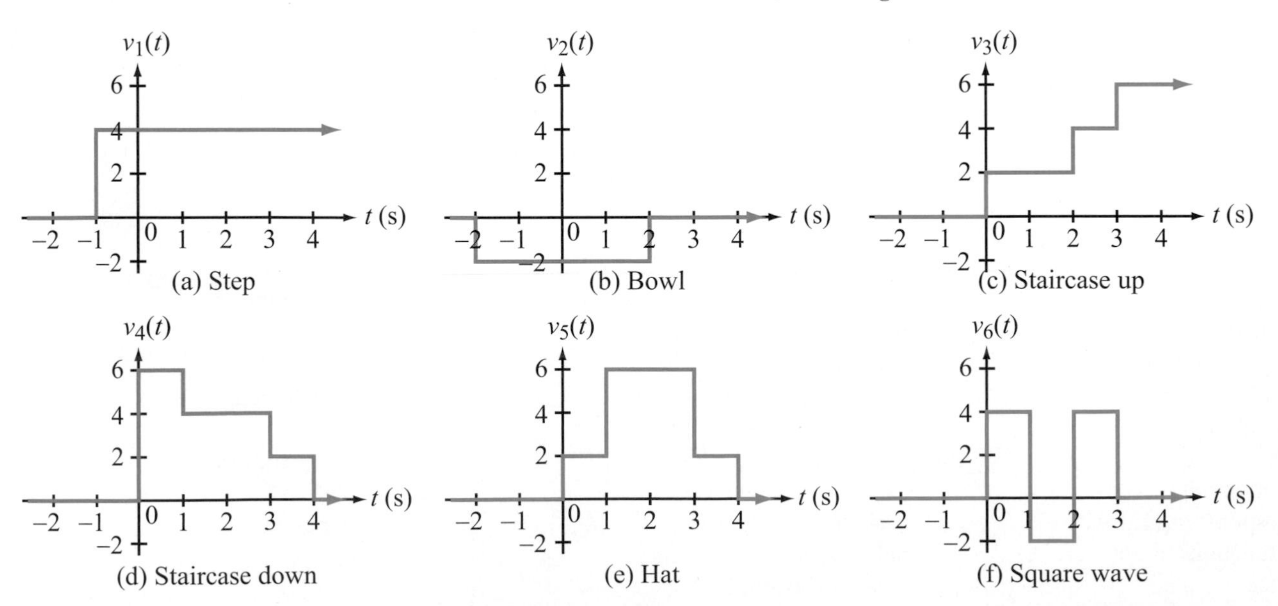

Figure P5.2: Waveforms for Problem 5.2.

*Answer(s) in Appendix E.

5.6 Provide plots for the following functions (over a time span and with a time scale that will appropriately display the shape of the associated waveform):

(a) $v_1(t) = 100e^{-2t}\ u(t)$

(b) $v_2(t) = -10e^{-0.1t}\ u(t)$

(c) $v_3(t) = -10e^{-0.1t}\ u(t-5)$

(d) $v_4(t) = 10(1 - e^{-10^3 t})\ u(t)$

(e) $v_5(t) = 10e^{-0.2(t-4)}\ u(t)$

(f) $v_6(t) = 10e^{-0.2(t-4)}\ u(t-4)$

***5.7** After opening a certain switch at $t = 0$ in a circuit containing a capacitor, the voltage across the capacitor started decaying exponentially with time. Measurements indicate that the voltage was 7.28 V at $t = 1$ s and 0.6 V at $t = 6$ s. Determine the initial voltage at $t = 0$ and the time constant of the voltage waveform.

Section 5-2: Capacitors

5.8 After plotting the voltage waveform, obtain expressions and generate plots for $i(t)$, $p(t)$, and $w(t)$ for a 0.2-mF capacitor. The voltage waveforms are given by

(a) $v_1(t) = 5r(t) - 5r(t-2)$ V

(b) $v_2(t) = 10u(-t) + 10u(t) - 5r(t-2) + 5r(t-4)$ V

(c) $v_3(t) = 15u(-t) + 15e^{-0.5t}\ u(t)$ V

(d) $v_4(t) = 15[1 - e^{-0.5t}]\ u(t)$ V

5.9 In response to a change introduced by a switch at $t = 0$, the current flowing through a 100-μF capacitor, defined in accordance with the passive sign convention, was observed to be

$$i(t) = -0.4e^{-0.5t} \text{ mA} \qquad (\text{for } t > 0).$$

If the final energy stored in the capacitor (at $t = \infty$) is 0.2 mJ, determine $v(t)$ for $t \geq 0$.

5.10 The voltage $v(t)$ across a 20-μF capacitor is given by the waveform shown in Fig. P5.10.

(a) Determine and plot the corresponding current $i(t)$.

(b) Specify the time interval(s) during which power transfers into the capacitor and that (those) during which it transfers out of the capacitor.

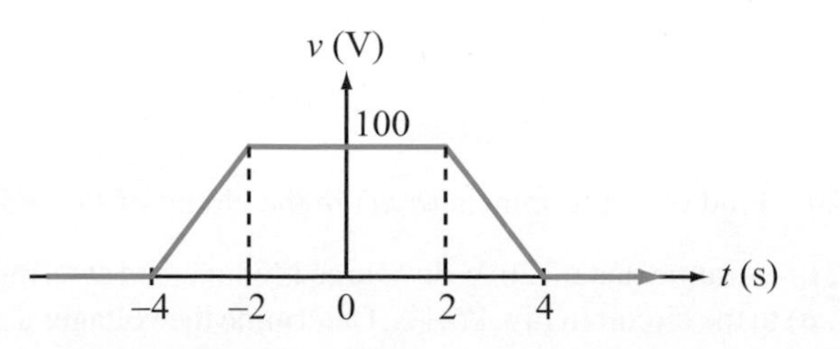

Figure P5.10: Waveform for Problems 5.10 and 5.11.

(c) At what instant in time is the power transfer into the capacitor a maximum? And at what instant is the power transfer out of the capacitor a maximum?

(d) What is the maximum amount of energy stored in the capacitor, and when does it occur?

5.11 Suppose the waveform shown in Fig. P5.10 is the current $i(t)$ through a 0.2-mF capacitor (rather than the voltage) and its peak value is 100 μA. Given that the initial voltage on the capacitor was zero at $t = -4$ s, determine and plot $v(t)$.

5.12 The current through a 40-μF capacitor is given by a rectangular pulse as

$$i(t) = 40 \text{ rect}\left(\frac{t-1}{2}\right) \text{ mA}.$$

If the capacitor was initially uncharged, determine $v(t)$, $p(t)$, and $w(t)$.

5.13 The voltage across a 0.2-mF capacitor was 20 V until a switch in the circuit was opened at $t = 0$, causing the voltage to vary with time as

$$v(t) = (60 - 40e^{-5t}) \text{ V} \qquad (\text{for } t > 0).$$

(a) Did the switch action result in an instantaneous change in $v(t)$?

(b) Did the switch action result in an instantaneous change in the current $i(t)$?

(c) How much initial energy was stored in the capacitor at $t = 0$?

(d) How much final energy will be stored in the capacitor (at $t = \infty$)?

5.14 Determine voltages v_1 to v_4 in the circuit of Fig. P5.14 under dc conditions.

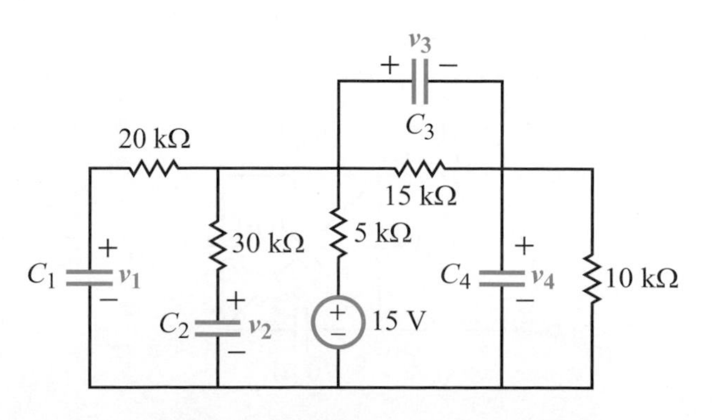

Figure P5.14: Circuit for Problem 5.14.

*5.15 Determine voltages v_1 to v_3 in the circuit of Fig. P5.15 under dc conditions.

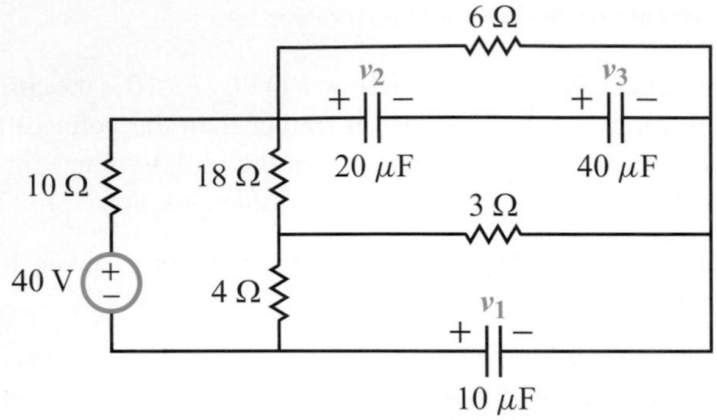

Figure P5.15: Circuit for Problem 5.15.

5.16 Determine the voltages across the two capacitors in the circuit of Fig. P5.16 under dc conditions.

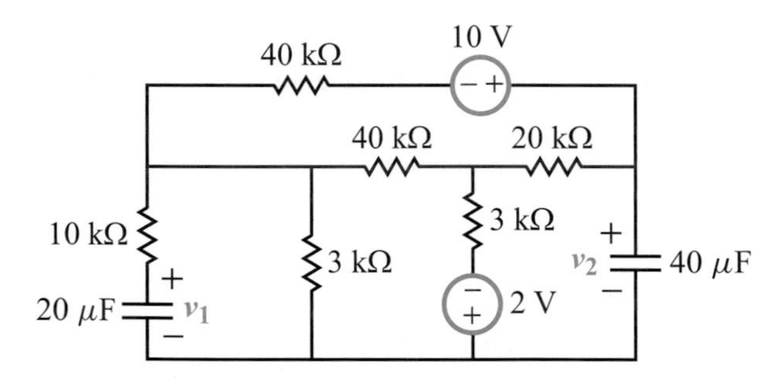

Figure P5.16: Circuit for Problem 5.16.

5.17 Reduce the circuit in Fig. P5.17 into a single equivalent capacitor at terminals (a, b). Assume that all initial voltages are zero at $t = 0$.

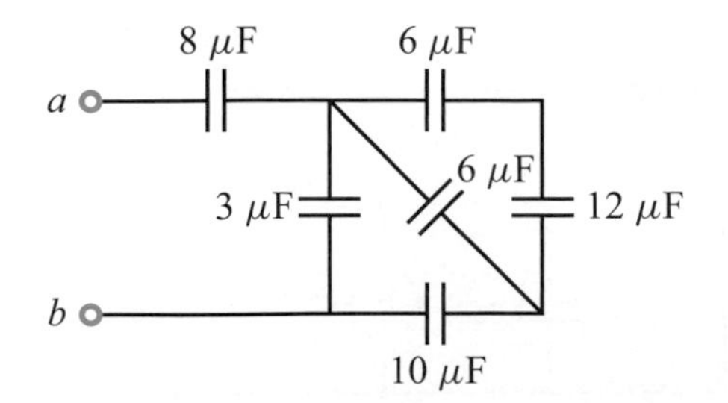

Figure P5.17: Circuit for Problems 5.17 and 5.21.

5.18 Reduce the circuit in Fig. P5.18 into a single equivalent capacitor at terminals (a, b). Assume that all initial voltages are zero at $t = 0$.

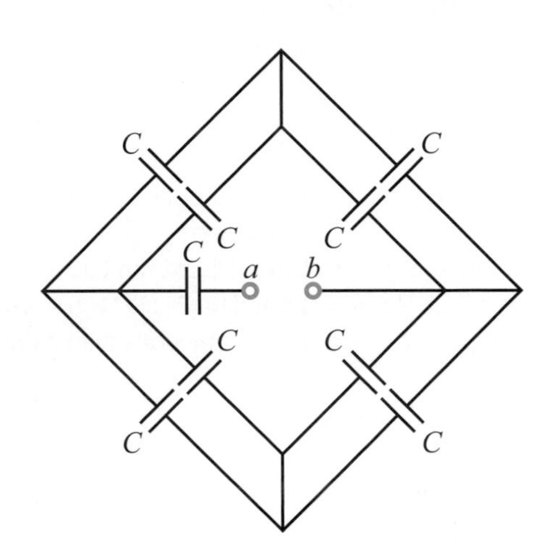

Figure P5.18: Circuit for Problem 5.18.

*5.19 For the circuit in Fig. P5.19, find C_{eq} at terminals (a, b) Assume all initial voltages to be zero.

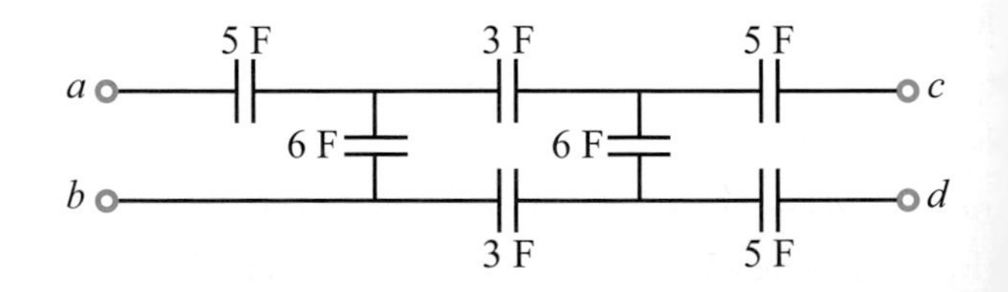

Figure P5.19: Circuit for Problems 5.19 and 5.20.

5.20 Find C_{eq} at terminals (c, d) in the circuit of Fig. P5.19

5.21 Assume that a 120-V dc source is connected at terminals (a, b) to the circuit in Fig. P5.17. Determine the voltages across all capacitors.

5.22 Determine (a) the amount of energy stored in each of the three capacitors shown in Fig. P5.22, (b) the equivalent capacitance at terminals (a, b), and (c) the amount of energy stored in the equivalent capacitor.

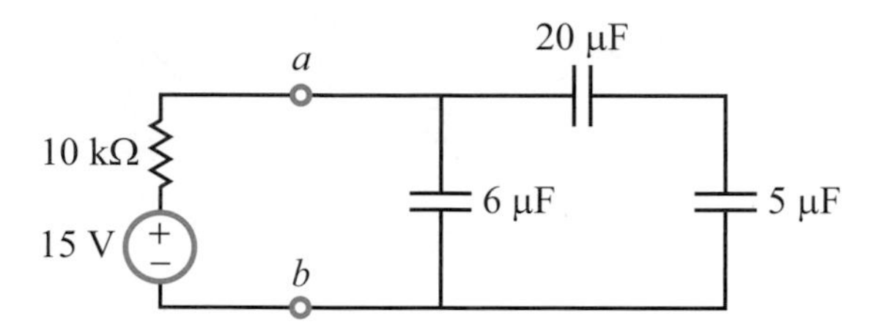

Figure P5.22: Circuit for Problem 5.22.

Section 5-3: Inductors

5.23 After plotting the current waveform, obtain expressions and generate plots for $v(t)$, $p(t)$, and $w(t)$ for a 0.5-mH inductor. The current waveforms are given by

(a) $i_1(t) =$
$0.2r(t-2) - 0.2r(t-4) - 0.2r(t-8) + 0.2r(t-10)$ A

(b) $i_2(t) = 2u(-t) + 2e^{-0.4t}\ u(t)$ A

(c) $i_3(t) = -4(1 - e^{-0.4t})\ u(t)$ A

5.24 The current $i(t)$ passing through a 0.1-mH inductor is given by the waveform shown in Fig. P5.24.

(a) Determine and plot the corresponding voltage $v(t)$ across the inductor.

(b) Specify the time interval(s) during which power is transferred into the inductor and that (those) during which power transfers out of the inductor. Also specify the amount of energy transferred in each case.

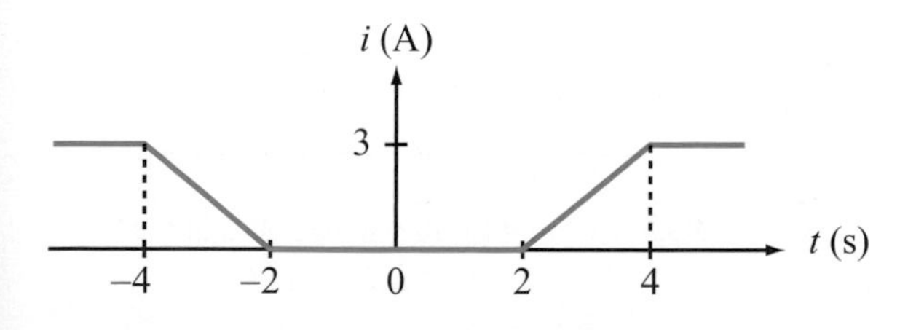

Figure P5.24: Current waveform for Problem 5.24.

5.25 Activation of a switch at $t = 0$ in a certain circuit caused the voltage across a 20-mH inductor to exhibit the voltage response

$$v(t) = 4e^{-0.2t}\text{ mV} \qquad (\text{for } t \geq 0).$$

Determine $i(t)$ for $t \geq 0$ given that the energy stored in the inductor at $t = \infty$ is 0.64 mJ.

5.26 The waveform shown in Fig. P5.26 represents the voltage across a 0.2-H inductor for $t \geq 0$. If the current flowing through the inductor is -20 mA at $t = 0$, determine the current $i(t)$ for $t \geq 0$.

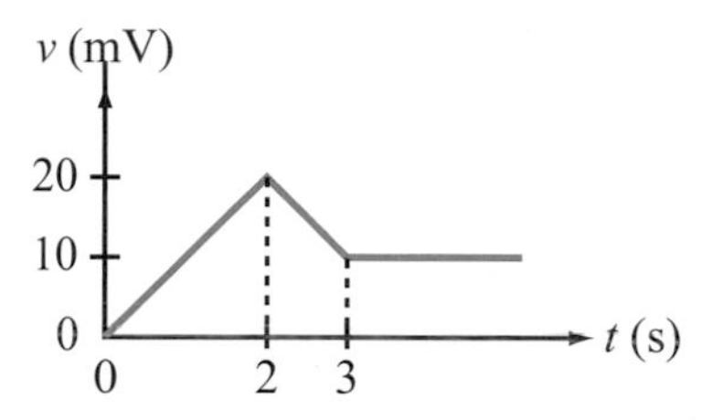

Figure P5.26: Voltage waveform for Problem 5.26.

5.27 The waveform shown in Fig. P5.27 represents the voltage across a 50-mH inductor. Determine the corresponding current waveform. Assume $i(0) = 0$.

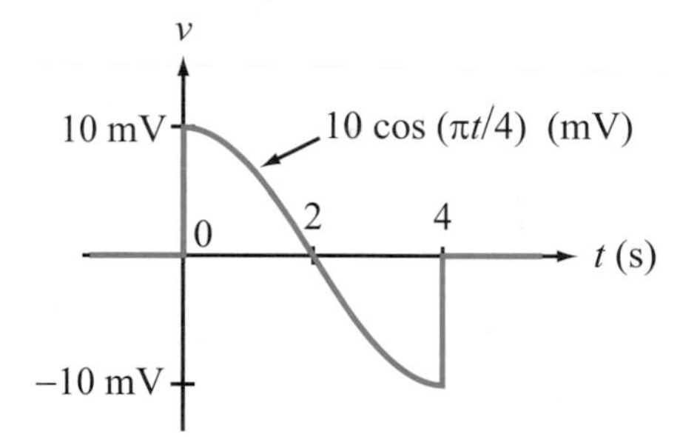

Figure P5.27: Voltage waveform for Problem 5.27.

5.28 For the circuit in Fig. P5.28, determine the voltage across C and the currents through L_1 and L_2 under dc conditions.

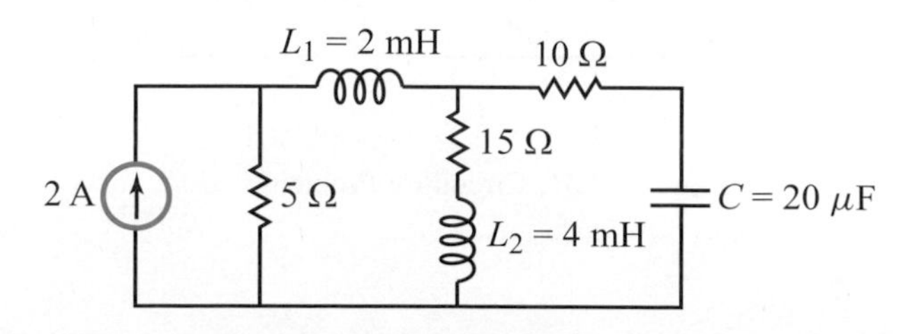

Figure P5.28: Circuit for Problem 5.28.

*5.29 For the circuit in Fig. P5.29, determine the voltages across C_1 and C_2 and the currents through L_1 and L_2 under dc conditions.

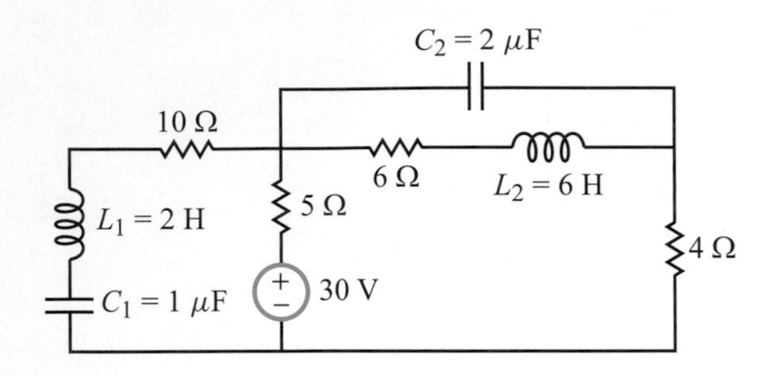

Figure P5.29: Circuit for Problem 5.29.

5.30 All elements in Fig. P5.30 are 10-mH inductors. Determine L_{eq}.

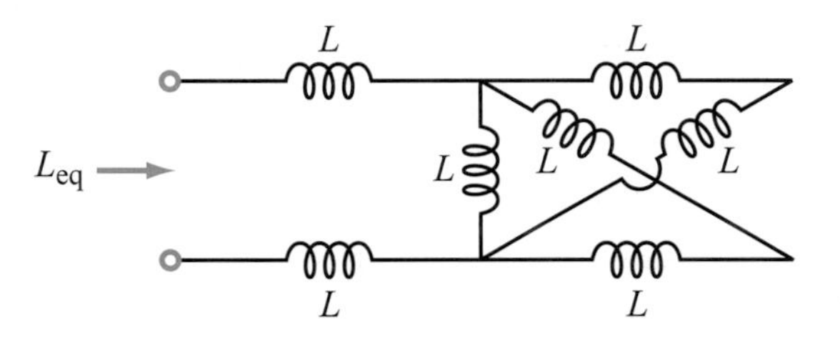

Figure P5.30: Circuit for Problem 5.30.

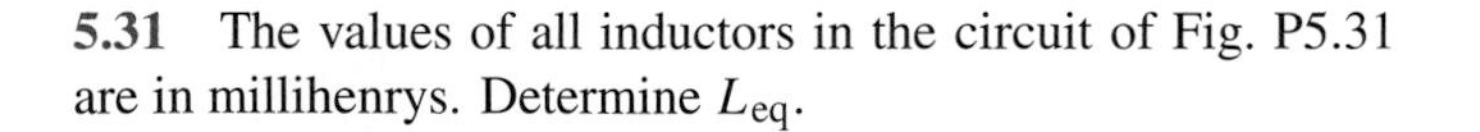

5.31 The values of all inductors in the circuit of Fig. P5.31 are in millihenrys. Determine L_{eq}.

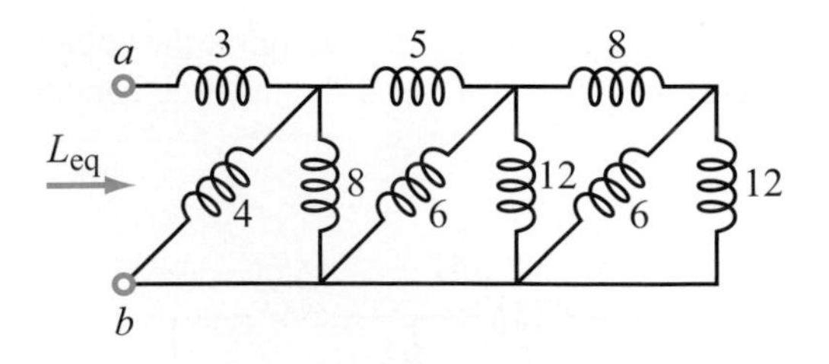

Figure P5.31: Circuit for Problem 5.31.

5.32 Determine L_{eq} at terminals (a, b) in the circuit of Fig. P5.32. All inductor values are in millihenrys.

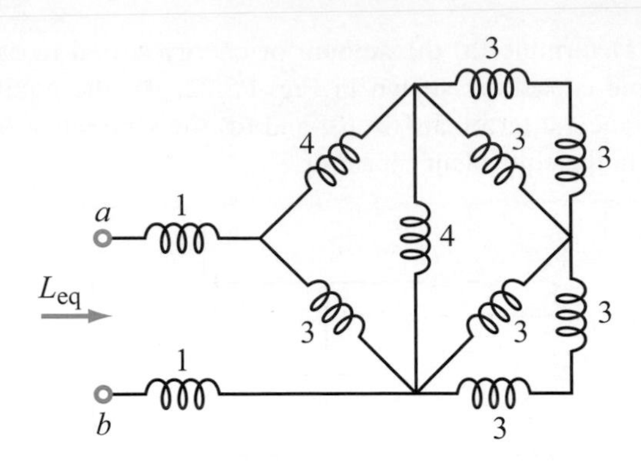

Figure P5.32: Circuit for Problem 5.32.

Section 5-4: Response of the RC Circuit

5.33 After having been in position 1 for a long time, the switch in the circuit of Fig. P5.33 was moved to position 2 at $t = 0$. Given that $V_0 = 12$ V, $R_1 = 30$ kΩ, $R_2 = 120$ kΩ, $R_3 = 60$ kΩ, and $C = 100\ \mu$F, determine:

(a) $i_C(0^-)$ and $v_C(0^-)$

(b) $i_C(0)$ and $v_C(0)$

*(c) $i_C(\infty)$ and $v_C(\infty)$

(d) $v_C(t)$ for $t \geq 0$

(e) $i_C(t)$ for $t \geq 0$

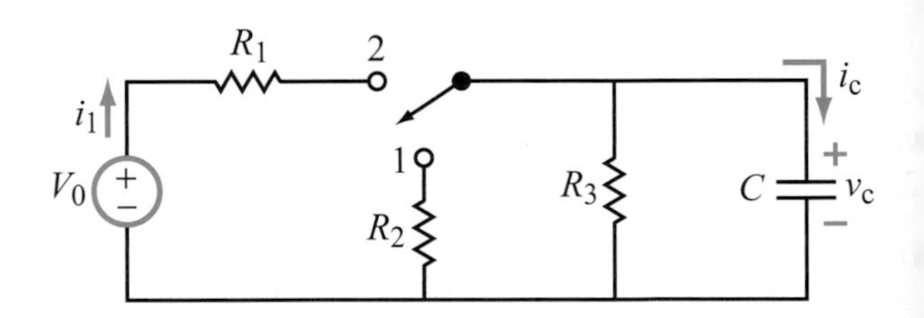

Figure P5.33: Circuit for Problems 5.33 and 5.34.

5.34 Repeat Problem 5.33, but with the switch having bee[n] in position 2 for a long time, and then moved to position 1 [a]t $t = 0$.

5.35 The circuit in Fig. P5.35 contains two switches, bo[th] of which had been open for a long time before $t = 0$. Switch 1 closes at $t = 0$, and switch 2 follows suit at $t = 5$ Determine and plot $v_C(t)$ for $t \geq 0$ given that $V_0 = 24$ $R_1 = R_2 = 16$ kΩ, and $C = 250\ \mu$F. Assume $v_C(0) = 0$.

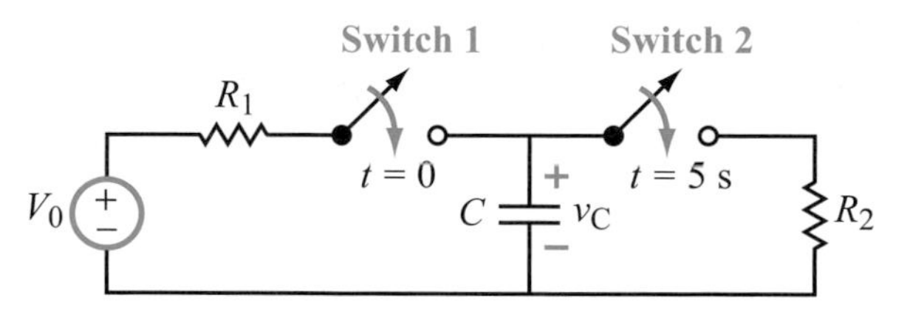

Figure P5.35: Circuit for Problem 5.35.

5.36 The circuit in Fig. P5.36 was in steady state until the switch was moved from terminal 1 to terminal 2 at $t = 0$. Determine $v(t)$ for $t \geq 0$ given that $I_0 = 21$ mA, $R_1 = 2$ kΩ, $R_2 = 3$ kΩ, $R_3 = 4$ kΩ, and $C = 50$ μF.

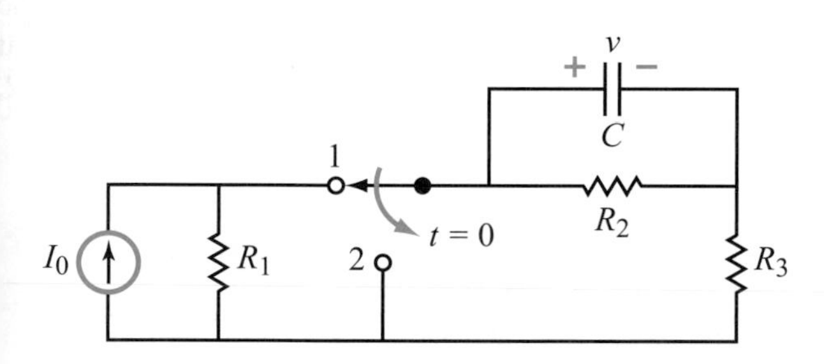

Figure P5.36: Circuit for Problem 5.36.

5.37 Prior to $t = 0$, capacitor C_1 in the circuit of Fig. P5.37 was uncharged. For $I_0 = 5$ mA, $R_1 = 2$ kΩ, $R_2 = 50$ kΩ, $C_1 = 3$ μ F, and $C_2 = 6$ μ F, determine:

(a) The equivalent circuit involving the capacitors for $t \geq 0$. Specify $v_1(0)$ and $v_2(0)$.

(b) $i(t)$ for $t \geq 0$.

(c) $v_1(t)$ and $v_2(t)$ for $t \geq 0$.

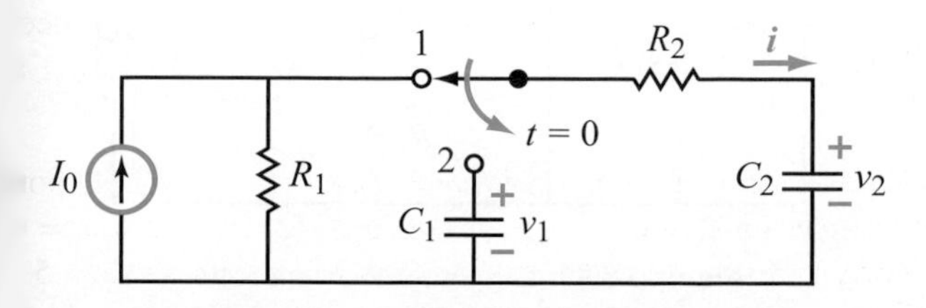

Figure P5.37: Circuit for Problem 5.37.

5.38 The switch in the circuit of Fig. P5.38 had been closed for a long time before it was opened at $t = 0$. Given that $V_s = 10$ V, $R_1 = 20$ kΩ, $R_2 = 100$ kΩ, $C_1 = 6$ μF, and $C_2 = 12$ μF, determine $i(t)$ for $t \geq 0$.

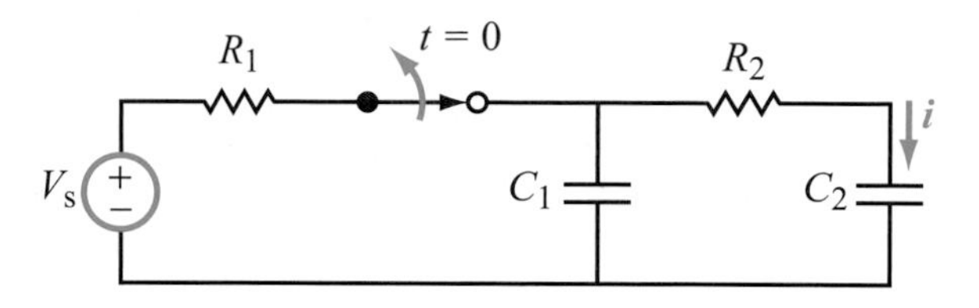

Figure P5.38: Circuit for Problem 5.38.

*__5.39__ The switch in the circuit of Fig. P5.39 had been in position 1 for a long time until it was moved to position 2 at $t = 0$. Determine $v(t)$ for $t \geq 0$, given that $I_0 = 6$ mA, $V_0 = 18$ V, $R_1 = R_2 = 4$ kΩ, and $C = 200$ μF.

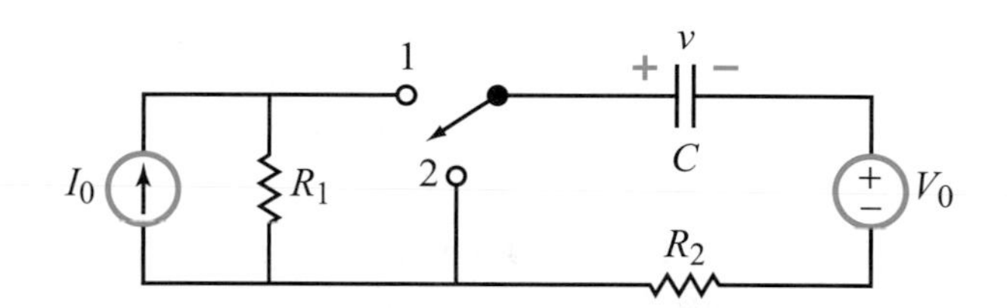

Figure P5.39: Circuit for Problems 5.39 and 5.40.

5.40 Repeat Problem 5.39, but reverse the switching sequence. [Switch starts in position 2 and is moved to position 1 at $t = 0$.]

5.41 Determine $i(t)$ for $t \geq 0$ where i is the current passing through R_3 in the circuit of Fig. P5.41. The element values are $v_s = 16$ V, $R_1 = R_2 = 2$ kΩ, $R_3 = 4$ kΩ, and $C = 25$ μF. Assume that the switch had been open for a long time prior to $t = 0$.

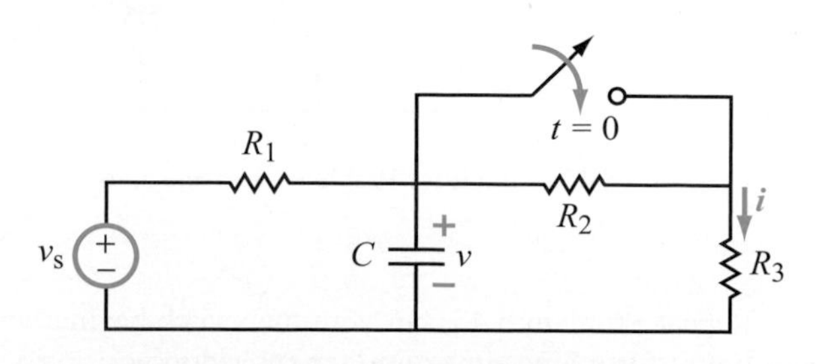

Figure P5.41: Circuit for Problems 5.41 to 5.43.

5.42 Repeat Problem 5.41, but start with the switch being closed prior to $t = 0$ and then opened at $t = 0$.

*__5.43__ Consider the circuit in Fig. P5.41, but without the switch. If the source v_s represents a 12-V, 100-ms-long rectangular pulse that starts at $t = 0$ and the element values are $R_1 = 6$ kΩ, $R_2 = 2$ kΩ, $R_3 = 4$ kΩ, and $C = 15$ μF, determine the voltage response $v(t)$ for $t \geq 0$.

5.44 Given that in Fig. P5.44, $I_1 = 4$ mA, $I_2 = 6$ mA, $R_1 = 3$ kΩ, $R_2 = 6$ kΩ, and $C = 0.2$ mF, determine $v(t)$. Assume the switch was connected to terminal 1 for a long time before it was moved to terminal 2.

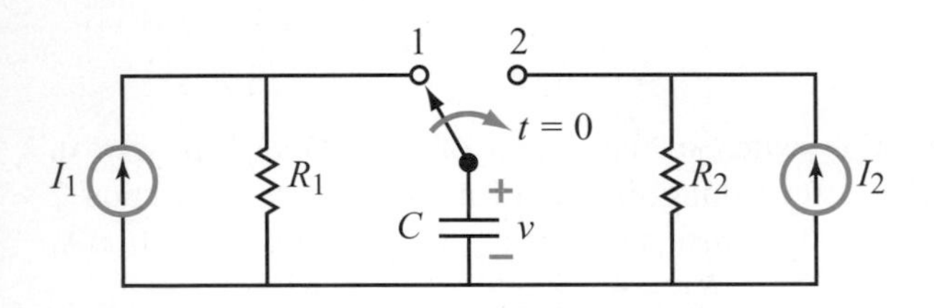

Figure P5.44: Circuit for Problem 5.44.

Section 5-5: Response of the RL Circuit

5.45 After having been in position 1 for a long time, the switch in the circuit of Fig. P5.45 was moved to position 2 at $t = 0$. Given that $V_0 = 12$ V, $R_1 = 30$ Ω, $R_2 = 120$ Ω, $R_3 = 60$ Ω, and $L = 0.2$ H, determine:

(a) $i_L(0^-)$ and $v_L(0^-)$

(b) $i_L(0)$ and $v_L(0)$

(c) $i_L(\infty)$ and $v_L(\infty)$

(d) $i_L(t)$ for $t \geq 0$

(e) $v_L(t)$ for $t \geq 0$

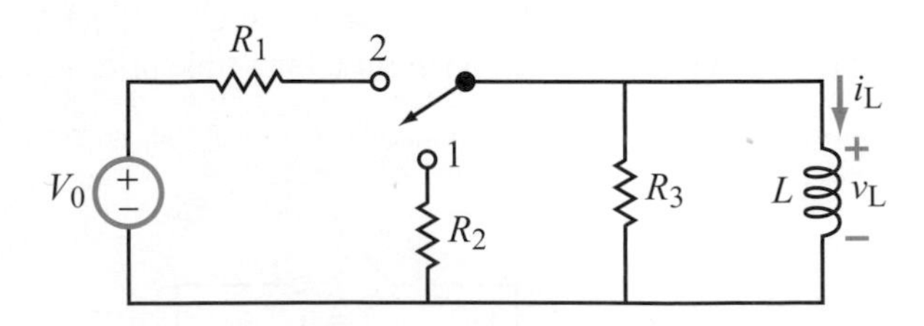

Figure P5.45: Circuit for Problems 5.45 and 5.46.

5.46 Repeat Problem 5.45, but with the switch having been in position 2 for a long time and then moved to position 1 at $t = 0$.

5.47 Determine $i(t)$ for $t \geq 0$ given that the circuit in Fig. P5.47 had been in steady state for a long time prior to $t = 0$. Also, $I_0 = 5$ A, $R_1 = 2$ Ω, $R_2 = 10$ Ω, $R_3 = 3$ Ω, $R_4 = 7$ Ω, and $L = 0.15$ H.

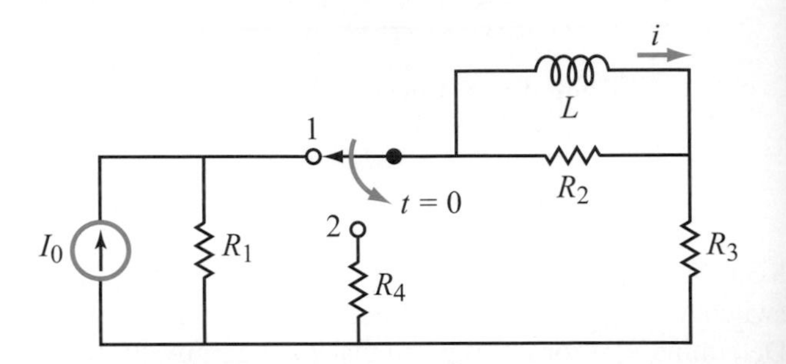

Figure P5.47: Circuit for Problem 5.47.

5.48 For the circuit in Fig. P5.48, determine $i_L(t)$ and plot it as a function of t for $t \geq 0$. The element values are $I_0 = 4$ A, $R_1 = 6$ Ω, $R_2 = 12$ Ω, and $L = 2$ H. Assume that $i_L = 0$ before $t = 0$.

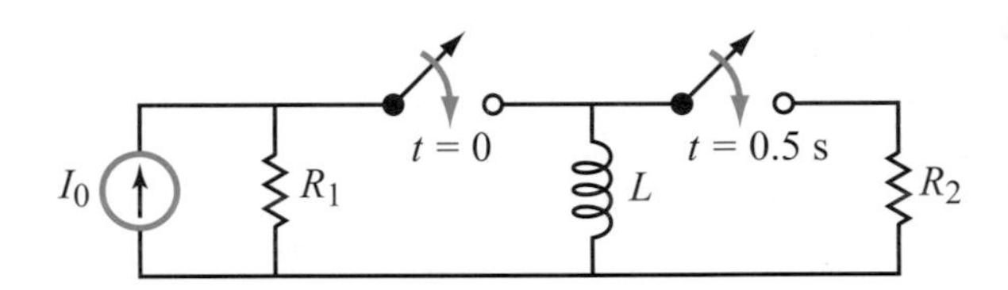

Figure P5.48: Circuit for Problem 5.48.

*__5.49__ After having been in position 1 for a long time, the switch in the circuit of Fig. P5.49 was moved to position 2 at $t = 0$. Determine $i_1(t)$ and $i_2(t)$ for $t \geq 0$, given that $I_0 = 6$ mA, $R_0 = 12$ Ω, $R_1 = 10$ Ω, $R_2 = 40$ Ω, $L_1 = 1$ H and $L_2 = 2$ H.

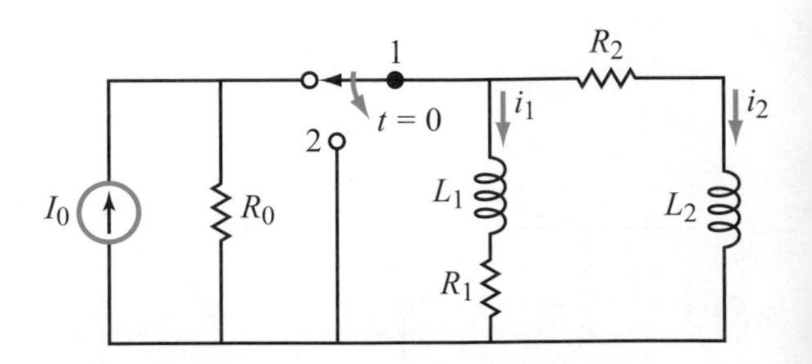

Figure P5.49: Circuit for Problem 5.49.

5.50 In the circuit of Fig. P5.50(a), $R_1 = R_2 = 20\ \Omega$, $R_3 = 10\ \Omega$, and $L = 2.5$ H. Determine $i(t)$ for $t \geq 0$ given that $v_s(t)$ is the step function described in Fig. P5.50(b).

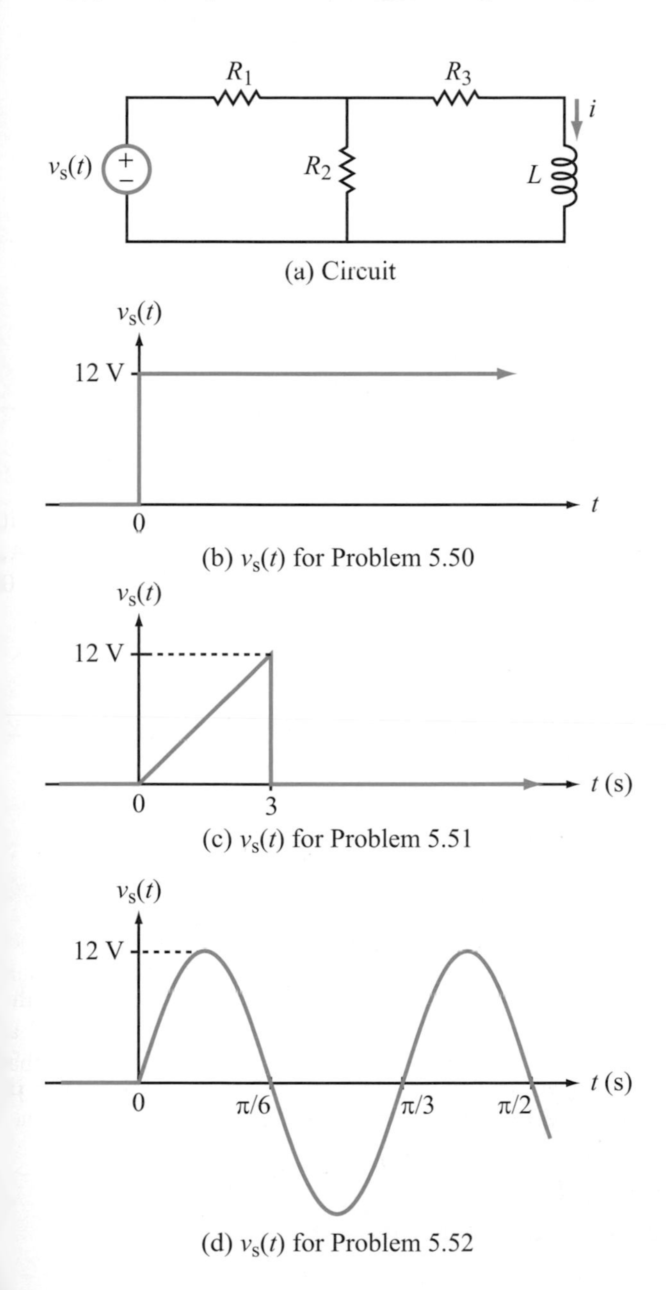

Figure P5.50: Circuit and excitation voltages for Problems 5.50 to 5.52.

5.51 Repeat Problem 5.50 for the triangular-source excitation given in Fig. P5.50(c).

$$\textit{Hint}: \quad \int x e^{ax}\, dx = \frac{e^{ax}}{a^2}\,(ax - 1).$$

5.52 Repeat Problem 5.50 for the sinusoidal-source excitation $v_s(t) = 12 \sin 6t$ V displayed in Fig. P5.50(d).

$$\textit{Hint}: \quad \int e^{ax} \sin bx\, dx = e^{ax}\,\frac{[a \sin bx - b \cos(bx)]}{a^2 + b^2}.$$

***5.53** The switch in the circuit of Fig. P5.53 was moved from position 1 to position 2 at $t = 0$, after it had been in position 1 for a long time. If $L = 80$ mH, determine $i(t)$ for $t \geq 0$.

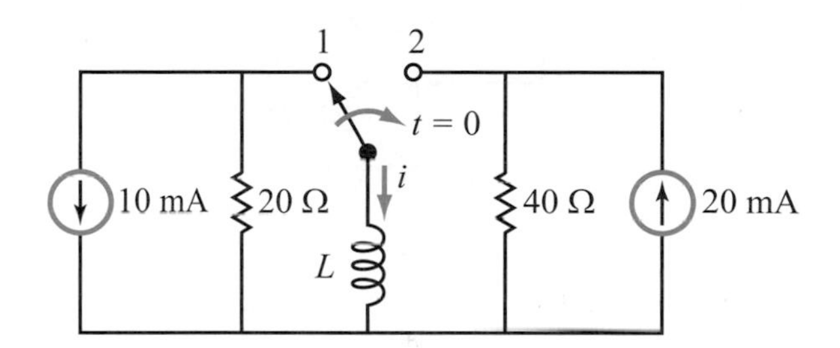

Figure P5.53: Circuit for Problems 5.53 and 5.54.

5.54 Repeat Problem 5.53, but with the switch having been in position 2 and then moved to position 1 at $t = 0$.

5.55 Determine $i(t)$ for $t \geq 0$ due to the rectangular-pulse excitation in the circuit of Fig. P5.55.

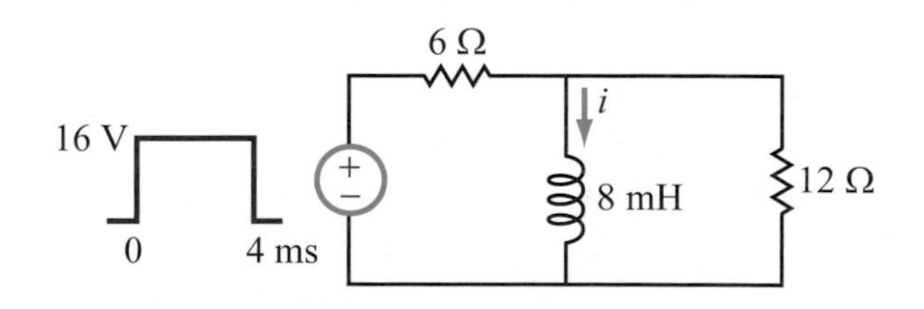

Figure P5.55: Circuit for Problem 5.55.

Section 5-6: RC Op-Amp Circuits

5.56 The input-voltage waveform shown in Fig. P5.56(a) is applied to the circuit in Fig. P5.56(b). Determine and plot the corresponding $v_{out}(t)$.

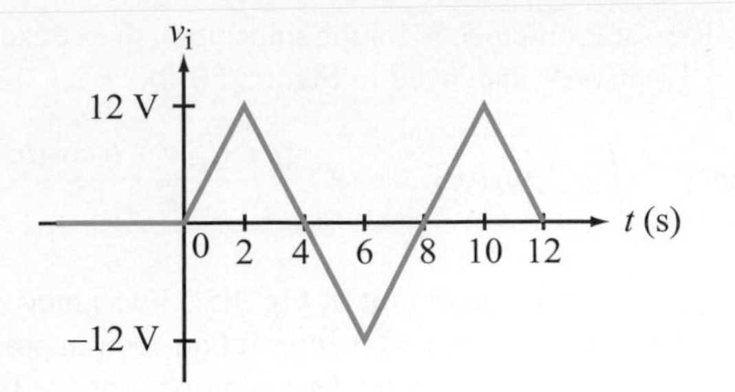

(a) Waveform of $v_i(t)$

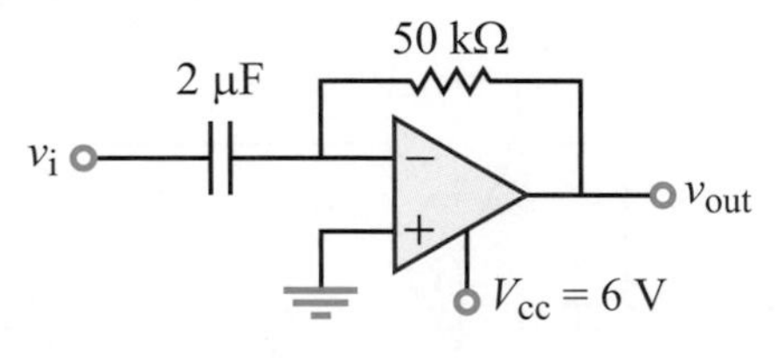

(b) Op-amp circuit

Figure P5.56: Waveform and circuit for Problem 5.56.

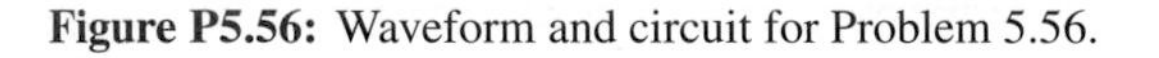

5.57 Relate v_{out} to v_i in the circuit of Fig. P5.57.

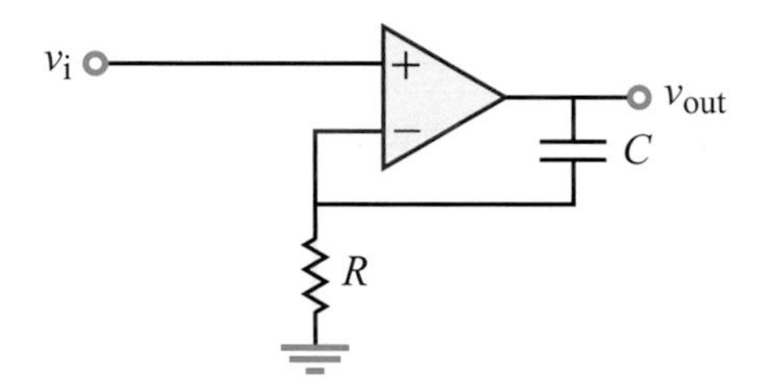

Figure P5.57: Circuit for Problem 5.57.

5.58 Develop the relationship between the output voltage v_{out} and the input voltage v_i for the circuit in Fig. P5.58.

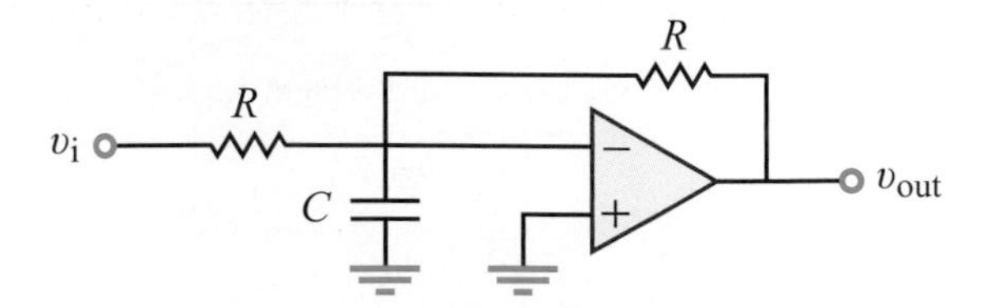

Figure P5.58: Circuit for Problem 5.58.

5.59 Relate v_{out} to v_i in the circuit of Fig. P5.59. Assume $v_C = 0$ at $t = 0$.

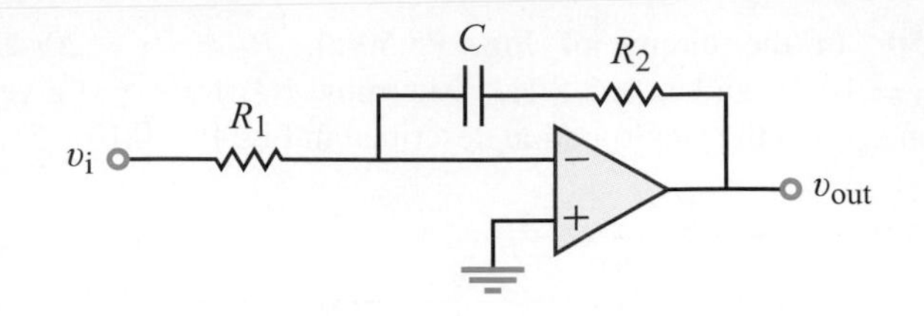

Figure P5.59: Circuit for Problem 5.59.

5.60 Relate $i_{out}(t)$ to $v_i(t)$ in the circuit of Fig. P5.60. Evaluate it for $v_C(0) = 3$ V, $R = 10$ kΩ, $C = 50$ μF, and $v_i(t) = 9\,u(t)$ V.

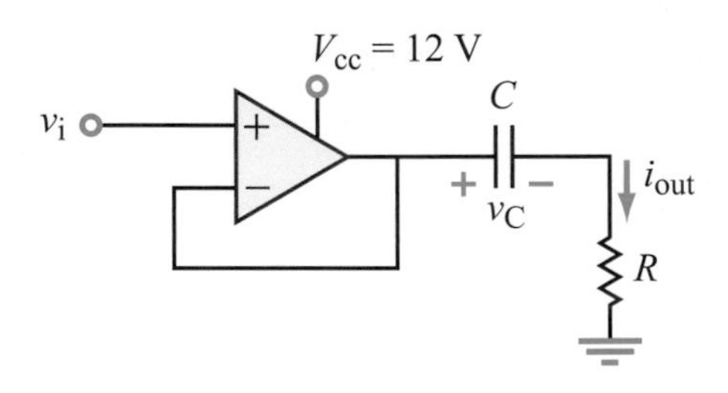

Figure P5.60: Circuit for Problem 5.60.

5.61 Design a circuit that can perform the following relationship between its output and input voltages:

$$v_{out} = -100 \int_0^t v_i \, dt,$$

with $v_{out}(0) = 0$ at $t = 0$. You are limited to one op-amp, one capacitor that does not exceed 0.1 F, and any resistor(s) of your choice.

5.62 The two-stage op-amp circuit in Fig. P5.62 is driven by an input step voltage given by $v_i(t) = 10\,u(t)$ mV. If $V_{cc} = 10$ V for both op amps and the two capacitors had no charge prior to $t = 0$, determine and plot:

(a) $v_{out_1}(t)$ for $t \geq 0$

(b) $v_{out_2}(t)$ for $t \geq 0$

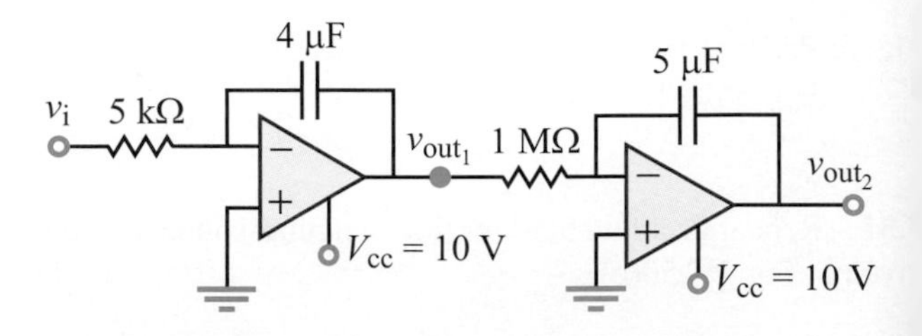

Figure P5.62: Two-stage op-amp circuit of Problem 5.62.

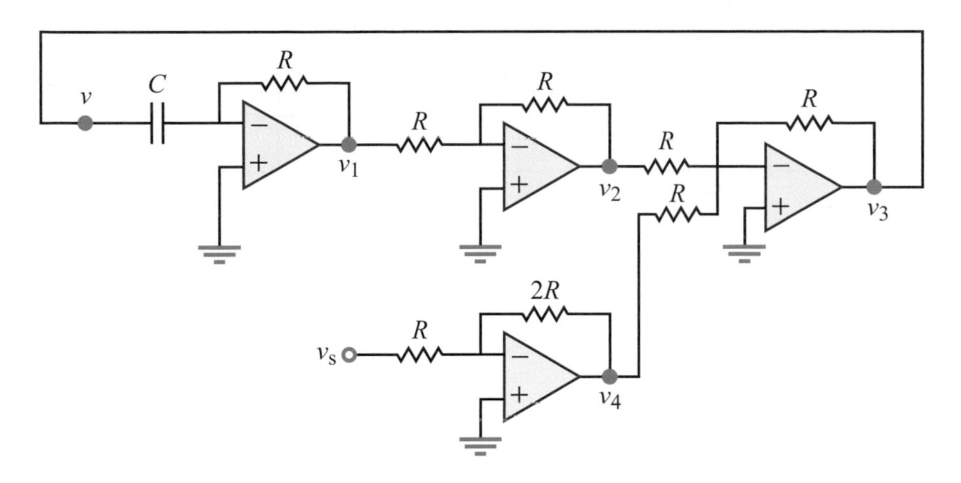

Figure P5.65: Circuit for Problem 5.65.

5.63 Design a single op-amp circuit that can perform the operation

$$v_{\text{out}} = -\int_0^t (5v_1 + 2v_2 + v_3)\,dt.$$

5.64 Design a single op-amp circuit that can perform the operation

$$i_{\text{out}} = -\int_0^t \left(\frac{v_1}{100} + \frac{v_2}{200} + \frac{v_3}{400}\right) dt.$$

5.65 Show that the op-amp circuit in Fig. P5.65 (in which $R = 10$ kΩ and $C = 20\ \mu$F) simulates the differential equation

$$\frac{dv}{dt} + 5v = 10v_s.$$

5.66 Design an op-amp circuit that can solve the differential equation

$$\frac{dv}{dt} + 0.2v = 4\sin 10t$$

with $v(0) = 0$. *Hint*: See Problem 5.65.

Sections 5-7 and 5-8: Parasitic Capacitance and Multisim Analysis

*__5.67__ In real transistors, both the MOSFET gain g and parasitic capacitances C_D^n and C_D^p depend on the size of the transistor. Assuming the functional relationships

$$g = 10^6 W, \qquad \textit{and} \qquad C_\text{D}^\text{n} = C_\text{D}^\text{p} = (2.5 \times 10^3)W^2,$$

where W is the transistor width in meters, how small should W be in order for the CMOS inverter to have a fall time of 1 ns? [The width of modern digital MOSFETs varies between 40 nm and 4 μm.]

5.68 Draw and simulate in Multisim the circuit in Fig. 5-41(a) of Example 5-15. Using the Grapher tool, plot $v_{\text{out}}(t)$ for $t \geq 0$.

5.69 Consider the circuit in Fig. P5.69. Switch S1 begins in the closed position and opens at $t = 0$. Switch S2 begins in the open position and toggles between the open and closed positions every 250 ps. Model this circuit in Multisim and plot v_0 and v_1 as a function of time until all nodes are discharged below 1 mV.

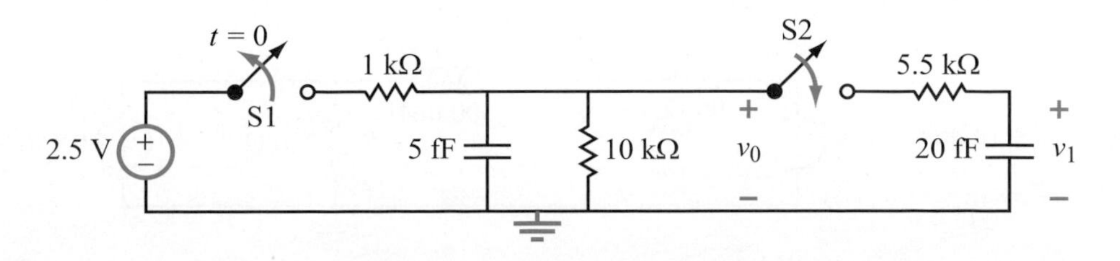

Figure P5.69: Circuit for Problem 5.69.

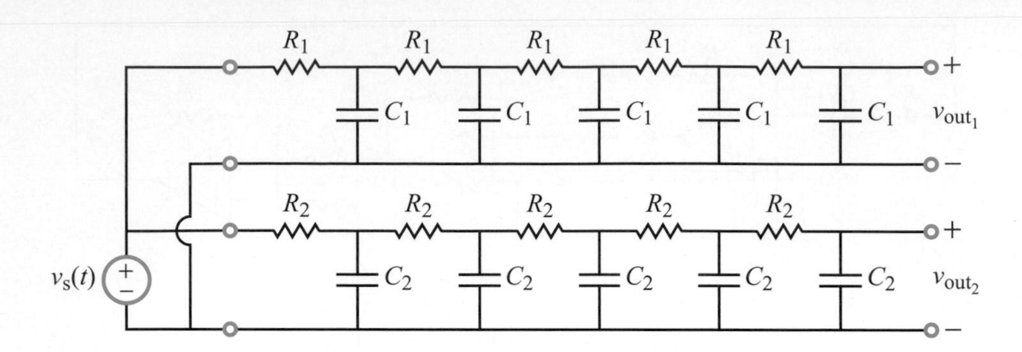

Figure P5.70: Circuit for Problem 5.70 with $R_1 = R_2 = 10\ \Omega$, $C_1 = 7$ pF, and $C_2 = 5$ pF.

5.70 A step voltage source $v_s(t)$ sends a signal down two transmission lines simultaneously (Fig. P5.70). In Multisim, the step voltage may be modeled as a 1-V square wave with a period of 10 ns. Model the circuit in Multisim and answer the following questions:

(a) If a detector registers a signal when the output voltage reaches 0.75 V, which signal arrives first?

(b) By how much?

Hint: When using cursors in the Grapher View, select a trace, then right-click on a cursor and select Set Y_Value, and enter 750 m. This will give you the exact time point at which that trace equals 0.75 V.

5.71 Consider the delta topology in Fig. P5.71. Use Multisim to generate response curves for v_a, v_b, and v_c. Apply Transient Analysis with TSTOP $= 3 \times 10^{-10}$ s.

5.72 Use Multisim to generate a plot for current $i(t)$ in the circuit in Fig. P5.72 from 0 to 15 ms.

5.73 Construct the integrator circuit shown in Fig. P5.73, using a 3-terminal virtual op amp. Print the output corresponding to each of the following input signals:

(a) $v_{in}(t)$ is a 0-to-1-V square wave with a period of 1 ms and a 50 percent duty cycle. Plot the output from 0 to 10 ms.

(b) $v_{in}(t) = -0.2t$ V. Plot the output from 0 to 50 ms.

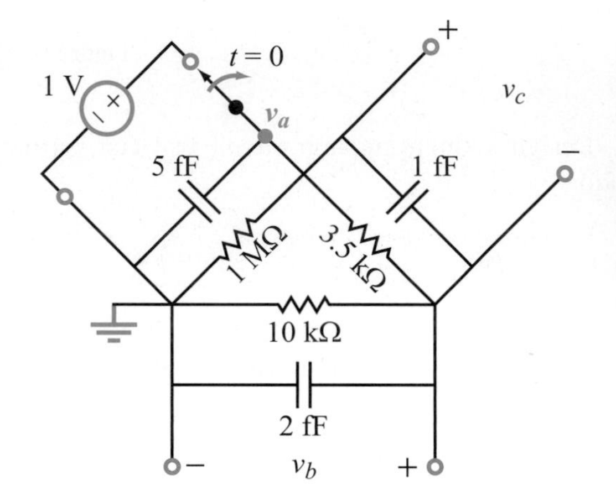

Figure P5.71: Circuit for Problem 5.71.

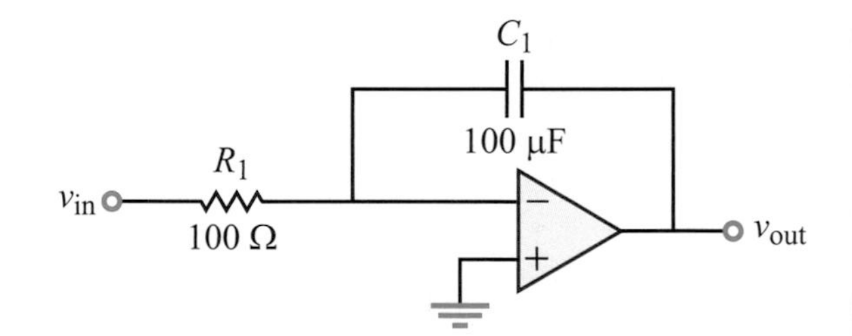

Figure P5.73: Circuit for Problem 5.73.

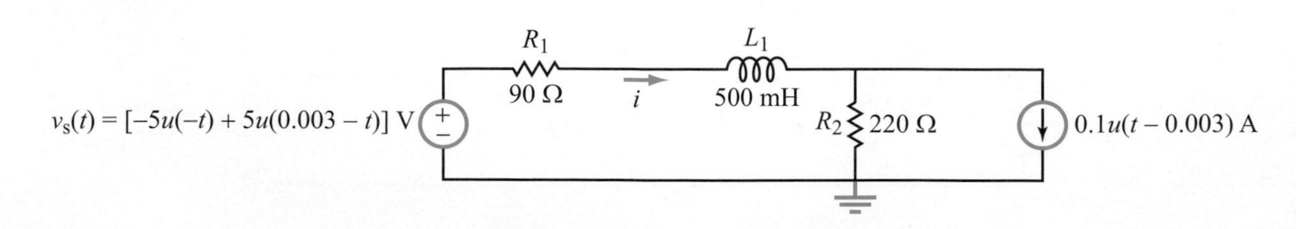

Figure P5.72: Circuit for Problem 5.72.

C H A P T E R

6

RLC Circuits

Chapter Contents

Objectives

Upon learning the material presented in this chapter, you should be able to:

1. Determine the initial and final conditions of a second-order circuit under dc conditions.
2. Develop the second-order differential equation for a series RLC circuit, a parallel RLC circuit, or any second-order circuit, and then solve it to determine the transient and steady-state components of its overall solution.
3. Establish whether the response of a second-order circuit is overdamped, critically damped, or underdamped.
4. Solve second-order op-amp circuits.
5. Describe the operation of a microelectromechanical accelerometer.
6. Apply Multisim to simulate the response of any second-order circuit.

Overview

The currents and voltages of the first-order RC and RL circuits we examined in the preceding chapter were characterized by first-order differential equations. A key provision of a ***first-order circuit*** is that it is reducible to a single series or parallel circuit containing a single capacitor or a single inductor, in addition to sources and resistors. If a circuit contains two capacitors, as in Fig. 6-1(a), and if the circuit architecture is such that it is not possible to replace the two capacitors with a single equivalent, then the circuit does not qualify as a first-order circuit. As will become evident later in this chapter, the two-capacitor circuit is a ***second-order circuit*** characterized by a second-order differential equation. The same is true for the two-inductor circuit in part (b) and for the series and parallel RLC circuits shown in parts (c) and (d) of the same figure.

A second-order circuit may contain any combination of two energy-storage elements (two capacitors, two inductors, or one of each), provided like elements cannot be replaced with a single-element equivalent.

A circuit containing more than two energy-storage elements may still qualify as a second-order circuit, but only if its elements can be combined together so that the equivalent circuit contains two, and only two, energy-storage elements.

The goal of the present chapter is to develop the mathematical tools needed for analyzing second-order circuits when excited by dc sources (in the form of step functions) or by rectangular pulses. The method of solution, which involves solving a second-order differential equation in the time domain, is relatively straightforward to apply. We limit our consideration to second-order circuits, because for third- and higher-order circuits, the differential-equation solution approach is rather cumbersome. In such cases, more robust methods are called for, such as the Laplace transform technique introduced in Chapter 10.

6-1 Initial and Final Conditions

The general form of the solution of the differential equation associated with a second-order circuit always includes a number of unknown constants. To determine the values of these constants, we usually match the solution to known values of the voltage or current under consideration. For a circuit where the solution we seek is for the time period following a sudden change (such as when a SPST switch is closed or opened or when a SPDT switch is moved from one terminal to another) we can analyze the circuit conditions at the beginning and at the end of that time period and then apply the results to match

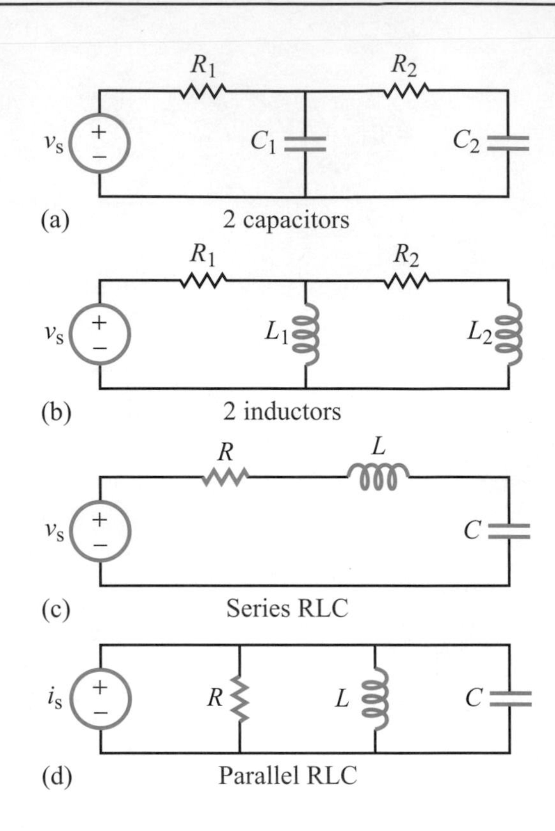

Figure 6-1: Examples of second-order circuits.

the solution of the differential equation. We call the process ***invoking initial and final conditions***.

Analyzing a circuit in its initial and final states relies on the following fundamental properties:

(a) The voltage υ_{C} across a capacitor cannot change instantaneously, and neither can the current i_{L} through an inductor.

If we denote:

$t = 0^-$: instant before sudden change,
$t = 0$: instant after sudden change,

then

$$\upsilon_{\mathrm{C}}(0) = \upsilon_{\mathrm{C}}(0^-) \tag{6.1a}$$

and

$$i_{\mathrm{L}}(0) = i_{\mathrm{L}}(0^-). \tag{6.1b}$$

We note that these constraints do not extend to the current i_{C} through the capacitor or to the voltage υ_{L} across an inductor, both of which *are* allowed to change instantaneously.

Table 6-1: Modes of behavior of a capacitor C and an inductor L. For each circuit, A represents the configuration before the sudden change (at $t = 0^-$) and B the configuration after the change.

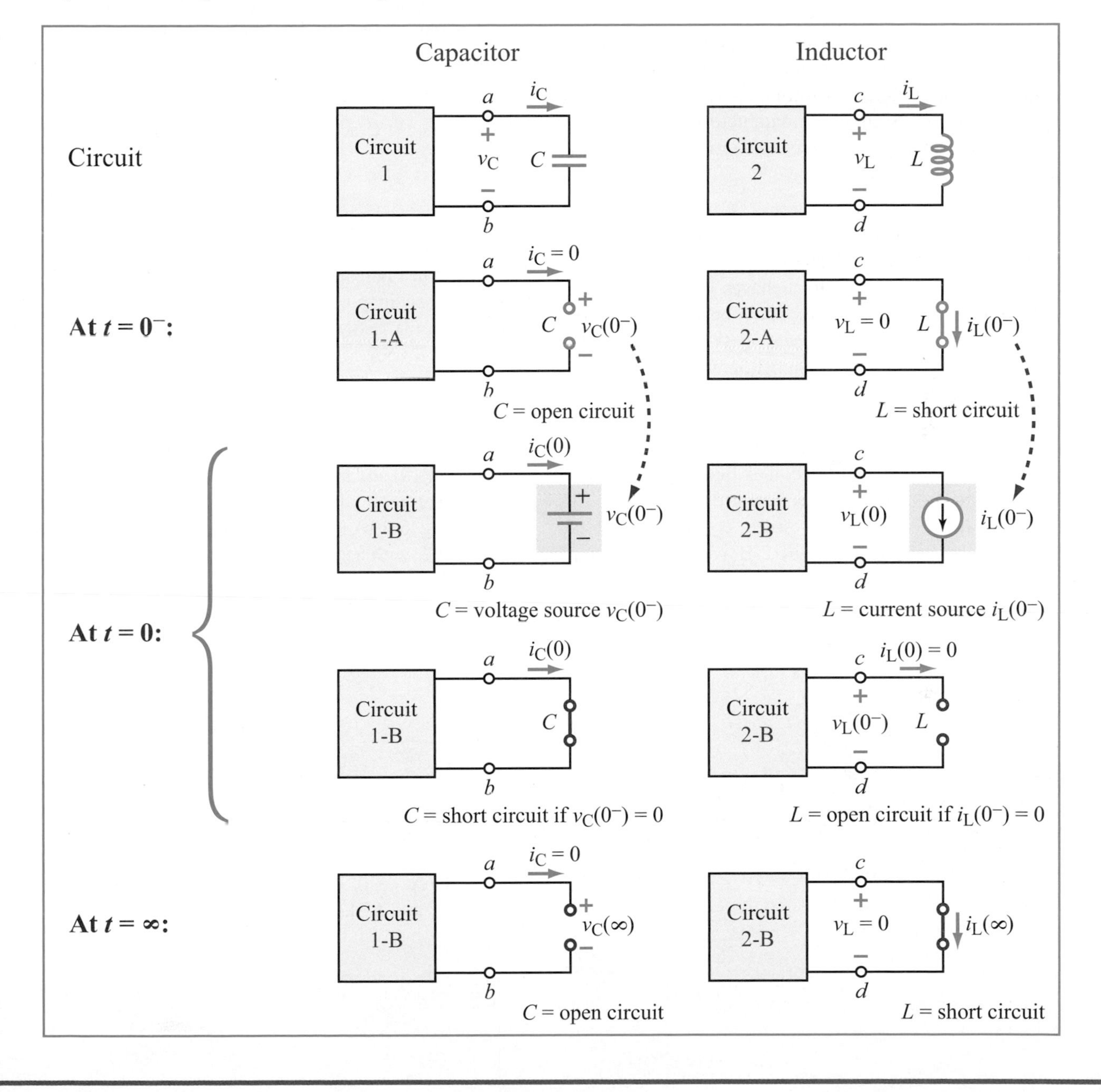

(b) In circuits containing dc sources, the steady-state condition of the circuit (after all transients have died out) is such that no currents flow through capacitors and no voltages exist across inductors, allowing us to represent capacitors as open circuits and inductors as short circuits.

Table 6-1 outlines the modes of behavior of a capacitor C contained in a circuit labeled Circuit 1 and an inductor L contained in Circuit 2 at three distinct instants in time. It is assumed that sudden changes in the circuits occur at $t = 0$, and that by $t = 0^-$, the circuits already had reached a steady state. To distinguish between the capacitor circuit configuration during $t \leq 0^-$ (before the sudden change) and its configuration at $t \geq 0$, we label them 1-A and 1-B, respectively. Similar designations apply to the inductor circuits.

- At $t = 0^-$, C acts like an open circuit and L like a short circuit. The capacitor voltage $v_C(0^-)$ is equal to the open-circuit voltage at terminals (a, b) for circuit configuration 1-A. Similarly, the inductor current $i_L(0^-)$ is the short-circuit current between terminals (c, d) in circuit configuration 2-A.

- At $t = 0$ (which is synonymous with $t = 0^+$), the circuits change to configurations 1-B for the capacitor and 2-B for the inductor. The capacitor behaves like a dc voltage source of magnitude $v_C(0) = v_C(0^-)$, because v_C cannot change instantaneously. Whereas $i_C(0^-)$ was equal to zero at $t = 0^-$ (because the capacitor was like an open circuit before the sudden change), now $i_C(0)$ is free to assume whatever value is dictated by Circuit 1-B—subject to C being represented by a voltage source $v_C(0^-)$. If the value of $v_C(0^-)$ is 0, C becomes like a short circuit. Similarly, the inductor behaves like a current source of magnitude $i_L(0^-)$, and the voltage $v_L(0)$ across it is determined by analyzing circuit 2-B with the inductor replaced with current source $i_L(0^-)$. If $i_L(0^-) = 0$, the inductor becomes like an open circuit.

- At $t > 0$ and before reaching a steady state at $t = \infty$, C and L behave "normally"; v_C, i_C, v_L, and i_L may all vary with time.

- At $t = \infty$, Circuits 1-B and 2-B are in a steady state, in which case C acts like an open circuit and L like a short circuit.

Example 6-1: Initial and Final Values

The circuit in Fig. 6-2(a) contains a dc source V_s and a switch that had been in position 1 for a long time prior to $t = 0$. Determine: (a) $v_C(0)$ and $i_L(0)$, (b) $i_C(0)$ and $v_L(0)$, and (c) $v_C(\infty)$ and $i_L(\infty)$.

Solution:

(a) To determine $v_C(0)$ and $i_L(0)$, we analyze the circuit configuration at $t = 0^-$ (before moving the switch), whereas to determine $i_C(0)$ and $v_L(0)$, we analyze the circuit configuration at $t = 0$. At $t = 0^-$, the circuit is equivalent to the arrangement shown in Fig. 6-2(b), in which C has been replaced with an open circuit and L with a short circuit. Because the circuit contains

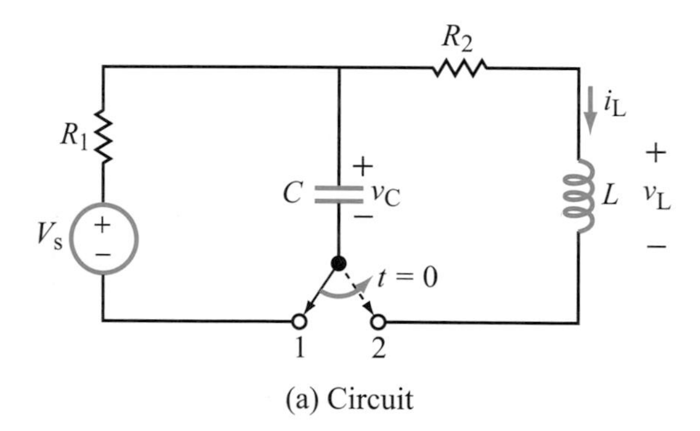

(a) Circuit

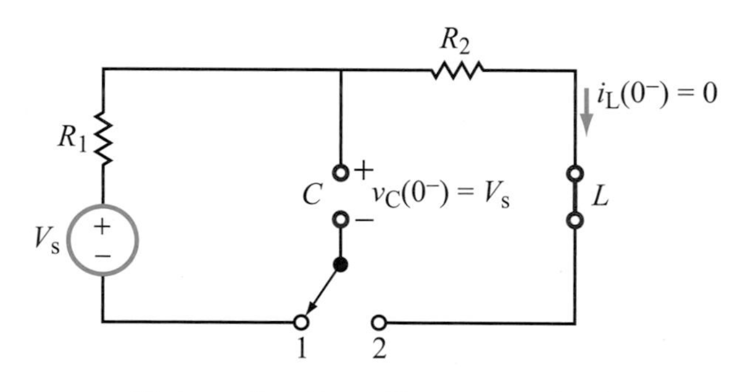

(b) At $t = 0^-$, C acts like an open circuit and L like a short circuit

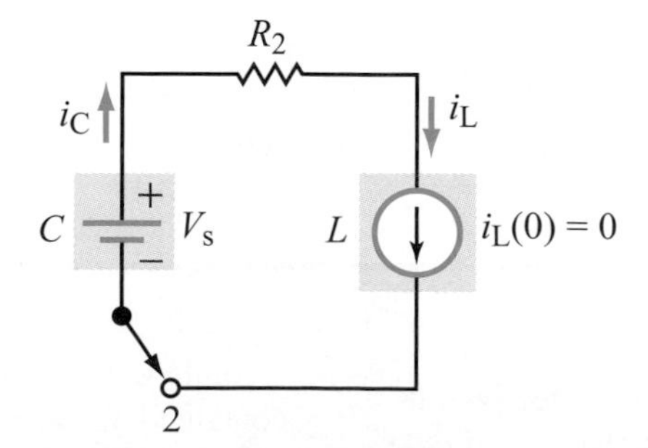

(c) At $t = 0$, C acts like a voltage source and L like a current source with zero current

Figure 6-2: Circuit of Example 6-1.

no closed loops, no current flows anywhere in the circuit. With no voltage drop across R_1, it follows that

$$v_C(0^-) = V_s.$$

Also,

$$i_L(0^-) = 0.$$

Time-continuity of v_C and i_L mandates that after moving the switch to terminal 2 we have

$$v_C(0) = v_C(0^-) = V_s$$

and

$$i_L(0) = i_L(0^-) = 0.$$

(b) The circuit in Fig. 6-2(c) depicts the state of the circuit at $t = 0$ (after moving the switch). The capacitor behaves like a dc voltage source of magnitude V_s, and the inductor behaves like a dc current source with zero current, which is equivalent to an open circuit. Even though (in general) there is no requirement disallowing a sudden change in i_C, in this case $i_C = i_L$ and $i_L(0) = 0$.

Consequently,

$$i_C(0) = 0.$$

With no voltage drop across R_2, the voltage across the inductor is

$$v_L(0) = v_C(0) = V_s.$$

(c) The analysis for v_C and i_L as $t \to \infty$ is totally straightforward; with no active sources remaining in the circuit that contains L and C, all of the energy that may have been stored in L and C will have dissipated completely by $t = \infty$, rendering the circuit inactive. Hence,

$$v_C(\infty) = 0,$$

and

$$i_L(\infty) = 0.$$

Example 6-2: Initial and Final Conditions

The circuit in Fig. 6-3(a) contains a dc voltage source and a step-function current source. The element values are $V_0 = 24$ V, $I_0 = 4$ A, $R_1 = 2\ \Omega$, $R_2 = 4\ \Omega$, $R_3 = 6\ \Omega$, $L = 0.2$ H, and $C = 8$ mF. Determine: (a) $v_C(0)$ and $i_L(0)$, (b) $i_C(0)$ and $v_L(0)$, and (c) $v_C(\infty)$ and $i_L(\infty)$.

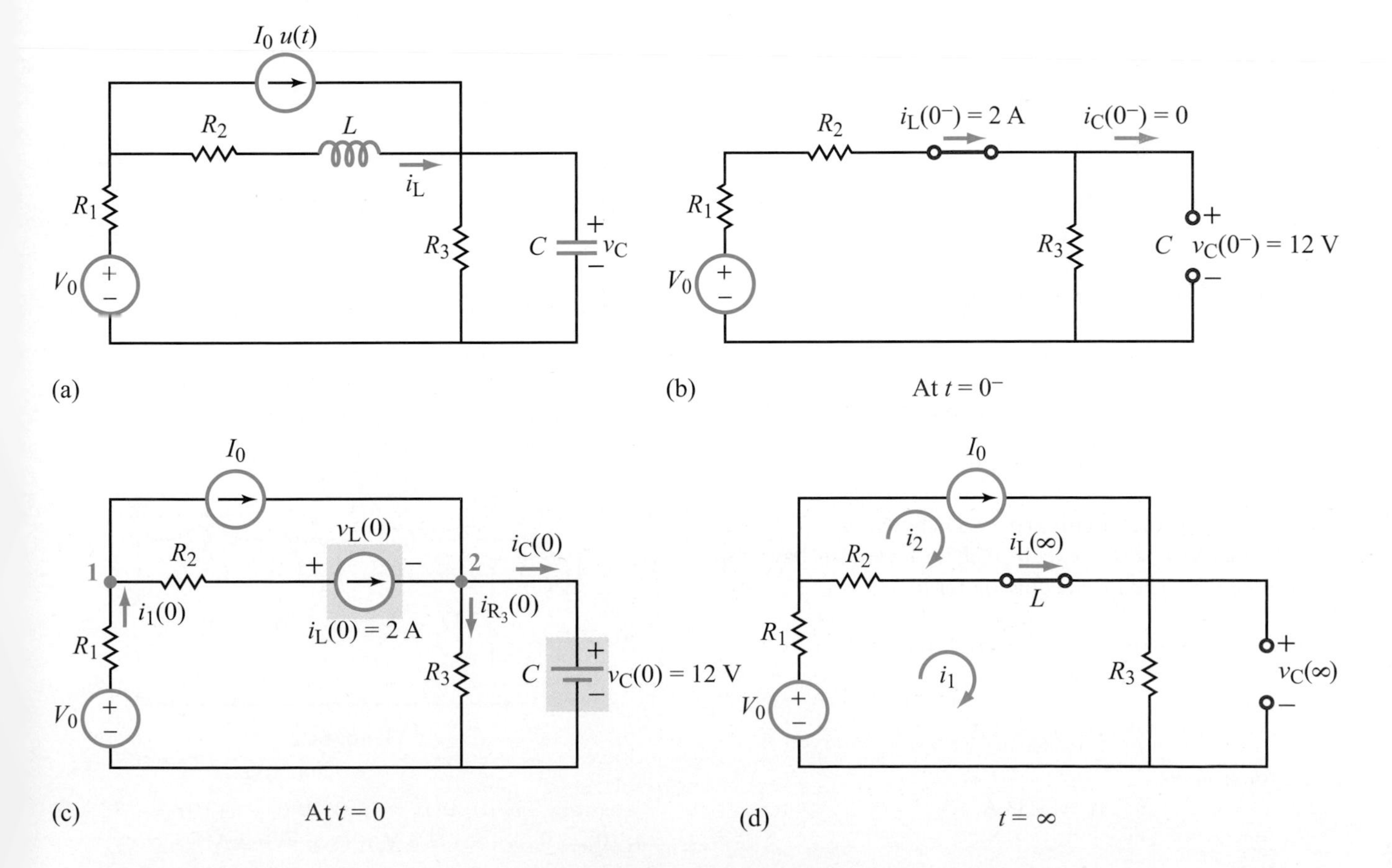

Figure 6-3: Circuit for Example 6-2.

Solution:

(a) To find v_C and i_L at $t = 0$, we have to determine their values at $t = 0^-$, and then invoke the requirement that neither the voltage across a capacitor nor the current through an inductor can change in zero time. The state of the circuit at $t = 0^-$ is shown in Fig. 6-3(b), wherein the inductor has been replaced with a short circuit, the capacitor replaced with an open circuit, and the current source is absent altogether. Since $i_C(0^-) = 0$,

$$i_L(0^-) = \frac{V_0}{R_1 + R_2 + R_3} = 2 \text{ A},$$

and

$$v_C(0^-) = i_L(0^-)\, R_3 = 12 \text{ V}.$$

Hence,

$$i_L(0) = i_L(0^-) = 2 \text{ A},$$

and

$$v_C(0) = v_C(0^-) = 12 \text{ V}.$$

(b) At $t = 0$, the state of the circuit is as shown in Fig. 6-3(c). Since

$$v_{R_3}(0) = v_C(0) = 12 \text{ V},$$

it follows that

$$i_{R_3}(0) = \frac{12}{6} = 2 \text{ A}.$$

Application of KCL at node 2 leads to

$$\begin{aligned} i_C(0) &= I_0 + i_L(0) - i_{R_3}(0) \\ &= 4 + 2 - 2 \\ &= 4 \text{ A}. \end{aligned}$$

Next, we need to determine $v_L(0)$. At node 1,

$$i_1(0) = I_0 + i_L(0) = 4 + 2 = 6 \text{ A}.$$

By applying KVL around the lower left loop, we find that

$$v_L(0) = -8 \text{ V}.$$

(c) The state of the circuit at $t = \infty$ shown in Fig. 6-3(d) resembles that at $t = 0^-$, except that now we also have the current source I_0. The mesh equation for loop 1 is

$$-V_0 + R_1 i_1 + R_2(i_1 - i_2) + R_3 i_1 = 0,$$

and for loop 2,

$$i_2 = I_0 = 4 \text{ A}.$$

Solving for i_1 gives

$$i_1 = 3.33 \text{ A},$$

which leads to

$$i_L(\infty) = i_1 - I_0 = 3.33 - 4 = -0.67 \text{ A}$$

and

$$v_C(\infty) = i_1 R_3 = 3.33 \times 6 = 20 \text{ V}.$$

Review Question 6-1: Determination of initial circuit conditions after a sudden change relies on two fundamental properties of capacitors and inductors. What are they?

Review Question 6-2: Under dc steady-state conditions, does a capacitor resemble an open circuit or a short circuit? What does an inductor resemble?

Review Question 6-3: What role do initial and final values play in the solution of a second-order circuit?

Exercise 6-1: For the circuit in Fig. E6.1, determine $v_C(0)$, $i_L(0)$, $v_L(0)$, $i_C(0)$, $v_C(\infty)$, and $i_L(\infty)$.

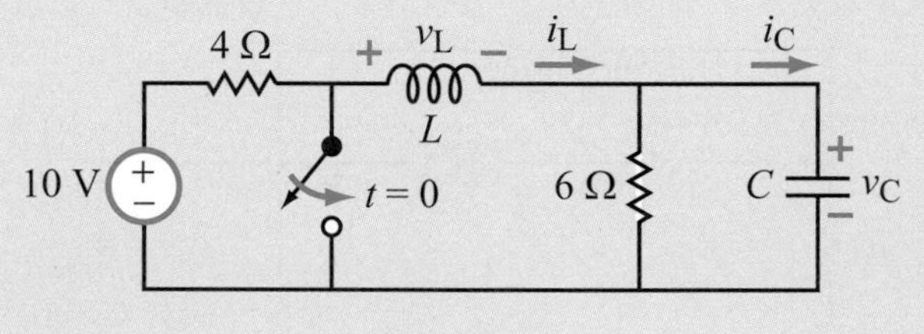

Figure E6.1

Answer: $v_C(0) = 6$ V, $i_L(0) = 1$ A, $v_L(0) = -6$ V, $i_C(0) = 0$, $v_C(\infty) = 0$, $i_L(\infty) = 0$. (See)

Exercise 6-2: For the circuit in Fig. E6.2, determine $v_C(0)$, $i_L(0)$, $v_L(0)$, $i_C(0)$, $v_C(\infty)$, and $i_L(\infty)$.

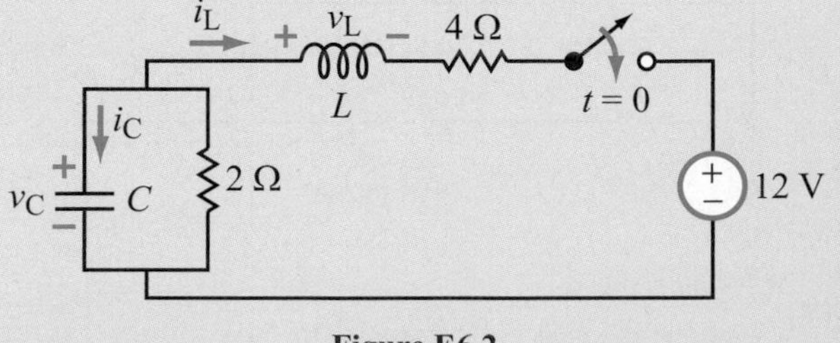

Figure E6.2

Answer: $v_C(0) = 0$, $i_L(0) = 0$, $v_L(0) = -12$ V, $i_C(0) = 0$, $v_C(\infty) = 4$ V, $i_L(\infty) = -2$ A. (See)

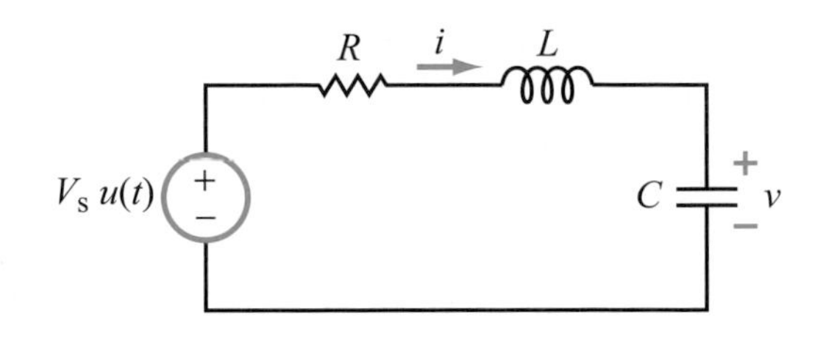

Figure 6-4: Series RLC circuit excited by a step-function input voltage.

6-2 General Solution

Consider the series RLC circuit shown in Fig. 6-4. The source υ_s is given by

$$\upsilon_s = V_s\, u(t), \tag{6.2}$$

where V_s is a constant and $u(t)$ is the unit step function. The KVL loop equation is given by

$$Ri + L\,\frac{di}{dt} + \upsilon = V_s \qquad (\text{for } t \geq 0). \tag{6.3}$$

By incorporating the relation

$$i = C\,\frac{d\upsilon}{dt} \tag{6.4}$$

and rearranging terms, Eq. (6.3) becomes

$$\frac{d^2\upsilon}{dt^2} + \frac{R}{L}\,\frac{d\upsilon}{dt} + \frac{1}{LC}\,\upsilon = \frac{V_s}{LC}. \tag{6.5}$$

For convenience, we rewrite Eq. (6.5) in the abbreviated form as

$$\upsilon'' + a\upsilon' + b\upsilon = c, \tag{6.6}$$

where

$$a = \frac{R}{L}, \qquad b = \frac{1}{LC}, \qquad \text{and} \qquad c = \frac{V_s}{LC}. \tag{6.7}$$

The second-order differential equation given by Eq. (6.6) is specific to the capacitor voltage of the series RLC circuit of Fig. 6-4, but the form of the equation is equally applicable to any current or voltage in any second-order circuit. The same is true for the general form of the solution of the differential equation.

6-2.1 Solution Outline

Given a second-order circuit, the procedure for obtaining analytical expressions for its currents and voltages consists of three basic steps.

Step 1: Apply KVL and KCL to the nodes, branches and loops and then combine them together to end up with a second-order differential equation in only one variable for each voltage or current of interest. The differential equation given by Eq. (6.6) is one such example.

Step 2: Solve the differential equation.

The general solution of a second-order differential equation of the form given by Eq. (6.6) consists of two components:

$$\upsilon(t) = \upsilon_p(t) + \upsilon_h(t), \tag{6.8}$$

where $\upsilon_p(t)$ is the ***particular solution*** (so named because it is related to the particular forcing function on the right-hand side of Eq. (6.6)) and $\upsilon_h(t)$ is the ***homogeneous solution*** (a solution of the source-free equation) given by

$$\upsilon_h'' + a\upsilon_h' + b\upsilon_h = 0 \quad (\text{source-free}). \tag{6.9}$$

Because a and b are not functions of the source exciting the circuit, the solution $\upsilon_h(t)$ of Eq. (6.9) is independent of whether the voltage source is dc, ac, or some other function of time. Accordingly, the homogeneous solution of the differential equation often is called the ***natural response*** of the circuit. In contrast, the particular solution $\upsilon_p(t)$ is strongly dependent on the forcing function, and in fact, its functional form usually mimics the forcing function.

Instead of using the terms *particular* and *homogeneous*—which come from the language of mathematics—we prefer to use ***steady state*** and ***transient***, respectively, because the terminology is more meaningful in the language of electronics. Accordingly, we will rewrite Eq. (6.8) using the subscripts "ss" and "t" as

$$\upsilon(t) = \upsilon_{ss}(t) + \upsilon_t(t). \tag{6.10}$$

The methods of solution for arriving at explicit expressions for $\upsilon_{ss}(t)$ and $\upsilon_t(t)$ are outlined in Sections 6-3 and 6-4. Usually the total expression for $\upsilon(t)$ may contain up to three unknown constants.

Step 3: Invoke initial and final conditions to determine the values of the unknown constants.

6-2.2 Role of the Excitation Source

In the differential equation given by Eq. (6.6), the forcing function (represented by the constant c on the right-hand side of the equation) is directly proportional to the voltage source V_s. The transient response $\upsilon_t(t)$ is independent of c—and hence of V_s—because $\upsilon_t(t)$ is the solution of Eq. (6.6) with c set equal to zero. The steady-state response, on the other hand, is a solution of the complete equation and takes on a functional form similar to that of the forcing function. In the present case, because V_s is a dc source, c is a constant, and so is v_{ss}. In fact, for a dc circuit,

$$v_{ss} = v(\infty) = \text{constant.} \tag{6.11}$$

This is because (as we shall see later in Section 6-3) $v_t(t)$ always decays down to zero as $t \to \infty$, leaving behind v_{ss} as the only term remaining in Eq. (6.10). Furthermore, since $v(\infty)$ is a constant, it follows that $v'(\infty) = v''(\infty) = 0$. Consequently, at $t = \infty$, the first two terms of Eq. (6.6) become zero, and it reduces to

$$v(\infty) = \frac{c}{b}. \tag{6.12}$$

In this chapter, our examination will be limited to second-order circuits excited by dc sources—usually in combination with SPST and SPDT switches. We will develop the circuit response to the step-function waveform, and we will synthesize the response to a rectangular pulse—just as we had done previously in Chapter 5 in connection with first-order circuits.

But what about circuits driven by ac sources, or sources with other waveforms? Can we apply the same method of solution to those circuits?

The simple answer is: Yes, we can, but we usually do not. Analyzing a second-order circuit by solving its second-order differential equation is a perfectly viable approach, but when the circuit contains time-varying sources, we usually employ other mathematical techniques because they are easier to apply.

Periodic Waveforms

Among periodic excitations, the sinusoidally time-varying waveform is by far the most widely used in electronic circuits and systems. Hence, ac circuits are accorded two entire chapters (7 and 8) in this book. The preferred technique for analyzing ac circuits is the phasor-domain method; transformation from the time domain to the phasor domain converts a differential equation into a linear equation, thereby facilitating its solution considerably.

Non-sinusoidal periodic waveforms—such as the square wave, pulse train, and triangle wave—can be synthesized in the form of a Fourier series composed of many sinusoidal harmonics, as demonstrated in Chapter 11. The linearity property of electric circuits allows us to use the circuit response to a sine-wave excitation to synthesize the response to any other periodic waveform.

Nonperiodic Waveforms

Even though the differential equation given by Eq. (6.6) can be solved in the time domain for any nonperiodic excitation, including the ramp and exponential waveforms, we usually resort to the use of more robust solution techniques in such situations. The Laplace Transform technique of Chapter 10 is particularly noteworthy in that regard.

Review Question 6-4: The natural response of a second-order circuit does not depend on the voltage and current sources in the circuit. Why not?

Review Question 6-5: What does the steady-state response of a circuit have in common with the voltage or current source exciting the circuit?

6-3 Natural Response of the Series RLC Circuit

In reaction to a sudden change in its configuration, a circuit may respond by exhibiting voltages and currents that contain transient components and steady-state components. If the sudden change results in a circuit containing no active sources, the circuit response will consist of a transient component only (also called the ***natural response***). An example of such a situation is the circuit shown in Fig. 6-5(a) for which the action of the switch at $t = 0$ leads to the source-free circuit shown in Fig. 6-5(d). To determine the capacitor voltage $v(t)$ for $t \geq 0$, we follow the three-step procedure outlined in Section 6-2.1.

Step 1: Obtain Differential Equation for $v(t)$

The second-order differential equation for $v(t)$ of the series RLC circuit was derived earlier in connection with the circuit in Fig. 6-4. For the source-free case corresponding to the state of the circuit shown in Fig. 6-5(d), the differential equation is given by Eq. (6.6) with $c = 0$, namely

$$v'' + av' + bv = 0, \tag{6.13}$$

with

$$a = \frac{R}{L} \quad \text{and} \quad b = \frac{1}{LC}. \tag{6.14}$$

The source-free equation given by Eq. (6.13) applies specifically to the transient component $v_t(t)$, but because the circuit in Fig. 6-5(d) contains no active sources, the steady-state component v_{ss} is zero, and $v(t)$ becomes synonymous with $v_t(t)$.

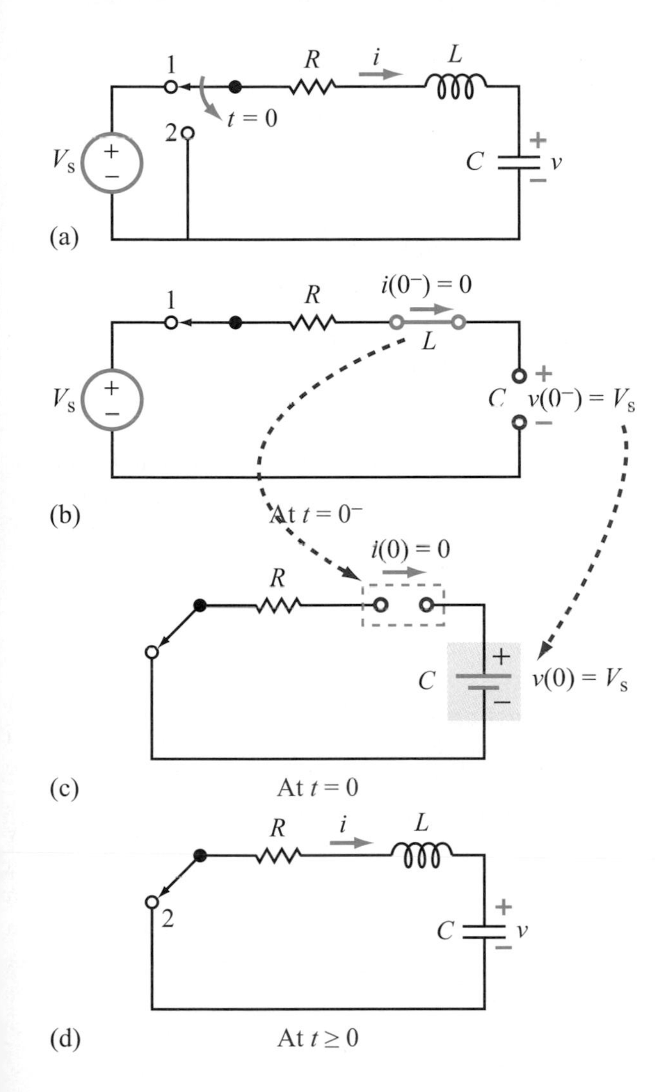

Figure 6-5: Series RLC circuit with SPDT switch. As the circuit condition progresses from $t = 0^-$ to $t = 0$, v across the capacitor remains unchanged as V_s, and i through the inductor remains unchanged at zero current, which is equivalent to an open circuit.

Step 2: Solve Differential Equation

When differentiated, the exponential function e^{st} replicates itself (within a multiplying factor), so it is often offered as a candidate solution when solving homogeneous differential equations. Thus, we will assume that

$$v(t) = Ae^{st}, \tag{6.15}$$

where A and s are constants to be determined later. To ascertain that Eq. (6.15) is indeed a viable solution of Eq. (6.13), we insert the expression for $v(t)$ and its first and second derivatives in Eq. (6.13). The result is

$$s^2 Ae^{st} + asAe^{st} + bAe^{st} = 0, \tag{6.16}$$

which simplifies to

$$s^2 + as + b = 0. \tag{6.17}$$

Hence, the proposed solution given by Eq. (6.15) is indeed an acceptable solution, so long as Eq. (6.17) is satisfied.

The quadratic equation given by Eq. (6.17) is known as the ***characteristic equation*** of the differential equation. It has two roots:

$$s_1 = -\frac{a}{2} + \sqrt{\left(\frac{a}{2}\right)^2 - b}, \tag{6.18a}$$

$$s_2 = -\frac{a}{2} - \sqrt{\left(\frac{a}{2}\right)^2 - b}. \tag{6.18b}$$

Since the values of a and b are governed by the values of only the passive components in the circuit, so are the values of s_1 and s_2. Strictly speaking, the unit of s_1 and s_2 is 1/second, but it is customary to add the dimensionless neper to the units of quantities that appear in exponential functions. Hence, s_1 and s_2 are measured in ***nepers/second*** (Np/s).

The existence of two distinct roots implies that Eq. (6.13) has two viable solutions, one in terms of $e^{s_1 t}$ and another in terms of $e^{s_2 t}$. Hence, we should generalize the form of our solution to

$$v(t) = A_1 e^{s_1 t} + A_2 e^{s_2 t} \qquad (\text{for } t \geq 0), \tag{6.19}$$

where A_1 and A_2 are to be determined next.

Step 3: Invoke Initial and Final Conditions

To determine A_1 and A_2, we need to know the values of $v(0)$ and $v'(0)$, where $v'(0)$ is the time derivative of $v(t)$ evaluated at $t = 0$. Since $i(t) = C\, dv/dt = C\, v'(t)$, the second requirement is equivalent to needing to know $i(0)$.

> As a general rule, to find v_C across a capacitor or a current i_L though an inductor at $t = 0$, we do so by analyzing the circuit configuration as it existed at $t = 0^-$ and then carrying their values forward to $t = 0$. For the other pair, namely $i_C(0)$ and $v_L(0)$, we need to perform the analysis on the circuit configuration specific to $t = 0$ (after the sudden change and after carrying forward the values of v_C and i_L).

By time $t = 0^-$, the RLC circuit had been in steady state for a long time. Hence, C acts like an open circuit (Fig. 6-5(b)), allowing no current to flow through the loop. Consequently,

$$v(0^-) = V_s \qquad \text{and} \qquad i(0^-) = 0. \tag{6.20}$$

Since neither the voltage across a capacitor nor the current through an inductor can change instantaneously, it follows that at $t = 0$ (Fig. 6-5(c))

$$v(0) = v(0^-) = V_s \tag{6.21a}$$

and

$$i(0) = i(0^-) = 0. \tag{6.21b}$$

Moreover, because $i = C\,dv/dt$,

$$v'(0) = \left.\frac{dv}{dt}\right|_{t=0} = \frac{1}{C}\,i(0) = 0. \tag{6.22}$$

Matching Eq. (6.21a) to Eq. (6.19), when evaluated at $t = 0$, gives

$$v(0) = A_1 + A_2 = V_s. \tag{6.23}$$

Similarly, matching Eq. (6.22) to the derivative of Eq. (6.19), also at $t = 0$, leads to

$$v'(0) = (s_1 A_1 e^{s_1 t} + s_2 A_2 e^{s_2 t})\big|_{t=0} = 0, \tag{6.24}$$

which simplifies to

$$s_1 A_1 + s_2 A_2 = 0. \tag{6.25}$$

Simultaneous solution of Eqs. (6.23) and (6.25) gives

$$A_1 = \left(\frac{s_2}{s_2 - s_1}\right) V_s \tag{6.26a}$$

and

$$A_2 = -\left(\frac{s_1}{s_2 - s_1}\right) V_s. \tag{6.26b}$$

Inserting Eqs. (6.26a and b) in Eq. (6.19) provides the final expression for $v(t)$ as

$$v(t) = \frac{V_s}{s_2 - s_1}\,(s_2 e^{s_1 t} - s_1 e^{s_2 t}) \qquad (\text{for } t \geq 0). \tag{6.27}$$

The solution given by Eq. (6.27) is valid so long as the roots s_1 and s_2 are real and distinct. If s_1 and s_2 are complex or if $s_1 = s_2$, the proposed form of the solution would require certain modifications. To facilitate our discussion of such cases, we start by rewriting the expressions given by Eqs. (6.18a and b) for s_1 and s_2 in terms of a new pair of parameters, α and ω_0, as follows:

$$s_1 = -\frac{a}{2} + \sqrt{\left(\frac{a}{2}\right)^2 - b} = -\alpha + \sqrt{\alpha^2 - \omega_0^2}, \tag{6.28a}$$

$$s_2 = -\frac{a}{2} - \sqrt{\left(\frac{a}{2}\right)^2 - b} = -\alpha - \sqrt{\alpha^2 - \omega_0^2}. \tag{6.28b}$$

where

$$\alpha = \frac{a}{2} = \frac{R}{2L}, \tag{6.29a}$$

$$\omega_0 = \sqrt{b} = \frac{1}{\sqrt{LC}}. \tag{6.29b}$$

The parameter α is called the ***damping coefficient***, and it has the unit Np/s as do s_1 and s_2. For reasons that will become apparent later, ω_0 is called the ***resonant frequency***, and it is measured in radians per second (rad/s).

The quantity $(\alpha^2 - \omega_0^2)$ inside the square root in Eq. (6.28) plays a critically important role, because its value and its sign (+ or −) determine the character of the transient response of the circuit. The circuit exhibits markedly different responses depending on whether $\alpha > \omega_0$, $\alpha = \omega_0$, or $\alpha < \omega_0$. We will shortly examine these three conditions individually.

> **Review Question 6-6:** Can either a or b in Eq. (6.14) be negative? If not, can either s_1 or s_2 of Eq. (6.18) have a positive real value? To what conclusion would that lead you in terms of the behavior of the natural response as $t \to \infty$?
>
> **Review Question 6-7:** To determine $v_C(0)$ in a series RLC circuit, is it necessary to analyze the circuit at $t = 0^-$, or only at $t = 0$? Which circuit condition(s) are needed to determine $i_C(0)$?

6-3.1 Overdamped Response ($\alpha > \omega_0$)

For the series RLC circuit, the condition $\alpha > \omega_0$ corresponds to

$$R > 2\sqrt{\frac{L}{C}} \qquad (\text{overdamped}), \tag{6.30}$$

In that case, the quantity $(\alpha^2 - \omega_0^2) > 0$, but it is smaller than α. Consequently, s_1 and s_2 are negative, real, and distinct (different from one another), and Eq. (6.19) retains its form, namely

$$v(t) = A_1 e^{s_1 t} + A_2 e^{s_2 t} \qquad (\text{for } t \geq 0) \quad (\text{overdamped response}). \tag{6.31}$$

This is called an ***overdamped response***, a reference to the shape of the waveform of $v(t)$ when compared with the waveforms of the other two conditions. Illustrative examples of all three responses are shown in Fig. 6-6.

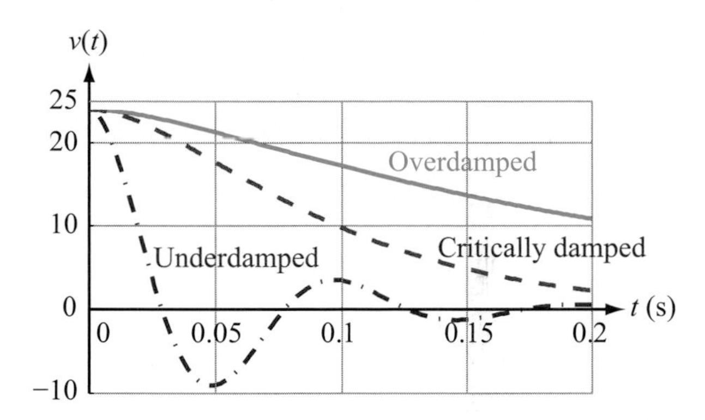

Figure 6-6: Voltage response of the series RLC circuit of Fig. 6-5(a) under overdamped, critically damped, and underdamped conditions.

Example 6-3: Overdamped Response

Determine $v(t)$ for the circuit in Fig. 6-5(a), given that $V_s = 24$ V, $R = 12\ \Omega$, $L = 0.3$ H, and $C = 0.01$ F.

Solution: The numerical values of α and ω_0 are

$$\alpha = \frac{R}{2L} = \frac{12}{2 \times 0.3} = 20 \text{ Np/s}$$

and

$$\omega_0 = \frac{1}{\sqrt{LC}} = \frac{1}{\sqrt{0.3 \times 10^{-2}}} = 18.26 \text{ rad/s}.$$

Since $\alpha > \omega_0$, the circuit will exhibit an overdamped response with roots given by

$$\begin{aligned} s_1 &= -\alpha + \sqrt{\alpha^2 - \omega_0^2} \\ &= -20 + \sqrt{(20)^2 - (18.26)^2} = -11.84 \text{ Np/s} \end{aligned}$$

and

$$\begin{aligned} s_2 &= -\alpha - \sqrt{\alpha^2 - \omega_0^2} \\ &= -20 - \sqrt{(20)^2 - (18.26)^2} = -28.16 \text{ Np/s}. \end{aligned}$$

The constants A_1 and A_2 associated with the overdamped response are given by Eq. (6.26) as

$$A_1 = \frac{s_2}{s_2 - s_1}\, V_s = \frac{-28.16}{-28.16 + 11.84} \times 24 = 41.4 \text{ V}$$

and

$$A_2 = \frac{-s_1}{s_2 - s_1}\, V_s = \frac{11.84}{-28.16 + 11.84} \times 24 = -17.4 \text{ V}.$$

Incorporating the values of s_1, s_2, A_1, and A_2 in Eq. (6.31) leads to

$$\begin{aligned} v(t) &= (A_1 e^{s_1 t} + A_2 e^{s_2 t})\, u(t) \\ &= (41.4 e^{-11.84t} - 17.4 e^{-28.16t}) \text{ V} \\ &\qquad \text{for } t \ge 0. \end{aligned}$$

A time plot of $v(t)$ is displayed in Fig. 6-7.

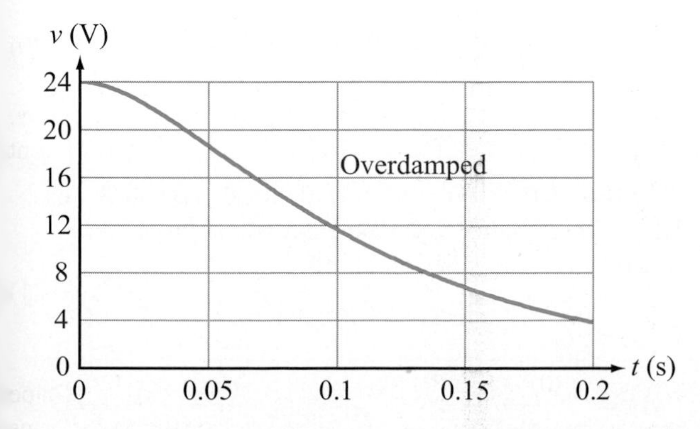

Figure 6-7: Overdamped response of the series RLC circuit of Example 6-3.

Example 6-4: Current-Source Excitation

Determine $v(t)$ for the circuit in Fig. 6-8(a), given that $I_s = 2$ A, $R_s = 10\ \Omega$, $R_1 = 1.81\ \Omega$, $R_2 = 0.2\ \Omega$, $L = 5$ mH, and $C = 5$ mF.

Solution: Figures 6-8(b), (c), and (d) depict the state of the circuit at $t = 0^-$, $t = 0$, and $t > 0$, respectively.

At $t = 0^-$: The capacitor in Fig. 6-8(b) has been replaced with an open circuit and the inductor with a short circuit. Also, the current source I_s (including its associated resistor R_s) has been converted into a voltage source $V_s = I_s R_s$ in series with a resistor R_s. With C as an open circuit, no current flows through the circuit, and therefore, no voltage drop occurs across R_s or R_2. Consequently,

$$v(0^-) = V_s = I_s R_s = 20 \text{ V}.$$

We also note that no current flows through L,

$$i_L(0^-) = 0.$$

At $t = 0$: Carrying the information on $v(0^-)$ and $i_L(0^-)$ forward to $t = 0$, we replace C in Fig. 6-8(c) with a voltage source V_s and L with a current source of zero magnitude, which is equivalent to an open circuit. Hence,

$$v(0) = v(0^-) = 20 \text{ V},$$

and

$$i_C(0) = -i_L(0) = 0.$$

Equivalently,

$$v'(0) = \frac{i_C(0)}{C} = 0.$$

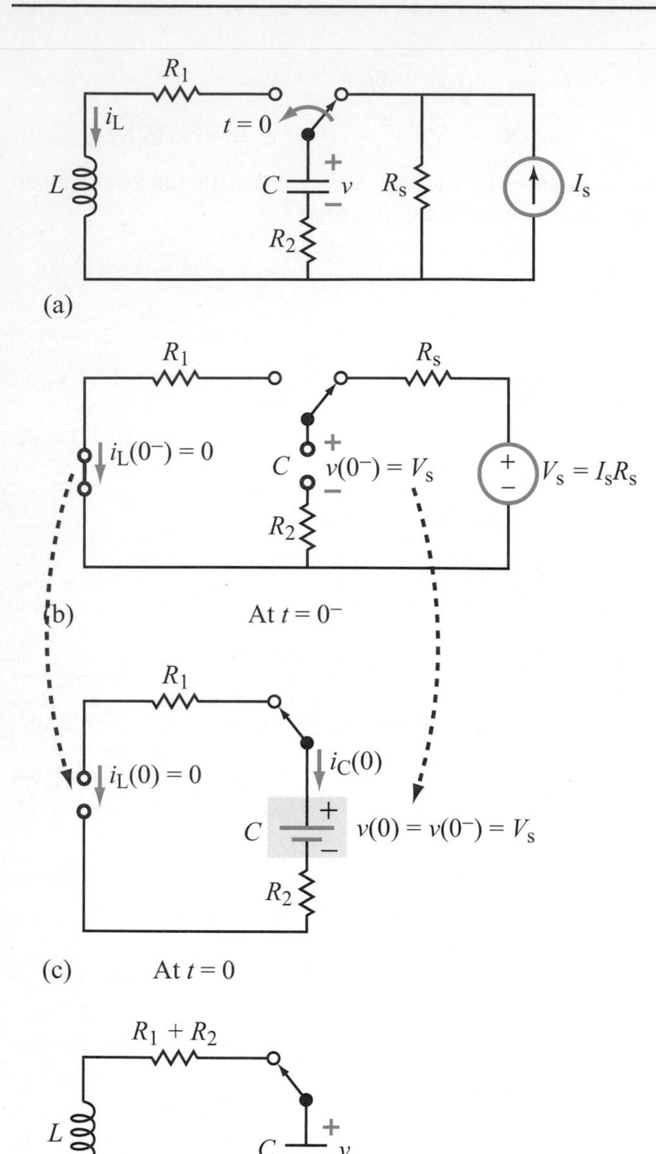

Figure 6-8: Example 6-4.

At $t > 0$: The circuit in Fig. 6-8(d) is a source-free series RLC circuit with

$$R = R_1 + R_2 = 1.81 + 0.2 = 2.01 \ \Omega.$$

The parameters α and ω_0 are given by

$$\alpha = \frac{R}{2L} = \frac{2.01}{2 \times 5 \times 10^{-3}} = 201 \text{ Np/s}$$

and

$$\omega_0 = \frac{1}{\sqrt{LC}} = \frac{1}{\sqrt{5 \times 10^{-3} \times 5 \times 10^{-3}}} = 200 \text{ rad/s}.$$

Since $\alpha > \omega_0$, the response is overdamped and given by Eq. (6.31) as

$$v(t) = A_1 e^{s_1 t} + A_2 e^{s_2 t},$$

with

$$s_1 = -\alpha + \sqrt{\alpha^2 - \omega_0^2}$$
$$= -201 + \sqrt{(201)^2 - (200)^2} = -181 \text{ Np/s}$$

and

$$s_2 = -\alpha - \sqrt{\alpha^2 - \omega_0^2} = -221 \text{ Np/s}.$$

Application of initial conditions: The constants A_1 and A_2 are determined by matching the expression for $v(t)$ to the initial conditions determined earlier, namely $v(0) = 20$ V and $v'(0) = 0$. Hence,

$$A_1 + A_2 = 20,$$
$$(s_1 A_1 e^{s_1 t} + s_2 A_2 e^{s_2 t})\big|_{t=0} = 0,$$

or

$$s_1 A_1 + s_2 A_2 = 0.$$

Simultaneous solution of the two equations leads to

$$A_1 = \frac{20 s_2}{s_2 - s_1} = \frac{20 \times (-221)}{-221 + 181} = 110.4 \text{ V}$$

and

$$A_2 = \frac{-20 s_1}{s_2 - s_1} = \frac{-20 \times (-181)}{-221 + 181} = -90.4 \text{ V}.$$

Incorporating the values of s_1, s_2, A_1, and A_2, we have

$$v(t) = (110.4 e^{-181t} - 90.4 e^{-221t}) \text{ V} \qquad (\text{for } t \geq 0).$$

Exercise 6-3: After interchanging the locations of L and C in Fig. 6-8(a), repeat Example 6-4 to determine $v_C(t)$ across C.

Answer: $v(t) = 9.8(e^{-221t} - e^{-181t})$ V. (See ◎)

Exercise 6-4: For the overdamped response, use Eq. (6.31) to develop general expressions for A_1 and A_2 in terms of s_1, s_2, $v(0)$, and $v'(0)$.

Answer:

$$A_1 = \frac{v'(0) - s_2 \, v(0)}{s_1 - s_2}, \qquad A_2 = -\frac{v'(0) - s_1 \, v(0)}{s_1 - s_2}.$$

(See ◎)

6-3.2 Critically Damped Response ($\alpha = \omega_0$)

When

$$R = 2\sqrt{\frac{L}{C}} \quad \text{(critically damped)}, \tag{6.32}$$

$\alpha = \omega_0$, and

$$s_1 = s_2 = -\alpha. \tag{6.33}$$

Repeated roots are problematic because Eq. (6.19) becomes

$$\begin{aligned} v(t) &= A_1 e^{-\alpha t} + A_2 e^{-\alpha t} \\ &= (A_1 + A_2)e^{-\alpha t} \\ &= A_3 e^{-\alpha t}, \end{aligned} \tag{6.34}$$

where $A_3 = A_1 + A_2$. A solution containing a single constant cannot simultaneously satisfy the initial condition on both the voltage across the capacitor and the current through the inductor. For this ***critically damped*** case, we introduce two new constants: B_1 and B_2. We use the modified form

$$v(t) = B_1 e^{-\alpha t} + B_2 t e^{-\alpha t} = (B_1 + B_2 t)e^{-\alpha t} \quad \text{(for } t \geq 0) \quad \text{(critically damped)}, \tag{6.35}$$

which contains a term that involves $e^{-\alpha t}$ and a second term that involves the product $(te^{-\alpha t})$. As will be demonstrated in Example 6-5, the expression given by Eq. (6.35) will indeed satisfy all initial conditions of the critically damped RLC circuit. Figure 6-9 displays a typical example of its response.

The critically damped case represents a *resonant* condition at which ω_0 is exactly equal to the damping coefficient α.

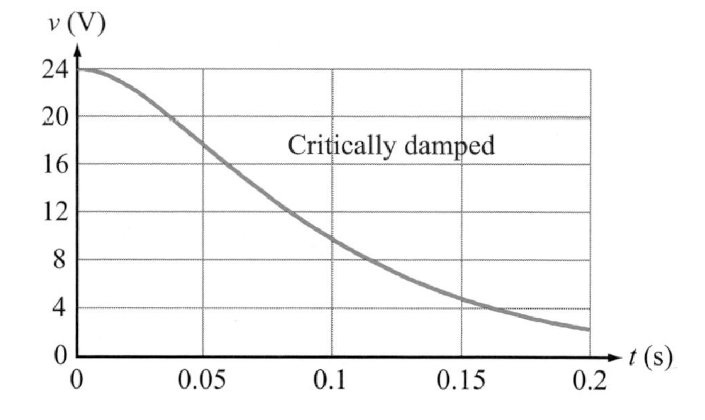

Figure 6-9: Critically damped response of the series RLC circuit of Example 6-5.

Example 6-5: Critically Damped Response

Repeat Example 6-3, keeping $V_s = 24$ V, $R = 12\ \Omega$, and $L = 0.3$ H but changing the capacitance to $C = 8.33$ mF.

Solution: The parameters α and ω_0 are given by

$$\alpha = \frac{R}{2L} = \frac{12}{2 \times 0.3} = 20 \text{ Np/s}$$

and

$$\omega_0 = \frac{1}{\sqrt{LC}} = \frac{1}{\sqrt{0.3 \times 8.33 \times 10^{-3}}} = 20 \text{ rad/s}.$$

Hence, because $\alpha = \omega_0$, the response is critically damped, and it is given by Eq. (6.35) as

$$v(t) = (B_1 + B_2 t)e^{-\alpha t} = (B_1 + B_2 t)e^{-20t}.$$

The initial conditions given by Eqs. (6.21a) and (6.22) state that at $t = 0$

$$v(0) = V_s = 24 \text{ V} \quad \text{and} \quad v'(0) = 0.$$

Application of these conditions to the expression of $v(t)$ gives

$$B_1 = 24 \text{ V} \quad \text{and} \quad [B_2 e^{-20t} - 20(B_1 + B_2 t)e^{-20t}]\Big|_{t=0} = 0,$$

which yields

$$B_2 - 20B_1 = 0 \quad \text{or} \quad B_2 = 20B_1.$$

Incorporating the values of B_1 and B_2 into the expression for $v(t)$ leads to

$$v(t) = 24(1 + 20t)e^{-20t} \text{ V} \quad (\text{for } t \geq 0).$$

The time profile of $v(t)$ is displayed in Fig. 6-9.

Exercise 6-5: The switch in Fig. E6.5 is moved to position 2 after it had been in position 1 for a long time. Determine (a) $v_C(0)$ and $i_C(0)$ and (b) $i_C(t)$ for $t \geq 0$.

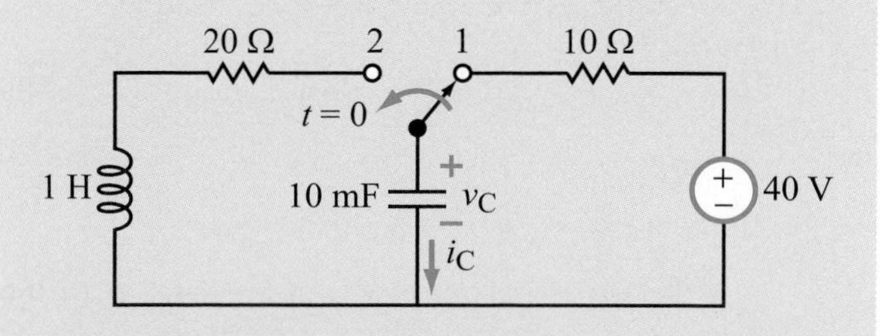

Figure E6.5

Answer: (a) $v_C(0) = 40$ V, $i_C(0) = 0$, (b) $i_C(t) = [-40te^{-10t}]$ A. (See 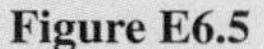)

Exercise 6-6: The circuit in Fig. E6.6 is a replica of the circuit in Fig. E6.5 but with the capacitor and inductor interchanged in location. Determine (a) $i_L(0)$ and $v_L(0)$ and (b) $i_L(t)$ for $t \geq 0$.

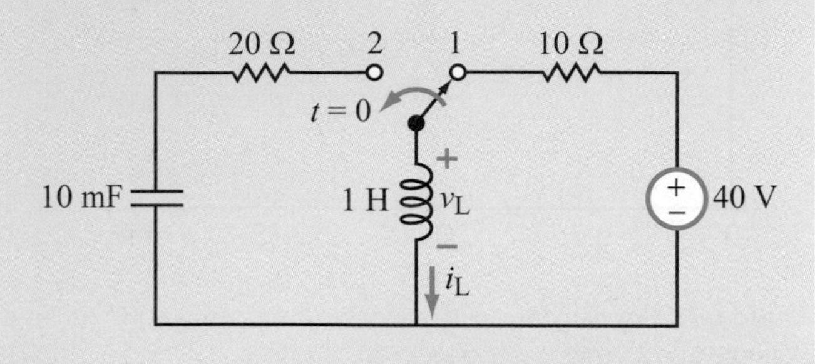

Figure E6.6

Answer: (a) $i_L(0) = 4$ A, $v_L(0) = -80$ V, (b) $i_L(t) = [4(1 - 10t)e^{-10t}]$ A. (See)

Exercise 6-7: For the critically damped response given by Eq. (6.35), develop general expressions for B_1 and B_2 in terms of α, $v(0)$, and $v'(0)$.

Answer: $B_1 = v(0)$, $B_2 = v'(0) + \alpha\, v(0)$. (See)

6-3.3 Underdamped Response ($\alpha < \omega_0$)

If $\alpha < \omega_0$, corresponding to

$$R < 2\sqrt{\frac{L}{C}} \qquad \text{(underdamped)}, \tag{6.36}$$

we introduce the ***damped natural frequency*** ω_d defined as

$$\omega_d^2 = \omega_0^2 - \alpha^2. \tag{6.37}$$

Since $\alpha < \omega_0$, it follows that $\omega_d > 0$. In terms of ω_d, the expressions for the roots s_1 and s_2 given by Eq. (6.28) become

$$s_1 = -\alpha + \sqrt{\alpha^2 - \omega_0^2} = -\alpha + \sqrt{-\omega_d^2}$$
$$= -\alpha + j\omega_d \tag{6.38a}$$

and

$$s_2 = -\alpha - \sqrt{\alpha^2 - \omega_0^2}$$
$$= -\alpha - j\omega_d, \tag{6.38b}$$

where $j = \sqrt{-1}$. The fact that s_1 and s_2 are complex conjugates of one another will prove central to the form of the solution. Inserting the expressions for s_1 and s_2 into Eq. (6.19) gives

$$v(t) = A_1 e^{-\alpha t} e^{j\omega_d t} + A_2 e^{-\alpha t} e^{-j\omega_d t}. \tag{6.39}$$

The Euler identity

$$e^{\pm j\theta} = \cos\theta \pm j\sin\theta \tag{6.40}$$

allows us to expand Eq. (6.39) as follows:

$$\begin{aligned} v(t) &= A_1 e^{-\alpha t}(\cos\omega_d t + j\sin\omega_d t) \\ &\quad + A_2 e^{-\alpha t}(\cos\omega_d t - j\sin\omega_d t) \\ &= e^{-\alpha t}[(A_1 + A_2)\cos\omega_d t + j(A_1 - A_2)\sin\omega_d t]. \end{aligned} \tag{6.41}$$

Next, by introducing a new pair of constants, $D_1 = A_1 + A_2$ and $D_2 = j(A_1 - A_2)$, we have

$$v(t) = e^{-\alpha t}[D_1\cos\omega_d t + D_2\sin\omega_d t] \qquad (\text{for } t \geq 0) \quad \text{(underdamped)}. \tag{6.42}$$

The negative exponential $e^{-\alpha t}$ signifies that $v(t)$ has a damped waveform with a ***time constant*** $\tau = 1/\alpha$, and the sine and cosine terms signify that $v(t)$ is oscillatory with an angular frequency ω_d and a corresponding ***time period***

$$T = \frac{2\pi}{\omega_d}. \tag{6.43}$$

Since ω_d is a measure of the oscillation associated with the damped natural response of the circuit, it is appropriate that it be called the ***damped natural frequency*** of the circuit.

The oscillatory behavior of the underdamped response is visible in the time waveform of Fig. 6-10.

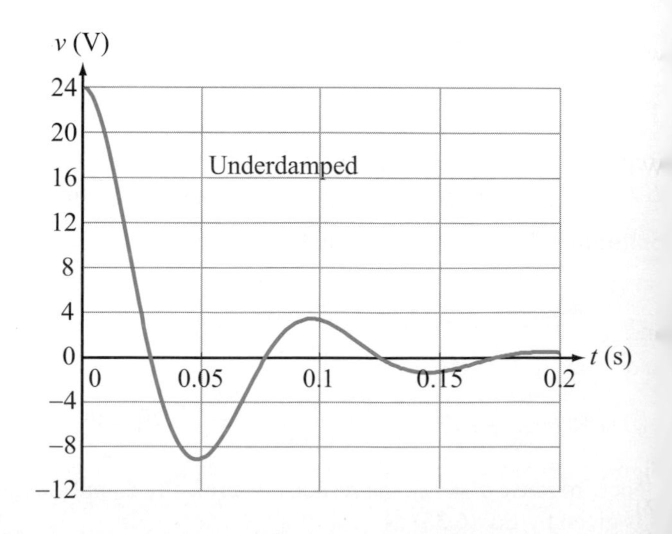

Figure 6-10: Underdamped response of the series RLC circuit c Example 6-6.

Example 6-6: Underdamped Response

Repeat Example 6-3, keeping $V_s = 24$ V, $R = 12\ \Omega$, and $L = 0.3$ H but changing the capacitance to $C = 0.72$ mF.

Solution: For the specified values,

$$\alpha = \frac{R}{2L} = \frac{12}{2 \times 0.3} = 20 \text{ Np/s},$$

and

$$\omega_0 = \frac{1}{\sqrt{LC}} = \frac{1}{\sqrt{0.3 \times 0.72 \times 10^{-3}}} = 68 \text{ rad/s}.$$

Since $\alpha < \omega_0$, the voltage response is given by Eq. (6.42), namely

$$v(t) = e^{-\alpha t}[D_1 \cos \omega_d t + D_2 \sin \omega_d t]$$

with

$$\omega_d = \sqrt{\omega_0^2 - \alpha^2} = \sqrt{(68)^2 - (20)^2} = 65 \text{ rad/s}.$$

Initial conditions require

$$v(0) = D_1 = V_s = 24 \text{ V}$$

and

$$v'(0) = \{-\alpha[D_1 \cos \omega_d t + D_2 \sin \omega_d t] + [-\omega_d D_1 \sin \omega_d t + \omega_d D_2 \cos \omega_d t]\} e^{-\alpha t}\Big|_{t=0} = 0,$$

which yields

$$-\alpha D_1 + \omega_d D_2 = 0.$$

With $D_1 = 24$ V, $\alpha = 20$ (Np/s), and $\omega_d = 65$ (rad/s), we have

$$D_2 = \frac{\alpha}{\omega_d} D_1 = \frac{20 \times 24}{65} = 7.4 \text{ V}$$

and

$$v(t) = e^{-20t}[24 \cos 65t + 7.4 \sin 65t] \text{ V} \qquad (\text{for } t \geq 0).$$

Figure 6-10 shows a time plot of $v(t)$ which exhibits an exponential decay (due to e^{-20t}) in combination with the oscillatory behavior associated with the sine and cosine functions.

Review Question 6-8: What specific feature distinguishes the waveform of the underdamped response from those of the overdamped and critically damped responses?

Review Question 6-9: Why is ω_0 called the resonant frequency?

Exercise 6-8: Repeat Example 6-4, changing the value of R_1 to 1.7 Ω.

Answer:

$$v(t) = e^{-190t}(20 \cos 62.45t + 60.85 \sin 62.45t) \text{ V}.$$

(See CD)

Exercise 6-9: For the underdamped response given by Eq. (6.42), develop general expressions for D_1 and D_2 in terms of α, ω_d, $v(0)$, and $v'(0)$.

Answer:

$$D_1 = v(0), \qquad D_2 = \frac{v'(0) + \alpha\, v(0)}{\omega_d}.$$

(See CD)

6-4 Total Response of the Series RLC Circuit

After moving the switch from position 1 to position 2, both circuits in Fig. 6-11 become series RLC circuits, but they differ in one important aspect. After $t = 0$, Circuit 1 will contain no sources, and hence, its response consists of only one component: the transient response $v_t(t)$. In contrast, Circuit 2 will continue to contain a source, namely V_{s_2}, after $t = 0$. Consequently, the response of Circuit 2 will include not only a transient component but also a steady-state component v_{ss},

$$v(t) = v_{ss} + v_t(t). \tag{6.44}$$

In a circuit with dc sources, the steady-state component is a constant given by

$$v_{ss} = v(\infty). \tag{6.45}$$

The validity of Eq. (6.45) is based on the fact that the transient response $v_t(t)$ always decays to zero as $t \to \infty$. That was evident for all three types of responses we considered in Section 6-3.

By combining Eq. (6.45) with the transient solutions we derived in the preceding section for the three types of damped conditions, the ***total response*** for each is listed here.

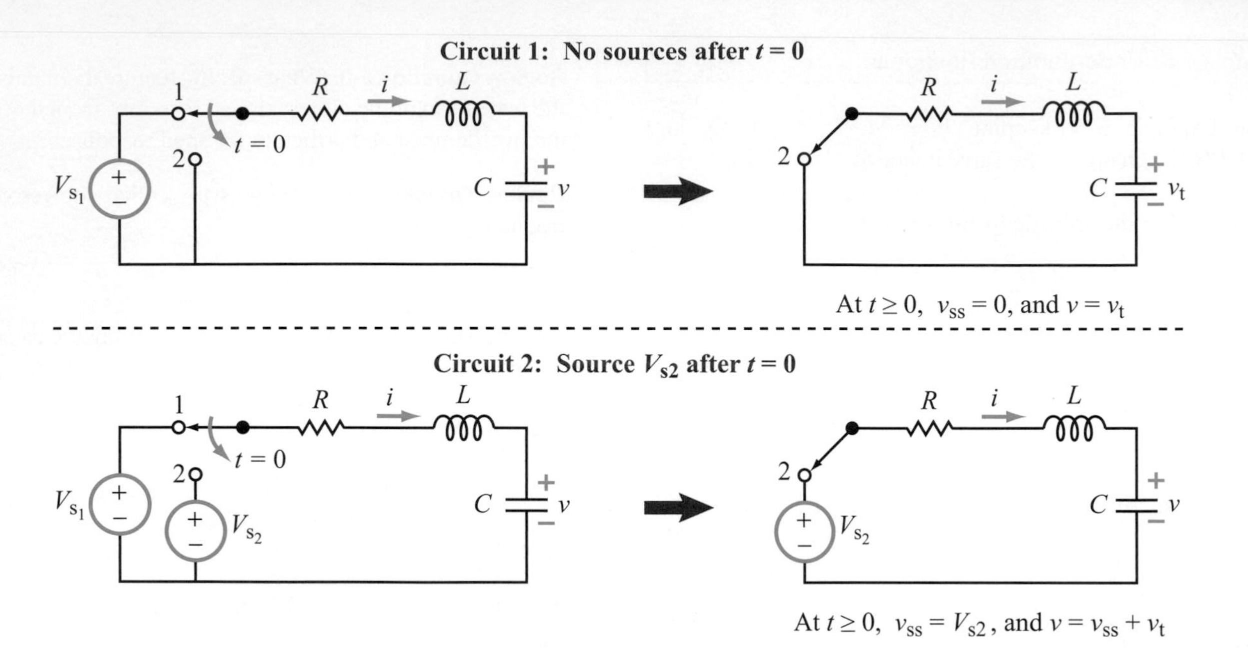

Figure 6-11: Circuit 1 will exhibit a transient response after $t = 0$, whereas the response of Circuit 2 also will feature a steady-state component.

Series RLC

Overdamped ($\alpha > \omega_0$)

$$v(t) = [v(\infty) + A_1 e^{s_1 t} + A_2 e^{s_2 t}] \qquad (t \geq 0), \quad (6.46a)$$

Critically Damped ($\alpha = \omega_0$)

$$v(t) = [v(\infty) + (B_1 + B_2 t)e^{-\alpha t}] \qquad (t \geq 0), \quad (6.46b)$$

Underdamped ($\alpha < \omega_0$)

$$v(t) = [v(\infty) + e^{-\alpha t}(D_1 \cos \omega_d t + D_2 \sin \omega_d t)] \qquad (t \geq 0). \quad (6.46c)$$

The pairs of constants (A_1 and A_2, B_1 and B_2, and D_1 and D_2) are determined by matching the expression for $v(t)$ to the initial conditions, namely to $v(0)$ and $v'(0)$ (as illustrated next by Exercise 6.10 and in the examples that follow). The general expressions and auxiliary relations for the series and parallel RLC circuits are provided in summary form in Table 6-2.

Time Shift to T_0

We should remember that should the sudden change in the circuit occur at $t = T_0$ instead of at $t = 0$, the only changes that need to be made are

(1) t should be replaced with $(t - T_0)$ everywhere on the right-hand side of Eq. (6.46)

(2) $v(0)$ and $v'(0)$ should be replaced with $v(T_0)$ and $v'(T_0)$, respectively, in the expressions for the constants in Exercise 6.10 and Table 6-2.

Exercise 6-10: Develop expressions for the unknown constants in Eq. (6.46) in terms of $v(0)$, $v(\infty)$, and $v'(0)$, where $v'(0)$ is defined as $dv/dt|_{t=0}$.

Answer:
Overdamped case

$$A_1 = \frac{v'(0) - s_2[v(0) - v(\infty)]}{s_1 - s_2},$$

$$A_2 = -\left[\frac{v'(0) - s_1[v(0) - v(\infty)]}{s_1 - s_2}\right]$$

Critically damped case

$$B_1 = v(0) - v(\infty), \qquad B_2 = v'(0) + \alpha[v(0) - v(\infty)]$$

Underdamped case

$$D_1 = v(0) - v(\infty), \qquad D_2 = \frac{v'(0) + \alpha[v(0) - v(\infty)]}{\omega_d}$$

(See ◎)

Table 6-2: Step response of RLC circuits for $t \geq 0$.

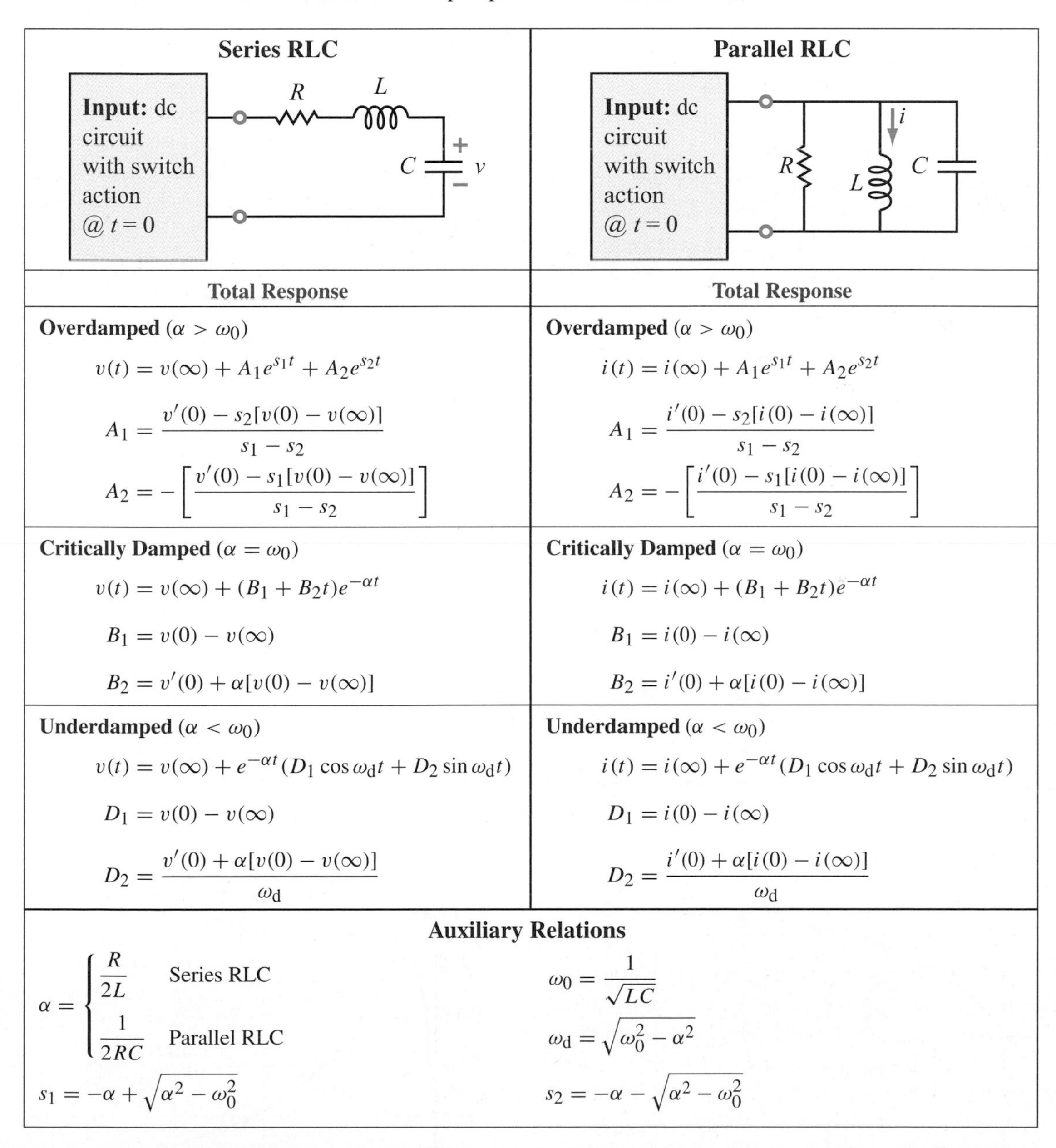

Series RLC	Parallel RLC
Total Response	**Total Response**
Overdamped ($\alpha > \omega_0$) $v(t) = v(\infty) + A_1 e^{s_1 t} + A_2 e^{s_2 t}$ $A_1 = \dfrac{v'(0) - s_2[v(0) - v(\infty)]}{s_1 - s_2}$ $A_2 = -\left[\dfrac{v'(0) - s_1[v(0) - v(\infty)]}{s_1 - s_2}\right]$	**Overdamped** ($\alpha > \omega_0$) $i(t) = i(\infty) + A_1 e^{s_1 t} + A_2 e^{s_2 t}$ $A_1 = \dfrac{i'(0) - s_2[i(0) - i(\infty)]}{s_1 - s_2}$ $A_2 = -\left[\dfrac{i'(0) - s_1[i(0) - i(\infty)]}{s_1 - s_2}\right]$
Critically Damped ($\alpha = \omega_0$) $v(t) = v(\infty) + (B_1 + B_2 t)e^{-\alpha t}$ $B_1 = v(0) - v(\infty)$ $B_2 = v'(0) + \alpha[v(0) - v(\infty)]$	**Critically Damped** ($\alpha = \omega_0$) $i(t) = i(\infty) + (B_1 + B_2 t)e^{-\alpha t}$ $B_1 = i(0) - i(\infty)$ $B_2 = i'(0) + \alpha[i(0) - i(\infty)]$
Underdamped ($\alpha < \omega_0$) $v(t) = v(\infty) + e^{-\alpha t}(D_1 \cos \omega_d t + D_2 \sin \omega_d t)$ $D_1 = v(0) - v(\infty)$ $D_2 = \dfrac{v'(0) + \alpha[v(0) - v(\infty)]}{\omega_d}$	**Underdamped** ($\alpha < \omega_0$) $i(t) = i(\infty) + e^{-\alpha t}(D_1 \cos \omega_d t + D_2 \sin \omega_d t)$ $D_1 = i(0) - i(\infty)$ $D_2 = \dfrac{i'(0) + \alpha[i(0) - i(\infty)]}{\omega_d}$

Auxiliary Relations

$$\alpha = \begin{cases} \dfrac{R}{2L} & \text{Series RLC} \\[2ex] \dfrac{1}{2RC} & \text{Parallel RLC} \end{cases} \qquad \omega_0 = \frac{1}{\sqrt{LC}} \qquad \omega_d = \sqrt{\omega_0^2 - \alpha^2}$$

$$s_1 = -\alpha + \sqrt{\alpha^2 - \omega_0^2} \qquad s_2 = -\alpha - \sqrt{\alpha^2 - \omega_0^2}$$

Example 6-7: Overdamped RLC Circuit

Given that in the circuit of Fig. 6-12(a) $V_s = 16$ V, $R = 64\ \Omega$, $L = 0.8$ H, and $C = 2$ mF, determine $v(t)$ and $i(t)$ for $t \geq 0$. The capacitor had no charge prior to $t = 0$.

Solution: We begin by establishing the damping condition of the circuit. From the definitions for α and ω_0 given by Eq. (6.29), we have

$$\alpha = \frac{R}{2L} = \frac{64}{2 \times 0.8} = 40 \text{ Np/s}$$

and

$$\omega_0 = \frac{1}{\sqrt{LC}} = \frac{1}{\sqrt{0.8 \times 2 \times 10^{-3}}} = 25 \text{ rad/s}.$$

Hence, $\alpha > \omega_0$, which means that the circuit will exhibit an overdamped response after the switch is closed. The applicable expression for $v(t)$ is given by Eq. (6.46a) as

$$v(t) = [v(\infty) + A_1 e^{s_1 t} + A_2 e^{s_2 t}]. \tag{6.47}$$

Using Eq. (6.28), the characteristic roots are

$$s_1 = -\alpha + \sqrt{\alpha^2 - \omega_0^2} = -40 + \sqrt{40^2 - 25^2} = -8.8 \text{ Np/s}$$

and

$$s_2 = -\alpha - \sqrt{\alpha^2 - \omega_0^2} = -71.2 \text{ Np/s}.$$

At $t = \infty$, the capacitor becomes like an open circuit, allowing no current to flow through the circuit. Consequently,

$$v(\infty) = V_s = 16 \text{ V}.$$

At $t = 0^-$, the capacitor was uncharged. Hence,

$$v(0) = v(0^-) = 0. \tag{6.48}$$

We also need the initial condition on the time derivative of $v(t)$ evaluated at $t = 0$. We obtain that from the knowledge that the current through L (which also is the current through C) cannot change instantaneously. Thus,

$$v'(0) = \frac{1}{C}\, i(0) = 0. \tag{6.49}$$

The expressions for A_1 and A_2 are given in Exercise 6.10 (as well as in Table 6-2) as

$$A_1 = \frac{v'(0) - s_2[v(0) - v(\infty)]}{s_1 - s_2} = \frac{0 + 71.2(0 - 16)}{-8.8 + 71.2} = -18.25 \text{ V}$$

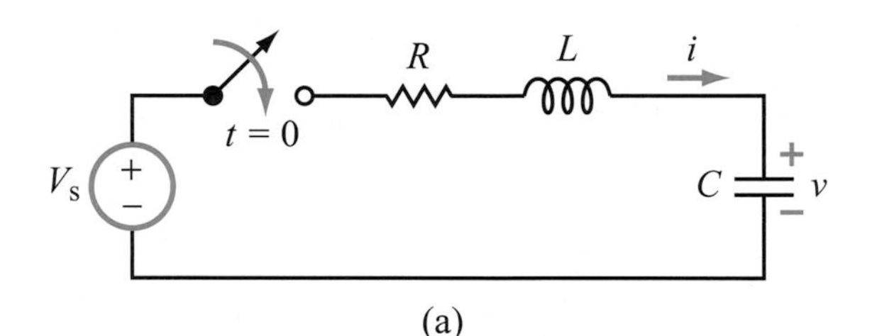

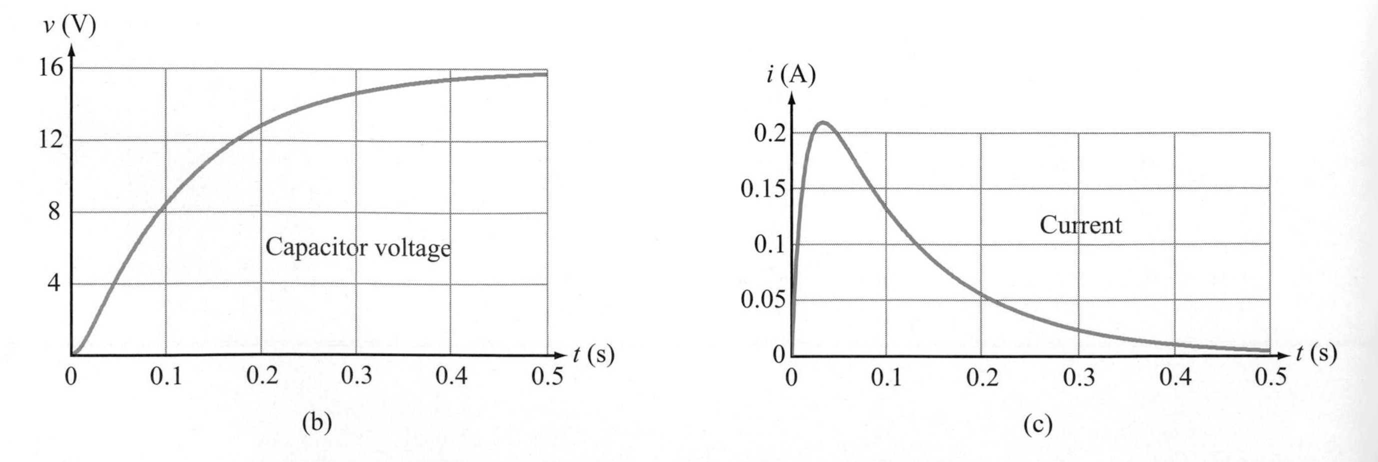

Figure 6-12: Example 6-7 (a) circuit, (b) $v(t)$, and (c) $i(t)$.

and

$$A_2 = -\left[\frac{v'(0) - s_1[v(0) - v(\infty)]}{s_1 - s_2}\right]$$
$$= -\left[\frac{0 + 8.8(0 - 16)}{-8.8 + 71.2}\right]$$
$$= 2.25 \text{ V}.$$

The total response $v(t)$ then is given by

$$v(t) = [16 - 18.25e^{-8.8t} + 2.25e^{-71.2t}] \text{ V} \quad (\text{for } t \geq 0), \tag{6.50}$$

and the associated current is

$$i(t) = C\,\frac{dv}{dt}$$
$$= 2 \times 10^{-3}[18.25 \times 8.8e^{-8.8t} - 2.25 \times 71.2e^{-71.2t}]$$
$$= 0.32(e^{-8.8t} - e^{-71.2t}) \text{ A} \quad (\text{for } t \geq 0). \tag{6.51}$$

The waveforms of $v(t)$ and $i(t)$ are displayed in Figs. 6-12(b) and (c), respectively.

Review Question 6-10: If $R = 0$ in the series RLC circuit of Fig. 6-12, how would that affect the nature of the response of $v(t)$?

Review Question 6-11: For the circuit in Fig. 6-12, how much (net) energy gets transferred into L and C between $t = 0$ and $t = \infty$?

Exercise 6-11: The circuit in Fig. E6.11 was already in a steady state when the switch was opened at $t = 0$. Determine $v_C(0)$, $i_C(0)$, α, ω_0, and $v_C(t)$ for $t \geq 0$.

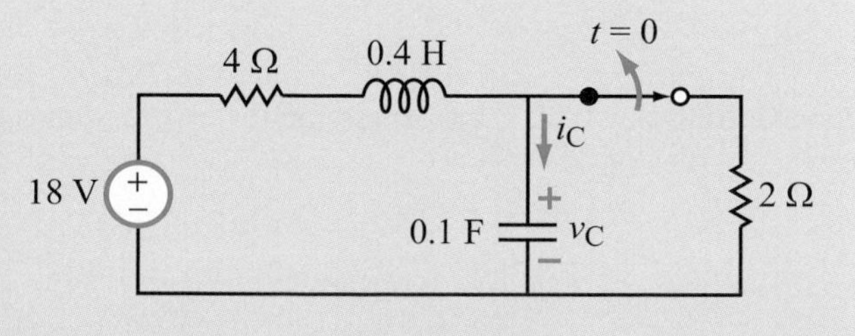

Figure E6.11

Answer: $v_C(0) = 6$ V, $i_C(0) = 3$ A, $\alpha = 5$ Np/s, $\omega_0 = 5$ rad/s, $v_C(t) = [18 - (12 + 30t)e^{-5t}]$ V. (See CD)

Example 6-8: Rectangular-Pulse Excitation

The switch in the circuit of Fig. 6-13(a) was in position 1 for a long time before it was moved to position 2 at $t = 0$ and then back to position 1 at $t = 20$ ms. If $V_s = 12$ V, $R = 40\ \Omega$, $L = 0.8$ H, and $C = 2$ mF, determine the waveforms of $v(t)$ and $i(t)$ for $t \geq 0$.

Solution: From Eq. (6.29),

$$\alpha = \frac{R}{2L} = \frac{40}{2 \times 0.8} = 25 \text{ Np/s},$$

and

$$\omega_0 = \frac{1}{\sqrt{LC}} = \frac{1}{\sqrt{0.8 \times 2 \times 10^{-3}}} = 25 \text{ rad/s}.$$

Since $\alpha = \omega_0$, the circuit will exhibit a critically damped response. We will divide the solution into two time segments.

Time Segment 1: $0 \leq t \leq 20$ ms.

The general expression for the critically damped response of the series RLC circuit is given by Eq. (6.46b) as

$$v_1(t) = v_1(\infty) + (B_1 + B_2 t)e^{-\alpha t}. \tag{6.52}$$

Even though we know that the switch will be moved back to position 1 at $t = 20$ ms, when we evaluate the constants in Eq. (6.52) for Time Segment 1, we do so as if the state of the circuit shown in Fig. 6-13(b) is to remain the same until $t = \infty$. Since the circuit is "unaware" of the change that will be taking place at $t = 20$ ms, its reaction to the change at $t = 0$ presumes that the new condition of the circuit will continue indefinitely. Hence, the voltage across the capacitor at $t = \infty$ would have been

$$v_1(\infty) = V_s = 12 \text{ V}. \tag{6.53}$$

At $t = 0^-$, the RLC circuit contains no active sources, so both $v_1(0^-)$ and $i_1(0^-)$ are zero. Moreover, since neither the voltage across C nor the current through L can change instantaneously, it follows that

$$v_1(0) = v_1(0^-) = 0$$

and

$$v_1'(0) = \frac{1}{C}\,i(0) = \frac{1}{C}\,i(0^-) = 0.$$

Application of the expressions for B_1 and B_2 available in Table 6-2 gives

$$B_1 = v(0) - v(\infty) = 0 - 12 = -12 \text{ V} \tag{6.54a}$$

and

$$B_2 = v'(0) + \alpha[v(0) - v(\infty)]$$
$$= 0 + 25[0 - 12] = -300 \text{ V/s}. \tag{6.54b}$$

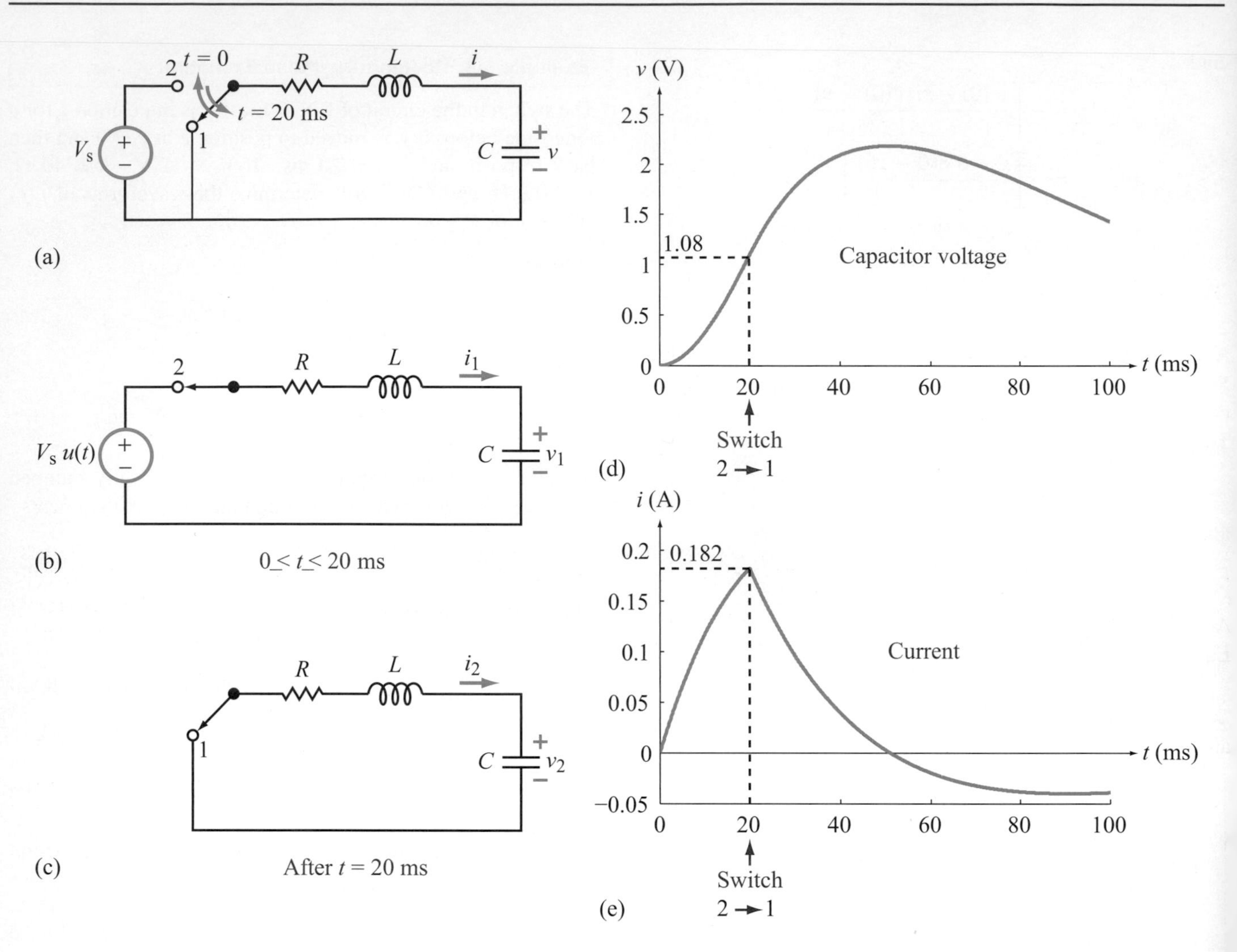

Figure 6-13: Example 6-8.

Consequently, $v_1(t)$ is given by

$$v_1(t) = 12 - (12 + 300t)e^{-25t} \text{ V} \qquad (6.55)$$
$$(\text{for } 0 \le t \le 20 \text{ ms}).$$

The associated current is

$$i_1(t) = C\,\frac{dv_1}{dt} = 2 \times 10^{-3}\,\frac{d}{dt}[12 - (12 + 300t)e^{-25t}]$$
$$= 15te^{-25t} \text{ A} \quad (\text{for } 0 \le t \le 20 \text{ ms}). \quad (6.56)$$

Time Segment 2: $t \ge 20$ ms.

After moving the switch back to position 1 at $t = 20$ ms, the circuit no longer has any active sources, and yet it is part of a closed circuit (Fig. 6-13(c)). This allows the capacitor and inductor to dissipate their stored energies through the resistor. Hence, at $t = \infty$,

$$v_2(\infty) = 0.$$

Upon shifting t by 0.02 s, the expression for $v_2(t)$ assumes the form

$$v_2(t) = [B_3 + B_4(t - 0.02)]e^{-25(t-0.02)} \text{ V}$$
$$(\text{for } t \ge 20 \text{ ms}), \qquad (6.57)$$

where constants B_3 and B_4 are so labeled to avoid confusion with B_1 and B_2 of the earlier time segment. The associated current is

$$i_2(t) = C\,\frac{dv_2}{dt}$$
$$= 2 \times 10^{-3}\,\frac{d}{dt}\{[B_3 + B_4(t - 0.02)]e^{-25(t-0.02)}\}$$
$$= [(2B_4 - 50B_3) - 50B_4(t - 0.02)]$$
$$\cdot\, e^{-25(t-0.02)} \times 10^{-3}\text{ A} \qquad (\text{for } t \geq 20\text{ ms}). \tag{6.58}$$

Across the juncture between Time Segment 1 and Time Segment 2, neither the voltage can change (as mandated by the capacitor) nor can the current (as mandated by the inductor). Thus,

$$v_1(t = 20\text{ ms}) = v_2(t = 20\text{ ms}), \tag{6.59a}$$

and

$$i_1(t = 20\text{ ms}) = i_2(t = 20\text{ ms}). \tag{6.59b}$$

Application of Eqs. (6.59a and b) to the expressions given by Eqs. (6.55) through (6.58) gives

$$12 - (12 + 300 \times 0.02)e^{-25\times 0.02} = B_3$$

and

$$15 \times 0.02e^{-25\times 0.02} = (2B_4 - 50B_3) \times 10^{-3},$$

whose joint solution leads to

$$B_3 = 1.08\text{ V} \qquad \text{and} \qquad B_4 = 118.04\text{ V/s}.$$

Consequently,

$$v_2(t) = [1.08 + 118.04(t - 0.02)]e^{-25(t-0.02)}\text{ V}$$
$$(\text{for } t \geq 20\text{ ms}), \tag{6.60a}$$

and

$$i_2(t) = [0.182 - 5.90(t - 0.02)]e^{-25(t-0.02)}\text{ A}$$
$$(\text{for } t \geq 20\text{ ms}). \tag{6.60b}$$

The waveforms of $v(t)$ and $i(t)$ are displayed in Figs. 6-13(d) and (e), respectively.

Example 6-9: Underdamped Response

The switch in the circuit of Fig. 6-14(a) was opened at $t = 0$ after it had been closed for a long time. If $V_{s_1} = 20$ V, $V_{s_2} = 24$ V, $R_1 = 40\ \Omega$, $R_2 = R_3 = 20\ \Omega$, $R_4 = 10\ \Omega$, $L = 0.8$ H, and $C = 2$ mF, determine $v_C(t)$ for $t \geq 0$.

Solution: Consider the state of the circuit at $t = 0^-$ (before opening the switch), as depicted by Fig. 6-14(b). The mesh current equations for the indicated loops are

$$-V_{s_1} + R_1 I_1 + R_2(I_1 - I_2) = 0$$

and

$$R_2(I_2 - I_1) + R_3 I_2 + V_{s_2} + R_4 I_2 = 0.$$

After substituting the given values for the sources and the resistors, simultaneous solution of the two equations leads to

$$I_1 = 0.2\text{ A}, \qquad \text{and} \qquad I_2 = -0.4\text{ A}.$$

Hence,

$$v_C(0^-) = I_2 R_4 = -0.4 \times 10 = -4\text{ V} \tag{6.61a}$$

and

$$i_L(0^-) = I_1 = 0.2\text{ A}. \tag{6.61b}$$

Next, we consider Fig. 6-14(d), which depicts the circuit configuration at $t > 0$ (after opening the switch). To simplify the analysis, we use source transformation to convert the circuit into its Thévenin equivalent, as shown in Fig. 6-14(e), where

$$R_{eq} = (R_2 + R_3) \parallel R_4 = \frac{(R_2 + R_3)R_4}{R_2 + R_3 + R_4} = 8\ \Omega$$

and

$$V_{eq} = \frac{V_{s_2}}{R_2 + R_3} \times R_{eq} = 4.8\text{ V}.$$

Now we are ready to analyze the series RLC circuit of Fig. 6-14(e). To that end, we compute α and ω_0 as

$$\alpha = \frac{R_{eq}}{2L} = \frac{8}{2 \times 0.8} = 5\text{ Np/s}$$

$$\omega_0 = \frac{1}{\sqrt{LC}} = \frac{1}{\sqrt{0.8 \times 2 \times 10^{-3}}} = 25\text{ rad/s}.$$

Since $\alpha < \omega_0$, the capacitor voltage v_C will exhibit an underdamped oscillatory response of the form given by Eq. (6.46c) as

$$v_C(t) = \{v(\infty) + e^{-\alpha t}[D_1 \cos\omega_d t + D_2 \sin\omega_d t]\}, \tag{6.62}$$

where

$$\omega_d = \sqrt{\omega_0^2 - \alpha^2} = \sqrt{25^2 - 5^2} = 24.5\text{ rad/s}.$$

It is evident from the circuit in Fig. 6-14(e) that

$$v_C(\infty) = -V_{eq} = -4.8\text{ V}.$$

To determine D_1 and D_2, we need to know the values of $v_C(0)$ and $v_C'(0)$. In view of Eq. (6.61a),

$$v_C(0) = v_C(0^-) = -4\text{ V}.$$

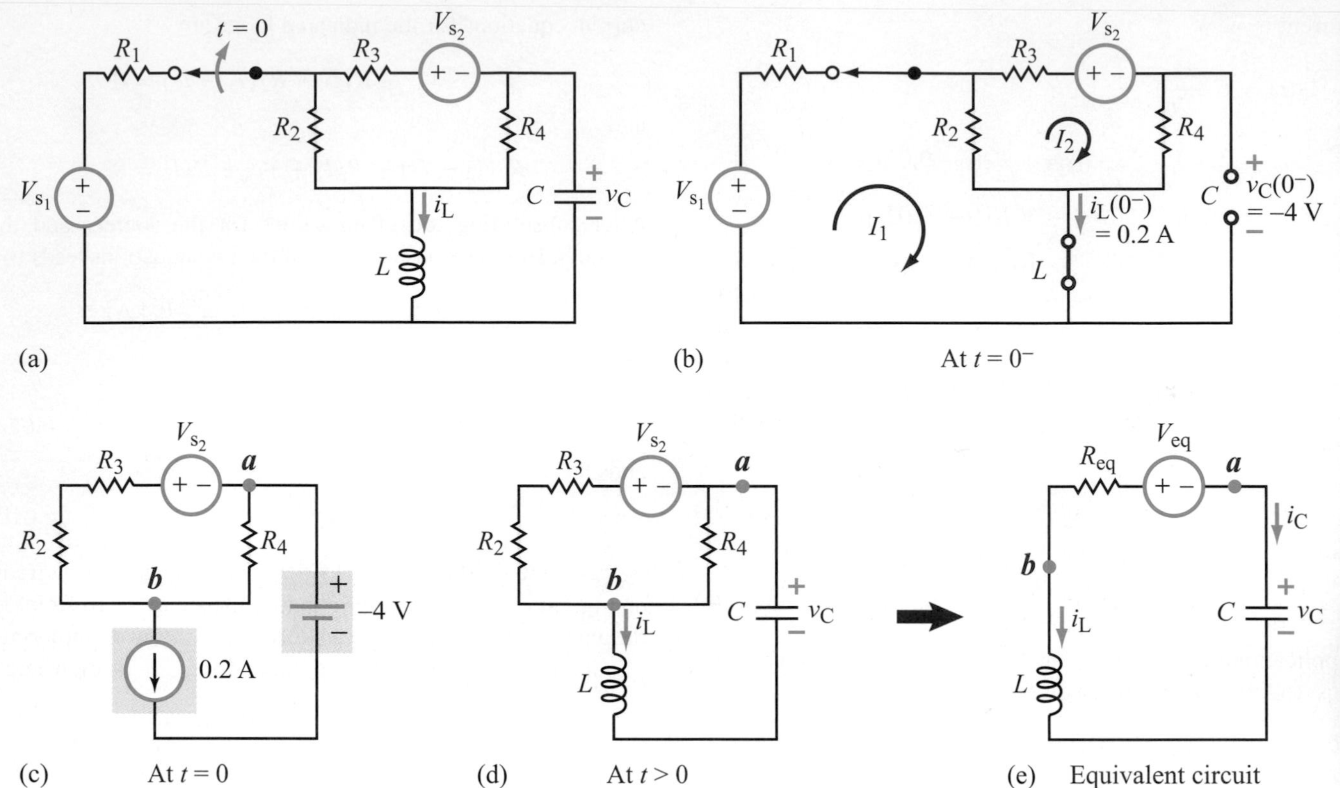

Figure 6-14: Circuit for Example 6-9.

From Fig. 6-14(e), it is evident that $i_C = -i_L$. Hence, we can use Eq. (6.61b) to determine

$$v'_C(0) = \frac{1}{C}\, i_C(0) = -\frac{1}{C}\, i_L(0)$$
$$= -\frac{1}{C}\, i_L(0^-) = -\frac{0.2}{2 \times 10^{-3}} = -100 \text{ V/s}.$$

From Table 6-2, the expressions for D_1 and D_2 are given by

$$D_1 = v_C(0) - v_C(\infty) = -4 + 4.8 = 0.8 \text{ V} \tag{6.63a}$$

and

$$D_2 = \frac{v'_C(0) + \alpha[v_C(0) - v_C(\infty)]}{\omega_d} = \frac{-100 + 5[-4 + 4.8]}{24.5} = -3.92 \text{ V}. \tag{6.63b}$$

With all unknown quantities accounted for, we have

$$v_C(t) = \{-4.8 + e^{-5t}[0.8 \cos 24.5t - 3.92 \sin 24.5t]\} \text{ V} \quad (\text{for } t \geq 0). \tag{6.64}$$

The waveform of $v_C(t)$ is displayed in Fig. 6-15.

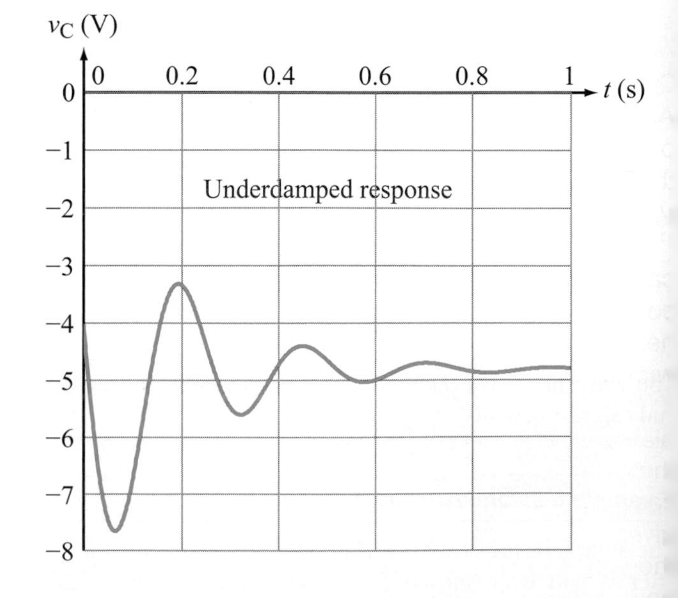

Figure 6-15: Plot of $v_C(t)$ as given by Eq. (6.64).

Technology Brief 12: Micromechanical Sensors and Actuators

Energy is stored in many different forms in the world around us. The conversion of energy from one form to another is called ***transduction***. Each of our five senses, for example, transduces a specific form of energy into electrochemical signals: tactile transducers on the skin convert mechanical and thermal energy; the eye converts electromagnetic energy; smell and taste receptors convert chemical energy; and our ears convert the mechanical energy of pressure waves. Any device, whether natural or man-made, that converts energy signals from one form to another is a ***transducer***.

Most modern man-made systems are designed to manipulate signals (i.e., information) using electrical energy. Computation, communication, and storage of information are examples of functions performed mostly with electrical circuits. Most systems also perform a fourth class of signal manipulation: the transduction of energy from the environment surrounding them into electrical signals that their circuits can use in support of their intended application. If the transducer converts external signals into electrical signals, it is called a ***sensor***. The charge-coupled device (CCD) chip on your camera is a sensor that converts electromagnetic energy (light) into electrical signals that can be processed, stored, and communicated by your camera circuits. Some transducers perform the reverse function, namely to convert a circuit's electrical signal into an environmental excitation. Such a transducer then is called an ***actuator***. The components that deploy the airbag in your car are actuators: given the right signal from the car's microcontroller, the actuators convert electrical energy into mechanical energy and the airbag is released and inflated.

Micro- and nanofabrication technology have begun to revolutionize many aspects of sensor and actuator design. Humans increasingly are able to embed transducers at very fine scales all over their environment. This is leading to big changes, as our computational elements are becoming increasingly aware of their environment. Shipping containers that track their own acceleration profiles (allowing the customer to detect falling and damage), laptops that scan fingerprints for routine login, cars that detect collisions, and even office suites that modulate energy consumption based on human activity are all examples of this transduction revolution. In this technology brief, we will focus on a specific type of microscale transducers that lend themselves to direct integration with silicon ICs. Collectively, devices of this type are called ***Microelectromechanical Systems (MEMS)*** or ***Microsystems Technologies (MST)***; the two names are used interchangeably.

A Capacitive Sensor: The MEMS Accelerometer

According to Eq. (5.21), the capacitance C of a parallel plate capacitor varies directly with A, the effective area of overlap between its two conducting plates, and inversely with d, the spacing between the plates. By capitalizing on these two attributes, capacitors can be made into motion sensors that can measure velocity and acceleration along x, y, and z.

Figure TF12-1 illustrates two mechanisms for translating motion into a change of capacitance. The first generally is called the ***gap-closing mode***, while the second one is called the ***overlap mode***. In the gap-closing mode, A remains constant, but if a vertical force is applied onto the upper plate, causing it to be displaced from its nominal position at height d above the lower plate to a new position $(d - z)$, then the value of capacitance C_z will change in accordance with the expression given in Fig. TF12-1(a). The sensitivity of C_z to the vertical displacement is given by dC_z/dz.

The overlap mode (Fig. TF12-1(b)) is used to measure horizontal motion. If a horizontal force causes one of the plates to shift by a distance y from its nominal position (where nominal position corresponds to a 100 percent overlap), the decrease in effective overlap area will lead to a corresponding change in the magnitude of capacitance C_y. In this case, d remains constant, but the width of the overlapped areas changes from w to $(w - y)$. The expression for C_y given in Fig. TF12-1(b) is reasonably accurate (even though it ignores the effects of the ***fringing electric field*** between the edges of the two plates) so long as $y \ll w$. To measure and amplify changes in capacitance, the capacitor can be integrated into an appropriate op-amp circuit whose output voltage is proportional to C. As we shall see shortly, a combination of three capacitors, one to sense vertical motion and two to measure horizontal motion along orthogonal axes, can provide complete information on both the velocity and acceleration vectors associated with the applied force. The capacitor configurations shown in Fig TF12-1 illustrate the basic concept of how a capacitor is used to measure motion, although more complex capacitor geometries also are possible, particularly for sensing angular motion.

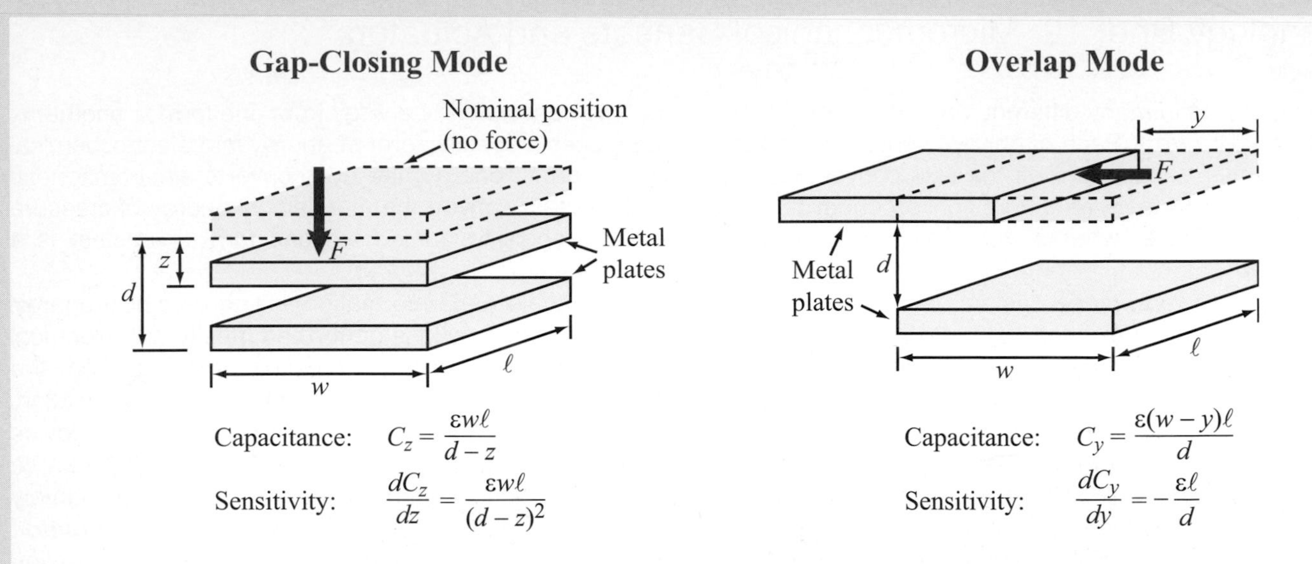

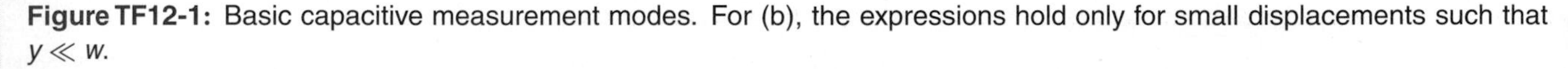

Figure TF12-1: Basic capacitive measurement modes. For (b), the expressions hold only for small displacements such that $y \ll w$.

To convert the capacitor-accelerometer concept into a practical sensor—such as the automobile accelerometer that controls the release of the airbag—let us consider the arrangement shown in Fig. TF12-2(b). The lower plate is fixed to the body of the vehicle, and the upper plate sits on a plane at a height d above it. The upper plate is attached to the body of the vehicle through a spring with a ***spring constant*** k. When no horizontal force is acting on the upper plate, its position is such that it provides a 100 percent overlap with the lower plate, in which case the capacitance will be a maximum at $C_y = \epsilon W\ell/d$. If the vehicle accelerates in the y-direction with acceleration a_y, the acceleration force F_{acc} will generate an opposing spring force F_{sp} of equal magnitude. Equating the two forces leads to an expression relating the displacement y to the acceleration a_y, as shown in the figure. Furthermore, the capacitance C_y is directly proportional to the overlap area $\ell(w - y)$ and therefore is proportional to the acceleration a_y. Thus, by measuring C_y, the accelerometer determines the value of a_y. A similar overlap-mode capacitor attached to the vehicle along the x-direction can be used to measure a_x. Through a similar analysis for the gap-closing mode capacitor shown in Fig TF12-2(a), we can arrive at a functional relationship that can be used to determine the vertical acceleration a_z by measuring capacitance C_z.

If we designate the time when the ignition starts the engine as $t = 0$, we then can set the initial conditions on both the velocity u of the vehicle and its acceleration a as zero at $t = 0$. That is, $u(0) = a(0) = 0$. The capacitor accelerometers measure continuous-time waveforms $a_x(t)$, $a_y(t)$, and $a_z(t)$. Each waveform then can be used by an op-amp integrator circuit to calculate the corresponding velocity waveform. For u_x, for example,

$$u_x(t) = \int_0^t a_x(t)\, dt,$$

and similar expressions apply to u_y and u_z.

Figure TF12-3 shows the Analog Devices ADXL202 accelerometer which uses the gap-closing mode to detec accelerations on a tiny micromechanical capacitor structure that works on the same principle described above, althoug slightly more complicated geometrically. Commercial accelerometers, however, make use of negative feedback t prevent the plates from physically moving. When an acceleration force attempts to move the plate, an electric negative feedback circuit applies a voltage across the plates to generate an electrical force between the plates that counteract the acceleration force exactly, thereby preventing any motion by the plate. The magnitude of the applied voltag

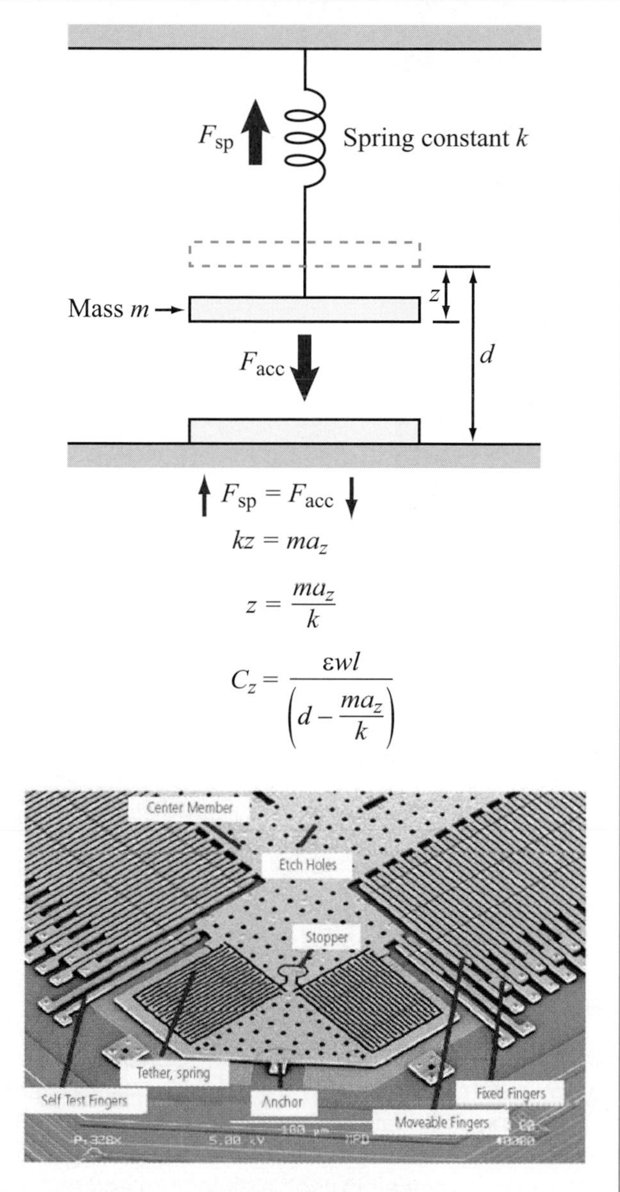

(a) The ADXL202 accelerometer employs many gap-closing capacitor sensors to detect acceleration. (Courtesy Analog Devices.)

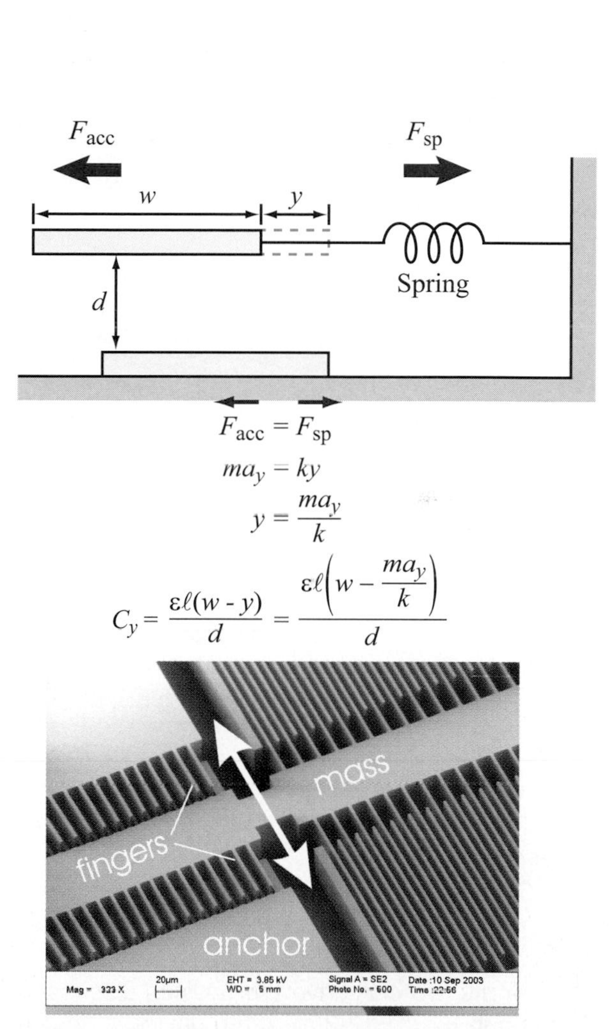

(b) A silicon sensor that uses overlap mode fingers. The white arrow shows the direction of motion of the moving mass and its fingers in relation to the fixed anchors. Note that the moving fingers move into and out of the fixed fingers on either side of the mass during motion. (Courtesy of the Adriatic Research Institute.)

Figure TF12-2: Adding a spring to a movable plate capacitor makes an accelerometer.

becomes a measure of the acceleration force that the capacitor plate is subjected to. Because of their small size and low power consumption, chip-based microfabricated silicon accelerometers are used in most modern cars to activate the release mechanism of airbags. They also are used heavily in many toy applications to detect position, velocity and acceleration. The Nintendo Wii, for example, uses the newer ADXL330 accelerometer in each remote to detect orientation and acceleration. Incidentally, a condenser microphone operates much like the device shown in

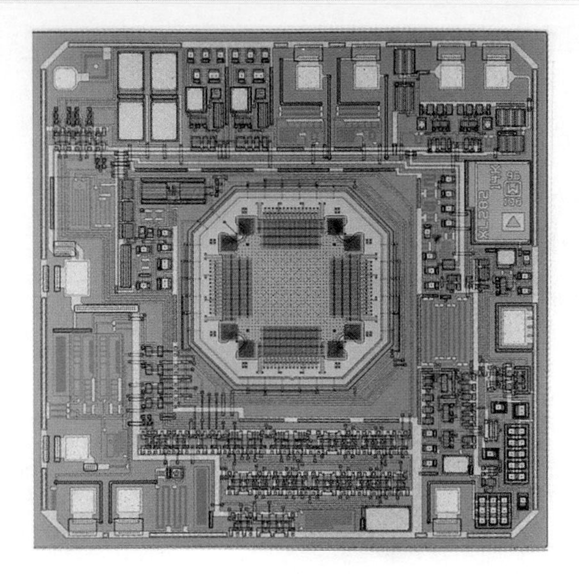

Figure TF12-3: The complete ADXL202 accelerometer chip. The center region holds the micromechanical sensor; the majority of the chip space is used for the electronic circuits that measure the capacitance change, provide feedback, convert the measurement into a digital signal, and perform self-tests. (Courtesy of Analog Devices.)

Fig. TF12-2(a): as air pressure waves (sound) hit the spring-mounted plate, it moves and the change in capacitance can be read and recorded.

A Capacitive Actuator: MEMS Electrostatic Resonators

Not surprisingly, we can drive the devices discussed previously in reverse to obtain actuators. Consider again the configuration in Fig. TF12-2(a). If the device is not experiencing any external forces and we apply a voltage V across the two plates, an attractive force F will develop between the plates. This is because charges of opposite polarity on the two plates give rise to an electrostatic force between them. This, in fact, is true for all capacitors. In the case of our actuator, however, we replace the normally stiff, dielectric material with air (since air is itself a dielectric) and attach it to a spring as before. With this modification, an applied potential generates an electrostatic force that moves the plates.

This basic idea can be applied to a variety of applications. The most successful one to date is the ***Digital Light Projector (DLP)*** system (see Technology Brief 6 on page 106) that drives most digital projectors used today. In the DLP hundreds of thousands of capacitor actuators are arranged in a 2-D array on a chip, with each actuator corresponding to a pixel on an image displayed by the projector. One capacitive plate of each pixel actuator (which is mirror smooth and can reflect light exceedingly well) is connected to the chip via a spring. In order to brighten or darken a pixel, a voltage is applied between the plates, causing the mirror to move into or out of the path of the projected light. These same devices have been used for many other applications, including microfluidic valves and tiny force sensors used to measure forces as small as a zeptonewton (1 zeptonewton $= 10^{-21}$ newtons).

6-5 The Parallel RLC Circuit

Having completed our examination of the series RLC circuit (Fig. 6-16(a)), we now turn our attention to the parallel RLC circuit shown in Fig. 6-16(b). As we will see shortly, the current $i(t)$ flowing through the inductor in the parallel RLC circuit is characterized by a second-order differential equation identical in form with that for the voltage $\upsilon(t)$ across the capacitor of the series RLC circuit. Accordingly, we will take advantage of this correspondence between the series and parallel RLC circuits by adapting the solutions we obtained in the preceding section for the series circuit to the solutions we seek in this section for the parallel circuit.

Application of KCL to the circuit in Fig. 6-16(b) gives

$$i_R + i + i_C = I_s \qquad (\text{for } t \geq 0). \qquad (6.65)$$

In terms of υ, the voltage common to all three passive elements, the equation becomes

$$\frac{\upsilon}{R} + i + C\,\frac{d\upsilon}{dt} = I_s. \qquad (6.66)$$

Using $\upsilon = L\,di/dt$ and rearranging the terms leads to

$$\frac{d^2i}{dt^2} + \frac{1}{RC}\frac{di}{dt} + \frac{1}{LC}\,i = \frac{I_s}{LC}, \qquad (6.67)$$

which can be rewritten in the abbreviated form

$$i'' + ai' + bi = c, \qquad (6.68)$$

where

$$a = \frac{1}{RC}, \qquad b = \frac{1}{LC}, \qquad \text{and} \qquad c = \frac{I_s}{LC}. \qquad (6.69)$$

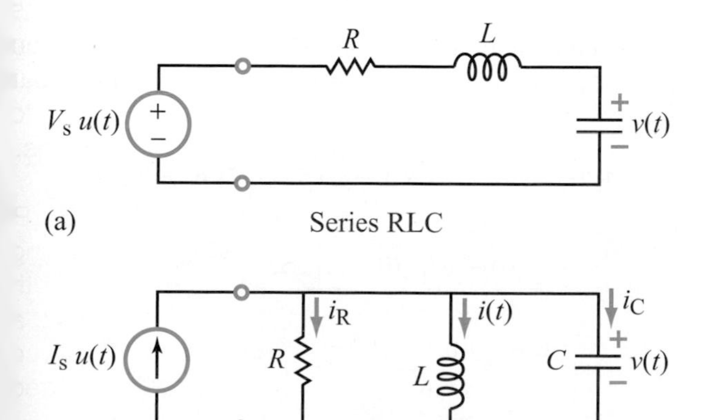

Figure 6-16: The differential equation for $\upsilon(t)$ of the series RLC circuit shown in (a) is identical in form with that of the current $i(t)$ in the parallel RLC circuit in (b).

Also, $i' = di/dt$. and $i'' = d^2i/dt^2$. Comparison of Eq. (6.68) with Eq. (6.6) for the capacitor voltage of the series RLC circuit reveals that the two differential equations are identical in form, albeit the constants a and c have different expressions in the two cases. By replacing υ with i in the set of expressions given by Eq. (6.46), we obtain the following general solution for $i(t)$.

Parallel RLC

Overdamped ($\alpha > \omega_0$)

$$i(t) = [i(\infty) + A_1e^{s_1t} + A_2e^{s_2t}] \qquad (\text{for } t \geq 0), \qquad (6.70a)$$

Critically Damped ($\alpha = \omega_0$)

$$i(t) = [i(\infty) + (B_1 + B_2t)e^{-\alpha t}] \qquad (\text{for } t \geq 0), \qquad (6.70b)$$

Underdamped ($\alpha < \omega_0$)

$$i(t) = [i(\infty) + e^{-\alpha t}(D_1 \cos \omega_d t + D_2 \sin \omega_d t)] \qquad (\text{for } t \geq 0), \qquad (6.70c)$$

In this situation s_1, s_2, ω_0, and ω_d retain the same expressions given earlier by Eqs. (6.28a), (6.28b), (6.29b), and (6.37), respectively. However, α is now given by

$$\alpha = \frac{1}{2RC} \qquad (\text{parallel RLC}). \qquad (6.71)$$

A summary of all relevant expressions is available in Table 6-2.

The constant $i(\infty)$ is the steady-state value of the inductor current at $t = \infty$, and the pairs of constants (A_1 and A_2, B_1 and B_2, and D_1 and D_2) are to be determined by matching the expressions for $i(t)$ to initial and final values.

Example 6-10: Parallel RLC Circuit

Determine $i_L(t)$ in the circuit of Fig. 6-17(a) for $t \geq 0$ given that $I_s = 0.5$ A, $V_0 = 12$ V, $R_1 = 60\ \Omega$, $R_2 = 30\ \Omega$, $L = 0.2$ H, and $C = 500\ \mu$F.

Solution: The circuit in Fig. 6-17(b) represents the steady-state condition of the circuit at $t = 0^-$ (prior to moving the switch). Given that I_s flows entirely through the short circuit representing the inductor, it follows that

$$i_L(0^-) = I_s = 0.5 \text{ A}$$

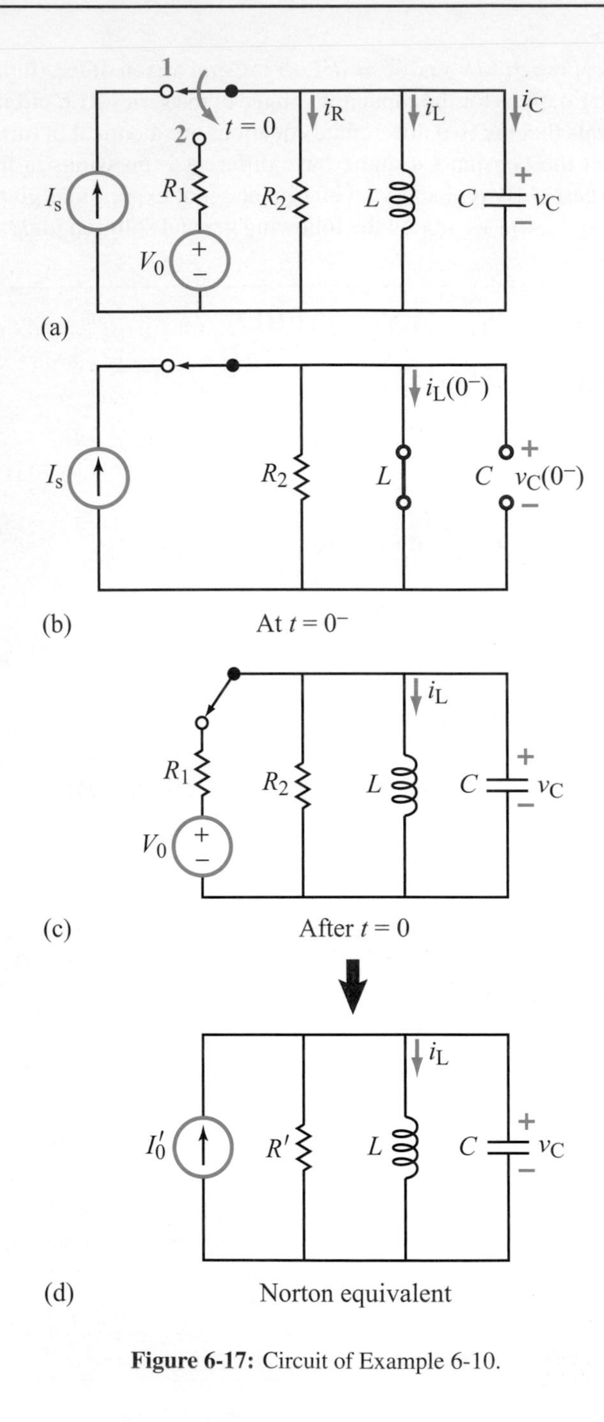

Figure 6-17: Circuit of Example 6-10.

and

$$v_C(0^-) = 0.$$

Since i_L through an inductor cannot change instantaneously (nor can v_C across a capacitor) these conditions are equally applicable at $t = 0$. Consequently,

$$i_L(0) = i_L(0^-) = 0.5 \text{ A}$$

and

$$i'_L(0) = \frac{1}{L}\, v_L(0) = \frac{1}{L}\, v_C(0) = 0.$$

After moving the switch ($t > 0$), the circuit assumes the configuration shown in Fig. 6-17(d), in which

$$I'_0 = \frac{V_0}{R_1} = \frac{12}{60} = 0.2 \text{ A}$$

and

$$R' = R_1 \parallel R_2 = \frac{R_1 R_2}{R_1 + R_2} = 20\ \Omega.$$

In view of the circuit in Fig. 6-17(d), the expressions for α and ω_0 are given by

$$\alpha = \frac{1}{2R'C} = \frac{1}{2 \times 20 \times 500 \times 10^{-6}} = 50 \text{ Np/s}$$

and

$$\omega_0 = \frac{1}{\sqrt{LC}} = \frac{1}{\sqrt{0.2 \times 500 \times 10^{-6}}} = 100 \text{ rad/s}.$$

Since $\alpha < \omega_0$, the circuit will exhibit an underdamped response with a damped natural frequency ω_d given by

$$\omega_d = \sqrt{\omega_0^2 - \alpha^2} = \sqrt{100^2 - 50^2} = 86.6 \text{ rad/s}.$$

The expression for $i_L(t)$ then is given by

$$i_L(t) = [i_L(\infty) + e^{-\alpha t}(D_1 \cos \omega_d t + D_2 \sin \omega_d t)] \quad (\text{for } t \geq 0).$$

At $t = \infty$, the inductor behaves like a short circuit, forcing I'_0 to flow through it exclusively. Hence,

$$i_L(\infty) = I'_0 = 0.2 \text{ A}.$$

The only remaining unknowns are D_1 and D_2, which we shall determine by applying the expressions given in Table 6-2, namely

$$D_1 = i_L(0) - i_L(\infty) = (0.5 - 0.2) \text{ A} = 0.3 \text{ A}$$

and

$$D_2 = \frac{i'_L(0) + \alpha[i_L(0) - i_L(\infty)]}{\omega_d} = \frac{0 + 50(0.5 - 0.2)}{86.6} = 0.17 \text{ A}.$$

The final expression for $i_L(t)$ then is given by

$$i_L(t) = [0.2 + e^{-50t}(0.3 \cos 86.6t + 0.17 \sin 86.6t)] \text{ A} \quad (\text{for } t \geq 0).$$

Exercise 6-12: Determine the initial and final values for i_L in the circuit of Fig. E6.12 and provide an expression for $i_L(t)$.

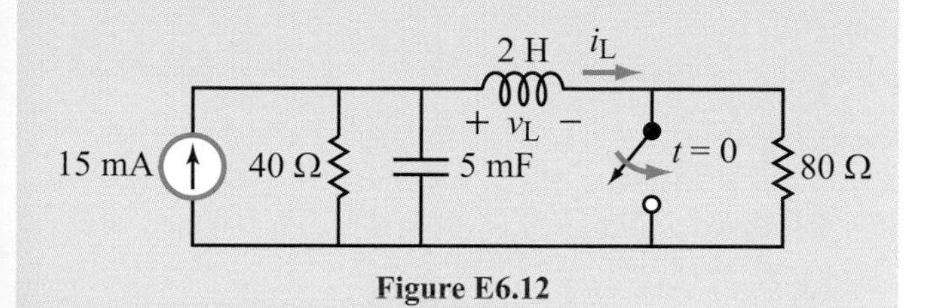

Figure E6.12

Answer: $i_L(0) = 5$ mA, $v_L(0) = 0.4$ V, $i_L(\infty) = 15$ mA, $\alpha = 2.5$ Np/s, $\omega_0 = 10$ rad/s, $\omega_d = 9.68$ rad/s, $i_L(t) = \{15 - [10\cos 9.68t - 18.08 \sin 9.68t]e^{-2.5t}\}$ mA. (See)

Exercise 6-13: In the parallel RLC circuit shown in Fig. 6-16(b), how much energy will be stored in L and C at $t = \infty$?

Answer: $w_L = \frac{1}{2}LI_s^2$, $w_C = 0$. (See)

6-6 General Solution for Any Second-Order Circuit

ccording to the material covered in the preceding sections, eries and parallel RLC circuits share a common set of haracteristics. An RLC circuit is characterized by a resonant requency ω_0 and a damping coefficient α, and when driven by a udden dc excitation, the circuit exhibits a response that decays xponentially as $e^{-\alpha t}$. It may or may not contain an oscillatory ariation, depending on whether ω_0 is or is not larger than α n magnitude, respectively. These characteristics arise from he interplay between energy storage and dissipation. During he operation of the RLC circuit, energy is exchanged between he two storage elements—the capacitor and the inductor— hrough the resistor. Dissipation is governed by $e^{-\alpha t}$, which e can redefine as $e^{-t/\tau}$ with $\tau = 1/\alpha$. In this alternative orm, the decay rate is specified by the time constant τ. If is short (rapid decay) in comparison with the duration of a ngle oscillation period T (where $T = 2\pi/\omega$) it means that nergy burns away too quickly to generate an oscillation. This the overdamped case. On the other hand, if τ is sufficiently ng (slow decay) in comparison with T, energy will move ack and forth between L and C, generating an oscillation. ith every cycle, however, the resistance will burn off some the remaining energy, resulting in an underdamped response that decays and oscillates simultaneously. If $R = 0$, the circuit will oscillate forever at the resonant frequency ω_0 (see Exercise 6-14).

Building on the experience we gained from our examination of the series and parallel RLC circuits, we now will extend the method of solution to any second-order circuit, including those containing op amps. For a circuit containing only dc sources (or no independent sources at all), we seek to find the circuit response $x(t)$ for $t \geq 0$, where $x(t)$ is a voltage or current of interest in the circuit and $t = 0$ is the instant at which the circuit experiences a sudden change (usually caused by a switch). To that end, we propose the following solution outline.

Step 1: Develop a second-order differential equation for $x(t)$ for $t \geq 0$. Express the equation in the general form

$$x'' + ax' + bx = c, \tag{6.72}$$

where a, b, and c are constants.

Step 2: Determine the values of α and ω_0 from

$$\alpha = \frac{a}{2} \quad \text{and} \quad \omega_0 = \sqrt{b}. \tag{6.73}$$

Step 3: Determine whether the response $x(t)$ is overdamped, critically damped, or underdamped and write down the expression corresponding to that case from the following general solution:

General Solution

Overdamped ($\alpha > \omega_0$)

$$x(t) = [x(\infty) + A_1 e^{s_1 t} + A_2 e^{s_2 t}] \quad (\text{for } t \geq 0), \tag{6.74a}$$

Critically Damped ($\alpha = \omega_0$)

$$x(t) = [x(\infty) + (B_1 + B_2 t)e^{-\alpha t}] \quad (\text{for } t \geq 0), \tag{6.74b}$$

Underdamped ($\alpha < \omega_0$)

$$x(t) = [x(\infty) + e^{-\alpha t}(D_1 \cos \omega_d t + D_2 \sin \omega_d t)] \quad (\text{for } t \geq 0), \tag{6.74c}$$

Table 6-3: General solution for second-order circuits for $t \geq 0$.

$x(t)$ = unknown variable (voltage or current)
Differential equation: $x'' + ax' + bx = c$
Initial conditions: $x(0)$ and $x'(0)$
Final condition: $x(\infty) = \frac{c}{b}$
$\alpha = \frac{a}{2}$ $\qquad \omega_0 = \sqrt{b}$
Overdamped Response $\alpha > \omega_0$
$x(t) = [x(\infty) + A_1 e^{s_1 t} + A_2 e^{s_2 t}]\, u(t)$
$s_1 = -\alpha + \sqrt{\alpha^2 - \omega_0^2}$ $\qquad s_2 = -\alpha - \sqrt{\alpha^2 - \omega_0^2}$
$A_1 = \frac{x'(0) - s_2[x(0) - x(\infty)]}{s_1 - s_2}$ $\qquad A_2 = -\left[\frac{x'(0) - s_1[x(0) - x(\infty)]}{s_1 - s_2}\right]$
Critically Damped $\alpha = \omega_0$
$x(t) = [x(\infty) + (B_1 + B_2 t)e^{-\alpha t}]\, u(t)$
$B_1 = x(0) - x(\infty)$ $\qquad B_2 = x'(0) + \alpha[x(0) - x(\infty)]$
Underdamped $\alpha < \omega_0$
$x(t) = x(\infty) + [D_1 \cos \omega_d t + D_2 \sin \omega_d t]e^{-\alpha t}\, u(t)$
$D_1 = x(0) - x(\infty)$ $\qquad D_2 = \frac{x'(0) + \alpha[x(0) - x(\infty)]}{\omega_d}$
$\omega_d = \sqrt{\omega_0^2 - \alpha^2}$

where

$$s_1 = -\alpha + \sqrt{\alpha^2 - \omega_0^2}, \tag{6.75a}$$

$$s_2 = -\alpha - \sqrt{\alpha^2 - \omega_0^2}, \tag{6.75b}$$

and

$$\omega_d = \sqrt{\omega_0^2 - \alpha^2}. \tag{6.75c}$$

[The three expressions given by Eq. (6.74) represent the circuit response to a sudden change that occurred at $t = 0$. Had the sudden change occurred at $t = T_0$ instead, the expressions would continue to apply, but t will need to be replaced with $(t - T_0)$ everywhere on the right-hand side (only) of those expressions.]

Step 4: Evaluate the circuit to determine $x(\infty)$ at $t = \infty$. Alternatively, we apply the form of Eq. (6.12), namely

$$x(\infty) = \frac{c}{b}. \tag{6.75d}$$

Step 5: Apply initial conditions for $x(t)$ and $x'(t)$ at $t = 0$ (or at $t = T_0$ if the sudden change occurred at T_0) to determine the remaining unknown constants.

This procedure is highlighted in Table 6-3 and demonstrated through Examples 6-11 through 6-13.

Example 6-11: RLC Circuit with a Short-Circuit Switch

The switch in the circuit of Fig. 6-18(a) had been open for a long time before it was closed at $t = 0$. Determine $i_L(t)$ for $t \geq 0$. The circuit elements have the following values: $V_0 = 24$ V, $R_1 = 4\ \Omega$, $R_2 = 8\ \Omega$, $R_3 = 12\ \Omega$, $L = 2$ H, and $C = 0.2$ F.

Solution: Figures 6-18(b), (c), and (d) depict the state of the circuit at $t = 0^-$, $t \geq 0$, and $t = \infty$, respectively.

Step 1: Obtain differential equation for $\boldsymbol{i_L(t)}$

After closing the switch, node 1 gets connected to node 2, and R_2 becomes inconsequential to the rest of the circuit because it is connected in parallel with a short circuit. At node 2 of the circuit in Fig. 6-18(c), KCL gives

$$-i_1 + i_L + i_C = 0. \tag{6.76}$$

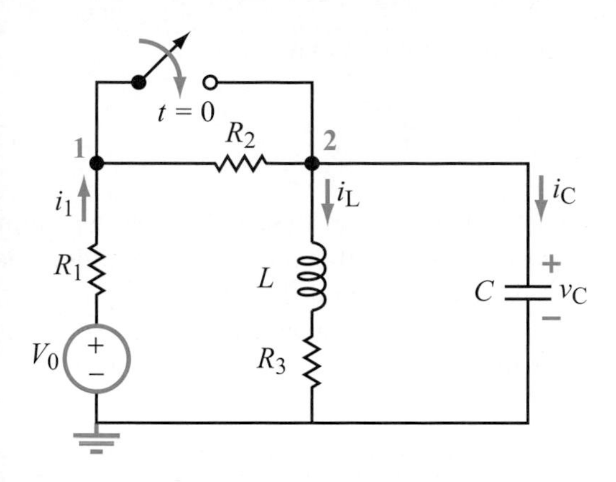

(a) Circuit with switch

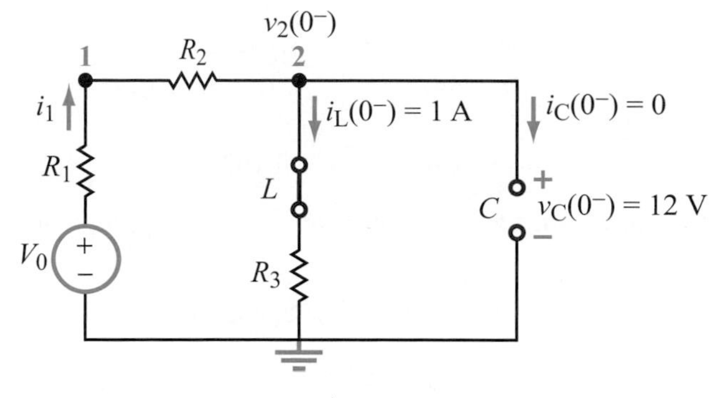

(b) At $t = 0^-$: $i_L(0^-) = V_0/(R_1 + R_2 + R_3) = 1$ A, and $v_C(0^-) = i_L(0^-)\,R_3 = 12$ V.

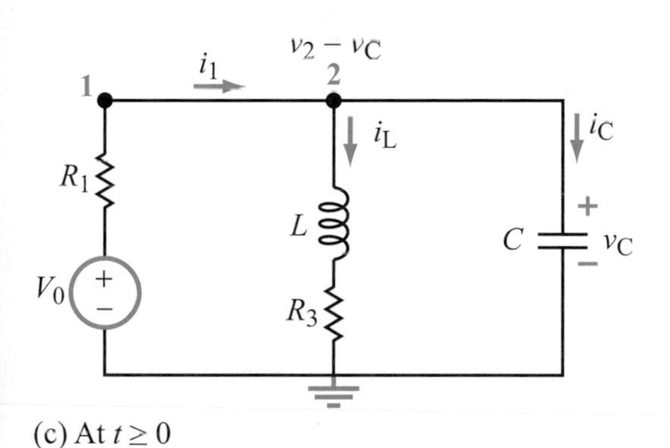

(c) At $t \geq 0$

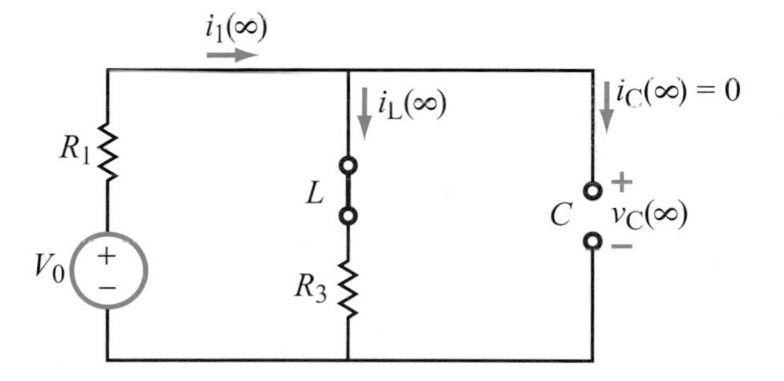

(d) At $t = \infty$: $i_L(\infty) = V_0/(R_1 + R_3) = 1.5$ A.

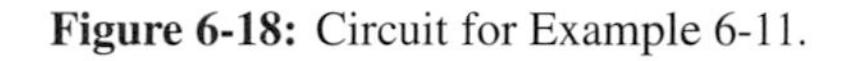

Figure 6-18: Circuit for Example 6-11.

n terms of the node voltage v_C,

$$-i_1 = \frac{v_C - V_0}{R_1} \tag{6.77a}$$

nd

$$i_C = C\,\frac{dv_C}{dt}. \tag{6.77b}$$

ence,

$$\frac{v_C}{R_1} + i_L + C\,\frac{dv_C}{dt} = \frac{V_0}{R_1}. \tag{6.78}$$

he voltage v_C is equal to the sum of the voltages across L nd R_3,

$$v_C = L\,\frac{di_L}{dt} + i_L R_3. \tag{6.79}$$

ubstituting Eq. (6.79) in Eq. (6.78) leads to

$$\frac{1}{R_1}\left(L\,\frac{di_L}{dt} + i_L R_3\right) + i_L + C\,\frac{d}{dt}\left(L\,\frac{di_L}{dt} + i_L R_3\right) = \frac{V_0}{R_1}. \tag{6.80}$$

After carrying out the differentiation in the third term and rearranging the terms, we have

$$\frac{d^2 i_L}{dt^2} + \left(\frac{L + R_1 R_3 C}{R_1 LC}\right)\frac{di_L}{dt} + \left(\frac{R_1 + R_3}{R_1 LC}\right) i_L = \frac{V_0}{R_1 LC}. \tag{6.81}$$

For convenience, we rewrite Eq. (6.81) in the compact form

$$i_L'' + a i_L' + b i_L = c, \tag{6.82}$$

where

$$a = \frac{L + R_1 R_3 C}{R_1 LC} = \frac{2 + 4 \times 12 \times 0.2}{4 \times 2 \times 0.2} = 7.25,$$

$$b = \frac{R_1 + R_3}{R_1 LC} = \frac{4 + 12}{4 \times 2 \times 0.2} = 10,$$

and

$$c = \frac{V_0}{R_1 LC} = \frac{24}{4 \times 2 \times 0.2} = 15.$$

Step 2: Determine $\boldsymbol{\alpha}$ and $\boldsymbol{\omega_0}$

$$\alpha = \frac{a}{2} = \frac{7.25}{2} = 3.625,$$

and

$$\omega_0 = \sqrt{b} = \sqrt{10} = 3.162.$$

Step 3: Determine damping condition and select appropriate expression

Since $\alpha > \omega_0$, the response is overdamped, and

$$i_L(t) = [i_L(\infty) + (A_1 e^{s_1 t} + A_2 e^{s_2 t})]$$

with

$$s_1 = -\alpha + \sqrt{\alpha^2 - \omega_0^2} = -1.85 \text{ Np/s},$$

and

$$s_2 = -\alpha - \sqrt{\alpha^2 - \omega_0^2} = -5.40 \text{ Np/s}.$$

Step 4: Determine $\boldsymbol{i_L(\infty)}$

From the circuit in Fig. 6-18(d), $i_C = 0$ (open-circuit capacitor) and

$$i_L(\infty) = \frac{V_0}{R_1 + R_3} = \frac{24}{4 + 12} = 1.5 \text{ A}. \tag{6.83}$$

Step 5: Invoke initial conditions

With C acting like an open circuit at $t = 0^-$ (Fig. 6-18(b)),

$$I_L(0^-) = i_1(0^-) = \frac{V_0}{R_1 + R_2 + R_3} = 1 \text{ A}.$$

Since i_L cannot change in zero time,

$$i_L(0) = i_L(0^-) = 1 \text{ A}. \tag{6.84}$$

We need one additional relationship involving A_1 and A_2, which can be provided by the initial condition on i'_L. From the circuit in Fig. 6-18(b) at $t = 0^-$, we have

$$\begin{aligned} v_C(0^-) &= i_L(0^-)\ R_3 \\ &= 1 \times 12 \\ &= 12 \text{ V}. \end{aligned}$$

As we transition from $t = 0^-$ (before closing the switch) to $t = 0$ (after closing the switch), neither i_L nor v_C can change, which means that the voltage $v_2(0)$ at node 2 will continue to be 12 V and the current i_L through R_3 will continue to be 1 A. Hence, the voltage $v_L(0)$ has to be

$$\begin{aligned} v_L(0) &= v_2(0) - i_L(0)\ R_3 \\ &= 12 - 1 \times 12 \\ &= 0. \end{aligned}$$

Since $v_L = L\, di_L/dt$, it follows that

$$i'_L(0) = 0. \tag{6.85}$$

The expressions for A_1 and A_2 in Table 6-3 are given in terms of x, the variable associated with the second-order differential equation. In the present case, our differential equation is given by Eq. (6.82) with $i_L(t)$ as the unknown variable. Hence, by setting $x = i_L$ in the expressions for A_1 and A_2, we have

$$\begin{aligned} A_1 &= \frac{i'_L(0) - s_2[i_L(0) - i_L(\infty)]}{s_1 - s_2} \\ &= \frac{0 + 5.4(1 - 1.5)}{-1.85 + 5.4} \\ &= -0.76 \text{ A} \end{aligned}$$

and

$$\begin{aligned} A_2 &= -\left[\frac{i'_L(0) - s_1[i_L(0) - i_L(\infty)]}{s_1 - s_2}\right] \\ &= -\left[\frac{0 + 1.85(1 - 1.5)}{-1.85 + 5.4}\right] \\ &= 0.26 \text{ A}, \end{aligned}$$

and the final solution is then given by

$$i_L(t) = [1.5 - 0.76e^{-1.85t} + 0.26e^{-5.4t}] \text{ A} \quad (\text{for } t \ge 0).$$

Exercise 6-14: Develop an expression for $i_C(t)$ in the circuit of Fig. E6.14 for $t \ge 0$.

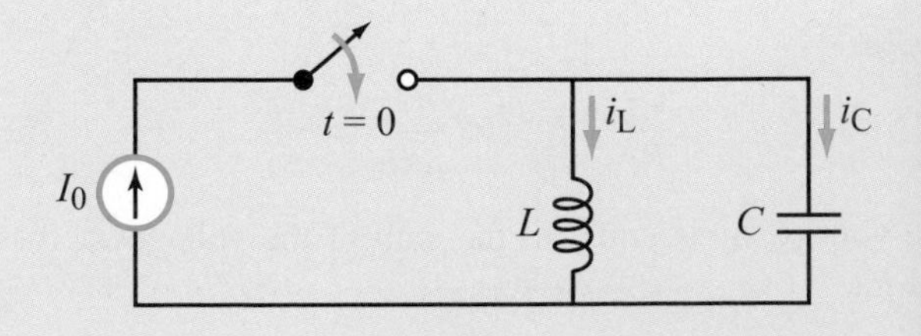

Figure E6.14

Answer: $i_C(t) = I_0 \cos \omega_0 t$ with $\omega_0 = 1/\sqrt{LC}$. This is an LC *oscillator* circuit in which dc energy provided by the current source is converted into ac energy in the LC circuit. (See)

Example 6-12: Two-Inductor Circuit

Determine $i_1(t)$ and $i_2(t)$ in the circuit of Fig. 6-19 for $t \geq 0$. The component values are $V_s = 1.4$ V, $R_1 = 0.4\ \Omega$, $R_2 = 0.3\ \Omega$, $L_1 = 0.1$ H, and $L_2 = 0.2$ H.

Solution: We designate i_x and i_y as the mesh currents in the two loops, as shown. We will analyze the circuit in terms of i_x and i_y and then use the solutions to determine i_1 and i_2.

For $t \geq 0$, the mesh equations are given by:

$$-V_s + R_1 i_x + L_1 \frac{d}{dt}(i_x - i_y) = 0 \qquad (i_x \text{ loop})$$

and

$$L_1 \frac{d}{dt}(i_y - i_x) + R_2 i_y + L_2 \frac{di_y}{dt} = 0, \qquad (i_y \text{ loop})$$

which can be rearranged and rewritten in the forms

$$R_1 i_x + L_1 i'_x - L_1 i'_y = V_s \qquad (i_x \text{ loop}) \qquad (6.86)$$

and

$$-L_1 i'_x + R_2 i_y + (L_1 + L_2) i'_y = 0. \qquad (i_y \text{ loop}) \qquad (6.87)$$

Step 1: Develop a differential equation in $\boldsymbol{i_x}$ alone

Take the time derivative of all terms in the i_y-loop equation:

$$-L_1 i''_x + R_2 i'_y + (L_1 + L_2) i''_y = 0. \qquad (6.88)$$

To convert Eq. (6.88) into a differential equation in i_x alone, we need to develop expressions for i'_y and i''_y in terms of i_x and its derivatives. By isolating i'_y in Eq. (6.86), we have

$$i'_y = \frac{R_1}{L_1}\, i_x + i'_x - \frac{V_s}{L_1}. \qquad (6.89)$$

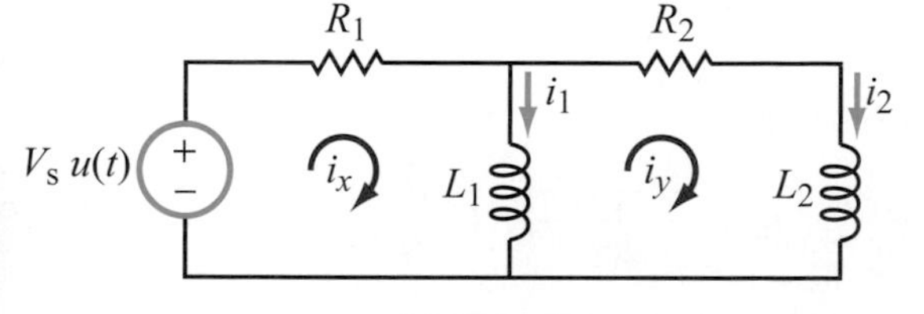

(a) Circuit

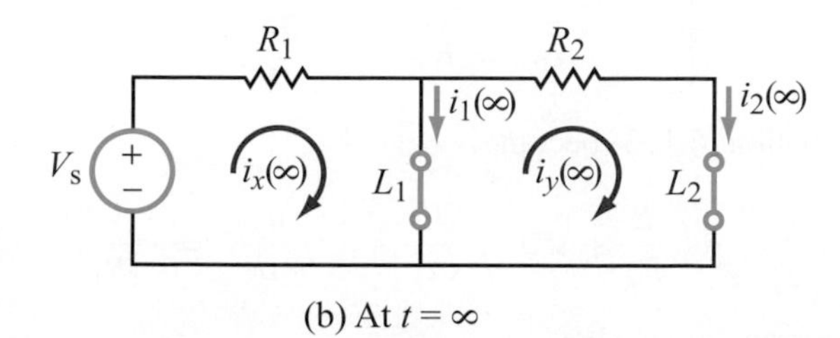

(b) At $t = \infty$

Figure 6-19: Circuit for Example 6-12.

To obtain an expression for i''_y, we simply take the derivative of Eq. (6.89),

$$i''_y = \frac{R_1}{L_1}\, i'_x + i''_x. \qquad (6.90)$$

After inserting Eqs. (6.89) and (6.90) into Eq. (6.88) and rearranging terms, we have

$$i''_x + \left[\frac{(R_1 + R_2)L_1 + R_1 L_2}{L_1 L_2}\right] i'_x + \left(\frac{R_1 R_2}{L_1 L_2}\right) i_x = \frac{R_2 V_s}{L_1 L_2}, \qquad (6.91)$$

which can be rewritten in the compact form

$$i''_x + a i'_x + b i_x = c, \qquad (6.92)$$

where

$$a = \frac{(R_1 + R_2)L_1 + R_1 L_2}{L_1 L_2} = 7.5,$$

$$b = \frac{R_1 R_2}{L_1 L_2} = 6, \qquad \text{and} \qquad c = \frac{R_2 V_s}{L_1 L_2} = 21.$$

Step 2: Evaluate $\boldsymbol{\alpha}$, $\boldsymbol{\omega_0}$, $\boldsymbol{s_1}$, and $\boldsymbol{s_2}$

$$\alpha = \frac{a}{2} = \frac{7.5}{2} = 3.75 \text{ Np/s},$$

$$\omega_0 = \sqrt{b} = \sqrt{6} = 2.45 \text{ rad/s},$$

$$s_1 = -\alpha + \sqrt{\alpha^2 + \omega_0^2}$$

$$= -3.75 + \sqrt{(3.75)^2 - 6} = -0.91 \text{ Np/s},$$

and

$$s_2 = -3.75 - \sqrt{(3.75)^2 - 6} = -6.6 \text{ Np/s}.$$

Step 3: Write expression for $\boldsymbol{i_x(t)}$

Since $\alpha > \omega_0$, i_x will exhibit an overdamped response given by

$$i_x(t) = [i_x(\infty) + A_1 e^{s_1 t} + A_2 e^{s_2 t}]$$

$$= [i_x(\infty) + A_1 e^{-0.91t} + A_2 e^{-6.6t}]. \qquad (6.93)$$

Step 4: Evaluate final condition

At $t = \infty$, the inductors in the circuit behave like short circuits (Fig. 6-19(b)), in which case the current generated by V_s will flow entirely through L_1. Hence,

$$i_x(\infty) = \frac{V_s}{R_1} = \frac{1.4}{0.4} = 3.5 \text{ A},$$

and

$$i_y(\infty) = 0.$$

The expression for $i_x(t)$ becomes

$$i_x(t) = 3.5 + A_1 e^{-0.91t} + A_2 e^{-6.6t}. \qquad (6.94)$$

Step 5: Invoke initial conditions

Before $t = 0$, the circuit contained no sources. Hence,

$$i_1(0) = i_1(0^-) = 0,$$

and

$$i_2(0) = i_2(0^-) = 0,$$

which implies that

$$i_x(0) = i_x(0^-) = 0, \quad (6.95)$$

and

$$i_y(0) = i_y(0^-) = 0. \quad (6.96)$$

At $t = 0$ with no currents flowing through either loop, the voltages across L_1 and L_2 are both equal to V_s. That is

$$i'_1(0) = \frac{1}{L_1}\, \upsilon_{L_1}(0) = \frac{V_s}{L_1}, \quad (6.97a)$$

and

$$i'_2(0) = \frac{1}{L_2}\, \upsilon_{L_2}(0) = \frac{V_s}{L_2}, \quad (6.97b)$$

Consequently,

$$i'_x(0) = i'_1(0) + i'_2(0) = \frac{V_s}{L_1} + \frac{V_s}{L_2} = 21. \quad (6.98)$$

Now that we know the values of $i_x(0)$, $i'_x(0)$, and $i_x(\infty)$, we can apply the general expressions for A_1 and A_2 in Table 6-3 to get

$$A_1 = \frac{i'_x(0) - s_2[i_x(0) - i_x(\infty)]}{s_1 - s_2} = \frac{21 + 6.6(0 - 3.5)}{-0.91 + 6.6} = -0.36 \text{ A}$$

and

$$A_2 = -\left[\frac{i'_x(0) - s_1[i_x(0) - i_x(\infty)]}{s_1 - s_2}\right] = -\left[\frac{21 + 0.91(0 - 3.5)}{-0.91 + 6.6}\right] = -3.14 \text{ A}.$$

The final expression for $i_x(t)$ then is given by

$$i_x(t) = [3.5 - 0.36e^{-0.91t} - 3.14e^{-6.6t}] \text{ A}. \quad (6.99)$$

Repetition of Steps 1 through 4 for i_y requires that we start by taking the time derivative of the i_x-loop equation (Eq. (6.86)) and then using the i_y-loop equation (Eq. (6.87)) to generate expressions for i'_x and i''_x. The procedure leads to

$$i_y(t) = 1.23(e^{-0.91t} - e^{-6.6t}) \text{ A}. \quad (6.100)$$

Finally, the solutions for $i_1(t)$ and $i_2(t)$ are

$$i_1(t) = i_x(t) - i_y(t) = [3.5 - 1.59e^{-0.91t} - 1.91e^{-6.6t}] \text{ A}, \quad (6.101a)$$

and

$$i_2(t) = i_y(t) = 1.23(e^{-0.91t} - e^{-6.6t}) \text{ A} \quad (6.101b)$$

(for $t \geq 0$).

Exercise 6-15: For the circuit in Fig. E6.15, determine $i_C(t)$ for $t \geq 0$.

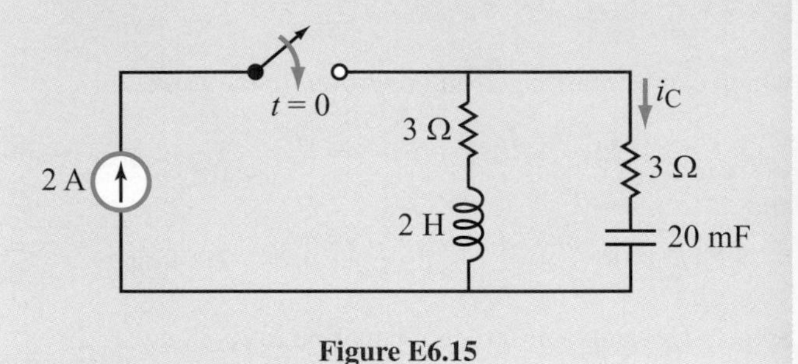

Figure E6.15

Answer: $i_C(t) = 2e^{-1.5t} \cos 4.77t$ A. (See [CD icon])

Example 6-13: Second-Order Op-Amp Circuit

Determine $i_L(t)$ in the op-amp circuit of Fig. 6-20(a) for $t \geq 0$. Assume $V_s = 1$ mV, $R_1 = 10$ kΩ, $R_2 = 1$ MΩ, $R_3 = 100$ Ω, $L = 5$ H, and $C = 1$ μF.

Solution: KCL at node υ_n gives

$$i_1 + i_n + i_2 + i_3 = 0,$$

or equivalently,

$$\frac{\upsilon_n - V_s}{R_1} + i_n + \frac{\upsilon_n - \upsilon_{out}}{R_2} + C\,\frac{d}{dt}(\upsilon_n - \upsilon_{out}) = 0. \quad (6.103)$$

Since $\upsilon_n = \upsilon_p = 0$, $i_n = 0$, and

$$\upsilon_{out} = R_3 i_L + L\,\frac{di_L}{dt}, \quad (6.104)$$

Equation (6.103) becomes

$$\frac{R_3}{R_2}\, i_L + \left(\frac{L}{R_2} + R_3 C\right)\frac{di_L}{dt} + LC\,\frac{d^2 i_L}{dt^2} = -\frac{V_s}{R_1}. \quad (6.105)$$

Rearranging, we have

$$i''_L + a i'_L + b i_L = c, \quad (6.106)$$

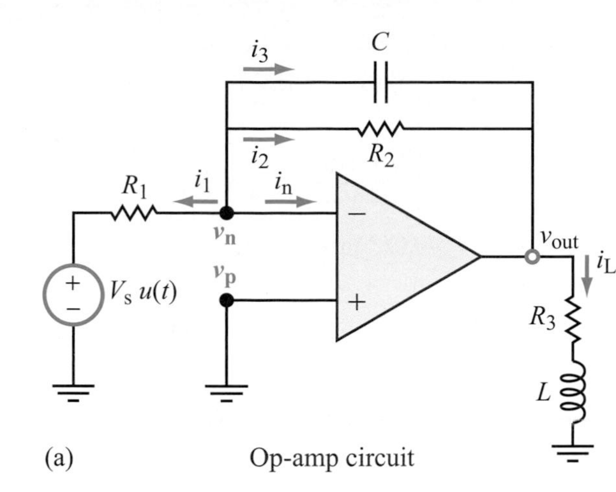

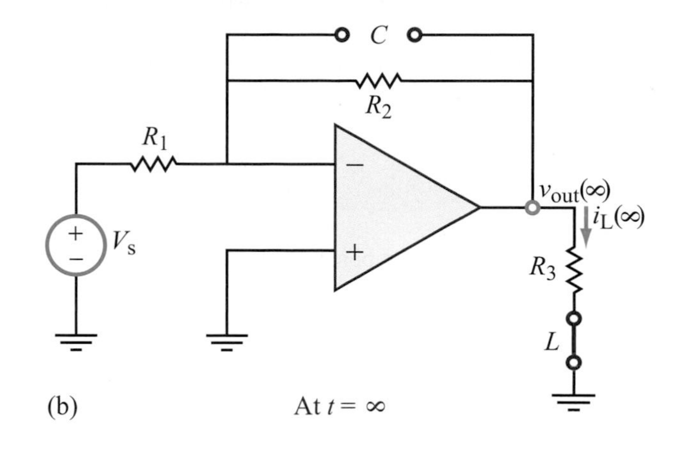

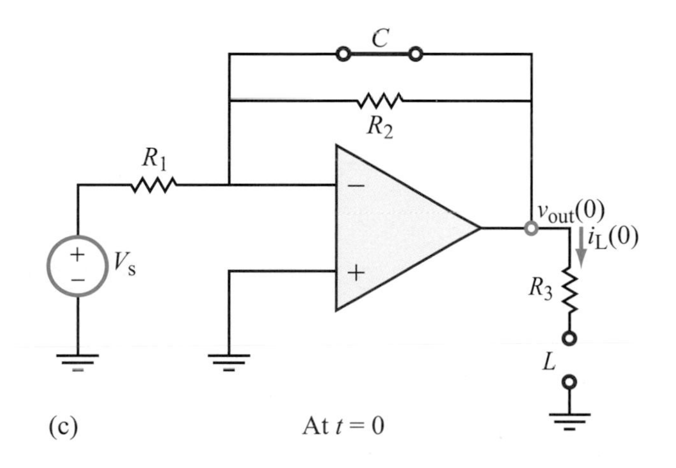

Figure 6-20: Op-amp circuit of Example 6-13.

where

$$a = \frac{L + R_2 R_3 C}{R_2 LC} = 21,$$

$$b = \frac{R_3}{R_2 LC} = 20,$$

and

$$c = \frac{-V_s}{R_1 LC} = -0.02.$$

The damping behavior of i_L is determined by how the magnitude of α compares with that of ω_0:

$$\alpha = \frac{a}{2} = 10.5 \text{ Np/s},$$

and

$$\omega_0 = \sqrt{b} = \sqrt{20} = 4.47 \text{ rad/s}.$$

Since $\alpha > \omega_0$, i_L will exhibit an overdamped response given by

$$i_L(t) = [i_L(\infty) + A_1 e^{s_1 t} + A_2 e^{s_2 t}]\, u(t),$$

with

$$s_1 = -\alpha + \sqrt{\alpha^2 - \omega_0^2} = -1.0$$

and

$$s_2 = -\alpha - \sqrt{\alpha^2 - \omega_0^2} = -20.$$

At $t = \infty$, the circuit assumes the equivalent configuration shown in Fig. 6-20(b), which is an inverting amplifier with an output voltage

$$v_{out}(\infty) = -\frac{R_2}{R_1} V_s.$$

Hence,

$$i_L(\infty) = \frac{v_{out}(\infty)}{R_3} = -\frac{R_2 V_s}{R_1 R_3} = -1 \text{ mA}.$$

The expression for $i_L(t)$ becomes

$$i_L(t) = [-10^{-3} + A_1 e^{-t} + A_2 e^{-20t}]. \qquad (6.107)$$

To determine the values of A_1 and A_2, we examine initial conditions for i_L and i'_L. At $t = 0^-$, there were no active sources in the circuit, and since i_L cannot change instantaneously, it follows that

$$i_L(0) = i_L(0^-) = 0,$$

which means that the inductor behaves like an open circuit at $t = 0$, as depicted in Fig. 6-20(c). Also, since the voltage υ_C across the capacitor was zero before $t = 0$, it has to remain at zero at $t = 0$, which is why it has been replaced with a short circuit in Fig. 6-20(c). Consequently, $\upsilon_{out}(0) = 0$, $\upsilon_L(0) = 0$, and

$$i'_L(0) = \frac{1}{L}\,\upsilon_L(0) = 0.$$

From Table 6-3 with $x = i_L$,

$$\begin{aligned} A_1 &= \frac{i'_L(0) - s_2[i_L(0) - i_L(\infty)]}{s_1 - s_2} \\ &= \frac{0 + 20(0+1)}{-1+20} \times 10^{-3} && (6.108) \\ &= 1.05 \text{ mA}, && (6.109) \end{aligned}$$

and

$$\begin{aligned} A_2 &= -\left[\frac{i'_L(0) - s_1[i_L(0) - i_L(\infty)]}{s_1 - s_2}\right] \\ &= -\left[\frac{0 + 1(0+1)}{-1+20}\right] \times 10^{-3} && (6.110) \\ &= -0.053 \text{ mA}. && (6.111) \end{aligned}$$

The final expression for $i_L(t)$ then is given by

$$i_L(t) = -[1 - 1.05e^{-t} + 0.053e^{-20t}] \text{ mA} \qquad (\text{for } t \geq 0).$$

Review Question 6-12: A circuit contains two capacitors and three inductors, in addition to resistors and sources. Under what circumstance is it a second-order circuit?

Review Question 6-13: Suppose $a = 0$ in Eq. (6.72). What type of response will $x(t)$ have in that case?

6-7 Application Note: Micromechanical Transducers

In Section 6-6, we noted that for any circuit containing two energy-storage elements (capacitors and/or inductors) in addition to dissipative elements in the form of resistors, its voltages and currents are characterized by second-order differential equations of the type given by Eq. (6.72) as

$$x'' + ax' + bx = c. \qquad (6.102)$$

In fact, this equation also is applicable to a wide variety of non-electrical systems, so long as (1) they contain two energy storage media or mechanisms, (2) energy can transfer between them, and (3) when the energy transfer takes place, it occurs through a dissipative third medium. A guitar string provides one such example. When plucked, a pulse of force is applied to the string, causing it to vibrate at an angular frequency ω_d. The applied force provides energy to the string in two forms: as kinetic energy as the string moves back and forth and as potential energy because of the tension induced in the string. In this case, the same medium (the string) serves as the storage medium for both forms of energy.

In addition to the back-and-forth transfer between the two forms of energy, the system experiences dissipation by the air resistance surrounding the string. By moving the air molecules, the string transfers kinetic energy to them, resulting in a small rise in air temperature. Thus, the guitar string does contain two forms of energy storage and a dissipative medium, thereby satisfying the basic conditions that underlie Eq. (6.102). When plucked, a guitar string exhibits an underdamped response; it vibrates at some angular frequency ω_d. The amplitude of its vibrations decays exponentially with time, just like the underdamped voltage behavior displayed in Fig. 6-10 for the series RLC circuit. If we were to dip the guitar string in molasses and then try to pluck it, the string movement would exhibit an overdamped response with no oscillations (and no sound generated by it).

Systems that are characterized by a differential equation of the form of Eq. (6.102) are called ***resonators***. The two critical parameters defining the resonance character of a resonator are $\alpha = a/2$ and $\omega_0 = \sqrt{b}$. Examples of mechanical resonators include bridges, pendulums, and most musical instruments.

Example 6-14: Resonant Frequency of MEMS Accelerometer

In Technology Brief 12 on page 269, we examined how a capacitor-spring combination can be used to measure the acceleration of the body to which it is attached. The diagram shown in Fig. 6-21 depicts a situation in which the upper plate of the capacitor is under the influence of an ***external force*** $\mathbf{F}_{\text{ext}}$, which is directed along the $+z$ direction, defined to be the downward direction. Had the plate been in perfect isolation, the external force would have imparted an ***acceleration force*** $\mathbf{F}_{\text{acc}}$ on the plate of the same magnitude and direction as $\mathbf{F}_{\text{ext}}$. Because the plate is connected to the spring, stretching the spring by a distance z (from its unstretched position) induces a ***spring force*** $\mathbf{F}_{\text{sp}}$ given by

$$\mathbf{F}_{\text{sp}} = -\hat{\mathbf{z}}kz, \tag{6.112}$$

where k is the spring constant and $\hat{\mathbf{z}}$ is a unit vector pointed along the $+z$ direction. Known as ***Hooke's law***, the spring force is proportional to z, which is the distance by which it is stretched, and it contains a negative sign to denote that when the spring is stretched out (downward) the downward-defined spring force is actually upward, and vice versa.

The movement of the plate induces another opposition force known as a ***drag*** or ***damper force***, which is a resistance force due to the air molecules that the plate is pushing along its path. The drag force is given by

$$\mathbf{F}_{\text{drag}} = -\hat{\mathbf{z}}\beta u, \tag{6.113}$$

where β is an air-resistance constant and u is the downward velocity of the upper plate.

Show that the differential equation for the displacement z is of the form given by Eq. (6.102). Determine the associated resonator parameters α and ω_0, given that the capacitor accelerometer is a ***MEMS*** (microelectromechanical system) design with an upper plate whose mass is 10 μg, $k = 9$ N/m, and $\beta = 0.06$ (N·s/m).

Solution: The acceleration force on the plate is equal to the net force acting on it,

$$\mathbf{F}_{\text{acc}} = \mathbf{F}_{\text{ext}} + \mathbf{F}_{\text{sp}} + \mathbf{F}_{\text{drag}} \tag{6.114}$$

or, equivalently,

$$\hat{\mathbf{z}}ma = \hat{\mathbf{z}}F_{\text{ext}} - \hat{\mathbf{z}}kz - \hat{\mathbf{z}}\beta u, \tag{6.115}$$

where m is the mass of the upper plate and a is its downward acceleration. The velocity u and acceleration a are related to z by

$$u = \frac{dz}{dt} = z', \quad \text{and} \quad a = \frac{d^2z}{dt^2} = z''. \tag{6.116}$$

Hence, Eq. (6.115) simplifies to

$$mz'' + \beta z' + kz = F_{\text{ext}}, \tag{6.117}$$

which can be simplified further to

$$z'' + az' + bz = c, \tag{6.118}$$

where

$$a = \frac{\beta}{m}, \qquad b = \frac{k}{m}, \qquad \text{and} \qquad c = \frac{F_{\text{ext}}}{m}. \tag{6.119}$$

Applying Eq. (6.73), we have

$$\alpha = \frac{a}{2} = \frac{\beta}{2m} = \frac{0.06}{2 \times 10^{-8}} = 3 \times 10^6 \text{ Np/s},$$

and

$$\omega_0 = \sqrt{b} = \sqrt{\frac{k}{m}} = \sqrt{\frac{9}{10^{-8}}} = 30 \text{ krad/s},$$

where we converted the value of m to 10^{-8} kg to adhere to the mks system.

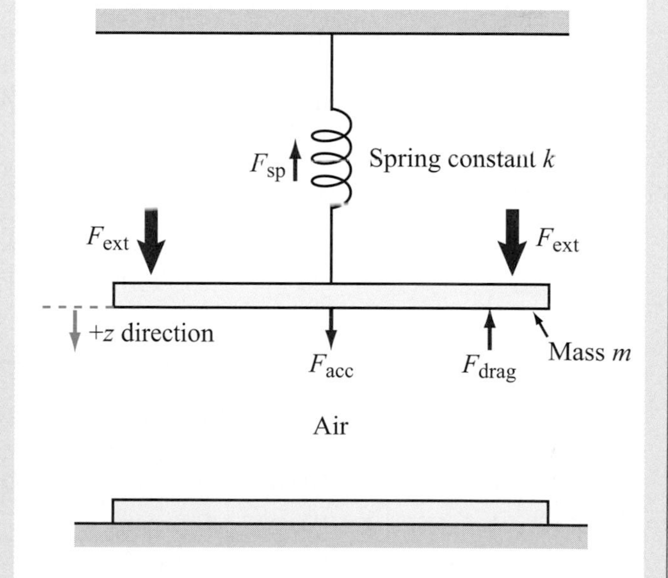

Figure 6-21: Capacitor accelerometer. Note that the $+z$ direction is downward. The arrows shown in the figure depict the directions of forces when an external force tries to move the upper plate downward.

Review Question 6-14: Under what conditions can non-electrical systems be modeled by a second-order differential equation of the form given by Eq. (6.102)?

Review Question 6-15: What are the two energy-storage modes in a vibrating guitar string? What is the primary dissipative mechanism?

Review Question 6-16: What is Hooke's law?

Exercise 6-16: What is the resonant frequency ω_0 and damping coefficient α of a MEMS capacitive accelerometer with $k = 0.1$ N/m, $m = 1$ ng $= 10^{-12}$ kg, and $\beta = 0.05$ (N·s/m)?

Answer: $\omega_0 = 316$ krad/s, $\alpha = 2.5 \times 10^{10}$ Np/s. (See)

6-8 Multisim Analysis of Circuit Response

Understanding the behavior of even a simple RLC circuit is sometimes a challenging task for electrical and computer engineering students. In reaction to a sudden change, a circuit gives rise to voltage and current variations that depend on the circuit topology, the initial conditions of its components, and the values of those components. In this section, we first will describe how to use Multisim to analyze the response of the series RLC circuit we discussed earlier in Section 6-3. The procedure is intended to demonstrate the steps one would follow to analyze any circuit with Multisim. As an example of a real-world application of the RLC-circuit response, we will then examine how such a circuit is used in radio frequency identification (RFID) technology.

6-8.1 The Series RLC Circuit

Using the now (hopefully) familiar schematic tools, draw a series RLC circuit including a switch in the Multisim Schematic Capture window. Use the parts and component values listed in Table 6-4, and add an oscilloscope as shown in Fig. 6-22. The scope is used for both L_1 and C_1, so that we

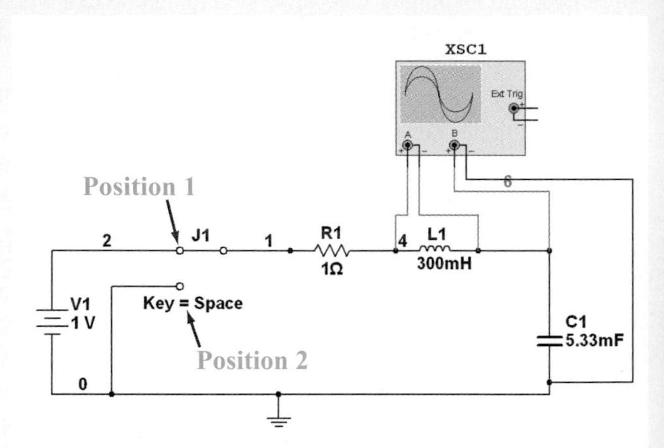

Figure 6-22: Multisim screen with RLC circuit.

Table 6-4: Component values for the circuit in Fig. 6-22.

Component	Group	Family	Quantity	Description
1	Basic	Resistor	1	1-Ω resistor
300 m	Basic	Inductor	1	300-mH inductor
5.3 3m	Basic	Capacitor	1	5.33-mF capacitor
SPDT	Basic	Switch	1	Single-pole double-throw (SPDT) switch
DC_POWER	Sources	Power_Sources	1	1-V dc source

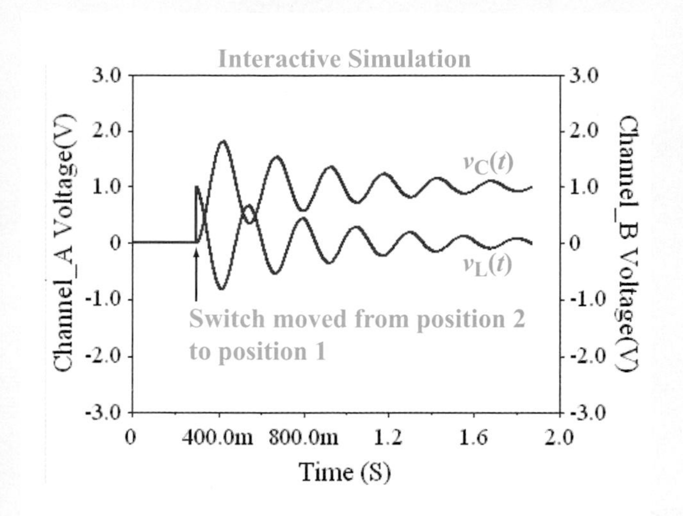

Figure 6-23: Voltage responses to moving the switch in the RLC circuit from position 2 to position 1.

may compare the voltages across them on the same screen. Make sure that before starting the interactive simulation, the initial condition of the switch is in position 2, so that the dc voltage source is not connected directly to the RLC circuit. Upon starting the simulation, you should see no voltage across any of the three components. After hitting the space bar to move the switch (Fig. 6-23), $v_L(t)$ initially will jump in level to 1 V and then exhibit an underdamped oscillatory response as a function of time. In contrast, $v_C(t)$ will exhibit an oscillatory behavior that will dampen out with time to assume a final value of 1 V.

A note on the Interactive Simulation settings is appropriate here. When you run an Interactive Simulation, Multisim numerically solves for the solution to the circuit at successive points in time. The resolution of this time step can be modified under Simulate → Interactive Simulation Settings. Both the ***maximum time step*** (TMAX) and the ***initial time step*** can be changed. Normally, there is no reason to do this, and Multisim's defaults will work well. However, when using the virtual instruments, sometimes time points are generated too quickly, and this makes it difficult for the user to observe the behavior. Conversely, the resolution may be too small so that the progression of time in the Interactive Simulation becomes annoyingly slow. When generating the traces in Fig. 6-23, for example, it may be difficult to see the damped behavior directly on the scope window because it scrolls by too fast. In that case, it can be helpful to reduce both the maximum and initial time steps (10 to 100× reduction usually works fine). This forces the computer to simulate more data points and slows it down, allowing you to see the trace appear more slowly. The drawback of this tweak is that you also use up more memory (and filespace).

Exercise 6-17: Given the component values in the Multisim circuit of Fig. 6-22, what are the values of ω_0 and α for the circuit response?

Answer: (See)

Exercise 6-18: Is the natural response for the circuit in Fig. 6-22 over-, under-, or critically damped? You can determine this both graphically (from the oscilloscope) and mathematically by comparing ω_0 and α.

Answer: (See)

Exercise 6-19: Modify the value of R in the circuit of Fig. 6-22 to obtain a critically damped response.

Answer: (See)

6-8.2 RFID Circuit

Radio frequency identification (RFID) circuits are fast becoming ubiquitous in many mass-consumer applications, ranging from tracking parcels and shipments to "smart" ID badges. Most systems in use today rely on a transceiver (usually hand-held) that can remotely interrogate one or more RFID tags (ranging in size from a few millimeters to a few centimeters). Some tags reply with only a serial number, while others are connected to miniature sensors and return values for temperature, humidity, acceleration, position, etc. The key to the widespread success of these RFID tags is that they do not require batteries to operate! If the transceiver is in close proximity to the tag (usually within a fraction of a meter), the radio-frequency power it transmits is sufficient to activate the RFID tag. The RFID tag uses an RLC circuit to harvest this power and communicate back to the transceiver (Fig. 6-24). The essential elements of the RFID communication system are shown in the circuit of Fig. 6-25.

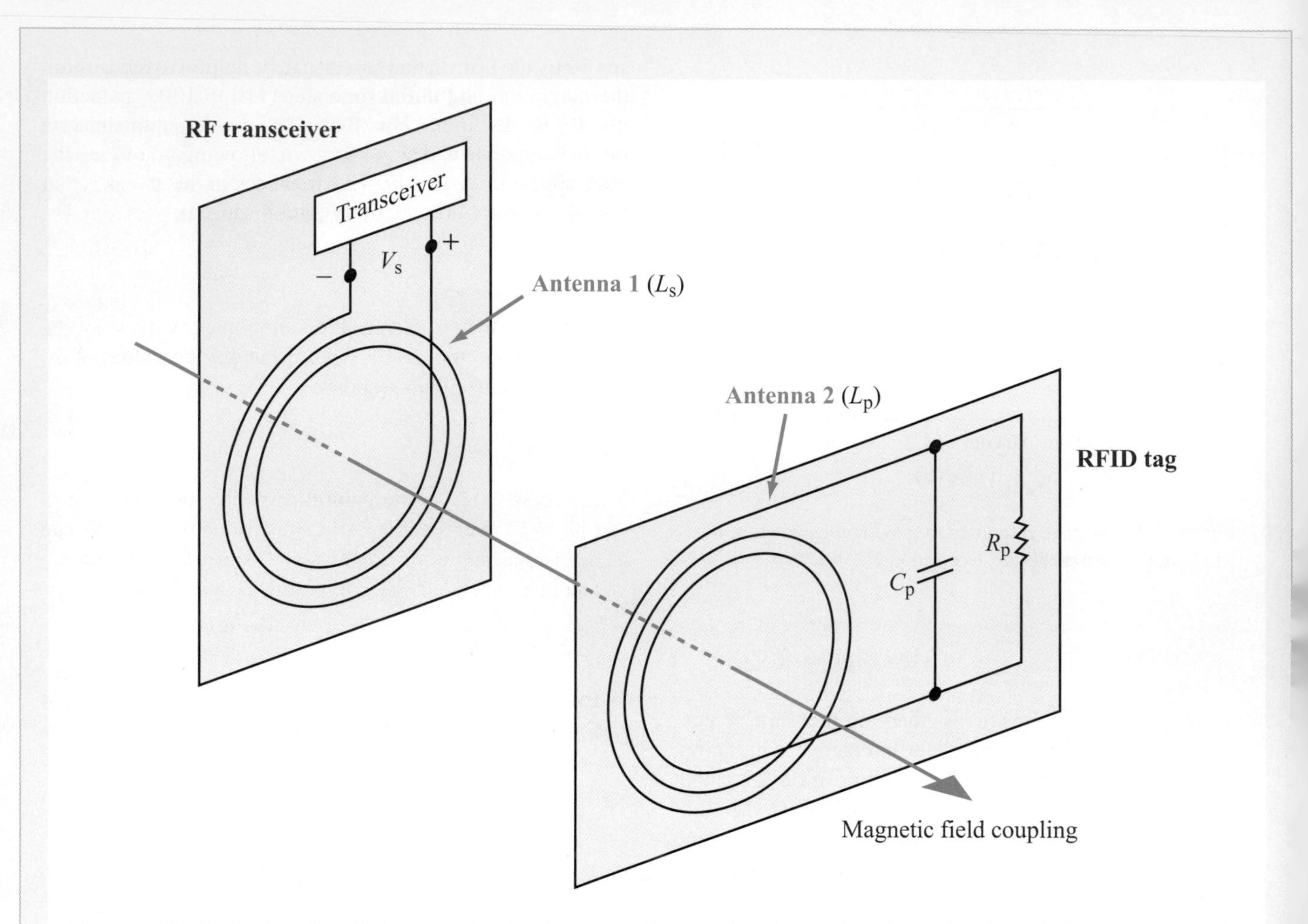

Figure 6-24: Illustration of an RFID transceiver in close proximity to an RFID tag. Note that the RFID tag will only couple to the transceiver when the two inductors are aligned along the magnetic field (shown in blue).

[An actual RFID circuit is more sophisticated, but the basic principle of operation is the same.] In transmit mode—with the SPDT switch connected to terminal T—the transceiver circuit consists of an ac voltage source υ_s connected in series with an inductor L_s. By moving the switch to terminal R, the transceiver circuit becomes a receiver with output voltage $\upsilon_{out}(t)$. In transmit mode, υ_s generates a current through L_s, which induces a magnetic field around it. If inductor L_p of the RFID tag is close to L_s, the magnetic field generated by L_s will induce a current through L_p. This current becomes the power source in the RFID-tag circuit and is the mechanism for building up the voltage across C_p to some maximum value V_C.

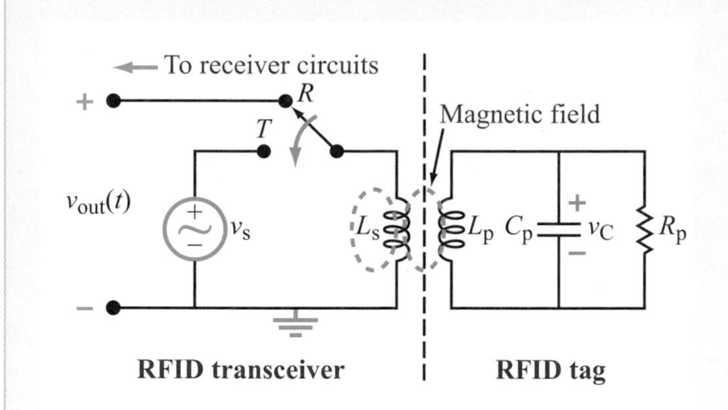

Figure 6-25: Basic elements of the RFID.

When the switch is moved from transmit mode to receive mode, v_s stops delivering power to L_s. The current through L_p, however, cannot change to zero instantaneously. The RLC circuit will react to the sudden change with an oscillatory underdamped response characterized by a damped natural frequency ω_d, whose value is governed by the choice of values for R_p, L_p, and C_p of the RFID tag. This oscillation frequency becomes part of the ID of that particular tag. In the same way that magnetic coupling served to transfer power from L_s to L_p during the transmit mode, it also serves to transfer information in the opposite direction—from L_p to L_s—during the receive mode. Since

$$v_{out}(t) = L_s \frac{di_{L_s}}{dt},$$

the output voltage recorded after moving the switch to the receive mode provides the *reply* by the RFID tag to the earlier excitation introduced by v_s during the transmit mode. [Real RFID transceivers transmit a few bits of data by superimposing digital bits onto the oscillations.]

To illustrate the operation of the RFID tag, we can simulate the process in Multisim. Using the parts listed in Table 6-5, we can build the circuit shown in the Multisim window of Fig. 6-26. To simulate magnetic coupling between inductors L_s and L_p, we use transformer T_1, which represents two closely coupled inductors sharing a common magnetic field. In Multisim, we set the inductance of each of the two transformer units to 1 mH and the coupling coefficient to 1. The circuit uses an oscilloscope to monitor $v_{out}(t)$. The oscilloscope trace is displayed in Fig. 6-27. Note that when the switch is moved from transmit to receive mode, $v_{out}(t)$ exhibits an immediate response that then decays exponentially with time. You may also want to plot $v_C(t)$ and $i_C(t)$ to examine the voltage and current experienced by the RFID tag itself during transmit and receive periods.

Review Question 6-17: How does the transmitter in the RFID system transfer power to the RLC circuit?

Review Question 6-18: How does the transceiver elicit a reply from the RFID tag?

Table 6-5: Parts for the Multisim circuit in Fig. 6-26.

Component	Group	Family	Quantity	Description
TS_IDEAL	Basic	Transformer	1	1 mH:1 mH ideal transformer
1 k	Basic	Resistor	1	1-kΩ resistor
1 μ	Basic	Capacitor	1	1-μF capacitor
SPDT	Basic	Switch	1	SPDT switch
AC_CURRENT	Sources	Signal_Current_Source	1	1 mA, 5.033 kHz

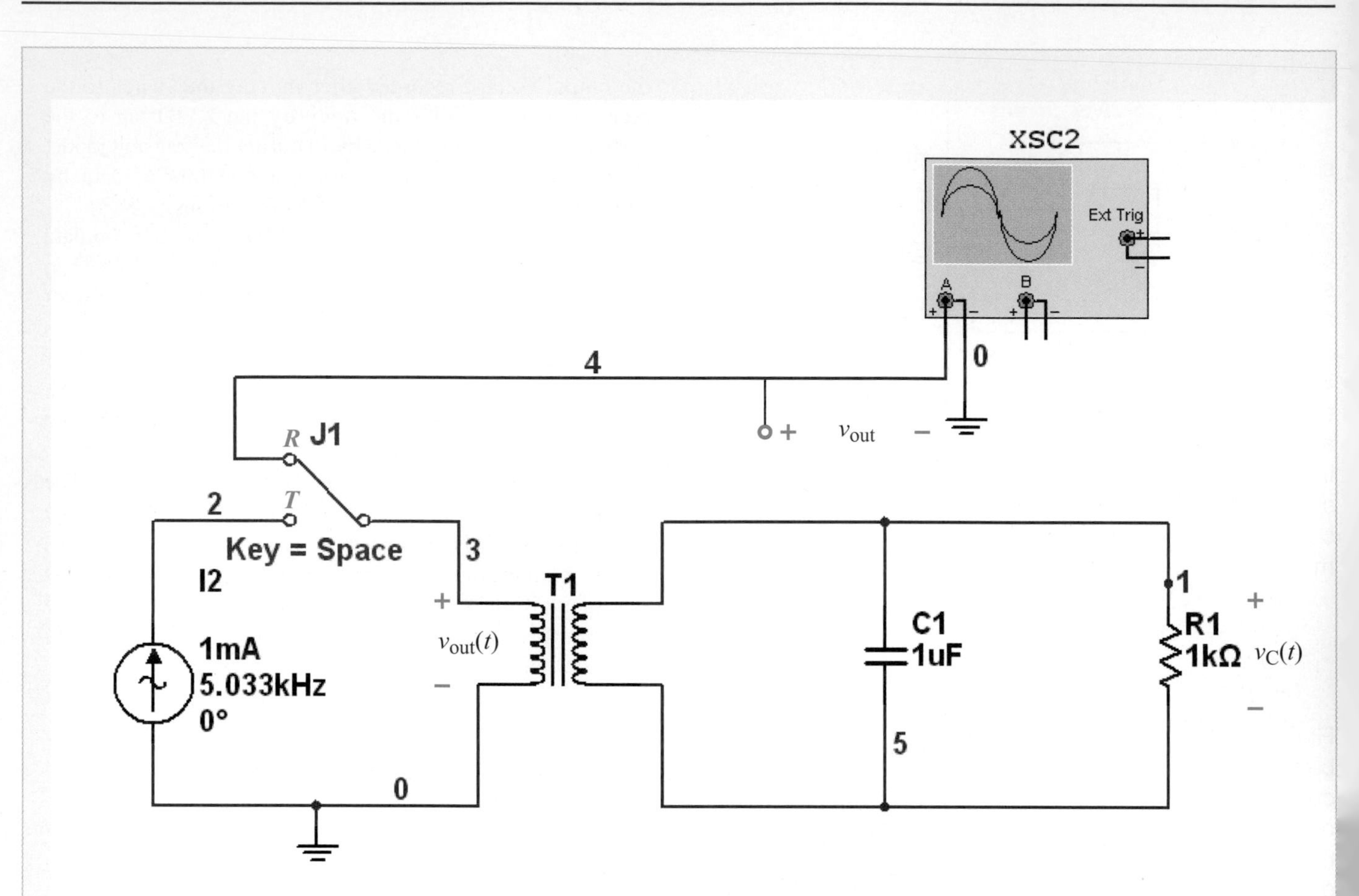

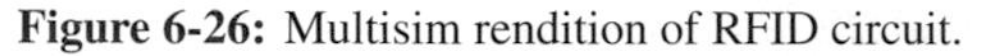
Figure 6-26: Multisim rendition of RFID circuit.

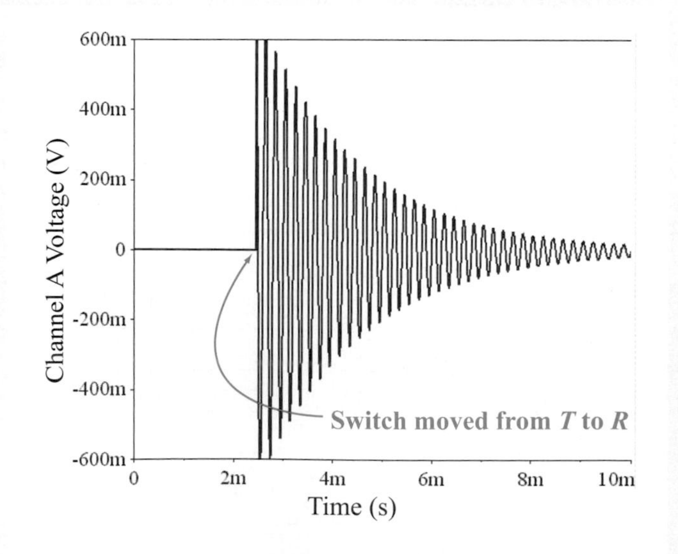

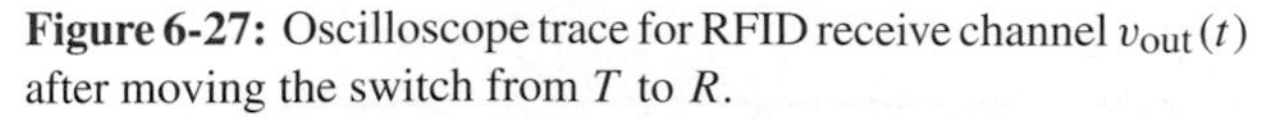
Figure 6-27: Oscilloscope trace for RFID receive channel $v_{out}(t)$ after moving the switch from T to R.

Exercise 6-20: Calculate ω_0, α, and ω_d for the RLC circuit in Fig. 6-26. How do ω_0 and ω_d compare with the angular frequency of the current source? This result (as we will learn later when we study resonant circuits in Chapter 9) is not at all by coincidence.

Answer: (See)

Exercise 6-21: Ideally, we would like the response of the RFID tag to take a very long time to decay down to zero in order to contain as many digital bits as possible. What determines the decay time? Change the values of some of the components in Fig. 6-26 to decrease the damping coefficient by a factor of 2.

Answer: (See)

Technology Brief 13: Touchscreens and Active Digitizers

Touchscreen is the common name given to a wide variety of technologies that allow computer displays to directly sense information from the user. In older systems, this usually meant the display could detect and pinpoint where a user touched the screen surface; newer systems can detect touch location as well as the associated pressure at multiple locations simultaneously, with very high resolution. This has led to a surge of applications in mobile computing, cell phones, personal digital assistants (PDA), and consumer appliances. One of the most common modern interactive screens is the tablet-style notebook computer, which uses a pen or stylo to input pressure-sensitive information, thereby allowing a user to write realistically and directly onto the screen.

Numerous technologies have been developed since the invention of the electronic touch interface in 1971 by Samuel C. Hurst. Some of the earlier technologies were susceptible to dust, damage from repeat use, and poor transparency. These issues largely have been resolved over the years (even for older technologies) as experience and advanced material selection have led to improved devices. With the explosion of consumer interest in portable, interactive electronics, newer technologies have emerged that are more suitable for these applications. Figure TF13-1 summarizes the general categories of touchscreens in use today. Historically, touchscreens were manufactured separately from displays and added as an extra layer of the display. More recently, display companies have begun to manufacture sensing technology directly into the displays; some of the newer technologies reflect this.

Resistive

Resistive touchscreens are perhaps the simplest to understand. A thin, flexible membrane is separated from a plastic base by insulating spacers. Both the thin membrane and the plastic base are coated on the inside with a transparent conductive film (indium tin oxide (ITO) often is used). When the membrane is touched, the two conductive surfaces come into contact. Detector circuits at the edges of the screen can detect this change in resistance between the two membranes and pinpoint the location on the *X*–*Y* plane. Older designs of this type were susceptible to membrane damage (from repeated flexing) and suffered from poor transparency.

Capacitive

Capacitive touchscreens employ a single thin, transparent conductive film (usually ITO) on a plastic or glass base. The conductive film is coated with another thin, transparent insulator for protection. Since the human body stores charge, a finger tip moved close to the surface of the film effectively forms a capacitor with the film as one of the plates and the finger as the other. The protective coating and the air form the intervening dielectric insulator. This capacitive coupling changes how a current flowing across the film surface is distributed; by placing electrodes at the screen corners and applying an ac electric signal, the location of the finger capacitance can be calculated precisely. A variant of this idea is to divide the sensing area into many smaller squares (just like pixels on the display) and to sense the change in capacitance across each of them continuously and independently. Capacitive technologies are much more resistant to wear and tear (since they are not flexed) than resistive touchscreen and are somewhat more transparent (80 to 85 percent transparency) since they have fewer films. These types of screens can be used to detect metal objects as well, so pens with conductive tips can be used on writing interfaces.

Pressure

Touch also can be detected mechanically. Pressure sensors can be placed at the corners of the display screen or even the entire display assembly, so whenever the screen is depressed, the four corners will experience different stresses depending on the (X, Y) position of the pressure point. Pressure screens benefit from high resistance to wear and tear and no losses in transparency (since there is no need to add layers over the display screen).

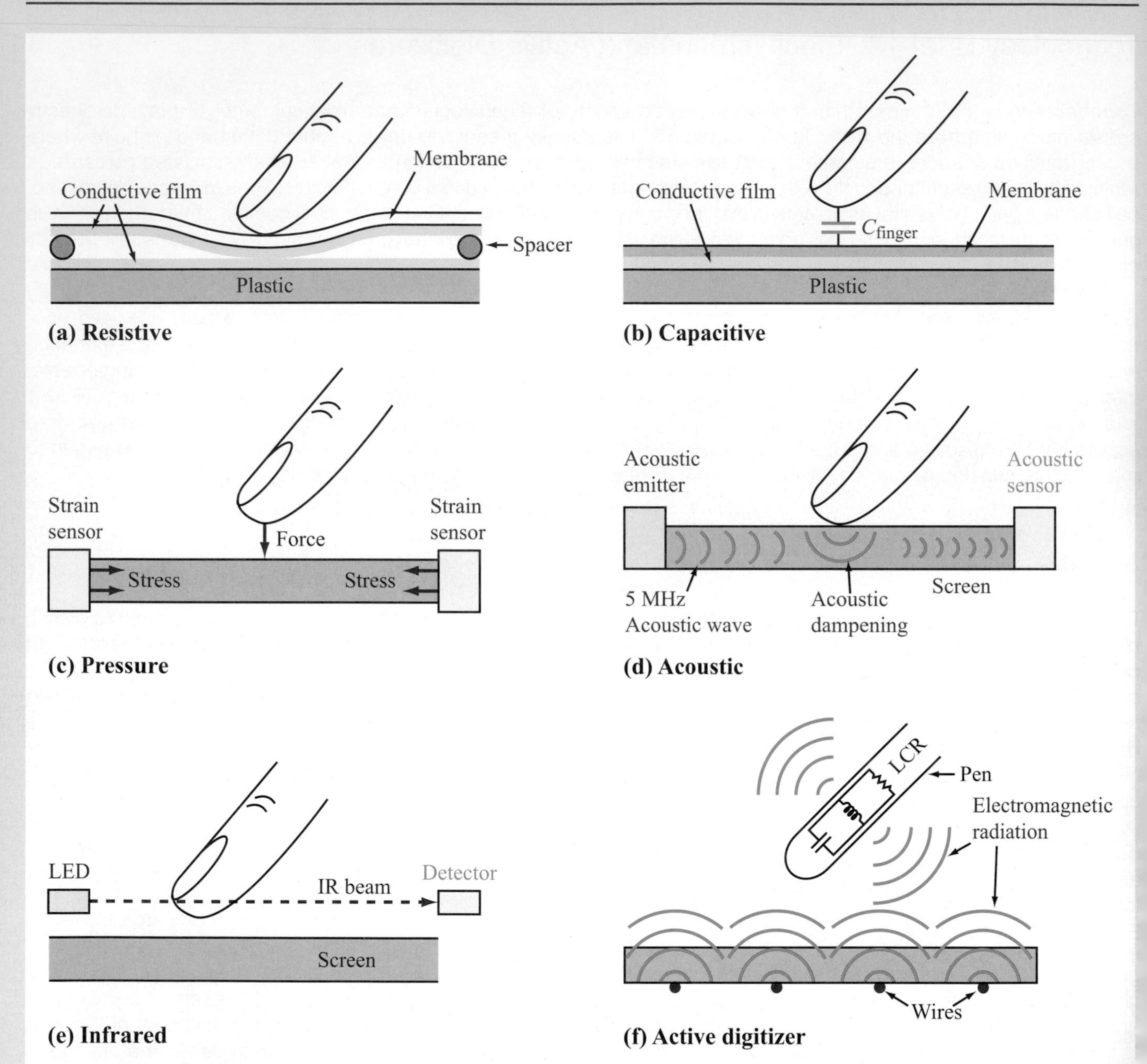

Figure TF13-1: Touchscreen technologies: (a) resistive, (b) capacitive, (c) pressure/strain sensor, (d) acoustic, (e) infrared, and (f) active digitizer.

Acoustic

A completely different way to detect touch relies on the transmission of high-frequency acoustic energy across the surface of the display material. Bursts of 5 MHz tones are launched by acoustic actuators from two corners of the screen. Acoustic reflectors all along the edges of the screen re-direct the incoming waves to the sensors. Any time an object comes into contact with the screen, it dampens or absorbs some fraction of the energy traveling across the material. The exact (X, Y) position can be calculated from the energy hitting the acoustic sensors. The contact force can be calculated as well, because the acoustic energy is dampened more or less depending on how hard the screen is pressed. Acoustic touchscreens also can detect multiple touches simultaneously, whereas resistive, capacitive (except those that use sensing "pixels"), and pressure touchscreens cannot do so easily.

Infrared

One of the oldest and least used technologies is the infrared touchscreen. This technology relies on infrared emitters (usually IR diodes; see Technology Brief 5 on page 96) aligned along two adjoining edges of the screen and infrared detectors aligned across from the emitters at the other two edges. The position of a touch event can be determined through a process based on which light paths are interrupted. The detection of multiple simultaneous touch events is possible. Infrared screens are somewhat bulky, prone to damage or interference from dust and debris, and need special modifications to work in daylight. They largely have been displaced by newer technologies.

Electromagnetic Resonance

The newest technology in widespread use is the electromagnetic resonance detection scheme used by most modern tablet PCs, developed by The Wacom Company. Strictly speaking, tablet PC screens are not touchscreens; they are called ***active digitizers*** because they can detect the presence and location of the tablet pen as it approaches the screen (even without contact). In this scheme, a very thin wire grid is integrated within the display screen (which usually is a flat-profile LCD display, see Technology Brief 6: Displays on page 106). The pen itself contains a simple RLC resonator (see Section 6-1) with no power supply. The wire grid alternates between two modes (transmit and receive) every $\sim$ 20 milliseconds. The grid essentially acts as an antenna. During the transmit mode, an ac signal is applied to the grid and part of that signal is emitted into the air around the display. As the pen approaches the grid, some energy from the grid travels across to the pen's resonator which begins to oscillate. In receive mode, the grid is used to "listen" for ac signals at the resonator frequency; if those signals are present, the grid can pinpoint where they are across the screen. A tuning fork provides a good analogy. Imagine a surface vibrating at a musical note; if a tuning fork designed to vibrate at that note comes very close to that surface, it will begin to oscillate at the same frequency. Even if we were to stop the surface vibrations, the tuning fork will continue to make a sound for a little while longer (as the resonance dies down). In a similar way, the laptop screen continuously transmits a signal and listens for the pen's electromagnetic resonance. Functions (such as buttons and pressure information) can be added to the pen by having the buttons change the capacitance value of the LCR when pressed; in this way, the resonance frequency will shift (see Section 6-2), and the shift can be detected by the grid and interpreted as a button press.

Chapter 6 Relationships

Step response of series and parallel RLC circuits (See Table 6-2)

Hook's law $\mathbf{F}_{\text{sp}} = -\hat{\mathbf{z}}kz$

General Solution for Second Order Circuits:
(see details in Table 6-3)

Differential equation $x'' + ax' + bx = c$

General Solution for Second Order Circuits (cont'd.):

Overdamped Response ($\alpha > \omega_0$)
$x(t) = [x(\infty) + A_1 e^{s_1 t} + A_2 e^{s_2 t}]\, u(t)$

Critically Damped Response ($\alpha > \omega_0$)
$x(t) = [x(\infty) + (B_1 + B_2 t)e^{-\alpha t}]\, u(t)$

Underdamped Response ($\alpha > \omega_0$)
$x(t) = x(\infty) + [D_1 \cos \omega_{\text{d}} t + D_2 \sin \omega_{\text{d}} t]e^{-\alpha t}\, u(t)$

CHAPTER HIGHLIGHTS

- Under dc steady-state conditions, a capacitor behaves like an open circuit, and an inductor behaves like a short circuit.
- Second-order circuits include series and parallel RLC circuits, as well as any circuit containing two passive energy-storage elements (capacitors and inductors).
- The response of a second-order circuit (containing dc sources) to a sudden change consists of a transient component (which decays to zero as $t \to \infty$) and a steady-state component that has a constant value.
- The transient response may be overdamped, critically damped, or underdamped, depending on the values of the circuit elements.
- The general solution for second-order circuits is applicable to circuits containing op-amps.
- Multisim can be used to simulate the response of any second-order circuit.

GLOSSARY OF IMPORTANT TERMS

Provide definitions or explain the meaning of the following terms:

characteristic equation
critically damped response
damped natural frequency
damping coefficient
initial condition
final condition
homogeneous solution
Hooke's law
MEMS
natural response
overdamped response
particular solution
resonant frequency
RFID
second-order circuit
steady-state response
time constant
transient response
underdamped response

PROBLEMS

Section 6-1: Initial and Final Conditions

6.1 The SPST switch in the circuit of Fig. P6.1 closes at $t = 0$, after it had been open for a long time. Draw the configurations that appropriately represent the state of the circuit at $t = 0^-$, $t = 0$, and $t = \infty$, and use them to determine (a) $v_{\text{C}}(0)$ and $i_{\text{L}}(0)$, (b) $i_{\text{C}}(0)$ and $v_{\text{L}}(0)$, and (c) $v_{\text{C}}(\infty)$ and $i_{\text{L}}(\infty)$.

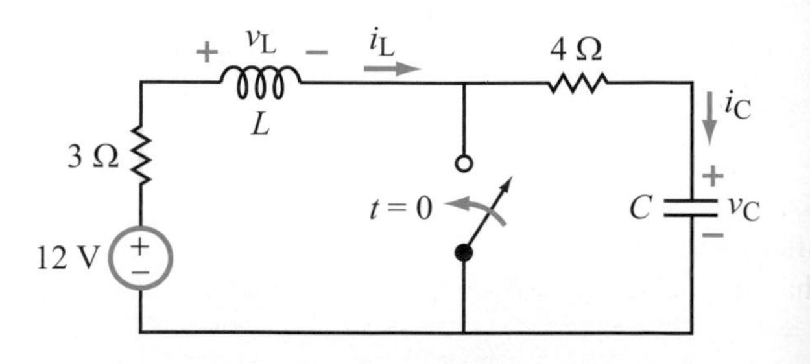

Figure P6.1: Circuit for Problem 6.1.

.2 The SPST switch in the circuit of Fig. P6.2 opens at $t = 0$, fter it had been closed for a long time. Draw the configurations hat appropriately represent the state of the circuit at $t = 0^-$, $= 0$, and $t = \infty$ and use them to determine (a) $v_C(0)$ and $_L(0)$, (b) $i_C(0)$ and $v_L(0)$, and (c) $v_C(\infty)$ and $i_L(\infty)$.

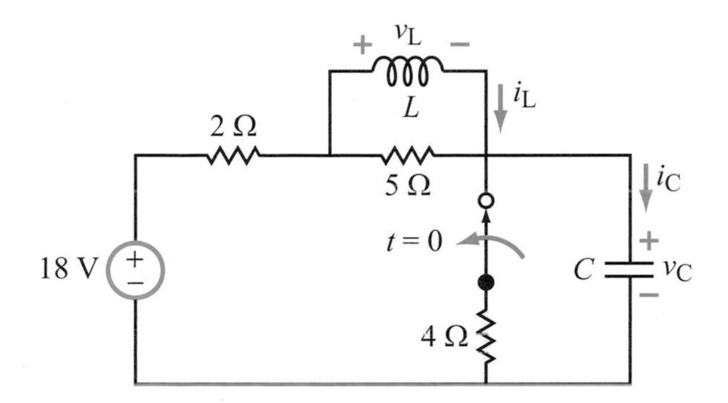

Figure P6.2: Circuit for Problem 6.2.

.3 The SPST switch in the circuit of Fig. P6.3 opens at $t = 0$, fter it had been closed for a long time. Draw the configurations hat appropriately represent the state of the circuit at $t = 0^-$, $= 0$, and $t = \infty$ and use them to determine (a) $v_C(0)$ and $_L(0)$, (b) $i_C(0)$ and $v_L(0)$, and (c) $v_C(\infty)$ and $i_L(\infty)$.

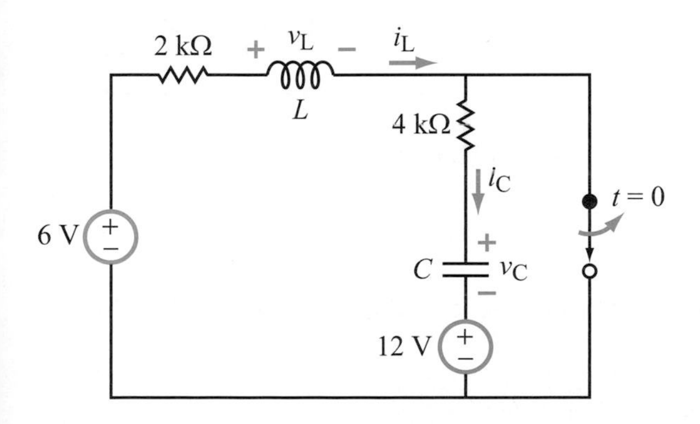

Figure P6.3: Circuit for Problem 6.3.

.4 The SPST switch in the circuit of Fig. P6.4 opens at $t = 0$, fter it had been closed for a long time. Draw the configurations hat appropriately represent the state of the circuit at $t = 0^-$, $= 0$, and $t = \infty$ and use them to determine (a) $v_C(0)$ and $_L(0)$, (b) $i_C(0)$ and $v_L(0)$, and (c) $v_C(\infty)$ and $i_L(\infty)$.

*Answer(s) in Appendix E.

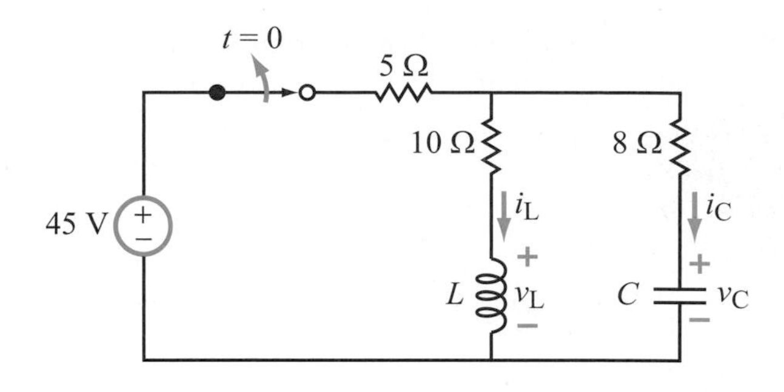

Figure P6.4: Circuit for Problem 6.4.

6.5 The SPST switch in the circuit of Fig. P6.5 closes at $t = 0$, after it had been opened for a long time. Draw the configurations that appropriately represent the state of the circuit at $t = 0^-$, $t = 0$, and $t = \infty$ and use them to determine (a) $v_C(0)$ and $i_L(0)$, (b) $i_C(0)$ and $v_L(0)$, and (c) $v_C(\infty)$ and $i_L(\infty)$.

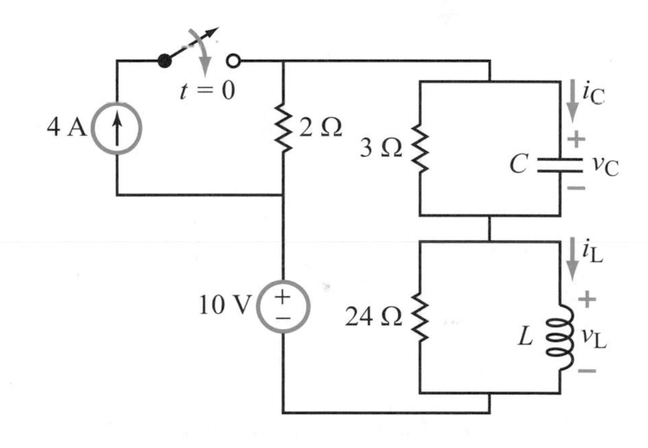

Figure P6.5: Circuit for Problems 6.5 and 6.6.

6.6 Repeat Problem 6.5, but start with a closed switch that opens at $t = 0$.

***6.7** For the circuit in Fig. P6.7, determine $i_1(0)$ and $i_2(0)$.

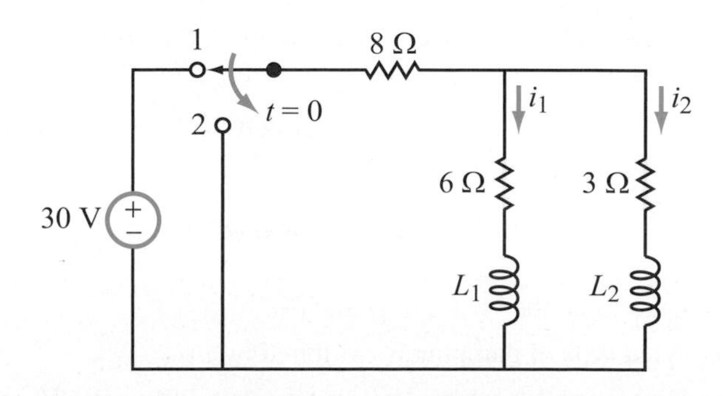

Figure P6.7: Circuit for Problem 6.7.

6.8 For the circuit of Fig. P6.8, determine (a) $i_{C_1}(0)$, $i_{R_1}(0)$, $i_{C_2}(0)$, and $i_{R_2}(0)$ and (b) $v_{C_1}(\infty)$ and $v_{C_2}(\infty)$.

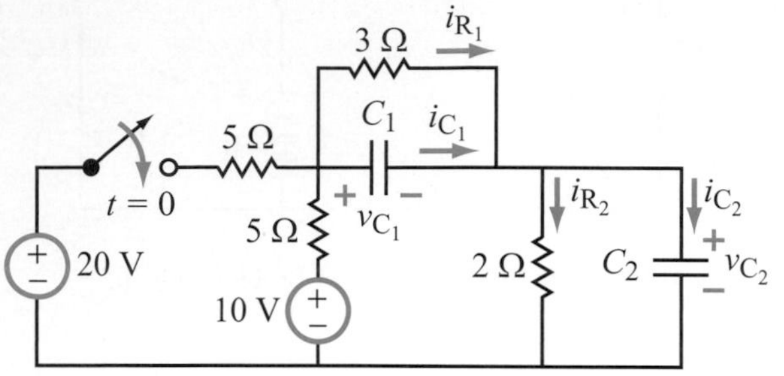

Figure P6.8: Circuit for Problem 6.8.

Section 6-3: Natural Response of the Series RLC Circuit

6.9 Determine $v_C(t)$ in the circuit of Fig. P6.9 and plot its waveform for $t \geq 0$, given that $V_0 = 12$ V, $R_1 = 0.4\ \Omega$, $R_2 = 1.2\ \Omega$, $L = 0.1$ H, and $C = 0.4$ F. Use a time scale that appropriately captures the shape of the waveform in your plot.

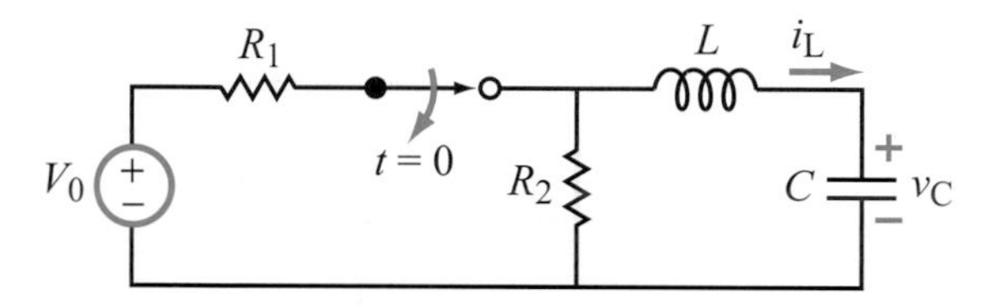

Figure P6.9: Circuit for Problems 6.9 to 6.11.

6.10 Determine $i_L(t)$ in the circuit of Fig. P6.9 and plot its waveform for $t \geq 0$, given that $V_0 = 12$ V, $R_1 = 0.4\ \Omega$, $R_2 = 1.2\ \Omega$, $L = 0.1$ H, and $C = 0.1$ F. Use a time scale that appropriately captures the shape of the waveform in your plot.

*__6.11__ In the circuit of Fig. P6.9, $V_0 = 12$ V, $R_1 = 0.4\ \Omega$, $R_2 = 1.2\ \Omega$, and $L = 0.1$ H. What should the value of C be in order for $i_L(t)$ to exhibit a critically damped response? Provide an expression for $i_L(t)$ and plot its waveform for $t \geq 0$.

6.12 The voltage v in a certain circuit is described by the differential equation

$$3v'' + 24v' + 75v = 0.$$

(a) Determine the values of α and ω_0.

(b) What type of damping is exhibited by $v(t)$?

(c) Determine $v(t)$ for $t \geq 0$, given that $v(0) = 10$ V and $v'(0) = 50$ V/s.

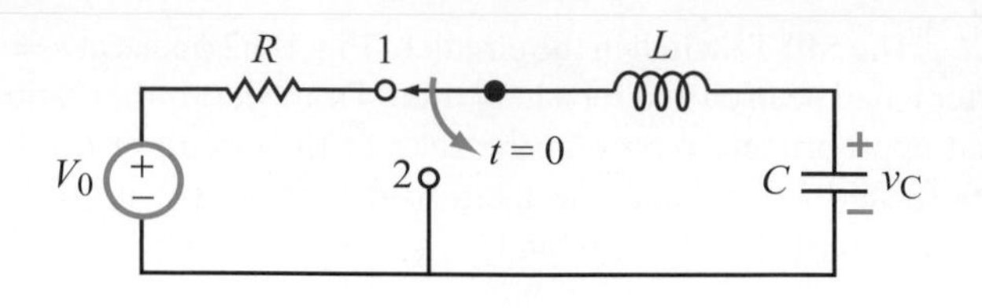

Figure P6.13: Circuit for Problem 6.13.

6.13 In the circuit of Fig. P6.13, the switch is moved from position 1 to position 2 at $t = 0$. Provide an expression for $v_C(t)$ for $t \geq 0$.

6.14 A series RLC circuit exhibits the following voltage and current responses:

$$v_C(t) = (6\cos 4t - 3\sin 4t)e^{-2t}\ u(t)\ \text{V},$$

$$i_C(t) = -(0.24\cos 4t + 0.18\sin 4t)e^{-2t}\ u(t)\ \text{A}.$$

Determine α, ω_0, R, L, and C.

*__6.15__ Determine $i_C(t)$ in the circuit of Fig. P6.15 for $t \geq 0$.

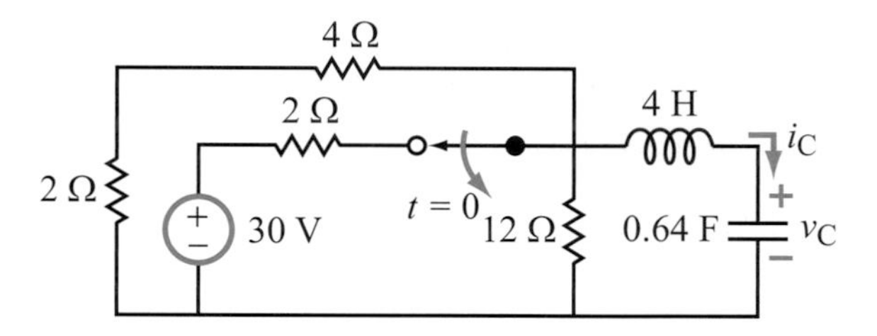

Figure P6.15: Circuit for Problem 6.15.

6.16 Determine $v_C(t)$ in the circuit of Fig. P6.16 for $t \geq 0$.

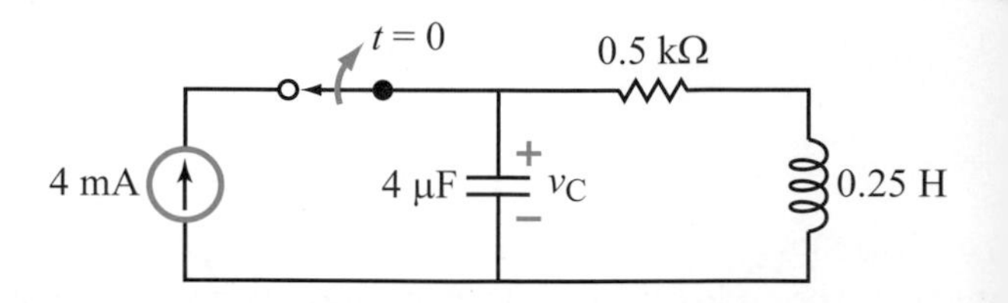

Figure P6.16: Circuit for Problem 6.16.

6.17 Determine $i_C(t)$ in the circuit of Fig. P6.17 for $t \geq 0$.

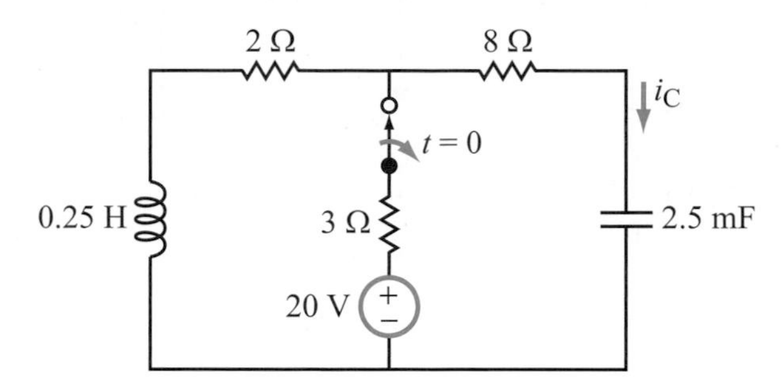

Figure P6.17: Circuit for Problem 6.17.

6.18 The circuit in Fig. 6-5(a) exhibits the response

$$v(t) = (12 + 36t)e^{-3t} \text{ V} \quad (\text{for } t \geq 0).$$

If $R = 12\ \Omega$, determine the values of V_s, L, and C.

*__6.19__ Determine $i_C(t)$ in the circuit of Fig. P6.19 and plot its waveform for $t \geq 0$.

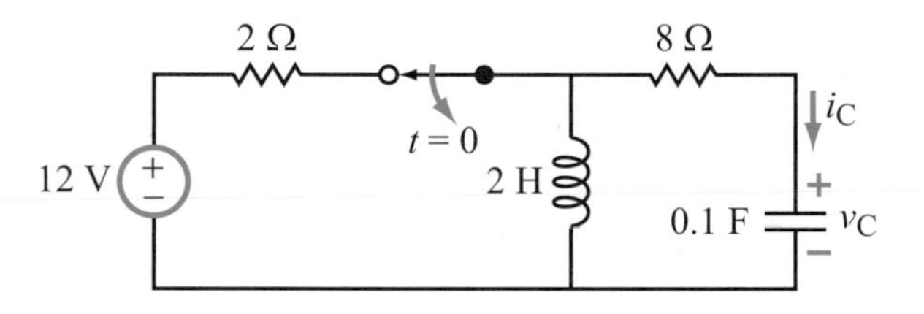

Figure P6.19: Circuit for Problems 6.19 and 6.20.

6.20 Repeat Problem 6.19, retaining the same values for all elements in the circuit except C. Choose the value of C so that the response of $i_C(t)$ is critically damped.

Section 6-4: Total Response of the Series RLC Circuit

6.21 Determine $i_C(t)$ in the circuit of Fig. P6.21 and plot its waveform for $t \geq 0$, given that $L = 0.05$ H. Use a time scale that appropriately captures the shape of the waveform in your plot.

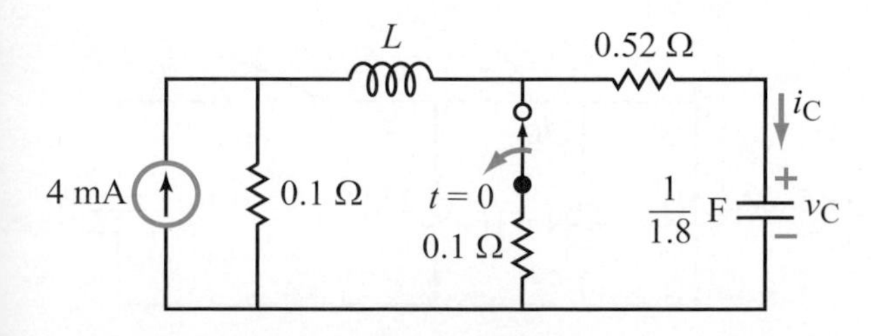

Figure P6.21: Circuit for Problem 6.21 and 6.22.

6.22 Choose the value of the inductor in the circuit of Fig. P6.21 so that v_C exhibits a critically damped response and determine $v_C(t)$ for $t \geq 0$.

6.23 Determine $i_C(t)$ in the circuit of Fig. P6.23 and plot its waveform for $t \geq 0$, given that $V_s = 24$ V, $R_1 = 2\ \Omega$, $R_2 = 4\ \Omega$, $L = 0.4$ H, and $C = \frac{10}{24}$ F.

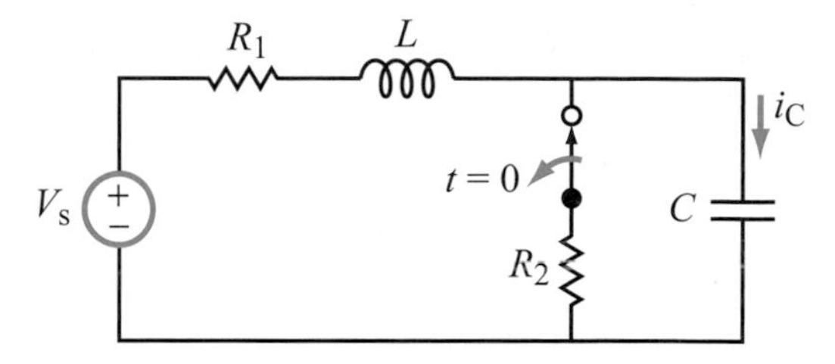

Figure P6.23: Circuit for Problems 6.23 and 6.24.

6.24 Repeat Problem 6.23 with the elements retaining their values, except change C to $\frac{10}{29}$ F.

6.25 Determine $i_L(t)$ in the circuit of Fig. P6.25, given that the switch was moved to position 2 at $t = 0$ (after it had been in position 1 for a long time) and then back to position 1 at $t = 0.5$ s. The element values are $V_s = 36$ V, $R_1 = 4\ \Omega$, $R_2 = 8\ \Omega$, $L = 0.8$ H, and $C = \frac{5}{24}$ F. Assume the capacitor had no charge prior to $t = 0$.

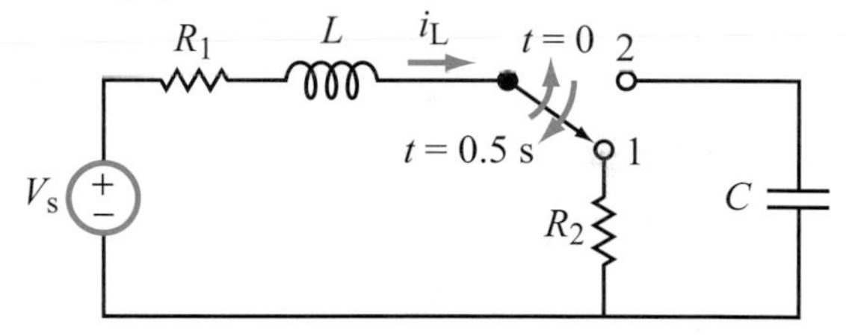

Figure P6.25: Circuit for Problem 6.25.

6.26 In the circuit of Fig. P6.26:

(a) What is the value of $v_C(\infty)$?

(b) How long does it take after $t = 0$ for v_C to reach 0.99 of its final value? [*Hint*: After solving for $v_C(t)$, step through values of t over the range $2 \leq t \leq 2.5$ to determine the value that satisfies the stated condition.]

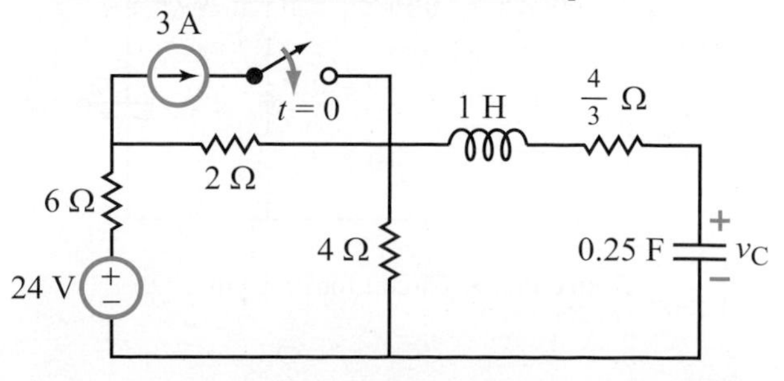

Figure P6.26: Circuit for Problem 6.26.

*6.27 Choose the value of C in the circuit of Fig. P6.27 so that $v_C(t)$ has a critically damped response for $t \geq 0$. Plot the waveform of $v_C(t)$.

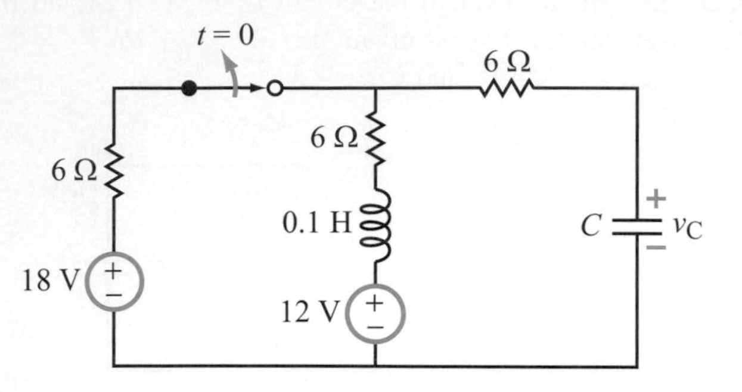

Figure P6.27: Circuit for Problem 6.27.

6.28 After having been open for a long time, the switch in the circuit of Fig. P6.28 was closed at $t = 0$ and then reopened at $t = 0.4$ s. Determine $i_L(t)$ and plot its waveform for $t \geq 0$.

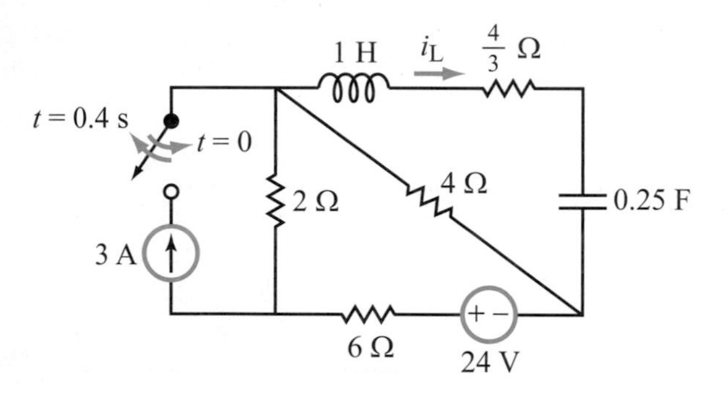

Figure P6.28: Circuit for Problem 6.28.

6.29 Determine $i_L(t)$ in the circuit of Fig. P6.29 and plot its waveform for $t \geq 0$.

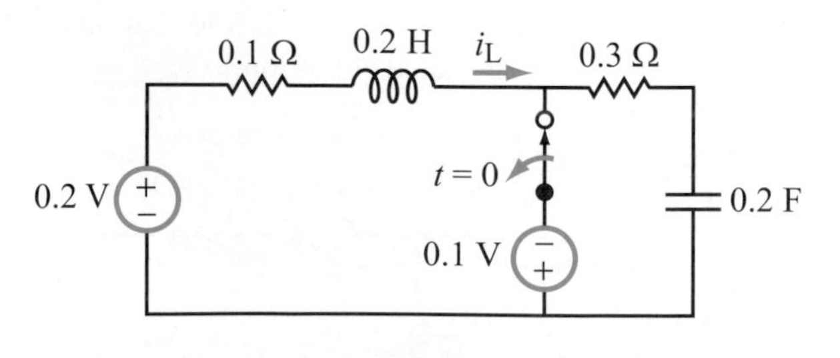

Figure P6.29: Circuit for Problem 6.29.

6.30 Determine $i_C(t)$ and $i_L(t)$ in the circuit of Fig. P6.30 for $t \geq 0$.

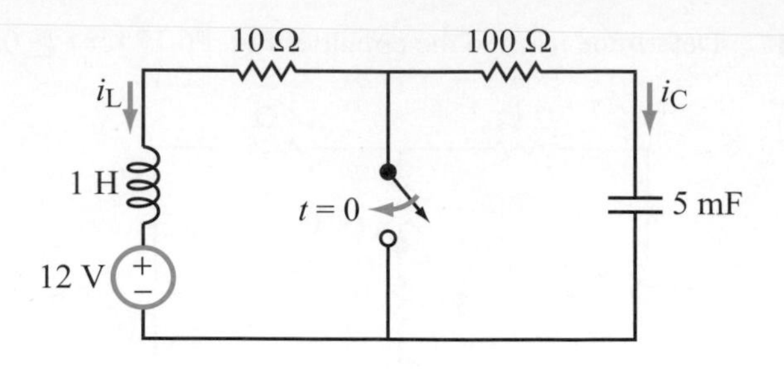

Figure P6.30: Circuit for Problem 6.30.

Section 6-5: The Parallel RLC Circuit

6.31 Determine $i_L(t)$ and $i_C(t)$ in the circuit of Fig. P6.31 and plot both waveforms for $t \geq 0$. The SPDT switch was moved from position 1 to position 2 at $t = 0$.

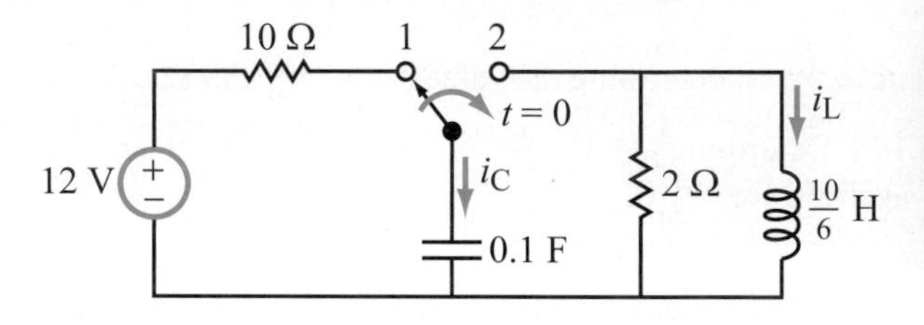

Figure P6.31: Circuit for Problem 6.31.

6.32 Determine $i_L(t)$ in the circuit of Fig. P6.32 and plot its waveform for $t \geq 0$.

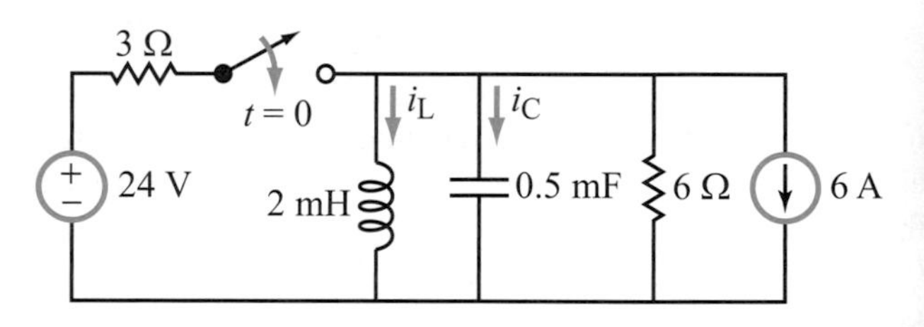

Figure P6.32: Circuit for Problems 6.32 and 6.34.

*6.33 Determine $i_L(t)$ in the circuit of Fig. P6.33 and plot its waveform for $t \geq 0$. The capacitor had no charge on it prior to $t = 0$.

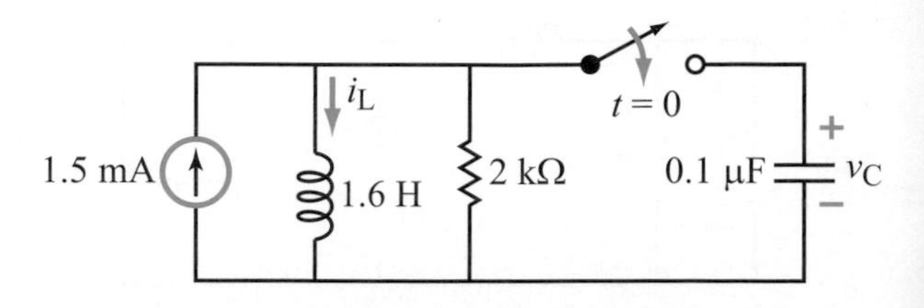

Figure P6.33: Circuit for Problem 6.33.

6.34 Determine $i_C(t)$ in the circuit of Fig. P6.32 for $t \geq 0$.

6.35 Determine $i_L(t)$ in the circuit of Fig. P6.35 and plot its waveform for $t \geq 0$.

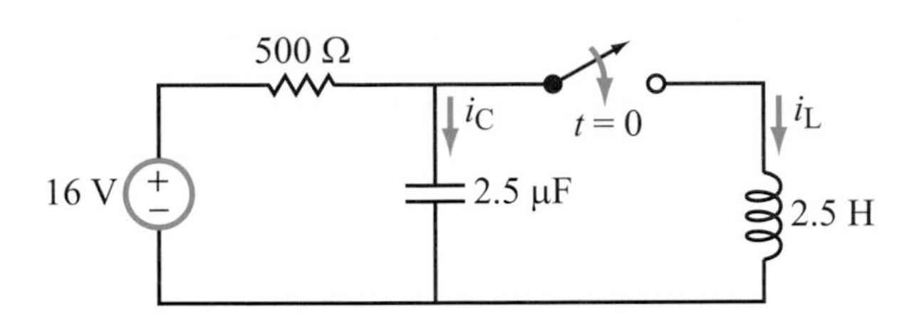

Figure P6.35: Circuit for Problems 6.35 and 6.36.

6.36 Determine $i_C(t)$ in the circuit of Fig. P6.35 and plot its waveform for $t \geq 0$.

6.37 Determine $i_L(t)$ in the circuit of Fig. P6.37 and plot its waveform for $t \geq 0$.

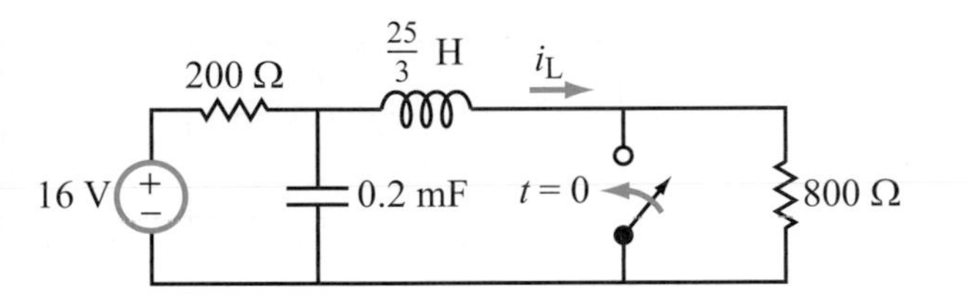

Figure P6.37: Circuit for Problem 6.37.

6.38 The switch in the circuit of Fig. P6.38 was closed at $t = 0$ and then reopened at $t = 1$ ms. Determine $i_L(t)$ and $v_C(t)$ for $t \geq 0$. Assume the capacitor had no charge prior to $t = 0$.

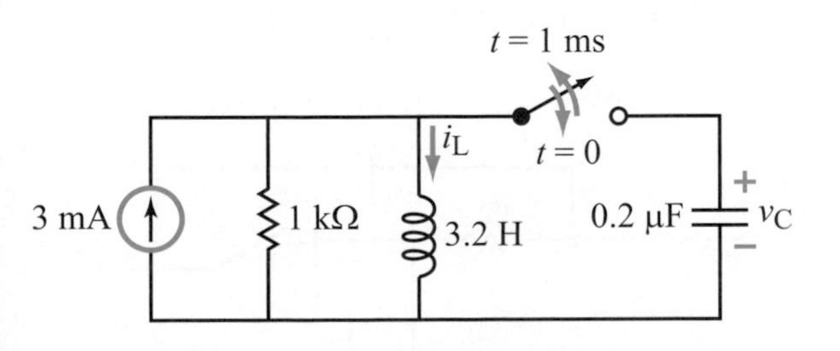

Figure P6.38: Circuit for Problem 6.38.

6.39 After closing the switch in the circuit of Fig. P6.39 at $t = 0$, it was reopened at $t = 1$ ms. Determine $i_C(t)$ and plot its waveform for $t \geq 0$. Assume no energy was stored in either L or C prior to $t = 0$.

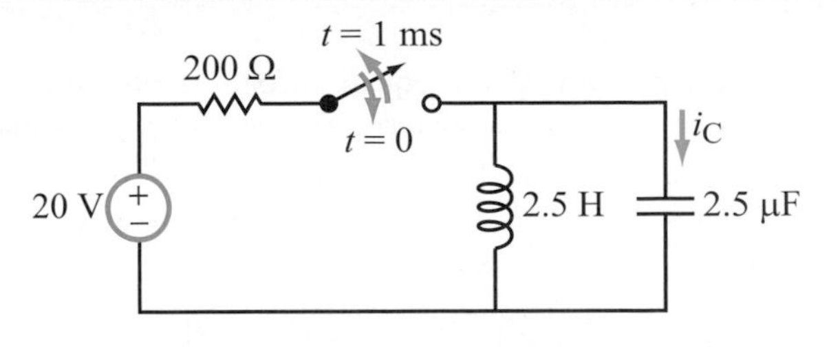

Figure P6.39: Circuit for Problem 6.39.

6.40 Determine the current responses $i_L(t)$ and $i_C(t)$ to a rectangular-current pulse as shown in Fig. P6.40, given that $I_s = 10$ mA and $R = 499.99\ \Omega$. Plot the waveforms of $i_L(t)$, $i_C(t)$, and $i_s(t)$ on the same scale.

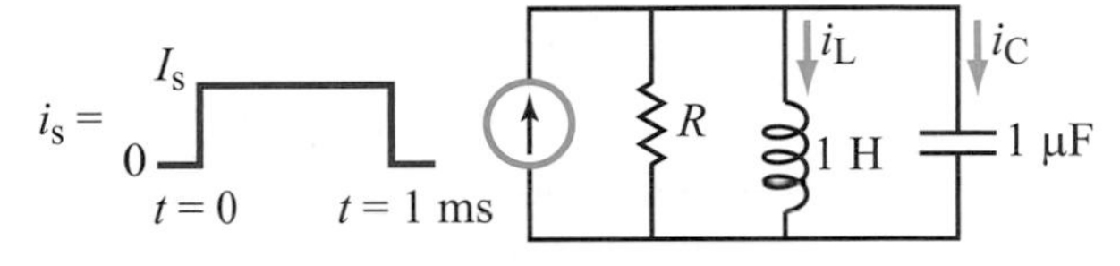

Figure P6.40: Circuit for Problem 6.40.

Section 6-6: General Solution for Any Second-Order Circuit

6.41 The voltage in a certain circuit is described by the differential equation

$$v'' + 5v' + 6v = 144 \qquad (\text{for } t \geq 0).$$

Determine $v(t)$ for $t \geq 0$, given that $v(0) = 16$ V and $v'(0) = 9.6$ V/s.

6.42 The current in a certain circuit is described by the differential equation

$$i'' + \sqrt{24}\, i' + 6i = 18 \qquad (\text{for } t \geq 0).$$

Determine $i(t)$ for $t \geq 0$, given that $i(0) = -2$ A and $i'(0) = 8\sqrt{6}$ A/s.

6.43 For the circuit in Fig. P6.43:

(a) Determine $i_L(0)$ and $v_L(0)$.

(b) Derive the differential equation for $i_L(t)$ for $t \geq 0$.

(c) Solve the differential equation and obtain an explicit expression for $i_L(t)$, given that $V_s = 12$ V, $R_s = 3\ \Omega$, $R_1 = 0.5\ \Omega$, $R_2 = 1\ \Omega$, $L = 2$ H, and $C = 2$ F.

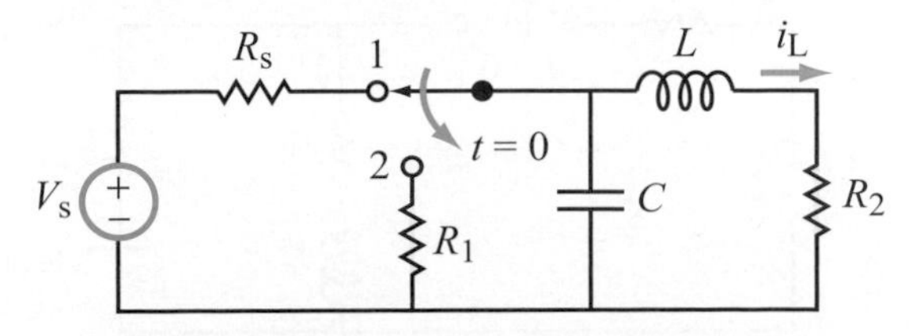

Figure P6.43: Circuit for Problem 6.43.

6.44 Develop a differential equation for $i_L(t)$ in the circuit of Fig. P6.44. Solve it to determine $i_L(t)$ for $t \geq 0$ subject to the following element values: $I_s = 36\ \mu$A, $R_s = 100$ kΩ, $R = 100\ \Omega$, $L = 10$ mH, and $C = 10\ \mu$F.

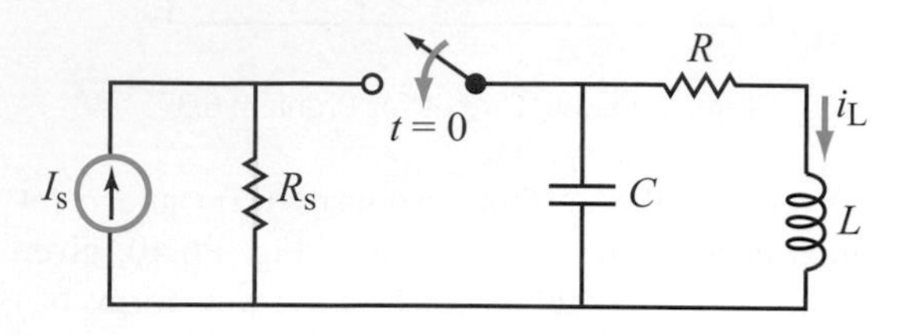

Figure P6.44: Circuit for Problem 6.44.

*__6.45__ Develop a differential equation for v_C in the circuit of Fig. P6.45. Solve it to determine $v_C(t)$ for $t \geq 0$. The element values are $I_s = 0.2$ A, $R_s = 30\ \Omega$, $R_1 = 10\ \Omega$, $R_2 = 20\ \Omega$, $R_3 = 20\ \Omega$, $L = 4$ H, and $C = 5$ mF.

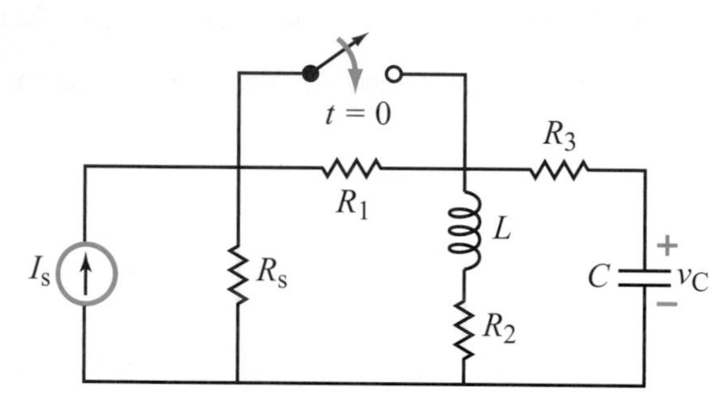

Figure P6.45: Circuit for Problem 6.45.

6.46 Develop a differential equation for i_L in the circuit of Fig. P6.46. Solve it for $t \geq 0$. The switch was closed at $t = 0$ and then reopened at $t = 0.5$ s, and the element values are $V_s = 18$ V, $R_s = 1\ \Omega$, $R_1 = 5\ \Omega$, $R_2 = 2\ \Omega$, $L = 2$ H, and $C = \frac{1}{17}$ F.

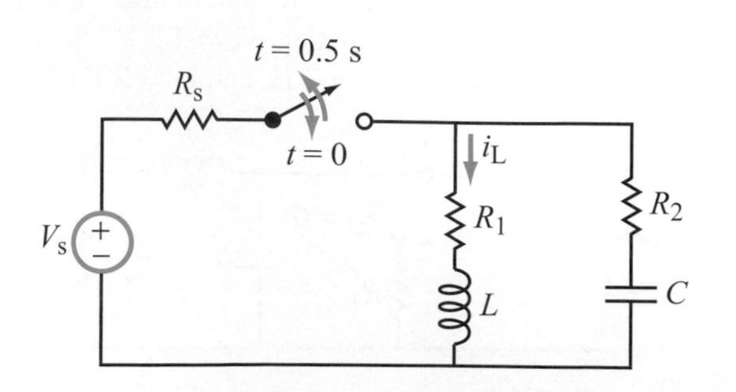

Figure P6.46: Circuit for Problem 6.46.

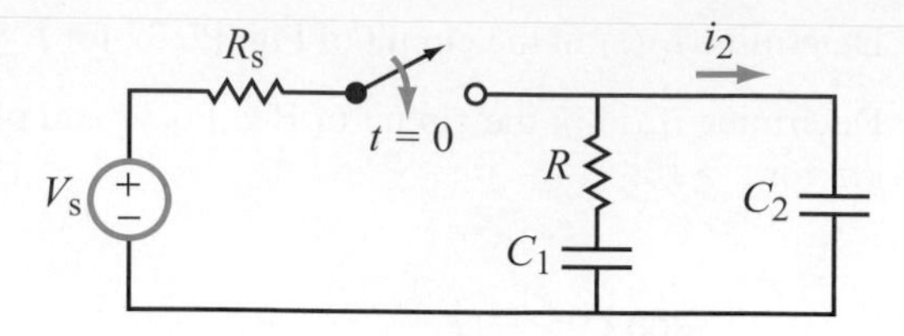

Figure P6.47: Circuit for Problems 6.47 and 6.48.

6.47 Determine i_2 in the circuit of Fig. P6.47 for $t \geq 0$, given that $V_s = 10$ V, $R_s = 0.1$ MΩ, $R = 1$ MΩ, $C_1 = 1\ \mu$F, and $C_2 = 2\ \mu$F.

6.48 Repeat Problem 6.47, but this time assume that the switch had been closed for a long time and then opened at $t = 0$.

*__6.49__ The op-amp circuit shown in Fig. P6.49 is called a multiple-feedback bandpass filter. If $v_{in} = A\,u(t)$, determine $v_{out}(t)$ for $t \geq 0$ for $A = 6$ V, $R_1 = 10$ kΩ, $R_2 = 5$ kΩ, $R_f = 50$ kΩ, and $C_1 = C_2 = 1\ \mu$F.

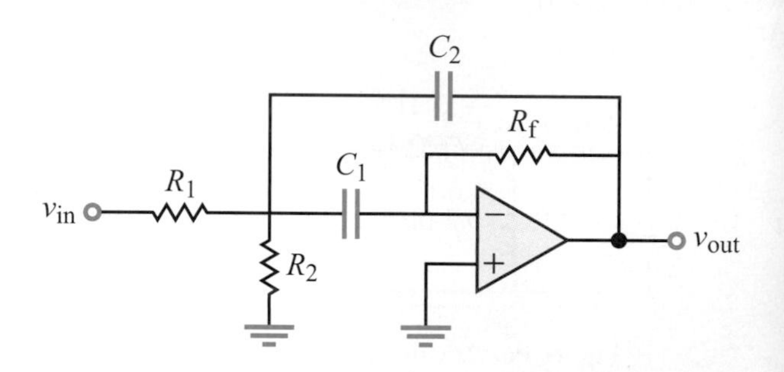

Figure P6.49: Circuit for Problem 6.49.

6.50 The op-amp circuit shown in Fig. P6.50 is called a two-pole low-pass filter. If $v_{in} = A\,u(t)$, determine $v_{out}(t)$ for $t \geq 0$ for $A = 2$ V, $R_1 = 5$ kΩ, $R_2 = 10$ kΩ, $R_3 = 12$ kΩ, $R_4 = 20$ kΩ, $C_1 = 100\ \mu$F, and $C_2 = 200\ \mu$F.

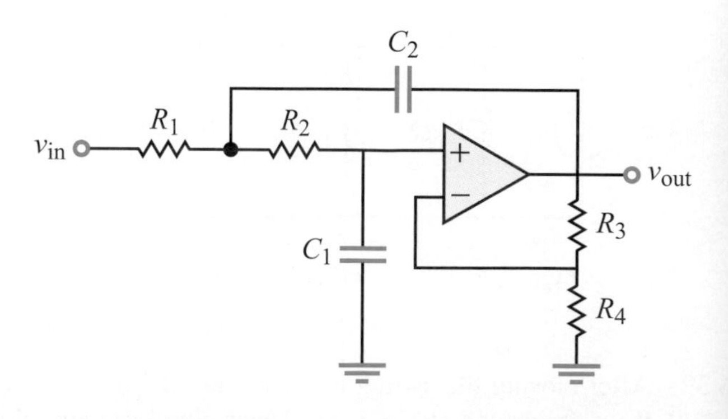

Figure P6.50: Circuit for Problem 6.50.

Section 6-8: Multisim

6.51 Using Multisim, draw the circuit in Fig. 6-5(a) using the component values given in Example 6-3. Use the Transient Analysis tool to obtain a plot of $v(t)$ for $0 < t < 0.2$ s.

6.52 Using Multisim, draw the circuit in Fig. E6.5. Use the Transient Analysis tool to obtain a plot of $i_C(t)$ for $0 < t < 1$ s.

6.53 Using Multisim, draw the circuit in Figure E6.5. Use the Transient Analysis tool to obtain three plots of $i_C(t)$ for (a) an underdamped response, (b) a critically damped response, and (c) an overdamped response. To obtain the three desired responses, adjust the value of the 20-Ω resistor as needed.

6.54 Adjust the values of the source and the components in Fig. 6-26 such that the RLC circuit is excited and oscillates at a frequency of 1 MHz and the oscillation envelope decays to 10 percent of its initial value after 12 oscillations once the circuit is switched to "listen" mode.

6.55 Build the circuit shown in Fig. P6.55 in Multisim and then plot the voltage $v_C(t)$ from 0 to 200 ms using Transient Analysis.

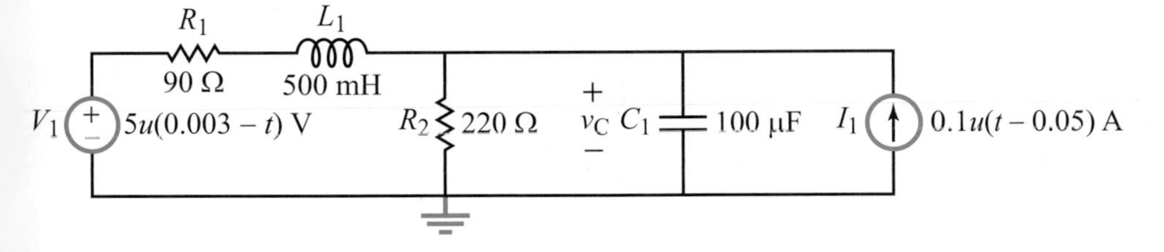

Figure P6.55: Circuit for Problem 6.55.

6.56 Build the active second-order circuit shown in Fig. P6.56. Plot the signal v_{out} from 0 to 5 ms, and note how long it takes before the amplitude of the oscillations drops below 1 V. Change the value of R_2 to 100 kΩ and repeat the simulation. (You may need to readjust your timescale.)

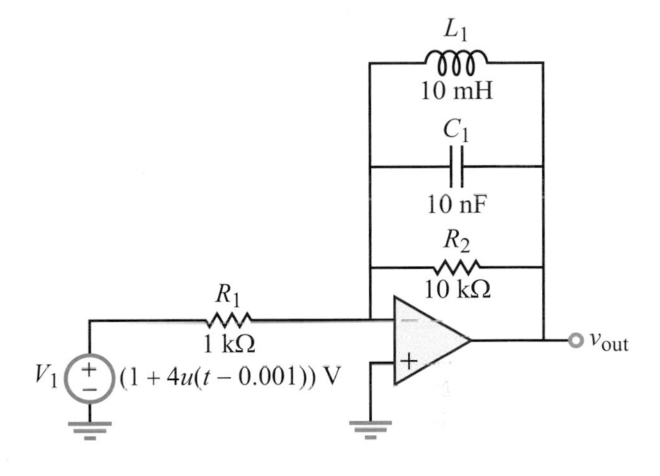

Figure P6.56: Circuit for Problem 6.56.

C H A P T E R

7

ac Analysis

Chapter Contents

Objectives

Upon learning the material presented in this chapter, you should be able to:

1. Relate the parameters of a sinusoidal function to the geometry of the associated waveform.
2. Manipulate complex algebra with relative ease.
3. Transform time-varying sinusoidal functions to the phasor domain and vice versa.
4. Analyze any linear circuit in the phasor domain.
5. Determine the impedance of any passive element, or the combination of elements connected in series or in parallel.
6. Perform Y–Δ transformations, source transformations, current division and voltage division, and determine Thévenin and Norton equivalent circuits, all in the phasor domain.
7. Apply nodal analysis, mesh analysis, and other analysis techniques, all in the phasor domain.
8. Design simple RC phase-shift circuits.
9. Design a dc power-supply circuit.
10. Use Multisim to analyze ac circuits.

Overview

From solar illumination to radio and cell-phone transmissions, we are surrounded by electromagnetic (EM) waves all of the time. *EM waves are composed of sinusoidally varying electric and magnetic fields*, and the fundamental parameter that distinguishes one EM wave from another is the wave's frequency f (or equivalently, its wavelength $\lambda = c/f$, where $c = 3 \times 10^8$ m/s is the velocity of light in a vacuum). The frequency of red light, for example, is 4.3×10^{14} Hz, and one of the frequencies assigned to cell-phone traffic is 1,900 MHz (1.9×10^9 Hz). Both are EM waves—and so are X-rays, infrared waves, and microwaves—but they oscillate at different frequencies (see Technology Brief 8 on page 158).

The term "ac" (***alternating current***) is associated with electric circuits whose currents and voltages vary sinusoidally with time, just like EM waves. In fact, ac circuits and EM waves are not only similar, but they also are connected directly: an ac current with an oscillation frequency f radiates EM waves of the same frequency. The radiated waves can couple signals from one part of the circuit to another through the air space they share. The coupling may serve as an intentional means of communication, as in the case of ***radio frequency identification*** (RFID) circuits (Section 6-8.2), or it may introduce unwelcome signals that interfere with the intended operation of the circuit. Mitigation of such undesirable consequences is part of a subdiscipline of electrical engineering called ***electromagnetic compatibility***.

This and the next two chapters will be devoted to the study of ac circuits, which are far more prevalent than dc circuits and offer a much broader array of practical applications. In our study, we will assume that all currents and voltages are confined to the discrete elements in the circuit and to the connections between them, allowing us to ignore EM-compatibility issues altogether.

7-1 Sinusoidal Signals

The voltage between two points in a circuit (or the current flowing through a branch) is said to have a ***sinusoidal waveform*** if its time variation is given by a sinusoidal function. *The term sinusoid includes both sine and cosine functions.* For example, the expression

$$v(t) = V_{\mathrm{m}} \cos \omega t \tag{7.1}$$

describes a sinusoidal voltage $v(t)$ that has an ***amplitude*** V_{m} and an ***angular frequency*** ω. The amplitude defines the maximum or ***peak value*** that $v(t)$ can reach, and $-V_{\mathrm{m}}$ is its lowest negative value. The ***argument*** of the cosine function ωt is measured either in degrees or in radians, with

$$\pi \text{ (rad)} \simeq 3.1416 \text{ (rad)} = 180^\circ. \tag{7.2}$$

Since ωt is measured in radians, the unit for ω is (rad/s). Figure 7-1(a) displays a plot of $v(t)$ as a function of ωt. The familiar cosine function starts at its maximum value (at $\omega t = 0$), decreases to zero at $\omega t = \pi/2$, goes into negative territory for half of a cycle, and completes its first cycle at $\omega t = 2\pi$. Occasionally, we may want to display a sinusoidal signal as a function of t, instead of ωt. We note that the angular frequency ω is related to the ***oscillation frequency*** (or simply the ***frequency***) f of the signal by

$$\omega = 2\pi f \qquad \text{(rad/s)}, \tag{7.3}$$

with f measured in hertz (Hz), which is equivalent to cycles/second. A sinusoidal voltage with a frequency of 100 Hz makes 100 oscillations in 1 s, each of duration $1/100 = 0.01$ s. The duration of a cycle is its ***period*** T. Thus,

$$T = \frac{1}{f} \qquad \text{(s)}. \tag{7.4}$$

By combining Eqs. (7.1), (7.3), and (7.4), $v(t)$ can be rewritten as

$$v(t) = V_{\mathrm{m}} \cos \frac{2\pi t}{T}, \tag{7.5}$$

which is displayed in Fig. 7-1(b) as a function of t. We observe that the cyclical pattern of the waveform repeats itself every T seconds. That is,

$$v(t) = v(t + nT) \tag{7.6}$$

for any integer value of n.

Sinusoidal waveforms can be expressed in terms of either sine or cosine functions.

To avoid confusion, we will adopt the cosine form as our reference standard throughout this and the next chapter.

This means that we will always express voltages and currents in terms of cosine functions, so if a voltage (or current) waveform is given in terms of a sine function, we should first convert it to a cosine form with a positive amplitude before proceeding with our circuit analysis. Conversion from sine to cosine form is realized through the application of Eq. (7.7a) of Table 7-1. For example,

$$\begin{aligned} i(t) &= 6 \sin(\omega t + 30^\circ) \\ &= 6 \cos(\omega t + 30^\circ - 90^\circ) \\ &= 6 \cos(\omega t - 60^\circ). \end{aligned} \tag{7.8}$$

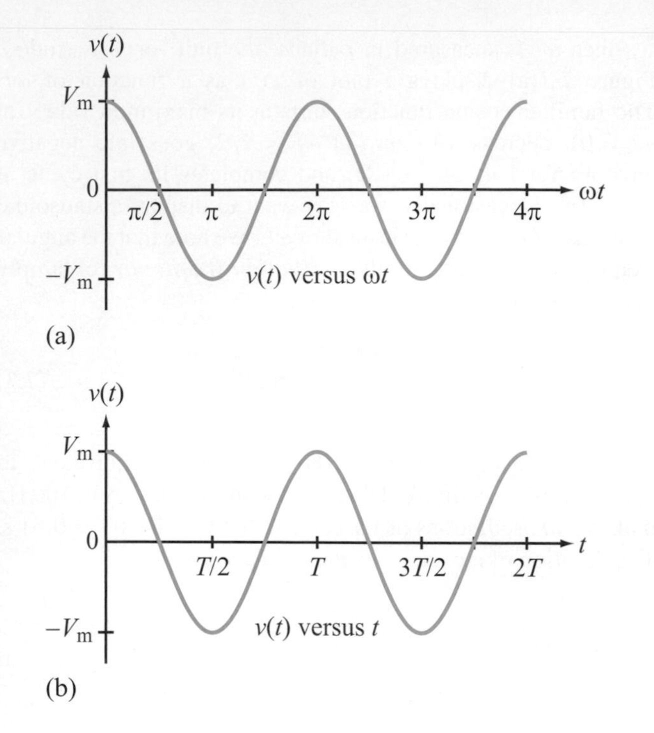

Figure 7-1: The function $v(t) = V_{\text{m}} \cos \omega t$ plotted as a function of (a) ωt and (b) t.

In addition to ωt, the argument of the cosine function contains a constant angle of $-60°$. A ***cosine-referenced*** sinusoidal function generally takes the form

$$v(t) = V_{\text{m}} \cos(\omega t + \phi), \tag{7.9}$$

where ϕ is called its ***phase angle***. For $i(t)$ of Eq. (7.8), $\phi = -60°$.

The angle ϕ may assume any positive or negative value, but we usually add or subtract multiples of 2π radians (or equivalently, multiples of 360°) so that the remainder is between $-180°$ and $+180°$. The magnitude and sign (+ or −) of ϕ will then determine, respectively, how much and in what direction the waveform of $v(t)$ is shifted along the time axis, relative to the reference waveform corresponding to $v(t)$ with $\phi = 0$. Figure 7-2 displays three waveforms:

$$v_1(t) = V_{\text{m}} \cos\left(\frac{2\pi t}{T} - \frac{\pi}{4}\right), \tag{7.10a}$$

$$v_2(t) = V_{\text{m}} \cos \frac{2\pi t}{T} \quad \text{(Reference waveform with } \phi = 0\text{)}, \tag{7.10b}$$

and

$$v_3(t) = V_{\text{m}} \cos\left(\frac{2\pi t}{T} + \frac{\pi}{4}\right). \tag{7.10c}$$

We observe that waveform $v_3(t)$, which is shifted backwards in time relative to the reference waveform $v_2(t)$, attains its peak value before $v_2(t)$ does. Consequently, waveform $v_3(t)$ is said

Table 7-1: Useful trigonometric identities (additional relations are given in Appendix D).

$\sin x = \pm\cos(x \mp 90°)$	(7.7a)
$\cos x = \pm\sin(x \pm 90°)$	(7.7b)
$\sin x = -\sin(x \pm 180°)$	(7.7c)
$\cos x = -\cos(x \pm 180°)$	(7.7d)
$\sin(-x) = -\sin x$	(7.7e)
$\cos(-x) = \cos x$	(7.7f)
$\sin(x \pm y) = \sin x \cos y \pm \cos x \sin y$	(7.7g)
$\cos(x \pm y) = \cos x \cos y \mp \sin x \sin y$	(7.7h)
$2 \sin x \sin y = \cos(x - y) - \cos(x + y)$	(7.7i)
$2 \sin x \cos y = \sin(x + y) + \sin(x - y)$	(7.7j)
$2 \cos x \cos y = \cos(x + y) + \cos(x - y)$	(7.7k)

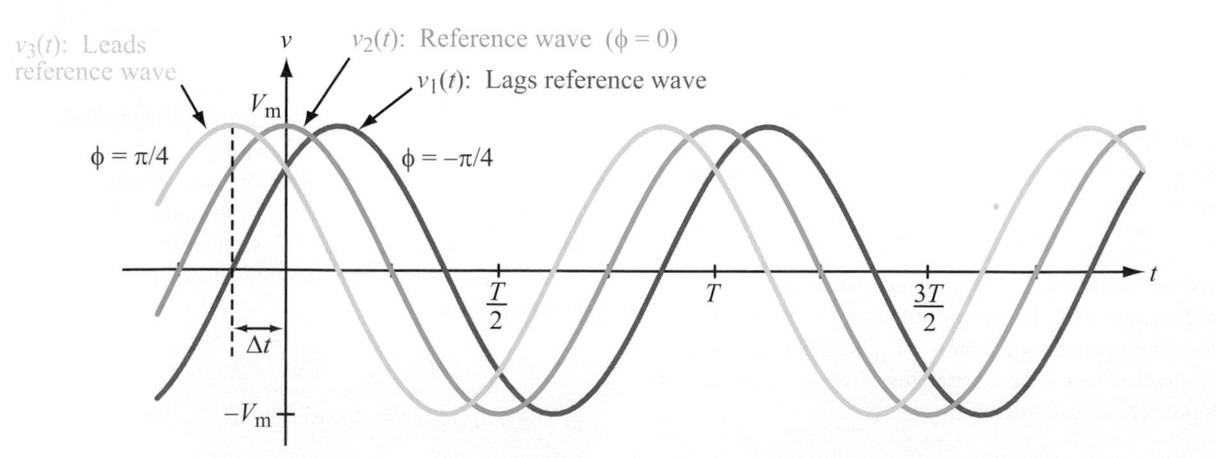

Figure 7-2: Plots of $v(t) = V_{\text{m}} \cos[(2\pi t/T) + \phi]$ for three different values of ϕ.

to ***lead*** $v_2(t)$ by a ***phase lead*** of $\pi/4$. Similarly, waveform $v_1(t)$ ***lags*** $v_2(t)$ by a ***phase lag*** of $\pi/4$. A cosine function with a negative phase angle ϕ takes longer to reach a specified reference level (such as the peak value) than it takes the zero-phase angle function to reach that level, signifying a phase lag. When ϕ is positive, it signifies a phase lead. A phase angle of 2π corresponds to a time shift along the time axis equal to one full period T. Proportionately, a phase angle of ϕ (in radians) corresponds to a time shift Δt given by

$$\Delta t = \left(\frac{\phi}{2\pi}\right) T. \tag{7.11}$$

We generalize our discussion of phase lead and lag by stating that:

Given two sinusoidal functions with the same angular frequency ω, and both expressed in standard cosine form as

$$v_1(t) = V_1 \cos(\omega + \phi_1)$$

and

$$v_2(t) = V_2 \cos(\omega + \phi_2),$$

the relevant terminology is:

v_2 **leads** v_1 **by** $(\phi_2 - \phi_1)$,

v_2 **lags** v_1 **by** $(\phi_1 - \phi_2)$,

v_1 and v_2 are **in-phase** if $\phi_2 = \phi_1$,

v_1 and v_2 are **in phase-opposition** if $\phi_2 = \phi_1 \pm 180°$.

Example 7-1: Voltage Waveform

A sampling oscilloscope is used to measure a voltage signal $v(t)$. The measurements reveal that $v(t)$ is periodic with an amplitude of 10 V, its maxima are separated by 20 ms, and one of its maxima occurs at $t = 1.2$ ms. Determine the functional form of $v(t)$.

Solution: Given that $V_m = 10$ V and

$$T = 20 \text{ ms} = 2 \times 10^{-2} \text{ s},$$

$v(t)$ is given by

$$v(t) = 10\cos\left(\frac{2\pi t}{2 \times 10^{-2}} + \phi\right) = 10\cos(100\pi t + \phi) \text{ V}.$$

Application of $v(t = 1.2 \text{ ms}) = 10$ V gives

$$10 = 10\cos(100\pi \times 1.2 \times 10^{-3} + \phi),$$

which requires the argument of the cosine to be a multiple of 2π,

$$0.12\pi + \phi = 2n\pi, \qquad n = 0, \pm 1, \pm 2, \ldots$$

The smallest value of ϕ in the range $[-180°, 180°]$ that satisfies the preceding equation corresponds to $n = 0$, and is given by

$$\phi = -0.12\pi = -21.6°.$$

Hence,

$$v(t) = 10\cos(100\pi t - 21.6°) \text{ V}.$$

Example 7-2: Phase Lead/Lag

Given the current waveforms

$$i_1(t) = -8\cos(\omega t - 30°) \text{ A}$$

and

$$i_2(t) = 12\sin(\omega t + 45°) \text{ A},$$

does $i_1(t)$ lead $i_2(t)$, or the other way around, and by how much?

Solution: Standard cosine format requires that the sinusoidal functions be cosines and that the amplitudes have positive values. Application of Eq. (7.7d) of Table 7-1 allows us to remove the negative sign preceding the amplitude of $i_1(t)$,

$$\begin{aligned} i_1(t) &= -8\cos(\omega t - 30°) = 8\cos(\omega t - 30° + 180°) \\ &= 8\cos(\omega t + 150°) \text{ A}. \end{aligned}$$

Application of Eq. (7.7a) to $i_2(t)$ leads to

$$\begin{aligned} i_2(t) &= 12\sin(\omega t + 45°) = 12\cos(\omega t + 45° - 90°) \\ &= 12\cos(\omega t - 45°) \text{ A}. \end{aligned}$$

Hence, $\phi_1 = 150°$, $\phi_2 = -45°$, and

$$\Delta\phi = \phi_2 - \phi_1 = -195°.$$

The concept of phase lead/lag requires that $\Delta\phi$ be within the range $[-180°, 180°]$. Addition of 360° to $\Delta\phi$ converts it to 165°, which means that i_2 leads i_1 by 165°.

Review Question 7-1: A sinusoidal waveform is characterized by three parameters. What are they, and what does each one of them specify?

Review Question 7-2: Waveforms $v_1(t)$ and $v_2(t)$ have the same angular frequency, but $v_1(t)$ leads $v_2(t)$. Will the peak value of $v_1(t)$ occur sooner or later than that of $v_2(t)$? Explain.

Exercise 7-1: Provide an expression for a 100-V, 60-Hz voltage that exhibits a minimum at $t = 0$.

Answer: $v(t) = 100\cos(120\pi t + 180°)$ V. (See CD)

Exercise 7-2: Given two current waveforms:

$$i_1(t) = 3\cos\omega t$$

and

$$i_2(t) = 3\sin(\omega t + 36°),$$

does $i_2(t)$ lead or lag $i_1(t)$, and by what phase angle?

Answer: $i_2(t)$ lags $i_1(t)$ by 54°. (See CD)

7-2 Review of Complex Numbers

A ***complex number*** $\mathbf{z}$ may be written in the ***rectangular form***

$$\mathbf{z} = x + jy, \tag{7.12}$$

where x and y are the ***real*** ($\Re\mathfrak{e}$) and ***imaginary*** ($\Im\mathfrak{m}$) parts of $\mathbf{z}$, respectively, and $j = \sqrt{-1}$. That is,

$$x = \Re\mathfrak{e}(\mathbf{z}) \qquad \text{and} \qquad y = \Im\mathfrak{m}(\mathbf{z}). \tag{7.13}$$

Alternatively, $\mathbf{z}$ may be written in ***polar form*** as

$$\mathbf{z} = |\mathbf{z}|e^{j\theta} = |\mathbf{z}|\angle\theta \tag{7.14}$$

where $|\mathbf{z}|$ is the magnitude of $\mathbf{z}$, θ is its phase angle, and the form $\angle\theta$ is a useful shorthand representation commonly used in numerical calculations. By applying ***Euler's identity***

$$e^{j\theta} = \cos\theta + j\sin\theta, \tag{7.15}$$

we can convert $\mathbf{z}$ from polar form, as in Eq. (7.14), into rectangular form, as in Eq. (7.12) by

$$\mathbf{z} = |\mathbf{z}|e^{j\theta} = |\mathbf{z}|\cos\theta + j|\mathbf{z}|\sin\theta, \tag{7.16}$$

which leads to the relations

$$x = |\mathbf{z}|\cos\theta, \qquad y = |\mathbf{z}|\sin\theta, \tag{7.17}$$

$$|\mathbf{z}| = \sqrt[+]{x^2 + y^2}, \quad \text{and} \quad \theta = \tan^{-1}(y/x). \tag{7.18}$$

The two forms of $\mathbf{z}$ are illustrated graphically in Fig. 7-3. Because in the complex plane a complex number assumes the form of a vector, it is represented by a bold letter in this book.

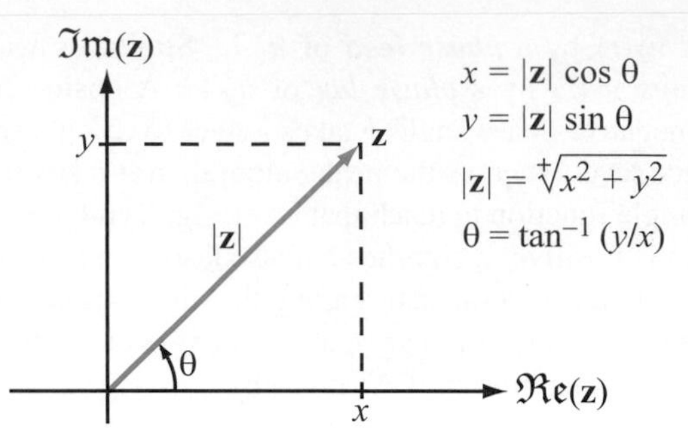

Figure 7-3: Relation between rectangular and polar representations of a complex number $\mathbf{z} = x + jy = |\mathbf{z}|e^{j\theta}$.

When using Eq. (7.18), care should be taken to ensure that θ is in the proper quadrant by noting the signs of x and y individually, as illustrated in Fig. 7-4. Complex numbers $\mathbf{z}_2$ and $\mathbf{z}_4$ point in opposite directions and their phase angles θ_2 and θ_4 differ by 180°, despite the fact that (y/x) has the same value in both cases. Also note that, since $|\mathbf{z}|$ is a positive quantity, only the positive root in Eq. (7.18) is applicable. This is denoted by the + sign above the square-root sign.

The ***complex conjugate*** of $\mathbf{z}$, denoted with a star superscript (or asterisk), is obtained by replacing j (wherever it appears) with $-j$, so that

$$\mathbf{z}^* = (x + jy)^* = x - jy = |\mathbf{z}|e^{-j\theta} = |\mathbf{z}|\angle{-\theta}. \tag{7.19}$$

The magnitude $|\mathbf{z}|$ is equal to the positive square root of the product of $\mathbf{z}$ and its complex conjugate:

$$|\mathbf{z}| = \sqrt[+]{\mathbf{z}\mathbf{z}^*}. \tag{7.20}$$

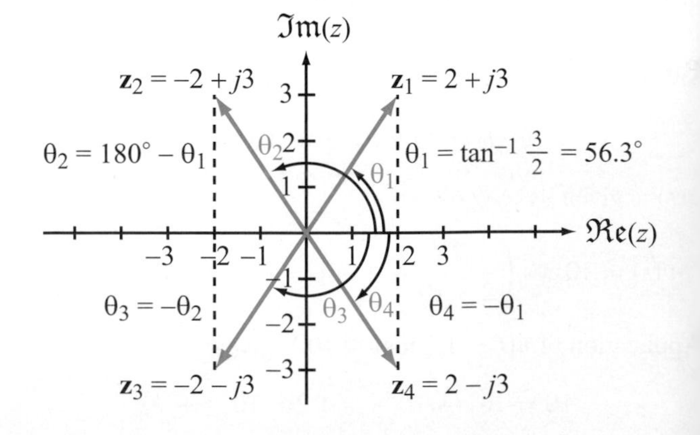

Figure 7-4: Complex numbers $\mathbf{z}_1$ to $\mathbf{z}_4$ have the same magnitud $|\mathbf{z}| = \sqrt[+]{2^2 + 3^2} = 3.61$, but their polar angles depend on the polaritie of their real and imaginary components.

We now highlight some of the properties of complex algebra that we will likely encounter in future sections.

Equality: If two complex numbers $\mathbf{z}_1$ and $\mathbf{z}_2$ are given by

$$\mathbf{z}_1 = x_1 + jy_1 = |\mathbf{z}_1|e^{j\theta_1} \tag{7.21a}$$

and

$$\mathbf{z}_2 = x_2 + jy_2 = |\mathbf{z}_2|e^{j\theta_2}, \tag{7.21b}$$

then $\mathbf{z}_1 = \mathbf{z}_2$ if and only if (*iff*) $x_1 = x_2$ and $y_1 = y_2$, or equivalently, $|\mathbf{z}_1| = |\mathbf{z}_2|$, and $\theta_1 = \theta_2$.

Addition:

$$\mathbf{z}_1 + \mathbf{z}_2 = (x_1 + x_2) + j(y_1 + y_2). \tag{7.22}$$

Multiplication:

$$\begin{aligned}\mathbf{z}_1\mathbf{z}_2 &= (x_1 + jy_1)(x_2 + jy_2)\\ &= (x_1x_2 - y_1y_2) + j(x_1y_2 + x_2y_1),\end{aligned} \tag{7.23a}$$

or

$$\begin{aligned}\mathbf{z}_1\mathbf{z}_2 &= |\mathbf{z}_1|e^{j\theta_1} \cdot |\mathbf{z}_2|e^{j\theta_2} = |\mathbf{z}_1||\mathbf{z}_2|e^{j(\theta_1+\theta_2)}\\ &= |\mathbf{z}_1||\mathbf{z}_2|[\cos(\theta_1 + \theta_2) + j\sin(\theta_1 + \theta_2)].\end{aligned} \tag{7.23b}$$

Division: For $\mathbf{z}_2 \neq 0$,

$$\begin{aligned}\frac{\mathbf{z}_1}{\mathbf{z}_2} &= \frac{x_1 + jy_1}{x_2 + jy_2} = \frac{(x_1 + jy_1)}{(x_2 + jy_2)} \cdot \frac{(x_2 - jy_2)}{(x_2 - jy_2)}\\ &= \frac{(x_1x_2 + y_1y_2) + j(x_2y_1 - x_1y_2)}{x_2^2 + y_2^2},\end{aligned} \tag{7.24a}$$

or

$$\begin{aligned}\frac{\mathbf{z}_1}{\mathbf{z}_2} &= \frac{|\mathbf{z}_1|e^{j\theta_1}}{|\mathbf{z}_2|e^{j\theta_2}} = \frac{|\mathbf{z}_1|}{|\mathbf{z}_2|}e^{j(\theta_1-\theta_2)}\\ &= \frac{|\mathbf{z}_1|}{|\mathbf{z}_2|}[\cos(\theta_1 - \theta_2) + j\sin(\theta_1 - \theta_2)].\end{aligned} \tag{7.24b}$$

Powers: For any positive integer n,

$$\begin{aligned}\mathbf{z}^n &= (|\mathbf{z}|e^{j\theta})^n\\ &= |\mathbf{z}|^n e^{jn\theta} = |\mathbf{z}|^n(\cos n\theta + j\sin n\theta),\end{aligned} \tag{7.25}$$

$$\begin{aligned}\mathbf{z}^{1/2} &= \pm|\mathbf{z}|^{1/2}e^{j\theta/2}\\ &= \pm|\mathbf{z}|^{1/2}[\cos(\theta/2) + j\sin(\theta/2)].\end{aligned} \tag{7.26}$$

Useful Relations:

$$-1 = e^{j\pi} = e^{-j\pi} = 1\angle 180^\circ, \tag{7.27a}$$

$$j = e^{j\pi/2} = 1\angle 90^\circ, \tag{7.27b}$$

$$-j = -e^{j\pi/2} = e^{-j\pi/2} = 1\angle -90^\circ, \tag{7.27c}$$

$$\sqrt{j} = (e^{j\pi/2})^{1/2} = \pm e^{j\pi/4} = \frac{\pm(1 + j)}{\sqrt{2}}, \tag{7.27d}$$

$$\sqrt{-j} = \pm e^{-j\pi/4} = \frac{\pm(1 - j)}{\sqrt{2}}. \tag{7.27e}$$

For quick reference, the preceding properties of complex numbers are summarized in Table 7-2. Note that if a complex number is given by $(a + jb)$ and $b = 1$, it can be written either as $(a + j1)$ or simply as $(a + j)$. Thus, j is synonymous with $j1$.

Example 7-3: Working with Complex Numbers

Given two complex numbers:

$$\mathbf{V} = 3 - j4$$

and

$$\mathbf{I} = -(2 + j3),$$

(a) Express **V** and **I** in polar form, and find (b) **VI**, (c) **VI***, (d) **V**/**I**, and (e) $\sqrt{\mathbf{I}}$.

Solution:

(a)

$$\begin{aligned}|\mathbf{V}| &= \sqrt[+]{\mathbf{V}\mathbf{V}^*} = \sqrt[+]{(3 - j4)(3 + j4)} = \sqrt[+]{9 + 16} = 5,\\ \theta_V &= \tan^{-1}(-4/3) = -53.1^\circ,\\ \mathbf{V} &= |\mathbf{V}|e^{j\theta_V} = 5e^{-j53.1^\circ} = 5\angle -53.1^\circ,\end{aligned}$$

and

$$|\mathbf{I}| = \sqrt[+]{2^2 + 3^2} = \sqrt[+]{13} = 3.61.$$

Since $\mathbf{I} = (-2 - j3)$ is in the third quadrant in the complex plane [Fig. 7-5],

$$\theta_I = -180^\circ + \tan^{-1}\left(\tfrac{3}{2}\right) = -123.7^\circ,$$

and

$$\mathbf{I} = 3.61\angle -123.7^\circ.$$

Table 7-2: Properties of complex numbers.

Euler's Identity: $e^{j\theta} = \cos\theta + j\sin\theta$	
$\sin\theta = \dfrac{e^{j\theta} - e^{-j\theta}}{2j}$	$\cos\theta = \dfrac{e^{j\theta} + e^{-j\theta}}{2}$
$\mathbf{z} = x + jy = \lvert\mathbf{z}\rvert e^{j\theta}$	$\mathbf{z}^* = x - jy = \lvert\mathbf{z}\rvert e^{-j\theta}$
$x = \mathfrak{Re}(\mathbf{z}) = \lvert\mathbf{z}\rvert\cos\theta$	$\lvert\mathbf{z}\rvert = \sqrt[+]{\mathbf{z}\mathbf{z}^*} = \sqrt[+]{x^2 + y^2}$
$y = \mathfrak{Im}(\mathbf{z}) = \lvert\mathbf{z}\rvert\sin\theta$	$\theta = \tan^{-1}(y/x)$
$\mathbf{z}^n = \lvert\mathbf{z}\rvert^n e^{jn\theta}$	$\mathbf{z}^{1/2} = \pm\lvert\mathbf{z}\rvert^{1/2} e^{j\theta/2}$
$\mathbf{z}_1 = x_1 + jy_1$	$\mathbf{z}_2 = x_2 + jy_2$
$\mathbf{z}_1 = \mathbf{z}_2$ iff $x_1 = x_2$ and $y_1 = y_2$	$\mathbf{z}_1 + \mathbf{z}_2 = (x_1 + x_2) + j(y_1 + y_2)$
$\mathbf{z}_1\mathbf{z}_2 = \lvert\mathbf{z}_1\rvert\lvert\mathbf{z}_2\rvert e^{j(\theta_1+\theta_2)}$	$\dfrac{\mathbf{z}_1}{\mathbf{z}_2} = \dfrac{\lvert\mathbf{z}_1\rvert}{\lvert\mathbf{z}_2\rvert} e^{j(\theta_1-\theta_2)}$
$-1 = e^{j\pi} = e^{-j\pi} = 1\angle\pm 180^\circ$	
$j = e^{j\pi/2} = 1\angle 90^\circ$	$-j = e^{-j\pi/2} = 1\angle -90^\circ$
$\sqrt{j} = \pm e^{j\pi/4} = \pm\dfrac{(1+j)}{\sqrt{2}}$	$\sqrt{-j} = \pm e^{-j\pi/4} = \pm\dfrac{(1-j)}{\sqrt{2}}$

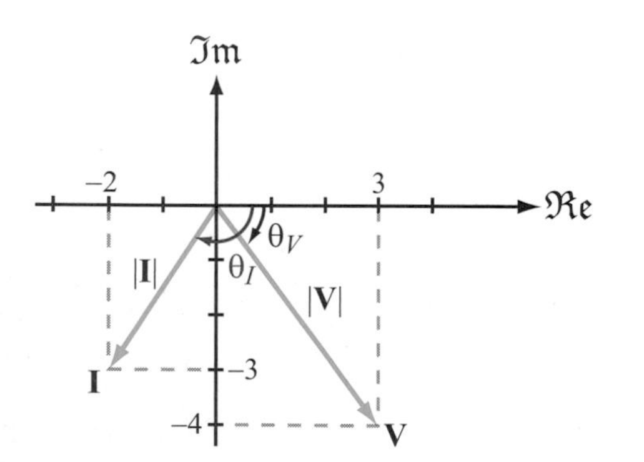

Figure 7-5: Complex numbers **V** and **I** in the complex plane (Example 7-3).

Alternatively, whenever the real part of a complex number is negative, we can factor out a (-1) multiplier and then use Eq. (7.27a) to replace it with a phase angle of either $+180^\circ$ or -180°, as needed. In the case of **I**, the process is as follows:

$$\begin{aligned}\mathbf{I} &= -2 - j3 = -(2 + j3)\\ &= e^{\pm j180^\circ} \cdot \sqrt[+]{2^2 + 3^2}\; e^{j\tan^{-1}(3/2)}\\ &= 3.61 e^{j57.3^\circ} e^{\pm j180^\circ}.\end{aligned}$$

Since our preference is to end up with a phase angle within the range between -180° and $+180^\circ$, we will choose -180°. Hence,

$$\mathbf{I} = 3.61 e^{-j123.7^\circ}.$$

(b)

$$\begin{aligned}\mathbf{VI} &= (5\angle -53.1^\circ)(3.61\angle -123.7^\circ)\\ &= (5 \times 3.61)\angle(-53.1^\circ - 123.7^\circ) = 18.05\angle -176.8^\circ.\end{aligned}$$

(c)

$$\begin{aligned}\mathbf{VI}^* &= 5e^{-j53.1^\circ} \times 3.61 e^{j123.7^\circ}\\ &= 18.05 e^{j70.6^\circ}.\end{aligned}$$

(d)

$$\begin{aligned}\frac{\mathbf{V}}{\mathbf{I}} &= \frac{5e^{-j53.1^\circ}}{3.61 e^{-j123.7^\circ}}\\ &= 1.39 e^{j70.6^\circ}.\end{aligned}$$

(e)

$$\begin{aligned}\sqrt{\mathbf{I}} &= \sqrt{3.61 e^{-j123.7^\circ}}\\ &= \pm\sqrt{3.61}\; e^{-j123.7^\circ/2} = \pm 1.90 e^{-j61.85^\circ}.\end{aligned}$$

Review Question 7-3: If $\mathbf{Z}$ is a complex number that lies in the first quadrant in the complex plane, its complex conjugate $\mathbf{Z}^*$ will lie in which quadrant?

Review Question 7-4: If two complex numbers have the same magnitude, are they necessarily equal to each other?

Exercise 7-3: Express the following complex functions in polar form:

$$\mathbf{z}_1 = (4 - j3)^2$$

and

$$\mathbf{z}_2 = (4 - j3)^{1/2}.$$

Answer: $\mathbf{z}_1 = 25\angle{-73.7^\circ}$, $\mathbf{z}_2 = \pm\sqrt{5}\,\angle{-18.4^\circ}$. (See)

Exercise 7-4: Show that $\sqrt{2j} = \pm(1 + j)$. (See)

7-3 Phasor Domain

A ***domain transformation*** is a mathematical process that converts a set of variables from their domain into a corresponding set of variables defined in another domain.

We will explore how currents and voltages defined in the time domain are transformed into their counterparts in the phasor domain, and why such a transformation facilitates the analysis of ac circuits.

The KVL and KCL equations characterizing an ac circuit containing capacitors and inductors take the form of integro-differential equations with forcing functions (representing the real sources in the circuit) that vary sinusoidally with time. The phasor-analysis technique allows us to transform the equations from the time domain to the phasor domain, as a result of which the integro-differential equations get converted into linear equations with no sinusoidal functions. After solving for the desired variable—such as a particular voltage or current—in the phasor domain, conversion back to the time domain provides the same solution that we would have obtained had we solved the integro-differential equations entirely in the time domain. The procedure involves multiple steps, but it avoids the complexity of solving differential equations containing sinusoidal functions.

7-3.1 Time-Domain/Phasor-Domain Correspondence

Transformation from the time domain to the phasor domain entails transforming all time-dependent quantities in the circuit, which in effect transforms the entire circuit from the time domain to an equivalent circuit in the phasor domain. The quantities involved in the transformation include all currents and voltages, all sources, and all capacitors and inductors. The values of capacitors and inductors do not change per se, but their i–v relationships will undergo a transformation because they involve differentiation or integration with respect to t.

Any cosinusoidally ***time-varying function*** $x(t)$, representing a voltage or a current, can be expressed in the form

$$x(t) = \mathfrak{Re}[\mathbf{X}e^{j\omega t}], \tag{7.28}$$

where $\mathbf{X}$ is a ***time-independent*** function called the ***phasor equivalent*** or ***phasor counterpart*** of $x(t)$. Thus, $x(t)$ is defined in the time domain, while its counterpart $\mathbf{X}$ is defined in the phasor domain.

To distinguish phasor quantities from their time-domain counterparts, phasors are always represented by bold letters in this book.

In general, the phasor-domain quantity $\mathbf{X}$ is complex, consisting of a magnitude $|\mathbf{X}|$ and a phase angle ϕ by

$$\mathbf{X} = |\mathbf{X}|e^{j\phi}. \tag{7.29}$$

Using this expression in Eq. (7.28) gives

$$\begin{aligned} x(t) &= \mathfrak{Re}[|\mathbf{X}|e^{j\phi}e^{j\omega t}] \\ &= \mathfrak{Re}[|\mathbf{X}|e^{j(\omega t+\phi)}] \\ &= |\mathbf{X}|\cos(\omega t + \phi). \end{aligned} \tag{7.30}$$

Application of the $\mathfrak{Re}$ operator allows us to transform a function from the phasor domain to the time domain. The reverse operation, namely to specify the phasor-domain equivalent of a time function, can be ascertained by comparing the two sides of Eq. (7.30). Thus, for a voltage $v(t)$ with phasor counterpart $\mathbf{V}$, the correspondence between the two domains is as follows:

Time Domain		**Phasor Domain**	
$v(t) = V_0 \cos \omega t$	$\longleftrightarrow$	$\mathbf{V} = V_0$	(7.31a)
$v(t) = V_0 \cos(\omega t + \phi)$	$\longleftrightarrow$	$\mathbf{V} = V_0 e^{j\phi}$.	(7.31b)

If $\phi = -\pi/2$,

$$v(t) = V_0 \cos(\omega t - \pi/2) \longleftrightarrow \mathbf{V} = V_0 e^{-j\pi/2}. \tag{7.32}$$

Application of Eq. (7.7a) of Table 7-1 to the cosine of $v(t)$ and Eq. (7.27c) to $e^{-j\pi/2}$ gives

$$v(t) = V_0 \sin \omega t \longleftrightarrow \mathbf{V} = -jV_0, \tag{7.33}$$

Technology Brief 14: Noise-Cancellation Headphones

Noise-cancellation headphones are a class of devices that use ***active noise-control*** technology to reduce the level of environmental noise reaching a listener's ear. They were invented by Amar Bose in 1978, based on the concepts developed in the mid-20th century (most notably by Paul Lueg, who developed a system for cancelling noise in air ducts using loudspeakers). The primary advantage of such systems is the ability to selectively reduce noise without having to use heavy and expensive sound padding. Beyond the commercial headphones used by airline passengers, specialized active noise-control systems have been in use by pilots and heavy equipment operators for several decades. In small enclosed environments, active noise-control systems can employ microphone and speaker arrays to lower the amount of ambient noise experienced by the listener. Examples of this include noise-cancellation systems used to dampen engine noise in cockpits, active mufflers for industrial exhaust stacks, noise reduction around large fans and, recently, systems for reducing road and traffic noise in automobile interiors.

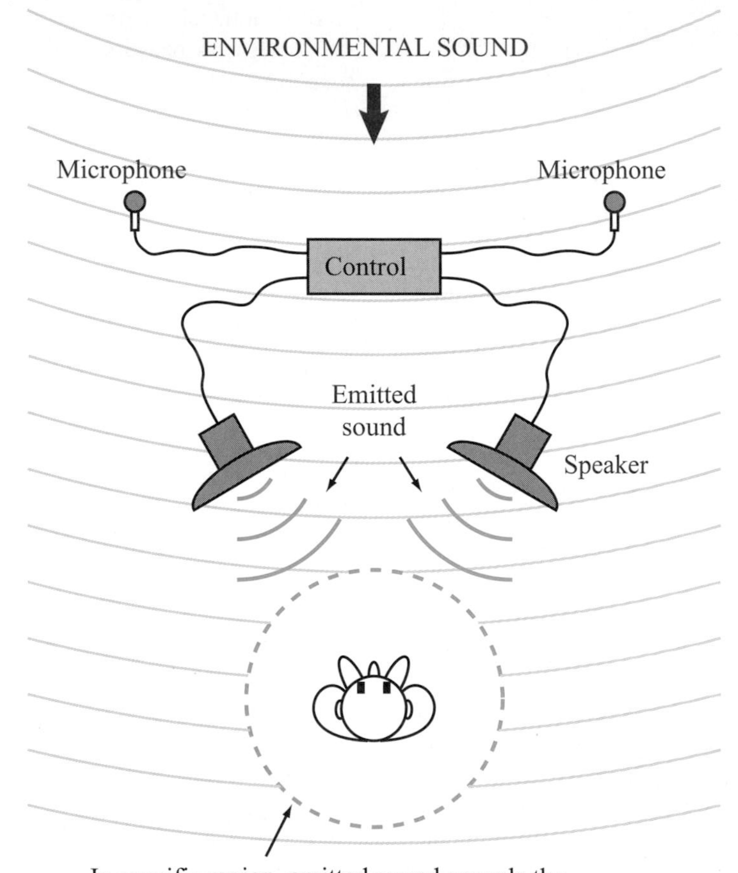

Figure TF14-1: Active noise control. A basic active noise-control system uses a set of microphones to sense incoming ambient sound; the microphone signals are fed into control circuits which drive a set of speakers. The control circuit generates exactly the signals required to cancel out the incoming ambient sound in a specific region.

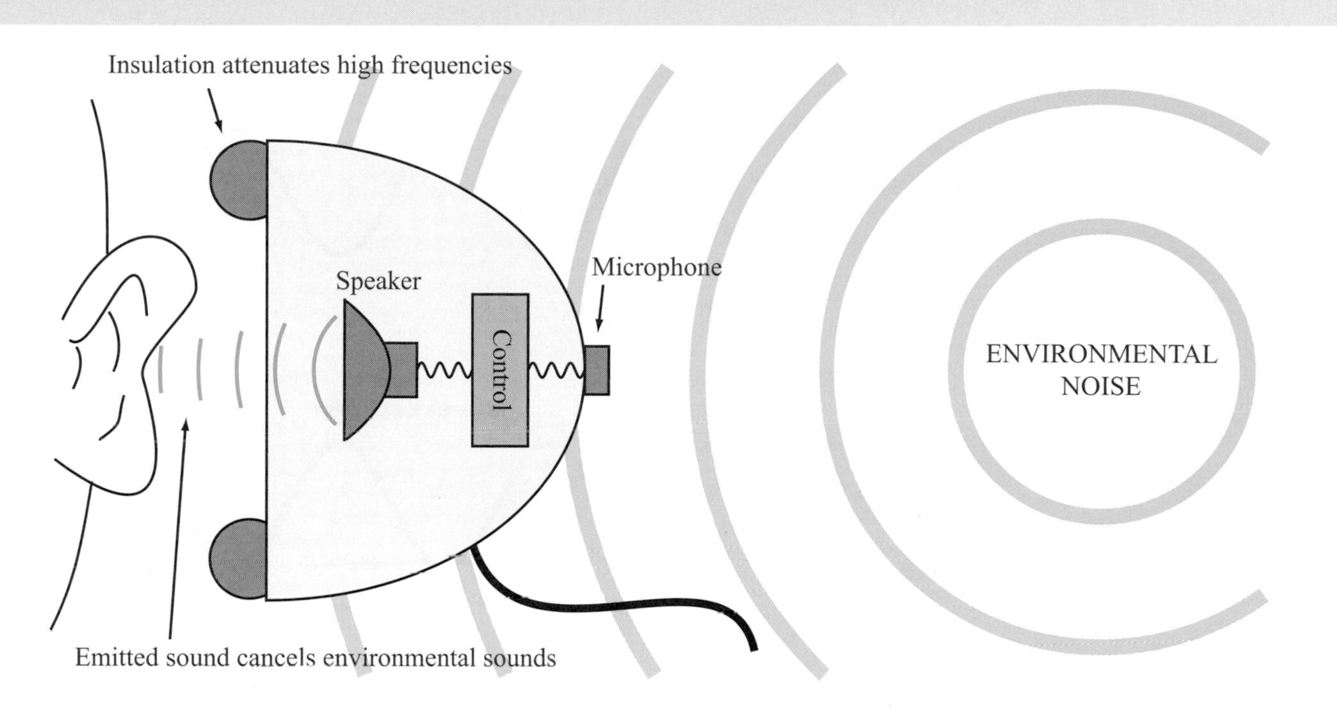

Figure TF14-2: Noise-cancellation headphones employ the active noise-control principle to eliminate noise around the ear. Sound insulation is used commonly to remove high-frequency noise signals, while the active system removes low-frequency sound.

Active Noise Control

In its most basic form, active noise control consists of measuring the sound levels at certain points in the environment and then using that data to emit noise from speakers whose frequency, phase shift, and amplitude are selected in order to cancel out the incoming environmental noise (Fig. TF14-1). In noise-cancellation headphones, small microphones outside the headphones measure the incoming ambient noise, and the measured signal is then fed to circuitry that produces output noise in the headset that cancels out the ambient sound (Fig. TF14-2). The general phenomenon whereby one waveform is added to another to cancel it out is called ***destructive interference***. Consider a vibrating wave traveling along a one-dimensional string (Fig. TF14-3), which is analogous to a sound pressure wave moving through the air. If we superimpose a second traveling wave onto the string (perhaps by waving the end up and down with a second hand), the two waves will overlap and the result will be the sum of the two individual waves (superposition). If we precisely time the second wave so that it is the exact mirror of the first (i.e., it is phase-shifted by 180° from the first wave), the two waves will cancel out exactly, and the string will not vibrate. This (in principle) is what active noise control aims to do, even though (in reality) the technology faces a number of limitations.

Limitations

In order to truly cancel out all ambient noise, the emitted noise would have to exactly match the ambient noise in both space and time for all audible frequencies across a three-dimensional volume (such as the interior of a car or an airplane cabin). This is very difficult to accomplish in real environments. High frequencies are the hardest to match, because to correctly cancel them, the system would need to employ large arrays of microphones and speakers. Moreover,

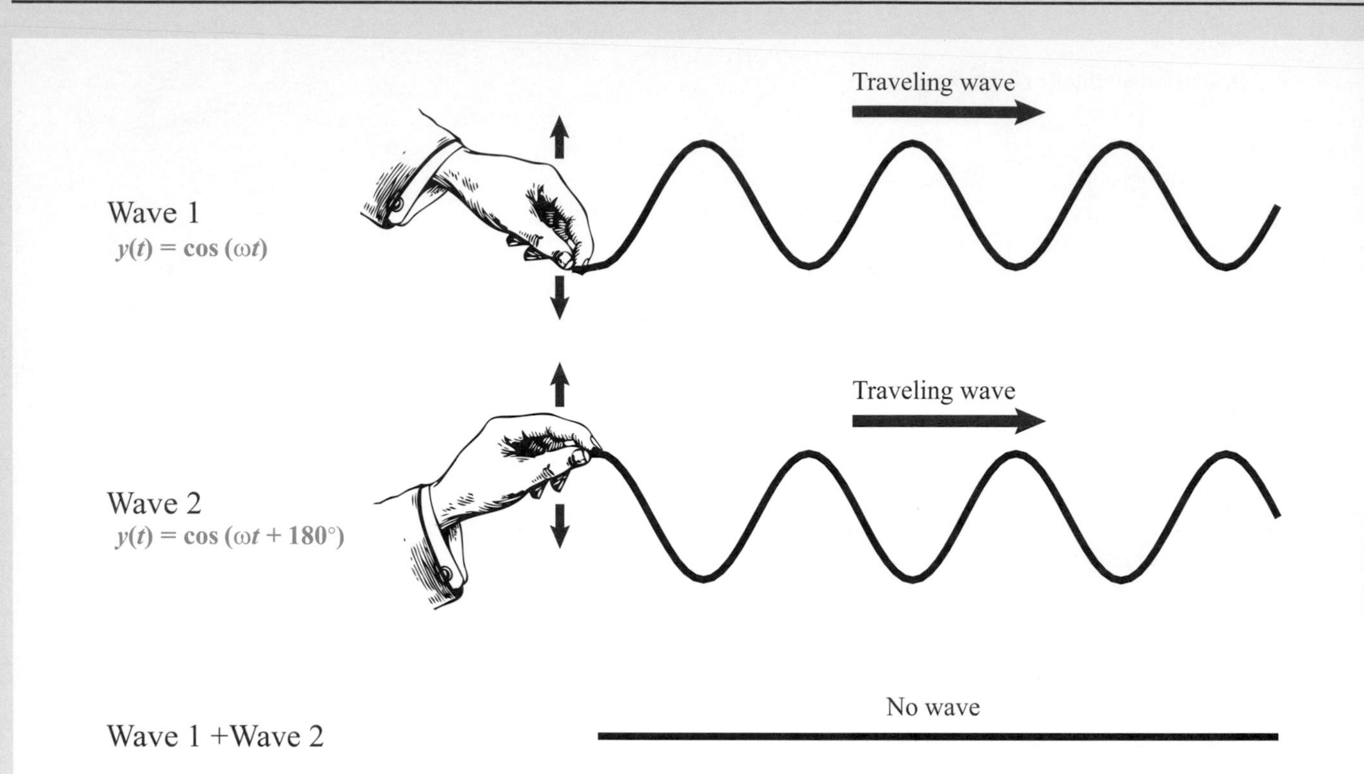

Figure TF14-3: Destructive interference via the superposition of two waves. A string can be agitated so as to generate traveling waves along its length. In order to cancel out a wave traveling along the string (Wave 1), we simultaneously can generate a second wave of the same amplitude and frequency with a 180° phase shift (Wave 2). Because the vibration of the string will be the result of the two waves superimposed, the two waves cancel out, and no vibration occurs. This is analogous to what active control systems do: Microphones sense ambient waves that are then canceled out by emitting an appropriately phase-shifted wave from the speakers.

objects within the environment will reflect, absorb, and emit sound—further complicating the signals required to cance the ambient sound. The situation is somewhat easier for headphones, as the area of interest is simply the user's ear However, most commercial noise-cancellation headphones do not attempt to cancel high frequencies. Padding anc passive layers are instead used to absorb the high frequencies and the system actively cancels out only the lowe frequencies (e.g., the airplane engine hum). In general, noise cancellation only works well for sound that is periodic Noise that is random or has very fast changes is very hard to mask, because the system cannot compute what the interfering signal should be instantaneously.

Feedforward versus Feedback Control

Active noise-control systems provide an interesting comparison between feedback control (which we examined i Chapter 4) and feedforward control (Fig. TF14-4). Consider again Fig. TF14-1. If the microphones of this system ar positioned relatively far away from the speakers, they sample the incoming ambient sound signal and send it ahead t the control circuit, which then drives the speakers. There is no microphone at the speaker location and thus no way t measure the "output" of the system (i.e., there is no microphone that measures how well the system is canceling th sound at the listener). This is an example of ***feedforward*** control. If we were to move the microphones very close to the speakers (or, better yet right next to the listener), the microphones continuously would report how well the speakers were canceling the sound. If the control system is doing poorly, the microphones will detect some sound, and the control circuit can attempt to correct for this. In such a configuration, the system is operating with ***feedback*** control. Figure TF14-4 illustrates both of these control configurations. Some sophisticated active noise-control systems use both modes simultaneously: They have distant microphones as well as microphones near the listener. In general, feedforward systems are less practical to implement in consumer systems, and feedback systems tend to be less stable.

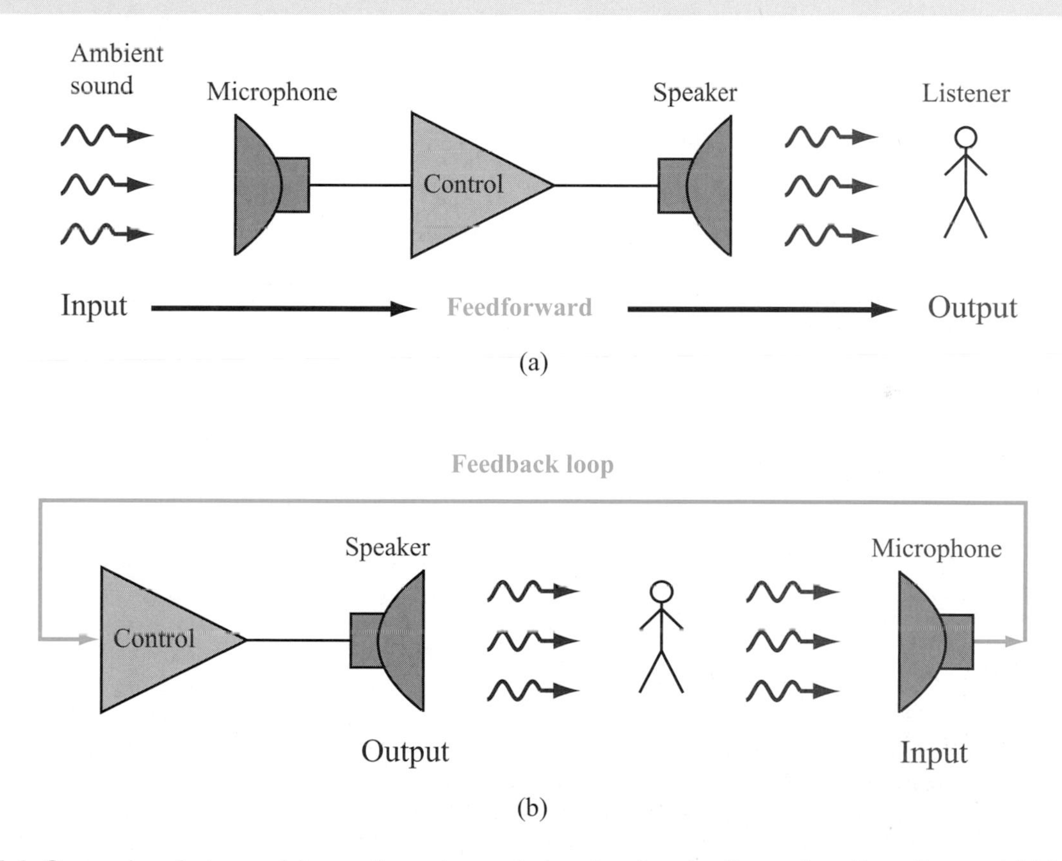

Figure TF14-4: Comparison between (a) an active noise-control system in a feedforward configuration, and (b) in a feedback configuration.

which can be generalized to

$$v(t) = V_0 \sin(\omega t + \phi) \quad \longleftrightarrow \quad \mathbf{V} = V_0 e^{j(\phi - \pi/2)}. \tag{7.34}$$

Occasionally, voltage and current time functions may encounter differentiation or integration. For example, consider a current $i(t)$ with a corresponding phasor $\mathbf{I}$,

$$i(t) = \mathfrak{Re}[\mathbf{I} e^{j\omega t}], \tag{7.35}$$

where $\mathbf{I}$ may be complex but, by definition, not a function of time. The derivative di/dt is given by

$$\begin{aligned} \frac{di}{dt} &= \frac{d}{dt}[\mathfrak{Re}(\mathbf{I} e^{j\omega t})] \\ &= \mathfrak{Re}\left[\frac{d}{dt}(\mathbf{I} e^{j\omega t})\right] \\ &= \mathfrak{Re}[j\omega \mathbf{I} e^{j\omega t}], \end{aligned} \tag{7.36}$$

where in the second step we interchanged the order of the two operators, $\mathfrak{Re}$ and d/dt, which is justified by the fact that the two operators are independent of one another, meaning that *taking the real part* of a quantity has no influence on taking its time derivative, and vice versa. We surmise from Eq. (7.36) that

$$\frac{di}{dt} \quad \longleftrightarrow \quad j\omega \mathbf{I}, \tag{7.37}$$

or:

> Differentiation of a time function $i(t)$ in the time domain is equivalent to multiplication of its phasor counterpart $\mathbf{I}$ by $j\omega$ in the phasor domain.

Similarly,

$$\begin{aligned} \int i \, dt &= \int \mathfrak{Re}[\mathbf{I} e^{j\omega t}] \, dt \\ &= \mathfrak{Re}\left[\int \mathbf{I} e^{j\omega t} \, dt\right] \\ &= \mathfrak{Re}\left[\frac{\mathbf{I}}{j\omega} e^{j\omega t}\right], \end{aligned} \tag{7.38}$$

or

$$\int i \, dt \quad \longleftrightarrow \quad \frac{\mathbf{I}}{j\omega}, \tag{7.39}$$

which states that:

> Integration of $i(t)$ in the time domain is equivalent to dividing its phasor $\mathbf{I}$ by $j\omega$ in the phasor domain.

Table 7-3 provides a summary of some time functions and their phasor-domain counterparts.

7-3.2 Impedance of Circuit Elements

Resistors

The v–i relationship for a resistor R is

$$v_R = R i_R. \tag{7.40}$$

The time-domain quantities v_R and i_R are related to their phasor-domain counterparts by

$$v_R = \mathfrak{Re}[\mathbf{V}_R e^{j\omega t}] \tag{7.41a}$$

and

$$i_R = \mathfrak{Re}[\mathbf{I}_R e^{j\omega t}]. \tag{7.41b}$$

Inserting these expressions into Eq. (7.40) gives

$$\begin{aligned} \mathfrak{Re}[\mathbf{V}_R e^{j\omega t}] &= R\, \mathfrak{Re}[\mathbf{I}_R e^{j\omega t}] \\ &= \mathfrak{Re}[R \mathbf{I}_R e^{j\omega t}]. \end{aligned} \tag{7.42}$$

Upon combining both sides under the same real-part ($\mathfrak{Re}$) operator, we have

$$\mathfrak{Re}[(\mathbf{V}_R - R\mathbf{I}_R) e^{j\omega t}] = 0. \tag{7.43a}$$

Through a somewhat similar treatment that uses a sine reference—rather than a cosine reference—to define sinusoidal functions, we can obtain the result

$$\mathfrak{Im}[(\mathbf{V}_R - R\mathbf{I}_R) e^{j\omega t}] = 0, \tag{7.43b}$$

Table 7-3: Time-domain sinusoidal functions $x(t)$ and their cosine-reference phasor-domain counterparts $\mathbf{X}$, where $x(t) = \mathfrak{Re}\,[\mathbf{X} e^{j\omega t}]$.

$x(t)$		$\mathbf{X}$
$A \cos \omega t$	$\longleftrightarrow$	A
$A \cos(\omega t + \phi)$	$\longleftrightarrow$	$A e^{j\phi}$
$-A \cos(\omega t + \phi)$	$\longleftrightarrow$	$A e^{j(\phi \pm \pi)}$
$A \sin \omega t$	$\longleftrightarrow$	$A e^{-j\pi/2} = -jA$
$A \sin(\omega t + \phi)$	$\longleftrightarrow$	$A e^{j(\phi - \pi/2)}$
$-A \sin(\omega t + \phi)$	$\longleftrightarrow$	$A e^{j(\phi + \pi/2)}$
$\frac{d}{dt}(x(t))$	$\longleftrightarrow$	$j\omega \mathbf{X}$
$\frac{d}{dt}[A \cos(\omega t + \phi)]$	$\longleftrightarrow$	$j\omega A e^{j\phi}$
$\int x(t) \, dt$	$\longleftrightarrow$	$\frac{1}{j\omega} \mathbf{X}$
$\int A \cos(\omega t + \phi) \, dt$	$\longleftrightarrow$	$\frac{1}{j\omega} A e^{j\phi}$

which, for the sake of expediency, we simply state without taking the steps to prove it. In view of Eqs. (7.43a) and (7.43b), both the real and imaginary components of the quantity inside the square bracket are zero. Hence, the quantity itself is zero, and since $e^{j\omega t} \neq 0$, it follows that

$$\mathbf{V}_\mathrm{R} - R\mathbf{I}_\mathrm{R} = 0. \tag{7.44}$$

In the phasor domain:

> The ***impedance*** **Z** of a circuit element is defined as the ratio of the phasor voltage across it to the phasor current entering through its plus terminal,

$$\mathbf{Z} = \frac{\mathbf{V}}{\mathbf{I}} \quad (\Omega), \tag{7.45}$$

and the unit of **Z** is the ohm (Ω). For a resistor, Eq. (7.44) gives

$$\mathbf{Z}_\mathrm{R} = \frac{\mathbf{V}_\mathrm{R}}{\mathbf{I}_\mathrm{R}} = R. \tag{7.46}$$

Thus, for a resistor the impedance is entirely real, and the form of the v–i relationship is the same in both the time and phasor domains.

Inductors

In the time domain, the voltage v_L across an inductor L is related to i_L by

$$v_\mathrm{L} = L\,\frac{di_\mathrm{L}}{dt}. \tag{7.47}$$

Phasors $\mathbf{V}_\mathrm{L}$ and $\mathbf{I}_\mathrm{L}$ are related to their time-domain counterparts by

$$v_\mathrm{L} = \mathfrak{Re}[\mathbf{V}_\mathrm{L} e^{j\omega t}] \tag{7.48a}$$

and

$$i_\mathrm{L} = \mathfrak{Re}[\mathbf{I}_\mathrm{L} e^{j\omega t}]. \tag{7.48b}$$

Consequently,

$$\begin{aligned}\mathfrak{Re}[\mathbf{V}_\mathrm{L} e^{j\omega t}] &= L\,\frac{d}{dt}[\mathfrak{Re}(\mathbf{I}_\mathrm{L} e^{j\omega t})] \\ &= \mathfrak{Re}[j\omega L \mathbf{I}_\mathrm{L} e^{j\omega t}],\end{aligned} \tag{7.49}$$

which leads to

$$\mathbf{V}_\mathrm{L} = j\omega L \mathbf{I}_\mathrm{L} \tag{7.50}$$

and

$$\mathbf{Z}_\mathrm{L} = \frac{\mathbf{V}_\mathrm{L}}{\mathbf{I}_\mathrm{L}} = j\omega L. \tag{7.51}$$

According to Eq. (7.51), $\mathbf{Z}_\mathrm{L}$ is positive and entirely imaginary (no real component); $\mathbf{Z}_\mathrm{L} \to 0$ as $\omega \to 0$ (dc); and $\mathbf{Z}_\mathrm{L} \to \infty$ as $\omega \to \infty$. Consequently:

> In the phasor domain, an inductor behaves like a short circuit at dc and like an open circuit at very high frequencies.

Capacitors

Since for a capacitor

$$i_\mathrm{C} = C\,\frac{dv_\mathrm{C}}{dt}, \tag{7.52}$$

it follows that in the phasor domain,

$$\mathbf{I}_\mathrm{C} = j\omega C \mathbf{V}_\mathrm{C} \tag{7.53}$$

and

$$\mathbf{Z}_\mathrm{C} = \frac{\mathbf{V}_\mathrm{C}}{\mathbf{I}_\mathrm{C}} = \frac{1}{j\omega C}. \tag{7.54}$$

Because $\mathbf{Z}_\mathrm{L}$ and $\mathbf{Z}_\mathrm{C}$ are, respectively, directly and inversely proportional to ω, $\mathbf{Z}_\mathrm{L}$ and $\mathbf{Z}_\mathrm{C}$ assume inverse roles as ω approaches zero and ∞.

> In the phasor domain, a capacitor behaves like an open circuit at dc and like a short circuit at very high frequencies.

We note that the impedance of a resistor is purely real, that of an inductor is purely imaginary and positive, and that of a capacitor is purely imaginary and negative (because $1/j\omega C = -j/\omega C$). Table 7-4 provides a summary of the v–i properties for R, L, and C.

Example 7-4: Phasor Quantities

Determine the phasor-domain counterparts of the following quantities:

(a) $v_1(t) = 10\cos(2 \times 10^4 t + 53^\circ)$ V,

(b) $v_2(t) = -6\sin(3 \times 10^3 t - 15^\circ)$ V,

(c) $L = 0.4$ mH at 1 kHz,

(d) $C = 2\ \mu$F at 1 MHz.

Table 7-4: Summary of v–i properties for R, L, and C.

Property	R	L	C
v–i	$v = Ri$	$v = L \frac{di}{dt}$	$i = C \frac{dv}{dt}$
V–I	$\mathbf{V} = R\mathbf{I}$	$\mathbf{V} = j\omega L\mathbf{I}$	$\mathbf{V} = \frac{\mathbf{I}}{j\omega C}$
Z	R	$j\omega L$	$\frac{1}{j\omega C}$
dc equivalent	R	Short circuit	Open circuit
High-frequency equivalent	R	Open circuit	Short circuit
Frequency response	$\lvert\mathbf{Z}_R\rvert$: R (constant vs. ω)	$\lvert\mathbf{Z}_L\rvert$: ωL (vs. ω)	$\lvert\mathbf{Z}_C\rvert$: 1/ωC (vs. ω)

Solution:
(a) Since $v_1(t)$ is already in cosine format,

$$\mathbf{V}_1 = 10e^{j53^\circ} = 10\angle 53^\circ \text{ V}.$$

(b) To determine the phasor $\mathbf{V}_2$ corresponding to $v_2(t)$, we should either convert the expression for $v_2(t)$ to standard cosine format or apply the transformation for a sine function given in Table 7-3. We choose the first option,

$$\begin{aligned} v_2(t) &= -6\sin(3 \times 10^3 t - 15^\circ) \\ &= -6\cos(3 \times 10^3 t - 15^\circ - 90^\circ) \\ &= -6\cos(3 \times 10^3 t - 105^\circ) \text{ V}. \end{aligned}$$

To convert the amplitude from -6 to $+6$, we use Eq. (7.7d) of Table 7-1, namely

$$-\cos(x) = \cos(x \pm 180^\circ).$$

We can either add or subtract 180° from the argument of the cosine. Since the argument has a negative phase angle (-105°), it is more convenient to add 180°. Hence,

$$\begin{aligned} v_2(t) &= 6\cos(3 \times 10^3 t - 105^\circ + 180^\circ) \\ &= 6\cos(3 \times 10^3 t + 75^\circ) \text{ V}, \end{aligned}$$

and

$$\mathbf{V}_2 = 6e^{j75^\circ} = 6\angle 75^\circ \text{ V}.$$

(c)

$$\mathbf{Z}_L = j\omega L = j2\pi \times 10^3 \times 0.4 \times 10^{-3} = j2.5\ \Omega.$$

(d)

$$\mathbf{Z}_C = \frac{-j}{\omega C} = \frac{-j}{2\pi \times 10^6 \times 2 \times 10^{-6}} = -j0.08\ \Omega.$$

Review Question 7-5: Why is the phasor domain useful for analyzing ac circuits?

Review Question 7-6: Differentiation in the time domain corresponds to what mathematical operation in the phasor domain?

Review Question 7-7: The unit for inductance is the henry (H). What is the unit for the impedance $\mathbf{Z}_L$ of an inductor?

Review Question 7-8: What type of circuit is equivalent to the behavior of (a) an inductor at dc and (b) a capacitor at very high frequencies?

Exercise 7-5: Determine the phasor counterparts of the following waveforms:

(a) $i_1(t) = 2\sin(6 \times 10^3 t - 30°)$ A,

(b) $i_2(t) = -4\sin(1000t + 136°)$ A

Answer: (a) $\mathbf{I}_1 = 2\angle{-120°}$ A, (b) $\mathbf{I}_2 = 4\angle{-134°}$ A. (See ◎)

Exercise 7-6: Obtain the time-domain waveforms (in standard cosine format) corresponding to the following phasors at angular frequency $\omega = 3 \times 10^4$ rad/s:

(a) $\mathbf{V}_1 = (-3 + j4)$ V

(b) $\mathbf{V}_2 = (3 - j4)$ V

Answer: (a) $v_1(t) = 5\cos(3 \times 10^4 t + 126.87°)$ V, (b) $v_2(t) = 5\cos(3 \times 10^4 t - 53.13°)$ V. (See ◎)

Exercise 7-7: At $\omega = 10^6$ rad/s, the phasor voltage across and current through a certain element are given by $\mathbf{V} = 4\angle{-20°}$ V and $\mathbf{I} = 2\angle{70°}$ A. What type of element is it?

Answer: Capacitor with $C = 0.5\ \mu$F. (See ◎)

7-4 Phasor-Domain Analysis

In the time domain, Kirchhoff's voltage law states that the algebraic sum of all voltages v_1 to v_n around a closed path containing n elements is zero,

$$v_1(t) + v_2(t) + \cdots + v_n(t) = 0. \quad (7.55)$$

If $\mathbf{V}_1$ to $\mathbf{V}_n$ are respectively the phasor-domain counterparts of v_1 to v_n, then

$$\mathfrak{Re}[\mathbf{V}_1 e^{j\omega t}] + \mathfrak{Re}[\mathbf{V}_2 e^{j\omega t}] + \cdots + \mathfrak{Re}[\mathbf{V}_n e^{j\omega t}] = 0, \quad (7.56)$$

or equivalently,

$$\mathfrak{Re}[(\mathbf{V}_1 + \mathbf{V}_2 + \cdots + \mathbf{V}_n)e^{j\omega t}] = 0. \quad (7.57)$$

Since $e^{j\omega t} \neq 0$, it follows that

$$\mathbf{V}_1 + \mathbf{V}_2 + \cdots + \mathbf{V}_n = 0, \quad (7.58)$$

which states that KVL is equally applicable in the phasor domain.

Similarly, KCL at a node leads to

$$\mathbf{I}_1 + \mathbf{I}_2 + \cdots + \mathbf{I}_n = 0, \quad (7.59)$$

where $\mathbf{I}_1$ to $\mathbf{I}_n$ are the phasor counterparts of i_1 to i_n. The fact that KCL and KVL are valid in the phasor domain is highly significant, because it implies that the analysis tools we developed earlier on the basis of these two laws also are valid in the phasor domain. These include the nodal and mesh analysis methods, the Thévenin and Norton techniques, and several others. Revisiting these tools and learning to apply them to ac circuits is the subject of future sections in this chapter. However, we will now introduce the basic elements of the phasor analysis process through a simple example.

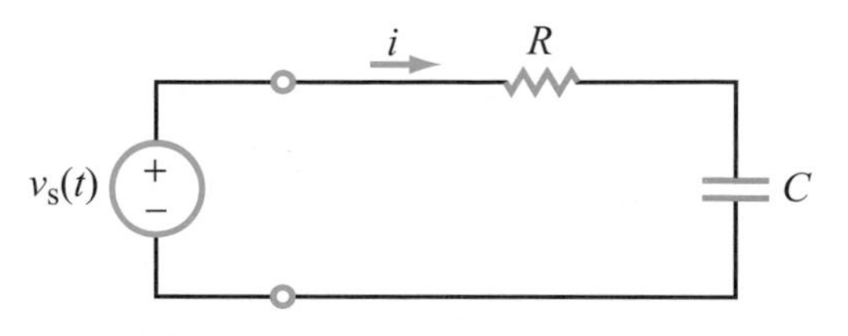

Figure 7-6: RC circuit connected to an ac source.

The phasor analysis method consists of five steps. To assist us in presenting it, we will use the RC circuit shown in Fig. 7-6. The voltage source is given by

$$v_s = 12\sin(\omega t - 45°) \text{ V}, \quad (7.60)$$

with $\omega = 10^3$ rad/s, $R = \sqrt{3}$ kΩ, and $C = 1\ \mu$F. Application of KVL generates the following loop equation:

$$Ri + \frac{1}{C}\int i\, dt = v_s \qquad \text{(time domain)}. \quad (7.61)$$

Our goal is to obtain a solution for $i(t)$. In general, $i(t)$ consists of a transient response, obtained by solving Eq. (7.61) with v_s set equal to zero (as we had done previously in Chapter 5), and a steady-state response that involves the sinusoidal function $v_s(t)$. Our interest at present is in the sinusoidal response, which we can obtain by solving Eq. (7.61) in the time domain, but the method of solution is somewhat cumbersome—even for such a simple circuit—on account of the sinusoidal voltage source. Alternatively, we can obtain the desired solution by applying the phasor technique, which avoids dealing with sine and cosine functions altogether.

Step 1: Adopt Cosine Reference

All voltages and currents with known sinusoidal functions should be expressed in the standard cosine format (Section 7-1). For our RC circuit, $v_s(t)$ is the only time-varying quantity with an explicit expression, and since $v_s(t)$ is given in terms of a sine function, we need to convert it into a cosine by applying Eq. (7.7a) of Table 7-1:

$$\begin{aligned} v_s(t) &= 12\sin(\omega t - 45°) \\ &= 12\cos(\omega t - 45° - 90°) \\ &= 12\cos(\omega t - 135°) \text{ V}. \end{aligned} \quad (7.62)$$

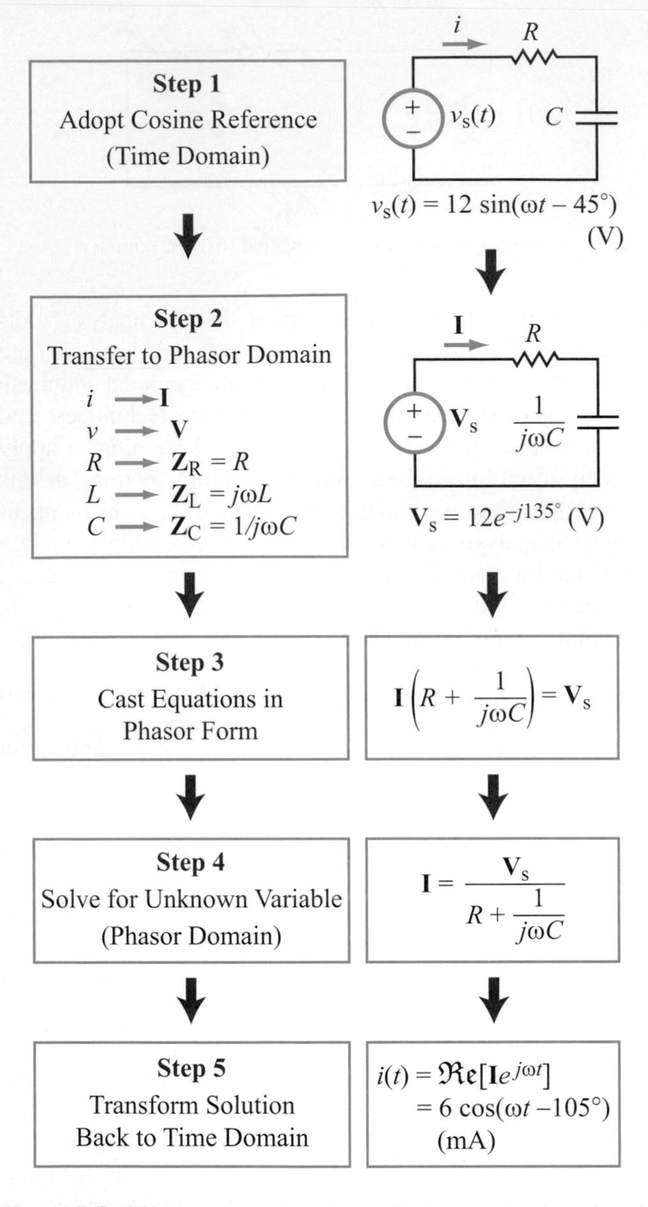

Figure 7-7: Five-step procedure for analyzing ac circuits using the phasor-domain technique.

In accordance with Table 7-3, the phasor equivalent of $v_s(t)$ is

$$\mathbf{V}_s = 12e^{-j135^\circ} \text{ V}. \tag{7.63}$$

Step 2: Transform Circuit to Phasor Domain

The current $i(t)$ in Eq. (7.61) is related to its phasor counterpart $\mathbf{I}$ by

$$i(t) = \mathfrak{Re}[\mathbf{I}e^{j\omega t}]. \tag{7.64}$$

As yet, we do not have an explicit expression for either $i(t)$ or $\mathbf{I}$, but we will obtain those expressions later on in Steps 4 and 5. Step 2 in Fig. 7-7 shows the RC circuit in the phasor domain, with loop current $\mathbf{I}$, impedance $\mathbf{Z}_R = R$ representing the resistance and impedance $\mathbf{Z}_C = 1/j\omega C$ representing the capacitor. The voltage source is represented by its phasor $\mathbf{V}_s$.

Step 3: Cast KCL and/or KVL Equations in Phasor Domain

For the circuit in Step 2 of Fig. 7-7, its loop equation is given by

$$\mathbf{Z}_R\mathbf{I} + \mathbf{Z}_C\mathbf{I} = \mathbf{V}_s, \tag{7.65}$$

which is equivalent to

$$\left(R + \frac{1}{j\omega C}\right)\mathbf{I} = 12e^{-j135^\circ}. \tag{7.66}$$

This equation also could have been obtained by transforming Eq. (7.61) from the time domain to the phasor domain, which entails replacing i with $\mathbf{I}$, $\int i\,dt$ with $\mathbf{I}/j\omega$, and v_s with $\mathbf{V}_s$.

Step 4: Solve for Unknown Variable

Solving Eq. (7.66) for $\mathbf{I}$ gives

$$\mathbf{I} = \frac{12e^{-j135^\circ}}{R + \frac{1}{j\omega C}} = \frac{j12\omega Ce^{-j135^\circ}}{1 + j\omega RC}. \tag{7.67}$$

Using the specified values, namely $R = \sqrt{3}$ kΩ, $C = 1$ μF, and $\omega = 10^3$ rad/s, Eq. (7.67) becomes

$$\mathbf{I} = \frac{j12 \times 10^3 \times 10^{-6}e^{-j135^\circ}}{1 + j10^3 \times \sqrt{3} \times 10^3 \times 10^{-6}} = \frac{j12e^{-j135^\circ}}{1 + j\sqrt{3}} \text{ mA}. \tag{7.68}$$

In preparation for the next step, we should convert the expression for $\mathbf{I}$ into polar form ($Ae^{j\theta}$, where A is a positive real number). To that end, we should replace j in the numerator with $e^{j\pi/2}$ and convert the denominator into polar form:

$$1 + j\sqrt{3} = \sqrt{1+3}\,e^{j\phi} = 2e^{j\phi}, \tag{7.69}$$

where

$$\phi = \tan^{-1}\left(\frac{\sqrt{3}}{1}\right) = 60^\circ. \tag{7.70}$$

Hence,

$$\mathbf{I} = \frac{12e^{-j135^\circ} \cdot e^{j90^\circ}}{2e^{j60^\circ}} = 6e^{j(-135^\circ+90^\circ-60^\circ)} = 6e^{-j105^\circ} \text{ mA}. \tag{7.71}$$

Step 5: Transform Solution Back to Time Domain

To return to the time domain, we apply the fundamental relation between a sinusoidal function and its phasor counterpart, namely

$$\begin{aligned} i(t) &= \mathfrak{Re}[\mathbf{I}e^{j\omega t}] \\ &= \mathfrak{Re}[6e^{-j105^\circ} e^{j\omega t}] \\ &= 6\cos(\omega t - 105^\circ) \text{ mA.} \end{aligned} \quad (7.72)$$

This concludes our demonstration of the five-step procedure of the phasor-domain analysis technique. The procedure is equally applicable for solving any linear ac circuit.

Example 7-5: RL Circuit

The voltage source of the circuit shown in Fig. 7-8(a) is given by

$$v_s(t) = 15\sin(4 \times 10^4 t - 30^\circ) \text{ V}.$$

Also, $R = 3\ \Omega$ and $L = 0.1$ mH. Obtain an expression for the voltage across the inductor.

Solution:

Step 1: Convert $\boldsymbol{v_s(t)}$ to the cosine reference

$$\begin{aligned} v_s(t) &= 15\sin(4 \times 10^4 t - 30^\circ) \\ &= 15\cos(4 \times 10^4 t - 30^\circ - 90^\circ) \\ &= 15\cos(4 \times 10^4 t - 120^\circ) \text{ V}, \end{aligned}$$

and its corresponding phasor $\mathbf{V}_s$ is given by

$$\mathbf{V}_s = 15e^{-j120^\circ} \text{ V}.$$

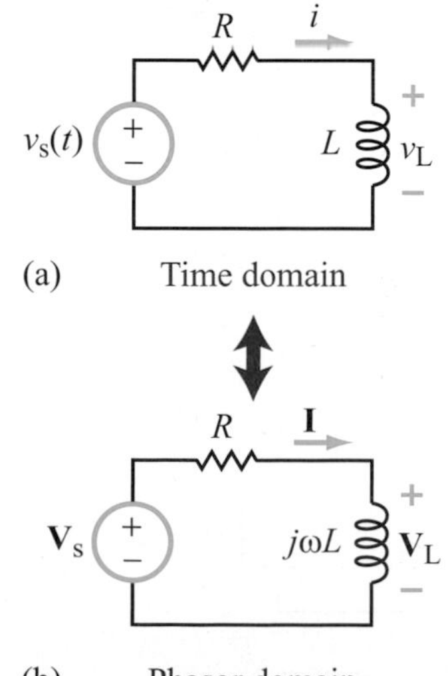

Figure 7-8: RL circuit of Example 7-5.

Step 2: Transform circuit to the phasor domain

Phasor-domain circuit is shown in Fig. 7-8(b).

Step 3: Cast KVL in phasor domain

$$R\mathbf{I} + j\omega L\mathbf{I} = \mathbf{V}_s.$$

Step 4: Solve for unknown variable

$$\begin{aligned} \mathbf{I} = \frac{\mathbf{V}_s}{R + j\omega L} &= \frac{15e^{-j120^\circ}}{3 + j4 \times 10^4 \times 10^{-4}} \\ &= \frac{15e^{-j120^\circ}}{3 + j4} = \frac{15e^{-j120^\circ}}{5e^{j53.1^\circ}} = 3e^{-j173.1^\circ} \text{ A.} \end{aligned}$$

The phasor voltage across the inductor is related to $\mathbf{I}$ by

$$\begin{aligned} \mathbf{V}_L &= j\omega L\mathbf{I} \\ &= j4 \times 10^4 \times 10^{-4} \times 3e^{-j173.1^\circ} \\ &= j12e^{-j173.1^\circ} \\ &= 12e^{-j173.1^\circ} \cdot e^{j90^\circ} = 12e^{-j83.1^\circ} \text{ V}, \end{aligned}$$

where we replaced j with e^{j90°.

Step 5: Transform solution to the time domain

The corresponding time-domain voltage is

$$\begin{aligned} v_L(t) &= \mathfrak{Re}[\mathbf{V}_L e^{j\omega t}] \\ &= \mathfrak{Re}[12e^{-j83.1^\circ} e^{j4 \times 10^4 t}] \\ &= 12\cos(4 \times 10^4 t - 83.1^\circ) \text{ V}. \end{aligned}$$

Exercise 7-8: Repeat the analysis of the circuit in Example 7-5 for $v_s(t) = 20\cos(2 \times 10^3 t + 60^\circ)$ V, $R = 6\ \Omega$, and $L = 4$ mH.

Answer: $v_L(t) = 16\cos(2 \times 10^3 t + 96.9^\circ)$ V. (See)

7-5 Impedance Transformations

Voltage division, current division, and the Y–Δ transformation are among the many analysis tools we developed in Chapter 2 in connection with circuits composed solely of sources and resistors. All of these tools are based on two fundamental laws: KCL and KVL. Having established in the preceding section that KCL and KVL also are valid in the phasor domain, it follows that these simplification and transformation techniques can be used in the phasor domain as well. The fundamental difference between the two cases is that in Chapter 2 we dealt with resistors, and with voltages and currents expressed in the time domain, whereas in the phasor domain the circuit quantities are impedances and phasors. Thus, once an ac circuit has been transformed into the phasor domain, we can apply the same techniques of Chapters 2 and 3, but we do so using complex algebra.

In this and the next section, we will illustrate how impedance and source transformations are executed in the phasor domain. Before we do so, however, we should expand our definition of impedance to encompass more than the impedance of a single element. The three passive elements, R, L, and C, are measured in ohms, henrys, and farads. *Their corresponding impedances* $\mathbf{Z}_R$, $\mathbf{Z}_L$, *and* $\mathbf{Z}_C$ *are all measured in ohms,* and are given by

$$\mathbf{Z}_R = R, \qquad \mathbf{Z}_L = j\omega L, \qquad \mathbf{Z}_C = \frac{-j}{\omega C}. \tag{7.73}$$

Consider the three series combinations shown in Fig. 7-9. Application of KVL to the circuits on the left-hand side and to their counterparts leads to

$$\mathbf{Z}_1 = \mathbf{Z}_{R_1} + \mathbf{Z}_{L_1} = R_1 + j\omega L_1, \tag{7.74a}$$

$$\mathbf{Z}_2 = \mathbf{Z}_{R_2} + \mathbf{Z}_{C_2} = R_2 - \frac{j}{\omega C_2}, \tag{7.74b}$$

and

$$\mathbf{Z}_3 = \mathbf{Z}_{L_3} + \mathbf{Z}_{C_3} = j\left(\omega L_3 - \frac{1}{\omega C_3}\right). \tag{7.74c}$$

From these three simple examples, we observe that an impedance $\mathbf{Z}$ is, in general, a complex quantity composed of a real part and an imaginary part. We usually use the symbol R to represent its real part and we call it its ***resistance***, and we use the symbol X to represent its imaginary part and we call it its ***reactance***. Thus,

$$\mathbf{Z} = R + jX. \tag{7.75}$$

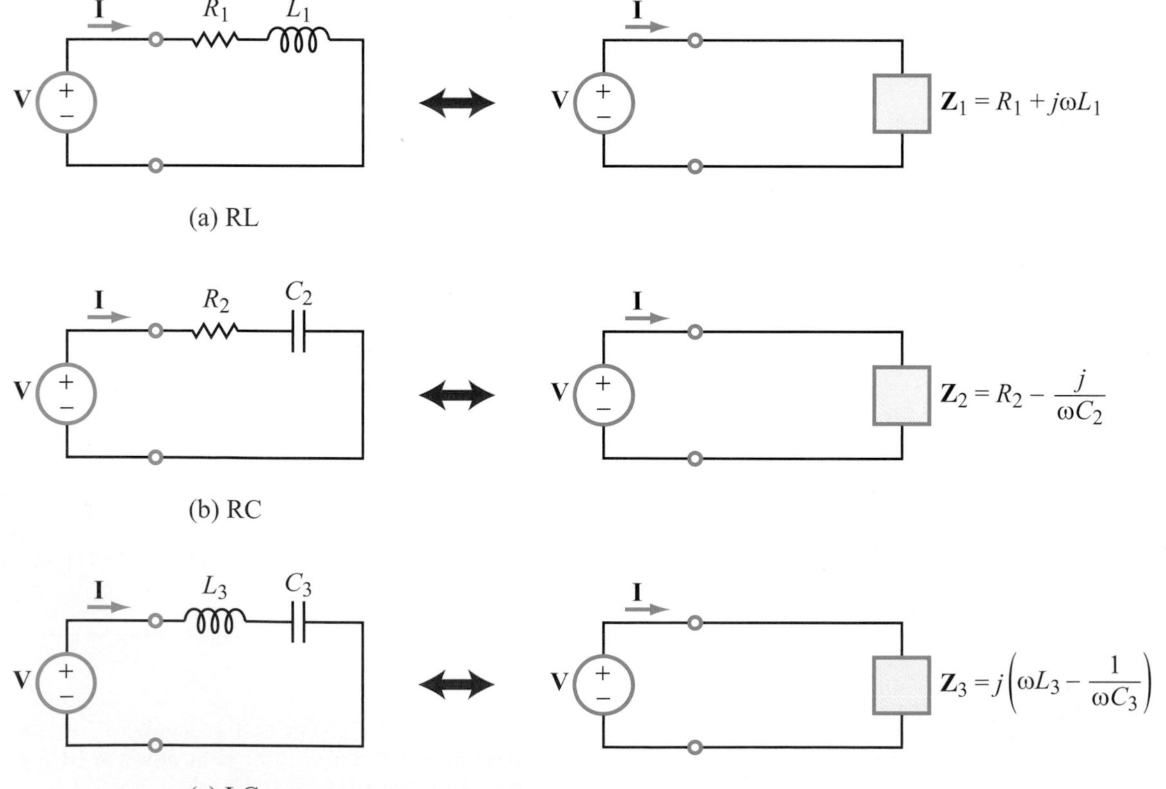

Figure 7-9: Three different, two-element, series combinations.

Impedances $\mathbf{Z}_1$ and $\mathbf{Z}_2$ have reactances with opposite polarities. When X is positive, as in $\mathbf{Z}_1$, we call $\mathbf{Z}$ an ***inductive impedance***, and when X is negative, we call it a ***capacitive impedance***. Impedance $\mathbf{Z}_2$ is capacitive. Impedance $\mathbf{Z}_3$ is purely imaginary, and it may be inductive or capacitive depending on how the magnitude of ωL compares with that of $1/\omega C$.

Occasionally, we may need to express $\mathbf{Z}$ in polar form

$$\mathbf{Z} = |\mathbf{Z}|e^{j\theta}, \tag{7.76}$$

where its magnitude $|\mathbf{Z}|$ and phase angle θ are related to components R and X of the rectangular form by

$$|\mathbf{Z}| = \sqrt[+]{R^2 + X^2}, \quad \text{and} \quad \theta = \tan^{-1}\left(\frac{X}{R}\right). \tag{7.77}$$

The inverse relationships are given by

$$R = \mathfrak{Re}[\mathbf{Z}] = \mathfrak{Re}[|\mathbf{Z}|e^{j\theta}] = |\mathbf{Z}|\cos\theta \tag{7.78a}$$

and

$$X = \mathfrak{Im}[\mathbf{Z}] = \mathfrak{Im}[|\mathbf{Z}|e^{j\theta}] = |\mathbf{Z}|\sin\theta. \tag{7.78b}$$

In Chapter 2, we defined the conductance G as the reciprocal of R, namely $G = 1/R$. The phasor analogue of G is the ***admittance*** $\mathbf{Y}$, defined as

$$\mathbf{Y} = \frac{1}{\mathbf{Z}} = G + jB, \tag{7.79}$$

where $G = \mathfrak{Re}[\mathbf{Y}]$ is called the ***conductance*** of $\mathbf{Y}$ and $B = \mathfrak{Im}[\mathbf{Y}]$ is called its ***susceptance***. The unit for $\mathbf{Y}$, G, and B is the siemen (S).

7-5.1 Impedances in Series and in Parallel

The three in-series examples of Fig. 7-9 consisted each of only two impedances. By extension, we can assert that:

> N impedances connected in series (sharing the same phasor current) can be combined into a single equivalent impedance $\mathbf{Z}_{eq}$ whose value is equal to the algebraic sum of the individual impedances.

$$\mathbf{Z}_{eq} = \sum_{i=1}^{N} \mathbf{Z}_i \qquad \text{(impedances in series)}. \tag{7.80}$$

The phasor voltage across any individual impedance $\mathbf{Z}_i$ is a proportionate fraction $(\mathbf{Z}_i/\mathbf{Z}_{eq})$ of the phasor voltage across the entire group.

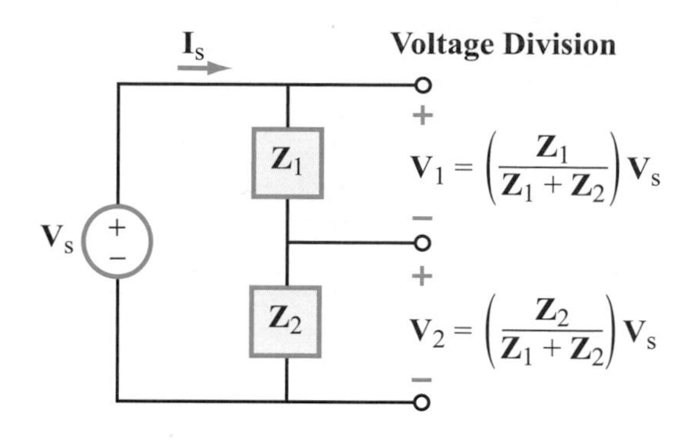

Figure 7-10: Voltage division among two impedances in series.

This is a statement of ***voltage division***, which for the two-impedance circuit of Fig. 7-10, assumes the form

$$\mathbf{V}_1 = \left(\frac{\mathbf{Z}_1}{\mathbf{Z}_1 + \mathbf{Z}_2}\right)\mathbf{V}_s, \qquad \mathbf{V}_2 = \left(\frac{\mathbf{Z}_2}{\mathbf{Z}_1 + \mathbf{Z}_2}\right)\mathbf{V}_s. \tag{7.81}$$

Similarly:

> N admittances connected in parallel between a pair of nodes, all sharing the same voltage, can be combined into a single, equivalent admittance $\mathbf{Y}_{eq}$, whose value is equal to the algebraic sum of the individual admittances.

$$\mathbf{Y}_{eq} = \sum_{i=1}^{N} \mathbf{Y}_i \qquad \text{(admittances in parallel)}. \tag{7.82}$$

The phasor current flowing through any individual admittance $\mathbf{Y}_i$ is a proportionate fraction $(\mathbf{Y}_i/\mathbf{Y}_{eq})$ of the phasor current flowing through the entire group.

The ***current division*** analogue of Eq. (7.81), defining how current splits up among two admittances connected in parallel (Fig. 7-11), is

$$\mathbf{I}_1 = \left(\frac{\mathbf{Y}_1}{\mathbf{Y}_1 + \mathbf{Y}_2}\right)\mathbf{I}_s, \qquad \mathbf{I}_2 = \left(\frac{\mathbf{Y}_2}{\mathbf{Y}_1 + \mathbf{Y}_2}\right)\mathbf{I}_s. \tag{7.83}$$

Since $\mathbf{Z}_1 = 1/\mathbf{Y}_1$ and $\mathbf{Z}_2 = 1/\mathbf{Y}_2$, Eq. (7.83) can be rewritten in terms of impedances as

$$\mathbf{I}_1 = \left(\frac{\mathbf{Z}_2}{\mathbf{Z}_1 + \mathbf{Z}_2}\right)\mathbf{I}_s, \qquad \mathbf{I}_2 = \left(\frac{\mathbf{Z}_1}{\mathbf{Z}_1 + \mathbf{Z}_2}\right)\mathbf{I}_s. \tag{7.84}$$

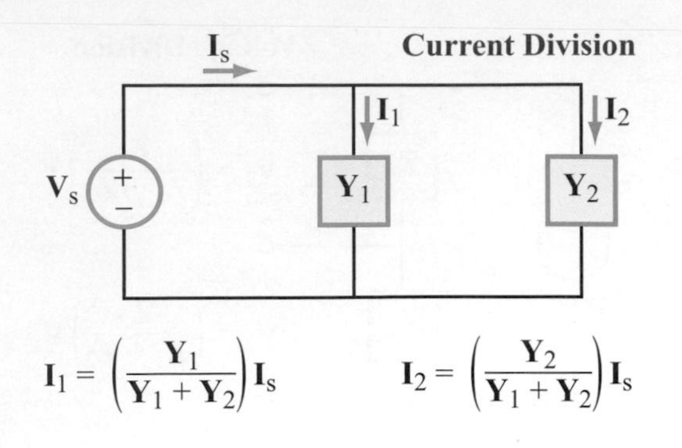

Figure 7-11: Current division among two admittances in parallel.

Example 7-6: Input Impedance

The circuit in Fig. 7-12(a) is connected to a source given by

$$v_s(t) = 16\cos 10^6 t \text{ V}.$$

Determine (a) the input impedance of the circuit, given that $R_1 = 2$ kΩ, $R_2 = 4$ kΩ, $L = 3$ mH, and $C = 1$ nF, and (b) the voltage $v_2(t)$ across R_2.

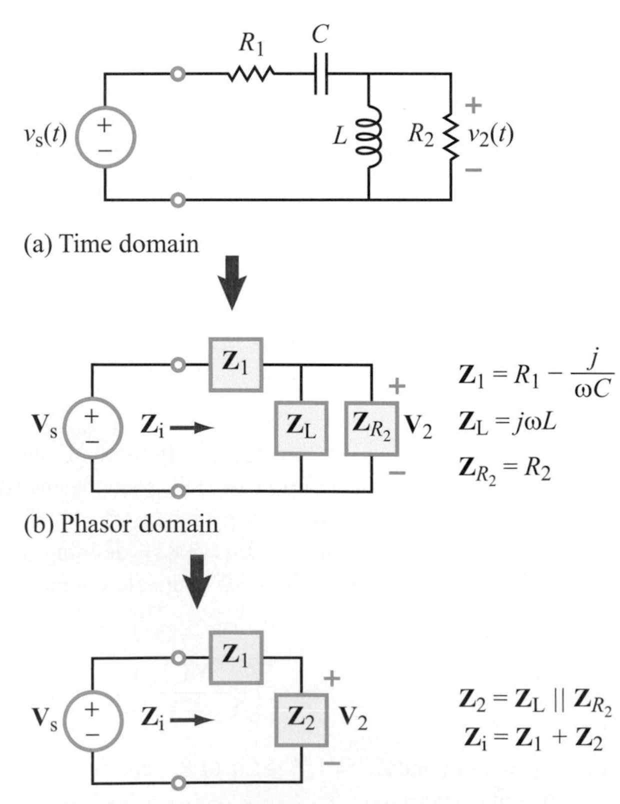

Figure 7-12: Circuit for Example 7-6.

Solution:

(a) The phasor-domain equivalent circuit is shown in Fig. 7-12(b), where

$$\mathbf{V}_s = 16,$$

$$\mathbf{Z}_1 = R_1 - \frac{j}{\omega C} = 2 \times 10^3 - \frac{j}{10^6 \times 10^{-9}} = (2 - j1) \text{ k}\Omega,$$

$$\mathbf{Z}_L = j\omega L = j \times 10^6 \times 3 \times 10^{-3} = j3 \text{ k}\Omega,$$

and

$$\mathbf{Z}_{R_2} = R_2 = 4 \text{ k}\Omega.$$

The parallel combination of $\mathbf{Z}_L$ and $\mathbf{Z}_{R_2}$ is denoted $\mathbf{Z}_2$ in Fig. 7-12(c), and it is given by

$$\mathbf{Z}_2 = \frac{\mathbf{Z}_L \mathbf{Z}_{R_2}}{\mathbf{Z}_L + \mathbf{Z}_{R_2}} = \frac{j3 \times 10^3 \times 4 \times 10^3}{(4 + j3) \times 10^3} = \frac{j12 \times 10^3}{4 + j3}.$$

A useful "trick" for converting the expression for $\mathbf{Z}_2$ into the form $(a + jb)$ is to multiply the numerator and denominator by the complex conjugate of the denominator:

$$\mathbf{Z}_2 = \frac{j12 \times 10^3}{4 + j3} \times \frac{4 - j3}{4 - j3} = \frac{36 + j48}{16 + 9} \times 10^3 = (1.44 + j1.92) \text{ k}\Omega.$$

The input impedance $\mathbf{Z}_i$ is equal to the sum of $\mathbf{Z}_1$ and $\mathbf{Z}_2$,

$$\mathbf{Z}_i = \mathbf{Z}_1 + \mathbf{Z}_2 = (2 - j1 + 1.44 + j1.92) \times 10^3 = (3.44 + j0.92) \text{ k}\Omega.$$

(b) By voltage division,

$$\mathbf{V}_2 = \frac{\mathbf{Z}_2 \mathbf{V}_s}{\mathbf{Z}_1 + \mathbf{Z}_2} = \frac{(1.44 + j1.92) \times 10^3 \times 16}{(3.44 + j0.92) \times 10^3} = 10.8 e^{j38.2^\circ} \text{ V}.$$

Transforming $\mathbf{V}_2$ to its time-domain counterpart leads to

$$v_2(t) = \mathfrak{Re}[\mathbf{V}_2 e^{j\omega t}] = \mathfrak{Re}[10.8 e^{j38.2^\circ} e^{j10^6 t}] = 10.8\cos(10^6 t + 38.2^\circ) \text{ V}.$$

Example 7-7: Current Division

The circuit in Fig. 7-13(a) is connected to a source

$$v_s(t) = 4\sin(10^7 t + 15°) \text{ V}.$$

Determine (a) the input admittance $\mathbf{Y}_i$, given that $R_1 = 10\ \Omega$, $R_2 = 30\ \Omega$, $L = 2\ \mu$H, and $C = 10$ nF, and (b) the current $i_2(t)$ flowing through R_2.

Solution:

(a) We start by converting $v_s(t)$ to cosine format:

$$v_s(t) = 4\sin(10^7 t + 15°) = 4\cos(10^7 t + 15° - 90°)$$
$$= 4\cos(10^7 t - 75°) \text{ V}.$$

The corresponding phasor voltage is

$$\mathbf{V}_s = 4e^{-j75°} \text{ V},$$

and the impedances shown in Fig. 7-13(b) are given by

$$\mathbf{Z}_{R_1} = R_1 = 10\ \Omega,$$

$$\mathbf{Z}_C = \frac{-j}{\omega C} = \frac{-j}{10^7 \times 10^{-8}} = -j10\ \Omega,$$

and

$$\mathbf{Z}_a = R_2 + j\omega L$$
$$= 30 + j10^7 \times 2 \times 10^{-6} = (30 + j20)\ \Omega.$$

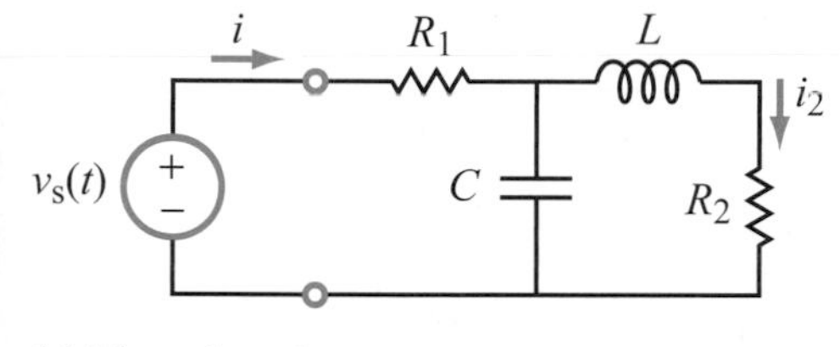

(a) Time domain

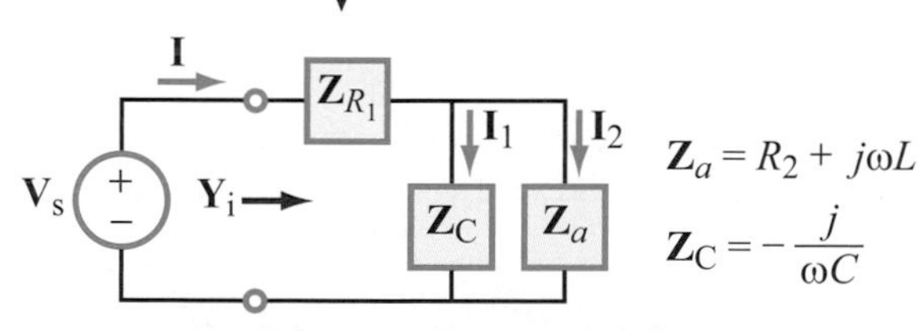

(b) Phasor domain

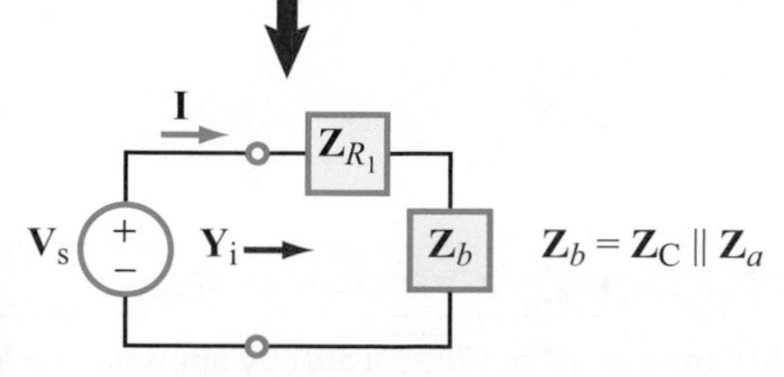

(c) Combining impedances

Figure 7-13: Circuit for Example 7-7.

In Fig. 7-13(c), $\mathbf{Z}_b$ represents the parallel combination of $\mathbf{Z}_C$ and $\mathbf{Z}_a$,

$$\mathbf{Z}_b = \mathbf{Z}_C \parallel \mathbf{Z}_a$$
$$= \frac{(-j10)(30 + j20)}{-j10 + 30 + j20} = \frac{20 - j30}{3 + j1}$$
$$= \frac{(20 - j30)}{(3 + j1)} \frac{(3 - j1)}{(3 - j1)} = (3 - j11)\ \Omega.$$

The input impedance is

$$\mathbf{Z}_i = \mathbf{Z}_{R_1} + \mathbf{Z}_b = 10 + 3 - j11 = (13 - j11)\ \Omega,$$

and its reciprocal is

$$\mathbf{Y}_i = \frac{1}{\mathbf{Z}_i}$$
$$= \frac{1}{13 - j11} \times \frac{13 + j11}{13 + j11}$$
$$= \frac{13 + j11}{169 + 121} = (4.5 + j3.8) \times 10^{-2} \text{ S}.$$

(b) The current $\mathbf{I}$ is given by

$$\mathbf{I} = \frac{\mathbf{V}_s}{\mathbf{Z}_i} = \frac{4e^{-j75°}}{13 - j11} = \frac{4e^{-j75°}}{17.03e^{-j40.2°}} = 0.235e^{-j34.8°} \text{ A}.$$

By current division in Fig. 7-13(b),

$$\mathbf{I}_2 = \frac{\mathbf{Z}_C}{\mathbf{Z}_a + \mathbf{Z}_C}\, \mathbf{I}$$
$$= \frac{-j10}{30 + j20 - j10} \times 0.235e^{-j34.8°}$$
$$= \frac{2.35e^{-j34.8°} \cdot e^{-j90°}}{31.6e^{j18.4°}} = 7.4 \times 10^{-2} e^{-j143.2°} \text{ A}.$$

The corresponding current in the time domain is

$$i_2(t) = \mathfrak{Re}[\mathbf{I}_2 e^{j\omega t}]$$
$$= \mathfrak{Re}[7.4 \times 10^{-2} e^{-j143.2°} e^{j10^7 t}]$$
$$= 7.4 \times 10^{-2} \cos(10^7 t - 143.2°) \text{ A}.$$

Review Question 7-9: The rule for adding two in-series capacitors is different from that for adding two in-series resistors, but the rule for adding the impedances of those two in-series capacitors is the same as the rule for adding two in-series resistors. Does this pose a contradiction? Explain.

Review Question 7-10: Is it possible to construct a circuit composed solely of capacitors and inductors such that the impedance of the overall combination has a non-zero real part? Explain.

Exercise 7-9: Determine the input impedance at $\omega = 10^5$ rad/s for each of the circuits in Fig. E7.9.

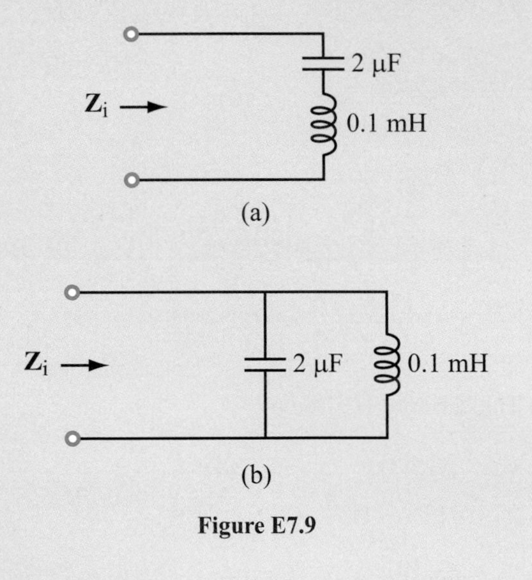

Figure E7.9

Answer: (a) $\mathbf{Z}_i = j5\ \Omega$, (b) $\mathbf{Z}_i = -j10\ \Omega$. (See)

7-5.2 Y–Δ Transformation

The Y–Δ transformation outlined in Section 2-5 allows us to replace a Y circuit connected to three nodes with a Δ circuit, or vice versa, without altering the voltages at the three nodes or the currents entering them. The same principle applies to impedances, as do the relationships between impedances $\mathbf{Z}_1$ to $\mathbf{Z}_3$ of the Y circuit (Fig. 7-14) and impedances $\mathbf{Z}_a$ to $\mathbf{Z}_c$ of the Δ circuit.

Δ →Y Transformation:

$$\mathbf{Z}_1 = \frac{\mathbf{Z}_b \mathbf{Z}_c}{\mathbf{Z}_a + \mathbf{Z}_b + \mathbf{Z}_c}, \quad (7.85a)$$

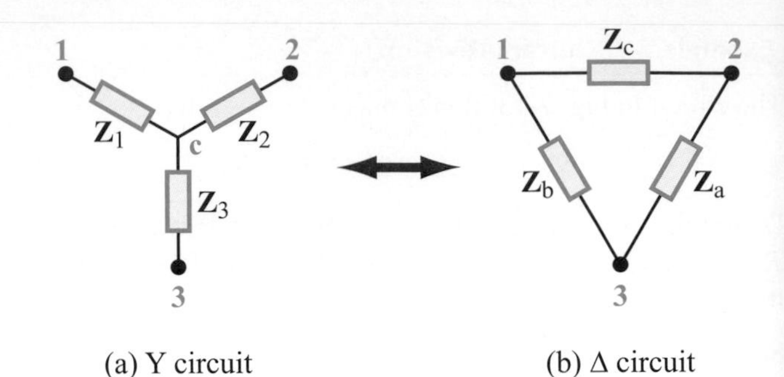

Figure 7-14: Y–Δ equivalent circuits.

$$\mathbf{Z}_2 = \frac{\mathbf{Z}_a \mathbf{Z}_c}{\mathbf{Z}_a + \mathbf{Z}_b + \mathbf{Z}_c}, \quad (7.85b)$$

$$\mathbf{Z}_3 = \frac{\mathbf{Z}_a \mathbf{Z}_b}{\mathbf{Z}_a + \mathbf{Z}_b + \mathbf{Z}_c}. \quad (7.85c)$$

Y→ **Δ** Transformation:

$$\mathbf{Z}_a = \frac{\mathbf{Z}_1\mathbf{Z}_2 + \mathbf{Z}_2\mathbf{Z}_3 + \mathbf{Z}_1\mathbf{Z}_3}{\mathbf{Z}_1}, \quad (7.86a)$$

$$\mathbf{Z}_b = \frac{\mathbf{Z}_1\mathbf{Z}_2 + \mathbf{Z}_2\mathbf{Z}_3 + \mathbf{Z}_1\mathbf{Z}_3}{\mathbf{Z}_2}, \quad (7.86b)$$

$$\mathbf{Z}_c = \frac{\mathbf{Z}_1\mathbf{Z}_2 + \mathbf{Z}_2\mathbf{Z}_3 + \mathbf{Z}_1\mathbf{Z}_3}{\mathbf{Z}_3}. \quad (7.86c)$$

Balanced Circuits:

If the Y circuit is balanced (all of its impedances are equal), so will be the Δ circuit, and vice versa. Accordingly:

$$\mathbf{Z}_1 = \mathbf{Z}_2 = \mathbf{Z}_3 = \frac{\mathbf{Z}_a}{3}, \quad \text{if } \mathbf{Z}_a = \mathbf{Z}_b = \mathbf{Z}_c, \quad (7.87a)$$

$$\mathbf{Z}_a = \mathbf{Z}_b = \mathbf{Z}_c = 3\mathbf{Z}_1, \quad \text{if } \mathbf{Z}_1 = \mathbf{Z}_2 = \mathbf{Z}_3. \quad (7.87b)$$

Example 7-8: Applying Y–Δ Transformation

(a) Simplify the circuit in Fig. 7-15(a) by applying the Y–Δ transformation so as to determine the current **I**. (b) Determine the corresponding $i(t)$, given that the oscillation frequency of the voltage source is 1 MHz.

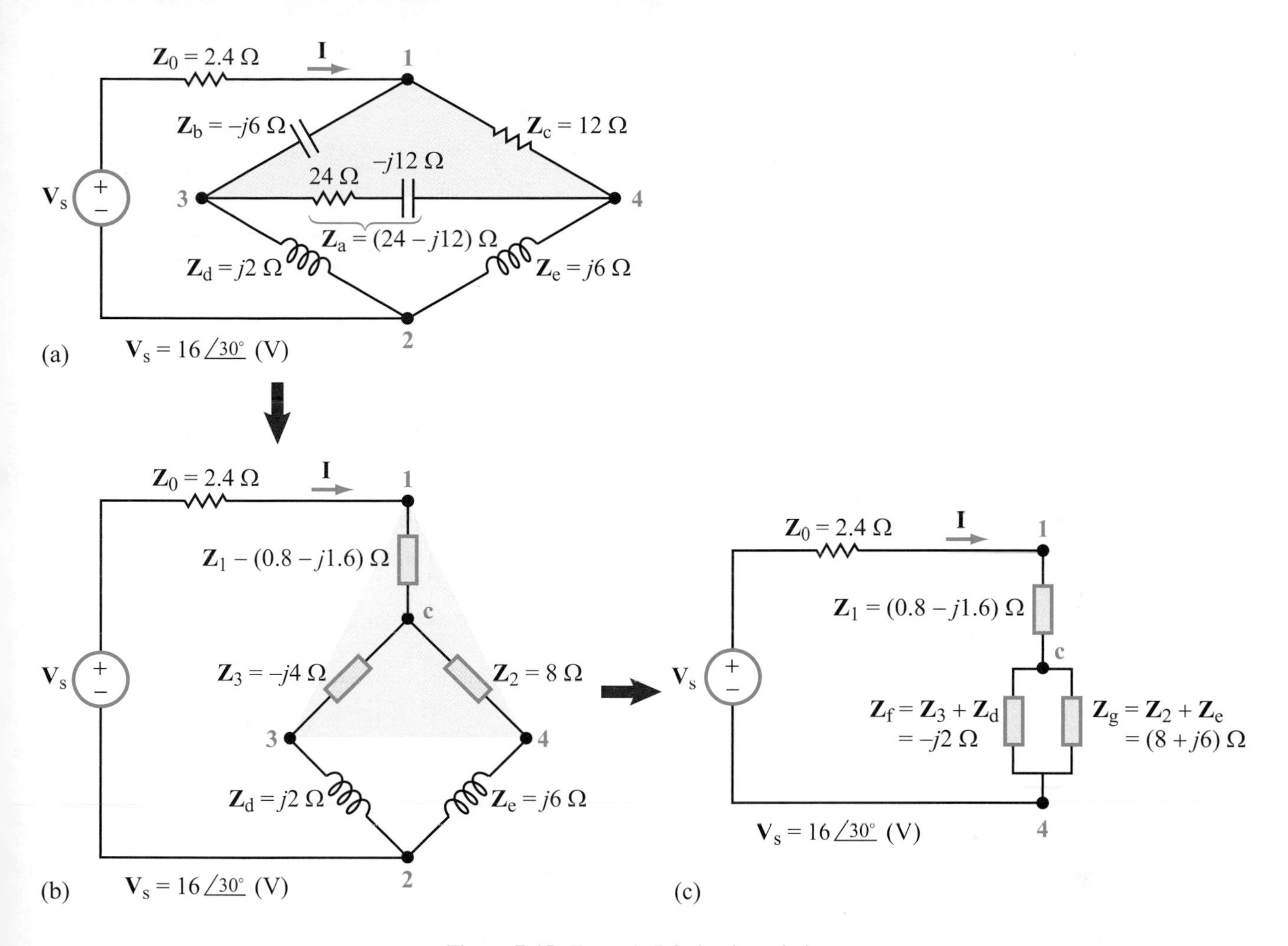

Figure 7-15: Example 7-8 circuit evolution.

Solution:

(a) The Δ circuit connected to nodes 1, 2, and 3 can be replaced with a Y circuit, as shown in Fig. 7-15(b), with impedances

$$\begin{aligned}\mathbf{Z}_1 &= \frac{\mathbf{Z}_b\mathbf{Z}_c}{\mathbf{Z}_a+\mathbf{Z}_b+\mathbf{Z}_c}\\ &= \frac{-j6\times 12}{24-j12-j6+12} = \frac{-j72}{36-j18} = (0.8-j1.6)\ \Omega,\end{aligned}$$

$$\begin{aligned}\mathbf{Z}_2 &= \frac{\mathbf{Z}_a\mathbf{Z}_c}{\mathbf{Z}_a+\mathbf{Z}_b+\mathbf{Z}_c}\\ &= \frac{(24-j12)\times 12}{36-j18} = 8\ \Omega,\end{aligned}$$

and

$$\begin{aligned}\mathbf{Z}_3 &= \frac{\mathbf{Z}_b\mathbf{Z}_a}{\mathbf{Z}_a+\mathbf{Z}_b+\mathbf{Z}_c}\\ &= \frac{-j6(24-j12)}{36-j18} = -j4\ \Omega.\end{aligned}$$

In Fig. 7-15(c), $\mathbf{Z}_f$ represents the series combination of $\mathbf{Z}_3$ and $\mathbf{Z}_d$,

$$\mathbf{Z}_f = \mathbf{Z}_3+\mathbf{Z}_d = -j4+j2 = -j2\ \Omega.$$

Similarly,

$$\mathbf{Z}_g = \mathbf{Z}_2+\mathbf{Z}_e = (8+j6)\ \Omega.$$

Impedances $\mathbf{Z}_f$ and $\mathbf{Z}_g$ are connected in parallel, and their combination is in series with $\mathbf{Z}_0$ and $\mathbf{Z}_1$. Hence,

$$\begin{aligned}\mathbf{I} &= \frac{\mathbf{V}_s}{\mathbf{Z}_0+\mathbf{Z}_1+(\mathbf{Z}_f \parallel \mathbf{Z}_g)}\\ &= \frac{16e^{j30^\circ}}{2.4+(0.8-j1.6)+\dfrac{-j2\times(8+j6)}{-j2+8+j6}}.\end{aligned}$$

After a few steps of complex algebra, we obtain the result

$$\mathbf{I} = 3.06\angle 76.55^\circ \text{ A}.$$

(b)

$$\begin{aligned} i(t) &= \Re\mathfrak{e}[\mathbf{I}e^{j\omega t}] \\ &= \Re\mathfrak{e}[3.06e^{j76.55^\circ} e^{j2\pi \times 10^6 t}] \\ &= 3.06\cos(2\pi \times 10^6 t + 76.55^\circ) \text{ A}. \end{aligned}$$

Exercise 7-10: Convert the Y-impedance circuit in Fig. E7.10 into a Δ-impedance circuit.

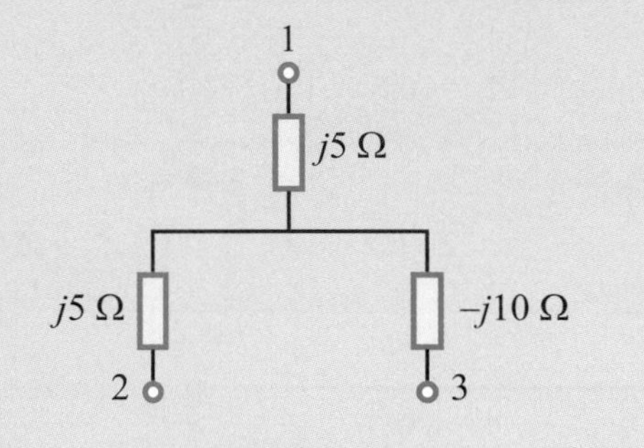

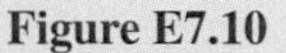
Figure E7.10

Answer:

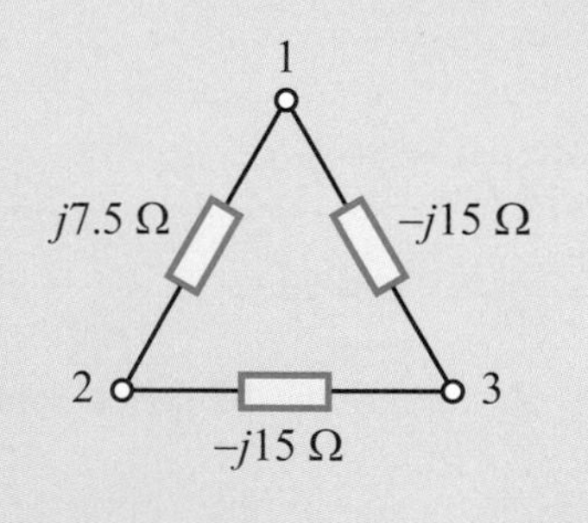

(See)

7-6 Equivalent Circuits

Having examined in the preceding section how phasor-domain circuits can be simplified by applying impedance transformations, we now extend our review of the rules of circuit equivalency to circuits containing voltage and current sources.

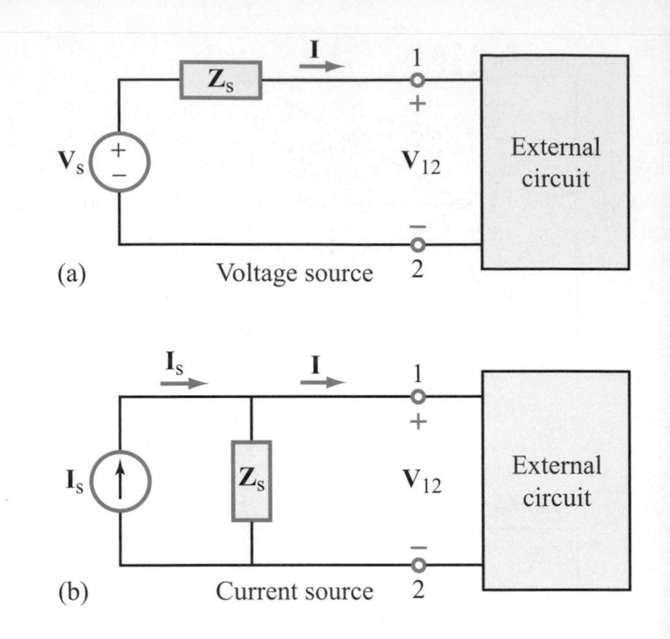

Figure 7-16: Source-transformation equivalency.

7-6.1 Source Transformation

Section 2-4.4 provides an outline of the ***source-transformation principle*** as it applies to resistive circuits. Its phasor-domain analogue is diagrammed in Fig. 7-16; from the vantage point of the external circuit:

> A voltage source $\mathbf{V}_s$ in series with a source impedance $\mathbf{Z}_s$ is equivalent to the combination of a current source $\mathbf{I}_s = \mathbf{V}_s/\mathbf{Z}_s$, in parallel with a shunt impedance $\mathbf{Z}_s$.

Equivalence implies that both input circuits would deliver the same current $\mathbf{I}$ and voltage $\mathbf{V}_{12}$ to the external circuit.

7-6.2 Thévenin Equivalent Circuit

When restated for the phasor domain, ***Thévenin's theorem*** of Section 3-5.1 becomes:

> A linear circuit can be represented at its output terminals by an equivalent circuit consisting of a series combination of a voltage source $\mathbf{V}_{Th}$ and an impedance $\mathbf{Z}_{Th}$, where $\mathbf{V}_{Th}$ is the open-circuit voltage at those terminals (no load) and $\mathbf{Z}_{Th}$ is the equivalent impedance between the same terminals when all independent sources in the circuit have been deactivated.

Technology Brief 15: Night-Vision Imaging

Either in movies, on television, or when playing video games, most people have seen images taken through night-vision goggles or imaging systems. These are usually monochromatic (single color), green-tinted images of indoor or nighttime environments, such as those shown in Figs. TF15-1 and TF15-2. Historically, two approaches have been pursued to "see in the dark;" one that relies on measuring self-emitted ***thermal energy*** by the scene and another that focuses on ***intensifying*** the light reflected by the scene when illuminated by very weak sources, such as the moon or the stars. We will explore each of the two approaches briefly.

Figure TF15-1: A night-vision image taken with military-grade goggles.

Figure TF15-2: A full-color thermal-infrared image of a soldier.

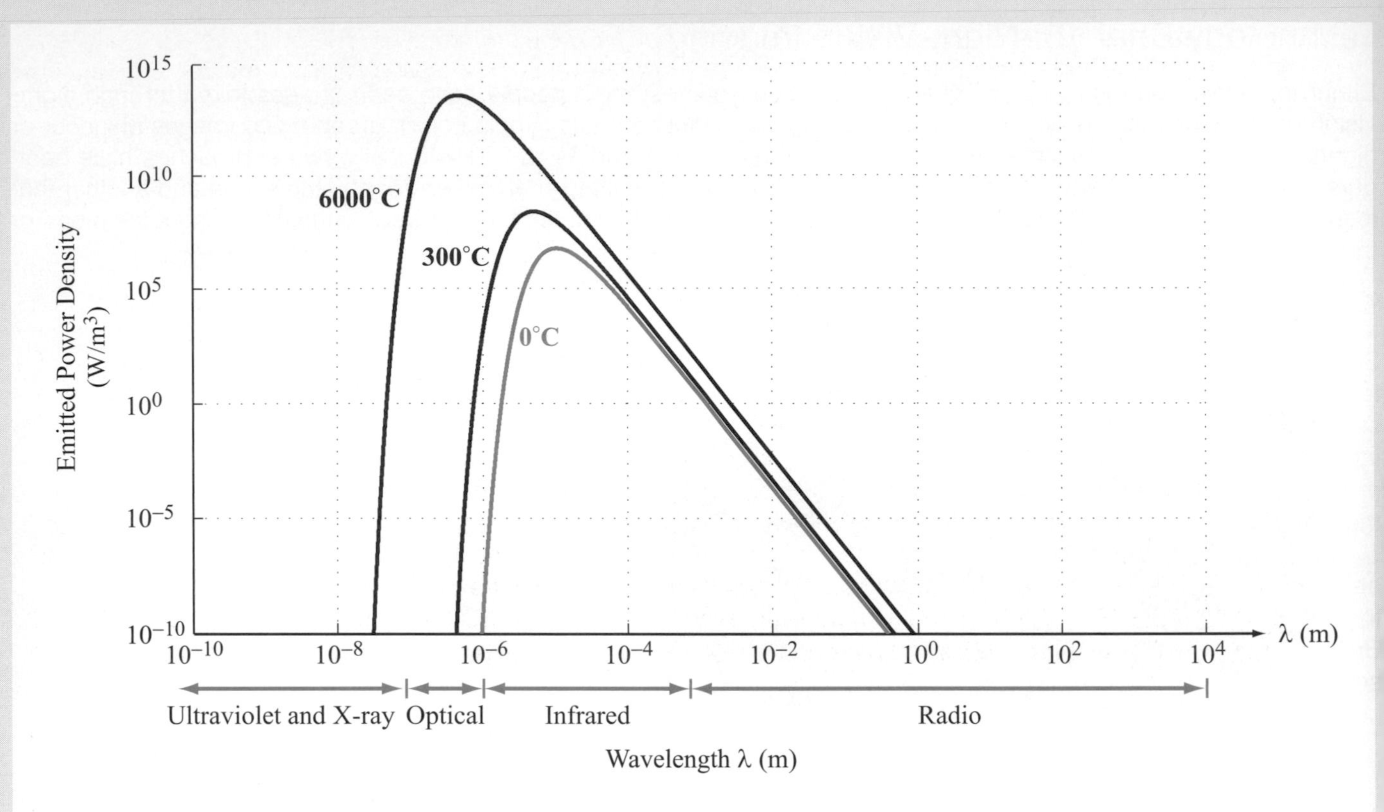

Figure TF15-3: Spectra of power density emitted by ideal blackbodies at 0°C, 300°C, and 6000°C.

Thermal-Infrared Imaging

The visible spectrum extends from the violet (wavelength $\lambda \simeq 0.38$ μm) to the red ($\simeq 0.78$ μm). As noted in Technolog Brief 8 on page 158, the spectral region next to the visible is the ***infrared (IR)***, and it is subdivided into the ***near-IR*** ($\simeq 0.7$ to 1.3 μm), ***mid-IR*** (1.3 to 3 μm), and ***thermal-IR*** (3 to 30 μm). Infrared waves cannot be perceived by humans because our eyes are not sensitive to EM waves outside of the visible spectrum. In the visible spectrum, we see o image a scene by detecting the light reflected by it, but in the thermal-IR region, we image a scene without an externa source of energy, because the scene itself is the source. All material media emit electromagnetic energy all of the time—with hotter objects emitting more than cooler objects. The amount of energy emitted by an object and the shape of its emission spectrum depend on the object's temperature and its material properties. Most of the emitted energ occurs over a relatively narrow spectral range, as illustrated in Fig. TF15-3, which is centered around a peak valu that is highly temperature dependent. For a high-temperature object like the sun (≈ 6000°C), the peak value occurs a about 0.5 μm (red-orange color), whereas for a terrestrial object, the peak value occurs in the thermal-IR region.

Through a combination of lenses and a 2-D array of infrared detectors, the energy emitted by a scene can b focused onto the array, thereby generating an image of the scene. The images sometimes are displayed with a rainbo coloring—with hotter objects displayed in red and cooler objects in blue.

In the near- and mid-IR regions, the imaging process is based on reflection—just as in the visible. Interestingly, th sensor chips used in commercial digital cameras are sensitive not only to visible light but to near-IR energy as wel To avoid image blur caused by the IR energy, the camera lens usually is coated with an IR-blocking film that filter out the IR energy but passes visible light with near-perfect transmission. TV remote controls use near-IR signals t communicate with TV sets, so if an inexpensive digital camera with no IR-blocking coating is used to image an activated TV remote control in the dark, the image will show a bright spot at the tip of the remote control. Some cameras are now making use of this effect to offer IR-based night-vision recording. These cameras emit IR energy from LEDs mounted near the lens, so upon reflection by a nighttime scene, the digital camera is able to record an image "in the dark."

Image Intensifier

A second approach to nighttime imaging is to build sensors with much greater detection sensitivity than the human eye. Such sensors are called ***image intensifiers***. Greater sensitivity means that fewer photons are required in order to detect and register an input signal against the random "noise" in the receiver (or the brain in the case of vision). Some animals can see in the dark (but not in total darkness) because their eye receptors and neural networks require fewer numbers of photons than humans to generate an image under darker conditions. ***Image intensifiers*** work by a simple principle (Fig. TF15-4). Incident photons (of which there are relatively few in a dark scene) are focused through lenses and onto a thin plate of ***gallium arsenide*** material. This material emits one electron every time a photon hits it. Importantly, these electrons are emitted at the locations where the photons hit the plate, preserving the shape of the light image. These photoelectrons then are accelerated by a high voltage (~ 5000 V) onto a ***microchannel plate (MCP)***. The MCP is a plate that emits 10,000 new electrons every time one electron impacts its surface. In essence, it is an amplifier with a current gain of 10,000. These secondary electrons again are accelerated—this time onto phosphors that glow when impacted with electrons. This works on the same principle as the cathode ray tube (see Technology Brief 6 on page 106). The phosphors are arranged in arrays and form pixels on a display, allowing the image to be seen by the naked eye.

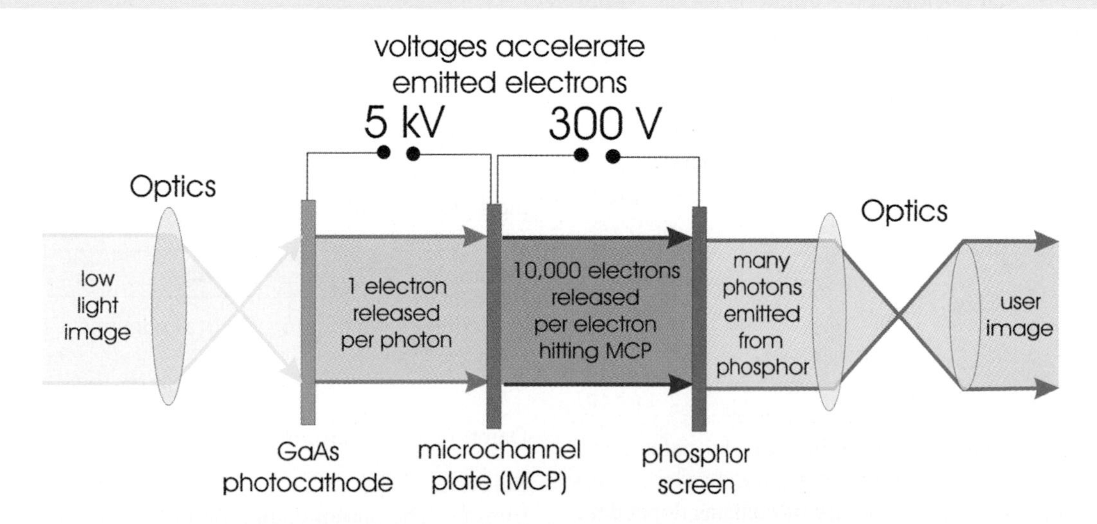

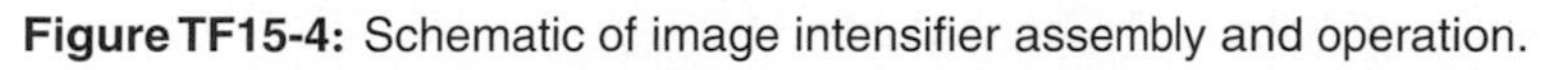
Figure TF15-4: Schematic of image intensifier assembly and operation.

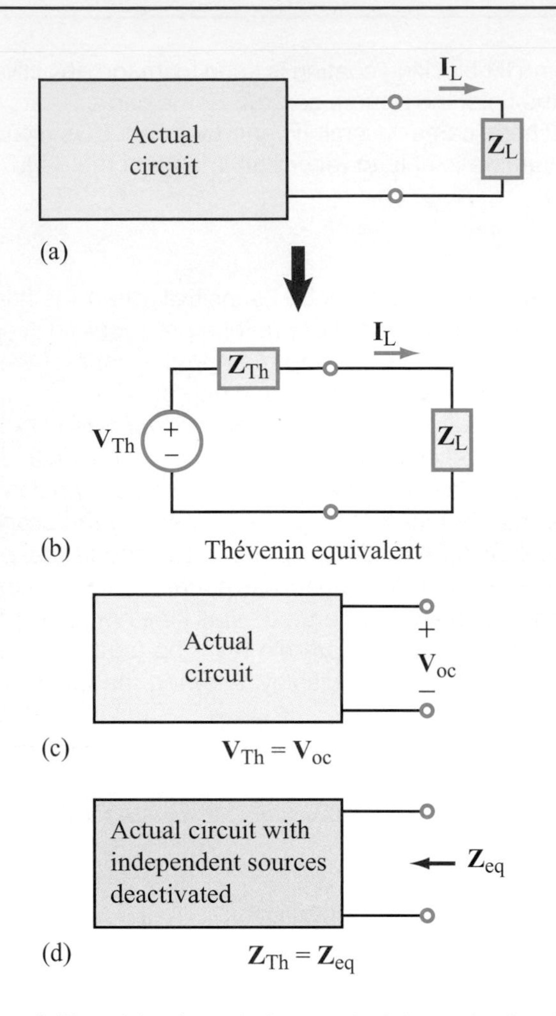

Figure 7-17: Thévenin-equivalent method for a circuit with no dependent sources.

Equivalence implies that if a load $\mathbf{Z}_\mathrm{L}$ is connected at the output terminals of any actual circuit (as portrayed in Fig. 7-17(a)) thereby inducing a current $\mathbf{I}_\mathrm{L}$ to flow through it, the Thévenin equivalent circuit (Fig. 7-17(b)) would deliver the same current $\mathbf{I}_\mathrm{L}$ when connected to the same load impedance $\mathbf{Z}_\mathrm{L}$. For the equivalence to hold, the voltage $\mathbf{V}_\mathrm{Th}$ and impedance $\mathbf{Z}_\mathrm{Th}$ of the Thévenin circuit have to be related to the actual circuit by (Figs. 7-17(c) and (d)):

$$\mathbf{V}_\mathrm{Th} = \mathbf{V}_\mathrm{oc} \tag{7.88a}$$

and

$$\mathbf{Z}_\mathrm{Th} = \mathbf{Z}_\mathrm{eq}. \tag{7.88b}$$

Application of Eq. (7.88a) to determine $\mathbf{V}_\mathrm{Th}$ by calculating or measuring the open-circuit voltage $\mathbf{V}_\mathrm{oc}$ is always a valid approach, whether or not the actual circuit contains dependent sources. That is not so for Eq. (7.88b). The equivalent-impedance method cannot be used to determine $\mathbf{Z}_\mathrm{Th}$ if the

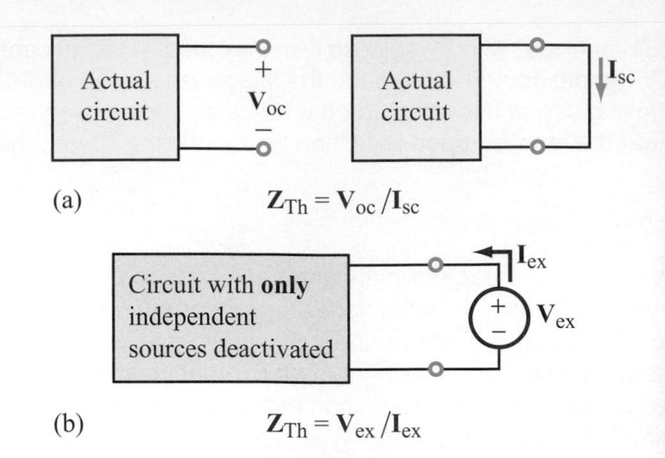

Figure 7-18: The (a) open-circuit/short-circuit method and (b) the external-source method are both suitable for determining $\mathbf{Z}_\mathrm{Th}$, whether or not the circuit contains dependent sources.

circuit contains dependent sources. Alternative approaches include the following.

Open-Circuit/Short-Circuit Method

$$\mathbf{Z}_\mathrm{Th} = \frac{\mathbf{V}_\mathrm{oc}}{\mathbf{I}_\mathrm{sc}}, \tag{7.89}$$

where $\mathbf{I}_\mathrm{sc}$ is the short-circuit current at the circuit's output terminals (Fig. 7-18(a)).

External-Source Method

$$\mathbf{Z}_\mathrm{Th} = \frac{\mathbf{V}_\mathrm{ex}}{\mathbf{I}_\mathrm{ex}}, \tag{7.90}$$

where $\mathbf{I}_\mathrm{ex}$ is the current generated by an external source $\mathbf{V}_\mathrm{ex}$ connected at the circuit's terminals (as shown in Fig. 7-18(b)) after deactivating all independent sources in the circuit.

For the sake of completeness, we should remind the reader that a Thévenin equivalent circuit always can be transformed into a Norton equivalent circuit—or vice versa—by applying the source-transformation method of Section 7-6.1.

Example 7-9: Thévenin Circuit

The circuit shown in Fig. 7-19(a) contains a sinusoidal source given by

$$v_\mathrm{s}(t) = 10\cos 10^5 t \text{ V}.$$

Determine the Thévenin equivalent circuit.

Solution:
Step 1: The phasor counterpart of $v_\mathrm{s}(t)$ is

$$\mathbf{V}_\mathrm{s} = 10 \text{ V}.$$

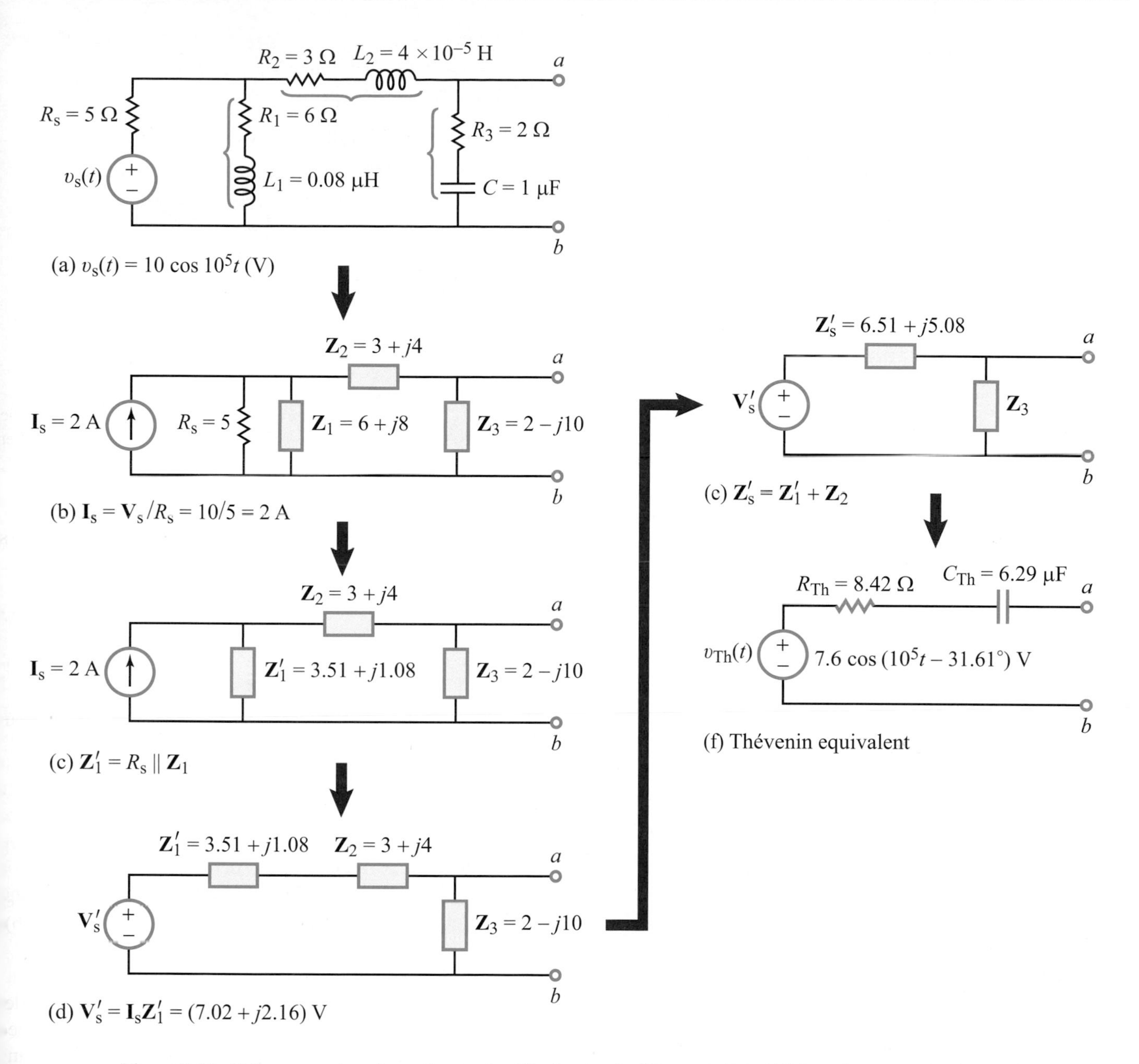

Figure 7-19: Using source transformation to simplify the circuit of Example 7-9. (All impedances are in ohms.)

Figure 7-19(b) displays the circuit in the phasor domain, in addition to having replaced the series combination ($\mathbf{V}_s$, R_s) with the parallel combination ($\mathbf{I}_s$, R_s), where

$$\mathbf{I}_s = \frac{\mathbf{V}_s}{R_s} = \frac{10}{5} = 2 \text{ A}.$$

Step 2: Combining R_s with $\mathbf{Z}_1$ in parallel gives

$$\mathbf{Z}_1' = R_s \parallel \mathbf{Z}_1 = \frac{5(6 + j8)}{5 + 6 + j8} = (3.51 + j1.08)\ \Omega.$$

Step 3: Converting back to a voltage source in series with $\mathbf{Z}_1'$ leads to the circuit in Fig. 7-19(d), with

$$\mathbf{V}_s' = \mathbf{I}_s \mathbf{Z}_1' = 2(3.51 + j1.08) = (7.02 + j2.16) \text{ V}.$$

Step 4: Combining $\mathbf{Z}_1'$ with $\mathbf{Z}_2$ in series leads to the circuit in Fig. 7-19(e), where

$$\mathbf{Z}_s' = \mathbf{Z}_1' + \mathbf{Z}_2 = (3.51 + j1.08) + (3 + j4) = (6.51 + j5.08)\ \Omega.$$

Step 5: Application of voltage division provides

$$\mathbf{V}_{\text{Th}} = \mathbf{V}_{\text{oc}} = \frac{\mathbf{V}'_{\text{s}}\mathbf{Z}_3}{\mathbf{Z}'_{\text{s}} + \mathbf{Z}_3} = \frac{(7.02 + j2.16)(2 - j10)}{(6.51 + j5.08) + (2 - j10)} = 7.6\angle{-31.61^\circ} \text{ V}.$$

Hence,

$$v_{\text{Th}}(t) = \Re\mathfrak{e}[\mathbf{V}_{\text{Th}} e^{j\omega t}] = \Re\mathfrak{e}[7.6 e^{-j31.61^\circ} e^{j10^5 t}] = 7.6\cos(10^5 t - 31.61^\circ) \text{ V}.$$

Step 6: Suppressing the source $\mathbf{V}'_{\text{s}}$ in Fig. 7-19(e) reduces the circuit at terminals (a, b) to $\mathbf{Z}'_{\text{s}}$ in parallel with $\mathbf{Z}_3$, leading to

$$\mathbf{Z}_{\text{Th}} = \mathbf{Z}'_{\text{s}} \parallel \mathbf{Z}_3 = \frac{(6.51 + j5.08)(2 - j10)}{(6.51 + j5.08) + (2 - j10)} = (8.42 - j1.59)\ \Omega.$$

Step 7: The impedance $\mathbf{Z}_{\text{Th}}$ is capacitive because the sign of the imaginary component is negative. Hence, it is equivalent to

$$\mathbf{Z}_{\text{Th}} = R_{\text{Th}} - \frac{j}{\omega C_{\text{Th}}}.$$

Matching the two expressions gives

$$R_{\text{Th}} = 8.42\ \Omega, \qquad C_{\text{Th}} = \frac{1}{1.59\omega} = 6.29\ \mu\text{F}.$$

The time-domain Thévenin equivalent circuit is shown in Fig. 7-19(f).

Review Question 7-11: In the phasor domain, is the Thévenin equivalent method valid for circuits containing dependent sources? If yes, what methods are amenable to finding $\mathbf{Z}_{\text{Th}}$ of such circuits?

Review Question 7-12: If $\mathbf{Z}_{\text{Th}}$ of a certain circuit is purely imaginary, what would be your expectation about whether or not the circuit contains resistors?

Exercise 7-11: Determine $\mathbf{V}_{\text{Th}}$ and $\mathbf{Z}_{\text{Th}}$ for the circuit in Fig. E7.11 at terminals (a, b).

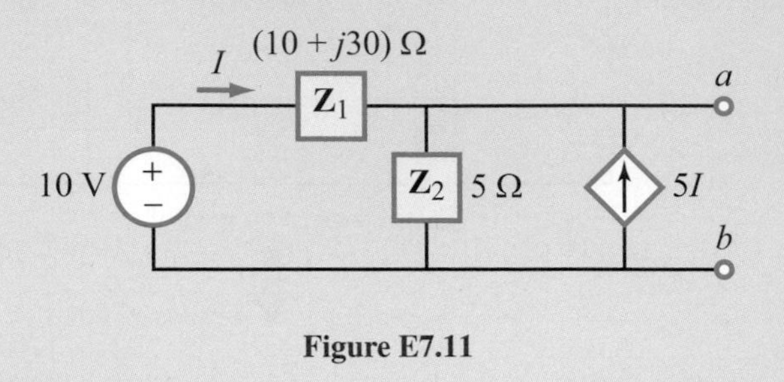

Figure E7.11

Answer: $\mathbf{V}_{\text{Th}} = 6\angle{-36.9^\circ}$ V, $\mathbf{Z}_{\text{Th}} = (2.6 + j1.8)\ \Omega$. (See ◎)

7-7 Phasor Diagrams

Consider the following sinusoidal signal $v_{\text{s}}(t)$ and its phasor counterpart $\mathbf{V}_{\text{s}}$:

$$v_{\text{s}}(t) = V_0 \cos(\omega t + \phi) \quad \longleftrightarrow \quad \mathbf{V}_{\text{s}} = V_0\angle{\phi}. \tag{7.91}$$

The time-domain voltage $v_{\text{s}}(t)$ is characterized by three attributes: the amplitude V_0, the angular frequency ω, and the phase angle ϕ. In contrast, its counterpart in the phasor domain $\mathbf{V}_{\text{s}}$ is specified by only two attributes, V_0 and ϕ. This may suggest that ω becomes irrelevant when we analyze a circuit in the phasor domain, but that certainly is not true if the circuit contains capacitors and/or inductors. Whereas ω does not appear explicitly in the expressions for phasor currents and voltages, it is integral to the definitions of the capacitor impedance $\mathbf{Z}_{\text{C}}$ and inductor impedance $\mathbf{Z}_{\text{L}}$, which in turn define the **I**–**V** relationships for those two elements as

$$\mathbf{Z}_{\text{C}} = \frac{\mathbf{V}_{\text{C}}}{\mathbf{I}_{\text{C}}} = \frac{1}{j\omega C} = \frac{1}{\omega C}\angle{-90^\circ} \tag{7.92a}$$

and

$$\mathbf{Z}_{\text{L}} = \frac{\mathbf{V}_{\text{L}}}{\mathbf{I}_{\text{L}}} = j\omega L = \omega L\angle{90^\circ}. \tag{7.92b}$$

In fact, the value of ω (relative to the values of L of C) can drastically change the behavior of a circuit:

At dc, $\mathbf{Z}_{\text{C}} \to \infty$ (open circuit) and $\mathbf{Z}_{\text{L}} \to 0$ (short circuit); and conversely, as $\omega \to \infty$, $\mathbf{Z}_{\text{C}} \to 0$ and $\mathbf{Z}_{\text{L}} \to \infty$.

A ***phasor diagram*** is a useful graphical tool for examining the relationships among the various currents and voltages in a circuit. Before considering multi-element circuits, however, we will start by examining the phasor diagrams for R, L and C

Resistor

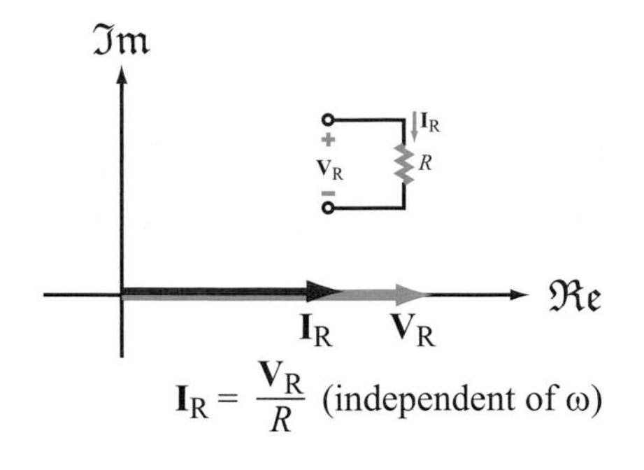

Capacitor

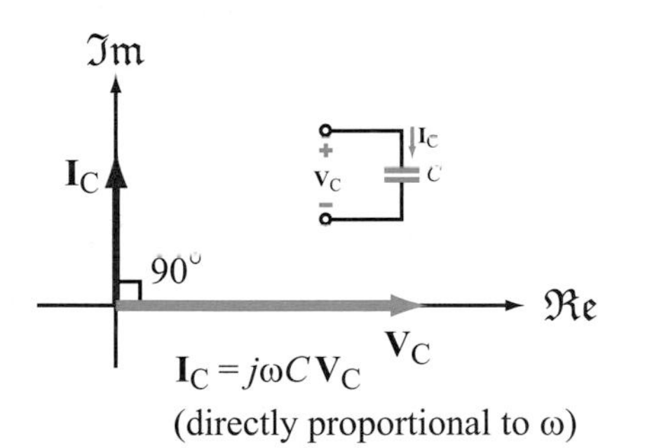

Inductor

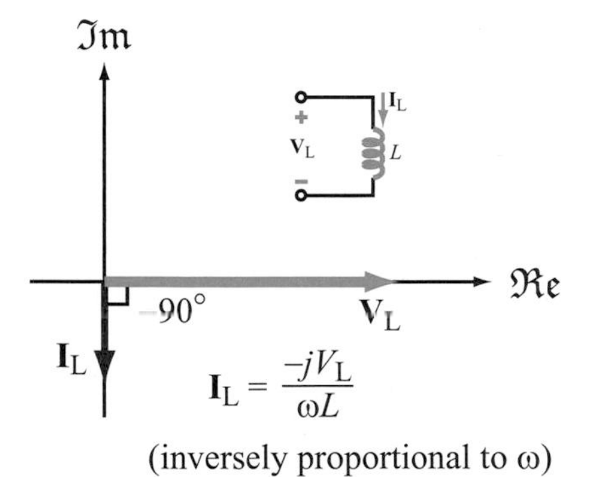

Figure 7-20: Phasor diagrams for R, L, and C.

ndividually. Figure 7-20 displays the phasor diagrams for **I** nd **V** for all three elements, with **V** chosen as a reference by electing its phase angle to be zero. Each phasor quantity is isplayed in the complex plane in terms of its magnitude and hase angle. For the resistor, $\mathbf{V}_R$ and $\mathbf{I}_R$ always line up along ne same direction because they are always in-phase. Since $\mathbf{V}_R$ vas chosen to be purely real, so is $\mathbf{I}_R$.

Next, we consider the capacitor. In view of Eq. (7.92a),

$$\mathbf{I}_C = \frac{\mathbf{V}_C}{\mathbf{Z}_C} = j\omega C\mathbf{V}_C = \omega C\mathbf{V}_C\angle 90^\circ, \qquad (7.93a)$$

which ***positions the vector*** $\mathbf{I}_C$ ***ahead of*** $\mathbf{V}_C$ ***by*** 90°. For the inductor,

$$\mathbf{I}_L = \frac{\mathbf{V}_L}{j\omega L} = \frac{-j\mathbf{V}_L}{\omega L} = \frac{\mathbf{V}_L}{\omega L}\angle{-90^\circ}. \qquad (7.93b)$$

Consequently, $\mathbf{I}_L$ ***lags*** $\mathbf{V}_L$ ***by*** 90°.

For individual elements, the relationship between **I** and **V** is straightforward; given the position of either one of them in the complex plane, we can place the other one in accordance with the phase-angle shift appropriate to that element.

For a multi-element circuit, we can draw either a ***relative phasor diagram*** or an ***absolute phasor diagram***. In the case of the relative phasor diagram, we usually choose a specific current or voltage and designate it as our reference phasor by arbitrarily assigning it a phase angle of 0°.

The goal then is to use the phasor diagram to examine the relationships between and among the various currents and voltages in the circuit—which includes their magnitudes and ***relative*** phase angles—rather than to establish their absolute phase angles. In principle, it does not matter much which specific phasor voltage or current is selected as the reference, but in practice, we usually choose a phasor current or voltage that is common to lots of elements in the circuit. By way of illustration, Example 7-10 examines a series RLC circuit by displaying its phasor diagram twice, once with the current flowing through the loop as reference, and a second time with the voltage source as reference. The former results in a *relative phasor diagram*, whereas the latter results in an *absolute phasor diagram*.

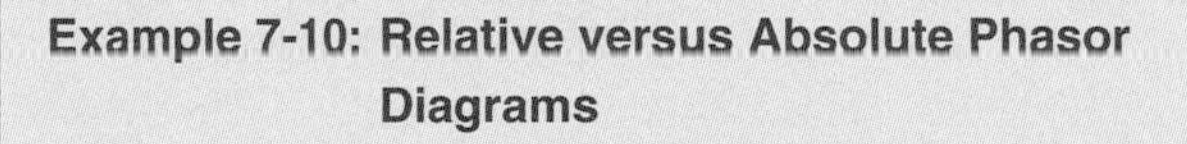
Example 7-10: Relative versus Absolute Phasor Diagrams

The circuit in Fig. 7-21(a) is driven by a voltage source given by

$$v_s(t) = 20\cos(500t + 30^\circ)\ \text{V}.$$

Generate: (a) a relative phasor diagram by selecting the phasor current **I** as a reference, and (b) an absolute phasor diagram by selecting the phasor voltage source as a reference.

Solution: Figure 7-21(b) displays the phasor-domain circuit with its RLC elements represented by their respective impedances.

(a) Relative Phasor Diagram

Selecting **I** as the reference phasor means that we assign it an unknown magnitude I_0 and a phase angle of 0°:

$$\mathbf{I} = I_0\angle 0^\circ.$$

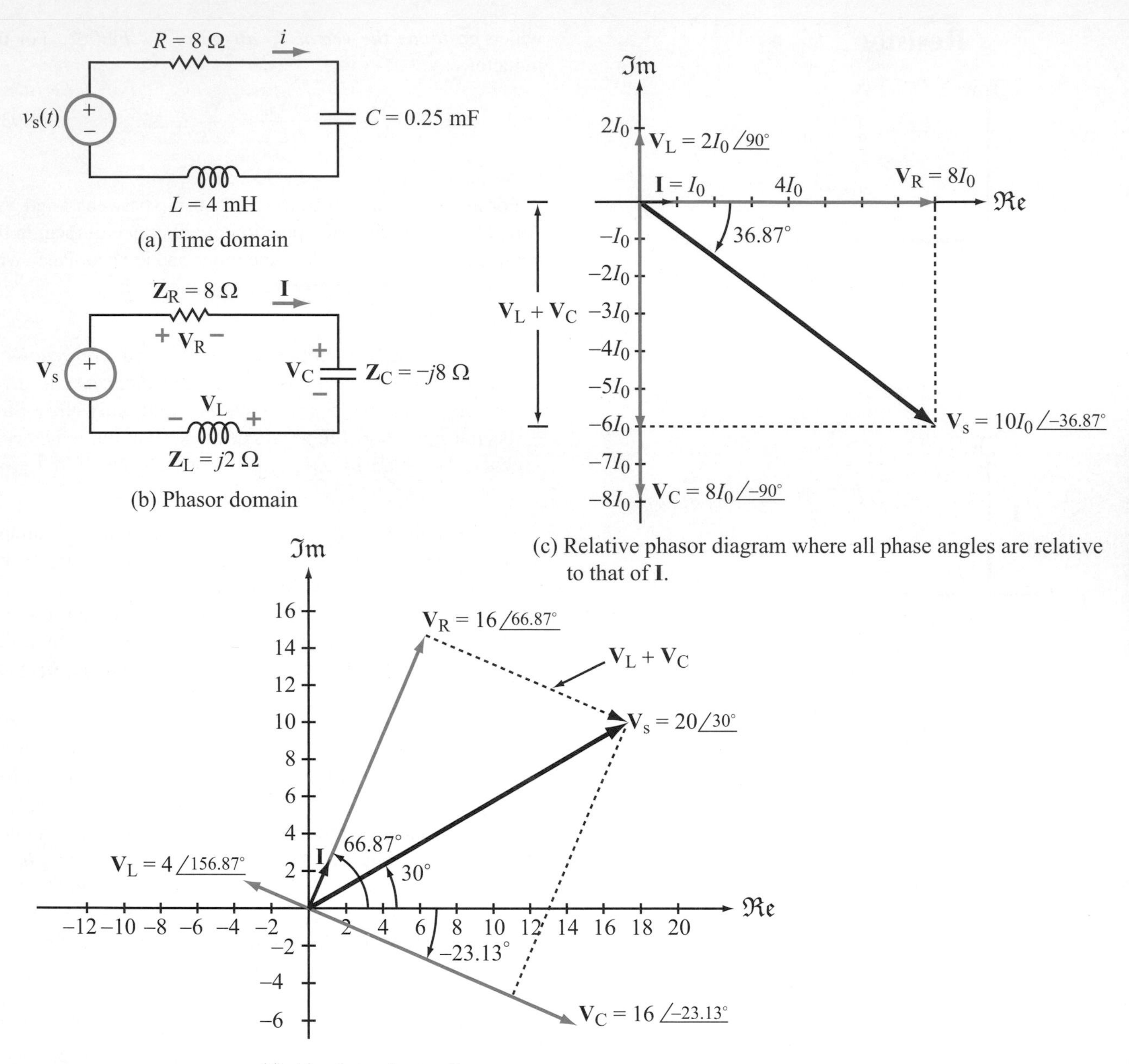

Figure 7-21: Circuit and phasor diagrams for Example 7-10. The true phase angle of **I** is 66.87°, so if the relative phasor diagram in (c) were to be rotated counterclockwise by that angle and the scale adjusted to incorporate the fact $I_0 = 2$, the diagram would coincide with the absolute phasor diagram in (d).

Because the true phase angle of **I** actually may not be zero, the vectors we will draw in the complex plane of the phasor diagram all will be shifted in orientation by exactly the same amount (namely by the true phase angle of **I**) so even though they may not have the correct orientations, they all will bear the correct *relative* orientations to one another.

We deduce from the functional form of $v_s(t)$ that $\omega = 500$ rad/s. In terms of **I**, the voltages across R, C, and L are

$$\mathbf{V}_R = R\mathbf{I} = 8I_0\angle 0°,$$

$$\mathbf{V}_C = \frac{\mathbf{I}}{j\omega C} = \frac{-jI_0}{500 \times 2.5 \times 10^{-4}} = -j8I_0 = 8I_0\angle -90°,$$

and

$$\mathbf{V}_L = j\omega L\mathbf{I} = j500 \times 4 \times 10^{-3} I_0 = j2I_0 = 2I_0\angle 90°,$$

and the sum of all three gives

$$\begin{aligned}\mathbf{V}_s = \mathbf{V}_R + \mathbf{V}_C + \mathbf{V}_L &= 8I_0 - j8I_0 + j2I_0 \\ &= (8 - j6)I_0 \\ &= \sqrt{8^2 + 6^2}\, I_0 e^{j\phi} \\ &= 10I_0\angle\phi,\end{aligned}$$

where

$$\phi = -\tan^{-1}\frac{6}{8} = -36.87°.$$

Figure 7-21(c) displays the relative phasor diagram of the RLC circuit with $\mathbf{I}$ as a reference; the magnitudes of $\mathbf{V}_R$, $\mathbf{V}_C$, $\mathbf{V}_L$, and $\mathbf{V}_s$ are all measured in units of I_0, and their orientations are relative to that of $\mathbf{I}$.

(b) Absolute Phasor Diagram

The phasor counterpart of $v_s(t)$ is

$$\mathbf{V}_s = 20\angle 30° \text{ V},$$

and the application of KVL around the loop leads to

$$\begin{aligned}\mathbf{I} = \frac{\mathbf{V}_s}{R + j\omega L - \dfrac{j}{\omega C}} &= \frac{20e^{j30°}}{8 + j2 - j8} \\ &= \frac{20e^{j30°}}{8 - j6} \\ &= \frac{20e^{j30°}}{10e^{-j36.87°}} \\ &= 2e^{j66.87°} \text{ A},\end{aligned}$$

which states that the ***true*** phase angle of $\mathbf{I}$ is 66.87°. Given $\mathbf{I}$, we easily can calculate $\mathbf{V}_R$, $\mathbf{V}_C$, and $\mathbf{V}_L$. The phasor diagram shown in Fig. 7-21(d) is identical with that in Fig. 7-21(c), except that all vectors have been rotated in a counterclockwise direction by 66.87°.

Review Question 7-13: For a capacitor, what is the phase angle of its phasor voltage, relative to that of its phasor current?

Review Question 7-14: What is the difference between a *relative* phasor diagram and an *absolute* phasor diagram?

Exercise 7-12: Establish the relative phasor diagram for the circuit in Fig. E7.12 with $\mathbf{V}$ as the reference phasor.

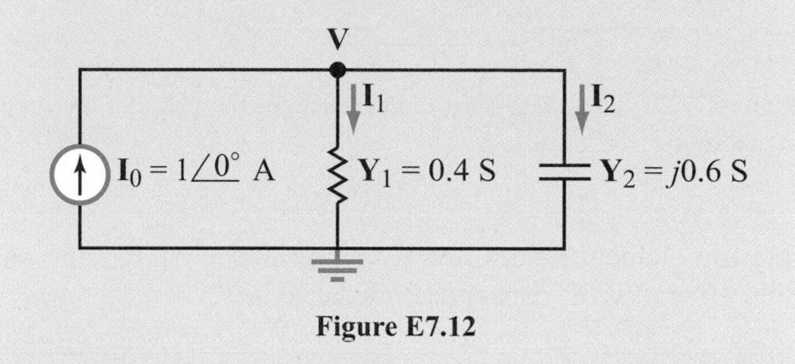

Figure E7.12

Answer:

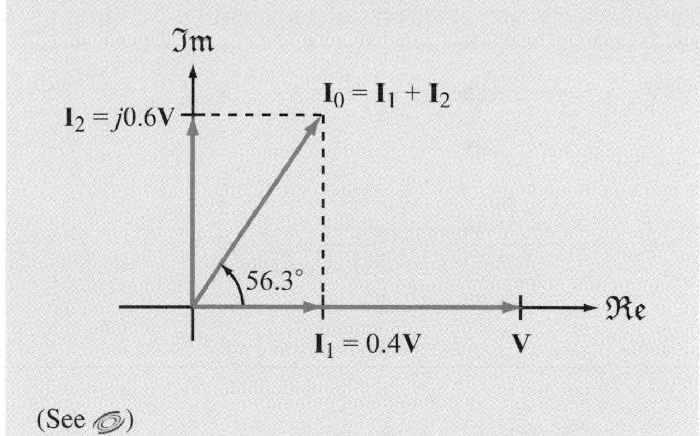

(See CD)

7-8 Phase-Shift Circuits

In certain communication and signal-processing applications, we often need to shift the phase of an ac signal by adding (or subtracting) a phase angle of a specified value, ϕ. Thus, if the input voltage in Fig. 7-22 is

$$v_{in}(t) = V_1 \cos\omega t, \tag{7.94}$$

the function of the ***phase-shift circuit*** is to provide an output voltage given by

$$v_{out}(t) = V_2 \cos(\omega t + \phi). \tag{7.95}$$

The amplitude V_2 of the output voltage is related to V_1 (the amplitude of the input voltage) and to the configuration of the phase-shift circuit. RC circuits can be designed as phase shifters, with any specified positive or negative value of ϕ:

$$\begin{aligned}&v_{out} \text{ will lead } v_{in} \quad \text{if } 0 \le \phi \le 180°, \\ &v_{out} \text{ will lag } v_{in} \quad \text{if } -180° \le \phi \le 0°.\end{aligned}$$

To illustrate the process, let us consider the simple RC circuit shown in Fig. 7-23(a). The input signal is given by

$$v_{in}(t) = 10\cos 10^6 t \quad \text{V},$$

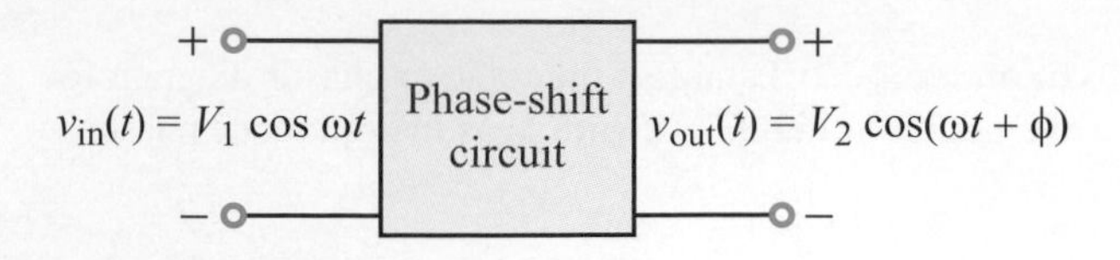

Figure 7-22: The phase-shift circuit changes the phase of the input signal by ϕ.

and the element values are $R = 2\ \Omega$ and $C = 0.2\ \mu$F. At $\omega = 10^6$ rad/s, the capacitor impedance is

$$\mathbf{Z}_C = \frac{-j}{\omega C} = \frac{-j}{10^6 \times 0.2 \times 10^{-6}} = -j5\ \Omega.$$

By voltage division in the phasor domain (Fig. 7-23(b)),

$$\mathbf{V}_{out1} = \frac{\mathbf{V}_{in} R}{R - \dfrac{j}{\omega C}} = \frac{\omega RC}{\sqrt{1+\omega^2 R^2 C^2}}\ \mathbf{V}_{in}\angle\phi_1, \tag{7.96a}$$

$$\mathbf{V}_{out2} = \frac{\mathbf{V}_{in}\left(\dfrac{-j}{\omega C}\right)}{R - \dfrac{j}{\omega C}} = \frac{1}{\sqrt{1+\omega^2 R^2 C^2}}\ \mathbf{V}_{in}\angle\phi_2, \tag{7.96b}$$

and the phase angles ϕ_1 and ϕ_2 are given by

$$\phi_1 = \tan^{-1}\left(\frac{1}{\omega RC}\right) \tag{7.97a}$$

and

$$\phi_2 = \phi_1 - 90° = \tan^{-1}\left(\frac{1}{\omega RC}\right) - 90°. \tag{7.97b}$$

For $\omega = 10^6$ rad/s, $R = 2\ \Omega$, $C = 0.2\ \mu$F, and $\mathbf{V}_{in} = 10$ V,

$$\mathbf{V}_{out1} = 3.71\angle 68.2° = (1.38 + j3.45)\text{ V}$$

and

$$\mathbf{V}_{out2} = 9.28\angle -21.8° = (8.62 - j3.45)\text{ V}.$$

The phase angle ϕ_1 associated with $\mathbf{V}_{out1}$ is 68.2°, and the angle ϕ_2 associated with $\mathbf{V}_{out2}$ is −21.8°. As shown in the complex plane of Fig. 7-23(c), the angular separation between $\mathbf{V}_{out1}$ and $\mathbf{V}_{out2}$ is exactly 90°. Also, if we were to add $\mathbf{V}_{out1}$ and $\mathbf{V}_{out2}$ in the complex plane, their imaginary parts would cancel out and their real parts would add up to 10 V (the amplitude of $\mathbf{V}_{in}$).

In the time domain,

$$\begin{aligned} v_{out1}(t) &= \Re\mathfrak{e}[\mathbf{V}_{out1} e^{j\omega t}] \\ &= 3.716 \cos(10^6 t + 68.2°)\text{ V} \end{aligned} \tag{7.98}$$

and

$$\begin{aligned} v_{out2}(t) &= \Re\mathfrak{e}[\mathbf{V}_{out2} e^{j\omega t}] \\ &= 9.285 \cos(10^6 t - 21.8°)\text{ V}. \end{aligned} \tag{7.99}$$

Figure 7-23(a) provides a comparison of the waveform of the input signal $v_{in}(t)$ with that of $v_{out2}(t)$, the voltage across the capacitor. We note that because v_{out2} lags v_{in}, it always crosses the time axis later than v_{in} by a time delay Δt. If we denote t_0 as the time when $v_{in}(t)$ crosses the time axis and t_2 as the time when $v_{out2}(t)$ does, then

$$10^6 t_0 = \frac{\pi}{2}$$

and

$$10^6 t_2 + \phi_2 = \frac{\pi}{2},$$

where

$$\phi_2 = -21.8° \times \left(\frac{\pi}{180°}\right) = -0.38\text{ radians}.$$

Now that all quantities are in the same units, we can determine the time delay from

$$\Delta t_2 = t_2 - t_0 = -\phi_2 \times 10^{-6} = 0.38\ \mu\text{s}.$$

By the same argument, v_{out1} leads v_{in} by 68.2°, and it crosses the time axis *sooner* than does $v_{in}(t)$ by

$$\Delta t_1 = 68.2° \times \frac{\pi}{180°} \times 10^{-6} = 1.19\ \mu\text{s}.$$

From the foregoing analysis, we conclude that for the simple RC circuit, we can use v_{out1} as our output if we want to add a positive phase angle to the input v_{in}, and we can use v_{out2} as our output if we want to add a negative phase angle to v_{in}. Moreover, by adjusting the values of R and C (at a specific value of ω), we can change ϕ_1 to any value between 0 and 90°, and similarly, we can change ϕ_2 to any value between 0 and −90° (but not independently); as was noted earlier in connection with Fig. 7-23(c), the absolute values of ϕ_1 and ϕ_2 always add up to 90°. Another consideration that we should be aware of is that the magnitudes of v_{out1} and v_{out2} are linked to the magnitudes of ϕ_1 and ϕ_2 through the choices we make for R, C, and ω. For example, as ϕ_1 approaches 90°, v_{out1} approaches zero, so we can indeed phase-shift the input signal by an angle close to 90° but the magnitude of the output signal will be too small to be useful. To overcome this limitation or to introduce phase-shift angles greater than 90°, we can use circuits with more than two elements, such as the cascaded circuit of Example 7-11.

To generate a phase lead at the output, the cascading arrangement should be as that shown in Fig. 7-24, but to generate a phase lag, the locations of R and C should be interchanged.

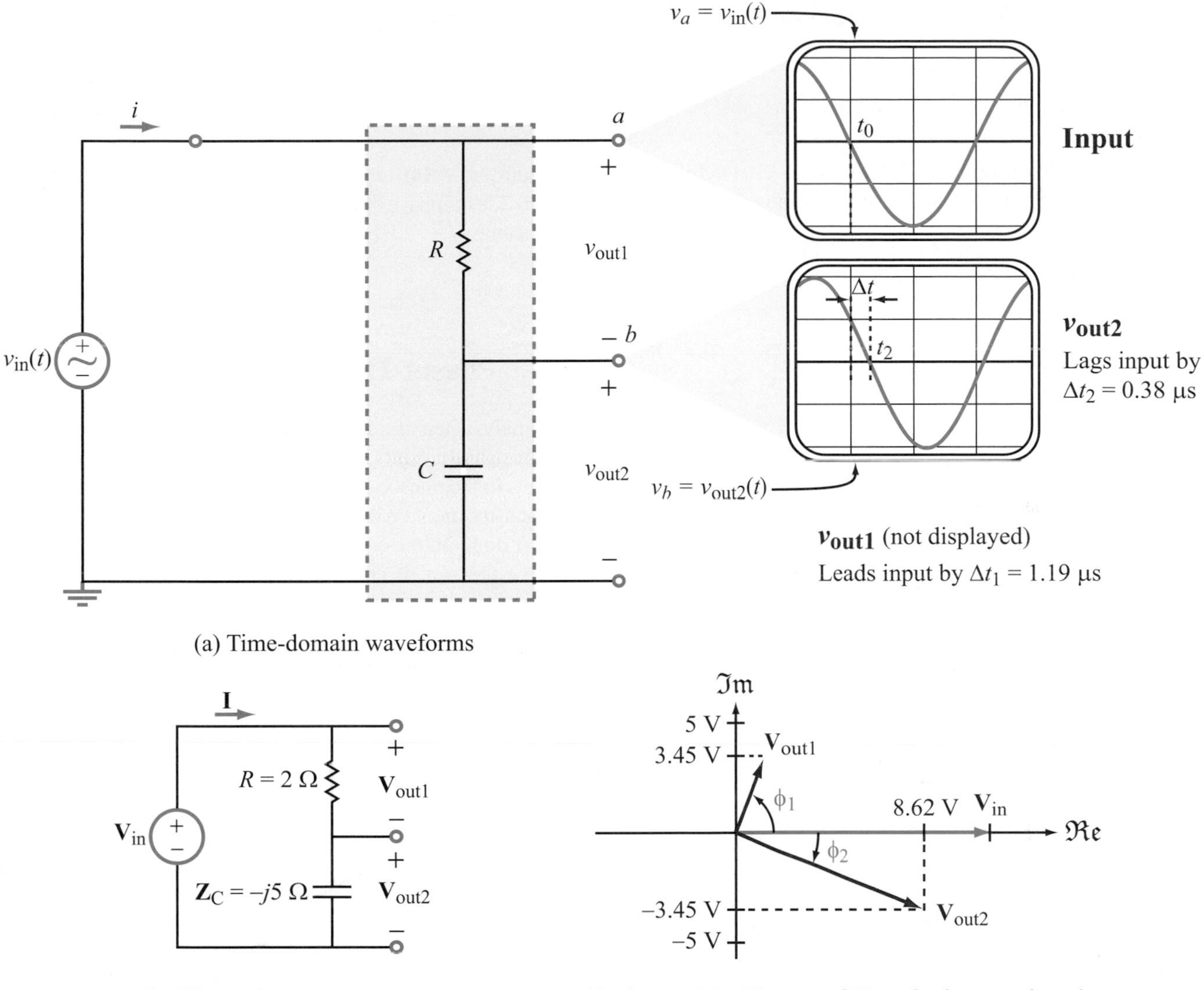

Figure 7-23: RC phase-shift circuit: the phase of v_{out1} (across R) leads the phase of $v_{in}(t)$, whereas the phase of v_{out2} (across C) lags the phase of $v_{in}(t)$.

Example 7-11: Cascaded Phase-Shifter

The circuit in Fig. 7-24 uses cascaded phase-shifters to produce an output signal $v_{out}(t)$ whose phase is 120° ahead of the input signal $v_s(t)$. If $\omega = 10^3$ (rad/s) and $C = 1\ \mu$F, determine R and the ratio of the amplitude of v_{out} to that of v_s.

Solution: Application of nodal analysis at nodes $\mathbf{V}_1$ and $\mathbf{V}_2$ in the phasor domain gives

$$\frac{\mathbf{V}_1 - \mathbf{V}_s}{\mathbf{Z}_C} + \frac{\mathbf{V}_1}{R} + \frac{\mathbf{V}_1 - \mathbf{V}_2}{\mathbf{Z}_C} = 0 \tag{7.100}$$

and

$$\frac{\mathbf{V}_2 - \mathbf{V}_1}{\mathbf{Z}_C} + \frac{\mathbf{V}_2}{R} + \frac{\mathbf{V}_2}{R + \mathbf{Z}_C} = 0, \tag{7.101}$$

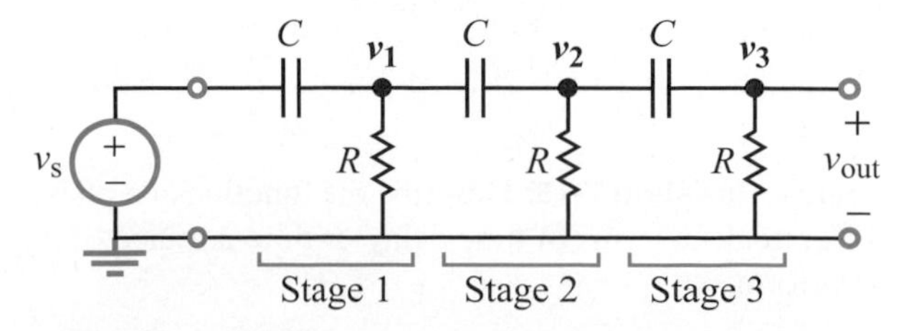

Figure 7-24: Three-stage, cascaded, RC phase-shifter (Example 7-11).

where $\mathbf{Z}_C = 1/j\omega C$. Moreover, through voltage division, $\mathbf{V}_3$ is related to $\mathbf{V}_2$ by

$$\mathbf{V}_3 = \left(\frac{R}{R + \mathbf{Z}_C}\right) \mathbf{V}_2. \tag{7.102}$$

Simultaneous solution of Eqs. (7.100) and (7.101), followed with several steps of algebra, leads to the expressions

$$\frac{\mathbf{V}_1}{\mathbf{V}_s} = \frac{x[(x^2 - 1) - j3x]}{(x^3 - 5x) + j(1 - 6x^2)}, \tag{7.103}$$

$$\frac{\mathbf{V}_2}{\mathbf{V}_s} = \frac{x^2(x - j1)}{(x^3 - 5x) + j(1 - 6x^2)}, \tag{7.104}$$

and

$$\frac{\mathbf{V}_3}{\mathbf{V}_s} = \frac{x^3}{(x^3 - 5x) + j(1 - 6x^2)}, \tag{7.105}$$

where

$$x = \omega RC. \tag{7.106}$$

The magnitude and phase of $\mathbf{V}_3$ (both relative to those of $\mathbf{V}_s$) are

$$\left|\frac{\mathbf{V}_3}{\mathbf{V}_s}\right| = \frac{x^3}{[(x^3 - 5x)^2 + (1 - 6x^2)^2]^{1/2}}, \tag{7.107a}$$

and

$$\phi_3 = -\tan^{-1}\left(\frac{1 - 6x^2}{x^3 - 5x}\right). \tag{7.107b}$$

To satisfy the stated requirement, we set $\phi_3 = 120°$ and solve for x:

$$\tan 120° = -1.732 = -\left(\frac{1 - 6x^2}{x^3 - 5x}\right),$$

which leads to

$$x = 1.1815. \tag{7.108}$$

Given that $\omega = 10^3$ rad/s and $C = 1\ \mu$F, it follows that

$$R = \frac{x}{\omega C} = \frac{1.1815}{10^3 \times 10^{-6}} = 1.1815\ \text{k}\Omega \simeq 1.2\ \text{k}\Omega.$$

With $x = 1.1815$, Eq. (7.107a) gives

$$\left|\frac{\mathbf{V}_3}{\mathbf{V}_s}\right| = 0.194.$$

Review Question 7-15: Describe the function of a phase-shift circuit in terms of time delay or time advance of the waveform.

Review Question 7-16: When is it necessary to use multiple stages to achieve the desired phase shift?

Exercise 7-13: Repeat Example 7-11, but use only two stages of RC phase shifters.

Answer: $R \simeq 2.2\ \text{k}\Omega$; $|\mathbf{V}_{out}/\mathbf{V}_s| = 0.63$. (See)

Exercise 7-14: Design a two-stage RC phase shifter that provides a phase shift of negative 120° at $\omega = 10^4$ rad/s. Assume $C = 1\ \mu$F.

Answer: $R \simeq 220\ \Omega$. (See)

7-9 Phasor-Domain Analysis Techniques

The analysis techniques introduced in Chapter 3 in connection with resistive circuits are all equally applicable for analyzing ac circuits in the phasor domain. The only fundamental difference is that after transferring the circuit from the time domain to the phasor domain, the operations conducted in the phasor domain involve the use of complex algebra, as opposed to just real numbers. Otherwise, the circuit laws and methods of solution are identical.

At this stage, instead of repeating the details of these various techniques, a more effective approach would be to illustrate their implementation procedures through concrete examples. Examples 7-12 through 7-16 are designed to do just that.

Example 7-12: Nodal Analysis

Apply the nodal-analysis method to determine $i_L(t)$ in the circuit of Fig. 7-25(a). The sources are given by:

$$\upsilon_{s_1}(t) = 12 \cos 10^3 t \text{ V},$$

$$\upsilon_{s_2}(t) = 6 \sin 10^3 t \text{ V}.$$

Solution: We first will demonstrate how to solve this problem using the standard nodal-analysis method, and then we will solve it again by applying the by-inspection method.

Nodal-Analysis Method

Our first step is to transform the given circuit to the phasor domain. Accordingly,

$$\mathbf{Z}_C = \frac{1}{j\omega C} = \frac{-j}{10^3 \times 0.25 \times 10^{-3}} = -j4\ \Omega,$$

$$\mathbf{Z}_L = j\omega L = j10^3 \times 10^{-3} = j1\ \Omega,$$

$$\mathbf{V}_{s_1} = 12 \text{ V},$$

and

$$\mathbf{V}_{s_2} = -j6 \text{ V},$$

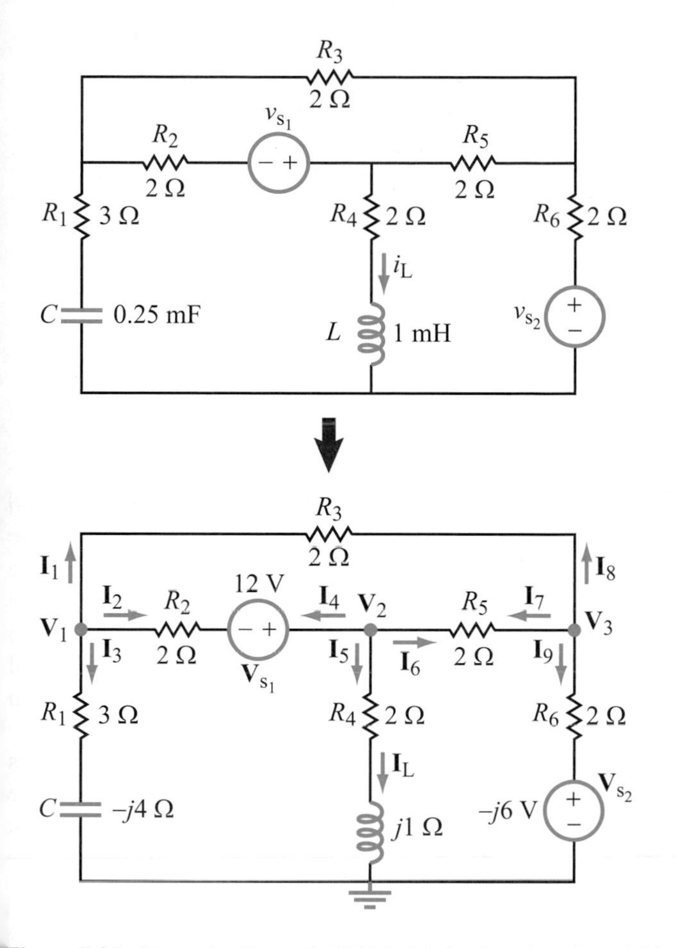

Figure 7-25: Circuit for Example 7-12 in (a) the time domain and (b) the phasor domain.

where for $\mathbf{V}_{s_2}$ we used the property given in Table 7-3, namely that the phasor counterpart of $\sin \omega t$ is $-j$. Using these values, we generate the phasor-domain circuit given in Fig. 7-25(b) in which we selected one of the extraordinary nodes as a ground node and assigned phasor voltages $\mathbf{V}_1$ to $\mathbf{V}_3$ to the other three.

Our plan is to write the voltage node equations at nodes 1 to 3, solve them simultaneously to find $\mathbf{V}_1$ to $\mathbf{V}_3$, and then use the value of $\mathbf{V}_3$ to obtain $\mathbf{I}_L$. The final step will involve transforming $\mathbf{I}_L$ to the time domain to obtain $i_L(t)$.

At node 1, KCL requires that

$$\mathbf{I}_1 + \mathbf{I}_2 + \mathbf{I}_3 = 0. \tag{7.109}$$

In terms of node voltages $\mathbf{V}_1$ to $\mathbf{V}_3$,

$$\mathbf{I}_1 = \frac{\mathbf{V}_1 - \mathbf{V}_3}{R_3} = \frac{\mathbf{V}_1 - \mathbf{V}_3}{2},$$

$$\mathbf{I}_2 = \frac{\mathbf{V}_1 - \mathbf{V}_2 + \mathbf{V}_{s_1}}{R_2} = \frac{\mathbf{V}_1 - \mathbf{V}_2 + 12}{2},$$

and

$$\mathbf{I}_3 = \frac{\mathbf{V}_1}{R_1 + \mathbf{Z}_C} = \frac{\mathbf{V}_1}{3 - j4}.$$

Inserting the expressions for $\mathbf{I}_1$ to $\mathbf{I}_3$ in Eq. (7.109) and then rearranging the terms leads to

$$\left(\frac{1}{2} + \frac{1}{2} + \frac{1}{3 - j4}\right)\mathbf{V}_1 - \frac{1}{2}\mathbf{V}_2 - \frac{1}{2}\mathbf{V}_3 = -6. \tag{7.110}$$

The coefficient of $\mathbf{V}_1$ can be simplified as follows:

$$\begin{aligned}\frac{1}{2} + \frac{1}{2} + \frac{1}{3 - j4} = 1 + \frac{1}{3 - j4} &= \frac{3 - j4 + 1}{3 - j4} \\ &= \frac{4 - j4}{3 - j4} \times \frac{3 + j4}{3 + j4} \\ &= \frac{(12 + 16) + j(16 - 12)}{9 + 16} \\ &= 1.12 + j0.16. \end{aligned} \tag{7.111}$$

Inserting Eq. (7.111) in Eq. (7.110) and multiplying all terms by 2 leads to the following simplified algebraic equation for node 1:

$$(2.24 + j0.32)\mathbf{V}_1 - \mathbf{V}_2 - \mathbf{V}_3 = -12 \qquad \text{(node 1)}. \tag{7.112}$$

Similarly, at node 2,

$$\frac{\mathbf{V}_2 - \mathbf{V}_1 - 12}{2} + \frac{\mathbf{V}_2}{2 + j1} + \frac{\mathbf{V}_2 - \mathbf{V}_3}{2} = 0,$$

which can be simplified to

$$-\mathbf{V}_1 + (2.8 - j0.4)\mathbf{V}_2 - \mathbf{V}_3 = 12 \qquad \text{(node 2)}, \tag{7.113}$$

and at node 3,

$$\frac{\mathbf{V}_3 - \mathbf{V}_2}{2} + \frac{\mathbf{V}_3 - \mathbf{V}_1}{2} + \frac{\mathbf{V}_3 + j6}{2} = 0,$$

or

$$-\mathbf{V}_1 - \mathbf{V}_2 + 3\mathbf{V}_3 = -j6 \qquad \text{(node 3)}. \tag{7.114}$$

Equations (7.112) to (7.114) now are ready to be cast in matrix form:

$$\begin{bmatrix} (2.24 + j0.32) & -1 & -1 \\ -1 & (2.8 - j0.4) & -1 \\ -1 & -1 & 3 \end{bmatrix} \begin{bmatrix} \mathbf{V}_1 \\ \mathbf{V}_2 \\ \mathbf{V}_3 \end{bmatrix} = \begin{bmatrix} -12 \\ 12 \\ -j6 \end{bmatrix}. \tag{7.115}$$

Matrix inversion, either manually or by MATLAB® software, provides the solution:

$$\mathbf{V}_1 = -(4.72 + j0.88) \text{ V}, \tag{7.116a}$$

$$\mathbf{V}_2 = (2.46 - j0.89) \text{ V}, \tag{7.116b}$$

and

$$\mathbf{V}_3 = -(0.76 + j2.59) \text{ V}. \quad (7.116c)$$

Hence,

$$\mathbf{I}_L = \frac{\mathbf{V}_2}{2+j1} = \frac{2.46 - j0.89}{2+j1}$$
$$= 0.81 - j0.85 = 1.17e^{-j46.5^\circ} \text{ A},$$

and its corresponding time-domain counterpart is

$$i_L(t) = \mathfrak{Re}[\mathbf{I}_L e^{j1000t}]$$
$$= \mathfrak{Re}[1.17e^{-j46.4^\circ} e^{j1000t}]$$
$$= 1.17\cos(1000t - 46.5^\circ) \text{ A}. \quad (7.117)$$

By-Inspection Method

Implementation of the nodal-analysis by-inspection method requires that the circuit contain no dependent sources and that all independent sources in the circuit be current sources. The first condition is valid for the circuit in Fig. 7-25(b), but the second one is not. However, both voltage sources in Fig. 7-25(b) have in-series resistors associated with them, so we easily can transform them into current sources. The resultant circuit is shown in Fig. 7-26, in which not only have the voltage sources been replaced with equivalent current sources, but all impedances have also been replaced with their equivalent admittances ($\mathbf{Y} = 1/\mathbf{Z}$). For the 3-node case, the phasor-domain equivalent of Eq. (3.26) is given by

$$\begin{bmatrix} \mathbf{Y}_{11} & \mathbf{Y}_{12} & \mathbf{Y}_{13} \\ \mathbf{Y}_{21} & \mathbf{Y}_{22} & \mathbf{Y}_{23} \\ \mathbf{Y}_{31} & \mathbf{Y}_{32} & \mathbf{Y}_{33} \end{bmatrix} \begin{bmatrix} \mathbf{V}_1 \\ \mathbf{V}_2 \\ \mathbf{V}_3 \end{bmatrix} = \begin{bmatrix} \mathbf{I}_{t_1} \\ \mathbf{I}_{t_2} \\ \mathbf{I}_{t_3} \end{bmatrix}, \quad (7.118)$$

where

$\mathbf{Y}_{kk} =$ sum of all admittances connected to node k

$\mathbf{Y}_{k\ell} = \mathbf{Y}_{\ell k} =$ ***negative*** of admittance(s) connecting nodes k and ℓ, with $k \neq \ell$

$\mathbf{V}_k =$ phasor voltage at node k

$\mathbf{I}_{t_k} =$ total of phasor current sources entering node k (a negative sign applies to a current source leaving the node).

For the circuit in Fig. 7-26,

$$\mathbf{Y}_{11} = \mathbf{Y}' + \mathbf{Y}_2 + \mathbf{Y}_3$$
$$= (\mathbf{Y}' + 0.5 + 0.5) \text{ S}, \quad (7.119)$$

where $\mathbf{Y}'$ is the sum of $\mathbf{Y}_1$ and $\mathbf{Y}_C$. The rule for adding two in-series admittances is the same as that for adding two in-parallel impedances:

$$\mathbf{Y}' = \mathbf{Y}_1 \parallel \mathbf{Y}_C = \frac{\frac{1}{3} \times j\frac{1}{4}}{\frac{1}{3} + j\frac{1}{4}} = (0.12 + j0.16) \text{ S}.$$

Hence,

$$\mathbf{Y}_{11} = (1.12 + j0.16) \text{ S}.$$

Similarly,

$$\mathbf{Y}_{22} = \mathbf{Y}'' + 0.5 + 0.5$$
$$= (\mathbf{Y}_4 \parallel \mathbf{Y}_L) + 1$$
$$= \frac{0.5 \times (-j1)}{0.5 - j1} + 1 = (1.4 - j0.2) \text{ S},$$
$$\mathbf{Y}_{33} = 0.5 + 0.5 + 0.5 = 1.5 \text{ S}.$$

Also, $\mathbf{Y}_{12} = \mathbf{Y}_{21} = \mathbf{Y}_{13} = \mathbf{Y}_{31} = \mathbf{Y}_{23} = \mathbf{Y}_{32} = -0.5$ S, $\mathbf{I}_{t_1} = -6$ A, $\mathbf{I}_{t_2} = 6$ A, and $\mathbf{I}_{t_3} = -j3$ A. Entering the values

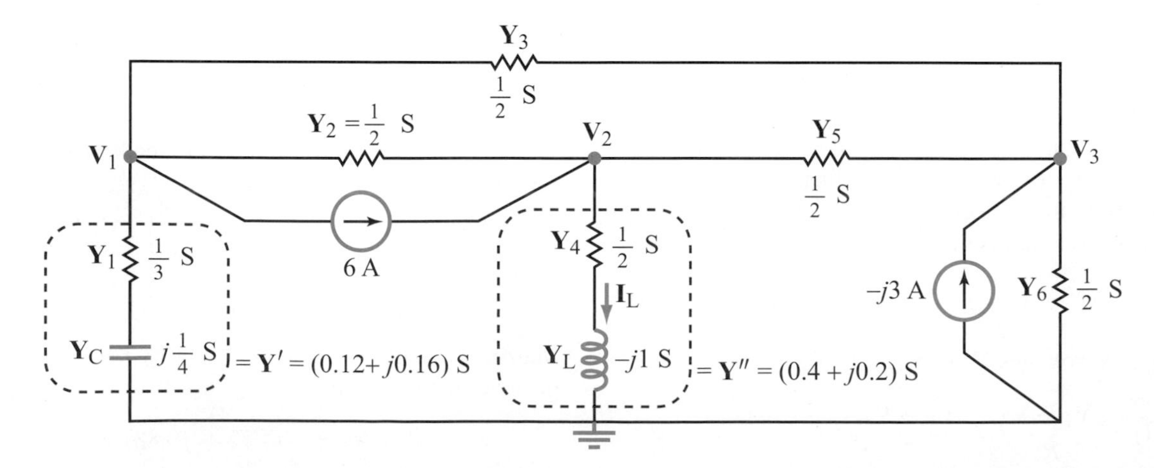

Figure 7-26: Equivalent of the circuit in Fig. 7-25, after source transformation of voltage sources into current sources and replacement of passive elements with their equivalent admittances.

of all of these quantities in Eq. (7.118) gives

$$\begin{bmatrix} (1.12+j0.16) & -0.5 & -0.5 \\ -0.5 & (1.4-j0.2) & -0.5 \\ -0.5 & -0.5 & 1.5 \end{bmatrix} \begin{bmatrix} \mathbf{V}_1 \\ \mathbf{V}_2 \\ \mathbf{V}_3 \end{bmatrix} = \begin{bmatrix} -6 \\ 6 \\ -j3 \end{bmatrix}. \tag{7.120}$$

Multiplication of both sides of Eq. (7.120) by a factor of 2 would produce exactly the matrix equation given by Eq. (7.115), as expected.

Exercise 7-15: Write down the node-voltage matrix equation for the circuit in Fig. E7.15.

Answer:

$$\begin{bmatrix} (2+j2) & -(2+j2) \\ -(2+j2) & (2-j2) \end{bmatrix} \begin{bmatrix} \mathbf{V}_1 \\ \mathbf{V}_2 \end{bmatrix} = \begin{bmatrix} 2-4e^{j60^\circ} \\ 4e^{j60^\circ} \end{bmatrix}.$$

(See ◎)

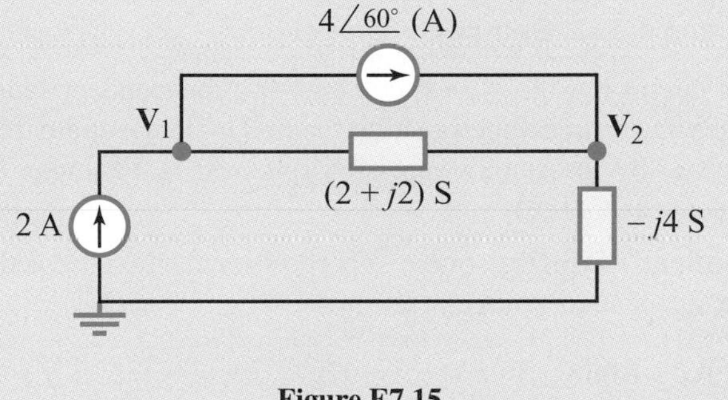

Figure E7.15

Example 7-13: Circuit with a Supernode

The circuit in Fig. 7-27, which is already in the phasor domain, contains two independent voltage sources, both oscillating at an angular frequency $\omega = 2 \times 10^3$ rad/s, and both characterized by a phase angle of 0°. Determine $i_L(t)$.

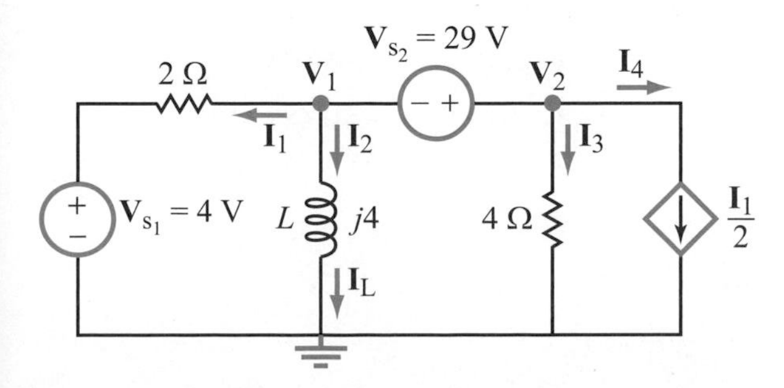

Figure 7-27: Phasor-domain circuit containing a supernode and a dependent source (Example 7-13).

Solution: Because nodes $\mathbf{V}_1$ and $\mathbf{V}_2$ are connected by a voltage source, their combination constitutes a supernode. When we apply KCL to a supernode, we simply sum all the currents leaving both of its nodes as if the two nodes are one,

$$\mathbf{I}_1 + \mathbf{I}_2 + \mathbf{I}_3 + \mathbf{I}_4 = 0,$$

or

$$\frac{\mathbf{V}_1 - 4}{2} + \frac{\mathbf{V}_1}{j4} + \frac{\mathbf{V}_2}{4} + \frac{\mathbf{I}_1}{2} = 0, \tag{7.121}$$

and we also incorporate the auxiliary equation relating the two nodes, namely

$$\mathbf{V}_2 - \mathbf{V}_1 = 29. \tag{7.122}$$

From the circuit, the current $\mathbf{I}_1$ in Eq. (7.121) is given by

$$\mathbf{I}_1 = \frac{\mathbf{V}_1 - 4}{2}. \tag{7.123}$$

Using Eqs. (7.122) and (7.123) in Eq. (7.121) and then solving for $\mathbf{V}_1$, leads to

$$\mathbf{V}_1 = -(4 + j1) \text{ V},$$

which in turn gives

$$\mathbf{I}_L = \frac{\mathbf{V}_1}{j4} = -\frac{(4+j1)}{j4} = (-0.25 + j1) = 1.03\angle 104^\circ \text{ A}.$$

With $\omega = 2 \times 10^3$ rad/s, the inductor current in the time domain is given by

$$i_L(t) = \mathfrak{Re}[\mathbf{I}_L e^{j\omega t}] = \mathfrak{Re}[1.03 e^{j104^\circ} e^{j2\times 10^3 t}]$$
$$= 1.03\cos(2\times 10^3 t + 104^\circ) \text{ A}.$$

Example 7-14: Mesh Analysis

Apply the mesh-analysis method to determine $i_L(t)$ in the circuit of Fig. 7-25 with υ_{s_1} and υ_{s_2} given by the expressions in the statement of Example 7-12.

Solution: The phasor-domain version of the circuit is shown in Fig. 7-28 with mesh currents $\mathbf{I}_1$ to $\mathbf{I}_3$. Since the circuit has no dependent sources and no independent current sources, it is suitable for application of the mesh-analysis by-inspection method. For a three-loop circuit, the phasor-domain parallel of Eq. (3.29) assumes the form:

$$\begin{bmatrix} \mathbf{Z}_{11} & \mathbf{Z}_{12} & \mathbf{Z}_{13} \\ \mathbf{Z}_{21} & \mathbf{Z}_{22} & \mathbf{Z}_{23} \\ \mathbf{Z}_{31} & \mathbf{Z}_{32} & \mathbf{Z}_{33} \end{bmatrix} \begin{bmatrix} \mathbf{I}_1 \\ \mathbf{I}_2 \\ \mathbf{I}_3 \end{bmatrix} = \begin{bmatrix} \mathbf{V}_{t_1} \\ \mathbf{V}_{t_2} \\ \mathbf{V}_{t_3} \end{bmatrix}, \tag{7.124}$$

where

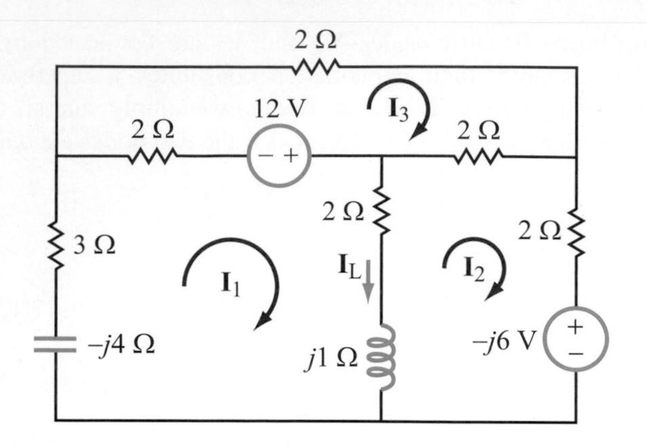

Figure 7-28: Circuit for Example 7-14.

$\mathbf{Z}_{kk}$ = sum of all impedances in loop k

$\mathbf{Z}_{k\ell}$ = $\mathbf{Z}_{\ell k}$ = ***negative*** of impedance(s) shared by loop k and ℓ, with $k \neq \ell$

$\mathbf{I}_k$ = phasor current of loop k

$\mathbf{V}_{t_k}$ = total of phasor voltage sources contained in loop k, with the polarity defined as positive if $\mathbf{I}_k$ flows from (−) to (+) through the source.

In view of these definitions, the matrix equation for the circuit in Fig. 7-28 is given by

$$\begin{bmatrix} (7-j3) & -(2+j1) & -2 \\ -(2+j1) & (6+j1) & -2 \\ -2 & -2 & 6 \end{bmatrix} \begin{bmatrix} \mathbf{I}_1 \\ \mathbf{I}_2 \\ \mathbf{I}_3 \end{bmatrix} = \begin{bmatrix} 12 \\ j6 \\ -12 \end{bmatrix}. \quad (7.125)$$

Matrix inversion leads to

$$\mathbf{I}_1 = (0.43 + j0.86) \text{ A},$$

$$\mathbf{I}_2 = (-0.38 + j1.71) \text{ A},$$

and

$$\mathbf{I}_3 = (-1.98 + j0.86) \text{ A}.$$

The current $\mathbf{I}_\text{L}$ through the inductor is given by

$$\begin{aligned} \mathbf{I}_\text{L} = \mathbf{I}_1 - \mathbf{I}_2 &= (0.43 + j0.86) - (-0.38 + j1.71) \\ &= 0.81 - j0.85 = 1.17e^{-j46.4^\circ} \text{ A}, \end{aligned} \quad (7.126)$$

and its time-domain counterpart is

$$\begin{aligned} i_\text{L}(t) = \mathfrak{Re}[\mathbf{I}_\text{L} e^{j\omega t}] &= \mathfrak{Re}[1.17e^{-j46.4^\circ} e^{j1000t}] \\ &= 1.17\cos(1000t - 46.4^\circ) \text{ A}. \end{aligned} \quad (7.127)$$

Exercise 7-16: Write down the mesh-current matrix equation for the circuit in Fig. E7.16.

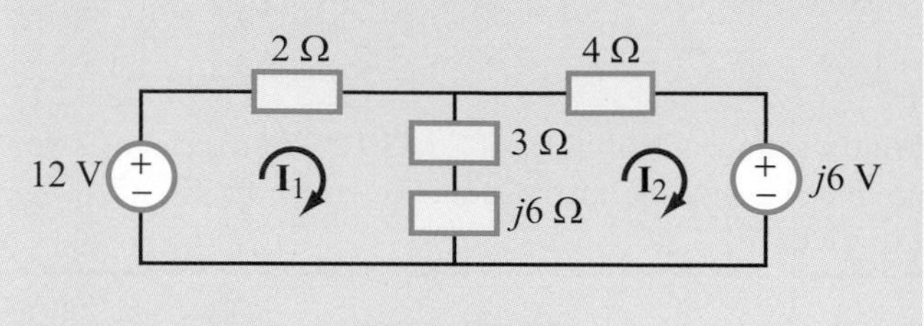

Figure E7.16

Answer:

$$\begin{bmatrix} (5+j6) & -(3+j6) \\ -(3+j6) & (7+j6) \end{bmatrix} \begin{bmatrix} \mathbf{I}_1 \\ \mathbf{I}_2 \end{bmatrix} = \begin{bmatrix} 12 \\ -j6 \end{bmatrix}.$$

(See 💿)

Example 7-15: Source Superposition

The circuit in Fig. 7-29(a) contains two independent sources. Apply the source-superposition method to demonstrate that $\mathbf{I}_\text{L}$ is given by the same expression obtained in Example 7-14, namely Eq. (7.126).

Solution: With the source-superposition method, we activate one independent source at a time.

Source 1 Alone: In part (b) of Fig. 7-29, only the 12-V source is active, and the other source has been replaced with a short circuit. The loop currents are designated $\mathbf{I}'_1$ through $\mathbf{I}'_3$, and the corresponding current through the inductor is $\mathbf{I}'_\text{L}$. Application of the mesh-current by-inspection method gives the matrix equation

$$\begin{bmatrix} (7-j3) & -(2+j1) & -2 \\ -(2+j1) & (6+j1) & -2 \\ -2 & -2 & 6 \end{bmatrix} \begin{bmatrix} \mathbf{I}'_1 \\ \mathbf{I}'_2 \\ \mathbf{I}'_3 \end{bmatrix} = \begin{bmatrix} 12 \\ 0 \\ -12 \end{bmatrix}, \quad (7.128)$$

whose inversion leads to

$$\mathbf{I}'_1 = (0.79 + j0.52) \text{ A},$$

$$\mathbf{I}'_2 = (-0.36 + j0.48) \text{ A},$$

and

$$\mathbf{I}'_3 = (-1.86 + j0.33) \text{ A}.$$

Hence,

$$\begin{aligned} \mathbf{I}'_\text{L} &= \mathbf{I}'_1 - \mathbf{I}'_2 \\ &= (0.79 + j0.52) - (-0.36 + j0.48) \\ &= (1.15 + j0.04) \text{ A}. \end{aligned} \quad (7.129)$$

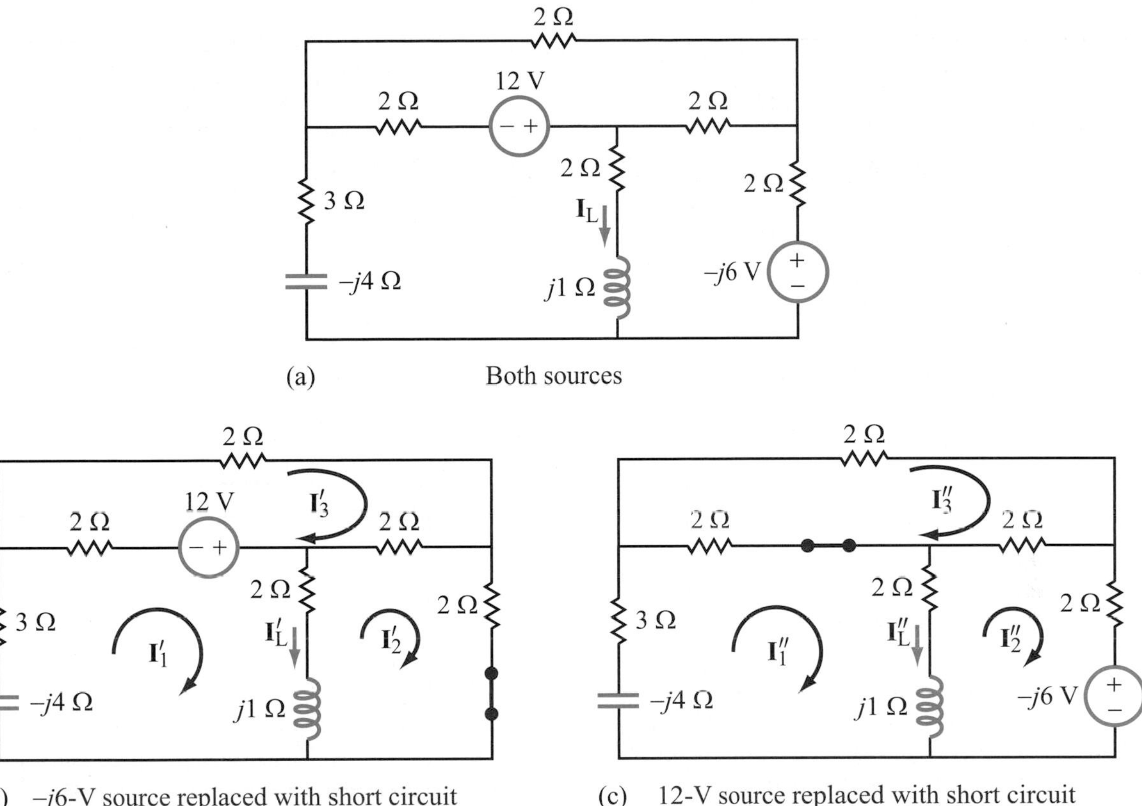

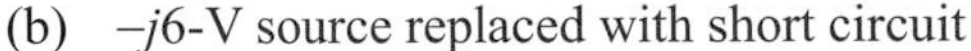
(b) –j6-V source replaced with short circuit

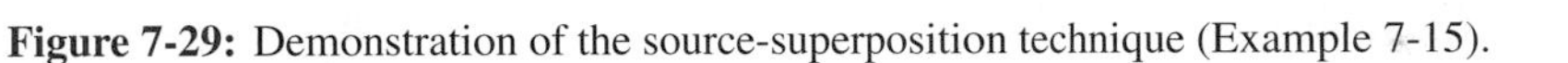
Figure 7-29: Demonstration of the source-superposition technique (Example 7-15).

Source 2 Alone: Deactivation of the 12-V source and reactivation of the $-6j$-V source produces the circuit shown in part (c) of Fig. 7-29. Now the loop currents are $\mathbf{I}''_1$, $\mathbf{I}''_2$, and $\mathbf{I}''_3$, and their matrix equation is

$$\begin{bmatrix} (7-j3) & -(2+j1) & -2 \\ -(2+j1) & (6+j1) & -2 \\ -2 & -2 & 6 \end{bmatrix} \begin{bmatrix} \mathbf{I}''_1 \\ \mathbf{I}''_2 \\ \mathbf{I}''_3 \end{bmatrix} = \begin{bmatrix} 0 \\ j6 \\ 0 \end{bmatrix}. \tag{7.130}$$

The solution of Eq. (7.130) is

$$\mathbf{I}''_1 = (-0.36 + j0.34) \text{ A},$$
$$\mathbf{I}''_2 = (-0.02 + j1.23) \text{ A},$$
$$\mathbf{I}''_3 = (-0.13 + j0.53) \text{ A},$$

and

$$\mathbf{I}''_L = \mathbf{I}''_1 - \mathbf{I}''_2 = -0.36 + j0.34 - (-0.02 + j1.23)$$
$$= (-0.34 - j0.89) \text{ A}.$$

Total Solution: Given $\mathbf{I}'_L$ due to source 1 alone and $\mathbf{I}''_L$ due to source 2 alone, the total current due to both sources simultaneously is

$$\mathbf{I}_L = \mathbf{I}'_L + \mathbf{I}''_L = (1.15 + j0.04) + (-0.34 - j0.89)$$
$$= (0.81 - j0.85) \text{ A}, \tag{7.131}$$

which is identical with the expression given by Eq. (7.126).

Example 7-16: Thévenin Approach

For the circuit of Fig. 7-30, (a) obtain its Thévenin equivalent at terminals (a, b), as if the inductor were an external load, and (b) then use the Thévenin circuit to determine $\mathbf{I}_L$.

Solution:

(a) We will apply the open-circuit/short-circuit method to determine the values of $\mathbf{V}_{Th}$ and $\mathbf{Z}_{Th}$ of the Thévenin equivalent circuit.

Open-Circuit Voltage: With the inductor replaced with an open circuit in Fig. 7-30(b), the matrix equation for loop currents $\mathbf{I}_1$ and $\mathbf{I}_2$ is

$$\begin{bmatrix} (9-j4) & -4 \\ -4 & 6 \end{bmatrix} \begin{bmatrix} \mathbf{I}_1 \\ \mathbf{I}_2 \end{bmatrix} = \begin{bmatrix} 12+j6 \\ -12 \end{bmatrix}, \tag{7.132}$$

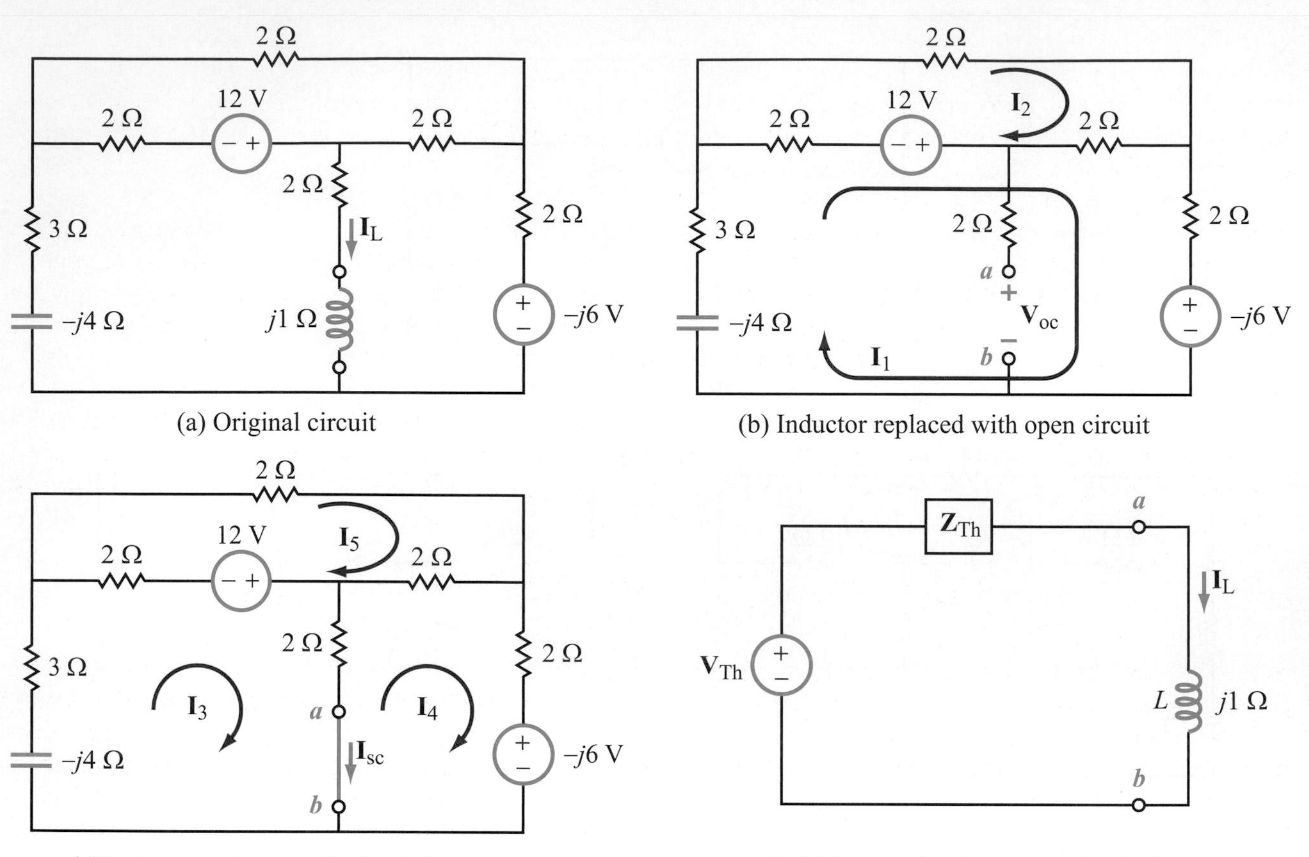

(a) Original circuit

(b) Inductor replaced with open circuit

(c) Inductor replaced with short circuit

(d) Thévenin circuit connected to inductor

Figure 7-30: After determining the open-circuit voltage in part (b) and the short-circuit current in part (c), the Thévenin equivalent circuit is connected to the inductor to determine $\mathbf{I}_L$.

and its inversion gives

$$\mathbf{I}_1 = (0.02 + j0.96) \text{ A} \quad \text{and} \quad \mathbf{I}_2 = (-1.98 + j0.64) \text{ A}.$$

With $\mathbf{I}_1$ and $\mathbf{I}_2$ known, application of KVL around the loop containing the $-j6$ V source leads to

$$\begin{aligned} \mathbf{V}_{Th} = \mathbf{V}_{oc} &= 2(\mathbf{I}_1 - \mathbf{I}_2) + 2\mathbf{I}_1 - j6 \\ &= 4I_1 - 2I_2 - j6 = (4.06 - j3.44) \text{ V}. \end{aligned} \quad (7.133)$$

Short-Circuit Current: In part (c) of Fig. 7-30, the inductor has been replaced with a short circuit. The matrix equation for loop currents $\mathbf{I}_3$ to $\mathbf{I}_5$ is given by

$$\begin{bmatrix} (7-j4) & -2 & -2 \\ -2 & 6 & -2 \\ -2 & -2 & 6 \end{bmatrix} \begin{bmatrix} \mathbf{I}_3 \\ \mathbf{I}_4 \\ \mathbf{I}_5 \end{bmatrix} = \begin{bmatrix} 12 \\ j6 \\ -12 \end{bmatrix}. \quad (7.134)$$

Solution of Eq. (7.134) gives

$$\mathbf{I}_3 = (0.44 + j0.95) \text{ A},$$

$$\mathbf{I}_4 = (-0.53 + j1.60) \text{ A},$$

and

$$\mathbf{I}_5 = (-2.03 + j0.85) \text{ A},$$

from which we have

$$\begin{aligned} \mathbf{I}_{sc} = \mathbf{I}_3 - \mathbf{I}_4 &= (0.44 + j0.95) - (-0.53 + j1.60) \\ &= (0.97 - j0.65) \text{ A}. \end{aligned} \quad (7.135)$$

Given $\mathbf{V}_{oc}$ and $\mathbf{I}_{sc}$, it follows that

$$\mathbf{Z}_{Th} = \frac{\mathbf{V}_{oc}}{\mathbf{I}_{sc}} = \frac{4.06 - j3.44}{0.97 - j0.65} = (4.53 - j0.51)\ \Omega. \quad (7.136)$$

(b) Having established $\mathbf{V}_{Th}$ and $\mathbf{Z}_{Th}$, we now connect the Thévenin equivalent circuit to the inductor at terminals (a, b) as shown in Fig. 7-30(d). The current $\mathbf{I}_L$ is simply

$$\begin{aligned} \mathbf{I}_L = \frac{\mathbf{V}_{Th}}{\mathbf{Z}_{Th} + j1} &= \frac{4.06 - j3.44}{4.53 - j0.51 + j1} \\ &= (0.80 - j0.85) \text{ A}. \end{aligned} \quad (7.137)$$

Technology Brief 16: Crystal Oscillators

Circuits that produce well-defined ac oscillations are fundamental to many applications: frequency generators for radio transmitters, filters for radio receivers, and processor clocks, among many. An ***oscillator*** is a circuit that takes a dc input and produces an ac output at a desired frequency. Temperature stability, long lifetime, and little frequency drift over time are important considerations when designing oscillators.

In Exercise 6.14 in Chapter 6, we saw how a circuit consisting of an inductor and a capacitor will resonate at a specific natural frequency $\omega_0 = 1/\sqrt{LC}$. In these circuits, energy is stored in an electric field (capacitor) and a magnetic field (inductor). Once energy is introduced into the circuit (for example, by applying an initial voltage to the capacitor), it will begin to flow back and forth between the two components; this constant conversion gives rise to oscillations in voltage and current at the resonant frequency. In an ideal circuit with no dissipation (no resistor), the oscillations will continue at this one frequency forever.

Making oscillating circuits from individual inductor and capacitor components, however, is relatively impractical and yields devices with poor reproducibility, high temperature drift (i.e., the resonant frequency changes with the temperature surrounding the circuit), and poor overall lifetime. Since the early part of the 20th century, resonators have been made in a completely different way, namely by using tiny, mechanically resonating pieces of quartz glass.

Quartz Crystals and Piezoelectricity

In 1880, the Curie brothers demonstrated that certain crystals—such as ***quartz***, topaz, and tourmaline—become electrically polarized when subjected to mechanical stress. That is, such a crystal exhibits a voltage across it if compressed, and a voltage of opposite polarity if stretched. The converse property, namely that if a voltage is applied across a crystal it will change its shape (compress or stretch), was predicted a year later by Gabriel Lippman (who received the 1908 Nobel prize in physics for producing the first color photographic plate). Collectively, these bidirectional properties of crystals are known as ***piezoelectricity***. Piezoelectric crystals are used in microphones to convert

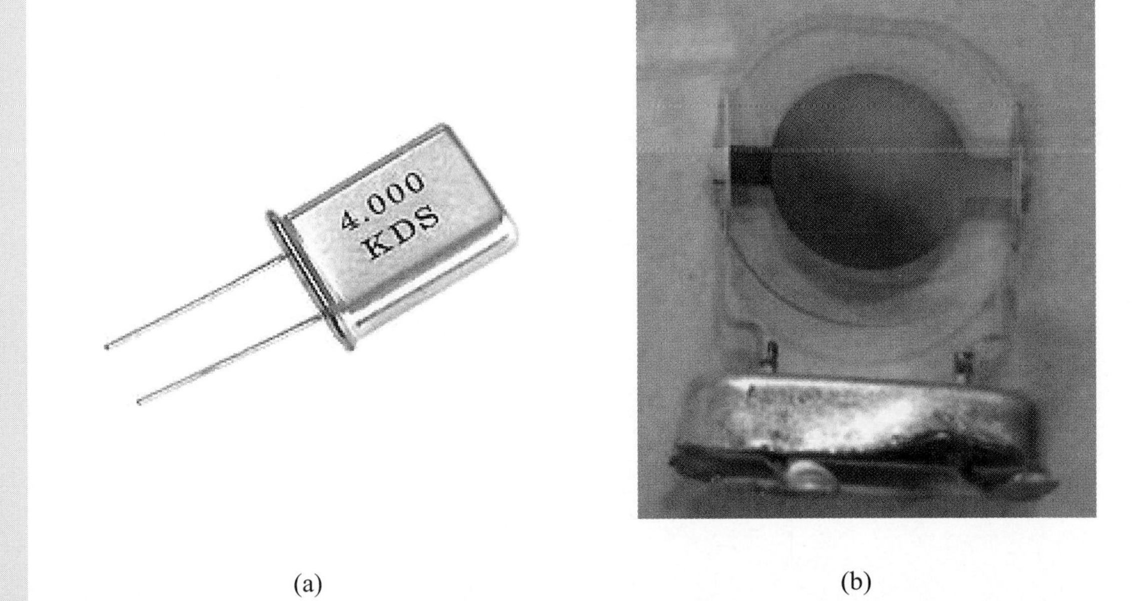

Figure TF16-1: (a) An HC49 crystal package (crystal inside); (b) an opened HC49 crystal package showing the transparent quartz plate and the metal electrodes.

mechanical vibrations of the crystal surface, caused by acoustic waves, into electrical signals, and the converse is used in loudspeakers. Piezoelectricity can also be applied to make a quartz crystal to resonate. If a voltage of the proper polarity is applied across one of the principal axes of the crystal, it will shrink along the direction of that axis. Upon removing the voltage, the crystal will try to restore its shape to its original unstressed state by stretching itself, but its stored compression energy is sufficient to allow it to stretch beyond the unstressed state, thereby generating a voltage whose polarity is opposite of that of the original voltage that was used to compress it. This induced voltage will cause it to shrink, and the process will continue back and forth until the energy initially introduced by the external voltage source is totally dissipated. The behavior of the crystal is akin to an underdamped RLC circuit.

In addition to crystals, some metals and ceramics are also used for making oscillators. Because the resonant frequency can be chosen by specifying the type of material and its shape, such oscillators are easy to manufacture in large quantities, and their oscillation frequencies can be designed with a high degree of precision. Moreover, quartz crystals have good temperature performance, which means that they can be used in many applications without the need for temperature compensation, including in clocks, radios, and cellphones.

Crystal Equivalent Circuit and Oscillator Design

The electrical behavior of a quartz crystal can be modeled as a series RLC circuit (L_S, C_S, R_S) in parallel with a shunt capacitor (C_O). The RLC circuit models the fundamental oscillator behavior with dissipation. The shunt capacitor is mostly due to the capacitance between the two plates that actuate the quartz crystal. Figure TF16-2 shows the circuit symbol and equivalent circuit with sample values for a 5-MHz commercial crystal.

The crystal is, of course, not sufficient to produce a continuous oscillating waveform; we need to excite the circuit and keep it running. A common way to do this is to insert the crystal in the positive feedback path of an amplifier

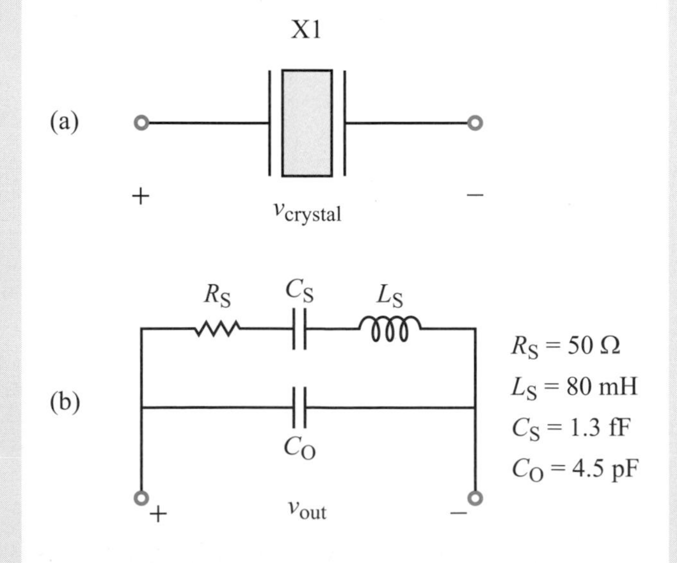

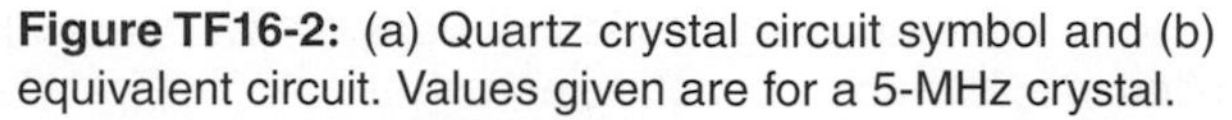
Figure TF16-2: (a) Quartz crystal circuit symbol and (b) equivalent circuit. Values given are for a 5-MHz crystal.

(Fig. TF16-3). The amplifier, of course, is supplied with dc power (V_{CC}^{+} and V_{CC}^{-}). Note that no input signal is applied to the circuit. Initially, the output generates no oscillations; however, any noise at v_{out} that is at the resonant frequency of X1 will be fed back to the input and amplified. This positive feedback will quickly ramp up the output so that it is oscillating at the resonant frequency of the crystal. A negative feedback loop is also commonly used to control the overall gain and prevent the circuit from clipping the signal against the op amp's supply voltages V_{CC}^{+} and V_{CC}^{-}.

In order to oscillate continuously, a circuit must meet the following two ***Barkhausen criteria***:

(a) The gain of the circuit must be greater than 1. (This makes sense, for otherwise the signal neither get amplified nor establish a resonating condition.)

(b) The phase shift from the input to the output, then across the feedback loop to the input must be 0. (This also makes sense, since if there is non-zero phase shift, the signals will destructively interfere and the oscillator will not be able to start up.)

Advances in Resonators and Clocks

As good as quartz resonators are, even the best among them will drift in frequency by 0.01 ppm per year as a result of aging of the crystal. If the oscillator is being used to keep time (as in your digital watch), this dictates how many seconds (or fractions thereof) the clock will lose per year. Put differently, this drift puts a hard limit on how long a clock can run without calibration. The same phenomenon limits how well independent clocks can stay synchronized with each other. Atomic clocks provide an extra level of precision by basing their oscillations on atomic transitions; these clocks are accurate to about 10^{-9} seconds per day. Recently, a chip-scale version of an atomic clock was demonstrated by the National Institute for Standards and Technology (NIST); it consumes 75 mW and was the size of a grain of rice (10 mm^3). Other recent efforts for making oscillators for communication have focused on replacing the quartz crystal with a micromechanical resonator of the type discussed in Technology Brief 12: Sensors and Actuators on page 269.

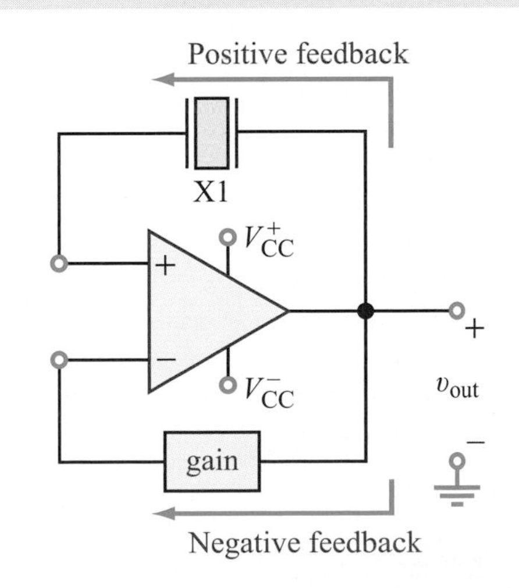

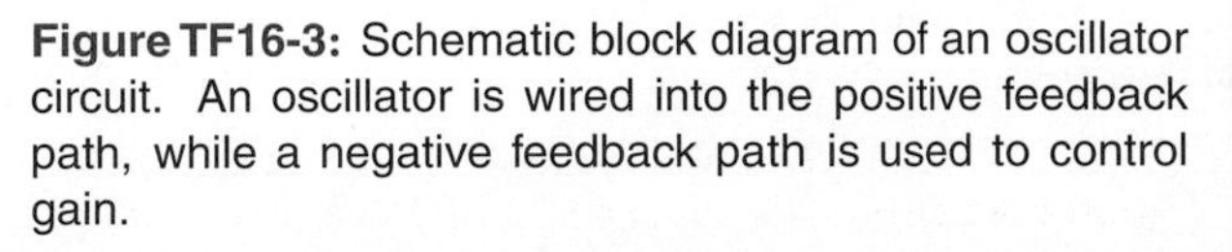

Figure TF16-3: Schematic block diagram of an oscillator circuit. An oscillator is wired into the positive feedback path, while a negative feedback path is used to control gain.

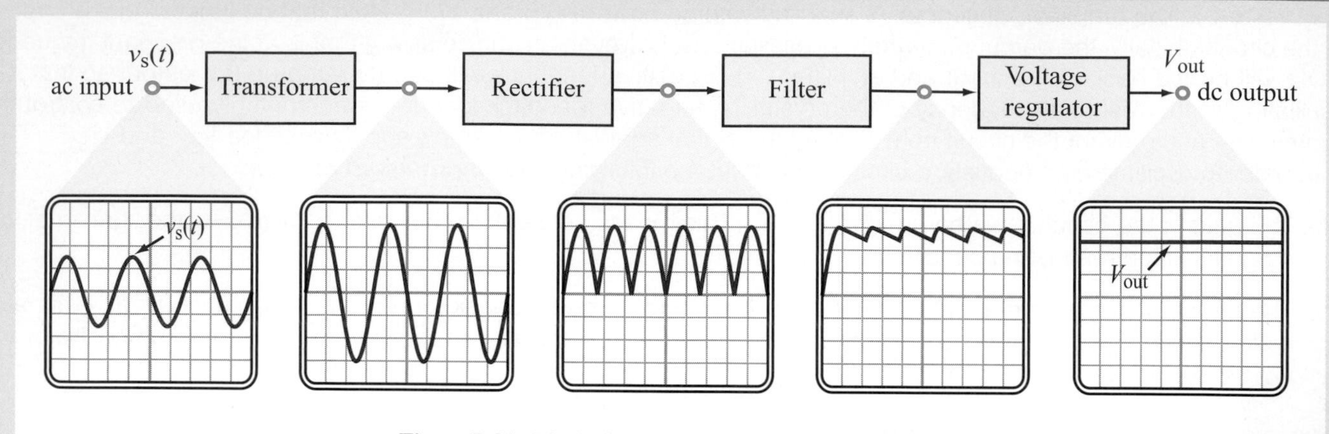

Figure 7-31: Block diagram of a basic dc power supply.

7-10 Application Note: Power-Supply Circuits

Systems composed of one or more electronic circuits usually contain power-supply circuits that convert the ac power available from the wall outlet into dc power, thereby providing the internal dc voltages required for proper operation of the electronic circuits. Most dc power supplies consist of the four subsystems diagrammed in Fig. 7-31. The input is an ac voltage $v_s(t)$ of amplitude V_s and angular frequency ω, and the final output is a dc voltage V_{out}. Our plan in this section is to describe the operation of each of the intermediate stages, and then connect them all together.

7-10.1 Transformers

A transformer consists of two inductors called ***windings***, that are in close proximity to each other but not connected electrically. The two windings are called the ***primary*** and the ***secondary***, as shown in Fig. 7-32. Even though the two windings are isolated electrically—meaning that no current flows between them—when an ac voltage is applied to the primary, it creates a magnetic flux that permeates both windings through a common ***core***, inducing an ac voltage in the secondary.

The ***transformer*** gets its name from the fact that it is used to transform currents, voltages, and impedances between its primary and secondary circuits.

The key parameter that determines the relationships between the primary and the secondary is the ***turns ratio*** N_1/N_2, where N_1 is the number of turns in the primary coil and N_2 is the number of turns in the secondary. An additionally important attribute is the direction of the primary winding, relative to that of the secondary, around the common magnetic core. The relative directions determine the voltage polarity and current direction at the secondary, relative to those at the primary. To distinguish between the two cases, a dot usually is placed at one or the other end of each winding, as shown in Fig. 7-32. For the ideal transformer, voltage v_2 at the secondary side is related to voltage v_1 at the primary side by

$$\frac{v_2}{v_1} = \frac{N_2}{N_1}, \tag{7.138}$$

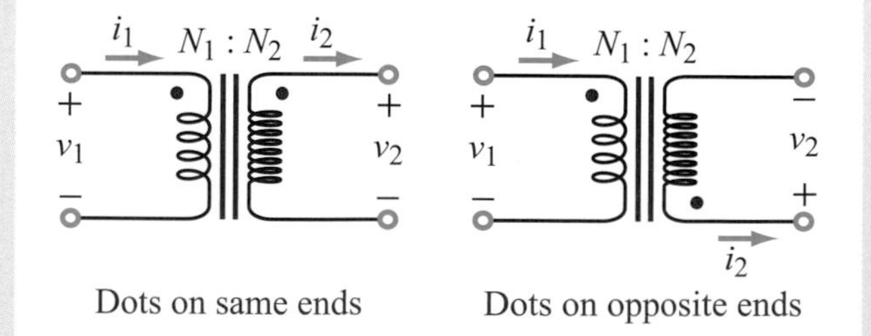

Figure 7-32: Schematic symbol for an ideal transformer. Note the reversal of the voltage polarity and current direction when the dot location at the secondary was moved from the top end of the coil to the bottom end. For both configurations:

$$\frac{v_2}{v_1} = \frac{N_2}{N_1} \qquad \frac{i_2}{i_1} = \frac{N_1}{N_2}$$

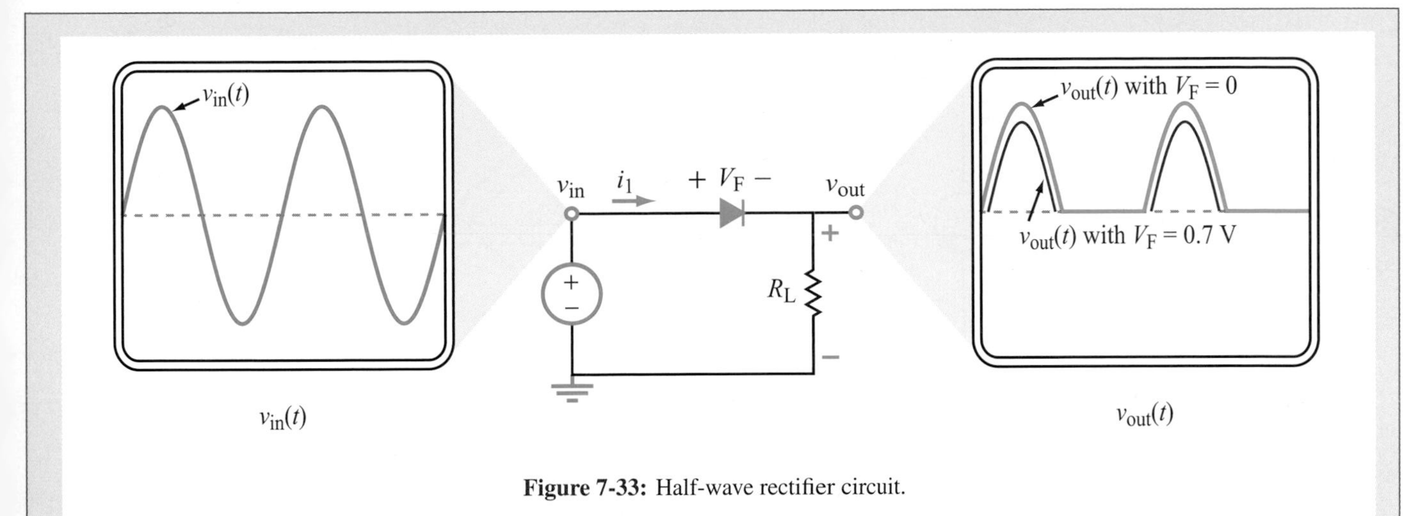

Figure 7-33: Half-wave rectifier circuit.

where *the polarities of* v_1 *and* v_2 *are defined such that their (+) terminals are at the ends with the dots.* In an ideal transformer, no power is lost in the core, so all of the power supplied by a source to its primary coil is transferred to the load connected at its secondary side. Thus, $p_1 = p_2$, and since $p_1 = i_1 v_1$ and $p_2 = i_2 v_2$, it follows that

$$\frac{i_2}{i_1} = \frac{N_1}{N_2}, \qquad (7.139)$$

with i_1 *always defined in the direction towards the dot* on the primary side and i_2 *defined in the direction away from the dot* on the secondary side. If $N_2/N_1 > 1$, the transformer is called a ***step-up transformer*** because it transforms v_1 to a higher voltage, and if $N_2/N_1 < 1$, it is called a ***step-down transformer***. Most office and household electronic gadgets (such as telephones, clocks, radios, and answering machines) require dc voltages that are on the order of volts (or at most a few tens of volts) and certainly much smaller than the voltage level available at the wall outlet. The transformer part of power supplies built into such gadgets is invariably a step-down transformer.

Review Question 7-17: In a transformer, how are the voltage polarities and current directions defined relative to the dots on the primary and secondary windings?

Review Question 7-18: For an ideal transformer, how is power p_2 related to power p_1?

7-10.2 Rectifiers

A rectifier is a diode circuit that converts an ac waveform into one that is either always positive or always negative, depending on the direction(s) of the diode(s). Power supplies usually use a ***bridge rectifier***, but to appreciate how it functions, we will first consider the simple single-diode rectifier circuit shown in Fig. 7-33. As discussed in Section 2-7.2, a diode is modeled by a practical response that allows current to flow through it in the direction shown in the figure if and only if the voltage across it is greater than a threshold value known as the ***forward-bias voltage*** V_F. That is, for the circuit in Fig. 7-33, the output voltage across the load resistor is given by

$$v_{out} = \begin{cases} v_{in} - V_F & \text{if } v_{in} \geq V_F, \\ 0 & \text{if } v_{in} \leq V_F. \end{cases} \qquad (7.140)$$

For an ideal diode with $V_F = 0$, the output waveform is identical with the input waveform for the half cycles during which v_{in} is positive, and the output is zero when v_{in} is negative. In the case of a real diode with $V_F \simeq 0.7$ V, the peak amplitude of the output is smaller than that of the input by 0.7 V. Because the output waveform essentially replicates only the positive half cycles of the input waveform (with a negative shift equal to V_F), the circuit of Fig. 7-33 is called a ***half-wave rectifier***.

Next, we consider the ***bridge-rectifier*** circuit of Fig. 7-34. The bridge rectifier uses four diodes. During the positive half cycle of $v_{in}(t)$, two of the diodes conduct, and the other

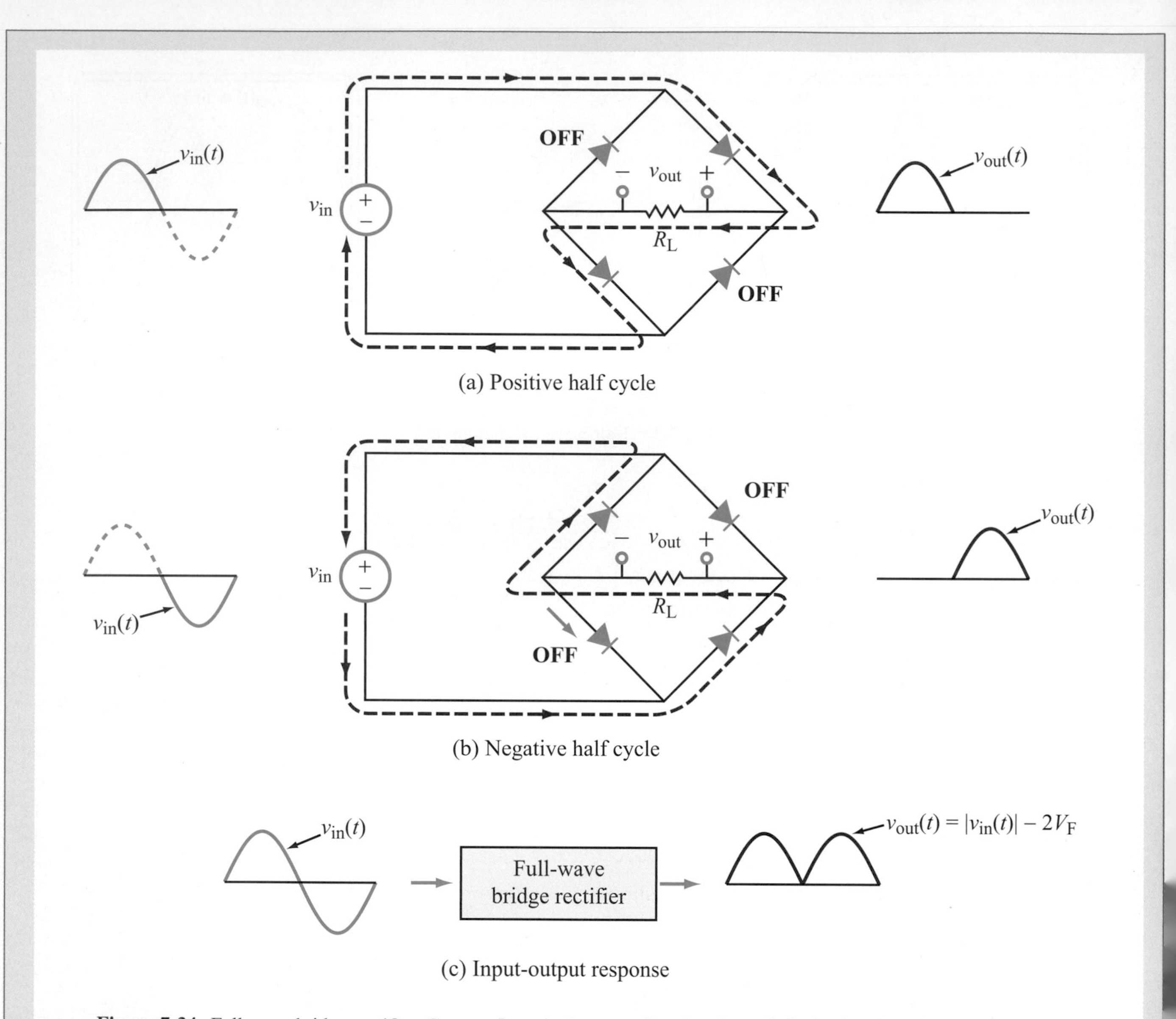

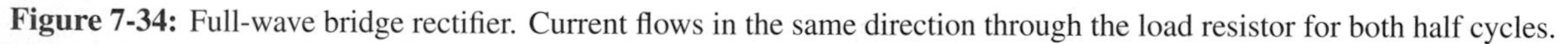

Figure 7-34: Full-wave bridge rectifier. Current flows in the same direction through the load resistor for both half cycles.

two are OFF. The reverse happens during the second half cycle, but the direction of the current through R_L is the same during both half cycles. Consequently, the output waveform essentially is equivalent to taking the absolute value of the input waveform (if V_F is so small relative to the peak value as to be neglected).

Exercise 7-17: Suppose the input voltage in the circuit of Fig. 7-34 is a 10-V amplitude square wave. What would the output look like?

Answer: 8.6-V dc. (See CD)

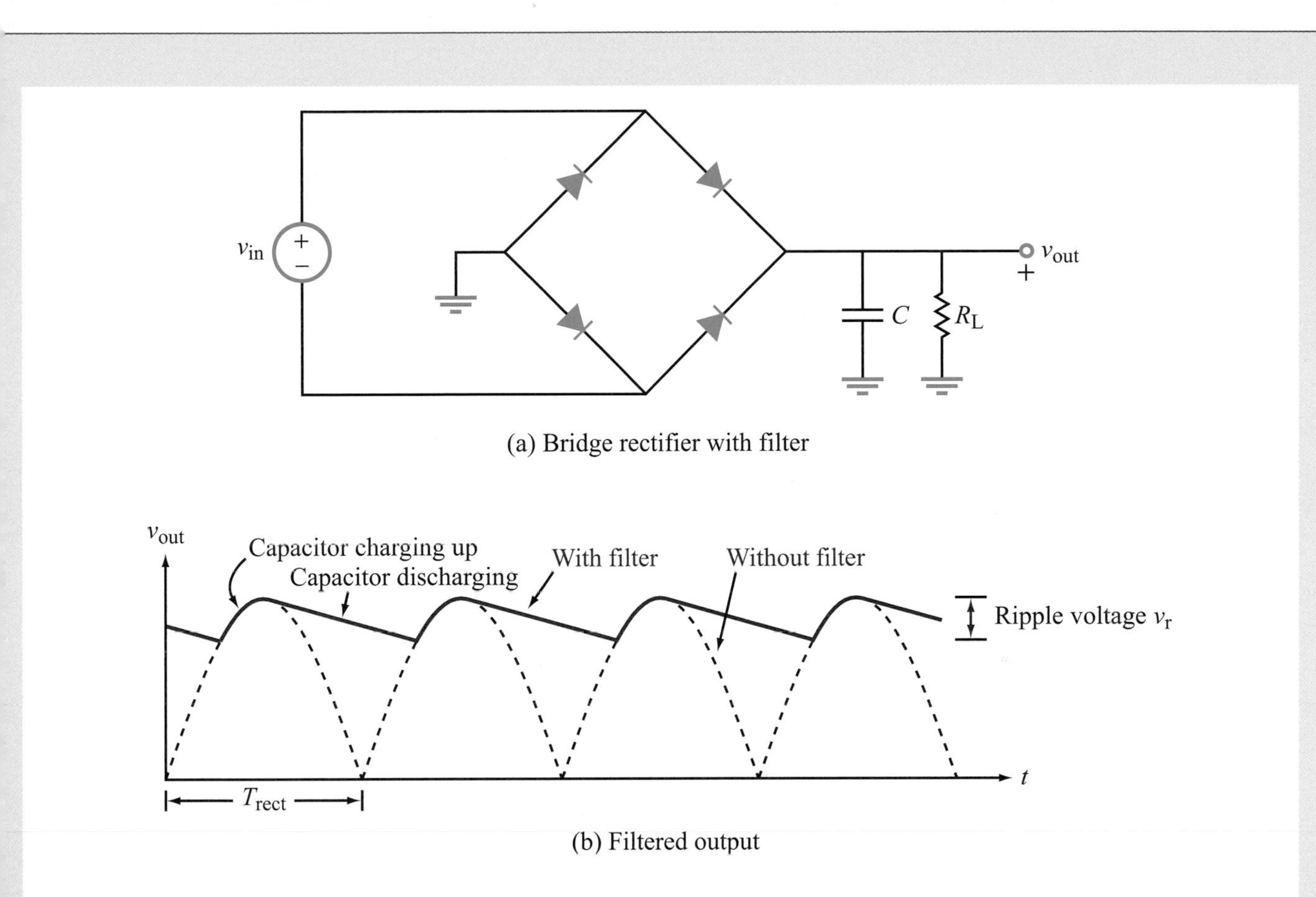

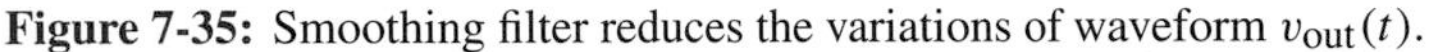
Figure 7-35: Smoothing filter reduces the variations of waveform $v_{out}(t)$.

7-10.3 Smoothing Filters

So far, we have examined two of the four subcircuits of the dc power supply. The transformer serves to adjust the amplitude of the ac signal to a level close to the desired dc voltage level of the final output. The bridge rectifier converts the ac signal into an all-positive waveform. Next, we need to reduce the variations of the full-wave rectified waveform to bring it to as close to a constant level as possible. We accomplish this goal by subjecting the full-wave rectified waveform to a smoothing filter. This is accomplished easily by adding a capacitor C in parallel with the load resistor. The modified circuit is shown in Fig. 7-35(a), and the associated output waveform is displayed in Fig. 7-35(b). The capacitor is a storage device that goes through partial charging-up and discharging-down cycles. During the charging-up period, the ***upswing time constant*** of the circuit is given by

$$\begin{aligned} \tau_{up} &= (2R_D \parallel R_L)C \\ &\simeq 2R_DC \qquad \text{if } R_L \gg R_D, \end{aligned} \tag{7.141}$$

where R_D is the diode resistance. Typically, R_D is on the order of ohms and R_L is on the order of kiloohms, so the approximation given by Eq. (7.141) is quite reasonable. In the absence of the capacitor in the circuit, R_D usually is ignored because it is in series with a much larger resistance, R_L. Adding a capacitor, however, creates an RC circuit in which R is the parallel combination of R_D and R_L, placing R_D in a controlling position.

During the discharging period, the diode turns off, and the capacitor discharges through R_L alone. Consequently, the ***downswing time constant*** involves R_L and C only,

$$\tau_{dn} = R_LC. \tag{7.142}$$

For a specified value of the diode resistance R_D, we can choose the values of R_L and C so that τ_{up} is short and τ_{dn} is long—both relative to the period of the rectified waveform—thereby realizing a fast response on the upswing part and a very slow response on the downswing part. In practice, it is possible to generate an approximately constant dc voltage with a ripple component on the order of 1 to 10 percent of its average value (Fig. 7-35(b)).

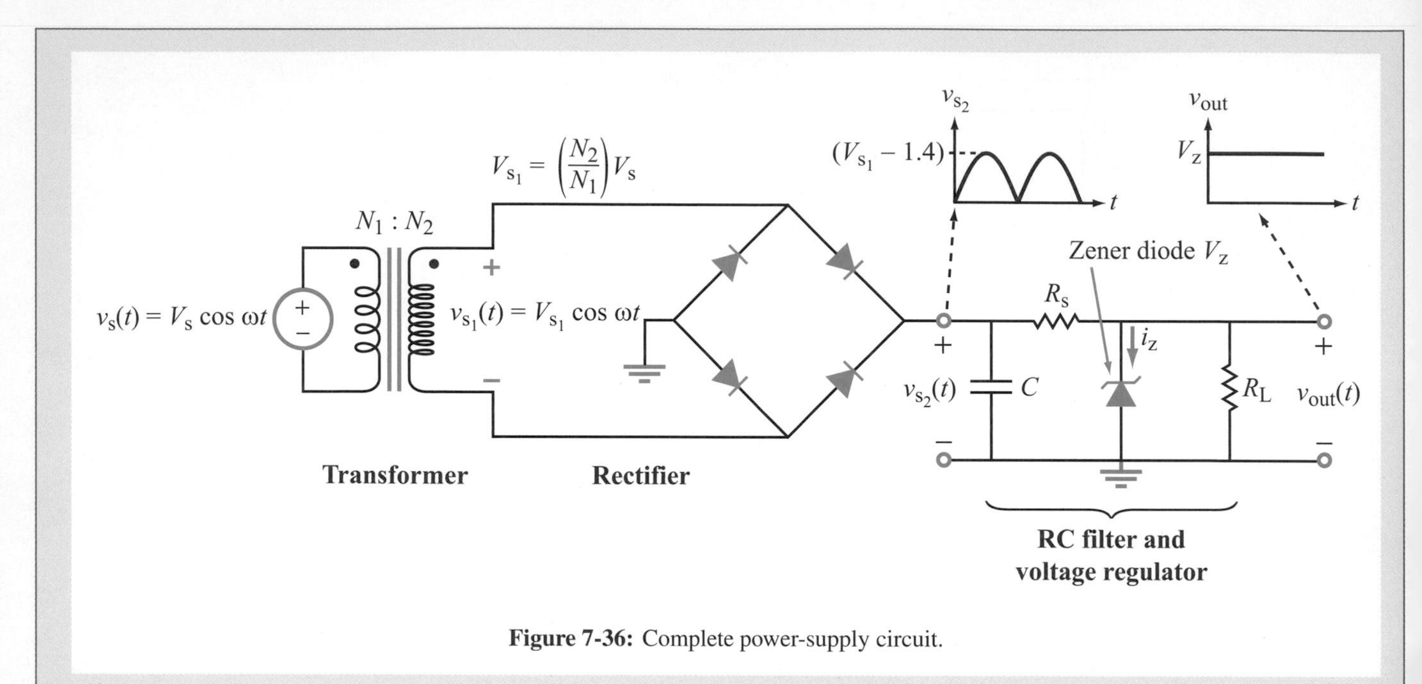

Figure 7-36: Complete power-supply circuit.

Example 7-17: Filter Design

If the bridge rectifier circuit of Fig. 7-35(a) has a 60-Hz ac input signal, determine the values of R_L and C that would result in $\tau_{up} = T_{rect}/12$ and $\tau_{dn} = 12T_{rect}$, where T_{rect} is the period of the *rectified* waveform. Assume $R_D = 5\ \Omega$.

Solution: If the frequency of the original ac signal is 60 Hz, the frequency of the rectified waveform is 120 Hz. Hence, the period of the rectified waveform is

$$T_{rect} = \frac{1}{120} = 8.33 \text{ ms},$$

and the corresponding design specifications are

$$\tau_{up} = \frac{T_{rect}}{12} = 0.69 \text{ ms}, \quad \text{and} \quad \tau_{dn} = 12T_{rect} = 100 \text{ ms}.$$

Application of Eq. (7.141) leads to

$$\tau_{up} \simeq 2R_D C$$

or

$$C = \frac{\tau_{up}}{2R_D} = \frac{0.69 \times 10^{-3}}{2 \times 5} = 69\ \mu\text{F}.$$

With the value of C known, application of Eq. (7.142) gives

$$R_L = \frac{\tau_{dn}}{C} = \frac{100 \times 10^{-3}}{69 \times 10^{-6}} = 1.45\ \text{k}\Omega.$$

7-10.4 Voltage Regulator

The circuit shown in Fig. 7-36 includes all of the power-supply subcircuits we have discussed thus far, plus two additional elements, namely series resistance R_s and a ***zener diode***. When operated in reverse breakdown, the zener diode maintains the voltage across it at a constant level V_z—so long as the current i_z passing through it remains between certain limits. Since the diode is connected in parallel with R_L, the output voltage becomes equal to the ***zener voltage*** V_z, and the effective time constant of the smoothing filter becomes $\tau = R_s C$. It is worth noting that the addition of the zener diode reduces the ***peak-to-peak ripple voltage*** V_r (Fig. 7-35(b)) at the output of the RC filter by about an order of magnitude. An approximate expression for the peak-to-peak ripple voltage with the zener diode in place is given by

$$V_r = \frac{[(V_{s_1} - 1.4) - V_z]T_{rect}}{R_s C} \times \frac{(R_z \parallel R_L)}{R_s + (R_z \parallel R_L)}, \quad (7.143)$$

where V_{s_1} is the amplitude of the ac signal at the output of the transformer (Fig. 7-36), the factor 1.4 V accounts for the voltage drop across a pair of diodes in the rectifier, V_z is the manufacturer-rated zener voltage for the specific model used in the circuit, T_{rect} is the period of the rectified waveform, and R_z is the manufacturer specified value of the ***zener-diode resistance***.

Example 7-18: Power-Supply Design

A power supply with the circuit configuration shown in Fig. 7-36 has the following specifications: the input voltage is 60 Hz with an rms amplitude $V_{\text{rms}} = 110$ V where $V_{\text{rms}} = V_s/\sqrt{2}$ (the rms value of a sinusoidal function is discussed in Chapter 8), $N_1/N_2 = 5$, $C = 2$ mF, $R_s = 50\ \Omega$, $R_L = 1\ \text{k}\Omega$, $V_z = 24$ V, and $R_z = 20\ \Omega$. Determine v_{out}, the ripple voltage, and the ripple fraction relative to v_{out}.

Solution:
At the secondary side of the transformer,

$$v_{s_1}(t) = \left(\frac{N_2}{N_1}\right)(V_s \cos 377t)$$
$$= \frac{1}{5} \times 110\sqrt{2} \cos 377t = 31.11 \cos 377t \text{ V}.$$

Hence, $V_{s_1} = 31.11$ V, which is greater than the zener voltage $V_z = 24$ V.

The output voltage is then

$$v_{\text{out}} = V_z = 24 \text{ V}.$$

In Example 7-17, we established that $T_{\text{rect}} = 8.33$ ms. Also,

$$R_z \parallel R_L = \frac{20 \times 1000}{20 + 1000} = 19.6\ \Omega.$$

Application of Eq. (7.143) gives

$$V_r = \frac{[(V_{s_1} - 1.4) - V_z]T_{\text{rect}}}{R_s C} \times \frac{(R_z \parallel R_L)}{R_s + (R_z \parallel R_L)}$$
$$= \frac{[(31.11 - 1.4) - 24]}{50 \times 2 \times 10^{-3}} (8.33 \times 10^{-3}) \times \frac{19.6}{50 + 19.6}$$
$$= 0.13 \text{ V (peak-to-peak)}.$$

Hence,

$$\text{Ripple fraction} = \frac{(V_r/2)}{V_z} = \frac{0.13/2}{24} = 0.0027,$$

which represents a relative variation of less than ± 0.3 percent.

7-11 Multisim Analysis of ac Circuits

Even though we usually treat the wires in a circuit as ideal short circuits, in reality a wire has a small but non-zero resistance. Also, as noted earlier in Section 5-7.1, when two wires are in close proximity to one another, they form a non-zero capacitor. A pair of parallel wires on a circuit board is modeled as a distributed transmission line with each small length segment ℓ represented by a series resistance R and a shunt capacitance C, as depicted by the circuit model shown in Fig. 7-37. For a parallel-wire segment of length ℓ, R and C are given by

$$R = \frac{2\ell}{\pi a^2 \sigma} \quad \text{(low-frequency approximation)} \quad (a\sqrt{f\sigma} \leq 500), \tag{7.144a}$$

$$R = \sqrt{\frac{\pi f \mu}{\sigma}} \left(\frac{\ell}{\pi a}\right) \quad \text{(high-frequency approximation)} \quad (a\sqrt{f\sigma} \geq 1250), \tag{7.144b}$$

and

$$C = \frac{\pi \epsilon \ell}{\ln(d/a)} \quad \text{for } (d/2a)^2 \gg 1, \tag{7.144c}$$

where a is the wire radius, d is the separation between the wires, f is the frequency of the signal propagating along the wires, μ and σ are respectively the magnetic permeability and conductivity of the wire material, and ϵ is the permittivity of the material between the two wires. Note that R represents the resistance of both wires. There is actually a third distributed element to consider in the general case of a transmission line: the distributed inductance. This inductance is placed in series with the resistance R of each segment. It arises because current flowing through the transmission-line wires gives rise to a magnetic field around the wires and, hence, an inductance (as discussed in Section 5-3). However, modeling the behavior of a transmission line with all three components is rather complex. So, for the purposes of this section, we will ignore the inductance altogether so that we may illustrate the performance of an RC transmission line using Multisim. Keeping this in mind, the distributed model shown in Fig. 7-37 allows us to represent the wires by a series of cascaded RC circuits. For the model to faithfully represent the behavior of the real two-wire configuration, each RC stage should represent a physical length ℓ that is no longer than a fraction (≈ 10 percent) of the distance that the signal travels during one period of the signal frequency. Thus, ℓ should be on the order of

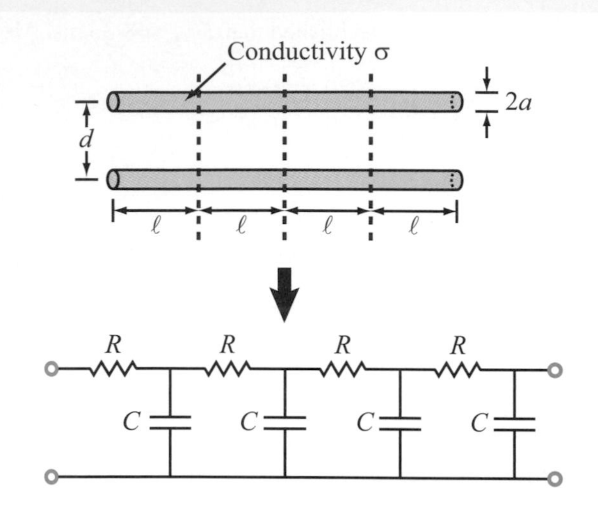

Figure 7-37: Distributed impedance model of two-wire transmission line.

$$\ell \leq \frac{u_{\mathrm{p}} T}{10} \simeq \frac{c}{10 f}, \tag{7.145}$$

where u_{p} is the signal velocity along the wires, which is on the order of the velocity of light by $c = 3 \times 10^8$ m/s, and the period T is related to the frequency f by $T = 1/f$. For example, if the signal frequency is 1 GHz ($= 10^9$ Hz), then ℓ should be on the order of

$$\ell \simeq \frac{c}{10 f} = \frac{3 \times 10^8}{10 \times 10^9} = 3 \text{ cm},$$

and if the total length of the parallel wires is $\ell_{\mathrm{t}} = 15$ cm, then their transmission-line equivalent circuit should consist of n sections with

$$n = \frac{\ell_{\mathrm{t}}}{\ell} = \frac{15 \text{ cm}}{3 \text{ cm}} = 5.$$

We will now use Multisim to simulate such a transmission line.

Example 7-19: Transmission-Line Simulation

A pair of parallel wires made of a conducting material with conductivity $\sigma = 1.9 \times 10^5$ S/m is used to carry a 1-GHz square-wave signal between two circuits on a circuit board. The wires are 15 cm in length and separated by 1 mm, and their radii are 0.1 mm. (a) Develop a transmission-line equivalent model for the wires and (b) use Multisim to evaluate the voltage response along the transmission line.

Solution:

(a) With $\ell = 3$ cm (to satisfy Eq. (7.145)), application of Eqs. (7.144b and c) gives

$$R = \sqrt{\frac{\pi f \mu}{\sigma}} \left(\frac{\ell}{\pi a}\right) = \sqrt{\frac{\pi \times 10^9 \times 4\pi \times 10^{-7}}{1.9 \times 10^5}} \left(\frac{3 \times 10^{-2}}{\pi \times 10^{-4}}\right) = 13.76\ \Omega$$

and

$$C = \frac{\pi \epsilon \ell}{\ln(d/a)} = \frac{\pi \times (10^{-9}/36\pi) \times 3 \times 10^{-2}}{\ln(10)} = 3.6 \times 10^{-13}\text{ F} = 0.36\text{ pF}.$$

(b) To use Multisim, we need to select values for R and C—from the libraries of available values—that are approximately equal to those we calculated. The selected values are less critical to the simulation than the value of their product, because it is the product $RC = 13.76 \times 0.36 \times 10^{-12} \simeq 5 \times 10^{-12}$ s that determines the time constant of the voltage response. Hence, we select

$$R = 10\ \Omega \quad \text{and} \quad C = 0.5\text{ pF},$$

and we draw the 5-stage circuit shown in Fig. 7-38. The square wave is generated by a pulse generator that alternates between 0 and 1 V. Its pulses are 500 ps long and the pulse period is 1000 ps (or equivalently, 1 ns, which is the period corresponding to a frequency $f = 1$ GHz). The Rise Time and Fall Time should be set to 1 ps. Figure 7-39 displays V(1) at node 1, which represents the pulse-generator voltage waveform, and the voltages at nodes 2, 3, 4, 5, and 6 corresponding to the outputs of the five RC stages.

During the charging-up period, it takes longer for the nodes further away from the pulse generator to reach the steady-state voltage of 1 V than it does for those closer to the generator. The same pattern applies during the discharge period. In addition to the parallel-wire configuration, the distributed transmission-line concept is equally applicable to other transmission media, including the shielded cable commonly used for the transmission of audio, video, and digital data

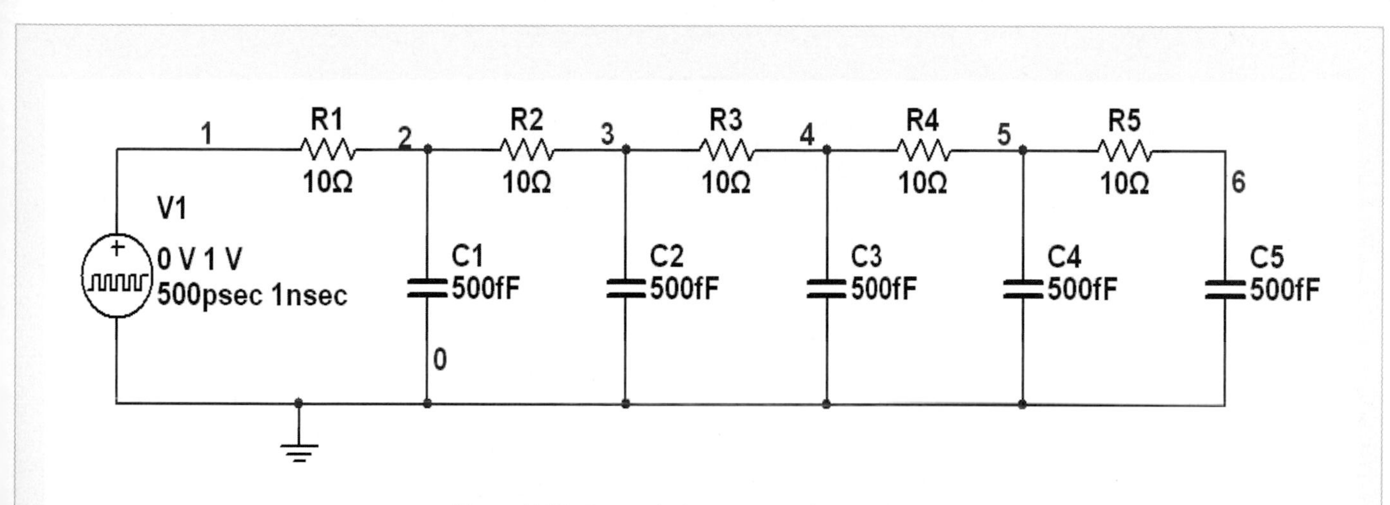

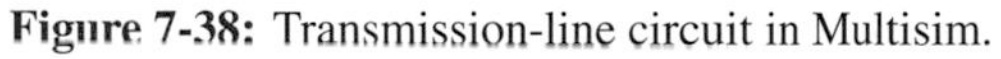
Figure 7-38: Transmission-line circuit in Multisim.

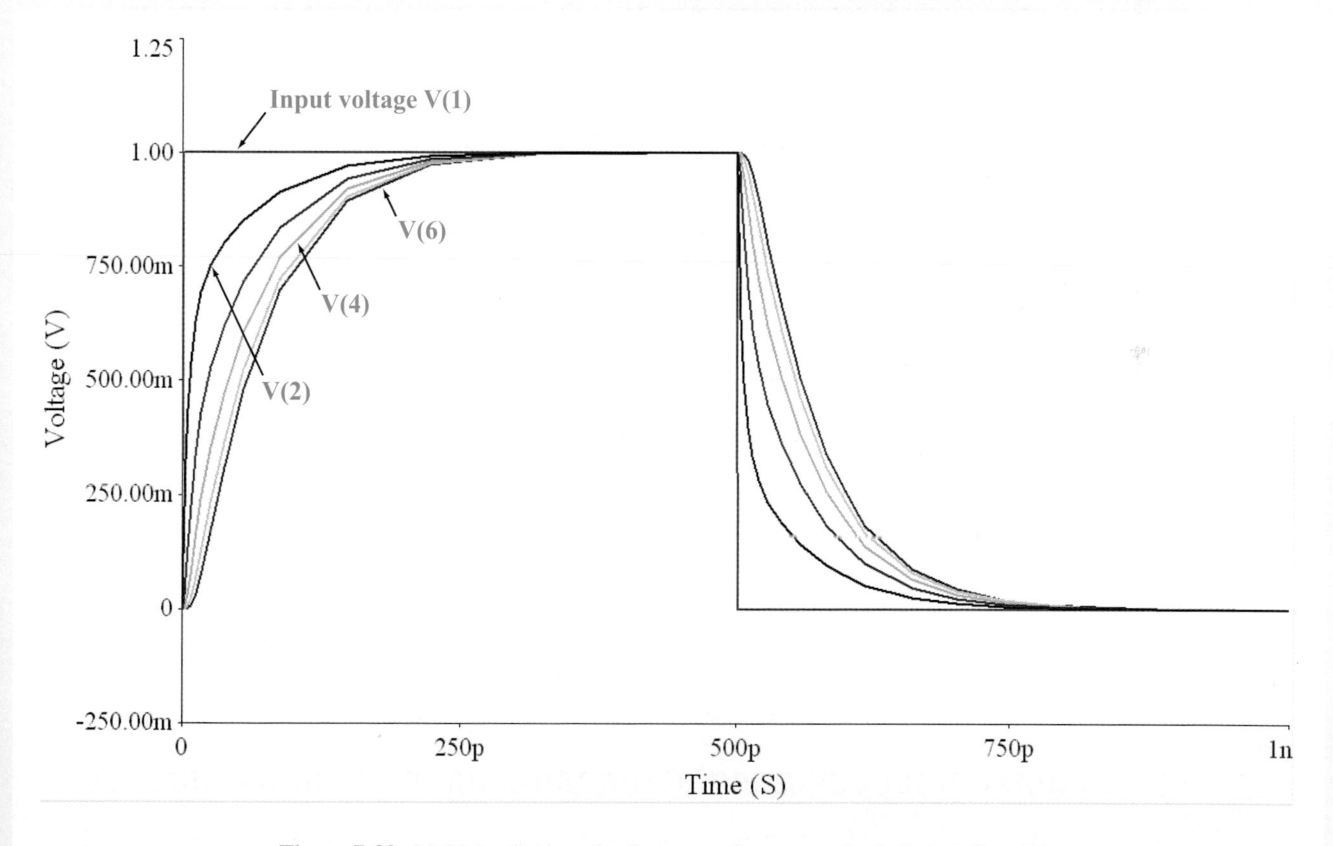

Figure 7-39: Multisim display of voltage waveforms at nodes 1, 2, 3, 4, 5, and 6.

between different circuits. If a digital signal with *logic* 0 = 0 V and *logic* 1 = 1 V is to be transmitted along a coaxial cable or some other transmission line, it may be of interest to simulate the process using Multisim to determine how long it takes to charge the different nodes along the line up to 1 V. This is also known as *propagating* the "logic 1" down the transmission line. The Logic Analyzer (Simulate → Instruments → Logic Analyzer) is used to visualize a large number of logic levels at once. (See the Multisim Tutorial for a detailed explanation on how to use the Logic Analyzer Instrument.) An example is shown in Fig. 7-40. The circuit uses 1-MΩ resistors, 5-fF capacitors, and a

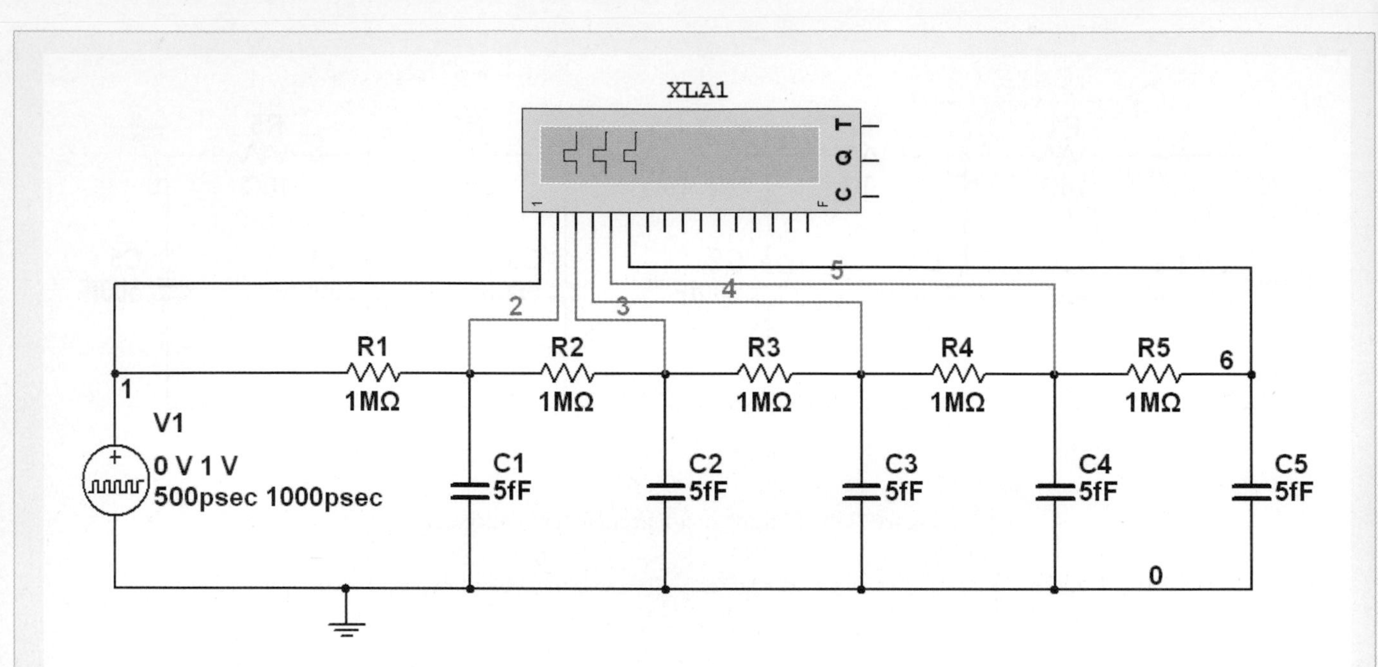

Figure 7-40: Using the Logic Analyzer to measure time delay in Multisim.

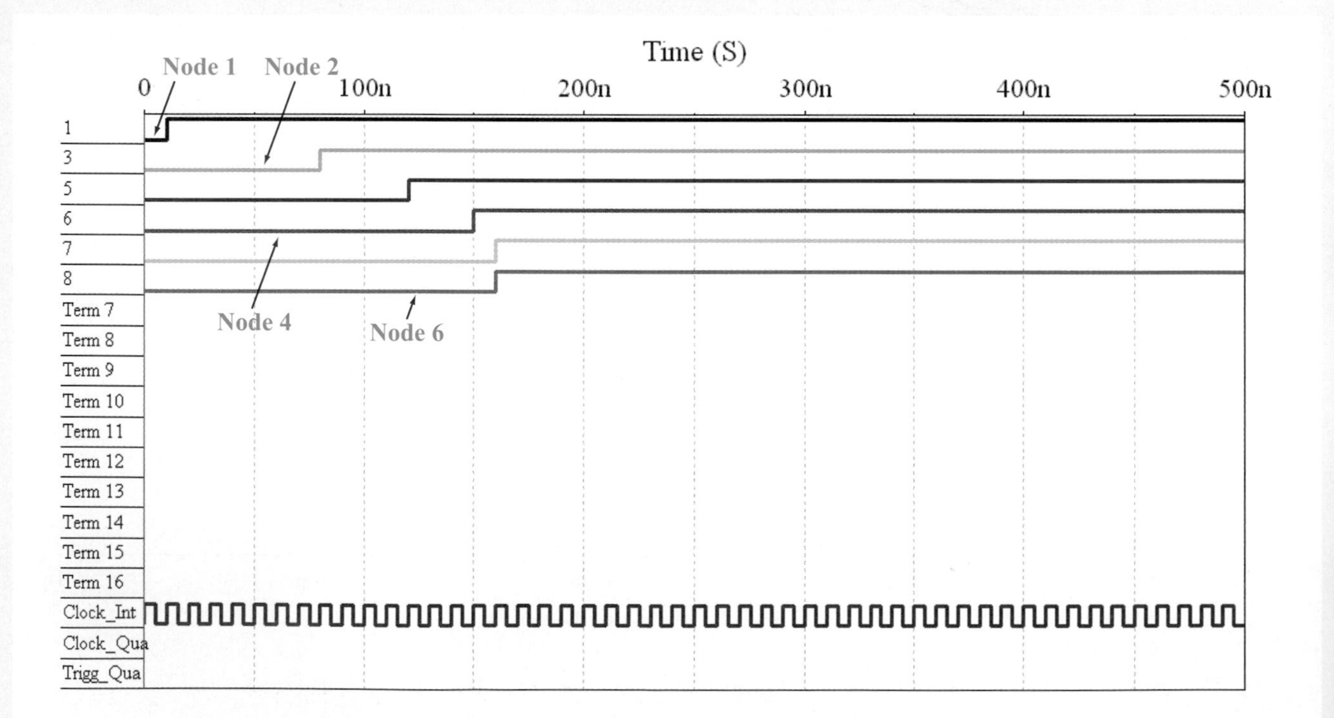

Figure 7-41: Logic Analyzer readout at nodes 1, 2, 3, 4, 5, and 6.

pulse generator. The pulse length is set at 500 ps and the pulse period at 1000 ps (= 1 ns). The circuit nodes are wired to the logic analyzer. In Fig. 7-41, we can observe how long it takes each node to charge up sufficiently to register as a *logic* 1. Note that the logic analyzer's cursor can be used to read out the exact time points.

Example 7-20: Measuring Phase Shift

Run a Transient Analysis on the Multisim circuit in Fig. 7-40 after replacing the pulse generator with a 1-V amplitude, 10-MHz ac source. The goal is to determine the phase of node 2, relative to the phase of node 1 (the voltage source). Select a Start Time of 2.7 μs and an End Time (TSTOP) of 3.0 μs, and set TSTEP and TMAX to 1e-10 seconds so as to generate smooth-looking curves. [We did not choose a Start Time of 0 s simply because it takes the circuit a few microseconds to reach its steady-state solution.]

Solution: Figure 7-42 shows the traces of selected nodes V(1), V(2), and V(6) on Grapher View. Clicking on the Show/Hide Cursors button enables the measurement cursor, which can be used to quantify the amplitude (vertical axis) and time (horizontal axis) for each curve. To measure the phase shift between nodes V(2) and V(1), two cursors are needed.

Step 1: Place cursor 1 slightly to the left of a maximum of the V(1) trace.

Step 2: Click on the trace for V(1) to select it. White triangles will appear on the V(1) trace.

Step 3: Right-click the cursor itself and select Go to next Y_Max=>. On row x1, at column V(1), the value in the table should be 2.7250 μs.

Step 4: Repeat the process using cursor 2 to select the nearby maximum of the V(2) trace. The entry in row x2, at column V(2), should be 2.7312 μs.

The time difference between the two values is

$$\Delta t = 2.7312\ \mu\text{s} - 2.7250\ \mu\text{s} = 0.0062\ \mu\text{s}.$$

Given that $f = 10$ MHz, the period is

$$T = \frac{1}{f} = \frac{1}{10^7} = 10^{-7} = 0.1\ \mu\text{s}.$$

Application of Eq. (7.11) gives

$$\phi = 2\pi\left(\frac{\Delta t}{T}\right) = 360^\circ \times \left(\frac{0.0062}{0.1}\right) = 22.3^\circ.$$

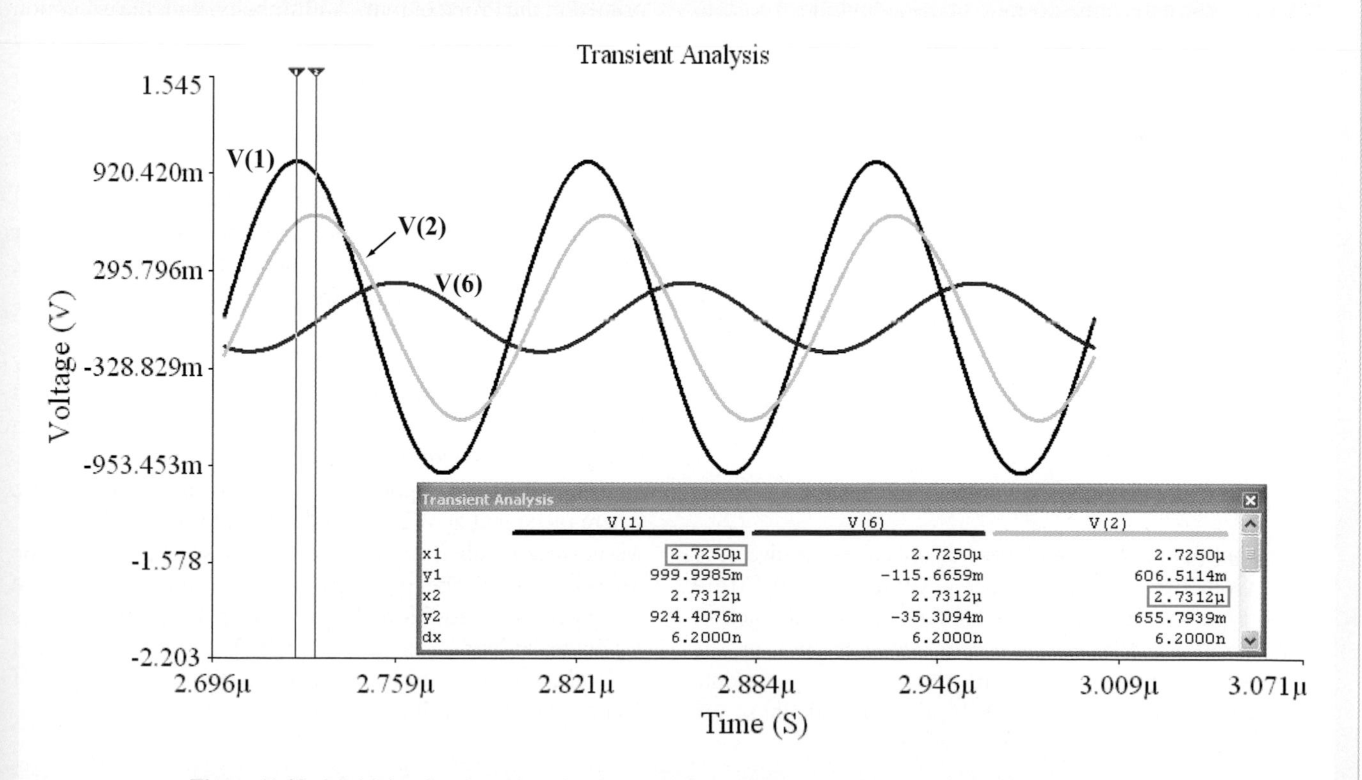

Figure 7-42: Multisim Grapher Plot of voltage nodes V(1), V(2), and V(6) in the circuit of Fig. 7-38.

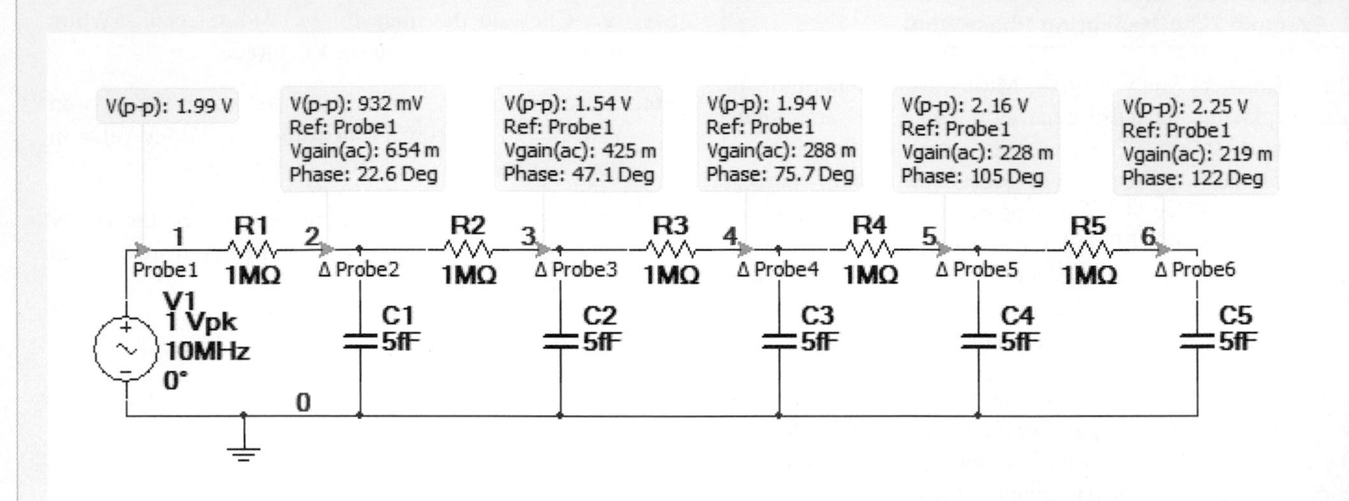

Figure 7-43: Using Measurement Probes to determine phase and amplitude of signal at various points on transmission line.

We also can determine the ratio of the amplitude of V(2) to that of V(1). The ratio of y2 in column V(2) to y1 in column V(1) gives

$$\frac{V(2)}{V(1)} = \frac{0.656}{1} \simeq 66 \text{ percent.}$$

Exercise 7-18: Determine the amplitude and phase of V(6) in the circuit of Example 7-20, relative to those of V(1).

Answer: (See ◎)

Additional Method to Measure Amplitude and Phase

Let us continue working with the transmission-line circuit of the previous two examples. Place a Measurement Probe (of the type we introduced in Chapters 2 and 3) at each of the appropriate nodes in the circuit. Double-click on the Probe, and under the Parameters tab, select the appropriate parameters so that only V(p-p), Vgain(ac), and Phase are printed in the Probe output. Additionally, with the exception of Probe 1 (located right above V1), at the top of the Probe Properties window, check Use reference probe, and select Probe 1. Note that "phase" here refers to the phase difference between the voltage at the specific probe and the reference probe. So if a particular signal is leading the reference node, then the phase will appear *negative*, and if a particular signal is lagging the reference node, then the phase will appear *positive*. This is the opposite of how we are taught to think of phase, so keep this at the front of your mind when using this approach.

Run the Interactive Simulation by pressing F5 (or any of the appropriate buttons or toggles, which you should know by now) and the result should resemble that shown in Fig. 7-43. We can see that the Phase at Node 2 is 22.6°, which of course is opposite to what we see in Fig. 7-42, where the signal at V(2) is *behind* V(1) by 22.3°. However, we must remember that the phase values are flipped in the Measurement Probe readings, so the values actually are in agreement. Additionally, we see in Fig. 7-43 that the Vgain(ac) at Node 2 is "654m" (which corresponds to 65.4 percent), which is very nearly in agreement with the value of 66 percent obtained in Example 7-20.

Chapter 7 Relationships

Trigonometric identities	Table 7-1
Euler's identity	$e^{j\theta} = \cos\theta + j\sin\theta$
Properties of complex numbers	Table 7-2
Time domain/phasor domain correspondence	Table 7-3
Impedance	$\mathbf{Z}_R = R$ $\mathbf{Z}_C = 1/j\omega C$ $\mathbf{Z}_L = j\omega L$
Impedances in series	$\mathbf{Z}_{eq} = \sum_{i=1}^{N} \mathbf{Z}_i$
Admittances in parallel	$\mathbf{Y}_{eq} = \sum_{i=1}^{N} \mathbf{Y}_i$
Y–Δ transformation	Section 7-5.2
Transformer	$\frac{v_2}{v_1} = \frac{N_2}{N_1}$ $\frac{i_2}{i_1} = \frac{N_1}{N_2}$
Wire resistance	$R = \frac{2\ell}{\pi a^2 \sigma}$ for $(a\sqrt{f\sigma} \leq 500)$ $R = \sqrt{\frac{\pi f \mu}{\sigma}} \left(\frac{\ell}{\pi a}\right)$ for $(a\sqrt{f\sigma} \geq 1250)$
Wire capacitor	$C = \frac{\pi \epsilon \ell}{\ln(d/a)}$ for $(d/2a)^2 \gg 1$

CHAPTER HIGHLIGHTS

- A sinusoidal waveform is characterized by three independent parameters: its amplitude, its angular frequency, and its phase angle.
- Complex algebra is used extensively in the phasor domain to analyze ac circuits. Hence, it behooves every student taking a course in circuit analysis to become proficient in using complex numbers.
- By transforming an ac circuit from the time domain to the phasor domain, its integro-differential equation gets transformed into a linear equation. After solving the linear equation, the solution is then transformed back to the time domain.
- Voltages and currents in the time domain have phasor counterparts in the phasor domain; resistors, capacitors, and inductors are transformed into impedances.
- The rules for combining impedances (when connected in series or in parallel) are the same as those for resistors in resistive circuits. The same is true for Y–Δ transformations.
- All of the techniques of circuit analysis are equally applicable in the phasor domain.
- A phase shifter is a circuit that can modify the phase angle of a sinusoidal waveform.
- An ac waveform can be converted into dc by subjecting it to a four-step process that includes a transformer, bridge rectifier, smoothing filter, and voltage regulator.
- Multisim is very useful for analyzing an ac circuit and evaluating its response as a function of frequency.

GLOSSARY OF IMPORTANT TERMS

Provide definitions or explain the meaning of the following terms:

admittance
amplitude
angular frequency
bridge rectifier
complex conjugate
electromagnetic compatibility
Euler's identity
impedance
oscillation frequency
period (of a cycle)
phase angle
phase lag
phase lead
phase-shift circuit
phasor diagram
phasor domain
polar form
reactance
rectangular form
rectifier
ripple
sinusoidal waveform
susceptance
transformer
voltage regulator
zener diode

PROBLEMS

Section 7-1: Sinusoidal Signals

7.1 Express the sinusoidal waveform

$$v(t) = -4\sin(8\pi \times 10^3 t - 45°) \text{ V}$$

in standard cosine form and then determine its amplitude, frequency, period, and phase angle.

7.2 Express the current waveform

$$i(t) = -0.2\cos(6\pi \times 10^9 t + 60°) \text{ mA}$$

in standard cosine form and then determine the following:

(a) Its amplitude, frequency, and phase angle.

(b) $i(t)$ at $t = 0.1$ ns.

*__7.3__ A 4-kHz sinusoidal voltage waveform $v(t)$, with a 12-V amplitude, was observed to have a value of 6 V at $t = 1$ ms. Determine the functional form of $v(t)$.

7.4 Two waveforms, $v_1(t)$ and $v_2(t)$, have identical amplitudes and oscillate at the same frequency, but $v_2(t)$ lags $v_1(t)$ by a phase angle of 60°. If

$$v_1(t) = 4\cos(2\pi \times 10^3 t + 30°) \text{ V},$$

write the expression appropriate for $v_2(t)$ and plot both waveforms over the time span from −1 ms to +1 ms.

7.5 Waveforms $v_1(t)$ and $v_2(t)$ are given by:

$$v_1(t) = -4\sin(6\pi \times 10^4 t + 30°) \text{ V},$$

$$v_2(t) = 2\cos(6\pi \times 10^4 t - 30°) \text{ V}.$$

Does $v_2(t)$ lead or lag $v_1(t)$, and by what phase angle?

*Answer(s) in Appendix E.

7.6 A phase angle of 120° was added to a 3-MHz signal, causing its waveform to shift by Δt along the time axis. In what direction did it shift and by how much?

7.7 Provide an expression for a 24-V signal that exhibits adjacent minima at $t = 1.04$ ms and $t = 2.29$ ms.

7.8 A multiplier circuit has two input ports, designated v_1 and v_2, and one output port whose voltage v_{out} is equal to the product of v_1 and v_2. Assume

$$v_1 = 10\cos 2\pi f_1 t \text{ V},$$

$$v_2 = 10\cos 2\pi f_2 t \text{ V}.$$

(a) Obtain an expression for v_{out} in terms of the sum and difference frequencies, $f_s = f_1 + f_2$ and $f_d = f_1 - f_2$.

(b) Plot its waveform over the time interval [0, 2 s], given that $f_1 = 3$ Hz and $f_2 = 2$ Hz.

*__7.9__ Provide an expression for a 12-V signal that exhibits a maximum at $t = 2.5$ ms, followed by an adjacent minimum at $t = 12.5$ ms.

Section 7-2: Complex Numbers

7.10 Express the following complex numbers in polar form.

(a) $\mathbf{z}_1 = 3 + j4$

(b) $\mathbf{z}_2 = -6 + j8$

***(c)** $\mathbf{z}_3 = -6 - j4$

(d) $\mathbf{z}_4 = j2$

***(e)** $\mathbf{z}_5 = (2 + j)^2$

(f) $\mathbf{z}_6 = (3 - j2)^3$

(g) $\mathbf{z}_7 = (-1 + j)^{1/2}$

7.11 Express the following complex numbers in rectangular form:

(a) $\mathbf{z}_1 = 2e^{j\pi/6}$

(b) $\mathbf{z}_2 = -3e^{-j\pi/4}$

*(c) $\mathbf{z}_3 = \sqrt{3}\, e^{-j3\pi/4}$

(d) $\mathbf{z}_4 = -j^3$

(e) $\mathbf{z}_5 = -j^{-4}$

*(f) $\mathbf{z}_6 = (2 + j)^2$

(g) $\mathbf{z}_7 = (3 - j2)^3$

7.12 Complex numbers $\mathbf{z}_1$ and $\mathbf{z}_2$ are given by:

$$\mathbf{z}_1 = 6 - j4,$$
$$\mathbf{z}_2 = -2 + j1.$$

(a) Express $\mathbf{z}_1$ and $\mathbf{z}_2$ in polar form.

(b) Determine $|\mathbf{z}_1|$ by applying Eq. (7.20) to the given expression.

(c) Determine the product $\mathbf{z}_1\mathbf{z}_2$ in polar form.

(d) Determine the ratio $\mathbf{z}_1/\mathbf{z}_2$ in polar form.

(e) Determine $\mathbf{z}_1^2$ and compare it with $|\mathbf{z}_1|^2$.

(f) Determine $\mathbf{z}_1/(\mathbf{z}_1 - \mathbf{z}_2)$ in polar form.

7.13 For the complex number $\mathbf{z} = 1 + j$, show that

$$\mathbf{z}^2 - |\mathbf{z}|^2 = -2(1 - j).$$

7.14 If $\mathbf{z} = -8 + j6$, determine the following quantities:

(a) $|\mathbf{z}|^2$

(b) $\mathbf{z}^2$, in polar form

(c) $1/\mathbf{z}$, in polar form

(d) $\mathbf{z}^{-3}$, in polar form

(e) $\mathfrak{Re}(1/\mathbf{z}^2)$

(f) $\mathfrak{Im}(\mathbf{z}^*)$

(g) $\mathfrak{Im}[(\mathbf{z}^*)^2]$

(h) $\mathfrak{Re}[(\mathbf{z}^*)^{-1/2}]$

7.15 Complex numbers $\mathbf{z}_1$ and $\mathbf{z}_2$ are given by

$$\mathbf{z}_1 = 2\angle{-60^\circ},$$
$$\mathbf{z}_2 = 5\angle{45^\circ}.$$

Determine in polar form:

(a) $\mathbf{z}_1\mathbf{z}_2$

(b) $\mathbf{z}_1/\mathbf{z}_2$

(c) $\mathbf{z}_1\mathbf{z}_2^*$

(d) $\mathbf{z}_1^2$

(e) $\sqrt{\mathbf{z}_2}$

(f) $\sqrt{\mathbf{z}_2^*}$

(g) $\mathbf{z}_1(\mathbf{z}_2 - \mathbf{z}_1)^*$

(h) $\mathbf{z}_2^*/(\mathbf{z}_1 + \mathbf{z}_2)$

7.16 Given $\mathbf{z} = 1.2 - j2.4$, determine the value of:

(a) $\ln \mathbf{z}$

(b) $e^{\mathbf{z}}$

(c) $\ln(\mathbf{z}^*)$

(d) $\exp(\mathbf{z}^* + 1)$

7.17 Simplify the following expressions into the form $(a + jb)$, where a and b are real numbers:

(a) $\sqrt{j} + \sqrt{-j}$

(b) $\sqrt{j}\sqrt{-j}$

(c) $\dfrac{(1+j)^2}{(1-j)^2}$

7.18 Simplify the following expressions and express the result in polar form:

(a) $\mathbf{A} = \dfrac{5e^{-j30^\circ}}{2 + j3} - j4$

(b) $\mathbf{B} = \dfrac{(-20\angle{45^\circ})(3 - j4)}{(2 - j)} + (2 + j)$

(c) $\mathbf{C} = \dfrac{j4}{(3 + j2) - 2(1 - j)} + \dfrac{1}{1 + j4}$

(d) $\mathbf{D} = \begin{vmatrix} (2 - j) & -(3 + j4) \\ -(3 + j4) & (2 + j) \end{vmatrix}$

(e) $\mathbf{E} = \begin{vmatrix} 5\angle{30^\circ} & -2\angle{45^\circ} \\ 2\angle{45^\circ} & 4\angle{60^\circ} \end{vmatrix}$

Sections 7-3 to 7-5: Phasor Domain and Impedance Transformations

7.19 Determine the phasor counterparts of the following sinusoidal functions:

(a) $v_1(t) = 4\cos(377t - 30^\circ)$ V

*(b) $v_2(t) = -2\sin(8\pi \times 10^4 t + 18^\circ)$ V

(c) $v_3(t) = 3\sin(1000t + 53^\circ) - 4\cos(1000t - 17^\circ)$ V

7.20 Determine the instantaneous time functions corresponding to the following phasors:

(a) $\mathbf{I}_1 = 6e^{j60^\circ}$ A at $f = 60$ Hz

(b) $\mathbf{I}_2 = -2e^{-j30^\circ}$ A at $f = 1$ kHz

(c) $\mathbf{I}_3 = j3$ A at $f = 1$ MHz

(d) $\mathbf{I}_4 = -(3 + j4)$ A at $f = 10$ kHz

(e) $\mathbf{I}_5 = -4\angle{-120^\circ}$ A at $f = 3$ MHz

7.21 Show that the instantaneous time function corresponding to the phasor $\mathbf{V} = 4e^{j60°} + 6e^{-j60°}$ V is given by

$$v(t) = 5.29\cos(\omega t - 19.1°) \text{ V}.$$

7.22 Determine the impedances of the following elements:

(a) $R = 1$ kΩ at 1 MHz

(b) $L = 30\ \mu$H at 1 MHz

(c) $C = 50\ \mu$F at 1 kHz

7.23 The function $v(t)$ is the sum of two sinusoids,

$$v(t) = 4\cos(\omega t + 30°) + 6\cos(\omega t + 60°) \text{ V}. \qquad (1)$$

(a) Apply the necessary trigonometric identities from Table 7-1 to show that

$$v(t) = 9.67\cos(\omega t + 48.1°) \text{ V}. \qquad (2)$$

(b) Transform the expression given by Eq. (1) to the phasor domain, simplify it into a single term, and then transform it back to the time domain to show that the result is identical with the expression given by Eq. (2).

7.24 Use phasors to simplify each of the following expressions into a single term [*Hint*: See Problem 7.23]:

(a) $v_1(t) = 12\cos(6t + 30°) - 6\cos(6t - 45°)$

(b) $v_2(t) = -3\sin(1000t - 15°) - 6\sin(1000t + 15°) + 12\cos(1000t - 60°)$

(c) $v_3(t) = 2\cos(377t + 60°) - 2\cos(377t - 60°)$

(d) $v_4(t) = 10\cos 800t + 10\sin 800t$

***7.25** The current source in the circuit of Fig. P7.25 is given by

$$i_s(t) = 12\cos(2\pi \times 10^4 t - 60°) \text{ mA}.$$

Apply the phasor-domain analysis technique to determine $i_C(t)$, given that $R = 20\ \Omega$ and $C = 1\ \mu$F.

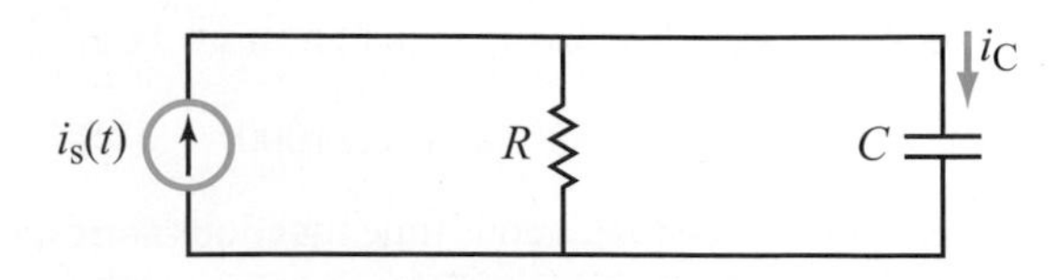

Figure P7.25: Circuit for Problems 7.25 and 7.26.

7.26 Repeat Problem 7.25, after replacing the capacitor with a 0.5-mH inductor and then calculating the current through it.

7.27 Determine the equivalent impedance:

(a) $\mathbf{Z}_1$ at 1000 Hz (Fig. P7.27(a))

***(b)** $\mathbf{Z}_2$ at 500 Hz (Fig. P7.27(b))

(c) $\mathbf{Z}_3$ at $\omega = 10^6$ rad/s (Fig. P7.27(c))

(d) $\mathbf{Z}_4$ at $\omega = 10^5$ rad/s (Fig. P7.27(d))

(e) $\mathbf{Z}_5$ at $\omega = 2000$ rad/s (Fig. P7.27(e))

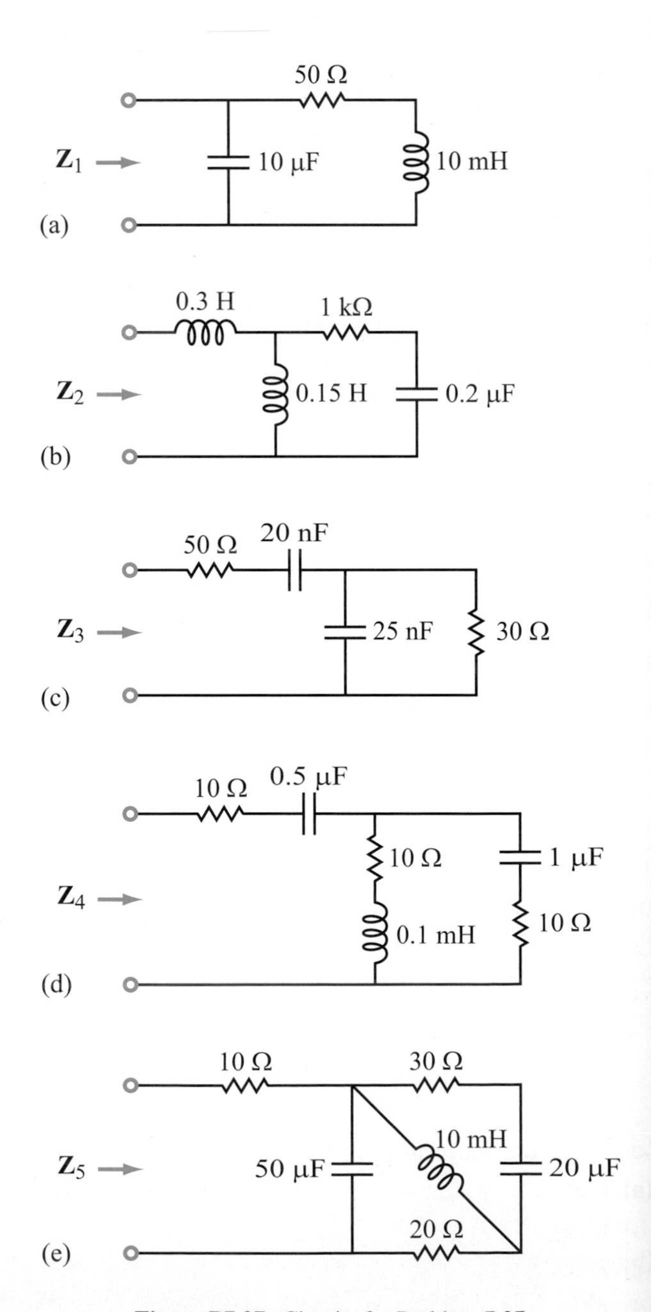

Figure P7.27: Circuits for Problem 7.27.

7.28 The voltage source in the circuit of Fig. P7.28 is given by

$$v_s(t) = 12\cos 10^4 t \text{ V}.$$

(a) Transform the circuit to the phasor domain and then determine the equivalent impedance $\mathbf{Z}$ at terminals (a, b). [*Hint*: Application of Δ–Y transformation should prove helpful.]

(b) Determine the phasor $\mathbf{I}$, corresponding to $i(t)$.

(c) Determine $i(t)$.

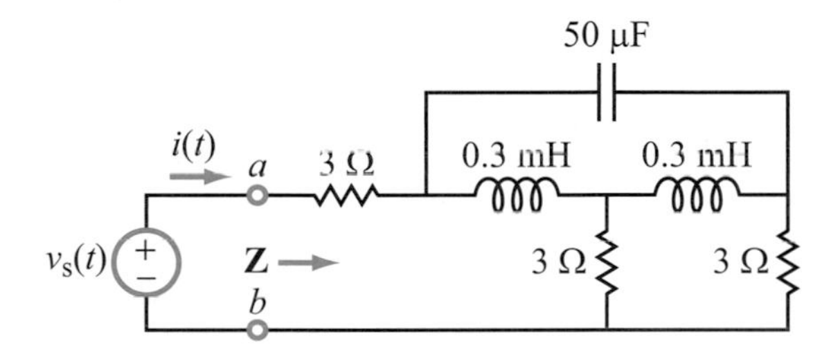

Figure P7.28: Circuit for Problem 7.28.

7.29 The circuit in Fig. P7.29 is in the phasor domain. Determine the following:

(a) The equivalent input impedance $\mathbf{Z}$ at terminals (a, b).

(b) The phasor current $\mathbf{I}$, given that $\mathbf{V}_s = 25\angle 45^\circ$ V.

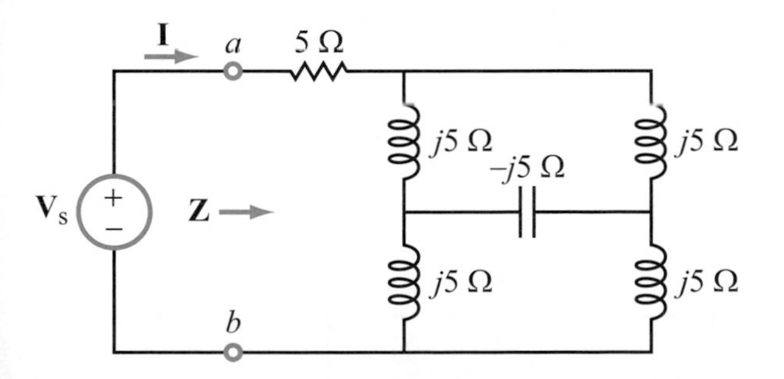

Figure P7.29: Circuit for Problem 7.29.

7.30 Use the phasor domain circuit in Fig. P7.30.

(a) Determine the value of $\mathbf{Z}_x$ that would make the input impedance $\mathbf{Z}$ purely real.

(b) Specify what type of element would be needed to realize that condition, and what its magnitude should be if $\omega = 6250$ rad/s.

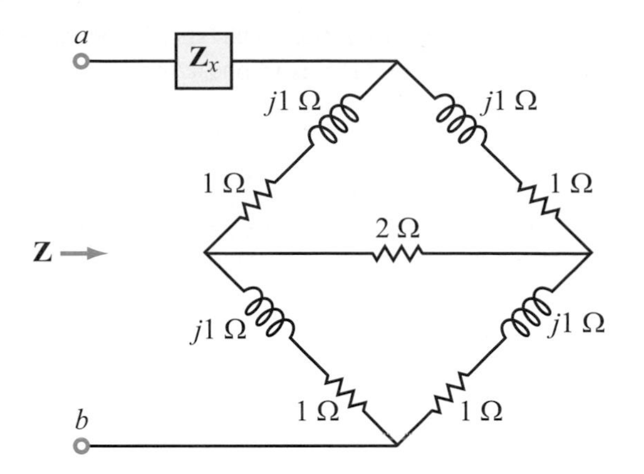

Figure P7.30: Circuit for Problem 7.30.

7.31 In response to an input signal voltage $v_s(t) = 24\cos 2000\pi t$, the input current in the circuit of Fig. P7.31 was measured as $i(t) = 6\cos(2000\pi t - 60^\circ)$ mA. Determine the equivalent input impedance $\mathbf{Z}$ of the circuit.

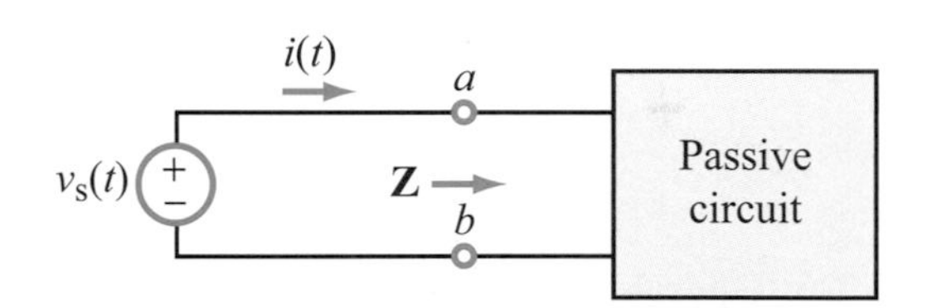

Figure P7.31: Configuration for Problem 7.31.

7.32 At $\omega = 400$ rad/s, the input impedance of the circuit in Fig. P7.32 is $\mathbf{Z} = (74 + j72)$ Ω. What is the value of L?

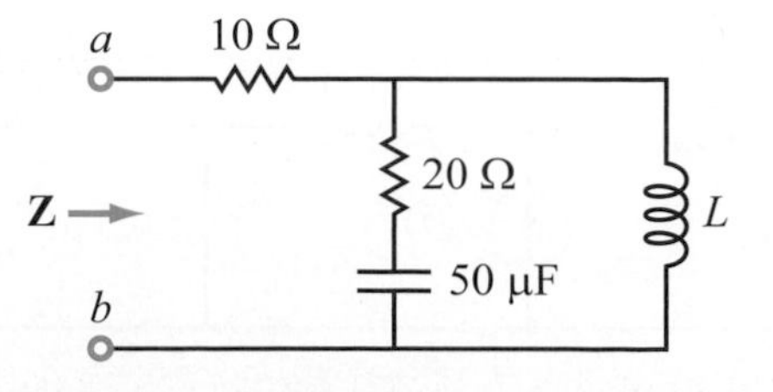

Figure P7.32: Circuit for Problem 7.32.

*7.33 In the circuit of Fig. P7.33, what should the value of L be at $\omega = 10^4$ rad/s so that $i(t)$ is in-phase with $v_s(t)$?

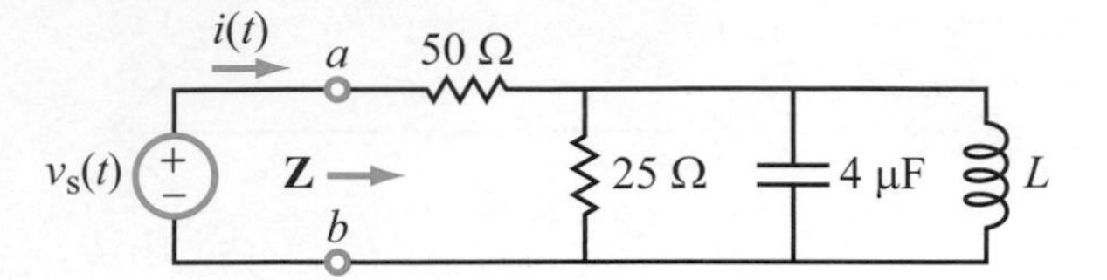

Figure P7.33: Circuit for Problem 7.33.

7.34 At what angular frequency ω is the current $i(t)$ in the circuit of Fig. P7.34 in-phase with the source voltage $v_s(t)$?

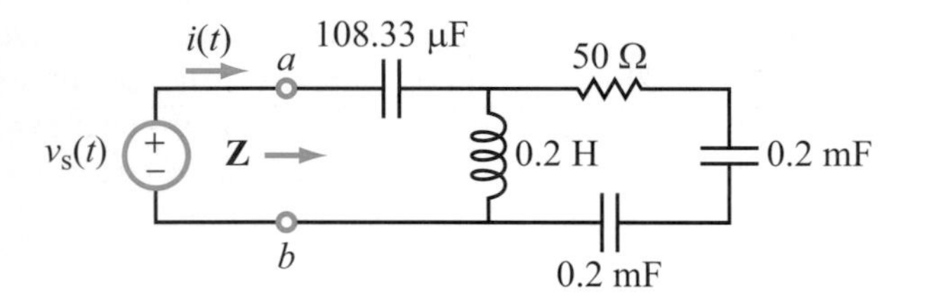

Figure P7.34: Circuit for Problem 7.34.

Sections 7-6: Equivalent Circuits

7.35 Your objective is to obtain a Thévenin equivalent for the circuit shown in Fig. P7.35, given that $i_s(t) = 3\cos 4 \times 10^4 t$ A. To that end:

(a) Transform the circuit to the phasor domain.

(b) Apply the source-transformation technique to obtain the Thévenin equivalent circuit at terminals (a, b).

(c) Transform the phasor-domain Thévenin circuit back to the time domain.

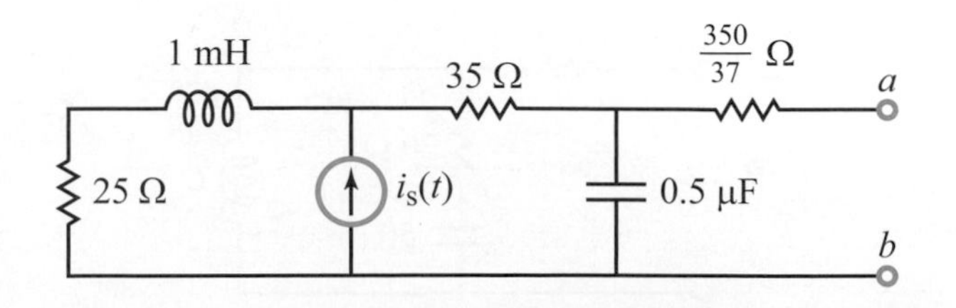

Figure P7.35: Circuit for Problem 7.35.

7.36 The input circuit shown in Fig. P7.36 contains two sources, given by

$$i_s(t) = 2\cos 10^3 t \text{ A},$$
$$v_s(t) = 8\sin 10^3 t \text{ V}.$$

This input circuit is to be connected to a load circuit that provides optimum performance when the impedance $\mathbf{Z}$ of the input circuit is purely real. The circuit includes a "matching" element whose type and magnitude should be chosen to realize that condition. What should those attributes be?

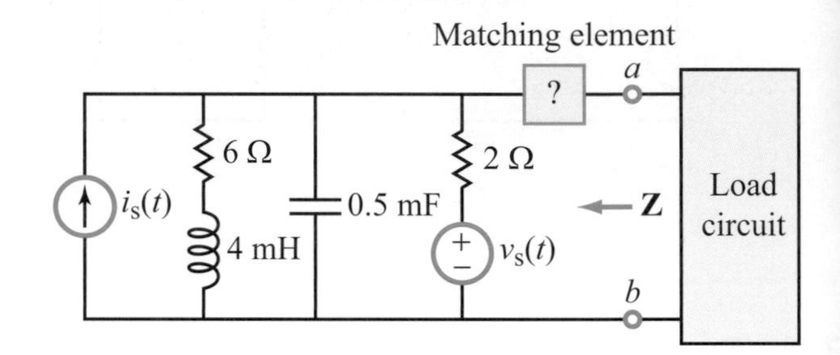

Figure P7.36: Circuit for Problem 7.36.

*7.37 Determine the Thévenin equivalent of the circuit in Fig. P7.37 at terminals (a, b), given that

$$v_s(t) = 12\cos 2500t \text{ V},$$
$$i_s(t) = 0.5\cos(2500t - 30°) \text{ A}.$$

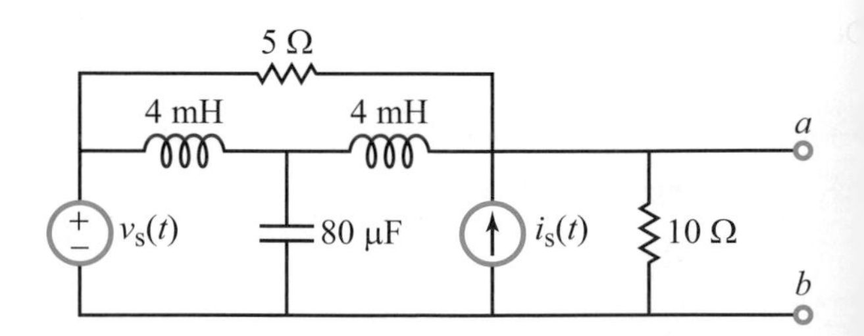

Figure P7.37: Circuit for Problem 7.37.

7.38 The circuit in Fig. P7.38 is in the phasor domain. Determine its Thévenin equivalent at terminals (a, b).

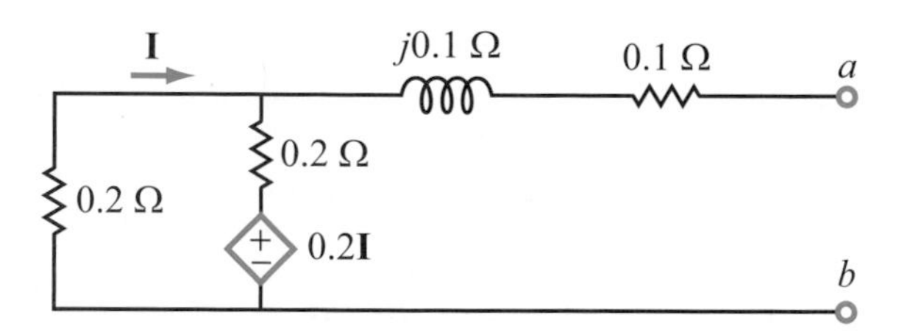

Figure P7.38: Circuit for Problem 7.38.

7.39 As we will learn in Chapter 8, to maximize the transfer of power from an input circuit to a load $\mathbf{Z}_L$, it is necessary to choose $\mathbf{Z}_L$ such that it is equal to the complex conjugate of the impedance of the input circuit. For the circuit in Fig. P7.39, such a condition translates into requiring $\mathbf{Z}_L = \mathbf{Z}^*_{Th}$. Determine $\mathbf{Z}_L$ such that it satisfies this condition.

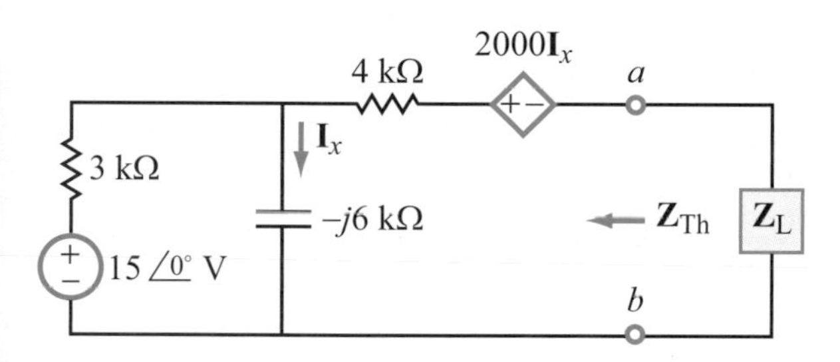

Figure P7.39: Circuit for Problem 7.39.

7.40 The phasor current $\mathbf{I}_L$ in the circuit of Fig. P7.40 was measured to be

$$\mathbf{I}_L = \left(\frac{78}{41} - j\frac{36}{41}\right) \text{ mA.}$$

Determine $\mathbf{Z}_L$.

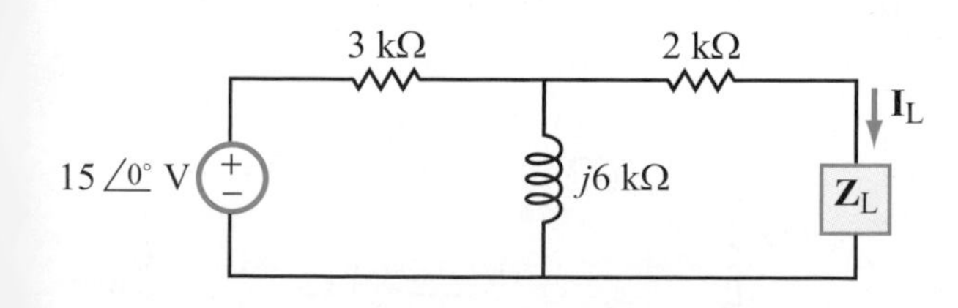

Figure P7.40: Circuit for Problem 7.40.

Sections 7-7 and 7-8: Phasor Diagrams and Phase Shifters

7.41 For the circuit in Fig. P7.41:

(a) Apply current division to express $\mathbf{I}_C$ and $\mathbf{I}_R$ in terms of $\mathbf{I}_s$.

(b) Using $\mathbf{I}_s$ as reference, generate a relative phasor diagram showing $\mathbf{I}_C$, $\mathbf{I}_R$, and $\mathbf{I}_s$ and demonstrate that the vector sum $\mathbf{I}_R + \mathbf{I}_C = \mathbf{I}_s$ is satisfied.

(c) Analyze the circuit to determine $\mathbf{I}_s$ and then generate the absolute phasor diagram with $\mathbf{I}_C$, $\mathbf{I}_R$, and $\mathbf{I}_s$ drawn according to their true phase angles.

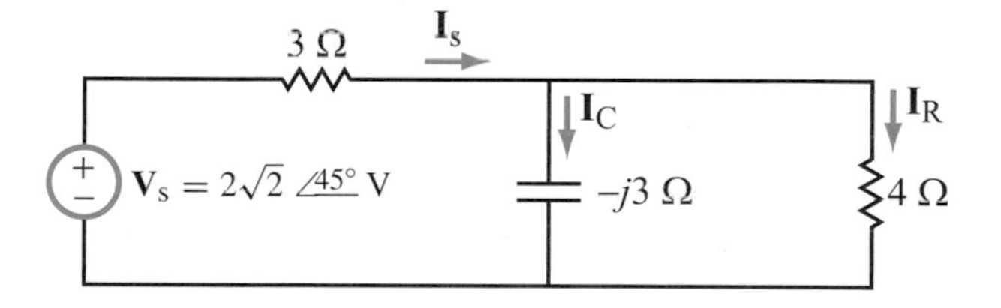

Figure P7.41: Circuit for Problem 7.41.

7.42 For the circuit in Fig. P7.42:

(a) Apply current division to express $\mathbf{I}_1$ and $\mathbf{I}_2$ in terms of $\mathbf{I}_s$.

(b) With $\mathbf{I}_s$ as reference, generate a relative phasor diagram showing that the vector sum $\mathbf{I}_1 + \mathbf{I}_2 = \mathbf{I}_s$ is indeed satisfied.

(c) Analyze the circuit to determine $\mathbf{I}_s$ and then generate the absolute phasor diagram for the three currents.

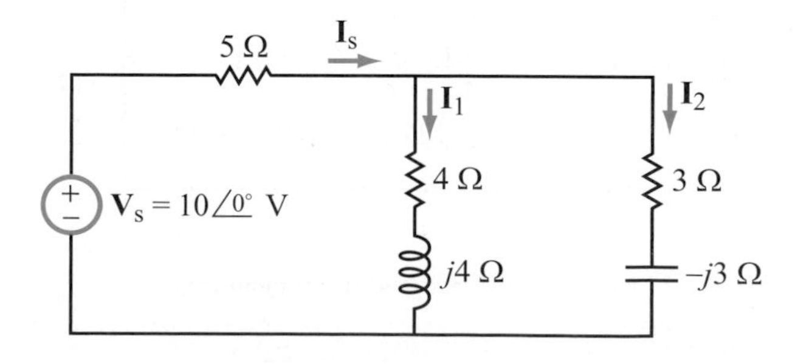

Figure P7.42: Circuit for Problem 7.42.

7.43 Design a two-stage 1-MHz RC phase-shift circuit whose output voltage is 120° behind that of the input signal. All capacitors are 1 nF each. Determine the values of the resistors and the amplitude ratio of the output voltage to that of the input.

7.44 A two-stage RC circuit provides a phase shift lead of 120°. What is the ratio of the output-voltage amplitude to that of the input?

*__7.45__ The element values of a single-stage phase-shift circuit are $R = 40\ \Omega$ and $C = 5\ \mu$F. At what frequency f is $\phi_1 = -\phi_2$, where ϕ_1 and ϕ_2 are the phase angles of the output voltages across R and C, respectively?

Section 7-9: Analysis Techniques

7.46 Apply nodal analysis in the phasor domain to determine $i_x(t)$ in the circuit of Fig. P7.46.

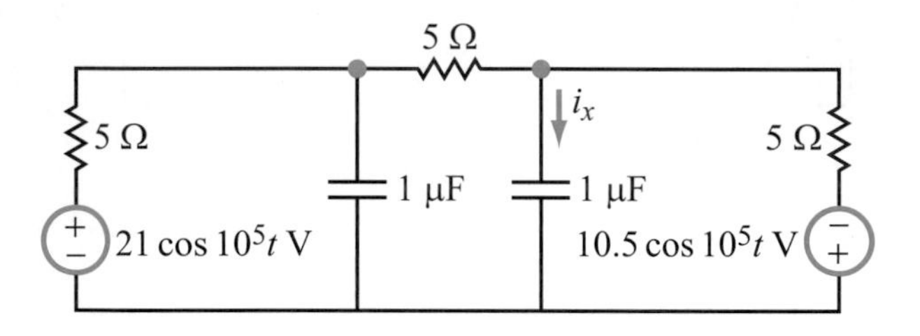

Figure P7.46: Circuit for Problem 7.46.

7.47 Apply nodal analysis in the phasor domain to determine $i_C(t)$ in the circuit of Fig. P7.47.

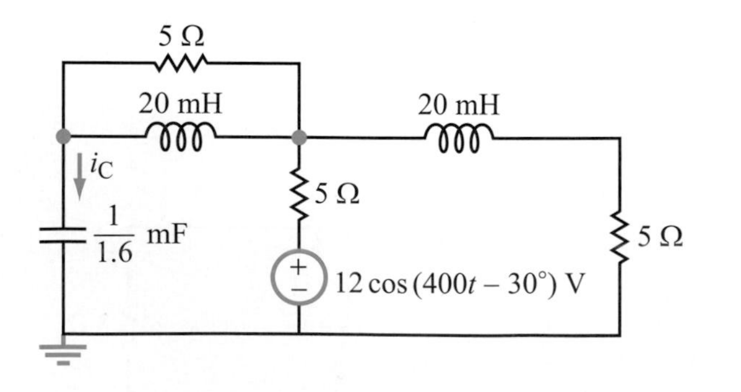

Figure P7.47: Circuit for Problem 7.47.

7.48 The circuit in Fig. P7.48 contains a supernode between nodes $\mathbf{V}_1$ and $\mathbf{V}_2$. Apply the supernode method to determine $\mathbf{V}_1$, $\mathbf{V}_2$, and $\mathbf{V}_3$, and then calculate $\mathbf{I}_C$.

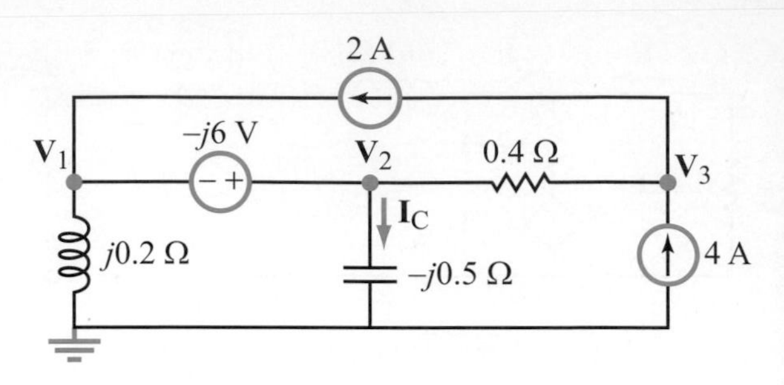

Figure P7.48: Circuit for Problem 7.48.

7.49 Apply the by-inspection method to develop a node-voltage matrix equation for the circuit in Fig. P7.49, and then use MATLAB® software to solve for $\mathbf{V}_1$ and $\mathbf{V}_2$.

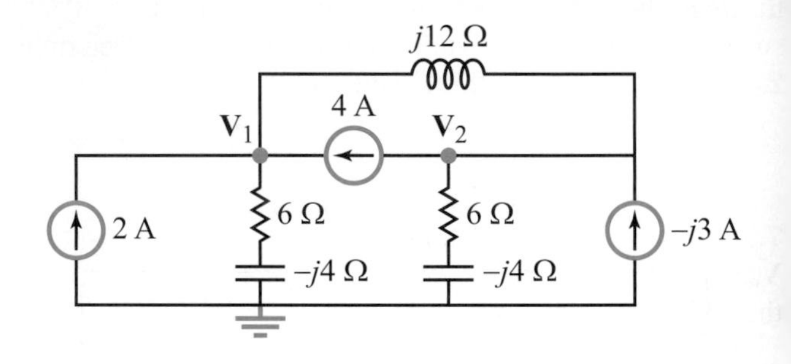

Figure P7.49: Circuit for Problem 7.49.

7.50 With $\mathbf{I}_s = 12\angle 120°$ V in the circuit of Fig. P7.50, apply the by-inspection method to develop a node-voltage matrix equation and then use MATLAB® software to solve for $\mathbf{I}_x$.

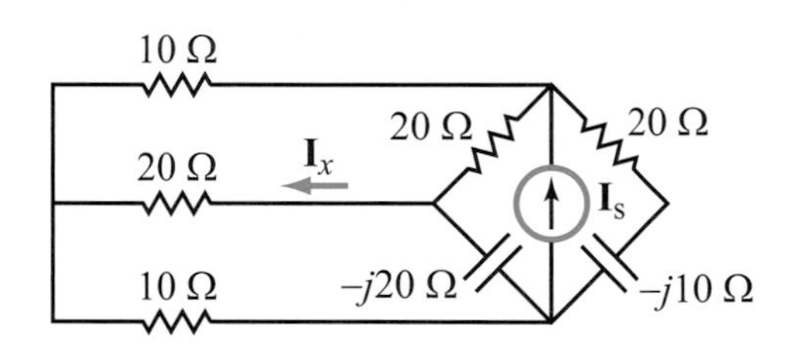

Figure P7.50: Circuit for Problem 7.50.

*__7.51__ Apply nodal analysis to determine $\mathbf{I}_C$ in the circuit of Fig. P7.51.

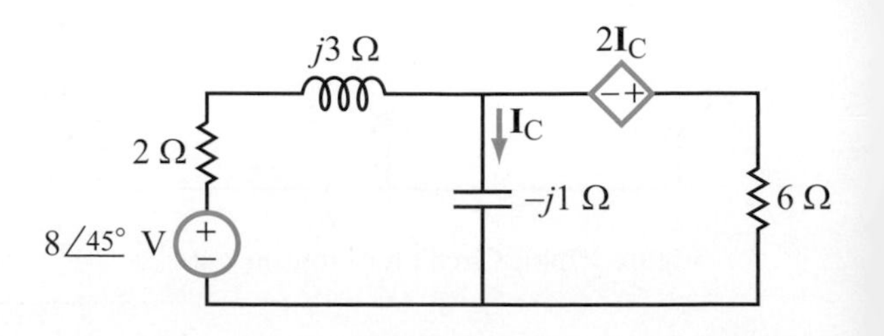

Figure P7.51: Circuits for Problems 7.51 and 7.52.

7.52 Apply mesh analysis to determine $\mathbf{I}_C$ in the circuit of Fig. P7.51.

7.53 Apply mesh analysis to determine $i_L(t)$ in the circuit of Fig. P7.53.

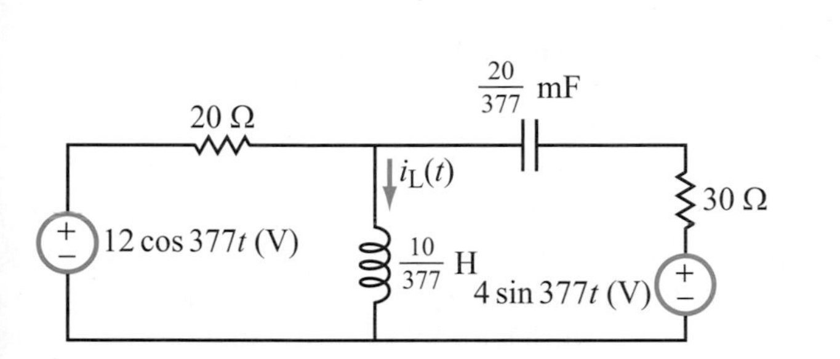

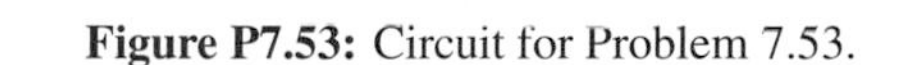

Figure P7.53: Circuit for Problem 7.53.

7.54 Use mesh analysis to obtain an expression for the phasor $\mathbf{V}_{out}$ in the circuit of Fig. P7.54, in terms of $\mathbf{V}_s$ and R, given that $R = \omega L = 1/\omega C$.

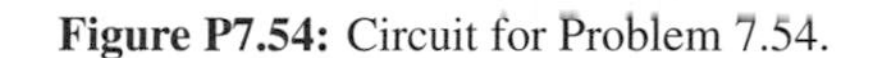

Figure P7.54: Circuit for Problem 7.54.

7.55 Apply the by-inspection method to develop a mesh-current matrix equation for the circuit in Fig. P7.55 and then use MATLAB software to solve for $\mathbf{I}_1$, $\mathbf{I}_2$, and $\mathbf{I}_3$.

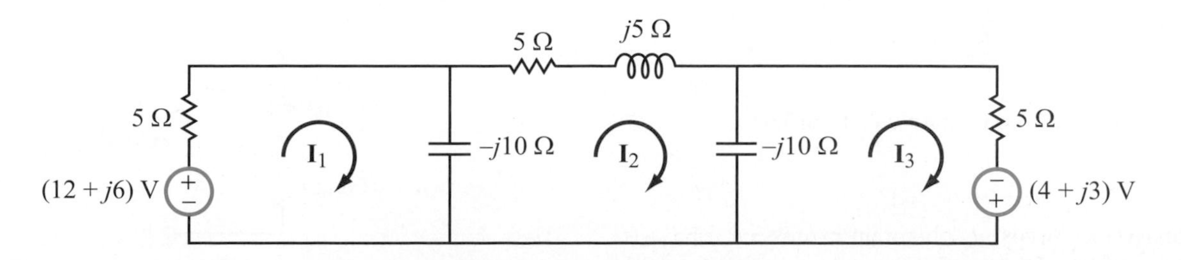

Figure P7.55: Circuit for Problem 7.55.

7.56 Use any analysis technique of your choice to determine $i_C(t)$ in the circuit of Fig. P7.56.

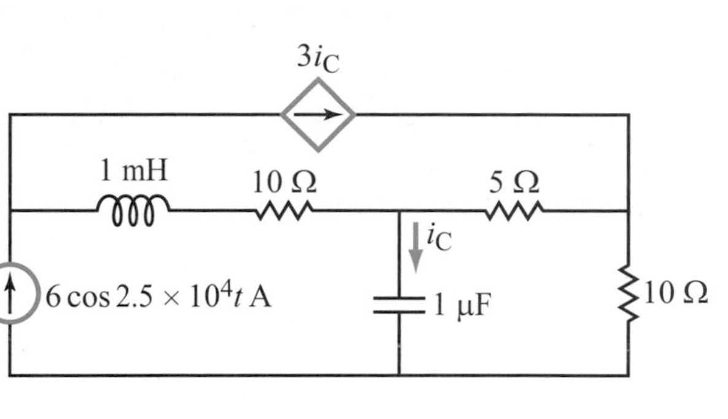

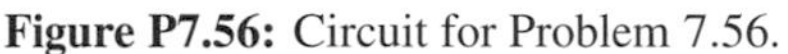

Figure P7.56: Circuit for Problem 7.56.

*__7.57__ Determine $i_x(t)$ in the circuit of Fig. P7.57, given that $v_s(t) = 6\cos 5 \times 10^5 t$ V.

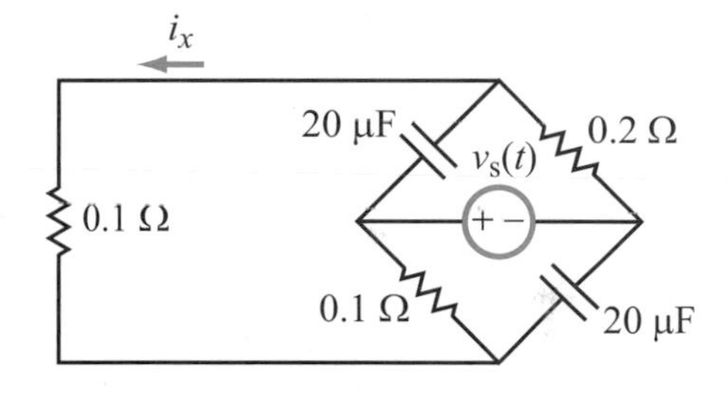

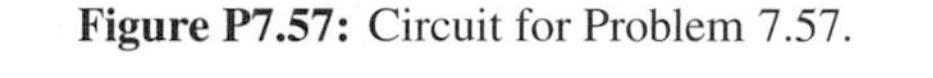

Figure P7.57: Circuit for Problem 7.57.

7.58 The input signal in the op-amp circuit of Fig. P7.58 is given by

$$v_{in}(t) = V_0 \cos \omega t.$$

Assuming the op amp is operating within its linear range, obtain an expression for $v_{out}(t)$ by applying the phasor-domain technique and then evaluate it for $\omega RC = 1$.

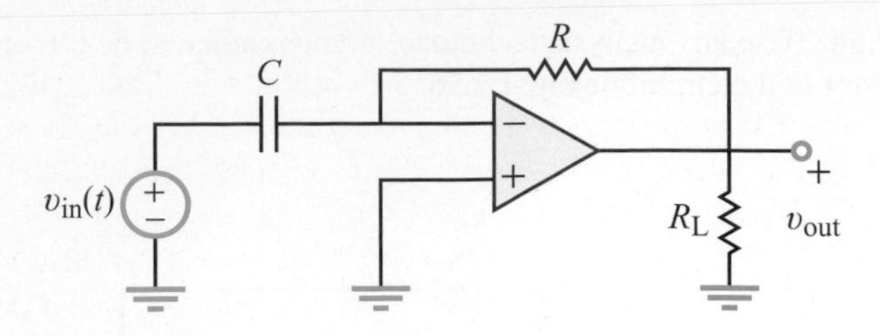

Figure P7.58: Op-amp circuit for Problem 7.58.

7.59 The input signal in the op-amp circuit of Fig. P7.59 is given by

$$v_{in}(t) = 0.5 \cos 2000t \text{ V}.$$

Obtain an expression for $v_{out}(t)$ and then evaluate it for $R_1 = 2$ kΩ, $R_2 = 10$ kΩ, and $C = 0.1$ μF.

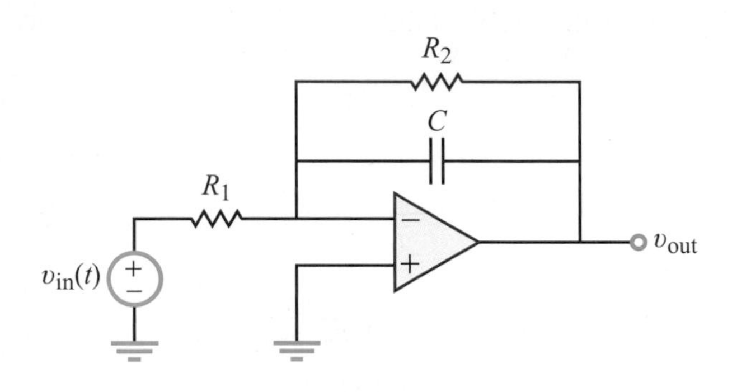

Figure P7.59: Op-amp circuit for Problem 7.59.

7.60 For $v_i(t) = V_0 \cos \omega t$, obtain an expression for $v_{out}(t)$ in the circuit of Fig. P7.60 and then evaluate it for $V_0 = 4$ V, $\omega = 400$ rad/s, $R = 5$ kΩ, and $C = 2.5$ μF.

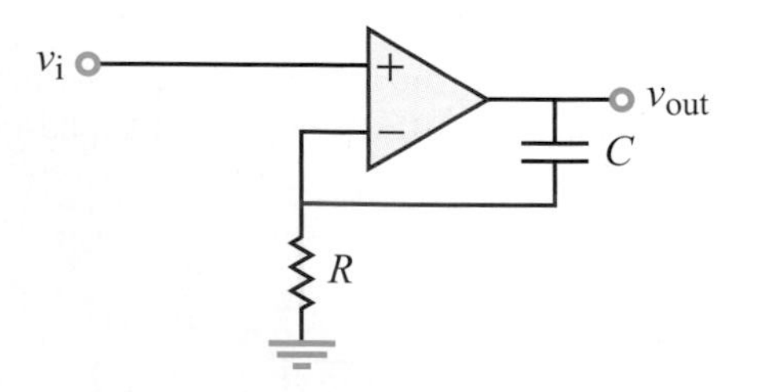

Figure P7.60: Circuit for Problem 7.60.

*__7.61__ For $v_i(t) = V_0 \cos \omega t$, obtain an expression for $v_{out}(t)$ in the circuit of Fig. P7.61 and then evaluate it for $V_0 = 2$ V, $\omega = 377$ rad/s, $R_1 = 2$ kΩ, $R_2 = 10$ kΩ, and $C = 0.5$ μF.

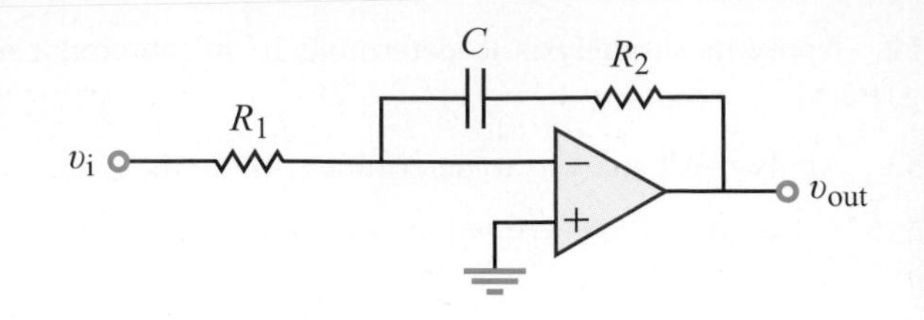

Figure P7.61: Circuit for Problem 7.61.

Section 7-10: Power-Supply Circuits

7.62 The signal voltage at the input of a half-wave rectifier circuit is given by $v_{in}(t) = A \cos(377t + 30°)$ V. Determine and plot the waveform of $v_{out}(t)$. Calculate the fraction of a full period over which $v_{out} = 0$ for each of the following values of A (assume $V_F = 0.7$ V):

(a) $A = 0.5$ V

(b) $A = 5$ V

7.63 A bridge rectifier is driven by a 1-kHz input signal with an amplitude of 10 V. The smoothing filter at the rectifier output uses a 1-μF capacitor in parallel with a load resistor R_L. If $R_D = 5$ Ω:

(a) What should R_L be so that $\tau_{dn}/\tau_{up} = 2500$?

(b) How does τ_{dn} compare with the period of the rectified waveform?

(c) What is the approximate peak value of the output waveform?

7.64 A power supply with the circuit configuration shown in Fig. 7-36 has the following specifications: $v_s = 24 \cos(2\pi \times 10^3 t + 30°)$ V, $N_2/N_1 = 2$, $C = 0.1$ mF, $R_s = 50$ Ω, $R_L = 20$ kΩ, $R_z = 20$ Ω, and $V_z = 42$ V. Determine v_{out} and the peak-to-peak ripple voltage.

Section 7-11: Multisim Analysis

7.65 Use the Network Analyzer (see Appendix C) in Multisim to determine the equivalent impedance $\mathbf{Z}_{eq}$ of the circuit in Fig. P7.65. Using the Network Analyzer, plot $\mathbf{Z}_{eq}$ from 1 kHz to 1 MHz and provide a hand calculation demonstrating that the simulated results are correct.

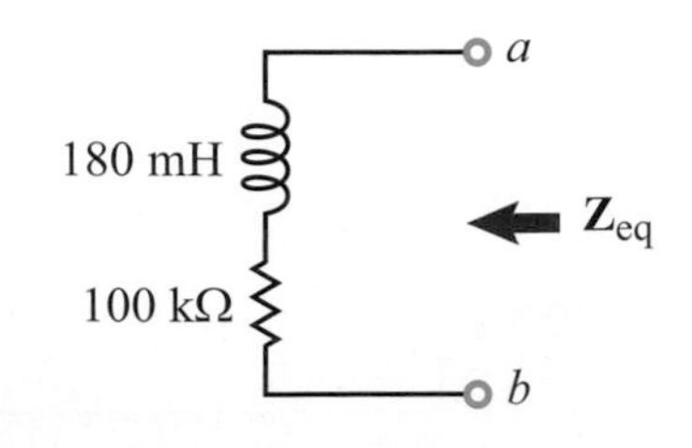

Figure P7.65: Circuit for Problem 7.65.

7.66 Use the Network Analyzer (see Appendix C) in Multisim to determine the equivalent impedance $\mathbf{Z}_{eq}$ of the circuit in Fig. P7.66. Using the Network Analyzer, plot the real and imaginary parts of $\mathbf{Z}_{eq}$ from 100 Hz to 100 kHz and provide a hand calculation demonstrating that the simulated results are correct.

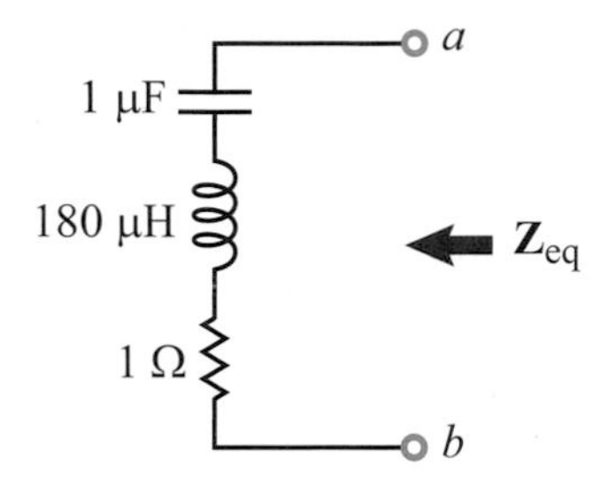

Figure P7.66: Circuit for Problem 7.66.

7.67 A 1-V, 100-MHz voltage source $v_s(t)$ sends a signal down two transmission lines simultaneously, as depicted by Fig. P7.67. Model this circuit in Multisim with $R_1 = R_2 = 10\ \Omega$, $C_1 = 7$ pF, and $C_2 = 5$ pF and answer the following questions.

(a) What is the phase shift between $v_s(t)$ and the two output nodes, v_{out_1} and v_{out_2}?

(b) What is the amplitude ratio for v_{out_1}/v_s and v_{out_2}/v_s?

7.68 Phase-shift circuits have many uses. They can be the fundamental component of an oscillator (a circuit which produces a repetitive electronic signal). The circuit shown in Fig. P7.68 is a ***phase-shift oscillator***. While a detailed analysis is too complex for this text, Multisim allows us to easily create and analyze this circuit. Using the 3-terminal virtual op-amp component, construct the phase-shift oscillator shown in the figure and plot the output from 0 to 1.5 ms in Transient Analysis. Determine the frequency and amplitude of the oscillations as well as the DC offset. Note that you may need to decrease the maximum time step (TMAX) in order to get a clear plot.

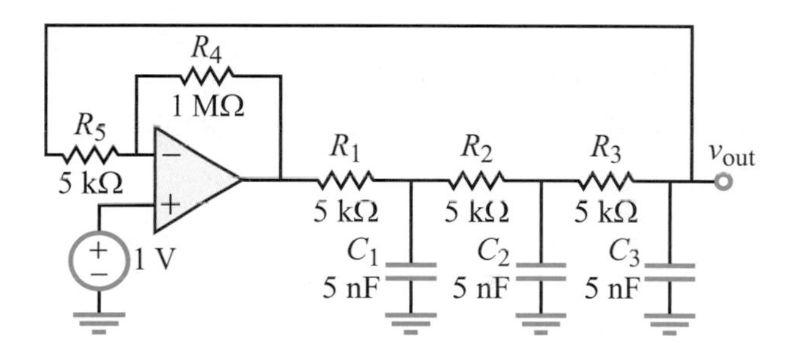

Figure P7.68: Circuit for Problem 7.68.

7.69 Using a Multisim tool or analysis of your choice, find the phase and magnitude of the voltage at each node in the circuit in Fig. P7.69.

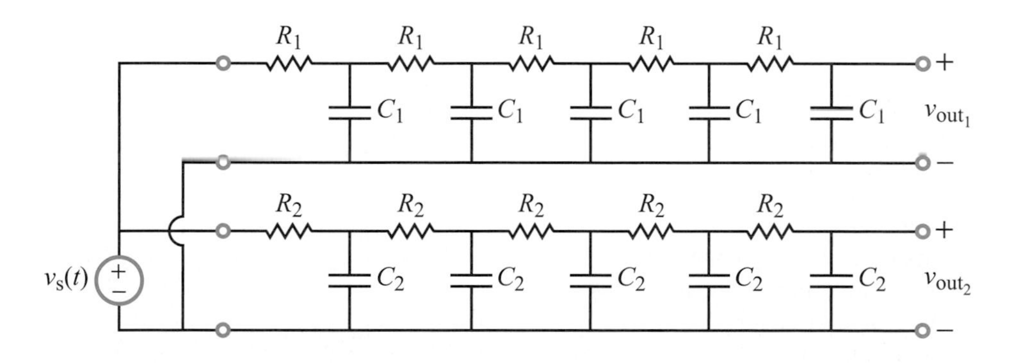

Figure P7.67: Circuit for Problem 7.67.

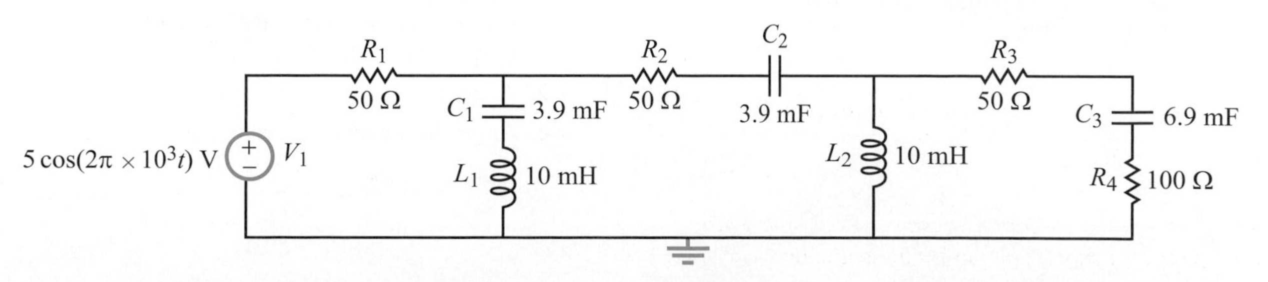

Figure P7.69: Circuit for Problem 7.69.

CHAPTER

8

ac Power

Chapter Contents

Objectives

Upon learning the material presented in this chapter, you should be able to:

1. Calculate the average and rms value of a periodic waveform.
2. Determine the complex power, average real power, and reactive power for any complex load with known input voltage or current.
3. Determine the power factor for a complex load and evaluate the improvement realized by compensating the load through the addition of a shunt capacitor.
4. Choose the load impedance so as to maximize the transfer of power from the input circuit to the load.
5. Analyze three-phase circuits.
6. Apply Multisim to measure power.

)verview

he power absorbed by a resistor R when a current i passes rough it is given by $p = i^2 R$. If i is time varying, we usually esignate it $i(t)$, call it the instantaneous current, and call the rresponding power the ***instantaneous power***

$$p(t) = i^2(t)\, R \qquad \text{(W)}. \tag{8.1}$$

Usually, we are interested in the ***average power*** P consumed y a given circuit—or by a collection of circuits, as in an entire usehold—and since ac signals are periodic in time with an gular frequency ω and a time period $T = 2\pi/\omega$, we define $_{\text{av}}$ to be the average value of $p(t)$ over one (or more) complete riod(s). For an ac current given by

$$i(t) = I_{\text{m}} \cos \omega t, \tag{8.2}$$

e average power consumed by a resistor R is

$$P_{\text{av}} = \frac{1}{2} I_{\text{m}}^2 R \qquad \text{(W)}. \tag{8.3}$$

is result, which we will derive in Section 8-2, is somewhat triguing, primarily because P_{av} is independent of ω and its pression contains a factor of 1/2. The explanations are fairly raightforward and will be covered later. The more important int we wish to make at this time is that had our intent been to scuss ac power in resistive circuits only, the discussion would t have required more than just a few pages and perhaps no ore than one or two examples. Instead, we are devoting this tire chapter to ac power, because real circuits contain more an just resistors; they contain capacitors and inductors, both which cannot consume power but can store it and then release

The current through a resistor is always in-phase with the ltage across it. This phase attribute is responsible, in part, r the functional form of the expression for P_{av} given by . (8.3). The expression generally is not valid when the load rcuit contains reactive elements (capacitors and inductors) ther alone or in combination with resistive elements. So, for e general case, we need to develop a formulation appropriate r any complex load—from the purely resistive to the purely active. That defines one of the objectives of the present apter.

When we transform a circuit from the time domain to the asor domain, voltages and currents are assigned phasor unterparts, and passive elements become impedances. What out power? Is there a phasor power **P**, corresponding to $p(t)$? e answer is: Not exactly. We will introduce a quantity **S** hich we will call ***complex power***, but **S** is not the phasor unterpart of $p(t)$. In fact, we assign it the symbol **S** (rather an **P**) to avoid the possible misinterpretation that it bears a e-to-one correspondence to $p(t)$. As we will see in Section 8-3, **S** consists of a real part and an imaginary part with the real part representing the real average power ***consumed*** by the circuit and the imaginary part representing the average power ***stored*** by the circuit.

Towards the end of Chapter 3, we posed the question: *When an input circuit is connected to a resistive load, under what condition(s) is the power transferred from the circuit to the load a maximum?* Through the application of Thévenin's theorem, we demonstrated that the transferred power is a maximum when the load resistance is equal to the Thévenin resistance of the input circuit. In the present chapter, we re-pose the question, but we generalize the load to a complex load $\mathbf{Z}_{\text{L}} = R_{\text{L}} + jX_{\text{L}}$, composed of a resistive part R_{L} and a reactive part X_{L}. In view of the fact that $\mathbf{Z}_{\text{L}}$ consists of two parts, we should expect the answer to consist of two conditions (not just one) and it does. The details are given in Section 8-5.

8-1 Periodic Waveforms

Even though the focus of this chapter is on the ac power carried by sinusoidally time-varying signals, we will preface our examination by first reviewing some of the important properties shared by all periodic waveforms, including sinusoids.

Mathematically, a periodic waveform $x(t)$ with period T satisfies the ***periodicity property***

$$x(t) = x(t + nT) \tag{8.4}$$

for any integer value of n. The periodicity property simply states that the waveform of $x(t)$ repeats itself every T seconds. Figure 8-1 displays the waveforms of three typical (and unrelated) periodic functions. In part (a), $\upsilon(t)$ is a sine wave; in (b) $i(t)$ is a sawtooth with a clipped top; and part (c) displays a function given by $p(t) = P_{\text{m}} \cos^2 \omega t$.

8-1.1 Instantaneous and Average Values

Each of the three waveforms shown in Fig. 8-1 describes the exact variation of its magnitude as a function of time. Consequently, the time function $\upsilon(t)$, for example, is referred to as the ***instantaneous voltage***. Similarly, $i(t)$ is the ***instantaneous current***, and $p(t)$ is the ***instantaneous power***. Often times, however, we may be interested in specifying an attribute of the waveform that conveys useful information about it, and yet it is much simpler to use than the complete waveform. When ac circuits are concerned, two attributes of particular interest are the ***average value*** of the waveform and its ***root-mean-square (rms) value***. The latter is introduced in the next subsection, so for the present we will pursue only the former.

The average value of a periodic function $x(t)$ with period T is given by

$$X_{av} = \frac{1}{T}\int_0^T x(t)\,dt. \tag{8.5}$$

We note that X_{av} is obtained by integrating $x(t)$ over a complete period T and then normalizing the integrated value by dividing it by T. The limits of integration are from 0 to T, but the definition is equally valid for any two limits so long as the upper limit is greater than the lower limit by exactly T, such as from T_0 to $T_0 + T$.

The voltage waveform shown in Fig. 8-1(a) is given by

$$v(t) = V_m \sin\frac{2\pi t}{T}. \tag{8.6}$$

Application of Eq. (8.5) gives

$$V_{av} = \frac{1}{T}\int_0^T V_m \sin\frac{2\pi t}{T}\,dt$$

$$= \frac{V_m}{T}\left(-\frac{T}{2\pi}\right)\cos\frac{2\pi t}{T}\Bigg|_0^T = -\frac{V_m}{2\pi}[1-1] = 0. \tag{8.7}$$

The fact that the *average value of a sine wave is zero* is not at all surprising; it is clear from the characteristic symmetry of its waveform that the area under the curve (integrand) during the first half of any cycle is equal (but opposite in polarity) to the area under the curve during the second half of the cycle, so the net sum of the two is exactly zero. In contrast, the lack of symmetry in the waveforms of $i(t)$ and $p(t)$ in Figs. 8-1(b) and (c) (between that part of the waveform above the t-axis and the part below it) is an obvious indicator that their average values are not only non-zero but also positive.

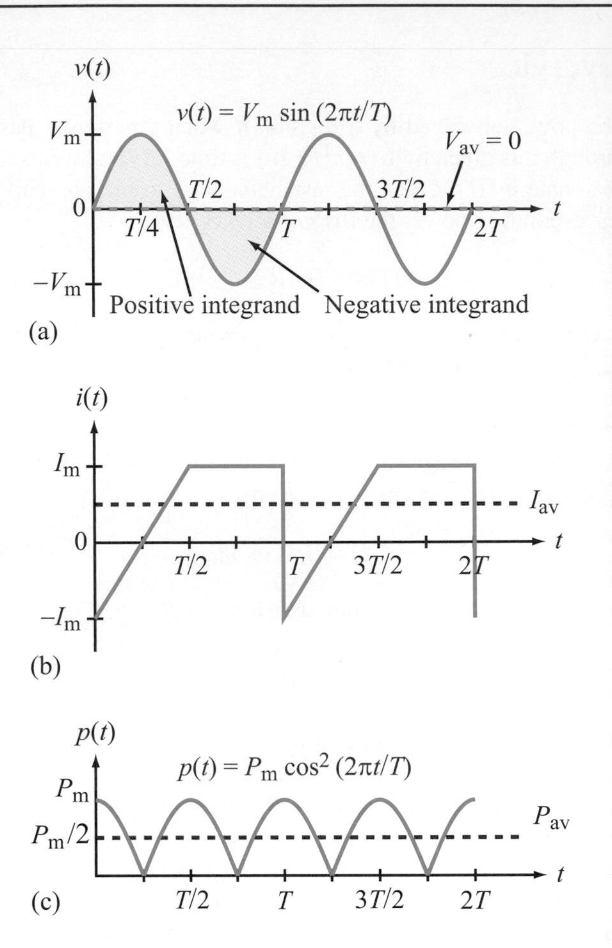

Figure 8-1: Examples of three periodic waveforms.

Example 8-1: Average Values

Determine the average values of the waveforms displayed in parts (b) and (c) of Fig. 8-1.

Solution: During the first half of the first cycle, $i(t)$ is described by a linear ramp of the form

$$i(t) = at + b \qquad \text{for } 0 \le t \le \frac{T}{2}.$$

Its slope is

$$a = \frac{2I_m}{\left(\frac{T}{2}\right)} = \frac{4I_m}{T},$$

and its intercept at $t = 0$ is

$$b = -I_m.$$

Hence,

$$i(t) = \begin{cases} \left(\frac{4t}{T} - 1\right) I_m & \text{for } 0 \le t \le \frac{T}{2}, \\ I_m & \text{for } \frac{T}{2} \le t < T. \end{cases} \tag{8.8}$$

By Eq. (8.5), the average value of $i(t)$ is

$$I_{av} = \frac{1}{T}\int_0^T i(t)\,dt = \frac{1}{T}\left[\int_0^{T/2}\left(\frac{4t}{T} - 1\right) I_m\,dt + \int_{T/2}^{T} I_m\,dt\right]$$

$$= \frac{I_m}{2}. \tag{8.9}$$

We also can obtain the same result by adding up the areas bounded by one cycle of the waveform of $i(t)$ and then dividing the net total by T, as in

$$I_{av} = \frac{1}{T}\left[-\frac{1}{2}I_m \times \frac{T}{4} + \frac{1}{2}I_m \times \frac{T}{4} + I_m \times \frac{T}{2}\right] = \frac{I_m}{2}.$$

To determine P_{av} of $p(t)$, we apply Eq. (8.5) to the co function:

$$P_{av} = \frac{1}{T}\int_0^T P_m \cos^2\left(\frac{2\pi t}{T}\right) dt.$$

The integration is facilitated by applying the trigonometric relation

$$\cos^2 x = \frac{1}{2} + \frac{1}{2}\cos 2x,$$

which leads to the final result

$$P_{av} = \frac{P_m}{2}.$$

We should take note for future reference of the fact that *the average value of* $\cos^2 \omega t$ *is* $1/2$. In fact, it is easy to show that

$$\frac{1}{T}\int_0^T \cos^2\left(\frac{2\pi nt}{T} + \phi_1\right) dt = \frac{1}{T}\int_0^T \sin^2\left(\frac{2\pi nt}{T} + \phi_2\right) dt = \frac{1}{2} \quad (8.10)$$

for any values of ϕ_1 and ϕ_2 and any integer value of n equal to or greater than 1. That is, *the average values of* $\cos^2(n\omega t)$ *and* $\sin^2(n\omega t)$ *are both* $1/2$, irrespective of whether or not their arguments are shifted by a constant phase angle, so long as the averaging process is performed over a complete period $T = 2\pi/\omega$.

8-1.2 Root-Mean-Square (rms) Value

For a periodic current waveform $i(t)$ flowing through a resistor R, the average power absorbed by the resistor is

$$P_{av} = \frac{1}{T}\int_0^T p(t)\, dt = \frac{1}{T}\int_0^T i^2(t)\, R\, dt. \quad (8.11)$$

We would like to introduce a new attribute of $i(t)$, called its ***effective value***, I_{eff}, defined such that the average power P_{av} delivered by $i(t)$ to resistor R is equivalent to what a dc current I_{eff} would deliver to R, namely $I_{eff}^2 R$. That is,

$$I_{eff}^2 R = P_{av} = \frac{1}{T}\int_0^T i^2(t)\, R\, dt. \quad (8.12)$$

Solving for I_{eff} gives

$$I_{eff} = \sqrt{\frac{1}{T}\int_0^T i^2(t)\, dt} \quad (8.13)$$

According to Eq. (8.13), I_{eff} is obtained by taking the square ***root*** of the ***mean*** (average value) of the ***square*** of $i(t)$. The three terms characterizing the operation are coupled together to form ***root-mean-square*** (***rms*** for short) and I_{eff} is relabeled I_{rms}. Even though the idea to define an effective or rms value is introduced in connection with a periodic current waveform, the definition is equally applicable to a periodic voltage waveform as well as to any other periodic waveform. For a periodic waveform $x(t)$, its rms value therefore is defined as

$$X_{rms} = X_{eff} = \sqrt{\frac{1}{T}\int_0^T x^2(t)\, dt}. \quad (8.14)$$

Example 8-2: rms Values

Determine the rms values of (a) $v(t) = V_m \sin(2\pi t/T + \phi)$ and (b) $i(t)$ of Fig. 8-1(b).

Solution:

(a) Application of Eq. (8.14) to $v(t)$ gives

$$V_{rms} = \left[\frac{1}{T}\int_0^T V_m^2 \sin^2\left(\frac{2\pi t}{T} + \phi\right) dt\right]^{1/2} \quad (8.15)$$

In view of Eq. (8.10),

$$V_{rms} = \frac{V_m}{\sqrt{2}}. \quad (8.16)$$

Hence:

For any sinusoidal function, its rms value is equal to its maximum value (its amplitude) divided by $\sqrt{2}$.

(b) From Eq. (8.8) of Example 8-1, $i(t)$ is given by

$$i(t) = \begin{cases} \left(\frac{4t}{T} - 1\right) I_m & \text{for } 0 \le t \le \frac{T}{2}, \\ I_m & \text{for } \frac{T}{2} \le t < T. \end{cases}$$

Its rms value therefore is given by

$$I_{\text{rms}} = \left\{ \frac{1}{T} \left[\int_0^{T/2} \left(\frac{4t}{T} - 1 \right)^2 I_{\text{m}}^2 \, dt + \int_{T/2}^{T} I_{\text{m}}^2 \, dt \right] \right\}^{1/2},$$

which leads to

$$I_{\text{rms}} = \frac{2I_{\text{m}}}{\sqrt{6}} = 0.82 I_{\text{m}}.$$

Review Question 8-1: What is the average value of a sinusoidal waveform?

Review Question 8-2: Why is Eq. (8.10) true, irrespective of the values of ϕ_1 and ϕ_2? Explain in terms of a diagram.

Review Question 8-3: What does rms stand for and how does it relate to its definition?

Exercise 8-1: Determine the average and rms values of the waveform $v(t) = 12 + 6\cos 400t$ V.

Answer: $V_{\text{av}} = 12$ V, $V_{\text{rms}} = 12.73$ V. (See)

Exercise 8-2: Determine the average and rms value of the waveform $i(t) = 8\cos 377t - 4\sin(377t - 30°)$ A.

Answer: $I_{\text{av}} = 0$, $I_{\text{rms}} = 7.48$ A. (See)

8-2 Average Power

The circuit configuration shown in Fig. 8-2 consists of an active ac circuit supplying power to a passive load. The load circuit is not restricted in terms of either its architecture or the combination of resistors, capacitors, and inductors it may contain. The instantaneous voltage across the load is $v(t)$ and the corresponding instantaneous current flowing into it—whose direction is defined in accordance with the passive sign convention—is $i(t)$. Since this is an ac circuit, all of its currents and voltages oscillate sinusoidally at the same angular frequency ω. The general functional forms for $v(t)$ and $i(t)$ are given by

$$v(t) = V_{\text{m}} \cos(\omega t + \phi_v), \tag{8.17a}$$

and

$$i(t) = I_{\text{m}} \cos(\omega t + \phi_i), \tag{8.17b}$$

where V_{m} and I_{m} are the amplitudes of $v(t)$ and $i(t)$, and ϕ_v and ϕ_i are their phase angles, respectively. Our objective is to relate the average power absorbed by the load P_{av} to the parameters of $v(t)$ and $i(t)$.

The instantaneous power flowing into the load circuit is

$$\begin{aligned} p(t) &= v(t)\, i(t) \\ &= V_{\text{m}} I_{\text{m}} \cos(\omega t + \phi_v) \cos(\omega t + \phi_i). \end{aligned} \tag{8.18}$$

By applying the trigonometric identity

$$\cos x \cos y = \frac{1}{2}\cos(x - y) + \frac{1}{2}\cos(x + y), \tag{8.19}$$

$p(t)$ can be cast in the form

$$p(t) = \frac{V_{\text{m}} I_{\text{m}}}{2} \cos(\phi_v - \phi_i) + \frac{V_{\text{m}} I_{\text{m}}}{2} \cos(2\omega t + \phi_v + \phi_i). \tag{8.20}$$

Before proceeding to find the average value of $p(t)$, let us briefly examine the significance of the two terms of Eq. (8.20). The first term is a constant, as it contains no dependence on t, and the second term is sinusoidal, but its angular frequency is 2ω. Thus:

$p(t)$ is the sum of a dc-like term and an ac term that oscillates at a frequency twice that of $i(t)$ and $v(t)$.

This behavior is evident in the waveforms of $v(t)$, $i(t)$, and $p(t)$ displayed in Fig. 8-3. The angular frequency $\omega = 2\pi$ corresponds to $f = 60$ Hz, and the phase angles were arbitrarily chosen as $\phi_v = 30°$ and $\phi_i = -30°$. The waveform patterns elicit the following observations:

(a) The voltage $v(t)$ oscillates symmetrically relative to the t-axis, with a peak-to-peak variation extending from -4 V to $+4$ V. The current $i(t)$ exhibits a similar pattern between -3 A and $+3$ A.

(b) The waveforms of $v(t)$ and $i(t)$ are separated from each other by a time shift Δt, corresponding to the difference in phase angle between them. Given that $\phi_v = +30°$ and $\phi_i = -30°$, $v(t)$ leads $i(t)$ by 60°. A complete period

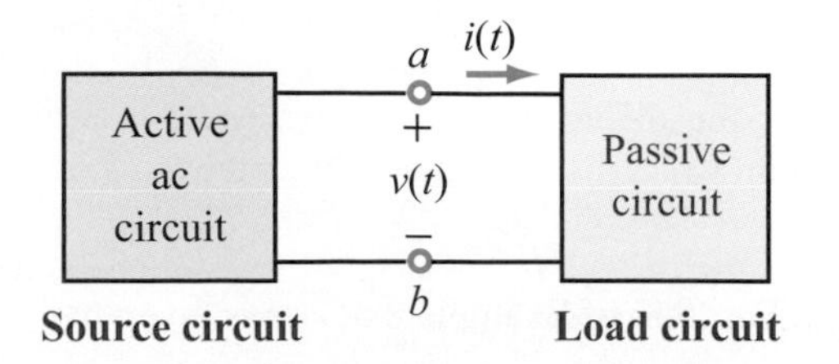

Figure 8-2: Passive load circuit connected to an input source at terminals (a, b).

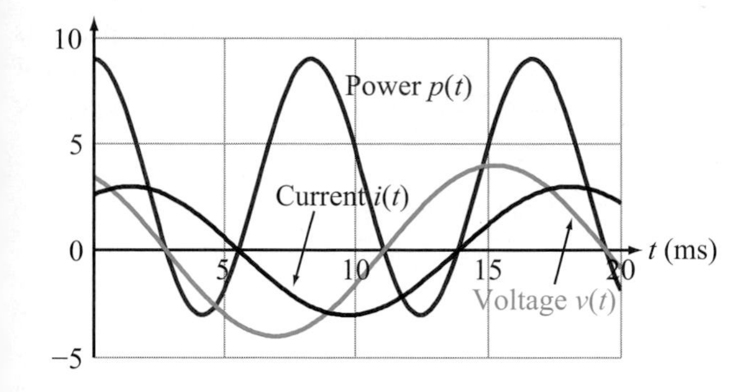

Figure 8-3: Waveforms for a 60-Hz circuit with $v(t) = 4\cos(377t + 30°)$ V, $i(t) = 3\cos(377t - 30°)$ A, and $p(t) = v(t)\, i(t)$. The waveform of $i(t)$ is shifted by 60° behind that of $v(t)$, and the oscillation frequency of $p(t)$ is twice that of $v(t)$ or $i(t)$.

($T = 1/f = 1/60 = 16.67$ ms) corresponds to a total phase angle of 360°. Hence, $v(t)$ leads $i(t)$ by

$$\Delta t = \left(\frac{60°}{360°}\right) \times 16.67 \text{ ms}$$
$$= 2.78 \text{ ms}.$$

(c) The waveform of the power $p(t)$ is not symmetrical with respect to the t-axis. Also, it traces twice as many cycles per unit time, in comparison with the waveforms of $v(t)$ or $i(t)$.

Returning to the task at hand, we now apply Eq. (8.5) to the expression of $p(t)$ given by Eq. (8.20) to determine P_{av}, the average power delivered to the load:

$$P_{\text{av}} = \frac{1}{T}\int_0^T p(t)\, dt$$
$$= \frac{1}{T}\int_0^T \frac{V_{\text{m}} I_{\text{m}}}{2}[\cos(\phi_v - \phi_i) + \cos(2\omega t + \phi_v + \phi_i)]\, dt. \tag{8.21}$$

For any sinusoidal function with $\omega = 2\pi/T$, it is fairly straightforward to show that for integer values of n equal to or greater than 1 and any constant angle θ

$$\frac{1}{T}\int_0^T \cos(n\omega t + \theta)\, dt = 0 \quad (n = 1, 2, \ldots). \tag{8.22}$$

Thus, the average value over a period $T = 2\pi/\omega$ of a sinusoidal function of angular frequency ω or integer multiple of ω is zero. In view of Eq. (8.22), the integral of the second term in Eq. (8.21) is zero. Consequently, the expression for P_{av} simplifies to

$$P_{\text{av}} = \frac{V_{\text{m}} I_{\text{m}}}{2}\cos(\phi_v - \phi_i) \qquad \text{(W)}. \tag{8.23}$$

According to Eq. (8.16), the rms value of a sinusoidal voltage waveform is related to its amplitude by $V_{\text{rms}} = V_{\text{m}}/\sqrt{2}$, and a similar relationship holds for $i(t)$. Hence,

$$P_{\text{av}} = V_{\text{rms}} I_{\text{rms}} \cos(\phi_v - \phi_i) \qquad \text{(W)}. \tag{8.24}$$

The quantity $(\phi_v - \phi_i)$ is called the ***power factor angle*** and plays a critical role with respect to P_{av}. For a purely resistive load R, $v(t)$ and $i(t)$ are in-phase, which means that $\phi_v = \phi_i$. Consequently,

$$P_{\text{av}} = V_{\text{rms}} I_{\text{rms}} = \frac{V_{\text{rms}}^2}{R} \quad \text{(purely resistive load)}. \tag{8.25}$$

The orthogonal state is when the load is purely reactive, in which case $(\phi_v - \phi_i) = \pm 90°$, with the (+) sign corresponding to an inductive load (because v_{L} leads i_{L} by 90°) and the (−) sign corresponding to a capacitive load (v_{C} lags i_{C} by 90°). In either case,

$$P_{\text{av}} = V_{\text{rms}} I_{\text{rms}} \cos 90° = 0 \quad \text{(purely reactive load)}. \tag{8.26}$$

A purely reactive load can store power and then release it, but the net average power it absorbs is zero.

Review Question 8-4: How is the rms value related to the amplitude of a sinusoidal signal?

Review Question 8-5: How much average power is consumed by a reactive load? Explain.

Exercise 8-3: The voltage across and current through a certain load are given by

$$v(t) = 8\cos(754t - 30^\circ) \text{ V},$$
$$i(t) = 0.2\sin 754t \text{ A}.$$

What is the average power consumed by the load, and by how far in time is $i(t)$ shifted relative to $v(t)$?

Answer: $P_{av} = 0.4$ W; $\Delta t = 1.39$ ms. (See)

8-3 Complex Power

The correspondence between the instantaneous voltage $v(t)$ and instantaneous current $i(t)$ and their respective phasors (**V** and **I**) is embodied by the relationships

$$v(t) = V_m \cos(\omega t + \phi_v) \quad \longleftrightarrow \quad \mathbf{V} = V_m e^{j\phi_v} \qquad (8.27a)$$

and

$$i(t) = I_m \cos(\omega t + \phi_i) \quad \longleftrightarrow \quad \mathbf{I} = I_m e^{j\phi_i}. \qquad (8.27b)$$

In the time domain, in general it is not possible to combine all of the elements of a passive load circuit into a single equivalent element, but it is possible to do so in the phasor domain. A passive ac circuit always can be represented by an equivalent impedance **Z**, as shown in Fig. 8-4, and it has to satisfy the condition

$$\mathbf{Z} = \frac{\mathbf{V}}{\mathbf{I}} = \frac{V_m}{I_m}\, e^{j(\phi_v - \phi_i)} \qquad (\Omega), \qquad (8.28)$$

where **V** and **I** are the phasor voltage and current at its input terminals. The ***complex power*** **S** is a phasor quantity defined in terms of **V** and **I**, but it is not simply the product of **V** and **I**. The definition of **S** is constructed such that the real part of **S** is exactly equal to P_{av}, the real average power absorbed by the load **Z**. To that end, **S** is defined as

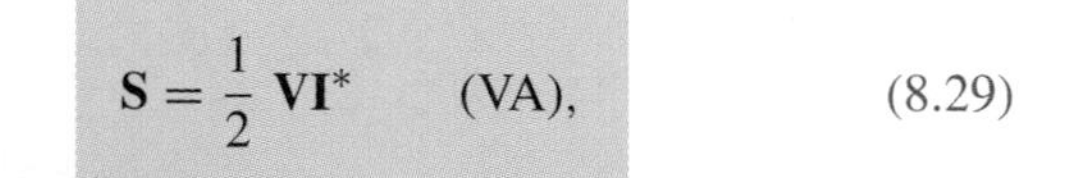

$$\mathbf{S} = \frac{1}{2}\,\mathbf{V}\mathbf{I}^* \qquad \text{(VA)}, \qquad (8.29)$$

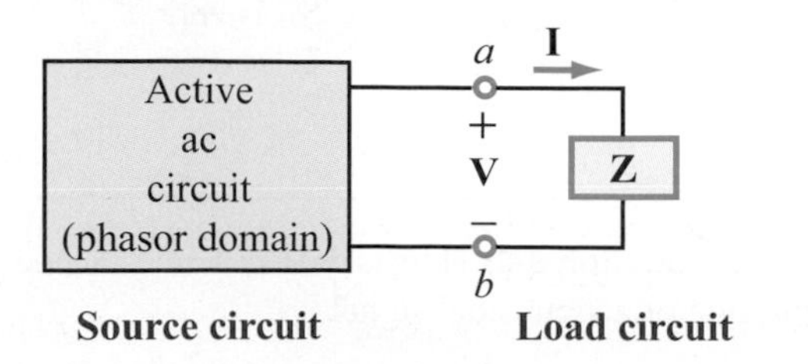

Figure 8-4: Source circuit connected to an impedance **Z** of a load circuit.

where $\mathbf{I}^*$ is the complex conjugate of **I**, realized by replacing j with $-j$ everywhere in **I**. Upon inserting the expressions for **V** and **I** given by Eqs. (8.27a and b) into Eq. (8.29) (after replacing $j\phi_i$ with $-j\phi_i$), we obtain the result

$$\begin{aligned} \mathbf{S} &= \frac{1}{2}(V_m e^{j\phi_v})(I_m e^{-j\phi_i}) \\ &= \frac{1}{2}\, V_m I_m e^{j(\phi_v - \phi_i)} \\ &= \frac{1}{2}\, V_m I_m \cos(\phi_v - \phi_i) + j\frac{1}{2}\, V_m I_m \sin(\phi_v - \phi_i). \end{aligned} \qquad (8.30)$$

For the sake of consistency, we introduce the rms phasor voltage and current as

$$\mathbf{V}_{rms} = \frac{\mathbf{V}}{\sqrt{2}} = \frac{V_m}{\sqrt{2}}\, e^{j\phi_v} \qquad (8.31a)$$

and

$$\mathbf{I}_{rms} = \frac{\mathbf{I}}{\sqrt{2}} = \frac{I_m}{\sqrt{2}}\, e^{j\phi_i}, \qquad (8.31b)$$

and we rewrite Eqs. (8.29) and (8.30) in terms of rms quantities as

$$\mathbf{S} = \mathbf{V}_{rms}\mathbf{I}^*_{rms} \qquad \text{(VA)}, \qquad (8.32)$$

and

$$\mathbf{S} = V_{rms} I_{rms} \cos(\phi_v - \phi_i) + j\, V_{rms} I_{rms} \sin(\phi_v - \phi_i). \qquad (8.33)$$

We note that the real part of **S** (first term) is equal to the expression for P_{av} given by Eq. (8.24). The second term is called the ***reactive power*** Q:

$$Q = V_{rms} I_{rms} \sin(\phi_v - \phi_i) \qquad \text{(VAR)}. \qquad (8.34)$$

Hence,

$$\mathbf{S} = P_{av} + jQ \qquad \text{(VA)}, \qquad (8.35)$$

and conversely,

$$P_{av} = \mathfrak{Re}[\mathbf{S}] \qquad \text{(average absorbed power)} \qquad (8.36a)$$

and

$$Q = \mathfrak{Im}[\mathbf{S}] \qquad \text{(peak exchanged power)}. \qquad (8.36b)$$

Whereas P_{av} represents real dissipated power, Q represents the peak amount of power exchanged (back and forth) between the source circuit and the load circuit.

During a single oscillation cycle of duration T:

$P_{av}T$ = energy dissipated in the load,
QT = energy transferred to the load and then returned to the source.

The three quantities—$\mathbf{S}$, P_{av}, and Q—are each a product of a voltage and a current and therefore should be measured in watts. However, to help distinguish between them, only P_{av} retains the unit of watt, and the other two have been assigned artificially different units. $\mathbf{S}$ has been given the unit ***volt-ampere*** (VA) and Q the unit ***volt-ampere reactive (VAR)***.

8-3.1 Complex Power for a Load

So far, we have expressed $\mathbf{S}$ in terms of $\mathbf{V}$ and $\mathbf{I}$, but $\mathbf{V}$ and $\mathbf{I}$ are linked to one another through the impedance of the load circuit $\mathbf{Z}$ (Fig. 8-4). In general, $\mathbf{Z}$ has a real, resistive component R, and an imaginary, reactive component X:

$$\mathbf{Z} = R + jX.$$

We should recall from Chapter 7 that the reactive component is inductive if $X > 0$ and capacitive if $X < 0$. In terms of $\mathbf{Z}$,

$$\mathbf{V} = \mathbf{ZI}, \tag{8.37}$$

and the expression for $\mathbf{S}$ given by Eq. (8.29) becomes

$$\begin{aligned} \mathbf{S} &= \frac{1}{2}\,\mathbf{VI}^* \\ &= \frac{1}{2}|\mathbf{I}|^2\mathbf{Z} \\ &= I_{\text{rms}}^2(R + jX), \end{aligned} \tag{8.38}$$

From this, we deduce that

$$P_{\text{av}} = \mathfrak{Re}[\mathbf{S}] = \frac{1}{2}|\mathbf{I}|^2 R = I_{\text{rms}}^2 R \quad \text{(W)} \tag{8.39a}$$

and

$$Q = \mathfrak{Im}[\mathbf{S}] = \frac{1}{2}|\mathbf{I}|^2 X = I_{\text{rms}}^2 X \quad \text{(VAR)}. \tag{8.39b}$$

The relationship between $\mathbf{S}$ and its components P_{av} and Q is illustrated graphically in Fig. 8-5(a) for an impedance with an inductive component ($X > 0$). A similar illustration is contained in Fig. 8-5(b) for an impedance with a capacitive component ($X < 0$). The vector $\mathbf{S}$ lies in quadrant 1 $(0 < (\phi_v - \phi_i) \leq 90^\circ)$ if X is inductive and in quadrant 4 $(-90^\circ \leq (\phi_v - \phi_i) < 0)$ if X is capacitive. [If $\mathbf{S}$ were to lie in quadrants 2 or 3, P_{av} would be negative, indicating that the load is actually a source supplying power, not consuming it.]

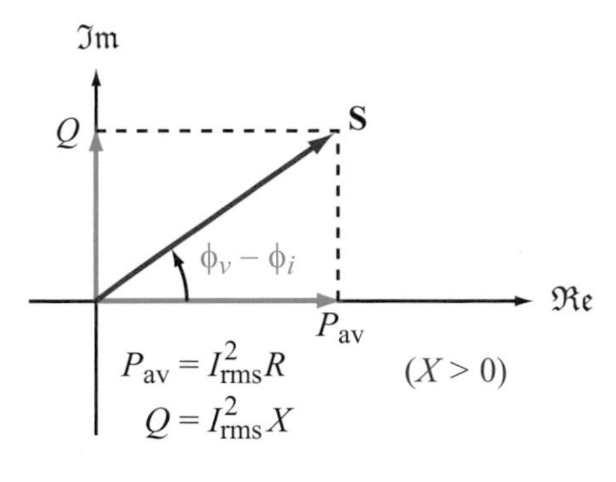

(a) Inductive load

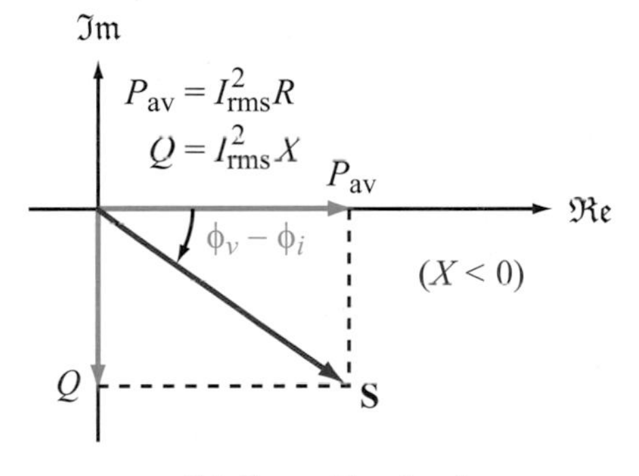

(b) Capacitive load

Figure 8-5: Complex power $\mathbf{S}$ lies in quadrant 1 for an inductive load and in quadrant 4 for a capacitive load.

8-3.2 Conservation of Complex Power

In a circuit containing n elements, energy conservation requires that the sum of the complex powers associated with all n elements be equal to zero:

$$\sum_{i=1}^{n} \mathbf{S}_i = 0.$$

Since $\mathbf{S}_i$ is complex, it follows that both the real and imaginary components of the sum have to individually be equal to zero, which, in view of Eq. (8.35), leads to

$$\sum_{i=1}^{n} P_{\text{av}_i} = 0, \qquad \sum_{i=1}^{n} Q_i = 0. \tag{8.40}$$

Keeping in mind that P_{av_i} has a positive (+) value if the ith element is a resistor and a negative (−) value if it is a generator of power, the first summation in Eq. (8.40) states that the power consumed by the resistors is equal to the (real) power generated by the sources in the circuit. Similarly, the summation over Q_i states that there is no net exchange of reactive power between the sources and the reactive elements in the circuit.

Example 8-3: RL Load

An input circuit consisting of a source $v_s = 10\cos 10^5 t$ V in series with a source resistance $R_s = 100\ \Omega$ is connected to an RL load circuit, as shown in Fig. 8-6(a). If $R = 300\ \Omega$ and $L = 3$ mH, determine: $\mathbf{I}$, P_{av}, Q, $\mathbf{S}$, and ϕ_v.

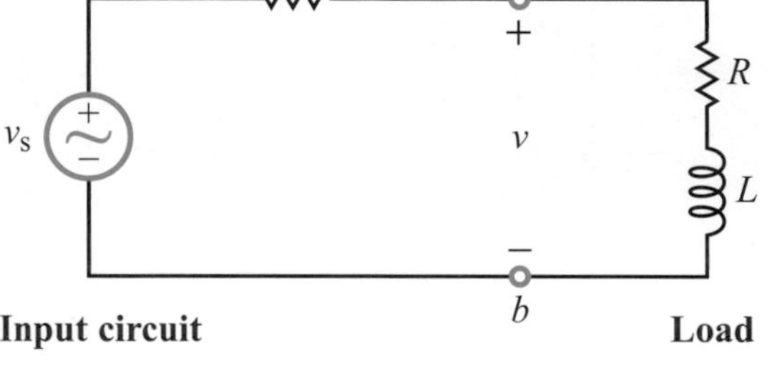

(a) Circuit

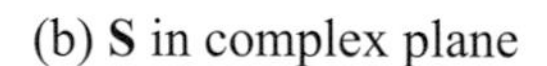

(b) **S** in complex plane

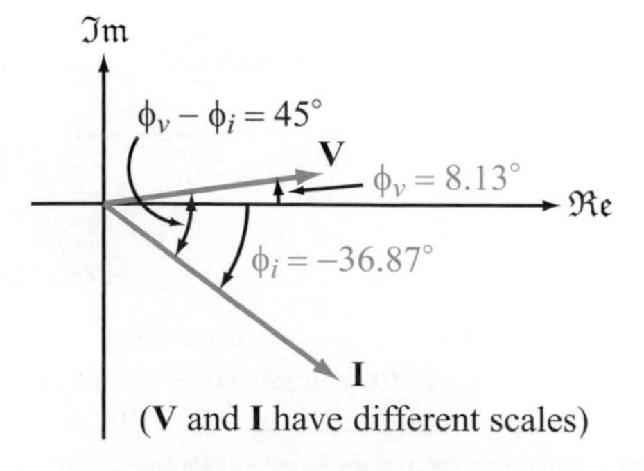

(c) **V** and **I** in complex plane

Figure 8-6: Example 8-3.

Solution: From the expression for v_s, we deduce that $\mathbf{V}_s = 10$ V and $\omega = 10^5$ rad/s. Hence, the load impedance is

$$\mathbf{Z} = R + j\omega L$$
$$= 300 + j10^5 \times 3 \times 10^{-3} = (300 + j300)\ \Omega.$$

The phasor current $\mathbf{I}$ corresponding to $i(t)$ is given by

$$\mathbf{I} = \frac{\mathbf{V}_s}{R_s + \mathbf{Z}}$$
$$= \frac{10}{100 + 300 + j300}$$
$$= \frac{10}{400 + j300} = 20e^{-j36.87°} \text{ mA}.$$

Given that $I_m = 20$ mA and $R = X = 300\ \Omega$,

$$P_{av} = I_{rms}^2 R$$
$$= \frac{I_m^2 R}{2} = \frac{(20 \times 10^{-3})^2}{2} \times 300 = 60 \text{ mW},$$
$$Q = I_{rms}^2 X$$
$$= \frac{(20 \times 10^{-3})^2}{2} \times 300 = 60 \text{ mVAR},$$

and their combination specifies $\mathbf{S}$ as

$$\mathbf{S} = 60 + j60$$
$$= 84.85e^{j45°} \text{ mVA}.$$

According to Eq. (8.30), the phase angle of $\mathbf{S}$ is equal to $\phi_v - \phi_i$ Hence,

$$45° = \phi_v - (-36.87°),$$

which yields

$$\phi_v = 8.13°.$$

We also can determine $\mathbf{V}$ independently by applying voltage division to the circuit,

$$\mathbf{V} = \frac{\mathbf{V}_s \mathbf{Z}}{R_s + \mathbf{Z}}$$
$$= \frac{10(300 + j300)}{400 + j300} = 8.48e^{j8.13°} \text{ V},$$

which confirms the value we found earlier for ϕ_v. Figure 8-6(b) and (c) provide graphical renditions of $\mathbf{S}$, $\mathbf{V}$, and $\mathbf{I}$ in the complex plane.

Example 8-4: Capacitive Load

The current source $i_s(t) = 20\cos(10^3 t + 30°)$ mA and associated shunt resistance $R_s = 400\ \Omega$, as shown in Fig. 8-7(a), provide ac power to the load circuit to the right of terminals (a, b). If $R_1 = 200\ \Omega$, $R_2 = 2\ \text{k}\Omega$, and $C = 1\ \mu\text{F}$, determine: (a) **I**, **V**, **S**, P_{av}, and Q for the entire load circuit (to the right of terminals (a, b)), (b) $\mathbf{S}_C$ for the capacitor alone, and (c) $\mathbf{S}_s$ for the current source.

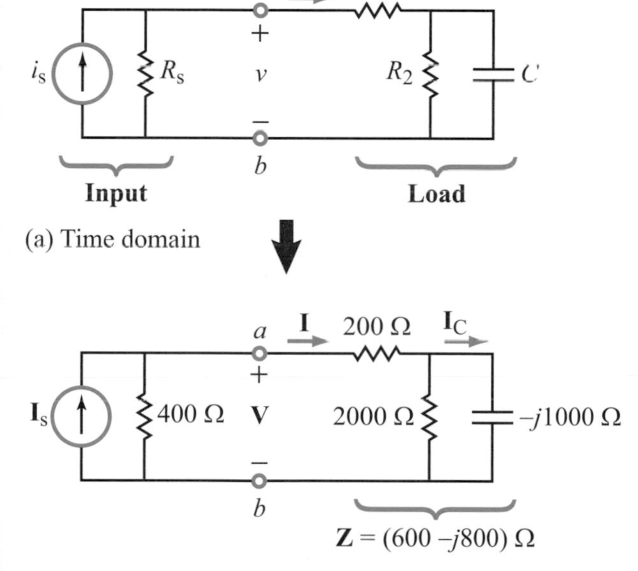

Figure 8-7: Circuit for Example 8-4.

Solution:
(a) In the phasor domain,

$$\mathbf{I}_s = 20e^{j30°} \text{ mA},$$

$$\mathbf{Z}_C = \frac{-j}{\omega C} = \frac{-j}{10^3 \times 10^{-6}} = -j1000\ \Omega,$$

and the impedance **Z** of the load circuit is

$$\mathbf{Z} = R_1 + R_2 \parallel \mathbf{Z}_C$$
$$= 200 + \frac{2000 \times (-j1000)}{2000 - j1000} = (600 - j800)\ \Omega.$$

Current division in the phasor-domain circuit of Fig. 8-7(b) yields

$$\mathbf{I} = \frac{\mathbf{I}_s R_s}{R_s + \mathbf{Z}} = \frac{20 \times 10^{-3} e^{j30°} \times 400}{400 + (600 - j800)}$$
$$= 6.25e^{j68.66°} \text{ mA},$$

and the phasor voltage at terminals (a, b) is

$$\mathbf{V} = \mathbf{IZ} = 6.25 \times 10^{-3} e^{j68.66} \times (600 - j800)$$
$$= 6.25e^{j15.53°} \text{ V}.$$

Given **I** and **V**, the complex power **S** is

$$\mathbf{S} = \frac{1}{2}\mathbf{VI}^*$$
$$= \frac{1}{2} \times 6.25e^{j15.53°} \times 6.25 \times 10^{-3} e^{-j68.66°}$$
$$= 19.53e^{-j53.13°} \text{ mVA},$$

with real and imaginary components given by

$$P_{av} = \mathfrak{Re}[\mathbf{S}]$$
$$= 19.53 \times 10^{-3} \cos(-53.13°)$$
$$= 11.72 \text{ mW}$$

and

$$Q = \mathfrak{Im}[\mathbf{S}]$$
$$= 19.53 \times 10^{-3} \sin(-53.13°)$$
$$= -15.62 \text{ mVAR}.$$

(b) The phasor current $\mathbf{I}_C$ flowing through C is related to **I** by

$$\mathbf{I}_C = \frac{R_2 \mathbf{I}}{R_2 + \mathbf{Z}_C} = \frac{2000 \times 6.25 \times 10^{-3} e^{j68.66°}}{2000 - j1000}$$
$$= 5.59e^{j95.23°} \quad \text{(mA)},$$

and the corresponding voltage $\mathbf{V}_C$ across the capacitor is

$$\mathbf{V}_C = \mathbf{I}_C \mathbf{Z}_C = 5.59e^{j95.23°} \times 10^{-3} \times (-j1000)$$
$$= 5.59e^{j5.23°} \text{ V},$$

where we used the identity

$$-j = e^{-j90°}.$$

The complex power associated with the capacitor is

$$\begin{aligned}\mathbf{S}_{\mathrm{C}} &= \frac{1}{2}\,\mathbf{V}_{\mathrm{C}}\mathbf{I}_{\mathrm{C}}^{*}\\ &= \frac{1}{2}\,5.59e^{j5.23^\circ} \times 5.59 \times 10^{-3}e^{-j95.23^\circ}\\ &= 15.62e^{-j90^\circ}\\ &= 0 - j15.62 \text{ mVA}.\end{aligned}$$

As expected, the real part of $\mathbf{S}_{\mathrm{C}}$ (representing the amount of power *dissipated* in the capacitor) is zero, and the imaginary part is exactly equal to Q of the overall load circuit (the capacitor is the only element in the load circuit capable of *exchanging* power back and forth with the input circuit).

(c) Recall that for any device, **S** represents the complex power transferred into the device, and it is defined such that the current direction through the device is from the (+) terminal to the (−) terminal of the voltage across it. For the current source $\mathbf{I}_{\mathrm{s}}$, it flows through itself from the (−) terminal of **V** to the (+) terminal of **V**, in exact opposition to the definition of **S**. Hence,

$$\begin{aligned}\mathbf{S}_{\mathrm{s}} &= -\frac{1}{2}\,\mathbf{V}\mathbf{I}_{\mathrm{s}}^{*}\\ &= -\frac{1}{2} \times 6.25e^{j15.53^\circ} \times 20e^{-j30^\circ} \times 10^{-3}\\ &= -62.5e^{-j14.47^\circ} \text{ mVA}\\ &= -62.5\cos(-14.47^\circ) - j62.5\sin(-14.47^\circ)\\ &= (-60.52 + j15.62) \text{ mVA}.\end{aligned}$$

The real part of $\mathbf{S}_{\mathrm{s}}$ represents the real average power generated by $\mathbf{I}_{\mathrm{s}}$ and is equal in magnitude to the average power dissipated in the three resistors in the circuit. The imaginary part of $\mathbf{S}_{\mathrm{s}}$ is equal in magnitude and opposite in sign to $\mathbf{S}_{\mathrm{C}}$.

Review Question 8-6: What are the two components of the complex power **S**, what type of power do they represent, and what units are assigned to them?

Review Question 8-7: If **S** lies in quadrant 2 in the complex plane, what does that tell you about the load?

Exercise 8-4: The current flowing into a load is given by $i(t) = 2\cos 2500t$ A. If the load is known to consist of a series of two passive elements, and $\mathbf{S} = (10 - j8)$ VA, determine the identities of the elements and their values.

Answer: $R = 5\ \Omega$, $C = 100\ \mu$F. (See ◎)

8-4 The Power Factor

Several power-related terms were introduced in the preceding two sections, including the complex power **S**, the real average power P_{av}, and the reactive power Q. We plan to introduce two additional terms in this section, so lest this apparent profusion of terms contribute to any possible confusion, we have prepared a summary of all relevant terms and expressions in the form of Table 8-1. This is intended to provide the reader easy access to and greater clarity about the interrelationships among the various power quantities.

In terms of the complex quantities **V** and **I** (representing the phasor voltage across a load circuit and the associated current into it) the complex power **S** transferred to the load circuit (Fig. 8-8) is given by

$$\mathbf{S} = P_{\mathrm{av}} + jQ, \tag{8.41}$$

with

$$P_{\mathrm{av}} = V_{\mathrm{rms}} I_{\mathrm{rms}} \cos(\phi_v - \phi_i) \tag{8.42a}$$

and

$$Q = V_{\mathrm{rms}} I_{\mathrm{rms}} \sin(\phi_v - \phi_i). \tag{8.42b}$$

For reasons that we will discuss in the next subsection, the magnitude of **S** is called the ***apparent power*** S, and it is given by

$$S = |\mathbf{S}| = \sqrt{P_{\mathrm{av}}^2 + Q^2} = V_{\mathrm{rms}} I_{\mathrm{rms}}, \tag{8.43}$$

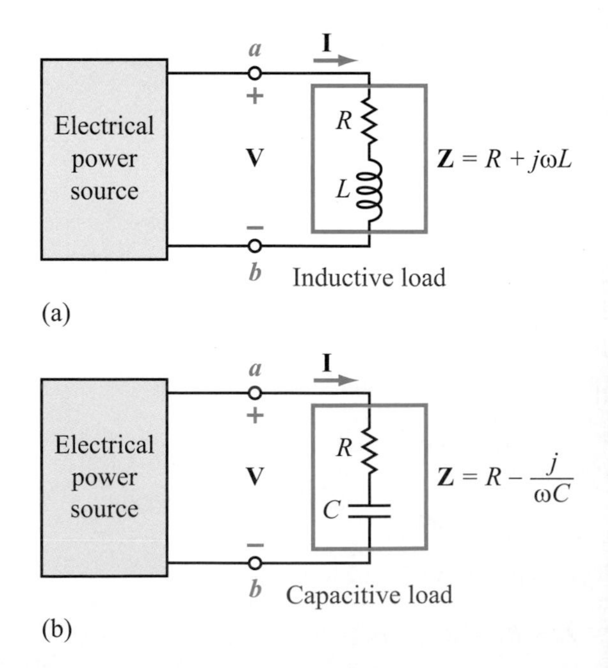

Figure 8-8: Inductive and capacitive loads connected to an electrical source.

Table 8-1: Summary of power-related quantities.

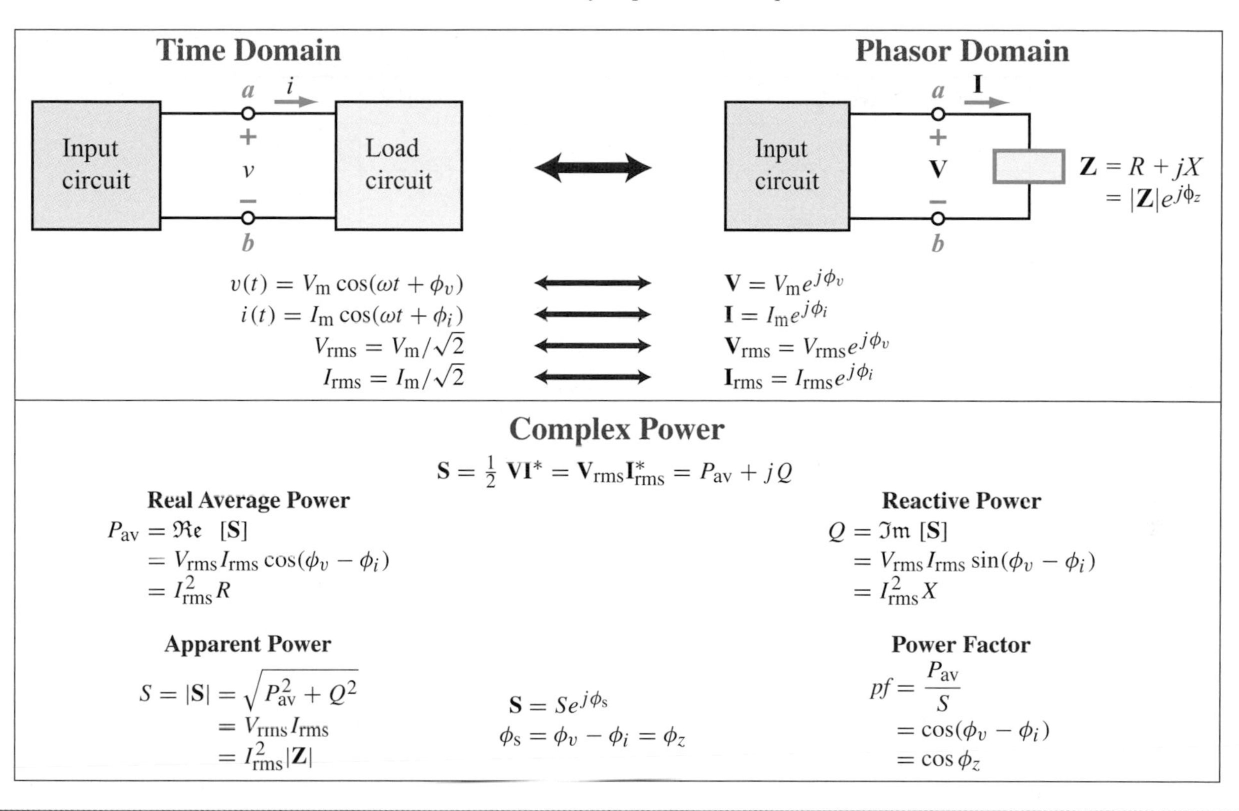

$$v(t) = V_m \cos(\omega t + \phi_v) \longleftrightarrow \mathbf{V} = V_m e^{j\phi_v}$$

$$i(t) = I_m \cos(\omega t + \phi_i) \longleftrightarrow \mathbf{I} = I_m e^{j\phi_i}$$

$$V_{rms} = V_m/\sqrt{2} \longleftrightarrow \mathbf{V}_{rms} = V_{rms} e^{j\phi_v}$$

$$I_{rms} = I_m/\sqrt{2} \longleftrightarrow \mathbf{I}_{rms} = I_{rms} e^{j\phi_i}$$

Complex Power

$$\mathbf{S} = \tfrac{1}{2}\,\mathbf{V}\mathbf{I}^* = \mathbf{V}_{rms}\mathbf{I}^*_{rms} = P_{av} + jQ$$

Real Average Power

$$P_{av} = \Re\mathfrak{e}\,[\mathbf{S}] = V_{rms} I_{rms} \cos(\phi_v - \phi_i) = I^2_{rms} R$$

Reactive Power

$$Q = \Im\mathfrak{m}\,[\mathbf{S}] = V_{rms} I_{rms} \sin(\phi_v - \phi_i) = I^2_{rms} X$$

Apparent Power

$$S = |\mathbf{S}| = \sqrt{P^2_{av} + Q^2} = V_{rms} I_{rms} = I^2_{rms}|\mathbf{Z}|$$

$$\mathbf{S} = Se^{j\phi_s}$$
$$\phi_s = \phi_v - \phi_i = \phi_z$$

Power Factor

$$pf = \frac{P_{av}}{S} = \cos(\phi_v - \phi_i) = \cos\phi_z$$

and the ratio of P_{av} to S is called the ***power factor*** pf, and is given by

$$pf = \frac{P_{av}}{S} = \cos(\phi_v - \phi_i). \tag{8.44}$$

The argument of the cosine $(\phi_v - \phi_i)$ is called the ***power factor angle***. This angle is equal to the phase angle of the load impedance ϕ_z. To verify this statement, we first express $\mathbf{Z}$ in polar form,

$$\mathbf{Z} = R + jX = |\mathbf{Z}|e^{j\phi_z}, \tag{8.45}$$

with

$$|\mathbf{Z}| = \sqrt[+]{R^2 + X^2}, \qquad \text{and} \qquad \phi_z = \tan^{-1}\left(\frac{X}{R}\right). \tag{8.46}$$

Then, we write $\mathbf{Z}$ as the ratio of $\mathbf{V}$ to $\mathbf{I}$; namely

$$\mathbf{Z} = \frac{\mathbf{V}}{\mathbf{I}} = \frac{|\mathbf{V}|e^{j\phi_v}}{|\mathbf{I}|e^{j\phi_i}} = \frac{V_m}{I_m}\, e^{j(\phi_v - \phi_i)}. \tag{8.47}$$

Equality of two complex numbers mandates that they have equal magnitudes and equal phase angles. Hence, equating Eq. (8.45) to Eq. (8.47) leads to

$$\phi_z = \phi_v - \phi_i, \qquad |\mathbf{Z}| = \frac{V_m}{I_m}. \tag{8.48}$$

In view of Eq. (8.48), the expression for the power factor can be rewritten as

$$pf = \cos\phi_z. \tag{8.49a}$$

We also should note that the phase angle of the complex power $\mathbf{S}$ is equal to the phase angle of $\mathbf{Z}$. That is,

$$\phi_s = \phi_z. \tag{8.49b}$$

Inductive Load

An inductive load, such as a series RL circuit, has an impedance

$$\mathbf{Z}_{ind} = R + j\omega L. \tag{8.50}$$

As both components of $\mathbf{Z}_{ind}$ are positive quantities, ϕ_z is positive. Since R cannot be negative, the range of ϕ_z is $0 \le \phi_z \le 90°$ with $0°$ corresponding to a purely resistive load and $90°$ corresponding to a purely inductive load.

Table 8-2: Power factor leading and lagging relationships for a load $\mathbf{Z} = R + jX$.

Load Type	$\phi_z = \phi_v - \phi_i$	I-V Relationship	*pf*
Purely Resistive ($X = 0$)	$\phi_z = 0$	**I** in-phase with **V**	1
Inductive ($X > 0$)	$0 < \phi_z \leq 90°$	**I** lags **V**	lagging
Purely Inductive ($X > 0$ and $R = 0$)	$\phi_z = 90°$	**I** lags **V** by 90°	lagging
Capacitive ($X < 0$)	$-90° \leq \phi_z < 0$	**I** leads **V**	leading
Purely Capacitive ($X < 0$ and $R = 0$)	$\phi_z = -90°$	**I** leads **V** by 90°	leading

Capacitive Load

The equivalent circuit of a capacitive load is a series RC circuit with

$$\mathbf{Z}_{\text{cap}} = R - \frac{j}{\omega C}. \tag{8.51}$$

Consequently, ϕ_z is negative and its range is $-90° \leq \phi_z \leq 0$, with $-90°$ corresponding to a purely capacitive load.

Because $\cos(-\theta) = \cos\theta$ for any angle θ between $-90°$ and $+90°$, the power factor (Eq. (8.49a)) is insensitive to the sign of ϕ_z, and therefore, it cannot differentiate between an inductive load and a capacitive load. To qualify *pf* with such information:

The load is said to have a ***leading pf*** or a ***lagging pf***, depending on whether the current **I** leads or lags the voltage **V** (see Table 8-2).

8-4.1 Power Factor Significance

Most industrial loads involve the use of large motors or other inductive machinery that require the supply of tens of kilowatts of power, typically at 440 V rms. Household appliances (such as refrigerators and air conditioners) also contain inductive coils, and most are designed to operate at either 110 V rms or 220 V rms. Thus, most loads to which an electrical source has to supply power have an RL equivalent circuit of the type shown in Fig. 8-8(a). From the perspective of an energy supplier (such as the electric power company) the load has two important attributes: S and P_{av}. The amount of power the company has to supply is S, but it can charge for only P_{av}, because P_{av} is the only real power consumed by the load. The company appears to supply S—hence, the name apparent power—but it gets paid for a fraction of that, and the power factor is that fraction. For two loads—one purely resistive with $\mathbf{Z}_1 = R$ and the second inductive with $\mathbf{Z}_2 = R + j\omega L$—with both requiring the same voltage **V** and consuming the same power P_{av}, the inductive load will require the transmission of a larger current to it than would the purely resistive load. This point is demonstrated numerically through Example 8-5.

Example 8-5: ac Motor

The equivalent circuit of a dishwasher motor is characterized by an impedance $\mathbf{Z} = (20 + j20)\ \Omega$. The household voltage is 110 V rms. Determine: (a) *pf*, S, and P_{av} and (b) the current that the electric company would have supplied to the motor had it been purely resistive and consumed the same amount of power.

Solution:

(a) We will treat the phase of the voltage as our reference by setting it arbitrarily equal to zero. Thus,

$$\mathbf{V}_{\text{rms}} = V_{\text{rms}}\angle 0° = 110\ \text{V}.$$

This is justified by the fact that none of the quantities of interest require knowledge of the values of ϕ_v and ϕ_i individually; it is the difference $(\phi_v - \phi_i)$ that counts. The corresponding current is

$$\mathbf{I}_{\text{rms}} = \frac{\mathbf{V}_{\text{rms}}}{\mathbf{Z}} = \frac{110}{20 + j20} = \frac{110}{20\sqrt{2}\, e^{j45°}} = 3.9\angle{-45°}\ \text{A},$$

from which we deduce that $I_{\text{rms}} = 3.9$ A and $\phi_z = 45°$. The quantities of interest are then given by

$$S = V_{\text{rms}} I_{\text{rms}} = 110 \times 3.9 = 427.8\ \text{VA},$$

$$P_{\text{av}} = S\cos\phi_z = 429\cos 45° = 302.5\ \text{W},$$

and

$$pf = \frac{P_{\text{av}}}{S} = 0.707.$$

(b) A purely resistive load that consumes 302.5 W at 110 V rms must have a current of

$$I_{\rm rms} = \frac{P_{\rm av}}{V_{\rm rms}} = \frac{302.5}{110} = 2.75 \text{ A}.$$

For the same amount of consumed power, the power supplier has to provide 3.89 A to an inductive load with a power factor of 0.707, compared with only 2.75 A to a purely resistive load with $pf = 1$.

8-4.2 Power Factor Compensation

Raising the power factor of an inductive load (such as an electric drill or a compressor) is highly desirable, not only for the energy supplier but also ultimately for its customers as well. Redesigning the load circuit itself to raise its power factor to a value closer to 1, however, may not be practical, primarily because its motor or other inductive components were presumably selected to meet certain operational specifications that may be incompatible with a higher power factor. This problem of partial incompatibility raises the following question: can we raise the *pf* of a load (as seen by the generator circuit) while keeping it the same as far as the inductive load itself is concerned? The answer is yes, and the solution is fairly straightforward: it entails adding a shunt capacitor across the inductive load, as shown in Fig. 8-9(b). Without the capacitor (Fig. 8-9(a)), the inductor load requires a voltage $\mathbf{V}_{\rm L}$ across it and a current $\mathbf{I}_{\rm L}$ through it. The source current $\mathbf{I}_{\rm s}$ is equal to $\mathbf{I}_{\rm L}$. The presence of the shunt capacitor does not change $\mathbf{V}_{\rm L}$, and by virtue of the load impedance $\mathbf{Z}_{\rm L}$, the current $\mathbf{I}_{\rm L} = \mathbf{V}_{\rm L}/\mathbf{Z}_{\rm L}$ also remains unchanged. In other words, the capacitor exercises no influence on the inductive load, but it does change the overall load circuit as far as the generator is concerned. The new load circuit—which we will call the ***compensated load circuit***—consists of the parallel combination of C and the original RL circuit. Because of the new current $\mathbf{I}_{\rm C}$, the source current becomes

$$\mathbf{I}'_{\rm s} = \mathbf{I}_{\rm L} + \mathbf{I}_{\rm C}. \tag{8.52}$$

Had C and the RL load been purely resistive, both $\mathbf{I}_{\rm C}$ and $\mathbf{I}_{\rm L}$ would have been real and of the same sign, resulting in a larger source current rather than smaller. Fortunately, $\mathbf{I}_{\rm C}$ and $\mathbf{I}_{\rm L}$ are phasor quantities, and their imaginary components have opposite polarities [actually, $\mathbf{I}_{\rm C}$ is purely imaginary]. With $\mathbf{V}_{\rm L}$ chosen to serve as the phase reference, Fig. 8-10(a) illustrates

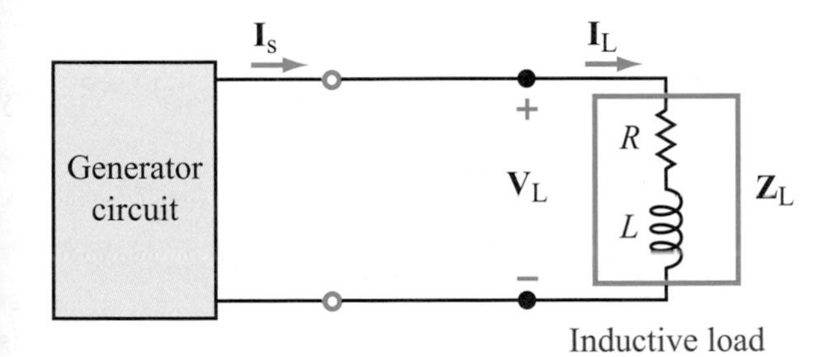

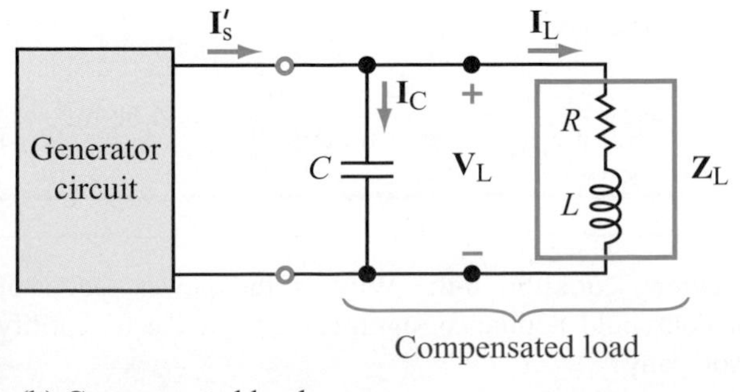

Figure 8-9: Adding a shunt capacitor across an inductive load reduces the current supplied by the generator.

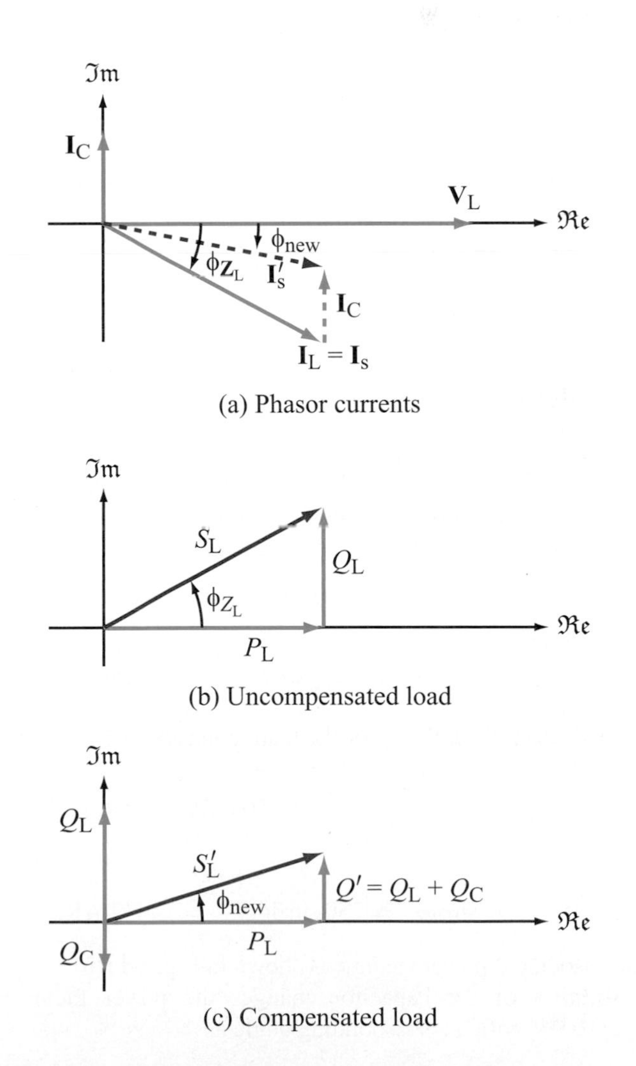

Figure 8-10: Comparison of source currents and power triangles for the compensated and uncompensated circuits.

how the vector sum of $\mathbf{I}_L$ (the current into the RL circuit) and $\mathbf{I}_C$ leads to a vector $\mathbf{I}'_s$, whose length (or equivalently, its magnitude) is shorter than the length it was before adding the capacitor. In terms of the power factor,

$$pf = \begin{cases} \cos\phi_{\mathbf{Z}_L} & \text{for the RL circuit alone,} \\ \cos\phi_{new} & \text{for the compensated circuit,} \end{cases} \tag{8.53}$$

where ϕ_{new} is the phase angle between $\mathbf{I}'_s$ and $\mathbf{V}_L$ in the compensated load circuit.

Another approach to demonstrate how the addition of the capacitor improves the power factor is by comparing the power triangle of the RL circuit alone with that of the compensated load circuit that includes the capacitor. The two triangles are diagrammed in parts (b) and (c) of Fig. 8-10, in which P_L and Q_L represent the consumed and reactive powers associated with the RL load, and Q_C is associated with the capacitor C. The capacitor introduces reactive power Q_C, and since Q_C is negative, the net sum

$$Q' = Q_L + Q_C \tag{8.54}$$

is smaller than Q_L alone, thereby reducing the phase angle from $\phi_{\mathbf{Z}_L}$ to ϕ_{new}, where

$$\phi_{new} = \tan^{-1}\left(\frac{Q'}{P_L}\right). \tag{8.55}$$

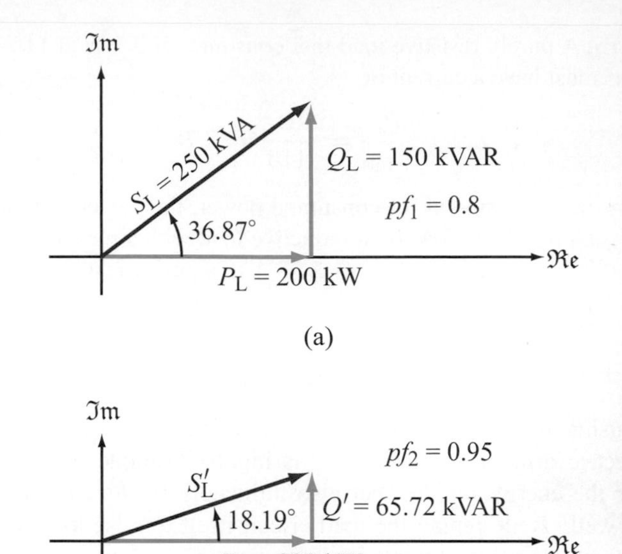

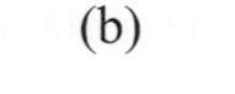

Figure 8-11: Power triangles for Example 8-6.

Example 8-6: *pf* Compensation

A 60-Hz electric generator supplies a 220-V rms to a load that consumes 200 kW at $pf = 0.8$ lagging. By adding a shunt capacitor C, the power factor of the overall circuit was improved to 0.95 lagging. Determine the value of C.

Solution: A power factor of 0.8 corresponds to a phase angle $\phi_{\mathbf{Z}_L}$ given by

$$\phi_{\mathbf{Z}_L} = \cos^{-1}(pf_1) = \cos^{-1}(0.8) = 36.87°.$$

The values of S_L and Q_L for the load alone are

$$S_L = \frac{P_L}{pf_1} = \frac{200 \times 10^3}{0.8} = 250 \text{ kVA}$$

and

$$Q_L = S_L \sin\phi_{\mathbf{Z}_L} = 250 \sin 36.87° = 150 \text{ kVAR}.$$

The associated power triangle is shown in Fig. 8-11(a).

Addition of the capacitor changes the power factor to $pf_2 = 0.95$, with a corresponding angle as

$$\phi_{new} = \cos^{-1}(pf_2) = \cos^{-1}(0.95) = 18.19°.$$

The consumed power P_L does not change, but from Fig. 8-11(b), the new reactive power is now

$$Q' = 200 \tan\phi_{new} = 200 \tan 18.19° = 65.72 \text{ kVAR}.$$

Using the value of Q_L we determined earlier, the reactive power introduced by the capacitor is

$$Q_C = Q' - Q_L = (65.74 - 150) = -84.26 \text{ kVAR}.$$

With $\mathbf{Z}_C = 1/j\omega C$, the complex power of C is

$$\mathbf{S}_C = \mathbf{V}_{L_{rms}} \mathbf{I}^*_{C_{rms}} = \mathbf{V}_{L_{rms}} \frac{\mathbf{V}^*_{L_{rms}}}{\mathbf{Z}^*_C} = -j|\mathbf{V}_{L_{rms}}|^2 \omega C.$$

Hence, $P_C = 0$, and

$$Q_C = -|\mathbf{V}_{L_{rms}}|^2 \omega C.$$

Solving for C gives

$$C = \frac{-Q_C}{2\pi f V^2_{rms}} = \frac{84.26 \times 10^3}{2\pi \times 60 \times (220)^2} = 4.62 \text{ mF}.$$

Review Question 8-8: Why is the power factor of a household appliance significant to an electric utility company?

Review Question 8-9: What is *pf* compensation, and why is it used?

Exercise 8-5: At 60 Hz, the impedance of a RL load is $\mathbf{Z}_L = (50 + j50)\ \Omega$. (a) What is the value of the power factor of $\mathbf{Z}_L$. (b) What will be the new power factor if a capacitance $C = \frac{1}{12\pi}$ mF is added in parallel with the RL load?

Answer: (a) $pf_1 = 0.707$, (b) $pf_2 = 1$. (See CD)

8-5 Maximum Power Transfer

Consider the network configuration shown in Fig. 8-12 in which an ac source circuit is represented by its Thévenin equivalent circuit, composed of a phasor voltage $\mathbf{V}_s$ and a source impedance

$$\mathbf{Z}_s = R_s + jX_s. \tag{8.56}$$

Similarly, the load is represented by its impedance $\mathbf{Z}_L$ with

$$\mathbf{Z}_L = R_L + jX_L. \tag{8.57}$$

In Section 3-6, we established that for a purely resistive circuit, the power transferred from the source circuit to the load is a maximum when $R_L = R_S$. The question we now pose is: What are the equivalent conditions for an ac circuit with complex impedances?

To answer the question, we start by writing down the expression for $\mathbf{I}_L$ (the current flowing into the load) namely

$$\mathbf{I}_L = \frac{\mathbf{V}_s}{\mathbf{Z}_s + \mathbf{Z}_L} = \frac{\mathbf{V}_s}{(R_s + R_L) + j(X_s + X_L)}. \tag{8.58}$$

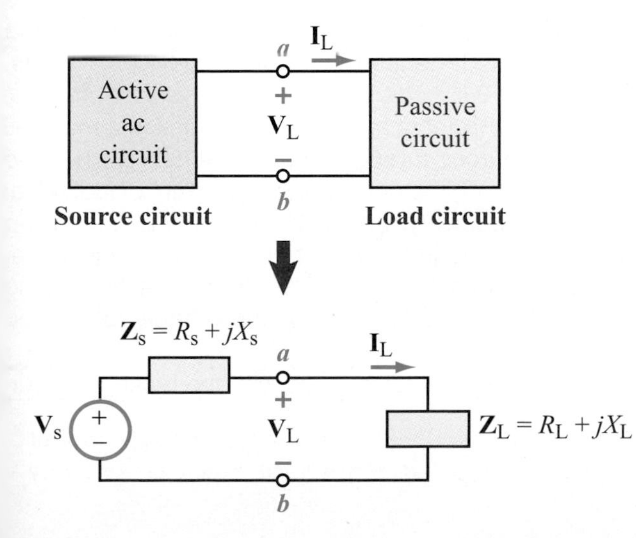

Figure 8-12: Replacing the source and load circuits with their respective Thévenin equivalents.

From Eq. (8.39a), the average power transferred to (consumed by) the load is

$$\begin{aligned} P_{av} &= \frac{1}{2}\,|\mathbf{I}_L|^2 R_L \\ &= \frac{1}{2}\,\mathbf{I}_L \times \mathbf{I}_L^* R_L \\ &= \frac{1}{2}\,\frac{\mathbf{V}_s}{(R_s + R_L) + j(X_s + X_L)} \\ &\quad \times \frac{\mathbf{V}_s^*}{(R_s + R_L) - j(X_s + X_L)} \cdot R_L \\ &= \frac{1}{2}\,\frac{|\mathbf{V}_s|^2 R_L}{(R_s + R_L)^2 + (X_s + X_L)^2}. \end{aligned} \tag{8.59}$$

The load parameters R_L and X_L represent orthogonal dimensions in the complex plane. Hence, the values of R_L and X_L that maximize P_{av} can be obtained by performing independent maximization processes: one by setting $\partial P_{av}/\partial R_L = 0$ and another by setting $\partial P_{av}/\partial X_L = 0$. For R_L,

$$\begin{aligned} \frac{\partial P_{av}}{\partial R_L} &= \frac{1}{2}\,|\mathbf{V}_s|^2 \\ &\quad \cdot \left[\frac{(R_s + R_L)^2 + (X_s + X_L)^2 - 2R_L(R_s + R_L)}{[(R_s + R_L)^2 + (X_s + X_L)^2]^2} \right]. \end{aligned} \tag{8.60}$$

The right-hand side of Eq. (8.60) is equal to zero if its numerator is equal to zero (because the other alternative, namely setting the denominator equal to infinity, produces a solution in which $\mathbf{Z}_L$ and $\mathbf{Z}_s$ are open circuits corresponding to no power transfer to the load). That is,

$$(R_s + R_L)^2 + (X_s + X_L)^2 - 2R_L(R_s + R_L) = 0,$$

which simplifies to

$$R_s^2 - R_L^2 + (X_s + X_L)^2 = 0. \tag{8.61}$$

Similarly, the partial derivative of P_{av} with respect to X_L is

$$\frac{\partial P_{av}}{\partial X_L} = \frac{1}{2}\,|\mathbf{V}_s|^2 R_L \left[\frac{-2(X_s + X_L)}{(R_s + R_L)^2 + (X_s + X_L)^2} \right], \tag{8.62}$$

which when set equal to zero yields

$$X_L = -X_s. \tag{8.63}$$

Incorporating Eq. (8.63) in Eq. (8.61) gives

$$R_L = R_s. \tag{8.64}$$

Technology Brief 17: Bandwidth, Data Rate, and Communication

In Section 9-4.1, we defined the ***bandwidth*** B of a resonant circuit as the frequency span over which power transfer through the circuit is greater than half of the maximum level possible (Fig. 9-15(b)). This common half-power (or −3-dB) definition for B can be extended to many devices, circuits, and transmission channels. But how does the everyday use of the word ***bandwidth*** refer to the data rate of a transmission channel, such as the rate at which your internet connection can download data?

Signal and Noise in Communication Channels

Every circuit (including switches, amplifiers, filters, phase shifters, rectifiers, etc.) and every transmission medium (air, wires, and optical fibers) operate with acceptable performance over a specific range of frequencies, outside of which ac signals are severely damped. The actual span of this operational frequency range is dictated by the physical characteristics of the circuit or transmission medium. One such example is the ***coaxial cable*** commonly used to connect a TV to a "cable network" or to an outside antenna. The coaxial cable is a high-fidelity transmission medium—causing negligible distortion or attenuation of the signal passing through it—so long as the carrier frequency of the signal is not much higher than about 10 GHz (because higher frequencies are highly attenuated and the shape of the signal waveform gets distorted). The "cutoff" frequency of a typical coaxial cable is determined by the cable's distributed capacitance, inductance, and resistance, which are governed in turn by the geometry of the cable, the conductivity of its inner and outer conductors, and the permittivity of the insulator that separates them. The ***MOSFET*** offers another example; in Section 5-7 we noted that the switching speed of a MOSFET circuit is limited by parasitic capacitances, setting an upper limit on the switching frequency that a given MOSFET circuit can handle. A circuit with a maximum switching speed of 100 ps, for example, cannot respond to frequencies greater than 1/100 ps (or 10 GHz) without distorting the output waveform in some significant way. Our third example is the ***earth's atmosphere***. According to Fig. TF8-2 (in Technology Brief 8: The Electromagnetic Spectrum on page 158), the transmission spectrum for the atmosphere is characterized by a limited set of ***transmission windows***, with each window extending over a specific range of frequencies.

The overall effective bandwidth B of a communication system is determined by the operational bandwidths of its constituent circuits and the transmission spectra of the cables or other transmission media it uses. As we will see shortly, the channel capacity (or data rate) of the system is directly proportional to B, but it also is influenced by the intensity and character of the ***noise*** in the system. Noise is random power self-generated by all real devices, circuits, and transmission media. In fact, *any material at a temperature greater than* 0 K *(which includes all physical materials, since no material can exist at exactly* 0 K*) emits noise power all of the time.* Random motions of electrons in a material generate mini-currents that function like a random distribution of power sources throughout the material. Such noise sources exist in both passive and active devices, and models exist for describing the noise power spectrum of individual types of devices, circuits, and even entire systems (such as a cell-phone receiver). Figure TF17-1 illustrates how the noise generated by a circuit modifies the waveform of the signal passing through it. The input is an ideal sine wave, whereas the output consists of the same sine wave but with a fluctuating component added to it. The fluctuating component represents the noise generated by the circuit, which is random in polarity because the voltage associated with the noise fluctuates randomly between positive and negative values.

If we know both the power P_S carried by the signal and the average power P_N associated with the noise, we then can determine the ***signal-to-noise ratio (SNR)***:

$$\text{SNR} = \frac{P_S}{P_N}\,.$$

The bandwidth B and the SNR (Fig. TF17-2), jointly determine the highest data rate that can be transmitted reliably through a circuit or a communication system.

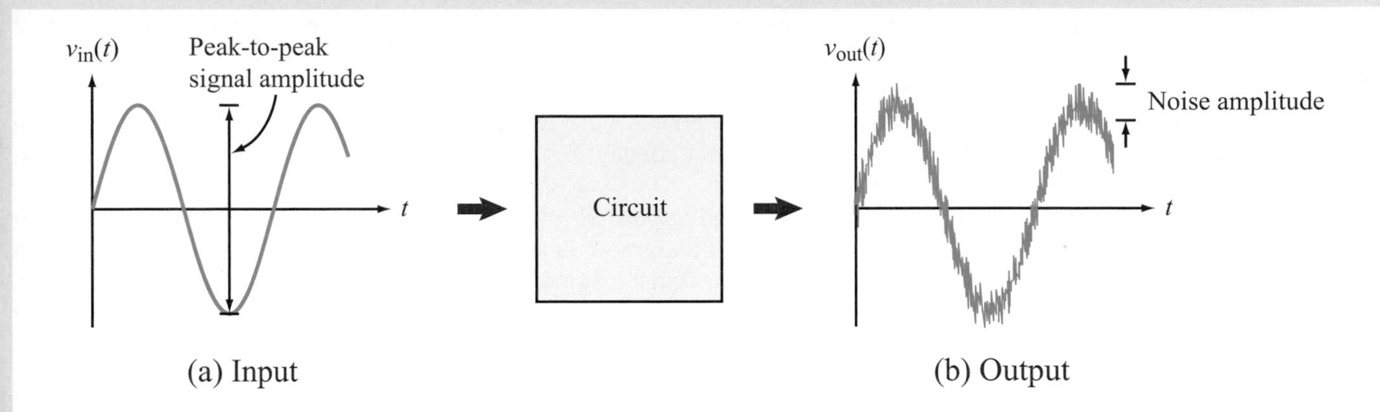

Figure TF17-1: The noise generated by the circuit adds a fluctuating component to the input signal.

Shannon-Hartley Theorem

In the late 1940s and early 1950s, ***Claude Shannon***, building on earlier work by ***Harry Nyquist*** and ***Ralph Hartley***, developed a complete theory that established the limits of information transfer in a communication system. This seminal work represents the foundation of ***information theory*** and underlies all of the subsequent developments that shaped today's ***information revolution***, including the Internet, cell phones, satellite communications, and much more.

The ***Shannon-Hartley theorem*** defines how much data can be transferred through a channel (with no error) in terms of the bandwidth B and the SNR. It states that

$$C = B \ln\left(1 + \frac{P_S}{P_N}\right),$$

where C is the ***channel capacity*** (or ***data rate***) in bits/second, B is the bandwidth in Hz, and P_S/P_N is the SNR. As an example, let us consider a communication channel with $B = 100$ MHz, $P_S = 1$ mW and $P_N = 1$ μW. The corresponding SNR is 1000, and the corresponding value of C is 690.87×10^6 bits/s or $\simeq 691$ Mb/s. By way of comparison, a 100GBE

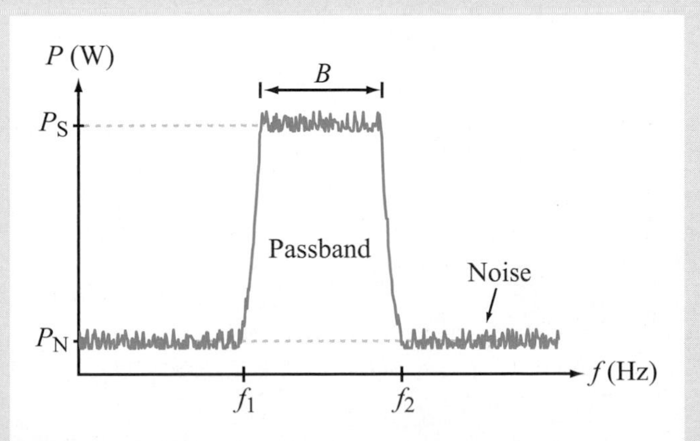

Figure TF17-2: Typical spectral response of a communication system with bandwidth B. The signal-to-noise ratio is given by P_S/P_N, where P_S and P_N are the average power levels of the signal and noise, respectively.

Ethernet connection can operate at 100 Gbps (or approximately two orders of magnitude faster), a 100 Base-T Ethernet connection can manage a maximum rate of only 100 Mbps, 802.11 Wifi networks are rated at 54 Mbps, and the Bluetooth 2.0 protocol used by cell phones is limited to 2.1 Mbps. The channel capacity of a conventional telephone used to support audio transmissions is only 56 kbps. We should note at this juncture that when people use the term "bandwidth" in everyday speech, they really mean channel capacity; B and C are directly proportional to one another, but they obviously are not the same quantity.

According to the expression for C, for a sufficiently high bandwidth, it is possible to achieve reasonably high data-transfer rates even when $\text{SNR} < 1$! The implication of this statement is that information can be transmitted reliably on channels whose noise levels exceed that of the signal, provided the signal is spread across a wide frequency spectrum. For example, with a bandwidth of 1 GHz, it is possible to transfer error-free data at a rate of 95 Mbps, even when SNR is only 0.1 (that is, with the signal power an order of magnitude smaller than the noise power). This is (in part) the basis for ***ultra-wideband*** communication schemes. In the U.S., the 3.1 to 10.6 GHz band recently was reserved for this use.

The development and proof of the Shannon-Hartley theorem is beyond the scope of this book, but a simple understanding of its implications helps explain the technology drive towards higher-frequency circuits and systems. Given an ever-present amount of noise in our electrical systems and transmission channels (even the atmosphere generates noise), increasing the data rate beyond current levels will require either a boost in signal power or an increase in signal bandwidth. The former of these two options (boosting signal power) poses a number of technical challenges and may introduce certain undesirable consequences (such as health effects in the case of a cell phone, for example), so the preferred option is to extend the range of the carrier-frequency into the upper part of the microwave spectrum. Shannon's foundational work also has impacted the development of many related disciplines, including encryption, encoding, jamming, efficient use of frequency space, and even quantum-level information manipulations.

The conditions on X_L and R_L can be combined into

$$\mathbf{Z}_L = \mathbf{Z}_s^* \qquad \text{(Maximum power transfer)}, \qquad (8.65)$$

where $\mathbf{Z}_s^* = (R_s - jX_s)$ is the complex conjugate of $\mathbf{Z}_s$. When the condition represented by Eq. (8.65) is true, the load is said to be ***matched*** to the source.

According to the result encapsulated by Eq. (8.65):

The average power transferred to (consumed by) an ac load is a maximum when its impedance $\mathbf{Z}_L$ is equal to $\mathbf{Z}_s^*$, which is the complex conjugate of the Thévenin impedance of the source circuit.

Under the conditions of maximum power transfer represented by Eqs. (8.63) and (8.64), the expression for P_{av} given by Eq. (8.59) reduces to

$$P_{av}(\max) = \frac{1}{8}\,\frac{|\mathbf{V}_s|^2}{R_L}. \qquad (8.66)$$

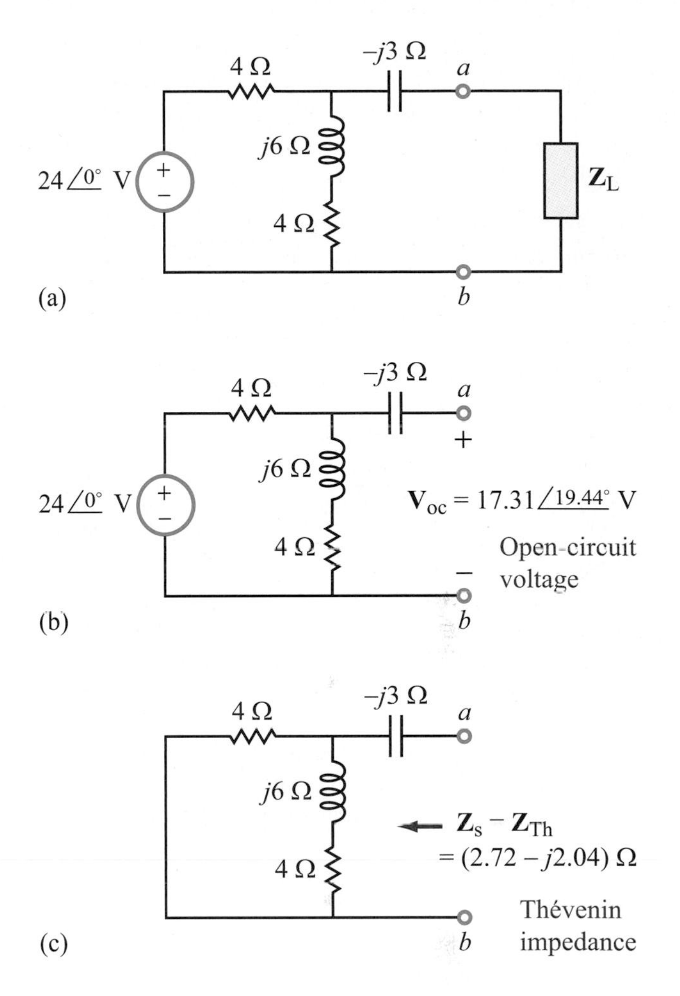

Figure 8-13: Circuit for Example 8-7.

Example 8-7: Maximum Power

Determine the maximum amount of power that can be consumed by the load $\mathbf{Z}_L$ in the circuit of Fig. 8-13.

Solution: We start by determining the Thévenin equivalent of the circuit to the left of terminals (a, b). In Fig. 8-13(b), the load has been removed so as to calculate the open-circuit voltage. Voltage division yields

$$\begin{aligned}\mathbf{V}_s &= \mathbf{V}_{oc} \\ &= \frac{(4 + j6)}{4 + 4 + j6} \times 24 \\ &= 17.31\angle 19.44^\circ \text{ V},\end{aligned}$$

where $\mathbf{V}_s$ is the Thévenin voltage of the source circuit to the left of terminals (a, b). The Thévenin impedance of the source circuit, $\mathbf{Z}_s$, is obtained by calculating the impedance at terminals (a, b), as shown in Fig. 8-13(c), after deactivating the 24-V voltage source,

$$\begin{aligned}\mathbf{Z}_s &= 4 \parallel (4 + j6) - j3 \\ &= \frac{4(4 + j6)}{4 + 4 + j6} - j3 \\ &= (2.72 - j2.04)\ \Omega.\end{aligned}$$

For maximum transfer of power to the load, the load impedance should be

$$\begin{aligned}\mathbf{Z}_L &= \mathbf{Z}_s^* \\ &= (2.72 + j2.04)\ \Omega,\end{aligned}$$

and the corresponding value of P_{av} is

$$\begin{aligned}P_{av}(\max) &= \frac{|\mathbf{V}_s|^2}{8R_L} \\ &= \frac{(17.31)^2}{8 \times 2.72} \\ &= 13.77 \text{ W}.\end{aligned}$$

Review Question 8-10: To achieve maximum transfer of power from a source circuit to a load, how should the impedance of the load be related to that of the source circuit?

Review Question 8-11: Suppose that a certain passive circuit—containing resistors, capacitors, and inductors—is connected to a square-wave voltage source. What procedure would you use to analyze the voltages and currents in the circuit?

Exercise 8-6: The "maximum power transfer" question is stated as follows: Given $\mathbf{Z}_s$, what should $\mathbf{Z}_L$ be so as to maximize power transfer to $\mathbf{Z}_L$? Suppose we were to reverse the question to: Given $\mathbf{Z}_L$, choose $\mathbf{Z}_s$ to maximize power transfer to $\mathbf{Z}_L$. (a) What would the answer be in that case? (b) What would be the corresponding expression for $P_{av}(\max)$?

Answer: (a) $\mathbf{Z}_s = 0 - jX_L$, (b) $P_{av}(\max) = \dfrac{1}{2}\dfrac{|\mathbf{V}_s|^2}{R_L}$. (See CD)

8-6 Application Note: Three-Phase Circuits

Have you ever wondered why a house is wired for electricity with all outlets having three wires, one of which is referred to as the ***neutral wire***? Electric power generation and distribution circuits in the industrialized world are a special case of alternating current circuits, and they can be analyzed handily using the techniques presented in this and the preceding chapter. Although there are many national and regional variants of the circuits described in this section, we will provide enough information to understand the fundamentals.

At most large electrical plants, power is generated simultaneously on three different circuits, all having the same ac frequency f but with different phases. Figure 8-14(a) is a representative cross-sectional view of a typical ***three-phase ac generator***. The generator consists of a rotating electromagnet (called the ***rotor***) and three separate stationary coils distributed evenly around a circular tube (called the ***stator***). The rotor is driven by an external force, such as a turbine powered by steam or gas. The three coils are arranged 120° apart over the circumference of the stator. As the electromagnet rotates, its magnetic field induces a sinusoidal voltage at the terminals of each of the three coils. If the coils are identical in shape and number of turns, the three induced phasor voltages ($\mathbf{V}_1$ to $\mathbf{V}_3$) all will have the same amplitude, and their time-domain counterpart ($v_1(t)$ to $v_3(t)$) will vary sinusoidally at the same frequency of $f = \omega/2\pi$, where ω is the angular rotation frequency of the rotor. However, because the coils physically are distributed 120° apart, the voltages induced in adjacent coils will be shifted in phase by 120° relative to one another. By designating $\mathbf{V}_1$ in Fig. 8-14(a) as the reference voltage with a zero phase, the phase of $\mathbf{V}_2$ will be either 120° or −120°, relative to the phase of $\mathbf{V}_1$ and depending on the relative directions of the two windings. If all windings are the same, then the phases of $\mathbf{V}_1$, $\mathbf{V}_2$, and $\mathbf{V}_3$ are 0, 120°, and 240° (Fig. 8-14(b)), respectively, and their waveforms are shifted in time accordingly (Fig. 8-14(c)). We will refer to this arrangement as a ***balanced three-phase source*** characterized by:

$$\mathbf{V}_1 = V_0\angle 0^\circ, \tag{8.67a}$$

$$\mathbf{V}_2 = V_0\angle 120^\circ, \tag{8.67b}$$

and

$$\mathbf{V}_3 = V_0\angle 240^\circ, \tag{8.67c}$$

with magnitude V_0. We should note that for a balanced three-phase source

$$\mathbf{V}_1 + \mathbf{V}_2 + \mathbf{V}_3 = 0, \tag{8.68}$$

which can be verified numerically by inserting Eq. (8.67) into Eq. (8.68) or graphically by summing the three vectors in Fig. 8-14(b).

In the wiring configuration of Fig. 8-14(a), which is redrawn diagrammatically in Fig. 8-15(a), the three voltage sources share a common terminal called ***neutral terminal n*** and a common wire—namely the ***neutral wire*** referred to earlier. This configuration (which is the most common in North America) is called a ***Y-source configuration***. Alternatively, the three sources can be wired in the form of a ***Δ-source configuration***—without a neutral wire, as shown in Fig. 8-15(b). In either case, each source is represented by an ideal voltage source in series with a complex ***coil impedance*** given by

$$\mathbf{Z}_c = R_c + j\omega L_c. \tag{8.69}$$

The impedance is inductive, because it is associated with a magnetic coil. In most cases, $\mathbf{Z}_c$ is much smaller than the impedances of the loads connected to the generator, allowing us to ignore $\mathbf{Z}_c$ altogether, but we will retain $\mathbf{Z}_c$ for the present.

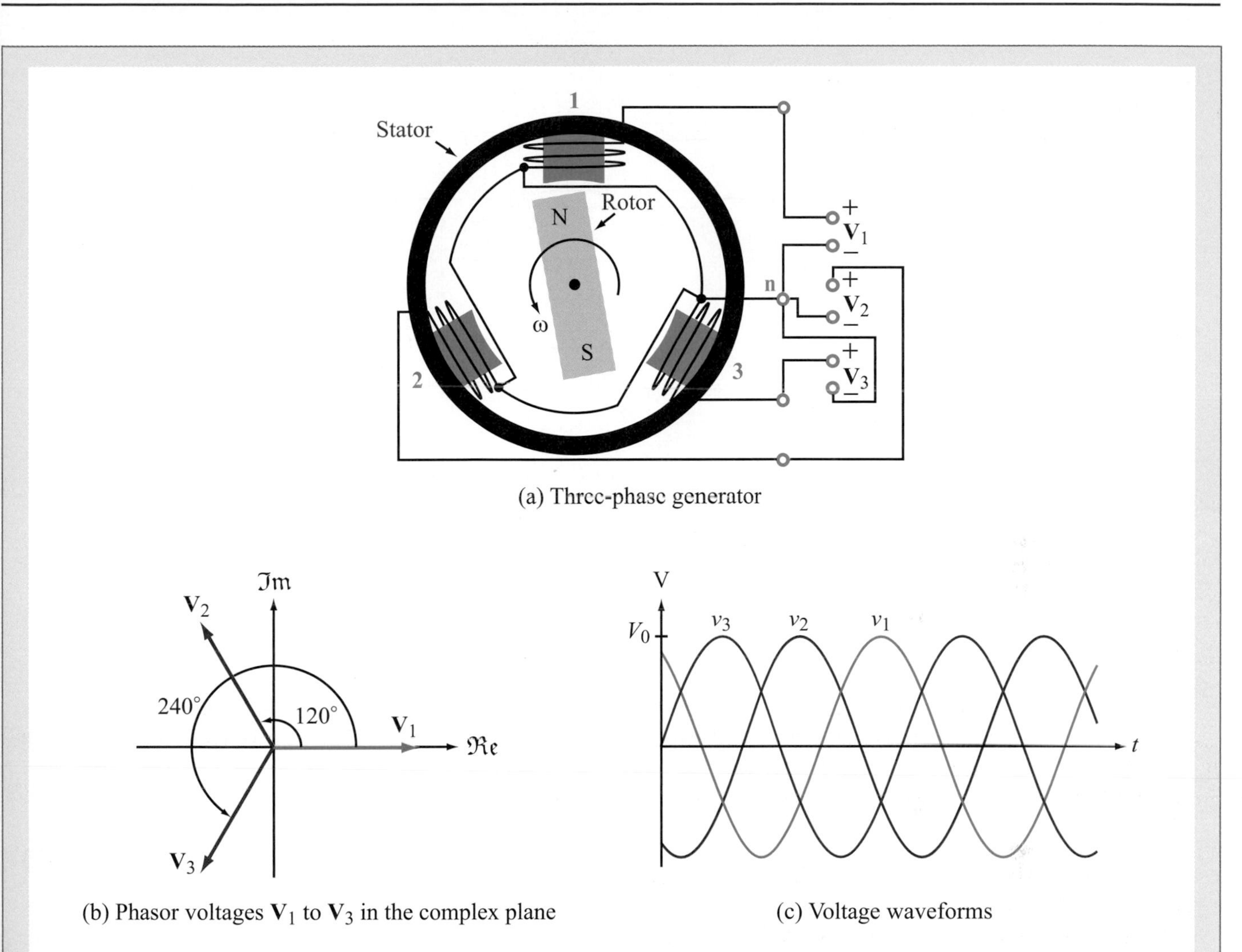

Figure 8-14: Three-phase ac generator and associated voltage waveforms.

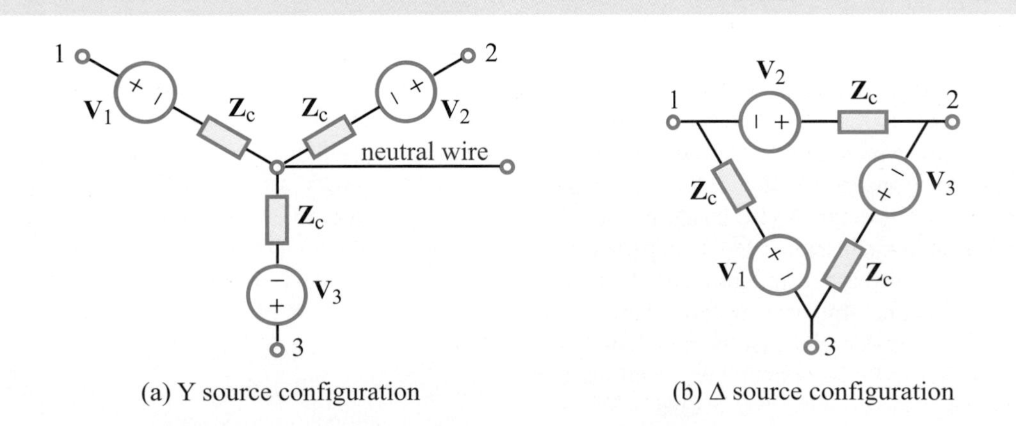

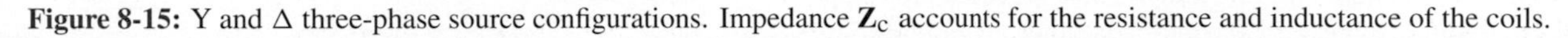

Figure 8-15: Y and Δ three-phase source configurations. Impedance $\mathbf{Z}_c$ accounts for the resistance and inductance of the coils.

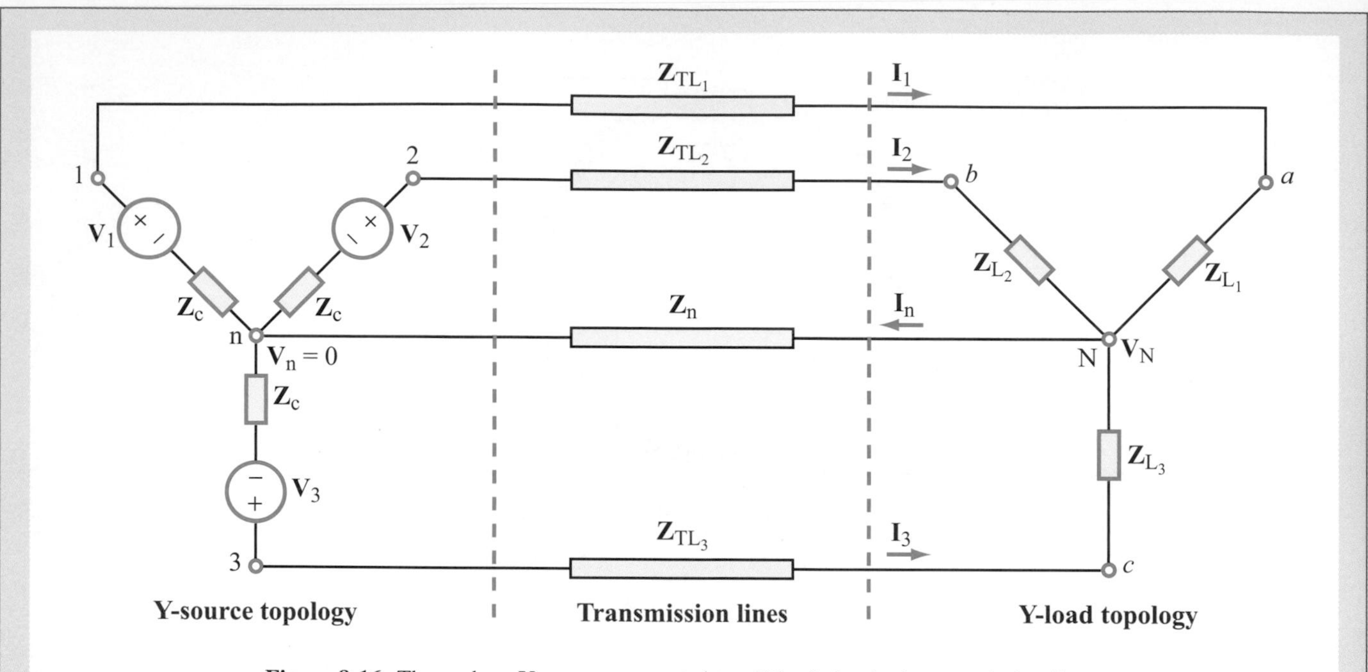

Figure 8-16: Three-phase Y source connected to a Y load circuit via transmission lines.

The three-phase generator is equivalent to three separate ***single-phase generators***, each one of which can be connected to a separate load. Transmission of three-phase power from the generator to loads is more efficient than three separate, single-phase transmissions. The three loads connected to the three sources may be arranged in either a Y- or a Δ-load configuration (with the Δ configuration being more common). Hence, the source–load connections may assume any one of four possible combinations: Y–Y, Y–Δ, Δ–Y, Δ–Δ. Since by applying equivalent-circuit transformations any one of the four connection configurations can be converted into any one of the other three, we will limit our treatment to only one configuration, namely the Y–Y connection shown in Fig. 8-16.

The Y-load network is connected to the Y-source circuit through four wires, usually referred to as ***transmission lines***. The transmission lines (which may or may not be identical) are characterized by impedances $\mathbf{Z}_{\text{TL}_1}$, $\mathbf{Z}_{\text{TL}_2}$, and $\mathbf{Z}_{\text{TL}_3}$ connecting sources $\mathbf{V}_1$ through $\mathbf{V}_3$ to loads $\mathbf{Z}_{\text{L}_1}$ to $\mathbf{Z}_{\text{L}_3}$. In addition, a fourth transmission line of impedance $\mathbf{Z}_n$ connects node n of the source configuration to node N of the load configuration. In the simplest scheme, each load may represent one residential customer, as depicted by the illustration in Fig. 8-17. Commercial operations may connect to more than one branch. For a single residence, a step-down center-tapped transformer is used to ***step-down*** the single-phase ac voltage to a level manageable by household appliances. The power carried to the house from the transformer is called ***three-wire single phase***, with the middle wire assuming the role of the neutral wire. It is single phase because the two 120-V rms voltages at the secondary side of the transformer have the same frequency and phase. Instead of providing a return path to the generating station through an actual wire, the earth ground is used to provide the feedback path for electrons. This is accomplished by using a cable to connect the middle wire at the transformer output to ground (Fig. 8-17). This is done to prevent charging up machinery to dangerous levels, as well as for the discharging of high-voltage events (like lightning) and to prevent current-related heating of wires due to unbalanced loads.

Example 8-8: Balanced Y–Y Network

With reference to the network shown in Fig. 8-16:

(a) Develop a node voltage equation for $\mathbf{V}_N$ (the voltage at node N) with node n treated as (the ground) reference.

(b) A network is said to be ***balanced*** if its source voltages are balanced and if it has identical transmission lines and identical loads. Determine currents $i_1(t)$ to $i_3(t)$ and $i_n(t)$ for a balanced network given that $V_0 = 120\sqrt{2}$ V, $f = 60$ Hz, $\mathbf{Z}_\text{c} = (0.1 + j0.2)\ \Omega$, $\mathbf{Z}_{\text{TL}_1} = \mathbf{Z}_{\text{TL}_2} = \mathbf{Z}_{\text{TL}_3} = (0.9 + j0.8)\ \Omega$, and $\mathbf{Z}_{\text{L}_1} = \mathbf{Z}_{\text{L}_2} = \mathbf{Z}_{\text{L}_3} = (29 + j9)\ \Omega$.

Solution: **(a)** Relative to node n (i.e., with $\mathbf{V}_n = 0$), the node equation at node N is

$$\mathbf{I}_n - \mathbf{I}_1 - \mathbf{I}_2 - \mathbf{I}_3 = 0, \tag{8.70}$$

or equivalently,

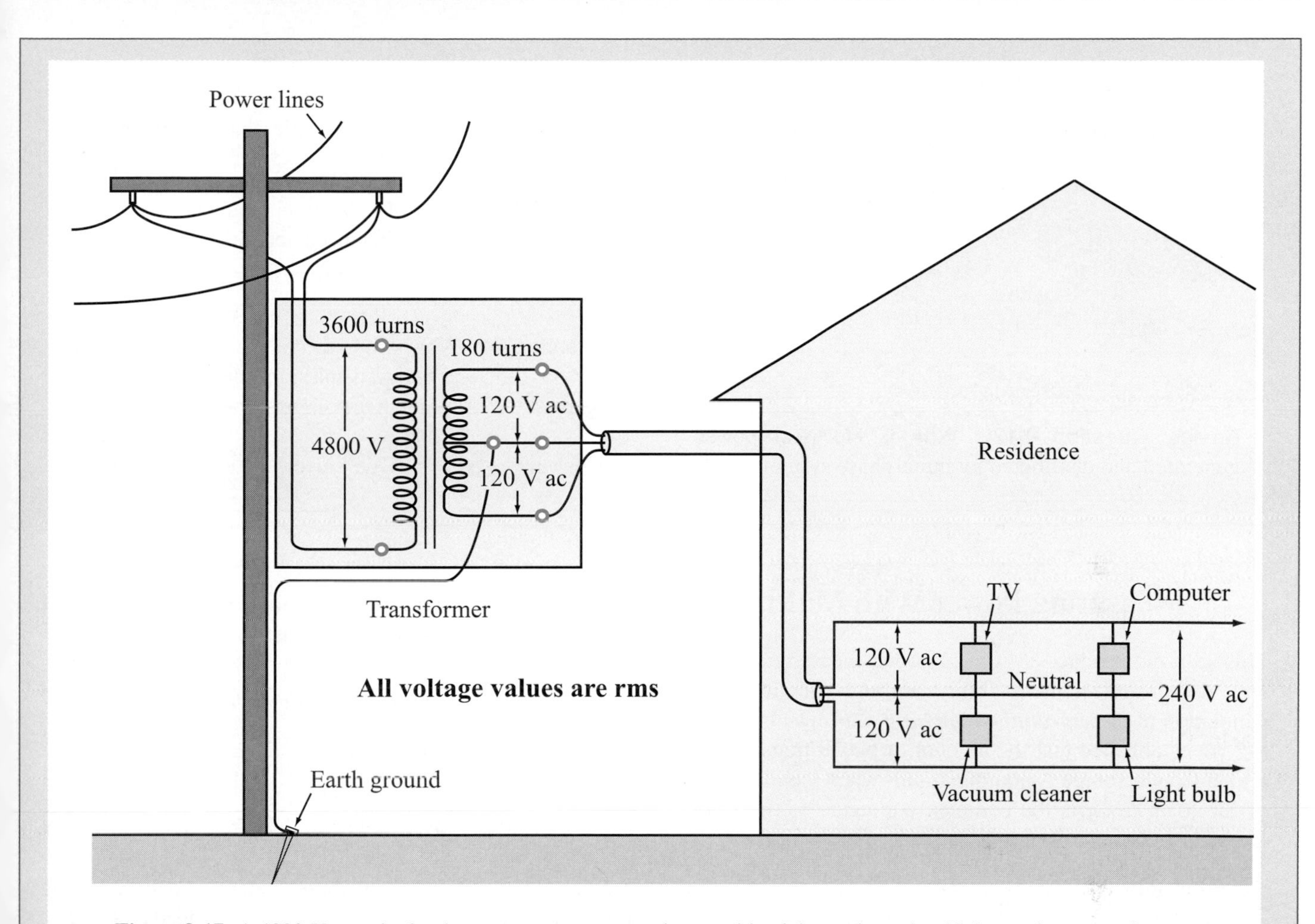

Figure 8-17: A 4800-V rms single-phase ac source connected to a residential user through a 20:1 step-down transformer.

$$\frac{\mathbf{V}_N}{\mathbf{Z}_n} - \frac{(\mathbf{V}_1 - \mathbf{V}_N)}{\mathbf{Z}_c + \mathbf{Z}_{\text{TL}_1} + \mathbf{Z}_{\text{L}_1}} - \frac{(\mathbf{V}_2 - \mathbf{V}_N)}{\mathbf{Z}_c + \mathbf{Z}_{\text{TL}_2} + \mathbf{Z}_{\text{L}_2}} - \frac{(\mathbf{V}_3 - \mathbf{V}_N)}{\mathbf{Z}_c + \mathbf{Z}_{\text{TL}_3} + \mathbf{Z}_{\text{L}_3}} = 0. \tag{8.71}$$

(b) For a balanced network, the denominators of terms 2 to 4 in Eq. (8.71) become identical, which we shall denote as $\mathbf{Z}_0$:

$$\begin{aligned}\mathbf{Z}_0 &= \mathbf{Z}_c + \mathbf{Z}_{\text{TL}_1} + \mathbf{Z}_{\text{L}_1}\\ &= (0.1 + j0.2) + (0.9 + j0.8) + (29 + j9)\\ &= (30 + j10)\ \Omega.\end{aligned} \tag{8.72}$$

The node–voltage equation given by Eq. (8.71) simplifies to

$$\mathbf{V}_N\left(\frac{1}{\mathbf{Z}_n} + \frac{3}{\mathbf{Z}_0}\right) = \frac{\mathbf{V}_1 + \mathbf{V}_2 + \mathbf{V}_3}{\mathbf{Z}_0}. \tag{8.73}$$

According to Eq. (8.68), for a balanced source, $\mathbf{V}_1 + \mathbf{V}_2 + \mathbf{V}_3 = 0$. Hence,

$$\mathbf{V}_N = 0 \qquad \text{(balanced source).} \tag{8.74}$$

Consequently,

$$\mathbf{I}_n = \frac{\mathbf{V}_N}{\mathbf{Z}_n} = 0,$$

$$\mathbf{I}_1 = \frac{\mathbf{V}_1 - \mathbf{V}_N}{\mathbf{Z}_0} = \frac{120\sqrt{2}}{30 + j10} = 5.4e^{-j18.4^\circ},$$

$$\mathbf{I}_2 = \frac{\mathbf{V}_2 - \mathbf{V}_N}{\mathbf{Z}_0} = \frac{120\sqrt{2}e^{j120^\circ}}{30 + j10} = 5.4e^{j101.6^\circ},$$

and

$$\mathbf{I}_3 = \frac{\mathbf{V}_3 - \mathbf{V}_N}{\mathbf{Z}_0} = \frac{120\sqrt{2}e^{j240^\circ}}{30 + j10} = 5.4e^{j221.6^\circ}.$$

The corresponding time-domain currents are

$$i_1(t) = \mathfrak{Re}[\mathbf{I}_1 e^{j\omega t}] = 5.4\cos(2\pi ft - 18.4°) \text{ A},$$

$$i_2(t) = \mathfrak{Re}[\mathbf{I}_2 e^{j\omega t}] = 5.4\cos(2\pi ft + 101.6°) \text{ A},$$

and

$$i_3(t) = \mathfrak{Re}[\mathbf{I}_3 e^{j\omega t}] = 5.4\cos(2\pi ft + 221.6°) \text{ A},$$

with $f = 60$ Hz.

Review Question 8-12: Why is electrical power generated and distributed by three-phase systems?

Review Question 8-13: Is the power coming into a residential unit (Fig. 8-17) single phase or two phase? Why?

Review Question 8-14: What is the magnitude of the return current I_n in a balanced network?

Exercise 8-7: What would the value of $|\mathbf{I}_1|$ of Example 8-8 be if the coil and transmission-line impedances are ignored? What percentage of error does that represent?

Answer: $|\mathbf{I}_1| = 5.6$ A, % error = 4 percent. (See)

8-7 Measuring Power With Multisim

This section introduces Multisim power-measurement tools and demonstrates their ability through an interactive simulation of an ***impedance-matching network***. In Section 8-5 we established that the amount of power transferred to a load from a source is at a maximum when the impedance of the load $\mathbf{Z}_{\text{Load}}$ is the complex conjugate of the source impedance $\mathbf{Z}_{\text{Source}}$. That is,

$$\mathbf{Z}_{\text{Load}} = \mathbf{Z}^*_{\text{Source}}. \tag{8.75}$$

Consider the circuit shown in Fig. 8-18. The circuit is supplied by a realistic source composed of an ideal voltage source $\mathbf{V}_s$ in series with a source resistance R_s. The load is a series RL circuit. In the phasor domain:

$$\mathbf{Z}_{\text{Source}} = R_s, \qquad \mathbf{Z}_{\text{Load}} = R_L + j\omega L_L. \tag{8.76}$$

For the general case where $L_L \neq 0$ and $R_s \neq R_L$, the load would not be matched to the source, and power transfer would not be a maximum. By inserting a matching network in between the source and the load and selecting the values of its components appropriately, we can match the source to the load, thereby realizing the maximum transfer of power from the source to the circuit segment to the right of terminals (a, b), which includes the matching network and the load. If the load has an inductor, the matching network should have a capacitor, and vice versa.

For the circuit to the right of terminals (a, b), which includes both the matching network and the load,

$$\mathbf{Z}_{\text{Load+Match}} = (R_M + R_L) + j\left(\omega L_L - \frac{1}{\omega C_M}\right). \tag{8.77}$$

For the maximum transfer of power at terminals (a, b) towards the load, it is necessary that

$$\mathbf{Z}_{\text{Source}} = \mathbf{Z}^*_{\text{Load+Match}}, \tag{8.78}$$

which can be satisfied by selecting R_M and C_M as

$$R_M = R_s - R_L \qquad \text{and} \qquad C_M = \frac{1}{\omega^2 L_L}, \tag{8.79}$$

provided $R_s \geq R_L$. Under these matched conditions, the impedance of the capacitor cancels out the impedance of L_L, and the source is matched to the combination of the matching network and load. This means that the power transferred from the source to this combination is a maximum, but it does not mean that the power transferred to the load alone is a maximum. In fact, if the values of R_s and R_L cannot be changed, power transfer to the load is a maximum when $R_M = 0$ and $C_M = 1/(\omega^2 L_L)$.

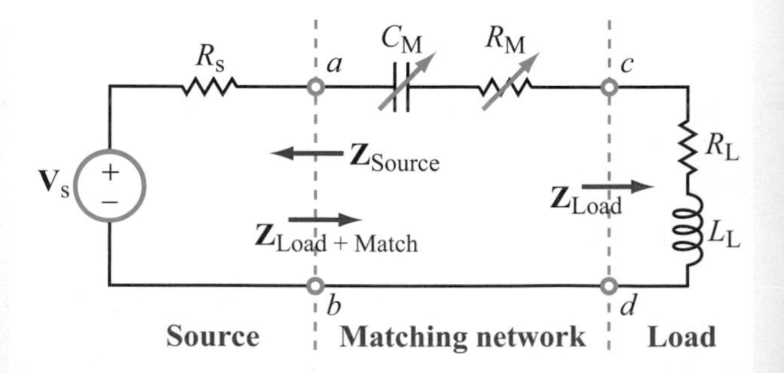

Figure 8-18: Matching network in between the source and the load.

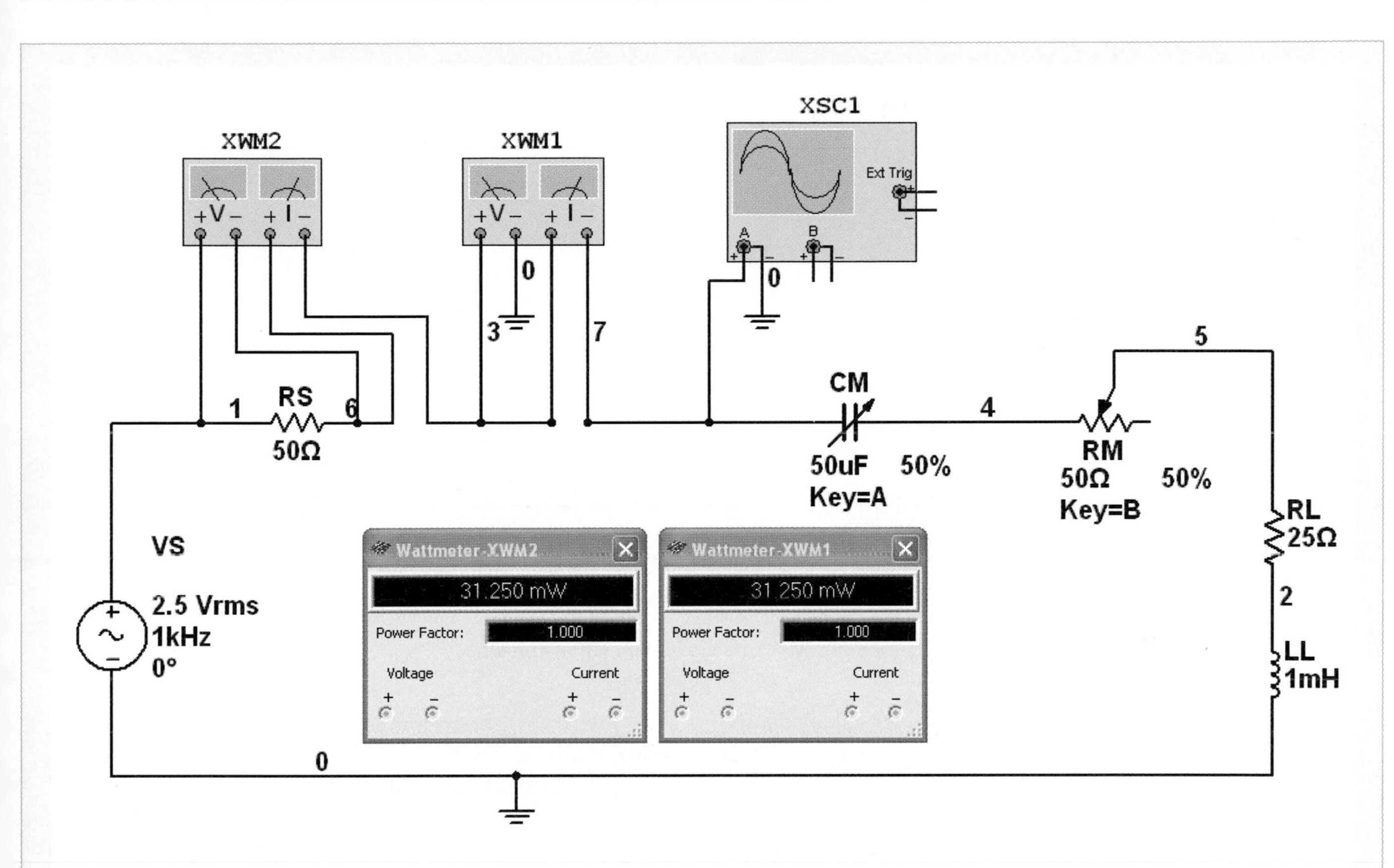

Figure 8-19: Multisim simulation of matching network (CM, RM) in between the source and the load, and wattmeter displays for maximum power transfer.

We also should note that the value of C_M required to achieve the matching condition is a function of ω. Thus, if the value of C_M is selected so as to match the circuit at a given frequency, the circuit will cease to remain matched if ω is changed to a significantly different value. To serve its intended function with significant flexibility, the matching network usually is configured to include a potentiometer and an adjustable capacitor, allowing for manual tuning of R_M and C_M to satisfy Eq. (8.79) at any specified value of ω (within a certain range).

The circuit in Fig. 8-18 can be simulated and analyzed by Multisim, as shown in Fig. 8-19. For variable components, you can choose which keys will shift the component values by double-clicking the component and selecting the desired key letter under Values → Key. Measurement instruments XWM1 and XWM2 are wattmeters configured to measure the average power dissipated by a component or circuit:

$$P_{av} = \frac{1}{2}\Re\mathfrak{e}[\mathbf{VI}^*],$$

where **V** is the phasor voltage across the component or circuit and **I** is the phasor current flowing into its positive voltage terminal. In Fig. 8-19, XWM2 measures the current through R_s and the voltage across it, and XWM1 measures the voltage at node 7 (relative to the ground terminal) and the current through the loop at node 7. Thus, XWM2 measures the average power dissipated in R_s, and XWM1 measures the average power delivered by the source to the matching network and load combined. To match the load to the source in the circuit of Fig. 8-19, we should select

$$R_M = R_s - R_L = 50 - 25 = 25\ \Omega,$$

and

$$C_M = \frac{1}{\omega^2 L_L} = \frac{1}{(2\pi \times 10^3)^2 \times 10^{-3}} = 25.33\ \mu\text{F}.$$

In Fig. 8-19, R_M is a 50-Ω potentiometer set at 50 percent of its maximum value (or 25 Ω), and C_M is a variable 50-μF capacitor, also set at 50 percent of its maximum value (which is very close to the required value of 25.33 μF). The wattmeter displays confirm that the average powers reported by XWM1 and XWM2 are indeed equal.

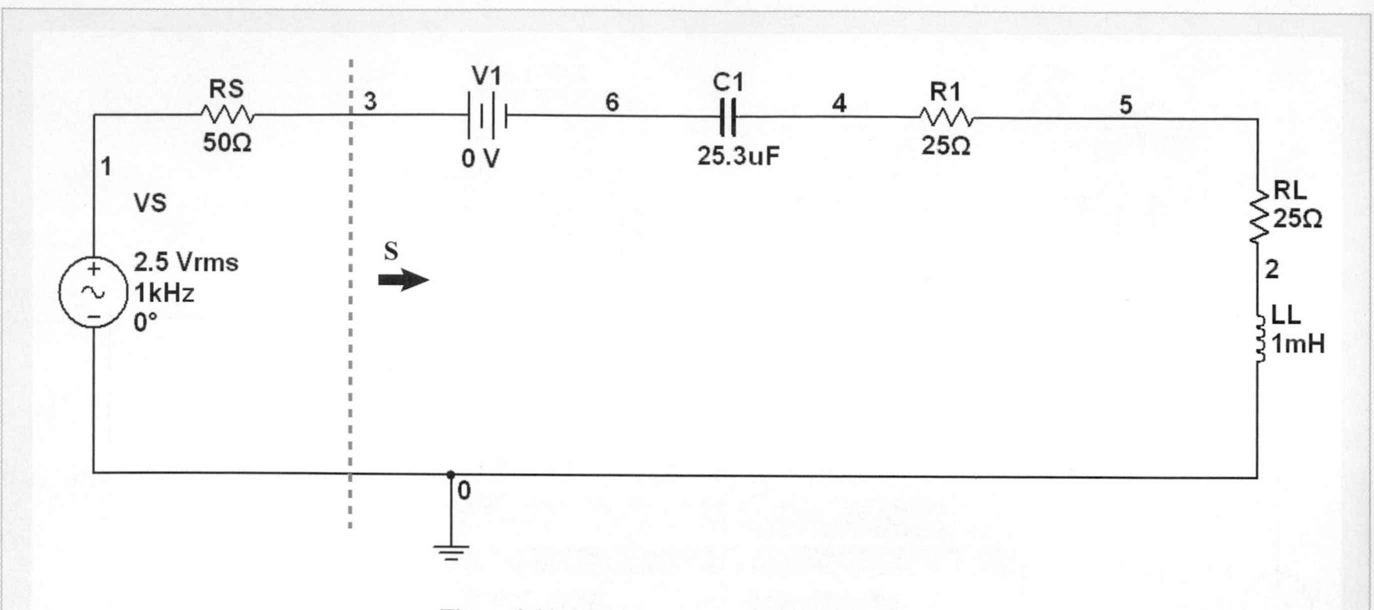

Figure 8-20: Multisim circuit without instruments.

It is important to note that the wattmeter calculates the average power by measuring the voltage and current at a sampling rate specified by the Maximum Time Step (TMAX) in the Interactive Simulation Settings. The default value is 10^{-5} s, which means that the voltage and current are sampled at a time spacing of 10^{-5} s. At 1 kHz, the period is 10^{-3} s. Hence at a time spacing of 10^{-5} s each cycle gets sampled 100 times, which is quite adequate for generating a reliable measurement of the average power. At higher oscillation frequencies, however, the period is much shorter necessitating that TMAX be selected such that TMAX $\leq 10^{-2}/f$ where f is the oscillation frequency in Hz. Thus, at $f = 1$ MHz, for example, TMAX should be set at 10^{-8} s.

Another method for measuring average power in Multisim is to use the Analysis functions to plot the complex power across any section of a circuit. Figure 8-20 is a Multisim reproduction of the circuit in Fig. 8-19 but with no instruments and fixed-value components. Note that to perform the AC Analysis Simulation properly, the AC Analysis Magnitude value of the VS source must be changed to $2.5 * \text{sqrt}(2) = 3.5355$ V. We can plot the magnitude and phase of the complex power **S** across terminals (3,0) in Fig. 8-20 by performing AC Analysis in Multisim. Under Simulate → Analyses → AC Analysis, set FSTART to 1 Hz and FSTOP to 1 MHz. Make sure to include at least 10 points per decade to produce a good plot. Under Output, enter the following expression: 0.5*(real(I(v1)),-imag(I(v1)))*V(3). Note that this expression is equivalent to $\mathbf{S} = \frac{1}{2}\mathbf{I}^*\mathbf{V}$ (Eq. (8.29)). (The expression (real(X),-imag(X)) gives us the complex conjugate of any complex number X; we need to do this because Multisim does not have a complex conjugate function). Figure 8-21 shows a plot of the AC Analysis output. As expected, the phase of **S** goes to 0 at 1 kHz (since it is at this frequency that the inductor and capacitor reactances cancel each other out).

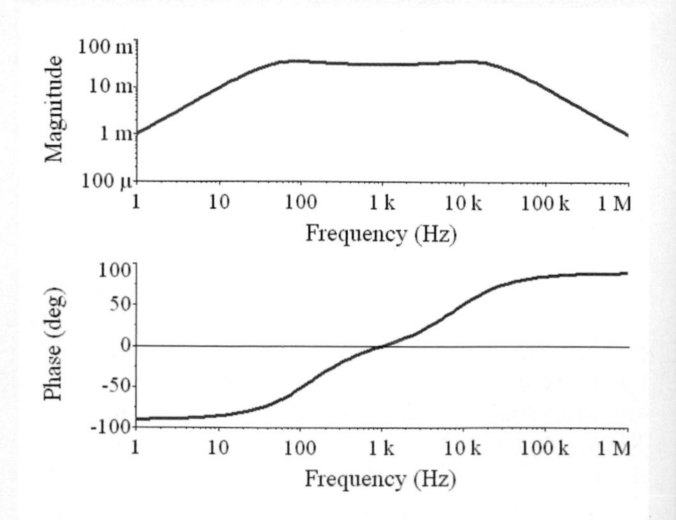

Figure 8-21: Spectral plots of the magnitude and phase of the complex power **S** at terminals (3,0) in Fig. 8-20.

Review Question 8-15: How is power measured in Multisim? Why must all four terminals of the wattmeter be used to obtain a power measurement?

Review Question 8-16: Assuming the values of $\mathbf{V}_s$, R_s, R_L, and L_L are fixed, what values of R_M and C_M lead to maximum transfer of power from the source to R_L?

Exercise 8-8: Use Multisim to simulate the circuit in Fig. 8-19. Connect Channel B of the oscilloscope across the voltage source $\mathbf{V}_s$. Vary C_M over its full range, noting the phase difference between the two channels of the oscilloscope at $C_M = 0$, $C_M = 25\ \mu$F, and $C_M = 50\ \mu$F.

Answer: (See CD)

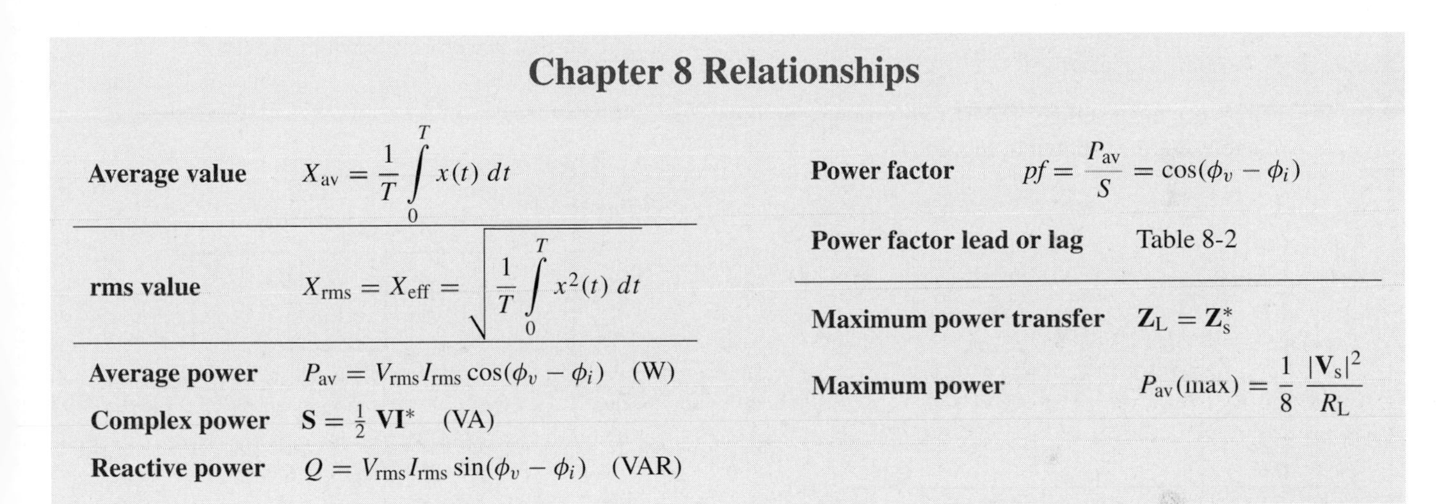

Chapter 8 Relationships

Average value	$X_{av} = \frac{1}{T}\int_0^T x(t)\, dt$
rms value	$X_{rms} = X_{eff} = \sqrt{\frac{1}{T}\int_0^T x^2(t)\, dt}$
Average power	$P_{av} = V_{rms} I_{rms} \cos(\phi_v - \phi_i)$ (W)
Complex power	$\mathbf{S} = \frac{1}{2}\mathbf{V}\mathbf{I}^*$ (VA)
Reactive power	$Q = V_{rms} I_{rms} \sin(\phi_v - \phi_i)$ (VAR)
Power factor	$pf = \frac{P_{av}}{S} = \cos(\phi_v - \phi_i)$
Power factor lead or lag	Table 8-2
Maximum power transfer	$\mathbf{Z}_L = \mathbf{Z}_s^*$
Maximum power	$P_{av}(\max) = \frac{1}{8}\frac{\lvert\mathbf{V}_s\rvert^2}{R_L}$

CHAPTER HIGHLIGHTS

- The rms value of a periodic waveform is obtained by squaring the expression describing the waveform, integrating it over a complete period, dividing by the period, and then taking its square root.
- Even though the average values of the sinusoidal voltage across and current through a load are both zero, the average power consumed by the load is not zero, unless the load is purely reactive (no resistors).
- Power is characterized by several attributes, including the complex power $\mathbf{S}$, the average power P_{av}, and reactive power Q.
- The power factor *pf* is the ratio of the average real power P_{av} consumed by the load to S (the magnitude of the complex power) which incorporates the reactive power Q through $S = [P_{av}^2 + Q^2]^{1/2}$.
- An RL load—typical for machinery that includes coils (such as in a refrigerator, a compressor, and a can opener)—can be compensated by adding a shunt capacitor, causing its *pf* to increase, and in turn reducing the amount of current that has to be supplied by the electrical power source.
- The power transferred from an input source circuit with Thévenin impedance $\mathbf{Z}_s = R_s + jX_s$ to a complex load with impedance $\mathbf{Z}_L = R_L + jX_L$ is at a maximum when $\mathbf{Z}_L = \mathbf{Z}_s^*$.
- Transmission of three-phase power from the generator to loads is more efficient than three separate, single-phase transmissions.
- Multisim can be used to measure the magnitude and phase of complex power as a function of frequency.

GLOSSARY OF IMPORTANT TERMS

Provide definitions or explain the meaning of the following terms:

apparent power
average power
average value
balanced three-phase source
compensated load
complex power
impedance matching network
instantaneous power
matched load
periodicity property
power factor
power factor angle
power factor compensation
reactive power
root-mean-square (rms) value
rotor
stator
VAR

PROBLEMS

Section 8-1: Periodic Waveforms

*8.1 Determine (a) the average and (b) rms values of the periodic voltage waveform shown in Fig. P8.1.

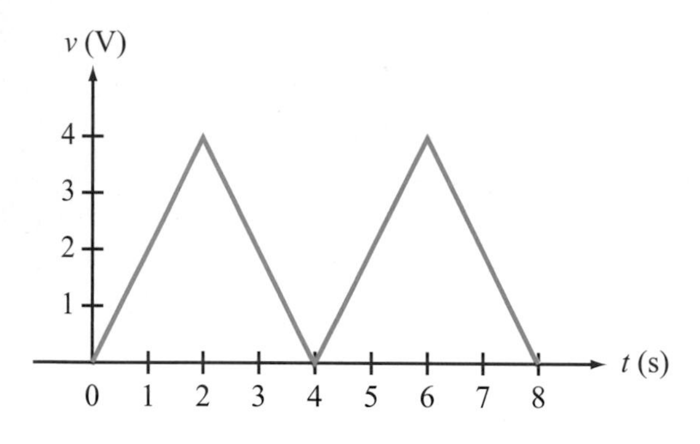

Figure P8.1: Waveform for Problem 8.1.

8.2 Determine (a) the average and (b) rms values of the periodic voltage waveform shown in Fig. P8.2.

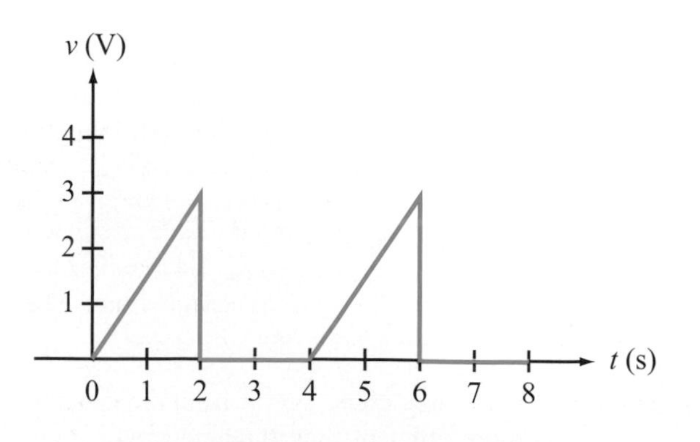

Figure P8.2: Waveform for Problem 8.2.

8.3 Determine (a) the average and (b) rms values of the periodic current waveform shown in Fig. P8.3.

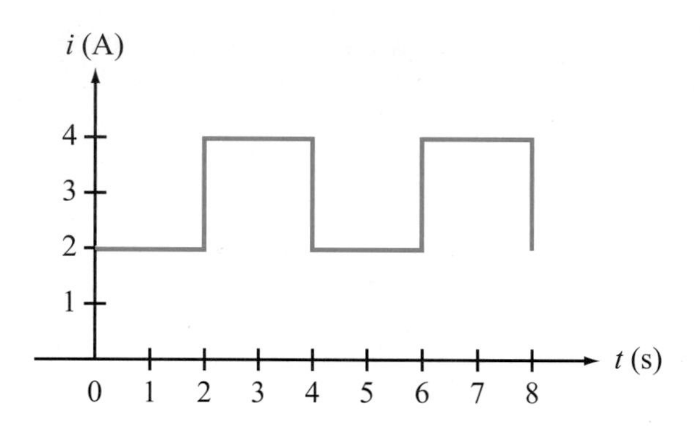

Figure P8.3: Waveform for Problem 8.3.

8.4 Determine (a) the average and (b) rms values of the periodic current waveform shown in Fig. P8.4.

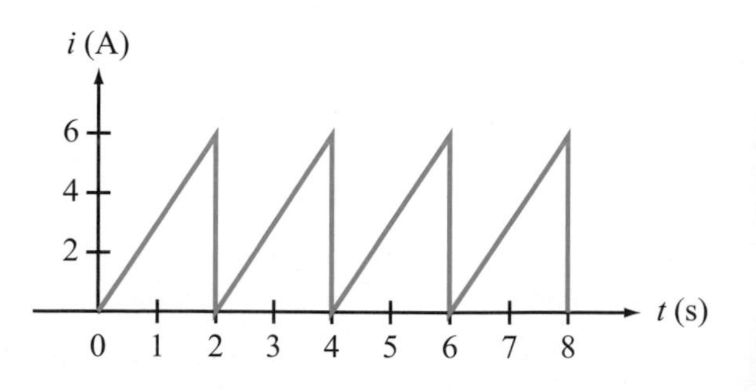

Figure P8.4: Waveform for Problem 8.4.

*Answer(s) in Appendix E.

8.5 Determine (a) the average and (b) rms values of the periodic voltage waveform shown in Fig. P8.5.

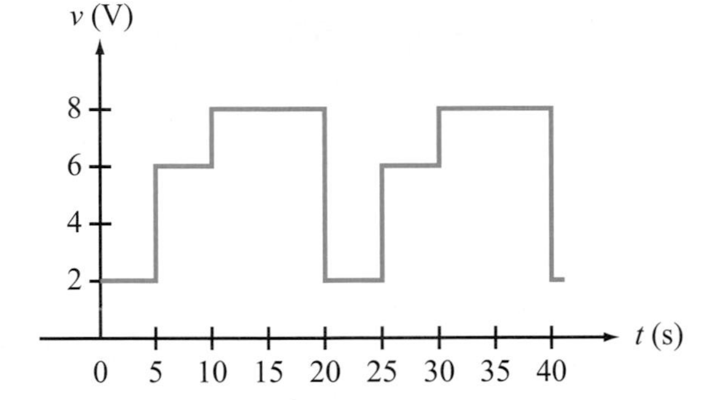

Figure P8.5: Waveform for Problem 8.5.

8.6 Determine (a) the average and (b) rms values of the periodic current waveform shown in Fig. P8.6.

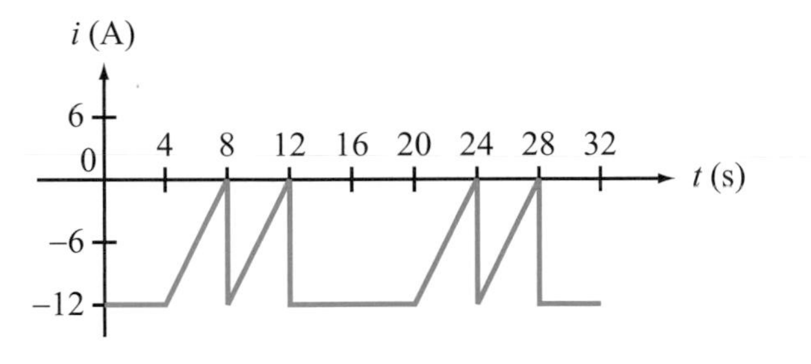

Figure P8.6: Waveform for Problem 8.6.

8.7 Determine (a) the average and (b) rms values of the periodic voltage waveform shown in Fig. P8.7.

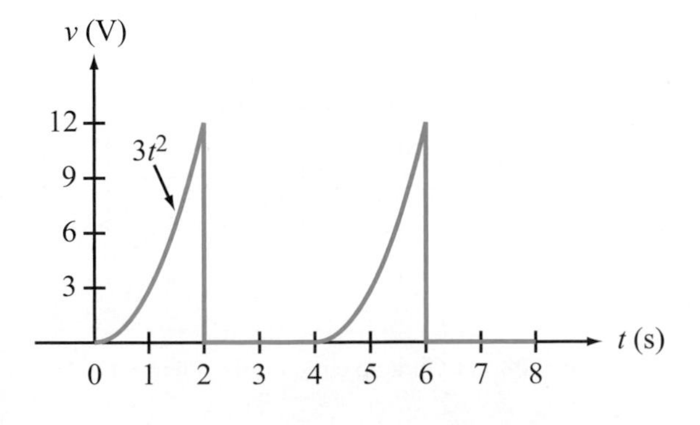

Figure P8.7: Waveform for Problem 8.7.

8.8 Determine the average and rms values of the following periodic waveforms:

(a) $v(t) = |12\cos(\omega t + \theta)|$ V

(b) $v(t) = 4 + 6\cos(2\pi f t + \phi)$ V

(c) $v(t) = 2\cos\omega t - 4\sin(\omega t + 30^\circ)$ V

(d) $v(t) = 9\cos\omega t \sin(\omega t + 30^\circ)$ V

Section 8-2 and 8-3: Average and Complex Power

8.9 Determine the complex power, apparent power, average power absorbed, reactive power, and power factor (including whether it is leading or lagging) for a load circuit whose voltage and current at its input terminals are given by:

(a) $v(t) = 100\cos(377t - 30^\circ)$ V,
$i(t) = 2.5\cos(377t - 60^\circ)$ A.

(b) $v(t) = 25\cos(2\pi \times 10^3 t + 40^\circ)$ V,
$i(t) = 0.2\cos(2\pi \times 10^3 t - 10^\circ)$ A.

(c) $\mathbf{V}_{\text{rms}} = 110\angle 60^\circ$ V, $\mathbf{I}_{\text{rms}} = 3\angle 45^\circ$ A.

(d) $\mathbf{V}_{\text{rms}} = 440\angle 0^\circ$ V, $\mathbf{I}_{\text{rms}} = 0.5\angle 75^\circ$ A.

(e) $\mathbf{V}_{\text{rms}} = 12\angle 60^\circ$ V, $\mathbf{I}_{\text{rms}} = 2\angle 30^\circ$ A.

8.10 In the circuit of Fig. P8.10, $v_s(t) = 60\cos 4000t$ V, $R_1 = 200\ \Omega$, $R_2 = 100\ \Omega$, and $C = 2.5\ \mu$F. Determine the average power absorbed by each passive element and the average power supplied by the source.

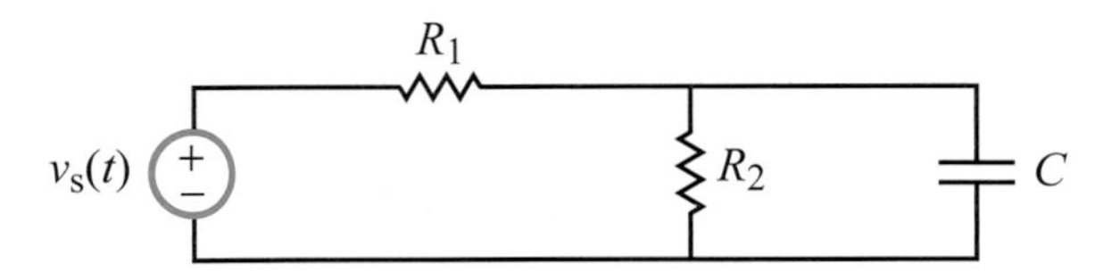

Figure P8.10: Circuit for Problem 8.10.

8.11 In the circuit of Fig. P8.11, $i_s(t) = 0.2\sin 10^5 t$ A, $R = 20\ \Omega$, $L = 0.1$ mH, and $C = 2\ \mu$F. Show that the sum of the complex powers for the three passive elements is equal to the complex power of the source.

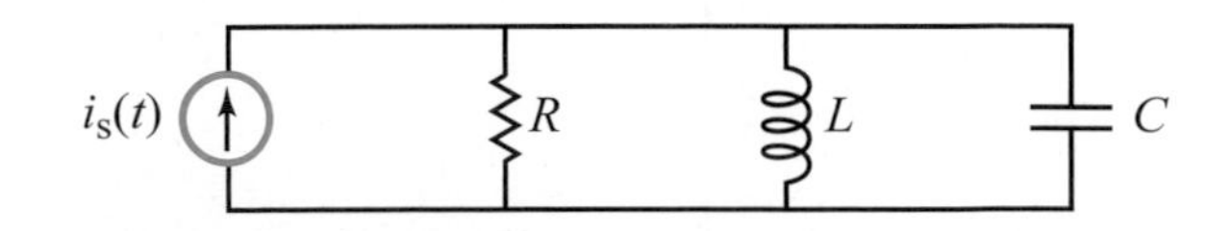

Figure P8.11: Circuit for Problem 8.11.

8.12 In the phasor-domain circuit of Fig. P8.12, $\mathbf{V}_s = 20$ V, $\mathbf{I}_s = 0.3\angle 30^\circ$ A, $R_1 = R_2 = 100$ Ω, $\mathbf{Z}_L = j50$ Ω, and $\mathbf{Z}_C = -j50$ Ω. Determine the complex power for each of the four passive elements and for each of the two sources. Verify that conservation of energy is satisfied.

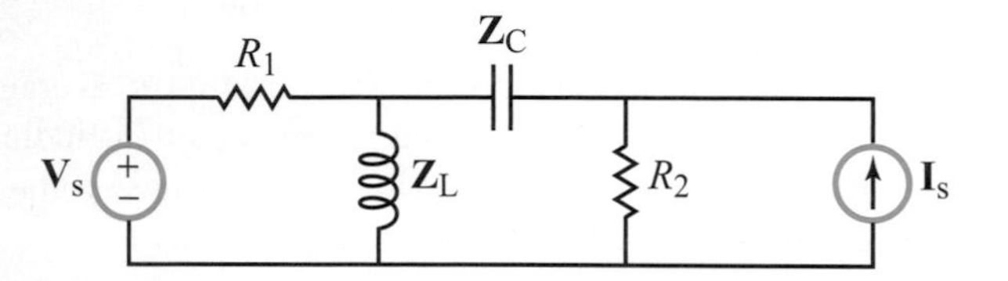

Figure P8.12: Circuit for Problem 8.12.

***8.13** Determine the average power dissipated in the load resistor R_L of the circuit in Fig. P8.13, given that $\mathbf{V}_s = 100$ V, $R_1 = 1$ kΩ, $R_2 = 0.5$ kΩ, $R_L = 2$ kΩ, $\mathbf{Z}_L = j0.8$ kΩ, and $\mathbf{Z}_C = -j4$ kΩ.

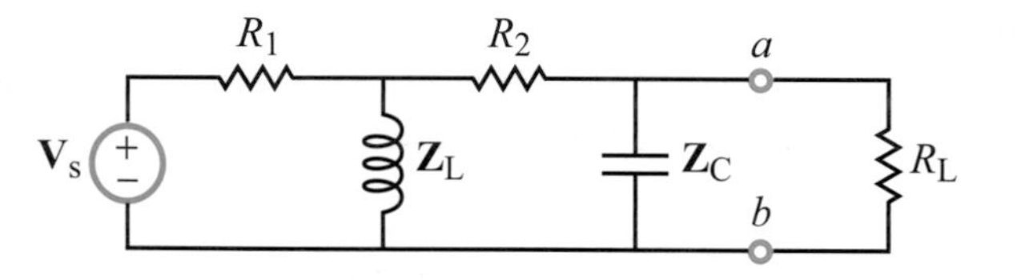

Figure P8.13: Circuit for Problem 8.13.

8.14 Determine $\mathbf{S}$ for the RL load in the circuit of Fig. P8.14, given that $\mathbf{I}_s = 4\angle 0^\circ$ A, $R_1 = 10$ Ω, $R_2 = 5$ Ω, $\mathbf{Z}_C = -j20$ Ω, $R = 10$ Ω, and $\mathbf{Z}_L = j20$ Ω.

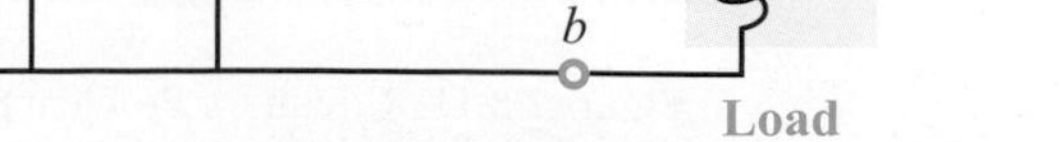

Figure P8.14: Circuit for Problem 8.14.

8.15 Determine the power dissipated in R_L of the circuit in Fig. P8.15.

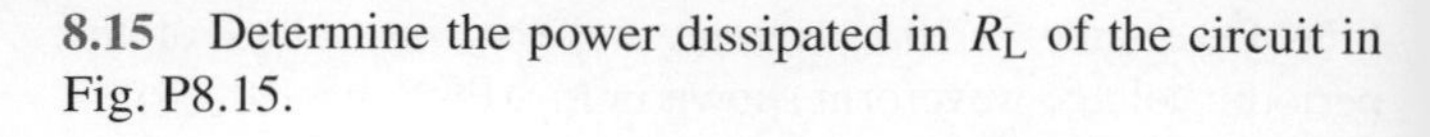

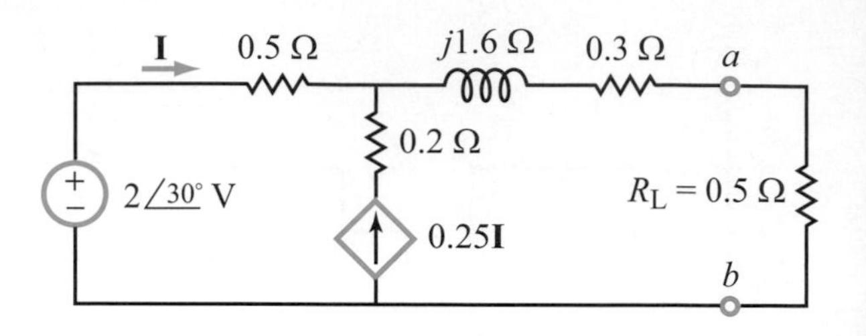

Figure P8.15: Circuit for Problem 8.15.

8.16 Determine the power dissipated in R_L of the circuit in Fig. P8.16.

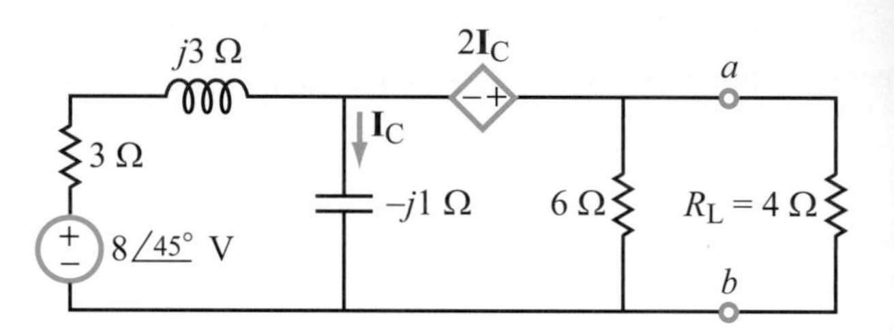

Figure P8.16: Circuit for Problem 8.16.

8.17 In the op-amp circuit of Fig. P8.17, $v_{in}(t) = V_0 \cos \omega t$ V, with $V_0 = 10$ V, $\omega RC = 1$, and $R_L = 10$ kΩ. Determine the power delivered to R_L.

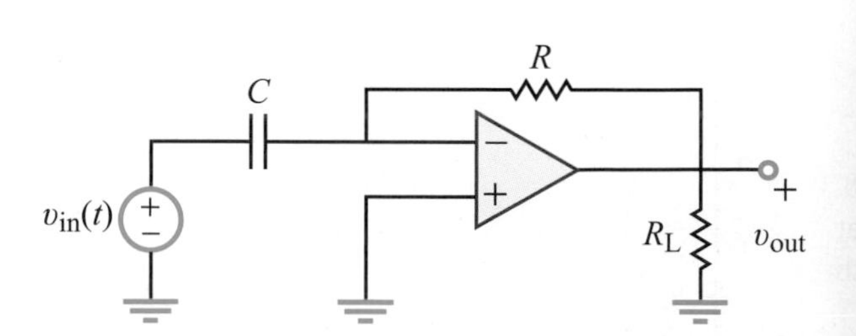

Figure P8.17: Op-amp circuit of Problem 8.17.

8.18 Determine the amount of power delivered to R_L in the circuit of Fig. P8.18, given that $v_{in}(t) = 0.5\cos 2000t$ V, $R_1 = 1$ kΩ, $R_2 = 10$ kΩ, $C = 0.1\ \mu$F, $R_L = 1$ kΩ, and $L = 0.2$ H.

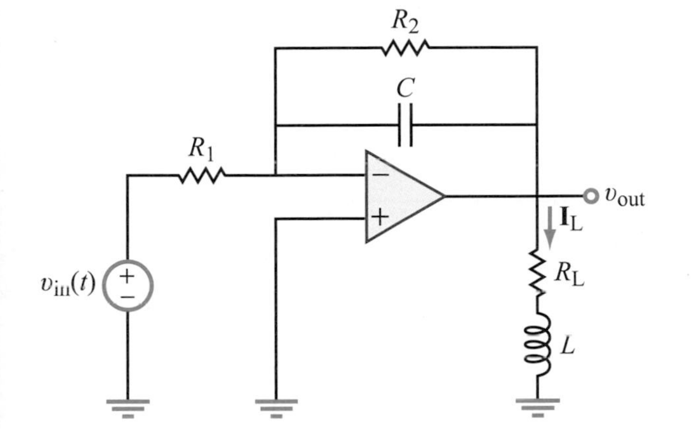

Figure P8.18: Op-amp circuit for Problem 8.18.

*8.19 Given that $v_s(t) = 2\cos 10^3 t$ V in the circuit of Fig. P8.19, determine the power delivered to R_L.

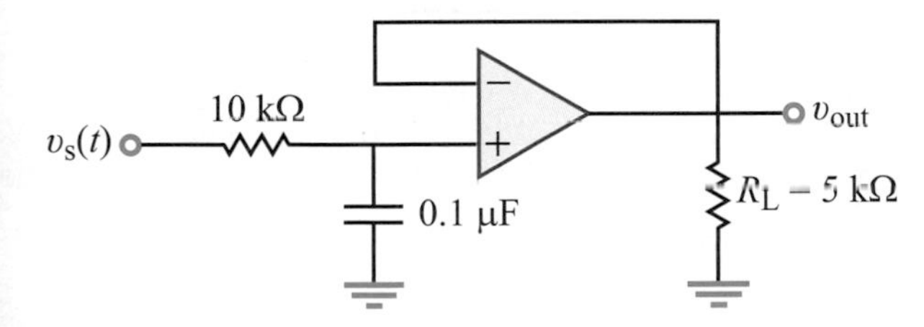

Figure P8.19: Circuit for Problem 8.19.

8.20 The apparent power entering a certain load **Z** is 250 VA at a power factor of 0.8 leading. If the rms phasor voltage of the source is 125 V at 1 MHz:

(a) Determine $\mathbf{I}_{rms}$ going into the load

(b) Determine **S** into the load

(c) Determine **Z**

(d) The equivalent impedance of the load circuit should be of the form $\mathbf{Z} = R + j\omega L$ or $\mathbf{Z} = R - j/\omega C$. Determine the value of L or C, whichever is applicable.

8.21 Voltage source $\mathbf{V}_s$ in the circuit of Fig. P8.21 supplies power to three load circuits with impedances $\mathbf{Z}_1$, $\mathbf{Z}_2$, and $\mathbf{Z}_3$. The following partial power information was deduced from measurements performed on the three load circuits:

Load $\mathbf{Z}_1$: 80 W at $pf = 0.8$ lagging

Load $\mathbf{Z}_2$: 60 VA at $pf = 0.7$ leading

Load $\mathbf{Z}_3$: 40 VA at $pf = 0.6$ leading

If $\mathbf{I}_{rms} = 0.4\angle 37^\circ$ A, determine:

(a) the rms value of $\mathbf{V}_s$ by applying the law of conservation of energy

(b) $\mathbf{Z}_1$, $\mathbf{Z}_2$, and $\mathbf{Z}_3$.

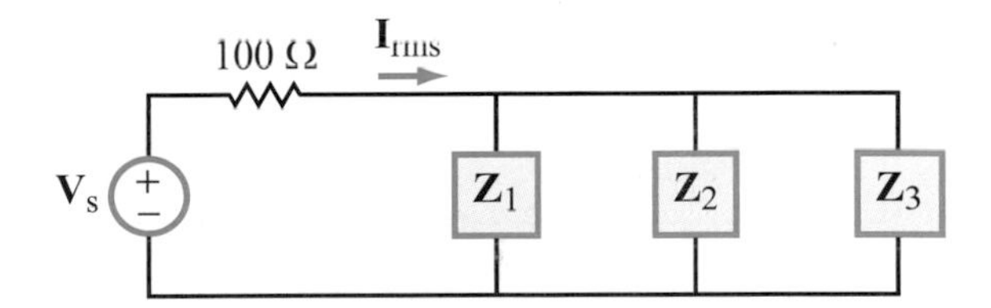

Figure P8.21: Circuit for Problem 8.21.

Section 8-4: Power Factor

8.22 The RL load in Fig. P8.22 is compensated by adding the shunt capacitance C so that the power factor of the combined (compensated) circuit is exactly unity. How is C related to R, L, and ω in that case?

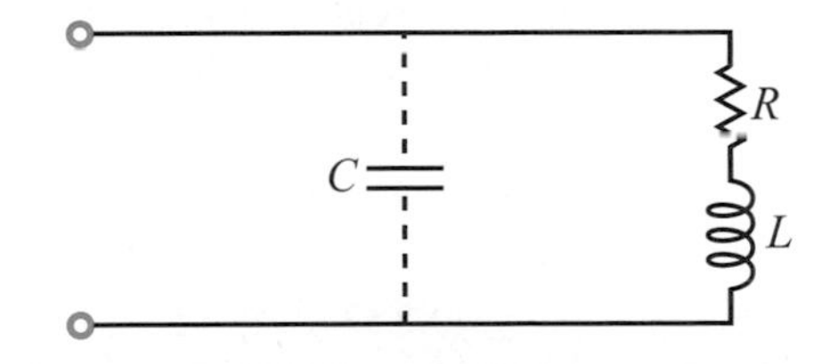

Figure P8.22: Circuit for Problem 8.22.

8.23 The generator circuit shown in Fig. P8.23 is connected to a distant load via a long coaxial transmission line. The overall circuit can be modeled as in Fig. P8.23(b), in which the transmission line is represented by an equivalent impedance $\mathbf{Z}_{line} = (5 + j2)$ Ω.

(a) Determine the power factor of voltage source $\mathbf{V}_s$.

(b) Specify the capacitance of a shunt capacitor C that would raise the power factor of the source to unity when connected between terminals (a, b). The source frequency is 1.5 kHz.

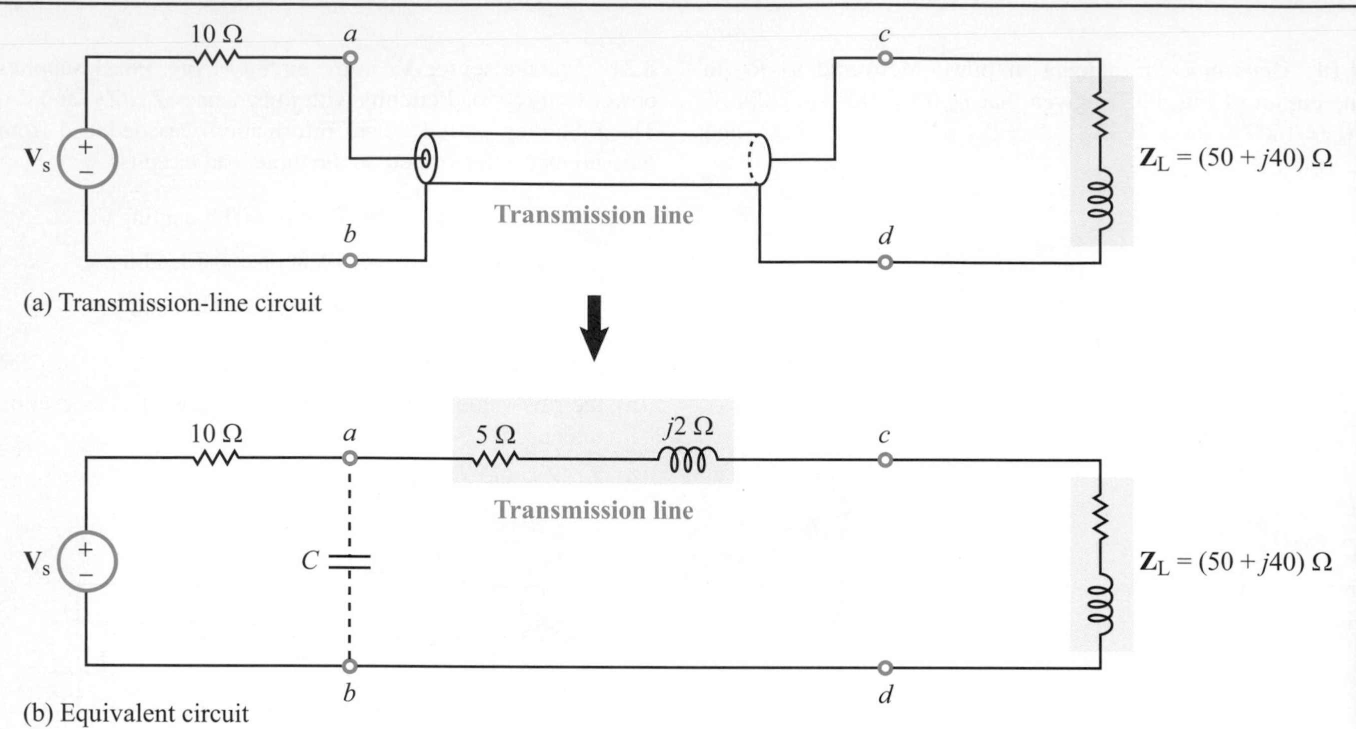

Figure P8.23: Circuit for Problem 8.23.

8.24 Source $\mathbf{V}_s$ in the circuit of Fig. P8.24 is connected to two industrial loads, with equivalent impedances $\mathbf{Z}_1$ and $\mathbf{Z}_2$, via two identical transmission lines, each characterized by an equivalent impedance $\mathbf{Z}_{\text{line}} = (0.5 + j0.3)\ \Omega$. If $\mathbf{Z}_1 = (8 + j12)\ \Omega$ and $\mathbf{Z}_2 = (6 + j3)\ \Omega$:

(a) Determine the power factors for $\mathbf{Z}_1$, $\mathbf{Z}_2$, and source $\mathbf{V}_s$.

(b) Specify the capacitance of a shunt capacitor C that would raise the power factor of the source to 0.95 when connected between terminals (a, b). The source frequency is 12 kHz.

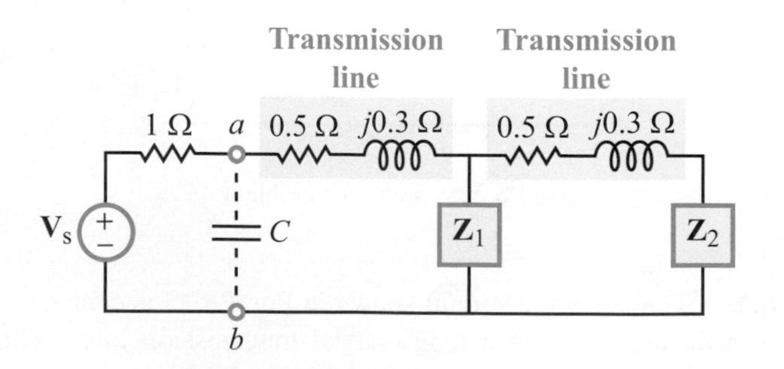

Figure P8.24: Circuit for Problem 8.24.

*__8.25__ Use the power information given for the circuit in Fig. P8.25 to determine:

(a) $\mathbf{Z}_1$ and $\mathbf{Z}_2$

(b) the rms value of $\mathbf{V}_s$.

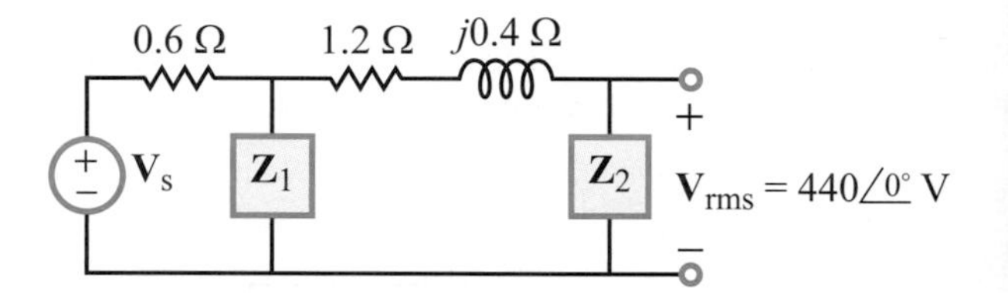

Load $\mathbf{Z}_1$: 24 kW @ pf = 0.66 leading
Load $\mathbf{Z}_2$: 18 kW @ pf = 0.82 lagging

Figure P8.25: Circuit for Problem 8.25.

Section 8-5: Maximum Power Transfer

8.26 For the circuit in Fig. P8.26, choose the load impedance $\mathbf{Z}_L$ so that the power dissipated in it is a maximum. How much power will that be?

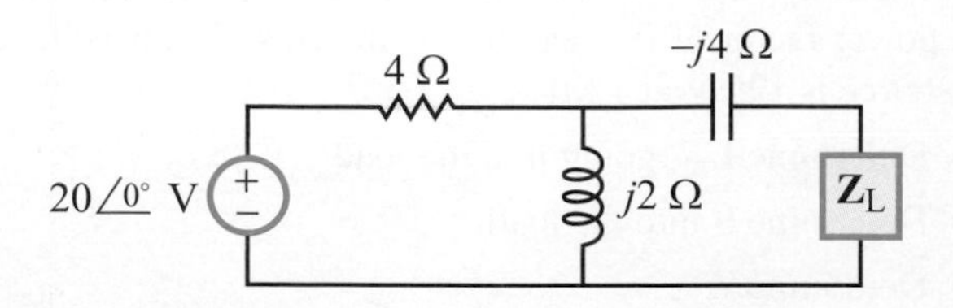

Figure P8.26: Circuit for Problem 8.26.

8.27 For the circuit in Fig. P8.27, choose the load impedance $\mathbf{Z}_L$ so that the power dissipated in it is a maximum. How much power will that be?

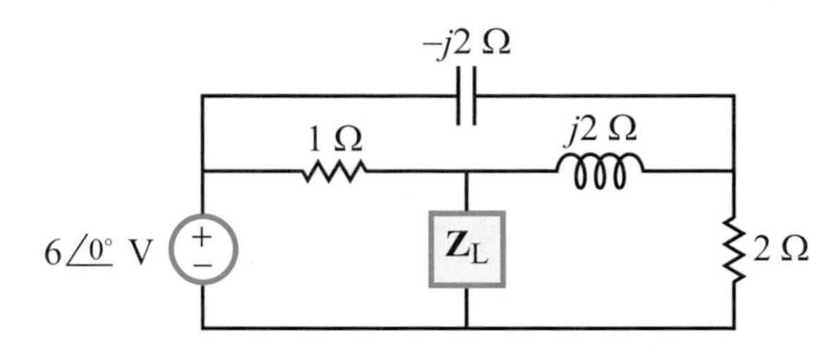

Figure P8.27: Circuit for Problem 8.27.

8.28 For the circuit in Fig. P8.28, choose the load impedance $\mathbf{Z}_L$ so that the power dissipated in it is a maximum. How much power will that be?

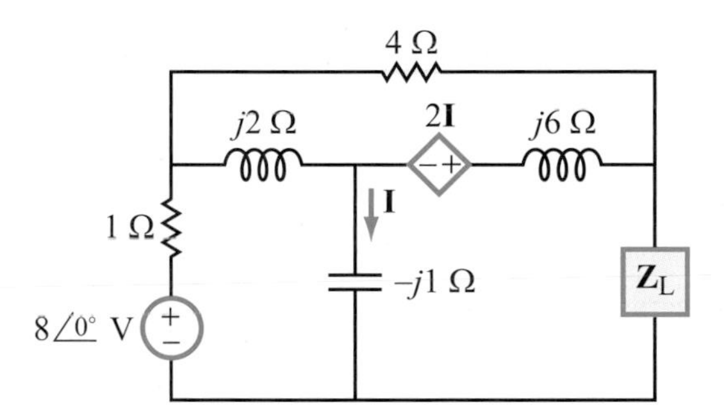

Figure P8.28: Circuit for Problem 8.28.

8.29 For the circuit in Fig. P8.29, choose the load impedance $\mathbf{Z}_L$ so that the power dissipated in it is a maximum. How much power will that be?

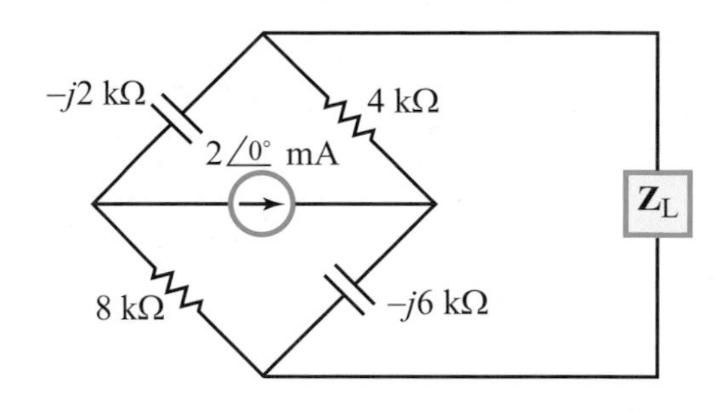

Figure P8.29: Circuit for Problem 8.29.

8.30 For the circuit in Fig. P8.30, choose the load impedance $\mathbf{Z}_L$ so that the power dissipated in it is a maximum. How much power will that be?

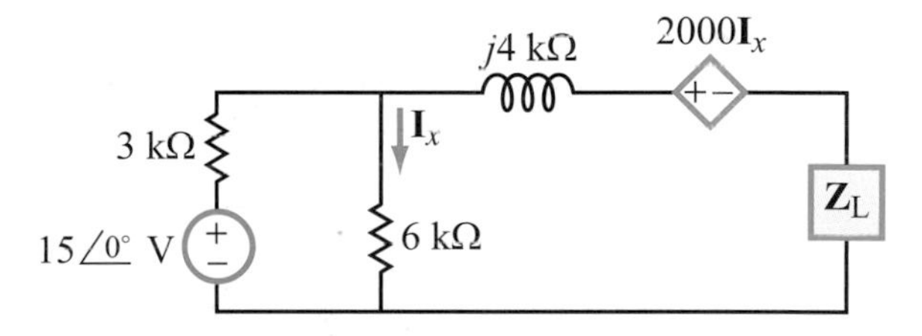

Figure P8.30: Circuit for Problem 8.30.

Sections 8-6 and 8-7: Three-Phase and Multisim

8.31 The circuit depicted in Fig. P8.31 is a 60-Hz Y–Δ network for which coil and transmission line impedances have been ignored. Determine:

(a) $\mathbf{I}_1$, $\mathbf{I}_2$, and $\mathbf{I}_3$ for the circuit as is.

(b) $\mathbf{I}_1$, $\mathbf{I}_2$, and $\mathbf{I}_3$ after applying $\Delta \rightarrow Y$ transformation to the load circuit.

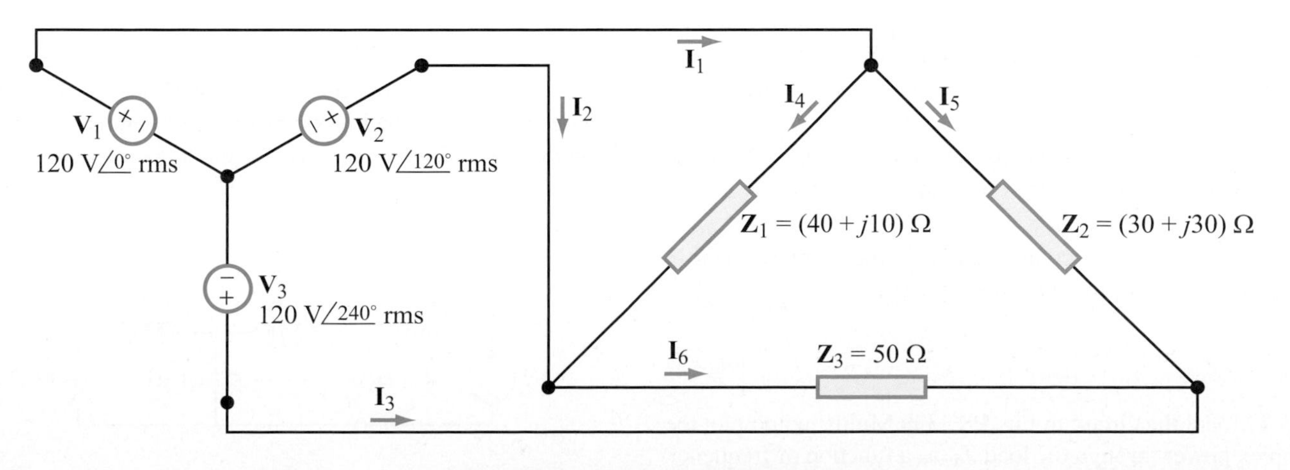

Figure P8.31: Circuit for Problem 8.31.

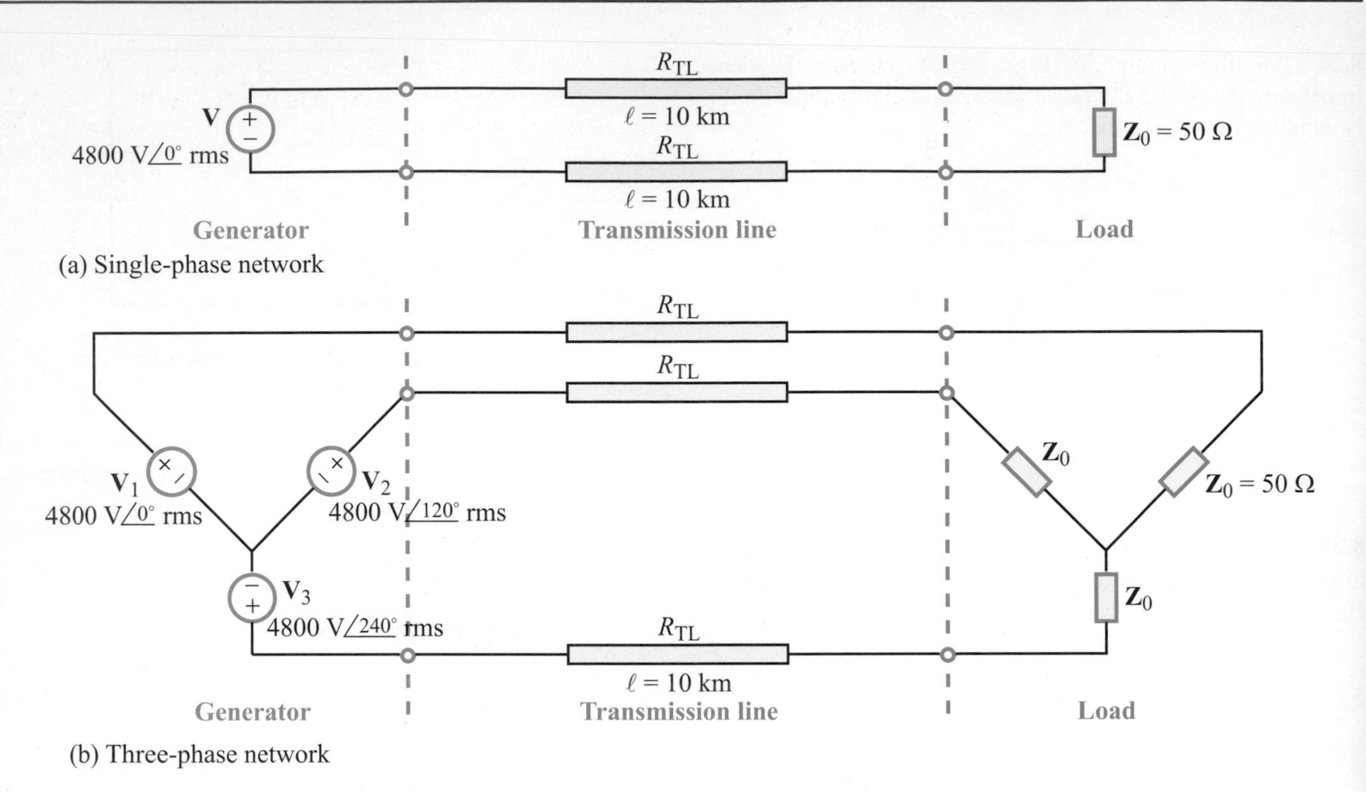

Figure P8.32: Networks for Problem 8.32. All loads are $\mathbf{Z}_0 = 50\ \Omega$.

8.32 The circuits shown in Figs. P8.32(a) and (b) represent a single-phase network and a three-phase network, respectively, both operating at 60 Hz.

(a) The single-phase network consists of a source **V** connected to a load impedance $\mathbf{Z}_0 = 50\ \Omega$ via a two-wire transmission line. The wires are made of copper, and each is 10 km in length and 0.8 cm in radius. Determine the average power consumed by the transmission lines.

(b) Determine the average power consumed by all of the transmission lines of the three-phase circuit.

(c) Which of the two configurations is the more efficient in terms of power loss per single phase?

8.33 Model the circuit in Fig. P8.33 in Multisim and plot the complex power through the load $\mathbf{Z}_L$ as a function of frequency from 1 kHz to 1 GHz. Assume $v_s(t)$ has an amplitude of 1 V.

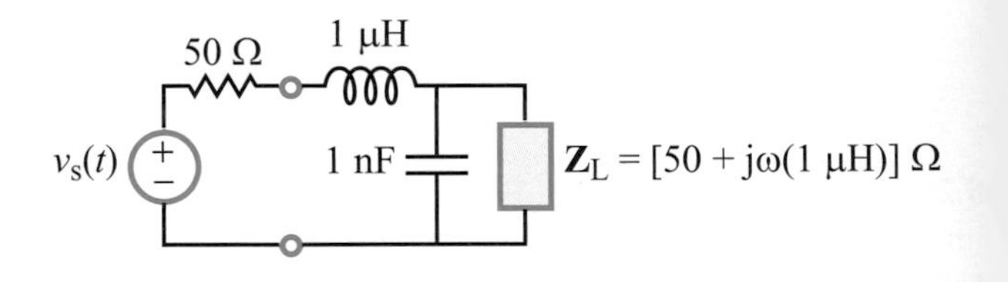

Figure P8.33: Circuit for Problem 8.33.

8.34 Model the circuit in Fig. P8.34 in Multisim and find the frequency at which the input impedance of the load circuit $\mathbf{Z}_{in}$ is purely real. Assume $v_s(t)$ has an amplitude of 1 V.

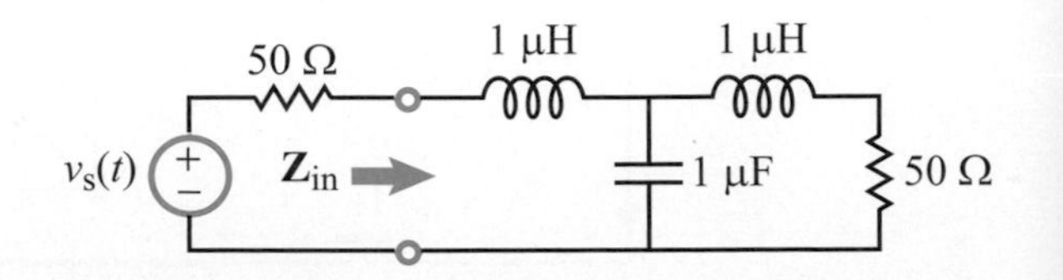

Figure P8.34: Circuit for Problems 8.34 and 8.37.

8.35 Model the circuit in Fig. P8.35 in Multisim and find the frequency at which the input impedance of the load circuit $\mathbf{Z}_{\text{in}}$ is purely real.

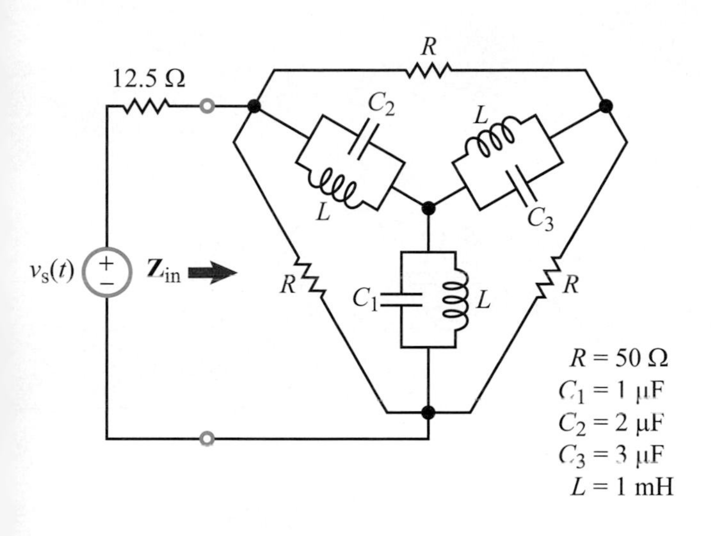

Figure P8.35: Circuit for Problem 8.35.

8.36 Model the circuit in Fig. P8.36 and use the wattmeter to determine the average power consumed by the load $\mathbf{Z}_{\text{L}}$. Also, perform an AC Analysis from 100 kHz to 1 GHz and show that the average power value given by the AC Analysis at 1 MHz matches the value provided by the wattmeter.

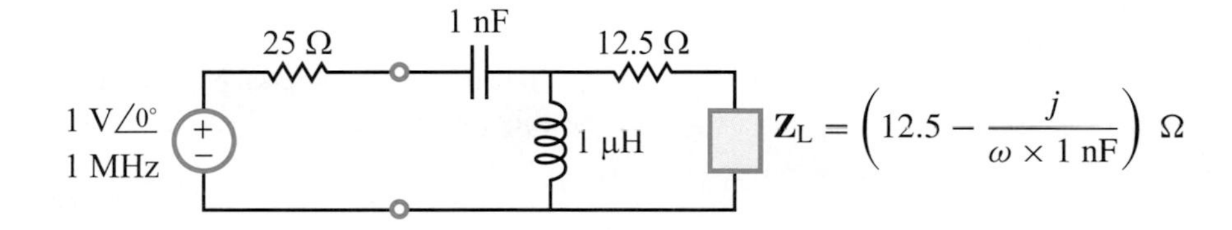

$$\mathbf{Z}_{\text{L}} = \left(12.5 - \frac{j}{\omega \times 1\text{ nF}}\right)\ \Omega$$

Figure P8.36: Circuit for Problem 8.36.

8.37 Plot the power factor and phase angle ϕ_z across the load $\mathbf{Z}_{\text{in}}$ in Fig. P8.34 using AC Analysis in Multisim from 1 kHz to 1 MHz. (See Multisim Demo 8.3 in the Tutorial for help on how to do this.)

CHAPTER 9

Frequency Response of Circuits and Filters

Chapter Contents

Objectives

Upon learning the material presented in this chapter, you should be able to:

1. Derive the transfer function of an ac circuit.
2. Generate magnitude and phase spectral plots.
3. Design first-order lowpass, highpass, bandpass, and bandreject filters.
4. Generate Bode plots for any transfer function.
5. Design active filters.
6. Describe amplitude modulation, frequency modulation, and the operation of a superheterodyne receiver.
7. Apply Multisim to generate spectral responses for passive and active circuits.

Overview

To avoid interference, every radio and TV transmission station is assigned a unique transmission frequency different from those assigned to other radio and TV stations in the area. At the receiver end, even though the antenna will intercept the signals transmitted by all sources within a certain distance, the receiver is able to select from among them the specific channel of interest, while rejecting all others. The selection process is based on the oscillation frequency of the desired signal, and it is realized by passing the intercepted signals through a narrow ***bandpass filter*** whose center frequency is aligned with the frequency of the desired channel. The bandpass filter is one of many different types of ***frequency selective circuits*** employed in analog and digital communication networks to manage the traffic of signals between multiple sources and multiple recipients. The behavior of an ac circuit as a function of the angular frequency ω is called its ***frequency response***. Building on the phasor-domain analysis tools we acquired in the preceding two chapters, we now are ready to develop and adopt a standard set of metrics and design methodologies for characterizing the frequency response of any resonant circuit and to apply them to various types of active and passive circuits.

9-1 The Transfer Function

At input terminals (a, b), the passive linear circuit represented by the block diagram in Fig. 9-1 has an input phasor voltage $\mathbf{V}_{\text{in}}$ and an associated input phasor current $\mathbf{I}_{\text{in}}$. A corresponding set of phasors, $\mathbf{V}_{\text{out}}$ and $\mathbf{I}_{\text{out}}$, exist at output terminals (c, d). The voltage gain of the circuit is defined as

$$\mathbf{H}(\omega) = \frac{\mathbf{V}_{\text{out}}(\omega)}{\mathbf{V}_{\text{in}}(\omega)}, \tag{9.1}$$

where all quantities are written explicitly as functions of the angular frequency ω simply to emphasize the notion that ω will play a central role in our forthcoming discussions. If the circuit contains capacitors and inductors, $\mathbf{V}_{\text{out}}$ likely will be a function of ω, and in the general case $\mathbf{V}_{\text{in}}$ may vary with ω also. The phasor $\mathbf{H}(\omega)$ is called the ***voltage transfer function*** of the circuit and carries a connotation broader than just another name for voltage gain. In fact, $\mathbf{H}(\omega)$ can be defined to convey

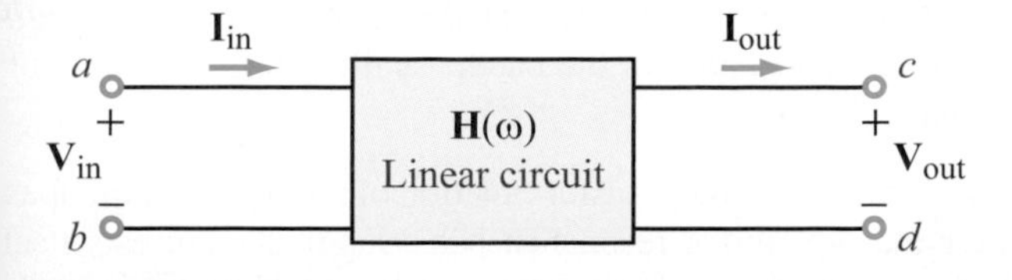

Figure 9-1: The voltage-gain transfer function is $\mathbf{H}(\omega) = \mathbf{V}_{\text{out}}(\omega)/\mathbf{V}_{\text{in}}(\omega)$.

the relationship between any input excitation and any output response. For example, we may define other transfer functions for the circuit in Fig. 9-1, such as:

$$\text{Current gain:} \quad \mathbf{H}_I(\omega) = \frac{\mathbf{I}_{\text{out}}(\omega)}{\mathbf{I}_{\text{in}}(\omega)}, \tag{9.2a}$$

$$\text{Transfer impedance:} \quad \mathbf{H}_{\mathbf{Z}}(\omega) = \frac{\mathbf{V}_{\text{out}}(\omega)}{\mathbf{I}_{\text{in}}(\omega)}, \tag{9.2b}$$

and

$$\text{Transfer admittance:} \quad \mathbf{H}_{\mathbf{Y}}(\omega) = \frac{\mathbf{I}_{\text{out}}(\omega)}{\mathbf{V}_{\text{in}}(\omega)}. \tag{9.2c}$$

In any case, because $\mathbf{H}(\omega)$ always is defined as the ratio of an output quantity to an input quantity, we may think of it as equal to the output generated by the circuit in response to a unity input $(1\angle 0^\circ)$.

As a complex quantity, the transfer function $\mathbf{H}(\omega)$ has a ***magnitude***—to which we will assign the symbol $M(\omega)$—and an associated ***phase angle*** $\phi(\omega)$,

$$\mathbf{H}(\omega) = M(\omega)\, e^{j\phi(\omega)}, \tag{9.3}$$

where by definition,

$$M(\omega) = |\mathbf{H}(\omega)|, \qquad \phi(\omega) = \tan^{-1}\left\{\frac{\mathfrak{Im}[\mathbf{H}(\omega)]}{\mathfrak{Re}[\mathbf{H}(\omega)]}\right\}. \tag{9.4}$$

9-1.1 Terminology

The voltage transfer functions most commonly encountered in electronic circuits are those belonging to ***lowpass***, ***highpass***, ***bandpass***, and ***bandreject filters***. To visualize the frequency response of a transfer function, we usually generate plots of its magnitude and phase angle as a function of frequency from $\omega = 0$ (dc) to $\omega = \infty$. Figure 9-2 displays typical magnitude responses for the four aforementioned types of filters. Each of the four filters is characterized by at least one ***passband*** and one ***stopband***. The lowpass filter allows low-frequency signals to pass through (essentially unimpeded) but blocks the transmission of high-frequency signals. The qualifiers *low* and *high* are relative to the ***corner frequency*** ω_c (Fig. 9-2(a)), which we shall define shortly. The high-pass filter exhibits the opposite behavior, blocking low-frequency signals while allowing high frequencies to go through. The bandpass filter (Fig. 9-2(c)) is transparent to signals whose frequencies are within a certain range centered at ω_0, but cuts off both very high and very low frequencies. The response of the bandreject filter provides the opposite function to that of the bandpass filter; it is transparent to low- and high-frequency signals and opaque to intermediate-frequency signals.

We often use the term "frequency" for both the angular frequency ω and the oscillation frequency $f = \omega/2\pi$. Because the impedances of inductors and capacitors are given by $j\omega L$

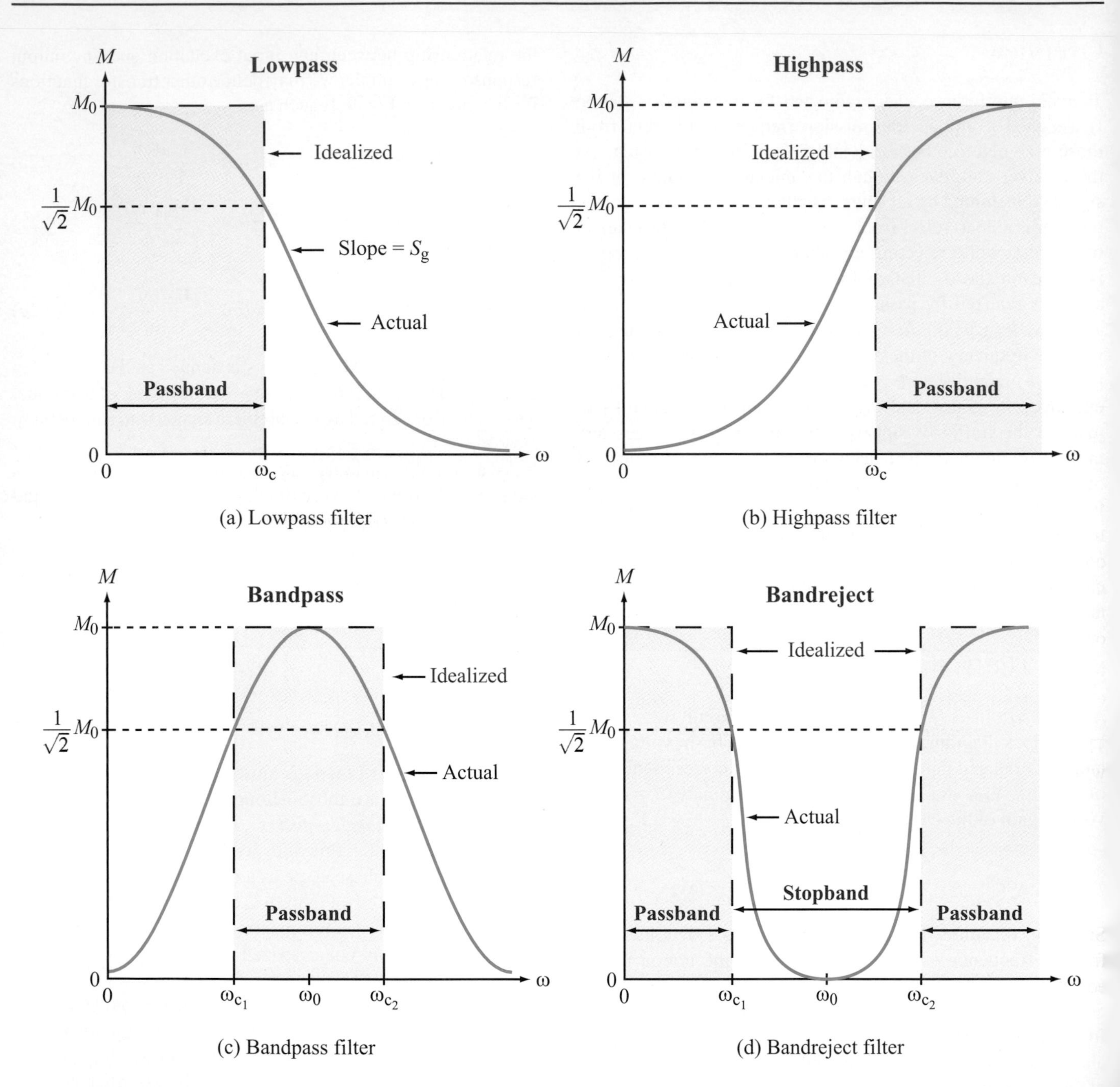

Figure 9-2: Typical magnitude spectral responses for the four types of filters.

and $1/j\omega C$, it is easier to analyze a circuit and plot its response as a function of ω, but if the circuit performance is specified in Hz, ω should be replaced with $2\pi f$ everywhere.

Gain Factor M_0

All four spectral plots shown in Fig. 9-2 exhibit smooth patterns as a function of ω, and each has a peak value M_0 in its passband. If M_0 occurs at dc, as in the case of the lowpass filter, it is called the ***dc gain***; if it occurs at $\omega = \infty$, it is called the ***high-frequency gain***; and for the bandpass filter, it is called simply the ***gain factor***.

In some cases, the transfer function of a lowpass or highpass filter may exhibit a resonance behavior that manifests itself in the form of a peaking pattern in the neighborhood of the resonant frequency of the circuit ω_0, as illustrated in Fig. 9-3. Obviously, the peak value at $\omega = \omega_0$ exceeds M_0, but we wil

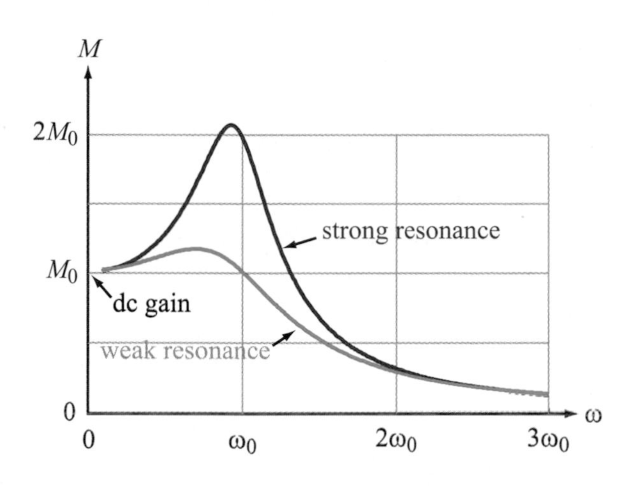

Figure 9-3: Resonant peak in the spectral response of a lowpass filter circuit.

continue to refer to M_0 as the dc gain of $M(\omega)$ because M_0 *is defined as the reference level in the passband of the transfer function*, whereas the behavior of $M(\omega)$ in the neighborhood of ω_0 is specific to that neighborhood.

Corner Frequency ω_c

The corner frequency ω_c is defined as the angular frequency at which $M(\omega)$ is equal to $1/\sqrt{2}$ of the reference peak value,

$$M(\omega_c) = \frac{M_0}{\sqrt{2}} = 0.707 M_0. \tag{9.5}$$

Since $M(\omega)$ is a voltage transfer function, $M^2(\omega)$ is the transfer function for power. The condition described by Eq. (9.5) is equivalent to

$$M^2(\omega_c) = \frac{M_0^2}{2} \quad \text{or} \quad P(\omega_c) = \frac{P_0}{2}. \tag{9.6}$$

Hence, ω_c also is called the ***half-power frequency***. The spectra of the lowpass and highpass filters shown in Fig. 9-2(a) and (b) have only one half-power frequency each, but the bandpass and bandreject responses have two half-power frequencies each, ω_{c_1} and ω_{c_2}. Even though the *actual* frequency response of a filter is a gently varying curve, it usually is approximated to that of an equivalent idealized response, as illustrated in Fig. 9-2. The idealized version for the lowpass filter has a rectangle-like envelope with a sudden transition at $\omega = \omega_c$. Accordingly, ω_c also is referred to as the ***cutoff frequency*** of the filter. This term also applies to the other three types of filters.

Bandwidth B

The filter *bandwidth* B *is defined as the range of* ω *corresponding to the filter's passband*:

$$B = \begin{cases} 0 \le \omega < \omega_c & \text{for lowpass filter,} \\ \omega > \omega_c & \text{for highpass filter,} \\ \omega_{c_1} < \omega < \omega_{c_2} & \text{for bandpass filter,} \\ \omega < \omega_{c_1} \text{ and } \omega > \omega_{c_2} & \text{for bandreject filter.} \end{cases} \tag{9.7}$$

The stopband of the bandreject filter extends from ω_{c_1} to ω_{c_2} (Fig. 9-2(d)).

Resonant Frequency ω_0

Resonance is a condition that occurs when the input impedance or input admittance of a circuit containing reactive elements is purely real, and the angular frequency at which it occurs is called the ***resonant frequency*** ω_0. Often (but not always) the transfer function $\mathbf{H}(\omega)$ also is purely real at $\omega = \omega_0$, and its magnitude is at its maximum or minimum value.

Let us consider the two circuits shown in Fig. 9-4. The input impedance of the RL circuit is simply

$$\mathbf{Z}_{\text{in}_1} = R + j\omega L. \tag{9.8}$$

Resonance corresponds to when the imaginary part of $\mathbf{Z}_{\text{in}_1}$ is zero, which occurs at $\omega = 0$. Hence, the resonant frequency of the RL circuit is

$$\omega_0 = 0 \qquad \text{(RL circuit)}. \tag{9.9}$$

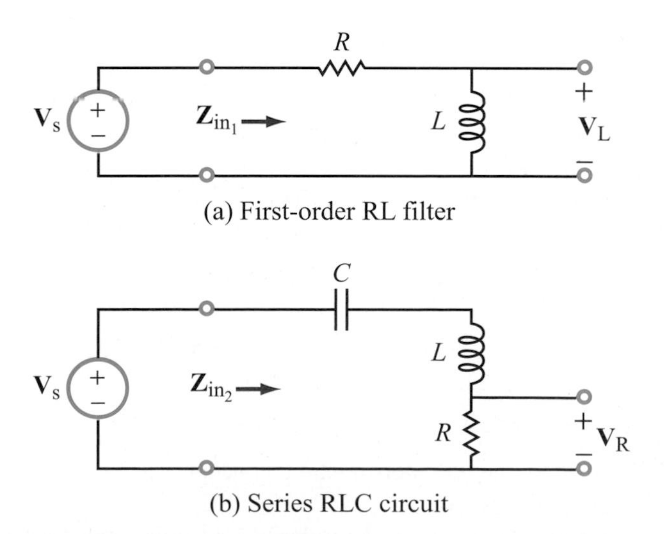

Figure 9-4: Resonance occurs when the imaginary part of the input impedance is zero. For the RL circuit, $\mathfrak{Im}\,[\mathbf{Z}_{\text{in}_1}] = 0$ when $\omega = 0$ (dc), but for the RLC circuit, $\mathfrak{Im}\,[\mathbf{Z}_{\text{in}_2}] = 0$ requires that $\mathbf{Z}_L = -\mathbf{Z}_C$ or, equivalently, $\omega^2 = 1/LC$.

When $\omega_0 = 0$ (dc) or ∞, the resonance is regarded as a ***trivial resonance*** because it occurs at the extreme ends of the spectrum. This usually happens when the circuit has either an inductor or a capacitor (but not both simultaneously). A circuit that exhibits only a trivial resonance, such as the RL circuit in Fig. 9-4(a), is not considered a resonator.

If the circuit contains at least one capacitor and at least one inductor, resonance can occur at intermediate values of ω. A case in point is the series RLC circuit shown in Fig. 9-4(b). Its input impedance is

$$\mathbf{Z}_{\text{in}_2} = R + j\left(\omega L - \frac{1}{\omega C}\right). \tag{9.10}$$

At resonance ($\omega = \omega_0$), the imaginary part of $\mathbf{Z}_{\text{in}_2}$ is equal to zero. Thus,

$$\omega_0 L - \frac{1}{\omega_0 C} = 0,$$

or

$$\omega_0 = \frac{1}{\sqrt{LC}} \qquad \text{(RLC circuit).} \tag{9.11}$$

So long as neither L nor C is zero or ∞, the transfer function $\mathbf{H}(\omega) = \mathbf{V}_\text{R}/\mathbf{V}_\text{s}$ will exhibit a two-sided spectrum with a peak at ω_0—similar in shape to that of the bandpass filter response shown in Fig. 9-2(c).

Roll-Off Rate S_g

Outside the passband, the rectangle-shaped idealized responses shown in Fig. 9-2 have infinite slopes, but of course, the actual responses have finite slopes. The steeper the slope, the more discriminating the filter is, and the closer it approaches the idealized response. Hence, the slope S_g outside the passband (called the ***gain roll-off rate***) is an important attribute of the filter response.

9-1.2 RC Circuit Example

To illustrate the transfer-function concept with a concrete example, let us consider the series RC circuit shown in Fig. 9-5(a). Voltage source $\mathbf{V}_\text{s}$ is designated as the input phasor, and on the output side, we have designated two voltage phasors, namely $\mathbf{V}_\text{R}$ and $\mathbf{V}_\text{C}$. We will examine the frequency responses of the transfer functions corresponding to each of those two output voltages.

Lowpass Filter

Application of voltage division gives

$$\mathbf{V}_\text{C} = \frac{\mathbf{V}_\text{s}\mathbf{Z}_\text{C}}{R + \mathbf{Z}_\text{C}} = \frac{\mathbf{V}_\text{s}/j\omega C}{R + \frac{1}{j\omega C}}. \tag{9.12}$$

The transfer function corresponding to $\mathbf{V}_\text{C}$ is

$$\mathbf{H}_\text{C}(\omega) = \frac{\mathbf{V}_\text{C}}{\mathbf{V}_\text{s}} = \frac{1}{1 + j\omega RC}, \tag{9.13}$$

where we have multiplied the numerator and denominator of Eq. (9.12) by $j\omega C$ to simplify the form of the expression. In terms of its magnitude $M_\text{C}(\omega)$ and phase angle $\phi_\text{C}(\omega)$, the transfer function is given by

$$\mathbf{H}_\text{C}(\omega) = M_\text{C}(\omega)\, e^{j\phi_\text{C}(\omega)}, \tag{9.14}$$

with

$$M_\text{C}(\omega) = |\mathbf{H}_\text{C}(\omega)| = \frac{1}{\sqrt{1 + \omega^2 R^2 C^2}} \tag{9.15a}$$

and

$$\phi_\text{C}(\omega) = -\tan^{-1}(\omega RC). \tag{9.15b}$$

Spectral plots for $M_\text{C}(\omega)$ and $\phi_\text{C}(\omega)$ are displayed in Fig. 9-5(b). It is clear from the plot of its magnitude that the expression given by Eq. (9.13) represents the transfer function of a lowpass filter with a dc gain factor $M_0 = 1$. At dc, the capacitor acts like an open circuit—allowing no current to flow through the loop—with the obvious consequence that $\mathbf{V}_\text{C} = \mathbf{V}_\text{s}$. At very high values of ω, the capacitor acts like a short circuit, in which case the voltage across it is approximately zero.

Application of Eq. (9.6) allows us to determine the corner frequency ω_c as follows

$$M_\text{C}^2(\omega_\text{c}) = \frac{1}{1 + \omega_\text{c}^2 R^2 C^2} = \frac{1}{2}, \tag{9.16}$$

which leads to

$$\omega_\text{c} = \frac{1}{RC}. \tag{9.17}$$

Highpass Filter

The output across R in Fig. 9-5(a) leads to

$$\mathbf{H}_\text{R}(\omega) = \frac{\mathbf{V}_\text{R}}{\mathbf{V}_\text{s}} = \frac{j\omega RC}{1 + j\omega RC}. \tag{9.18}$$

The magnitude and phase angle of $\mathbf{H}_\text{R}(\omega)$ are given by

$$M_\text{R}(\omega) = |\mathbf{H}_\text{R}(\omega)| = \frac{\omega RC}{\sqrt{1 + \omega^2 R^2 C^2}} \tag{9.19a}$$

and

$$\phi_\text{R}(\omega) = \frac{\pi}{2} - \tan^{-1}(\omega RC). \tag{9.19b}$$

Their spectral plots are displayed in Fig. 9-5(c).

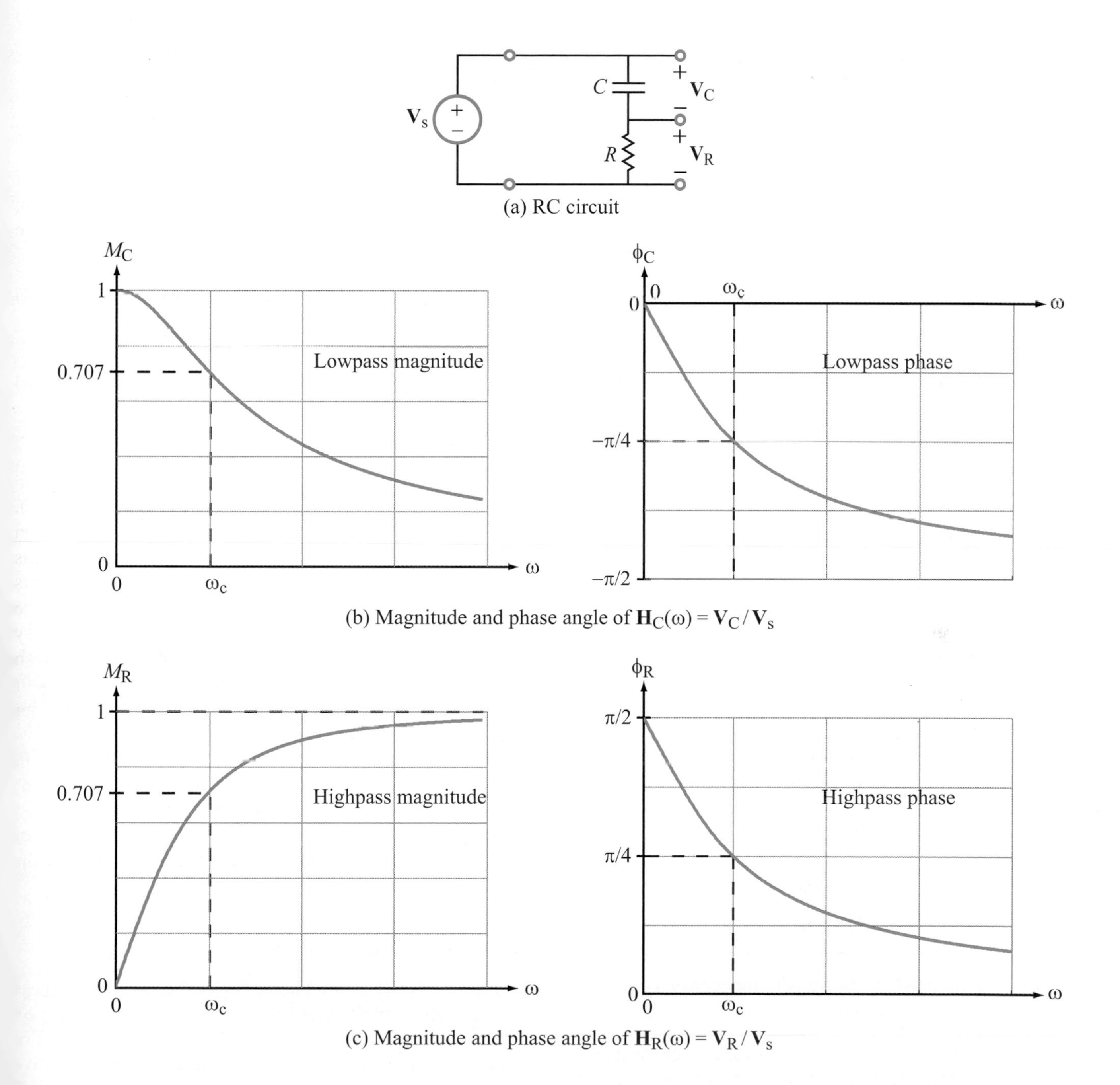

Figure 9-5: Lowpass and highpass transfer functions.

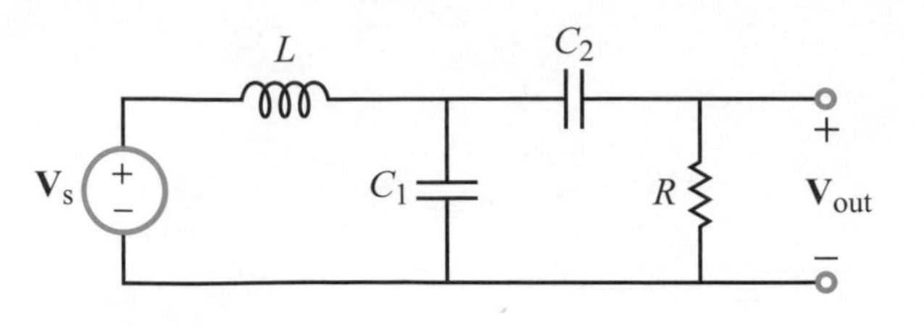

Figure 9-6: Circuit of Example 9-1.

Example 9-1: Resonant Frequency

For the circuit in Fig. 9-6, (a) obtain an expression for $\mathbf{H}(\omega) = \mathbf{V}_{\text{out}}/\mathbf{V}_{\text{s}}$ and (b) show that $\mathbf{H}(\omega)$ becomes purely real at $\omega_0 = 1/\sqrt{L(C_1 + C_2)}$.

Solution: (a) Application of KCL and KVL leads to

$$\mathbf{H}(\omega) = \frac{\mathbf{V}_{\text{out}}}{\mathbf{V}_{\text{s}}} = \frac{RZ_{C_1}}{Z_{C_1}Z_{C_2} + Z_{L}(Z_{C_1} + Z_{C_2}) + R(Z_{C_1} + Z_{L})},$$

where $Z_L = j\omega L$, $Z_{C_1} = 1/j\omega C_1$, and $Z_{C_2} = 1/j\omega C_2$. After a few steps of algebra aimed at transforming the expression into a form whose denominator is purely real, we end up with

$$\mathbf{H}(\omega) = \frac{\omega^2 R^2 C_2^2(1 - \omega^2 LC_1) + j\omega RC_2[1 - \omega^2 L(C_1 + C_2)]}{[1 - \omega^2 L(C_1 + C_2)]^2 + \omega^2 R^2 C_2^2(1 - \omega^2 LC_1)^2}.$$

(b) At

$$\omega = \omega_0 = \frac{1}{\sqrt{L(C_1 + C_2)}},$$

the imaginary part of the expression becomes equal to zero and the expression simplifies to

$$\mathbf{H}(\omega_0) = \frac{C_1 + C_2}{C_2}.$$

Review Question 9-1: Is the transfer function of a circuit always the same as its voltage gain?

Review Question 9-2: Is the gain factor M_0 always the peak value of $M(\omega)$?

Review Question 9-3: When is a circuit in a resonance condition?

Review Question 9-4: Why is the corner frequency also called the half-power frequency?

Exercise 9-1: A series RL circuit is connected to a voltage source $\mathbf{V}_{\text{s}}$. Obtain an expression for $\mathbf{H}(\omega) = \mathbf{V}_{\text{R}}/\mathbf{V}_{\text{s}}$, where $\mathbf{V}_{\text{R}}$ is the phasor voltage across R. Also, determine the corner frequency of $\mathbf{H}(\omega)$.

Answer: $\mathbf{H}(\omega) = R/(R + j\omega L)$, $\omega_c = R/L$. (See)

Exercise 9-2: Obtain an expression for the input impedance of the circuit in Fig. E9.2 and then use it to determine the resonant frequency.

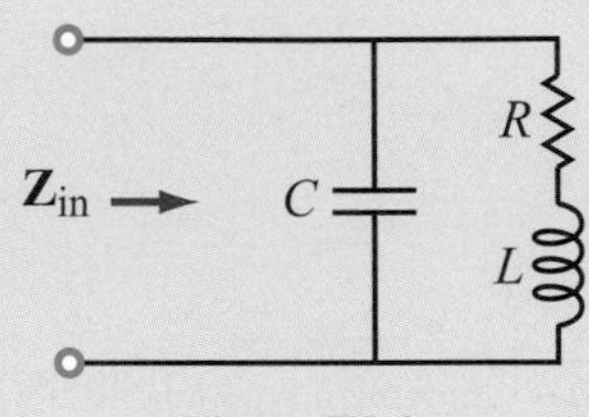

Figure E9.2

Answer: $\omega_0 = \sqrt{\dfrac{1}{LC} - \dfrac{R^2}{L^2}}$.

9-2 Scaling

When designing a resonant circuit such as a filter, it often is convenient to start by designing a ***prototype model*** in which the elements have values on the order of ohms, henrys, and farads and then to scale the prototype circuit into a ***practical circuit*** that not only contains elements with realistic values but also provides the specified frequency response. A circuit can be scaled in magnitude, in frequency, or both. ***Magnitude scaling*** changes the values of the elements in the circuit, but it does not modify its frequency response. ***Frequency scaling*** allows the designer to translate the frequency response into higher or lower frequency ranges while keeping the impedances of the circuit elements unchanged.

9-2.1 Magnitude Scaling

The transfer function of a circuit is based on the impedances of its elements. If all impedances are multiplied (scaled) by the same ***magnitude scaling factor*** K_{m}, the absolute level of the transfer function may or may not change, but its relative frequency response will remain the same. To distinguish between the prototype and scaled circuits, we shall:

(a) Denote elements and impedances of the prototype circuit with unprimed symbols:

$$\mathbf{Z}_{\text{R}} = R, \qquad \mathbf{Z}_{\text{L}} = j\omega L, \qquad \text{and} \qquad \mathbf{Z}_{\text{C}} = \frac{1}{j\omega C}. \quad (9.2C$$

(**b**) Denote elements and impedances of the scaled circuit with primed symbols:

$$\mathbf{Z}'_R = R', \qquad \mathbf{Z}'_L = j\omega L', \qquad \text{and} \qquad \mathbf{Z}'_C = \frac{1}{j\omega C'}. \tag{9.21}$$

Magnitude scaling by a factor K_m implies that:

$$\mathbf{Z}'_R = K_m \mathbf{Z}_R, \qquad \mathbf{Z}'_L = K_m \mathbf{Z}_L, \qquad \text{and} \qquad \mathbf{Z}'_C = K_m \mathbf{Z}_C, \tag{9.22}$$

which translates into the relations

$$\begin{aligned} R' &= K_m R, \\ L' &= K_m L, \\ C' &= \frac{C}{K_m}, \\ \omega &= \omega'. \end{aligned}$$

(Magnitude scaling only) (9.23)

Thus, resistor and inductor values scale by K_m, but capacitor values scale by $1/K_m$.

To illustrate with an example, consider the transfer function given by Eq. (9.18),

$$\mathbf{H}_R(\omega) = \frac{j\omega RC}{1 + j\omega RC} \tag{9.24a}$$

and its scaled version

$$\mathbf{H}'_R(\omega) = \frac{j\omega R'C'}{1 + j\omega R'C'}. \tag{9.24b}$$

Applying the recipe given by Eq. (9.23) leads to

$$\mathbf{H}'_R(\omega) = \mathbf{H}_R(\omega),$$

which means that the frequency response remains unchanged.

9-2.2 Frequency Scaling

To shift the profile of a transfer function along the ω-axis by a ***frequency scaling factor*** K_f while keeping its relative shape the same, we can replace ω in the transfer function of the prototype circuit with $\omega' = K_f \omega$ and scale the element values so that their impedances remain unchanged. For an inductor, $\mathbf{Z}_L = j\omega L$, so if ω is to be scaled up by K_f, L has to be scaled down by the same factor in order for $\mathbf{Z}_L$ to stay the same. Hence, the impedance condition requires that

$$\begin{aligned} R' &= R, \\ L' &= \frac{L}{K_f}, \\ C' &= \frac{C}{K_f}, \\ \omega' &= K_f \omega. \end{aligned}$$

(Frequency scaling only) (9.25)

9-2.3 Combined Magnitude and Frequency Scaling

To transfer the prototype circuit design into a realizable circuit, we often apply magnitude and frequency scaling simultaneously, in which case the relationships between the prototype and scaled circuits become

$$\begin{aligned} R' &= K_m R, \\ L' &= \frac{K_m}{K_f}\, L, \\ C' &= \frac{1}{K_m K_f}\, C, \\ \omega' &= K_f \omega. \end{aligned}$$

(Magnitude and frequency scaling) (9.26)

Example 9-2: Third-Order LP Filter

The order of a filter is a measure of how steep its response is as a function of ω. For example, the response of the third-order lowpass filter shown in Fig. 9-7 is much steeper than the response of the first-order LP filter shown earlier in Fig. 9-5(b). The circuit in Fig. 9-7 is a prototype model with a cutoff frequency of $\omega_c = 1$ rad/s. Develop a scaled version with a cutoff frequency of $\omega'_c = 10^6$ rad/s and a resistor value of 2 kΩ.

Solution: Based on the given information, the scaling factors are

$$K_m = \frac{R'}{R} = \frac{2\text{k}}{2} = 10^3 \qquad \text{and} \qquad K_f = \frac{\omega'_c}{\omega_c} = \frac{10^6}{1} = 10^6.$$

Application of Eq. (9.26) leads to

$$L'_1 = \frac{K_m}{K_f}\, L_1 = \frac{10^3}{10^6} \times 3 = 3 \text{ mH},$$

$$L'_2 = \frac{K_m}{K_f}\, L_2 = 1 \text{ mH},$$

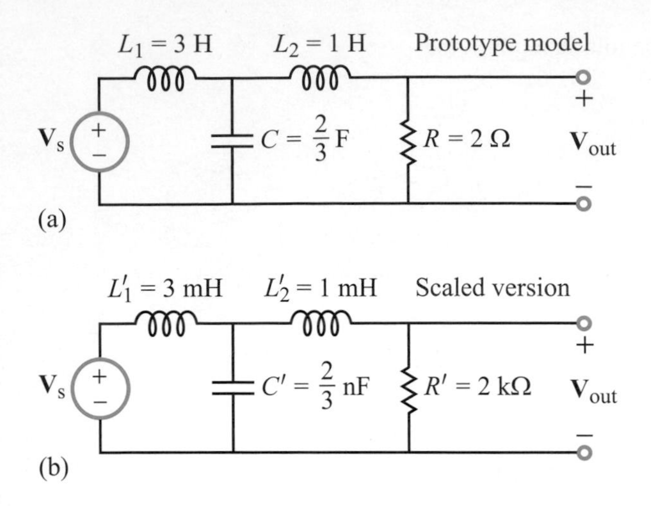

Figure 9-7: Prototype and scaled circuits of Example 9-2.

and

$$C' = \frac{1}{K_m K_f}\ C = \frac{1}{10^3 \times 10^6} \times \frac{2}{3} = \frac{2}{3} \text{ nF}.$$

The scaled circuit is displayed in Fig. 9-7(b).

Review Question 9-5: How is the scaling concept used in the design of resonant circuits and filters?

Review Question 9-6: What remains unchanged in (a) magnitude scaling alone and (b) frequency scaling alone?

Exercise 9-3: Determine (a) $\mathbf{Z}_{in}$ of the prototype circuit shown in Fig. E9.3 at $\omega = 1$ rad/s and (b) $\mathbf{Z}'_{in}$ of the same circuit after scaling it by $K_m = 1000$ and $K_f = 1000$.

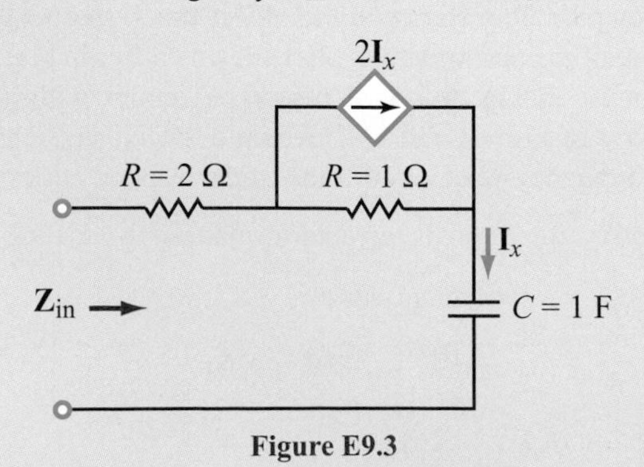

Figure E9.3

Answer: (a) $\mathbf{Z}_{in} = (1 - j)\ \Omega$, (b) $\mathbf{Z}'_{in} = (1 - j)$ kΩ. (See)

9-3 Bode Plots

In the late 1930s, inventor Hendrik Bode (pronounced Boh-dee) developed a graphical technique that has since become a standard tool for the analysis and design of resonant circuits, including filters, oscillators, and amplifiers. Bode's technique, which generates what we today call ***Bode plots*** or a ***Bode diagram***, relies on *using a logarithmic scale for ω and on expressing the magnitude of the transfer function in decibels (dB).* To make sure the reader is fully familiar with the properties of the dB operator, a quick review is in order.

9-3.1 The dB Scale

The ratio of the power P relative to a reference power level P_0—such as the output power generated by an amplifier, relative to the input power supplied by the source—is called ***relative*** or ***normalized power***. In many engineering applications, P/P_0 may vary over several orders of magnitude when plotted against a specific variable of interest, such as the frequency ω of the circuit. The dB scale originally was introduced as a logarithmic conversion tool to facilitate the generation of plots involving relative power, but its use has since been expanded to other physical quantities. The dB operator is intended as a scale converter of relative quantities, such as P/P_0, rather than of P itself, but it still can be applied to P by setting P_0 equal to a specified value, such as 1 watt or 1 mwatt, so long as P is expressed in the same units as P_0.

If G is defined as the power gain,

$$G = \frac{P}{P_0}, \tag{9.27}$$

then the corresponding gain in dB is defined as

$$G\ [\text{dB}] = 10 \log G = 10 \log \left(\frac{P}{P_0}\right) \qquad (\text{dB}). \tag{9.28}$$

The logarithm is in base 10. *The dB scale converts a power ratio to its logarithmic value and then multiplies it by 10.* Table 9-1 (left) provides a listing of some values of G and the corresponding values of G [dB]. Note that when G varies across six orders of magnitude, from 10^{-3} to 10^3, G [dB] varies from -30 dB to $+30$ dB. Also note that the dB value of 2 is $+3$ dB and the dB value of 0.5 is -3 dB.

Even though the scale originally was applied to power ratios, it now is used to express voltage and current ratios as well. If P and P_0 are the average powers absorbed by resistors of equal value and the corresponding phasor voltages across the resistor are $\mathbf{V}$ and $\mathbf{V}_0$, respectively, then

$$G\ [\text{dB}] = 10 \log \left(\frac{\frac{1}{2}|\mathbf{V}|^2 R}{\frac{1}{2}|\mathbf{V}_0|^2 R}\right) = 20 \log \left(\frac{|\mathbf{V}|}{|\mathbf{V}_0|}\right). \tag{9.29}$$

Table 9-1: Correspondence between power ratios in natural numbers and their dB values (left table) and between voltage or current ratios and their dB values (right table).

$\dfrac{P}{P_0}$	dB
10^N	$10N$ dB
10^3	30 dB
100	20 dB
10	10 dB
4	$\simeq 6$ dB
2	$\simeq 3$ dB
1	0 dB
0.5	$\simeq -3$ dB
0.25	$\simeq -6$ dB
0.1	-10 dB
10^{-N}	$-10N$ dB

$\left\lvert\dfrac{\mathbf{V}}{\mathbf{V}_0}\right\rvert$ or $\left\lvert\dfrac{\mathbf{I}}{\mathbf{I}_0}\right\rvert$	dB
10^N	$20N$ dB
10^3	60 dB
100	40 dB
10	20 dB
4	$\simeq 12$ dB
2	$\simeq 6$ dB
1	0 dB
0.5	$\simeq -6$ dB
0.25	$\simeq -12$ dB
0.1	-20 dB
10^{-N}	$-20N$ dB

Similarly,

$$G \text{ [dB]} = 20 \log \left(\frac{|\mathbf{I}|}{|\mathbf{I}_0|} \right). \tag{9.30}$$

Whereas the dB definition for power ratio includes a scaling factor of 10, the scaling factor for voltage and current is 20.

A useful property of the log operator is that *the log of the product of two numbers is equal to the sum of their logs.* That is if

$$G = XY \Rightarrow G \text{ [dB]} = X \text{ [dB]} + Y \text{ [dB]}. \tag{9.31}$$

This result follows from

$$G \text{ [dB]} = 10 \log(XY) = 10 \log X + 10 \log Y = X \text{ [dB]} + Y \text{ [dB]}.$$

By the same token, if:

$$G = \frac{X}{Y} \Rightarrow G \text{ [dB]} = X \text{ [dB]} - Y \text{ [dB]}. \tag{9.32}$$

Conversion of products and ratios into sums and differences will prove to be quite useful when constructing the frequency response of a resonant circuit.

Example 9-3: RL Highpass Filter

For the series RL circuit shown in Fig. 9-8(a):

(a) Obtain an expression for the transfer function $\mathbf{H} = \mathbf{V}_{\text{out}}/\mathbf{V}_{\text{s}}$ in terms of ω/ω_{c} where $\omega_{\text{c}} = R/L$.

(b) Determine the magnitude M [dB] $= 20 \log |\mathbf{H}|$ and plot it as a function of ω on a log scale with ω expressed in units of ω_{c}.

(c) Determine and plot the phase angle of $\mathbf{H}$.

Solution:

(a) Voltage division gives

$$\mathbf{V}_{\text{out}} = \frac{j\omega L \mathbf{V}_{\text{s}}}{R + j\omega L},$$

which leads to

$$\mathbf{H} = \frac{\mathbf{V}_{\text{out}}}{\mathbf{V}_{\text{s}}} = \frac{j\omega L}{R + j\omega L} = \frac{j(\omega/\omega_{\text{c}})}{1 + j(\omega/\omega_{\text{c}})}, \tag{9.33}$$

with $\omega_{\text{c}} = R/L$.

(b) The magnitude of $\mathbf{H}$ is given by

$$M = |\mathbf{H}| = \frac{(\omega/\omega_{\text{c}})}{|1 + j(\omega/\omega_{\text{c}})|} = \frac{(\omega/\omega_{\text{c}})}{\sqrt{1 + (\omega/\omega_{\text{c}})^2}}. \tag{9.34}$$

Since H is a voltage ratio, the appropriate dB scaling factor is 20, so

$$\begin{aligned} M \text{ [dB]} &= 20 \log M \\ &= 20 \log(\omega/\omega_{\text{c}}) - 20 \log[1 + (\omega/\omega_{\text{c}})^2]^{1/2} \\ &= 20 \log(\omega/\omega_{\text{c}}) - 10 \log[1 + (\omega/\omega_{\text{c}})^2]. \end{aligned} \tag{9.35}$$

In the Bode-diagram terminology introduced later in Section 9-3.2, the components of M [dB] are called ***factors***, so in the present case, M [dB] consists of two factors with the second one having a negative coefficient. A magnitude plot is displayed on semilog graph paper with the vertical axis in dB and the horizontal axis in (rad/s). If in the expression for M [dB], ω appears in a normalized format—as in $(\omega/\omega_{\text{c}})$—we may choose to express the horizontal axis in units of ω_{c}. Figure 9-8(b) contains individual plots for each of the two factors comprising M [dB] as well as a plot for their sum.

On semilog graph paper, the plot of $\log(\omega/\omega_{\text{c}})$ is a straight line that crosses the ω-axis at $(\omega/\omega_{\text{c}}) = 1$. This is because $\log 1 = 0$. At $(\omega/\omega_{\text{c}}) = 10$, $20 \log 10 = 20$ dB. Hence;

$$20 \log \left(\frac{\omega}{\omega_{\text{c}}} \right) \Rightarrow$$ straight line with slope = 20 dB/decade and ω-axis crossing at $\omega/\omega_{\text{c}} = 1$.

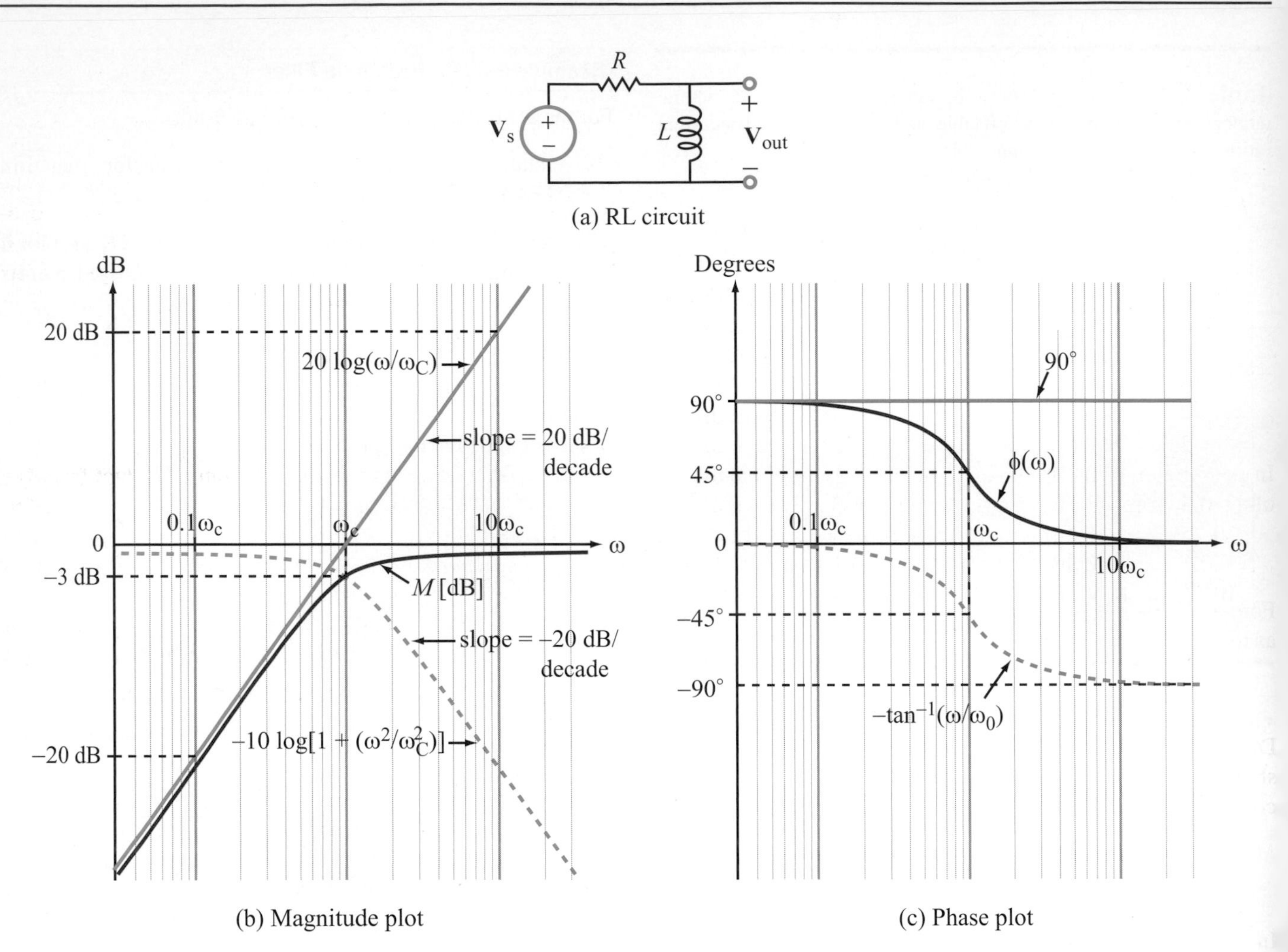

Figure 9-8: Magnitude and phase plots of $\mathbf{H} = \mathbf{V}_{\text{out}}/\mathbf{V}_{\text{s}}$.

The second factor has a nonlinear plot, with the following properties.

Low-Frequency Asymptote

$$\text{As } (\omega/\omega_c) \Rightarrow 0, \qquad -10\log\left[1+\left(\frac{\omega}{\omega_c}\right)^2\right] \Rightarrow 0.$$

High-Frequency Asymptote

$$\text{As } (\omega/\omega_c) \Rightarrow \infty, \qquad -10\log\left[1+\left(\frac{\omega}{\omega_c}\right)^2\right]$$
$$\Rightarrow -20\log\left(\frac{\omega}{\omega_c}\right).$$

The plot of M [dB] is obtained by graphically adding together the two plots of its individual factors (Fig. 9-8(b)). At low frequencies such that $(\omega/\omega_c \ll 1)$, M [dB] is dominated by its first factor; at $\omega/\omega_c = 1$, M [dB] $= -3$ dB; and at high frequencies $(\omega/\omega_c \gg 1)$, M [dB] $\rightarrow 0$, because its two factors cancel each other out. The overall profile is typical of the spectral response of a highpass filter with a cutoff frequency ω_c

(c) From Eq. (9.33), the phase angle of **H** is

$$\phi(\omega) = 90^\circ - \tan^{-1}\left(\frac{\omega}{\omega_c}\right). \tag{9.36}$$

The 90° component is contributed by j in the numerator and the second term is the phase angle of the denominator. The phase plot is displayed in Fig. 9-8(c).

Review Question 9-7: When is it helpful to use the dB scale?

Review Question 9-8: What is the scaling factor for power ratio? For current ratio?

Exercise 9-4: Convert the following voltage ratios to dB: (a) 20, (b) 0.03, (c) 6×10^6.

Answer: (a) 26.02 dB, (b) −30.46 dB, (c) 135.56 dB. (See)

Exercise 9-5: Convert the following dB values to voltage ratios: (a) 36 dB, (b) −24 dB, (c) −0.5 dB.

Answer: (a) 63.1, (b) 0.063, (c) 0.94. (See)

9-3.2 Poles and Zeros

In polar coordinates, the transfer function $\mathbf{H}(\omega)$ is composed of a magnitude $M(\omega)$ and a phase angle $\phi(\omega)$,

$$\mathbf{H}(\omega) = M(\omega)\, e^{j\phi(\omega)}. \tag{9.37}$$

For any circuit, the expression for $\mathbf{H}(\omega)$ in general can be cast as the product of multiple *factors* $\mathbf{A}_1(\omega)$ to $\mathbf{A}_n(\omega)$,

$$\mathbf{H}(\omega) = \mathbf{A}_1(\omega)\, \mathbf{A}_2(\omega) \ldots \mathbf{A}_n(\omega). \tag{9.38}$$

Discussion of the functional forms of $\mathbf{A}_1$ to $\mathbf{A}_n$ will follow shortly, but to clarify what we mean by Eq. (9.38), let us consider the simple example of a transfer function given by

$$\mathbf{H}(\omega) = 10\, \frac{1 + j\omega/\omega_z}{1 + j\omega/\omega_p}. \tag{9.39}$$

In this case,

$$\mathbf{A}_1 = 10, \tag{9.40a}$$

$$\mathbf{A}_2 = 1 + j\omega/\omega_z, \tag{9.40b}$$

and

$$\mathbf{A}_3 = \frac{1}{1 + j\omega/\omega_p}. \tag{9.40c}$$

The expression for $\mathbf{H}(\omega)$ was structured intentionally into a form—called the ***standard form***—in which the two terms involving ω each are written such that the real part is unity and the coefficient of ω in the imaginary part is defined as the reciprocal of an angular frequency. For the circuit represented by the transfer function given by Eq. (9.39), ω_z and ω_p are related to the circuit architecture and the element values of the circuit. The quantity ω_z (which has the same units as ω) is called a ***zero*** of $\mathbf{H}(\omega)$, because it appears in a factor contained in the numerator of $\mathbf{H}(\omega)$. Similarly, ω_p is called a ***pole*** because it is part of a factor contained in the denominator of $\mathbf{H}(\omega)$. If the numerator is a product of multiple factors, $\mathbf{H}(\omega)$ will have multiple zeros—one associated with each factor (except for frequency independent factors, such as $\mathbf{A}_1 = 10$). A transfer function also may have multiple poles if the denominator of $\mathbf{H}(\omega)$ is the product of multiple factors. Moreover, the factors may assume functional forms different from those given by $\mathbf{A}_1$ to $\mathbf{A}_3$ in Eq. (9.40).

To analyze the frequency response of the circuit, we need to extract from Eq. (9.38) explicit expressions for the magnitude $M(\omega)$ and the phase angle $\phi(\omega)$. For any two complex numbers, the phase angle of their product is equal to the sum of their individual phase angles. Application of this multiplication principle to Eq. (9.38) gives

$$\phi(\omega) = \phi_{\mathbf{A}_1}(\omega) + \phi_{\mathbf{A}_2}(\omega) + \cdots + \phi_{\mathbf{A}_n}(\omega), \tag{9.41}$$

where $\phi_{\mathbf{A}_1}(\omega)$ to $\phi_{\mathbf{A}_n}(\omega)$ are the phase angles of factors $\mathbf{A}_1$ to $\mathbf{A}_n$, respectively. Transformation from a product form (as in Eq. (9.38)), into a sum (as in Eq. (9.41)) allows us to generate a phase plot for each factor separately and then add them together graphically—rather than having to deal with a single complicated expression all at once. The dB conversion introduced by Bode accomplishes a similar transformation for the magnitude $M(\omega)$. Application of the log property described by Eq. (9.31) leads to

$$\begin{aligned} M\ [\text{dB}] &= 20\log|\mathbf{H}| \\ &= 20\log|\mathbf{A}_1| + 20\log|\mathbf{A}_2| + \cdots + 20\log|\mathbf{A}_n| \\ &= A_1\ [\text{dB}] + A_2\ [\text{dB}] + \cdots + A_n\ [\text{dB}] \end{aligned} \tag{9.42}$$

where

$$A_1\ [\text{dB}] = 20\log|\mathbf{A}_1|, \tag{9.43}$$

and a similar definition applies to the other factors. The transformations represented by Eqs. (9.41) and (9.42) constitute the basic framework for generating Bode diagrams. Our next step is to examine the possible functional forms that factors $\mathbf{A}_1$ to $\mathbf{A}_n$ may assume.

Standard form refers to an arrangement in which factors $\mathbf{A}_1$ to $\mathbf{A}_n$ of Eq. (9.38) each can assume any one of only seven possible functional forms. We will examine the general character of each of these factors individually (as if it were the only component of $\mathbf{H}(\omega)$) by considering two types of plots: ***exact plots*** based on the exact expression for $\mathbf{H}(\omega)$ and straight-line approximations—called ***Bode plots***—that are much easier to generate and yet provide reasonable accuracy. The symbol N (which we will call the ***order*** of a factor) is an integer equal to or greater than 1.

9-3.3 Functional Forms

Constant factor: $\mathbf{H} = K$

This is a frequency-independent constant that may be positive or negative.

Magnitude: M [dB] $= 20 \log |K|$
Phase: $\phi = 0°$ if $K > 0$, or $\pm 180°$ if $K < 0$
Bode plots: Same as exact plots; straight horizontal lines (Table 9-2)

Zero @ origin: $\mathbf{H} = (j\omega)^N$ (N = positive integer)

The name of this factor reflects the fact that $\mathbf{H} \to 0$ as $\omega \to 0$. N is its order, so for example, $(j\omega)^2$ is called a second-order zero @ origin.

Magnitude: M [dB] $= 20 \log |(j\omega)^N| = 20N \log \omega$
Phase: $\phi = (90N)°$
Bode plots: Same as exact
Magnitude: Straight line through $\omega = 1$ with slope $= 20N$ dB/decade
Phase: Constant level (Table 9-2)

Pole @ origin: $\mathbf{H} = 1/(j\omega)^N$

This factor is called a pole because $\mathbf{H} \to \infty$ when $\omega \to 0$. The function $1/(j\omega)^3$ is called a third-order pole @ origin, for example.

Magnitude: M [dB] $= 20 \log \left| \dfrac{1}{(j\omega)^N} \right| = -20N \log \omega$
Phase: $\phi = (-90N)°$
Bode plots: Same as exact (Table 9-2)

Magnitude and phase plots are identical with those of zero @ origin except for (−) sign in both cases.

Simple zero: $\mathbf{H} = (1 + j\omega/\omega_c)^N$

Standard form requires that the real part be 1 and the imaginary part be positive. The constant ω_c is the corner frequency of the simple-zero factor, and N is its order.

Magnitude:

$$M \text{ [dB]} = 20 \log |(1 + j\omega/\omega_c)^N| = 10N \log[1 + (\omega/\omega_c)^2] \simeq \begin{cases} 0 \text{ dB}, & \text{for } \omega/\omega_c \ll 1 \\ 20N \log(\omega/\omega_c), & \text{for } \omega/\omega_c \gg 1 \end{cases}$$

Phase:

$$\phi = N \tan^{-1}\left(\frac{\omega}{\omega_c}\right) \simeq \begin{cases} 0, & \text{for } \omega/\omega_c \ll 1 \\ (90N)°, & \text{for } \omega/\omega_c \gg 1 \end{cases}$$

Bode plots: Different from exact; for magnitude, the maximum difference is $3N$ dB and it occurs at $\omega/\omega_c = 1$
Magnitude: 0 dB horizontal line to $\omega = \omega_c$, followed by straight line with slope $= 20N$ dB/decade
Phase: 0° horizontal line to $\omega = 0.1\omega_c$; straight line connecting coordinates $[0.1\omega_c, 0]$ to $[10\omega_c, (90N)°]$; followed by horizontal line $(90N)°$

Figure 9-9 provides a comparison between the Bode approximation and the exact solution for both magnitude and phase. The corner frequency ω_c gets its name from the Bode magnitude plot, which turns the corner at ω_c.

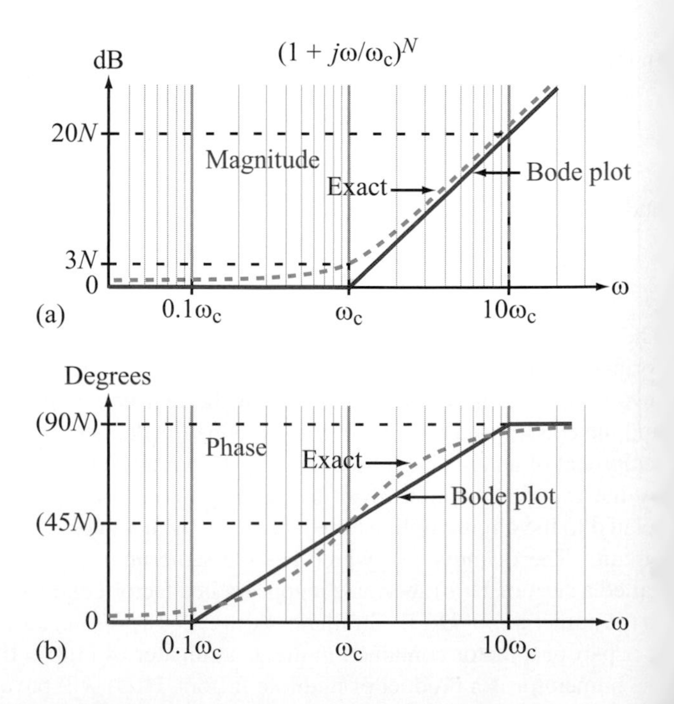

Figure 9-9: Comparison of exact plots with Bode approximations for a simple zero with a corner frequency ω_c.

Table 9-2: Bode straight-line approximations for magnitude and phase.

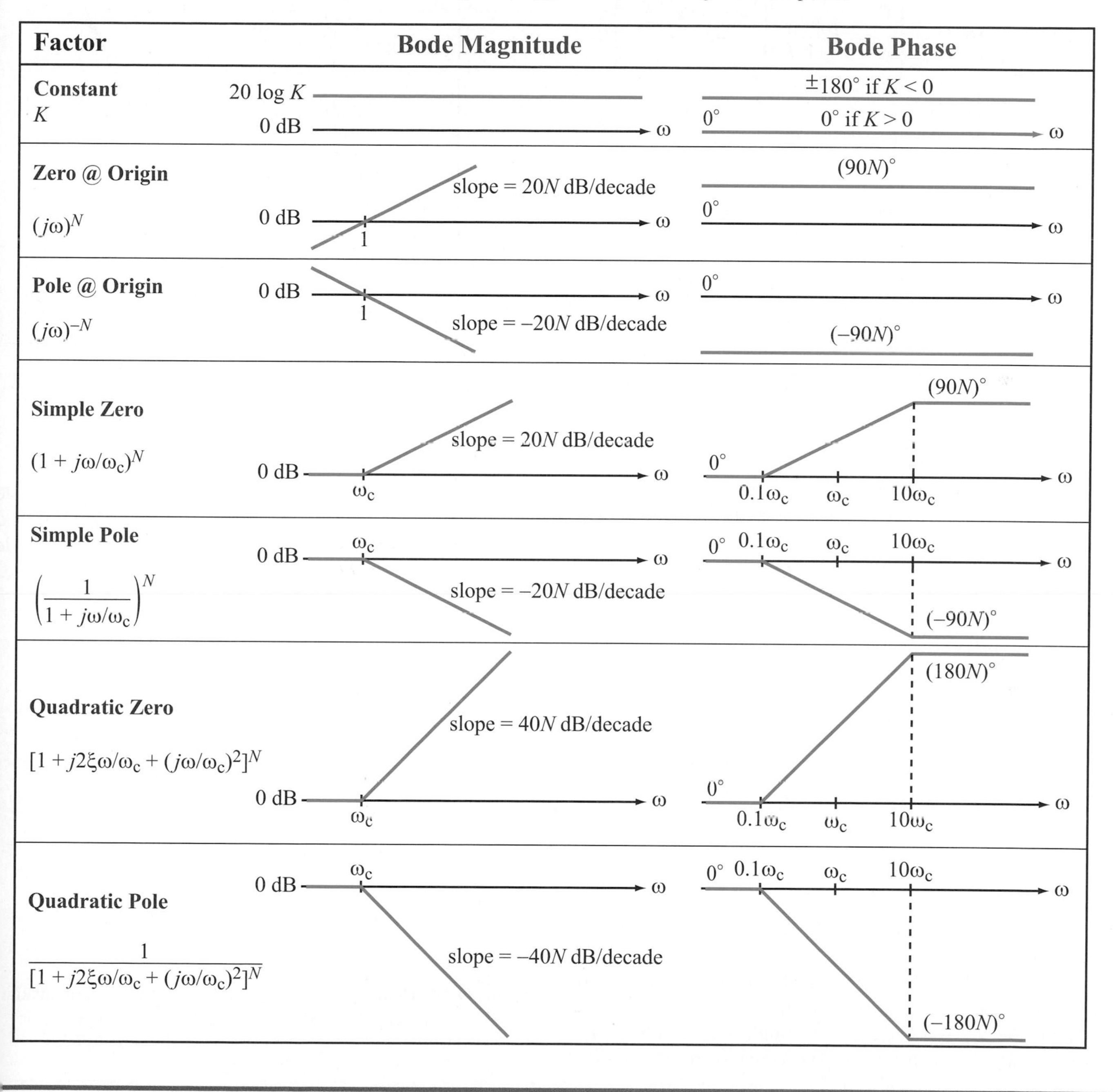

imple pole: $\mathbf{H} = 1/(1 + j\omega/\omega_c)^N$

Plots are mirror images (relative to ω-axis) of those for the mple zero (Table 9-2).

Quadratic zero:
$\mathbf{H} = [1 + j2\xi\omega/\omega_c + (j\omega/\omega_c)^2]^N$

N is the order of the quadratic zero, ω_c is its corner frequency, and ξ is its ***damping factor***.

Magnitude:

$$M\ [\text{dB}] = 10N \log\left\{\left[1-\left(\frac{\omega}{\omega_c}\right)^2\right]^2 + 4\xi^2\left(\frac{\omega}{\omega_c}\right)^2\right\}$$

$$\simeq \begin{cases} 0\ \text{dB} & \text{for } \omega/\omega_c \ll 1, \\ 40N\log(\omega/\omega_c) & \text{for } \omega/\omega_c \gg 1 \end{cases}$$

Phase: $\phi = N \tan^{-1}\left[\dfrac{2\xi(\omega/\omega_c)}{1-(\omega/\omega_c)^2}\right]$

$$\simeq \begin{cases} 0^\circ & \text{for } \omega/\omega_c \ll 1, \\ (180N)^\circ & \text{for } \omega/\omega_c \gg 1 \end{cases}$$

Bode plots: **Magnitude:** Same as simple zero except at twice the slope
Phase: Same as simple zero except at twice the slope and twice the level at $\omega/\omega_c \gg 1$

Figure 9-10 displays plots of the magnitude and phase of the quadratic-zero factor with $N = 1$ for three different values of the damping coefficient. Whereas the value of ξ has little influence on the shape of the plots when $\omega/\omega_c \ll 1$ or $\omega/\omega_c \gg 1$, it exercises significant influence when ω is in the neighborhood of ω_c. *In terms of the Bode approximation, the Bode plots (Table 9-2) for a quadratic factor of order N are identical with those for a simple factor of order 2N.*

Quadratic pole:
$\mathbf{H} = [1 + j2\xi\omega/\omega_c + (j\omega/\omega_c)^2]^{-N}$
Plots are mirror images, relative to ω-axis, as those for the quadratic zero.

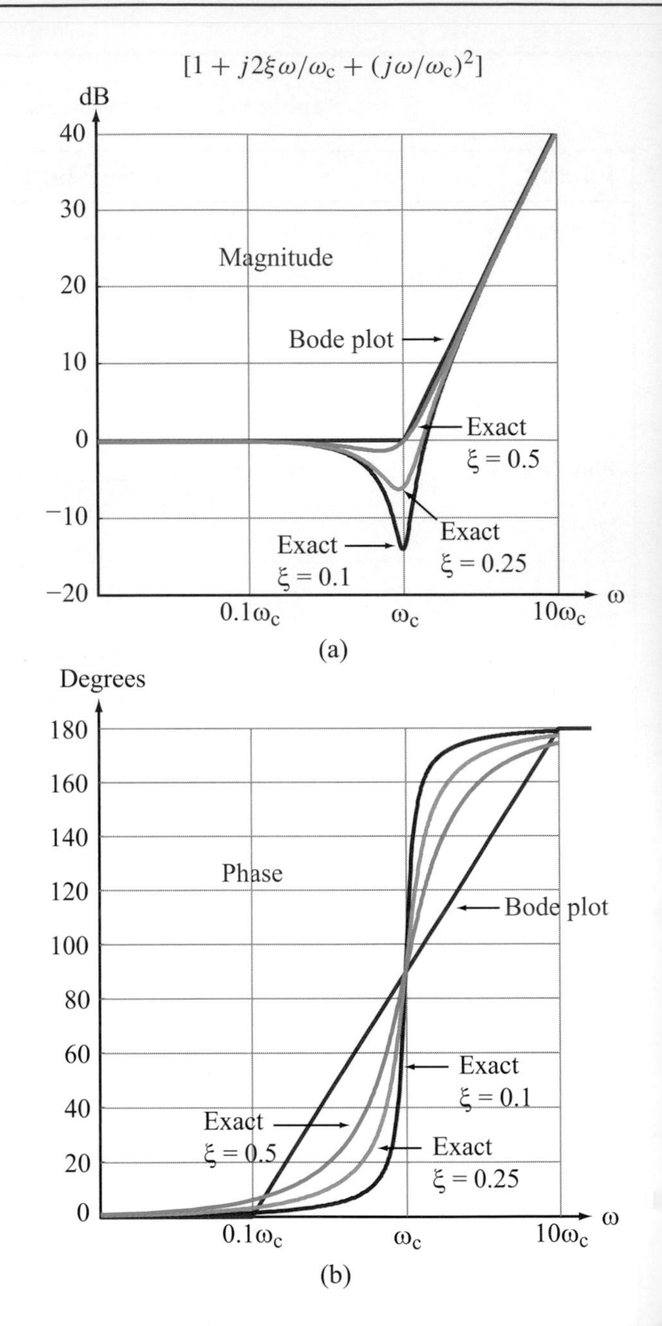

Figure 9-10: Comparison of exact plots with Bode approximation for a quadratic zero $[1 + j2\xi\omega/\omega_c + (j\omega/\omega_c)^2]$.

9-3.4 General Observations

A few general observations about the Bode plots shown in Table 9-2 are in order.

(1) For $N = 1$, the slope of the non-horizontal Bode magnitude line (called ***gain roll-off rate***) is 20 dB/decade for both the zero @ origin and simple-zero factors. The corresponding slope for their pole counterparts is −20 dB/decade.

(2) For $N = 1$, the slopes of the non-horizontal Bode magnitude lines for the quadratic zero and quadratic pole factors are 40 dB/decade and −40 dB/decade, respectively.

(3) The slopes of all Bode magnitude and phase lines are proportional to N. For example, the slope of the magnitude of a first-order simple-zero factor $(1 + j\omega/\omega_c)$ is 20 dB/decade, so the slope of a third-order simple-zero factor $(1 + j\omega/\omega_c)^3$ is 60 dB/decade.

Example 9-4: Bode Plots I

The voltage transfer function of a certain circuit is given by

$$\mathbf{H}(\omega) = \frac{(20 + j4\omega)^2}{j40\omega(100 + j2\omega)}.$$

(a) Rearrange the expression into standard form. (b) Generate Bode plots for the magnitude and phase of $\mathbf{H}(\omega)$.

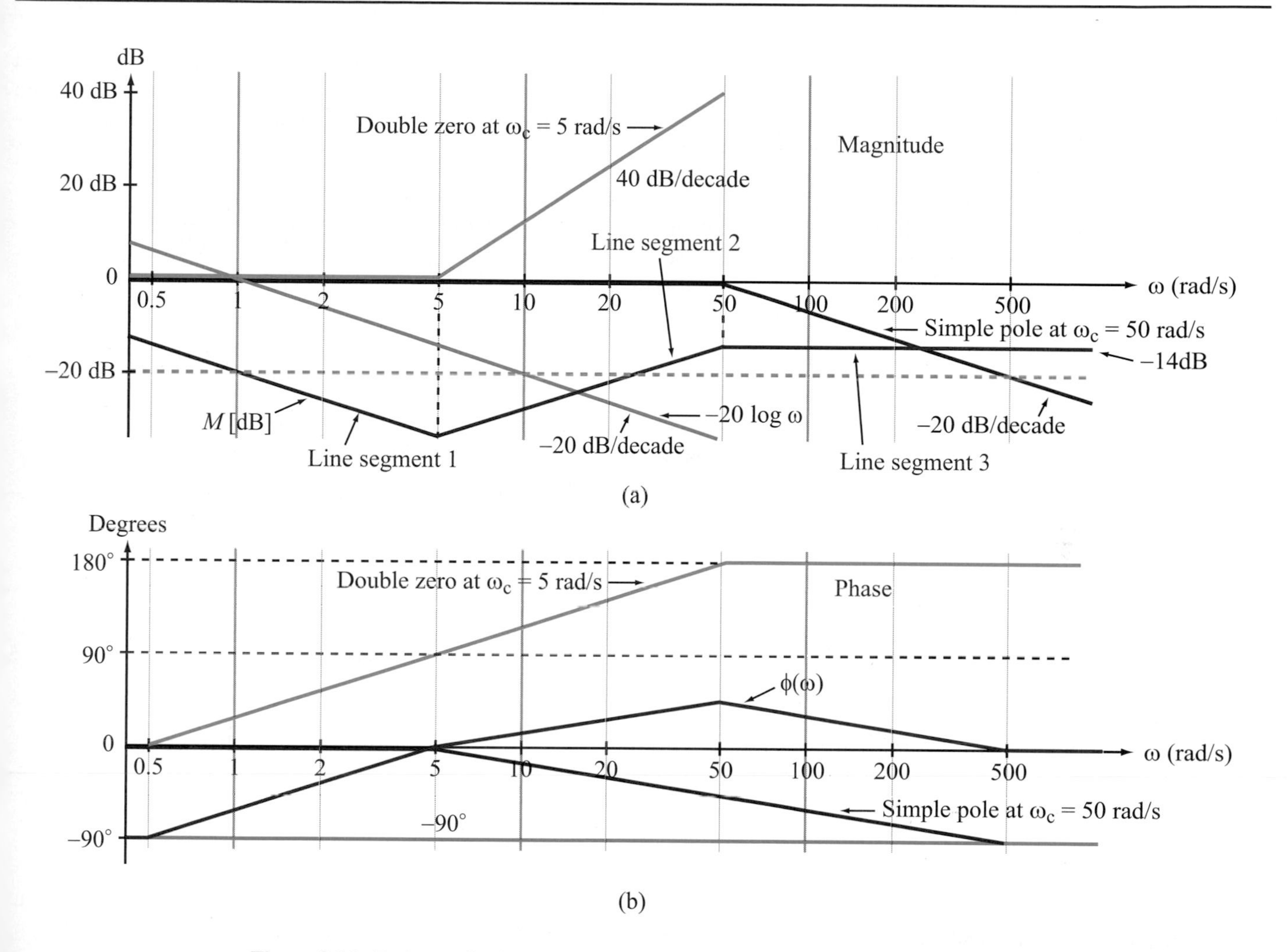

Figure 9-11: Bode amplitude and phase plots for the transfer function of Example 9-4.

Solution:

(a) By factoring out 20^2 from the factor in the numerator and 100 from the factor in the denominator, we have

$$\mathbf{H}(\omega) = \frac{400(1 + j\omega/5)^2}{j4000\omega(1 + j\omega/50)} = \frac{-j0.1(1 + j\omega/5)^2}{\omega(1 + j\omega/50)}.$$

The corner frequency of the double-zero factor given by $(1+j\omega/5)^2$ is $\omega_{c_1} = 5$ rad/s, and similarly, the corner frequency of the simple-pole factor given by $(1+j\omega/50)$ is $\omega_{c_2} = 50$ rad/s.

(b)

$$\begin{aligned} M\text{ [dB]} &= 20 \log |\mathbf{H}| \\ &= 20 \log 0.1 + 40 \log |1 + j\omega/5| \\ &\quad - 20 \log \omega - 20 \log |1 + j\omega/50| \\ &= -20 \text{ dB} + 40 \log |1 + j\omega/5| \\ &\quad - 20 \log \omega - 20 \log |1 + j\omega/50|. \end{aligned}$$

The Bode line-approximations for the four terms constituting M [dB] and their sum are shown in Fig. 9-11(a). The sum is obtained by graphically adding the line-approximations corresponding to the four individual terms. An alternative method for generating the Bode magnitude plot is to start by plotting the line-approximation of the term with the lowest corner frequency, and then to move forward along the ω-axis while sequentially changing the slope of the line as we encounter terms with higher corner frequencies. To illustrate the procedure, we labeled the three line segments of M [dB] in Fig. 9-11(a) as line segments 1, 2, and 3.

(1) The constant term is −20 dB (horizontal line with zero slope).

(2) The lowest-frequency term is the pole @ origin $(1/\omega)$. Its Bode line goes through 0 dB at $\omega = 1$ rad/s and has a slope of −20 dB/decade.

(3) The combination of (1) and (2) generates line segment 1, which goes through $\omega = 1$ at -20 dB.

(4) The term with the next higher corner frequency is the double zero with $\omega_c = 5$ rad/s. A double zero has a Bode line with a slope of $+40$ dB/decade. Hence, at $\omega_c = 5$ rad/s, we change the slope of line segment 1 by adding 40 dB/decade to its original slope of -20 dB/decade. This step generates line segment 2, with a slope of $+20$ dB/decade.

(5) Line segment 2 continues until we encounter the corner frequency of the next term, namely the simple pole with $\omega_c = 50$ rad/s. Adding a slope of -20 dB/decade leads to line segment 3, with a net slope of 0, and a constant level of -14 dB.

The phase of $\mathbf{H}(\omega)$ is given by

$$\phi = -90^\circ + 2\tan^{-1}\frac{\omega}{5} - \tan^{-1}\frac{\omega}{50}.$$

Bode plots for ϕ and its three components are shown in Fig. 9-11(b).

Example 9-5: Bode Plots II

Transfer function $\mathbf{H}(\omega)$ is given by

$$\mathbf{H}(\omega) = \frac{(j10\omega + 30)^2}{(300 - 3\omega^2 + j90\omega)}.$$

(a) Rearrange $\mathbf{H}(\omega)$ into standard form. (b) Generate Bode plots for its magnitude and phase.

Solution: **(a)** Upon reversing the order of the real and imaginary components in the numerator, factoring out 30^2 from it, and factoring out 300 from the denominator, we get

$$\mathbf{H}(\omega) = \frac{3(1 + j\omega/3)^2}{[1 + j3\omega/10 + (j\omega/10)^2]},$$

which consists of a constant factor $K = 3$, a zero factor with a corner frequency of 3 (rad/s), and a quadratic pole with a corner frequency of 10 rad/s.

(b)

$$\begin{aligned} M\text{ [dB]} &= 20\log|\mathbf{H}| \\ &= 20\log 3 + 40\log|1 + j\omega/3| \\ &\quad - 20\log|1 + j3\omega/10 + (j\omega/10)^2| \\ &= 9.5\text{ dB} + 40\log|1 + j\omega/3| \\ &\quad - 20\log|1 + j3\omega/10 + (j\omega/10)^2|, \end{aligned}$$

$$\phi = 2\tan^{-1}(\omega/3) - \tan^{-1}\left(\frac{3\omega/10}{1 - \omega^2/100}\right).$$

Bode plots of M [dB] and ϕ are shown in Fig. 9-12. We note that M [dB] exhibits a highpass filter-like response.

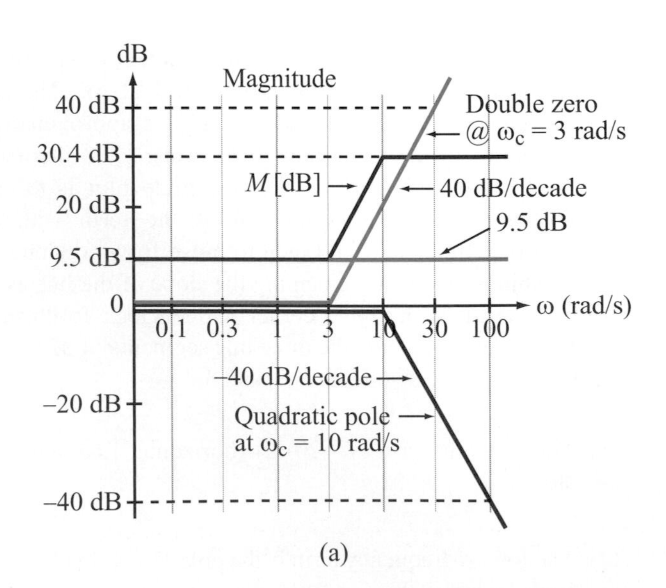

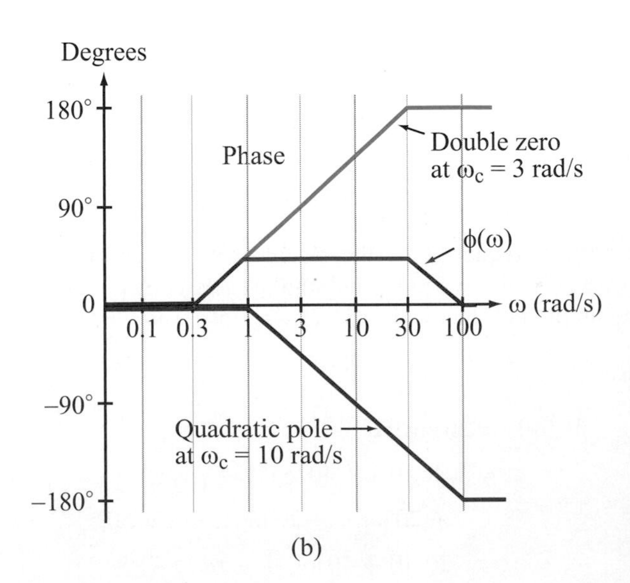

Figure 9-12: Bode magnitude and phase plots for Example 9-6.

Exercise 9-6: Generate a Bode magnitude plot for the transfer function

$\mathbf{H} = 10(100 + j\omega)(1000 + j\omega)/[(10 + j\omega)(10^4 + j\omega)]$.

Answer:

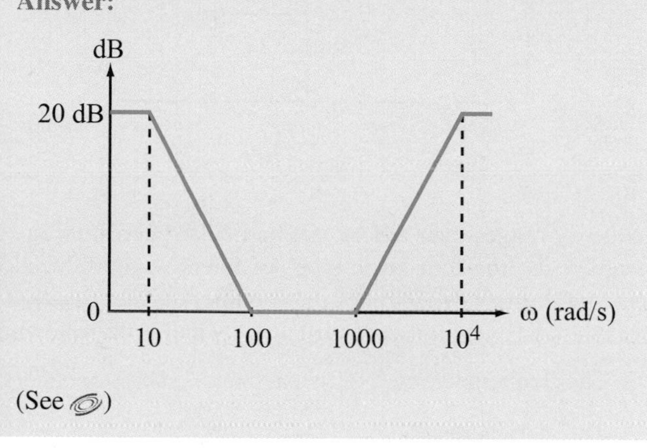

(See)

Example 9-6: Bandreject Filter

The Bode magnitude plot shown in Fig. 9-13 belongs to a bandreject filter that provides significant gain at low and high frequencies, but no gain to frequencies in the 10 to 50 (rad/s) range. Obtain the transfer function $\mathbf{H}(\omega)$.

Solution: The Bode plot consists of five segments.

The first segment, corresponding to $\omega \leq 1$ rad/s, is generated by a pole @ the origin that goes through $\omega = 3$ rad/s and has a slope of -40 dB/decade. The slope indicates that it is a double pole, so it must be given by

$$\mathbf{H}_1 = \left(\frac{1}{\omega/3}\right)^2 = \frac{9}{\omega^2}.$$

To verify the validity of our expression, let us convert it to dB as

$$M_1 \text{ [dB]} = 20 \log \frac{9}{\omega^2} = 20 \log 9 - 40 \log \omega$$
$$= 19.1 \text{ dB} - 40 \log \omega.$$

At $\omega = 1$ rad/s, M_1 [dB] $= 19.1$ dB, which matches the figure.

As we progress along the ω-axis, the second segment has a slope of only -20 dB/decade, which means that a simple-zero factor with a corner frequency of 1 rad/s has come into play. Hence,

$$\mathbf{H}_2 = (1 + j\omega).$$

At $\omega = 10$ rad/s, the slope becomes zero, signifying the introduction of another simple-zero factor given by

$$\mathbf{H}_3 = (1 + j\omega/10).$$

Similarly,

$$\mathbf{H}_4 = (1 + j\omega/50),$$

and

$$\mathbf{H}_5 = (1 + j\omega/200).$$

Hence,

$$\mathbf{H}(\omega) = \mathbf{H}_1\mathbf{H}_2\mathbf{H}_3\mathbf{H}_4\mathbf{H}_5$$
$$= \frac{9(1 + j\omega)(1 + \frac{j\omega}{10})(1 + \frac{j\omega}{50})(1 + \frac{j\omega}{200})}{\omega^2}.$$

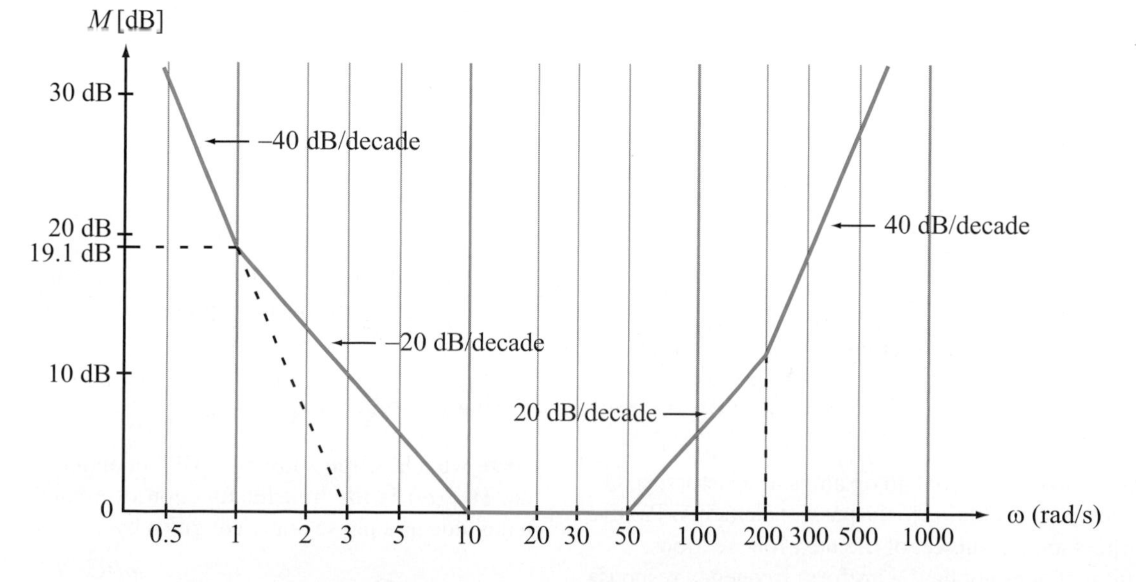

Figure 9-13: Bode plot of bandreject filter of Example 9-6.

Since we are not given any information about the phase pattern of $\mathbf{H}(\omega)$, the solution we obtained is correct within a multiplication factor of j^N, which can accommodate j (for $N = 1$), -1 (for $N = 2$), $-j$ (for $N = 3$), and 1 (for $N = 4$). The magnitude of $\mathbf{H}(\omega)$ is the same, regardless of the value of N.

Exercise 9-7: Determine the functional form of the transfer function whose Bode magnitude plot is shown in Fig. E9.7, given that its phase angle at dc is 90°.

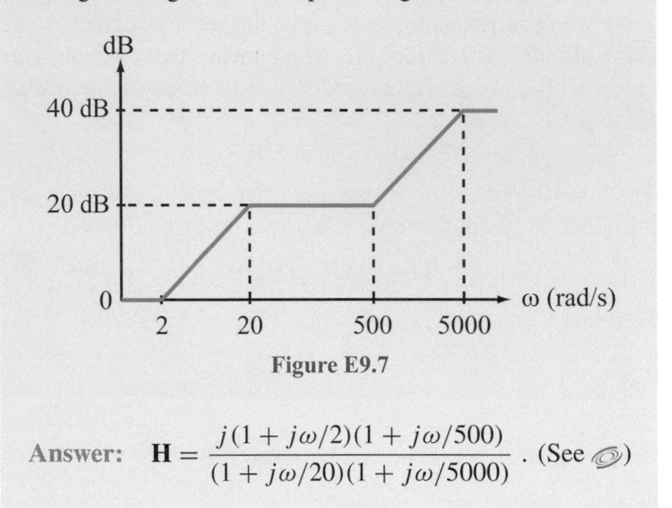

Figure E9.7

Answer: $\mathbf{H} = \dfrac{j(1 + j\omega/2)(1 + j\omega/500)}{(1 + j\omega/20)(1 + j\omega/5000)}$. (See)

Review Question 9-9: What does the term *standard form* of a transfer function refer to, and what purpose does it serve?

Review Question 9-10: For which of the seven standard factors are the Bode plots identical with the exact plots and for which are they different?

Review Question 9-11: What is the *gain roll-off rate*?

9-4 Passive Filters

Filters are of two types: passive and active.

Passive filters are resonant circuits that contain only passive elements, namely resistors, capacitors, and inductors.

In contrast, ***active filters*** contain op amps, transistors, and/or other active devices, in addition to the passive elements. Passive and active filters are the subject of the next four sections.

Any circuit that does not have a uniform frequency response is (by definition) a filter, simply because its output favors certain

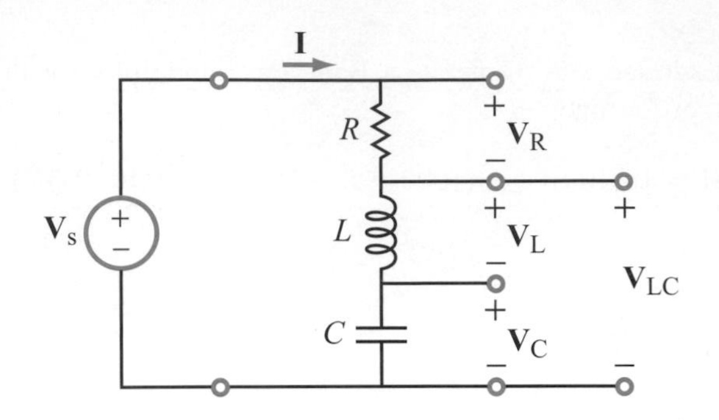

Figure 9-14: Series RLC circuit.

frequency ranges over others. Of particular interest to circuit designers are the four basic types of filters we introduced in Section 9-1. As we mentioned there, a filter transfer function is characterized by a number of attributes, including the following:

1. The frequency ranges of its passband(s) and stopband(s).
2. The gain factor M_0.
3. The gain roll-off rate S_g.

The objective of this section is to examine the basic properties of passive filters by analyzing their transfer functions. To that end, we will use the series RLC circuit shown in Fig. 9-14, in which we have designated four voltage outputs, namely $\mathbf{V}_R$, $\mathbf{V}_L$, and $\mathbf{V}_C$ across the individual elements, and $\mathbf{V}_{LC}$ across the combination of L and C. We will examine the frequency responses of the transfer functions corresponding to all four output voltages.

9-4.1 Bandpass Filter

The current $\mathbf{I}$ flowing through the loop in Fig. 9-15(a) is given by

$$\mathbf{I} = \frac{\mathbf{V}_s}{R + j\left(\omega L - \dfrac{1}{\omega C}\right)} = \frac{j\omega C\mathbf{V}_s}{(1 - \omega^2 LC) + j\omega RC}, \qquad (9.44)$$

where we multiplied the numerator and denominator by $j\omega C$ to simplify the form of the expression. The transfer function corresponding to $\mathbf{V}_R$ is

$$\mathbf{H}_{BP}(\omega) = \frac{\mathbf{V}_R}{\mathbf{V}_s} = \frac{R\mathbf{I}}{\mathbf{V}_s} = \frac{j\omega RC}{(1 - \omega^2 LC) + j\omega RC}, \qquad (9.45)$$

where we added the subscript "BP" in anticipation of the fact that $\mathbf{H}_{BP}(\omega)$ is the transfer function of a bandpass filter. Its magnitude and phase angle are given by

$$M_{BP}(\omega) = |\mathbf{H}_{BP}(\omega)| = \frac{\omega RC}{\sqrt{(1 - \omega^2 LC)^2 + \omega^2 R^2 C^2}}, \qquad (9.46)$$

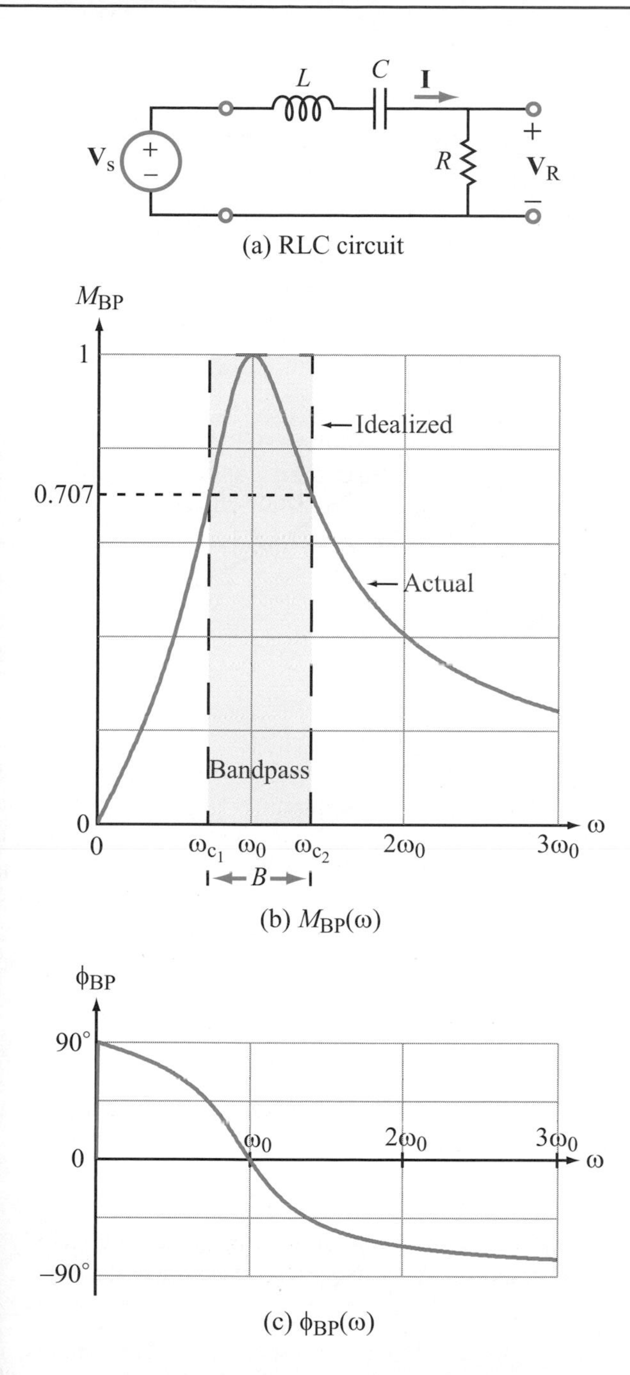

Figure 9-15: Series RLC bandpass filter.

nd

$$\phi_R(\omega) = 90° - \tan^{-1}\left[\frac{\omega RC}{1 - \omega^2 LC}\right]. \tag{9.47}$$

According to the spectral plot displayed in Fig. 9-15(b), M_{BP} goes to zero at both extremes of the frequency spectrum and exhibits a maximum across an intermediate range centered at ω_0. Hence, the circuit functions like a bandpass (BP) filter, allowing the transmission (through it) of signals whose angular frequencies are close to ω_0 and discriminating against those with frequencies that are far away from ω_0.

The general profile of $M_{BP}(\omega)$ can be discerned by examining the circuit of Fig. 9-15(a) at specific values of ω. At $\omega = 0$, the capacitor behaves like an open circuit, allowing no current to flow and no voltage to develop across R. At $\omega = \infty$, it is the inductor that acts like an open circuit, again allowing no current to flow. In the intermediate frequency range when the value of ω is such that $\omega L = 1/\omega C$, the impedances of L and C cancel each other out, reducing the total impedance of the RLC circuit to R and the current to $\mathbf{I} = \mathbf{V}_s/R$. Consequently, $\mathbf{V}_R = \mathbf{V}_s$, and $\mathbf{H}_{BP} = 1$. To note the significance of this specific condition, we call it the ***resonance condition***, and we refer to the frequency at which it occurs as the ***resonant frequency*** ω_0, which is given by

$$\omega_0 = \frac{1}{\sqrt{LC}}. \tag{9.48}$$

The phase plot in Fig. 9-15(c) conveys the fact that ϕ_{BP} is dominated by the phase of C at low frequencies and by the phase of L at high frequencies, and $\phi_{BP} = 0$ at $\omega = \omega_0$.

Filter Bandwidth

The ***bandwidth*** of the bandpass filter is defined as the frequency range extending between ω_{c_1} and ω_{c_2}, where ω_{c_1} and ω_{c_2} are the values of ω at which $M^2_{BP}(\omega) = 0.5$ or $M_{BP}(\omega) = 1/\sqrt{2} = 0.707$. As we will see shortly, M^2_{BP} is proportional to the power delivered to the resistor in the RLC circuit. At resonance, the power is at its maximum, and at ω_{c_1} and ω_{c_2}, the power delivered to R is equal to 1/2 of the maximum possible. That is why ω_{c_1} and ω_{c_2} also are referred to as the ***half-power frequencies*** (or the ***−3-dB frequencies*** on a dB scale). Thus,

$$M^2_{BP}(\omega) = \frac{1}{2} \qquad @ \ \omega_{c_1} \text{ and } \omega_{c_2}. \tag{9.49}$$

Upon inserting the expression for $M_{BP}(\omega)$ given by Eq. (9.46) and carrying out several steps of algebra, we obtain the solutions

$$\omega_{c_1} = -\frac{R}{2L} + \sqrt{\left(\frac{R}{2L}\right)^2 + \frac{1}{LC}}, \tag{9.50a}$$

and

$$\omega_{c_2} = \frac{R}{2L} + \sqrt{\left(\frac{R}{2L}\right)^2 + \frac{1}{LC}}. \tag{9.50b}$$

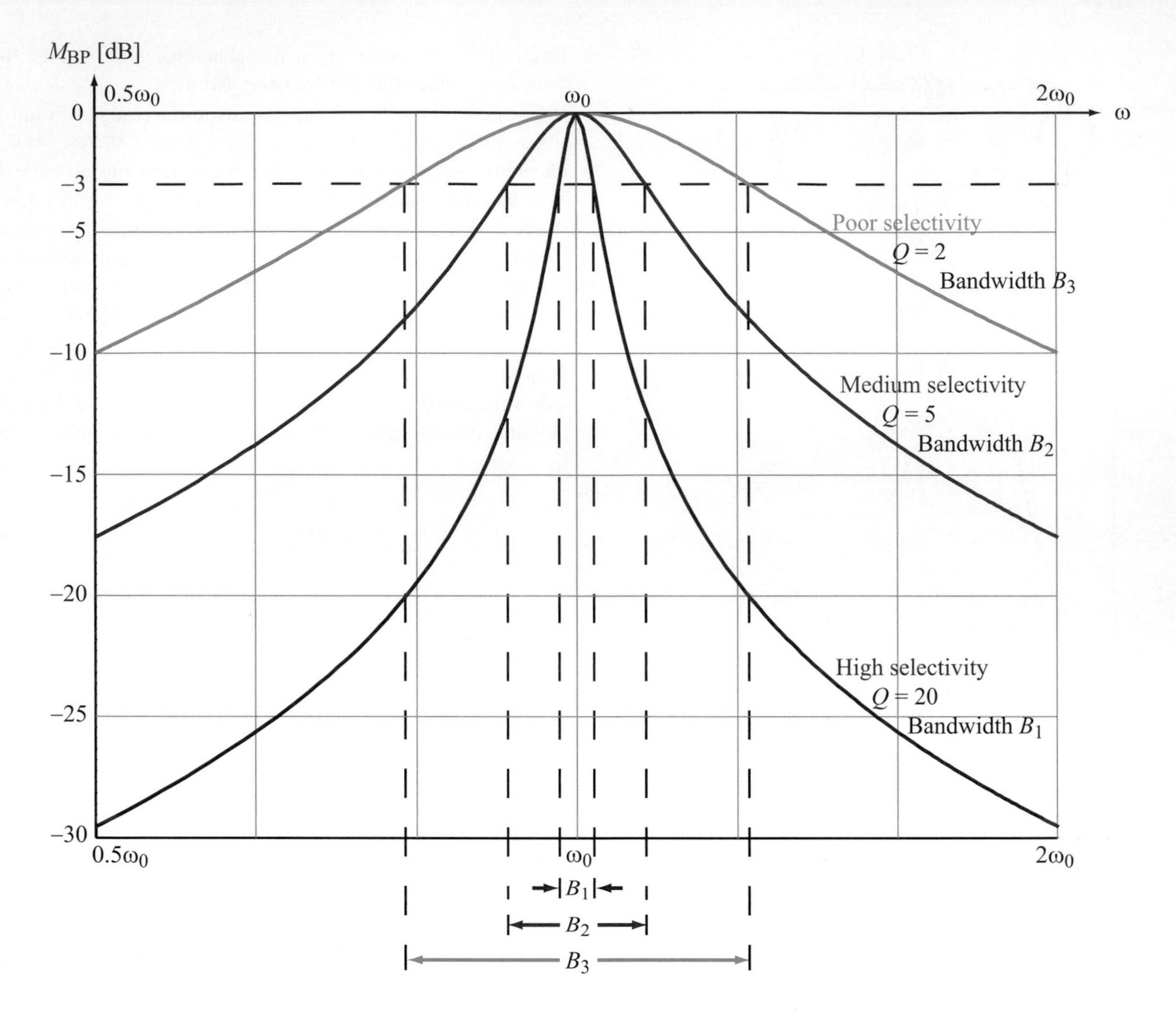

Figure 9-16: Examples of bandpass-filter responses.

The bandwidth then is given by

$$B = \omega_{c_2} - \omega_{c_1} = \frac{R}{L}. \tag{9.51}$$

It is worth noting that ω_0 is equal to the geometric mean of ω_{c_1} and ω_{c_2}:

$$\omega_0 = \sqrt{\omega_{c_1}\omega_{c_2}}. \tag{9.52}$$

Quality Factor

According to the foregoing discussion, the choice of values we make for R, L, and C will specify the overall shape of the transfer function completely, as well as its center frequency ω_0 and bandwidth B.

The ***quality factor*** of a circuit Q is an attribute commonly used to characterize the ***degree of selectivity*** of the circuit.

Figure 9-16 displays frequency responses for three circuits, al with the same ω_0. The high-Q circuit exhibits a sharp respons with a narrow bandwidth (relative to ω_0), the medium-Q circui has a broader pattern, and the low-Q circuit has a pattern wit limited selectivity.

For the bandpass-filter response, Q obviously is related t the ratio ω_0/B, but the formal definition of Q applies to an resonant circuit and is based on energy considerations, namel

$$Q = 2\pi \left(\frac{W_{\text{stor}}}{W_{\text{diss}}}\right)\bigg|_{\omega=\omega_0}, \tag{9.5}$$

where W_{stor} is the maximum energy that can be ***stored*** in the circuit at resonance ($\omega = \omega_0$), and W_{diss} is the ***energy dissipated*** by the circuit during a single period T. The factor 2π is an artificial multiplier introduced solely so that the expression for Q (that we will be deriving shortly) is simple in form and easy to remember.

In the RLC circuit of Fig. 9-15(a), the source is represented by the phasor voltage $\mathbf{V}_s$. In order to obtain expressions for the stored and dissipated energies at resonance, we need to (a) go back to the time domain and (b) specify ω as ω_0. For the sake of convenience and without loss of generality, we will assign the source the functional form

$$v_s(t) = V_0 \cos \omega_0 t \quad \longleftrightarrow \quad \mathbf{V}_s = V_0. \tag{9.54}$$

At resonance, $\omega_0 L = 1/\omega_0 C$, so the expressions for the total impedance of the circuit and the current flowing through it simplify to

$$\begin{aligned} \mathbf{Z} &= R + j\omega_0 L - j/\omega_0 C \\ &= R \qquad (@\ \omega_0 = 1/\sqrt{LC}) \end{aligned} \tag{9.55}$$

and

$$\mathbf{I} = \frac{\mathbf{V}_s}{\mathbf{Z}} = \frac{\mathbf{V}_s}{R} = \frac{V_0}{R}. \tag{9.56}$$

The time-domain current then is given by

$$i(t) = \mathfrak{Re}\left[\frac{V_0}{R}\, e^{j\omega_0 t}\right] = \frac{V_0}{R} \cos \omega_0 t. \tag{9.57}$$

At any instant in time, the instantaneous energies stored in the inductor and the capacitor are given by

$$w_L(t) = \frac{1}{2} L\, i_L^2(t) = \frac{V_0^2 L}{2R^2} \cos^2 \omega_0 t \qquad \text{(J)} \tag{9.58a}$$

and

$$\begin{aligned} w_C(t) &= \frac{1}{2} C\, v_C^2(t) = \frac{1}{2} C \left(\frac{1}{C}\int i\, dt\right)^2 \\ &= \frac{1}{2} C \left(\frac{V_0}{\omega_0 RC} \sin \omega_0 t\right)^2 \\ &= \frac{V_0^2 L}{2R^2} \sin^2 \omega_0 t \qquad \text{(J)}. \end{aligned} \tag{9.58b}$$

Even though both w_L and w_C vary with time, their sum is always a constant and equal to the maximum energy stored in the circuit,

$$\begin{aligned} W_{stor} = w_L(t) + w_C(t) &= \frac{V_0^2 L}{2R^2}[\cos^2 \omega_0 t + \sin^2 \omega_0 t] \\ &= \frac{V_0^2 L}{2R^2}. \end{aligned} \tag{9.59}$$

The energy dissipated by R during a single period is obtained by integrating the expression for the power p_R over a period $T = 1/f_0 = 2\pi/\omega_0$ so that

$$\begin{aligned} W_{diss} &= \int_0^T p_R\, dt = \int_0^T i^2 R\, dt \\ &= \int_0^{2\pi/\omega_0} \frac{V_0^2}{R} \cos^2 \omega_0 t\, dt = \frac{\pi V_0^2}{\omega_0 R}. \end{aligned} \tag{9.60}$$

Upon substituting Eqs. (9.59) and (9.60) into Eq. (9.53), we obtain the result

$$Q = \frac{\omega_0 L}{R}. \tag{9.61}$$

Using the relation given by Eq. (9.51), the expression for the quality factor becomes

$$Q = \frac{\omega_0}{B}, \tag{9.62}$$

which is dimensionless. Thus, for a bandpass filter, Q is the inverse of the bandwidth B normalized to the center frequency ω_0.

To highlight the role of Q, we can use the expressions for Q and ω_0 to rewrite Eqs. (9.46) and (9.47) for the magnitude and phase angle of $\mathbf{H}_{BP}(\omega)$ in the forms

$$M_{BP}(\omega) = \frac{(\omega/Q\omega_0)}{\{[1 - (\omega/\omega_0)^2]^2 + (\omega/Q\omega_0)^2\}^{1/2}}, \tag{9.63a}$$

and

$$\phi_{BP}(\omega) = 90^\circ - \tan^{-1}\left\{\frac{(\omega/\omega_0)}{Q[1 - (\omega/\omega_0)^2]}\right\}. \tag{9.63b}$$

Hence, the spectral response of the transfer function is specified completely by the combination of Q and ω_0.

Also, in view of Eq. (9.61), the expressions given by Eq. (9.50) for the half-power frequencies ω_{c_1} and ω_{c_2} can be rewritten as

$$\frac{\omega_{c_1}}{\omega_0} = -\frac{1}{2Q} + \sqrt{1 + \frac{1}{4Q^2}}, \tag{9.64a}$$

and

$$\frac{\omega_{c_2}}{\omega_0} = \frac{1}{2Q} + \sqrt{1 + \frac{1}{4Q^2}}. \tag{9.64b}$$

For a circuit with $Q > 10$, the expressions for ω_{c_1} and ω_{c_2} simplify to

$$\omega_{c_1} \simeq \omega_0 - \frac{B}{2}, \qquad \omega_{c_2} \simeq \omega_0 + \frac{B}{2}, \tag{9.65}$$

Table 9-3: Attributes of series and parallel RLC bandpass circuits.

	Series	Parallel
RLC Circuit	$\mathbf{V}_s$, L, C, R, $\mathbf{V}_R$	$\mathbf{I}_s$, L, C, R, $\mathbf{V}_R$
Transfer Function	$\mathbf{H} = \dfrac{\mathbf{V}_R}{\mathbf{V}_s}$	$\mathbf{H} = \dfrac{\mathbf{V}_R}{\mathbf{I}_s}$
Resonant Frequency, ω_0	$\dfrac{1}{\sqrt{LC}}$	$\dfrac{1}{\sqrt{LC}}$
Bandwidth, B	$\dfrac{R}{L}$	$\dfrac{1}{RC}$
Quality Factor, Q	$\dfrac{\omega_0}{B} = \dfrac{\omega_0 L}{R}$	$\dfrac{\omega_0}{B} = \dfrac{R}{\omega_0 L}$
Lower Half-Power Frequency, ω_{c_1}	$\left[-\dfrac{1}{2Q} + \sqrt{1 + \dfrac{1}{4Q^2}}\right]\omega_0$	$\left[-\dfrac{1}{2Q} + \sqrt{1 + \dfrac{1}{4Q^2}}\right]\omega_0$
Upper Half-Power Frequency, ω_{c_2}	$\left[\dfrac{1}{2Q} + \sqrt{1 + \dfrac{1}{4Q^2}}\right]\omega_0$	$\left[\dfrac{1}{2Q} + \sqrt{1 + \dfrac{1}{4Q^2}}\right]\omega_0$

Notes: (1) The expression for Q of the series RLC circuit is the inverse of that for Q of the parallel circuit. (2) For $Q \geq 10$, $\omega_{c_1} \simeq \omega_0 - \frac{B}{2}$, and $\omega_{c_2} \simeq \omega_0 + \frac{B}{2}$.

thereby forming a symmetrical bandpass centered at ω_0. Table 9-3 provides a summary of the salient features of the series RLC bandpass filter. For comparison, the table also includes the corresponding list for the parallel RLC circuit.

Example 9-7: Filter Design

(a) Design a series RLC bandpass filter with a center frequency $f_0 = 1$ MHz and a quality factor $Q = 20$, given that $L = 0.1$ mH.

(b) Determine the 10-dB bandwidth of the filter, which is defined as the bandwidth between frequencies at which the power level is 10 dB below the peak value.

Solution:

(a) Application of

$$\omega_0 = 2\pi f_0 = 2\pi \times 10^6 = \frac{1}{\sqrt{LC}} = \frac{1}{\sqrt{10^{-4}C}}$$

leads to $C = 0.25$ nF.

Solving Eq. (9.61) for R gives

$$R = \frac{\omega_0 L}{Q} = \frac{2\pi \times 10^6 \times 10^{-4}}{20} = 31.4\ \Omega.$$

(b) Voltage is proportional to M_{BP}, and power is proportional to M_{BP}^2. The definition for power in dB is

$$P\ [\text{dB}] = 10 \log P = 10 \log M_{\text{BP}}^2 = 20 \log M_{\text{BP}} = M_{\text{BP}}\ [\text{dB}].$$

We seek to find angular frequencies ω_a and ω_b corresponding to M_{BP} [dB] $= -10$ dB (Fig. 9-17). If

$$20 \log M_{\text{BP}} = -10 \text{ dB},$$

it follows that

$$M_{\text{BP}} = 10^{-0.5} = 0.316.$$

The expression for M_{BP} is given by Eq. (9.46) as

$$M_{\text{BP}} = \frac{\omega RC}{\sqrt{(1 - \omega^2 LC)^2 + \omega^2 R^2 C^2}}.$$

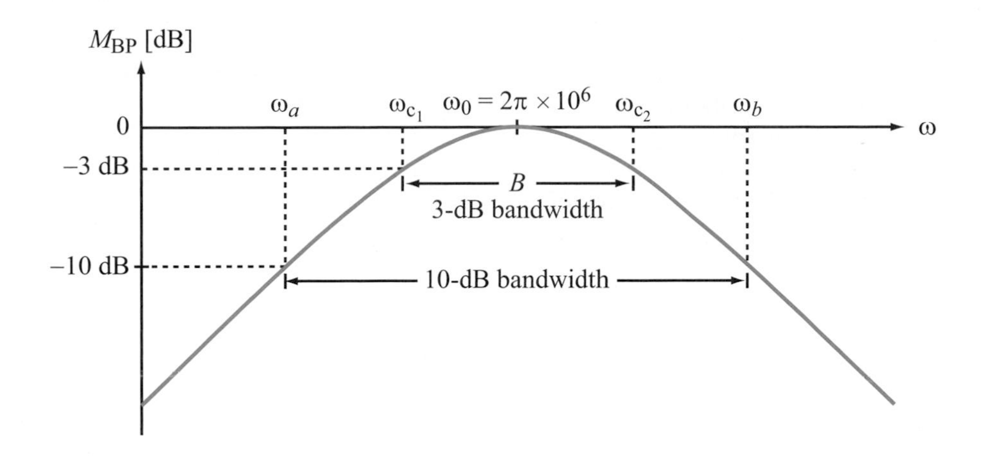

Figure 9-17: 10-dB bandwidth extends from ω_a to ω_b, corresponding to M_{BP} (dB) $= -10$ dB.

With $M_{BP} = 0.316$, $R = 31.4\ \Omega$, $L = 10^{-4}$ H, and $C = 0.25$ nF, solution of the expression yields

$$\frac{\omega_a}{\omega_0} = 0.93 \quad \text{and} \quad \frac{\omega_b}{\omega_0} = 1.08.$$

The corresponding bandwidth in Hz is

$$B_{10\text{ dB}} = (1.08 - 0.93) \times 1\text{ MHz} = 0.15\text{ MHz}.$$

Example 9-8: Two-Stage Bandpass Filter

Determine $\mathbf{H}(\omega) = \mathbf{V}_o/\mathbf{V}_s$ for the two-stage BP-filter circuit shown in Fig. 9-18. If $Q_1 = \omega_0 L/R$ is the quality factor of a single stage alone, what is Q_2 for the two stages in combination, given that $R = 2\ \Omega$, $L = 10$ mH, and $C = 1\ \mu$F?

Solution: For each stage alone,

$$\omega_0 = \frac{1}{\sqrt{LC}} = \frac{1}{\sqrt{10^{-2} \times 10^{-6}}} = 10^4 \text{ rad/s}$$

and

$$Q_1 = \frac{\omega_0 L}{R} = \frac{10^4 \times 10^{-2}}{2} = 50.$$

The loop equations for mesh currents $\mathbf{I}_1$ and $\mathbf{I}_2$ are

$$-\mathbf{V}_s + \mathbf{I}_1\left(j\omega L + \frac{1}{j\omega C} + R\right) - R\mathbf{I}_2 = 0$$

and

$$-R\mathbf{I}_1 + \mathbf{I}_2\left(2R + j\omega L + \frac{1}{j\omega C}\right) = 0.$$

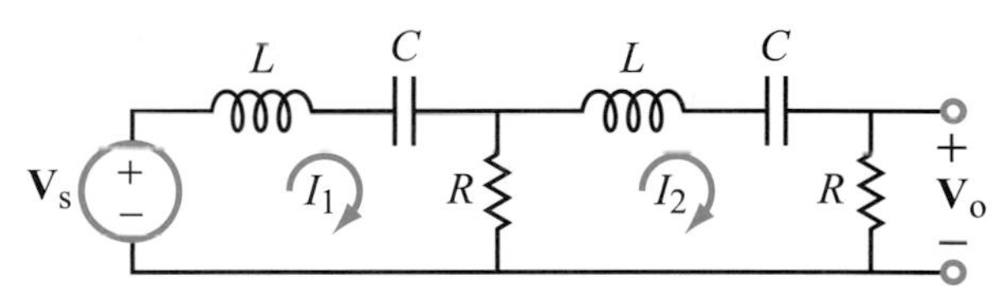

(a) Two-stage circuit

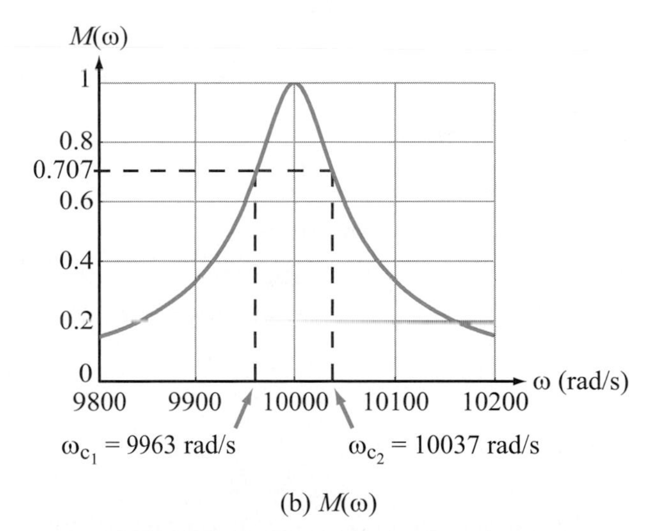

(b) $M(\omega)$

Figure 9-18: Two-stage RLC circuit of Example 9-8.

Simultaneous solution of the two equations leads to

$$\mathbf{H}(\omega) = \frac{\mathbf{V}_o}{\mathbf{V}_s}$$

$$= \frac{\omega^2 R^2 C^2}{\omega^2 R^2 C^2 - (1 - \omega^2 LC)^2 - j3\omega RC(1 - \omega^2 LC)}$$

$$= \frac{\omega^2 R^2 C^2[\omega^2 R^2 C^2 - (1 - \omega^2 LC)^2 + j3\omega RC(1 - \omega^2 LC)]}{[\omega^2 R^2 C^2 - (1 - \omega^2 LC)^2]^2 + 9\omega^2 R^2 C^2 (1 - \omega^2 LC)^2}.$$

Resonance occurs when the imaginary part of $\mathbf{H}(\omega)$ is zero, which is satisfied either when $\omega = 0$ (which is a trivial resonance) or when $\omega = 1/\sqrt{LC}$. Hence, the two-stage circuit has the same resonance frequency as a single-stage circuit.

Using the specified values of R, L, and C, we can calculate the magnitude $M(\omega) = |\mathbf{H}(\omega)|$ and plot it as a function of ω. The result is displayed in Fig. 9-18(b). From the spectral plot, we have

$$\omega_{c_1} = 9963 \text{ rad/s},$$

$$\omega_{c_2} = 10037 \text{ rad/s},$$

$$B_2 = \omega_{c_2} - \omega_{c_1} = 10037 - 9963 = 74 \text{ rad/s},$$

and

$$Q_2 = \frac{\omega_0}{B_2} = \frac{10^4}{74} = 135,$$

where B_2 is the bandwidth of the two-stage BP-filter response. The two-stage combination increases the quality factor from 50 to 135.

Exercise 9-8: Show that for the parallel RLC circuit shown in Fig. E9.8, the transfer-impedance transfer function $\mathbf{H_Z} = \mathbf{V_R}/\mathbf{I_s}$ exhibits a bandpass-filter response.

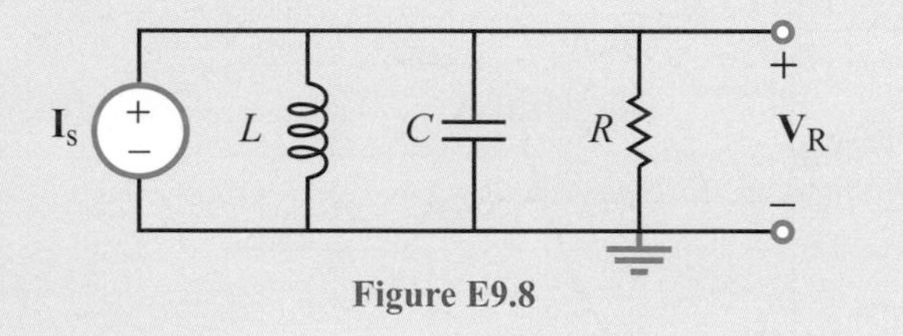

Figure E9.8

Answer:

$$\mathbf{H_Z} = \frac{\mathbf{V_R}}{\mathbf{I_s}} = \frac{j\omega L}{(1 - \omega^2 LC) + j\omega L/R}.$$

The functional form of $\mathbf{H_Z}(\omega)$ is identical with that given by Eq. (9.45) for the series RLC bandpass filter. Moreover, both circuits resonate at $\omega = 1/\sqrt{LC}$. (See)

9-4.2 Highpass Filter

Transfer function $\mathbf{H}_{\text{HP}}(\omega)$, corresponding to $\mathbf{V}_\text{L}$ in the circuit of Fig. 9-19(a), is given by

$$\mathbf{H}_{\text{HP}}(\omega) = \frac{\mathbf{V}_\text{L}}{\mathbf{V}_\text{s}} = \frac{j\omega L\mathbf{I}}{\mathbf{V}_\text{s}} = \frac{-\omega^2 LC}{(1 - \omega^2 LC) + j\omega RC} \tag{9.66}$$

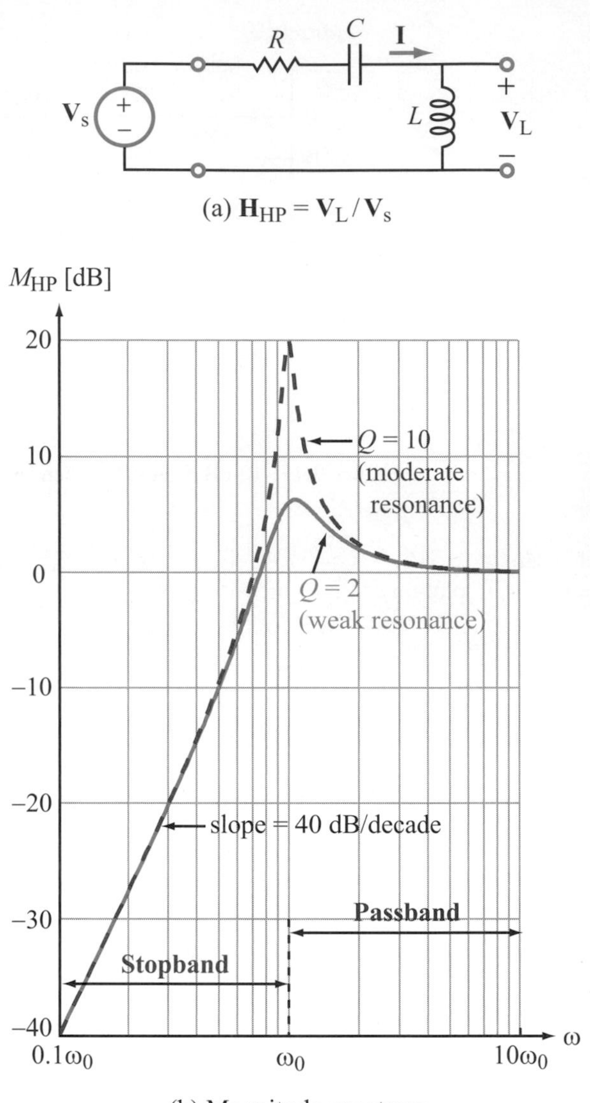

Figure 9-19: Plots of M_{HP} [dB] for $Q = 2$ (weak resonance) and $Q = 10$ (moderate resonance).

with magnitude and phase angle

$$M_{\text{HP}}(\omega) = \frac{\omega^2 LC}{[(1 - \omega^2 LC)^2 + \omega^2 R^2 C^2]^{1/2}} = \frac{(\omega/\omega_0)^2}{\{[1 - (\omega/\omega_0)^2]^2 + (\omega/Q\omega_0)^2\}^{1/2}} \tag{9.67a}$$

and

$$\phi_{\text{HP}}(\omega) = 180^\circ - \tan^{-1}\left[\frac{\omega RC}{1 - \omega^2 LC}\right] = 180^\circ - \tan^{-1}\left\{\frac{(\omega/\omega_0)}{Q[1 - (\omega/\omega_0)^2]}\right\}, \tag{9.67b}$$

where ω_0 and Q are defined by Eqs. (9.48) and (9.61), respectively. Figure 9-19(b) displays logarithmic plots of M_{HP} [dB] for two values of Q. Because $M_{\text{HP}}(\omega)$ has a quadratic zero, its slope in the stopband is 40 dB/decade.

Exercise 9-9: How should R be related to L and C so that the denominator of Eq. (9.66) becomes a simple pole of order 2? What will the value of Q be in that case?

Answer: $R = 2\sqrt{L/C}$, $Q = 1/2$. (See)

9-4.3 Lowpass Filter

The voltage across the capacitor in Fig. 9-20(a) generates a lowpass-filter transfer function given by

$$\mathbf{H}_{\text{LP}}(\omega) = \frac{\mathbf{V}_{\text{C}}}{\mathbf{V}_{\text{s}}} = \frac{(1/j\omega C)\mathbf{I}}{\mathbf{V}_{\text{s}}} = \frac{1}{(1-\omega^2 LC) + j\omega RC}, \tag{9.68}$$

with magnitude and phase angle given by

$$M_{\text{LP}}(\omega) = \frac{1}{[(1-\omega^2 LC)^2 + \omega^2 R^2 C^2]^{1/2}}$$
$$= \frac{1}{\{[1-(\omega/\omega_0)^2]^2 + (\omega/Q\omega_0)^2\}^{1/2}}, \tag{9.69a}$$

and

$$\phi_{\text{LP}}(\omega) = -\tan^{-1}\left(\frac{\omega RC}{1-\omega^2 LC}\right)$$
$$= -\tan^{-1}\left\{\frac{(\omega/\omega_0)}{Q[1-(\omega/\omega_0)^2]}\right\}. \tag{9.69b}$$

The spectral plots of M_{LP} [dB] shown in Fig. 9-20(b) are mirror images of the highpass-filter plots displayed in Fig. 9-19(b).

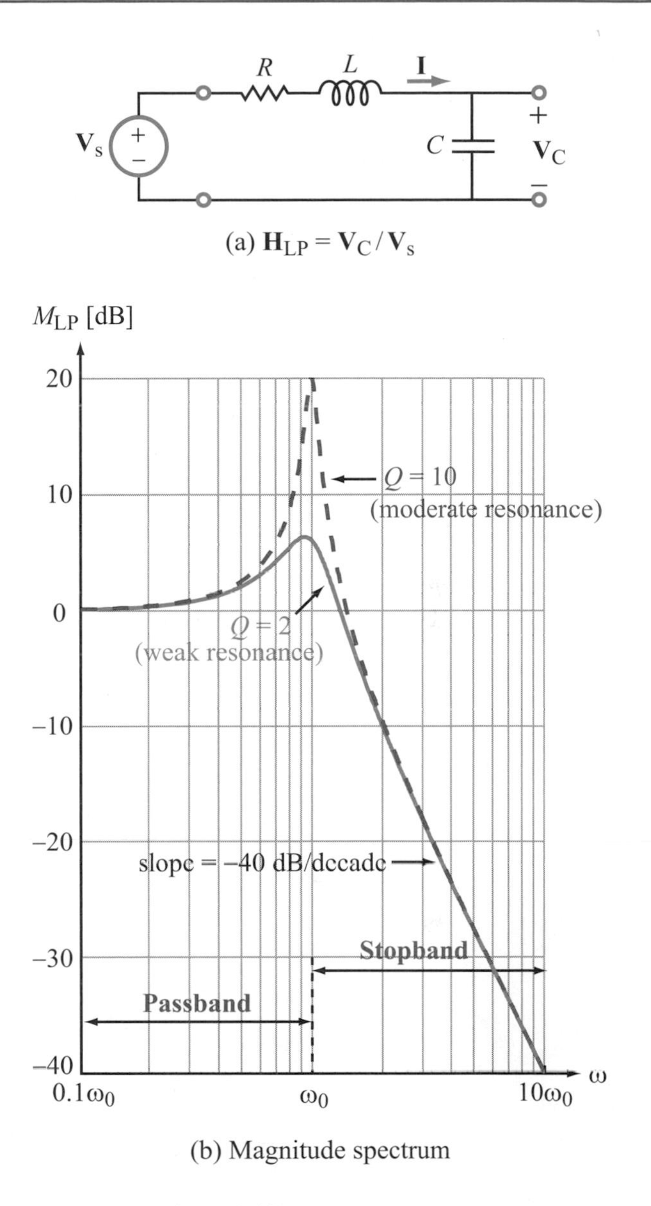

Figure 9-20: RLC lowpass filter.

9-4.4 Bandreject Filter

The output voltage across the combination of L and C in Fig. 9-21(a) generates a bandreject-filter transfer function and s equal to $\mathbf{V}_{\text{s}} - \mathbf{V}_{\text{R}}$:

$$\mathbf{H}_{\text{BR}}(\omega) = \frac{\mathbf{V}_{\text{L}} + \mathbf{V}_{\text{C}}}{\mathbf{V}_{\text{s}}} = \frac{\mathbf{V}_{\text{s}} - \mathbf{V}_{\text{R}}}{\mathbf{V}_{\text{s}}} = 1 - \mathbf{H}_{\text{BP}}(\omega), \tag{9.70}$$

where $\mathbf{H}_{\text{BP}}(\omega)$ is the bandpass-filter transfer function given by Eq. (9.45). The spectral response of $\mathbf{H}_{\text{BP}}$ passes all frequencies except for an intermediate band centered at ω_0, as shown in Fig. 9-21(b). The width of the stopband is determined by the values of ω_0 and Q.

Exercise 9-10: Is $M_{\text{BR}} = 1 - M_{\text{BP}}$?

Answer: No, because $M_{\text{BR}} = |\mathbf{H}_{\text{BR}}| = |1 - \mathbf{H}_{\text{BP}}| \neq 1 - |\mathbf{H}_{\text{BP}}| = 1 - M_{\text{BP}}$. (See)

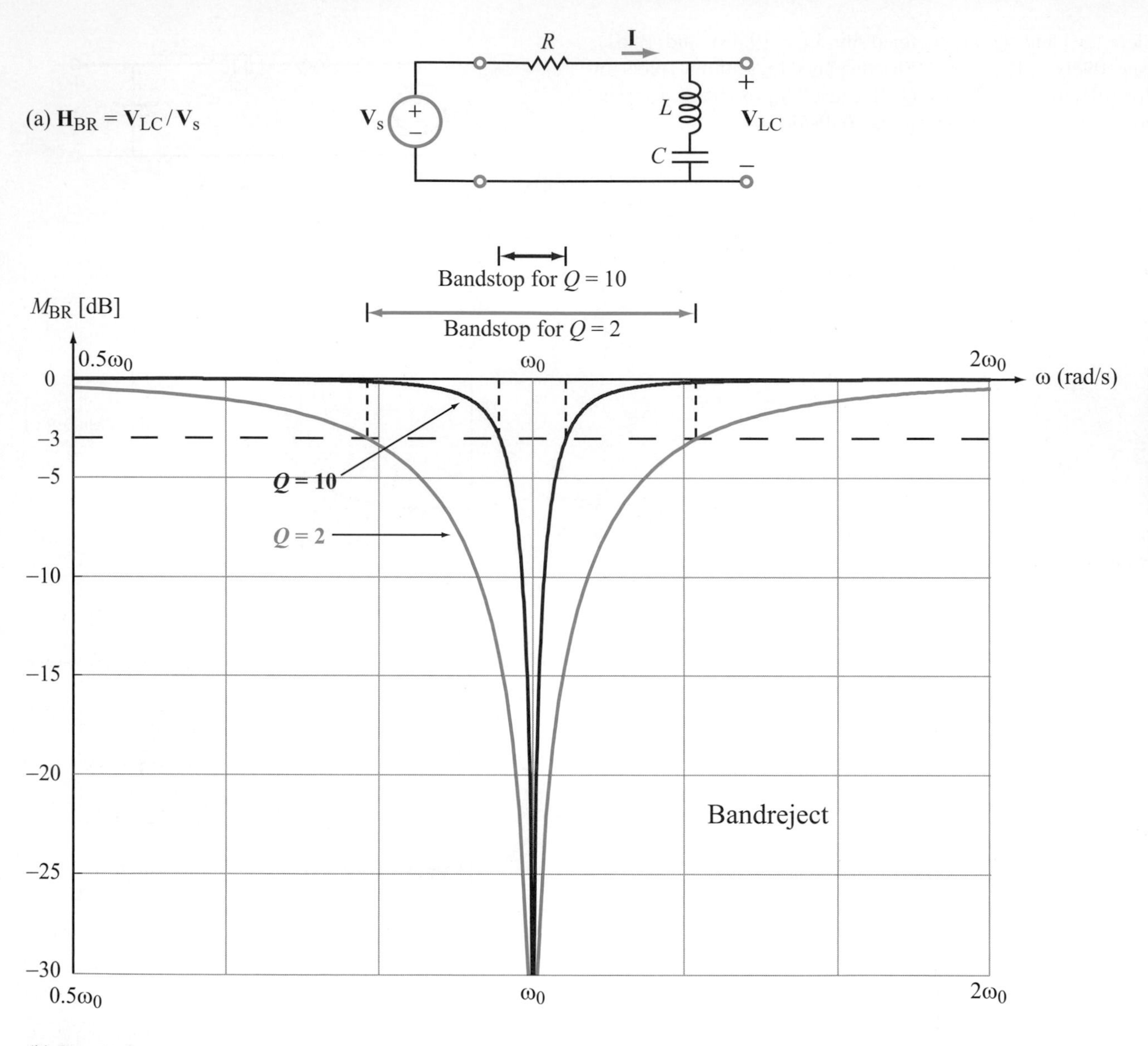

Figure 9-21: Bandreject filter.

9-5 Filter Order

In Section 9-3, we associated the term *order* with the power of ω, so a factor given by $(1 + j\omega/\omega_c)^2$ was called a second-order zero because the highest power of ω in the expression is 2. Similarly, $(1 + j\omega/\omega_c)^{-2}$ was called a second-order pole because the highest power of ω is also 2, but the expression appears in the denominator. The term order also is used to describe the overall filter response, which may be composed of the product of several zero and pole factors—each with its own order. Multiple, different characterizations have been used over the years to define the order of a filter, so to avoid ambiguity, we adopt the following definition.

> The order of a filter is equal to the absolute value of the highest power of ω in its transfer function when ω is in the filter's stopband(s).

Let us examine this definition for three circuit configurations

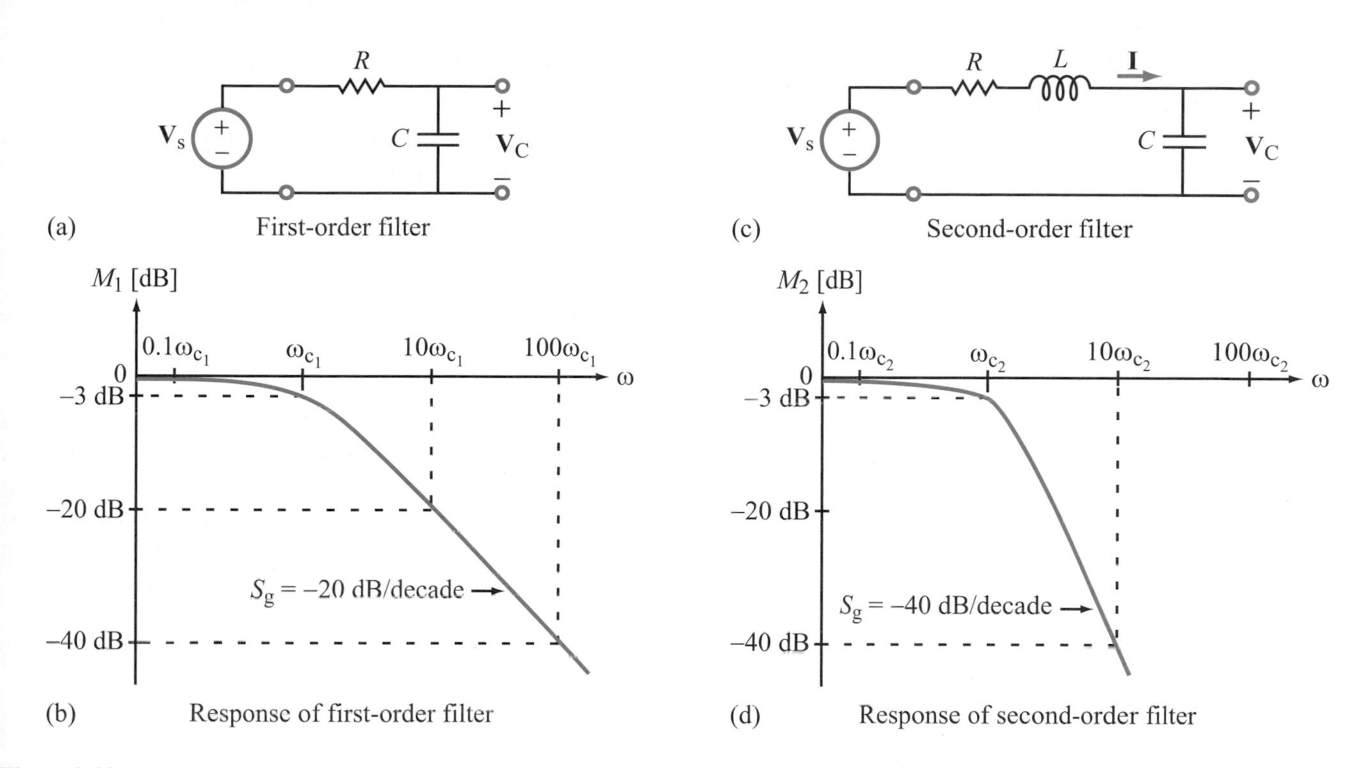

Figure 9-22: Comparison of magnitude responses of the first-order RC filter and the second-order RLC filter. The corner frequencies are given by $\omega_{c_1} = 1/RC$ and $\omega_{c_2} = 1.28/RC$.

9-5.1 First-Order Lowpass RC Filter

The transfer function of the RC circuit shown in Fig. 9-22(a) is given by

$$\begin{aligned} \mathbf{H}_1(\omega) = \frac{\mathbf{V}_C}{\mathbf{V}_s} &= \frac{1/j\omega C}{R + 1/j\omega C} \\ &= \frac{1}{1 + j\omega RC} \\ &= \frac{1}{1 + j\omega/\omega_{c_1}} \qquad \text{(first-order)}, \end{aligned} \tag{9.71}$$

where we multiplied both the numerator and denominator by $j\omega C$ so as to rearrange the expression into the ***standard form*** we discussed in Section 9-3.2. The expression given by Eq. (9.71) is a simple pole with a ***corner frequency*** given by

$$\omega_{c_1} = \frac{1}{RC} \qquad \text{(RC filter)}. \tag{9.72}$$

It is evident from the expression given by Eq. (9.71) that the highest order of ω is 1, and therefore the RC circuit is a first-order filter. Strict application of the definition for the order of a filter requires that we evaluate the power of ω when ω is in the stopband of the filter. In the present case, the stopband covers the range $\omega \geq \omega_{c_1}$. When ω is well into the stopband ($\omega/\omega_{c_1} \gg 1$), Eq. (9.71) simplifies to

$$\mathbf{H}_1(\omega) \simeq \frac{-j\omega_{c_1}}{\omega} \qquad (\text{for } \omega/\omega_{c_1} \gg 1), \tag{9.73}$$

which confirms the earlier conclusion that the RC circuit is first-order.

A circuit containing a single reactive element (capacitor or inductor) generates a first-order transfer function. Generally speaking, the order of a filter depends on the number of reactive elements contained in the circuit. The order may be smaller or equal to the number of reactive elements, but not greater.

The magnitude of $\mathbf{H}_1(\omega)$ is given by

$$M_1 = |\mathbf{H}_1(\omega)| = \frac{1}{|1 + j\omega/\omega_{c_1}|} = \frac{1}{\sqrt{1 + (\omega/\omega_{c_1})^2}}. \tag{9.74}$$

At $\omega = \omega_{c_1}$,

$$M_1(\omega_{c_1}) = \frac{1}{\sqrt{1+1}} = 0.707. \tag{9.75}$$

When expressed in dB, M_1 becomes

$$M_1 \text{ [dB]} = 20 \log M_1$$
$$= -10 \log[1 + (\omega/\omega_{c_1})^2]$$
$$= \begin{cases} 0 \text{ dB} & @ \ \omega = 0, \\ -3 \text{ dB} & @ \ \omega/\omega_{c_1} = 1, \\ -20 \log(\omega/\omega_{c_1}) & @ \ \omega/\omega_{c_1} \gg 1. \end{cases} \quad (9.76)$$

On the semilog scale of Fig. 9-22(b), M_1 [dB] starts out at 0 dB—corresponding to $M_1 = 1$ in natural units—decreases to -3 dB at $\omega = \omega_{c_1}$, and then its slope accelerates towards a steady-state value of -20 dB/decade at much greater values of ω. As noted earlier, the steepness (slope) of the transfer function after it has transitioned from its passband to its stopband is called its gain roll-off rate S_g. *For a first-order filter,* $S_g = -20$ *dB/decade.* To achieve a faster rate of decay, second- or higher-order filters are called for.

Example 9-9: Filter Transmission Spectrum

An RC lowpass filter uses a capacitor $C = 10\ \mu$F. (a) Specify R so that $\omega_{c_1} = 1$ krad/s. (b) The filter is considered acceptably transparent to a signal if the signal's voltage amplitude is reduced by no more than 12 dB as it passes through the filter. What is the filter's transmission spectrum according to this criterion?

Solution:

(a) Application of Eq. (9.72) leads to

$$R = \frac{1}{\omega_{c_1} C} = \frac{1}{10^3 \times 10^{-5}} = 100\ \Omega.$$

(b) If

$$M_1 \text{ [dB]} = 20 \log M_1 = 12 \text{ dB},$$

then

$$\log M_1 = -\frac{12}{20} = -0.6$$

and

$$M_1 = 10^{-0.6} = 0.25.$$

Equating this value of M_1 to the expression given by Eq. (9.74) leads to

$$M_1 = \frac{1}{\sqrt{1 + (\omega/\omega_{c_1})^2}} = 0.25,$$

which yields the solution

$$\frac{\omega}{\omega_{c_1}} = 3.85 \quad \text{or} \quad \omega = 3.85 \text{ krad/s}.$$

Hence, the transmission spectrum of the filter extends from 0 to 3.87 krad/s or equivalently from 0 to 616 Hz.

9-5.2 Second-Order Lowpass Filter

For the RLC circuit shown in Fig. 9-22(c), we determined in Section 9-4.3 that its transfer function is given by Eq. (9.68) as

$$\mathbf{H}_2(\omega) = \frac{\mathbf{V}_C}{\mathbf{V}_s} = \frac{1}{(1 - \omega^2 LC) + j\omega RC} \quad \text{(RLC filter)}. \quad (9.77)$$

The magnitude spectrum of the RLC lowpass filter was presented earlier in Fig. 9-20(b), where it was observed that the response may exhibit a resonance phenomenon in the neighborhood of $\omega_0 = 1/\sqrt{LC}$, and that it decays with $S_g = -40$ dB/decade in the stopband ($\omega \geq \omega_0$). This is consistent with the fact that the RLC circuit generates a second-order lowpass filter when the output voltage is taken across the capacitor. In terms of our definition for the order of a filter in the stopband ($\omega^2 \gg 1/LC$), Eq. (9.77) reduces to

$$\mathbf{H}_2(\omega) \simeq \frac{1}{\omega^2 LC} \quad (\text{for } \omega \gg \omega_0), \quad (9.78)$$

which assumes the form of a second-order pole.

The ripple-like effect exhibited by the RLC filter can be avoided through a judicious choice of the values of R, L, and C. By replacing the minus sign in Eq. (9.77) with j^2 and selecting R such that

$$R = 2\sqrt{\frac{L}{C}}, \quad (9.79)$$

the expression given by Eq. (9.77) can be converted into a perfect square:

$$\mathbf{H}_2(\omega) = \frac{1}{1 + j^2\omega^2 LC + j2\sqrt{LC}}$$
$$= \frac{1}{(1 + j\omega\sqrt{LC})^2}. \quad (9.80)$$

The constraint given by Eq. (9.79) allowed us to convert $\mathbf{H}_2(\omega)$ from a quadratic pole into a simple pole of second-order. Its magnitude response is displayed in Fig. 9-22(d).

The corner frequency of the RLC lowpass filter (ω_{c_2}) is determined by setting the magnitude of $\mathbf{H}_2(\omega)$ equal to $1/\sqrt{2}$. Thus,

$$|\mathbf{H}_2(\omega_{c_2})| = \frac{1}{1 + \omega_{c_2}^2 LC} = \frac{1}{\sqrt{2}},$$

which leads to

$$\omega_{c_2} = \left\{ \frac{\sqrt{2} - 1}{LC} \right\}^{1/2} = \frac{0.64}{\sqrt{LC}}. \quad (9.81)$$

From Eq. (9.79), $L = R^2C/4$. When used in Eq. (9.81), the expression for ω_{c_2} becomes

$$\omega_{c_2} = \frac{1.28}{RC} \quad \text{(RLC filter)}. \tag{9.82}$$

The foregoing analysis warrants the following observations:

1. The RC lowpass filter is first-order, its corner frequency is $\omega_{c_1} = 1/RC$, and its gain roll-off rate is $S_g = -20$ dB/decade.

2. By adding a series inductor whose value is specified by $L = R^2C/4$, the filter becomes second-order, its corner frequency shifts upward to $1.28/RC$, and its slope becomes twice as steep.

9-5.3 RLC Bandpass Filter

We just concluded that the RLC lowpass filter is second-order. The RLC circuit also behaves like a bandpass filter when the output voltage is taken across R instead of C. Does it then follow that the RLC bandpass filter is second-order?

To answer the question, we start by examining the expression for the transfer function when ω is in the stopbands of the filter. The expression for $\mathbf{H}_{\text{BP}}(\omega)$ is given by Eq. (9.45) as

$$\mathbf{H}_{\text{BP}}(\omega) = \frac{j\omega RC}{1 - \omega^2 LC + j\omega RC}, \tag{9.83}$$

and the spectral plot of its magnitude is shown in Fig. 9-15(b). For $\omega \lll \omega_0$ and $\omega \gg \omega_0$ (where $\omega_0 = 1/\sqrt{LC}$), the expression simplifies to

$$\mathbf{H}_{\text{BP}}(\omega) \simeq \begin{cases} j\omega RC & (\text{for } \omega \ll \omega_0 \text{ and } \omega \ll RC), \\ \dfrac{-jR}{\omega L} & (\text{for } \omega \gg \omega_0). \end{cases} \tag{9.84}$$

At the low-frequency end, $\mathbf{H}_{\text{BP}}(\omega)$ reduces to a first-order zero @ origin, and at the high-frequency end, it reduces to a first-order pole @ origin. Hence, the RLC bandpass filter is first-order, not second-order.

Review Question 9-12: How does S_g of a third-order filter compare with that of a first-order filter?

Review Question 9-13: When is a series RLC circuit a first-order circuit, and when is it a second-order circuit?

Exercise 9-11: What is the order of the two-stage bandpass-filter circuit shown in Fig. 9-18(a)?

Answer: Second-order. (See ◎)

Exercise 9-12: Determine the order of $\mathbf{H}(\omega) = \mathbf{V}_{\text{out}}/\mathbf{V}_{\text{s}}$ for the circuit in Fig. E9.12.

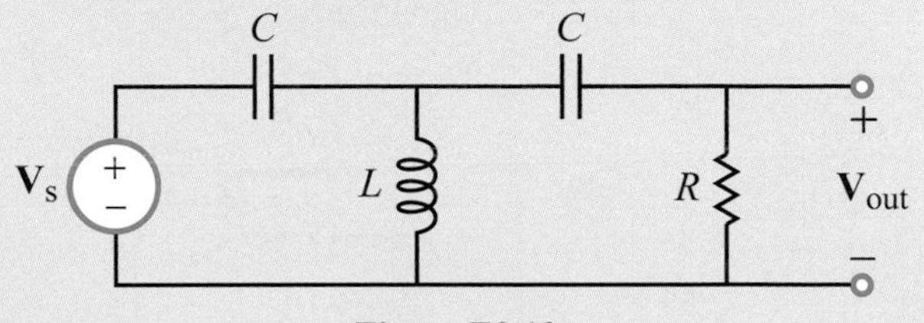

Figure E9.12

Answer:

$$\mathbf{H}(\omega) = \frac{j\omega^3 RLC^2}{\omega^2 LC - (1 - \omega^2 LC)(1 + j\omega RC)},$$

which describes a highpass filter. In the stopband (very small values of ω), $\mathbf{H}(\omega)$ varies as ω^3. Hence, it is third-order. (See ◎)

9-6 Active Filters

The four basic types of filters we examined in earlier sections (lowpass, highpass, bandpass, and bandreject) are all relatively easy to design, but they do have a number of drawbacks. Passive elements cannot generate energy, so the power gain of a passive filter cannot exceed 1. Active filters (by comparison) can be designed to provide significant gain in addition to realizing the specified filter performance. A second drawback of passive filters has to do with inductors. Whereas capacitors and resistors can be fabricated easily in planar form on machine-assembled printed circuit boards, inductors generally are more expensive to fabricate and more difficult to integrate into the rest of the circuit, because they are bulky and three-dimensional in shape. In contrast, op-amp circuits can be designed to function as filters without the use of inductors. The intended operating-frequency range is an important determining factor with regard to what type of filter is best to design and use. Op amps generally do not perform reliably at frequencies above 1 MHz, so their use as filters is limited to lower frequencies. Fortunately, inductor size becomes less of a problem above 1 MHz (because $\mathbf{Z}_{\text{L}} = j\omega L$ necessitating a smaller value for L, and consequently a physically smaller inductor), so passive filters are the predominant type used at the higher frequencies.

One of the major assets of op-amp circuits is that they easily can be cascaded together (both in series and in parallel) to

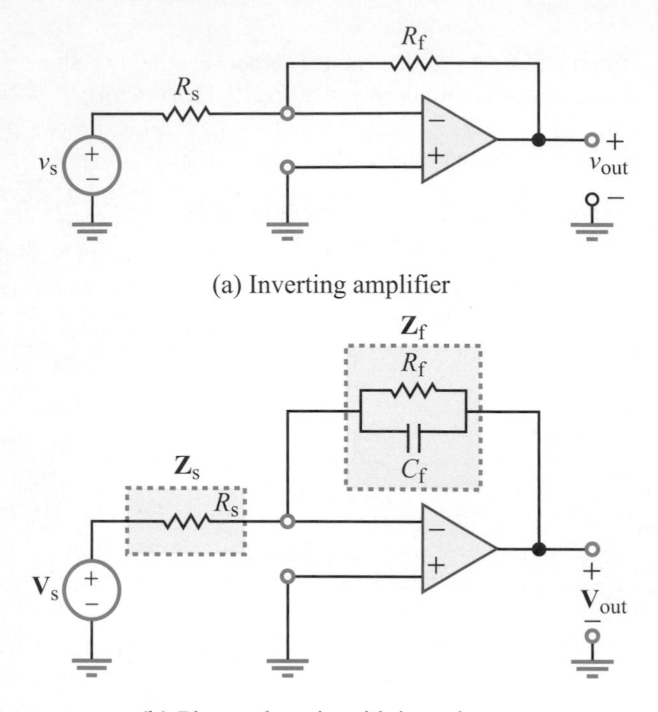

Figure 9-23: Inverting amplifier functioning like a lowpass filter.

realize the intended function. Moreover, by inserting buffer circuits (see Section 4-7) between successive stages, impedance mismatch and loading problems can be minimized or avoided altogether.

9-6.1 Single-Pole Lowpass Filter

Consider the circuit shown in Fig. 9-23(a), which essentially is a replica of the inverting amplifier circuit of Fig. 4-9(a) and for which the input and output voltages are related by

$$v_{out} = -\frac{R_f}{R_s}\, v_s. \tag{9.85}$$

Let us now transform the circuit into the phasor domain and generalize it by replacing resistors R_s and R_f with impedances $\mathbf{Z}_s$ and $\mathbf{Z}_f$, respectively, as shown in Fig. 9-23(b). Further, let us retain $\mathbf{Z}_s$ as R_s, but specify $\mathbf{Z}_f$ as the parallel combination of a resistor R_f and a capacitor C_f. By analogy with Eq. (9.85), the equivalent relationship for the circuit in Fig. 9-23(b) is

$$\mathbf{V}_{out} = -\frac{\mathbf{Z}_f}{\mathbf{Z}_s}\, \mathbf{V}_s, \tag{9.86}$$

with

$$\mathbf{Z}_s = R_s \tag{9.87a}$$

and

$$\mathbf{Z}_f = R_f \parallel \left(\frac{1}{j\omega C_f}\right) = \frac{R_f}{1 + j\omega R_f C_f}. \tag{9.87b}$$

The transfer function of the circuit, which we soon will recognize as that of a lowpass filter, is given by

$$\begin{aligned}\mathbf{H}_{LP}(\omega) = \frac{\mathbf{V}_{out}}{\mathbf{V}_s} = -\frac{\mathbf{Z}_f}{\mathbf{Z}_s} &= -\frac{R_f}{R_s}\left(\frac{1}{1 + j\omega R_f C_f}\right)\\ &= G_{LP}\left(\frac{1}{1 + j\omega/\omega_{LP}}\right),\end{aligned} \tag{9.88}$$

where

$$G_{LP} = -\frac{R_f}{R_s}, \qquad \omega_{LP} = \frac{1}{R_f C_f}. \tag{9.89}$$

The expression for G_{LP} is the same as that of the original inverting amplifier, and ω_{LP} is the cutoff frequency of the lowpass filter. Except for the gain factor, the expression given by Eq. (9.88) is identical in form with Eq. (9.71), which is the transfer function of the RC lowpass filter. A decided advantage of the active lowpass filter over its passive counterpart is that ω_{LP} is independent of both the input resistance R_s and any non-zero load resistance R_L that may be connected across the op amp's output terminals.

9-6.2 Single-Pole Highpass Filter

If in the inverting amplifier circuit we were to specify the input and feedback impedances as

$$\mathbf{Z}_s = R_s - \frac{j}{\omega C_s} \qquad \text{and} \qquad \mathbf{Z}_f = R_f, \tag{9.90}$$

as shown in Fig. 9-24, we would obtain the highpass-filter transfer function given by

$$\begin{aligned}\mathbf{H}_{HP}(\omega) = \frac{\mathbf{V}_{out}}{\mathbf{V}_s} = -\frac{\mathbf{Z}_f}{\mathbf{Z}_s} &= -\frac{R_f}{R_s - j/\omega C_s}\\ &= G_{HP}\left[\frac{j\omega/\omega_{HP}}{1 + j\omega/\omega_{HP}}\right],\end{aligned} \tag{9.91}$$

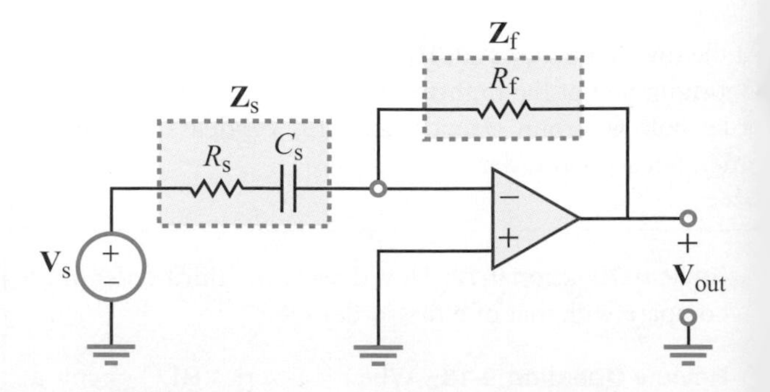

Figure 9-24: Single-pole active highpass filter.

Technology Brief 18: Electrical Engineering and the Audiophile

The reproduction of high-quality music with sufficient fidelity to sound like a live performance in one's living room was one of the technological hallmarks of the 20th century. In these days of iPods and online music distribution, good music is increasingly accessible to many people. The price of good quality tuners, amplifiers, and speakers continues to drop, and driven mostly by demand for home entertainment audio/video systems, audio equipment is increasingly "user-friendly." The reproduction of theater-quality or live-performance sound in a confined space is challenging enough to be a profession unto itself. It also can be a very rewarding technical hobby for the well-versed electrical engineer. In this Technology Brief, we will cover some of the basics of audio equipment and relate them directly to the concepts taught in this book. Several good audiophile websites exist with more in-depth treatments of these (and other) topics; beyond the audiophile community, the sub-field of audio engineering has an extensive academic and professional literature to consult.

The Basics

Reproduced sound starts out as an analog (e.g., the vinyl record) or digital (e.g., the mp3 file) recording. How that recording is made from real sound with high fidelity is beyond the scope of this Brief (and is a large component of the audio engineering profession). That recording is converted into an electrical signal that is first amplified and then transmitted via cables to speakers. Figure TF18-1 shows a schematic of the process.

The ***audible spectrum*** of the human ear extends from about 20 Hz to 20 kHz, although the frequency response may vary among different individuals depending on age and other factors. An audio signal is a superposition of many sinusoids oscillating at different frequencies—each with its own individual amplitude. When we say a sound has a lot of ***bass***, for example, we mean that the low-frequency segment of its spectrum (20 to 100 Hz) has a large amplitude when compared with higher-frequency components. Conversely, very shrill or high-pitched sounds have large-amplitude components in the high-frequency range (10 kHz to 20 kHz). When converting an electrical recording back into the original sound that generated it in the first place, the reproduction fidelity is determined by the degree of distortion that the spectrum undergoes during the playback process. In practice, minimizing ***spectral distortion*** can be quite a challenge!

Each component of the sound-reproduction system shown in Fig. TF18-1 is characterized by its own transfer function relating its output to its input, and since each of these components is equivalent to a circuit composed of resistive and reactive elements, its transfer function is bound to exhibit a non-uniform spectral response. The amplifier, for example,

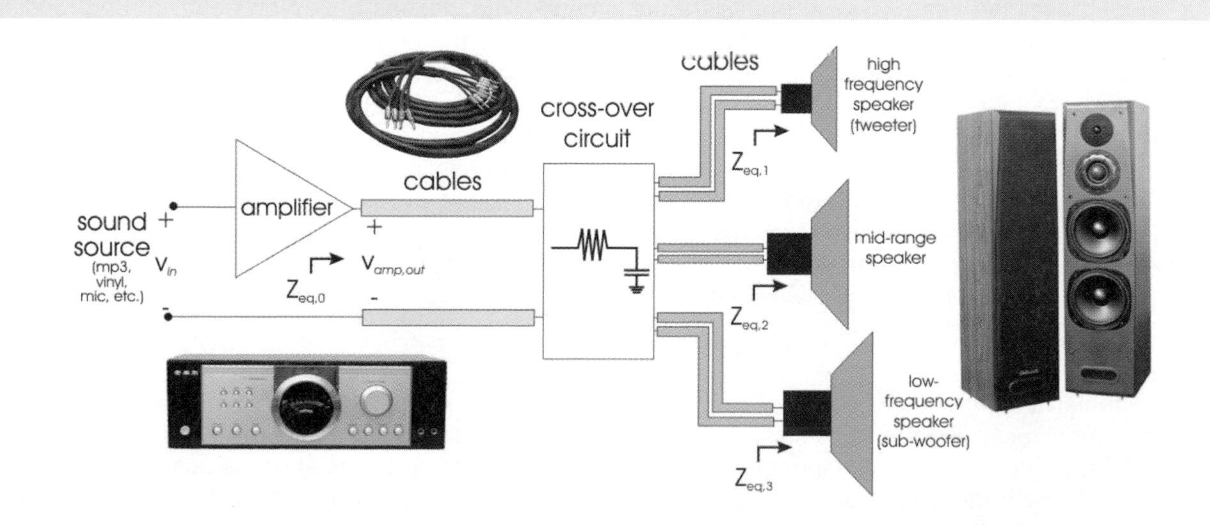

Figure TF18-1: Schematic of a basic audio-reproduction system.

may act like a filter, favoring parts of the audible spectrum over others. The cables, which behave (electrically) like the RC transmission line of Fig. 7-37, will favor low-frequency spectral components over high-frequency components. Thus, unless the components of an audio-reproduction system are well designed in order to generate a transfer function with a nearly flat spectral response over the audible range, the reproduced sound will exhibit a distorted spectrum when compared with the original spectrum. While there are many objective metrics by which to judge the fidelity of audio equipment, every listener processes a given sound differently, introducing a subjective component into the experience. A great way to appreciate the concepts introduced in this Technology Brief is to walk into a high-end audio-systems store with three favorite CDs and then to listen to them on many different amplifier-speaker combinations.

Recording

The specific method used for recording the sound will itself affect the harmonics of the signal. Purely analog methods have long been touted to be superior to digital methods since they do not encode the signal into discrete digital levels. Whereas this may have been true in the past, modern ADC/DAC systems can convert data with very high resolution and little distortion. Likewise, storage media are now large enough that recordings are of such a high resolution that it is debatable whether anyone can really "hear" the difference between a digital recording and an analog recording played on the same system. Of course, data files often are compressed to save storage space and this can introduce distortion and loss of signal fidelity. Online commercial mp3 music distributors often offer music files with different compression options. This means that the discriminating audiophile can download a high-resolution, uncompressed file while someone downloading a lecture will download a compressed, space-saving file.

Amplifiers

It takes quite a bit of power to drive speakers to produce sound in a room. The function of an amplifier is to boost the audio signal's power high enough to drive the speakers. In doing so, the amplifier must (1) keep frequency distortion to a minimum and (2) introduce as little noise as possible into the signal. In order to keep frequency distortion to a minimum, the amplifier's response must be as uniform as possible; in other words, signals of different frequencies and different amplitudes must be amplified with exactly the same gain. Consider again the MOSFET amplifier in Fig. 4-28. Assuming the MOSFET has a perfectly linear response (i.e., $i_{DS} = gV_{GS}$), an input signal will be amplified by the gain g independent of frequency or amplitude. In actuality, a MOSFET (or any other amplifier device) really does not have a uniform response. To address this non-uniform frequency response, many different transistor–amplifier topologies have been developed over the years. These amplifiers are grouped into classes based on behavior and topology; the principal differences lie in circuit complexity, power consumption, and the degree of fidelity with which the circuit reproduces an input signal. Audio-amplifier circuit topologies are categorized by letter—currently from A to G. Although a description of each class lies beyond the scope of this discussion, very succinct overviews can be found in many places online. In order to reduce the noise during the amplification step, two-amp stages often are used. The first stage is called the pre-amplifier. Pre-amps have very good noise characteristics and amplify the signal partway (this mid-level signal is called the ***line signal***). Often, this is simply an amplification of the voltage level. The power amp then boosts this signal (which does not have much noise) to a level high enough to drive speakers; this usually requires significant current amplification to provide enough overall power to the speakers.

Cables

The cables that transfer a signal between sources, amplifiers, crossovers, and speakers can themselves distort the signal. Cables behave exactly like the transmission line of Fig. 7-37; the distributed resistance and capacitance act like a filter with an associated frequency response. In general, cables should be (1) as short as possible, (2) properly impedance-matched to both the output of the amplifier and the input of the speakers, and (3) properly terminated so the cable connections to the equipment do not introduce capacitances. All three of these objectives easily are accomplished using industry-standard cables and connectors. In some modern systems, transmission of audio signals between non-speaker components (e.g., from a tuner to an amp or from an amp to a TV) is often performed in digital form so as to eliminate both noise issues and frequency distortion.

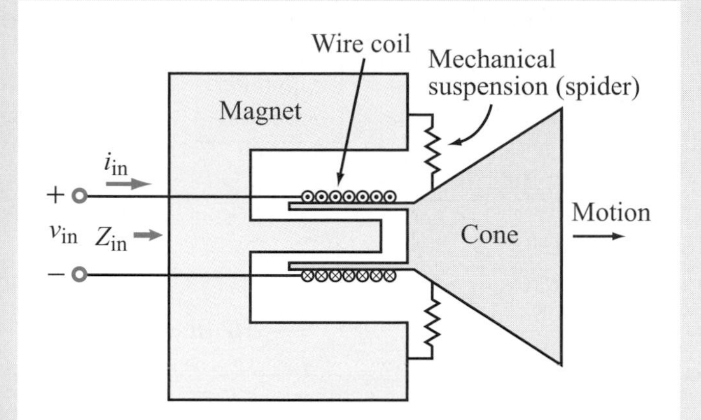

Figure TF18-2: Conceptual illustration of an electromagnetic speaker transducer. The current from the amplifier runs through a coil that induces an electromagnetic force on the cone in proportion to the amplitude of the input signal. The cone motion produces pressure waves and, hence, sound.

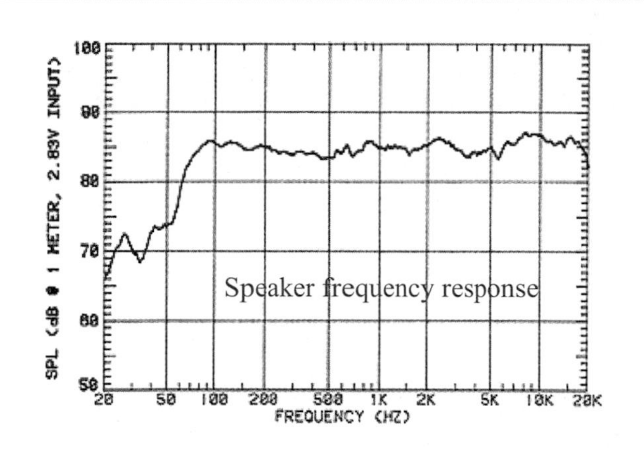

Figure TF18-3: Frequency response of a good consumer-quality speaker consisting of a tweeter and woofer.

Speakers

A speaker is any electro-mechanical device or transducer that converts an electrical signal into sound. Electromagnetic transducers are the most commonly used type for consumer audio applications (Fig. TF18-2), although several other technologies (such as electrostatic speakers) exist as well. The principal metric when choosing a speaker is arguably its frequency response (Fig. TF18-3). Ideally, a speaker will provide a very flat response. This means that signals at different frequencies recreated into sound all at the same audio level. Generally speaking, very small speakers have difficulty reproducing very low frequencies (i.e., bass); a deep drum or baseline may be lost entirely when listening through a small speaker. The most common method for obtaining a nice flat frequency response is to drive several speakers together—each with a different but complementary frequency response. When listened to as a group, the frequency response is close to flat. For example, ***tweeters*** are small speakers intended for reproducing high-frequency sound, while ***woofers*** only reproduce the lowest frequencies. A common entry-level speaker consists of a tweeter, a mid-range speaker, and a woofer all housed together. With appropriate crossover circuits the ensemble can exhibit a good response.

The equivalent impedance of a speaker depends on whether multiple transducers are connected in parallel or in series (within one speaker housing) and whether crossover circuits are used. The transducer impedance itself has a resistance (from the resistance of the wires) and an inductance (from the many turns of the wires that drive the magnet). Commercial audio amplifiers are designed to match the speaker impedance over a certain range, typically from 2 to 8 Ω.

Crossover circuits

As we noted earlier, most speakers cannot handle the entire range of frequencies in the audio range. In order to split the signal for use by the different speakers (such as a tweeter, a mid-range speaker, and a sub-woofer), passive filters are used. The signal is applied to a set of filters that produce three outputs: one output contains only low frequencies in some range, a second contains mid-range frequencies and a third output contains only high-frequency harmonics. In this way, each speaker receives a dedicated signal that contains only the frequencies it can reproduce properly. Designing crossovers can be an involved process that takes into account many variables, including the amount of current in the input signal, the input impedances of all of the speakers, and the frequency range of each speaker. Without careful design, the crossover circuit can provide too much signal power to one speaker and too little to another, thereby distorting the overall frequency response heard by the listener.

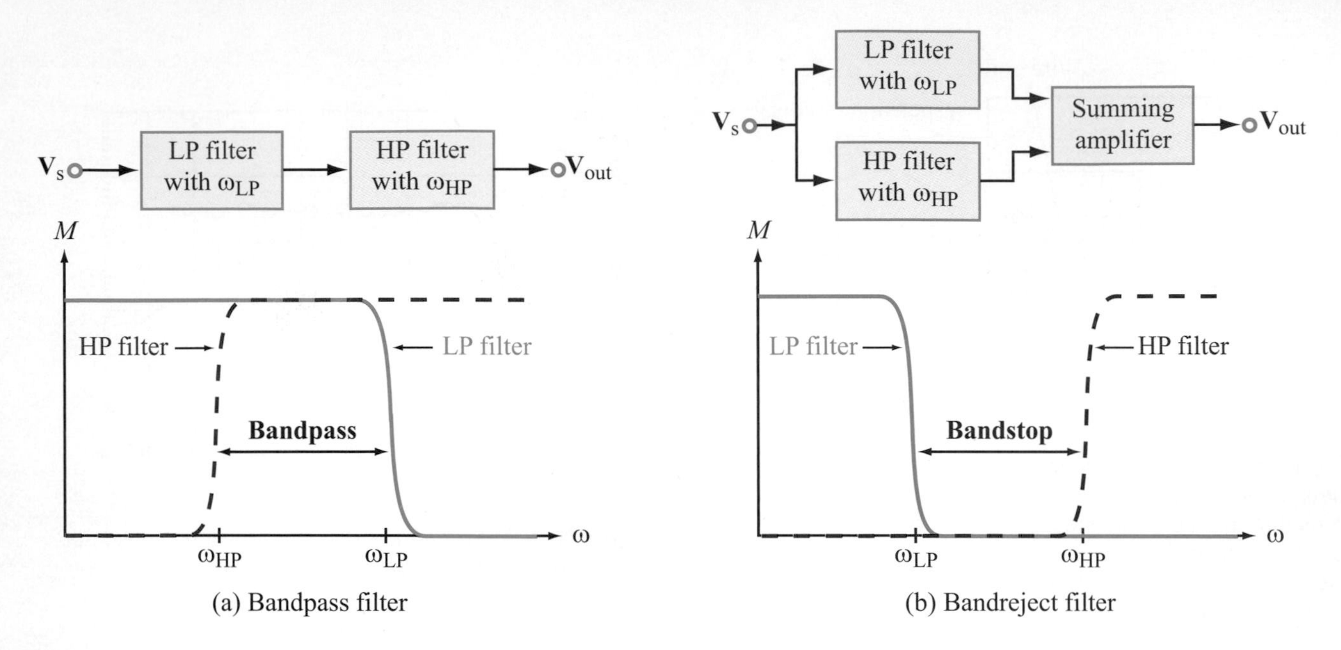

Figure 9-25: (a) In-series cascade of a lowpass and a highpass filter generates a bandpass filter; (b) in-parallel cascading generates a bandreject filter.

where

$$G_{HP} = -\frac{R_f}{R_s} \quad \text{and} \quad \omega_{HP} = \frac{1}{R_s C_s}. \tag{9.92}$$

The expression given by Eq. (9.91) represents a first-order highpass filter with a cutoff frequency ω_{HP} and a gain factor G_{HP}.

Review Question 9-14: What are the major advantages of active filters over their passive counterparts?

Review Question 9-15: Are active filters used mostly at frequencies below 1 MHz or above 1 MHz?

Exercise 9-13: Choose values for R_s and R_f in the circuit of Fig. 9-23(b) so that the gain magnitude is 10 and the corner frequency is 10^3 rad/s given that $C_f = 1\ \mu$F.

Answer: $R_s = 100\ \Omega$, $R_f = 1\ \text{k}\Omega$. (See)

9-7 Cascaded Active Filters

The active lowpass and highpass filters we examined thus far—as well as other op-amp configurations that provide these functions—can be regarded as basic building blocks that easily can be cascaded together to create second- or higher-order lowpass and highpass filters or to design bandpass and bandreject filters (Fig. 9-25).

The cascading approach allows the designer to work with each stage separately and then combine all of the stages together to achieve the desired specifications.

Moreover, inverting or noninverting amplifier stages can be added to the filter cascade to adjust the gain or polarity of the output signal, and buffer circuits can be inserted inbetween stages to provide impedance isolation, if necessary. Throughout the multistage process, it is prudent to compare the positive and negative peak values of the voltage at the output of every stage with the op amp's power-supply voltages V_{CC} and $-V_{CC}$ to make sure that the op amp will not go into saturation mode.

Example 9-10: Third-Order Lowpass Filter

For the three-stage active filter shown in Fig. 9-26, generate dB plots for M_1, M_2, and M_3, where $M_1 = |\mathbf{V}_1/\mathbf{V}_s|$, $M_2 = |\mathbf{V}_2/\mathbf{V}_s|$, and $M_3 = |\mathbf{V}_3/\mathbf{V}_s|$.

Solution: Since all three stages have the same values for R_f and C_f, they have the same cutoff frequency give by

$$\omega_{LP} = \frac{1}{R_f C_f} = \frac{1}{10^4 \times 10^{-9}} = 10^5 \text{ rad/s}.$$

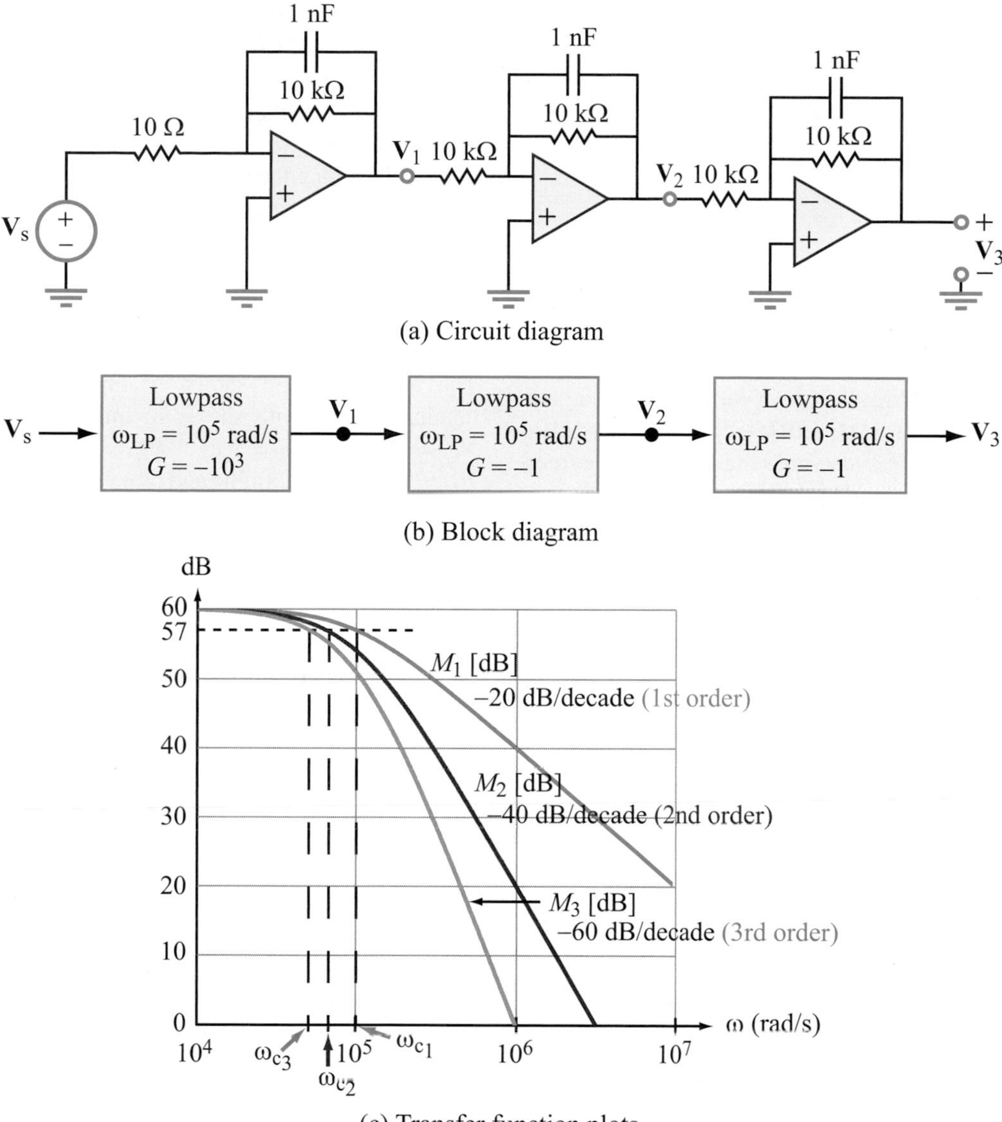

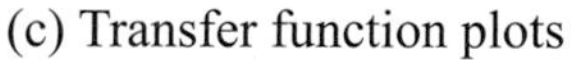
(c) Transfer function plots

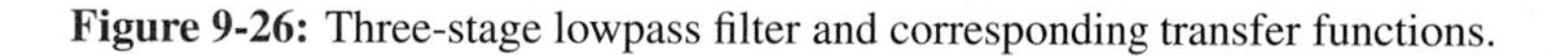
Figure 9-26: Three-stage lowpass filter and corresponding transfer functions.

The input resistance of the first stage is 10 Ω, but the input resistances of the second and third stages are 10 kΩ. Hence,

$$G_1 = -\frac{10\text{k}}{10} = -10^3 \qquad \text{and} \qquad G_2 = G_3 = -\frac{10\text{k}}{10\text{k}} = -1.$$

Transfer function M_1 therefore is given by

$$M_1 = \left|\frac{\mathbf{V}_1}{\mathbf{V}_s}\right| = \left|\frac{G_1}{1 + j\omega/\omega_{\text{LP}}}\right| = \frac{10^3}{\sqrt{1 + (\omega/10^5)^2}}$$

and

$$M_1\ [\text{dB}] = 20 \log\left[\frac{10^3}{\sqrt{1 + (\omega/10^5)^2}}\right]$$
$$= 60\ \text{dB} - 10 \log[1 + (\omega/10^5)^2].$$

The transfer function corresponding to $\mathbf{V}_2$ is

$$M_2 = \left|\frac{\mathbf{V}_2}{\mathbf{V}_1} \cdot \frac{\mathbf{V}_1}{\mathbf{V}_s}\right| = \left|\frac{G_1}{1 + j\omega/\omega_{\text{LP}}}\right| \left|\frac{G_2}{1 + j\omega/\omega_{\text{LP}}}\right|$$
$$= \frac{10^3}{1 + (\omega/10^5)^2}$$

and

$$M_2\ [\text{dB}] = 20\log\left[\frac{10^3}{1+(\omega/10^5)^2}\right] = 60\ \text{dB} - 20\log[1+(\omega/10^5)^2].$$

Similarly,

$$M_3\ [\text{dB}] = 60\ \text{dB} - 30\log[1+(\omega/10^5)^2].$$

The three-stage process is shown in Fig. 9-26(b) in block-diagram form, and spectral plots of M_1 [dB], M_2 [dB], and M_3 [dB] are displayed in Fig. 9-26(c). We note that the gain roll-off rate S_g is −20 dB for M_1 [dB], −40 dB for M_2 [dB], and −60 dB for M_3 [dB]. We also note that the −3 dB corner frequencies are not the same for the three stages.

Review Question 9-16: Why is it more practical to cascade multiple stages of active filters than to cascade multiple stages of passive filters?

Review Question 9-17: What determines the gain factors of the highpass and lowpass op-amp filters?

Exercise 9-14: What are the values of the corner frequencies associated with M_1, M_2, and M_3 of Example 9-10?

Answer: $\omega_{c_1} = 10^5$ rad/s, $\omega_{c_2} = 0.64\omega_{c_1} = 6.4\times10^4$ rad/s, $\omega_{c_3} = 0.51\omega_{c_1} = 5.1\times10^4$ rad/s. (See ◎)

Analogy to AND and OR Gates

An ***AND logic gate*** has two inputs (whose logic states can each be either 0 or 1) and one output. Its output state is 1 if and only if both input states are 1. Otherwise, its output state is zero. Hence, *the output of an AND gate is equal to the product of its input states.* The cascaded bandpass filter diagrammed in Fig. 9-25(a) is analogous to an AND gate. The filter consists of an idealized lowpass filter with cutoff frequency ω_{LP} ***connected in series*** with an idealized highpass filter with cutoff frequency ω_{HP}. *The frequency of the input signal has to be in the passband of both filters in order for the signal to make it to the output.* The filter passband is defined by the bandwidth extending from ω_{HP} to ω_{LP}, as illustrated in Fig. 9-25(a). The combination of the two filters when cascaded in series is equivalent to an AND gate, because the final output is proportional to the product of their transfer functions, $\mathbf{H}_{LP}\mathbf{H}_{HP}$. The process is illustrated through Example 9-11.

In contrast, the bandreject filter (Fig. 9-25(b)) is analogous to an ***OR gate***, for which the state of its output is 1 if either one or both of its inputs has a state of 1. The cascade configuration of the bandreject filter consists of a lowpass filter ***connected in parallel*** with a highpass filter—thereby offering the input signal to pass through either or both of them—and then their outputs are added by a summing amplifier. Example 9-12 provides more details.

Example 9-11: Bandpass Filter

The block diagram shown in Fig. 9-27(a) is a two-stage bandpass filter with the following elements: $R_{s_1} = 1$ kΩ, $R_{f_1} = 10$ kΩ, $C_{f_1} = 9$ nF, $R_{s_2} = 12$ kΩ, $C_{s_2} = 9$ nF, and $R_{f_2} = 96$ kΩ. Determine and plot the magnitude of the transfer function and obtain the values of ω_0, ω_{c_1}, ω_{c_2}, B, and Q.

Solution: The first stage is a lowpass filter with a transfer function given by Eq. (9.88) as

$$\mathbf{H}_{LP}(\omega) = G_{LP}\left(\frac{1}{1+j\omega/\omega_{LP}}\right)$$

with

$$G_{LP} = -\frac{R_{f_1}}{R_{s_1}} = -\frac{10^4}{10^3} = -10$$

and

$$\omega_{LP} = \frac{1}{R_{f_1}C_{f_1}} = \frac{1}{10^4\times9\times10^{-9}} = 11.11\ \text{krad/s}.$$

The transfer function of the highpass filter in the second stage is characterized by Eq. (9.91) as

$$\mathbf{H}_{HP}(\omega) = G_{HP}\left(\frac{j\omega/\omega_{HP}}{1+j\omega/\omega_{HP}}\right)$$

with

$$G_{HP} = -\frac{R_{f_2}}{R_{s_2}} = -\frac{96\times10^3}{12\times10^3} = -8,$$

and

$$\omega_{HP} = \frac{1}{R_{s_2}C_{s_2}} = \frac{1}{12\times10^3\times9\times10^{-9}} = 9.26\ \text{krad/s}.$$

The combined transfer function is then given by

$$\mathbf{H}(\omega) = \mathbf{H}_{LP}\mathbf{H}_{HP} = G_1G_2\left[\frac{j\omega/\omega_{HP}}{(1+j\omega/\omega_{LP})(1+j\omega/\omega_{HP})}\right], \quad (9.93)$$

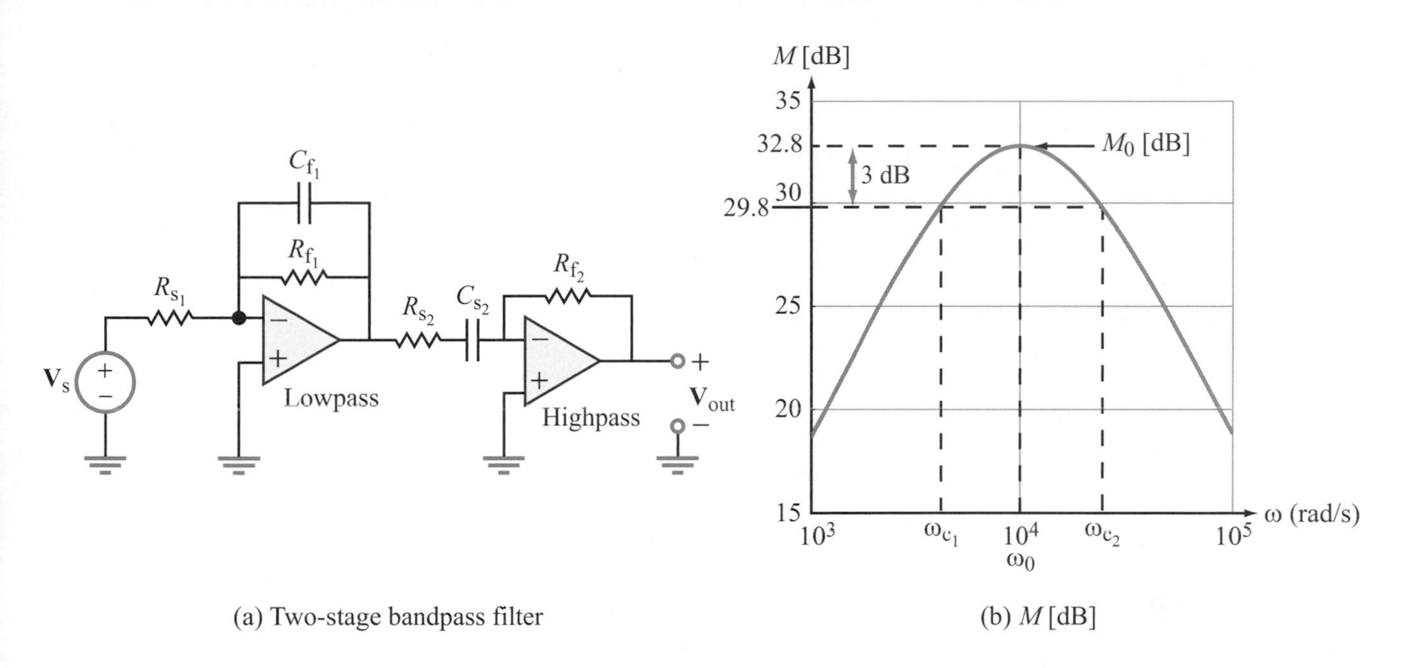

(a) Two-stage bandpass filter

(b) M [dB]

Figure 9-27: Active bandpass filter of Example 9-11.

and its magnitude is

$$M = |\mathbf{H}(\omega)|$$
$$= \left| \frac{80\omega/(9.26 \times 10^3)}{[1 + j\omega/(11.11 \times 10^3)][1 + j\omega/(9.26 \times 10^3)]} \right|$$
$$= \frac{80\omega/(9.26 \times 10^3)}{\{[1 + (\omega/11.11 \times 10^3)^2][1 + (\omega/9.26 \times 10^3)^2]\}^{1/2}} \tag{9.94}$$

The angular frequency ω_0 of a bandpass filter is defined as the frequency at which the transfer function is a maximum. For a high-Q filter, ω_0 is approximately midway between the lower and upper cutoff frequencies ω_{c_1} and ω_{c_2}, but ω_0 may be significantly closer to ω_1 or ω_2 if Q is not very large. We do not yet know the value of Q for the present filter, but we can generate an approximate estimate from knowledge of the values of ω_{LP} and ω_{HP}. The estimated bandwidth is

$$B \text{ (est)} = \omega_{LP} - \omega_{HP} = (11.11 - 9.26)\text{k} = 1.85 \text{ krad/s}$$

and if we assume that ω_0 is midway between ω_{HP} and ω_{LP}, then

$$\omega_0 \text{ (est)} = \left(9.26 + \frac{1.85}{2}\right) \text{k} = 10.185 \text{ krad/s},$$

and

$$Q \text{ (est)} = \frac{\omega_0 \text{ (est)}}{B \text{ (est)}} = \frac{10.185}{1.85} = 5.5.$$

Since Q is not greater than 10, the estimated values of B, ω_0, and Q are not likely to be very accurate, so we should return to the expression for M given by Eq. (9.94) and use it to determine the exact values of ω_0, ω_{c_1}, and ω_{c_2}. We can do so by calculating (or plotting) M as a function of ω to identify: (a) M_0 and ω_0, the maximum value of $M(\omega)$ and the corresponding value of ω at which it occurs, respectfully, and (b) ω_{c_1} and ω_{c_2}, the corner frequencies at which $M = M_0/\sqrt{2}$ (or -3 dB below the peak on a dB scale). According to the spectral plot of M [dB] shown in Fig. 9-27(b),

$$M_0 \text{ [dB]} = 32.8 \text{ dB}, \qquad \omega_0 \text{ (exact)} = 10.14 \text{ krad/s},$$
$$\omega_{c_1} = 4.19 \text{ krad/s}, \qquad \text{and} \qquad \omega_{c_2} = 24.56 \text{ krad/s}.$$

Hence,

$$B \text{ (exact)} = \omega_{c_2} - \omega_{c_1} = 20.37 \text{ krad/s},$$

and

$$Q \text{ (exact)} = \frac{\omega_0}{B} = \frac{10.14}{20.37} \simeq 0.51.$$

The obvious conclusion is that our estimated values for B and Q are way off in comparison with their exact counterparts. We assumed that the corner frequencies ω_{LP} and ω_{HP} associated with functions $\mathbf{H}_{LP}$ and $\mathbf{H}_{HP}$, respectively, are good estimates of the corner frequencies ω_{c_1} and ω_{c_2} of the product of the two functions. That was obviously a poor assumption.

Technology Brief 19: Smart Dust, Sensor Webs, and Ubiquitous Computing

Suppose it were possible to fit all of the parts of a sensor or computer into a small volume—comparable in size to a grain of sand. Also, suppose that many such grains of sand could be connected wirelessly to a vast, distributed, nearly imperceptible communications web. Were such a technological capability possible, would we embed microsensors and computational devices in our everyday surroundings? How would we use them? These questions pose not only some serious technological challenges, but also a number of ethical and societal issues as well.

The realization of ***ubiquitous computing*** (also known as ***ambient intelligence***, ***embedded computing***, and ***pervasive computing***) will require fundamental developments in several specialties within electrical engineering and computer science. These include extreme miniaturization of computer hardware, development of seamless computer networks across many size scales, interfaces for massively parallel computation, and special security protocols, among many others. The idea of miniaturizing a computer into a volume as small as a grain of sand is not as radical as it might seem. As of 2007, conventional commercial transistors could be fabricated in sizes as small as 45 nm (i.e., 45×10^{-9} m!). Similarly, extensive miniaturization has been applied to communication and sensor hardware, as well as to power sources. The so-called ***smart-dust*** device shown in Fig. TF19-1 is only a few millimeters in size, yet it contains all of the elements of a very simple computer. A tiny processor powered by a microbattery reads data from a

Figure TF19-1: A Smart-Dust mote (circa 2001) mounted on a microbattery. (Courtesy of Prof. Kristopher J. Pister of the University of California at Berkeley.) Note that the device is only a few millimeters across.

simple sensor. When queried by an external source, the processor can communicate using reflected light pulses or very low-power radio communication. Processors (such as the ones in this device) use many different strategies to conserve power; for example, they run at very slow clock speeds (as compared with desktop computers), they have ultra-low power (<1 nW) sleep modes, and the transistors are operated at extremely low voltages. Many obstacles remain before these systems become truly ***smart dust***, however. A principal challenge is power; even the most advanced battery technologies can store no more than 2 J/mm^3. Many strategies are being explored, including wireless transmission of power, higher efficiency of solar cells, scavenging power from mechanical vibrations in the environment, and even the extreme miniaturization of combustion engines! The miniaturization of communication components also remains a challenge. Efficient radio communication requires the use of antennas with dimensions comparable in size with the wavelength of the ac radio signal, so in order to communicate through sub-millimeter-long antennas, we should use electromagnetic waves whose wavelengths are shorter than a millimeter, which corresponds to the frequency range above 30 GHz. Communication technology is not developed as well in this part of the microwave spectrum as at lower frequencies, nor is the atmosphere as transparent. To circumvent these limitations, other communication modes have been explored, including the use of lasers and infrared devices. Other challenges include packaging, power delivery, and micro-sensor design and fabrication.

In addition to miniaturizing the hardware, we also face the challenge of having to design and program the sensor networks. As more and more of our everyday technological environment acquires the ability to sense, compute, and communicate (Fig. TF19-2), interesting problems develop. How should this communication and computation be organized? Should information be passed on to dedicated communication backbones (such as the high-bandwidth optical fiber that connects the Internet to buildings and houses), or can the information "hop" from one mobile mote to another until it reaches its target? How will the information be ***managed*** and what type of human interfaces will be needed? These are but a few of the many questions that will need to be addressed by the next generation of engineers and scientists.

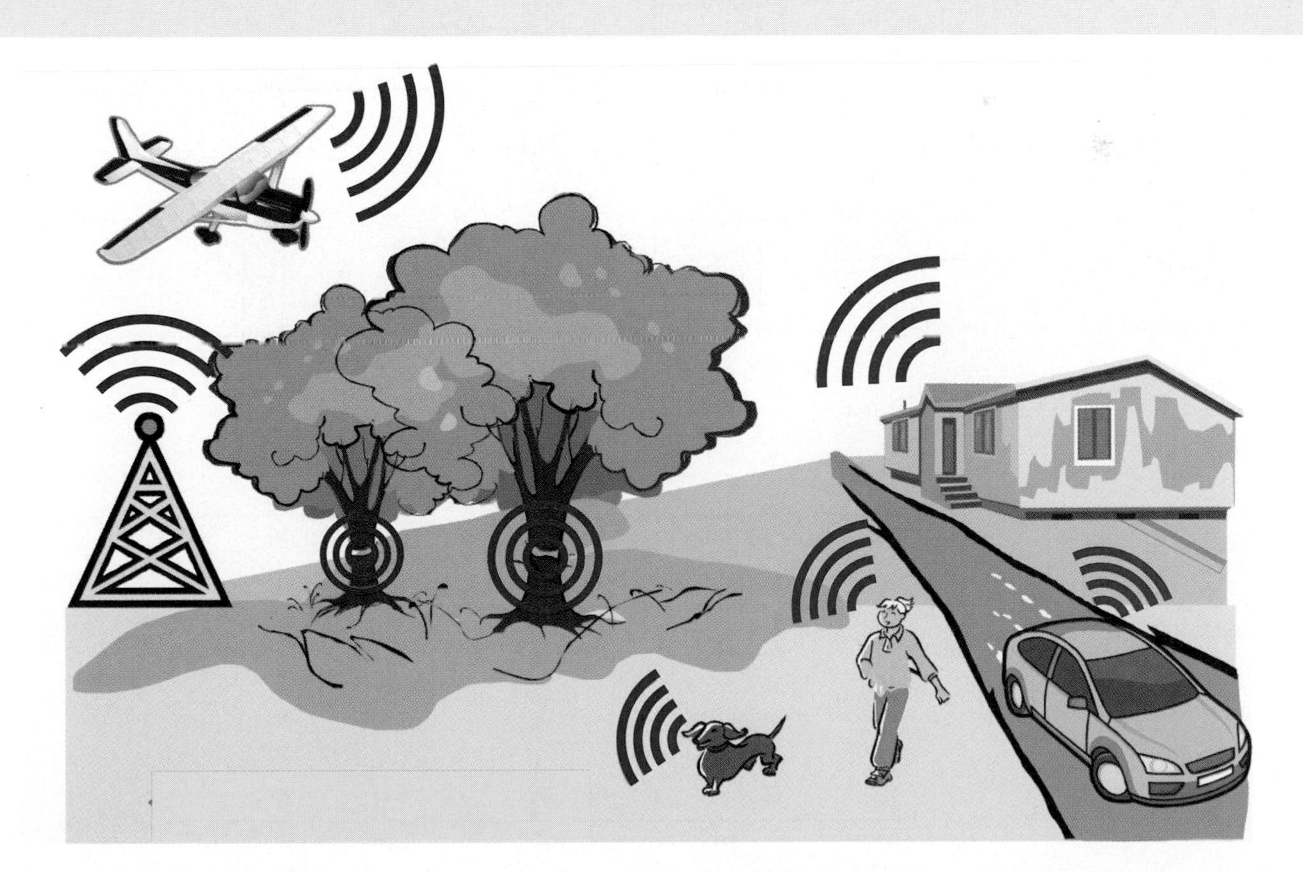

Figure TF19-2: Our environment will become increasingly aware as more and more sensing, computation, and communication systems are embedded everywhere.

Example 9-12: Bandreject Filter

Design a bandreject filter with the specifications: (a) Gain $= -50$, (b) bandstop extends from 20 kHz to 40 kHz, and (c) gain roll-off rate $= -40$ dB/decade along both boundaries of the bandstop.

Solution: The specified roll-off rate requires the use of two identical lowpass filters with $\omega_{LP} = 2\pi \times 2 \times 10^4 = 4\pi \times 10^4$ rad/s, and two identical highpass filters with $\omega_{HP} = 8\pi \times 10^4$ rad/s. To minimize performance variations among identical pairs, identical resistors will be used in all four units (Fig. 9-28(b)), which means that they all will have unity gain. The overall gain of -50 will be provided by the summing amplifier.

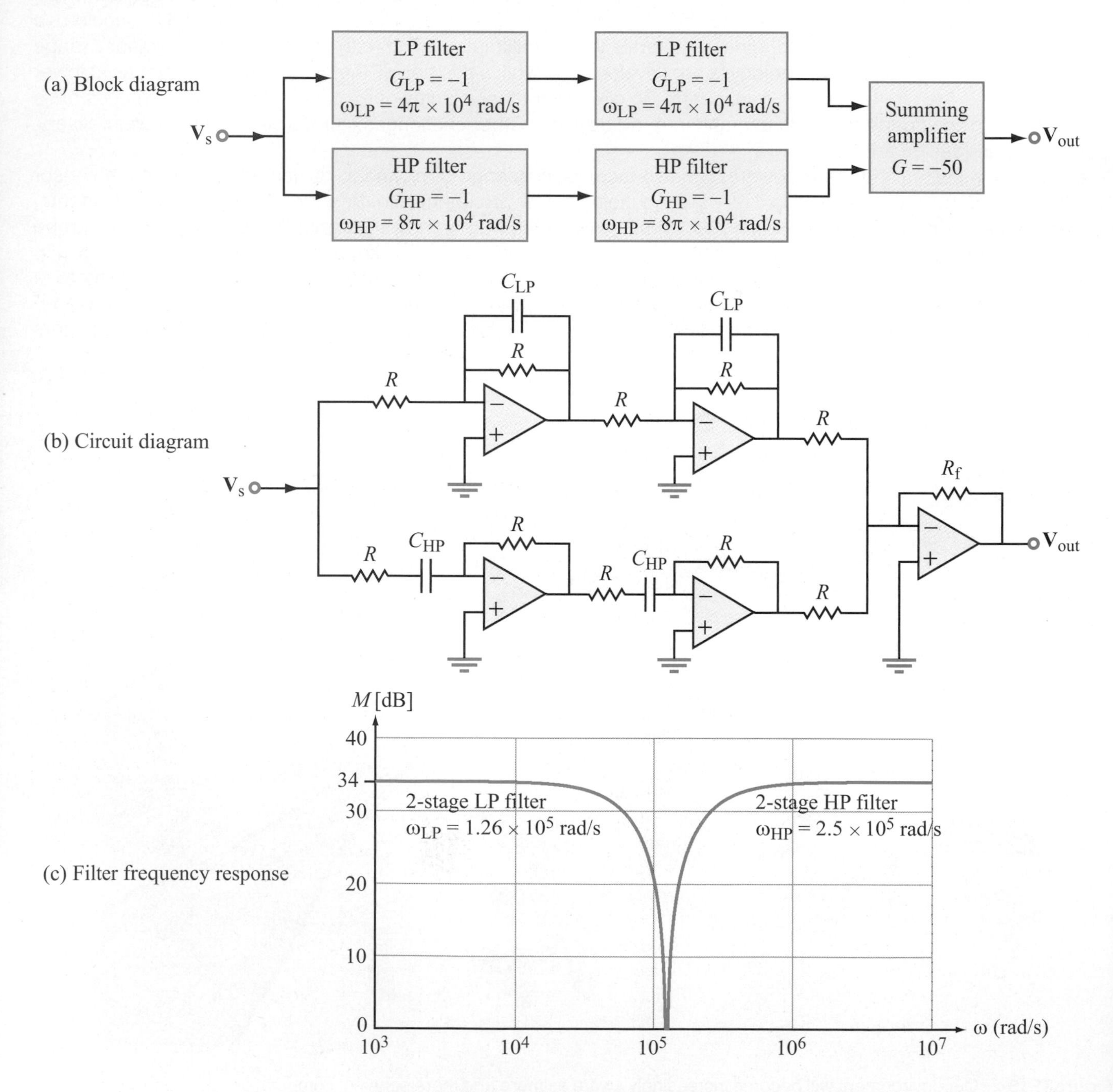

Figure 9-28: Bandreject filter of Example 9-12.

Somewhat arbitrarily, we select $R = 1$ kΩ. From

$$\omega_{LP} = 4\pi \times 10^4 = \frac{1}{RC_{LP}} \Rightarrow C_{LP} = 7.96 \text{ nF} \simeq 8 \text{ nF},$$

$$\omega_{HP} = 8\pi \times 10^4 = \frac{1}{RC_{HP}} \Rightarrow C_{HP} \simeq 4 \text{ nF}.$$

The value of R_f is specified by the gain of the summing amplifier as

$$G = -50 = -\frac{R_f}{R} \Rightarrow R_f = 50 \text{ k}\Omega.$$

The transfer function of the bandreject filter is given by

$$\begin{aligned} \mathbf{H}(\omega) &= G[\mathbf{H}_{LP}^2 + \mathbf{H}_{HP}^2] \\ &= -50\left[\left(\frac{1}{1+j\omega RC_{LP}}\right)^2 + \left(\frac{j\omega RC_{HP}}{1+j\omega RC_{HP}}\right)^2\right] \\ &= -50\left[\left(\frac{1}{1+j\omega/4\pi \times 10^4}\right)^2 + \left(\frac{j\omega/8\pi \times 10^4}{1+j\omega/8\pi \times 10^4}\right)^2\right]. \end{aligned}$$

The spectrum of M [dB] $= 20 \log |\mathbf{H}|$ is displayed in Fig. 9-28(c).

Exercise 9-15: The bandreject filter of Example 9-12 uses two lowpass-filter stages and two highpass-filter stages. If three stages of each were used instead, what would the expression for $\mathbf{H}(\omega)$ be in that case?

Answer:

$$\mathbf{H}(\omega) = 50\left[\left(\frac{1}{1+j\omega/4\pi \times 10^4}\right)^3 + \left(\frac{j\omega/8\pi \times 10^4}{1+j\omega/8\pi \times 10^4}\right)^3\right].$$

(See ◎)

9-8 Application Note: Modulation and the Superheterodyne Receiver

9-8.1 Modulation

In the language of electronic communication, the term ***signal*** refers to the information to be communicated between two different locations or between two different circuits, and the term ***carrier*** refers to the sinusoidal waveform that *carries* the information. The latter is of the form

$$v_c(t) = A \cos 2\pi f_c t, \tag{9.95}$$

where A is its amplitude and f_c is its ***carrier frequency***. The sinusoid can be used to carry information by varying its amplitude—in which case A becomes $A(t)$—while keeping f_c constant. In the example shown in Fig. 9-29(a), multiplication of the signal waveform by the sinusoidal carrier generates an ***amplitude-modulated (AM)*** carrier whose envelope is identical with the signal waveform. Alternatively, we can apply ***frequency modulation (FM)*** by keeping A constant and varying f_c in a fashion that mimics the variation of the signal waveform, as illustrated by Fig. 9-29(b). FM usually requires more bandwidth than AM, but it is also more immune to noise and interference, thereby delivering a higher-quality sound than AM. Many other types of modulation techniques also are available, including phase modulation and pulse-code modulation.

9-8.2 The Superheterodyne Receiver

Let us assume the signal $v_s(t)$ in Fig. 9-29(a) is an audio signal and the carrier frequency $f_c = 1$ MHz. Let us also assume that the signal was used to generate an amplitude-modulated waveform, which was then fed into a transmit antenna. After propagating through the air along many different directions (as dictated by the antenna radiation pattern), part of the AM waveform was intercepted by a receive antenna connected to an AM receiver. Prior to 1918, the receiver would have been a ***tuned-radio frequency receiver*** or a ***regenerative receiver***, both of which suffered from poor frequency selectivity and low immunity to noise. In either case, the receiver would have ***demodulated*** the AM signal by suppressing the carrier and preserving the envelope, thereby retrieving the original signal $v_s(t)$ (or more realistically, some distorted version of $v_s(t)$). To overcome the shortcomings of such receivers, ***Edwin Armstrong*** introduced the ***heterodyne receiver*** in 1918 by proposing the addition of a receiver stage to convert the carrier frequency of the AM signal f_c to a fixed lower frequency (now called the ***intermediate frequency*** f_{IF}) before detection (demodulation). [Armstrong also invented frequency modulation in 1935.] The superheterodyne concept proved to be one of the foundational enablers of 20th-century radio transmission. It is still in use in most AM and FM analog receivers, although it slowly is getting supplanted by the concept of software radio (Section 9-8.4).

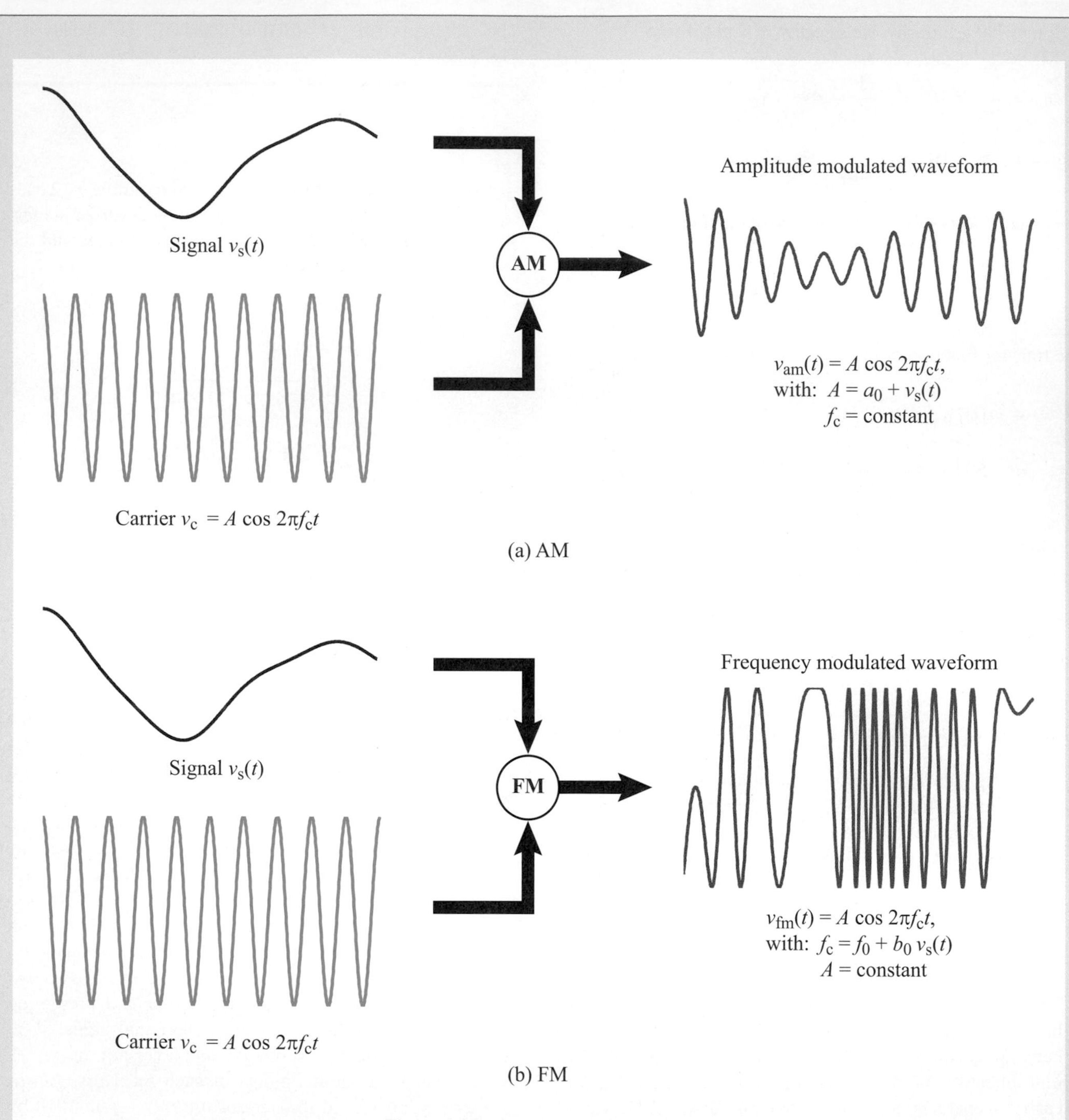

Figure 9-29: Overview of AM and FM.

Figure 9-30 shows a basic block diagram of a superheterodyne receiver. The ***tuner*** is a bandpass filter whose center frequency can be adjusted to allow the intended signal at $f_c = 1$ MHz (for example) to pass through, while rejecting signals at other carrier frequencies. After amplification by the ***radio-frequency (RF)*** amplifier, the AM signal either can be demodulated directly (which is what receivers did prior to 1918) or it can be converted into an IF signal by ***mixing*** it with another locally generated sinusoidal signal provided by a ***local oscillator***. As will be explained in

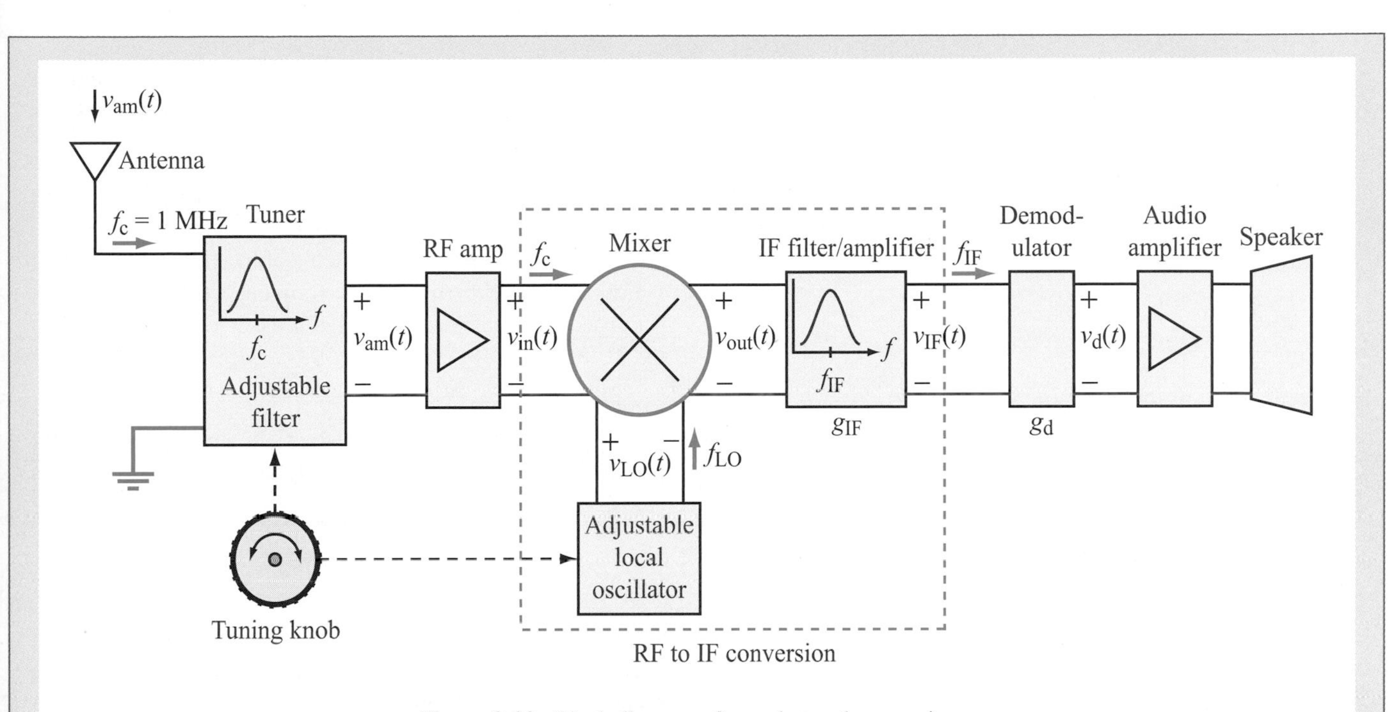

Figure 9-30: Block diagram of superheterodyne receiver.

Section 9-8.3, the ***mixer*** is a device that multiplies the two signals available at its input and generates an output signal whose frequency is

$$f_{\text{IF}} = f_{\text{LO}} - f_{\text{c}}, \tag{9.96}$$

where f_{LO} is the local-oscillator frequency. The frequency conversion given by Eq. (9.96) assumes that $f_{\text{LO}} \geq f_{\text{c}}$; otherwise, $f_{\text{IF}} = f_{\text{c}} - f_{\text{LO}}$ if $f_{\text{LO}} < f_{\text{c}}$. It is important to note that frequency conversion changes the carrier frequency of the AM waveform from f_{c} to f_{IF}, but the audio signal $\upsilon_{\text{s}}(t)$ remains unchanged; it merely is getting carried by a different carrier frequency.

The diagram in Fig. 9-30 indicates that the tuning knob controls the center of the adjustable tuner as well as the local oscillator frequency. By *synchronizing* these two frequencies to each other, the IF frequency remains always a constant. This is an important feature of the superheterodyne receiver, because it insures that the same ***IF filter/amplifier*** can be used to provide high-selectivity filtering and high-gain amplification, regardless of the carrier frequency of the AM signal. In the ***AM radio band***, the carrier frequency of the audio signals transmitted by an AM radio station may be at any frequency between 530 and 1610 kHz. Because of the built-in synchronization between the tuner and the local oscillator, the IF frequency of an AM receiver is always at 455 kHz, which is the standard IF for AM radio. Similarly, the standard IF for FM radio is 10 MHz, and the standard IF for television is 45 MHz.

It is impractical to design and manufacture high-performance components at every frequency in the radio spectrum. By designating certain frequencies as IF standards, industry was able to develop devices and systems that operate with very high performance at those frequencies. Consequently, frequency conversion to an IF band is very prevalent not only in radio and TV receivers but also in radar sensors, satellite communication systems and transponders, among others.

9-8.3 Frequency Conversion

Regardless of the specific type of modulation used in a modern communication system, it will employ one or more steps of frequency conversion, whereby the carrier frequency is changed from an initial frequency f_1 to a new frequency f_2. If f_2 is higher than f_1, it is called ***up-conversion***, and the reverse is called ***down-conversion***. In the AM example of Fig. 9-30, $f_1 = f_{\text{c}} = 1$ MHz and $f_2 = f_{\text{IF}} = 455$ kHz. To explain how the conversion takes place, consider the general case of two signals given by

$$\upsilon_{\text{in}}(t) = A(t)\ \cos 2\pi f_{\text{c}} t \tag{9.97a}$$

and

$$\upsilon_{\text{LO}}(t) = A_{\text{LO}} \cos 2\pi f_{\text{LO}} t, \tag{9.97b}$$

where $A(t)$ represents the audio signal waveform $v_s(t)$ (Fig. 9-29(a)) and A_{LO} is a constant amplitude associated with the local oscillator signal.

A ***mixer*** is a diode circuit that has two inputs and one output with its output voltage $v_{out}(t)$ being equal to the product of its input voltages:

$$\begin{aligned} v_{out}(t) &= v_{in}(t) \times v_{LO}(t) \\ &= A(t)\ A_{LO} \cos 2\pi f_c t \cos 2\pi f_{LO} t. \end{aligned} \tag{9.98}$$

Application of the trigonometric identity

$$\cos x \cos y = \tfrac{1}{2}[\cos(x+y) + \cos(x-y)] \tag{9.99}$$

leads to

$$\begin{aligned} v_{out}(t) = &\frac{A(t)\ A_{LO}}{2} \cos[2\pi(f_c + f_{LO})t] \\ &+ \frac{A(t)\ A_{LO}}{2} \cos[2\pi(f_{LO} - f_c)t]. \end{aligned} \tag{9.100}$$

Let us consider the case where $f_c = 1$ MHz and $f_{LO} = 1.445$ MHz. The expression for $v_{out}(t)$ becomes

$$v_{out}(t) = A'(t)\ \cos 2\pi f_s t + A'(t)\ \cos 2\pi f_d t, \tag{9.101}$$

where

$$A'(t) = \frac{A(t)\ A_{LO}}{2}, \tag{9.102}$$

and f_s and f_d are the sum and difference frequencies:

$$f_s = f_c + f_{LO} = 2.445 \text{ MHz} \tag{9.103a}$$

and

$$f_d = f_{LO} - f_c = 0.445 \text{ MHz}. \tag{9.103b}$$

Thus, $v_{out}(t)$ consists of two signal components with markedly different carrier frequencies. By selecting a narrow IF filter/amplifier in Fig. 9-30 with a center frequency $f_{IF} = f_d$, only the difference-frequency component of $v_{out}(t)$ will make it through the filter. Consequently, its output is given by

$$v_{IF}(t) = g_{IF}\ A'(t)\ \cos 2\pi f_{IF} t,$$

where g_{IF} is the voltage gain factor of the IF filter/amplifier. Demodulation, which is a low-frequency filtering process, removes the IF carrier, leaving behind a detected signal given by

$$v_d(t) = g_d g_{IF}\ A'(t),$$

where g_d is a demodulator constant. Since $A'(t)$ is directly proportional to the original audio signal $v_s(t)$, $v_d(t)$ becomes (ideally) a replica of $v_s(t)$.

9-8.4 Software Radio

The recent increase in the speed of digital circuits has made it possible to perform all of the functions of a superheterodyne receiver directly in the digital domain. Low-cost FM receivers thus consist of little more than an antenna connected to the input pin of a digital chip. The chip converts the input signal into digital format and then performs all of the mixing, filtering, amplifying, and demodulating functions by direct computation. This digital approach (which has become known as ***software radio***) already has made some inroads into the entertainment market.

Review Question 9-18: What are the advantages of FM over AM?

Review Question 9-19: What is the fundamental contribution of the superheterodyne receiver, and why is it significant?

Review Question 9-20: What does a mixer do?

9-9 Spectral Response with Multisim

The AC Analysis and Parameter Sweep tools are very useful when analyzing the frequency response of a circuit. The Network Analyzer, first introduced in Chapter 8, also provides a convenient way to evaluate the frequency response of a circuit using Multisim. These tools are illustrated in the next three examples.

Example 9-13: RLC Circuit

Design a series RLC bandpass filter with a center frequency of 10 MHz and $Q = 50$. Use Multisim to generate magnitude and phase plots covering the range from 8 to 12 MHz.

Solution: The specified filter can be designed with an infinite number of different combinations of R, L, and C.

We will choose a realistic value for L, namely 0.1 mH, which will dictate that C be

$$C = \frac{1}{\omega_0^2 L} = \frac{1}{(2\pi \times 10^7)^2 \times 10^{-4}} = 2.53 \text{ pF}.$$

Next, we select the value of R to satisfy the requirement on Q. From Table 9-3, we obtain

$$R = \frac{\omega_0 L}{Q} = \frac{2\pi \times 10^7 \times 10^{-4}}{50} = 125.7\ \Omega.$$

With all three elements specified, we construct the Multisim circuit shown in Fig. 9-31. Before performing AC Analysis, we should double-click on the ac source and change its value to 1 V (rms). Next, we select Simulate → Analyses → AC Analysis, and then we set FSTART to 8 MHz and FSTOP to 12 MHz. With both the Sweep Type and Vertical Scale set to Linear, the number of points set to 1000, and the variable selected for analysis is V(3), AC Analysis generates the plots displayed in Fig. 9-32. The magnitude plot exhibits a peak at 10 MHz, and the phase goes through 0° at that frequency. To verify that the circuit has a $Q = 50$, we use the cursors to establish the locations at which the vertical value of the curve is $1/\sqrt{2} = 0.707$ V. The separation between the two cursors (labeled "dx" in the cursor box) is 200.0699 kHz. This is the half-power bandwidth B. The quality factor is

$$Q = \frac{\omega_0}{B} = \frac{10^7}{200.0699 \times 10^3} = 49.98,$$

which is approximately equal to the specified value.

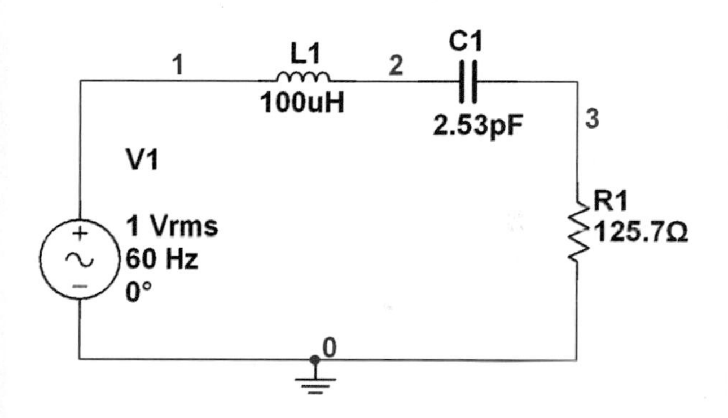

Figure 9-31: A series RLC filter implemented in Multisim.

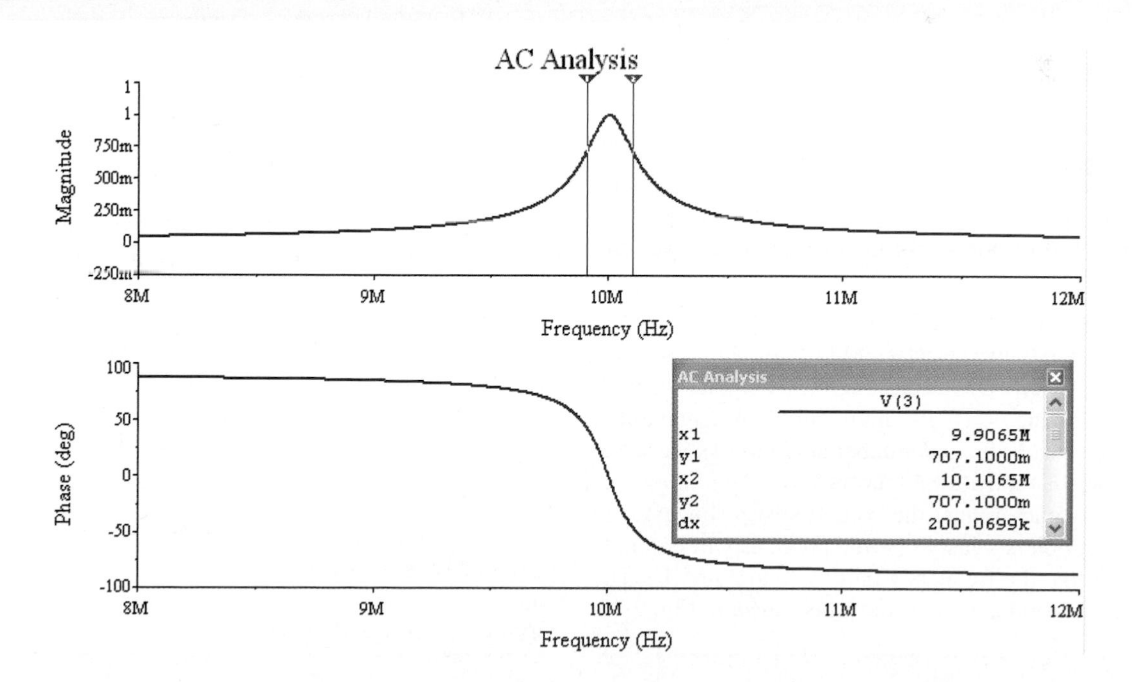

Figure 9-32: Magnitude and phase plots for the circuit of Fig. 9-31 generated by AC Analysis for Example 9-13.

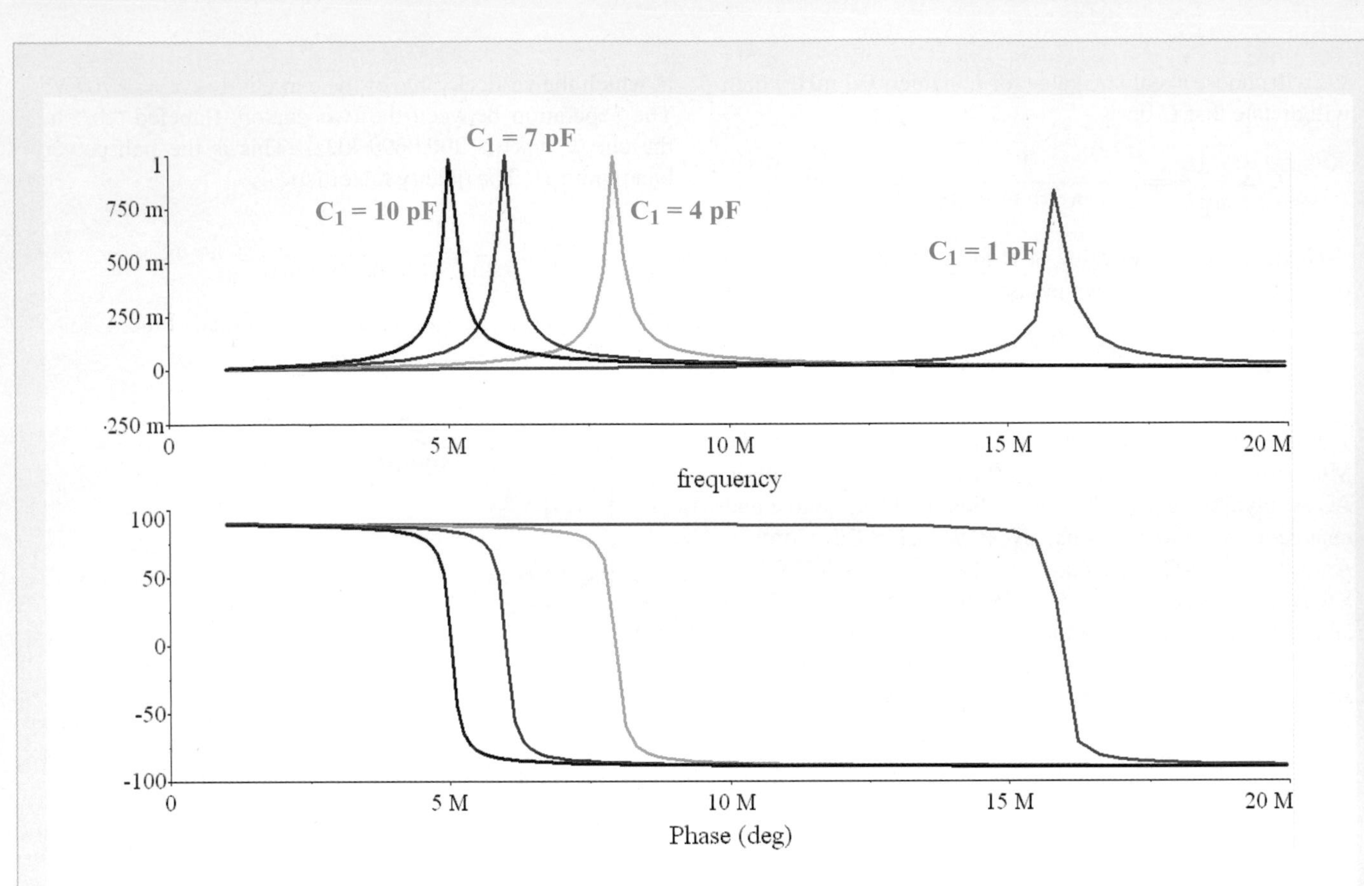

Figure 9-33: AC analysis plots for the circuit in Fig. 9-31 generated with the Parameter Sweep tool in Example 9-14. The capacitance was varied from 1 to 10 pF.

Example 9-14: Parameter Sweep

Apply Parameter Sweep to the circuit in Fig. 9-31 to generate spectral responses for C1 = 1 pF, 4 pF, 7 pF, and 10 pF.

Solution: Starting with the circuit in Fig. 9-31, we set V1 = 1 V. Next, we select Simulate → Analyses → Parameter Sweep. Upon selecting the parameter we wish to vary (capacitance C1), its minimum (1 pF), maximum (10 pF), step size (3 pF), and number of points (4), we select AC Analysis in the More Options box. This allows us to set the frequency range, the type of sweep (linear), and the number of points—just as we did previously in Example 9-13, except that the frequency range is 0 to 20 MHz. The Simulate command generates the plots shown in Fig. 9-33.

Example 9-15: Bode Plots

Reproduce the circuit of Fig. 9-26(a) in Multisim and generate Bode plots corresponding to the outputs of the three stages.

Solution: The circuit is reproduced in Fig. 9-34(a), and part (b) displays the results. In order to generate these plots, we use AC Analysis with
FSTART $= 10^4$ (rad/s)$/2\pi = 1.592$ kHz
and
FSTOP $= 10^7/2\pi = 1.592$ MHz.
The number of points was set to 200,
Sweep Type = Decade,
and
Vertical Scale = Decibel.

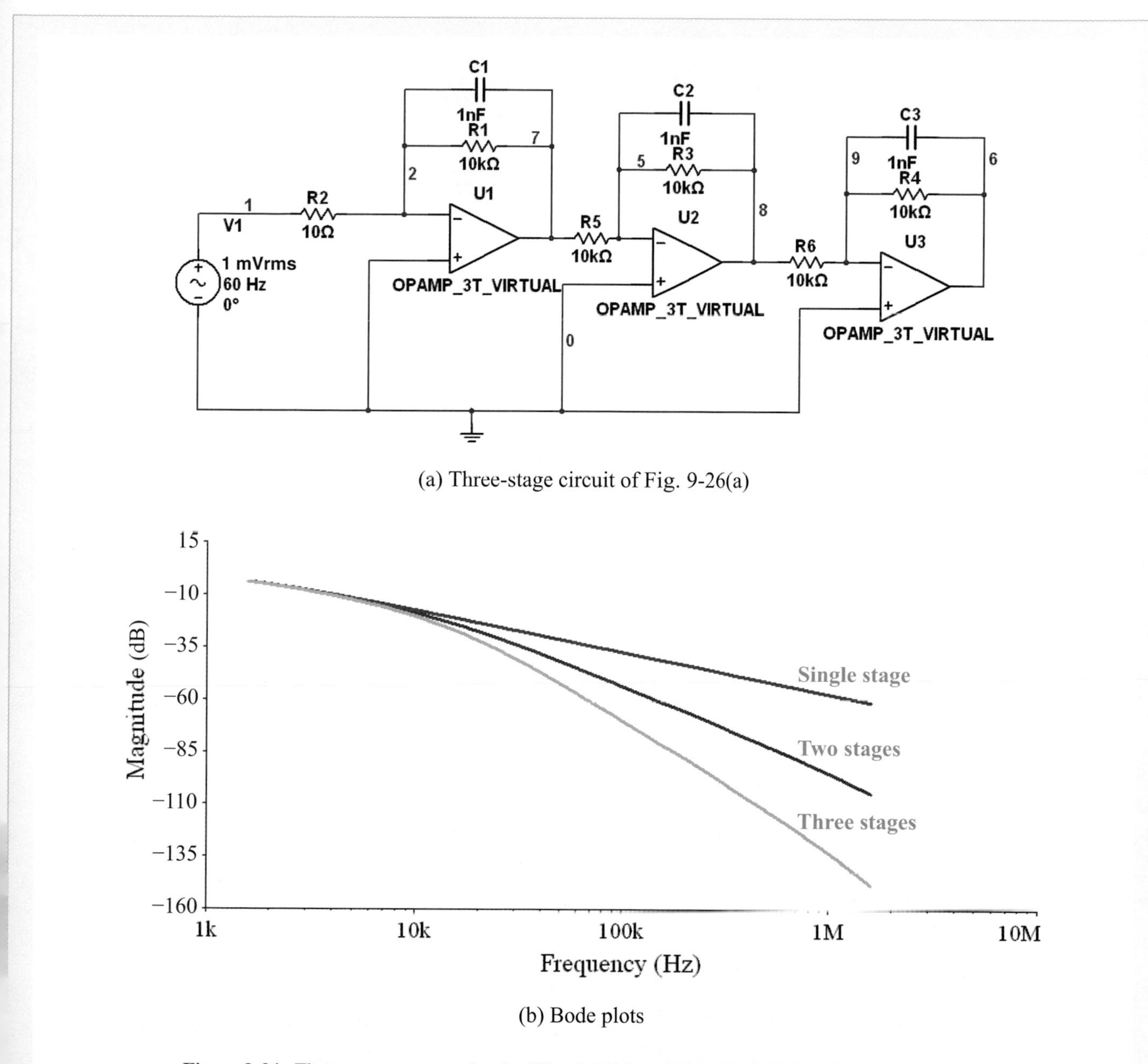

(a) Three-stage circuit of Fig. 9-26(a)

(b) Bode plots

Figure 9-34: Three-stage op-amp circuit of Fig. 9-26(a) reproduced in Multisim for Example 9-15.

Example 9-16: Bode Plotter Instrument

Use the Bode Plotter Instrument to generate magnitude and phase plots of the circuit in Fig. 9-31 over the frequency range of 8 to 12 MHz.

Solution: Go to Simulate → Instruments → Bode Plotter. Connect the "IN" terminals across the V1 source and connect the "OUT" terminals across the resistor R1. Bring up the Bode Plotter Instrument window. With the Magnitude Mode selected, set the horizontal scale to Lin (for linear), set I (initial frequency) to 8 MHz, and set F (final frequency) to 12 MHz. For the vertical scale, leave it on Log, set I to −50 dB, and set F to 5 dB. Select the Phase mode and for the vertical scale set I to −100 deg and F to 100 deg. Run the Interactive Simulation by pressing F5 or the appropriate button or toggle switch on the toolbar. In the Magnitude and Phase mode, you will generate plots similar to those shown in Fig. 9-35(a) and (b), respectively.

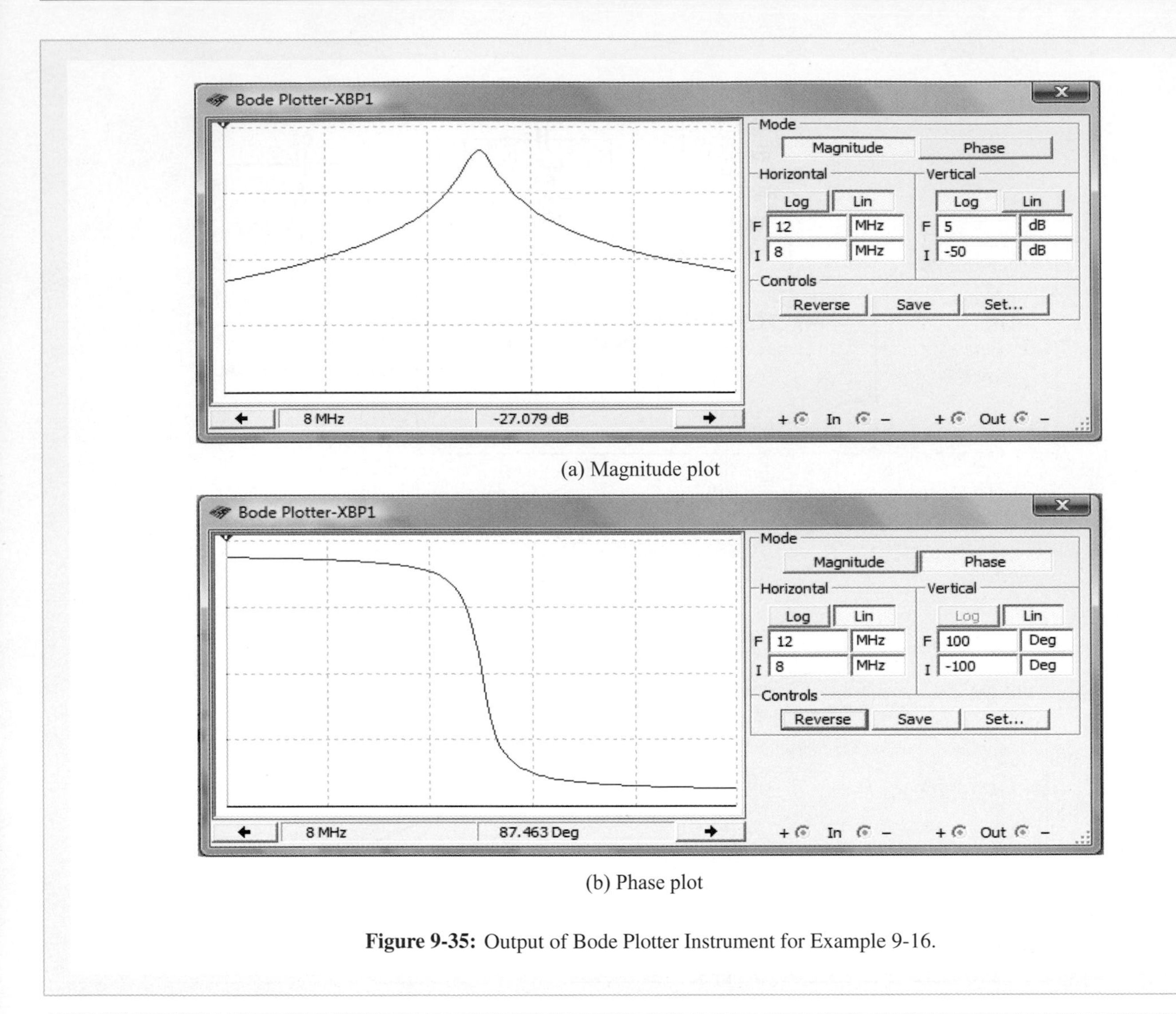

(a) Magnitude plot

(b) Phase plot

Figure 9-35: Output of Bode Plotter Instrument for Example 9-16.

Chapter 9 Relationships

Resonant Frequency ω_0

$$\Im\{\mathbf{Z}_{\text{in}}(\omega)\} = 0 \qquad @\ \omega = \omega_0$$

Magnitude and Frequency Scaling

$$R' = K_{\text{m}} R, \qquad L' = \frac{K_{\text{m}}}{K_{\text{f}}}\, L$$

$$C' = \frac{1}{K_{\text{m}} K_{\text{f}}}\, C, \qquad \omega' = K_{\text{f}}\omega$$

dB Scale

If $G = XY$ ➡ G [dB] $= X$ [dB] $+ Y$ [dB]

If $G = \dfrac{X}{Y}$ ➡ G [dB] $= X$ [dB] $- Y$ [dB]

Series and Parallel Bandpass RLC Filters

$$\omega_0 = \sqrt{\omega_{\text{c}_1}\omega_{\text{c}_2}} = \frac{1}{\sqrt{LC}}$$

$$Q = \frac{\omega_0 L}{R} \qquad \text{(series)}$$

$$Q = \frac{R}{\omega_0 L} \qquad \text{(parallel)}$$

Active Filters

Sections 9-6 and 9-7

CHAPTER HIGHLIGHTS

- The transfer function of a circuit is the ratio of a phasor output voltage or current to a phasor input voltage or current.
- The transfer function is characterized by magnitude and phase plots describing the spectral response of the circuit.
- At the resonant frequency ω_0, the input impedance of the circuit is purely real.
- The Bode diagram technique uses straight-line approximations on a semilog-scale to display the magnitude and phase spectra of the transfer function.
- The quality factor Q of a bandpass filter defines the degree of frequency selectivity of the filter.
- The order of a filter defines the gain roll-off rate of the magnitude spectrum in the stopband.
- Active filters are used primarily at frequencies below 1 MHz, whereas passive filters are better suited at higher frequencies.
- Active filters can provide power gain, and they easily can be cascaded in series or in parallel to generate the desired frequency response.
- In a superheterodyne receiver, the RF frequency is converted into an IF frequency for amplification and filtering prior to demodulation.
- Parameter sweep can be used in Multisim to compare the circuit response for different values of a key parameter.

GLOSSARY OF IMPORTANT TERMS

Provide definitions or explain the meaning of the following terms:

3-dB frequency
active filter
amplitude modulation
AM radio band
AND gate
bandpass filter
bandreject filter
bandwidth
Bode plot
carrier frequency
corner frequency
cutoff frequency
damping factor
dc gain
down-conversion
filter order
frequency conversion
frequency modulation
frequency response
frequency scaling
gain factor
gain roll-off rate
half-power frequency
high-frequency gain
highpass filter
IF
local oscillator
lowpass filter
magnitude scaling
magnitude response
mixer
OR gate
passband
passive filter
phase response
pole @ origin factor
pole factor
quadratic-pole factor
quadratic-zero factor
quality factor
resonant frequency
RF
simple-pole factor
simple-zero factor
software radio
stopband
transfer function
tuned radio frequency receiver
tuner
up-conversion
zero factor
zero @ origin factor

PROBLEMS

Section 9-1: Transfer Function

9.1 Determine the resonant frequency of the circuit shown in Fig. P9.1, given that $R = 100\ \Omega$, $L = 5$ mH, and $C = 1\ \mu$F.

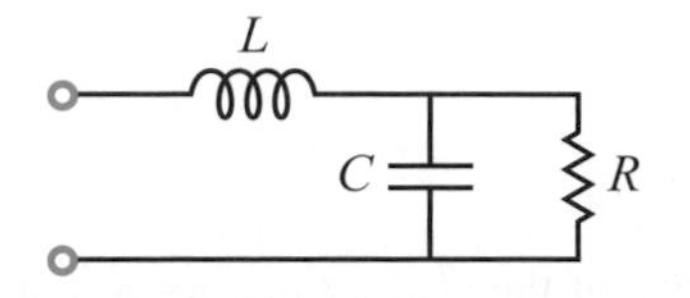

Figure P9.1: Circuit for Problem 9.1.

*Answer(s) in Appendix E.

9.2 Determine the resonant frequency of the circuit shown in Fig. P9.2, given that $R = 100\ \Omega$, $L = 5$ mH, and $C = 1\ \mu$F.

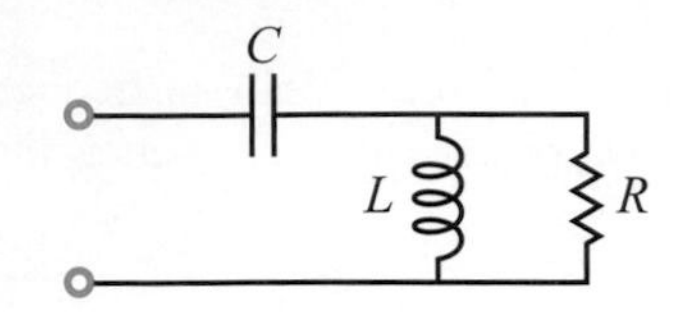

Figure P9.2: Circuit for Problem 9.2.

9.3 For the circuit shown in Fig. P9.3, determine (a) the transfer function $\mathbf{H} = \mathbf{V}_o/\mathbf{V}_i$ and (b) the frequency ω_o at which $\mathbf{H}$ is purely real.

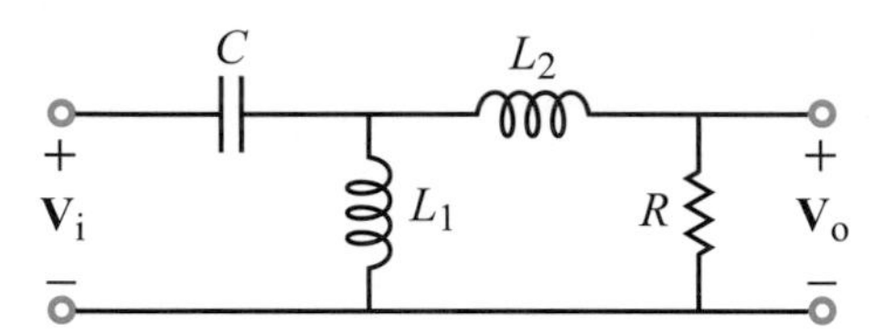

Figure P9.3: Circuit for Problem 9.3.

9.4 For the circuit shown in Fig. P9.4, determine (a) the transfer function $\mathbf{H} = \mathbf{V}_o/\mathbf{V}_i$ and (b) the frequency ω_o at which $\mathbf{H}$ is purely real.

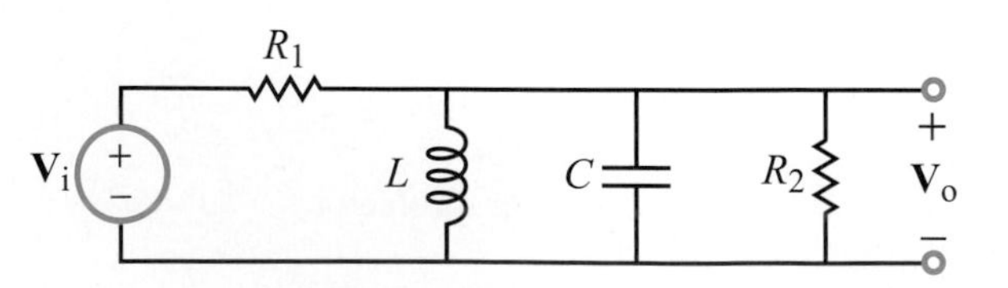

Figure P9.4: Circuit for Problem 9.4.

Section 9-2: Scaling

*__9.5__ What values of the scaling factors K_m and K_f should be applied to scale a circuit containing a 1-F capacitor and 4-H inductor into one containing 1 μF and 10 mH, respectively?

9.6 The corner frequency of the highpass-filter circuit shown in Fig. P9.6 is approximately 1 Hz. Scale the circuit up in frequency by a factor of 10^5 while keeping the values of the inductors unchanged.

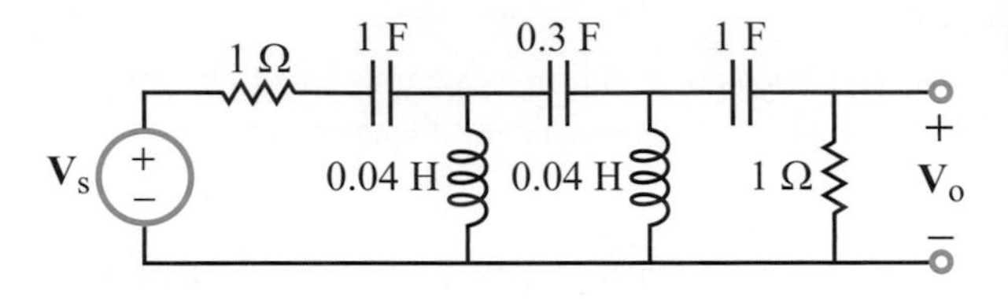

Figure P9.6: Circuit for Problem 9.6.

9.7 For the circuit shown in Fig. P9.7:

(a) Obtain an expression for the input impedance $\mathbf{Z}_{in}(\omega)$.

(b) If $R_1 = R_2 = 1\ \Omega$, $C = 1$ F, and $L = 5$ H, at what angular frequency is $\mathbf{Z}_{in}$ purely real?

(c) Scale the circuit by $K_m = 20$ and write down the new expression for the input impedance.

(d) Is the value of ω at which the input impedance of the scaled circuit the same or different from the answer of part (b)?

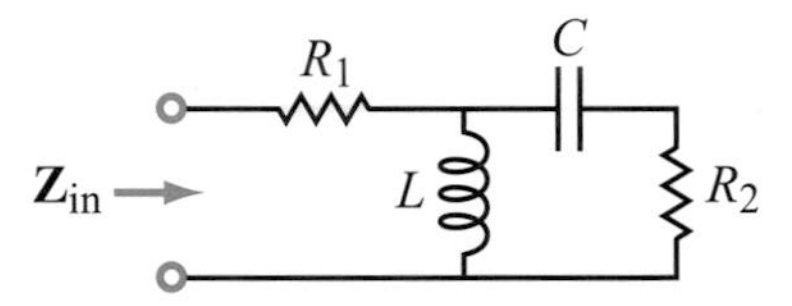

Figure P9.7: Circuit for Problem 9.7.

9.8 For the circuit shown in Fig. P9.8:

(a) Obtain an expression for the input impedance $\mathbf{Z}_{in}(\omega)$.

(b) If $R_1 = 1\ \Omega$, $R_2 = R_3 = 2\ \Omega$, $L = 1$ H, and $C = 1$ F, at what angular frequency is $\mathbf{Z}_{in}$ purely real?

(c) Redraw the circuit after scaling it by $K_m = 10^3$ and $K_f = 10^5$. Specify the new element values.

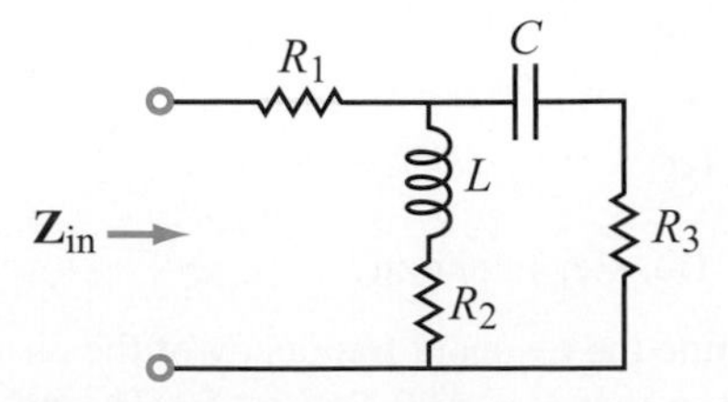

Figure P9.8: Circuit for Problem 9.8.

9.9 Circuit (b) in Fig. P9.9 is a scaled version of circuit (a). The scaling process may have involved magnitude or frequency scaling, or both simultaneously. If $R_1 = 1$ kΩ gets scaled to $R'_1 = 10$ kΩ, supply the impedance values of the other elements in the scaled circuit.

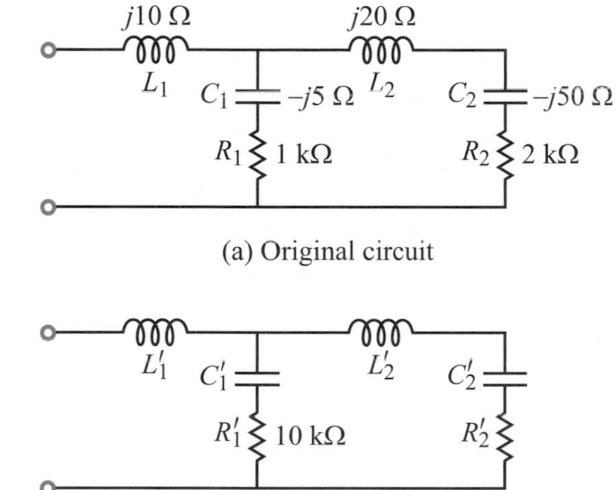

Figure P9.9: Circuits for Problem 9.9.

Section 9-3: Bode Plots

9.10 Convert the following power ratios to dB.

(a) 3×10^2

*(b) 0.5×10^{-2}

(c) $\sqrt{2000}$

(d) $(360)^{1/4}$

*(e) $6e^3$

(f) $2.3 \times 10^3 + 60$

(g) $24(3 \times 10^7)$

(h) $4/(5 \times 10^3)$

9.11 Convert the following voltage ratios to dB.

(a) 2×10^{-4}

(b) 3000

(c) $\sqrt{30}$

(d) $6/(5 \times 10^4)$

9.12 Convert the following dB values to voltage ratios.

(a) 46 dB

(b) 0.4 dB

*(c) −12 dB

(d) −66 dB

9.13 Generate Bode magnitude and phase plots for the following voltage transfer functions.

(a) $\mathbf{H}(\omega) = \dfrac{j100\omega}{10 + j\omega}$

(b) $\mathbf{H}(\omega) = \dfrac{0.4(50 + j\omega)^2}{(j\omega)^2}$

(c) $\mathbf{H}(\omega) = \dfrac{(40 + j80\omega)}{(10 + j50\omega)}$

(d) $\mathbf{H}(\omega) = \dfrac{(20 + j5\omega)(20 + j\omega)}{j\omega}$

(e) $\mathbf{H}(\omega) = \dfrac{30(10 + j\omega)}{(200 + j2\omega)(1000 + j2\omega)}$

(f) $\mathbf{H}(\omega) = \dfrac{j100\omega}{(100 + j5\omega)(100 + j\omega)^2}$

(g) $\mathbf{H}(\omega) = \dfrac{(200 + j2\omega)}{(50 + j5\omega)(1000 + j\omega)}$

9.14 Generate Bode magnitude and phase plots for the following voltage transfer functions.

(a) $\mathbf{H}(\omega) = \dfrac{4 \times 10^4(60 + j6\omega)}{(4 + j2\omega)(100 + j2\omega)(400 + j4\omega)}$

(b) $\mathbf{H}(\omega) = \dfrac{(1 + j0.2\omega)^2(100 + j2\omega)^2}{(j\omega)^3(500 + j\omega)}$

(c) $\mathbf{H}(\omega) = \dfrac{8 \times 10^{-2}(10 + j10\omega)}{j\omega(16 - \omega^2 + j4\omega)}$

(d) $\mathbf{H}(\omega) = \dfrac{4 \times 10^4\omega^2(100 - \omega^2 + j50\omega)}{(5 + j5\omega)(200 + j2\omega)^3}$

(e) $\mathbf{H}(\omega) = \dfrac{j5 \times 10^3\omega(20 + j2\omega)}{(2500 - \omega^2 + j20\omega)}$

(f) $\mathbf{H}(\omega) = \dfrac{512(1 + j\omega)(4 + j40\omega)}{(256 - \omega^2 + j32\omega)^2}$

(g) $\mathbf{H}(\omega) = \dfrac{j(10 + j\omega) \times 10^8}{(20 + j\omega)^2(500 + j\omega)(1000 + j\omega)}$

9.15 Determine the voltage transfer function $\mathbf{H}(\omega)$ corresponding to the Bode magnitude plot shown in Fig. P9.15.

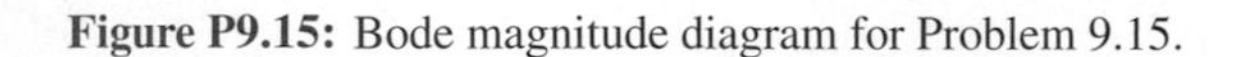

Figure P9.15: Bode magnitude diagram for Problem 9.15.

9.16 Determine the voltage transfer function $\mathbf{H}(\omega)$ corresponding to the Bode magnitude plot shown in Fig. P9.16. The phase of $\mathbf{H}(\omega)$ is 90° at $\omega = 0$.

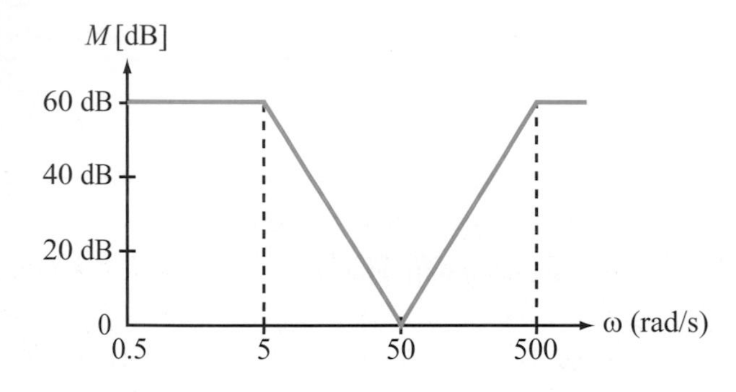

Figure P9.16: Bode magnitude plot for Problem 9.16.

*__9.17__ Determine the voltage transfer function $\mathbf{H}(\omega)$ corresponding to the Bode magnitude plot shown in Fig. P9.17. The phase of $\mathbf{H}(\omega)$ is 180° at $\omega = 0$.

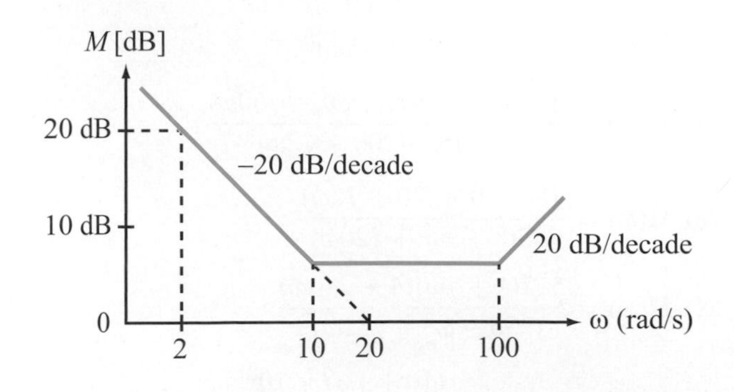

Figure P9.17: Bode magnitude plot for Problem 9.17.

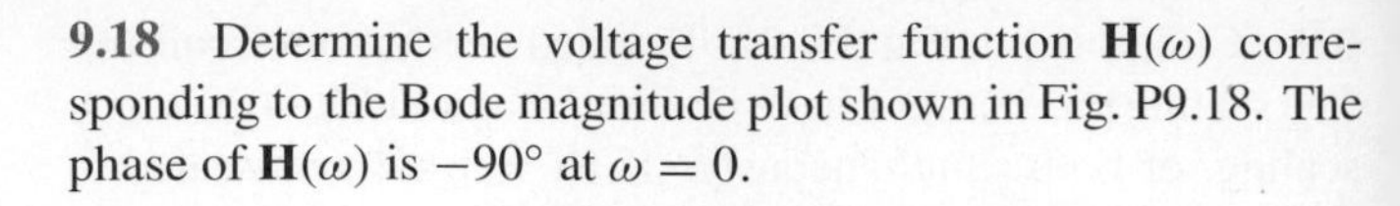

9.18 Determine the voltage transfer function $\mathbf{H}(\omega)$ corresponding to the Bode magnitude plot shown in Fig. P9.18. The phase of $\mathbf{H}(\omega)$ is −90° at $\omega = 0$.

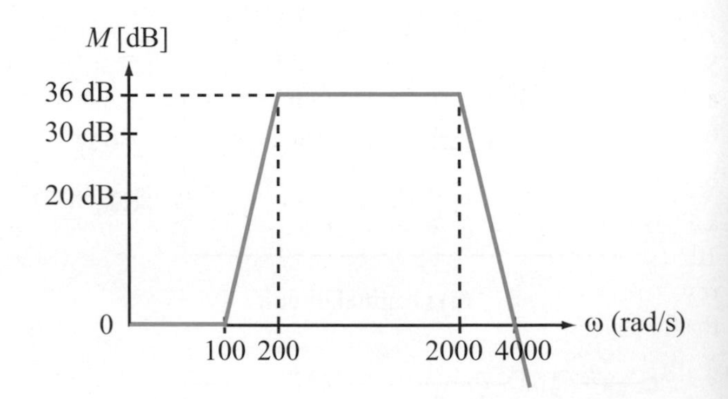

Figure P9.18: Bode magnitude plot for Problem 9.18.

9.19 Determine the voltage transfer function $\mathbf{H}(\omega)$ corresponding to the Bode magnitude plot shown in Fig. P9.19. The phase of $\mathbf{H}(\omega)$ is 0° at $\omega = 0$.

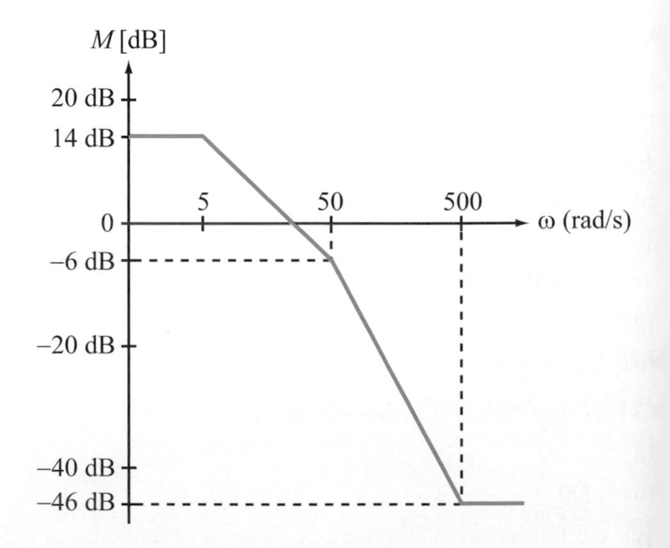

Figure P9.19: Bode magnitude plot for Problem 9.19.

Sections 9-4 and 9-5: Passive Filters

9.20 The element values of a series RLC bandpass filter are $R = 5\ \Omega$, $L = 20$ mH, and $C = 0.5\ \mu$F.

(a) Determine ω_0, Q, B, ω_{c_1}, and ω_{c_2}.

(b) Is it possible to double the magnitude of Q by changing the values of L and/or C while keeping ω_0 and R unchanged? If yes, propose such values, and if no, why not?

9.21 A series RLC bandpass filter has half-power frequencies at 1 kHz and 10 kHz. If the input impedance at resonance is 6 Ω, what are the values of R, L, and C?

9.22 A series RLC circuit is driven by an ac source with a phasor voltage $\mathbf{V}_s = 10\angle 30^\circ$ V. If the circuit resonates at 10^3 rad/s and the average power absorbed by the resistor at resonance is 2.5 W, determine the values of R, L, and C, given that $Q = 5$.

***9.23** The element values of a parallel RLC circuit are $R = 100\ \Omega$, $L = 10$ mH, and $C = 0.4$ mF. Determine ω_0, Q, B, ω_{c_1}, and ω_{c_2}.

9.24 Design a parallel RLC filter with $f_0 = 4$ kHz, $Q = 100$, and an input impedance of 25 kΩ at resonance.

9.25 For the circuit shown in Fig. P9.25:

(a) Obtain an expression for $\mathbf{H}(\omega) = \mathbf{V}_o/\mathbf{V}_i$ in standard form.

(b) Generate spectral plots for the magnitude and phase of $\mathbf{H}(\omega)$, given that $R_1 = 1\ \Omega$, $R_2 = 2\ \Omega$, $C_1 = 1\ \mu$F, and $C_2 = 2\ \mu$F.

(c) Determine the cutoff frequency ω_c and the slope of the magnitude (in dB) when $\omega/\omega_c \ll 1$ and when $\omega/\omega_c \gg 1$.

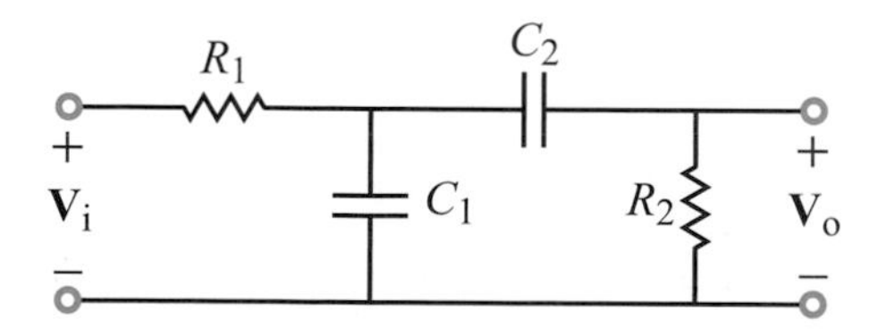

Figure P9.25: Circuit for Problem 9.25.

9.26 For the circuit shown in Fig. P9.26:

(a) Obtain an expression for $\mathbf{H}(\omega) = \mathbf{V}_o/\mathbf{V}_i$ in standard form.

(b) Generate spectral plots for the magnitude and phase of $\mathbf{H}(\omega)$, given that $R_1 = 1\ \Omega$, $R_2 = 2\ \Omega$, $L_1 = 1$ mH, and $L_2 = 2$ mH.

(c) Determine the cutoff frequency ω_c and the slope of the magnitude (in dB) when $\omega/\omega_c \ll 1$ and when $\omega/\omega_c \gg 1$.

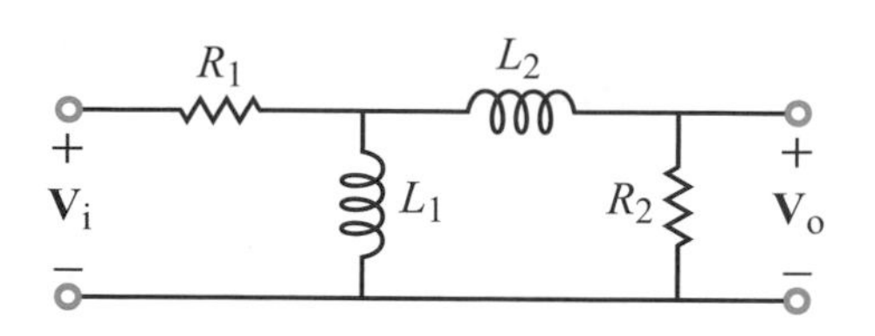

Figure P9.26: Circuit for Problem 9.26.

9.27 For the circuit shown in Fig. P9.27:

(a) Obtain an expression for $\mathbf{H}(\omega) = \mathbf{V}_o/\mathbf{V}_i$ in standard form.

(b) Generate spectral plots for the magnitude and phase of $\mathbf{H}(\omega)$, given that $R = 100\ \Omega$, $L = 0.1$ mH, and $C = 1\ \mu$F.

(c) Determine the cutoff frequency ω_c and the slope of the magnitude (in dB) when $\omega/\omega_c \gg 1$.

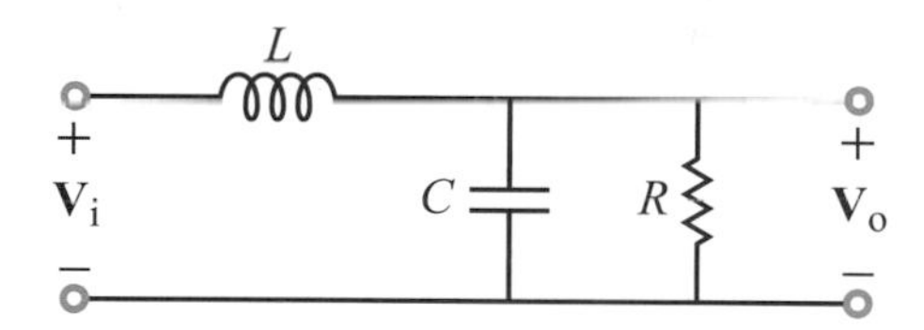

Figure P9.27: Circuit for Problem 9.27.

9.28 For the circuit shown in Fig. P9.28:

(a) Obtain an expression for $\mathbf{H}(\omega) = \mathbf{V}_o/\mathbf{V}_i$ in standard form.

(b) Generate spectral plots for the magnitude and phase of $\mathbf{H}(\omega)$, given that $R = 10\ \Omega$, $L = 1$ mH, and $C = 10\ \mu$F.

(c) Determine the cutoff frequency ω_c and the slope of the magnitude (in dB) when $\omega/\omega_c \ll 1$.

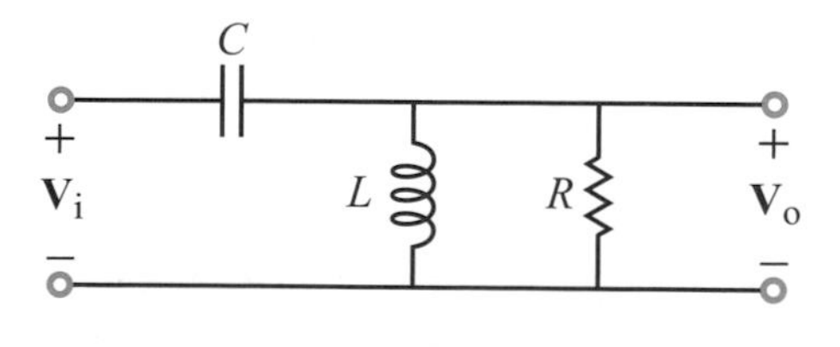

Figure P9.28: Circuit for Problem 9.28.

9.29 For the circuit shown in Fig. P9.29:

***(a)** Obtain an expression for $\mathbf{H}(\omega) = \mathbf{V}_o/\mathbf{V}_i$ in standard form.

(b) Generate spectral plots for the magnitude and phase of $\mathbf{H}(\omega)$, given that $R = 50\ \Omega$ and $L = 2$ mH.

(c) Determine the cutoff frequency ω_c and the slope of the magnitude (in dB) when $\omega/\omega_c \ll 1$.

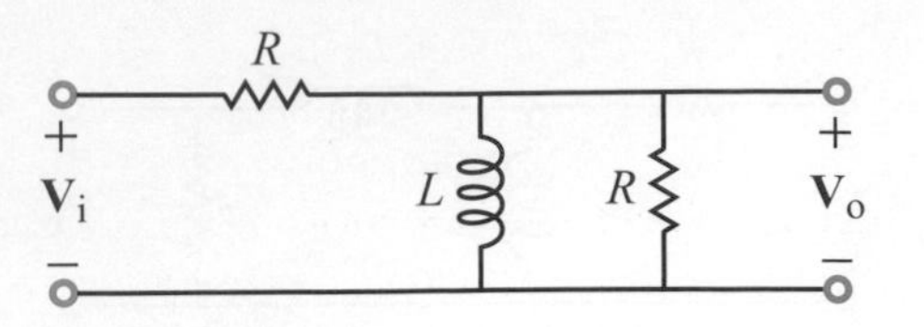

Figure P9.29: Circuit for Problem 9.29.

9.30 For the circuit shown in Fig. P9.30:

(a) Obtain an expression for $\mathbf{H}(\omega) = \mathbf{V}_o/\mathbf{V}_i$ in standard form.

(b) Generate spectral plots for the magnitude and phase of $\mathbf{H}(\omega)$, given that $R = 50\ \Omega$ and $L = 2$ mH.

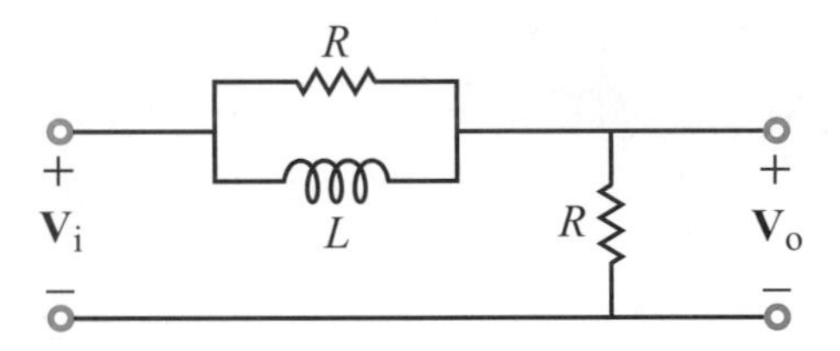

Figure P9.30: Circuit for Problem 9.30.

Sections 9-6 and 9-7: Active Filters

9.31 For the op-amp circuit of Fig. P9.31:

(a) Obtain an expression for $\mathbf{H}(\omega) = \mathbf{V}_o/\mathbf{V}_s$ in standard form.

(b) Generate spectral plots for the magnitude and phase of $\mathbf{H}(\omega)$, given that $R_1 = 1$ kΩ, $R_2 = 4$ kΩ, and $C = 1\ \mu$F.

(c) What type of filter is it? What is its maximum gain?

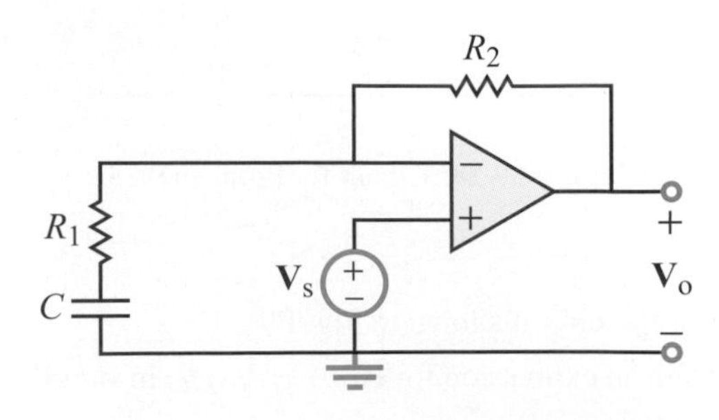

Figure P9.31: Circuit for Problem 9.31.

9.32 For the op-amp circuit of Fig. P9.32:

(a) Obtain an expression for $\mathbf{H}(\omega) = \mathbf{V}_o/\mathbf{V}_s$ in standard form.

(b) Generate spectral plots for the magnitude and phase of $\mathbf{H}(\omega)$, given that $R_1 = 99$ kΩ, $R_2 = 1$ kΩ, and $C = 0.1\ \mu$F.

(c) What type of filter is it? What is its maximum gain?

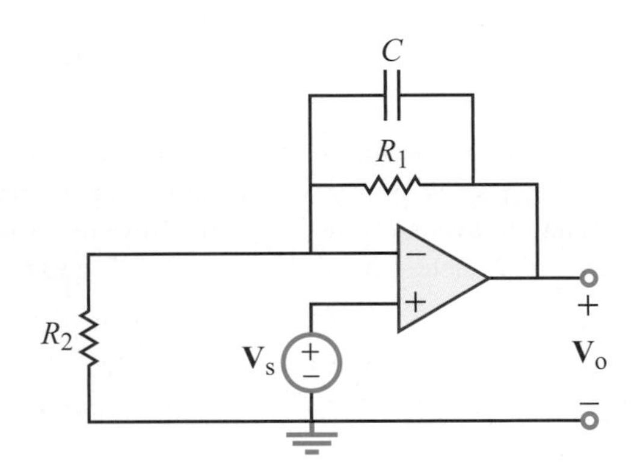

Figure P9.32: Circuit for Problem 9.32.

9.33 For the op-amp circuit of Fig. P9.33:

(a) Obtain an expression for $\mathbf{H}(\omega) = \mathbf{V}_o/\mathbf{V}_i$ in standard form.

(b) Generate spectral plots for the magnitude and phase of $\mathbf{H}(\omega)$, given that $R_1 = R_2 = 100\ \Omega$, $C_1 = 10\ \mu$F, and $C_2 = 0.4\ \mu$F.

(c) What type of filter is it? What is its maximum gain?

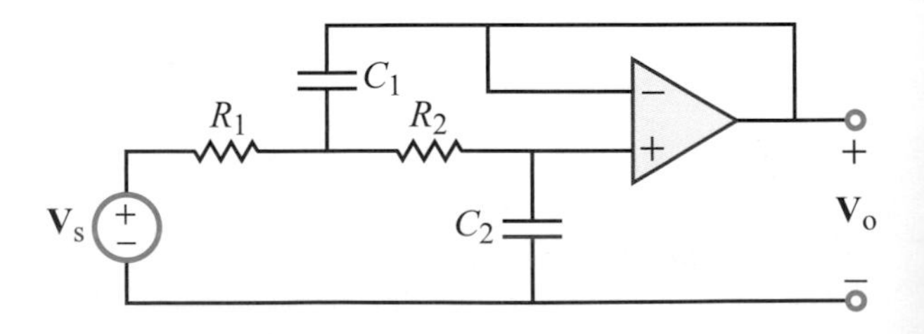

Figure P9.33: Circuit for Problems 9.33 and 9.34.

9.34 Repeat Problem 9.33 after interchanging the values o C_1 and C_2 to $C_1 = 0.4\ \mu$F and $C_2 = 10\ \mu$F.

9.35 For the op-amp circuit of Fig. P9.35:

*(a) Obtain an expression for $\mathbf{H}(\omega) = \mathbf{V}_o/\mathbf{V}_s$ in standard form.

(b) Generate spectral plots for the magnitude and phase of $\mathbf{H}(\omega)$, given that $R_1 = 1$ kΩ, $R_2 = 20$ Ω, $C_1 = 5$ μF, and $C_2 = 25$ nF.

(c) What type of filter is it? What is its maximum gain?

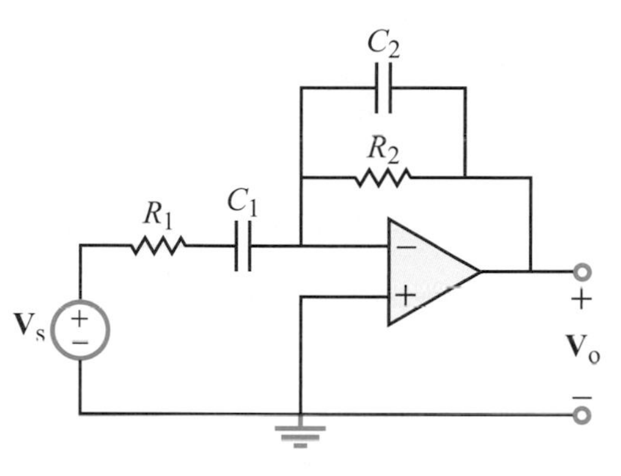

Figure P9.35: Circuit for Problem 9.35.

9.36 Design an active lowpass filter with a gain of 4, a corner frequency of 1 kHz, and a gain roll-off rate of −60 dB/decade.

9.37 Design an active highpass filter with a gain of 10, a corner frequency of 2 kHz, and a gain roll-off rate of 40 dB/decade.

9.38 The element values in the circuit of the second-order bandpass filter shown in Fig. P9.38 are $R_{f_1} = 100$ kΩ, $R_{s_1} = 10$ kΩ, $R_{f_2} = 100$ kΩ, $R_{s_2} = 10$ kΩ, $C_{f_1} = 3.98 \times 10^{-11}$ F, and $C_{s_2} = 7.96 \times 10^{-11}$ F. Generate a spectral plot for the magnitude of $\mathbf{H}(\omega) = \mathbf{V}_o/\mathbf{V}_s$. Determine the frequency locations of the maximum value of M [dB] and its half-power points.

Section 9-8: Superheterodyne Receiver

9.39 Using the circuit layout shown in Fig. 9-15, design a tuner that uses a variable inductor, a capacitor, and a resistor. The input impedance of the tuner should be 377 Ω at 1 MHz, and its bandwidth should be 2 percent.

***9.40** What range of frequencies should the local oscillator be able to provide to mix the FM radio range (87 to 102 MHz) down to 10 MHz?

Section 9-9: Multisim

9.41 Generate plots in Multisim for the magnitude and phase of the transfer function for a series bandpass filter with $L = 1$ mH, $f_0 = 1$ MHz, and $Q = 10$. Choose FSTART = 100 kHz and FSTOP = 10 MHz.

9.42 Perform a Parameter Sweep in Multisim for capacitor C_{s_2} of the two-stage bandpass filter shown in Fig. 9-27. Generate response plots from 10 Hz to 100 kHz for each of five equally spaced values of C_{s_2} starting at 1 nF and ending at 15 nF.

9.43 Use Multisim to generate spectral plots for the magnitudes and phases of voltages $\mathbf{V}_C$ and $\mathbf{V}_o$ in the circuit of Fig. P9.43. The circuit is a second-order passive lowpass filter followed by an active highpass filter. Use the following element values: $R_1 = 20.3$ Ω, $R_2 = R_3 = 1.592$ kΩ, $L_1 = 100$ nH, $C_1 = C_2 = 1$ nF, and $v_s(t) = \cos 2\pi f t$ V.

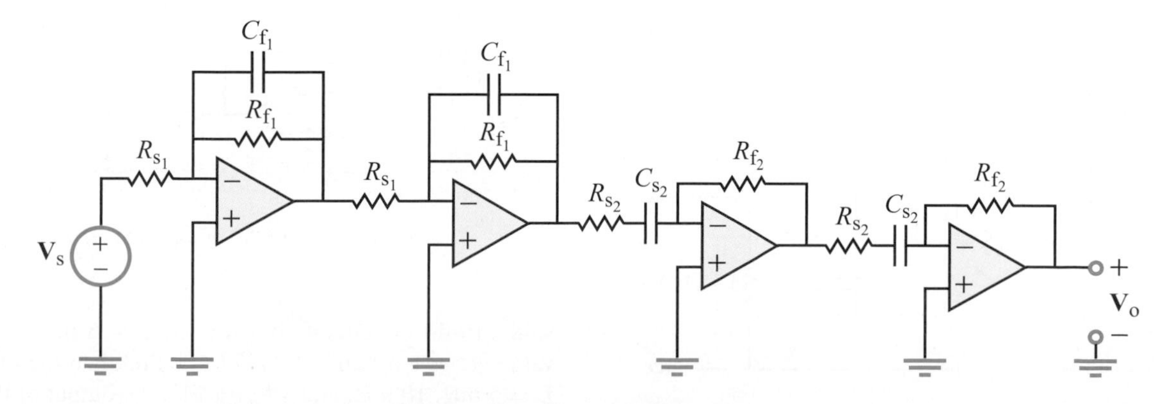

Figure P9.38: Circuit for Problem 9.38.

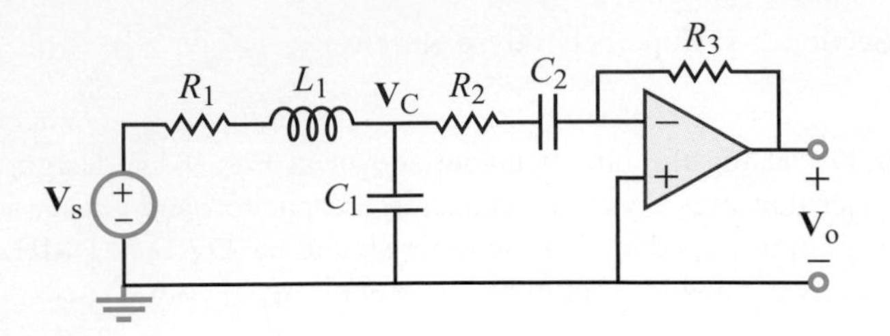

Figure P9.43: Circuit for Problem 9.43.

9.44 For the circuit in Fig. P9.44, use Multisim to generate spectral plots for the magnitude and phase of $\mathbf{H}(\omega) = \mathbf{V}_o/\mathbf{V}_i$ over the range from 100 Hz to 100 kHz. When performing the AC Analysis, use 200 points per decade. Determine the frequencies at which M [dB] is a maximum or a minimum.

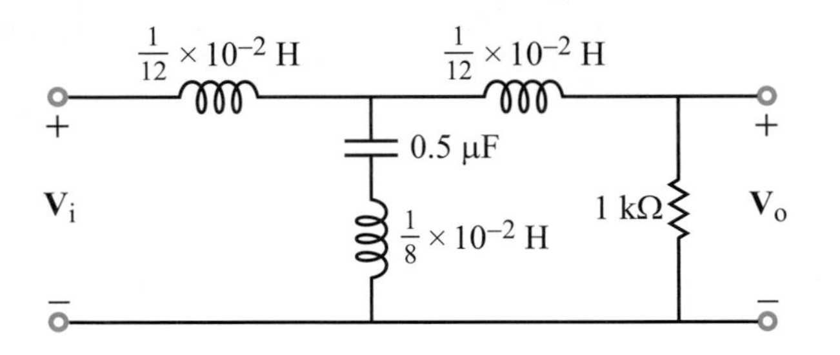

Figure P9.44: Circuit for Problem 9.44.

9.45 For the circuit in Fig. P9.45, use Multisim to generate spectral plots for the magnitude and phase of $\mathbf{H}(\omega) = \mathbf{V}_o/\mathbf{V}_i$ over the range from 1 to 15 kHz. When performing the AC Analysis, use 10^4 points in linear scan. Determine the frequencies at which M [dB] is a maximum or a minimum.

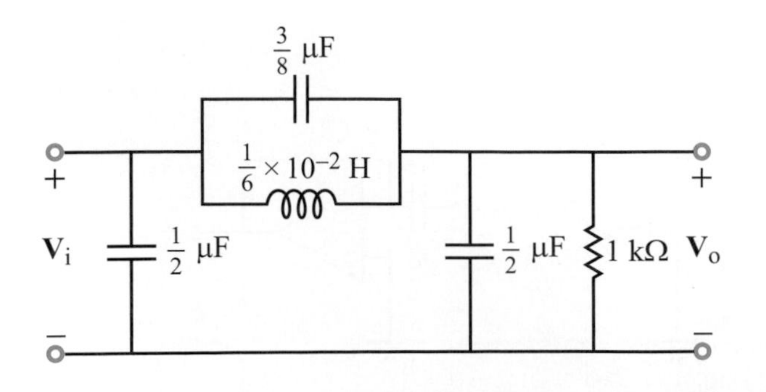

Figure P9.45: Circuit for Problem 9.45.

9.46 For the circuit in Fig. P9.46, use Multisim to generate spectral plots for the magnitude and phase of $\mathbf{H}(\omega) = \mathbf{V}_o/\mathbf{V}_i$ over the range from 100 Hz to 10 kHz. When performing the AC Analysis, use 200 points per decade. Determine the frequencies at which M [dB] is a maximum or a minimum.

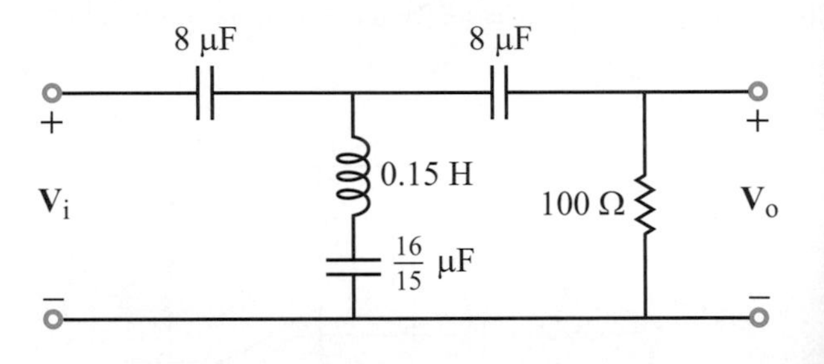

Figure P9.46: Circuit for Problem 9.46.

9.47 Figure P9.47 depicts a band-stop filter composed of a high-pass filter, a low-pass filter, and a summing amplifier. Construct it in Multisim using the values $R = 5$ kΩ, $C_{LP} = 26$ nF, and $C_{HP} = 1$ nF. Using Multisim's AC Analysis, plot the transfer function from 100 Hz to 100 kHz for the high-pass component alone, the low-pass component alone, and the overall filter all on the same graph. Find the frequency where minimum gain occurs for the overall filter.

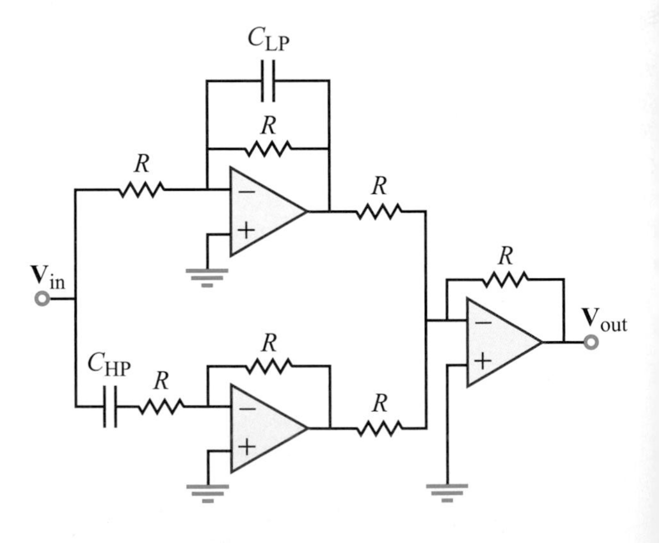

Figure P9.47: Circuit for Problem 9.47.

9.48 Build the circuit shown in Fig. 9-15 in Multisim wit values $C = 1$ pF and $R = 377$ Ω. Simulate the circuit wit $L = 5$ mH, 10 mH, and 15 mH. Plot the output of the filter a the three tunings on the same plot from 100 kHz to 100 MHz

Putting It All Together: Computer Architecture and the Digital Camera

This book covers many topics in circuit analysis and design, so it is only natural to wonder how they all fit together and how they are used in practice. Even though we have attempted to highlight the answers to these questions through the various Technology Briefs and Application Notes, we felt it would be helpful to pull much of these concepts together to show you how the various disciplines in electrical engineering come together to make the systems that power your world. To this end, we have chosen the digital camera as a platform with which to illustrate where all of the book's concepts become relevant in standard practice.

What Do Cameras, Phones, and Computers Have in Common?

The increasing pace of technological miniaturization (Technology Briefs 1 and 2 on pages 10 and 20, respectively) is enabling ever more functions to be packaged into single products. Digital cameras, mobile phones, and personal data assistants (PDAs) are examples of how extremely small sensing, computation, and communication elements enable the creation of indispensable technology. If you think about it a bit, you may begin to see that all of these devices are actually very similar. Even to non-engineers, the line between a cell phone, a computer, a PDA and a digital camera is starting to blur. With extreme miniaturization, even the distinction between environment and computer blurs (Technology Brief 19 on page 442).

In broad terms, all engineered systems perform the functions shown in Fig. P-1. The system converts energy to and from its environment into electrical signals; this is known as ***transduction***. As we saw in Technology Brief 12 on page 269, if environmental signals are converted into electrical signals, this is known as ***sensing***. If the conversion is from the system's electrical signals into the environment, it is usually termed ***actuation***. The internal manipulation of the electrical signals (or, rather, the information contained in the signals) by the system is known as ***computation***. Some of this information must be stored for later use; this is known as generally as ***storage*** or sometimes as ***memory***. All of the energy required to perform these transduction, computation, and storage functions must come from some source of ***energy***. Some of the signals sensed and generated by the system are meant explicitly for communicating with other systems; this is known as ***communication***.

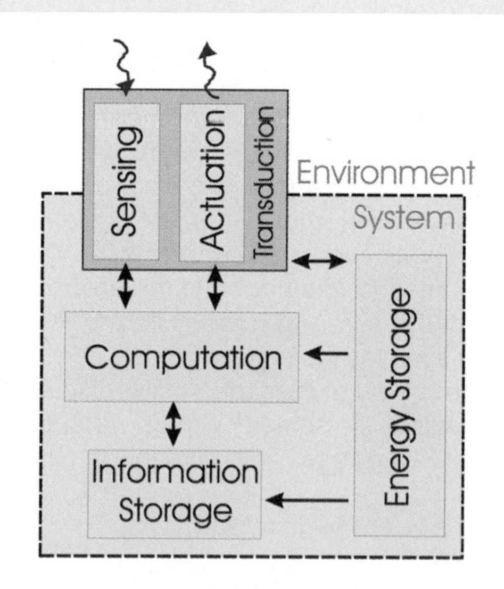

Figure P-1: A schematic of the functions and energy flow in a system from an electrical-engineering perspective. This very general model applies equally well to a digital camera, a laptop computer, or a human.

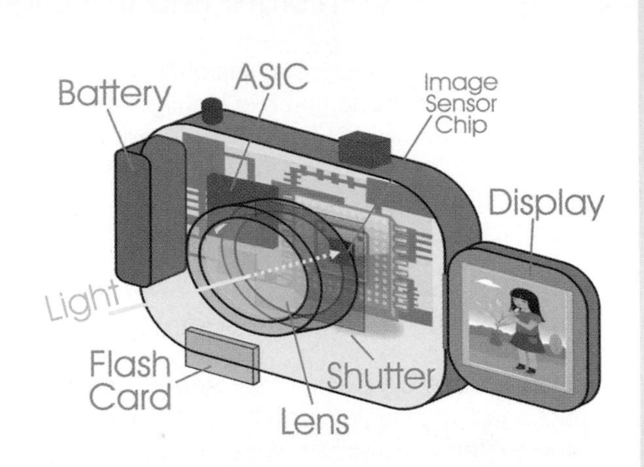

Figure P-2: A conceptual diagram of a digital camera. Light enters the camera through a series of lenses and strikes the CMOS or CCD imaging sensor. This sensor converts the light signals to electrical signals and sends the information to the ASIC, which modifies the image data as necessary and then stores or communicates it to the user. An LED display shows the user the image striking the sensor and allows data entry through a touchscreen.

Let us consider two example systems: a mobile phone and you. A mobile phone has a set of sophisticated integrated circuits (ICs, as described in Technology Brief 7 on page 135) that sense an electromagnetic signal from the environment (radio waves, Technology Brief 8 on page 158) through an antenna, then perform many computations on that signal. The resulting electrical signal is used to actuate a device that produces pressure waves in the air (i.e., sound). Some of this information, like voicemail or images, can be stored in memory (Technology Briefs 9 and 10 on pages 163 and 190, respectively). All of these functions are possible only if there is a battery connected to provide energy (Technology Brief 11 on page 199). Modern mobile phones transduce many other signals as well: inertial sensors detect motion and position (Technology Brief 12 on page 269), radio waves are used to provide very accurate position through the GPS system, and onboard optical sensors allow the user to take pictures with the phone. You are not so different (but vastly more sophisticated, of course). You transduce signals from the environment through your sensory organs which generate signals for your neural system. Your neural system processes these signals and performs computational operations. These computations result in transductions back into the environment through your limbs. None of this is possible without the energy stored in your body as fat and carbohydrates. You often use the same transducers that you use for sensing to communicate with other systems (i.e., your friends).

This analogy between man and machine is very significant. As we discuss the digital camera example in the next few pages, it is well worth asking to what extent the same concepts we apply to analyzing electrical systems can be applied to biological ones (Technology Brief 22 on page 536). With this in mind, let us take a look at the digital camera.

The Digital Camera

Figure P-2 shows a conceptual mechanical view of where things are placed in a hypothetical camera; the specifics vary from manufacturer to manufacturer. Figure P-3 shows a simplified block diagram of the various components in a hypothetical digital camera; for each component, the relevant Technology Brief or chapter section is highlighted in red. A digital camera (like a mobile phone or a PDA) is essentially a computer with specialized input/output capabilities;

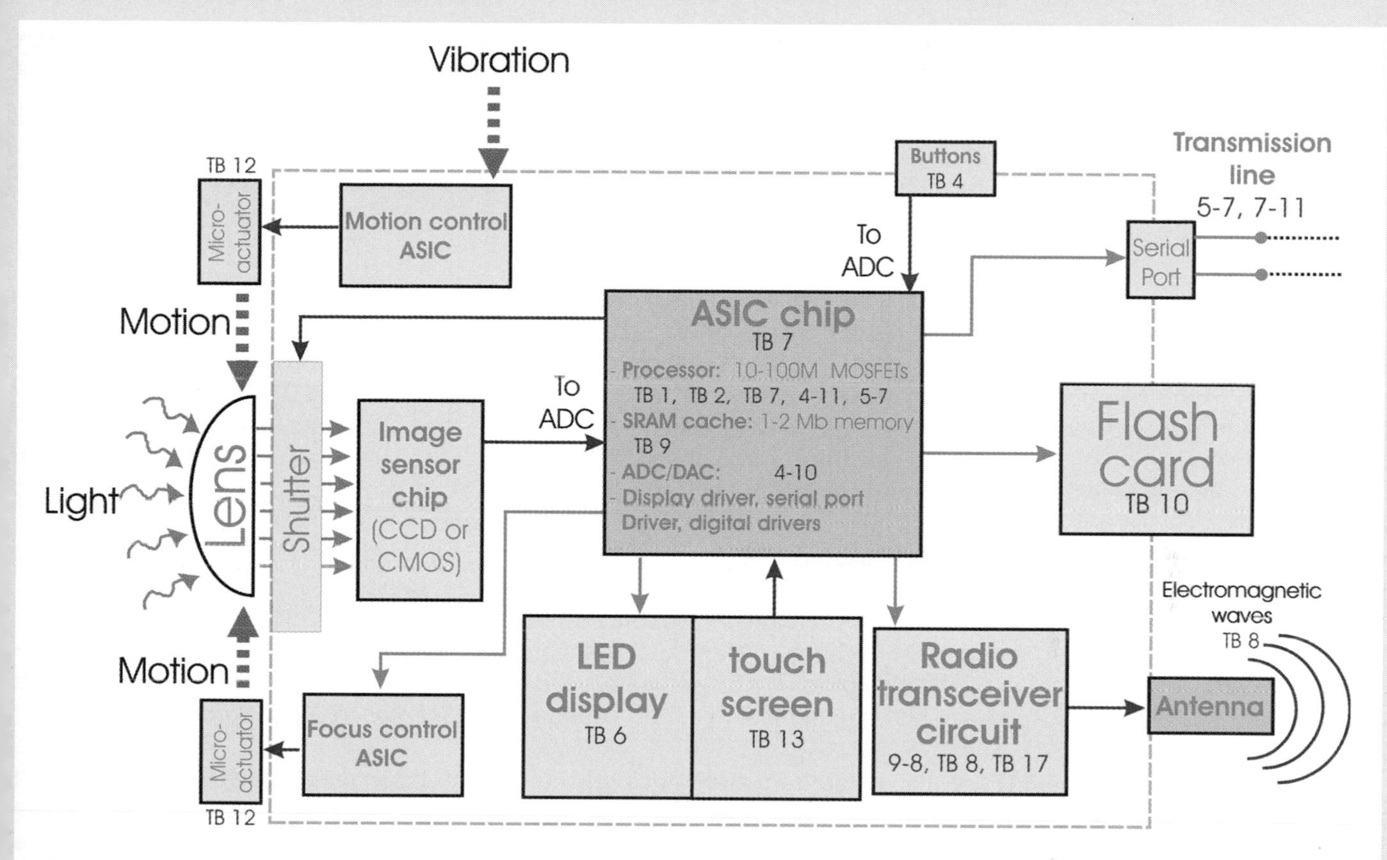

Figure P-3: A functional block diagram of the various components of the digital camera. Red arrows indicate analog electrical signals, while blue arrows indicate digital electrical signals. Broken arrows indicate transduction is occurring between a component and the outside environment. Relevant sections in the book are highlighted in red text.

detailed version of Fig. P-3, including connection specifications, processor details and the like, could be termed the computer architecture. Light enters the camera through an optical system that consists of one or more motorized lenses and mirrors. There are many subtleties to how these multiple-lens systems are built that affect both the quality of the image and the user interface. The optical system projects the light onto an imaging sensor. These sensors are custom-built integrated circuits that have light-sensitive pixels (numbering in the millions of pixels, or megapixels, in modern cameras). The imaging chips convert the light signal into electrical signals. These electrical signals are then run through an analog-to-digital converter (ADC) and sent to the computational element, an ***application-specific integrated circuit (ASIC)***. The ASIC can perform various operations with the digital image data. The data can be altered to perform color correction or other operations. Data can be stored locally (usually on a hard drive or flash), communicated over a wire to a computer, communicated wirelessly, and/or displayed on the camera's LED screen for the user to see.

Transduction: Optics and Motion Compensation (Technology Brief 12, Section 6-7)

Various configurations of lenses, mirrors, and apertures guide the light from the field of view onto the imaging sensor. Most of that detail is beyond the scope of this overview. Of note are cameras with motion compensation. During the motion-compensation operation, an on-board microfabricated inertial sensor (an accelerometer; see Technology Brief

12 on page 269) senses the changes in acceleration of the camera. The vibration information then is used to compute control signals that are sent to a microfabricated actuator (Technology Brief 12 on page 269) that moves the optical lens system in real time to counteract the artifacts caused by vibration or high frequency motion of the camera (i.e., shaking or vehicle motion).

Transduction: Image Sensor

The image sensor is a crucial component in the digital camera. This device consists of a chip, usually manufactured in silicon in a process similar to the one that makes ICs (Technology Brief 7 on page 135). When an image is focused onto the chip, millions of individual pixels record the light levels at their location and encode the levels into electrical signals. Currently, there are two competing technologies of sensors: ***charge-coupled devices (CCD)*** and ***complementary metal oxide semiconductor (CMOS)*** detectors. In CCD chips, each light-sensitive pixel is a capacitor where the two conducting elements of the capacitor are built from silicon on the chip using IC fabrication (Technology Brief 7 on page 135). Its operation is basically governed by the same physical property that makes solar cells possible: the photoelectric effect. Because silicon is a semiconductor, photons striking the silicon surface generate electrons. As photons hit each silicon capacitor, the generated electrons build up charge in the capacitor proportional to the intensity of the light at that location. To take a picture, the CCD array is exposed to an image for a set period of time (the exposure time). Once this is done, the charge in each pixel is dumped, in sequence, through an amplifier. This amplifier generates a voltage output proportional to the amount of charge moving out of each capacitor; it is, in effect, a current-to-voltage amplifier (Example 4-2). The output of the amplifier is a sequence of voltage pulses; each pulse represents the amount of charge on one of the CCD pixels. This voltage pulse must be sent through an analog-to-digital converter before it can be processed by the ASIC.

CMOS sensors (Section 4-11) are also made from silicon using IC fabrication techniques. However, the standard CMOS sensor has CMOS transistors built into each pixel. Recall from Section 5-7 that MOSFET transistors have associated parasitic capacitances; when light strikes the silicon, these capacitances will charge up as in a CCD. Unlike a CCD, the transistors sample the charge locally on each pixel and generate a digital "1" or "0" depending on the amount of charge. This eliminates the need for the external ADC used for CCDs. Also, the transistors enable the use of row and column addressing of the pixels (as in the dynamic memory of Technology Brief 9 on page 163), which also speeds up the device (because there is no need to wait for a sequence of voltage pulses as in a CCD). This technology only became commercially viable as transistor sizes decreased below 0.25 μm (Technology Briefs 1 and 2 on pages 10 and 20, respectively). CMOS sensors are increasingly displacing CCDs in the commercial market although they still suffer from noise and dynamic-range limitations when compared with CCDs.

In order to obtain color images, different pixels are coated with different filter materials that pass through only certain colors. For example, the common Bayer filter method coats a pixel in either a red, green, or blue filter. These red, green, or blue pixels are then tiled together. The ASIC can then reconstruct the image from the different color pixels. More sophisticated methods include using three different sensor arrays (one for each color), using more color filter types or even performing three exposures in succession onto the same sensor chip with a different color filter each time.

Transduction: LED Screen (Technology Briefs 5 and 6)

Most commercial digital cameras now display the image seen through the sensor using a miniature LED display (Technology Briefs 5 and 6 on pages 96 and 106, respectively). These allow the user to choose camera settings playback movies or images, modify the images, or control communications. The screen is driven by an ASIC.

Computation: ASIC (Technology Brief 7, Sections 4-11, 5-7, and 11-10)

The application specific integrated circuit (ASIC) is the brains of the camera. The ASIC chip usually has a microprocessor, on-chip memory, ADC and DAC circuits, and communication circuits. Like any IC, the ASIC chip is made with silicon fabrication processes (Technology Brief 7 on page 135). The ASIC contains thousands or millions of MOSFET transistors (Sections 4-11 and 5-7) and increasingly, mixed signal circuits (Section 11-10) are used for some of the ADC and DAC functions (such as the ADC required for the CCD, described earlier). These circuits are designed and analyzed in SPICE simulators like Multisim 10.

Once the image has been converted to digital format (by the ADC) the ASIC performs computations on it. These include interpolating colors from the different color pixels, smoothing, color correction of the data, and compression (for example, compressing the image into the common JPEG format). The ASIC also generates the signals for communicating image data to the LED screen, over an antenna (Technology Brief 8 on page 158), or over a transmission line to a computer (Sections 5-7 and 7-11).

Storage: Primary Memory and Secondary Memory (Technology Briefs 6, 7, and 21)

As you might expect, the microprocessor that runs the camera needs local memory to perform its computational operations. Moreover, image data must be stored somewhere for later use. The first requirement is dealt with by primary memory (Technology Brief 7 on page 135); most cameras have SRAM built into the microprocessor ASIC to act as a cache and stand-alone SRAM to store larger data sets.

Once images are ready to be stored, they are copied into a secondary storage device by the ASIC. Before the advent of high density non-volatile memory like Flash cards (Technology Brief 7 on page 135), many cameras used hard drives or optical (CD/DVD) drives (Technology Brief 21 on page 509). Flash cards have made huge inroads into the camera market, precisely because they have no moving parts, are portable and small, and are of sufficient density to store many high-quality pictures. High-performance cameras and cameras that record moving video continue to use optical drives and/or hard drives, although this is likely to change as storage technologies change (Technology Briefs 6 and 7 on pages 106 and 135, respectively).

Communication: Wired and Wireless Interfaces (Technology Brief 8, Sections 4-7, 5-7, 7-11, and 9-8)

Most digital cameras can communicate with external computers using a ***serial port***. A serial port is the name given to the collection of wires (usually two or three) through which two computers can exchange digital information. Several standard serial port types exist, each with its own protocol for exchanging information; the ***Universal Serial Bus (USB)*** and the FireWire standard are quite common. To communicate, the camera's ASIC generates digital pulses that pass through a buffer (Section 4-7) with a high voltage-to-current gain. The buffer is connected to the serial wires and sends a signal along 10 to 20 cm of wire to the computer; this wire behaves like a transmission line (Sections 5-7 and 7-11). The signal received at the computer's port is processed by the computer serial driver IC, which can then respond. As standard serial ports can exchange information at 60 MB/s, the camera can download image information rather quickly this way; for a more in-depth discussion of data transmission limits as they relate to circuit characteristics, see Technology Brief 17 on page 384.

Cameras increasingly communicate wirelessly, either because they form part of a mobile phone or because they have their own antennas. Wireless standards, other than those used for mobile phones, vary somewhat. For our purposes, it is sufficient to know that the camera's ASIC outputs the information digitally to a radio circuit, which converts the digital bits into ac signals that are launched into free space by the camera's antenna (Technology Brief 8 on page 158, Section 9-8).

Energy: Batteries (Technology Brief 11)

Energy sources have not kept pace with miniaturization (Technology Brief 11 on page 199). Most of the weight in mobile phones, PDAs and digital cameras is associated with the batteries. Despite the development of extremely low-power circuits, energy density and power density (Technology Brief 11 on page 199) requirements continue to place fundamental limitations on miniaturization. Energy technologies in the form of fuel cells, advanced supercapacitors, miniature combustion engines, and even energy scavenging from the environment are being pursued aggressively by researchers all over the world.

C H A P T E R

10

Laplace Transform Analysis Technique

Chapter Contents

Objectives

Upon learning the material presented in this chapter, you should be able to:

1. Find the Laplace transform of a time-domain function, if it exists.
2. Apply the properties of the Laplace transform to account for differentiation, integration, or time shift.
3. Apply the partial-fraction-expansion method so as to cast the Laplace transform function in a form amenable to easy transformation to the time domain.
4. Transform a circuit from the time domain to the **s**-domain.
5. Analyze the circuit in the **s**-domain and then transfer the solution to the time domain.
6. Determine the output of a circuit by applying the convolution-integral method.

Overview

In Chapters 5 and 6, we examined the transient response to a sudden change in RC, RL, and RLC circuits. The excitation sources were dc voltage and current sources in combination with SPST and SPDT switches. Voltage and current responses in such circuits are characterized by exponential functions of the form $e^{-\alpha t}$, where α is a damping coefficient that defines the rate at which the response transitions from its initial value immediately after the sudden change to its final value at $t = \infty$. The time-domain solution method employed in Chapters 5 and 6 is quite satisfactory, so long as the forcing function is a dc source and the differential equation describing the voltages and currents in the circuit is not higher than second order. If either of these two conditions is violated, a more robust method of solution is called for. The Laplace transform analysis technique introduced in this chapter is suited perfectly to deal with a wide range of circuits and any type of realistic forcing function, including pulses and sinusoids. In fact, the Laplace transform reduces to the phasor transform when the circuit sources are time-harmonic sinusoidal functions (and if the Laplace transform analysis is confined to the steady-state component of the overall solution). The phasor-domain technique served us well in Chapters 7 through 9, but it does not account for the transient component of the circuit response. Because in most ac circuits of interest the transient component decays rapidly after connecting the sources to the circuit, the steady-state solution provided by the phasor-domain technique is all that is needed. However, if we were to seek a solution that incorporates both the transient and steady-state components, then the Laplace transform technique is the solution method of choice.

When applying the Laplace transform analysis technique, the process entails steps similar to those associated with the phasor transform technique.

Solution Procedure: Laplace Transform

Step 1: The circuit is transformed to the Laplace domain—also known as the **s**-domain.

Step 2: In the **s**-domain, application of KVL and KCL yields a set of algebraic equations.

Step 3: The equations are solved for the variable of interest.

Step 4: The **s**-domain solution is transferred back to the time domain.

After introducing the Laplace transform and exploring its properties in the early part of this chapter, we will demonstrate its capabilities by applying the four-step procedure to analyze several types of passive and active circuits.

10-1 Singularity Functions

The waveforms commonly encountered in electric circuits include a variety of continuous-time functions—such as the ramp, exponential, and sinusoid—as well as some discontinuous functions—most notably the step and impulse functions. The step function is used to describe mathematically the instantaneous action by a switch to connect or disconnect a source to the circuit, and the impulse function is a useful mathematical tool for describing a sudden action of very short duration or for sampling a continuous function at discrete points in time. An example of the latter is when an analog-to-digital converter (ADC) is used to convert a continuous time signal into a digital sequence.

Application of the Laplace transform to a function $f(t)$ involves performing an integration over t. If $f(t)$ is a discontinuous function, special care should be exercised to make sure the transforms of $f(t)$ and its derivatives are performed correctly. With that in mind, we now will describe briefly the salient features of the step and impulse functions.

10-1.1 Unit Step Function

In Section 5-1.1, we defined the ***unit step function*** $u(t)$ as

$$u(t) = \begin{cases} 0 & \text{for } t < 0, \\ 1 & \text{for } t > 0. \end{cases} \tag{10.1}$$

Whereas $u(t)$ is defined at $t < 0$ and $t > 0$, it is not defined at $t = 0$. That is, $u(t)$ does not have a unique value at $t = 0$; it makes a *discontinuous* jump from 0 at $t = 0^-$ to 1 at $t = 0^+$ (Fig. 10-1(a)).

A ***singularity function*** is a function that either itself is not finite everywhere or one (or more) of its derivatives is (are) not finite everywhere.

Because of the discontinuity at $t = 0$, the derivative of $u(t)$ is infinite at $t = 0$, thereby qualifying $u(t)$ as a singularity function.

Occasionally, it may prove convenient to model the unit step function as a ramp over an infinitesimal interval extending between $-\epsilon$ and $+\epsilon$, as shown in Fig. 10-1(b). In that case, we can define $u(t)$ as

$$u(t) = \begin{cases} 0 & \text{for } t \le -\epsilon \\ \lim\limits_{\epsilon \to 0} \left[\dfrac{1}{2} \left(\dfrac{t}{\epsilon} + 1 \right) \right] & \text{for } -\epsilon \le t \le \epsilon \\ 1 & \text{for } t \ge \epsilon, \end{cases} \tag{10.2}$$

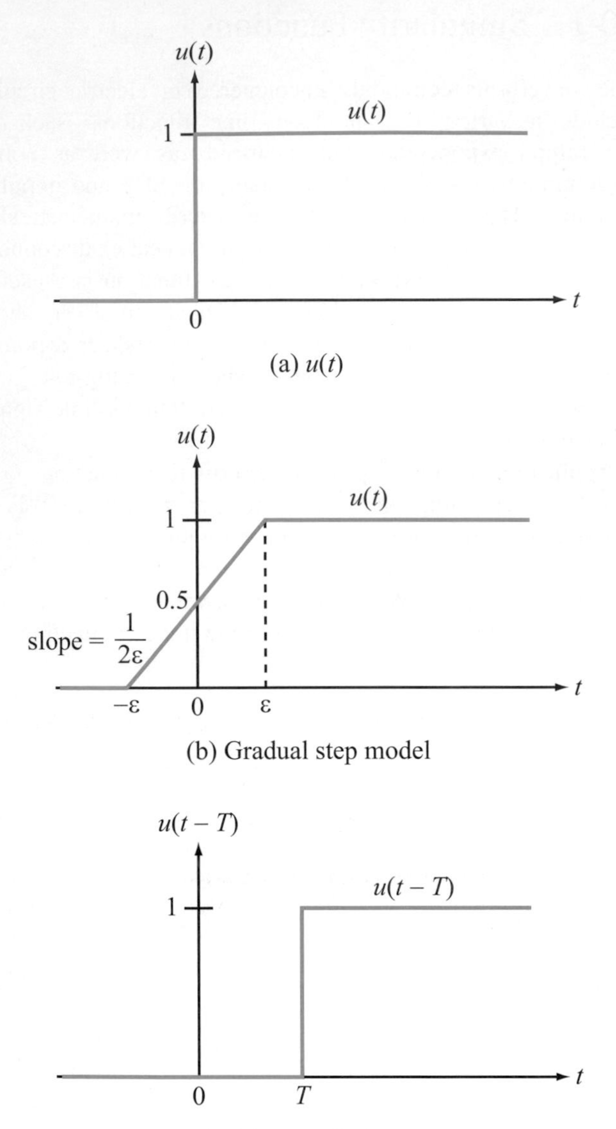

Figure 10-1: The unit step function.

which is a continuous function, but in the limit as $\epsilon \to 0$ its slope in the interval $(-\epsilon, \epsilon)$, is given by

$$u'(t) = \lim_{\epsilon \to 0} \frac{d}{dt}\left[\frac{1}{2}\left(\frac{t}{\epsilon}+1\right)\right] = \lim_{\epsilon \to 0}\left(\frac{1}{2\epsilon}\right) = \infty. \quad (10.3)$$

Thus, the slope of $u(t)$ still is not finite at $t = 0$, which is consistent with the formal definition of $u(t)$ as an *instantaneous* step function. Even though the more elaborate definition given by Eq. (10.2) does not add much information at this time, we will use it later to establish the connection between $u(t)$ and the impulse function $\delta(t)$.

In addition to $u(t)$, the ***time-shifted step function*** $u(t-T)$ will be used extensively in the present chapter. Recall from Chapter 5 that if the discontinuity occurs at $t = T$, then

$$u(t-T) = \begin{cases} 0 & \text{for } t < T \\ 1 & \text{for } t > T. \end{cases} \quad (10.4)$$

Function $u(t-T)$ is displayed in Fig. 10-1(c).

10-1.2 Unit Impulse Function

Graphically, the ***unit impulse function***—also known as the ***delta function*** $\delta(t)$—is represented by a vertical arrow, as shown in Fig. 10-2(a). If it is located at $t = T$, it is designated $\delta(t-T)$. For any fixed value T, the unit delta function is defined through the combination of two properties:

$$\delta(t-T) = 0 \qquad \text{for } t \neq T \quad (10.5a)$$

and

$$\int_{-\infty}^{\infty} \delta(t-T)\, dt = 1. \quad (10.5b)$$

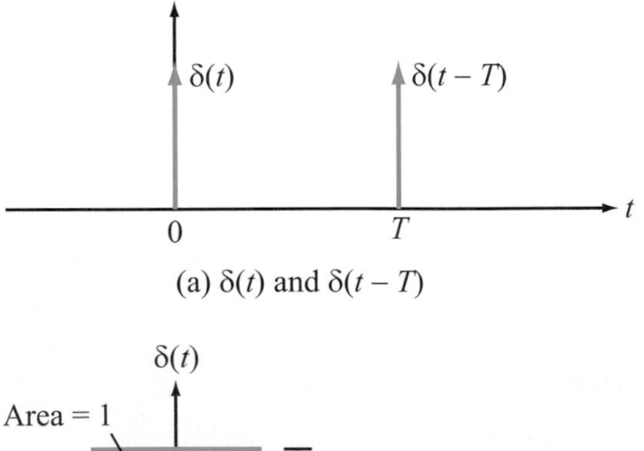

(a) δ(t) and δ(t − T)

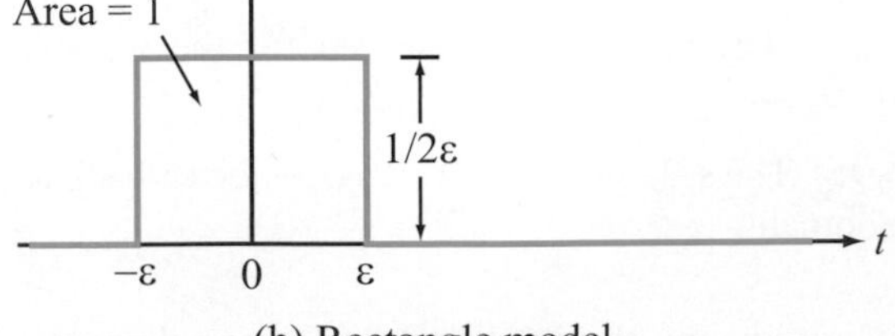

(b) Rectangle model

Figure 10-2: Unit impulse function.

The first property states that the unit impulse function $\delta(t-T)$ is zero everywhere, except at its own location $(t=T)$, and it is *not* defined at that location! The second property states that the total area under the unit impulse function is equal to 1, regardless of its location.

To visualize the meaning of the second property, we can represent the unit delta function by the graphical rectangle shown in Fig. 10-2(b) with the understanding that $\delta(t)$ is defined in the limit as $\epsilon \to 0$. The rectangle is of width $w = 2\epsilon$ and height $h = \frac{1}{2\epsilon}$, so its area is always unity, even as $\epsilon \to 0$.

Since $\delta(t-T)$ in the integral of Eq. (10.5b) is by definition equal to zero everywhere except over an infinitesimally narrow range surrounding $t = T$, Eq. (10.5b) can be reexpressed as

$$\int_{T-\epsilon}^{T+\epsilon} \delta(t-T)\, dt = 1. \tag{10.6}$$

The impulse function is related to the step function by

$$\delta(t-T) = \frac{d}{dt}\, u(t-T) = u'(t-T). \tag{10.7}$$

By differentiating the gradual-step waveform representing $u(t)$ in Fig. 10-1(b), we obtain the rectangle representation shown in Fig. 10-2(b) for $\delta(t)$, thereby demonstrating the validity of Eq. (10.7).

One of the most useful features of the impulse function is its ***sampling property***: For any function $f(t)$ that is continuous at $= T$,

$$\int_{-\infty}^{\infty} f(t)\, \delta(t-T)\, dt = f(T). \tag{10.8}$$

To show the validity of Eq. (10.8), we start by noting that because $\delta(t-T)$ is zero everywhere except over the region surrounding $t = T$, the contribution to the integral is limited to that region, in which case the limits can be shrunk, and $f(t)$ can be specified as $f(T)$ and removed from the integral, leading to

$$\int_{-\infty}^{\infty} f(t)\, \delta(t-T)\, dt = \int_{T-\epsilon}^{T+\epsilon} f(t)\, \delta(t-T)\, dt$$
$$= f(T) \int_{T-\epsilon}^{T+\epsilon} \delta(t-T)\, dt = f(T), \tag{10.9}$$

where in the final step we invoked the relationship given by Eq. (10.6).

Example 10-1: Periodic Sawtooth Waveform

Express the periodic sawtooth waveform shown in Fig. 10-3 in terms of step functions.

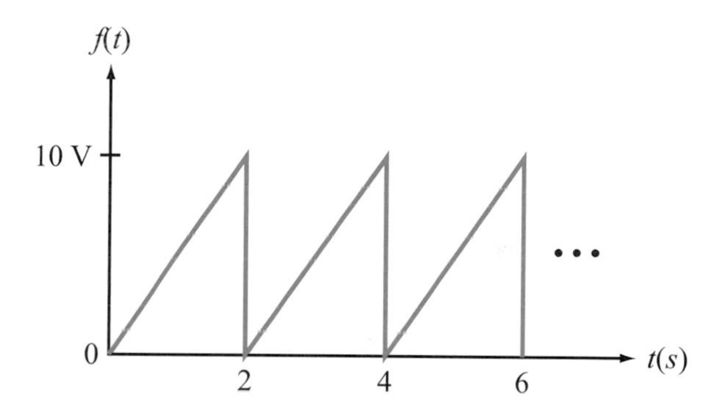

Figure 10-3: Periodic sawtooth waveform of Example 10-1.

Solution: The segment between $t = 0$ and $t = 2$ s is a ramp with a slope of 5 V/s. To effect a sudden drop from 10 V down to zero at $t = 2$ s, we need to add a negative ramp function at that point in time. Hence, for the first cycle,

$$f_1(t) = 5r(t) - 5r(t)\, u(t-2)$$
$$= 5t\, u(t) - 5t\, u(t-2) \text{ V}, \qquad 0 \le t \le 2 \text{ s}.$$

By extension, for the entire sawtooth waveform, we have

$$f(t) = \sum_{n=0}^{\infty} f_1(t - nT)$$
$$= \sum_{n=0}^{\infty} 5(t-2n)[u(t-2n) - u(t-2(n+1))] \text{ V}.$$

Review Question 10-1: What is the definition of a singularity function?

Review Question 10-2: How is $u(t)$ related to $\delta(t)$?

Review Question 10-3: Why is Eq. (10.8) called the *sampling property* of the impulse function?

10-2 Definition of the Laplace Transform

The symbol $\mathcal{L}[f(t)]$ is a shorthand notation for "the Laplace transform of function $f(t)$." Usually denoted $\mathbf{F(s)}$, the Laplace transform is defined by

$$\mathbf{F(s)} = \mathcal{L}[f(t)] = \int_{0^-}^{\infty} f(t)\, e^{-\mathbf{s}t}\, dt, \tag{10.10}$$

where $\mathbf{s}$ is a complex variable with a real part σ and an imaginary part ω given by

$$\mathbf{s} = \sigma + j\omega. \tag{10.11}$$

Given that the exponent $\mathbf{s}t$ has to be dimensionless, $\mathbf{s}$ has the unit of inverse second (which is the same as Hz or rad/s). Moreover, since $\mathbf{s}$ is a complex quantity, it often is termed ***complex frequency***.

In view of the definite limits on the integral in Eq. (10.10), the outcome of the integration will be an expression that depends on a single variable $\mathbf{s}$. The transform operation converts a function $f(t)$ defined in the time domain into a function $\mathbf{F(s)}$ defined in the $\mathbf{s}$-domain. Functions $f(t)$ and $\mathbf{F(s)}$ are called a ***Laplace transform pair***.

The ***uniqueness property*** of the Laplace transform states:

A given $f(t)$ has a unique $\mathbf{F(s)}$, and vice versa.

The uniqueness property can be expressed in symbolic form by

$$f(t) \longleftrightarrow \mathbf{F(s)}. \tag{10.12a}$$

The two-way arrow is a shorthand notation for the combination of the two statements

$$\mathcal{L}[f(t)] = \mathbf{F(s)}, \qquad \text{and} \qquad \mathcal{L}^{-1}[\mathbf{F(s)}] = f(t). \tag{10.12b}$$

The first statement asserts that $\mathbf{F(s)}$ is the Laplace transform of $f(t)$, and the second one asserts that the ***inverse Laplace transform*** ($\mathcal{L}^{-1}[\ ]$) of $\mathbf{F(s)}$ is $f(t)$.

Because the lower limit on the integral in Eq. (10.10) is 0^-, $\mathbf{F(s)}$ is called a ***one-sided transform***—in contrast with the ***two-sided transform*** for which the lower limit is $-\infty$. When we apply the Laplace transform technique to electric circuits, we select the start time for the circuit operation as $t = 0$, so the single-sided transform is suitable for our intended use, and we will adhere to it exclusively in this book. Moreover, *unless noted to the contrary, it will be assumed that $f(t)$ always is multiplied by an implicit invisible step function $u(t)$*. The inquisitive reader may ask why we use 0^- instead of simply 0, as our lower limit. We use it as a reminder that the integration should include initial conditions at $t = 0$, some of which may be in the form of step or impulse functions.

10-2.1 Convergence Condition

Depending on the functional form of $f(t)$, the Laplace transform integral given by Eq. (10.10) may or may not converge to a finite value. If it does not, the Laplace transform does not exist. Convergence requires that

$$\int_{0-}^{\infty} |f(t)\, e^{-\mathbf{s}t}|\, dt = \int_{0-}^{\infty} |f(t)||e^{-\sigma t}||e^{-j\omega t}|\, dt$$
$$= \int_{0^-}^{\infty} |f(t)|e^{-\sigma t}\, dt < \infty, \tag{10.13}$$

for some real value of σ. We used the fact that $|e^{-j\omega t}| = 1$ for any value of ωt and, since σ is real, $|e^{-\sigma t}| = e^{-\sigma t}$. If σ_c is the smallest value of σ for which the integral converges, then the ***region of convergence*** is $\sigma > \sigma_c$. Fortunately, this convergence issue is somewhat esoteric to circuit analysts and designers, because the waveforms of the excitation sources used in electric circuits are such that they do satisfy the convergence condition for all values of σ, and hence, their Laplace transforms do exist.

10-2.2 Inverse Laplace Transform

Equation (10.10) allows us to obtain Laplace transform $\mathbf{F(s)}$ corresponding to time function $f(t)$. The inverse process, denoted $\mathcal{L}^{-1}[\mathbf{F(s)}]$, allows us to perform an integration on $\mathbf{F(s)}$ to obtain $f(t)$ as

$$f(t) = \mathcal{L}^{-1}[\mathbf{F(s)}] = \frac{1}{2\pi j} \int_{\sigma-j\infty}^{\sigma+j\infty} \mathbf{F(s)}\, e^{\mathbf{s}t}\, d\mathbf{s}, \tag{10.14}$$

where $\sigma > \sigma_c$. The integration (which has to be performed in the two-dimensional complex plane) is rather cumbersome and to be avoided if an alternative approach is available for converting $\mathbf{F(s)}$ into $f(t)$. Fortunately, there is an alternative approach. Recall from our earlier discussion in the Overview of this chapter that the Laplace transform technique entails several steps, with the final step involving a transformation of the solution realized in the $\mathbf{s}$-domain to the time domain. Instead of applying Eq. (10.14), we can generate a table of Laplace transform pairs for all of the time functions commonly encountered in electric circuits, and then use it (sort of like a look-up table) to transform the $\mathbf{s}$-domain solution to the time domain. The validity of this approach is supported by the uniqueness property of the Laplace transform, which guarantees a one-to-one correspondence between every $f(t)$ and its corresponding $\mathbf{F(s)}$. The details of the inverse-transform process are covered in Section 10-4.

Example 10-2: Laplace Transforms of Singularity Functions

Determine the Laplace transforms of the waveforms displayed in Fig. 10-4.

Solution: **(a)** The step function in Fig. 10-4(a) is given by

$$f_1(t) = A\,u(t - T).$$

Application of Eq. (10.10) gives

$$\mathbf{F}_1(\mathbf{s}) = \int_{0^-}^{\infty} f_1(t)\,e^{-\mathbf{s}t}\,dt$$

$$= \int_{0^-}^{\infty} A\,u(t - T)\,e^{-\mathbf{s}t}\,dt$$

$$= A \int_{T}^{\infty} e^{-\mathbf{s}t}\,dt = -\frac{A}{\mathbf{s}}\,e^{-\mathbf{s}t}\bigg|_{T}^{\infty} = \frac{A}{\mathbf{s}}\,e^{-\mathbf{s}T}.$$

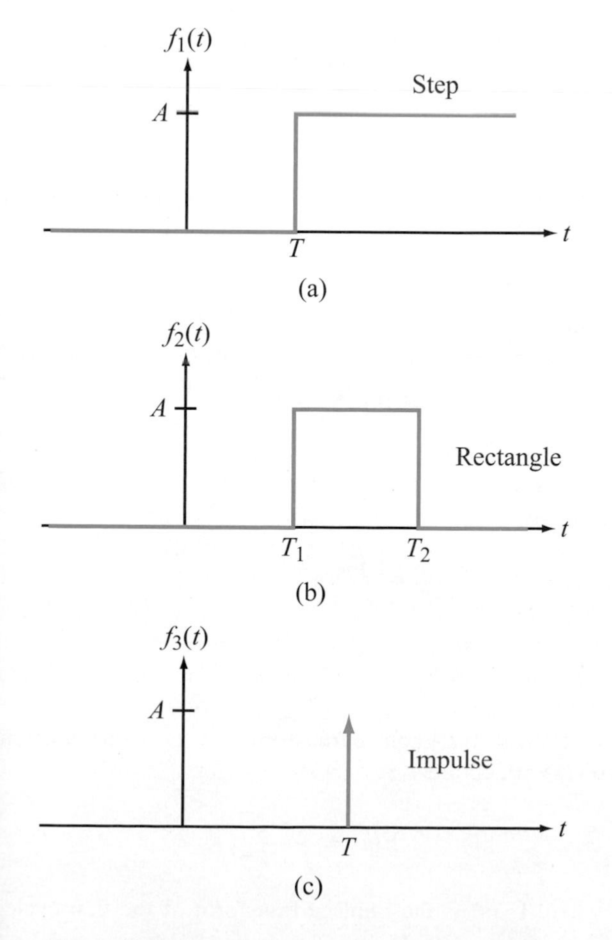

Figure 10-4: Singularity functions for Example 10-2.

For the special case where $A = 1$ and $T = 0$ (the step occurs at $t = 0$), the transform pair becomes

$$u(t) \quad \longleftrightarrow \quad \frac{1}{\mathbf{s}}. \tag{10.15}$$

(b) The rectangle function in Fig. 10-4(b) can be constructed as the sum of two step functions:

$$f_2(t) = A[u(t - T_1) - u(t - T_2)].$$

Its Laplace transform is

$$\mathbf{F}_2(\mathbf{s}) = \int_{0^-}^{\infty} A[u(t - T_1) - u(t - T_2)]e^{-\mathbf{s}t}\,dt$$

$$= A \int_{0^-}^{\infty} u(t - T_1)\,e^{-\mathbf{s}t}\,dt - A \int_{0^-}^{\infty} u(t - T_2)\,e^{-\mathbf{s}t}\,dt$$

$$= \frac{A}{\mathbf{s}}[e^{-\mathbf{s}T_1} - e^{-\mathbf{s}T_2}].$$

(c) The delta function in Fig. 10-4(c) is given by

$$f_3(t) = A\,\delta(t - T).$$

The corresponding Laplace transform is

$$\mathbf{F}_3(\mathbf{s}) = \int_{0^-}^{\infty} A\,\delta(t - T)\,e^{-\mathbf{s}t}\,dt$$

$$= A \int_{T-\epsilon}^{T+\epsilon} \delta(t - T)\,e^{-\mathbf{s}t}\,dt = Ae^{-\mathbf{s}T},$$

where we have used the procedure introduced earlier in connection with Eq. (10.9). For the special case where $A = 1$ and $T = 0$, the Laplace transform pair simplifies to

$$\delta(t) \quad \longleftrightarrow \quad 1. \tag{10.16}$$

Example 10-3: Laplace Transform of $\cos \omega t$

Obtain the Laplace transform for $[\cos \omega t]\, u(t)$.

Solution: The solution is facilitated by expressing $\cos \omega t$ in terms of complex exponentials (see Appendix D-1), namely

$$\cos \omega t = \frac{1}{2}[e^{j\omega t} + e^{-j\omega t}].$$

Use of this expression in Eq. (10.10) gives

$$\begin{aligned}\mathbf{F}(\mathbf{s}) &= \int_{0^-}^{\infty} \cos \omega t \; u(t)\; e^{-\mathbf{s}t}\; dt \\ &= \frac{1}{2}\left[\int_0^{\infty} e^{j\omega t} e^{-\mathbf{s}t}\; dt + \int_0^{\infty} e^{-j\omega t} e^{-\mathbf{s}t}\; dt\right] \\ &= \frac{1}{2}\left[\frac{e^{(j\omega-\mathbf{s})t}}{j\omega - \mathbf{s}} + \frac{e^{-(j\omega+\mathbf{s})t}}{-(j\omega+\mathbf{s})}\right]\Bigg|_0^{\infty} = \frac{\mathbf{s}}{\mathbf{s}^2+\omega^2}.\end{aligned}$$

Hence,

$$\cos \omega t \quad \leftrightarrow \quad \frac{\mathbf{s}}{\mathbf{s}^2+\omega^2}. \tag{10.17}$$

Review Question 10-4: Is the uniqueness property of the Laplace transform uni-directional or bi-directional? Why is that significant?

Review Question 10-5: Is convergence of the Laplace transform integral in doubt when applied to circuit analysis? If not, why not?

Exercise 10-1: Determine the Laplace transform for (a) $\sin \omega t$, (b) e^{-at}, and (c) $r(t-T)$ [see the ramp function in Chapter 5]. Assume that all waveforms are zero for $t < 0$.

Answer: (a) $\sin \omega t \leftrightarrow \dfrac{\omega}{\mathbf{s}^2+\omega^2}$, (b) $e^{-at} \leftrightarrow \dfrac{1}{\mathbf{s}+a}$, (c) $r(t-T) \leftrightarrow \dfrac{e^{-\mathbf{s}T}}{\mathbf{s}^2}$. (See CD)

Exercise 10-2: If $f(t)$ is the rectangular pulse shown in Fig. E10.2(a), determine $f'(t)$ and plot it.

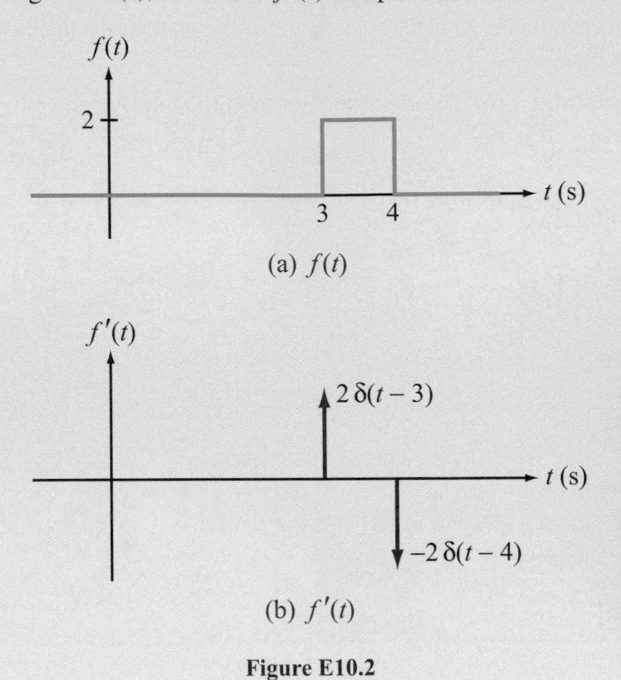

Figure E10.2

Answer: $f'(t) = 2\delta(t-3) - 2\delta(t-4)$. (See CD)

Exercise 10-3: Determine the Laplace transform corresponding to the sawtooth waveform shown in Fig. 10-3.

Answer:

$$\mathbf{F}(\mathbf{s}) = \mathbf{F}_1(\mathbf{s}) \sum_{n=0}^{\infty} e^{-2n\mathbf{s}} = \frac{\mathbf{F}_1(\mathbf{s})}{1-e^{-2\mathbf{s}}},$$

where

$$\begin{aligned}\mathbf{F}_1(\mathbf{s}) &= \int_0^2 5te^{-\mathbf{s}t}\; dt \\ &= \frac{5}{\mathbf{s}^2}\,[1-(2\mathbf{s}+1)e^{-2\mathbf{s}}].\end{aligned}$$

In general, the Laplace transform of a periodic function $f(t)$ with period T is

$$\mathbf{F}(\mathbf{s}) = \frac{\mathbf{F}_1(\mathbf{s})}{1-e^{-T\mathbf{s}}},$$

where $\mathbf{F}_1(\mathbf{s})$ is the Laplace transform of the first cycle. (See CD)

10-3 Properties of the Laplace Transform

The Laplace transform has a number of useful, universal properties that apply to any function $f(t)$, greatly facilitating the process of transforming a circuit from the t-domain to the $\mathbf{s}$-domain. To explain what we mean by a universal property, we borrow an example from the phasor transform. In Chapter 7 we established that differentiation in the time domain is equivalent to multiplication by $j\omega$ in the phasor domain. Thus, if $\mathbf{V}(\omega)$ is the phasor counterpart of $v(t)$, then

$$\frac{d}{dt}\,v(t) \;\Rightarrow\; j\omega\,\mathbf{V}(\omega). \tag{10.18}$$

This is a ***universal*** or ***operational property*** of the phasor transform because it applies to any $f(t)$, regardless of its specific functional form.

This section will conclude with a table outlining 13 universal properties of the Laplace transform which we will be making frequent reference to throughout this chapter. Some of these properties are intuitively obvious, while others may require some elaboration.

10-3.1 Time Scaling

If

$$f(t) \;\leftrightarrow\; \mathbf{F}(\mathbf{s}), \tag{10.19}$$

then the transform of the time-scaled function $f(at)$ is

$$f(at) \;\leftrightarrow\; \frac{1}{a}\,\mathbf{F}\left(\frac{\mathbf{s}}{a}\right), \qquad a > 0. \tag{10.20}$$

The ***time scaling*** property states that stretching the time axis by a factor a corresponds to shrinking the $\mathbf{s}$-axis and the amplitude of $\mathbf{F}(\mathbf{s})$ by the same factor, and vice versa.

To prove Eq. (10.20), we start with the standard definition of the Laplace transform given by Eq. (10.10) as

$$\mathcal{L}[f(at)] = \int_{0^-}^{\infty} f(at)\, e^{-\mathbf{s}t}\, dt. \tag{10.21}$$

In the integral, if we set $t' = at$ and $dt = \frac{1}{a}\,dt'$, we have

$$\begin{aligned}\mathcal{L}[f(at)] &= \frac{1}{a}\int_{0^-}^{\infty} f(t')\, e^{-(\mathbf{s}/a)t'}\, dt' \\ &= \frac{1}{a}\int_{0^-}^{\infty} f(t')\, e^{-\mathbf{s}'t'}\, dt', \qquad \text{with } \mathbf{s}' = \frac{\mathbf{s}}{a}.\end{aligned} \tag{10.22}$$

The definite integral is identical in form with the Laplace transform definition given by Eq. (10.10), except that the dummy variable is t' instead of t, and the coefficient of the exponent is $\mathbf{s}' = \mathbf{s}/a$ instead of just $\mathbf{s}$. Hence,

$$\mathcal{L}[f(at)] = \frac{1}{a}\,\mathbf{F}(\mathbf{s}') = \frac{1}{a}\,\mathbf{F}\left(\frac{\mathbf{s}}{a}\right), \qquad a > 0, \tag{10.23}$$

which proves the time-scaling property defined by Eq. (10.20).

10-3.2 Time Shift

If t is shifted by T along the time axis with $T \geq 0$, then

$$f(t-T)\,u(t-T) \;\leftrightarrow\; e^{-T\mathbf{s}}\,\mathbf{F}(\mathbf{s}), \qquad T \geq 0. \tag{10.24}$$

The validity of this property is demonstrated as follows:

$$\begin{aligned}\mathcal{L}[f(t-T)\,u(t-T)] &= \int_{0^-}^{\infty} f(t-T)\,u(t-T)\,e^{-\mathbf{s}t}\,dt \\ &= \int_{T}^{\infty} f(t-T)\,e^{-\mathbf{s}t}\,dt \\ &= \int_{0}^{\infty} f(x)\,e^{-\mathbf{s}(x+T)}\,dx \\ &= e^{-T\mathbf{s}}\int_{0}^{\infty} f(x)\,e^{-\mathbf{s}x}\,dx \\ &= e^{-T\mathbf{s}}\,\mathbf{F}(\mathbf{s}),\end{aligned} \tag{10.25}$$

where we made the substitutions $t - T = x$ and $dt = dx$ and then applied the definition for $\mathbf{F}(\mathbf{s})$ given by Eq. (10.10).

To illustrate the utility of the time-shift property, we consider the cosine function of Example 10-3, where it was shown that

$$\cos\omega t \;\leftrightarrow\; \frac{\mathbf{s}}{\mathbf{s}^2 + \omega^2}. \tag{10.26}$$

According to Eq. (10.24),

$$\cos\omega(t-T)\,u(t-T) \;\leftrightarrow\; e^{-T\mathbf{s}}\,\frac{\mathbf{s}}{\mathbf{s}^2 + \omega^2}. \tag{10.27}$$

Had we analyzed a circuit driven by a cosinusoidal voltage source that started at $t = 0$ then wanted to reanalyze it anew, but wanted to delay both the cosine function and the start time by T, Eq. (10.27) would provide an expedient solution to obtaining the transform of the delayed cosine function.

Exercise 10-4: Determine $\mathcal{L}[\sin\omega(t-T)\,u(t-T)]$.

Answer:
$$e^{-T\mathbf{s}}\,\frac{\omega}{\mathbf{s}^2+\omega^2}.$$
(See)

10-3.3 Frequency Shift

According to the time-shift property, if t is replaced with $(t-T)$ in the time domain, $\mathbf{F}(\mathbf{s})$ gets multiplied by $e^{-T\mathbf{s}}$ in the $\mathbf{s}$-domain. Within a $(-)$ sign, the converse is also true: if $\mathbf{s}$ is replaced with $(\mathbf{s}+a)$ in the $\mathbf{s}$-domain, $f(t)$, gets multiplied by e^{-at} in the time domain. Thus,

$$e^{-at}\,f(t) \leftrightarrow \mathbf{F}(\mathbf{s}+a). \tag{10.28}$$

Proof of Eq. (10.28) is part of Exercise 10-5.

Review Question 10-6: According to the time scaling property of the Laplace transform, "stretching the time axis corresponds to shrinking the $\mathbf{s}$-axis." What does that mean?

Review Question 10-7: Explain the similarities and differences between the time-shift and frequency-shift properties of the Laplace transform.

Exercise 10-5: (a) Prove Eq. (10.28) and (b) apply it to determine $\mathcal{L}[e^{-at}\cos\omega t]$.

Answer: (a) (See),
(b) $e^{-at}\cos\omega t \leftrightarrow \dfrac{\mathbf{s}+a}{(\mathbf{s}+a)^2+\omega^2}$. (See)

10-3.4 Time Differentiation

Differentiating $f(t)$ in the time domain is equivalent to (a) multiplying $\mathbf{F}(\mathbf{s})$ by $\mathbf{s}$ in the $\mathbf{s}$-domain and then (b) subtracting $f(0^-)$ from $\mathbf{s}\,\mathbf{F}(\mathbf{s})$:

$$f'=\frac{df}{dt} \leftrightarrow \mathbf{s}\,\mathbf{F}(\mathbf{s})-f(0^-). \tag{10.29}$$

To verify Eq. (10.29), we start with the standard definition for the Laplace transform:

$$\mathcal{L}[f'] = \int_{0^-}^{\infty} \frac{df}{dt}\,e^{-\mathbf{s}t}\,dt. \tag{10.30}$$

Integration by parts with

$$u=e^{-\mathbf{s}t}, \qquad du=-\mathbf{s}e^{-\mathbf{s}t}\,dt,$$
$$dv=\left(\frac{df}{dt}\right)dt, \qquad \text{and} \qquad v=f,$$

gives

$$\begin{aligned}
\mathcal{L}[f'] &= uv\big|_{0^-}^{\infty} - \int_{0^-}^{\infty} v\,du \\
&= e^{-\mathbf{s}t}\,f(t)\big|_{0^-}^{\infty} - \int_{0^-}^{\infty} -\mathbf{s}\,f(t)\,e^{-\mathbf{s}t}\,dt \\
&= -f(0^-)+\mathbf{s}\,\mathbf{F}(\mathbf{s}),
\end{aligned} \tag{10.31}$$

which is equivalent to Eq. (10.29).

Higher derivatives can be obtained by repeating the application of Eq. (10.29). For the second derivative of $f(t)$,

$$f''=\frac{d^2f}{dt^2} \leftrightarrow \mathbf{s}^2\,\mathbf{F}(\mathbf{s})-\mathbf{s}\,f(0^-)-f'(0^-), \tag{10.32}$$

where $f'(0^-)$ is the derivative of $f(t)$, evaluated at $t=0^-$.

Example 10-4: Second Derivative Property

Verify the second derivative property for $f(t)=\cos\omega t$ by (a) applying the transform equation to $f''(t)$ and (b) comparing it with the result obtained via Eq. (10.32).

Solution:

(a) The second derivative of $f(t)$ is

$$\begin{aligned}
f''(t) &= \frac{d^2}{dt^2}\cos\omega t \\
&= \frac{d}{dt}(-\omega\sin\omega t) \\
&= -\omega^2\cos\omega t.
\end{aligned}$$

The Laplace transform of $f''(t)$ is

$$\begin{aligned}\mathcal{L}[f''] &= \mathcal{L}[-\omega^2 \cos \omega t] \\ &= -\omega^2 \, \mathcal{L}[\cos \omega t] \\ &= \frac{-\omega^2 \mathbf{s}}{\mathbf{s}^2 + \omega^2},\end{aligned}$$

where we made use of Eq. (10.17).

(b) Application of Eq. (10.32) gives

$$\mathcal{L}[f''] = \mathbf{s}^2 \, \mathbf{F}(\mathbf{s}) - \mathbf{s} \, f(0^-) - f'(0^-).$$

For $f(t) = \cos \omega t$,

$$\mathbf{F}(\mathbf{s}) = \frac{\mathbf{s}}{\mathbf{s}^2 + \omega^2},$$

$$f(0^-) = 1,$$

and

$$f'(0^-) = -\omega \sin \omega t|_{t=0^-} = 0.$$

Hence,

$$\begin{aligned}\mathcal{L}[f''] &= \frac{\mathbf{s}^3}{\mathbf{s}^2 + \omega^2} - \mathbf{s} \\ &= \frac{-\omega^2 \mathbf{s}}{\mathbf{s}^2 + \omega^2}.\end{aligned}$$

10-3.5 Time Integration

Integration of $f(t)$ in the time domain is equivalent to dividing $\mathbf{F}(\mathbf{s})$ by $\mathbf{s}$ in the $\mathbf{s}$-domain:

$$\int_0^t f(t) \, dt \quad \leftrightarrow \quad \frac{1}{\mathbf{s}} \, \mathbf{F}(\mathbf{s}). \tag{10.33}$$

Application of the Laplace transform definition gives

$$\mathcal{L}\left[\int_0^t f(t) \, dt\right] = \int_{0^-}^{\infty} \left[\int_0^t f(x) \, dx\right] e^{-\mathbf{s}t} \, dt, \tag{10.34}$$

where (for the sake of clarity) we changed the dummy variable in the inner integral from t to x. Integration by parts with

$$u = \int_0^t f(x) \, dx, \qquad du = f(t) \, dt,$$

$$dv = e^{-\mathbf{s}t} \, dt, \qquad \text{and} \qquad v = -\frac{e^{-\mathbf{s}t}}{\mathbf{s}}$$

leads to

$$\begin{aligned}\mathcal{L}\left[\int_0^t f(t) \, dt\right] &= uv\big|_{0^-}^{\infty} - \int_{0^-}^{\infty} v \, du \\ &= \left[-\frac{e^{-\mathbf{s}t}}{\mathbf{s}} \int_0^t f(x) \, dx\right]\Bigg|_{0^-}^{\infty} + \frac{1}{\mathbf{s}} \int_{0^-}^{\infty} f(t) \, e^{-\mathbf{s}t} \, dt \\ &= \frac{1}{\mathbf{s}} \, \mathbf{F}(\mathbf{s}).\end{aligned} \tag{10.35}$$

Both limits on the first term on the right-hand side yield zero values.

10-3.6 Initial- and Final-Value Theorems

The relationship between $f(t)$ and $\mathbf{F}(\mathbf{s})$ is such that the initial value $f(0)$ and the final value $f(\infty)$ of $f(t)$ can be determined directly from the expression of $\mathbf{F}(\mathbf{s})$, provided certain conditions are satisfied (as discussed later in this subsection).

Consider the derivative property represented by Eq. (10.31) as

$$\mathcal{L}[f'] = \int_{0^-}^{\infty} \frac{df}{dt} \, e^{-\mathbf{s}t} \, dt = \mathbf{s} \, \mathbf{F}(\mathbf{s}) - f(0^-). \tag{10.36}$$

If we take the limit as $\mathbf{s} \to \infty$ while recognizing that $f(0^-)$ is independent of $\mathbf{s}$, we get

$$\lim_{\mathbf{s} \to \infty} \left[\int_{0^-}^{\infty} \frac{df}{dt} \, e^{-\mathbf{s}t} \, dt\right] = \lim_{\mathbf{s} \to \infty} [\mathbf{s} \, \mathbf{F}(\mathbf{s}) - f(0^-)]. \tag{10.37}$$

The integral on the left-hand side can be split into two integrals, one over the time segment $(0^-, 0^+)$—for which $e^{-\mathbf{s}t} = 1$, and another over the segment $(0^+, \infty)$:

$$\begin{aligned}\lim_{\mathbf{s} \to \infty} &\left[\int_{0^-}^{\infty} \frac{df}{dt} \, e^{-\mathbf{s}t} \, dt\right] \\ &= \lim_{\mathbf{s} \to \infty} \left[\int_{0^-}^{0^+} \frac{df}{dt} \, dt + \int_{0^+}^{\infty} \frac{df}{dt} \, e^{-\mathbf{s}t} \, dt\right] \\ &= f(0^+) - f(0^-).\end{aligned} \tag{10.38}$$

As $\mathbf{s} \to \infty$, the exponential function e^{-st} causes the integrand of the last term to vanish. Equating Eqs. (10.37) and (10.38) leads to

$$f(0^+) = \lim_{\mathbf{s}\to\infty} \mathbf{s}\,\mathbf{F}(\mathbf{s})$$

Initial-value theorem, (10.39)

which is known as the ***initial-value theorem***.

A similar treatment in which $\mathbf{s}$ is made to approach 0 (instead of ∞) in Eq. (10.37) leads to the ***final-value theorem***:

$$f(\infty) = \lim_{\mathbf{s}\to 0} \mathbf{s}\,\mathbf{F}(\mathbf{s})$$

Final-value theorem. (10.40)

We should note that Eq. (10.40) is useful for determining $f(\infty)$, so long as $f(\infty)$ exists. Otherwise, application of Eq. (10.40) may lead to an erroneous result. Consider, for example, $f(t) = \cos\omega t$, which does not have a unique value as $t \to \infty$. Yet application of Eq. (10.40) to Eq. (10.17) leads to $f(\infty) = 0$, which is incorrect.

Example 10-5: Initial and Final Values

Determine the initial and final values of a function $f(t)$ whose Laplace transform is given by

$$\mathbf{F}(\mathbf{s}) = \frac{25\mathbf{s}(\mathbf{s}+3)}{(\mathbf{s}+1)(\mathbf{s}^2+2\mathbf{s}+36)}.$$

Solution: Application of Eq. (10.39) gives

$$f(0^+) = \lim_{\mathbf{s}\to\infty} \mathbf{s}\,\mathbf{F}(\mathbf{s}) = \lim_{\mathbf{s}\to\infty} \frac{25\mathbf{s}^2(\mathbf{s}+3)}{(\mathbf{s}+1)(\mathbf{s}^2+2\mathbf{s}+36)}.$$

To avoid the problem of dealing with ∞, it often is more convenient to first apply the substitution $\mathbf{s} = 1/\mathbf{u}$, rearrange the function and then to find the limit as $\mathbf{u} \to 0$. That is,

$$\begin{aligned} f(0^+) &= \lim_{\mathbf{u}\to 0} \frac{25(1/\mathbf{u}+3)}{\mathbf{u}^2(1/\mathbf{u}+1)(1/\mathbf{u}^2+2/\mathbf{u}+36)} \\ &= \lim_{\mathbf{u}\to 0} \frac{25(1+3\mathbf{u})}{(1+\mathbf{u})(1+2\mathbf{u}+36\mathbf{u}^2)} \\ &= \frac{25(1+0)}{(1+0)(1+0+0)} = 25. \end{aligned}$$

To determine $f(\infty)$, we apply Eq. (10.40),

$$\begin{aligned} f(\infty) &= \lim_{\mathbf{s}\to 0} \mathbf{s}\,\mathbf{F}(\mathbf{s}) \\ &= \lim_{\mathbf{s}\to 0} \frac{25\mathbf{s}^2(\mathbf{s}+3)}{(\mathbf{s}+1)(\mathbf{s}^2+2\mathbf{s}+36)} = 0. \end{aligned}$$

Exercise 10-6: Determine the initial and final values of $f(t)$ if its Laplace transform is given by

$$\mathbf{F}(\mathbf{s}) = \frac{\mathbf{s}^2+6\mathbf{s}+18}{\mathbf{s}(\mathbf{s}+3)^2}.$$

Answer: $f(0^+) = 1$, $f(\infty) = 2$. (See ◎)

10-3.7 Frequency Differentiation

Given the definition of the Laplace transform, namely

$$\mathbf{F}(\mathbf{s}) = \mathcal{L}[f(t)] = \int_{0^-}^{\infty} f(t)\, e^{-\mathbf{s}t}\, dt, \tag{10.41}$$

if we take the derivative with respect to $\mathbf{s}$ on both sides, we have

$$\begin{aligned} \frac{d\,\mathbf{F}(\mathbf{s})}{d\mathbf{s}} &= \int_{0^-}^{\infty} \frac{d}{d\mathbf{s}}[f(t)\, e^{-\mathbf{s}t}]\, dt \\ &= \int_{0^-}^{\infty} [-t\, f(t)]e^{-\mathbf{s}t}\, dt = \mathcal{L}[-t\, f(t)], \end{aligned} \tag{10.42}$$

where we recognize the integral as the Laplace transform of the function $[-t\, f(t)]$. Rearranging Eq. (10.42) provides the ***frequency differentiation relation***

$$t\, f(t) \quad \leftrightarrow \quad -\frac{d\,\mathbf{F}(\mathbf{s})}{d\mathbf{s}} = -\mathbf{F}'(\mathbf{s}), \tag{10.43}$$

which states that multiplication of $f(t)$ by $-t$ in the time domain is equivalent to differentiating $\mathbf{F}(\mathbf{s})$ in the $\mathbf{s}$-domain.

Example 10-6: Applying Frequency Differentiation Property

Given that

$$\mathbf{F}(\mathbf{s}) = \mathcal{L}[e^{-at}] = \frac{1}{\mathbf{s}+a},$$

apply Eq. (10.43) to obtain the Laplace transform of te^{-at}.

Solution:

$$\begin{aligned} \mathcal{L}[te^{-at}] &= -\frac{d}{d\mathbf{s}}\,\mathbf{F}(\mathbf{s}) = -\frac{d}{d\mathbf{s}}\left[\frac{1}{\mathbf{s}+a}\right] \\ &= \frac{1}{(\mathbf{s}+a)^2}. \end{aligned}$$

10-3.8 Frequency Integration

If we integrate both sides of Eq. (10.41) from $\mathbf{s}$ to ∞, we get

$$\int_{\mathbf{s}}^{\infty} \mathbf{F}(\mathbf{s})\, d\mathbf{s} = \int_{\mathbf{s}}^{\infty} \left[\int_{0^-}^{\infty} f(t)\, e^{-\mathbf{s}t}\, dt \right] d\mathbf{s}. \qquad (10.44)$$

Since t and $\mathbf{s}$ are independent variables, we can interchange the order of the integration on the right-hand side of Eq. (10.44),

$$\int_{\mathbf{s}}^{\infty} \mathbf{F}(\mathbf{s})\, d\mathbf{s} = \int_{0^-}^{\infty} \left[\int_{\mathbf{s}}^{\infty} f(t)\, e^{-\mathbf{s}t}\, d\mathbf{s} \right] dt = \int_{0^-}^{\infty} \left[\frac{f(t)}{-t}\, e^{-\mathbf{s}t} \Big|_{\mathbf{s}}^{\infty} \right] dt$$

$$= \int_{0}^{\infty} \left[\frac{f(t)}{t} \right] e^{-\mathbf{s}t}\, dt = \mathcal{L}\left[\frac{f(t)}{t} \right]. \qquad (10.45)$$

This ***frequency integration property*** can be expressed as

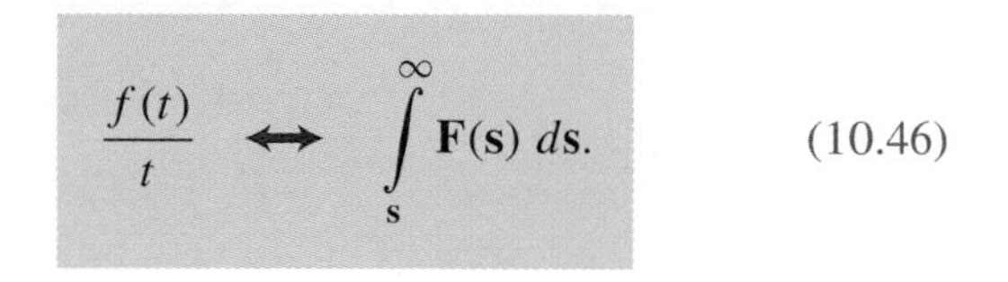

$$\frac{f(t)}{t} \leftrightarrow \int_{\mathbf{s}}^{\infty} \mathbf{F}(\mathbf{s})\, d\mathbf{s}. \qquad (10.46)$$

Table 10-1 provides a summary of the principal properties of the Laplace transform. The only property we have not yet examined is the convolution property, which we will do in Section 10-8. For easy reference, Table 10-2 contains a list of Laplace transform pairs that we are likely to encounter in future sections.

Table 10-1: Properties of the Laplace transform.

Property	$f(t)$		$\mathbf{F}(\mathbf{s}) = \mathcal{L}[f(t)]$
1. Multiplication by constant	$K\, f(t)$	$\leftrightarrow$	$K\, \mathbf{F}(\mathbf{s})$
2. Linearity	$K_1\, f_1(t) + K_2\, f_2(t)$	$\leftrightarrow$	$K_1\, \mathbf{F}_1(\mathbf{s}) + K_2\, \mathbf{F}_2(\mathbf{s})$
3. Time scaling	$f(at), \quad a > 0$	$\leftrightarrow$	$\frac{1}{a}\, \mathbf{F}\left(\frac{\mathbf{s}}{a}\right)$
4. Time shift	$f(t - T)\, u(t - T)$	$\leftrightarrow$	$e^{-T\mathbf{s}}\, \mathbf{F}(\mathbf{s})$
5. Frequency shift	$e^{-at}\, f(t)$	$\leftrightarrow$	$\mathbf{F}(\mathbf{s} + a)$
6. Time 1st derivative	$f' = \frac{df}{dt}$	$\leftrightarrow$	$\mathbf{s}\, \mathbf{F}(\mathbf{s}) - f(0^-)$
7. Time 2nd derivative	$f'' = \frac{d^2 f}{dt^2}$	$\leftrightarrow$	$\mathbf{s}^2 \mathbf{F}(\mathbf{s}) - \mathbf{s} f(0^-) - f'(0^-)$
8. Time integral	$\int_0^t f(t)\, dt$	$\leftrightarrow$	$\frac{1}{\mathbf{s}}\, \mathbf{F}(\mathbf{s})$
9. Frequency derivative	$t\, f(t)$	$\leftrightarrow$	$-\frac{d}{d\mathbf{s}}\, \mathbf{F}(\mathbf{s}) = -\mathbf{F}'(\mathbf{s})$
10. Frequency integral	$\frac{f(t)}{t}$	$\leftrightarrow$	$\int_{\mathbf{s}}^{\infty} \mathbf{F}(\mathbf{s})\, d\mathbf{s}$
11. Initial value	$f(0^+)$	=	$\lim_{\mathbf{s} \to \infty} \mathbf{s}\, \mathbf{F}(\mathbf{s})$
12. Final value	$f(\infty)$	=	$\lim_{\mathbf{s} \to 0} \mathbf{s}\, \mathbf{F}(\mathbf{s})$
13. Convolution	$f_1(t) * f_2(t)$	$\leftrightarrow$	$\mathbf{F}_1(\mathbf{s})\, \mathbf{F}_2(\mathbf{s})$

Table 10-2: Examples of Laplace transform pairs.
Note that $f(t) = 0$ for $t < 0^-$.

Laplace Transform Pairs			
	$f(t)$		$\mathbf{F(s)} = \mathcal{L}[f(t)]$
1	$\delta(t)$	$\longleftrightarrow$	1
1a	$\delta(t-T)$	$\longleftrightarrow$	$e^{-T\mathbf{s}}$
2	$u(t)$	$\longleftrightarrow$	$\frac{1}{\mathbf{s}}$
2a	$u(t-T)$	$\longleftrightarrow$	$\frac{e^{-T\mathbf{s}}}{\mathbf{s}}$
3	$e^{-at}\,u(t)$	$\longleftrightarrow$	$\frac{1}{\mathbf{s}+a}$
3a	$e^{-a(t-T)}\,u(t-T)$	$\longleftrightarrow$	$\frac{e^{-T\mathbf{s}}}{\mathbf{s}+a}$
4	$t\,u(t)$	$\longleftrightarrow$	$\frac{1}{\mathbf{s}^2}$
4a	$(t-T)\,u(t-T)$	$\longleftrightarrow$	$\frac{e^{-T\mathbf{s}}}{\mathbf{s}^2}$
5	$t^2\,u(t)$	$\longleftrightarrow$	$\frac{2}{\mathbf{s}^3}$
6	$te^{-at}\,u(t)$	$\longleftrightarrow$	$\frac{1}{(\mathbf{s}+a)^2}$
7	$t^2e^{-at}\,u(t)$	$\longleftrightarrow$	$\frac{2}{(\mathbf{s}+a)^3}$
8	$t^{n-1}e^{-at}\,u(t)$	$\longleftrightarrow$	$\frac{(n-1)!}{(\mathbf{s}+a)^n}$
9	$\sin\omega t\,u(t)$	$\longleftrightarrow$	$\frac{\omega}{\mathbf{s}^2+\omega^2}$
10	$\sin(\omega t+\theta)\,u(t)$	$\longleftrightarrow$	$\frac{\mathbf{s}\sin\theta+\omega\cos\theta}{\mathbf{s}^2+\omega^2}$
11	$\cos\omega t\,u(t)$	$\longleftrightarrow$	$\frac{\mathbf{s}}{\mathbf{s}^2+\omega^2}$
12	$\cos(\omega t+\theta)\,u(t)$	$\longleftrightarrow$	$\frac{\mathbf{s}\cos\theta-\omega\sin\theta}{\mathbf{s}^2+\omega^2}$
13	$e^{-at}\sin\omega t\,u(t)$	$\longleftrightarrow$	$\frac{\omega}{(\mathbf{s}+a)^2+\omega^2}$
14	$e^{-at}\cos\omega t\,u(t)$	$\longleftrightarrow$	$\frac{\mathbf{s}+a}{(\mathbf{s}+a)^2+\omega^2}$
15	$2e^{-at}\cos(bt-\theta)\,u(t)$	$\longleftrightarrow$	$\frac{e^{j\theta}}{\mathbf{s}+a+jb}+\frac{e^{-j\theta}}{\mathbf{s}+a-jb}$
16	$\frac{2t^{n-1}}{(n-1)!}\,e^{-at}\cos(bt-\theta)\,u(t)$	$\longleftrightarrow$	$\frac{e^{j\theta}}{(\mathbf{s}+a+jb)^n}+\frac{e^{-j\theta}}{(\mathbf{s}+a-jb)^n}$

Example 10-7: Laplace Transform

Obtain the Laplace transform of $f(t) = t^2 e^{-3t} \cos 4t$.

Solution: The given function is a product of three functions. We start with the cosine function which we will call $f_1(t)$:

$$f_1(t) = \cos 4t. \quad (10.47)$$

According to entry #11 in Table 10-2, the corresponding Laplace transform is

$$\mathbf{F}_1(\mathbf{s}) = \frac{\mathbf{s}}{\mathbf{s}^2 + 16}. \quad (10.48)$$

Next, we define

$$f_2(t) = e^{-3t} \cos 4t = e^{-3t} f_1(t), \quad (10.49)$$

and we apply the frequency-shift property (entry #5 in Table 10-1) to obtain

$$\mathbf{F}_2(\mathbf{s}) = \mathbf{F}_1(\mathbf{s}+3) = \frac{\mathbf{s}+3}{(\mathbf{s}+3)^2 + 16}, \quad (10.50)$$

where we replaced $\mathbf{s}$ with $(\mathbf{s}+3)$ everywhere in the expression of Eq. (10.48). Finally, we define

$$f(t) = t^2 f_2(t) = t^2 e^{-3t} \cos 4t, \quad (10.51)$$

and we apply the frequency derivative property (entry #9 in Table 10-1) twice.

$$\begin{aligned} \mathbf{F}(\mathbf{s}) = \mathbf{F}_2''(\mathbf{s}) &= \frac{d^2}{d\mathbf{s}^2}\left[\frac{\mathbf{s}+3}{(\mathbf{s}+3)^2+16}\right] \\ &= \frac{2(\mathbf{s}+3)[(\mathbf{s}+3)^2 - 48]}{[(\mathbf{s}+3)^2+16]^3}. \end{aligned} \quad (10.52)$$

Exercise 10-7: Obtain the Laplace transform of
(a) $f_1(t) = 2(2 - e^{-t})\, u(t)$ and
(b) $f_2(t) = e^{-3t} \cos(2t + 30°)\, u(t)$.

Answer: (a) $\mathbf{F}_1(\mathbf{s}) = \dfrac{2\mathbf{s}+4}{\mathbf{s}(\mathbf{s}+1)}$,

(b) $\mathbf{F}_2(\mathbf{s}) = \dfrac{0.866\mathbf{s} + 3.6}{\mathbf{s}^2 + 6\mathbf{s} + 13}$. (See CD)

10-4 Circuit Analysis Procedure

Now that we have learned how to transform a function $f(t)$, as defined in the time domain, to its Laplace counterpart $\mathbf{F}(\mathbf{s})$, as defined in the $\mathbf{s}$-domain, we shall demonstrate the basic steps of the Laplace transform technique by analyzing a relatively simple circuit. Figure 10-5 contains a series RLC circuit with no stored energy connected to a dc voltage source V_o via a SPST switch that closes at $t = 0$. Hence, the source should be represented as

$$v_s(t) = V_o\, u(t). \quad (10.53)$$

Fundamentally, the Laplace transfer technique consists of five steps.

Step 1: Apply KCL and/or KVL to obtain the integrodifferential equation(s) of the circuit.

Step 2: Define Laplace transform currents and voltages corresponding to the time-domain currents and voltages appearing in the integrodifferential equation(s) and then transform the equations to the $\mathbf{s}$-domain.

Step 3: Solve for the variable of interest in the $\mathbf{s}$-domain.

Step 4: Apply partial fraction expansion to express the solution of Step 3 into a sum of terms of appropriate form (as discussed later in Section 10-5).

Step 5: Transfer the solution back to the time domain with the help of Tables 10-1 and 10-2.

For the circuit in Fig. 10-5, KVL at $t \geq 0^-$ gives

$$Ri + \left[\frac{1}{C}\int_{0^-}^{t} i\, dt + v_C(0^-)\right] + L\frac{di}{dt} = V_o\, u(t). \quad (10.54)$$

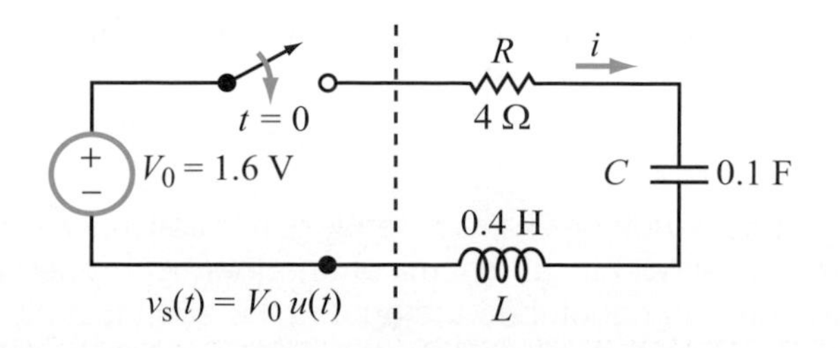

Figure 10-5: RLC circuit. The dc source (in combination with the switch) constitutes an input excitation $v_s(t) = V_o\, u(t)$.

Technology Brief 20: Carbon Nanotubes

Molecules made exclusively from carbon atoms can have very remarkable properties. Diamond, for example, occurs when carbon atoms are arranged in a specific, repeating three-dimensional arrangement (Fig. TF20-1(a)). This particular arrangement leads to a material known for its hardness, light dispersion (i.e., brilliance) and durability. Carbon atoms can be arranged into various other configurations, including flat hexagonal sheets known as ***graphene*** sheets (Fig. TF20-1(b)) and closed spheres known as ***buckyballs*** (Fig. TF20-1(c)), so named because they resemble the geodesic dome architecture developed by Buckminster Fuller. Carbon atoms can also be arranged into long tubes, called ***carbon nanotubes*** (CNTs) with single or multiple concentric walls (Fig. TF20-1(d)). Nanotubes can have radii ranging from 0.4 nm to tens of nanometers and may be several millimeters in length. Each of these arrangements leads to materials with interesting and still not entirely understood properties.

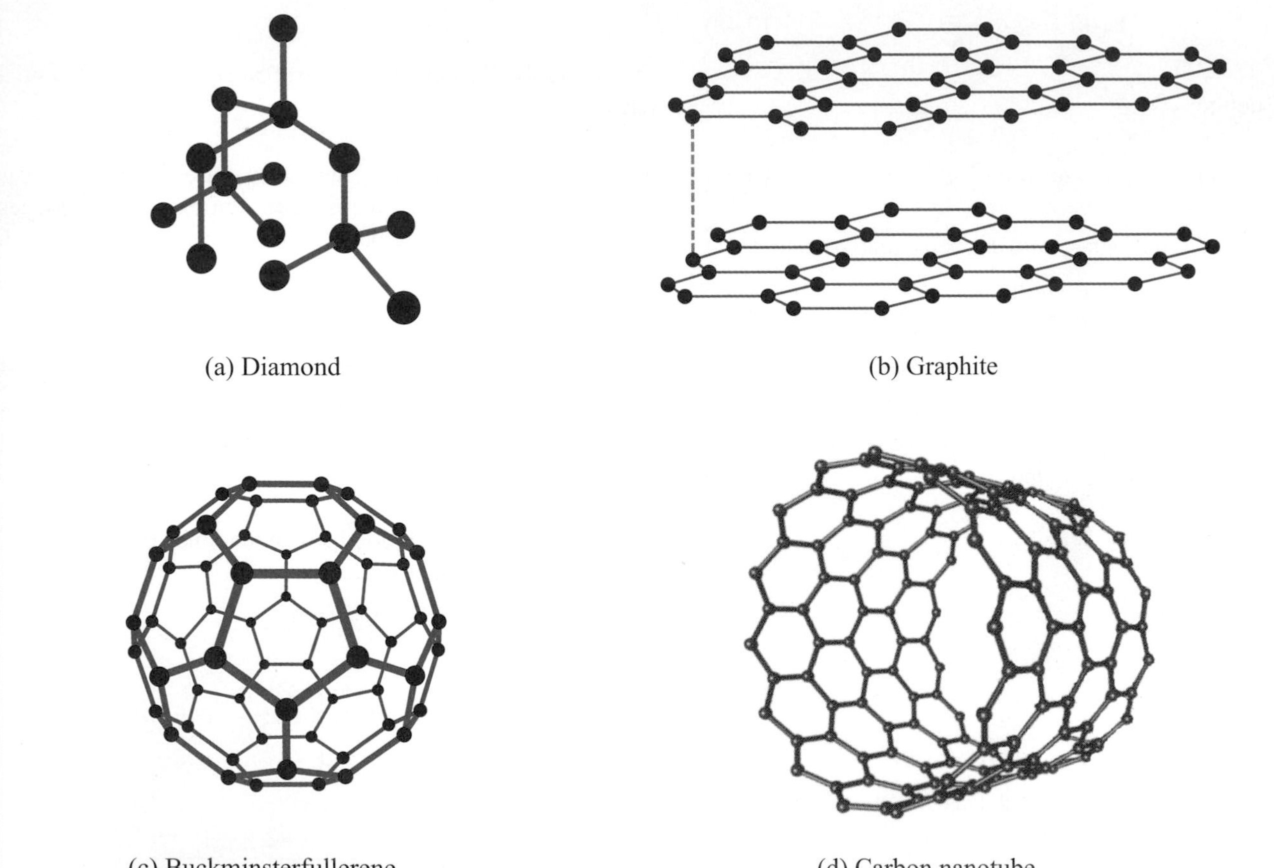

Figure TF20-1: Different configurations of carbon: (a) diamond, (b) graphene, (c) buckminsterfullerene spheres or buckyballs, and (d) carbon nanotube.

Carbon nanotubes have recently generated much interest in the academic and commercial worlds due to their unique mechanical, electrical, chemical and thermal properties. When pulled along the axis of the tube (i.e., under tensile load), carbon nanotubes are among the strongest materials, about 5 times stiffer than steel! Nanotubes can also be added to polymers to produce composite materials with great strength and durability; several companies are exploring such materials for aerospace, automotive and armor applications. For example, highly resistant automobile collision bumpers made from carbon nanotube composites recently went into production.

Nanotubes can also be modified to alter their electrical properties; they can be made conductive, like metals, or semiconductive, like silicon, via chemical modifications. As conductors, nanotubes can have theoretical current densities along the axis of the nanotube that are 1000 times higher than that of metals. Intensive research and development activities are carried out by numerous institutions aimed at building integrated circuits from combinations of conductive and semiconductive nanotubes. Carbon nanotubes also form the basis of several emerging sensor technologies (Fig. TF20-2).

By exploiting both the mechanical and electrical properties of nanotubes, it has been possible to develop new types of devices and circuits. One such example is shown in Fig. TF20-3, which displays a high-resolution image of the smallest radio transmitter built to date. It consists of a single, nanoscale carbon tube that vibrates mechanically at radio frequencies when placed in a small vacuum chamber. Amazingly, this single nanotube acts as a complete radio, including the antenna, tuner, amplifier, and demodulator!

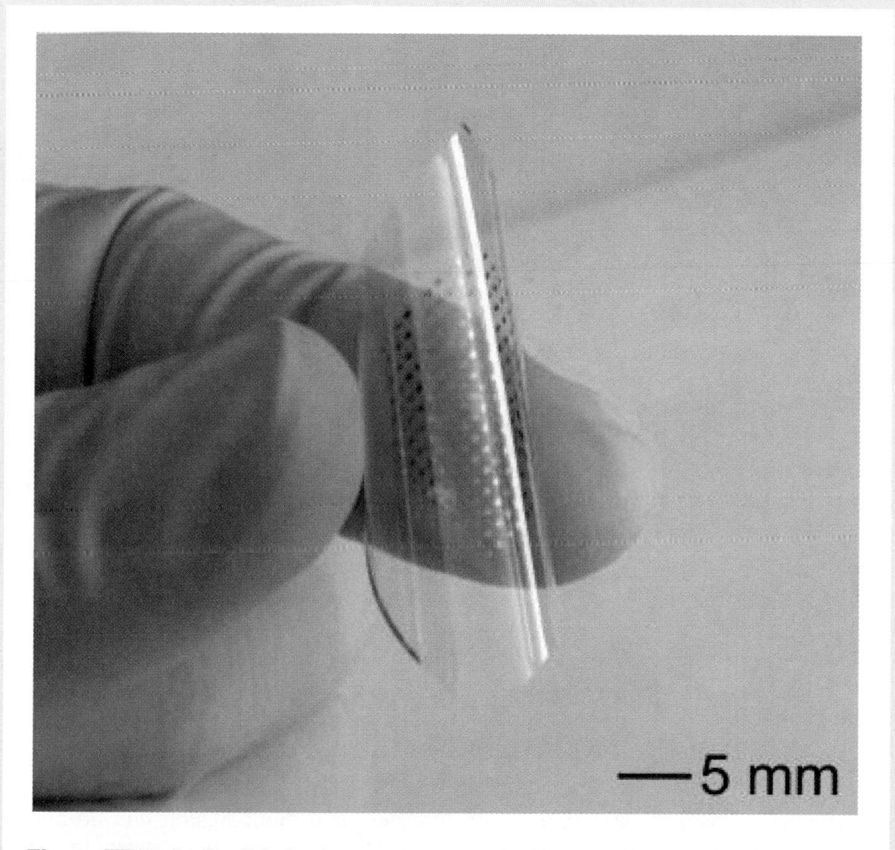

Figure TF20-2: Flexible hydrogen sensors can be made by transferring carbon nanotubes onto plastic substrates, then adding hydrogen-sensitive palladium nanoparticles to the ends. (Courtesy of Argonne National Laboratory.)

Carbon nanotubes may also enable more efficient organic photovoltaic solar cells. Organic photovoltaic devices (OPVs) are made from thin-films of organic semiconductors (as opposed to semiconductors made of silicon or gallium arsenide). Unlike silicon solar cells, these devices can be fabricated at low relative cost into large, flexible sheets. However, current OVPs are inherently poor in terms of how efficiently they can convert photons into electrons. Because carbon nanotubes can be made into good conductors, researchers have begun adding them to the OVP polymer mix to increase the internal conduction within the photoactive polymer so as to increase the photon-to-electron conversion efficiency. Nanotubes have also been proposed as a less expensive and higher conduction material for OVP electrodes.

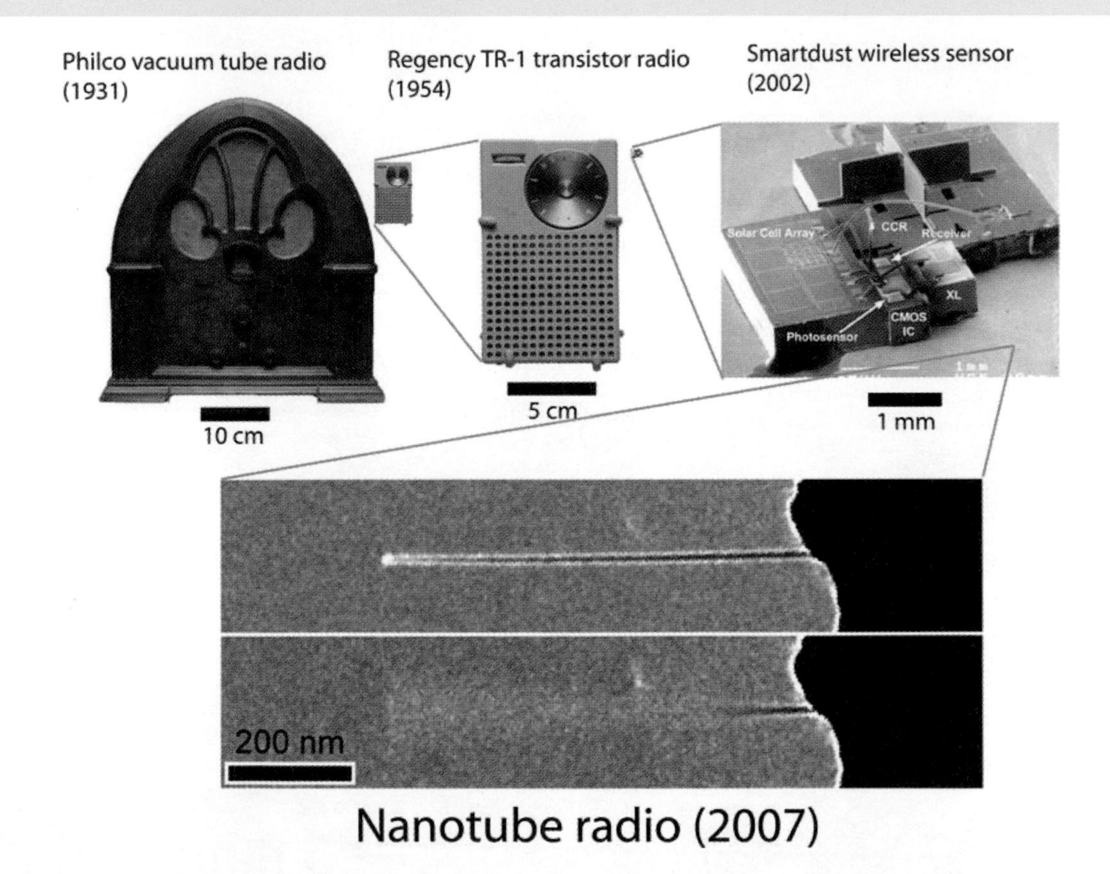

Figure TF20-3: A complete radio, including antenna, tuner, amplifier, and demodulator, can be built from a single carbon nanotube suspended from an electrode inside a vacuum chamber (bottom image shows the single nanotube when stationary and when oscillating during operation). (Courtesy Alex Zettl Research Group, Lawrence Berkeley National Laboratory and University of California at Berkeley.)

In terms of $\mathbf{I}(\mathbf{s})$, the Laplace counterpart of $i(t)$, the four terms of Eq. (10.54) have the following transforms:

$$R\, i(t) \leftrightarrow R\, \mathbf{I}(\mathbf{s}),$$

$$\frac{1}{C}\int_{0^-}^{t} i\, dt + \upsilon_C(0^-) \leftrightarrow \frac{1}{C}\left[\frac{\mathbf{I}(\mathbf{s})}{\mathbf{s}}\right] + \frac{\upsilon_C(0^-)}{\mathbf{s}},$$

(Time integral property),

$$L\,\frac{di}{dt} \leftrightarrow L[\mathbf{s}\,\mathbf{I}(\mathbf{s}) - i(0^-)] \quad \text{(Time derivative property)},$$

and

$$V_o\, u(t) \leftrightarrow \frac{V_o}{\mathbf{s}} \quad \text{(Transform of step function)}.$$

Hence, the $\mathbf{s}$-domain equivalent of Eq. (10.54) is

$$R\mathbf{I} + \frac{\mathbf{I}}{C\mathbf{s}} + L[\mathbf{s}\mathbf{I} - i(0^-)] = \frac{V_o}{\mathbf{s}}, \qquad (10.55)$$

where we have set $\upsilon_C(0^-) = 0$, because we are told that the circuit had no energy stored in it prior to $t = 0$. Moreover, from the circuit's history, we know that no current was flowing through L prior to closing the switch. Consequently, $i(0^-) = 0$. Solving for $\mathbf{I}(\mathbf{s})$ and then replacing R, L, C, and V_o with their numerical values, leads to

$$\mathbf{I}(\mathbf{s}) = \frac{V_o}{L\left[\mathbf{s}^2 + \frac{R}{L}\mathbf{s} + \frac{1}{LC}\right]} = \frac{4}{\mathbf{s}^2 + 10\mathbf{s} + 25} = \frac{4}{(\mathbf{s}+5)^2}. \qquad (10.56)$$

According to entry #6 in Table 10-2,

$$\mathcal{L}^{-1}\left[\frac{1}{(\mathbf{s}+a)^2}\right] = te^{-at}\, u(t).$$

Hence,

$$i(t) = 4te^{-5t}\, u(t). \qquad (10.57)$$

In this particular example, the expression for $\mathbf{I}(\mathbf{s})$ given by Eq. (10.56) matches one of the entries available in Table 10-2, but what should we do if it does not? We have two options:

(1) We can apply the inverse Laplace transform relation given by Eq. (10.14), which in general involves a rather cumbersome integration.

(2) We can apply the ***partial-fraction-expansion*** method to rearrange the expression for $\mathbf{I}(\mathbf{s})$ into a sum of terms, each of which has an appropriate match in Table 10-2. This latter approach is the subject of the next section.

10-5 Partial Fraction Expansion

Let us assume that after transforming the integrodifferential equation associated with a circuit of interest to the $\mathbf{s}$-domain and then solving it for the voltage or current whose behavior we wish to examine, we end up with an expression $\mathbf{F}(\mathbf{s})$. Our next step is to inverse transform $\mathbf{F}(\mathbf{s})$ to the time domain, thereby completing our solution. The degree of mathematical difficulty associated with the implementation of the inverse transformation depends on the mathematical form of $\mathbf{F}(\mathbf{s})$.

Consider, for example, the expression

$$\mathbf{F}(\mathbf{s}) = \frac{4}{\mathbf{s}+2} + \frac{6}{(\mathbf{s}^2+5)^2} + \frac{8}{\mathbf{s}^2+4\mathbf{s}+5}. \qquad (10.58)$$

The inverse transform $f(t)$ is given by

$$\begin{aligned} f(t) &= \mathcal{L}^{-1}[\mathbf{F}(\mathbf{s})] \\ &= \mathcal{L}^{-1}\left[\frac{4}{\mathbf{s}+2}\right] + \mathcal{L}^{-1}\left[\frac{6}{(\mathbf{s}+5)^2}\right] \\ &\quad + \mathcal{L}^{-1}\left[\frac{8}{\mathbf{s}^2+4\mathbf{s}+5}\right]. \end{aligned} \qquad (10.59)$$

By comparison with the entries in Table 10-2, we note:

(a) The first term in Eq. (10.59), $4/(\mathbf{s}+2)$, is functionally the same as entry #3 in Table 10-2 with $a = 2$. Hence,

$$\mathcal{L}^{-1}\left[\frac{4}{\mathbf{s}+2}\right] = 4e^{-2t}\, u(t). \qquad (10.60a)$$

(b) The second term, $6/(\mathbf{s}+5)^2$, is functionally the same as entry #6 in Table 10-2 with $a = 5$. Thus,

$$\mathcal{L}^{-1}\left[\frac{6}{(\mathbf{s}+5)^2}\right] = 6te^{-5t}\, u(t). \qquad (10.60b)$$

(c) The third term, $1/(\mathbf{s}^2+4\mathbf{s}+5)$, is similar but not identical in form with entry #13 in Table 10-2. However, it can be rearranged to assume the proper form:

$$\frac{1}{\mathbf{s}^2+4\mathbf{s}+5} = \frac{1}{(\mathbf{s}+2)^2+1}.$$

Consequently,

$$\mathcal{L}^{-1}\left[\frac{8}{(\mathbf{s}+2)^2+1}\right] = 8e^{-2t}\sin t\; u(t). \qquad (10.60c)$$

Combining the results represented by Eqs. (10.60a–c) gives:

$$f(t) = [4e^{-2t} + 6te^{-5t} + 8e^{-2t}\sin t]\, u(t). \qquad (10.61)$$

The preceding example demonstrated that the implementation of the inverse Laplace transform is a rather painless process—as long as the expression for $\mathbf{F(s)}$ is composed of a series of terms similar to those in Eq. (10.58). Usually, however, $\mathbf{F(s)}$ is not in the proper form, so we will need to reconfigure it before we can apply the inverse transform. At the most general level, $\mathbf{F(s)}$ is given by the ratio of a polynomial numerator $\mathbf{N(s)}$ to a polynomial denominator $\mathbf{D(s)}$ as

$$\mathbf{F(s)} = \frac{\mathbf{N(s)}}{\mathbf{D(s)}} = \frac{a_m \mathbf{s}^m + a_{m-1}\mathbf{s}^{m-1} + \cdots + a_1\mathbf{s} + a_0}{b_n \mathbf{s}^n + b_{n-1}\mathbf{s}^{n-1} + \cdots + b_1\mathbf{s} + b_0}, \tag{10.62}$$

where all of the a and b coefficients are real and the powers m and n are positive integers. The roots of $\mathbf{N(s)}$, namely the values of $\mathbf{s}$ at which $\mathbf{N(s)} = 0$, are called the ***zeros*** of $\mathbf{F(s)}$ and similarly the roots of $\mathbf{D(s)} = 0$ are called the ***poles*** of $\mathbf{F(s)}$. As we shall see shortly, the poles are critically important to the inverse transform process.

An equally important factor has to do with the magnitude of m relative to that of n:

(a) If $m < n$, $\mathbf{F(s)}$ is considered a ***proper rational function***, in which case $\mathbf{F(s)}$ can be expanded into a sum of partial fractions by applying the applicable recipe from among those outlined in Subsections 10-5.1 through 10-5.4.

(b) If $m \geq n$, the ratio $\mathbf{N(s)}/\mathbf{D(s)}$ is an ***improper rational function***, requiring a preparatory step of long division prior to the application of the partial-fraction-expansion recipes. Fortunately, circuit configurations leading to Laplace transforms characterized by improper rational functions are rarely (if ever) encountered in practice. Hence, we will limit all considerations in this chapter to proper rational functions.

10-5.1 Distinct Real Poles

Consider the $\mathbf{s}$-domain function

$$\mathbf{F(s)} = \frac{\mathbf{s}^2 - 4\mathbf{s} + 3}{\mathbf{s}(\mathbf{s}+1)(\mathbf{s}+3)}. \tag{10.63}$$

The poles of $F(\mathbf{s})$ are $\mathbf{s} = 0$, $\mathbf{s} = -1$, and $\mathbf{s} = -3$. All three poles are ***real*** and ***distinct***. By distinct, we mean that no two or more poles are the same. In $(\mathbf{s}+4)^2$, for example, the pole $\mathbf{s} = -4$ occurs twice, and therefore, it is not distinct. The highest power of $\mathbf{s}$ in the numerator of Eq. (10.63) is $m = 2$, and the highest power of $\mathbf{s}$ in the denominator is $n = 3$. Hence, $\mathbf{F(s)}$ is a proper rational function because $m < n$. Given these attributes, $\mathbf{F(s)}$ can be decomposed into partial fractions corresponding to the three factors in the denominator of $\mathbf{F(s)}$:

$$\mathbf{F(s)} = \frac{A_1}{\mathbf{s}} + \frac{A_2}{(\mathbf{s}+1)} + \frac{A_3}{(\mathbf{s}+3)}, \tag{10.64}$$

where A_1 to A_3 are ***expansion coefficients*** to be determined shortly. Equating the two functional forms of $\mathbf{F(s)}$, we have

$$\frac{A_1}{\mathbf{s}} + \frac{A_2}{(\mathbf{s}+1)} + \frac{A_3}{(\mathbf{s}+3)} = \frac{\mathbf{s}^2 - 4\mathbf{s} + 3}{\mathbf{s}(\mathbf{s}+1)(\mathbf{s}+3)}. \tag{10.65}$$

Associated with each expansion coefficient is a ***pole factor***; $\mathbf{s}$, $(\mathbf{s}+1)$, and $(\mathbf{s}+3)$ are the pole factors associated with A_1, A_2, and A_3, respectively. To determine the value of any expansion coefficient we multiply both sides of Eq. (10.65) by the pole factor of that expansion coefficient, and then we evaluate them at $\mathbf{s}$ = pole value of that pole factor. This procedure is called the ***residue method***.

To determine A_2, for example, we multiply both sides of Eq. (10.65) by $(\mathbf{s}+1)$, we reduce the expressions, and then we evaluate them at $\mathbf{s} = -1$:

$$\left\{(\mathbf{s}+1)\left[\frac{A_1}{\mathbf{s}} + \frac{A_2}{(\mathbf{s}+1)} + \frac{A_3}{(\mathbf{s}+3)}\right]\right\}\Bigg|_{\mathbf{s}=-1} = \left[\frac{(\mathbf{s}+1)(\mathbf{s}^2 - 4\mathbf{s} + 3)}{\mathbf{s}(\mathbf{s}+1)(\mathbf{s}+3)}\right]\Bigg|_{\mathbf{s}=-1}. \tag{10.66}$$

After reduction, the expression becomes

$$\left[\frac{A_1(\mathbf{s}+1)}{\mathbf{s}} + A_2 + \frac{A_3(\mathbf{s}+1)}{(\mathbf{s}+3)}\right]\Bigg|_{\mathbf{s}=-1} = \left[\frac{(\mathbf{s}^2 - 4\mathbf{s} + 3)}{\mathbf{s}(\mathbf{s}+3)}\right]\Bigg|_{\mathbf{s}=-1}. \tag{10.67}$$

We note that (a) the presence of $(\mathbf{s}+1)$ in the numerators of terms 1 and 3 on the left-hand side will force those terms to go to zero when evaluated at $\mathbf{s} = -1$, (b) the middle term has only A_2 in it, and (c) the reduction on the right-hand side of Eq. (10.67) eliminated the pole factor $(\mathbf{s}+1)$ from the expression. Consequently,

$$A_2 = \frac{(-1)^2 + 4 + 3}{(-1)(-1+3)} = -4.$$

Similarly,

$$A_1 = \mathbf{s}\,\mathbf{F(s)}|_{\mathbf{s}=0} = \left.\frac{\mathbf{s}^2 - 4\mathbf{s} + 3}{(\mathbf{s}+1)(\mathbf{s}+3)}\right|_{\mathbf{s}=0} = 1,$$

and

$$A_3 = (\mathbf{s}+3)\,\mathbf{F(s)}|_{\mathbf{s}=-3} = \left.\frac{\mathbf{s}^2 - 4\mathbf{s} + 3}{\mathbf{s}(\mathbf{s}+1)}\right|_{\mathbf{s}=-3} = 4.$$

Having established the values of A_1 to A_3, we now are ready to apply the inverse Laplace transform to Eq. (10.64):

$$\begin{aligned} f(t) &= \mathcal{L}^{-1}[\mathbf{F(s)}] \\ &= \mathcal{L}\left[\frac{1}{\mathbf{s}} - \frac{4}{\mathbf{s}+1} + \frac{4}{\mathbf{s}+3}\right] \\ &= [1 - 4e^{-t} + 4e^{-3t}]\,u(t). \end{aligned} \tag{10.68}$$

Building on this example, we can generalize the process to:

Distinct Real Poles

Given a proper rational function defined by

$$\mathbf{F}(\mathbf{s}) = \frac{\mathbf{N}(\mathbf{s})}{\mathbf{D}(\mathbf{s})} = \frac{\mathbf{N}(\mathbf{s})}{(\mathbf{s}+p_1)(\mathbf{s}+p_2)\dots(\mathbf{s}+p_n)}, \tag{10.69}$$

with distinct real poles $-p_1$ to $-p_n$, such that $p_i \neq p_j$ for all $i \neq j$, and $m < n$ (where m and n are the highest powers of $\mathbf{s}$ in $\mathbf{N}(\mathbf{s})$ and $\mathbf{D}(\mathbf{s})$, respectively), then $\mathbf{F}(\mathbf{s})$ can be expanded into the equivalent form:

$$\mathbf{F}(\mathbf{s}) = \frac{A_1}{\mathbf{s}+p_1} + \frac{A_2}{\mathbf{s}+p_2} + \dots + \frac{A_n}{\mathbf{s}+p_n} = \sum_{i=1}^{n} \frac{A_i}{(\mathbf{s}+p_i)}, \tag{10.70}$$

with expansion coefficients A_1 to A_n given by

$$A_i = (\mathbf{s}+p_i)\,\mathbf{F}(\mathbf{s})|_{\mathbf{s}=-p_i}, \quad i = 1, 2, \dots, n. \tag{10.71}$$

In view of entry #3 in Table 10-2, the inverse Laplace transform of Eq. (10.70) is

$$f(t) = \mathcal{L}^{-1}[\mathbf{F}(\mathbf{s})] = [A_1 e^{-p_1 t} + A_2 e^{-p_2 t} + \dots + A_n e^{-p_n t}]\, u(t). \tag{10.72}$$

Exercise 10-8: Apply the partial-fraction-expansion method to determine $f(t)$, given that its Laplace transform is

$$\mathbf{F}(\mathbf{s}) = \frac{10\mathbf{s}+16}{\mathbf{s}(\mathbf{s}+2)(\mathbf{s}+4)}.$$

Answer: $f(t) = [2 + e^{-2t} - 3e^{-4t}]\, u(t)$. (See)

10-5.2 Repeated Real Poles

We now will consider the case when $\mathbf{F}(\mathbf{s})$ is a proper rational function containing repeated real poles or a combination of distinct and repeated real poles. The partial-fraction-expansion method is outlined by the following steps.

Step 1. We are given a proper rational function $\mathbf{F}(\mathbf{s})$ composed of the product

$$\mathbf{F}(\mathbf{s}) = \mathbf{F}_1(\mathbf{s})\,\mathbf{F}_2(\mathbf{s}) \tag{10.73}$$

with

$$\mathbf{F}_1(\mathbf{s}) = \frac{\mathbf{N}(\mathbf{s})}{(\mathbf{s}+p_1)(\mathbf{s}+p_2)\dots(\mathbf{s}+p_n)} \tag{10.74}$$

and

$$\mathbf{F}_2(\mathbf{s}) = \frac{1}{(\mathbf{s}+p)^m}. \tag{10.75}$$

We note that $\mathbf{F}_1(\mathbf{s})$ is identical in form with Eq. (10.69) and contains only distinct real poles $-p_1$ to $-p_n$, thereby qualifying it for representation by a series of terms as in Eq. (10.70). The second function $\mathbf{F}_2(\mathbf{s})$ has an m-repeated pole at $\mathbf{s} = -p$, where m is a positive integer. Also, the repeated pole is not a pole of $\mathbf{F}_1(\mathbf{s})$; $p \neq p_i$ for $i = 1, 2, \dots, n$.

Step 2. Partial fraction representation for an m-repeated pole at $\mathbf{s} = -p$ consists of m terms as

$$\frac{B_1}{\mathbf{s}+p} + \frac{B_2}{(\mathbf{s}+p)^2} + \dots + \frac{B_m}{(\mathbf{s}+p)^m}. \tag{10.76}$$

Step 3. Partial fraction expansion for the combination of the product $\mathbf{F}_1(\mathbf{s})\,\mathbf{F}_2(\mathbf{s})$ is then given by

$$\begin{aligned}\mathbf{F}(\mathbf{s}) &= \frac{A_1}{\mathbf{s}+p_1} + \frac{A_2}{\mathbf{s}+p_2} + \dots + \frac{A_n}{\mathbf{s}+p_n} \\ &\quad + \frac{B_1}{\mathbf{s}+p} + \frac{B_2}{(\mathbf{s}+p)^2} + \dots + \frac{B_m}{(\mathbf{s}+p)^m} \\ &= \sum_{i=1}^{n} \frac{A_i}{\mathbf{s}+p_i} + \sum_{j=1}^{m} \frac{B_j}{(\mathbf{s}+p)^j}.\end{aligned} \tag{10.77}$$

Step 4. Expansion coefficients A_1 to A_n are determined by applying Eq. (10.71):

$$A_i = (\mathbf{s}+p_i)\,\mathbf{F}(\mathbf{s})|_{\mathbf{s}=-p_i}, \quad i = 1, 2, \dots, n. \tag{10.78}$$

Repeated Real Poles

Expansion coefficients B_1 to B_m are determined through a procedure that involves multiplication by $(\mathbf{s}+p)^m$, differentiation with respect to $\mathbf{s}$, and evaluation at $\mathbf{s}=-p$:

$$B_j = \left\{ \frac{1}{(m-j)!} \frac{d^{m-j}}{d\mathbf{s}^{m-j}} [(\mathbf{s}+p)^m \, \mathbf{F}(\mathbf{s})] \right\}\Bigg|_{\mathbf{s}=-p}, \quad j = 1, 2, \ldots, m. \tag{10.79}$$

For the m, $m-1$, and $m-2$ terms, Eq. (10.79) reduces to

$$B_m = (\mathbf{s}+p)^m \, \mathbf{F}(\mathbf{s})|_{\mathbf{s}=-p}, \tag{10.80a}$$

$$B_{m-1} = \left\{ \frac{d}{d\mathbf{s}} [(\mathbf{s}+p)^m \, \mathbf{F}(\mathbf{s})] \right\}\Bigg|_{\mathbf{s}=-p}, \tag{10.80b}$$

$$B_{m-2} = \left\{ \frac{1}{2!} \frac{d^2}{d\mathbf{s}^2} [(\mathbf{s}+p)^m \, \mathbf{F}(\mathbf{s})] \right\}\Bigg|_{\mathbf{s}=-p}. \tag{10.80c}$$

Thus, the evaluation of B_m does not involve any differentiation, that of B_{m-1} involves differentiation with respect to $\mathbf{s}$ only once (and division by 1!), and that of B_{m-2} involves differentiation twice and division by 2!. In practice, it is easiest to start by evaluating B_m first and then evaluating the other expansion coefficients in descending order.

Step 5. Once the values of all of the expansion coefficients of Eq. (10.77) have been determined, transformation to the time domain is accomplished by applying entry #8 of Table 10-2,

$$\mathcal{L}^{-1}\left[\frac{(n-1)!}{(\mathbf{s}+a)^n}\right] = t^{n-1} e^{-at} \, u(t). \tag{10.81}$$

The result is

$$f(t) = \mathcal{L}^{-1}[\mathbf{F}(\mathbf{s})] = \left[\sum_{i=1}^{n} A_i e^{-p_i t} + \sum_{j=1}^{m} \frac{B_j t^{j-1}}{(j-1)!} e^{-pt} \right] u(t). \tag{10.82}$$

Example 10-8: Repeated Poles

Determine the inverse Laplace transform of

$$\mathbf{F}(\mathbf{s}) = \frac{\mathbf{N}(\mathbf{s})}{\mathbf{D}(\mathbf{s})} = \frac{\mathbf{s}^2+3\mathbf{s}+3}{\mathbf{s}^4+11\mathbf{s}^3+45\mathbf{s}^2+81\mathbf{s}+54}.$$

Solution: In theory, any polynomial with real coefficients can be expressed as a product of linear and quadratic factors (of the form $(s+p)$ and (s^2+as+b), respectively). The process involves long division, but it requires knowledge of the roots of the polynomial, which can be determined through the application of numerical techniques. In the present case, a random check reveals that $\mathbf{s}=-2$ and $\mathbf{s}=-3$ are roots of $\mathbf{D}(\mathbf{s})$. Given that $\mathbf{D}(\mathbf{s})$ is fourth order, it should have four roots, including possible duplicates.

Since $\mathbf{s}=-2$ is a root of $\mathbf{D}(\mathbf{s})$, we should be able to factor out $(\mathbf{s}+2)$ from it. Long division gives

$$\begin{aligned} \mathbf{D}(\mathbf{s}) &= \mathbf{s}^4+11\mathbf{s}^3+45\mathbf{s}^2+81\mathbf{s}+54 \\ &= (\mathbf{s}+2)(\mathbf{s}^3+9\mathbf{s}^2+27\mathbf{s}+27). \end{aligned}$$

Next, we factor out $(\mathbf{s}+3)$ by

$$\begin{aligned} \mathbf{D}(\mathbf{s}) &= (\mathbf{s}+2)(\mathbf{s}+3)(\mathbf{s}^2+6\mathbf{s}+9) \\ &= (\mathbf{s}+2)(\mathbf{s}+3)^3. \end{aligned}$$

Hence, $\mathbf{F}(\mathbf{s})$ has a distinct real pole at $\mathbf{s}=-2$ and a thrice repeated pole at $\mathbf{s}=-3$. The given expression can be rewritten as

$$\mathbf{F}(\mathbf{s}) = \frac{\mathbf{s}^2+3\mathbf{s}+3}{(\mathbf{s}+2)(\mathbf{s}+3)^3}.$$

Through partial fraction expansion, $\mathbf{F}(\mathbf{s})$ can be decomposed into

$$\mathbf{F}(\mathbf{s}) = \frac{A}{\mathbf{s}+2} + \frac{B_1}{\mathbf{s}+3} + \frac{B_2}{(\mathbf{s}+3)^2} + \frac{B_3}{(\mathbf{s}+3)^3},$$

with

$$A = (\mathbf{s}+2)\,\mathbf{F}(\mathbf{s})|_{\mathbf{s}=-2} = \frac{\mathbf{s}^2+3\mathbf{s}+3}{(\mathbf{s}+3)^3}\bigg|_{\mathbf{s}=-2} = 1,$$

$$B_3 = (\mathbf{s}+3)^3\,\mathbf{F}(\mathbf{s})|_{\mathbf{s}=-3} = \frac{\mathbf{s}^2+3\mathbf{s}+3}{\mathbf{s}+2}\bigg|_{\mathbf{s}=-3} = -3,$$

$$B_2 = \frac{d}{d\mathbf{s}} [(\mathbf{s}+3)^3\,\mathbf{F}(\mathbf{s})]\bigg|_{\mathbf{s}=-3} = 0,$$

and

$$B_1 = \frac{1}{2}\frac{d^2}{d\mathbf{s}^2} [(\mathbf{s}+3)^3\,\mathbf{F}(\mathbf{s})]\bigg|_{\mathbf{s}=-3} = -1.$$

Hence,

$$\mathbf{F}(\mathbf{s}) = \frac{1}{\mathbf{s}+2} - \frac{1}{\mathbf{s}+3} - \frac{3}{(\mathbf{s}+3)^3},$$

and application of Eq. (10.81) leads to

$$\mathcal{L}^{-1}[\mathbf{F}(\mathbf{s})] = \left[e^{-2t} - e^{-3t} - \frac{3}{2}\,t^2 e^{-3t} \right] u(t).$$

Review Question 10-8: What purpose does the partial-fraction-expansion method serve?

Review Question 10-9: When evaluating the expansion coefficients of a function containing repeated poles, is it more practical to start by evaluating the coefficient of the fraction with the lowest-order pole or that with the highest-order pole? Why?

Exercise 10-9: Determine the inverse Laplace transform of

$$\mathbf{F}(\mathbf{s}) = \frac{4\mathbf{s}^2 - 15\mathbf{s} - 10}{(\mathbf{s}+2)^3}.$$

Answer: $f(t) = (18t^2 - 31t + 4)e^{-2t}\, u(t)$. (See ◎)

10-5.3 Distinct Complex Poles

The Laplace transform of a certain circuit is given by

$$\mathbf{F}(\mathbf{s}) = \frac{4\mathbf{s}+1}{(\mathbf{s}+1)(\mathbf{s}^2+4\mathbf{s}+13)}. \qquad (10.83)$$

In addition to the simple-pole factor, the denominator includes a quadratic-pole factor with roots $\mathbf{s}_1$ and $\mathbf{s}_2$. Solution of $\mathbf{s}^2 + 4\mathbf{s} + 13 = 0$ gives

$$\mathbf{s}_1 = -2 + j3, \quad \text{and} \quad \mathbf{s}_2 = -2 - j3. \qquad (10.84)$$

The fact that the two roots are complex conjugates of one another is a consequence of the property that *for any physically realizable circuit, if it has any complex poles, those poles always appear in conjugate pairs.*

In view of Eq. (10.84), the quadratic factor is given by

$$\mathbf{s}^2 + 4\mathbf{s} + 13 = (\mathbf{s}+2-j3)(\mathbf{s}+2+j3), \qquad (10.85)$$

and $\mathbf{F}(\mathbf{s})$ now can be expanded into partial fractions as

$$\mathbf{F}(\mathbf{s}) = \frac{A}{\mathbf{s}+1} + \frac{\mathbf{B}_1}{\mathbf{s}+2-j3} + \frac{\mathbf{B}_2}{\mathbf{s}+2+j3}. \qquad (10.86)$$

Expansion coefficients $\mathbf{B}_1$ and $\mathbf{B}_2$ are printed in bold letters to signify the fact that they may be complex quantities. Determination of A, $\mathbf{B}_1$, and $\mathbf{B}_2$ follows the same factor-multiplication technique employed in Section 10-5.1 with

$$A = (\mathbf{s}+1)\,\mathbf{F}(\mathbf{s})\big|_{\mathbf{s}=-1} = \left.\frac{4\mathbf{s}+1}{\mathbf{s}^2+4\mathbf{s}+13}\right|_{\mathbf{s}=-1} = -0.3, \qquad (10.87a)$$

$$\begin{aligned}\mathbf{B}_1 &= (\mathbf{s}+2-j3)\,\mathbf{F}(\mathbf{s})\big|_{\mathbf{s}=-2+j3} \\ &= \left.\frac{4\mathbf{s}+1}{(\mathbf{s}+1)(\mathbf{s}+2+j3)}\right|_{\mathbf{s}=-2+j3} \\ &= \frac{4(-2+j3)+1}{(-2+j3+1)(-2+j3+2+j3)} \\ &= \frac{-7+j12}{-18-j6} = 0.73e^{-j78.2^\circ}, \end{aligned} \qquad (10.87b)$$

and

$$\begin{aligned}\mathbf{B}_2 &= (\mathbf{s}+2+j3)\,\mathbf{F}(\mathbf{s})\big|_{\mathbf{s}=-2-j3} \\ &= \left.\frac{4\mathbf{s}+1}{(\mathbf{s}+1)(\mathbf{s}+2-j3)}\right|_{\mathbf{s}=-2-j3} = 0.73e^{j78.2^\circ}. \end{aligned} \qquad (10.87c)$$

We observe that $\mathbf{B}_2 = \mathbf{B}_1^*$. In fact, *the expansion coefficients associated with conjugate poles are always conjugate pairs themselves.*

The inverse Laplace transform of Eq. (10.86) is

$$\begin{aligned} f(t) &= \mathcal{L}^{-1}[\mathbf{F}(\mathbf{s})] \\ &= \mathcal{L}^{-1}\left(\frac{-0.3}{\mathbf{s}+1}\right) + \mathcal{L}^{-1}\left(\frac{0.73e^{-j78.2^\circ}}{\mathbf{s}+2-j3}\right) \\ &\quad + \mathcal{L}^{-1}\left(\frac{0.73e^{j78.2^\circ}}{\mathbf{s}+2+j3}\right) \\ &= [-0.3e^{-t} + 0.73e^{-j78.2^\circ}e^{-(2-j3)t} \\ &\quad + 0.73e^{j78.2^\circ}e^{-(2+j3)t}]\, u(t). \end{aligned} \qquad (10.88)$$

Because complex numbers do not belong in the time domain, our initial reaction to their presence in the solution given by Eq. (10.88) is that perhaps an error was committed somewhere along the way. The truth is the solution is correct, but incomplete. Terms 2 and 3 are conjugate pairs, so by applying Euler's formula, they can be combined into a single term containing real quantities only:

$$\begin{aligned} &0.73e^{-j78.2^\circ}e^{-(2-j3)t} + 0.73e^{j78.2^\circ}e^{-(2+j3)t} \\ &= 0.73e^{-2t}[e^{j(3t-78.2^\circ)} + e^{-j(3t-78.2^\circ)}] \\ &= 2 \times 0.73e^{-2t}\cos(3t - 78.2^\circ) \\ &= 1.46e^{-2t}\cos(3t - 78.2^\circ). \end{aligned} \qquad (10.89)$$

Hence, the final time-domain solution is

$$f(t) = [-0.3e^{-t} + 1.46e^{-2t}\cos(3t - 78.2^\circ)]\, u(t). \qquad (10.90)$$

Exercise 10-10: Determine the inverse Laplace transform of

$$\mathbf{F(s)} = \frac{2\mathbf{s} + 14}{\mathbf{s}^2 + 6\mathbf{s} + 25}.$$

Answer: $f(t) = [2\sqrt{2}\, e^{-3t} \cos(4t - 45°)]\, u(t)$. (See)

10-5.4 Repeated Complex Poles

If the Laplace transform $\mathbf{F(s)}$ contains repeated complex poles, we can expand it into partial fractions by using a combination of the tools introduced in Sections 10-5.2 and 10-5.3. The process is illustrated in Example 10-9.

Example 10-9: Five-Pole Function

Determine the inverse Laplace transform of

$$\mathbf{F(s)} = \frac{108(\mathbf{s}^2 + 2)}{(\mathbf{s} + 2)(\mathbf{s}^2 + 10\mathbf{s} + 34)^2}.$$

Solution: The roots of

$$\mathbf{s}^2 + 10\mathbf{s} + 34 = 0$$

are

$$\mathbf{s}_1 = -5 - j3$$

and

$$\mathbf{s}_2 = -5 + j3.$$

Hence,

$$\mathbf{F(s)} = \frac{108(\mathbf{s}^2 + 2)}{(\mathbf{s} + 2)(\mathbf{s} + 5 + j3)^2(\mathbf{s} + 5 - j3)^2},$$

and its partial fraction expansion can be expressed as

$$\mathbf{F(s)} = \frac{A}{\mathbf{s} + 2} + \frac{\mathbf{B}_1}{\mathbf{s} + 5 + j3} + \frac{\mathbf{B}_2}{(\mathbf{s} + 5 + j3)^2} + \frac{\mathbf{B}_1^*}{\mathbf{s} + 5 - j3} + \frac{\mathbf{B}_2^*}{(\mathbf{s} + 5 - j3)^2},$$

where $\mathbf{B}_1^*$ and $\mathbf{B}_2^*$ are the complex conjugates of $\mathbf{B}_1$ and $\mathbf{B}_2$, respectively. Coefficients A, $\mathbf{B}_1$, and $\mathbf{B}_2$ are evaluated as follows:

$$A = (\mathbf{s} + 2)\, \mathbf{F(s)}|_{\mathbf{s}=-2} = \left.\frac{108(\mathbf{s}^2 + 2)}{(\mathbf{s}^2 + 10\mathbf{s} + 34)^2}\right|_{\mathbf{s}=-2} = 2,$$

$$\begin{aligned}
\mathbf{B}_2 &= (\mathbf{s} + 5 + j3)^2\, \mathbf{F(s)}|_{\mathbf{s}=-5-j3} \\
&= \left.\frac{108(\mathbf{s}^2 + 2)}{(\mathbf{s} + 2)(\mathbf{s} + 5 - j3)^2}\right|_{\mathbf{s}=-5-j3} \\
&= \frac{108[(-5 - j3)^2 + 2]}{(-5 - j3 + 2)(-5 - j3 + 5 - j3)^2} \\
&= 24 + j6 = 24.74e^{j14°},
\end{aligned}$$

and

$$\begin{aligned}
\mathbf{B}_1 &= \left.\frac{d}{d\mathbf{s}}[(\mathbf{s} + 5 + j3)^2\, \mathbf{F(s)}]\right|_{\mathbf{s}=-5-j3} \\
&= \left.\frac{d}{d\mathbf{s}}\left[\frac{108(\mathbf{s}^2 + 2)}{(\mathbf{s} + 2)(\mathbf{s} + 5 - j3)^2}\right]\right|_{\mathbf{s}=-5-j3} \\
&= \left[\frac{108(2\mathbf{s})}{(\mathbf{s} + 2)(\mathbf{s} + 5 - j3)^2} - \frac{108(\mathbf{s}^2 + 2)}{(\mathbf{s} + 2)^2(\mathbf{s} + 5 - j3)^2}\right. \\
&\qquad \left.\left. - \frac{2 \times 108(\mathbf{s}^2 + 2)}{(\mathbf{s} + 2)(\mathbf{s} + 5 - j3)^3}\right]\right|_{\mathbf{s}=-5-j3} \\
&= -(1 + j9) = 9.06e^{-j96.34°}.
\end{aligned}$$

The remaining constants are

$$\mathbf{B}_1^* = 9.06e^{j96.34°},$$

and

$$\mathbf{B}_2^* = 24.74e^{-j14°},$$

and the inverse Laplace transform is

$$\begin{aligned}
f(t) &= \mathcal{L}^{-1}[\mathbf{F(s)}] \\
&= \mathcal{L}^{-1}\left[\frac{2}{\mathbf{s} + 2} + \frac{9.06e^{-j96.34°}}{\mathbf{s} + 5 + j3} + \frac{9.06e^{j96.34°}}{\mathbf{s} + 5 - j3}\right. \\
&\qquad \left. + \frac{24.74e^{j14°}}{(\mathbf{s} + 5 + j3)^2} + \frac{24.74e^{-j14°}}{(\mathbf{s} + 5 - j3)^2}\right] \\
&= \big[2e^{-2t} \\
&\quad + 9.06(e^{-j96.34°}e^{-(5+j3)t} + e^{j96.34°}e^{-(5-j3)t}) \\
&\quad + 24.74t(e^{j14°}e^{-(5+j3)t} + e^{-j14°}e^{-(5-j3)t})\big]\, u(t) \\
&= [2e^{-2t} + 18.12e^{-5t}\cos(3t + 96.34°) \\
&\quad + 49.48te^{-5t}\cos(3t - 14°)]\, u(t).
\end{aligned}$$

Table 10-3: Transform pairs for four types of poles.

Pole	$\mathbf{F}(\mathbf{s})$	$f(t)$
1. Distinct real	$\dfrac{A}{\mathbf{s}+a}$	$Ae^{-at}\,u(t)$
2. Repeated real	$\dfrac{A}{(\mathbf{s}+a)^n}$	$A\,\dfrac{t^{n-1}}{(n-1)!}\,e^{-at}\,u(t)$
3. Distinct complex	$\left[\dfrac{Ae^{j\theta}}{\mathbf{s}+a+jb}+\dfrac{Ae^{-j\theta}}{\mathbf{s}+a-jb}\right]$	$2Ae^{-at}\cos(bt-\theta)\,u(t)$
4. Repeated complex	$\left[\dfrac{Ae^{j\theta}}{(\mathbf{s}+a+jb)^n}+\dfrac{Ae^{-j\theta}}{(\mathbf{s}+a-jb)^n}\right]$	$\dfrac{2At^{n-1}}{(n-1)!}\,e^{-at}\cos(bt-\theta)\,u(t)$

Example 10-10: Interesting Transform!

Determine the time-domain equivalent of the Laplace transform

$$\mathbf{F}(\mathbf{s}) = \frac{\mathbf{s}e^{-3\mathbf{s}}}{\mathbf{s}^2+4}.$$

Solution: We start by separating out the exponential $e^{-3\mathbf{s}}$ from the remaining polynomial fraction. We do so by defining

$$\mathbf{F}(\mathbf{s}) = e^{-3\mathbf{s}}\,\mathbf{F}_1(\mathbf{s}),$$

where

$$\mathbf{F}_1(\mathbf{s}) = \frac{\mathbf{s}}{\mathbf{s}^2+4} = \frac{\mathbf{s}}{(\mathbf{s}+j2)(\mathbf{s}-j2)} = \frac{\mathbf{B}_1}{\mathbf{s}+j2}+\frac{\mathbf{B}_2}{\mathbf{s}-j2}$$

with

$$\mathbf{B}_1 = (\mathbf{s}+j2)\,\mathbf{F}(\mathbf{s})\big|_{\mathbf{s}=-j2} = \left.\frac{\mathbf{s}}{\mathbf{s}-j2}\right|_{\mathbf{s}=-j2} = \frac{-j2}{-j4} = \frac{1}{2}$$

and

$$\mathbf{B}_2 = \mathbf{B}_1^* = \frac{1}{2}.$$

Hence,

$$\mathbf{F}(\mathbf{s}) = e^{-3\mathbf{s}}\,\mathbf{F}_1(\mathbf{s}) = \frac{e^{-3\mathbf{s}}}{2(\mathbf{s}+j2)}+\frac{e^{-3\mathbf{s}}}{2(\mathbf{s}-j2)}.$$

By invoking property #3a of Table 10-2, we obtain the inverse Laplace transform

$$\begin{aligned} f(t) &= \mathcal{L}^{-1}[\mathbf{F}(\mathbf{s})] \\ &= \mathcal{L}^{-1}\left[\frac{1}{2}\,\frac{e^{-3\mathbf{s}}}{\mathbf{s}+j2}+\frac{1}{2}\,\frac{e^{-3\mathbf{s}}}{\mathbf{s}-j2}\right] \\ &= \left[\frac{1}{2}(e^{-j2(t-3)}+e^{j2(t-3)})\right]u(t-3) \\ &= [\cos(2t-6)]\,u(t-3). \end{aligned}$$

We conclude this section with Table 10-3, which lists $\mathbf{F}(\mathbf{s})$ and its corresponding inverse transform $f(t)$ for all combinations of real versus complex and distinct versus repeated poles.

10-6 s-Domain Circuit Element Models

In Chapter 7, we applied the phasor-domain technique to analyze ac circuits under steady-state conditions. The s-domain technique can do the same, but it also can be used to analyze circuits excited by sources with any type of variation—including pulse, step, ramp, and exponential—and provides a complete solution that incorporates both the steady-state and transient components of the overall response. Despite the differences between the two techniques, they nevertheless share a number of similarities. One of these similarities has to do with the circuit elements. In the phasor domain, elements R, L, and C are represented by impedances $\mathbf{Z}_R$, $\mathbf{Z}_L = j\omega L$, and $\mathbf{Z}_C = 1/j\omega C$, respectively. A somewhat similar representation occurs in the s-domain with one additional important attribute:

The **s**-domain transformation of circuit elements incorporates initial conditions associated with any energy storage that may have existed in capacitors and inductors at $t = 0^-$.

Resistor in the s-Domain

Application of the Laplace transform to Ohm's law,

$$\mathcal{L}[v] = \mathcal{L}[Ri], \qquad (10.91)$$

leads to

$$\mathbf{V} = R\mathbf{I}, \qquad (10.92)$$

where by definition,

$$\mathbf{V} = \mathcal{L}[v] \quad \text{and} \quad \mathbf{I} = \mathcal{L}[i]. \qquad (10.93)$$

Hence the correspondence between the time and **s**-domains is

$$v = Ri \quad \longleftrightarrow \quad \mathbf{V} = R\mathbf{I}. \qquad (10.94)$$

Inductor in the s-Domain

For R, the form of the i–v relationship remained invariant under the transformation to the **s**-domain. That is not the case for L and C. Application of the Laplace transform to the i–v relationship of the inductor,

$$\mathcal{L}[v] = \mathcal{L}\left[L\,\frac{di}{dt}\right], \qquad (10.95)$$

gives

$$\mathbf{V} = L[\mathbf{sI} - i(0^-)], \qquad (10.96)$$

where $i(0^-)$ is the current that was flowing through the inductor at $t = 0^-$. The time-differentiation property (#6 in Table 10-1) was used in obtaining Eq. (10.96). The correspondence between the two domains is expressed as

$$v = L\,\frac{di}{dt} \quad \longleftrightarrow \quad \mathbf{V} = \mathbf{s}L\mathbf{I} - L\,i(0^-). \qquad (10.97)$$

In the **s**-domain, an inductor is represented by an impedance $\mathbf{Z}_L = \mathbf{s}L$, in series with a dc voltage source given by $L\,i(0^-)$ or—through source transformation—in parallel with a dc current source $i(0^-)/\mathbf{s}$, as shown in Table 10-4. Note that the current $\mathbf{I}$ flows from (−) to (+) through the dc voltage source (if $i(0^-)$ is positive).

Capacitor in the s-Domain

Similarly,

$$i = C\,\frac{dv}{dt} \quad \longleftrightarrow \quad \mathbf{I} = \mathbf{s}C\mathbf{V} - C\,v(0^-), \qquad (10.98)$$

where $v(0^-)$ is the initial voltage across the capacitor. The **s**-domain circuit models for the capacitor are available in Table 10-4.

Impedances $\mathbf{Z}_R$, $\mathbf{Z}_L$, and $\mathbf{Z}_C$ are defined in the **s**-domain in terms of voltage to current ratios under zero initial conditions $[i(0^-) = v(0^-) = 0]$:

$$\mathbf{Z}_R = R, \qquad \mathbf{Z}_L = \mathbf{s}L, \qquad \text{and} \qquad \mathbf{Z}_C = \frac{1}{\mathbf{s}C}. \qquad (10.99)$$

These expressions reduce to the phasor-domain expressions when $\mathbf{s} = j\omega$.

Exercise 10-11: Convert the circuit in Fig. E10.11 into the **s**-domain.

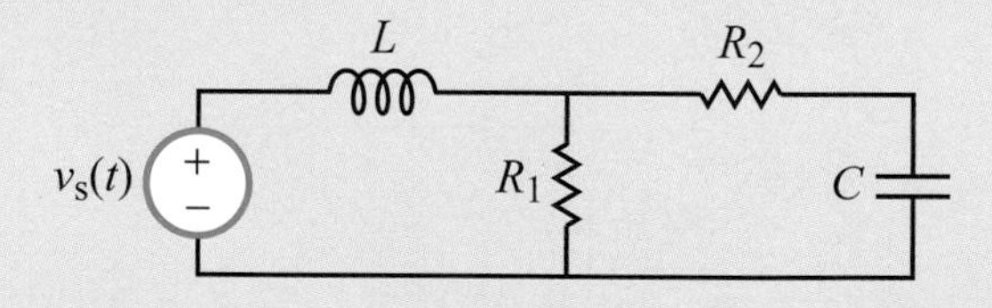

Figure E10.11

Answer:

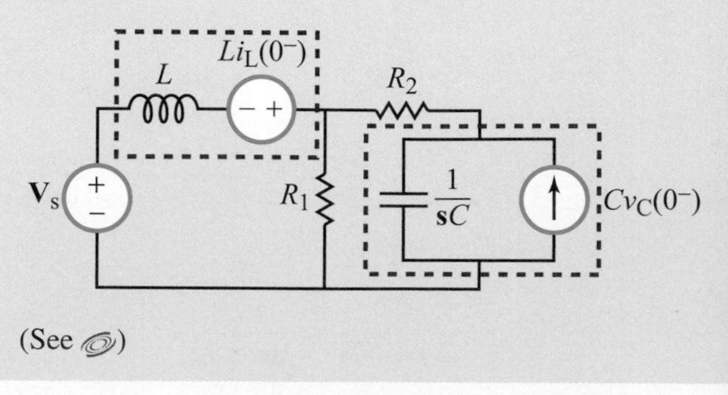

(See)

Table 10-4: Circuit models for R, L, and C in the **s**-domain.

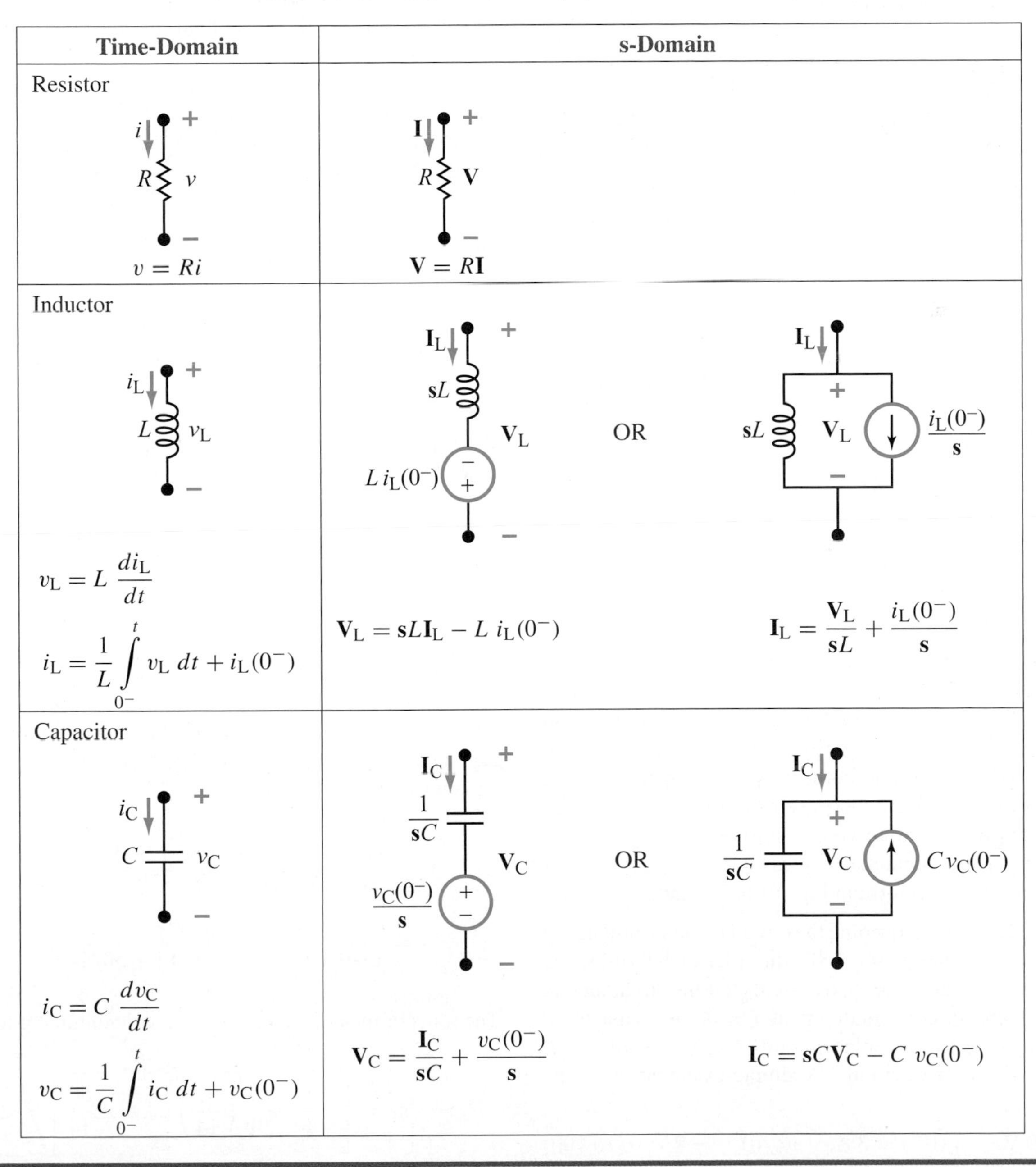

Time-Domain	s-Domain		
Resistor $v = Ri$	$\mathbf{V} = R\mathbf{I}$		
Inductor $v_L = L \dfrac{di_L}{dt}$ $i_L = \dfrac{1}{L}\displaystyle\int_{0^-}^{t} v_L \, dt + i_L(0^-)$	$\mathbf{V}_L = \mathbf{s}L\mathbf{I}_L - L\, i_L(0^-)$	OR	$\mathbf{I}_L = \dfrac{\mathbf{V}_L}{\mathbf{s}L} + \dfrac{i_L(0^-)}{\mathbf{s}}$
Capacitor $i_C = C \dfrac{dv_C}{dt}$ $v_C = \dfrac{1}{C}\displaystyle\int_{0^-}^{t} i_C \, dt + v_C(0^-)$	$\mathbf{V}_C = \dfrac{\mathbf{I}_C}{\mathbf{s}C} + \dfrac{v_C(0^-)}{\mathbf{s}}$	OR	$\mathbf{I}_C = \mathbf{s}C\mathbf{V}_C - C\, v_C(0^-)$

10-7 s-Domain Circuit Analysis

The circuit laws and analysis tools we used earlier in the time and phasor domains are equally applicable in the **s**-domain. They include KVL and KCL; voltage and current division; source transformation; source superposition; and Thévenin and Norton equivalent circuits. Execution of the **s**-domain analysis technique entails the following four steps.

Solution Procedure: s-Domain Technique

Step 1: Transform the circuit from the time domain to the **s**-domain.

Step 2: Apply KVL, KCL, and the other circuit tools to obtain an explicit expression for the voltage or current of interest.

Step 3: If necessary, expand the expression into partial fractions.

Step 4: Use the list of transform pairs given in Tables 10-2 and 10-3 and the list of properties in Table 10-1 (if needed) to transform the partial fraction to the time domain.

This process is illustrated through Examples 10-11 through 10-13, involving circuits excited by a variety of different waveforms.

Example 10-11: Interrupted Voltage Source

The circuit shown in Fig. 10-6(a) is driven by an input voltage source $\upsilon_{\text{in}}(t)$, and its output is taken across the 3-Ω resistor. The input waveform is depicted in Fig. 10-6(b). It starts out as a 15-V dc level that had existed for a long time prior to $t = 0$, then it experiences a momentary drop down to zero at $t = 0$, followed by a slow recovery towards its earlier level. The waveform of the output voltage is shown in Fig. 10-6(c). Analyze the circuit to obtain an expression for $\upsilon_{\text{out}}(t)$, in order to confirm that the waveform in Fig. 10-6(c) is indeed correct.

Solution: Before transforming the circuit to the **s**-domain, we always should evaluate it at $t = 0^-$, in order to determine the voltages across all capacitors and currents through all inductors at $t = 0^-$. The circuit condition at $t = 0^-$ is depicted in Fig. 10-6(d), where C is replaced with an open circuit and L is replaced with a short circuit. A simple examination of the circuit reveals:

$$\upsilon_C(0^-) = 9\text{ V}, \quad i_L(0^-) = 3\text{ A}, \quad \upsilon_{\text{out}}(0^-) = 9\text{ V}. \tag{10.100}$$

Next, we need to transform the circuit in Fig. 10-6(a) to the **s**-domain. The voltage waveform is given by

$$\upsilon_{\text{in}}(t) = \begin{cases} 15\text{ V} & \text{for } t \le 0^- \\ 15(1 - e^{-2t})\, u(t)\text{ V} & \text{for } t \ge 0. \end{cases} \tag{10.101}$$

With the help of Table 10-2, the corresponding **s**-domain function for $t \ge 0$ is given by

$$\mathbf{V}_{\text{in}}(\mathbf{s}) = \frac{15}{\mathbf{s}} - \frac{15}{\mathbf{s}+2}. \tag{10.102}$$

The **s**-domain circuit is shown in Fig. 10-6(e), in which L and C are represented by their **s**-domain models in accordance with Table 10-4, namely:

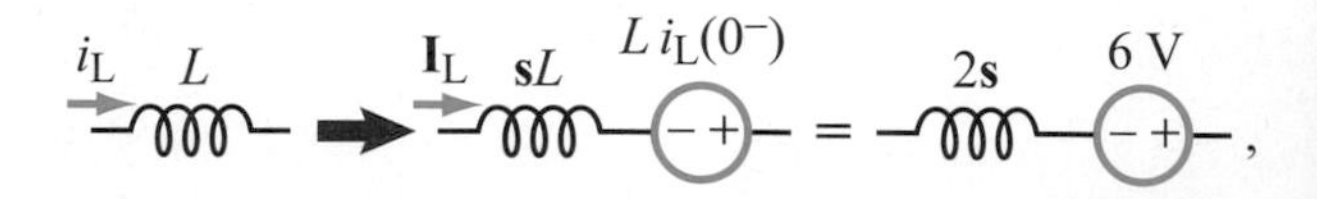

and

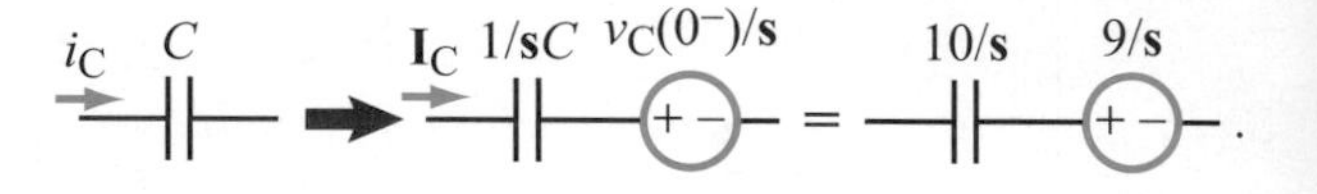

By inspection, the mesh-current equations for loops 1 and 2 are given by

$$\left(2 + 5 + \frac{10}{\mathbf{s}}\right)\mathbf{I}_1 - \left(5 + \frac{10}{\mathbf{s}}\right)\mathbf{I}_2 = \mathbf{V}_{\text{in}} - \frac{9}{\mathbf{s}}, \tag{10.103}$$

and

$$-\left(5 + \frac{10}{\mathbf{s}}\right)\mathbf{I}_1 + \left(8 + 2\mathbf{s} + \frac{10}{\mathbf{s}}\right)\mathbf{I}_2 = \frac{9}{\mathbf{s}} + 6. \tag{10.104}$$

After replacing $\mathbf{V}_{\text{in}}(\mathbf{s})$ with the expression given by Eq. (10.102), simultaneous solution of the two linear equations leads to

$$\begin{aligned}\mathbf{I}_2 &= \frac{42\mathbf{s}^3 + 162\mathbf{s}^2 + 306\mathbf{s} + 300}{\mathbf{s}(\mathbf{s}+2)(14\mathbf{s}^2 + 51\mathbf{s} + 50)} \\ &= \frac{42\mathbf{s}^3 + 162\mathbf{s}^2 + 306\mathbf{s} + 300}{14\mathbf{s}(\mathbf{s}+2)(\mathbf{s}^2 + 51\mathbf{s}/14 + 50/14)}.\end{aligned} \tag{10.105}$$

The roots of the quadratic term in the denominator are

$$\begin{aligned}\mathbf{s}_1 &= \left[-\frac{51}{14} - \sqrt{\left(\frac{51}{14}\right)^2 - 4 \times \frac{50}{14}}\right] \Big/ 2 \\ &= -1.82 - j0.5\end{aligned} \tag{10.106a}$$

and

$$\mathbf{s}_2 = -1.82 + j0.5. \tag{10.106b}$$

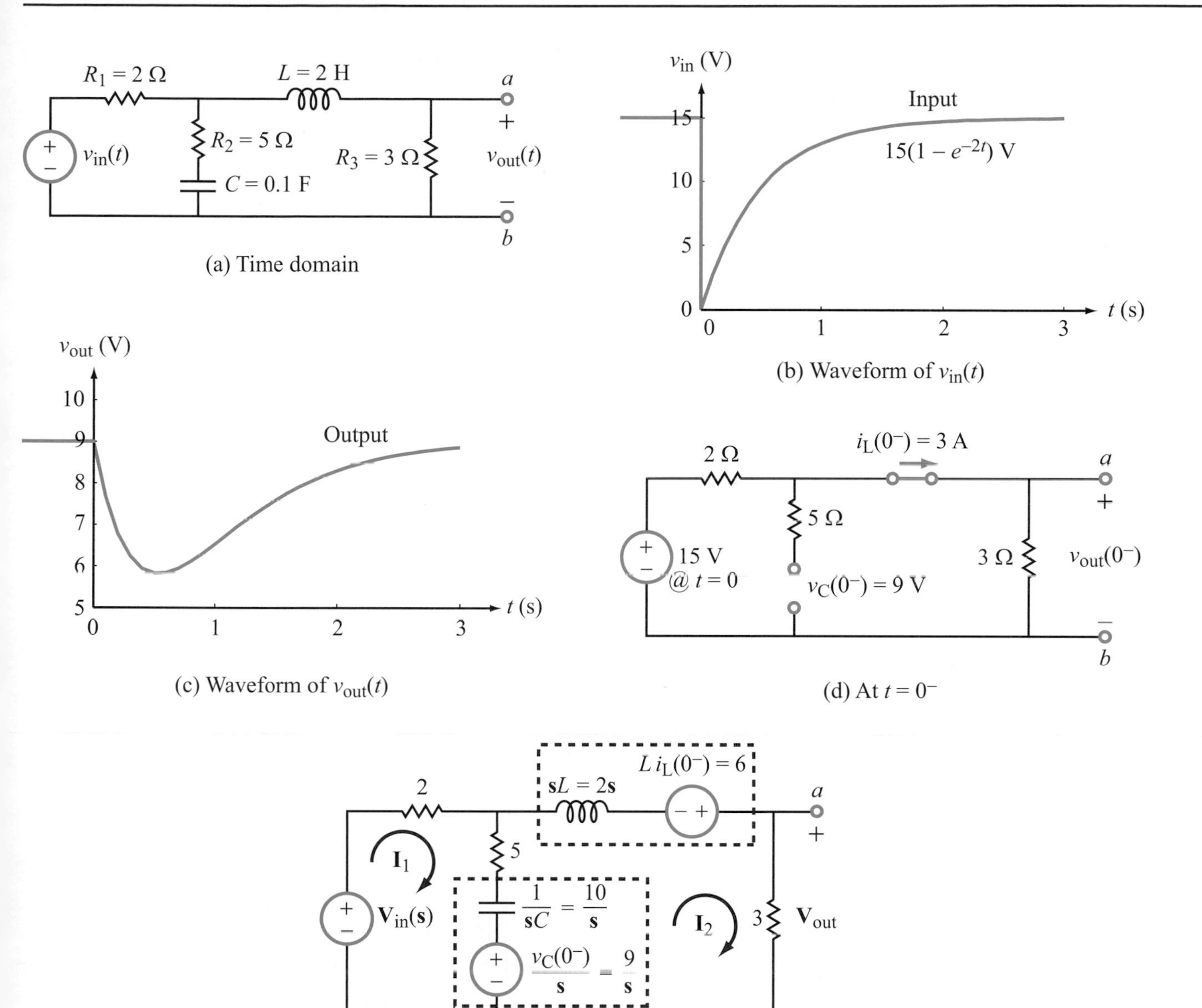

Figure 10-6: Circuit for Example 10-11.

Hence, Eq. (10.105) can be rewritten in the form

$$\mathbf{I}_2 = \frac{42\mathbf{s}^3 + 162\mathbf{s}^2 + 306\mathbf{s} + 300}{14\mathbf{s}(\mathbf{s}+2)(\mathbf{s}+1.82+j0.5)(\mathbf{s}+1.82-j0.5)}. \tag{10.107}$$

The expression for $\mathbf{I}_2$ now is ready for expansion in the form of partial fractions as

$$\mathbf{I}_2 = \frac{A_1}{\mathbf{s}} + \frac{A_2}{\mathbf{s}+2} + \frac{\mathbf{B}}{\mathbf{s}+1.82+j0.5} + \frac{\mathbf{B}^*}{\mathbf{s}+1.82-j0.5}, \tag{10.108}$$

with

$$A_1 = \mathbf{s}\mathbf{I}_2|_{\mathbf{s}=0} = \left.\frac{42\mathbf{s}^3 + 162\mathbf{s}^2 + 306\mathbf{s} + 300}{14(\mathbf{s}+2)(\mathbf{s}^2 + 51\mathbf{s}/14 + 50/14)}\right|_{\mathbf{s}=0} = 3, \tag{10.109a}$$

$$A_2 = (\mathbf{s}+2)\mathbf{I}_2|_{\mathbf{s}=-2} = \left.\frac{42\mathbf{s}^3 + 162\mathbf{s}^2 + 306\mathbf{s} + 300}{14\mathbf{s}(\mathbf{s}^2 + 51\mathbf{s}/14 + 50/14)}\right|_{\mathbf{s}=-2} = 0, \tag{10.109b}$$

$$\mathbf{B} = (\mathbf{s} + 1.82 + j0.5)\mathbf{I}_2|_{\mathbf{s}=-1.82-j0.5}$$
$$= \left.\frac{42\mathbf{s}^3 + 162\mathbf{s}^2 + 306\mathbf{s} + 300}{14\mathbf{s}(\mathbf{s}+2)(\mathbf{s} + 1.82 - j0.5)}\right|_{\mathbf{s}=-1.82-j0.5}$$
$$= 5.32e^{-j90°}. \qquad (10.109c)$$

Inserting the values of A_1, A_2, and $\mathbf{B}$ into Eq. (10.108) leads to

$$\mathbf{I}_2 = \frac{3}{\mathbf{s}} + \frac{5.32e^{-j90°}}{\mathbf{s} + 1.82 + j0.5} + \frac{5.32e^{j90°}}{\mathbf{s} + 1.82 - j0.5}. \qquad (10.110)$$

For the first term, entry #2 in Table 10-2 leads to the inverse Laplace transform

$$\frac{3}{\mathbf{s}} \leftrightarrow 3\,u(t),$$

and from Table 10-3,

$$\frac{Ae^{j\theta}}{\mathbf{s}+a+jb} + \frac{Ae^{-j\theta}}{\mathbf{s}+a-jb} \leftrightarrow 2Ae^{-at}\cos(bt - \theta)\,u(t). \qquad (10.111)$$

With $A = 5.32$, $\theta = -90°$, $a = 1.82$, and $b = 0.5$, the inverse Laplace transform corresponding to the expression given by Eq. (10.110) is

$$i_2(t) = [3 + 10.64e^{-1.82t}\cos(0.5t + 90°)]\,u(t)$$
$$= [3 - 10.64e^{-1.82t}\sin 0.5t]\,u(t)\text{ A}, \qquad (10.112)$$

and the corresponding output voltage is

$$\upsilon_{\text{out}}(t) = 3i_2(t)$$
$$= [9 - 31.92e^{-1.82t}\sin 0.5t]\,u(t)\text{ V}. \qquad (10.113)$$

The waveform of $\upsilon_{\text{out}}(t)$ shown in Fig. 10-6(c) was (indeed) generated using this expression.

Example 10-12: AC Source with a DC Bias

Repeat the analysis of the circuit shown in Fig. 10-6(a), but change the waveform of the voltage source into a 15-V dc level that has existed for a long time on top of which an ac signal is superimposed at $t = 0$, as shown in Fig. 10-7(a).

Solution:

$$\upsilon_{\text{in}}(t) = \begin{cases} 15\text{ V} & \text{for } t \le 0^-, \\ [15 + 5\sin 4t]\,u(t)\text{ V} & \text{for } t \ge 0. \end{cases} \qquad (10.114)$$

In view of entry #9 in Table 10-2, the s-domain counterpart of $\upsilon_{\text{in}}(t)$ is

$$\mathbf{V}_{\text{in}}(\mathbf{s}) = \frac{15}{\mathbf{s}} + \frac{5\omega}{\mathbf{s}^2 + \omega^2} = \frac{15}{\mathbf{s}} + \frac{80}{\mathbf{s}^2 + 16}, \qquad (10.115)$$

where we replaced ω with 4 rad/s. For $t \le 0^-$, the voltage waveform is the same as it was before in Example 10-11, namely 15 V, so initial conditions remain as before (Fig. 10-6(d)), as does the circuit configuration in the s-domain (Fig. 10-6(e)). The only quantity that has changed is the expression for $\mathbf{V}_{\text{in}}(\mathbf{s})$. Rewriting Eq. (10.103) with $\mathbf{V}_{\text{in}}(\mathbf{s})$ as given by Eq. (10.115) leads to

$$\left(7 + \frac{10}{\mathbf{s}}\right)\mathbf{I}_1 - \left(5 + \frac{10}{\mathbf{s}}\right)\mathbf{I}_2 = \mathbf{V}_{\text{in}} - \frac{9}{\mathbf{s}}$$
$$= \frac{15}{\mathbf{s}} + \frac{80}{\mathbf{s}^2+16} - \frac{9}{\mathbf{s}} = \frac{6\mathbf{s}^2 + 80\mathbf{s} + 96}{\mathbf{s}(\mathbf{s}^2 + 16)}, \qquad (10.116)$$

and Eq. (10.104) remains unchanged as

$$-\left(5 + \frac{10}{\mathbf{s}}\right)\mathbf{I}_1 + \left(8 + 2\mathbf{s} + \frac{10}{\mathbf{s}}\right)\mathbf{I}_2 = \frac{9}{\mathbf{s}} + 6. \qquad (10.117)$$

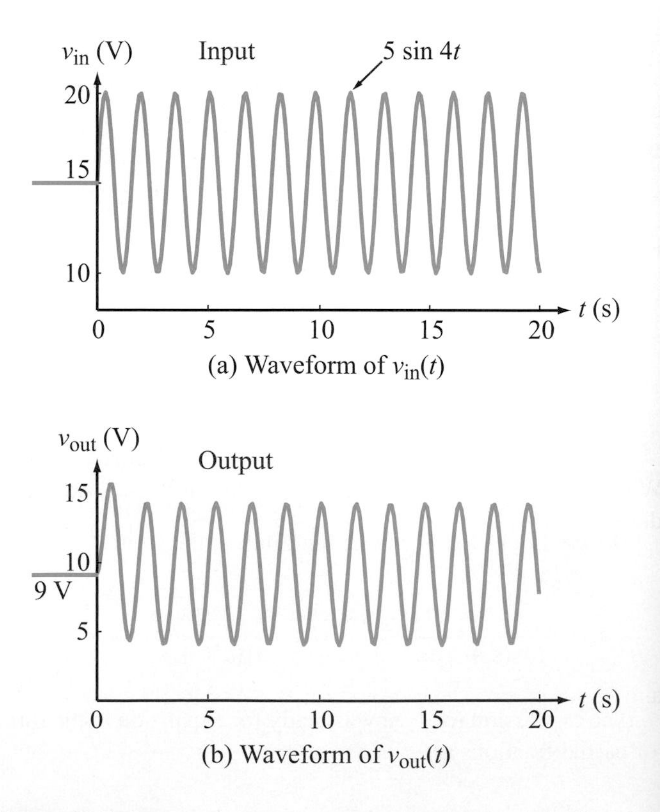

Figure 10-7: Input and output waveforms for the circuit in Fig. 10-6 (Example 10-12).

Simultaneous solution of Eqs. (10.116) and (10.117) leads to

$$\mathbf{I}_2 = \frac{42\mathbf{s}^4 + 153\mathbf{s}^3 + 1222\mathbf{s}^2 + 3248\mathbf{s} + 2400}{14\mathbf{s}(\mathbf{s}^2 + 16)(\mathbf{s}^2 + 51\mathbf{s}/14 + 50/14)} = \frac{42\mathbf{s}^4 + 153\mathbf{s}^3 + 1222\mathbf{s}^2 + 3248\mathbf{s} + 2400}{14\mathbf{s}(\mathbf{s} + j4)(\mathbf{s} - j4)(\mathbf{s} + 1.82 + j0.5)(\mathbf{s} + 1.82 - j0.5)}, \quad (10.118)$$

where we expanded the two quadratic terms in the denominator into a product of four simple-pole factors.

Partial fraction representation takes the form

$$\mathbf{I}_2 = \frac{A_1}{\mathbf{s}} + \frac{\mathbf{B}_1}{\mathbf{s} + j4} + \frac{\mathbf{B}_1^*}{\mathbf{s} - j4} + \frac{\mathbf{B}_2}{\mathbf{s} + 1.82 + j0.5} + \frac{\mathbf{B}_2^*}{\mathbf{s} + 1.82 - j0.5} \quad (10.119)$$

with

$$A_1 = \mathbf{s}\mathbf{I}_2|_{\mathbf{s}=0} = \left.\frac{42\mathbf{s}^4 + 153\mathbf{s}^3 + 1222\mathbf{s}^2 + 3248\mathbf{s} + 2400}{14(\mathbf{s} + j4)(\mathbf{s} - j4)(\mathbf{s} + 1.82 + j0.5)(\mathbf{s} + 1.82 - j0.5)}\right|_{\mathbf{s}=0} = 3, \quad (10.120a)$$

$$\mathbf{B}_1 = (\mathbf{s} + j4)\mathbf{I}_2|_{\mathbf{s}=-j4} = 0.834\angle 157.0^\circ, \quad (10.120b)$$

and

$$\mathbf{B}_2 = (\mathbf{s} + 1.82 + j0.5)\mathbf{I}_2|_{\mathbf{s}=-1.82-j0.5} = 0.79\angle 14.0^\circ. \quad (10.120c)$$

Hence, $\mathbf{I}_2$ becomes

$$\mathbf{I}_2 = \frac{3}{\mathbf{s}} + \frac{0.834e^{j157^\circ}}{\mathbf{s} + j4} + \frac{0.834e^{-j157^\circ}}{\mathbf{s} - j4} + \frac{0.79e^{j14^\circ}}{\mathbf{s} + 1.82 + j0.5} + \frac{0.79e^{-j14^\circ}}{\mathbf{s} + 1.82 - j0.5}. \quad (10.121)$$

With the help of item #3 in Table 10-3, conversion to the time domain gives

$$i_2(t) = [3 + 1.67\cos(4t - 157^\circ) + 1.576e^{-1.82t}\cos(0.5t - 14^\circ)]\, u(t) \text{ A}, \quad (10.122)$$

and the corresponding output voltage is

$$\upsilon_{\text{out}}(t) = 3\, i_2(t) = [9 + 5\cos(4t - 157^\circ) + 4.73e^{-1.82t}\cos(0.5t - 14^\circ)]\, u(t) \text{ V}. \quad (10.123)$$

The profile of $\upsilon_{\text{out}}(t)$ is shown in Fig. 10-7(b). In response to the introduction of the ac signal at the input at $t = 0$, the output consists of a dc term equal to 9 V, plus an oscillatory transient component (the last term in Eq. (10.123)) that decays down to zero over time, and a steady-state oscillatory component that continues indefinitely.

Another approach for analyzing the circuit is by applying the source-superposition method, in which case the circuit is analyzed twice—once with the 15-V dc component alone and a second time with only the ac component that starts at $t = 0$. The first solution would lead to the first term in Eq. (10.123). For the ac source, we can apply the phasor-analysis technique of Chapter 7, but the solution would have yielded the steady-state component only. The advantage of the Laplace transform technique is that it can provide a complete solution that automatically includes both the transient and steady-state components and would do so for any type of excitation.

Example 10-13: Circuit with a Switch

The circuit shown in Fig. 10-8(a) is a replica of the circuit in Fig. 6-18(a), which we analyzed earlier in Example 6-11 by solving a second-order differential equation in the time domain. The solution provided the following expression for the inductor current.

$$i_{\text{L}}(t) = [1.5 - 0.76e^{-1.85t} + 0.26e^{-5.4t}]\, u(t) \text{ A}. \quad (10.124)$$

Show that the same expression is realized by applying the Laplace transform technique.

Solution: We start by examining the state of the circuit at $t = 0^-$ (before closing the switch). Upon replacing L with a short circuit and C with an open circuit, as portrayed by the configuration in Fig. 10-8(b), we establish that

$$i_{\text{L}}(0^-) = 1 \text{ A} \quad \text{and} \quad \upsilon_{\text{C}}(0^-) = 12 \text{ V}. \quad (10.125)$$

For $t \geq 0$, the s-domain equivalent of the original circuit is shown in Fig. 10-8(c), where we have replaced R_2 with a short circuit, converted the dc source into its s-domain equivalent and, in accordance with the circuit models given in Table 10-4, converted L and C into impedances—each with its own appropriate voltage source. By inspection, the two mesh current equations are given by

$$(4 + 12 + 2\mathbf{s})\mathbf{I}_1 - (12 + 2\mathbf{s})\mathbf{I}_2 = \frac{24}{\mathbf{s}} + 2, \quad (10.126a)$$

$$-(12 + 2\mathbf{s})\mathbf{I}_1 + \left(12 + 2\mathbf{s} + \frac{5}{\mathbf{s}}\right)\mathbf{I}_2 = -2 - \frac{12}{\mathbf{s}}. \quad (10.126b)$$

Simultaneous solution of the two equations leads to

$$\mathbf{I}_1 = \frac{12\mathbf{s}^2 + 77\mathbf{s} + 60}{\mathbf{s}(4\mathbf{s}^2 + 29\mathbf{s} + 40)} \tag{10.127a}$$

and

$$\mathbf{I}_2 = \frac{8(\mathbf{s} + 6)}{4\mathbf{s}^2 + 29\mathbf{s} + 40}. \tag{10.127b}$$

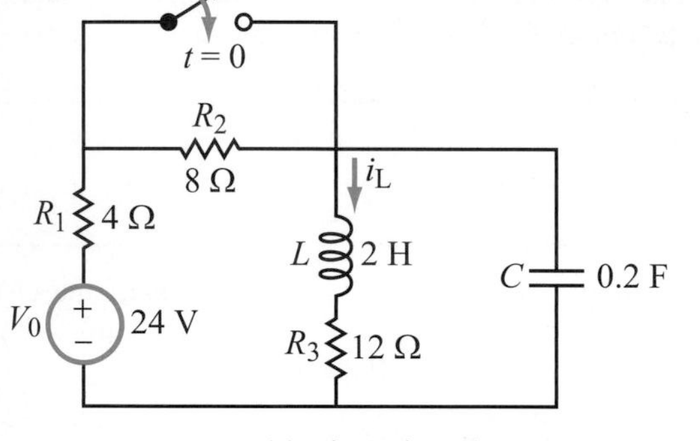

(a) Time-domain

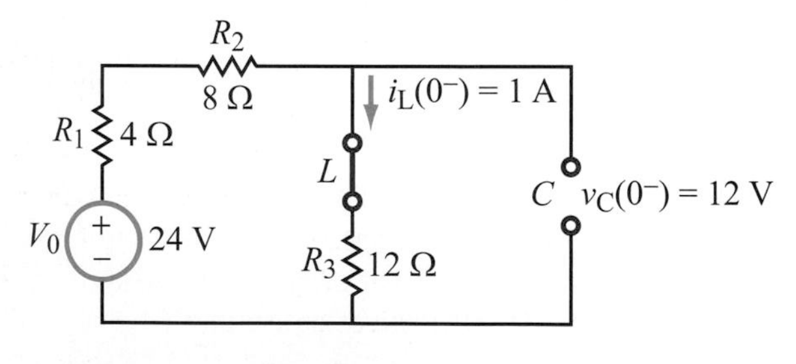

(b) Circuit at $t = 0^-$

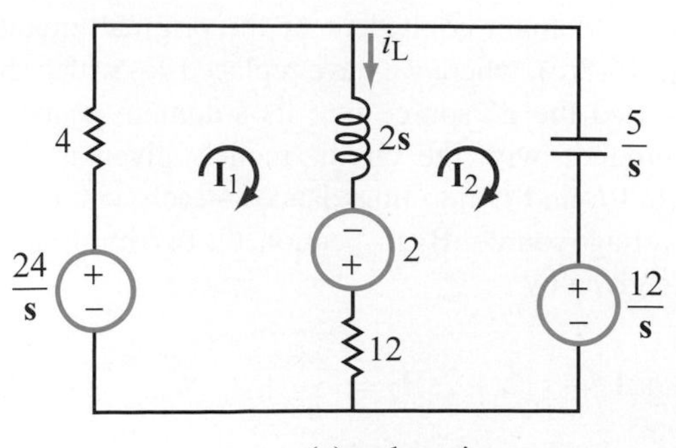

(c) s-domain

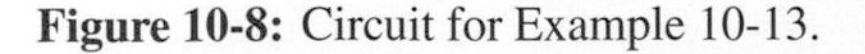
Figure 10-8: Circuit for Example 10-13.

The associated inductor current $\mathbf{I}_\mathrm{L}$ is

$$\begin{aligned} \mathbf{I}_\mathrm{L} &= \mathbf{I}_1 - \mathbf{I}_2 \\ &= \frac{4\mathbf{s}^2 + 29\mathbf{s} + 60}{\mathbf{s}(4\mathbf{s}^2 + 29\mathbf{s} + 40)} \\ &= \frac{4\mathbf{s}^2 + 29\mathbf{s} + 60}{4\mathbf{s}(\mathbf{s} + 1.85)(\mathbf{s} + 5.4)}, \end{aligned} \tag{10.128}$$

which can be represented by the partial fraction expansion

$$\mathbf{I}_\mathrm{L} = \frac{A_1}{\mathbf{s}} + \frac{A_2}{\mathbf{s} + 1.85} + \frac{A_3}{\mathbf{s} + 5.4}. \tag{10.129}$$

The values of A_1 through A_3 are obtained from

$$A_1 = \mathbf{s}\mathbf{I}_\mathrm{L}\big|_{\mathbf{s}=0} = \frac{60}{40} = 1.5, \tag{10.130a}$$

$$\begin{aligned} A_2 &= (\mathbf{s} + 1.85)\mathbf{I}_\mathrm{L}\big|_{\mathbf{s}=-1.85} \\ &= \frac{4\mathbf{s}^2 + 29\mathbf{s} + 60}{4\mathbf{s}(\mathbf{s} + 5.4)}\bigg|_{\mathbf{s}=-1.85} = -0.76, \end{aligned} \tag{10.130b}$$

and

$$A_3 = (\mathbf{s} + 5.4)\mathbf{I}_\mathrm{L}\big|_{\mathbf{s}=-5.4} = 0.26. \tag{10.130c}$$

Hence,

$$\mathbf{I}_\mathrm{L} = \frac{1.5}{\mathbf{s}} - \frac{0.76}{\mathbf{s} + 1.85} + \frac{0.26}{\mathbf{s} + 5.4}, \tag{10.131}$$

and the corresponding time-domain current is

$$i_\mathrm{L}(t) = [1.5 - 0.76e^{-1.85t} + 0.26e^{-5.4t}]\, u(t) \text{ A}, \tag{10.132}$$

which is identical with the expression given by Eq. (10.124).

Review Question 10-10: Under what circumstances do the **s**-domain circuit element models for a capacitor and an inductor resemble those in the phasor domain?

Review Question 10-11: In the **s**-domain, a capacitor is modeled by an impedance in series with a voltage source. The same is true for an inductor, except that the polarity of its associated voltage source is opposite that of the capacitor's. Why?

10-8 Transfer Function and Impulse Response

A linear circuit can be regarded as a linear system with input excitation $x(t)$ and output response $y(t)$, as shown in Fig. 10-9(a). Usually, $x(t)$ is a current or a voltage source, and $y(t)$ is a particular current or voltage of interest. The circuit is characterized by a yet-to-be-defined function called the ***unit impulse response*** of the circuit $h(t)$ or simply the ***impulse response***, for short. In the **s**-domain, the circuit is characterized by a ***transfer function*** $\mathbf{H}(\mathbf{s})$, defined as the ratio of the output $\mathbf{Y}(\mathbf{s})$ to the input $\mathbf{X}(\mathbf{s})$ *assuming that all initial conditions relating to currents and voltages in the circuit are zero at* $t = 0^-$. Thus,

$$\mathbf{H}(\mathbf{s}) = \frac{\mathbf{Y}(\mathbf{s})}{\mathbf{X}(\mathbf{s})}, \qquad (10.133)$$

with the aforementioned initial condition. For a linear circuit, $\mathbf{H}(\mathbf{s})$ does not depend on $\mathbf{X}(\mathbf{s})$, so an easy way by which to determine $\mathbf{H}(\mathbf{s})$ is to select an arbitrary excitation $\mathbf{X}(\mathbf{s})$,

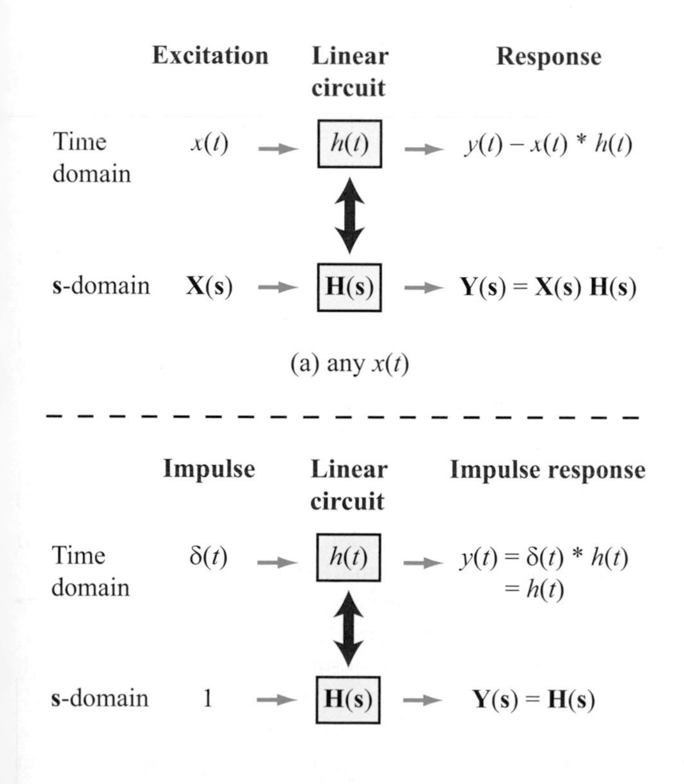

Figure 10-9: Correspondence between the time domain and the **s**-domain. In the **s**-domain, the output response $\mathbf{Y}(\mathbf{s})$ is equal to the product of the input excitation $\mathbf{X}(\mathbf{s})$ and the transfer function $\mathbf{H}(\mathbf{s})$, but in the time domain, $y(t)$ is equal to the convolution of $x(t)$ with the impulse response $h(t)$. If $x(t) = \delta(t)$, as in part (b), then $y(t) = h(t)$.

determine the corresponding response $\mathbf{Y}(\mathbf{s})$, and then form the ratio defined by Eq. (10.133). Of particular note is the excitation $\mathbf{X}(\mathbf{s}) = 1$, because then Eq. (10.133) simplifies to $\mathbf{H}(\mathbf{s}) = \mathbf{Y}(\mathbf{s})$. The inverse Laplace transform of 1 is the ***unit impulse function*** $\delta(t)$. Thus, when a circuit is excited by $x(t) = \delta(t)$, as illustrated by part (b) of Fig. 10-9, its **s**-domain output is equal to the circuit's transfer function $\mathbf{H}(\mathbf{s})$:

$$\mathbf{H}(\mathbf{s}) = \mathbf{Y}(\mathbf{s}), \qquad (\text{when } x(t) = \delta(t)). \qquad (10.134)$$

Once $\mathbf{H}(\mathbf{s})$ of a given circuit has been established, $y(t)$ can be determined readily for any excitation $x(t)$ by

Step 1. Transforming $x(t)$ to the **s**-domain to obtain $\mathbf{X}(\mathbf{s})$

Step 2. Multiplying $\mathbf{X}(\mathbf{s})$ by $\mathbf{H}(\mathbf{s})$ to obtain $\mathbf{Y}(\mathbf{s})$

Step 3. Expressing $\mathbf{Y}(\mathbf{s})$ in the form of a sum of partial fractions

Step 4. Transferring the partial-fraction expansion of $\mathbf{Y}(\mathbf{s})$ to the time domain to obtain $y(t)$

The process is straightforward, so long as $x(t)$ is transformable to the **s**-domain and $\mathbf{Y}(\mathbf{s})$ is transformable to the time domain (i.e., Steps 1 and 4). If $x(t)$ is some irregular or unusual waveform or if it consists of a series of experimental measurements generated by another circuit, it may not be possible to transform $x(t)$ to the **s**-domain analytically. Also, in some cases, the functional form of $\mathbf{Y}(\mathbf{s})$ may be so complicated that it may be very difficult to express it in a form amenable for transformation to the time domain. In such cases, an alternative approach is to determine $y(t)$ by operating entirely in the time domain. We do so by taking advantage of an important property of the Laplace transform. The property we refer to is the ***convolution property***, which states that if $\mathbf{H}(\mathbf{s})$ and $\mathbf{X}(\mathbf{s})$ are the Laplace transforms of, respectively, $h(t)$ and $x(t)$, then *multiplying* $\mathbf{X}(\mathbf{s})$ *by* $\mathbf{H}(\mathbf{s})$ *in the* **s**-*domain is equivalent to convolving* $x(t)$ *with* $h(t)$ *in the time domain.* Symbolically, the convolution property is expressed as

$$x(t) * h(t) \quad \longleftrightarrow \quad \mathbf{X}(\mathbf{s})\,\mathbf{H}(\mathbf{s}), \qquad (10.135)$$

and since $y(t) = \mathcal{L}^{-1}[\mathbf{Y}(\mathbf{s})] = \mathcal{L}^{-1}[\mathbf{X}(\mathbf{s})\,\mathbf{H}(\mathbf{s})]$, it follows that

$$y(t) = x(t) * h(t), \qquad (10.136)$$

where the asterisk * denotes "$x(t)$ is convolved with $h(t)$." We will explain what convolution means shortly, but before we do so, we should note that if $x(t) = \delta(t)$, it follows that

$$\mathbf{X}(\mathbf{s}) = \mathcal{L}[x(t)] = \mathcal{L}[\delta(t)] = 1, \qquad (10.137)$$

and consequently,

$$y(t) = \mathcal{L}^{-1}[\mathbf{Y(s)}] = \mathcal{L}^{-1}[\mathbf{H(s)\,X(s)}] = \mathcal{L}^{-1}[\mathbf{H(s)}] = h(t). \quad (10.138)$$

Accordingly, $h(t)$ is called the ***impulse response*** of the circuit, because it is equal to the output $y(t)$ when the circuit is excited by an impulse function $\delta(t)$ at its input. Its relation to $\mathbf{H(s)}$ is expressed symbolically by

$$h(t) \longleftrightarrow \mathbf{H(s)}. \quad (10.139)$$

(impulse response) (transfer function)

Example 10-14: Transfer Function

The output response of a system excited by a unit step function at $t = 0$ is given by

$$y(t) = [2t + 12te^{-3t}]\,u(t).$$

Determine (a) the transfer function of the system and (b) its impulse response.

Solution:

(a) The Laplace transform of a unit step function is

$$\mathbf{X(s)} = \frac{1}{\mathbf{s}},$$

and the Laplace transform of the output response is

$$\mathbf{Y(s)} = \frac{2}{\mathbf{s}^2} + \frac{12}{(\mathbf{s}+3)^2}.$$

The system transfer function then is given by

$$\mathbf{H(s)} = \frac{\mathbf{Y(s)}}{\mathbf{X(s)}} = \frac{2}{\mathbf{s}} + \frac{12\mathbf{s}}{(\mathbf{s}+3)^2}.$$

(b) The impulse response is obtained by transferring $\mathbf{H(s)}$ to the time domain. To do so, we need to have all terms in the expression for $\mathbf{H(s)}$ to be in proper form. The first term is indeed in proper form, because it belongs to transform pair 4 in Table 10-2, but the second term needs to be expanded into partial fractions. Let us define the second term as

$$\mathbf{H}_2\mathbf{(s)} = \frac{12}{(\mathbf{s}+3)^2} = \frac{B_1}{\mathbf{s}+3} + \frac{B_2}{(\mathbf{s}+3)^2},$$

with

$$B_2 = (\mathbf{s}+3)^2\; \mathbf{H}_2\mathbf{(s)}|_{\mathbf{s}=-3} = 12\mathbf{s}|_{\mathbf{s}=-3} = -36,$$

and

$$B_1 = \frac{d}{ds}(\mathbf{s}+3)^2\,\mathbf{H}_2\mathbf{(s)}\bigg|_{\mathbf{s}=-3} = 12.$$

Hence,

$$\mathbf{H}_2\mathbf{(s)} = \frac{12}{\mathbf{s}+3} - \frac{36}{(\mathbf{s}+3)^2},$$

and

$$\mathbf{H(s)} = \frac{2}{\mathbf{s}} + \frac{12}{\mathbf{s}+3} - \frac{36}{(\mathbf{s}+3)^2}.$$

The impulse response is therefore given by

$$h(t) = [2 + 12e^{-3t} - 36te^{-3t}]\,u(t).$$

Review Question 10-12: What does the term "impulse response" mean? What does it represent?

10-9 Convolution Integral

Now we shall return to the convolution relation given by Eq. (10.136), namely

$$y(t) = x(t) * h(t). \quad (10.140)$$

In the **s**-domain,

$$\mathbf{Y(s)} = \mathbf{X(s)\,H(s)}, \quad (10.141)$$

but in the time domain,

$$y(t) \neq x(t)\,h(t). \quad (10.142)$$

That is, the convolution operation,denoted by * in Eq. (10.140) is not the same as multiplication. Having stated what it is not, we will now proceed to define what it is. To that end, we will use the symbolic diagram process given in Fig. 10-10. In Step 1, we start with the result represented by Eq. (10.138), which states that for a linear system, $y(t) = h(t)$ when $x(t) = \delta(t)$. In the next step, we shift t at the input by a fixed amount λ by replacing $\delta(t)$ with $\delta(t-\lambda)$. Correspondingly, the output response becomes $h(t-\lambda)$. For this consequence to be true, the system has to be not only linear, but ***time-invariant*** as well. The ***time-invariance*** property of a system specifies that if input $x(t)$ causes output $y(t)$, then a delayed input $x(t-\lambda)$ causes an equally delayed output $y(t-\lambda)$. Moreover, if the time-shifted impulse function $\delta(t-\lambda)$ has an amplitude $x(\lambda)$, as shown in step 3 of Fig. 10-10, the output will scale by the same factor.

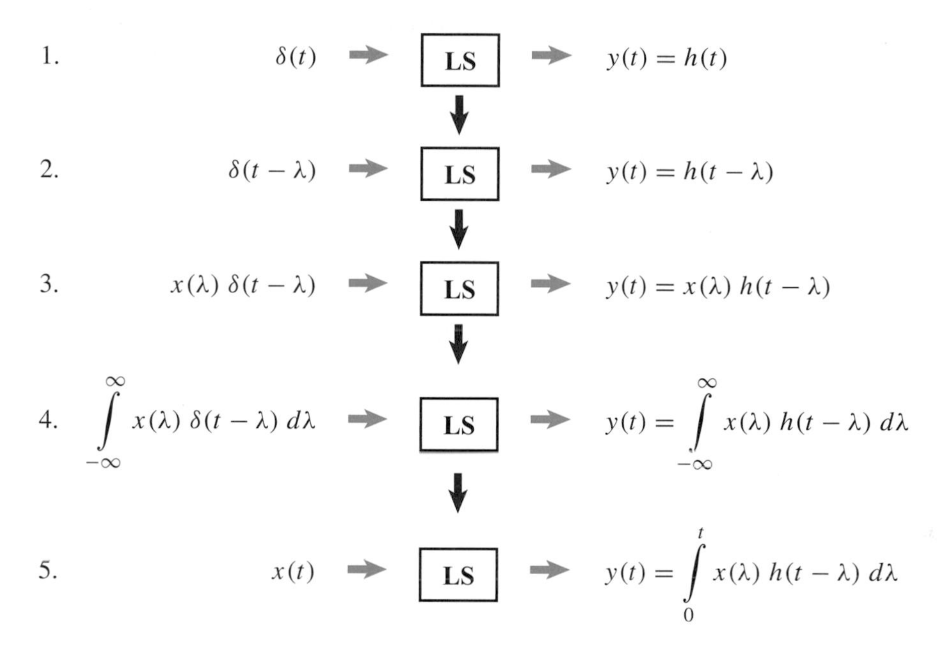

Figure 10-10: Derivation of the convolution integral for a linear time-invariant circuit. In the final step, the lower integration limit was changed to zero, presuming that excitation $x(t) = 0$ for $t < 0$, and the upper limit was changed to t because for a physically realizable circuit, the output $y(t)$ at time t cannot depend on what will occur at the input at future times later than t.

So far, $x(\lambda)$ is just a constant amplitude, so whether or not we write it as x or $x(\lambda)$ is irrelevant. However, we shortly will convert λ from a single-valued constant into a continuous dummy variable, at which time the notation $x(\lambda)$ will become indispensable.

According to Step 3 in the figure, an excitation $x_1(\lambda_1)$ at $t = \lambda_1$ will generate an output

$$y_1(t) = x_1(\lambda_1)\,h(t-\lambda_1), \qquad (10.143a)$$

and similarly, a second excitation of amplitude $x_2(\lambda_2)$ at $t = \lambda_2$ will generate

$$y_2(t) = x_2(\lambda_2)\,h(t-\lambda_2). \qquad (10.143b)$$

If the linear system is excited by both $x_1(\lambda_1)$ and $x_2(\lambda_2)$ simultaneously, the superposition theorem assures us that the output simply will be equal to the sum of $y_1(t)$ and $y_2(t)$. By extension, for a continuous input excitation $x(\lambda)$ extending over the λ-domain $[-\infty, \infty]$, the input and output of the system become definite integrals, as shown in Steps 4 and 5:

$$\text{Input:} \quad \int_{-\infty}^{\infty} x(\lambda)\,\delta(t-\lambda)\,d\lambda = x(t), \qquad (10.144)$$

$$\text{Output:} \quad y(t) = \int_{-\infty}^{\infty} x(\lambda)\,h(t-\lambda)\,d\lambda. \qquad (10.145)$$

The expression given by Eq. (10.145) is called a ***convolution integral***, defined as the product of two functions with one of them shifted by a constant amount t along the axis of the integration variable λ, integrated over the full range of possible values of λ. This stands in stark contrast with the simple multiplication on the right-hand side of Eq. (10.142). As a shorthand notation, the convolution integral is written using an asterisk *:

$$y(t) = x(t) * h(t) = \int_{-\infty}^{\infty} x(\lambda)\,h(t-\lambda)\,d\lambda. \qquad (10.146)$$

The integral given by Eq. (10.146) implies that the output $y(t)$ at time t depends on all excitations $x(t)$ occurring at the input, including those that will occur at times later than t. Since this cannot be true for a ***physically realizable system*** (such as an electric circuit) the upper integration limit should be replaced with t, instead of ∞. Also, if we choose our time scale such that no excitation exists before $t = 0$, we can replace the lower integration limit with zero. Hence, Eq. (10.146) becomes

$$y(t) = x(t) * h(t) = \int_{0}^{t} x(\lambda)\,h(t-\lambda)\,d\lambda. \qquad (10.147)$$

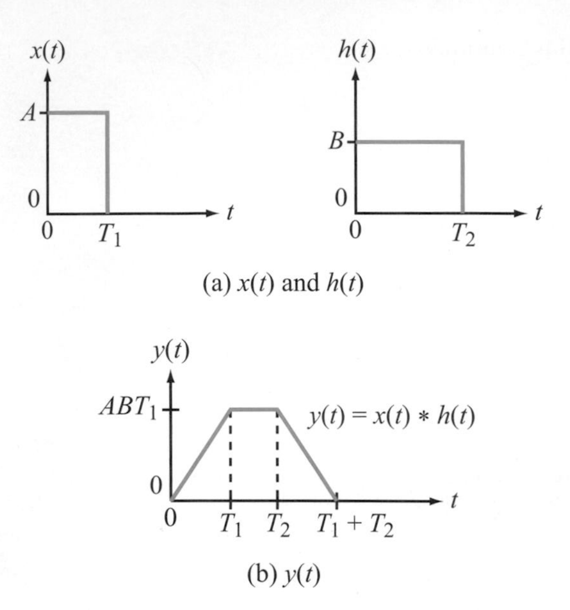

Figure 10-11: The convolution of the two rectangular waveforms in (a) is the pyramid shown in (b).

In Eq. (10.147), the convolution is realized by time-shifting the impulse response $h(t)$. Alternatively, the expression for $y(t)$ can be derived by time-shifting the excitation $x(t)$ instead, in which case the result would be

$$y(t) = x(t) * h(t) = \int_0^t x(t-\lambda)\, h(\lambda)\, d\lambda. \tag{10.148}$$

The transfer function $\mathbf{H}(\mathbf{s})$ defined by Eq. (10.133) represents a circuit with zero initial conditions. Hence, $x(t)$ and $h(t)$ are both zero for $t < 0$. If $x(t)$ and $h(t)$ are non-zero for all positive values of t, we denote that fact by inserting the unit step function $u(t)$ into their expressions, but if either or both of them are defined over a limited range of the positive t-axis, we use the delayed step function $u(t-T)$ instead. This means that the integrals in Eqs. (10.147) and (10.148) may contain one or more products of pairs of step functions. Whereas performing an integration involving a step function is straightforward, the process gets more complicated when the product of two step functions is involved. To facilitate the integration process, we will examine the simple example shown in Fig. 10-11. The rectangular waveforms of $x(t)$ and $h(t)$ are given by

$$x(t) = A[u(t) - u(t-T_1)] \tag{10.149a}$$

and

$$h(t) = B[u(t) - u(t-T_2)]. \tag{10.149b}$$

To evaluate the convolution integral given by Eq. (10.148), we need expressions for $x(t-\lambda)$ and $h(\lambda)$. Upon replacing the argument (t) in Eq. (10.149a) with $(t-\lambda)$ and replacing t in Eq. (10.149b) with λ, we have

$$x(t-\lambda) = A[u(t-\lambda) - u(t-T_1-\lambda)] \tag{10.150a}$$

and

$$h(\lambda) = B[u(\lambda) - u(\lambda - T_2)]. \tag{10.150b}$$

Inserting these expressions into Eq. (10.148) leads to

$$\begin{aligned} y(t) &= \int_0^t x(t-\lambda)\, h(\lambda)\, d\lambda \\ &= AB\left\{ \int_0^t u(t-\lambda)\, u(\lambda)\, d\lambda \right. \\ &\qquad - \int_0^t u(t-\lambda)\, u(\lambda-T_2)\, d\lambda \\ &\qquad - \int_0^t u(t-T_1-\lambda)\, u(\lambda)\, d\lambda \\ &\qquad \left. + \int_0^t u(t-T_1-\lambda)\, u(\lambda-T_2)\, d\lambda \right\}. \end{aligned} \tag{10.151}$$

We now will examine each term individually. Keeping in mind that the integration variable is λ and the upper integration limit is t, the difference $(t-\lambda)$ is never smaller than zero (over the range of integration). Also, T_1 and T_2 are both non-negative numbers.

Term 1:

Since over the range of integration both $u(\lambda)$ and $u(t-\lambda)$ are equal to 1, the integral in the first term simplifies to

$$\begin{aligned} \int_0^t u(t-\lambda)\, u(\lambda)\, d\lambda &= \left[\int_0^t d\lambda \right] u(t) \\ &= t\, u(t), \end{aligned} \tag{10.152}$$

where we have multiplied the integral by $u(t)$ to reflect the fact that the outcome of the integration is valid at any value of t greater than zero.

Term 2:

The unit step function $u(\lambda - T_2)$ is equal to zero, unless $\lambda > T_2$, requiring that the lower integration limit be replaced with T_2 and the outcome be multiplied by $u(t - T_2)$:

$$\int_0^t u(t-\lambda)\, u(\lambda - T_2)\, d\lambda = \left[\int_{T_2}^t d\lambda\right] u(t-T_2)$$
$$= (t - T_2)\, u(t - T_2). \quad (10.153)$$

Term 3:

The step function $u(t - T_1 - \lambda)$ is equal to zero, unless $\lambda < t - T_1$, requiring that the upper limit be replaced with $t - T_1$. Additionally, to satisfy the inequality, the smallest value that t can assume at the lower limit ($\lambda = 0$) is $t = T_1$. Consequently the outcome of the integration should be multiplied by $u(t - T_1)$:

$$\int_0^t u(t - T_1 - \lambda)\, u(\lambda)\, d\lambda = \left[\int_0^{t-T_1} d\lambda\right] u(t - T_1)$$
$$= (t - T_1)\, u(t - T_1). \quad (10.154)$$

Term 4:

To accommodate the product $u(t - T_1 - \lambda)\, u(\lambda - T_2)$, we need to (a) change the lower limit to T_2, (b) change the upper limit to $(t - T_1)$, and (c) multiply the outcome by $u(t - T_1 - T_2)$, leading to

$$\int_0^t u(t - T_1 - \lambda)\, u(\lambda - T_2)\, d\lambda$$
$$= \left[\int_{T_2}^{t-T_1} d\lambda\right] u(t - T_1 - T_2)$$
$$= (t - T_1 - T_2)\, u(t - T_1 - T_2). \quad (10.155)$$

Collecting the results given by Eqs. (10.152) through (10.155) leads to

$$y(t) = AB[t\, u(t) - (t - T_2)\, u(t - T_2) - (t - T_1)\, u(t - T_1) + t(t - T_1 - T_2)\, u(t - T_1 - T_2)]. \quad (10.156)$$

The waveform of $y(t)$ is displayed in Fig. 10-11(b).

Building on the experience gained from the preceding example, we can generalize the result to:

Convolution Integral

For functions $x(t)$ and $h(t)$ given by

$$x(t) = f_1(t)\, u(t - T_1) \quad (10.157a)$$

and

$$h(t) = f_2(t)\, u(t - T_2), \quad (10.157b)$$

where $f_1(t)$ and $f_2(t)$ are any constants or time-dependent functions and T_1 and T_2 are any non-negative numbers,

$$y(t) = x(t) * h(t)$$
$$= \int_0^t x(t - \lambda)\, h(\lambda)\, d\lambda$$
$$= \int_0^t f_1(t - \lambda)\, f_2(\lambda)\, u(t - T_1 - \lambda)\, u(\lambda - T_2)\, d\lambda$$
$$= \left[\int_{T_2}^{t-T_1} f_1(t - \lambda)\, f_2(\lambda)\, d\lambda\right] u(t - T_1 - T_2). \quad (10.158)$$

The convolution result represented by Eq. (10.158) will prove useful and efficient when evaluating the convolution integral analytically. As will be demonstrated through Examples 10-15 and 10-16, the convolution integral may be evaluated either analytically or through graphical integration.

Example 10-15: LP Filter Response to a Rectangular Pulse

Given the RC circuit shown in Fig. 10-12(a), determine the output response to a 1-s-long rectangular pulse. The pulse amplitude is 1 V.

Solution: We will offer three solution methods: (1) **s**-domain, (2) time-domain convolution by direct integration, and (3) time-domain convolution by graphical integration.

Method 1: **s**-domain

With $R = 0.5$ MΩ and $C = 1$ μF, the product is $RC = 0.5$ s. Voltage division in the **s**-domain (Fig. 10-12(b)) leads to

$$\mathbf{H}(\mathbf{s}) = \frac{\mathbf{V}_{\text{out}}(\mathbf{s})}{\mathbf{V}_{\text{in}}(\mathbf{s})} = \frac{1/\mathbf{s}C}{R + 1/\mathbf{s}C} = \frac{1/RC}{\mathbf{s} + 1/RC} = \frac{2}{\mathbf{s}+2}. \tag{10.159}$$

The rectangular pulse is given by

$$\upsilon_{\text{in}}(t) = [u(t) - u(t-1)] \text{ V}, \tag{10.160}$$

and with the help of Table 10-2, its **s**-domain counterpart is

$$\mathbf{V}_{\text{in}}(\mathbf{s}) = \left[\frac{1}{\mathbf{s}} - \frac{1}{\mathbf{s}}\, e^{-\mathbf{s}}\right] \text{V}. \tag{10.161}$$

Hence,

$$\begin{aligned}\mathbf{V}_{\text{out}}(\mathbf{s}) &= \mathbf{H}(\mathbf{s})\, \mathbf{V}_{\text{in}}(\mathbf{s}) \\ &= 2(1 - e^{-\mathbf{s}})\left[\frac{1}{\mathbf{s}(\mathbf{s}+2)}\right].\end{aligned} \tag{10.162}$$

In preparation for transformation to the time domain, we expand the function inside the square bracket into partial fractions as

$$\frac{1}{\mathbf{s}(\mathbf{s}+2)} = \frac{A_1}{\mathbf{s}} + \frac{A_2}{\mathbf{s}+2}, \tag{10.163}$$

with

$$A_1 = \mathbf{s}\left[\frac{1}{\mathbf{s}(\mathbf{s}+2)}\right]\bigg|_{\mathbf{s}=0} = \frac{1}{2},$$

and

$$A_2 = (\mathbf{s}+2)\left[\frac{1}{\mathbf{s}(\mathbf{s}+2)}\right]\bigg|_{\mathbf{s}=-2} = -\frac{1}{2}.$$

Incorporating these results gives

$$\mathbf{V}_{\text{out}}(\mathbf{s}) = \frac{1}{\mathbf{s}} - \frac{1}{\mathbf{s}+2} - \frac{1}{\mathbf{s}}\, e^{-\mathbf{s}} + \frac{1}{\mathbf{s}+2}\, e^{-\mathbf{s}}. \tag{10.164}$$

From Table 10-2, we deduce that

$$\begin{aligned} u(t) &\leftrightarrow \frac{1}{\mathbf{s}}, \\ e^{-2t}\, u(t) &\leftrightarrow \frac{1}{\mathbf{s}+2}, \\ u(t-1) &\leftrightarrow \frac{1}{\mathbf{s}}\, e^{-\mathbf{s}}, \end{aligned}$$

and

$$e^{-2(t-1)}\, u(t-1) \leftrightarrow \frac{1}{\mathbf{s}+2}\, e^{-\mathbf{s}}.$$

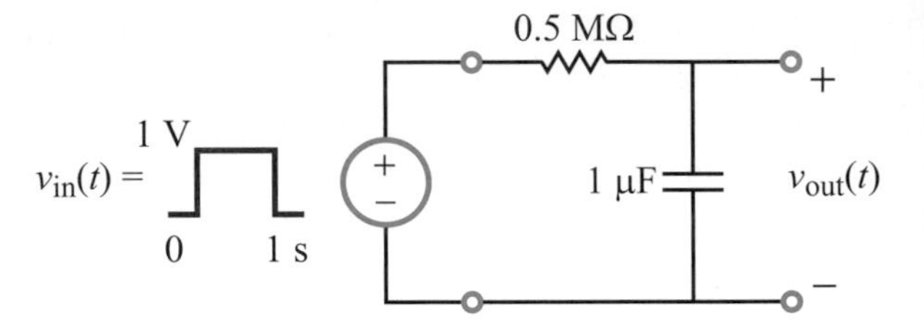

(a) RC lowpass filter

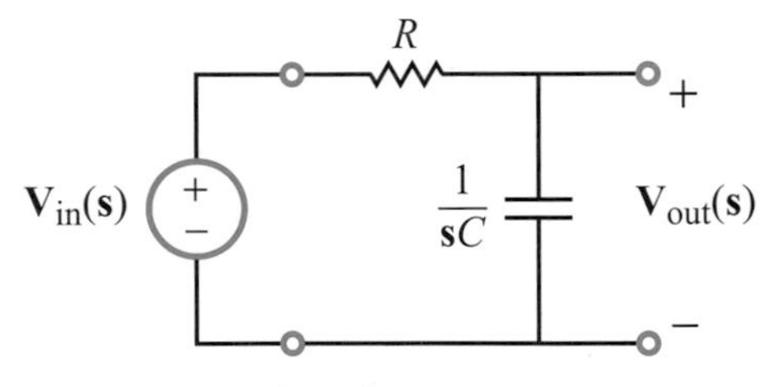

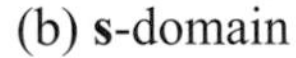
(b) **s**-domain

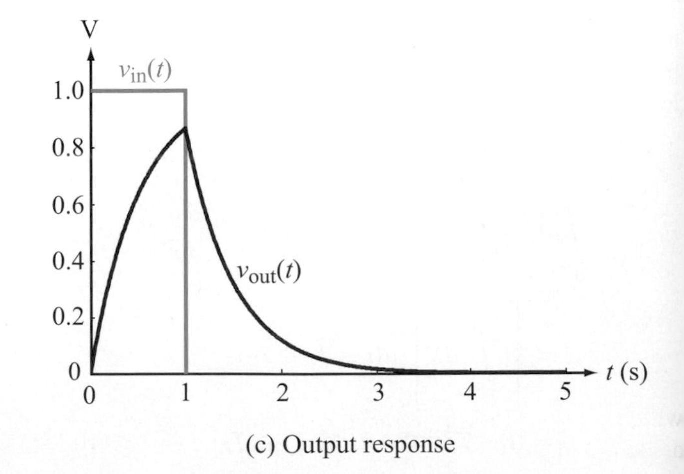

(c) Output response

Figure 10-12: Example 10-15.

Hence,

$$v_{out}(t) = \Big[[1 - e^{-2t}]\, u(t) - [1 - e^{-2(t-1)}]\, u(t-1) \Big] \text{ V}. \tag{10.165}$$

Figure 10-12(c) displays the temporal response of $v_{out}(t)$.

Method 2: Convolution by Direct Integration

In view of Eq. (10.159), the time-domain counterpart of $\mathbf{H}(\mathbf{s})$ is

$$h(t) = 2e^{-2t}\, u(t). \tag{10.166}$$

Application of Eq. (10.147) to the expressions for $v_{in}(t)$ and $h(t)$ given respectively by Eqs. (10.160) and (10.166) gives

$$\begin{aligned} v_{out}(t) &= v_{in}(t) * h(t) \\ &= \int_0^t v_{in}(\lambda)\, h(t-\lambda)\, d\lambda \\ &= \int_0^t [u(\lambda) - u(\lambda - 1)] \times 2e^{-2(t-\lambda)}\, u(t-\lambda)\, d\lambda \\ &= \int_0^t 2e^{-2(t-\lambda)}\, u(\lambda)\, u(t-\lambda)\, d\lambda \\ &\quad - \int_0^t 2e^{-2(t-\lambda)}\, u(\lambda - 1)\, u(t-\lambda)\, d\lambda. \end{aligned} \tag{10.167}$$

Upon application of the recipe described by Eq. (10.158), $v_{out}(t)$ becomes

$$\begin{aligned} v_{out}(t) &= \left[\int_0^t 2e^{-2(t-\lambda)}\, d\lambda \right] u(t) \\ &\quad - \left[\int_1^t 2e^{-2(t-\lambda)}\, d\lambda \right] u(t-1) \\ &= \frac{2}{2}\, e^{-2(t-\lambda)} \Big|_0^t\, u(t) - \frac{2}{2}\, e^{-2(t-\lambda)} \Big|_1^t\, u(t-1) \\ &= [1 - e^{-2t}]\, u(t) - [1 - e^{-2(t-1)}]\, u(t-1) \text{ V}, \end{aligned} \tag{10.168}$$

where we reintroduced the unit step functions $u(t)$ and $u(t-1)$ associated with the two integration terms. We note that the expression given by Eq. (10.168) is identical with the result obtained earlier by Method 1 (namely, Eq. (10.165)).

Method 3: Convolution by Graphical Integration

The convolution integral given by

$$v_{out}(t) = \int_0^t v_{in}(\lambda)\, h(t-\lambda)\, d\lambda \tag{10.169}$$

can be computed graphically at successive values of t. We start by plotting the rectangular-pulse excitation $v_{in}(t)$ and the circuit's impulse response $h(t)$, as shown in Fig. 10-13(a). To perform the convolution given by Eq. (10.169), we need to plot $v_{in}(\lambda)$ and $h(t-\lambda)$ along the λ-axis. Parts (b) through (e) show these plots at progressive values of t, starting at $t = 0$ and concluding at $t = 2$ s. In all cases, $v_{in}(\lambda)$ remains unchanged, but $h(t-\lambda)$ is obtained by "folding" the original function across the vertical axis to generate $h(-\lambda)$ and then shifting it to the right along the λ-axis by an amount t. The output voltage is equal to the integrated product of $v_{in}(\lambda)$ and $h(t-\lambda)$, which is equal to the shaded overlap areas in the figures. At $t = 0$ (Fig. 10-13(b)), no overlap exists; hence, $v_{out}(0) = 0$. Sliding $h(t-\lambda)$ by $t = 0.5$ s, as shown in Fig. 10-13(c), leads to $v_{out}(0.5) = 0.63$. Sliding $h(t-\lambda)$ further to the right leads to greater overlap, reaching a maximum at $t = 1$ s (Fig. 10-13(d)). Beyond $t = 1$ s, the overlap is smaller in area, as illustrated by part (e) of Fig. 10-13, which corresponds to a shift $t = 2$ s. If the values of $v_{out}(t)$ (determined through this graphical integration method at successive values of t) are plotted as a function of t, we would get the circuit response curve shown in Fig. 10-12(c).

Graphical Convolution Technique

Step 1: On λ-axis—display $x(\lambda)$ and $h(-\lambda)$, the latter being a folded image of $h(\lambda)$ about the vertical axis.

Step 2: Shift $h(-\lambda)$ to the right by a small increment t to obtain $h(t-\lambda)$.

Step 3: Determine the product of $x(\lambda)$ and $h(t-\lambda)$ and integrate it over the λ-domain from $\lambda = 0$ to $\lambda = t$ to get $y(t)$. The integration is equal to the area overlapped by the two functions.

Step 4: Repeat Steps 2 and 3 for each of many successive values of t to generate the complete response $y(t)$.

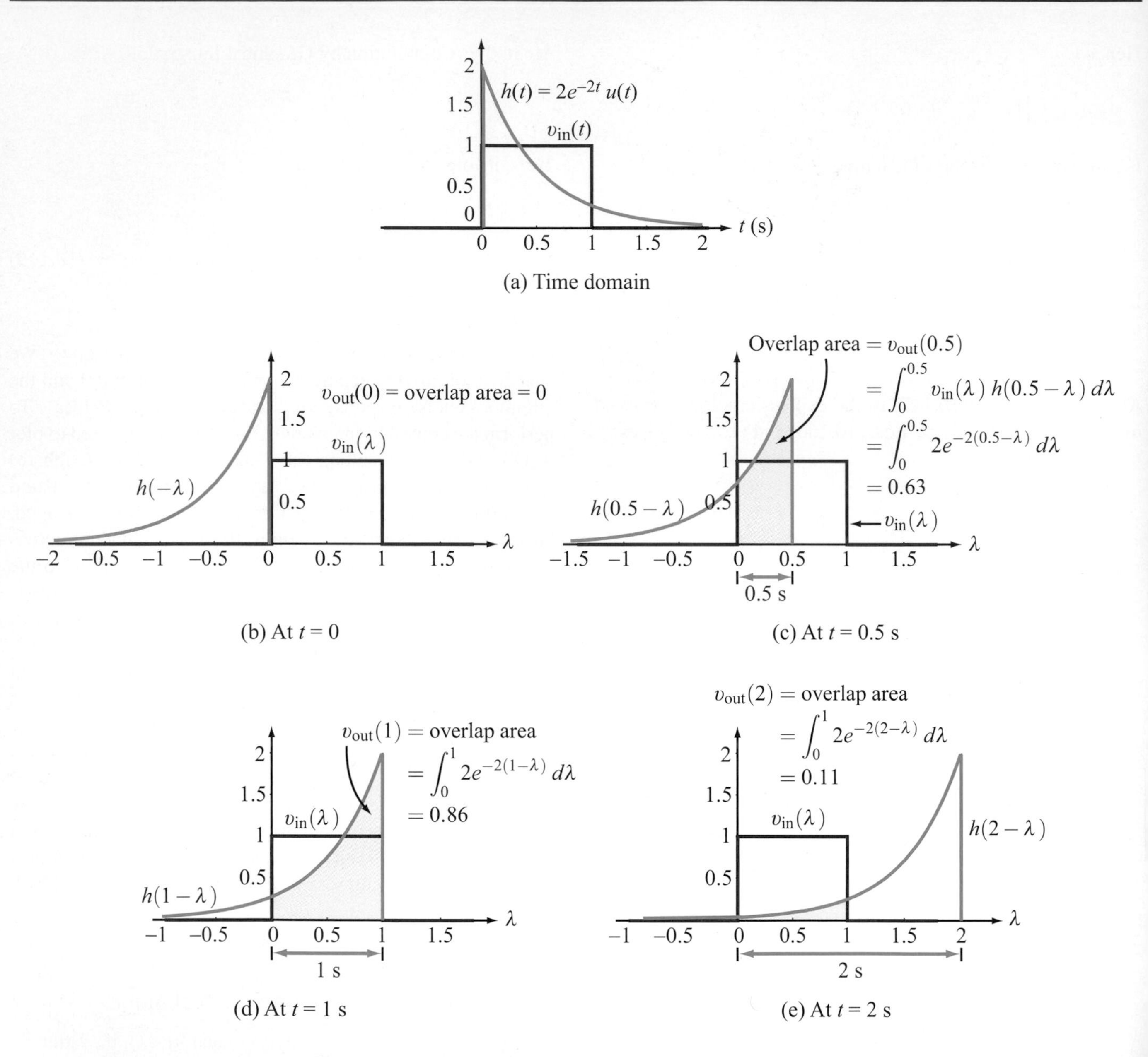

Figure 10-13: Graphical convolution solution for Example 10-15.

Example 10-16: Graphical Convolution

Given the waveforms shown in Fig. 10-14(a), apply the graphical convolution technique to determine the response $y(t) = x(t) * h(t)$.

Solution: Figure 10-14(b) shows waveforms $x(\lambda)$ and $h(-\lambda)$, plotted along the λ-axis. The waveform $h(-\lambda)$ is the mirror image of $h(\lambda)$ with respect to the vertical axis. In parts (c) through (e) of the same figure, $h(t-\lambda)$ is plotted for $t = 1$ s, 1.5 s, and 2 s, respectively. In each case, the shaded area is equal to $y(t)$. We note that when $t > 1$ s, one of the shaded areas contributes a positive number to $y(t)$ while the other contributes a negative number. The overall resultant response $y(t)$ generated by this process of sliding $h(t-\lambda)$ to the right is displayed in Fig. 10-14(f).

We note that when two functions with finite time widths T_1 and T_2 are convolved, the width of the resultant function will be equal to $(T_1 + T_2)$, regardless of the shapes of the two functions

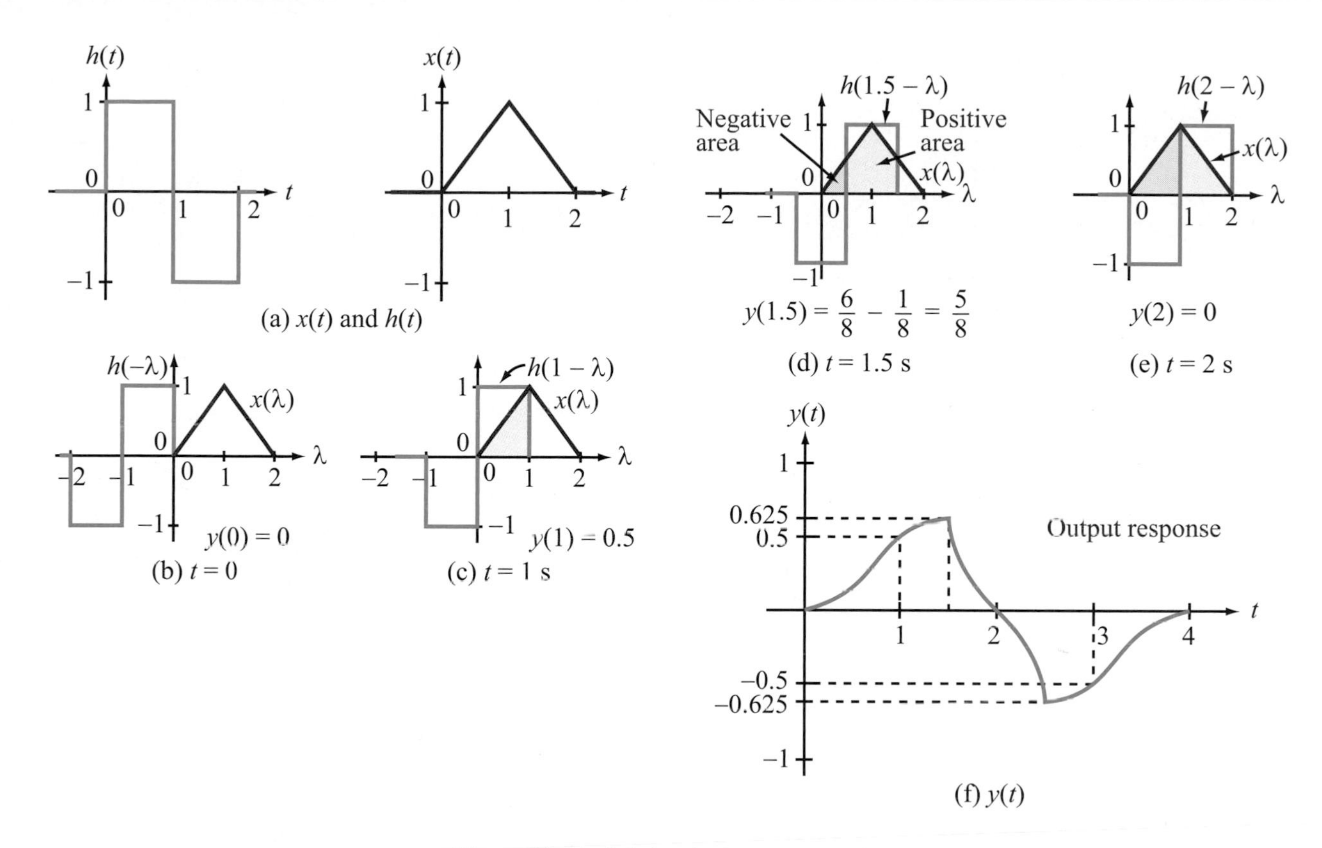

Figure 10-14: Solution for Example 10-16.

Review Question 10-13: Describe the process involved in the application of the graphical convolution technique.

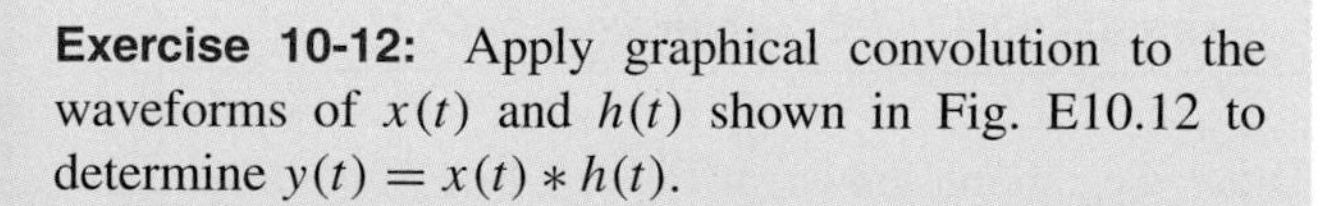

Exercise 10-12: Apply graphical convolution to the waveforms of $x(t)$ and $h(t)$ shown in Fig. E10.12 to determine $y(t) = x(t) * h(t)$.

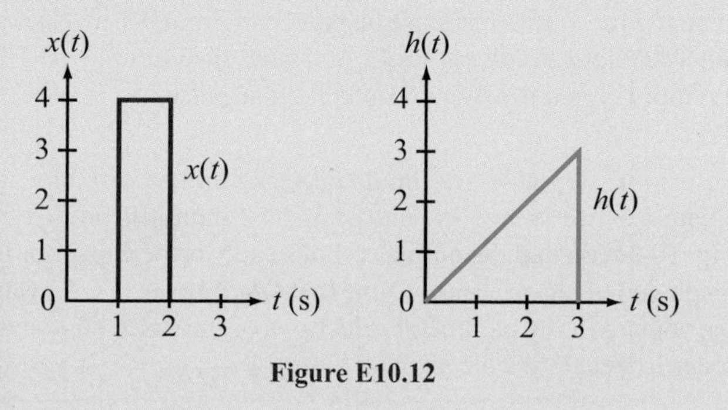

Figure E10.12

Answer:

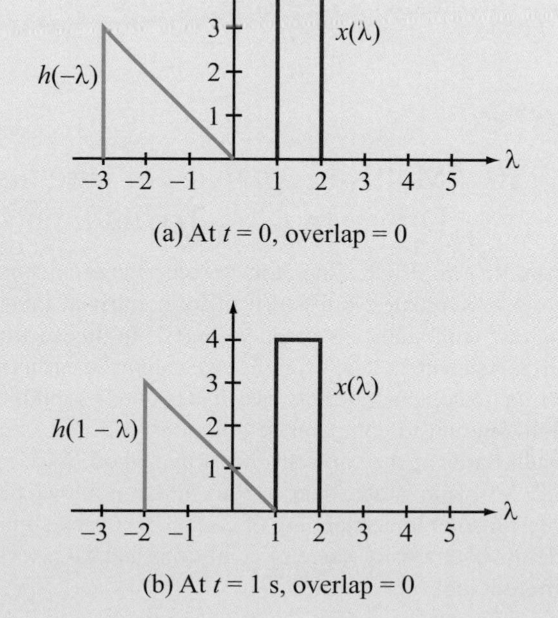

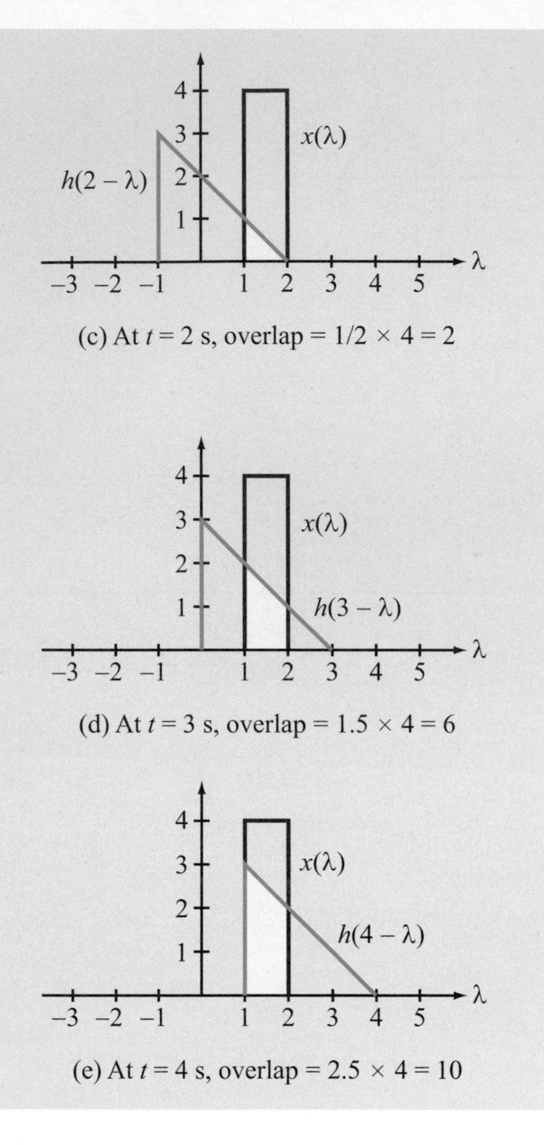

(c) At $t = 2$ s, overlap = $1/2 \times 4 = 2$

(d) At $t = 3$ s, overlap = $1.5 \times 4 = 6$

(e) At $t = 4$ s, overlap = $2.5 \times 4 = 10$

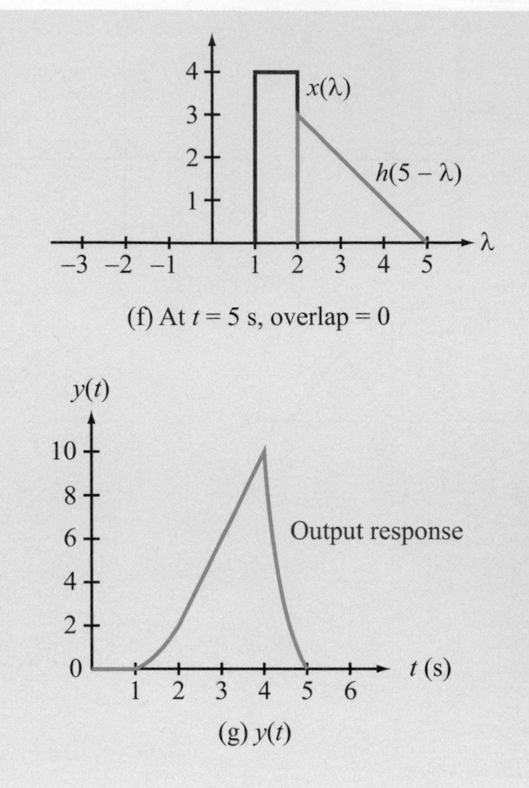

(f) At $t = 5$ s, overlap = 0

(g) $y(t)$

(See)

10-10 Multisim Analysis of Circuits Driven by Non-Trivial Inputs

The utility of SPICE simulators becomes most evident when trying to simulate circuits driven by non-trivial inputs—in contrast with sinusoids or dc voltages. In this section, we will revisit some of the examples we examined earlier in this chapter to demonstrate how easy it is to obtain solutions with Multisim and to compare the solutions with the analytical results based on the Laplace transform method. As a learning tool, Multisim is also very useful, in that it allows the user to test his/her understanding of core concepts by simulating circuits over a wide range of conditions and for a variety of different input waveforms.

Example 10-17: RC Circuit Response

Draw the circuit shown in Fig. 10-12(a) in Multisim and generate the output response displayed in Fig. 10-12(c) using the Transient Analysis tool. As stated in Example 10-15, the input signal is a 1-V, 1-s rectangular pulse.

Solution: By now, we should be very familiar with how to create a pulse source in Multisim. The circuit is shown in Fig. 10-15(a), and the output response across the capacitor is displayed in Fig. 10-15(b). Note that a ***delay time*** of 0.5 s was introduced in the parameter selections of the pulse source in order to generate a clearer plot.

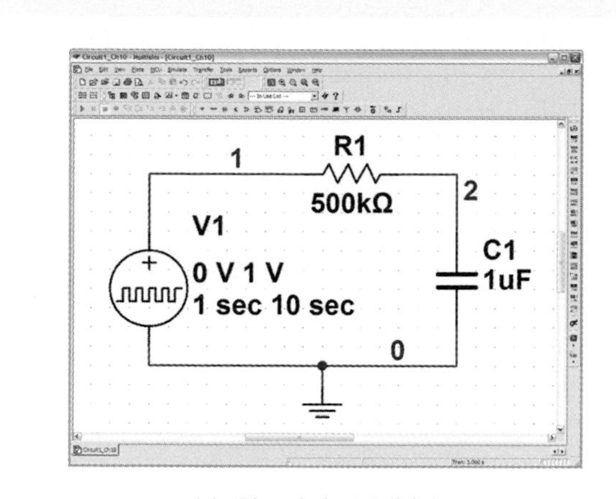

(a) Circuit in Multisim

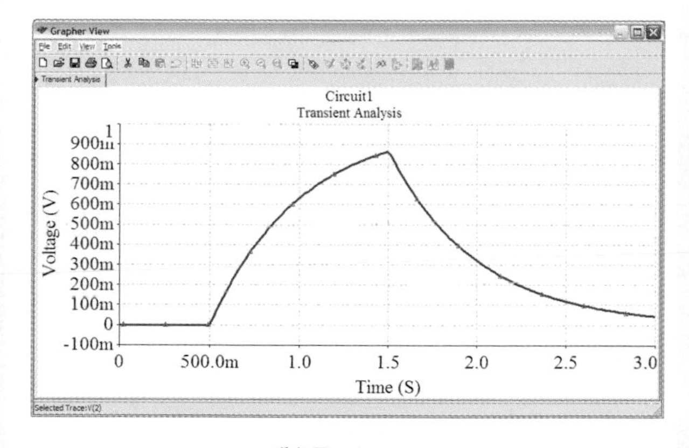

(b) Response

Figure 10-15: (a) RC circuit excited by a 1-V, 1-s rectangular pulse at 0.5 s, and (b) the corresponding response at node 2.

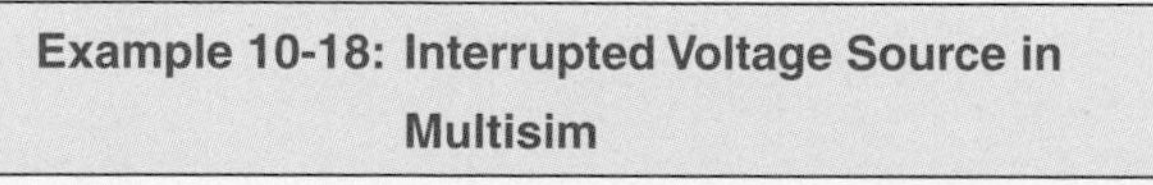

Example 10-18: Interrupted Voltage Source in Multisim

Draw the circuit shown in Fig. 10-6(a) in Multisim and then use Transient Analysis to generate a plot of the voltage across the 3-Ω resistor in response to an input excitation given by 15 V prior to $t = 0$, and $15(1 - e^{-2t})$ V afterwards.

Solution: The circuit is reproduced in Fig. 10-16(a). To model the exponential input voltage, we use the EXPONENTIAL_VOLTAGE source which can accommodate both rising and falling exponentials. Multisim divides the exponential voltage into two segments, with the first called the ***Rise*** segment and the second called the ***Fall*** segment, and this

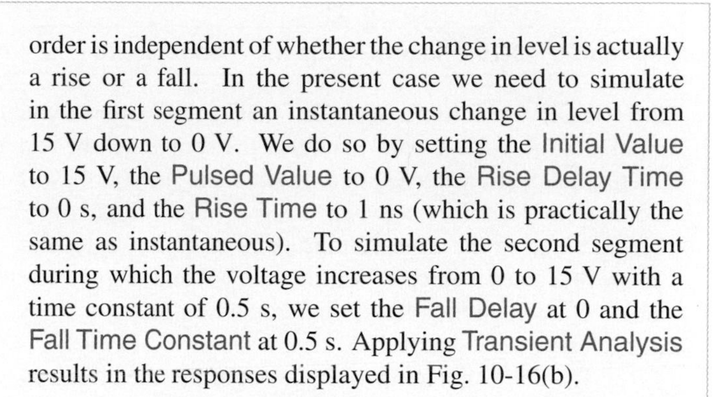

order is independent of whether the change in level is actually a rise or a fall. In the present case we need to simulate in the first segment an instantaneous change in level from 15 V down to 0 V. We do so by setting the Initial Value to 15 V, the Pulsed Value to 0 V, the Rise Delay Time to 0 s, and the Rise Time to 1 ns (which is practically the same as instantaneous). To simulate the second segment during which the voltage increases from 0 to 15 V with a time constant of 0.5 s, we set the Fall Delay at 0 and the Fall Time Constant at 0.5 s. Applying Transient Analysis results in the responses displayed in Fig. 10-16(b).

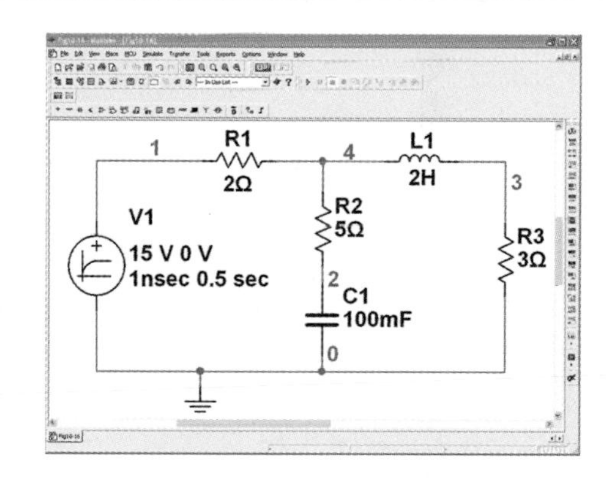

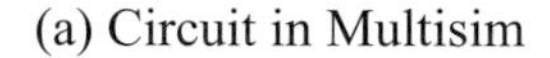

(a) Circuit in Multisim

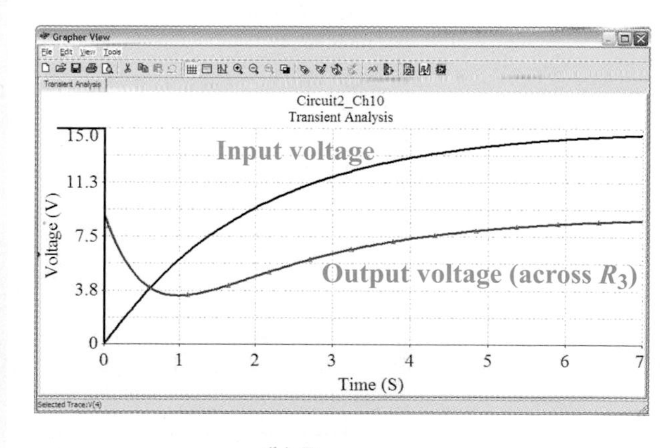

(b) Response

Figure 10-16: Multisim rendition of the circuit response to a sudden (but temporary) change in supply voltage level.

Example 10-19: Piecewise Linear Voltage Source

The PIECEWISE_LINEAR_VOLTAGE source allows you to define time-voltage pairs such that the source will be at a given voltage at the corresponding time and the source will "connect the dots" in between using a linear progression. Hence, entering the time-voltage pairs of (0,1), (1,1), and (2,4) will create a source which starts at 1 V and stays steady until 1 s, at which time it will increase with a slope of 3 V/s to reach a value of 4 V at 2 s.

Replace the Exponential voltage source in the circuit in Fig. 10-16(a) with a piecewise linear (PWL) voltage source with the time-voltage pairs: (0,1), (1,1), (2,4), (3,3), (7,−1), (8,5), and (9,5). Plot both the input and output in Transient Analysis from 0 to 10 s.

Solution: With the PIECEWISE_LINEAR_VOLTAGE source in place, double-click on it and then make sure the Value tab is selected. Click on Enter data points in table, and insert the time-voltage pairs shown in Fig. 10-17(a). Applying Transient Analysis results in the responses displayed in Fig. 10-17(b).

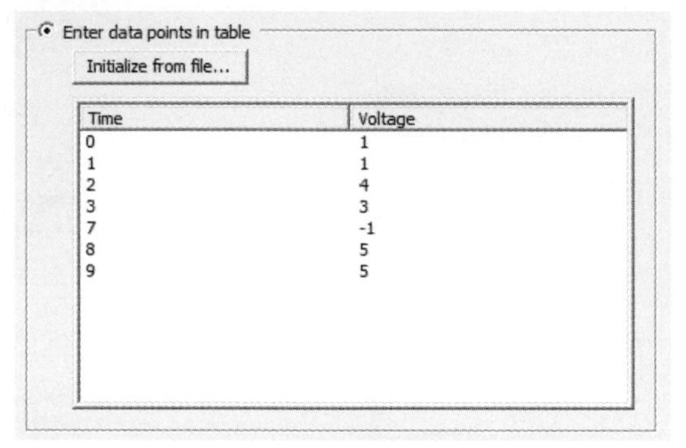

(a) Time-voltage pairs for circuit

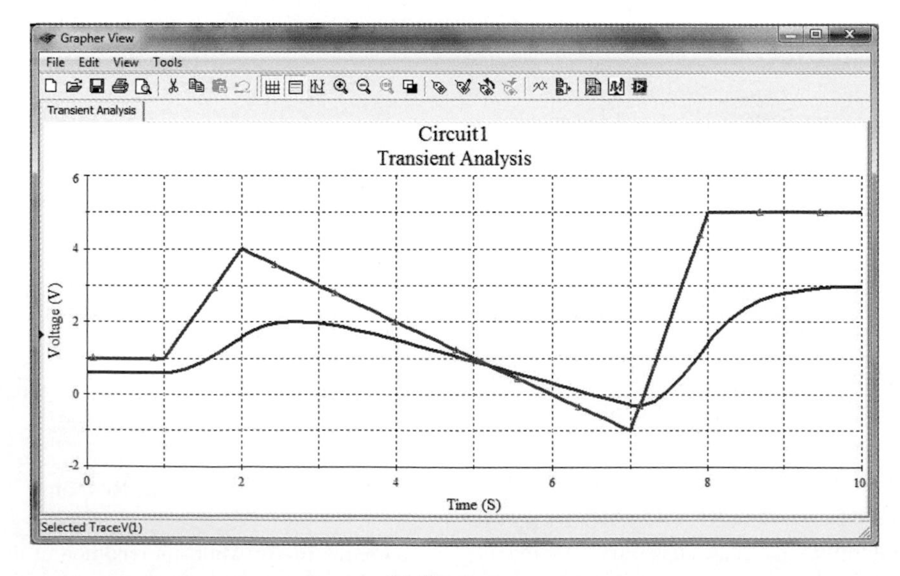

(b) Response

Figure 10-17: Multisim rendition of the circuit response to an arbitrary input signal produced by the PWL (Piecewise Linear) source.

Technology Brief 21: Hard Disk Drives (HDD)

Although invented in 1956, the ***hard disk drive (HDD)*** arguably is still the most commonly used data-storage device among non-volatile storage media available today. It is the availability of vast amounts of relatively inexpensive hard-drive space that has made search engines, webmail, and online games possible. Over the past 40 years, improvements in HDD technology have led to huge increases in storage density, which are simultaneous with the significant reduction in physical size. The term ***hard disk*** or ***hard drive*** evolved from common usage as a means to distinguish these devices from flexible (***floppy***) disk drives.

HDD Operation

Hard drives make use of magnetic material to read and write data. A non-magnetic disc ranging in diameter from 36 to 146 mm is coated with a thin film of magnetic material, such as an iron or cobalt alloy. When a strong magnetic field is applied across a small area of the disc, it causes the atoms in that area to align along the orientation of the field, providing the mechanism for writing bits of data onto the disc (Fig. TF21-1). Conversely, by detecting the aligned field, data can be read back from the disc. The hard drive is equipped with an arm that can be moved across the surface of the disc (Fig. TF21-2), and the disc itself is spun around to make all of the magnetic surface accessible to the writing or reading heads. Because writing onto or reading from the magnetized surface can be performed very rapidly (fraction of a microsecond), hard drives are spun at very high speeds (5000 to 15000 rpm) when directed to record or retrieve information. Amazingly, hard-drive heads usually hover at a height of about 25 nm above the surface of the magnetic disc while the disc is spinning at such high speeds! The extremely small gap between the head and the disc is maintained by having the head "ride" on a thin cushion of air trapped between the head and the surface of the spinning disc. To prevent accidental scratches, the disc is coated with carbon- or Teflon-like materials.

Hard drives are packaged carefully to prevent dust and other airborne particles from interfering with the drive's operation. In combination with the air motion caused by the spinning disc, a very fine air filter is used to keep dust out while maintaining the air pressure necessary to cushion the spinning discs. Hard drives intended for operation at high altitudes (or low air pressure) are sealed hermetically so as to make them airtight.

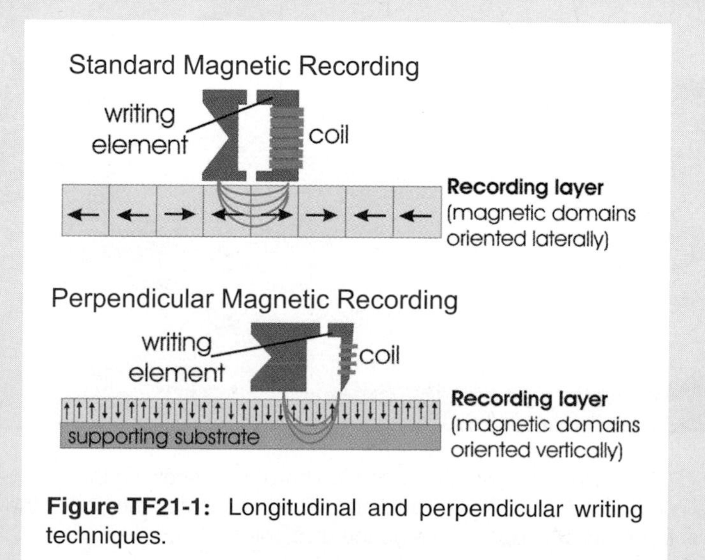

Figure TF21-1: Longitudinal and perpendicular writing techniques.

Technological Advances

Early hard drives performed read and write operations by using an inductor coil placed at the tip of the head. When electric current is made to flow through the coil, the coil induces a magnetic field which in turn aligns the atoms of the magnetic material (i.e., a ***write*** operation). The same coil also is used to detect the presence of aligned atoms, thereby providing the ***read*** operation. The many major developments that shaped the evolution of read/write heads over the past 50 years have introduced two major differences between the modern hard-drive heads and the original models. Instead of using the same head for both reading and writing, separate heads are now used for the two operations. Furthermore, the writing operation is now carried out with a lithographically defined ***thin-film*** head (see Technology Brief 7 on page 135), thereby reducing the ***feature size*** of the head by several orders of magnitude. The feature size is the area occupied by a single bit on the disc surface, which is determined in part by the size of the write head. Decreasing feature size leads to increased recording density. The read operation—housed separately next to the write head—uses a ***magnetoresistive*** material whose resistance changes when exposed to a magnetic field—even when the field intensity is exceedingly small. In modern hard drives, high magnetoresistive sensitivities are realized through the application of either the ***giant magnetoresistance (GMR)*** phenomena or the ***tunneling magnetoresistance (TMR)*** effect exhibited by certain materials. The 2007 Nobel prize in physics was awarded to Albert Fert and Peter Grünberg for their discovery of GMR. A consequence of the extremely small size of the magnetic bits (each bit in a 100-Gb/in^2 disc is about 40-nm long) is that temperature variations can lead to loss of information over time. One method developed to combat this issue is to use two magnetic layers separated by a thin (~ 1 nm) insulator, which increases the stability of the stored bit. Another recent innovation that is already in production involves the use of ***perpendicular magnetic recording (PMR)*** as illustrated in Fig. TF21-1. PMR makes it possible to align bits more compactly next to each other. It is estimated that PMR will enable densities on the order of 1000 Gb/in^2 in the next decade.

Figure TF21-2: Close-up of a disassembled hard drive showing the magnetic discs mounted on a spindle and an actuator arm. The head sits at the end of the arm and performs the read/write operations as the disc spins.

Chapter 10 Relationships

Singularity Functions

Unit Step Function

$$u(t-T) = \begin{cases} 0 & \text{for } t < T \\ 1 & \text{for } t > T \end{cases}$$

Unit Impulse Function

$$\delta(t-T) = 0 \qquad \text{for } t \neq T$$

$$\int_{-\infty}^{\infty} \delta(t-T)\, dt = 1$$

Laplace Transform

$$\mathbf{F}(\mathbf{s}) = \mathcal{L}[f(t)] = \int_{0^-}^{\infty} f(t)\, e^{-\mathbf{s}t}\, dt$$

Properties ➡ Table 10-1
Transform Pair ➡ Table 10-2

Time/s-Domain Equivalents

Resistor	$v = Ri$	⟷	$\mathbf{V} = R\mathbf{I}$
Inductor	$v = L\,\dfrac{di}{dt}$	⟷	$\mathbf{V} = \mathbf{s}L\mathbf{I} - L\,i(0^-)$
Capacitor	$i = C\,\dfrac{dv}{dt}$	⟷	$\mathbf{I} = \mathbf{s}C\mathbf{V} - C\,v(0^-)$

Convolution

Correspondence $x(t) * h(t)$ ⟷ $\mathbf{X}(\mathbf{s})\,\mathbf{H}(\mathbf{s})$

Integral $$y(t) = x(t) * h(t) = \int_0^t x(\lambda)\, h(t-\lambda)\, d\lambda$$

CHAPTER HIGHLIGHTS

- The Laplace transform analysis technique is similar to the phasor-domain technique (used in Chapters 7 through 9 to analyze ac circuits) in that it transforms the circuit to a new domain, solves for the quantity of interest in that domain, and then transforms the solution back to the time domain. However, the Laplace transform technique can be applied not only to ac circuits but to any other type of excitation as well.
- The time derivative of a step function that transitions from 0 to A at $t = T$ is an impulse function given by $A\,\delta(t-T)$.
- The Laplace transform has many useful properties that can facilitate the process of finding the Laplace transform of a time function.
- Under zero initial conditions, circuit elements R, L, and C transform to R, $\mathbf{s}L$, and $1/\mathbf{s}C$, respectively, in the $\mathbf{s}$-domain.
- The time-domain counterpart of the $\mathbf{s}$-domain transfer function of a system is called the impulse response of the system. It is equal to the output when the input is a unit impulse.
- The convolution-integral method allows us to determine the output response of a circuit by convolving the input excitation with the impulse response of the circuit. This approach is particularly useful when the excitation is in the form of experimental measurements that may be difficult to characterize in the form of a mathematical function.

GLOSSARY OF IMPORTANT TERMS

Provide definitions or explain the meaning of the following terms:

complex frequency
convergence condition
convolution
delta function
expansion coefficients
final-value theorem
frequency shift
improper rational function
impulse response
initial-value theorem
Laplace transform
partial fraction expansion
pole
pole factor
proper rational function
residue method
sampling property
singularity function
time invariance
time scaling
time shift
transfer function
uniqueness property
unit impulse function
unit step function
zero (of a polynomial)

PROBLEMS

Sections 10-1 to 10-3: Laplace Transform and Its Properties

*__10.1__ Express each of the waveforms in Fig. P10.1 in terms of step functions and then determine its Laplace transform. [Recall that the ramp function is related to the step function by $r(t-T) = (t-T)\, u(t-T)$.] Assume that all waveforms are zero for $t < 0$.

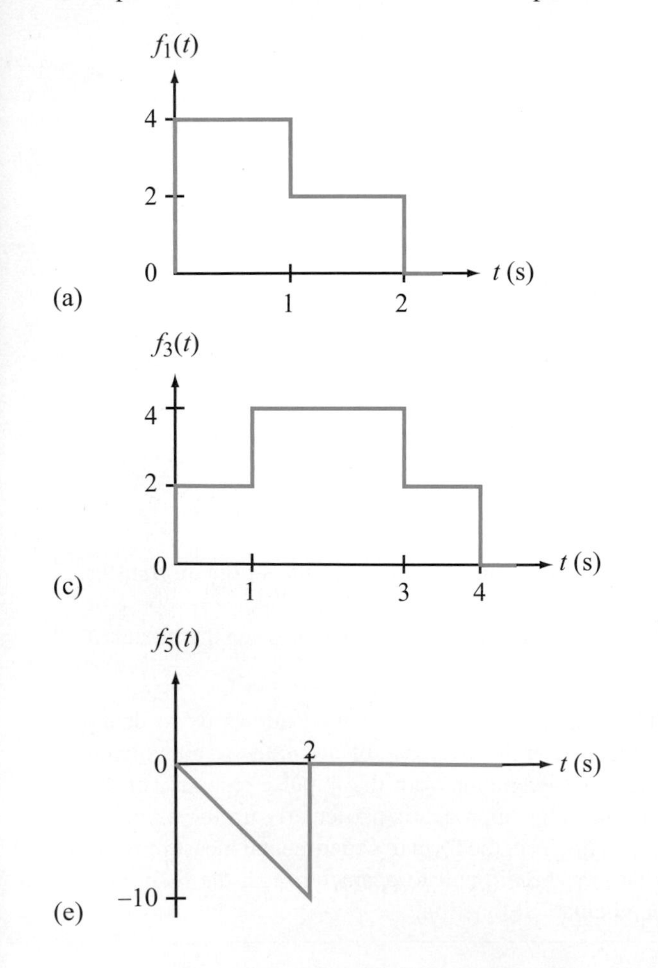

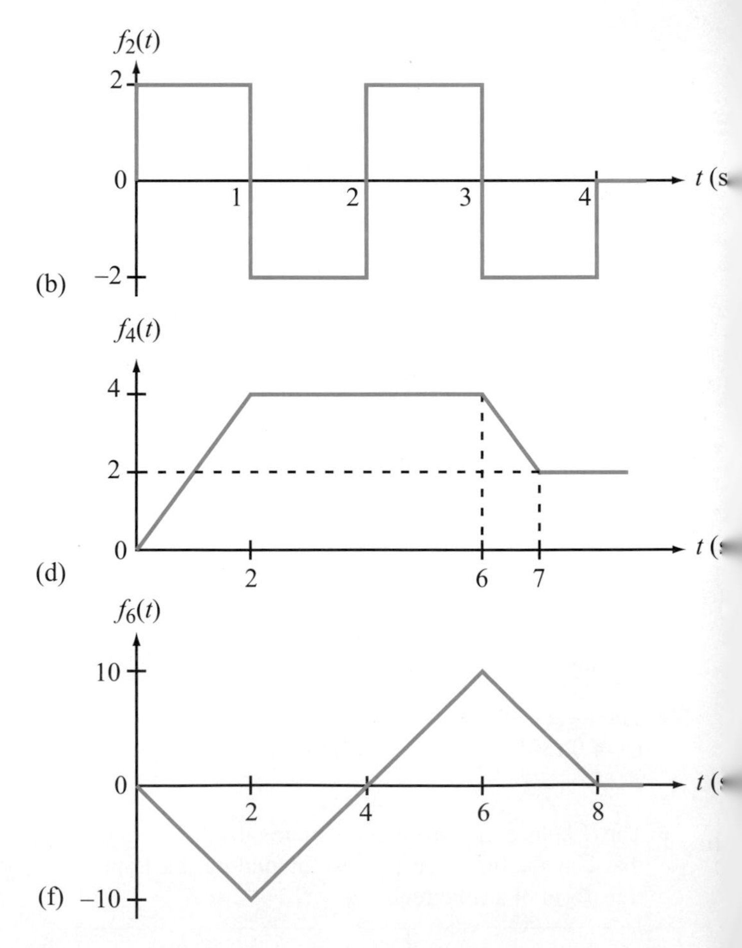

Figure P10.1: Waveforms for Problem 10.1.

*Answer(s) in Appendix E.

10.2 Determine the Laplace transform of each of the *periodic* waveforms shown in Fig. P10.2.

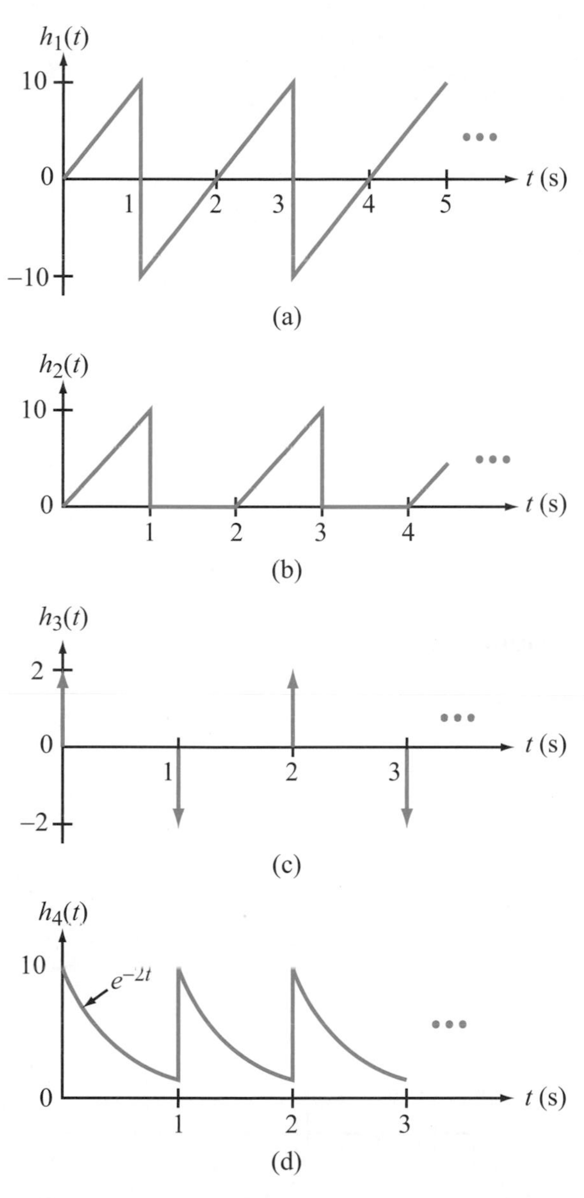

Figure P10.2: Periodic waveforms for Problem 10.2.

10.3 Evaluate each of the following integrals.

(a) $G_1 = \int_{-\infty}^{\infty} (3t^3 + 2t^2 + 1)[\delta(t) + 4\delta(t-2)]\, dt$

(b) $G_2 = \int_{-2}^{4} 4(e^{-2t} + 1)[\delta(t) - 2\delta(t-2)]\, dt$

(c) $G_3 = \int_{-20}^{20} 3(t\cos 2\pi t - 1)[\delta(t) + \delta(t-10)]\, dt$

10.4 Determine the Laplace transform of each of the following functions by applying the properties given in Tables 10-1 and 10-2.

(a) $f_1(t) = 4te^{-2t}\, u(t)$
(b) $f_2(t) = 10\cos(12t + 60^\circ)\, u(t)$
(c) $f_3(t) = 12e^{-3(t-4)}\, u(t-4)$
(d) $f_4(t) = 30(e^{-3t} + e^{3t})\, u(t)$
(e) $f_5(t) = 16e^{-2t}\cos 4t\, u(t)$
(f) $f_6(t) = 20te^{-2t}\sin 4t\, u(t)$

10.5 Determine the Laplace transform of each of the following functions by applying the properties given in Tables 10-1 and 10-2.

***(a)** $h_1(t) = 12te^{-3(t-4)}\, u(t-4)$
(b) $h_2(t) = 27t^2\sin(6t - 60^\circ)\, u(t)$
***(c)** $h_3(t) = 10t^3e^{-2t}\, u(t)$
(d) $h_4(t) = 5(t-6)\, u(t-3)$
(e) $h_5(t) = 10e^{-3t}\, u(t-4)$
(f) $h_6(t) = 4e^{-2(t-3)}\, u(t-4)$

10.6 Determine the Laplace transform of the following functions.

(a) $f_1(t) = 25\cos(4\pi t + 30^\circ)\, \delta(t)$
(b) $f_2(t) = 25\cos(4\pi t + 30^\circ)\, \delta(t - 0.2)$
(c) $f_3(t) = 10\, \dfrac{\sin 3t}{t}\, u(t)$
(d) $f_4(t) = \dfrac{d^2}{dt^2}\, [e^{-4t}\, u(t)]$
(e) $f_5(t) = \dfrac{d}{dt}\, [4te^{-2t}\cos(4\pi t + 30^\circ)\, u(t)]$
(f) $f_6(t) = e^{-3t}\cos(4t + 30^\circ)\, u(t)$
(g) $f_7(t) = t^2[u(t) - u(t-4)]$
(h) $f_8(t) = 10\cos(6\pi t + 30^\circ)\, \delta(t - 0.2)$

10.7 Determine $f(0^+)$ and $f(\infty)$, given that

$$\mathbf{F(s)} = \frac{4\mathbf{s}^2 + 28\mathbf{s} + 40}{\mathbf{s}(\mathbf{s}+3)(\mathbf{s}+4)}.$$

10.8 Determine $f(0^+)$ and $f(\infty)$, given that

$$\mathbf{F(s)} = \frac{\mathbf{s}^2 + 4}{2\mathbf{s}^3 + 4\mathbf{s}^2 + 10\mathbf{s}}.$$

***10.9** Determine $f(0^+)$ and $f(\infty)$, given that

$$\mathbf{F(s)} = \frac{12e^{-2\mathbf{s}}}{\mathbf{s}(\mathbf{s}+2)(\mathbf{s}+3)}.$$

10.10 Determine $f(0^+)$ and $f(\infty)$, given that

$$\mathbf{F}(\mathbf{s}) = \frac{19 - e^{-\mathbf{s}}}{\mathbf{s}(\mathbf{s}^2 + 5\mathbf{s} + 6)}.$$

Section 10-5: Partial Fraction Expansion

10.11 Obtain the inverse Laplace transform of each of the following functions by first applying the partial-fraction-expansion method.

(a) $\mathbf{F}_1(\mathbf{s}) = \dfrac{6}{(\mathbf{s}+2)(\mathbf{s}+4)}$

(b) $\mathbf{F}_2(\mathbf{s}) = \dfrac{4}{(\mathbf{s}+1)(\mathbf{s}+2)^2}$

(c) $\mathbf{F}_3(\mathbf{s}) = \dfrac{3\mathbf{s}^3 + 36\mathbf{s}^2 + 131\mathbf{s} + 144}{\mathbf{s}(\mathbf{s}+4)(\mathbf{s}^2 + 6\mathbf{s} + 9)}$

(d) $\mathbf{F}_4(\mathbf{s}) = \dfrac{2\mathbf{s}^2 + 4\mathbf{s} - 16}{(\mathbf{s}+6)(\mathbf{s}+2)^2}$

10.12 Obtain the inverse Laplace transform of each of the following functions.

(a) $\mathbf{F}_1(\mathbf{s}) = \dfrac{\mathbf{s}^2 + 17\mathbf{s} + 20}{\mathbf{s}(\mathbf{s}^2 + 6\mathbf{s} + 5)}$

(b) $\mathbf{F}_2(\mathbf{s}) = \dfrac{2\mathbf{s}^2 + 10\mathbf{s} + 16}{(\mathbf{s}+2)(\mathbf{s}^2 + 6\mathbf{s} + 10)}$

(c) $\mathbf{F}_3(\mathbf{s}) = \dfrac{4}{(\mathbf{s}+2)^3}$

(d) $\mathbf{F}_4(\mathbf{s}) = \dfrac{2(\mathbf{s}^3 + 12\mathbf{s}^2 + 16)}{(\mathbf{s}+1)(\mathbf{s}+4)^3}$

10.13 Obtain the inverse Laplace transform of each of the following functions.

(a) $\mathbf{F}_1(\mathbf{s}) = \dfrac{(\mathbf{s}+2)^2}{\mathbf{s}(\mathbf{s}+1)^3}$

(b) $\mathbf{F}_2(\mathbf{s}) = \dfrac{1}{(\mathbf{s}^2 + 4\mathbf{s} + 5)^2}$

(c) $\mathbf{F}_3(\mathbf{s}) = \dfrac{\sqrt{2}(\mathbf{s}+1)}{\mathbf{s}^2 + 6\mathbf{s} + 13}$

(d) $\mathbf{F}_4(\mathbf{s}) = \dfrac{-2(\mathbf{s}^2 + 20)}{\mathbf{s}(\mathbf{s}^2 + 8\mathbf{s} + 20)}$

10.14 Obtain the inverse Laplace transform of each of the following functions.

(a) $\mathbf{F}_1(\mathbf{s}) = 2 + \dfrac{4(\mathbf{s}-4)}{\mathbf{s}^2 + 16}$

(b) $\mathbf{F}_2(\mathbf{s}) = \dfrac{4}{\mathbf{s}} + \dfrac{4\mathbf{s}}{\mathbf{s}^2 + 9}$

(c) $\mathbf{F}_3(\mathbf{s}) = \dfrac{(\mathbf{s}+5)e^{-2\mathbf{s}}}{(\mathbf{s}+1)(\mathbf{s}+3)}$

(d) $\mathbf{F}_4(\mathbf{s}) = \dfrac{(1 - e^{-4\mathbf{s}})(24\mathbf{s} + 40)}{(\mathbf{s}+2)(\mathbf{s}+10)}$

(e) $\mathbf{F}_5(\mathbf{s}) = \dfrac{\mathbf{s}(\mathbf{s}-8)e^{-6\mathbf{s}}}{(\mathbf{s}+2)(\mathbf{s}^2 + 16)}$

(f) $\mathbf{F}_6(\mathbf{s}) = \dfrac{4\mathbf{s}(2 - e^{-4\mathbf{s}})}{\mathbf{s}^2 + 9}$

Sections 10-6 and 10-7: s-domain Analysis

*__10.15__ Determine $v(t)$ in the circuit of Fig. P10.15, given that $v_s(t) = 2u(t)$ V, $R_1 = 1\ \Omega$, $R_2 = 3\ \Omega$, $C = 0.3689$ F, and $L = 0.2259$ H.

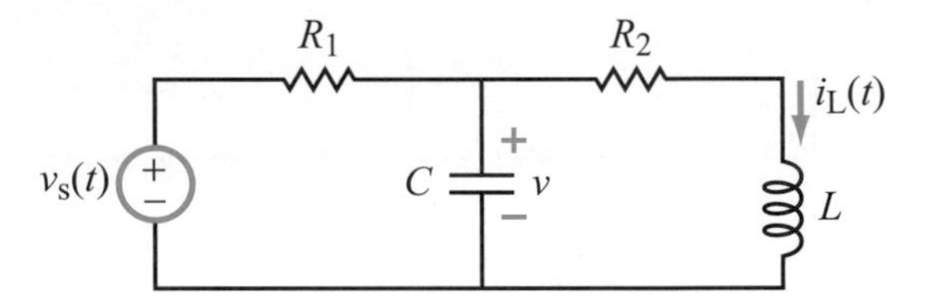

Figure P10.15: Circuit for Problems 10.15 and 10.16.

10.16 Determine $i_L(t)$ in the circuit in Fig. P10.15, given that $v_s(t) = 2u(t)$, $R_1 = 2\ \Omega$, $R_2 = 6\ \Omega$, $L = 2.215$ H, and $C = 0.0376$ F.

10.17 Determine $v_{out}(t)$ in the circuit in Fig. P10.17, given that $v_s(t) = 35u(t)$ V, $v_{C_1}(0^-) = 20$ V, $R_1 = 1\ \Omega$, $C_1 = 1$ F, $R_2 = 0.5\ \Omega$, and $C_2 = 2$ F.

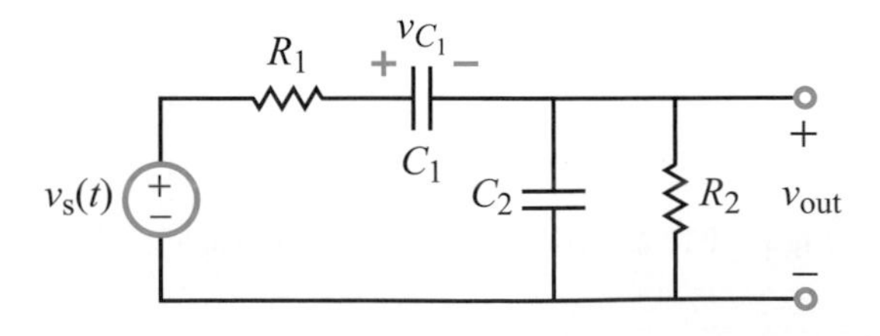

Figure P10.17: Circuit for Problem 10.17.

10.18 Determine $i_L(t)$ in the circuit of Fig. P10.18 for $t \geq 0$ given that the switch was opened at $t = 0$ after it had been closed for a long time, $v_s = 12$ mV, $R_0 = 5\ \Omega$, $R_1 = 10\ \Omega$, $R_2 = 20\ \Omega$, $L = 0.2$ H, and $C = 6$ mF.

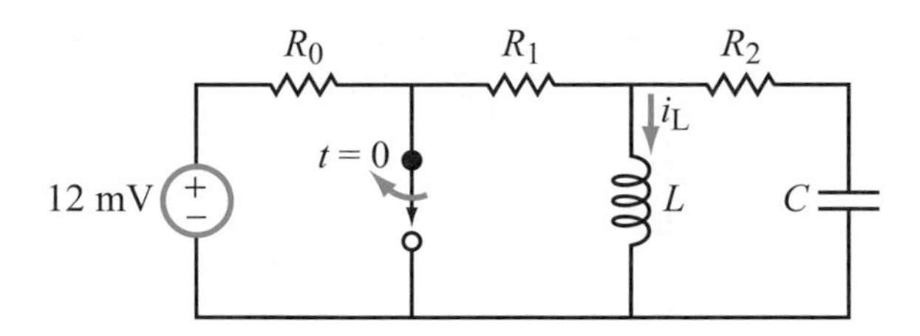

Figure P10.18: Circuit for Problems 10.18 and 10.19.

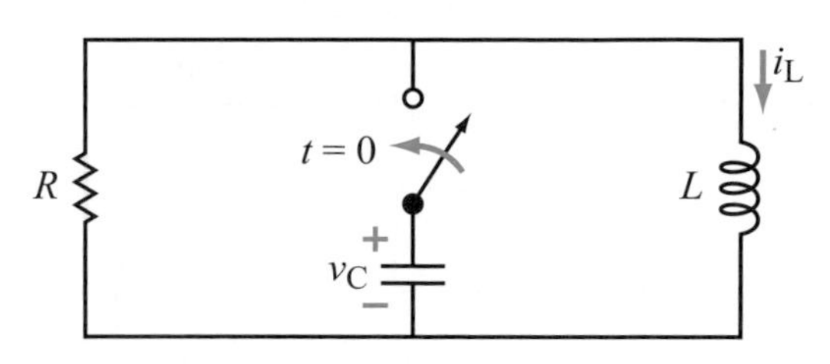

Figure P10.22: Circuit for Problem 10.22.

*10.19 Repeat Problem 10.18, but assume that the switch had been open for a long time and then closed at $t = 0$. Retain the dc source at 12 mV and the resistors at $R_0 = 5\ \Omega$, $R_1 = 10\ \Omega$, and $R_2 = 20\ \Omega$. Change L to 2 H and C to 0.4 F.

10.20 Determine $i_L(t)$ in the circuit of Fig. P10.20, given that $R_1 = 2\ \Omega$, $R_2 = 1/6\ \Omega$, $L = 1$ H, and $C = 1/13$ F. Assume no energy was stored in the circuit segment to the right of the switch prior to $t = 0$.

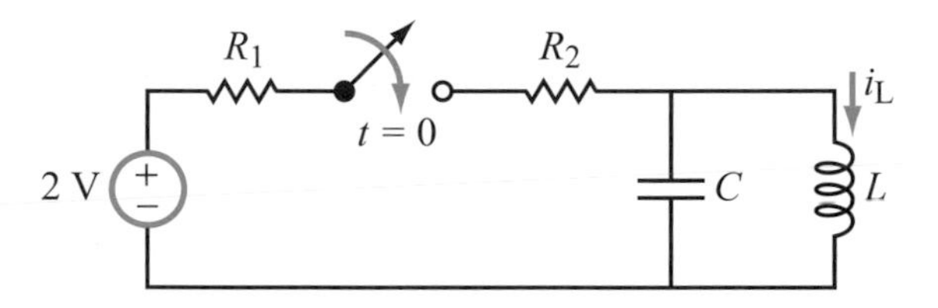

Figure P10.20: Circuit for Problem 10.20.

10.21 Determine $v_{C_2}(t)$ in the circuit of Fig. P10.21, given that $R = 200\ \Omega$, $C_1 = 1$ mF, and $C_2 = 5$ mF.

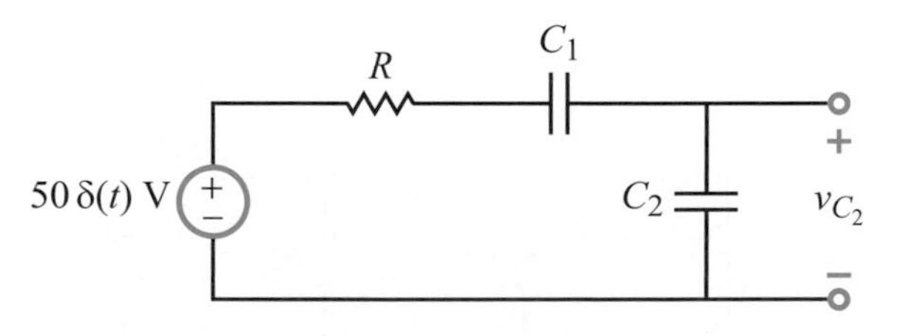

Figure P10.21: Circuit for Problem 10.21.

10.22 Determine $i_L(t)$ in the circuit of Fig. P10.22, given that before closing the switch $v_C(0^-) = 24$ V. Also, the element values are $R = 1\ \Omega$, $L = 0.8$ H, and $C = 0.25$ F.

10.23 Determine $v_{out}(t)$ in the circuit of Fig. P10.23, given that $v_s(t) = 11u(t)$ V, $R_1 = 2\ \Omega$, $R_2 = 4\ \Omega$, $R_3 = 6\ \Omega$, $L = 1$ H, and $C = 0.5$ F.

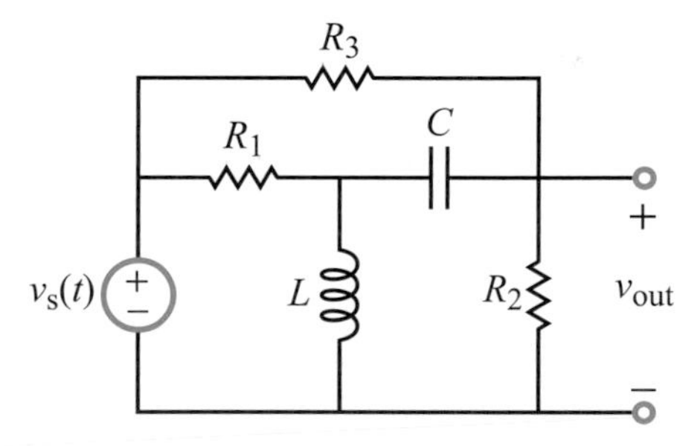

Figure P10.23: Circuit for Problem 10.23.

10.24 Determine $i_L(t)$ in the circuit of Fig. P10.24 for $t \geq 0$, given that $R = 3.5\ \Omega$, $L = 0.5$ H, and $C = 0.2$ F.

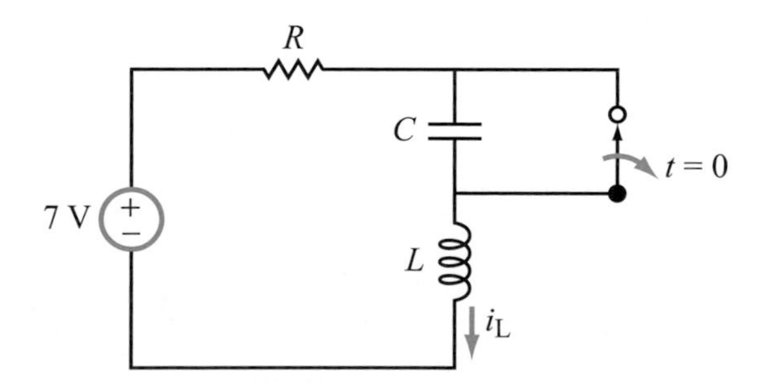

Figure P10.24: Circuit for Problem 10.24.

*10.25 Apply mesh-current analysis in the **s**-domain to determine $i_L(t)$ in the circuit of Fig. P10.25, given that $v_s(t) = 44u(t)$ V, $R_1 = 2\ \Omega$, $R_2 = 4\ \Omega$, $R_3 = 6\ \Omega$, $C = 0.1$ F, and $L = 4$ H.

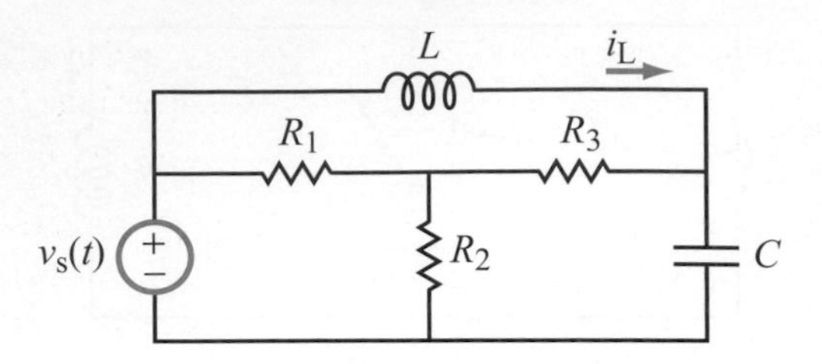

Figure P10.25: Circuit for Problem 10.25.

10.26 Determine $v_{out}(t)$ in the circuit of Fig. P10.26, given that $v_s(t) = 3u(t)$ V, $R_1 = 4\ \Omega$, $R_2 = 10\ \Omega$, and $L = 2$ H.

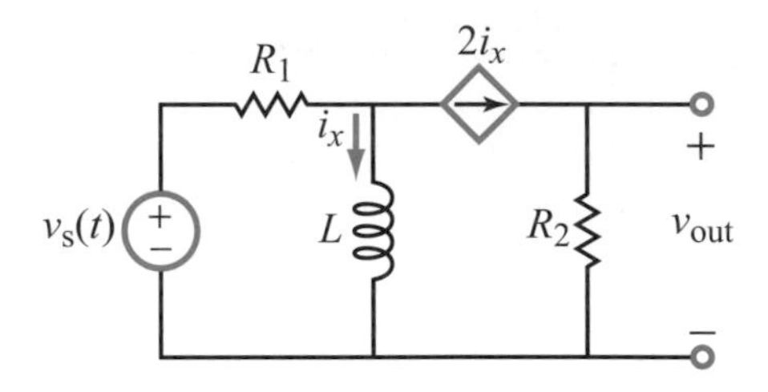

Figure P10.26: Circuit for Problems 10.26 and 10.27.

10.27 Repeat Problem 10.26 with $v_s(t) = 3\delta(t)$ V.

10.28 The voltage source in the circuit of Fig. P10.28 is, given by $v_s(t) = [10 - 5u(t)]$ V. Determine $i_L(t)$ for $t \geq 0$, given that $R_1 = 1\ \Omega$, $R_2 = 3\ \Omega$, $L = 2$ H, and $C = 0.5$ F.

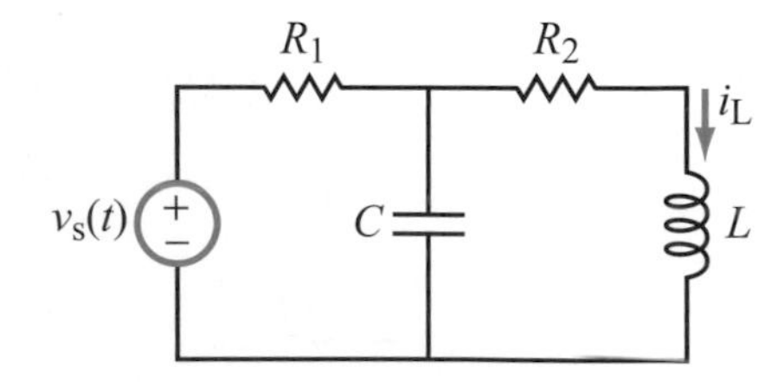

Figure P10.28: Circuit for Problems 10.28 and 10.32.

10.29 The current source in the circuit of Fig. P10.29 is given by $i_s(t) = [10u(t) + 20\delta(t)]$ mA. Determine $v_C(t)$ for $t \geq 0$, given that $R_1 = R_2 = 1$ kΩ and $C = 0.5$ mF.

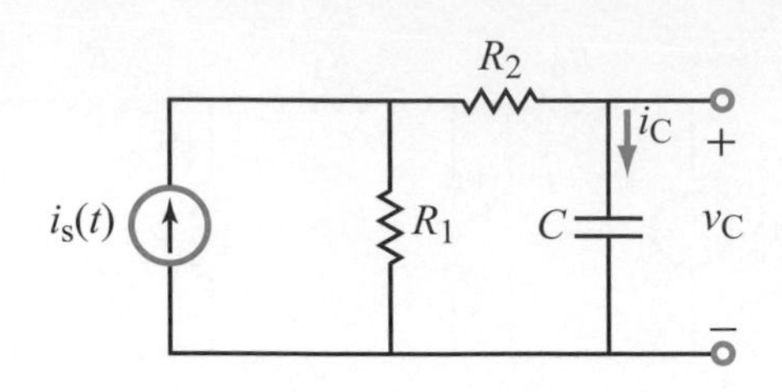

Figure P10.29: Circuit for Problems 10.29 and 10.31.

10.30 The circuit in Fig. P10.30 is excited by a 10-V, 1-long rectangular pulse. Determine $i(t)$, given that $R_1 = 1\ \Omega$, $R_2 = 2\ \Omega$, and $L = 1/3$ H.

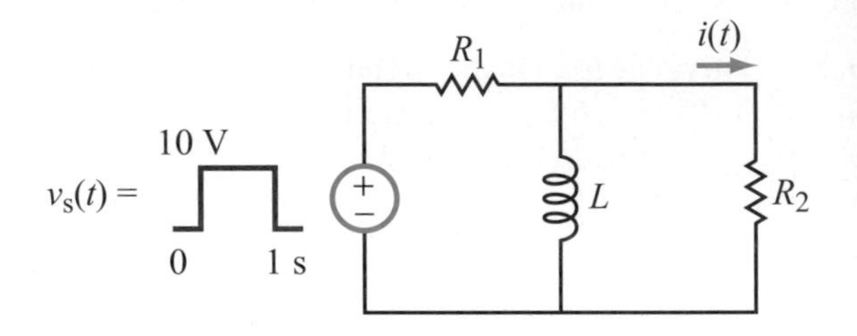

Figure P10.30: Circuit for Problems 10.30 and 10.58.

*__10.31__ Repeat Problem 10.29 after replacing the current source with a 10-mA, 2-s-long rectangular pulse.

10.32 Analyze the circuit shown in Fig. P10.28 to determine $i_L(t)$ in response to a voltage excitation $v_s(t)$ in the form of a 10-V rectangular pulse that starts at $t = 0$ and ends at $t = 5$ s. The element values are $R_1 = 1\ \Omega$, $R_2 = 3\ \Omega$, $L = 2$ H, and $C = 0.5$ F.

10.33 The current source in the circuit of Fig. P10.33 is given by $i_s(t) = 6e^{-2t}\ u(t)$ A. Determine $i_L(t)$ for $t \geq 0$, given that $R_1 = 10\ \Omega$, $R_2 = 5\ \Omega$, $L = 0.6196$ H, and $LC = (1/15)$ s.

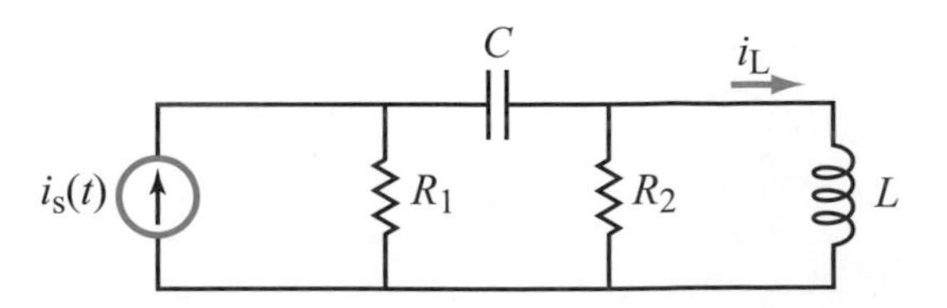

Figure P10.33: Circuit for Problems 10.33 and 10.34.

10.34 Given the current-source waveform displayed in Fig. P10.34, determine $i_L(t)$ in the circuit of Fig. P10.33, given that $R_1 = 10\ \Omega$, $R_2 = 5\ \Omega$, $L = 0.6196$ H, and $LC = (1/15)$ s.

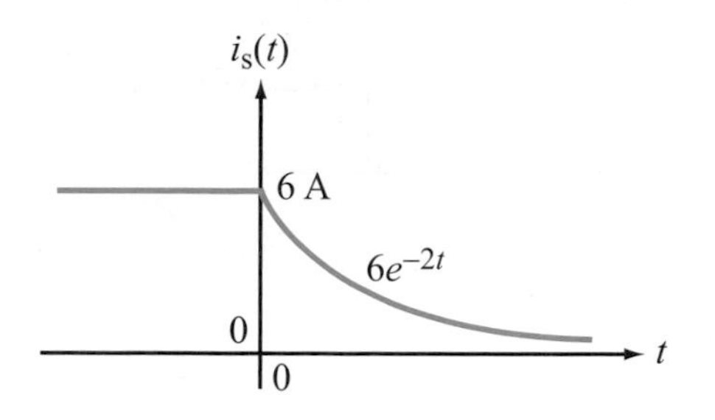

Figure P10.34: Waveform for Problem 10.34.

*10.35 The current source shown in the circuit of Fig. P10.35 is given by the displayed waveform. Determine $v_{out}(t)$ for $t \geq 0$, given that $R_1 = 1\ \Omega$, $R_2 = 0.5\ \Omega$, and $L = 0.5$ H.

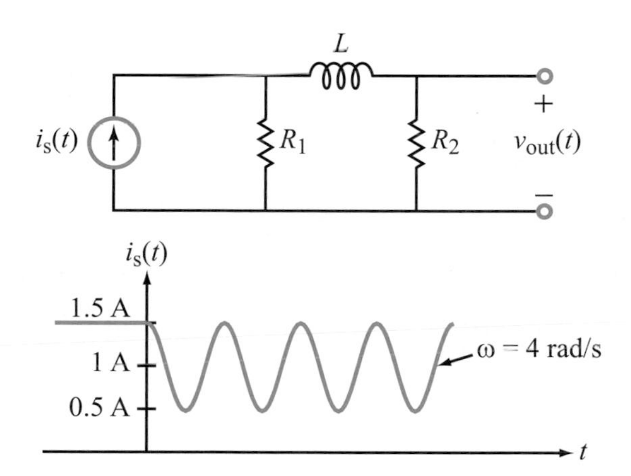

Figure P10.35: Circuit for Problem 10.35 and current waveform for Problems 10.35 and 10.36.

10.36 If the circuit shown in Fig. P10.36 is excited by the current waveform $i_s(t)$ shown in Fig. P10.35, determine $i(t)$ for $t \geq 0$, given that $R_1 = 10\ \Omega$, $R_2 = 5\ \Omega$, and $C = 0.02$ F.

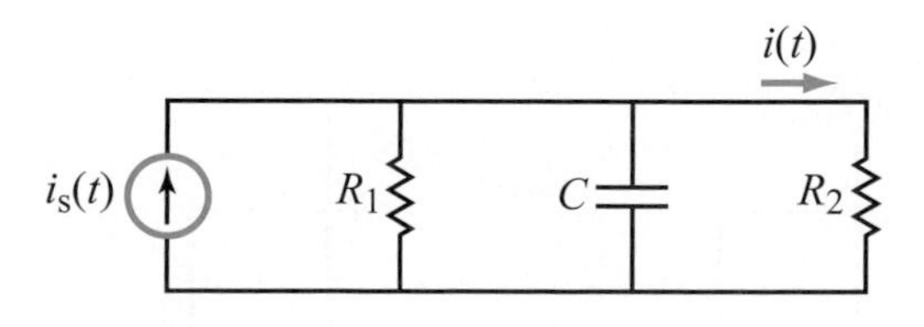

Figure P10.36: Circuit for Problems 10.36 to 10.38.

10.37 If the circuit shown in Fig. P10.36 is excited by current waveform $i_s(t) = 36te^{-6t}\ u(t)$ mA, determine $i(t)$ for $t \geq 0$, given that $R_1 = 2\ \Omega$, $R_2 = 4\ \Omega$, and $C = (1/8)$ F.

10.38 If the circuit shown in Fig. P10.36 is excited by a current waveform given by $i_s(t) = 9te^{-3t}\ u(t)$ mA, determine $i(t)$ for $t \geq 0$, given that $R_1 = 1\ \Omega$, $R_2 = 3\ \Omega$, and $C = 1/3$ F.

10.39 The circuit shown in Fig. P10.39 first was introduced in Problem 5.62. Then, a time-domain solution was sought for $v_{out_1}(t)$ and $v_{out_2}(t)$ for $t \geq 0$, given that $v_i(t) = 10u(t)$ mV, $V_{CC} = 10$ V for both op amps, and the two capacitors had no change prior to $t = 0$. Analyze the circuit and plot $v_{out_1}(t)$ and $v_{out_2}(t)$ using the Laplace transform technique.

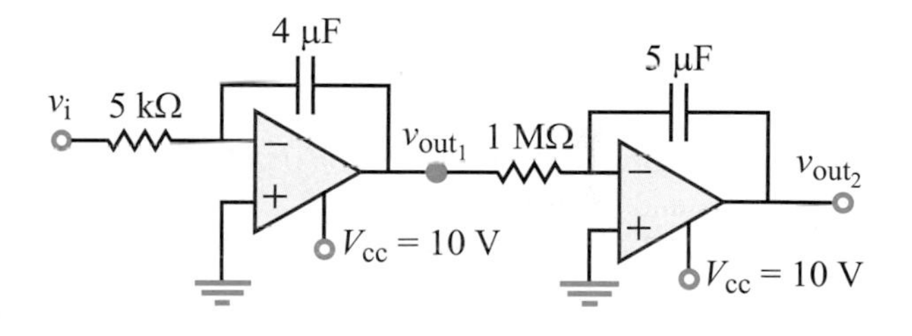

Figure P10.39: Circuit for Problems 10.39 and 10.40.

10.40 Repeat Problem 10.39 retaining all element values and conditions but changing the input voltage to $v_i(t) = 0.4te^{-2t}\ u(t)$.

Section 10-8: Transfer Function and Impulse Response

10.41 A system is characterized by a transfer function given by

$$\mathbf{H}(\mathbf{s}) = \frac{18\mathbf{s} + 10}{\mathbf{s}^2 + 6\mathbf{s} + 5}.$$

Determine the output response $y(t)$, if the input excitation is given by the following functions.

(a) $x_1(t) = u(t)$

*(b) $x_2(t) = 2t\ u(t)$

(c) $x_3(t) = 2e^{-4t}\ u(t)$

(d) $x_4(t) = [4 \cos 4t]\ u(t)$

10.42 When excited by a unit step function at $t = 0$, a system generates the output response

$$y(t) = [5 - 10t + 20 \sin 2t]\ u(t).$$

Determine (a) the system transfer function and (b) the impulse response.

10.43 For the circuit shown in Fig. P10.43, determine (a) $\mathbf{H(s)} = \mathbf{V_o}/\mathbf{V_i}$ and (b) $h(t)$, given that $R_1 = 1\ \Omega$, $R_2 = 2\ \Omega$, $C_1 = 1\ \mu$F, and $C_2 = 2\ \mu$F.

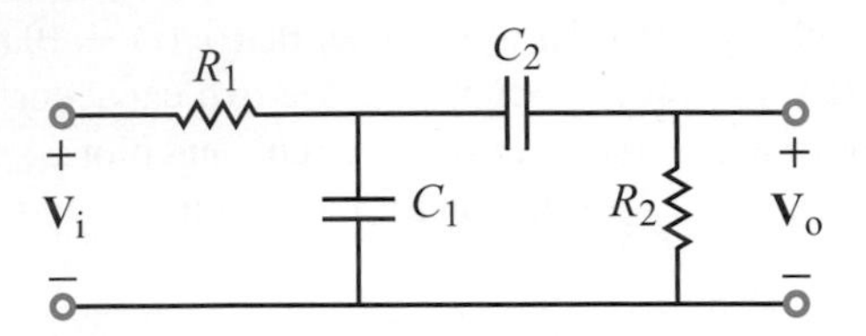

Figure P10.43: Circuit for Problem 10.43.

10.44 For the circuit shown in Fig. P10.44, determine (a) $\mathbf{H(s)} = \mathbf{V_o}/\mathbf{V_i}$ and (b) $h(t)$, given that $R_1 = 1\ \Omega$, $R_2 = 2\ \Omega$, $L_1 = 1$ mH, and $L_2 = 2$ mH.

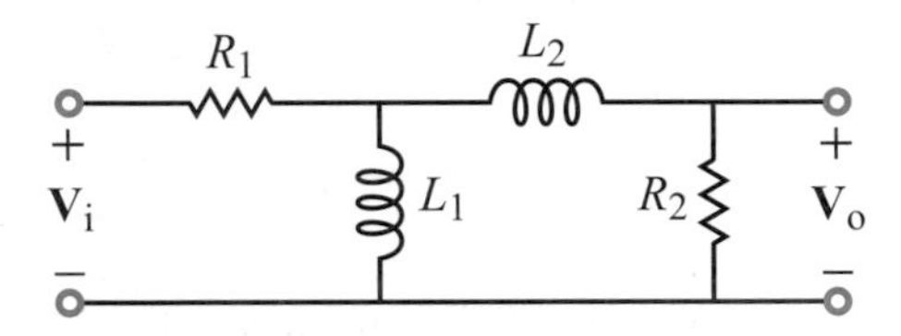

Figure P10.44: Circuit for Problem 10.44.

10.45 For the circuit shown in Fig. P10.45, determine (a) $\mathbf{H(s)} = \mathbf{V_o}/\mathbf{V_i}$ and (b) $h(t)$, given that $R = 5\ \Omega$, $L = 0.1$ mH, and $C = 1\ \mu$F.

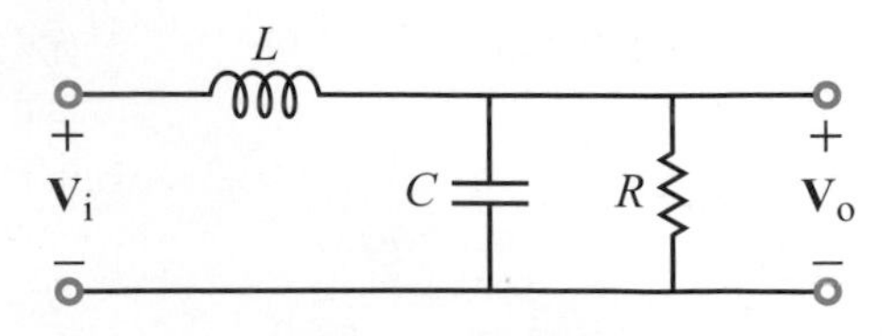

Figure P10.45: Circuit for Problem 10.45.

10.46 For the circuit shown in Fig. P10.46, determine (a) $\mathbf{H(s)} = \mathbf{V_o}/\mathbf{V_s}$ and (b) $h(t)$, given that $R_1 = 1$ kΩ, $R_2 = 4$ kΩ, and $C = 1\ \mu$F.

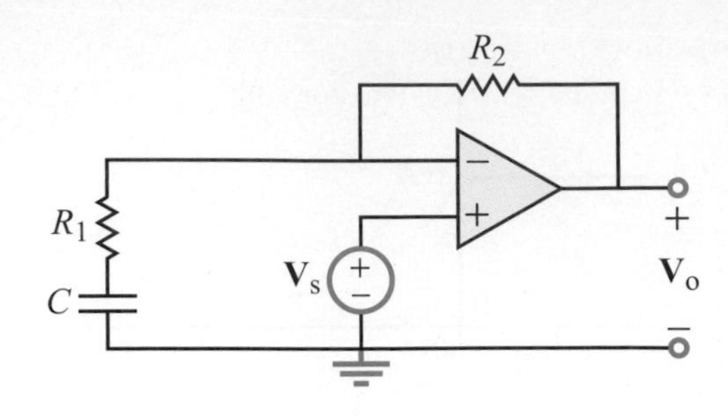

Figure P10.46: Op-amp circuit for Problem 10.46.

*__10.47__ For the circuit shown in Fig. P10.47, determine (a) $\mathbf{H(s)} = \mathbf{V_o}/\mathbf{V_s}$ and (b) $h(t)$, given that $R_1 = R_2 = 100\ \Omega$ and $C_1 = C_2 = 1\ \mu$F.

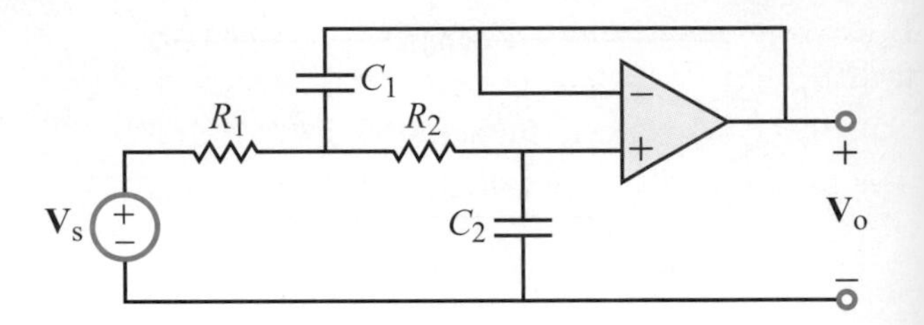

Figure P10.47: Op-amp circuit for Problem 10.47.

Section 10-9: Convolution Integral

10.48 Functions $x(t)$ and $h(t)$ are both rectangular pulses, as shown in Fig. P10.48. Apply graphical convolution to determine $y(t) = x(t) * h(t)$ for the following parameters.

(a) $A = 1$, $B = 1$, $T_1 = 2$ s, and $T_2 = 4$ s

(b) $A = 2$, $B = 1$, $T_1 = 4$ s, and $T_2 = 2$ s

(c) $A = 1$, $B = 2$, $T_1 = 4$ s, and $T_2 = 2$ s

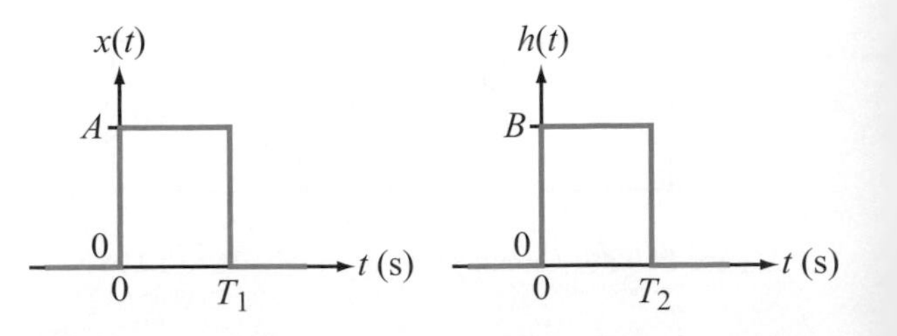

Figure P10.48: Waveforms of $x(t)$ and $h(t)$.

10.49 Apply graphical convolution to the waveforms of $x(t)$ and $h(t)$ shown in Fig. P10.49 to determine $y(t) = x(t) * h(t)$.

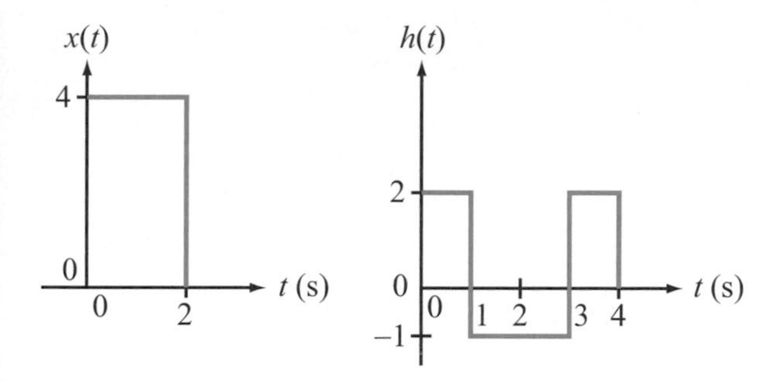

Figure P10.49: Waveforms for Problem 10.49.

10.50 Functions $x(t)$ and $h(t)$ have the waveforms shown in Fig. P10.50. Determine and plot $y(t) = x(t) * h(t)$ by each of the following techniques.

(a) Integrating the convolution analytically

(b) Integrating the convolution graphically

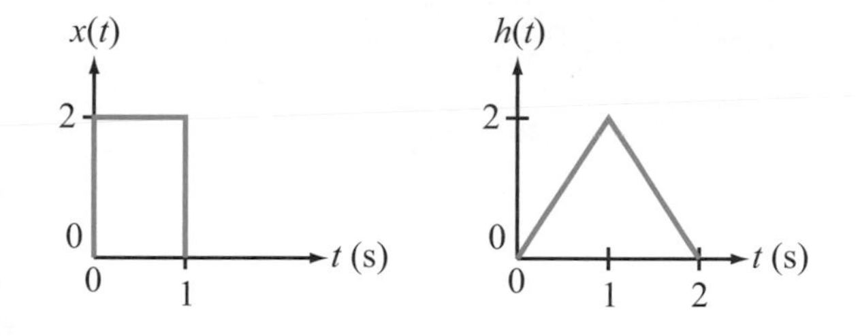

Figure P10.50: Waveforms for Problem 10.50.

10.51 Functions $x(t)$ and $h(t)$ have the waveforms shown in Fig. P10.51. Determine and plot $y(t) = x(t) * h(t)$ by each of the following techniques.

(a) Integrating the convolution analytically

(b) Integrating the convolution graphically

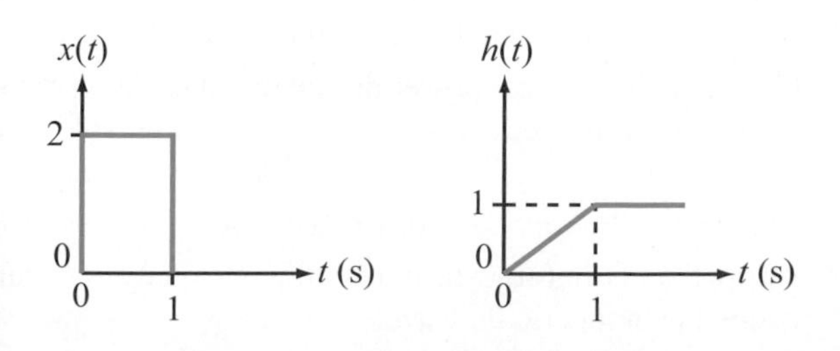

Figure P10.51: Waveforms for Problem 10.51.

10.52 Functions $x(t)$ and $h(t)$ are given by

$$x(t) = \begin{cases} 0 & \text{for } t < 0, \\ \sin \pi t & \text{for } 0 \le t \le 1 \text{ s}, \\ 0 & \text{for } t \ge 1 \text{ s}, \end{cases}$$

and

$$h(t) = u(t).$$

Determine $y(t) = x(t) * h(t)$.

Section 10-10: Multisim

10.53 Simulate the circuit in Example 10-11, but introduce a 1-V square-wave source as the input. What is the magnitude of the ripple in the output if the square wave has a period of 1 s and width of 0.5 s? Does the output ever reach 1 V?

10.54 Apply a 1-V, 1-Hz signal with a 10-V dc offset to the circuit in Fig. P10.54 and plot both $v_{\text{C}}(t)$ and $v_{\text{R}}(t)$ for $R = 1\ \Omega$ and $C = 1$ F. Why are the dc offsets different for the voltages across the two components?

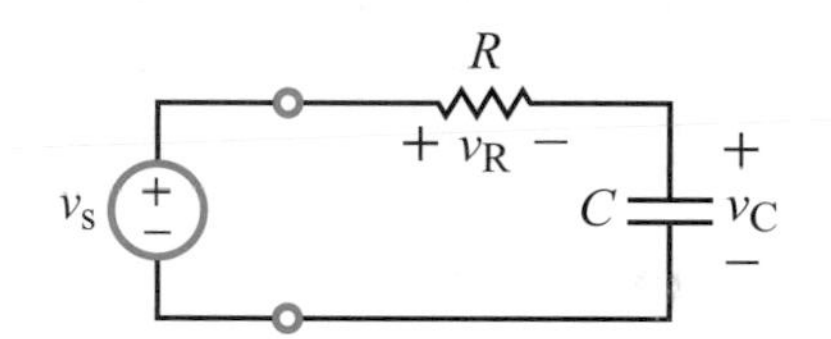

Figure P10.54: Circuit for Problems 10.54 and 10.56.

10.55 Repeat Example 10-18 but vary the time constant of the exponent from 0.1 s to 5 s (pick more than 3 points). Plot all responses on the same display.

10.56 In Multisim, apply input signal $v_{\text{s}}(t) = 5t$ V to the circuit shown in Fig. P10.54 and plot both $v_{\text{C}}(t)$ and $v_{\text{R}}(t)$ for 0 to 5 s. Find the point at which $v_{\text{C}}(t) = v_{\text{R}}(t)$. If we change the input signal to $v_{\text{s}}(t) = 10t$ V, will the point in time where $v_{\text{C}}(t) = v_{\text{R}}(t)$ change? Explain.

10.57 Using a Piecewise_Linear source in Multisim, build and simulate the circuit found in Fig. P10.30 (including the specified source). Plot $i(t)$ for 0 to 2 s. Use $R_1 = 1\ \Omega$, $R_2 = 2\ \Omega$, and $L = 1/3$ H.

CHAPTER 11

Fourier Analysis Technique

Chapter Contents

Objectives

Upon learning the material presented in this chapter, you should be able to:

1. Express a periodic function in terms of a Fourier series using the cosine/sine, the amplitude/phase, and the complex exponential representations.
2. Determine the line spectrum of a periodic waveform.
3. Utilize symmetry considerations in the evaluation of Fourier coefficients.
4. Explain the Gibbs phenomenon.
5. Analyze circuits excited by periodic waveforms.
6. Calculate the average power dissipated in or delivered by a component in a circuit excited by a periodic voltage or current.
7. Evaluate the Fourier transform of a nonperiodic waveform.
8. Apply the Fourier transform technique to analyze circuits excited by nonperiodic waveforms.
9. Generate the 2-D Fourier transform of a spatial image.
10. Use Multisim to model the behavior of the Sigma-Delta modulator.

Overview

First introduced in Chapter 7, the phasor-domain analysis technique has proven to be a potent—and easy to implement—tool for determining the steady-state response of circuits excited by sinusoidal waveforms. As a periodic function with a period T, a sinusoidal signal shares a distinctive property with all other members of the family of ***periodic functions***, namely the ***periodicity property*** given by Eq. (8.4) as

$$x(t) = x(t + nT), \tag{11.1}$$

where n is an integer. Given this natural connection between sinusoids and other periodic functions, can we somehow extend the phasor-domain solution technique to non-sinusoidal periodic excitations? The answer is yes, and the process for realizing it is facilitated by two enabling mechanisms: the ***Fourier theorem*** and the superposition principle. The Fourier theorem makes it possible to mathematically characterize any periodic excitation in the form of a sum of multiple sinusoidal harmonics. The superposition principle allows us to apply phasor analysis to calculate the circuit response due to each harmonic and then to add all of the responses together, thereby realizing the response to the original periodic excitation. The first half of this chapter aims to demonstrate the mechanics of the solution process as well as to explain the physics associated with the circuit response to the different harmonics.

The second half of the chapter is devoted to the ***Fourier transform***, which is useful particularly for analyzing circuits excited by nonperiodic waveforms, such as single pulses or step functions. As we will see in Section 11-7, the Fourier transform is sort of related to the Laplace transform of Chapter 10 and becomes identical with it under certain circumstances, but the two techniques are generally distinct (as are their conditions of applicability from the standpoint of circuit analysis).

11-1 Fourier Series Analysis Technique

By way of introducing the Fourier series analysis technique, let us consider the RL circuit shown in Fig. 11-1(a), which is excited by the square-wave voltage waveform shown in Fig. 11-1(b). The waveform amplitude is 3 V and its period $T = 2$ s. Our goal is to determine the output voltage response, $v_{\text{out}}(t)$. The solution procedure consists of three basic steps.

Step 1: Express the periodic excitation in terms of Fourier harmonics

According to the Fourier theorem (which we will introduce and examine in detail in Section 11-2), the waveform shown in Fig. 11-1(b) can be represented by the series

$$v_s(t) = \frac{12}{\pi}\left(\cos\omega_0 t - \frac{1}{3}\cos 3\omega_0 t + \frac{1}{5}\cos 5\omega_0 t - \cdots\right), \tag{11.2}$$

where $\omega_0 = 2\pi/T = 2\pi/2 = \pi$ (rad/s) is the ***fundamental angular frequency*** of the waveform. Since our present objective is to outline the solution procedure, we will accept it as a given that the infinite-series representation given by Eq. (11.2) is indeed equivalent to the square wave of Fig. 11-1(b). The series consists of cosine functions of the form $\cos m\omega_0 t$ with m assuming only odd values (1, 3, 5, etc.). Thus, the series contains only odd ***harmonics*** of ω_0. The coefficient of the mth harmonic is equal to $1/m$ (relative to the coefficient of the fundamental), and its polarity is positive if $m = 3, 7, \ldots$ and negative if $m = 5, 9, \ldots$. In view of these properties, we can

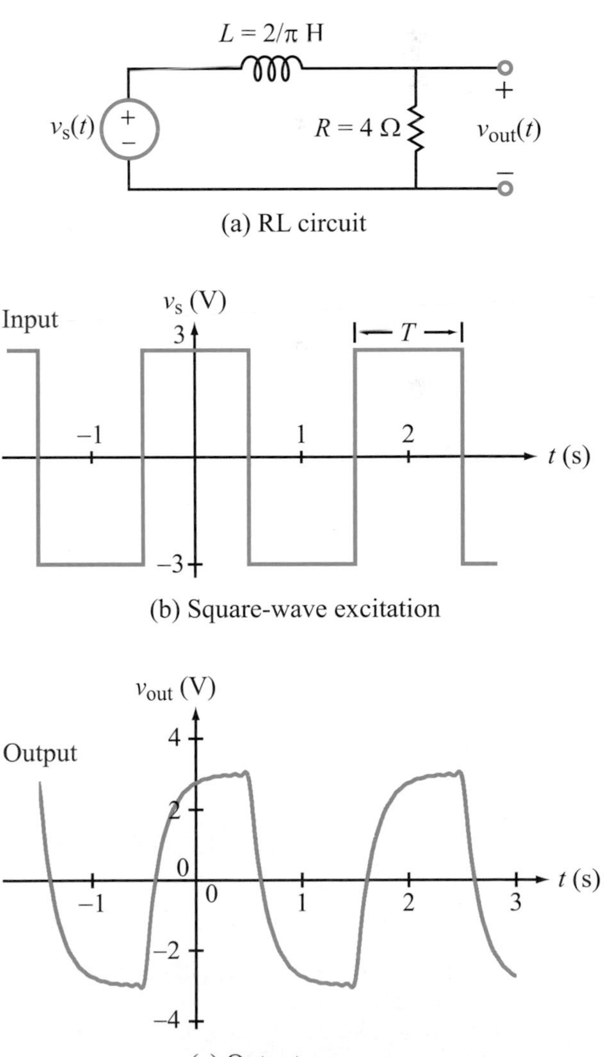

Figure 11-1: RL circuit excited by a square wave and corresponding output response.

replace m with $(2n-1)$ and cast $v_s(t)$ in the form

$$v_s(t) = \frac{12}{\pi}\sum_{n=1}^{\infty}(-1)^{n+1}\frac{1}{2n-1}\cos(2n-1)\pi t \text{ V}. \quad (11.3)$$

In terms of its first few components, $v_s(t)$ is given by

$$v_s(t) = v_{s_1}(t) + v_{s_2}(t) + v_{s_3}(t) + \cdots \quad (11.4)$$

with

$$v_{s_1}(t) = \frac{12}{\pi}\cos\omega_0 t \text{ V}, \quad (11.5a)$$

$$v_{s_2}(t) = -\frac{12}{3\pi}\cos 3\omega_0 t \text{ V}, \quad (11.5b)$$

$$v_{s_3}(t) = \frac{12}{5\pi}\cos 5\omega_0 t \text{ V}, \quad \text{etc.} \quad (11.5c)$$

In the phasor domain, the counterpart of $v_s(t)$ is given by:

$$\mathbf{V}_s(t) = \mathbf{V}_{s_1}(t) + \mathbf{V}_{s_2}(t) + \mathbf{V}_{s_3}(t) + \cdots \quad (11.6)$$

with

$$\mathbf{V}_{s_1} = \frac{12}{\pi} \text{ V} \quad (\text{for } \omega = \omega_0), \quad (11.7a)$$

$$\mathbf{V}_{s_2} = -\frac{12}{3\pi} \text{ V} \quad (\text{for } \omega = 3\omega_0), \quad (11.7b)$$

$$\mathbf{V}_{s_3} = \frac{12}{5\pi} \text{ V} \quad (\text{for } \omega = 5\omega_0), \quad \text{etc.} \quad (11.7c)$$

Phasor voltages $\mathbf{V}_{s_1}$, $\mathbf{V}_{s_2}$, $\mathbf{V}_{s_3}$, etc., are the counterparts of $v_{s_1}(t)$, $v_{s_2}(t)$, $v_{s_3}(t)$, etc., respectively.

Step 2: Determine output responses to input harmonics

For the circuit in Fig. 11-1(a), input voltage $\mathbf{V}_{s_1}$ acting alone would generate a corresponding output voltage $\mathbf{V}_{\text{out}_1}$. Keeping in mind that $\mathbf{V}_{s_1}$ corresponds to $v_{s_1}(t)$ at $\omega = \omega_0 = \pi$, voltage division gives

$$\mathbf{V}_{\text{out}_1} = \left(\frac{R}{R + j\omega_0 L}\right)\mathbf{V}_{s_1} = \frac{4}{4 + j\pi \times \frac{2}{\pi}} \cdot \frac{12}{\pi} = 3.42\angle{-26.56^\circ} \quad (11.8)$$

with a corresponding time-domain voltage

$$v_{\text{out}_1}(t) = \Re\mathfrak{e}[\mathbf{V}_{\text{out}_1} e^{j\omega_0 t}] = 3.42\cos(\omega_0 t - 26.56^\circ) \text{ V}. \quad (11.9)$$

Similarly, at $\omega = 3\omega_0 = 3\pi$,

$$\mathbf{V}_{\text{out}_2} = \frac{4}{4 + j3\pi \times \frac{2}{\pi}} \cdot \left(-\frac{12}{3\pi}\right) = -0.71\angle{-56.31^\circ} \text{ V}, \quad (11.10)$$

and

$$v_{\text{out}_2}(t) = \Re\mathfrak{e}[\mathbf{V}_{\text{out}_2} e^{j3\omega_0 t}] = -0.71\cos(3\omega_0 t - 56.31^\circ) \text{ V}. \quad (11.11)$$

In view of the harmonic pattern expressed in the form of Eq. (11.3), for the harmonic at angular frequency $\omega = (2n-1)\omega_0$,

$$\mathbf{V}_{\text{out}_n} = \frac{4}{4 + j(2n-1)\pi \times \frac{2}{\pi}} \cdot (-1)^{n+1}\frac{12}{\pi(2n-1)} = (-1)^{n+1}\frac{24}{\pi(2n-1)\sqrt{4+(2n-1)^2}} \cdot \angle{-\tan^{-1}[(2n-1)/2]} \text{ V}. \quad (11.12)$$

The corresponding time domain voltage is

$$v_{\text{out}_n}(t) = \Re\mathfrak{e}[\mathbf{V}_{\text{out}_n} e^{j(2n-1)\omega_0 t}] = (-1)^{n+1}\frac{24}{\pi(2n-1)\sqrt{4+(2n-1)^2}} \cdot \cos\left[(2n-1)\omega_0 t - \tan^{-1}\left(\frac{2n-1}{2}\right)\right] \text{ V}. \quad (11.13)$$

Step 3: Apply the superposition principle to determine $\boldsymbol{v}_{\textbf{out}}\boldsymbol{(t)}$

According to the superposition principle, if v_{out_1} is the output generated by a linear circuit when excited by an input voltage v_{s_1} acting alone and if similarly v_{out_2} is the output due to v_{s_2} acting alone, then the output due to the combination of v_{s_1} and v_{s_2} acting simultaneously is simply the sum of v_{out_1} and v_{out_2}. Moreover, the principle is extendable to any number of sources. In the present case, the square-wave excitation is equivalent to a series of sinusoidal sources $v_{s_1}, v_{s_2}, \ldots$ generating corresponding output voltages $v_{\text{out}_1}, v_{\text{out}_2}, \ldots$. Consequently,

$$\begin{aligned} v_{\text{out}}(t) &= \sum_{n=1}^{\infty} v_{\text{out}_n}(t) \\ &= \sum_{n=1}^{\infty}(-1)^{n+1}\frac{24}{\pi(2n-1)\sqrt{4+(2n-1)^2}} \cdot \cos\left[(2n-1)\omega_0 t - \tan^{-1}\left(\frac{2n-1}{2}\right)\right] \\ &= 3.42\cos(\omega_0 t - 26.56^\circ) \\ &\quad - 0.71\cos(3\omega_0 t - 56.31^\circ) \\ &\quad + 0.28\cos(5\omega_0 t - 68.2^\circ) + \cdots \text{ V}. \end{aligned} \quad (11.14)$$

We note that the fundamental component of $v_{out}(t)$ has the dominant amplitude and that the higher the harmonic is, the smaller is its amplitude. This allows us to approximate $v_{out}(t)$ by retaining only a few terms, such as up to $n = 10$, depending on the level of desired accuracy. The plot of $v_{out}(t)$ displayed in Fig. 11-1(c) (which is based on only the first 10 terms) is sufficiently accurate for most practical applications.

The foregoing three-step procedure (which equally is applicable to any linear circuit excited by any realistic periodic function) relied on the use of the Fourier theorem to express the square-wave pattern in terms of sinusoids. In the next section, we will examine the attributes of the Fourier theorem and how we may apply it to any periodic function.

Review Question 11-1: The Fourier series technique is applied to analyze circuits excited by what type of functions?

Review Question 11-2: How is the angular frequency of the nth harmonic related to that of the fundamental ω_0? How is ω_0 related to the period T of the periodic function?

Review Question 11-3: What steps constitute the Fourier series solution procedure?

11-2 Fourier Series Representation

In 1822, the French mathematician Jean Baptiste Joseph Fourier developed an elegant formulation for representing periodic functions in terms of a series of sinusoidal harmonics. The representation is known today as the ***Fourier series***, and the formulation is called the ***Fourier theorem***. To guarantee that a periodic function $f(t)$ has a realizable Fourier series, it should satisfy a set of conditions known as the ***Dirichlet conditions***. Fortunately, any periodic function generated by a real circuit will meet these conditions automatically and therefore, we are assured that its Fourier series does indeed exist.

The ***Fourier theorem*** states that a periodic function $f(t)$ of period T can be cast in the form

$$f(t) = a_0 + \sum_{n=1}^{\infty} (a_n \cos n\omega_0 t + b_n \sin n\omega_0 t), \quad \text{(cosine/sine representation)}, \tag{11.15}$$

where ω_0, the ***fundamental angular frequency*** of $f(t)$, is related to T by

$$\omega_0 = \frac{2\pi}{T}. \tag{11.16}$$

The summation is an infinite series whose first pair of terms (for $n = 1$) involve $\cos \omega_0 t$ and $\sin \omega_0 t$. Higher values of n involve sine and cosine functions at harmonic multiples of ω_0, namely $2\omega_0$, $3\omega_0$, etc. The constants a_0, a_n, and b_n, for $n = 1$ to ∞, are collectively called the ***Fourier coefficients*** of $f(t)$. Their values are determined by evaluating integral expressions involving $f(t)$:

$$a_0 = \frac{1}{T} \int_0^T f(t)\, dt, \tag{11.17a}$$

$$a_n = \frac{2}{T} \int_0^T f(t) \cos n\omega_0 t\, dt, \tag{11.17b}$$

$$b_n = \frac{2}{T} \int_0^T f(t) \sin n\omega_0 t\, dt. \tag{11.17c}$$

Even though the indicated limits of integration are from 0 to T, the expressions are equally valid if the lower limit is changed to t_0 and the upper limit to $(t_0 + T)$ for any value of t_0. In some cases, the evaluation is easier to perform by integrating from $-T/2$ to $T/2$.

Coefficient a_0 is equal to the time-average value of $f(t)$. It is called the ***dc component*** of $f(t)$, because the average values of the ac components are all zero.

11-2.1 Fourier Coefficients

To verify the validity of the expressions given by Eq. (11.17), we will make use of the trigonometric integral properties listed in Table 11-1.

dc Fourier Component a_0

Equation (8.5) in Chapter 8 states that the average value of a periodic function is obtained by integrating it over a complete period T and then dividing the integral by T. Applying the definition to Eq. (11.15) gives

Table 11-1: Trigonometric integral properties for any integers m and n. The integration period $T = 2\pi/\omega_0$.

Property	Integral
1	$\int_0^T \sin n\omega_0 t \; dt = 0$
2	$\int_0^T \cos n\omega_0 t \; dt = 0$
3	$\int_0^T \sin n\omega_0 t \sin m\omega_0 t \; dt = 0, \quad n \neq m$
4	$\int_0^T \cos n\omega_0 t \cos m\omega_0 t \; dt = 0, \quad n \neq m$
5	$\int_0^T \sin n\omega_0 t \cos m\omega_0 t \; dt = 0$
6	$\int_0^T \sin^2 n\omega_0 t \; dt = T/2$
7	$\int_0^T \cos^2 n\omega_0 t \; dt = T/2$

Note: All integral properties remain valid when the arguments $n\omega_0 t$ and $m\omega_0 t$ are phase shifted by a constant angle ϕ_0. Thus, Property 1, for example, becomes $\int_0^T \sin(n\omega_0 t + \phi_0)\; dt = 0$, and Property 5 becomes $\int_0^T \sin(n\omega_0 t + \phi_0)\cos(m\omega_0 t + \phi_0)\; dt = 0$.

$$\begin{aligned} \frac{1}{T}\int_0^T f(t)\; dt &= \frac{1}{T}\int_0^T a_0 \; dt \\ &\quad + \frac{1}{T}\int_0^T \left[\sum_{n=1}^{\infty} a_n \cos n\omega_0 t + b_n \sin n\omega_0 t\right] dt \\ &= a_0 + \frac{1}{T}\int_0^T a_1 \cos\omega_0 t \; dt + \frac{1}{T}\int_0^T a_2 \cos 2\omega_0 t \; dt + \cdots \\ &\quad + \frac{1}{T}\int_0^T b_1 \sin\omega_0 t \; dt + \frac{1}{T}\int_0^T b_2 \sin 2\omega_0 t \; dt + \cdots . \end{aligned} \tag{11.18}$$

According to Property 1 in Table 11-1, the average value of a sine function is zero, and the same is true for a cosine function (Property 2). Hence, all of the terms in Eq. (11.18) containing $\cos n\omega_0 t$ or $\sin n\omega_0 t$ will vanish, leaving behind

$$\frac{1}{T}\int_0^T f(t)\; dt = a_0, \tag{11.19}$$

which is identical with the definition given by Eq. (11.17a).

a_n Fourier Coefficients

Multiplication of both sides of Eq. (11.15) by $\cos m\omega_0 t$ (with m being any integer value equal to or greater than 1), followed with integration over $[0, T]$ yields

$$\begin{aligned} \int_0^T f(t) \; \cos m\omega_0 t \; dt &= \int_0^T a_0 \cos m\omega_0 t \; dt \\ &\quad + \int_0^T \sum_{n=1}^{\infty} a_n \cos n\omega_0 t \cos m\omega_0 t \; dt \\ &\quad + \int_0^T \sum_{n=1}^{\infty} b_n \sin n\omega_0 t \cos m\omega_0 t \; dt. \end{aligned} \tag{11.20}$$

On the right-hand side of Eq. (11.20):

(1) The term containing a_0 is equal to zero (Property 2 in Table 11-1).

(2) All terms containing b_n are equal to zero (Property 5).

(3) All terms containing a_n are equal to zero (Property 4), except when $m = n$, in which case Property 7 applies.

Hence, after eliminating all of the zero-valued terms and then setting $m = n$ in the two remaining terms, we have

$$\int_0^T f(t) \; \cos n\omega_0 t \; dt = a_n \frac{T}{2}, \tag{11.21}$$

which proves Eq. (11.17b).

b_n Fourier Coefficients

Similarly, if we were to repeat the preceding process, after multiplication of Eq. (11.15) by $\sin m\omega_0 t$ (instead of $\cos m\omega_0 t$), we would conclude with a result affirming the validity of Eq. (11.17c).

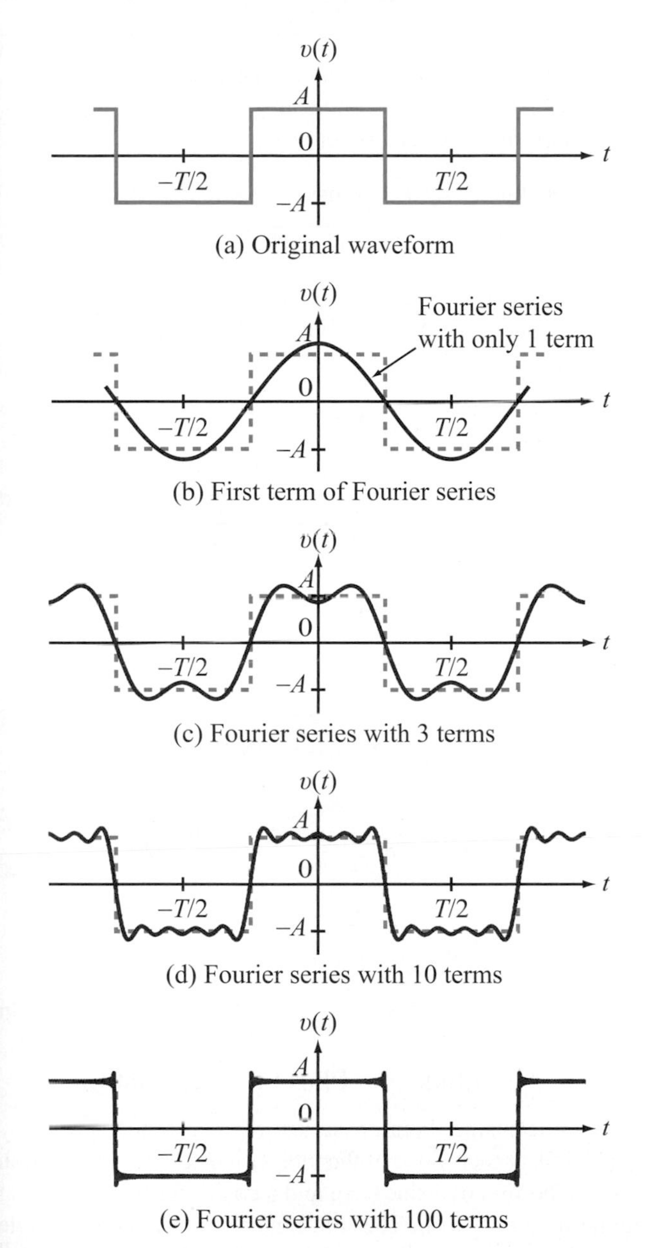

Figure 11-2: Comparison of the square-wave waveform with its Fourier series representation using only the first term (b), the sum of the first three (c), ten (d), and 100 terms (e).

To develop an appreciation for how the components of the Fourier series add up to represent the periodic waveform, let us consider the square-wave voltage waveform shown in Fig. 11-2(a). Over the period extending from $-T/2$ to $T/2$, $v(t)$ is given by

$$v(t) = \begin{cases} -A, & \text{for } -T/2 < t < -T/4, \\ A, & \text{for } -T/4 < t < T/4, \\ -A, & \text{for } T/4 < t < -T/2. \end{cases}$$

If we apply Eq. (11.17)—with integration limits $[-T/2, T/2]$—to evaluate the Fourier coefficients and then use them in Eq. (11.15), we end up with the series

$$\begin{aligned} v(t) &= \sum_{n=1}^{\infty} \frac{4A}{n\pi} \sin\left(\frac{n\pi}{2}\right) \cos\left(\frac{2n\pi t}{T}\right) \\ &= \frac{4A}{\pi} \cos\left(\frac{2\pi t}{T}\right) - \frac{4A}{3\pi} \cos\left(\frac{6\pi t}{T}\right) \\ &\quad + \frac{4A}{5\pi} \cos\left(\frac{10\pi t}{T}\right) - \cdots. \end{aligned}$$

Alone, the first term of the series provides a crude approximation of the square wave (Fig. 11-2(b)), but as we add more and more terms, the sum starts to better resemble the general shape of the square wave, as demonstrated by the waveforms in Figs. 11-2(c) to (e).

Example 11-1: Sawtooth Waveform

Express the sawtooth waveform shown in Fig. 11-3(a) in terms of a Fourier series, and then evaluate how well the original waveform is represented by a truncated series in which the summation stops when n reaches a specified truncation number $n_{\max}$. Generate plots for $n_{\max} = 1, 2, 10$, and 100.

Solution: The sawtooth waveform is characterized by a period $T = 4$ s and $\omega_0 = 2\pi/T = \pi/2$ (rad/s). Over the waveform's first cycle ($t = 0$ to $t = 4$ s), its amplitude variation is given by

$$f(t) = 5t \qquad (\text{for } 0 \leq t \leq 4 \text{ s}).$$

Application of Eq. (11.17) yields:

$$a_0 = \frac{1}{T} \int_0^T f(t)\, dt = \frac{1}{4} \int_0^4 5t\, dt = 10,$$

$$a_n = \frac{2}{T} \int_0^T f(t) \cos(n\omega_0 t)\, dt = \frac{2}{4} \int_0^4 5t \cos\left(\frac{n\pi}{2} t\right) dt = 0,$$

and

$$b_n = \frac{2}{T} \int_0^T f(t) \sin(n\omega_0 t)\, dt = \frac{2}{4} \int_0^4 5t \sin\left(\frac{n\pi}{2} t\right) dt = -\frac{20}{n\pi}.$$

Upon inserting these results into Eq. (11.15), we obtain the following ***complete*** Fourier series representation for the sawtooth waveform:

$$f(t) = 10 - \frac{20}{\pi} \sum_{n=1}^{\infty} \frac{1}{n} \sin\left(\frac{n\pi}{2} t\right).$$

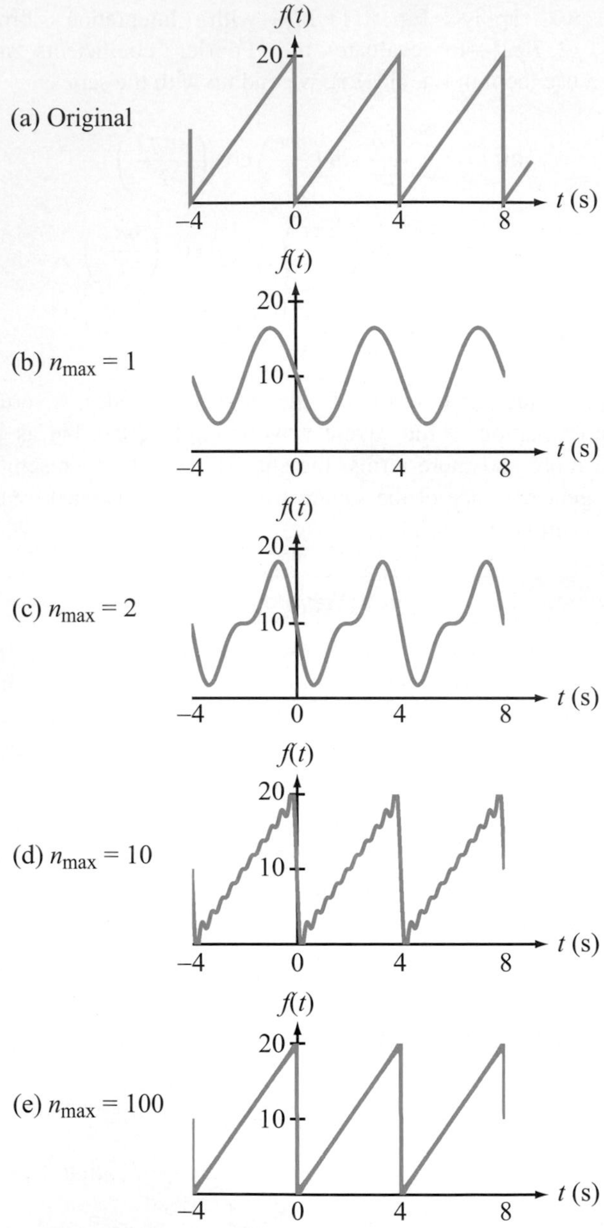

Figure 11-3: Sawtooth waveform: (a) original waveform, (b)–(e) representation by a truncated Fourier series with $n_{\max} = 1, 2, 10$, and 100, respectively.

The $n_{\max}$-***truncated series*** is identical in form with the complete series, except that the summation is terminated after the index n reaches $n_{\max}$. Figures 11-3(b) through (e) display the waveforms calculated using the truncated series with $n_{\max} = 1$, 2, 10, and 100. As expected, the addition of more terms improves the accuracy of the Fourier series representation, but even with only 10 terms (in addition to the dc component), the truncated series appears to provide a reasonable approximation of the original waveform.

Review Question 11-4: Is the Fourier series representation given by Eq. (11.15) applicable to a periodic function that starts at $t = 0$ (and is zero for $t < 0$)?

Review Question 11-5: What is a ***truncated*** series?

Exercise 11-1: Obtain the Fourier series representation for the waveform shown in Fig. E11.1.

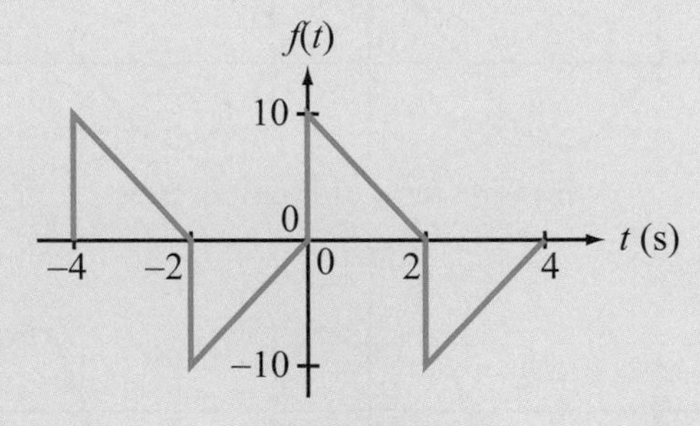

Figure E11.1

Answer:

$$f(t) = \sum_{n=1}^{\infty} \left[\frac{20}{n^2\pi^2} (1 - \cos n\pi) \cos \frac{n\pi t}{2} + \frac{10}{n\pi} (1 - \cos n\pi) \sin \frac{n\pi t}{2} \right].$$

(See ◎)

11-2.2 Amplitude and Phase Representation

In the sine/cosine Fourier series representation given by Eq. (11.15), at each value of the integer index n, the summation contains the sum of a sine term and a cosine term, with both at angular frequency $n\omega_0$. The sum can be converted into a single sinusoid as

$$a_n \cos n\omega_0 t + b_n \sin n\omega_0 t = A_n \cos(n\omega_0 t + \phi_n), \quad (11.22)$$

where A_n is called the ***amplitude of the nth harmonic*** and ϕ_n is its associated ***phase***. The relationships between (A_n, ϕ_n) and (a_n, b_n) are obtained by expanding the right-hand side of Eq. (11.22) in accordance with the trigonometric identity

$$\cos(x + y) = \cos x \cos y - \sin x \sin y. \quad (11.23)$$

Thus,

$$a_n \cos n\omega_0 t + b_n \sin n\omega_0 t = A_n \cos\phi_n \cos n\omega_0 t - A_n \sin\phi_n \sin n\omega_0 t. \quad (11.24)$$

Upon equating the coefficients of $\cos n\omega_0 t$ and $\sin n\omega_0 t$ on one side of the equation to their respective counterparts on the other side, we have

$$a_n = A_n \cos\phi_n, \qquad b_n = -A_n \sin\phi_n, \tag{11.25}$$

which can be combined to yield the relationships

$$A_n = \sqrt[+]{a_n^2 + b_n^2}, \qquad \phi_n = -\tan^{-1}\left(\frac{b_n}{a_n}\right). \tag{11.26}$$

The ambiguity associated with the sign of ϕ_n (when both a_n and b_n are negative) can be avoided by adopting the complex vector form

$$A_n \angle\phi_n = a_n - jb_n. \tag{11.27}$$

In view of Eq. (11.22), the cosine/sine Fourier series representation of $f(t)$ can be rewritten in the alternative ***amplitude/phase*** format

$$f(t) = a_0 + \sum_{n=1}^{\infty} A_n \cos(n\omega_0 t + \phi_n) \quad \text{(amplitude/phase representation)}. \tag{11.28}$$

Associated with each discrete frequency harmonic $n\omega_0$ is an amplitude A_n and a phase ϕ_n. A plot of A_n as a function of $n\omega_0$ is called the ***amplitude spectrum*** of $f(t)$, and similarly, a plot of ϕ_n versus $n\omega_0$ is its ***phase spectrum***. Because the plots are at discrete values along the ω-axis, they are called ***line spectra***. Example 11-2 provides an illustration.

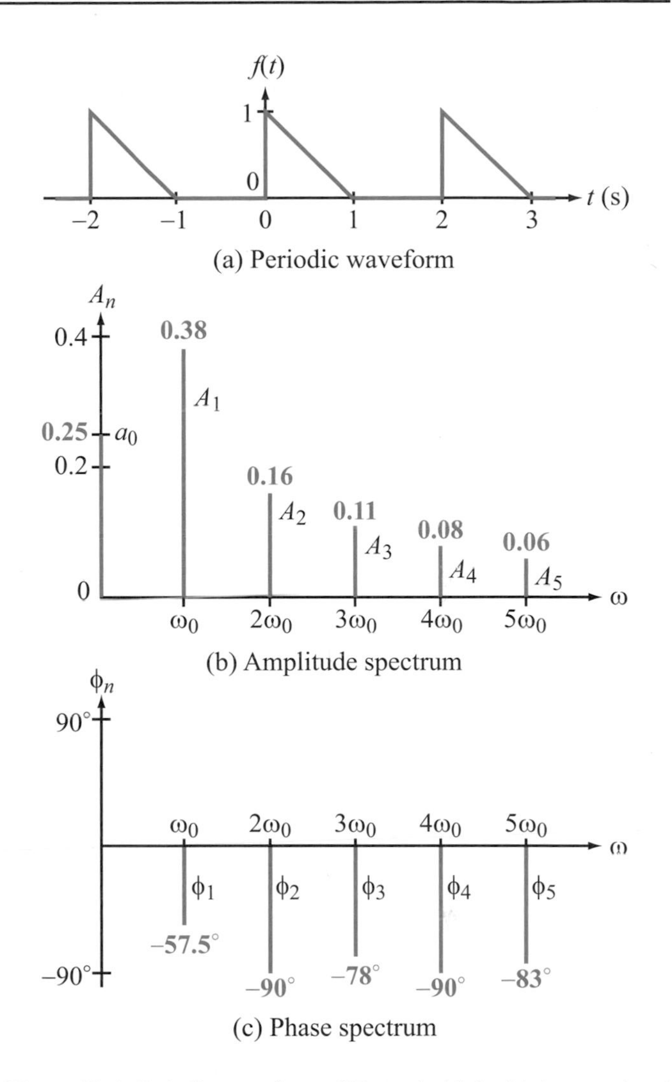

Figure 11-4: Periodic waveform of Example 11-2 with its associated line spectra.

Example 11-2: Line Spectra

Generate and plot the amplitude and phase spectra of the periodic waveform displayed in Fig. 11-4(a).

Solution: The periodic waveform has a period $T = 2$ s. Hence, $\omega_0 = 2\pi/T = 2\pi/2 = \pi$ rad/s, and the functional expression for $f(t)$ over its first cycle along the positive t-axis is

$$f(t) = \begin{cases} 1 - t & \text{for } 0 < t \le 1 \text{ s}, \\ 0 & \text{for } 1 \le t \le 2 \text{ s}. \end{cases}$$

The dc component of $f(t)$ is given by

$$a_0 = \frac{1}{T}\int_0^T f(t)\, dt = \frac{1}{2}\int_0^1 (1 - t)\, dt = 0.25,$$

which is equal to the area under a single triangle, divided by the period $T = 2$ s.

For the other Fourier coefficients, evaluation of the expressions given by Eqs. (11.17b and c) leads to

$$\begin{aligned} a_n &= \frac{2}{T}\int_0^T f(t)\cos n\omega_0 t\, dt \\ &= \frac{2}{2}\int_0^1 (1 - t)\cos n\pi t\, dt \\ &= \frac{1}{n\pi}\sin n\pi t\Big|_0^1 - \left(\frac{1}{n^2\pi^2}\cos n\pi t + \frac{t}{n\pi}\sin n\pi t\right)\Big|_0^1 \\ &= \frac{1}{n^2\pi^2}[1 - \cos n\pi] \end{aligned}$$

and

$$b_n = \frac{2}{T}\int_0^T f(t)\sin n\omega_0 t\,dt = \frac{2}{2}\int_0^1 (1-t)\sin n\pi t\,dt$$

$$= -\frac{1}{n\pi}\cos n\pi t\Big|_0^1 - \left(\frac{1}{n^2\pi^2}\sin n\pi t - \frac{t}{n\pi}\cos n\pi t\right)\Big|_0^1$$

$$= \frac{1}{n\pi}.$$

By Eq. (11.26), the harmonic amplitudes and phases are given by

$$A_n = \sqrt[+]{a_n^2 + b_n^2} = \left[\left(\frac{1}{n^2\pi^2}[1-\cos n\pi]\right)^2 + \left(\frac{1}{n\pi}\right)^2\right]^{1/2}$$

$$= \begin{cases} \left(\dfrac{4}{n^4\pi^4} + \dfrac{1}{n^2\pi^2}\right)^{1/2} & \text{for } n = \text{odd}, \\ \dfrac{1}{n\pi} & \text{for } n = \text{even}, \end{cases}$$

and

$$\phi_n = -\tan^{-1}\frac{b_n}{a_n} = -\tan^{-1}\left(\frac{n\pi}{[1-\cos n\pi]}\right)$$

$$= \begin{cases} -\tan^{-1}\left(\dfrac{n\pi}{2}\right) & \text{for } n = \text{odd}, \\ -90^\circ & \text{for } n = \text{even}. \end{cases}$$

The values of A_n and ϕ_n for the first three terms are

$$A_1 = 0.38, \qquad \phi_1 = -57.5^\circ,$$

$$A_2 = 0.16, \qquad \phi_2 = -90^\circ,$$

$$A_3 = 0.11, \qquad \phi_3 = -78^\circ.$$

Spectral plots of A_n and ϕ_n are shown in Figs. 11-4(b) and (c), respectively.

Exercise 11-2: Obtain the line spectra associated with the periodic function of Exercise 11.1.

Answer:

$$A_n = [1-\cos(n\pi)]\,\frac{20}{n^2\pi^2}\sqrt{1+\frac{n^2\pi^2}{4}}\;;$$

$$\phi_n = -\tan^{-1}\left(\frac{n\pi}{2}\right).$$

(See)

11-2.3 Symmetry Considerations

According to Eq. (11.17), determination of the Fourier coefficients involves evaluation of three definite integrals involving $f(t)$. If $f(t)$ exhibits symmetry properties, the evaluation process can be simplified significantly.

dc Symmetry

Since a_0 is equal to the average value of $f(t)$ which is proportional to the net area under the waveform over the span of a complete cycle, $a_0 = 0$ if the said area is zero. The waveform in Fig. 11-5(a) is an example of a dc-symmetrical function.

Even and Odd Symmetry

A waveform of a function $f(t)$ possesses ***even symmetry*** if it is symmetrical with respect to the vertical axis; the shape of the waveform on the left-hand side of the vertical axis is the mirror image of the waveform on the right-hand side. Mathematically, an ***even function*** satisfies the condition

$$f(t) = f(-t) \qquad \text{(even symmetry)}. \qquad (11.29)$$

The waveforms displayed in Figs. 11-5(b) and (c) exhibit even symmetry, as do the waveforms of $\sin^2 \omega t$ and $|\sin \omega t|$, among many others.

In contrast, the sine and square waves shown in Fig. 11-5(d) and (e) exhibit ***odd symmetry***; in each case, the shape of the waveform on the left-hand side of the vertical axis is the ***inverted*** mirror image of the waveform on the right-hand side. Thus, for an ***odd function***;

$$f(t) = -f(-t) \qquad \text{(odd symmetry)}. \qquad (11.30)$$

In the case of the square wave, were we to shift the waveform by $T/4$ to the left, it would switch from an odd function into an even function.

11-2.4 Even-Function Fourier Coefficients

Even symmetry allows us to simplify Eqs. (11.17) to the following equations:

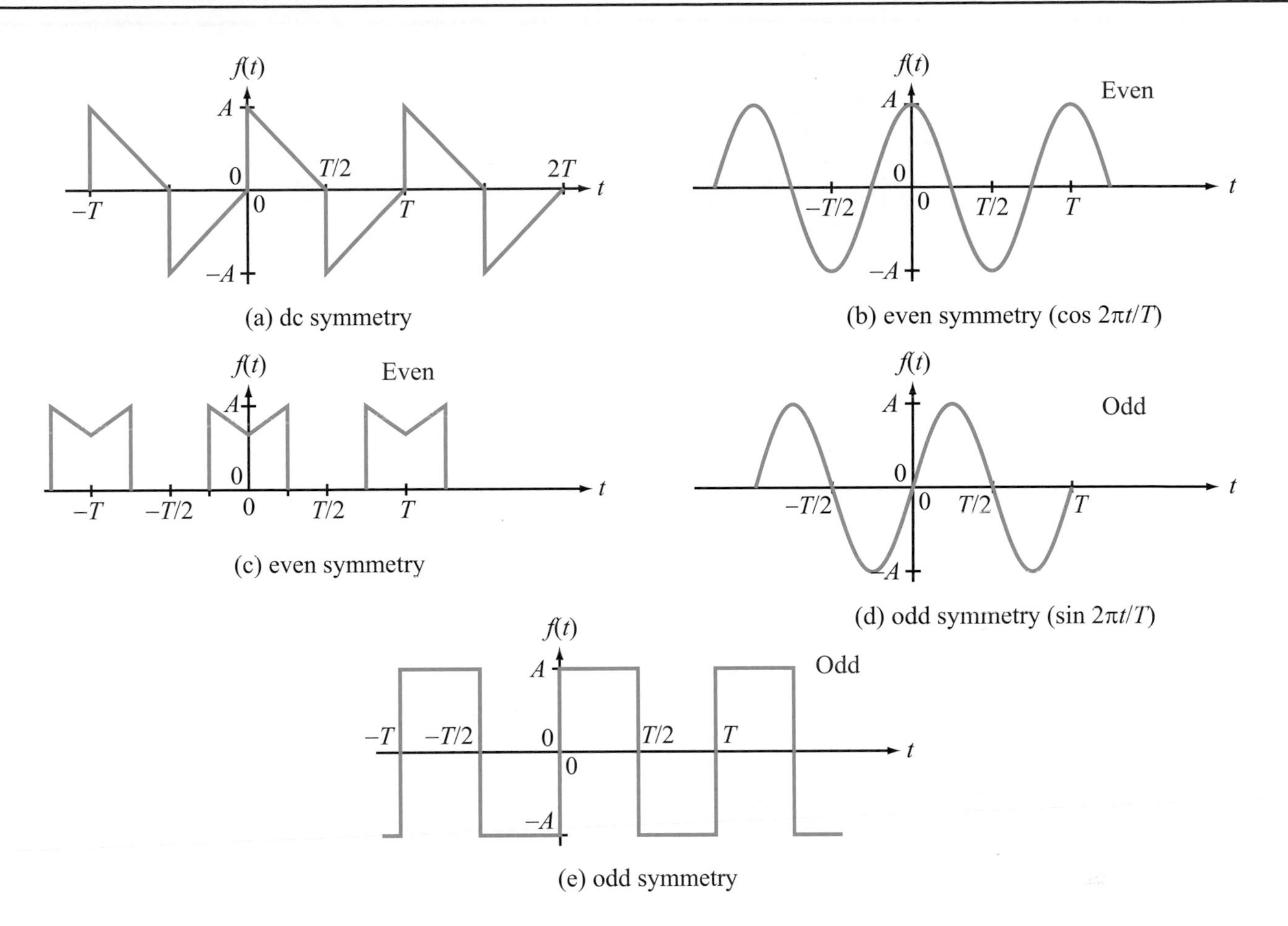

Figure 11-5: Waveforms with (a) dc symmetry, (b and c) even symmetry, and (d and e) odd symmetry.

Even Symmetry: $f(t) = f(-t)$

$$a_0 = \frac{2}{T} \int_0^{T/2} f(t)\, dt,$$

$$a_n = \frac{4}{T} \int_0^{T/2} f(t) \cos n\omega_0 t \, dt, \qquad (11.31)$$

$$b_n = 0,$$

$$A_n = |a_n|, \qquad \phi_n = \begin{cases} 0 & \text{if } a_n > 0, \\ 180^\circ & \text{if } a_n < 0. \end{cases}$$

The expressions for a_0 and a_n are the same as given earlier by Eq. (11.17a and b), except that the integration limits are now over half of a period and the integral has been multiplied by a factor of 2. The simplification is justified by the even symmetry of $f(t)$. As was stated in connection with Eq. (11.17), the only restriction associated with the integration limits is that the upper limit has to be greater than the lower limit by exactly T. Hence, by choosing the limits to be $[-T/2, T/2]$, and then recognizing that the integral of $f(t)$ over $[-T/2, 0]$ is equal to the integral over $[0, T/2]$, we justify the changes reflected in the expression for a_0. A similar argument applies to the expression for a_n based on the fact that multiplication of an even function $f(t)$ by $\cos n\omega_0 t$ (which itself is an even function) yields an even function.

The rationale for setting $b_n = 0$ for all n relies on the fact that multiplication of an even function $f(t)$ by $\sin n\omega_0 t$ (which is an odd function) yields an odd function, and integration of an odd function over $[-T/2, T/2]$ is always equal to zero. This is because the integral of an odd function over $[-T/2, 0]$ is equal in magnitude but opposite in sign to the integral over $[0, T/2]$.

11-2.5 Odd-Function Fourier Coefficients

In view of the preceding discussion, it follows that for a function with odd symmetry we have the following equations:

Table 11-2: Fourier series expressions for a select set of periodic waveforms.

Waveform	Fourier Series
1. Square Wave	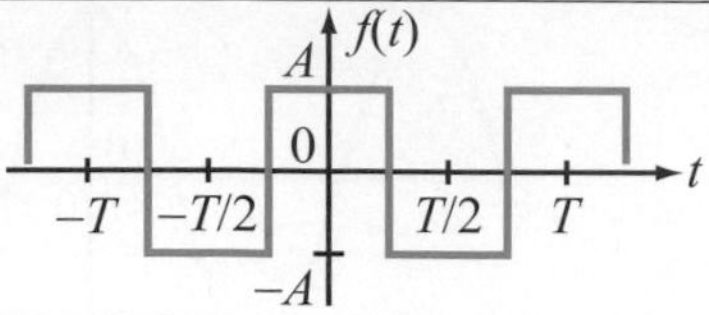 $f(t) = \sum_{n=1}^{\infty} \frac{4A}{n\pi} \sin\left(\frac{n\pi}{2}\right) \cos\left(\frac{2n\pi t}{T}\right)$
2. Time-Shifted Square Wave	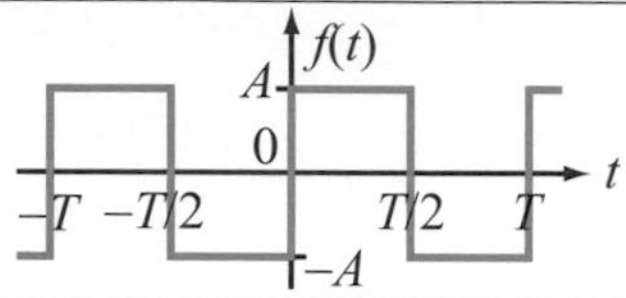 $f(t) = \sum_{\substack{n=1 \\ n=\text{odd}}}^{\infty} \frac{4A}{n\pi} \sin\left(\frac{2n\pi t}{T}\right)$
3. Pulse Train	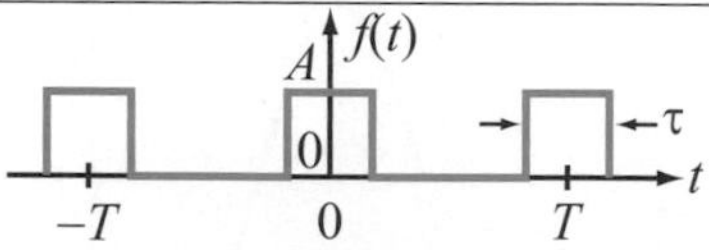 $f(t) = \frac{A\tau}{T} + \sum_{n=1}^{\infty} \frac{2A}{n\pi} \sin\left(\frac{n\pi\tau}{T}\right) \cos\left(\frac{2n\pi t}{T}\right)$
4. Triangular Wave	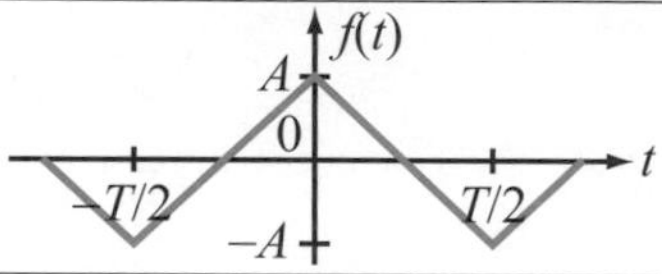 $f(t) = \sum_{\substack{n=1 \\ n=\text{odd}}}^{\infty} \frac{8A}{n^2\pi^2} \cos\left(\frac{2n\pi t}{T}\right)$
5. Shifted Triangular Wave	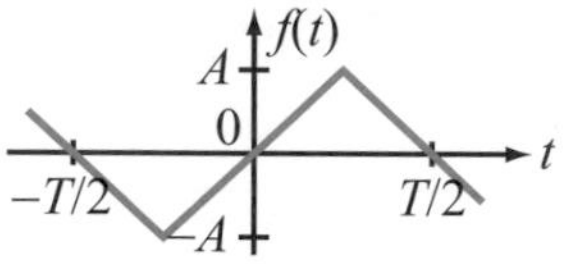 $f(t) = \sum_{\substack{n=1 \\ n=\text{odd}}}^{\infty} \frac{8A}{n^2\pi^2} \sin\left(\frac{n\pi}{2}\right) \sin\left(\frac{2n\pi t}{T}\right)$
6. Sawtooth	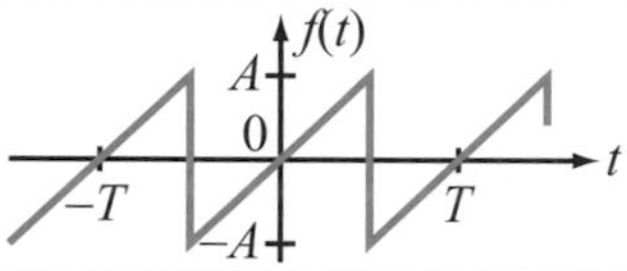 $f(t) = \sum_{n=1}^{\infty} (-1)^{n+1} \frac{2A}{n\pi} \sin\left(\frac{2n\pi t}{T}\right)$
7. Backward Sawtooth	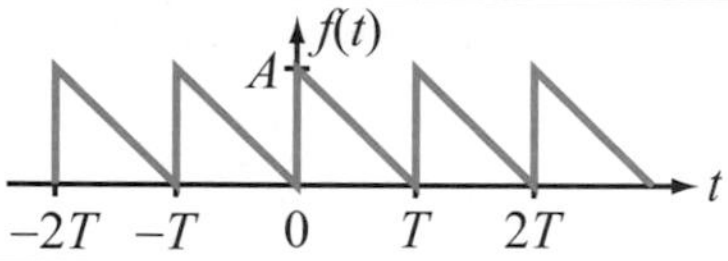 $f(t) = \frac{A}{2} + \sum_{n=1}^{\infty} \frac{A}{n\pi} \sin\left(\frac{2n\pi t}{T}\right)$
8. Full-Wave Rectified Sinusoid	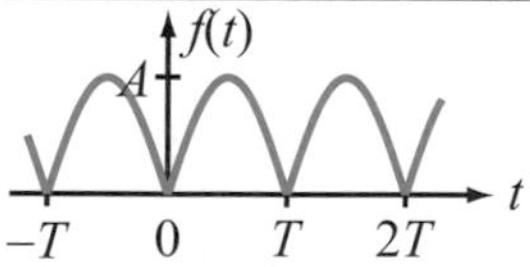 $f(t) = \frac{2A}{\pi} + \sum_{n=1}^{\infty} \frac{4A}{\pi(1-4n^2)} \cos\left(\frac{2n\pi t}{T}\right)$
9. Half-Wave Rectified Sinusoid	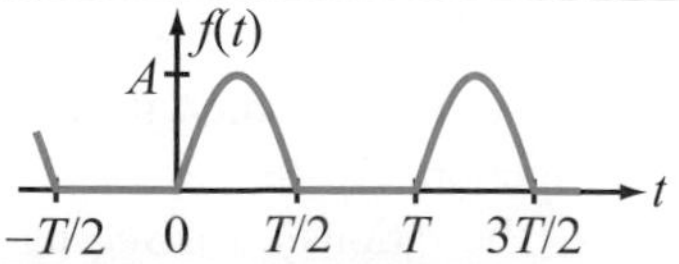 $f(t) = \frac{A}{\pi} + \frac{A}{2} \sin\left(\frac{2\pi t}{T}\right) + \sum_{\substack{n=2 \\ n=\text{even}}}^{\infty} \frac{2A}{\pi(1-n^2)} \cos\left(\frac{2n\pi t}{T}\right)$

Odd Symmetry: $f(t) = -f(-t)$

$$a_0 = 0 \qquad a_n = 0$$

$$b_n = \frac{4}{T} \int_0^{T/2} f(t) \sin n\omega_0 t \, dt \qquad (11.32)$$

$$A_n = |b_n| \qquad \phi_n = \begin{cases} -90^\circ & \text{if } b_n > 0, \\ 90^\circ & \text{if } b_n < 0. \end{cases}$$

Selected waveforms are displayed in Table 11-2, together with their corresponding Fourier series expressions.

Example 11-3: M-Periodic Waveform

Evaluate the Fourier coefficients of the M-periodic waveform shown in Fig. 11-6(a),

Solution: The M waveform is even-symmetrical, its period is $T = 4$ s, $\omega_0 = 2\pi/T = \pi/2$ rad/s, and its functional form

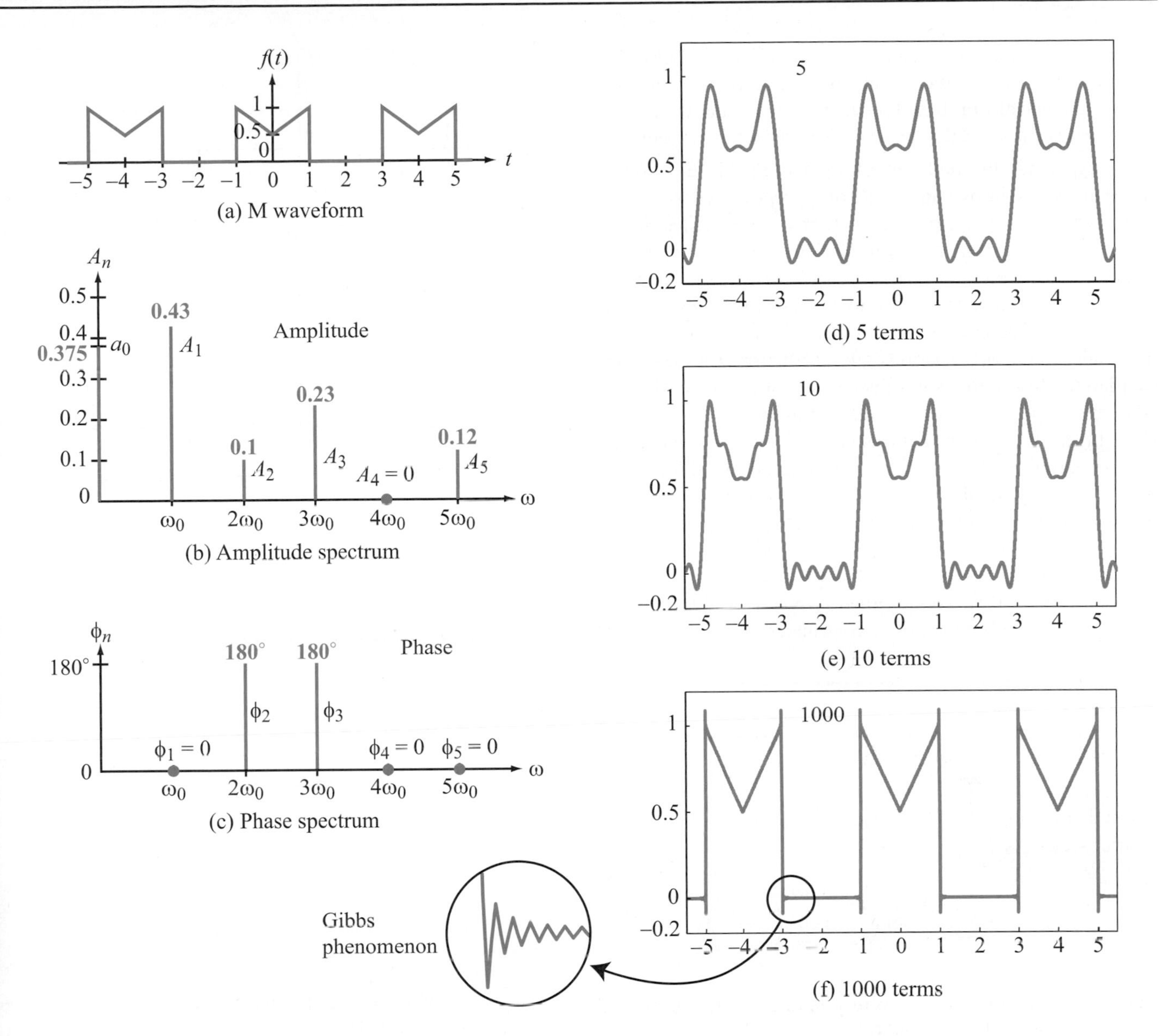

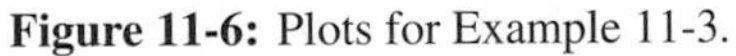

Figure 11-6: Plots for Example 11-3.

over the positive half period is:

$$f(t) = \begin{cases} \frac{1}{2}(1+t) & 0 \le t \le 1 \text{ s}, \\ 0 & 1 \le t \le 2 \text{ s}. \end{cases}$$

Application of Eq. (11.31) yields:

$$a_0 = \frac{2}{T} \int_0^{T/2} f(t)\, dt = \frac{2}{4} \int_0^1 \frac{1}{2}(1+t)\, dt = 0.375,$$

$$a_n = \frac{4}{T} \int_0^{T/2} f(t) \cos n\omega_0 t\, dt = \frac{4}{4} \int_0^1 \frac{1}{2}(1+t) \cos n\omega_0 t\, dt$$

$$= \frac{2}{n\pi} \sin \frac{n\pi}{2} + \frac{2}{n^2\pi^2} \left(\cos \frac{n\pi}{2} - 1 \right),$$

and

$$b_n = 0.$$

Since $b_n = 0$,

$$A_n = |a_n| \qquad \phi_n = \begin{cases} 0 & \text{if } a_n > 0, \\ 180^\circ & \text{if } a_n < 0. \end{cases}$$

Figures 11-6(b) and (c) display the amplitude and phase line spectra of the M-periodic waveform, and parts (d) through (f) display the waveforms based on the first five terms, the first ten terms, and the first 1000 terms of the Fourier series, respectively.

As expected, the addition of more terms in the Fourier series improves the overall fidelity of the reproduced waveform. However, no matter how many terms are included in the series representation, *the reproduction cannot duplicate the original M-waveform at points of discontinuity,* such as when the waveform jumps from zero to 1. *Discontinuities generate oscillations.* Increasing the number of terms (adding more harmonics) reduces the period of the oscillation. Ultimately, the oscillations fuse into a solid line, except at the discontinuities (see expanded view of the discontinuity at $t = -3$ s in Fig. 11-6(f)). As n approaches ∞, the Fourier series representation will reproduce the original waveform with perfect fidelity at all non-discontinuous points, but *at a point where the waveform jumps discontinuously between two different levels, the Fourier series will converge to a level half-way between them.* At $t = 1$ s, 3 s, 5 s, ..., the Fourier series will converge to 0.5. This oscillatory behavior of the Fourier series in the neighborhood of discontinuous points is called the ***Gibbs phenomenon***.

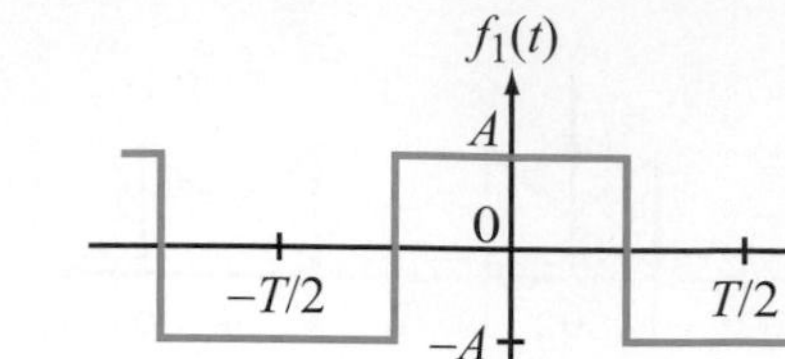

(a) $f_1(t)$

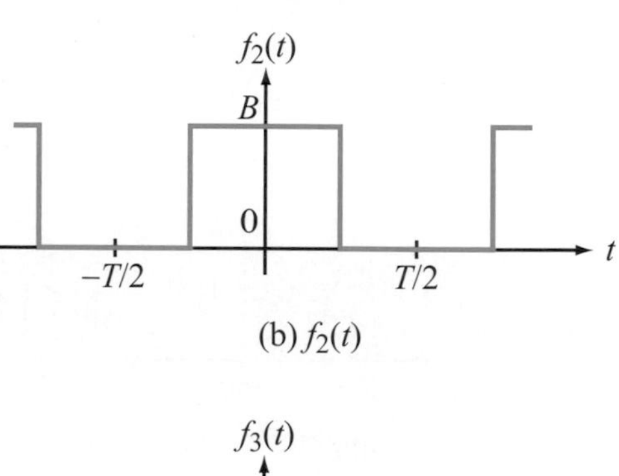

(b) $f_2(t)$

$f_3(t)$, A, 0, t, $-T/2$, $T/2$, $-A$

(c) $f_3(t)$

Figure 11-7: Waveforms for Example 11-4.

Example 11-4: Waveform Synthesis

Given that waveform $f_1(t)$ in Fig. 11-7(a) is represented by the Fourier series

$$f_1(t) = \sum_{n=1}^{\infty} \frac{4A}{n\pi} \sin\left(\frac{n\pi}{2}\right) \cos\left(\frac{2n\pi t}{T}\right),$$

generate the Fourier series corresponding to the waveforms displayed in Figs. 11-7(b) and (c).

Solution: Waveforms $f_1(t)$ and $f_2(t)$ are similar in shape and have the same period, but they also exhibit two differences: (1) the dc value of $f_1(t)$ is zero because it has dc symmetry, whereas the dc value of $f_2(t)$ is $B/2$, and (2) the peak-to-peak value of $f_1(t)$ is $2A$, compared with only B for $f_2(t)$. Mathematically, $f_2(t)$ is related to $f_1(t)$ by

$$f_2(t) = \frac{B}{2} + \left(\frac{B}{2A}\right) f_1(t)$$

$$= \frac{B}{2} + \sum_{n=1}^{\infty} \frac{2B}{n\pi} \sin\left(\frac{n\pi}{2}\right) \cos\left(\frac{2n\pi t}{T}\right).$$

Comparison of waveform $f_1(t)$ with waveform $f_3(t)$ reveals that the latter is shifted by $T/4$ along the t-axis relative to $f_1(t)$. That is,

$$f_3(t) = f_1\left(t - \frac{T}{4}\right)$$

$$= \sum_{n=1}^{\infty} \frac{4A}{n\pi} \sin\left(\frac{n\pi}{2}\right) \cos\left[\frac{2n\pi}{T}\left(t - \frac{T}{4}\right)\right].$$

Examination of the first few terms of $f_3(t)$ demonstrates that $f_3(t)$ can be rewritten in the simpler form

$$f_3(t) = \sum_{\substack{n=1 \\ n=\text{odd}}}^{\infty} \frac{4A}{n\pi} \sin\left(\frac{2n\pi t}{T}\right).$$

Review Question 11-6: What purpose is served by the symmetry properties of a periodic function?

Review Question 11-7: What distinguishes the phase angles ϕ_n of an even-symmetrical function from those of an odd-symmetrical function?

Review Question 11-8: What is the Gibbs phenomenon?

Exercise 11-3: (a) Does the waveform $f(t)$ shown in Fig. E11.3 exhibit either even or odd symmetry? (b) What is the value of a_0? (c) Does the function $g(t) = f(t) - a_0$ exhibit either even or odd symmetry?

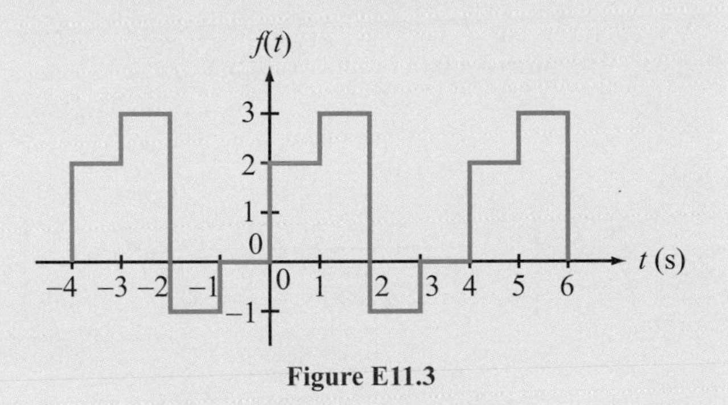

Figure E11.3

Answer: (a) Neither even nor odd symmetry, (b) $a_0 = 1$, (c) odd symmetry. (See)

11-3 Circuit Applications

Given the tools we developed in the preceding section for how to express a periodic function in terms of a Fourier series, we now will examine how to analyze linear circuits when excited by periodic voltage or current sources. The method of solution relies on the application of the phasor-domain technique that we introduced in Chapter 7 for analyzing circuits with cosinusoidal signals. A periodic function can be expressed as the sum of cosine and sine functions with coefficients a_n and b_n, and zero phase angles, or as the sum of only cosine functions with amplitudes A_n and phase angles ϕ_n. The latter form is amenable to direct application of the phasor-domain technique, whereas the former will require converting all $\sin n\omega_0 t$ terms into $\cos(n\omega_0 t - 90°)$ before implementation of the phasor-domain technique.

Even though the basic solution procedure was outlined earlier in Section 11-1, it is worth repeating it in a form that incorporates the concepts and terminology introduced in Section 11-2. To that end, we shall use $\upsilon_s(t)$ (or $i_s(t)$ if it is a current source) to denote the input excitation and $\upsilon_{out}(t)$ (or $i_{out}(t)$) to denote the output response for which we seek a solution.

Solution Procedure: Fourier Series Analysis Procedure

Step 1: Express $\upsilon_s(t)$ in terms of an amplitude/phase Fourier series as

$$\upsilon_s(t) = a_0 + \sum_{n=1}^{\infty} A_n \cos(n\omega_0 t + \phi_n) \tag{11.33}$$

with $A_n \angle \phi_n = a_n - jb_n$.

Step 2: Establish the generic transfer function of the circuit at frequency ω as

$$\mathbf{H}(\omega) = \mathbf{V}_{out} \qquad \text{when } \upsilon_s = 1 \cos \omega t. \tag{11.34}$$

Step 3: Write down the time-domain solution as

$$\upsilon_{out}(t) = a_0\, \mathbf{H}(\omega = 0) + \sum_{n=1}^{\infty} A_n \mathfrak{Re}\{\mathbf{H}(\omega = n\omega_0)\, e^{j(n\omega_0 t + \phi_n)}\}. \tag{11.35}$$

For each value of n, coefficient $A_n e^{j\phi_n}$ is associated with frequency harmonic $n\omega_0$. Hence, in Step 3, each harmonic amplitude is multiplied by its corresponding $e^{jn\omega_0 t}$ before application of the $\mathfrak{Re}\{\ \}$ operator.

Example 11-5: RC Circuit

Determine $\upsilon_{out}(t)$ when the circuit in Fig. 11-8(a) is excited by the voltage waveform shown in Fig. 11-8(b). The element values are $R = 20$ kΩ and $C = 0.1$ mF.

Solution: Step 1: The period of $\upsilon_s(t)$ is 4 s. Hence, $\omega_0 = 2\pi/4 = \pi/2$ rad/s, and by Eq. (11.17),

$$a_0 = \frac{1}{T}\int_0^T f(t)\, dt = \frac{1}{4}\int_0^1 10\, dt = 2.5 \text{ V},$$

$$a_n = \frac{2}{4}\int_0^1 10 \cos \frac{n\pi}{2} t\, dt = \frac{10}{n\pi} \sin \frac{n\pi}{2} \text{ V},$$

$$b_n = \frac{2}{4}\int_0^1 10 \sin \frac{n\pi}{2} t\, dt = \frac{10}{n\pi}\left(1 - \cos \frac{n\pi}{2}\right) \text{ V},$$

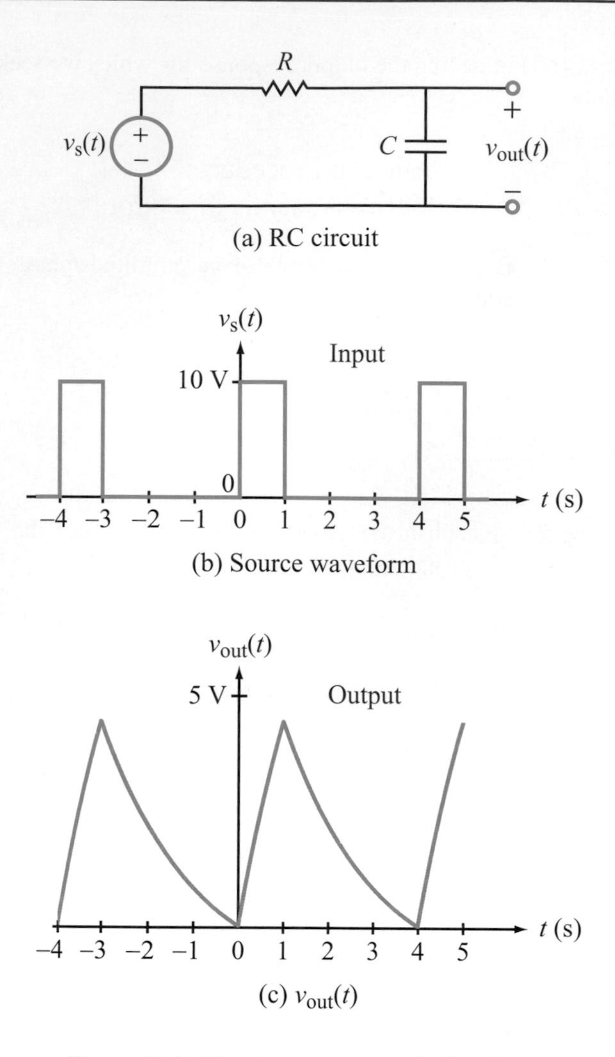

Figure 11-8: Circuit response to periodic pulses.

and

$$\begin{aligned}A_n\angle\phi_n &= a_n - jb_n\\&= \frac{10}{n\pi}\left[\sin\frac{n\pi}{2} - j\left(1-\cos\frac{n\pi}{2}\right)\right].\end{aligned}$$

The values of $A_n\angle\phi_n$ for the first four terms are

$$\begin{aligned}A_1\angle\phi_1 &= \frac{10\sqrt{2}}{\pi}\angle-45^\circ,\\A_2\angle\phi_2 &= \frac{10}{\pi}\angle-90^\circ,\\A_3\angle\phi_3 &= \frac{10\sqrt{2}}{3\pi}\angle-135^\circ,\\A_4\angle\phi_4 &= 0.\end{aligned}$$

Step 2: With $RC = 2\times10^4\times10^{-4} = 2$ s, the generic phasor-domain transfer function of the circuit is

$$\begin{aligned}\mathbf{H}(\omega) &= \mathbf{V}_{out} \qquad (\text{with } \mathbf{V}_s = 1)\\&= \frac{1}{1+j\omega RC}\\&= \frac{1}{\sqrt{1+\omega^2R^2C^2}}\,e^{-j\tan^{-1}(\omega RC)}\\&= \frac{1}{\sqrt{1+4\omega^2}}\,e^{-j\tan^{-1}(2\omega)}.\end{aligned} \tag{11.36}$$

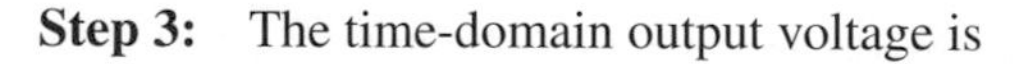

Step 3: The time-domain output voltage is

$$v_{out}(t) = 2.5 + \sum_{n=1}^{\infty} \mathfrak{Re}\left\{A_n \frac{1}{\sqrt{1+4n^2\omega_0^2}}\,e^{j[n\omega_0 t+\phi_n-\tan^{-1}(2n\omega_0)]}\right\}. \tag{11.37}$$

Using the values of $A_n\angle\phi_n$ determined earlier for the first four terms and replacing ω_0 with its numerical value of $\pi/2$ rad/s, the expression becomes

$$\begin{aligned}v_{out}(t) &= 2.5\\&\quad+\frac{10\sqrt{2}}{\pi\sqrt{1+\pi^2}}\cos\left[\frac{\pi t}{2} - 45^\circ - \tan^{-1}(\pi)\right]\\&\quad+\frac{10}{\pi\sqrt{1+4\pi^2}}\cos[\pi t - 90^\circ - \tan^{-1}(2\pi)]\\&\quad+\frac{10\sqrt{2}}{3\pi\sqrt{1+9\pi^2}}\cos\left[\frac{3\pi t}{2} - 135^\circ - \tan^{-1}(3\pi)\right]\cdots\\&= 2.5 + 1.37\cos\left(\frac{\pi t}{2} - 117^\circ\right) + 0.5\cos(\pi t - 171^\circ)\\&\quad+0.16\cos\left(\frac{3\pi t}{2} + 141^\circ\right)\cdots \text{ V}.\end{aligned}$$

The voltage response $v_{out}(t)$ is displayed in Fig. 11-8(c) and is computed using the series solution given by the preceding expression with $n_{max} = 1000$.

Example 11-6: Three-Stage Phase Shifter

In Chapter 7, we showed that the phasor-domain transfer function of the three-stage phase shifter shown in Fig. 11-9(a) is given by Eq. (7.105) as

$$\mathbf{H}(\omega) = \frac{\mathbf{V}_{\text{out}}}{\mathbf{V}_{\text{s}}} = \frac{x^3}{(x^3 - 5x) + j(1 - 6x^2)},$$

where $x = \omega RC$. Determine the output response to the periodic waveform shown in Fig. 11-9(b) given that $RC = 1$ s.

Solution: Step 1: With $T = 1$ s, $\omega_0 = 2\pi/T = 2\pi$ rad/s, and $v_s(t) = t$ over [0, 1],

$$a_0 = \frac{1}{T}\int_0^T v_s(t)\,dt = \int_0^1 t\,dt = 0.5,$$

$$a_n = \frac{2}{1}\int_0^1 t\cos 2n\pi t\,dt$$

$$= 2\left[\frac{1}{(2n\pi)^2}\cos 2n\pi t + \frac{t}{2n\pi}\sin 2n\pi t\right]\Bigg|_0^1 = 0,$$

$$b_n = \frac{2}{1}\int_0^1 t\sin 2n\pi t\,dt$$

$$= 2\left[\frac{1}{(2n\pi)^2}\sin 2n\pi t - \frac{t}{2n\pi}\cos 2n\pi t\right]\Bigg|_0^1$$

$$= -\frac{1}{n\pi},$$

and

$$A_n \angle\phi_n = 0 - jb_n$$

$$= 0 + j\,\frac{1}{n\pi}$$

$$= \frac{1}{n\pi}\angle 90^\circ \text{ V}.$$

Step 2: With $RC = 1$ and $x = \omega RC = \omega$, $\mathbf{H}(\omega)$ becomes

$$\mathbf{H}(\omega) = \frac{\omega^3}{(\omega^3 - 5\omega) + j(1 - 6\omega^2)}.$$

Step 5: With $\omega_0 = 2\pi$ rad/s, $\mathbf{H}(\omega = 0) = 0$, and $A_n = (1/n\pi)e^{j90^\circ}$, the time-domain voltage is obtained by multiplying each term in the summation by its corresponding $e^{jn\omega_0 t} = e^{j2n\pi t}$, and then taking the real part of the entire expression:

$$v_{\text{out}}(t) = \sum_{n=1}^{\infty} \mathfrak{Re}\left\{\frac{8n^2\pi^2}{[(2n\pi)((2n\pi)^2 - 5) + j(1 - 24n^2\pi^2)]} \cdot e^{j(2n\pi t + 90^\circ)}\right\}.$$

Evaluating the first few terms of $v_{\text{out}}(t)$ leads to

$$v_{\text{out}}(t) = 0.25\cos(2\pi t + 137^\circ) + 0.15\cos(4\pi t + 116^\circ) + 0.10\cos(6\pi t + 108^\circ) + \cdots.$$

A plot of $v_{\text{out}}(t)$ with 100 terms is displayed in Fig. 11-9(c).

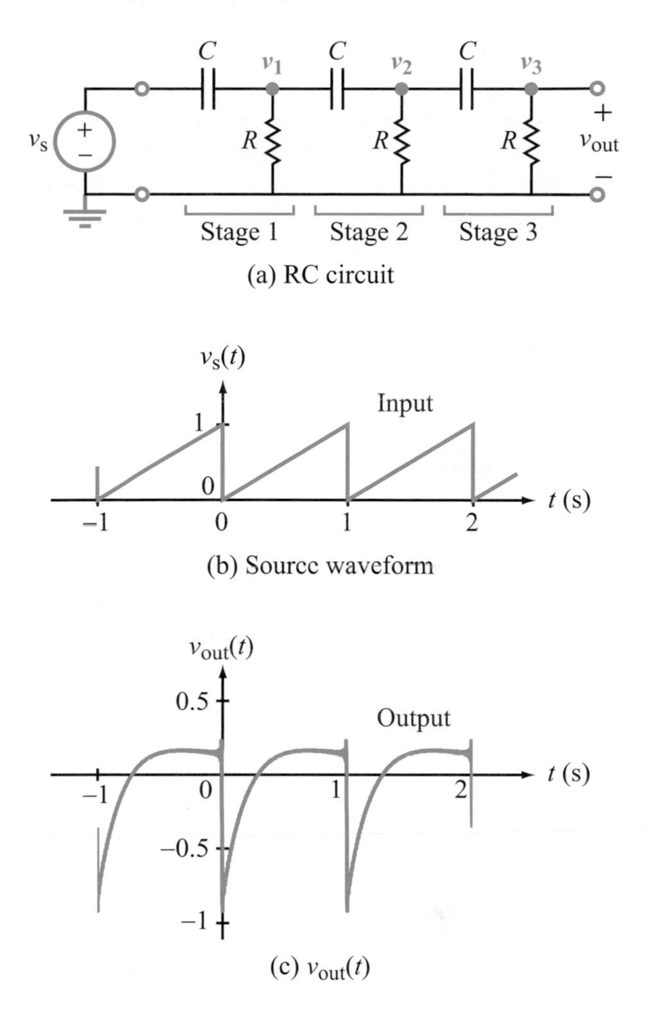

Figure 11-9: Circuit and plots for Example 11-6.

Technology Brief 22: Synthetic Biology

Whether amplifying, sensing, computing, or communicating, all of the circuits discussed in this book manipulate electric charge to process information. Voltage levels and current intensities all represent the collective properties of charges inside metals, insulators and semiconductors. The processing of information, however, can be accomplished in other media as well. Mechanical circuits (like the Babbage Engine, described in Technology Brief 1 on page 10), optical circuits (where computation is accomplished by manipulating light), and chemical circuits (where the operators are the reactants and products of chemical reactions) all have been demonstrated or in use for many years. Recently,

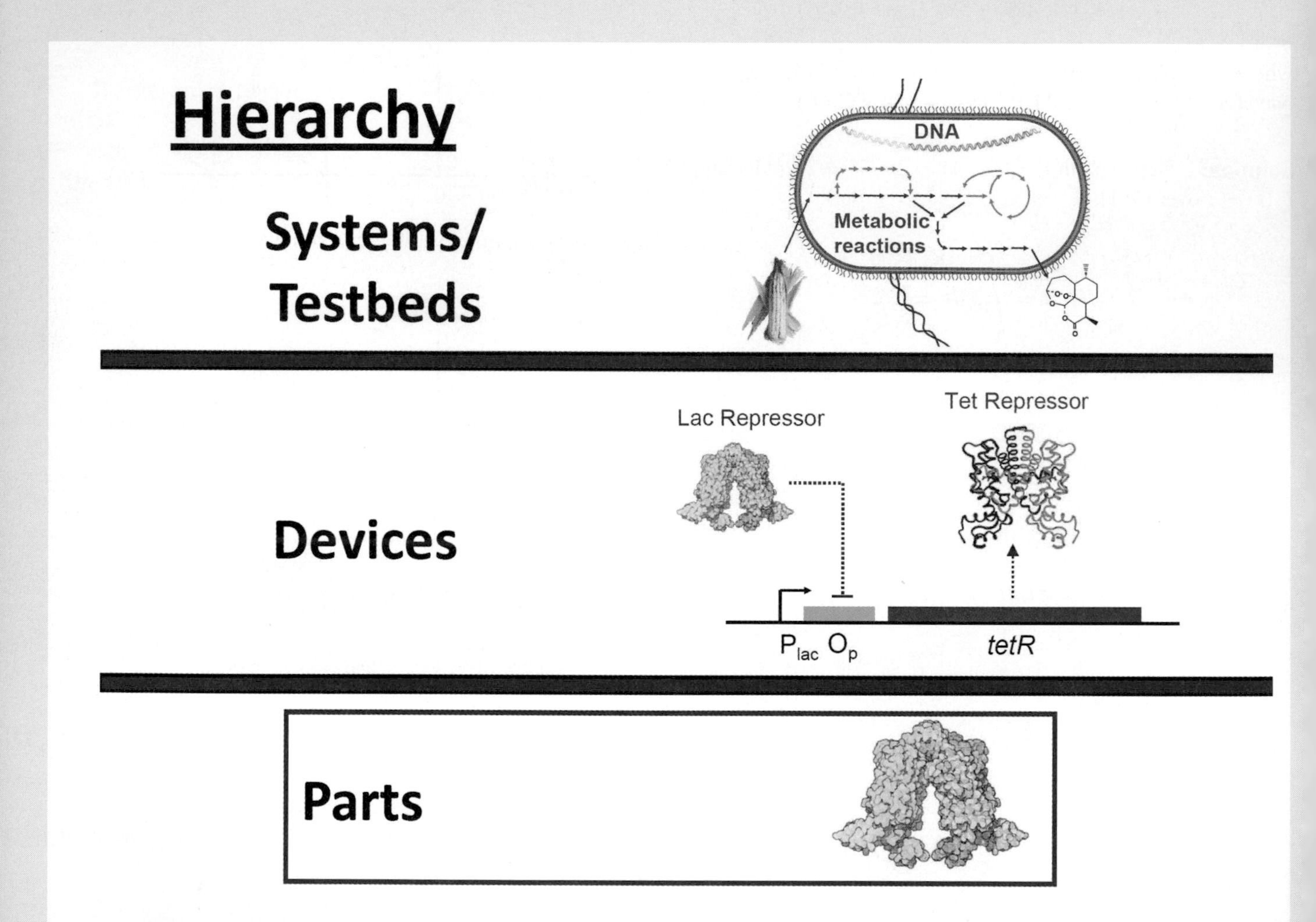

Figure TF22-1: Hierarchy of synthetic biological systems. In electric circuits, ***electrical components*** (resistors, capacitors, etc.) are assembled into ***circuits***, with each designed to perform a specific function. These circuits usually are combined into larger ***systems*** or ***architectures***. Synthetic biological systems have a similar hierarchy. ***Parts*** can be genes or proteins (a 3-D illustration of the LacR protein is shown here) which, when assembled into ***devices***, can be made to perform specific functions. The device shown above is a gene that produces the protein TetR, unless turned off by the LacR protein. This device is composed of five basic parts: the LacR protein, the TetR protein, the ***tetR*** gene (which contains instructions on how to make the TetR protein), the P_{lac} promoter (which turns on the ***tetR*** gene to make TetR protein) and the O_p repressor (which turns off the tetR gene when LacR is present). This is a chemical circuit with proteins and genes as components! Devices can then be assembled into a cell (in this case, the E. coli bacterium) to make ***systems*** that perform complex, multi-component tasks. For example, researchers are constructing a bacterium that senses tumor cells, penetrates them, and then kills the cancerous cells. (Courtesy of Wendell Lim, Associate Director of National Science Foundation's Synthetic Biology Engineering Research Center, University of California, Berkeley.)

engineers have begun to make synthetic information processing circuits inside biological cells. This new branch of engineering grew out of biochemical engineering and is called ***synthetic biology***. It promises to revolutionize the way we interact with biological systems.

In order to understand why synthetic biology is so powerful, and why it is so closely related to electrical engineering, we need to understand how biological cells process information. A cell, whether a free swimming bacterium or a human liver cell, is constantly transducing, storing and processing information from its environment. Cells produce molecules called ***proteins***, each of which can perform a specific function on a specific molecule. They can be thought of as little molecular robots. For example, certain proteins on a cell's membrane act as sensors, detecting the presence of molecules in the liquid around the cell. These surface proteins can change the state of other proteins inside the cell, which in turn affect other proteins, and so on. These chains of chemical reactions are called ***biochemical pathways***, and in this way the cell can adjust what molecules it produces based on what molecules are in its environment. If, say, the environment contains glucose, the cell's sensors can detect this and begin producing proteins that enable the cell to use glucose as fuel.

A key component of this regulation process depends on the ***genes*** the cell possesses. Although a discussion of genes is well beyond the scope of this book, for our purposes a cell contains a set of molecules called genes, which store descriptions of all the proteins it can make; a given gene will usually ***encode*** a single type of protein. In many ways, you can think of the genes as the cell's software. When we say a cell ***expresses*** a gene, we mean it has used the information in that gene to make the gene's protein. What is important here is that the biochemical pathways determine

Figure TF22-2: By introducing new genes into ***E. coli***, the bacteria were made sensitive to light. In this example, by a team from the University of Texas, Austin, and the University of California, San Francisco, a thin film of bacteria was grown in a Petri dish and exposed to patterns of light with the message "Hello World." Each bacterium in the film responded by changing color depending on the light level it was exposed to.

which genes are expressed. Thus, the surface proteins that detect glucose, for example, cause the cell to express the genes that encode the proteins that consume glucose. A single cell can make thousands of different proteins, which can interact with each other as well as with the cell's genes.

With the advent of modern molecular biology, our knowledge of the cell's pathways has expanded dramatically. In the latter half of the 20th century, biochemical engineers put this information to work by growing cells from many different species and modifying them to perform many useful functions. Waste water treatment, drug production, food additive production, and many other useful chemical processes are now carried out using cells.

Even more recently, synthetic biologists have begun to build ***information processing*** circuits into cells (Fig. TF22-1). These engineers hope to design components and circuits that perform many of the same functions you have studied in this book, using the cell's biochemical pathways! Amplification, logical functions, clocks, memory, multi-channel communication, sensing, and even rudimentary "software" programs are all being developed using cells, proteins, and genes instead of circuit boards and solid-state materials. If these efforts succeed, our ability to interact and guide the behavior of existing cells and to build entirely new types of cells with human-made programs will have a profound impact on the world of science and technology (Fig. TF22-2 and Fig. TF22-3). Along with these challenges comes a great responsibility to understand how our inventions can affect the natural world. What is very exciting is that synthetic biologists are realizing that many of the concepts electrical engineers developed for their electric circuits—noise, bandwidth, linear analysis, circuit diagrams, etc.—are proving useful for designing biological circuits!

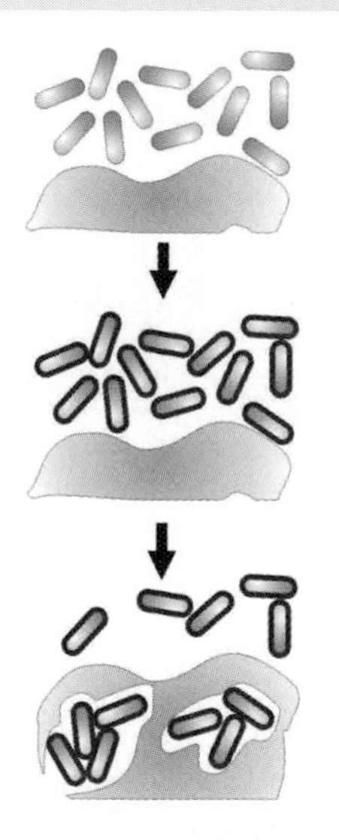

Figure TF22-3: Researchers at the University of California, Berkeley, and the University of California, San Francisco have devised synthetic pathways that may one day allow engineered bacteria to invade and kill tumors. (***top***) Modified ***E. coli*** bacteria at normal cell densities or oxygen conditions behave normally. (***middle***) Upon encountering low oxygen environments (associated with tumors) or high cell densities, the modified bacteria express invasin, a molecule that adheres to tumor cells and tricks the cells into absorbing the bacteria. (***bottom***) Once inside the tumor, the bacteria begin to invade and destroy it.

Review Question 11-9: What is the connection between the Fourier series solution method and the phasor-domain solution technique?

Review Question 11-10: Application of the Fourier series method in circuit analysis relies on which fundamental property of the circuit?

Exercise 11-4: The RL circuit shown in Fig. E11.4(a) is excited by the square-wave voltage waveform of Fig. E11.4(b). Determine $v_{\text{out}}(t)$.

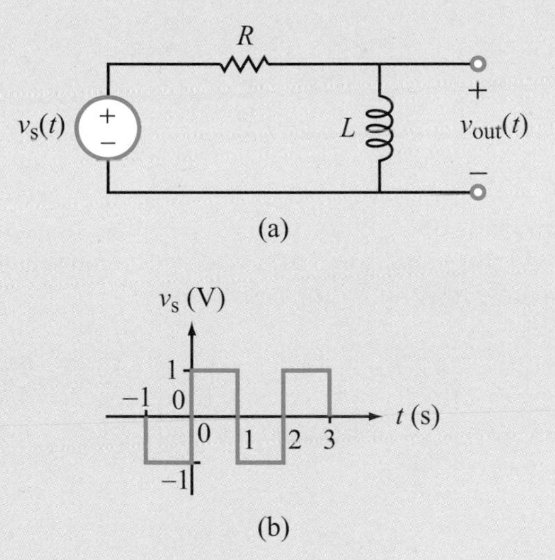

Figure E11.4

Answer:

$$v_{\text{out}}(t) = \sum_{\substack{n=1 \\ n=\text{odd}}}^{\infty} \frac{4L}{\sqrt{R^2 + n^2\pi^2 L^2}} \cos(n\pi t + \theta_n);$$

$$\theta_n = -\tan^{-1}\left(\frac{n\pi L}{R}\right).$$

(See)

11-4 Average Power

If a circuit is excited by a periodic voltage or current of period T and associated fundamental angular frequency $\omega_0 = 2\pi/T$, then every segment of the circuit will exhibit a voltage across it (Fig. 11-10), characterized by a Fourier series of the form

$$v(t) = V_{\text{dc}} + \sum_{n=1}^{\infty} V_n \cos(n\omega_0 t + \phi_{v_n}), \tag{11.38}$$

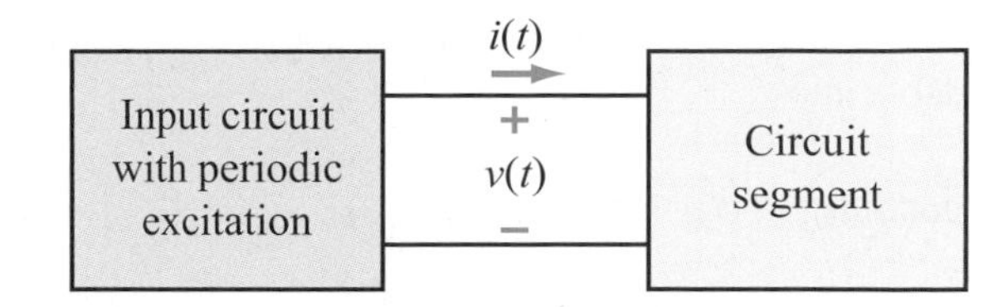

Figure 11-10: Voltage across and current into a circuit segment.

where V_{dc} is the average value of $v(t)$, V_n is the amplitude of the nth harmonic, and ϕ_{v_n} is the associated phase angle. Similarly, the current flowing into the circuit segment is also given by a Fourier series as

$$i(t) = I_{\text{dc}} + \sum_{m=1}^{\infty} I_m \cos(m\omega_0 t + \phi_{i_m}), \tag{11.39}$$

where similar definitions apply to I_{dc}, I_m and ϕ_{i_m} and we have used the integer index m instead of n in order to keep the two summations distinguishable from one another.

By denoting the current direction such that it is flowing into the (+) voltage terminal (Fig. 11-10), the passive sign convention stipulates that the product vi represents power flow into the circuit segment. Hence, the average power is given by

$$\begin{aligned}
P_{\text{av}} &= \frac{1}{T}\int_0^T vi\,dt \\
&= \frac{1}{T}\int_0^T V_{\text{dc}} I_{\text{dc}}\,dt \\
&\quad + \sum_{n=1}^{\infty} \frac{1}{T}\int_0^T V_n I_{\text{dc}} \cos(n\omega_0 t + \phi_{v_n})\,dt \\
&\quad + \sum_{m=1}^{\infty} \frac{1}{T}\int_0^T V_{\text{dc}} I_m \cos(m\omega_0 t + \phi_{i_m})\,dt \\
&\quad + \sum_{m=1}^{\infty}\sum_{n=1}^{\infty} \frac{1}{T}\int_0^T V_n I_m \cos(n\omega_0 t + \phi_{v_n}) \\
&\quad \cdot \cos(m\omega_0 t + \phi_{i_m})\,dt.
\end{aligned} \tag{11.40}$$

From Table 11-1, integral Property 2 states that

$$\int_0^T \cos(n\omega_0 t + \phi_0)\,dt = 0 \tag{11.41}$$

for any integer $n \geq 1$ and any constant phase angle ϕ_0. Consequently, terms 2 and 3 in Eq. (11.40) vanish. Moreover, the product of the two cosine functions inside the last term is expandable into

$$\begin{aligned}\cos(n\omega_0 t+\phi_{v_n})\cos(m\omega_0 t+\phi_{i_m}) &= \frac{1}{2}\cos(\text{sum of arguments}) \\ &\quad+\frac{1}{2}\cos(\text{difference between arguments}) \\ &= \frac{1}{2}\cos[(n+m)\omega_0 t+\phi_{v_n}+\phi_{i_m}] \\ &\quad+\frac{1}{2}\cos[(n-m)\omega_0 t+\phi_{v_n}-\phi_{i_m}]. \end{aligned} \tag{11.42}$$

Performing the integration over the two terms of Eq. (11.42) will yield zero values, except when $n = m$. Implementing these considerations in Eq. (11.40) leads to

$$P_{\text{av}} = V_{\text{dc}} I_{\text{dc}} + \frac{1}{2}\sum_{n=1}^{\infty} V_n I_n \cos(\phi_{v_n} - \phi_{i_n}). \tag{11.43}$$

We note that $\frac{1}{2} V_n I_n \cos(\phi_{v_n} - \phi_{i_n})$ represents the average power associated with frequency harmonic $n\omega_0$. Hence:

The total average power is equal to the dc power ($V_{\text{dc}} I_{\text{dc}}$) plus the sum of the average ac power associated with the fundamental frequency ω_0 and its harmonic multiples.

Example 11-7: ac Power Fraction

The periodic voltage across a certain circuit truncated to the first three ac terms of its Fourier series is given by

$$\begin{aligned} v(t) = 2 &+ 3\cos(4t+30^\circ) \\ &+ 1.5\cos(8t-30^\circ) \\ &+ 0.5\cos(12t-135^\circ) \text{ V}, \end{aligned}$$

and the associated current flowing into the (+) voltage terminal of the circuit is

$$\begin{aligned} i(t) &= 60 + 10\cos(4t-30^\circ) \\ &\quad + 5\cos(8t+15^\circ) + 2\cos 12t \text{ mA}. \end{aligned}$$

Determine the ac fraction of the average power.

Solution: Application of Eq. (11.43) yields

$$\begin{aligned} P_{\text{av}} &= 2\times 60 + \frac{3\times 10}{2}\cos(30^\circ+30^\circ) \\ &\quad + \frac{1.5\times 5}{2}\cos(-30^\circ-15^\circ) \\ &\quad + \frac{0.5\times 2}{2}\cos(-135^\circ) \\ &= 120 + 7.5 + 2.65 - 0.353 \\ &= 129.80 \text{ W}. \end{aligned}$$

The ac fraction is

$$\frac{7.5+2.65-0.353}{129.8} = 7.55\%.$$

Exercise 11-5: What will the expression given by Eq. (11.43) simplify to if the associated circuit segment is (a) purely resistive or (b) purely reactive?

Answer: (a) $P_{\text{av}} = V_{\text{dc}} I_{\text{dc}} + \frac{1}{2}\sum_{n=1}^{\infty} V_n I_n$; because $\phi_{v_n} = \phi_{i_n}$, (b) $P_{\text{av}} = 0$, because for a capacitor, $I_{\text{dc}} = 0$ and $\phi_{v_n} - \phi_{i_n} = -90^\circ$; and for an inductor, $V_{\text{dc}} = 0$ and $\phi_{v_n} - \phi_{i_n} = 90^\circ$. (See)

11-5 Fourier Transform

The Fourier series is a perfectly suitable construct for representing periodic functions, but what about nonperiodic functions? The pulse-train waveform shown in Fig. 11-11(a) consists of a sequence of rectangular pulses—with a width of $\tau = 2$ s. The period $T = 4$ s. In part (b) of the same figure, the individual pulses have the same shape as before, but T has been increased to 7 s. So long as T is finite, both waveforms are amenable to representation by the Fourier series, but what would happen if we let $T \to \infty$, ending up with the single pulse shown in Fig. 11-11(c)? Can we then represent the no-longer periodic pulse by a Fourier series? We will discover shortly that as $T \to \infty$, the summation $\sum_{n=1}^{\infty}$ in the Fourier series evolves into a continuous integral, which we call the ***Fourier transform***. Accordingly, when analyzing electric circuits:

We apply the Fourier series approach if the excitation is periodic in character, and we use the Fourier transform technique if the excitation is nonperiodic.

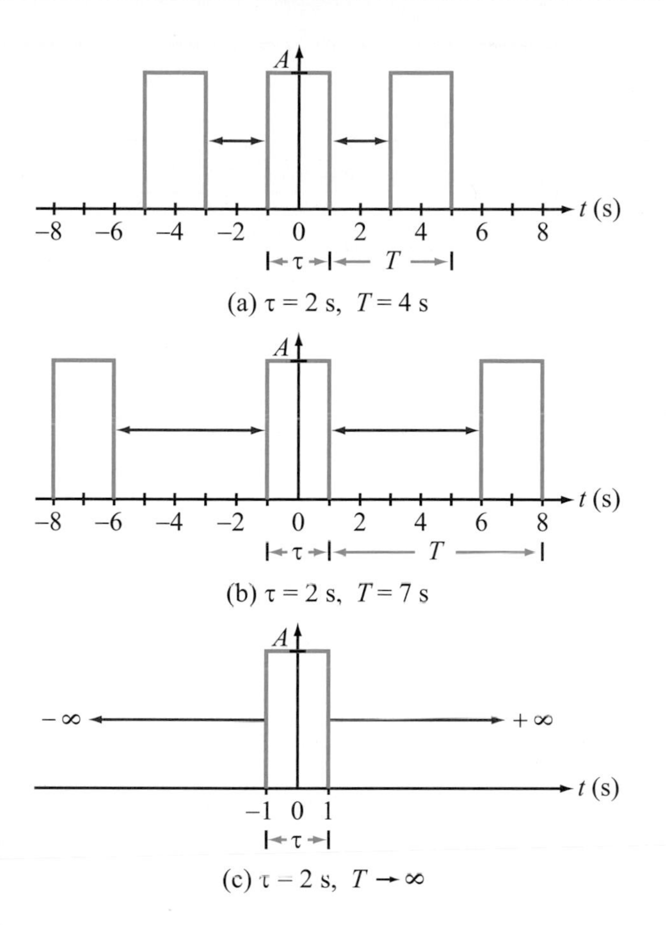

Figure 11-11: The single pulse in (c) is equivalent to a periodic pulse train with $T = \infty$.

Does that mean that we can use both the Laplace transform (Chapter 10) and the Fourier transform techniques to analyze circuits containing nonperiodic sources? If so, which of the two transforms should we use, and why? We will address these questions later (Section 11-7), after formally introducing the Fourier transform and discussing some of its salient features.

11-5.1 Exponential Fourier Series

According to Eq. (11.15), a periodic function of period T and corresponding fundamental frequency $\omega_0 = 2\pi/T$ can be represented by the series

$$f(t) = a_0 + \sum_{n=1}^{\infty} a_n \cos n\omega_0 t + b_n \sin n\omega_0 t. \tag{11.44}$$

Sine and cosine functions can be converted into complex exponentials via Euler's identity:

$$\cos n\omega_0 t = \frac{1}{2}(e^{jn\omega_0 t} + e^{-jn\omega_0 t}), \tag{11.45a}$$

$$\sin n\omega_0 t = \frac{1}{j2}(e^{jn\omega_0 t} - e^{-jn\omega_0 t}). \tag{11.45b}$$

Upon inserting Eqs. (11.45a and b) into Eq. (11.44), we have

$$\begin{aligned} f(t) &= \\ & a_0 + \sum_{n=1}^{\infty} \left[\frac{a_n}{2}(e^{jn\omega_0 t} + e^{-jn\omega_0 t}) + \frac{b_n}{j2}(e^{jn\omega_0 t} - e^{-jn\omega_0 t})\right] \\ &= a_0 + \sum_{n=1}^{\infty} \left[\left(\frac{a_n - jb_n}{2}\right) e^{jn\omega_0 t} + \left(\frac{a_n + jb_n}{2}\right) e^{-jn\omega_0 t}\right] \\ &= a_0 + \sum_{n=1}^{\infty} [\mathbf{c}_n e^{jn\omega_0 t} + \mathbf{c}_{-n} e^{-jn\omega_0 t}], \end{aligned} \tag{11.46}$$

where we introduced the complex coefficients:

$$\mathbf{c}_n = \frac{a_n - jb_n}{2}, \qquad \mathbf{c}_{-n} = \frac{a_n + jb_n}{2} = \mathbf{c}_n^*. \tag{11.47}$$

As the index n is incremented from 1 to ∞, the second term in Eq. (11.46) generates the series

$$\mathbf{c}_{-1} e^{-j\omega_0 t} + \mathbf{c}_{-2} e^{-j2\omega_0 t} + \cdots,$$

which also can be generated by $\mathbf{c}_n e^{jn\omega_0 t}$ with n incremented from -1 to $-\infty$. This equivalence allows us to express $f(t)$ in the compact ***exponential form*** as

$$f(t) = \sum_{n=-\infty}^{\infty} \mathbf{c}_n e^{jn\omega_0 t} \quad \text{(exponential representation)}, \tag{11.48}$$

where

$$c_0 = a_0, \tag{11.49}$$

and the range of n has been expanded to $[-\infty, \infty]$. For all coefficients $\mathbf{c}_n$ including c_0, it is easy to show that

$$\mathbf{c}_n = \frac{1}{T} \int_{-T/2}^{T/2} f(t)\, e^{-jn\omega_0 t}\, dt. \tag{11.50}$$

Table 11-3: Fourier series representations for a periodic function $f(t)$.

Cosine/Sine	Amplitude/Phase	Complex Exponential
$f(t) = a_0 + \sum_{n=1}^{\infty}(a_n \cos n\omega_0 t + b_n \sin n\omega_0 t)$	$f(t) = a_0 + \sum_{n=1}^{\infty} A_n \cos(n\omega_0 t + \phi_n)$	$f(t) = \sum_{n=-\infty}^{\infty} \mathbf{c}_n e^{jn\omega_0 t}$
$a_0 = \frac{1}{T}\int_0^T f(t)\, dt$	$A_n e^{j\phi_n} = a_n - jb_n$	$\mathbf{c}_n = \lvert\mathbf{c}_n\rvert e^{j\theta_n};\ \mathbf{c}_{-n} = \mathbf{c}_n^*$
$a_n = \frac{2}{T}\int_0^T f(t)\ \cos n\omega_0 t\ dt$	$A_n = \sqrt[+]{a_n^2 + b_n^2}$	$\lvert\mathbf{c}_n\rvert = A_n/2$
$b_n = \frac{2}{T}\int_0^T f(t)\ \sin n\omega_0 t\ dt$	$\phi_n = -\tan^{-1}\left(\frac{b_n}{a_n}\right)$	$\theta_n = \phi_n$

Even though the integration limits indicated in Eq. (11.50) are from $-T/2$ to $T/2$, they can be chosen arbitrarily so long as the upper limit exceeds the lower limit by exactly T.

For easy reference, Table 11-3 provides a summary of the relationships associated with all three Fourier series representations introduced in this chapter, namely the cosine/sine, amplitude/phase, and complex exponential.

Example 11-8: Pulse Train

Obtain the Fourier series exponential representation for the pulse-train waveform displayed in Fig. 11-11(a) in terms of the pulse width τ and the period T. Evaluate and plot the line spectrum of $|\mathbf{c}_n|$ for $A = 10$ and $\tau = 1$ s for each of the following values of T: 5 s, 10 s, and 20 s.

Solution: Over a period T extending from $-T/2$ to $T/2$,

$$f(t) = \begin{cases} A & \text{for } -\tau/2 \le t \le \tau/2, \\ 0 & \text{otherwise.} \end{cases}$$

With the integration domain chosen to be from $-T/2$ to $T/2$, Eq. (11.50) gives

$$\begin{aligned} \mathbf{c}_n &= \frac{1}{T}\int_{-T/2}^{T/2} f(t)\ e^{-jn\omega_0 t}\ dt = \frac{1}{T}\int_{-\tau/2}^{\tau/2} Ae^{-jn\omega_0 t}\ dt \\ &= \frac{A}{-jn\omega_0 T}\ e^{-jn\omega_0 t}\Big|_{-\tau/2}^{\tau/2} \\ &= \frac{2A}{n\omega_0 T}\left[\frac{e^{jn\omega_0\tau/2} - e^{-jn\omega_0\tau/2}}{2j}\right]. \end{aligned} \quad (11.51)$$

The quantity inside the square bracket matches the form of one of Euler's formulas, namely

$$\sin x = \frac{e^{jx} - e^{-jx}}{2j}. \quad (11.52)$$

Hence, Eq. (11.51) can be rewritten in the form

$$\begin{aligned} \mathbf{c}_n &= \frac{2A}{n\omega_0 T}\sin(n\omega_0\tau/2) \\ &= \frac{A\tau}{T}\frac{\sin(n\omega_0\tau/2)}{(n\omega_0\tau/2)} \\ &= \frac{A\tau}{T}\operatorname{sinc}(n\omega_0\tau/2), \end{aligned} \quad (11.53)$$

where in the last step we introduced the ***sinc function***, which is defined as

$$\operatorname{sinc}(x) = \frac{\sin x}{x}. \quad (11.54)$$

Among the important properties of the sinc function are the following:

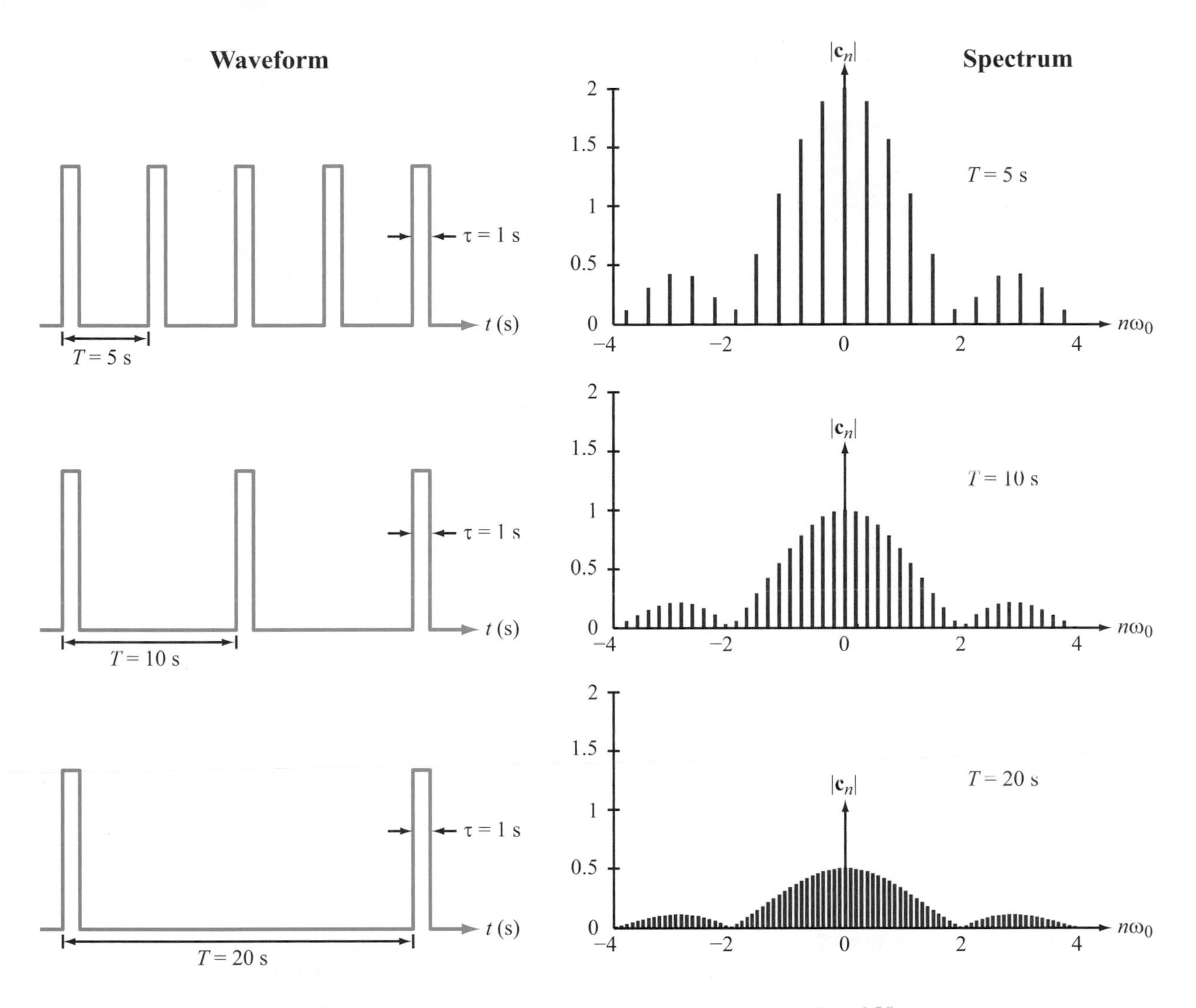

Figure 11-12. Line spectra for pulse trains with $T/\tau = 5$, 10, and 20.

(a) When its argument is zero, the sinc function is equal to 1,

$$\text{sinc}(0) = \left.\frac{\sin(x)}{x}\right|_{x=0} = 1. \qquad (11.55)$$

Verification of this property can be established by applying l'Hôpital's rule to Eq. (11.54) and then setting $x = 0$.

(b) Since $\sin(m\pi) = 0$ for any integer value of m, the same is true for the sinc function,

$$\text{sinc}(m\pi) = 0. \qquad (11.56)$$

(c) Because both $\sin x$ and x are odd functions, their ratio is an even function. Hence, the sinc function possesses even symmetry relative to the vertical axis. Consequently,

$$\mathbf{c}_n = \mathbf{c}_{-n}. \qquad (11.57)$$

Evaluation of Eq. (11.53) with $A = 10$ leads to the line spectra displayed in Fig. 11-12. The general shape of the envelope is dictated by the sinc function, exhibiting a symmetrical pattern with a peak at $n = 0$, a major lobe extending between $n = -T/\tau$ and $n = T/\tau$, and progressively smaller amplitude lobes on both sides. The density of spectral lines depends on the ratio of T/τ, so in the limit as $T \to \infty$, the line spectrum becomes a continuum.

11-5.2 Nonperiodic Waveforms

In Example 11-8, we noted that as the period $T \to \infty$ the periodic function becomes nonperiodic and the associated line spectrum evolves from one containing discrete lines into a continuum. We now will explore this evolution in mathematical terms, culminating in a definition for the Fourier transform of a nonperiodic function. To that end, we begin with the pair of expressions given by Eqs. (11.48) and (11.50), namely

$$f(t) = \sum_{n=-\infty}^{\infty} \mathbf{c}_n e^{jn\omega_0 t}, \tag{11.58a}$$

and

$$\mathbf{c}_n = \frac{1}{T} \int_{-T/2}^{T/2} f(t)\, e^{-jn\omega_0 t}\, dt. \tag{11.58b}$$

These two quantities form a complementary pair with $f(t)$ defined in the continuous time domain and $\mathbf{c}_n$ defined in the discrete frequency domain as $n\omega_0$, with $\omega_0 = 2\pi/T$. For a given value of T, the nth frequency harmonic is at $n\omega_0$ and the next harmonic after that is at $(n+1)\omega_0$. Hence, the ***spacing between adjacent harmonics*** is

$$\Delta\omega = (n+1)\omega_0 - n\omega_0 = \omega_0 = \frac{2\pi}{T}. \tag{11.59}$$

If we insert Eq. (11.58b) into Eq. (11.58a) and replace $1/T$ with $\Delta\omega/2\pi$, we get

$$f(t) = \sum_{n=-\infty}^{\infty} \left[\frac{1}{2\pi} \int_{-T/2}^{T/2} f(t)\, e^{-jn\omega_0 t}\, dt \right] e^{jn\omega_0 t}\, \Delta\omega. \tag{11.60}$$

As $T \to \infty$, $\Delta\omega \to d\omega$, $n\omega_0 \to \omega$, and the sum becomes a continuous integral:

$$f(t) = \frac{1}{2\pi} \int_{-\infty}^{\infty} \left[\int_{-\infty}^{\infty} f(t)\, e^{-j\omega t}\, dt \right] e^{j\omega t}\, d\omega. \tag{11.61}$$

Given this new arrangement, we now are ready to offer formal definitions for the Fourier transform $\mathbf{F}(\omega)$ and its inverse transform $f(t)$:

$$\mathbf{F}(\omega) = \mathcal{F}[f(t)] = \int_{-\infty}^{\infty} f(t)\, e^{-j\omega t}\, dt, \tag{11.62a}$$

and

$$f(t) = \mathcal{F}^{-1}[\mathbf{F}(\omega)] = \frac{1}{2\pi} \int_{-\infty}^{\infty} \mathbf{F}(\omega)\, e^{j\omega t}\, d\omega, \tag{11.62b}$$

where $\mathcal{F}[f(t)]$ is a shorthand notation for "the Fourier transform of $f(t)$," and similarly, $\mathcal{F}^{-1}[\mathbf{F}(\omega)]$ represents the inverse operation. Occasionally, we also may use the symbolic form

$$f(t) \longleftrightarrow \mathbf{F}(\omega).$$

Example 11-9: Rectangular Pulse

Determine the Fourier transform of the solitary rectangular pulse shown in Fig. 11-13(a) and then plot its ***amplitude spectrum*** $|\mathbf{F}(\omega)|$ for $A = 5$ and $\tau = 1$ s.

Solution: Application of Eq. (11.62a) with $f(t) = \text{rect}(t/\tau) = A$ over the integration interval $[-\tau/2, \tau/2]$ leads to

$$\mathbf{F}(\omega) = \int_{-\tau/2}^{\tau/2} Ae^{-j\omega t}\, dt = \frac{A}{-j\omega}\, e^{-j\omega t}\Big|_{-\tau/2}^{\tau/2}$$

$$= A\tau\, \frac{\sin \omega\tau/2}{(\omega\tau/2)} = A\tau \text{ sinc}\left(\frac{\omega\tau}{2}\right). \tag{11.63}$$

The ***frequency spectrum*** of $|\mathbf{F}(\omega)|$ is displayed in Fig. 11-13(b) for the specified values of $A = 5$ and $\tau = 1$ s. The ***nulls*** in the spectrum occur when the argument of the sinc function is a

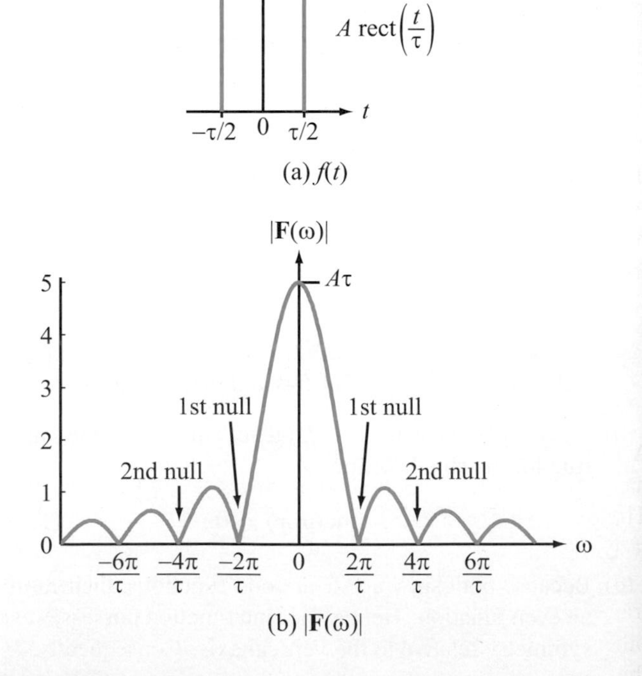

Figure 11-13: (a) Rectangular pulse of amplitude A and width τ; (b) frequency spectrum of $|\mathbf{F}(\omega)|$ for $A = 5$ and $\tau = 1$ s.

Review Question 11-11: For the cosine/sine and amplitude/phase Fourier series representations, the summation extends from $n = 1$ to $n = \infty$. What are the limits on the summation for the complex exponential representation?

Review Question 11-12: What is a sinc function, and what are its primary properties? Why is $\text{sinc}(0) = 1$?

Review Question 11-13: What is the functional form for the Fourier transform $\mathbf{F}(\omega)$ of a rectangular pulse of amplitude 1 and duration τ?

Exercise 11-6: For a single rectangular pulse of width τ, what is the spacing $\Delta\omega$ between first nulls? If τ is very wide, will its frequency spectrum be narrow and peaked or wide and gentle?

Answer: $\Delta\omega = 4\pi/\tau$. Wide τ leads to narrow spectrum. (See)

11-5.3 Convergence of the Fourier Integral

Not every function $f(t)$ has a Fourier transform. The Fourier transform $\mathbf{F}(\omega)$ exists if the Fourier integral given by Eq. (11.62a) converges to a finite number or to an equivalent expression, but as we shall discuss shortly, it also may exist even if the Fourier integral does not converge. Convergence depends on the character of $f(t)$ over the integration range $[-\infty, \infty]$. By character, we mean (1) whether or not $f(t)$ exhibits infinite discontinuities and (2) how $f(t)$ behaves as t approaches $\pm\infty$. As a general rule, the Fourier integral does converge if $f(t)$ has no discontinuities (i.e., it is single-valued) and the integral of its absolute magnitude is finite; so

$$\int_{-\infty}^{\infty} |f(t)|\, dt < \infty. \tag{11.64}$$

A function $f(t)$ still can have a Fourier transform—even if it has discontinuities as long as those discontinuities are finite. The step function $A\,u(t)$ exhibits a finite discontinuity at $t = 0$ if A is finite.

The stated conditions for the existence of the Fourier transform are sufficient (but not necessary) conditions. In other words, some functions may still have transforms even though their Fourier integrals do not converge. Among such functions are the constant $f(t) = A$ and the unit step function $f(t) = A\,u(t)$, both of which represent important excitation waveforms in linear circuits. To realize the Fourier transform of a function whose transform exists but its Fourier integral does not converge, we need to employ an indirect approach. The approach entails the following ingredients.

(a) If $f(t)$ is a function whose Fourier integral does not converge, we select a second function $f_\epsilon(t)$ whose functional form includes a parameter ϵ. If allowed to approach a certain limit, $f_\epsilon(t)$ becomes identical with $f(t)$.

(b) The choice of function $f_\epsilon(t)$ should be such that its Fourier integral does converge, and therefore, $f_\epsilon(t)$ has a definable Fourier transform $\mathbf{F}_\epsilon(\omega)$.

(c) By taking parameter ϵ in the expression for $\mathbf{F}_\epsilon(\omega)$ to its limit, $\mathbf{F}_\epsilon(\omega)$ reduces to the transform $\mathbf{F}(\omega)$ corresponding to the original function $f(t)$.

This procedure is illustrated through some of the examples presented in the next section.

11-6 Fourier Transform Pairs

In this section, we shall develop fluency in how to move back and forth between the time domain and the ω-domain. We will learn how to circumvent the convergence issues we noted earlier in Section 11-5.3, and in the process, we will identify a number of useful properties of the Fourier transform.

11-6.1 Linearity Property

If

$$f_1(t) \longleftrightarrow \mathbf{F}_1(\omega)$$

and

$$f_2(t) \longleftrightarrow \mathbf{F}_2(\omega),$$

then

$$K_1\, f_1(t) + K_2\, f_2(t) \longleftrightarrow K_1\, \mathbf{F}_1(\omega) + K_2\, \mathbf{F}_2(\omega) \quad \text{(linearity property)}, \tag{11.65}$$

where K_1 and K_2 are constants. Proof of Eq. (11.65) is ascertained easily through the application of Eq. (11.62a).

11-6.2 Fourier Transform of $\delta(t - t_0)$

By Eq. (11.62a), the Fourier transform of $\delta(t - t_0)$ is given by

$$\mathbf{F}(\omega) = \mathcal{F}[\delta(t - t_0)] = \int_{-\infty}^{\infty} \delta(t - t_0) e^{-j\omega t}\, dt$$

$$= e^{-j\omega t}\Big|_{t=t_0} = e^{-j\omega t_0}. \tag{11.66}$$

Hence,

$$\delta(t - t_0) \quad \leftrightarrow \quad e^{-j\omega t_0} \tag{11.67a}$$

and

$$\delta(t) \quad \leftrightarrow \quad 1. \tag{11.67b}$$

Thus, a unit impulse function $\delta(t)$ generates a constant of unit amplitude that extends over $[-\infty, \infty]$ in the ω-domain, as shown in Fig. 11-14(a),

11-6.3 Shift Properties

By Eq. (11.62b), the inverse Fourier transform of $\mathbf{F}(\omega) = \delta(\omega - \omega_0)$ is

$$f(t) = \mathcal{F}^{-1}[\delta(\omega - \omega_0)] = \frac{1}{2\pi} \int_{-\infty}^{\infty} \delta(\omega - \omega_0)\, e^{j\omega t}\, d\omega = \frac{e^{j\omega_0 t}}{2\pi}.$$

Hence,

$$e^{j\omega_0 t} \quad \leftrightarrow \quad 2\pi\, \delta(\omega - \omega_0), \tag{11.68a}$$

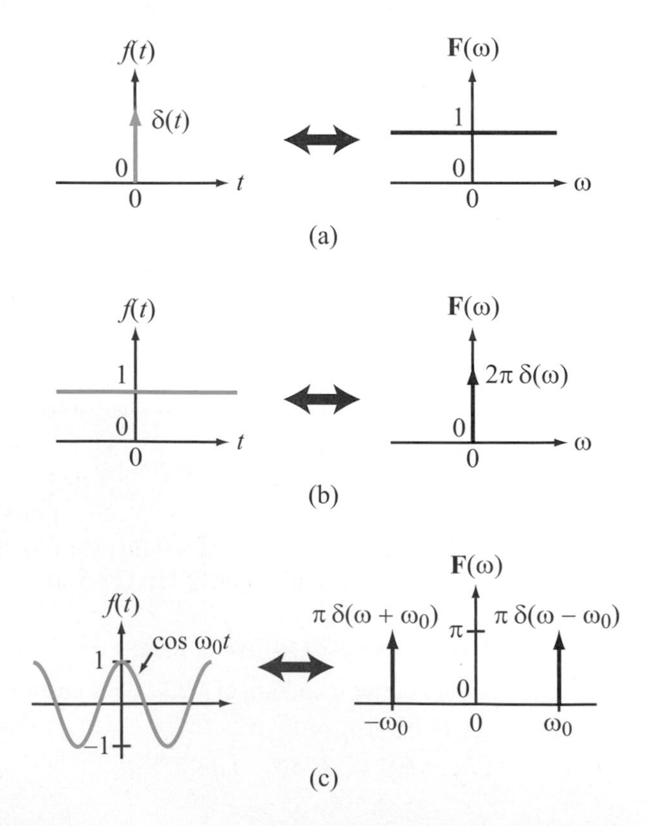

Figure 11-14: (a) The Fourier transform of $\delta(t)$ is 1 (b) the Fourier transform of 1 is $2\pi\, \delta(\omega)$, and (c) the Fourier transform of $\cos \omega_0 t$ is equal to two delta functions—one at ω_0 and another at $-\omega_0$.

$$1 \quad \leftrightarrow \quad 2\pi\, \delta(\omega). \tag{11.68b}$$

Comparison of the plots in Fig. 11-14(a) and (b) demonstrates the correspondence between the time domain and the ω-domain: an impulse $\delta(t)$ in the time domain generates a uniform spectrum in the frequency domain; conversely, a uniform (constant) waveform in the time domain generates an impulse $\delta(\omega)$ in the frequency domain.

It is straightforward to show that the result given by Eq. (11.68a) can be generalized to

$$e^{j\omega_0 t}\, f(t) \quad \leftrightarrow \quad \mathbf{F}(\omega - \omega_0)$$
$$\text{(frequency-shift property)}, \tag{11.69}$$

which is known as the ***frequency-shift property*** of the Fourier transform. It states that multiplication of a function $f(t)$ by $e^{j\omega_0 t}$ in the time domain corresponds to shifting the Fourier transform of $f(t)$ to $\mathbf{F}(\omega)$ by ω_0 along the ω-axis. The converse of the frequency-shift property is the ***time-shift property*** given by

$$f(t - t_0) \quad \leftrightarrow \quad e^{-j\omega t_0}\, \mathbf{F}(\omega)$$
$$\text{(time-shift property)}. \tag{11.70}$$

11-6.4 Fourier Transform of $\cos \omega_0 t$

By Euler's identity,

$$\cos \omega_0 t = \frac{e^{j\omega_0 t} + e^{-j\omega_0 t}}{2}.$$

In view of Eq. (11.68a),

$$\mathbf{F}(\omega) = \mathcal{F}\left[\frac{e^{j\omega_0 t}}{2} + \frac{e^{-j\omega_0 t}}{2}\right] = \pi\, \delta(\omega - \omega_0) + \pi\, \delta(\omega + \omega_0).$$

Hence,

$$\cos \omega_0 t \quad \leftrightarrow \quad \pi[\delta(\omega - \omega_0) + \delta(\omega + \omega_0)], \tag{11.71}$$

and similarly,

$$\sin \omega_0 t \quad \leftrightarrow \quad j\pi[\delta(\omega + \omega_0) - \delta(\omega - \omega_0)]. \tag{11.72}$$

As shown in Fig. 11-14(c), the Fourier transform of $\cos \omega_0 t$ consists of impulse functions at $\pm\omega_0$.

11-6.5 Fourier Transform of $Ae^{-at}\, u(t)$ with $a > 0$

The Fourier transform of an exponentially decaying function that starts at $t = 0$ is

$$\mathbf{F}(\omega) = \mathcal{F}[Ae^{-at}\, u(t)] = \int_{0}^{\infty} Ae^{-at} e^{-j\omega t}\, dt$$

$$= A \left. \frac{e^{-(a+j\omega)t}}{-(a+j\omega)} \right|_{0}^{\infty} = \frac{A}{a+j\omega}.$$

Hence,

$$Ae^{-at}\, u(t) \quad \longleftrightarrow \quad \frac{A}{a+j\omega}, \qquad \text{for } a > 0. \tag{11.73}$$

11-6.6 Fourier Transform of $u(t)$

The direct approach to finding $\mathbf{F}(\omega)$ for the unit step function leads to

$$\mathbf{F}(\omega) = \mathcal{F}[u(t)] = \int_{-\infty}^{\infty} u(t)\, e^{-j\omega t}\, dt$$

$$= \int_{0}^{\infty} e^{-j\omega t}\, dt = \left. \frac{e^{-j\omega t}}{-j\omega} \right|_{0}^{\infty} = \frac{j}{\omega}(e^{-j\infty} - 1),$$

which is problematic because $e^{-j\infty}$ does not converge. To avoid the convergence problem, we can pursue an alternative approach that involves the ***signum function***, which is defined by

$$\mathrm{sgn}(t) = u(t) - u(-t). \tag{11.74}$$

Shown graphically in Fig. 11-15(a), the signum function resembles a step-function waveform (with an amplitude of 2 units) that has been slid downward by 1 unit. Looking at the waveform, it is easy to see that one can generate a step function from the signum function as follows:

$$u(t) = \frac{1}{2} + \frac{1}{2}\, \mathrm{sgn}(t). \tag{11.75}$$

The corresponding Fourier transform is given by

$$\mathcal{F}[u(t)] = \mathcal{F}\left[\frac{1}{2}\right] + \frac{1}{2}\, \mathcal{F}[\mathrm{sgn}(t)]$$

$$= \pi\, \delta(\omega) + \frac{1}{2}\, \mathcal{F}[\mathrm{sgn}(t)], \tag{11.76}$$

where in the first term, we used the relationship given by Eq. (11.68b). Next, we will obtain $\mathcal{F}[\mathrm{sgn}(t)]$ by modeling the signum function as

$$\mathrm{sgn}(t) = \lim_{\epsilon \to 0} [e^{-\epsilon t}\, u(t) - e^{\epsilon t}\, u(-t)] \tag{11.77}$$

with $\epsilon > 0$. The shape of the modeled waveform is shown in Fig. 11-15(b) for a small value of ϵ.

Now we are ready to apply the formal definition of the Fourier transform given by Eq. (11.62a):

$$\mathcal{F}[\mathrm{sgn}(t)] = \int_{-\infty}^{\infty} \lim_{\epsilon \to 0} [e^{-\epsilon t}\, u(t) - e^{\epsilon t}\, u(-t)] e^{-j\omega t}\, dt$$

$$= \lim_{\epsilon \to 0} \left[\int_{0}^{\infty} e^{-(\epsilon + j\omega)t}\, dt - \int_{-\infty}^{0} e^{(\epsilon - j\omega)t}\, dt \right]$$

$$= \lim_{\epsilon \to 0} \left[\left. \frac{e^{-(\epsilon + j\omega)t}}{-(\epsilon + j\omega)} \right|_{0}^{\infty} - \left. \frac{e^{(\epsilon - j\omega)t}}{\epsilon - j\omega} \right|_{-\infty}^{0} \right]$$

$$= \lim_{\epsilon \to 0} \left[\frac{1}{\epsilon + j\omega} - \frac{1}{\epsilon - j\omega} \right] = \frac{2}{j\omega}. \tag{11.78}$$

Use of Eq. (11.78) in Eq. (11.76) gives

$$\mathcal{F}[u(t)] = \pi\, \delta(\omega) + \frac{1}{j\omega}.$$

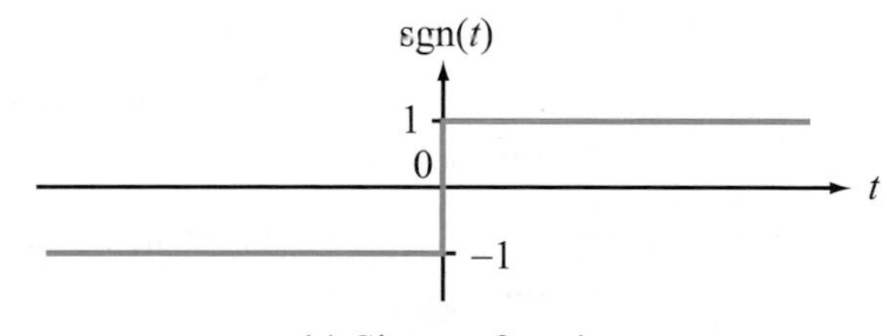

(a) Signum function

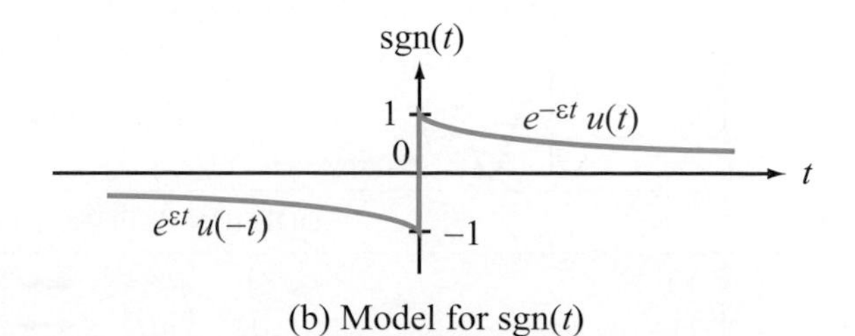

(b) Model for sgn(t)

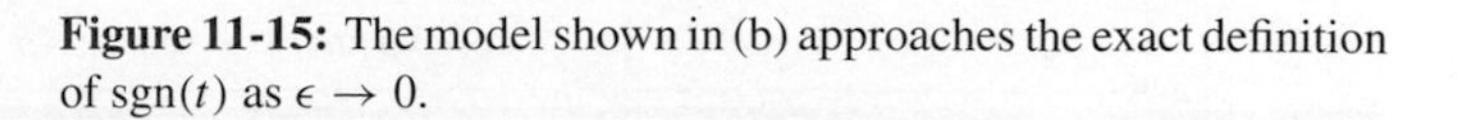

Figure 11-15: The model shown in (b) approaches the exact definition of $\mathrm{sgn}(t)$ as $\epsilon \to 0$.

Equivalently, the preceding result can be expressed in the form

$$u(t) \quad \leftrightarrow \quad \pi\,\delta(\omega) + \frac{1}{j\omega}. \tag{11.79}$$

Table 11-4 provides a list of commonly used time functions together with their corresponding Fourier transforms, and Table 11-5 offers a summary of the major properties of the Fourier transform—many of which resemble those we encountered earlier in Chapter 10 in connection with the Laplace transform.

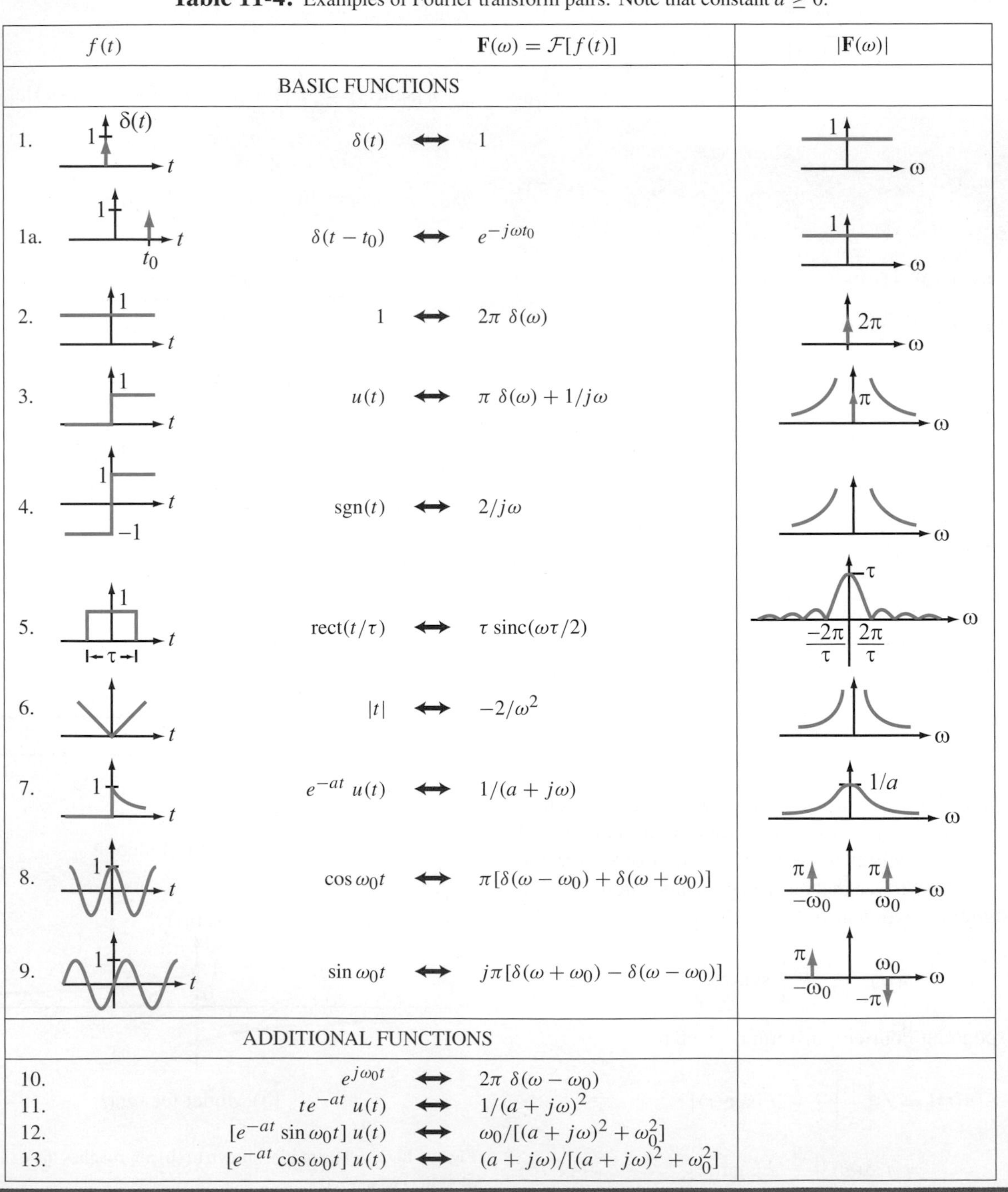

Table 11-4: Examples of Fourier transform pairs. Note that constant $a \geq 0$.

	$f(t)$		$\mathbf{F}(\omega) = \mathcal{F}[f(t)]$	$\lvert\mathbf{F}(\omega)\rvert$
	BASIC FUNCTIONS			
1.	$\delta(t)$	$\leftrightarrow$	1	
1a.	$\delta(t - t_0)$	$\leftrightarrow$	$e^{-j\omega t_0}$	
2.	1	$\leftrightarrow$	$2\pi\,\delta(\omega)$	
3.	$u(t)$	$\leftrightarrow$	$\pi\,\delta(\omega) + 1/j\omega$	
4.	$\text{sgn}(t)$	$\leftrightarrow$	$2/j\omega$	
5.	$\text{rect}(t/\tau)$	$\leftrightarrow$	$\tau\,\text{sinc}(\omega\tau/2)$	
6.	$\lvert t\rvert$	$\leftrightarrow$	$-2/\omega^2$	
7.	$e^{-at}\,u(t)$	$\leftrightarrow$	$1/(a + j\omega)$	
8.	$\cos\omega_0 t$	$\leftrightarrow$	$\pi[\delta(\omega - \omega_0) + \delta(\omega + \omega_0)]$	
9.	$\sin\omega_0 t$	$\leftrightarrow$	$j\pi[\delta(\omega + \omega_0) - \delta(\omega - \omega_0)]$	
	ADDITIONAL FUNCTIONS			
10.	$e^{j\omega_0 t}$	$\leftrightarrow$	$2\pi\,\delta(\omega - \omega_0)$	
11.	$te^{-at}\,u(t)$	$\leftrightarrow$	$1/(a + j\omega)^2$	
12.	$[e^{-at}\sin\omega_0 t]\,u(t)$	$\leftrightarrow$	$\omega_0/[(a + j\omega)^2 + \omega_0^2]$	
13.	$[e^{-at}\cos\omega_0 t]\,u(t)$	$\leftrightarrow$	$(a + j\omega)/[(a + j\omega)^2 + \omega_0^2]$	

Table 11-5: Major properties of the Fourier transform.

Property	$f(t)$		$\mathbf{F}(\omega) = \mathcal{F}[f(t)]$
1. Multiplication by a constant	$K\ f(t)$	$\leftrightarrow$	$K\ \mathbf{F}(\omega)$
2. Linearity	$K_1\ f_1(t) + K_2\ f_2(t)$	$\leftrightarrow$	$K_1\ \mathbf{F}_1(\omega) + K_2\ \mathbf{F}_2(\omega)$
3. Time scaling	$f(at)$	$\leftrightarrow$	$\frac{1}{\lvert a \rvert}\ \mathbf{F}\left(\frac{\omega}{a}\right)$
4. Time shift	$f(t - t_0)$	$\leftrightarrow$	$e^{-j\omega t_0}\ \mathbf{F}(\omega)$
5. Frequency shift	$e^{j\omega_0 t}\ f(t)$	$\leftrightarrow$	$\mathbf{F}(\omega - \omega_0)$
6. Time 1st derivative	$f' = \frac{df}{dt}$	$\leftrightarrow$	$j\omega\ \mathbf{F}(\omega)$
7. Time nth derivative	$\frac{d^n f}{dt^n}$	$\leftrightarrow$	$(j\omega)^n\ \mathbf{F}(\omega)$
8. Time integral	$\int_{-\infty}^{t} f(t)\ dt$	$\leftrightarrow$	$\frac{\mathbf{F}(\omega)}{j\omega}$
9. Frequency derivative	$t^n\ f(t)$	$\leftrightarrow$	$(j)^n \frac{d^n \mathbf{F}(\omega)}{d\omega^n}$
10. Modulation	$\cos\omega_0 t\ f(t)$	$\leftrightarrow$	$\frac{1}{2}[\mathbf{F}(\omega - \omega_0) + \mathbf{F}(\omega + \omega_0)]$
11. Convolution in t	$f_1(t) * f_2(t)$	$\leftrightarrow$	$\mathbf{F}_1(\omega)\ \mathbf{F}_2(\omega)$
12. Convolution in ω	$f_1(t)\ f_2(t)$	$\leftrightarrow$	$\frac{1}{2\pi}\ \mathbf{F}_1(\omega) * \mathbf{F}_2(\omega)$

Example 11-10: Fourier Transform Properties

Establish the validity of the time derivative and modulation properties of the Fourier transform (Properties 6 and 10 in Table 11-5).

Solution:

Time Derivative Property:

From Eq. (11.62b),

$$f(t) = \frac{1}{2\pi} \int_{-\infty}^{\infty} \mathbf{F}(\omega)\ e^{j\omega t}\ d\omega. \tag{11.80}$$

Differentiating both sides with respect to t gives

$$f'(t) = \frac{df}{dt} = \frac{1}{2\pi} \int_{-\infty}^{\infty} j\omega\ \mathbf{F}(\omega)\ e^{j\omega t}\ d\omega$$

$$= j\omega \left[\frac{1}{2\pi} \int_{-\infty}^{\infty} \mathbf{F}(\omega)\ e^{j\omega t}\ d\omega \right].$$

Hence, differentiating $f(t)$ in the time domain is equivalent to multiplying $\mathbf{F}(\omega)$ by $j\omega$ in the frequency domain as

$$f'(t) \leftrightarrow j\omega\ \mathbf{F}(\omega). \tag{11.81}$$

Time Modulation Property:

We start by multiplying both sides of Eq. (11.80) by $\cos\omega_0 t$, and for convenience, we change the dummy variable ω to ω':

$$\cos\omega_0 t\; f(t) = \frac{1}{2\pi}\int_{-\infty}^{\infty} \cos\omega_0 t\; \mathbf{F}(\omega')\, e^{j\omega' t}\, d\omega'.$$

Applying Euler's identity to $\cos\omega_0 t$ on the right-hand side leads to

$$\begin{aligned}\cos\omega_0 t\; f(t) &= \frac{1}{2\pi}\int_{-\infty}^{\infty}\left(\frac{e^{j\omega_0 t}+e^{-j\omega_0 t}}{2}\right)\mathbf{F}(\omega')\, e^{j\omega' t}\, d\omega' \\ &= \frac{1}{4\pi}\left[\int_{-\infty}^{\infty}\mathbf{F}(\omega')\, e^{j(\omega'+\omega_0)t}\, d\omega' \right. \\ &\qquad \left. + \int_{-\infty}^{\infty}\mathbf{F}(\omega')\, e^{j(\omega'-\omega_0)t}\, d\omega'\right].\end{aligned}$$

Upon making the substitution ($\omega = \omega'+\omega_0$) in the first integral and independently making the substitution ($\omega = \omega'-\omega_0$) in the second integral, we have

$$\begin{aligned}\cos\omega_0 t\; f(t) = \frac{1}{2}\left[\frac{1}{2\pi}\int_{-\infty}^{\infty}\mathbf{F}(\omega-\omega_0)\, e^{j\omega t}\, d\omega \right. \\ \left. + \frac{1}{2\pi}\int_{-\infty}^{\infty}\mathbf{F}(\omega+\omega_0)\, e^{j\omega t}\, d\omega\right],\end{aligned}$$

which can be cast in the abbreviated form

$$\cos\omega_0 t\; f(t) \;\longleftrightarrow\; \frac{1}{2}[\mathbf{F}(\omega-\omega_0)+\mathbf{F}(\omega+\omega_0)]. \tag{11.82}$$

Review Question 11-14: What is the Fourier transform of a dc voltage?

Review Question 11-15: "An impulse in the time domain is equivalent to an infinite number of sinusoids, all with equal amplitude." Is this a true statement? Can one construct an ideal impulse function?

Exercise 11-7: Use the entries in Table 11-4 to determine the Fourier transform of $u(-t)$.

Answer: $\mathbf{F}(\omega) = \pi\,\delta(\omega) - 1/j\omega$. (See ◎)

Exercise 11-8: Verify the Fourier transform expression for entry #10 in Table 11-4.

Answer: (See ◎)

11-6.7 Parseval's Theorem

If $f(t)$ represents the voltage across a 1-Ω resistor, then $f^2(t)$ represents the power dissipated in the resistor, and the integrated value of $f^2(t)$ over $[-\infty,\infty]$ represents the cumulative energy W expended in the resistor. Thus,

$$\begin{aligned}W &= \int_{-\infty}^{\infty} f^2(t)\, dt \\ &= \int_{-\infty}^{\infty} f(t)\left[\frac{1}{2\pi}\int_{-\infty}^{\infty}\mathbf{F}(\omega)\, e^{j\omega t}\, d\omega\right] dt,\end{aligned} \tag{11.83}$$

where one $f(t)$ was replaced with the inverse Fourier transform relationship given by Eq. (11.62b). By reversing the order of $f(t)$ and $\mathbf{F}(\omega)$, and reversing the order of integration, we have

$$\begin{aligned}W &= \frac{1}{2\pi}\int_{-\infty}^{\infty}\mathbf{F}(\omega)\left[\int_{-\infty}^{\infty} f(t)\, e^{j\omega t}\, dt\right] d\omega \\ &= \frac{1}{2\pi}\int_{-\infty}^{\infty}\mathbf{F}(\omega)\left[\int_{-\infty}^{\infty} f(t)\, e^{-j(-\omega)t}\, dt\right] d\omega \\ &= \frac{1}{2\pi}\int_{-\infty}^{\infty}\mathbf{F}(\omega)\,\mathbf{F}(-\omega)\, d\omega \\ &= \frac{1}{2\pi}\int_{-\infty}^{\infty}\mathbf{F}(\omega)\,\mathbf{F}^*(\omega)\, d\omega,\end{aligned} \tag{11.84}$$

where we used the ***reversal property*** of the Fourier transform (see Exercise 11-9) given as

$$\mathbf{F}(-\omega) = \mathbf{F}^*(\omega) \quad \text{(reversal property)}. \tag{11.85}$$

Technology Brief 23: Mapping the Entire World in 3-D

Mapping software has become increasingly indispensable in the 21st-century industrialized world. Giving verbal directions to someone's house has been replaced with directing them to MapQuest or Google Maps with the relevant address. Even more exciting, however, is the growing suite of 3-D ***virtual globe*** mapping software. Of these, arguably the most famous is the currently free Google Earth software. These packages allow the user to fly in virtual space around the world, into cities and remote areas and (in densely mapped areas) to view their own backyards, streets signs, and local landscapes. The tools are becoming an enabler for a new generation of armchair historians, archaeologists, demographers. They have already been used in search and rescue operations. How do these packages work? Where does this data come from? How is the world mapped?

Planes, Satellites, and Automobiles

Data for these packages is acquired by large, specialized companies that make use of satellites, aircraft, and (more recently) large fleets of specially equipped vans. The majority of data comes from several satellites orbiting earth financed by either national governments or private companies. For example, the U.S. National Aeronautics and Space Administration (NASA) has the long-standing Landsat 7 program which has 30-m imaging resolution and scans the earth in about 16 days. The European Space Agency's ERS and Envisat satellites perform similar functions. All of these satellites perform functions other than visual spectrum imaging; some have infrared sensors, radar sensors, temperature sensors, etc. Several commercial satellites are now in orbit whose primary function is to map the globe in high-resolution mode; these include Digital Globe's QuickBird and, in 2007, the WorldView-1 satellites. WorldView-1 revisits the same place approximately every 1.7 days and has a resolution of 50 cm—although not all data will be made publicly available. Most of these satellites maintain sun-synchronous orbits, which means that their orbits loop over or near the north and south poles and cross the equator twice on each loop. In this type of orbit, the satellite "visits" a given place at the same local time each visit, which is great for maintaining constant lighting for satellite images. Additionally, 2-D visual information is supplemented with digital elevation model (DEM) data collected by NASA's Shuttle Radar Topography Mission (SRTM). The SRTM (Fig. TF23-1) consisted of two radar antennas deployed on the space shuttle ***Endeavour*** during the 11-day mission of STS-99 in February 2000. A sample product is shown in Fig. TF23-2.

Figure TF23-1: The Shuttle Radar Topography Mission used an antenna located in the payload bay of the shuttle, and a second outboard antenna attached to the end of a 60-m mast. (Courtesy of NASA.)

Figure TF23-2: A shaded relief image of Mount St. Helens in the state of Washington. (Courtesy of NASA.)

Aircraft imaging complements the satellite data, although it is more expensive and available in limited areas. More recently, driven by growing demand for up-to-date mapping information, several companies have launched fleets of specially equipped vans with multiple cameras, laser distance sensors, and on-board computation to collect, merge, and store the data. Fig. TF23-3 shows one such vehicle developed by TeleAtlas. Hundreds of similar vehicles roam the earth; the cameras provide images over $360°$ around the vehicle, and a laser system measures important distances like bridge and building heights; GPS tracking hardware records the vehicle's position; and onboard computers synthesize everything and store it. As these vehicles visit more and more places, the 3-D map of the world continues to grow.

Imaging Software

All of this data is then compiled, corrected, and merged. This is not just a massive storage operation. Often, imagery comes from multiple sources that do not match exactly, there may be gaps between images and, very commonly, the color of the images must be corrected and made consistent. Fine-scale errors often are detectable with these map programs when data is incorrectly merged or have different dates; for example, pictures of a city might incorrectly show data from adjacent areas taken before and after major events, stitched together. Problems with incorporating 3-D topographical data with the visual information are common still. For public-accessible programs, not all data is taken at the same time nor with the same frequency; for example, Google Earth guarantees that image data is no more than three years old. More expensive commercial software is often more timely.

Beyond the compilation and merging of datasets, programs like Google Earth are increasingly integrating their software with both other software and mobile hardware. For example, Google Earth interfaces with both Wikipedia and the Google search engine as well as an increasing suite of information-providing programs. In a similar manner, some versions of commercial virtual globe programs can interface with GPS position-finding devices. Such programs take waypoints and tracks from the mobile GPS devices and merge them with available topographic, imaging, and other virtual globe datasets.

Figure TF23-3: A TeleAtlas van showing the imaging and laser equipment and the computation hardware inside the van.

The combination of Eqs. (11.83) and (11.84) can be written as

$$\int_{-\infty}^{\infty} f^2(t)\,dt = \frac{1}{2\pi}\int_{-\infty}^{\infty} |\mathbf{F}(\omega)|^2\,d\omega$$

(Parseval's theorem). (11.86)

Parseval's theorem states that the total energy in the time domain is equal to the total energy in the ω-domain.

Exercise 11-9: Verify the reversal property given by Eq. (11.85).

Answer: (See ◎)

11-7 Fourier versus Laplace

When a circuit is excited by a periodic voltage or current waveform (such as a square wave, a sequence of pulses, or any other repetitive pattern) the Fourier series technique is the method most proficient for analyzing the circuit performance. The solution procedure is not only straightforward to apply, but it also can be programmed easily into MATLAB® or MathScript.

The choice of solution method when the excitation is nonperiodic may be somewhat confusing or not so obvious. Beyond the time-domain differential-equation solution method, which in practice can accommodate only first- and second-order circuits, we have available to us two similar (but distinct) approaches, namely the Laplace transform technique of Chapter 10 and the Fourier transform technique introduced in the preceding two sections. Under what conditions should we choose the one transform technique over the other? The answer is: It depends on the properties of the excitation function, namely whether it is one-sided or two-sided and whether or not it has non-zero initial conditions.

A ***one-sided function*** $f(t)$ is defined over $[0, \infty]$ with $f(t) = 0$ for $t < 0$. In contrast, a ***two-sided function*** is defined over $[-\infty, \infty]$. Because the Laplace transform defined in Chapter 10 is one-sided, it is not applicable to two-sided functions, in which case the Fourier transform is the only viable option. Moreover, the concept of "initial conditions" has no relevant meaning in the world of two-sided functions. For one-sided functions, both transforms can be used if initial conditions are zero, but if they are not, the Laplace transform becomes the only practical choice because it is equipped to handle non-zero initial conditions, whereas the Fourier transform cannot (although the equivalent circuits of the capacitor and inductor can be modified in order to incorporate initial conditions).

In summary:

(a) If $f(t)$ is two-sided, the Fourier transform is the only option.

(b) If $f(t)$ is one-sided and has non-zero initial conditions, the Laplace transform is the only option.

(c) If $f(t)$ is one-sided and has zero initial conditions, either transform is equally applicable.

We also note that for one-sided functions with zero initial conditions, the Fourier transform $\mathbf{F}(\omega)$ is simply a special case of the Laplace transform $\mathbf{F}(\mathbf{s})$. Recall that $\mathbf{s} = \sigma + j\omega$, so when $\sigma = 0$,

$$\mathbf{F}(\omega) = \mathbf{F}(\mathbf{s})\big|_{\sigma=0}.$$

11-8 Circuit Analysis with Fourier Transform

As was mentioned earlier, the Fourier transform technique can be used to analyze circuits excited by either one-sided or two-sided nonperiodic waveforms—as long as the circuit has no initial conditions. The procedure (which is analogous to the Laplace transform technique) with $\mathbf{s}$ replaced by $j\omega$ is demonstrated through Example 11-11.

Example 11-11: RC Circuit

The RC circuit shown in Fig. 11-16(a) is excited by a voltage source $v_s(t)$. Apply Fourier analysis to determine $i_c(t)$ if: (a) $v_s = 10u(t)$, (b) $v_s(t) = 10e^{-2t}\,u(t)$, and (c) $v_s(t) = 10 + 5\cos 4t$ all measured in volts. The element values are $R_1 = 2$ kΩ, $R_2 = 4$ kΩ, and $C = 0.25$ mF.

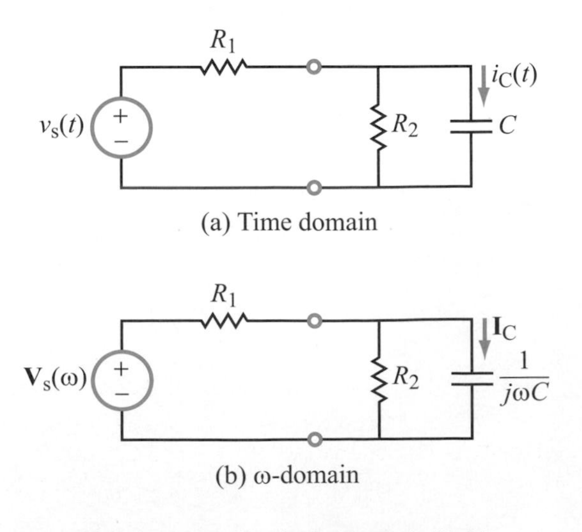

Figure 11-16: Circuits for Example 11-11.

Solution:

Step 1: Transfer Circuit to ω-Domain

In the frequency-domain circuit shown in Fig. 11-16(b), $\mathbf{V}_s(\omega)$ is the Fourier transform of $v_s(t)$.

Step 2: Determine $\mathbf{H}(\omega) = \mathbf{I}_C(\omega)/\mathbf{V}_s(\omega)$

Application of source transformation to the circuit in Fig. 11-16(b) followed with current division leads to

$$\mathbf{H}(\omega) = \frac{\mathbf{I}_C(\omega)}{\mathbf{V}_s(\omega)} = \frac{j\omega/R_1}{\dfrac{R_1+R_2}{R_1R_2C} + j\omega} = \frac{j0.5\omega \times 10^{-3}}{3+j\omega}. \quad (11.87)$$

Step 3: Solve for $\mathbf{I}_C(\omega)$ and $i_C(t)$

(a) $v_s(t) = 10u(t)$:

The corresponding Fourier transform per entry #3 in Table 11-4 is

$$\mathbf{V}_s(\omega) = 10\pi\ \delta(\omega) + \frac{10}{j\omega}.$$

The corresponding current is

$$\mathbf{I}_C(\omega) = \mathbf{H}(\omega)\ \mathbf{V}_s(\omega) = \frac{j5\pi\omega\ \delta(\omega) \times 10^{-3}}{3+j\omega} + \frac{5\times 10^{-3}}{3+j\omega}. \quad (11.88)$$

The inverse Fourier transform of $\mathbf{I}_C(\omega)$ is given by

$$i_C(t) = \frac{1}{2\pi}\int_{-\infty}^{\infty} \frac{j5\pi\omega\ \delta(\omega)\times 10^{-3}}{3+j\omega}\, e^{j\omega t}\, d\omega + \mathcal{F}^{-1}\left[\frac{5\times 10^{-3}}{3+j\omega}\right],$$

where we applied the formal definition of the inverse Fourier transform to the first term—because it includes a delta function—and the functional form to the second term—because we intend to use look-up entry #7 in Table 11-4. Accordingly,

$$i_C(t) = 0 + 5e^{-3t}\ u(t) \text{ mA}. \quad (11.89)$$

(b) $v_s(t) = 10e^{-2t}\ u(t)$:

By entry #7 in Table 11-4,

$$\mathbf{V}_s(\omega) = \frac{10}{2+j\omega}.$$

The corresponding current $\mathbf{I}_C(\omega)$ is given by

$$\mathbf{I}_C(\omega) = \mathbf{H}(\omega)\ \mathbf{V}_s(\omega) = \frac{j5\omega \times 10^{-3}}{(2+j\omega)(3+j\omega)}. \quad (11.90)$$

Application of partial fraction expansion (Section 10-5) gives

$$\mathbf{I}_C(\omega) = \frac{A_1}{2+j\omega} + \frac{A_2}{3+j\omega},$$

with

$$A_1 = (2+j\omega)\ \mathbf{I}_C(\omega)\big|_{j\omega=-2} = \left.\frac{j5\omega\times 10^{-3}}{3+j\omega}\right|_{j\omega=-2} = -10\times 10^{-3}$$

and

$$A_2 = (3+j\omega)\ \mathbf{I}_C(\omega)\big|_{j\omega=-3} = \left.\frac{j5\omega\times 10^{-3}}{2+j\omega}\right|_{j\omega=-3} = 15\times 10^{-3}.$$

Hence,

$$\mathbf{I}_C(\omega) = \left(\frac{-10}{2+j\omega} + \frac{15}{3+j\omega}\right)\times 10^{-3}$$

and

$$i_C(t) = (15e^{-3t} - 10e^{-2t})\ u(t) \text{ mA}. \quad (11.91)$$

(c) $v_s(t) = 10 + 5\cos 4t$:

By entries #2 and #8 in Table 11-4,

$$\mathbf{V}_s(\omega) = 20\pi\ \delta(\omega) + 5\pi[\delta(\omega-4) + \delta(\omega+4)],$$

and the capacitor current is

$$\mathbf{I}_C(\omega) = \mathbf{H}(\omega)\ \mathbf{V}_s(\omega) = \frac{j10\pi\omega\ \delta(\omega)\times 10^{-3}}{3+j\omega} + j2.5\pi\times 10^{-3}\left[\frac{\omega\ \delta(\omega-4)}{3+j\omega} + \frac{\omega\ \delta(\omega+4)}{3+j\omega}\right].$$

The corresponding time-domain current is obtained by applying Eq. (11.62b) as

$$\begin{aligned} i_C(t) &= \frac{1}{2\pi} \int_{-\infty}^{\infty} \frac{j10\pi\omega\, \delta(\omega) \times 10^{-3} e^{j\omega t}\, d\omega}{3+j\omega} \\ &+ \frac{1}{2\pi} \int_{-\infty}^{\infty} \frac{j2.5\pi\omega \times 10^{-3}}{3+j\omega}\, \delta(\omega-4)\, e^{j\omega t}\, d\omega \\ &+ \frac{1}{2\pi} \int_{-\infty}^{\infty} \frac{j2.5\pi\omega \times 10^{-3}}{3+j\omega}\, \delta(\omega+4)\, e^{j\omega t}\, d\omega \\ &= 0 + \frac{j5 \times 10^{-3} e^{j4t}}{3+j4} - \frac{j5 \times 10^{-3} e^{-j4t}}{3-j4} \\ &= 5 \times 10^{-3} \left(\frac{e^{j4t} e^{j36.9^\circ}}{5} + \frac{e^{-j4t} e^{-j36.9^\circ}}{5} \right) \\ &= 2\cos(4t + 36.9^\circ) \text{ mA}. \end{aligned} \quad (11.92)$$

Exercise 11-10: Determine $v_C(t)$, the voltage across the capacitor in Fig. 11-16(a) of Example 11-11, for each of the three voltage waveforms given in the example statement.

Answer: (a) $v_C(t) = \frac{10}{3} + \frac{20}{3}(1 - e^{-3t})\, u(t)$ V,
(b) $v_C(t) = 20(e^{-2t} - e^{-3t})\, u(t)$ V,
(c) $v_C(t) = \left[\frac{20}{3} + 2\cos(4t - 36.9^\circ)\right]$ V. (See)

11-9 Application Note: Phase Information

A sinusoidal waveform given by

$$v(t) = A\cos(\omega t + \phi) \quad (11.93)$$

is characterized by three parameters, namely its amplitude A, its angular frequency ω (with $\omega = 2\pi f$), and its phase angle ϕ. The role of A is straightforward; it determines the peak-to-peak swing of the waveform, and similarly, the role of ω also is easy to understand, as it defines the number of oscillations that the waveform goes through in 1 second. What about ϕ? At first glance, we might assign to ϕ a rather trivial role, because its only impact on the waveform is to specify in what direction and by how much the waveform is shifted in time relative to the waveform of $A\cos\omega t$. Whereas such an assignment may be quite reasonable in the case of the simple cosinusoidal waveform, we will demonstrate in this section that in the general case of a more elaborate waveform, the phase part of the waveform carries information that is equally important as that contained in the waveform's amplitude.

Let us consider the rectangular pulse shown in Fig. 11-17(a). According to entry #5 in Table 11-4, the pulse and its Fourier transform are given by

$$\text{rect}\left(\frac{t}{\tau}\right) \longleftrightarrow \mathbf{F}(\omega) = \tau \operatorname{sinc}\left(\frac{\omega\tau}{2}\right), \quad (11.94)$$

where τ is the pulse length and the sinc function is defined by Eq. (11.54). By defining $\mathbf{F}(\omega)$ as

$$\mathbf{F}(\omega) = |\mathbf{F}(\omega)| e^{j\phi(\omega)}, \quad (11.95)$$

we determine that the ***phase spectrum*** $\phi(\omega)$ can be ascertained from

$$e^{j\phi(\omega)} = \frac{\mathbf{F}(\omega)}{|\mathbf{F}(\omega)|} = \frac{\operatorname{sinc}(\omega\tau/2)}{|\operatorname{sinc}(\omega\tau/2)|}. \quad (11.96)$$

The quantity on the right-hand side of Eq. (11.96) is always equal to $+1$ or -1. Hence, $\phi(\omega) = 0$ when $[\sin(\omega\tau/2)]/(\omega\tau/2)$ is positive and $\pm 180^\circ$ when it is negative. The amplitude and phase spectra of the rectangular pulse are displayed in Figs. 11-17(b) and (c), respectively.

11-9.1 2-D Spatial Transform

When the independent variable is t, measured in seconds, the corresponding independent variable in the Fourier transform domain is ω, measured in radians/s. The (t, ω) correspondence has analogies in other domains, such as the spatial domain. In fact, in the case of planar images, we deal with two spatial dimensions—rather than just one—which we shall label x and y. Accordingly, the image intensity may vary with both x and y and should be generalized to $f(x, y)$. Moreover, since $f(x, y)$ is a function of two variables, so is its Fourier transform, which we will call the ***two-dimensional Fourier transform*** $\mathbf{F}(\omega_x, \omega_y)$, where ω_x and

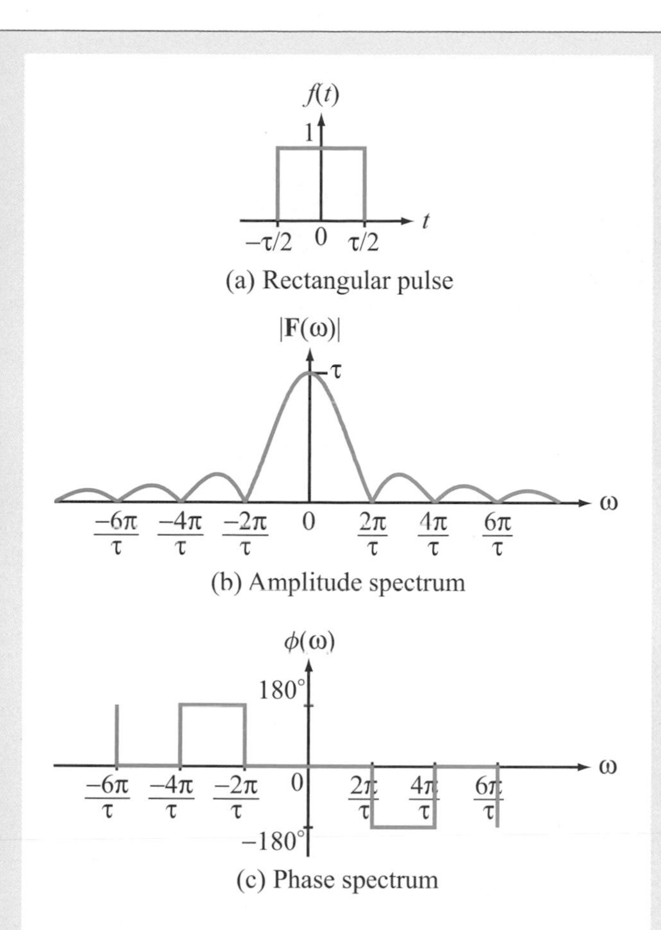

Figure 11-17: (a) Rectangular pulse, and corresponding (b) amplitude spectrum and (c) phase spectrum.

ω_y are called ***spatial frequencies***. If x and y are measured in meters, ω_x and ω_y will have units of radians/m. With digital images, x and y are measured in pixels, in which case ω_x and ω_y will have units of radians/pixel. Upon extending the Fourier transform definition given by Eq. (11.62) to the 2-D case—as well as replacing the time dimension with spatial dimensions—we have

$$\mathbf{F}(\omega_x, \omega_y) = \mathcal{F}[f(x, y)] = \int_{-\infty}^{\infty} f(x, y)\, e^{-j\omega_x x} e^{-j\omega_y y}\, dx\, dy, \tag{11.97a}$$

$$f(x, y) = \frac{1}{(2\pi)^2} \int_{-\infty}^{\infty} \mathbf{F}(\omega_x, \omega_y)\, e^{j\omega_x x} e^{j\omega_y y}\, d\omega_x\, d\omega_y. \tag{11.97b}$$

By way of an example, let us consider the white square shown in Fig. 11-18(a). If we assign an amplitude of 1 to the white part of the image and 0 to the black part, the variation across the image along the x-direction is analogous to that representing the time-domain pulse of Fig. 11-17(a), and the

(a) White square image

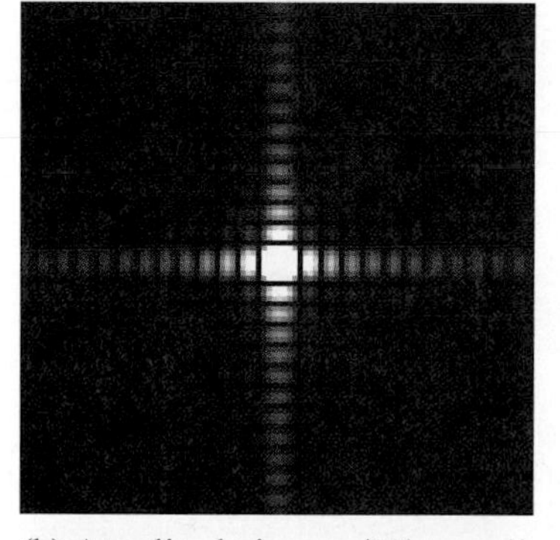

(b) Amplitude image $|\mathbf{F}(\omega_x,\omega_y)|$

(c) Phase image $\phi(\omega_x,\omega_y)$

Figure 11-18: (a) Grayscale image of a white square in a black background, (b) amplitude spectrum, and (c) phase spectrum.

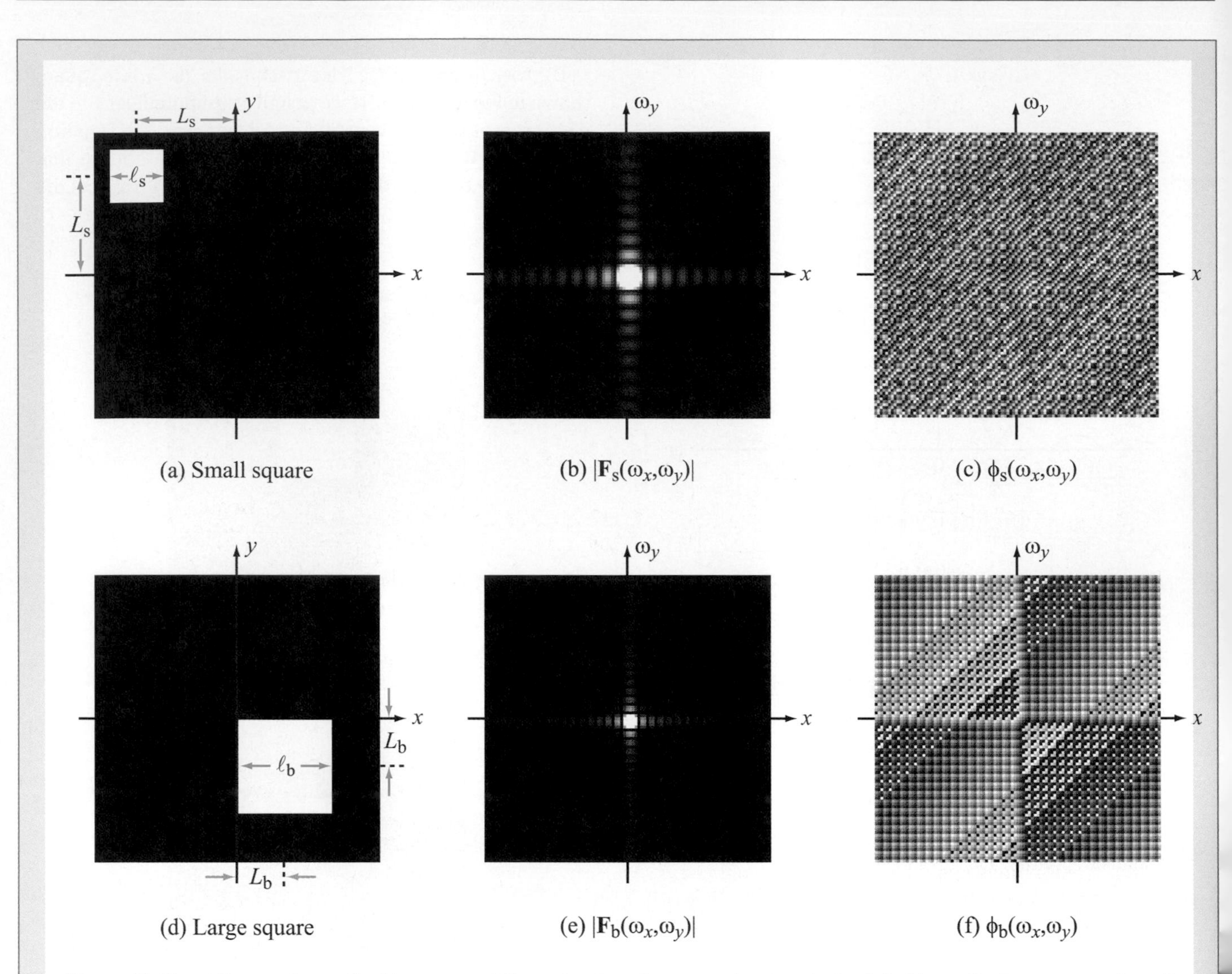

Figure 11-19: (a) Image of a small white square of dimension ℓ_s and center at $(x, y) = (-L_s, L_s)$, (b) amplitude spectrum of small square, (c) phase spectrum of small square, (d) image of a big white square of dimension ℓ_b and center at $(L_b, -L_b)$, (e) amplitude spectrum of large square, and (f) phase spectrum of large square.

same is true along y. Hence, the white square represents the superposition of two pulses, one along x and another along y, and is given by

$$f(x, y) = \text{rect}\left(\frac{x}{\ell}\right) \text{rect}\left(\frac{y}{\ell}\right), \tag{11.98}$$

where ℓ is the length of the square sides. Application of Eq. (11.97a) leads to

$$\mathbf{F}(\omega_x, \omega_y) = \ell^2 \,\text{sinc}\left(\frac{\omega_x \ell}{2}\right) \text{sinc}\left(\frac{\omega_y \ell}{2}\right). \tag{11.99}$$

The amplitude and phase spectra associated with the expression given by Eq. (11.99) are displayed in grayscale format in Fig. 11-18(b) and (c), respectively. For the amplitude spectrum, white represents the peak value of $|\mathbf{F}(\omega_x, \omega_y)|$ and black represents $|\mathbf{F}(\omega_x, \omega_y)| = 0$. The phase spectrum $\phi(\omega_x, \omega_y)$ varies between $-180°$ and $180°$, so the grayscale was defined such that white corresponds to $+180°$ and black to $-180°$. The tonal variations along ω_x and ω_y are equivalent to the patterns depicted in Figs. 11-17(b) and (c) for the rectangular pulse.

11-9.2 Amplitude and Phase Spectra

Next, we consider two white squares, of different sizes and at different locations, as shown in Fig. 11-19(a) and (d). The small square is of side ℓ_s and its center is at $(-L_s, +L_s)$, relative to the center of the image. In contrast, the big square is of side ℓ_b and located in the fourth quadrant with its center at $(+L_b, -L_b)$. Their corresponding functional expressions are

$$f_s(x, y) = \text{rect}\left(\frac{x + L_s}{\ell_s}\right) \text{rect}\left(\frac{x_y - L_s}{\ell_s}\right), \quad (11.100a)$$

$$f_b(x, y) = \text{rect}\left(\frac{x - L_b}{\ell_b}\right) \text{rect}\left(\frac{x_y + L_b}{\ell_b}\right). \quad (11.100b)$$

In view of property #4 in Table 11-5, the corresponding 2-D transforms are given by

$$\mathbf{F}_s(\omega_x, \omega_y) = \quad (11.101a)$$
$$\ell_s^2 e^{j\omega_x L_s} \text{sinc}\left(\frac{\omega_x \ell_s}{2}\right) e^{-j\omega_y L_s} \text{sinc}\left(\frac{\omega_y \ell_s}{2}\right),$$

$$\mathbf{F}_b(\omega_x, \omega_y) = \quad (11.101b)$$
$$\ell_b^2 e^{-j\omega_x L_b} \text{sinc}\left(\frac{\omega_x \ell_b}{2}\right) e^{j\omega_y L_b} \text{sinc}\left(\frac{\omega_y \ell_b}{2}\right).$$

Associated with $\mathbf{F}_s(\omega_x, \omega_y)$ are amplitude and phase spectra defined by

$$|\mathbf{F}_s(\omega_x, \omega_y)|$$

and

$$e^{j\phi_s(\omega_x, \omega_y)} = \frac{\mathbf{F}_s(\omega_x, \omega_y)}{|\mathbf{F}_s(\omega_x, \omega_y)|}, \quad (11.102)$$

and similar definitions apply to the amplitude and phase of $\mathbf{F}_b(\omega_x, \omega_y)$. The four 2-D spectra are displayed in Fig. 11-19.

11-9.3 Image Reconstruction

Figure 11-19 contains 2-D spectra of $|\mathbf{F}_s|$, ϕ_s, $|\mathbf{F}_b|$, and ϕ_b. If we were to apply the inverse Fourier transform to $|\mathbf{F}_s|e^{j\phi_s}$, we would reconstruct the original image of the small square, and similarly, application of the inverse transform to $|\mathbf{F}_b|e^{j\phi_b}$ would generate the image of the big square. Neither result would be a surprise, but what if we were to "mix" amplitude and phase spectra? That is, what would we get if in the reconstruction process we were to apply the inverse Fourier transform to $|\mathbf{F}_b|e^{j\phi_s}$, which contains the amplitude spectrum of the big square and the phase spectrum of the small square. Would we still obtain a square, what size would it be, and where will it be located? The result of such an experiment is displayed in Fig. 11-20(a). We observe that the dominant feature in the reconstructed image is still a square, but neither its size nor its location match those of the square in Fig. 11-19(d). In fact, the location corresponds to that of the small square in Fig. 11-19(a). Similarly, in Fig. 11-20(b) we display the image reconstructed by applying the inverse Fourier transform to $|\mathbf{F}_s|e^{j\phi_b}$. In both cases, the location of the square in the reconstructed image is governed primarily by the phase spectrum.

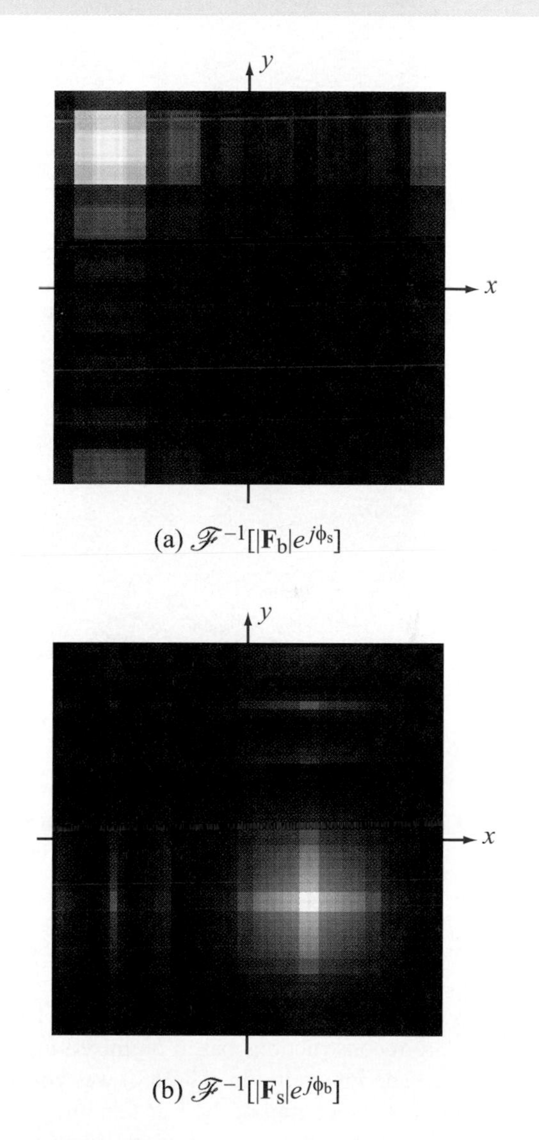

Figure 11-20: Reconstructed "mixed" images: (a) Although not sharp, the location and size of the small square are governed primarily by the phase information associated with the small square; (b) the phase information of the big square defines its location and approximate shape in the reconstructed image.

(a) Original Einstein image

(b) Original Mona Lisa image

(c) Reconstructed image based on Einstein amplitude and Mona Lisa phase

(d) Reconstructed image based on Mona Lisa amplitude and Einstein phase

Figure 11-21: In the reconstructed images, the phase spectrum exercised the dominant role, even more so than the amplitude spectrum. [After A. V. Oppenheim and J. S. Lim, "The importance of phase in signals," *Proceedings of the IEEE,* v. 69, no. 5, May 1981, pp. 529–541.]

Instead of squares, let us explore what happens when we use more complex images. Figures 11-21(a) and (b) are images of Albert Einstein and the Mona Lisa. The other two images are reconstructions based on mixed amplitude/phase spectra. The image in Fig. 11-21(c) was constructed using the amplitude spectrum of the Einstein image and the phase spectrum of the Mona Lisa image. Even though it contains the amplitude spectrum of the Einstein image, it fails to reproduce an image that resembles the original, but successfully reproduces an image with the likeness of the original Mona Lisa image. Thus, in this case, the phase spectrum has proven to be not only important, but even more important than the amplitude information. Further confirmation of the importance of phase information is evidenced by the image in Fig. 11-21(d), which reproduces a likeness of the Einstein image even though only the phase spectrum of the Einstein image was used in the reconstruction process.

11-9.4 Image Reconstruction Recipe

To perform a 2-D Fourier transform or an inverse transform on a black-and-white image, you can use MATLAB®, Mathematica, MathScript, or similar software. The process entails the following steps:

1. Your starting point has to be a digital image, so if your image is a hard copy print, you will need to scan it to convert it into digital format. Consisting of $M \times N$ pixels, the digital image is equivalent to an $M \times N$ matrix. Associated with an individual pixel (m, n) is a grayscale intensity $I(m, n)$.

2. In MATLAB® software, the command *fft(x)* generates the one-dimensional Fast Fourier transform of vector x, and similarly, *fft2(x,y)* generates the two-dimensional Fast Fourier transform of matrix (x, y). The outcome of the 2-D FFT is an $M \times N$ matrix whose elements we will designate as

$$\mathbf{F}(m,n) = A(m,n) + jB(m,n),$$

where $A(m,n)$ and $B(m,n)$ are the real and imaginary parts of $\mathbf{F}(m,n)$, respectively. In the frequency domain, coordinates (m,n) represent frequencies ω_x and ω_y, respectively.

3. The magnitude and phase matrices of the 2-D FFT can be generated from

$$|\mathbf{F}(m,n)| = [A^2(m,n) + B^2(m,n)]^{1/2},$$

$$\phi(m,n) = \tan^{-1}\left[\frac{B(m,n)}{A(m,n)}\right].$$

4. To display the matrices associated with $|\mathbf{F}(m,n)|$ and $\phi(m,n)$ as grayscale images, with zero frequency located at the center of the image (as opposed to having the dc component located at the bottom left corner of the matrix), it is necessary to apply the command ***fftshift*** to each of the images before displaying it.

5. Reconstruction back to the spatial domain (x, y) entails using the command ***ifft2*** on $\mathbf{F}(m,n)$, followed by ***fftshift***.

6. If two images are involved, $I_1(x,y)$ and $I_2(x,y)$, with corresponding 2-D FFTs $\mathbf{F}_1(m,n)$ and $\mathbf{F}_2(m,n)$, respectively, reconstruction of a mixed transform composed of the magnitude of one of the transforms and the phase of the other one will require a prerequisite step prior to applying *ifft2*. The artificial FFT composed of the magnitude of $\mathbf{F}_1(m,n)$ and the phase of $\mathbf{F}_2(m,n)$, for example, is given by

$$\mathbf{F}_3(m,n) = |\mathbf{F}_1(m,n)|\cos\phi_2(m,n) + j|\mathbf{F}_2(m,n)|\sin\phi_2(m,n).$$

11-10 Multisim: Mixed-Signal Circuits and the Sigma-Delta Modulator

Historically, circuit designers tended to fall into two broad classes: those who designed digital circuits and those who designed analog circuits. As a broad generalization, digital-circuit designers built logic gates, computational elements, memories, and so on, whereas analog designers tended to work on circuits that interfaced with the non-circuit world: amplifiers, drivers, radio frequency circuits, analog-to-digital converters, among others. Moreover, digital designers tended to have more comprehensive and powerful software design tools, mainly because digital circuits could be abstracted into modules that made hierarchical analysis possible. For example, a transistor could be modeled as a simple switch, several of these switches could be wired together as a simple logic gate, many logic gates could be wired together to make a counter or a memory, and so on, all of which can be readily modeled as "black boxes" in software. Analog circuits, by contrast, defied this type of compartmentalization due to feedback loops, non-linear behavior, and complex topologies; this made analog design almost an art form.

Advances in silicon fabrication technologies have now blurred the line between these two worlds considerably. A new generation of circuits, known as ***mixed-signal circuits***, contains elements of both worlds (Fig. TF20-1). This exciting area combines the power of analog designs with the scalability, modularity and computational power of digital circuits. Modern analog-to-digital conversion (ADC) and digital-to-analog conversion (DAC) circuits, cell phone communication circuits, software radio, internet routers, and audio synthesis circuits are all examples of mixed signal circuits. The advantages are numerous. Consider software radio, for example. We saw in Section 9-8 how a superheterodyne receiver works, which is a perfect example of a multi-stage analog circuit. But what if many of the functions of the superheterodyne receiver could be performed by digital circuits instead? What advantages might there be? One obvious advantage is the introduction of computational "intelligence" into the radio itself. If the receiver is designed, in part or in whole, with digital circuits, these circuits can be built around computation and memory. Programs can be loaded that allow the radio to change its power consumption, transmission patterns and protocols based on user or environmental parameters; this is often known as ***cognitive radio***. The integration of analog and digital circuits comes with certain drawbacks, however. Design remains a challenging, and highly paid, exercise. Design and testing software for mixed signal circuits is nowhere near as advanced as that for digital circuits. The fabrication of these circuits is often confined to specialized processes not compatible with standard, digital-processor fabrication methodologies (although this is rapidly changing).

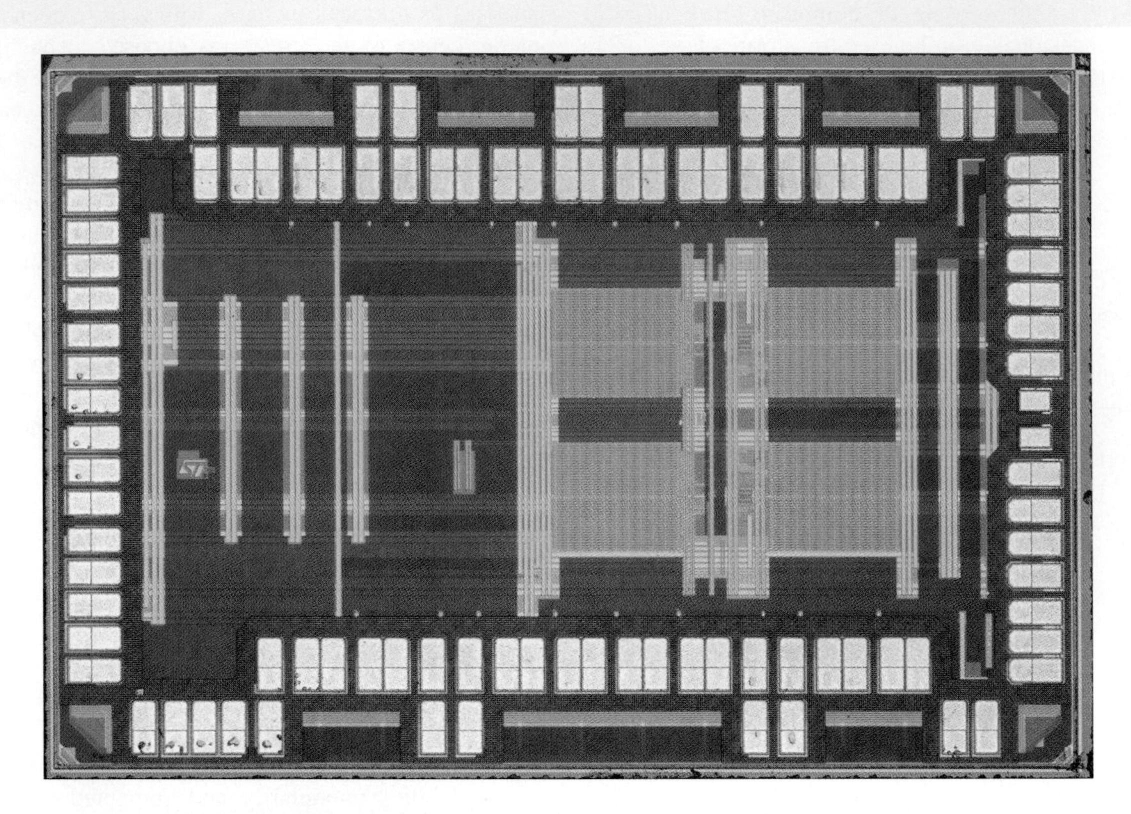

Figure 11-22: This mixed signal chip implements a highly reconfigurable RF receiver based on a down-converting Sigma-Delta A/D (courtesy Renaldi Winoto and Prof. Borivoje Nikolic, U.C. Berkeley)

11-10.1 The Sigma-Delta ($\Sigma\Delta$) Modulator and Analog-to-Digital Converters

So far in this book we have used Multisim to model amplifiers, digital circuits, filters, resonators, and circuits that employ feedback. In this section, we will put it all together and show you how to design a very useful circuit, the ***Sigma-Delta ($\Sigma\Delta$) Modulator***, which since its early development by Inose and Yasuda in 1962 has now become a standard tool for the design of inexpensive analog-to-digital converters (ADCs).

There are many ways to convert an analog waveform into a digital sequence of pulses. The classic ADC circuit takes a time-varying analog voltage, $v_{\text{in}}(t)$, and produces a corresponding time-varying digital output consisting of a number of bits (V_{out0}, V_{out1}, etc.). Figure 11-23(a) shows the process schematically. Here, a linearly increasing voltage is fed into a 4-bit ADC; as the input voltage changes with time, the four digital output bits change their state (either "0" or "1"). All of the pulses have the same duration and can change states instantaneously. With 4 binary bits, we can construct 16 different values (i.e., 0000, 0001, ..., 1111), so this 4-bit ADC converts any input voltage to one of 2^4 or, equivalently, 16 different digital values. Modern ADCs commonly have 12, 16 or even 24 output bits, giving

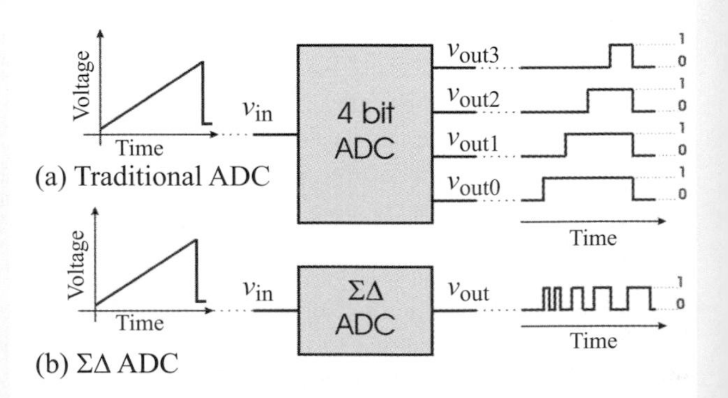

Figure 11-23: (a) A traditional 4-bit ADC converts an analog input voltage and produces 4 digital output bits; (b) a $\Sigma\Delta$ ADC generates a pulse train where the pulse duration is governed by the magnitude of the input voltage.

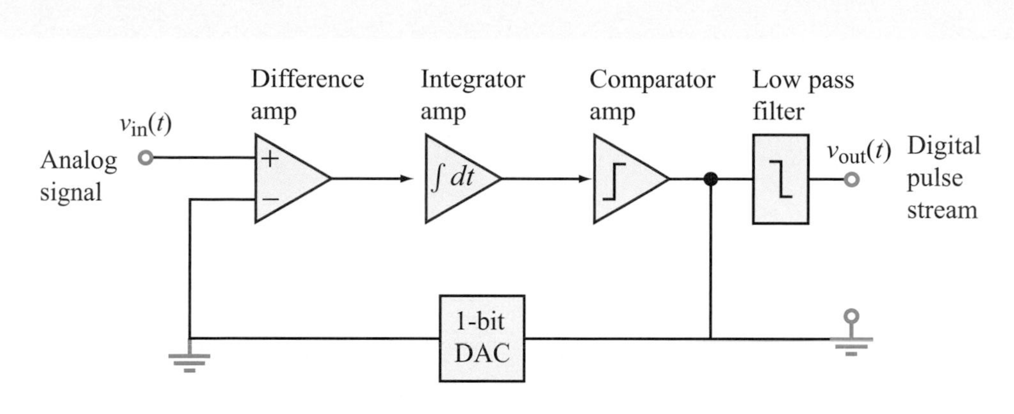

Figure 11-24: Block diagram of a ΣΔ modulator.

them very high resolution (e.g., $2^{24} = 16,777,216$ different values!). One usually trades off speed for resolution; the bits in a fast 12-bit ADC integrated circuit can change states once every 2 microseconds, which means that the ADC can measure the input voltage approximately 500,000 times per second.

Unlike conventional ADCs, the ΣΔ ***ADC*** generates an output consisting of a single digital bit. The ***duration*** of the voltage pulse, however, depends on the value of the input voltage (Fig. 11-23(b)), thereby encoding the magnitude of the input voltage into the duration of a single pulse, instead of encoding it into the binary states of several pulses (bits) of equal duration. The ΣΔ modulator is particularly attractive for designing and building inexpensive ADC circuits because it can (a) be made using digital components, which are less expensive to build and easier to test than analog components, and (b) the digital components can be re-programmed and modified by the user using firmware. Although invented in the 1960s, the ΣΔ modulator was not used commercially until digital CMOS processes became sufficiently fast (to produce time-varying output pulses faster than the changes exhibited by the input signal), small (enabling the mixing of digital with analog circuit components) and inexpensive to fabricate.

The entire circuit (Fig. 11-24) can be built from analog components introduced in this book, namely a subtractor, an integrator, a comparator, a 1-bit DAC and a low-pass filter. In real implementations, most of these components are replaced with digital substitutes, often re-programmable during operation. Thus, one could replace the analog filter with an adjustable digital filter, the integrator with a digital integrator, and so on. The only analog component in modern ΣΔs is usually the DAC.

11-10.2 How the ΣΔ Works

Table 11-6 shows the individual sub-circuits of the ΣΔ modulator and Fig. 11-25 displays the complete circuit, all drawn in Multisim. Our basic ΣΔ circuit takes an analog input, $v_{in}(t)$, subtracts from it a feedback signal, $v_{bit}(t)$, then integrates this signal, producing $v_{int}(t)$. The integrated signal is then compared with a reference voltage (in our case, 0 V) and produces $v_{out}(t)$. The output of the comparator, $v_{out}(t)$, can only have two values: V_{DD} or 0, where V_{DD} is the dc power supply voltage of the comparator. Hence, $v_{out}(t)$ is a time-varying digital signal. Note that $v_{out}(t)$ is also sent to a 1-bit DAC that converts the digital signal to an analog signal, $v_{bit}(t)$, which is fed back to the subtractor.

The overall functionality of the ΣΔ modulator is illustrated by Fig. 11-26 for an input signal composed of a 1-Hz sinusoid with an amplitude of 4 V. We observe that the corresponding output, $v_{out}(t)$, consists of a sequence of pulses whose durations are proportional to the instantaneous level of the input voltage, $v_{in}(t)$. Thus, the ΣΔ circuit encodes amplitude information contained in an analog signal into pulse-duration information in a digital sequence. After transmission of $v_{out}(t)$ through downstream digital circuits, the original information can be retrieved by measuring the durations of the pulses. This can be accomplished by a digital counter, either in hardware (using a counter circuit) or in software on a microcontroller. In hardware, transitions can be detected with ***Schmidt triggers*** or similar edge detectors, and the counter is made to "count" the duration between transitions.

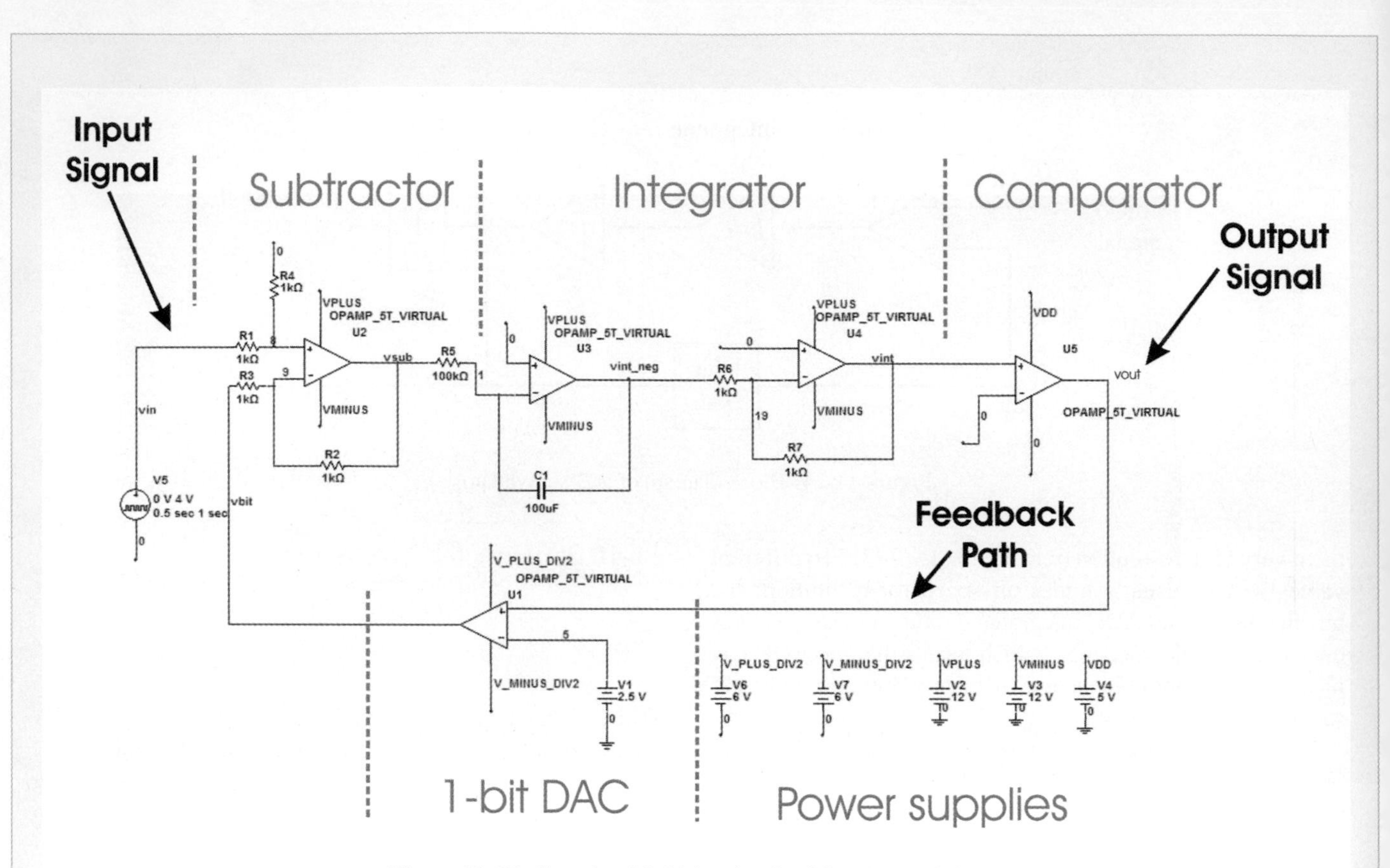

Figure 11-25: Complete Multisim circuit of the $\Sigma\Delta$ modulator.

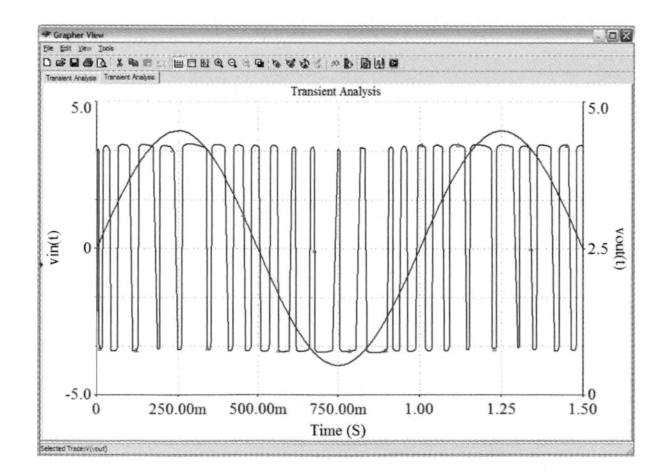

Figure 11-26: A 1-Hz sinusoidal ac signal, $v_{\text{in}}(t)$, blue trace, is converted to a series of pulses at the output, $v_{\text{out}}(t)$, red trace, by the Sigma Delta modulator. Note that the duration of the pulses is related to the instantaneous level of voltage $v_{\text{in}}(t)$.

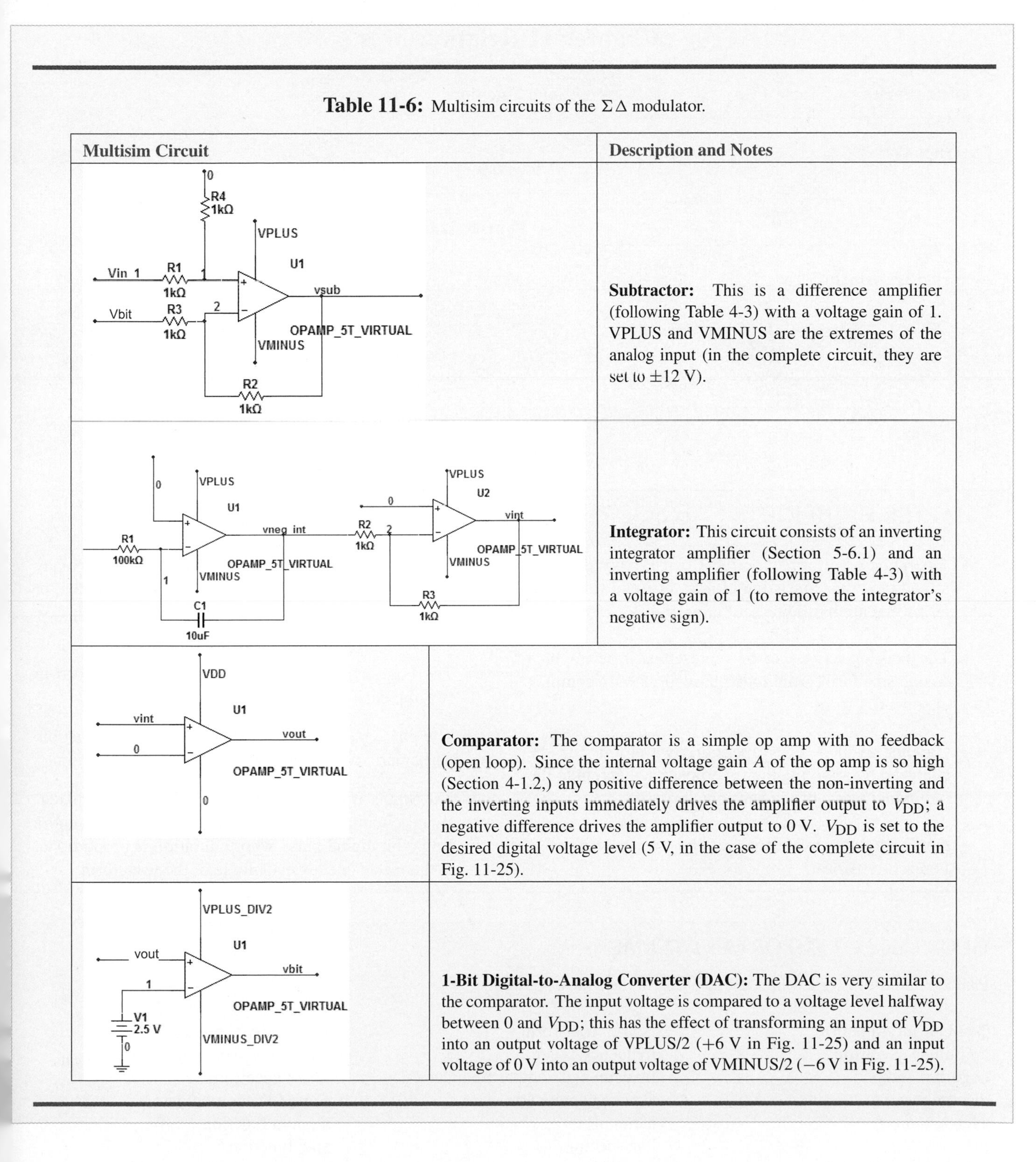

Table 11-6: Multisim circuits of the ΣΔ modulator.

Multisim Circuit	Description and Notes
(Subtractor circuit)	**Subtractor:** This is a difference amplifier (following Table 4-3) with a voltage gain of 1. VPLUS and VMINUS are the extremes of the analog input (in the complete circuit, they are set to ±12 V).
(Integrator circuit)	**Integrator:** This circuit consists of an inverting integrator amplifier (Section 5-6.1) and an inverting amplifier (following Table 4-3) with a voltage gain of 1 (to remove the integrator's negative sign).
(Comparator circuit)	**Comparator:** The comparator is a simple op amp with no feedback (open loop). Since the internal voltage gain A of the op amp is so high (Section 4-1.2,) any positive difference between the non-inverting and the inverting inputs immediately drives the amplifier output to V_{DD}; a negative difference drives the amplifier output to 0 V. V_{DD} is set to the desired digital voltage level (5 V, in the case of the complete circuit in Fig. 11-25).
(DAC circuit)	**1-Bit Digital-to-Analog Converter (DAC):** The DAC is very similar to the comparator. The input voltage is compared to a voltage level halfway between 0 and V_{DD}; this has the effect of transforming an input of V_{DD} into an output voltage of VPLUS/2 (+6 V in Fig. 11-25) and an input voltage of 0 V into an output voltage of VMINUS/2 (−6 V in Fig. 11-25).

Chapter 11 Relationships

Fourier Series Table 11-3

Average Power

$$P_{\text{av}} = V_{\text{dc}} I_{\text{dc}} + \frac{1}{2} \sum_{n=1}^{\infty} V_n I_n \cos(\phi_{v_n} - \phi_{i_n})$$

Fourier Transform

$$\mathbf{F}(\omega) = \mathcal{F}[f(t)] = \int_{-\infty}^{\infty} f(t)\, e^{-j\omega t}\, dt$$

$$f(t) = \mathcal{F}^{-1}[\mathbf{F}(\omega)] = \frac{1}{2\pi} \int_{-\infty}^{\infty} \mathbf{F}(\omega)\, e^{j\omega t}\, d\omega$$

sinc Function $\text{sinc}(x) = \dfrac{\sin x}{x}$

Properties of Fourier Transform Table 11-5

2-D Fourier Transform

$$\mathbf{F}(\omega_x, \omega_y) = \mathcal{F}[f(x,y)] = \int_{-\infty}^{\infty} f(x,y)\, e^{-j\omega_x x} e^{-j\omega_y y}\, dx\, dy$$

$$f(x,y) = \frac{1}{(2\pi)^2} = \int_{-\infty}^{\infty} \mathbf{F}(\omega_x, \omega_y)\, e^{j\omega_x x} e^{j\omega_y y}\, d\omega_x\, d\omega_y$$

CHAPTER HIGHLIGHTS

- A periodic waveform of period T can be represented by a Fourier series consisting of a dc term and sinusoidal terms that are harmonic multiples of $\omega_0 = 2\pi/T$.
- The Fourier series can be represented in terms of a cosine/sine form, amplitude/phase form, and a complex exponential form.
- Circuits excited by a periodic waveform can be analyzed by applying the superposition theorem to the individual terms of the harmonic series.
- Non-periodic waveforms can be represented by a Fourier transform.
- Upon transforming the circuit to the frequency domain, the circuit can be analyzed for the desired voltage or current of interest and then the result can be inverse transformed to the time domain.
- The Fourier transform technique can be extended to two-dimensional spatial images.
- The phase part of a signal contains vital information, particularly with regard to timing or spatial location.
- The Sigma-Delta modulator is an example of a mixed-signal circuit. It converts an analog waveform into a single-bit digital pulse whose duration is proportional to the instantaneous magnitude of the waveform.

GLOSSARY OF IMPORTANT TERMS

Provide definitions or explain the meaning of the following terms:

2-D Fourier transform
amplitude spectrum
cognitive radio
dc component
even symmetry
fft
fftshift
Fourier coefficient
Fourier series
Fourier transform
frequency spectrum
fundamental angular frequency
Gibbs phenomenon
harmonic
line spectra
mixed signal circuit
nulls
odd symmetry
periodicity property
periodic waveform
phase spectrum
Sigma-Delta modulator
signum function
sinc function
spatial frequency
truncated series

PROBLEMS

Sections 11-1 and 11-2: Fourier Series

For each of the waveforms in Problems 11.1 through 11.10:

(a) Determine if the waveform has dc, even, or odd symmetry.

(b) Obtain its cosine/sine Fourier series representation.

(c) Convert the representation to amplitude/phase format and plot the line spectra for the first five non-zero terms.

(d) Use MATLAB® software to plot the waveform using a truncated Fourier series representation with $n_{\max} = 100$.

11.1 Waveform in Fig. P11.1 with $A = 10$.

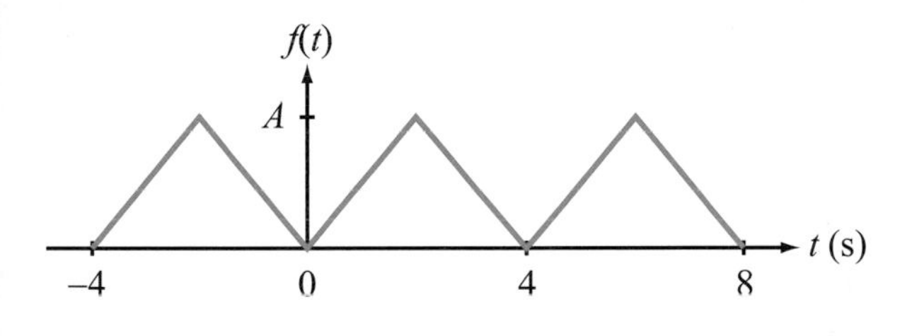

Figure P11.1: Waveform for Problem 11.1.

11.2 Waveform in Fig. P11.2 with $A = 4$.

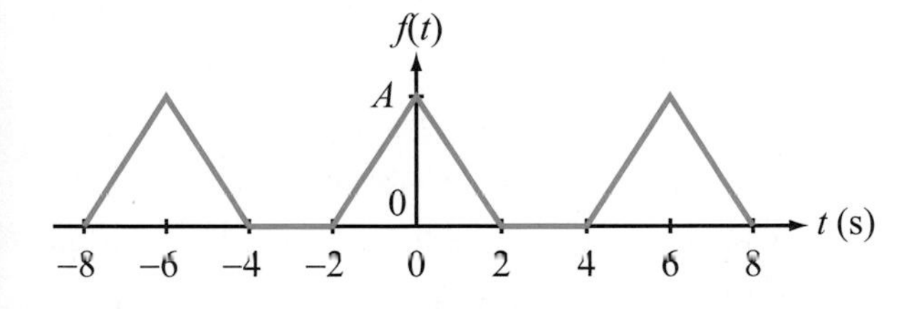

Figure P11.2: Waveform for Problem 11.2.

11.3 Waveform in Fig. P11.3 with $A = 6$.

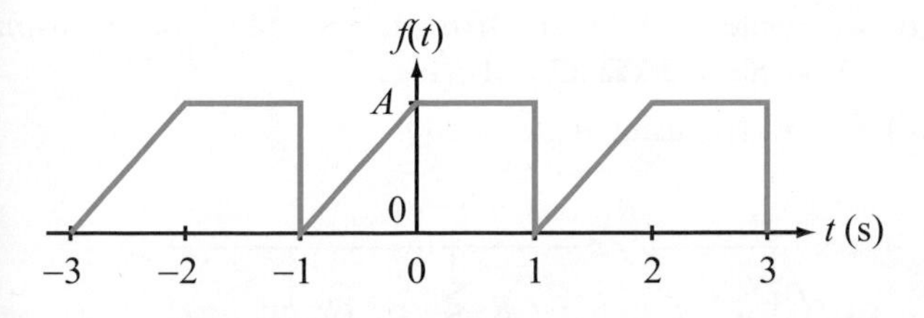

Figure P11.3: Waveform for Problem 11.3.

11.4 Waveform in Fig. P11.4 with $A = 10$.

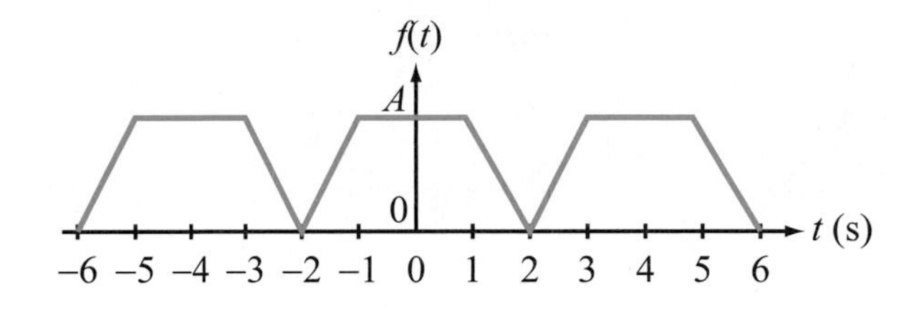

Figure P11.4: Waveform for Problem 11.4.

***11.5** Waveform in Fig. P11.5 with $A = 20$.

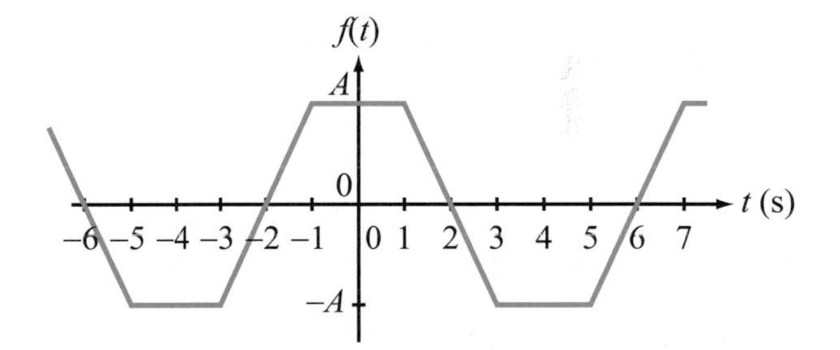

Figure P11.5: Waveform for Problem 11.5.

11.6 Waveform in Fig. P11.6 with $A = 100$.

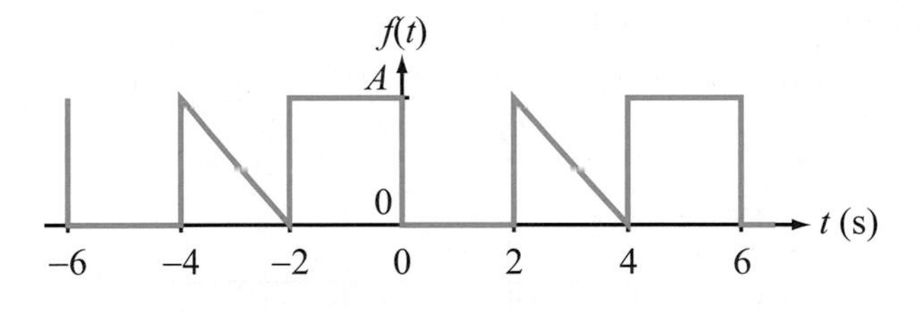

Figure P11.6: Waveform for Problem 11.6.

11.7 Waveform in Fig. P11.7 with $A = 4$.

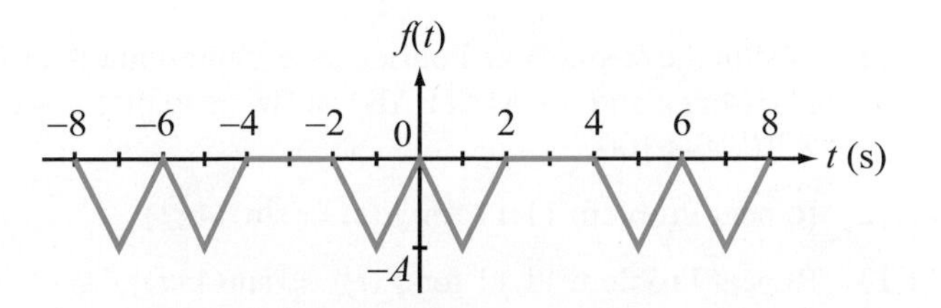

Figure P11.7: Waveform for Problem 11.7.

*Answer(s) in Appendix E.

11.8 Waveform in Fig. P11.8 with $A = 10$.

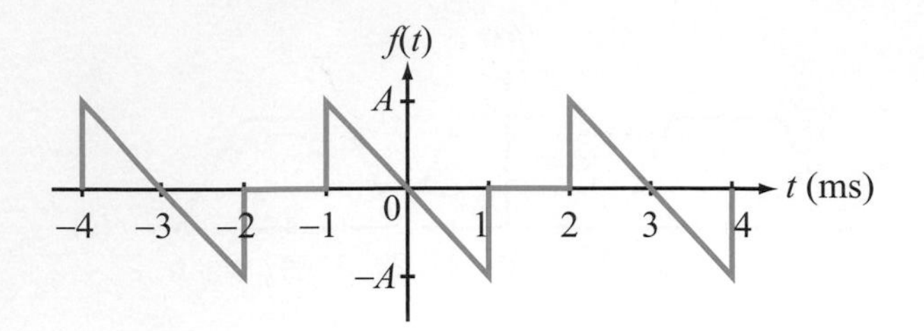

Figure P11.8: Waveform for Problem 11.8.

*__11.9__ Waveform in Fig. P11.9 with $A = 10$.

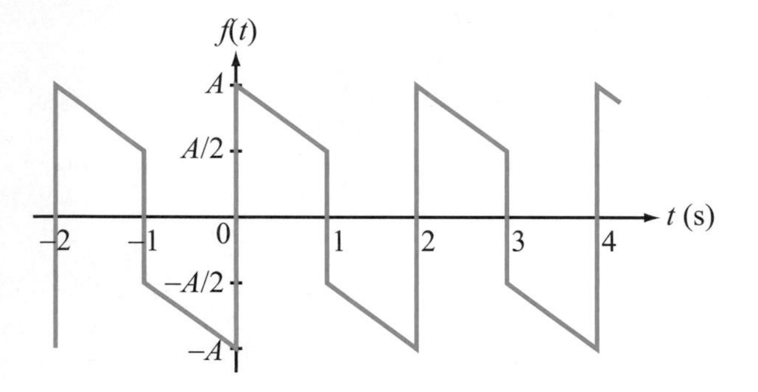

Figure P11.9: Waveform for Problem 11.9.

11.10 Waveform in Fig. P11.10 with $A = 20$.

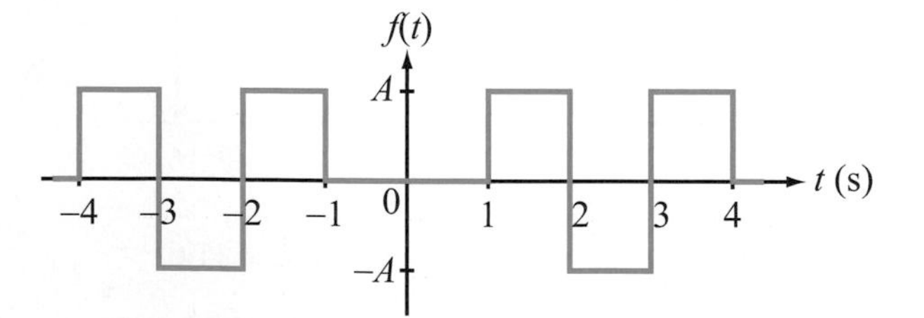

Figure P11.10: Waveform for Problem 11.10.

11.11 Obtain the cosine/sine Fourier series representation for $f(t) = \cos^2(4\pi t)$, and use MATLAB® software to plot it with $n_{\max} = 2$, 10, and 100.

11.12 Repeat Problem 11.11 for $f(t) = \sin^2(4\pi t)$.

11.13 Repeat Problem 11.11 for $f(t) = |\sin(4\pi t)|$.

11.14 Which of the six waveforms shown in Figs. P11.1 through P11.6 will exhibit the Gibbs oscillation phenomenon when represented by a Fourier series? Why?

11.15 Consider the sawtooth waveform shown in Fig. 11-3(a). Evaluate the Gibbs phenomenon in the neighborhood of $t = 4$ s by plotting the Fourier series representation with $n_{\max} = 100$ over the range between 3.99 s and 4.01 s (using expanded scales if necessary).

*__11.16__ The Fourier series of the periodic waveform shown in Fig. P11.16(a) is given by

$$f_1(t) = 10 - \frac{20}{\pi}\sum_{n=1}^{\infty}\frac{1}{n}\sin\left(\frac{n\pi t}{2}\right).$$

Determine the Fourier series of waveform $f_2(t)$ in Fig. P11.16(b).

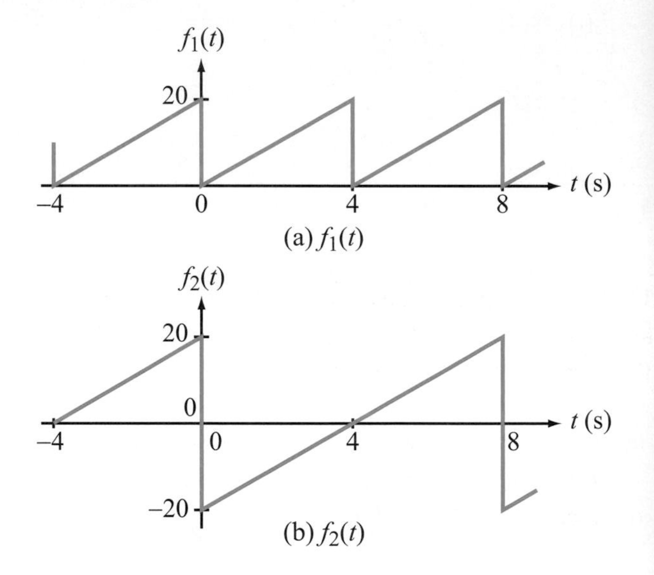

Figure P11.16: Waveforms of Problem 11.16.

Section 11-3: Circuit Applications

11.17 The voltage source $\upsilon_s(t)$ in the circuit of Fig. P11.17 generates a square wave (waveform #1 in Table 11-2) with $A = 10$ V and $T = 1$ ms.

(a) Derive the Fourier series representation of $\upsilon_{\text{out}}(t)$.

(b) Calculate the first five terms of $\upsilon_{\text{out}}(t)$ using $R_1 = R_2 = 2$ kΩ, $C = 1$ μF.

(c) Plot $\upsilon_{\text{out}}(t)$ using $n_{\max} = 100$.

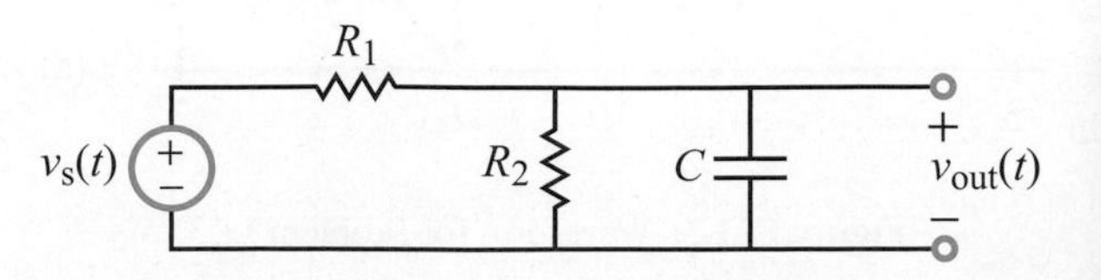

Figure P11.17: Circuit for Problem 11.17.

11.18 The current source $i_s(t)$ in the circuit of Fig. P11.18 generates a sawtooth wave (waveform in Fig. 11-3(a)) with a peak amplitude of 20 mA and a period $T = 5$ ms.

(a) Derive the Fourier series representation of $v_{out}(t)$.

(b) Calculate the first five terms of $v_{out}(t)$ using $R_1 = 500\ \Omega$, $R_2 = 2\ k\Omega$, and $C = 0.33\ \mu F$.

(c) Plot $v_{out}(t)$ and $i_s(t)$ using $n_{max} = 100$.

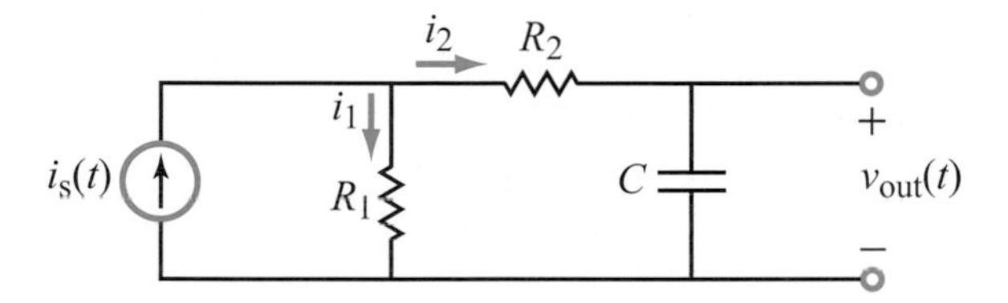

Figure P11.18: Circuit for Problem 11.18.

11.19 The current source $i_s(t)$ in the circuit of Fig. P11.19 generates a train of pulses (waveform #3 in Table 11-2) with $A = 6$ mA, $\tau = 1\ \mu s$, and $T = 10\ \mu s$.

(a) Derive the Fourier series representation of $i(t)$.

(b) Calculate the first five terms of $i(t)$ using $R = 1\ k\Omega$, $L = 1$ mH, and $C = 1\ \mu F$.

(c) Plot $i(t)$ and $i_s(t)$ using $n_{max} = 100$.

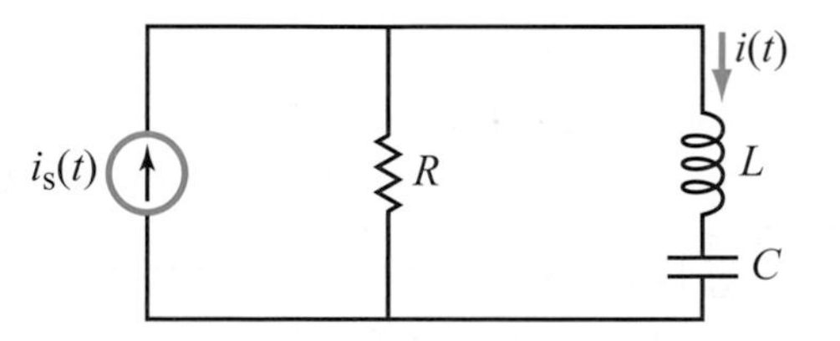

Figure P11.19: Circuit for Problem 11.19.

11.20 Voltage source $v_s(t)$ in the circuit of Fig. P11.20(a) has the waveform displayed in Fig. P11.20(b).

(a) Derive the Fourier series representation of $i(t)$.

(b) Calculate the first five terms of $i(t)$ using $R_1 = R_2 = 10\ \Omega$ and $L_1 = L_2 = 10$ mH.

(c) Plot $i(t)$ and $v_s(t)$ using $n_{max} = 100$.

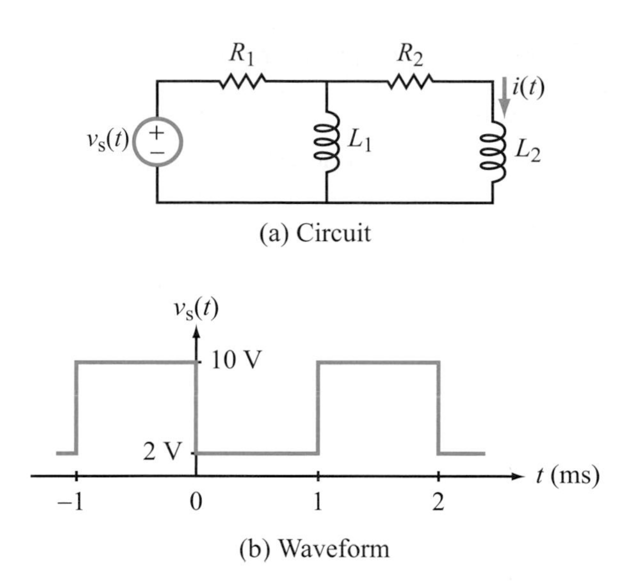

Figure P11.20: Circuit and waveform for Problem 11.20.

11.21 Determine the output voltage $v_{out}(t)$ in the circuit of Fig. P11.21, given that the input voltage $v_{in}(t)$ is a full-wave rectified sinusoid (waveform #8 in Table 11-2) with $A = 120$ V and $T = 1\ \mu s$.

(a) Derive the Fourier series representation of $v_{out}(t)$.

(b) Calculate the first five terms of $v_{out}(t)$ using $R = 1\ k\Omega$, $L = 1$ mH, and $C = 1$ nF.

(c) Plot $v_{out}(t)$ and $v_{in}(t)$ using $n_{max} = 100$.

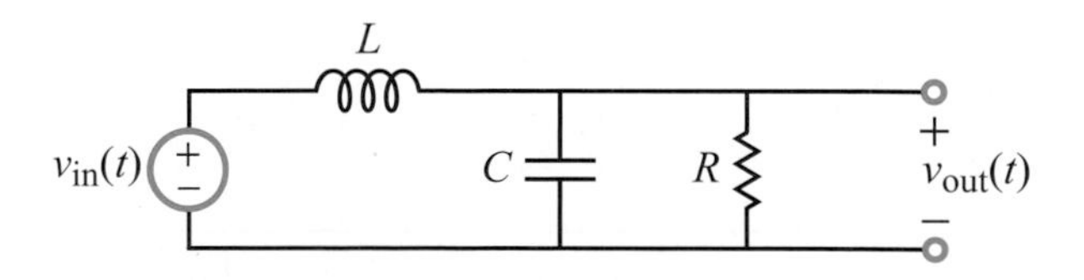

Figure P11.21: Circuit for Problem 11.21.

11.22

(a) Repeat Example 11-5, after replacing the capacitor with an inductor $L = 0.1$ H and reducing the value of R to $1\ \Omega$.

(b) Calculate the first five terms of $v_{out}(t)$.

(c) Plot $v_{out}(t)$ and $v_s(t)$ using $n_{max} = 100$.

11.23 Determine $v_{out}(t)$ in the circuit of Fig. P11.23, given that the input excitation is characterized by a triangular waveform (#4 in Table 11-2) with $A = 24$ V and $T = 20$ ms.

(a) Derive Fourier series representation of $v_{out}(t)$.

(b) Calculate first five terms of $v_{out}(t)$ using $R = 470\ \Omega$, $L = 10$ mH, and $C = 10\ \mu$F.

(c) Plot $v_{out}(t)$ and $v_s(t)$ using $n_{max} = 100$.

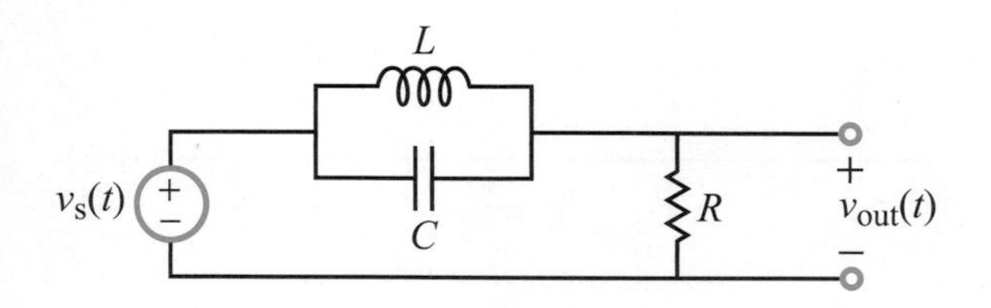

Figure P11.23: Circuit for Problem 11.23.

11.24 A backward-sawtooth waveform (#7 in Table 11-2) with $A = 100$ V and $T = 1$ ms is used to excite the circuit in Fig. P11.24.

(a) Derive Fourier series representation of $v_{out}(t)$.

(b) Calculate the first five terms of $v_{out}(t)$ using $R_1 = 1$ kΩ, $R_2 = 100\ \Omega$, $L = 1$ mH, and $C = 1\ \mu$F.

(c) Plot $v_{out}(t)$ and $v_s(t)$ using $n_{max} = 100$.

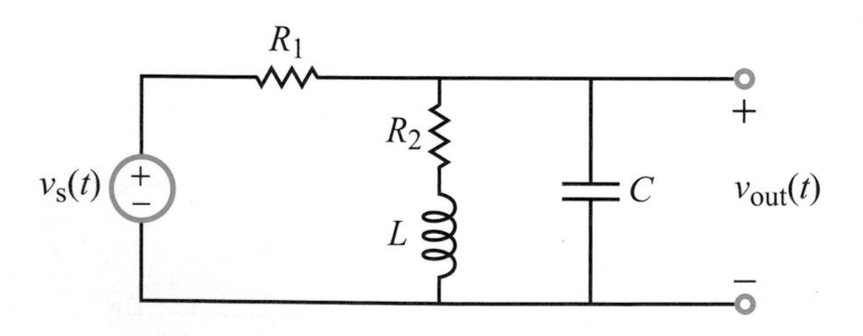

Figure P11.24: Circuit for Problem 11.24.

11.25 The circuit in Fig. P11.25 is excited by the source waveform shown in Fig. P11.20(b).

(a) Derive Fourier series representation of $i(t)$.

(b) Calculate the first five terms of $v_{out}(t)$ using $R_1 = R_2 = 100\ \Omega$, $L = 1$ mH, and $C = 1\ \mu$F.

(c) Plot $i(t)$ and $v_s(t)$ using $n_{max} = 100$.

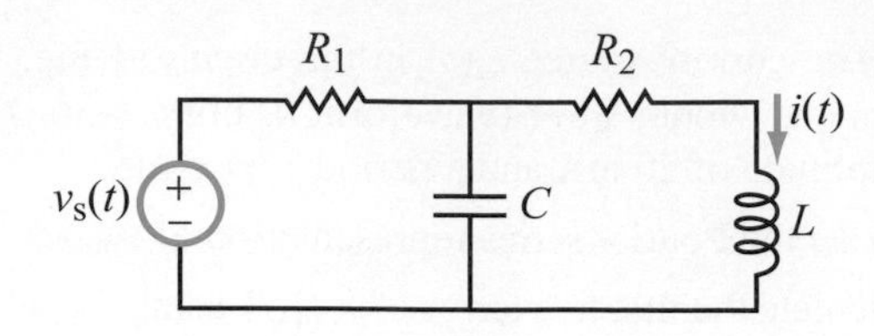

Figure P11.25: Circuit for Problem 11.25.

***11.26** The RC op-amp integrator circuit of Fig. P11.26 is excited by a square wave (waveform #1 in Table 11-2) with $A = 4$ V and $T = 2$ s.

(a) Derive Fourier series representation of $v_{out}(t)$.

(b) Calculate the first five terms of $v_{out}(t)$ using $R_1 = 1$ kΩ, $R_1 = 10$ kΩ, and $C = 10\ \mu$F.

(c) Plot $v_{out}(t)$ using $n_{max} = 100$.

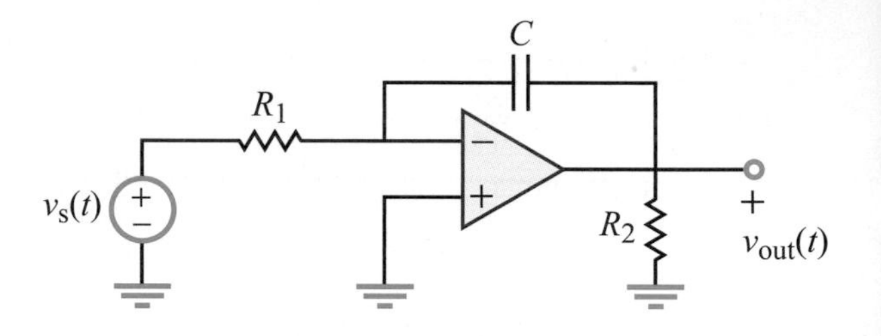

Figure P11.26: Circuit for Problem 11.26.

11.27 Repeat Problem 11.26 after interchanging the locations of the 1-kΩ resistor and the 10-μF capacitor.

Section 11-4: Average Power

11.28 The voltage across the terminals of a certain circuit and the current entering into its (+) voltage terminal are given by

$$v(t) = [4 + 12\cos(377t + 60°) - 6\cos(754t - 30°)]\text{ V},$$
$$i(t) = [5 + 10\cos(377t + 45°) + 2\cos(754t + 15°)]\text{ mA}.$$

Determine the average power consumed by the circuit, and the ac power fraction.

***11.29** The current flowing through a 2-kΩ resistor is given by

$$i(t) = [5 + 2\cos(400t + 30°) + 0.5\cos(800t - 45°)]\text{ mA}.$$

Determine the average power consumed by the resistor, as well as the ac power fraction.

11.30 The current flowing through a 10-kΩ resistor is given by a triangular waveform (#4 in Table 11-2) with $A = 4$ mA and $T = 0.2$ s.

(a) Determine the exact value of the average power consumed by the resistor.

(b) Using a truncated Fourier series representation of the waveform with only the first four terms, obtain an approximate value for the average power consumed by the resistor.

(c) What is the percentage of error in the value given in (b)?

11.31 The current source in the parallel RLC circuit of Fig. P11.31 is given by

$$i_s(t) = [10 + 5\cos(100t + 30^\circ) - \cos(200t - 30^\circ)] \text{ mA}.$$

Determine the average power dissipated in the resistor given that $R = 1$ kΩ, $L = 1$ H, and $C = 1$ μF.

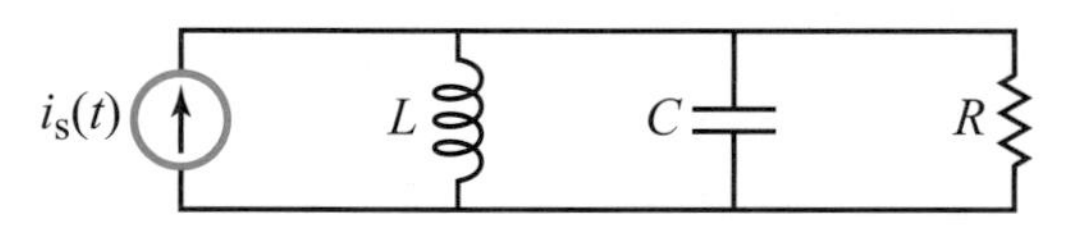

Figure P11.31: Circuit for Problem 11.31.

11.32 A series RC circuit is connected to a voltage source whose waveform is given by waveform #5 in Table 11-2, with $A = 12$ V and $T = 1$ ms. Using a truncated Fourier series representation composed of only the first three non-zero terms, determine the average power dissipated in the resistor, given that $R = 2$ kΩ and $C = 1$ μF.

Sections 11-5 and 11-6: Fourier Transform

For each of the waveforms in Problems 11.33 through 11.42, determine the Fourier transform.

11.33 Waveform in Fig. P11.33 with $A = 5$ and $T = 3$ s.

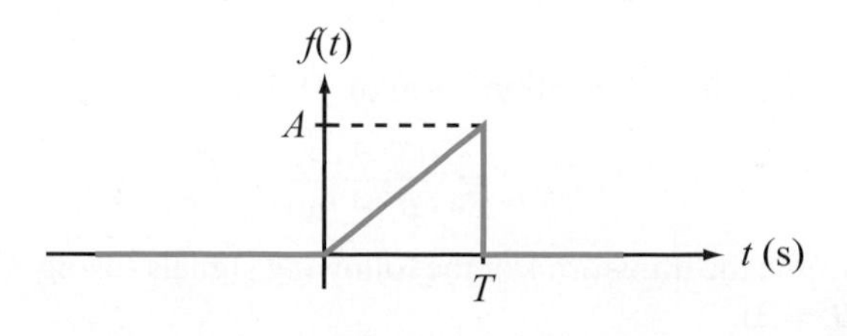

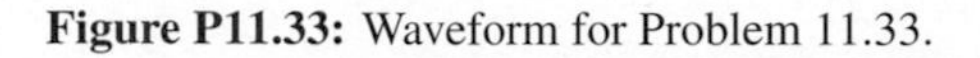

Figure P11.33: Waveform for Problem 11.33.

11.34 Waveform in Fig. P11.34 with $A = 10$ and $T = 6$ s.

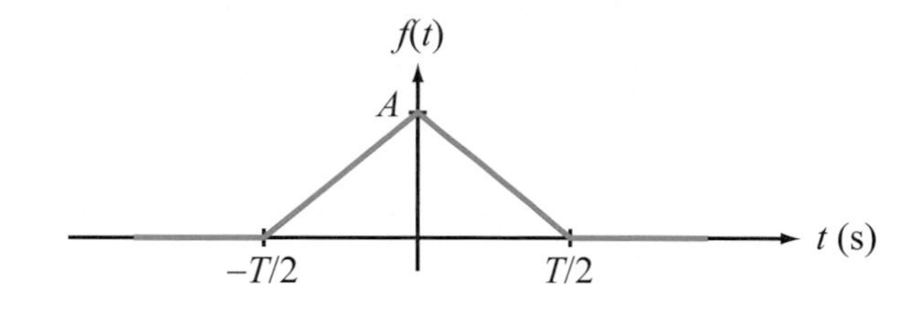

Figure P11.34: Waveform for Problem 11.34.

11.35 Waveform in Fig. P11.35 with $A = 12$ and $T = 3$ s.

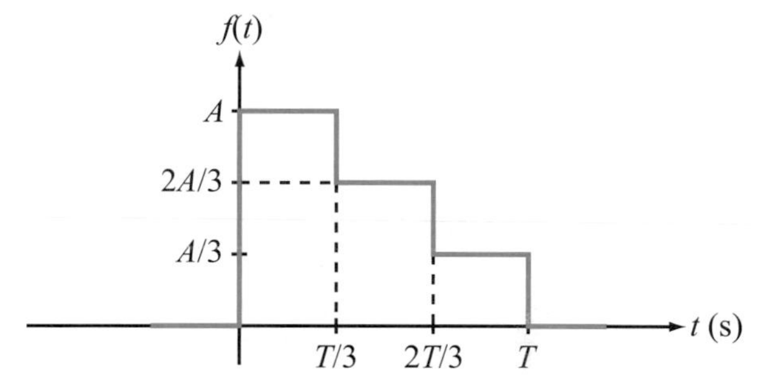

Figure P11.35: Waveform for Problem 11.35.

11.36 Waveform in Fig. P11.36 with $A = 2$ and $T = 12$ s.

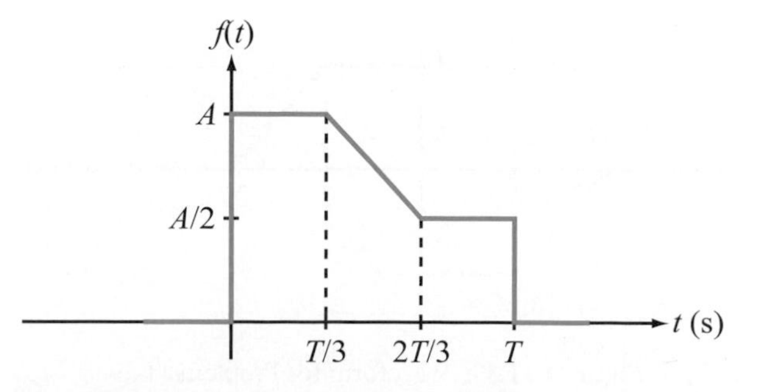

Figure P11.36: Waveform for Problem 11.36.

11.37 Waveform in Fig. P11.37 with $A = 1$ and $T = 3$ s.

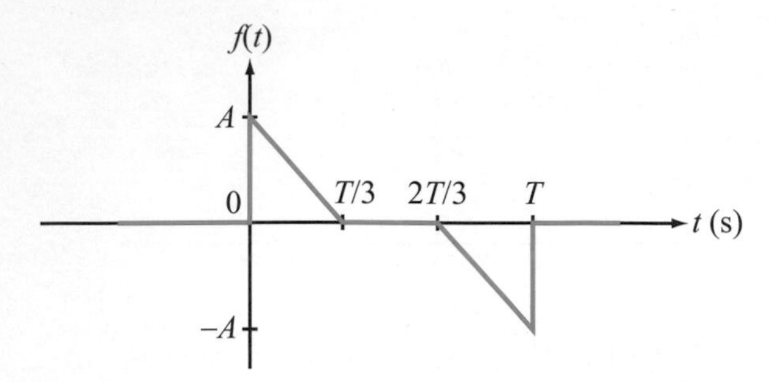

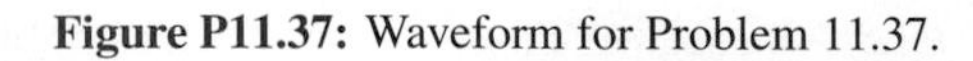

Figure P11.37: Waveform for Problem 11.37.

11.38 Waveform in Fig. P11.38 with $A = 1$ and $T = 2$ s.

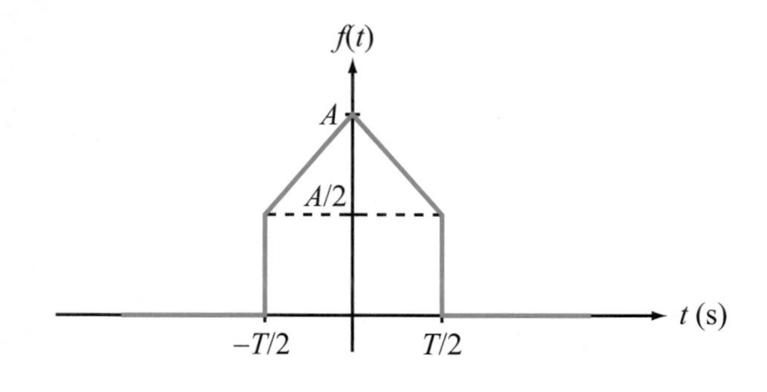

Figure P11.38: Waveform for Problem 11.38.

*__11.39__ Waveform in Fig. P11.39 with $A = 3$ and $T = 1$ s.

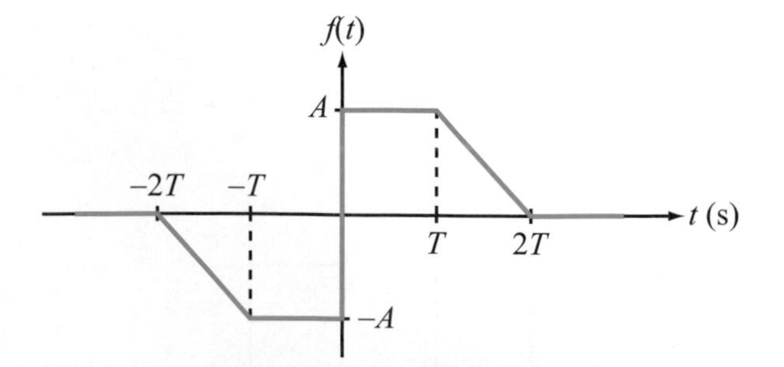

Figure P11.39: Waveform for Problem 11.39.

11.40 Waveform in Fig. P11.40 with $A = 5$, $T = 1$ s, and $\alpha = 10\text{ s}^{-1}$.

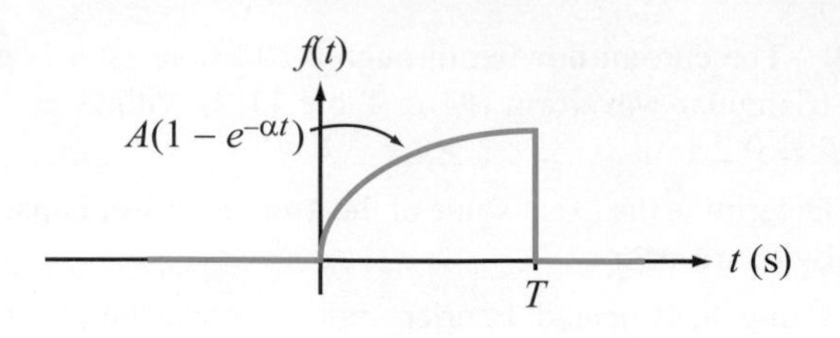

Figure P11.40: Waveform for Problem 11.40.

11.41 Waveform in Fig. P11.41 with $A = 10$ and $T = 2$ s.

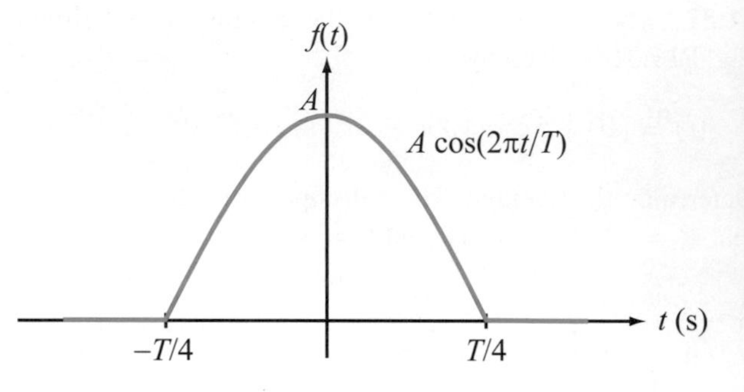

Figure P11.41: Waveform for Problem 11.41.

11.42 Find the Fourier transform of the following signals with $A = 2$, $\omega_0 = 5$ rad/s, $\alpha = 0.5\text{ s}^{-1}$, and $\phi_0 = \pi/5$.

(a) $f(t) = A\cos(\omega_0 t - \phi_0), \qquad -\infty \le t \le \infty$

(b) $g(t) = e^{-\alpha t}\cos(\omega_0 t)\, u(t)$

11.43 Find the Fourier transform of the following signals with $A = 3$, $B = 2$, $\omega_1 = 4$ rad/s, and $\omega_2 = 2$ rad/s.

(a) $f(t) = [A + B\sin(\omega_1 t)]\sin(\omega_2 t)$

(b) $g(t) = A|t|, \qquad |t| < (2\pi/\omega_1)$

11.44 Find the Fourier transform of the following signals with $\alpha = 0.5\text{ s}^{-1}$, $\omega_1 = 4$ rad/s, and $\omega_2 = 2$ rad/s.

(a) $f(t) = e^{-\alpha t}\sin(\omega_1 t)\cos(\omega_2 t)\, u(t)$

(b) $g(t) = te^{-\alpha t}, \qquad 0 \le t \le 10\alpha$

11.45 Using the definition of Fourier transform, prove that

$$\mathcal{F}[t\, f(t)] = j\,\frac{d}{d\omega}\,\mathcal{F}(\omega).$$

11.46 Let the Fourier transform of $f(t)$ be

$$F(\omega) = \frac{A}{(B + j\omega)}.$$

Determine the transforms of the following signals (using $A = 5$ and $B = 2$).

(a) $\mathcal{F}(3t - 2)$

*__(b)__ $t\, f(t)$

(c) $d\, f(t)/dt$

11.47 Let the Fourier transform of $f(t)$ be

$$F(\omega) = \frac{1}{(A + j\omega)} \, e^{-j\omega} + B.$$

Determine the Fourier transforms of the following signals (set $A = 2$ and $B = 1$).

(a) $f\left(\frac{5}{8}\,t\right)$

(b) $f(t)\cos(At)$

(c) $d^3 f/dt^3$

11.48 Prove the following two Fourier transform pairs.

(a) $\cos(\omega T)\; F(\omega) \quad \leftrightarrow \quad \frac{1}{2}[f(t - T) + f(t + T)]$

(b) $\sin(\omega T)\; F(\omega) \quad \leftrightarrow \quad \frac{1}{2j}[f(t + T) - f(t - T)]$

Section 11-8: Circuit Analysis with Fourier Transform

11.49 The circuit in Fig. P11.18 is excited by the source waveform shown in Fig. P11.33.

(a) Derive the expression for $v_{\text{out}}(t)$ using Fourier analysis.

(b) Plot $v_{\text{out}}(t)$ using $A = 5$, $T = 3$ ms, $R_1 = 500\ \Omega$, $R_2 = 2\ \text{k}\Omega$, and $C = 0.33\ \mu\text{F}$.

(c) Repeat (b) with $C = 0.33$ mF and comment on the results.

11.50 The circuit in Fig. P11.18 is excited by the source waveform shown in Fig. P11.34.

(a) Derive the expression for $v_{\text{out}}(t)$ using Fourier analysis.

(b) Plot $v_{\text{out}}(t)$ using $A = 5$, $T = 3$ s, $R_1 = 500\ \Omega$, $R_2 = 2\ \text{k}\Omega$, and $C = 0.33$ mF.

Section 11-10: Multisim

11.51 Design a Sigma-Delta converter that converts a sinusoidal voltage input with a magnitude always $\leq |1\ \text{V}|$ and generates a digital signal with 0–5-V range. No voltage into any op amp can exceed ± 20 V.

11.52 Design a Sigma-Delta converter that converts a sinusoidal current input with a magnitude always $\leq |1\ \text{mA}|$ and generates a digital signal with 0–5-V range. (Hint: The easiest way to do this is to add an additional op-amp buffer ahead of the subtractor input to convert the current signal into a voltage signal.) No voltage into any op amp can exceed ± 20 V.

APPENDIX

A

Symbols, Quantities, and Units

Symbol	Quantity	SI Unit	Abbreviation
A	Cross-sectional area	meter^2	m^2
A	Op-amp gain	dimensionless	—
B	Bandwidth	radians/second	rad/s
C	Capacitance	farad	F
d	Distance or spacing	meter	m
E	Electric field	volt/meter	V/m
F	Force	newton	N
$\mathbf{F}$	Laplace transform	(variable)	(variable)
$\mathbf{F}$	Fourier transform	(variable)	(variable)
f	Frequency	hertz	Hz
G	Conductance	siemen	S
G	Closed-loop gain	dimensionless	—
G	Power gain	dimensionless	—
g	MOSFET gain constant	amperes/volt	A/V
$\mathbf{H}$	Transfer function	(variable)	(variable)
h	Impulse response	(variable)	(variable)
I, i	Current	ampere	A

Symbol	Quantity	SI Unit	Abbreviation
k	Spring constant	newtons/meter	N/m
L	Inductance	henry	H
L, ℓ	Length	meter	m
N	Number of turns	dimensionless	—
P, p	Power	watt	W
P	Mechanical stress	newtons/meter^2	N/m^2
pf	Power factor	dimensionless	—
Q, q	Charge	coulomb	C
Q	Reactive power	volt·ampere reactive	VAR
Q	Quality factor	dimensionless	—
R	Resistance	ohm	Ω
S	Cross-sectional area	meter^2	m^2
$\mathbf{S}$	Complex power	volt·ampere	VA
$\mathbf{s}$	Complex frequency	radians/second	rad/s
T, t	Time	second	s
u	Velocity	meters/second	m/s
V, v	Voltage (potential difference)	volt	V
W	Width	meter	m
W, w	Energy	joule	J
X	Reactance part of impedance	ohm	Ω
$\mathbf{Y}$	Admittance	siemen	S
$\mathbf{Z}$	Impedance	ohm	Ω
α	Piezoresistive coefficient	meters^2/newton	m^2/N
α	Damping coefficient	nepers/second	Np/s
β	Common-emitter current gain	dimensionless	—
β	Air resistance constant	newtons·second/meter	N·s/m
ϵ	Permittivity	farads/meter	F/m
Λ	Magnetic flux linkage	weber	Wb
λ	Time shift	second	s
λ	Wavelength	meters	m
μ	Magnetic permeability	henrys/meter	H/m
ρ	Resistivity	ohms/meter	Ω/m
σ	Conductivity	siemens/meter	S/m
τ	Time constant or duration	second	s
ϕ	Phase	radians	rad
χ_e	Electrical susceptibility	dimensionless	—
ξ	Damping factor	dimensionless	—
ω	Angular frequency	radians/second	rad/s

APPENDIX B

Solving Simultaneous Equations

Electric circuits contain nodes and loops. Application of Kirchhoff's circuit laws usually generates a set of simultaneous equations with n unknowns. The two standard methods commonly used to determine the values of the unknown variables are Cramer's rule and matrix inversion. The latter is a standard tool in MATLAB®, MathScript, and similar software solvers. This Appendix provides a brief overview of some of these approaches.

B-1 Review of Cramer's Rule

Let us assume that the application of Kirchhoff's current and voltage laws to a certain circuit led to the following set of equations:

$$2(i_1 + i_2) - 10 + (3i_2 - i_1 - 4i_3) = 0 \quad \text{(B.1a)}$$

$$-3(i_1 + i_2) + 2(i_1 + 3i_3) = 0 \quad \text{(B.1b)}$$

$$i_1 - 5 - i_2 = 0 \quad \text{(B.1c)}$$

Our task is to solve the three independent, simultaneous, linear equations to determine the values of the three unknowns, i_1 to i_3. [Recall that ***independence*** means that none of the three equations can be generated through a linear combination of the other two.] One way to accomplish the specified task is to apply the method of elimination of variables. If we solve for i_1 in Eq. (B.1c), for example, and then use the expression $i_1 = (i_2 + 5)$ to replace i_1 in Eqs. (B.1a and b), we end up with two new equations containing only two variables, i_2 and i_3. Repeat of the substitution procedure leads to a single equation in only one unknown, which can be solved directly. Once that unknown has been determined, it is a straightforward process to solve for the values of the other two variables.

Such a solution method might prove effective for solving a simple set of three simultaneous equations, but what if the circuit we wish to analyze happens to contain a large number of variables? In that case, the more expeditious approach is to take advantage of Cramer's rule, whose implementation procedure is both systematic and straightforward.

Our review of Cramer's rule will initally use the set of three simultaneous equations given by Eq. (B.1) to demonstrate the mechanics of the solution procedure for a system of order 3. Afterwards, we will treat the general case of a ***system of order*** n (consisting of n independent equations in n unknowns).

B-1.1 System of Order 3

Step 1: Cast Equations in Standard Form

Before we can apply Cramer's rule, we need to regularize the simultaneous equations into a ***standard system*** of the following form:

$$a_{11}i_1 + a_{12}i_2 + a_{13}i_3 = b_1, \tag{B.2a}$$

$$a_{21}i_1 + a_{22}i_2 + a_{23}i_3 = b_2, \tag{B.2b}$$

$$a_{31}i_1 + a_{32}i_2 + a_{33}i_3 = b_3, \tag{B.2c}$$

where the a's are the coefficients of the variables, i_1 to i_3, and the b's are the unaffiliated constants. By expanding the bracketed quantities in Eq. (B.1) and collecting terms, we can convert the equations into the standard form defined by Eq. (B.2). Such a process leads to:

$$i_1 + 5i_2 - 4i_3 = 10, \tag{B.3a}$$

$$-i_1 - 3i_2 + 6i_3 = 0, \tag{B.3b}$$

$$i_1 - i_2 = 5. \tag{B.3c}$$

Note that $a_{11} = 1$, $a_{21} = -1$, and $a_{33} = 0$. The regularized set of three linear, simultaneous equations given by Eq. (B.3) is a system of order 3.

Step 2: General Solution

According to Cramer's rule, the solutions for i_1 to i_3 are given by

$$i_1 = \frac{\Delta_1}{\Delta}, \tag{B.4a}$$

$$i_2 = \frac{\Delta_2}{\Delta}, \tag{B.4b}$$

$$i_3 = \frac{\Delta_3}{\Delta}, \tag{B.4c}$$

where Δ is the value of the ***characteristic determinant*** of the system represented by Eq. (B.3), and Δ_1 to Δ_3 are the ***affiliated determinants*** for variables i_1 to i_3. The procedure for evaluating these determinants is covered in Steps 3 and 4. Before we do so, however, we should note that in view of the fact that Δ appears in the denominator in Eq. (B.4), *Cramer's rule cannot provide solutions for the unknown variables when* $\Delta = 0$. This is not surprising, because for any system of n unknowns, the condition $\Delta = 0$ occurs when one, or more, of the equations is not independent. This means that the system contains more unknowns than the available number of independent equations, in which case it has no unique solution.

Step 3: Evaluating the Characteristic Determinant

The characteristic determinant is composed of the a-coefficients of the 3×3 system of equations:

$$\Delta = \begin{vmatrix} a_{11} & a_{12} & a_{13} \\ a_{21} & a_{22} & a_{23} \\ a_{31} & a_{32} & a_{33} \end{vmatrix}. \tag{B.5}$$

Each element in the determinant has an ***address*** jk specified by its row number j and column number k. Thus, a_{12} is in the first row ($j = 1$) and second column ($k = 2$). For the system given by Eq. (B.3),

$$\Delta = \begin{vmatrix} 1 & 5 & -4 \\ -1 & -3 & 6 \\ 1 & -1 & 0 \end{vmatrix}. \tag{B.6}$$

To evaluate Δ, we ***expand*** it in terms of the elements of one of its rows. For simplicity, we will always perform the expansion using the top row. The expansion process converts Δ from a determinant of order 3 into the sum of 3 terms, each containing a determinant of order 2. Expanding Eq. (B.6) by its top row gives

$$\begin{aligned} \Delta &= a_{11}C_{11} + a_{12}C_{12} + a_{13}C_{13} \\ &= C_{11} + 5C_{12} - 4C_{13}, \end{aligned} \tag{B.7}$$

where C_{11}, C_{12}, and C_{13} are the ***cofactors*** of elements a_{11}, a_{12}, and a_{13}, respectively. The cofactor of any element a_{jk} located at the intersection of row j and column k is related to the minor determinant of that element by

$$C_{jk} = (-1)^{j+k} M_{jk}, \tag{B.8}$$

and the minor determinant M_{jk} is obtained by deleting from the parent determinant all elements contained in row j and column k. Hence, M_{11} is given by Δ after removal of the top row and the left column,

$$\begin{aligned} M_{11} &= \begin{vmatrix} a_{22} & a_{23} \\ a_{32} & a_{33} \end{vmatrix} \\ &= \begin{vmatrix} -3 & 6 \\ -1 & 0 \end{vmatrix}. \end{aligned} \tag{B.9}$$

For a determinant of order 2, expansion by the top row gives

$$\begin{aligned} M_{11} &= a_{22}M_{22} - a_{23}M_{23} \\ &= a_{22}a_{33} - a_{23}a_{32}, \end{aligned} \tag{B.10}$$

which is equivalent to diagonal multiplication of the upper-left and lower-right corners to get $a_{22}a_{33}$, followed with multiplication of the other two corners to get $a_{23}a_{32}$, and then subtracting the latter term from the former. Substituting the values of the coefficients we have

$$M_{11} = (-3) \times 0 - 6 \times (-1) = 6. \tag{B.11}$$

Similarly, M_{12} is obtained by removing from Δ the elements in row 1 and the elements in column 2,

$$\begin{aligned} M_{12} &= \begin{vmatrix} a_{21} & a_{23} \\ a_{31} & a_{33} \end{vmatrix} \\ &= a_{21}a_{33} - a_{23}a_{31} \\ &= (-1) \times 0 - 6 \times 1 = -6. \end{aligned} \tag{B.12}$$

Finally, M_{13} is obtained by removing from Δ row 1 and column 3,

$$\begin{aligned} M_{13} &= \begin{vmatrix} -1 & -3 \\ 1 & -1 \end{vmatrix} \\ &= (-1) \times (-1) - (-3) \times 1 = 4. \end{aligned} \tag{B.13}$$

Inserting the values of the three minor determinants in Eq. (B.7) gives

$$\begin{aligned} \Delta &= C_{11} + 5C_{12} - 4C_{13} \\ &= M_{11} - 5M_{12} - 4M_{13} \\ &= 6 - 5 \times (-6) - 4 \times 4 \\ &= 20. \end{aligned} \tag{B.14}$$

Step 4: Evaluating the Affiliated Determinants

The affiliated determinant Δ_1 for variable i_1 is obtained by replacing column 1 in the characteristic determinant Δ with a column comprised of the b's in Eq. (B.2). That is,

$$\Delta_1 = \begin{vmatrix} b_1 & a_{12} & a_{13} \\ b_2 & a_{22} & a_{23} \\ b_3 & a_{32} & a_{33} \end{vmatrix} = \begin{vmatrix} 10 & 5 & -4 \\ 0 & -3 & 6 \\ 5 & -1 & 0 \end{vmatrix}.$$

Evaluation of Δ_1 follows the same rules of expansion discussed earlier in Step 3 in connection with the evaluation of Δ. Hence

$$\Delta_1 = 10\begin{vmatrix} -3 & 6 \\ -1 & 0 \end{vmatrix} - 5\begin{vmatrix} 0 & 6 \\ 5 & 0 \end{vmatrix} - 4\begin{vmatrix} 0 & -3 \\ 5 & -1 \end{vmatrix}$$

$$= 10 \times 6 - 5 \times (-30) - 4 \times 15 = 150.$$

Application of Eq. (B.4a) gives

$$i_1 = \frac{\Delta_1}{\Delta} = \frac{150}{20} = 7.5.$$

Similarly, Δ_2 is obtained from Δ upon replacing column 2 with the b-column, and Δ_3 is obtained from Δ upon replacing column 3 with the b-column. The procedure leads to $\Delta_2 = 50$, $\Delta_3 = 50$, $i_2 = \Delta_2/\Delta = 50/20 = 2.5$, and $i_3 = \Delta_3/\Delta = 2.5$.

B-1.2 System of Order n

For a regularized system of linear simultaneous equations given by

$$a_{11}i_1 + a_{12}i_2 + a_{13}i_3 + \cdots + a_{1n}i_n = b_1, \tag{B.15a}$$

$$a_{21}i_1 + a_{22}i_2 + a_{23}i_3 + \cdots + a_{2n}i_n = b_2, \tag{B.15b}$$

$$\vdots \quad \vdots \quad \vdots \quad \vdots \quad \vdots \quad \vdots$$

$$a_{n1}i_1 + a_{n2}i_2 + a_{n3}i_3 + \cdots + a_{nn}i_n = b_n, \tag{B.15n}$$

the solution for any variable i_k of the system is

$$i_k = \frac{\Delta_k}{\Delta}, \tag{B.16}$$

where Δ is the characteristic determinant and Δ_k is the affiliated determinant for variable i_k. Analogous with the 3×3 system of Section 3-1, Δ is composed of the a-coefficients:

$$\Delta = \begin{vmatrix} a_{11} & a_{12} & a_{13} & \cdots & a_{1n} \\ a_{21} & a_{22} & a_{23} & \cdots & a_{2n} \\ \vdots & \vdots & \vdots & & \vdots \\ a_{n1} & a_{n2} & a_{n3} & \cdots & a_{nn} \end{vmatrix}, \tag{B.17}$$

and Δ_k is obtained from Δ by replacing column k with the b-column. For example, Δ_2 is given by

$$\Delta_2 = \begin{vmatrix} a_{11} & b_1 & a_{13} & \cdots & a_{1n} \\ a_{21} & b_2 & a_{23} & \cdots & a_{2n} \\ \vdots & \vdots & \vdots & & \vdots \\ a_{n1} & b_n & a_{n3} & \cdots & a_{nn} \end{vmatrix}. \tag{B.18}$$

To determine the value of a determinant of order n, we can carry out a process of successive expansion, analogous with that outlined in Step 3 of Section 3-1. The first step in the expansion process converts Δ from a determinant of order n into a sum of n terms, each containing a determinant of order $(n-1)$. Each of those new determinants can then be expanded into the sum of determinants of order $(n-2)$, and the process can be contined until it reaches a determinant of order 1, which consists of a single element.

B-2 Matrix Solution Method

The system of three simultaneous equations given by Eq. (B.2) can be cast in matrix form as

$$\begin{bmatrix} a_{11} & a_{12} & a_{13} \\ a_{21} & a_{22} & a_{23} \\ a_{31} & a_{32} & a_{33} \end{bmatrix} \begin{bmatrix} i_1 \\ i_2 \\ i_3 \end{bmatrix} = \begin{bmatrix} b_1 \\ b_2 \\ b_3 \end{bmatrix}, \tag{B.19}$$

or in symbolic form as

$$\mathbf{AI} = \mathbf{B}, \tag{B.20}$$

where

$$\mathbf{A} = \begin{bmatrix} a_{11} & a_{12} & a_{13} \\ a_{21} & a_{22} & a_{23} \\ a_{31} & a_{32} & a_{33} \end{bmatrix}, \tag{B.21a}$$

$$\mathbf{I} = \begin{bmatrix} i_1 \\ i_2 \\ i_3 \end{bmatrix}, \tag{B.21b}$$

$$\mathbf{B} = \begin{bmatrix} b_1 \\ b_2 \\ b_3 \end{bmatrix}. \tag{B.21c}$$

Matrix **A** is always a square matrix (same number of rows and columns), so long as the system of simultaneous equations contains the same number of independent equations as the number of unknowns. The solution for the unknown vector **I** is given by

$$\mathbf{I} = \mathbf{A}^{-1}\mathbf{B}, \tag{B.22}$$

where $\mathbf{A}^{-1}$ is the ***inverse*** of matrix **A**. The inverse of a square matrix is given by

$$\mathbf{A}^{-1} = \frac{\text{adj}\,\mathbf{A}}{\Delta}, \tag{B.23}$$

where adj **A** is the ***adjoint*** of **A** and Δ is the determinant of **A**. The adjoint of **A** is obtained from **A** by replacing each element a_{jk} with its cofactor C_{jk}, and then ***transposing*** the resultant matrix, wherein the rows and columns are interchanged. Thus,

$$\text{adj}\,\mathbf{A} = [C_{jk}]^T. \tag{B.24}$$

To illustrate the matrix solution method, let us return to the three simultaneous equations given by Eq. (B.3). Matrices **A** and **B** are given by

$$\mathbf{A} = \begin{bmatrix} 1 & 5 & -4 \\ -1 & -3 & 6 \\ 1 & -1 & 0 \end{bmatrix}, \tag{B.25a}$$

$$\mathbf{B} = \begin{bmatrix} 10 \\ 0 \\ 5 \end{bmatrix}. \tag{B.25b}$$

According to Eq. (B.24), adj **A** is given by

$$\text{adj}\,\mathbf{A} = \begin{bmatrix} C_{11} & C_{12} & C_{13} \\ C_{21} & C_{22} & C_{23} \\ C_{31} & C_{32} & C_{33} \end{bmatrix}^T \tag{B.26}$$

$$= \begin{bmatrix} C_{11} & C_{21} & C_{31} \\ C_{12} & C_{22} & C_{32} \\ C_{13} & C_{23} & C_{33} \end{bmatrix}. \tag{B.27}$$

Each cofactor is a 2×2 determinant. Application of the definition given by Eq. (B.8) leads to

$$\text{adj}\,\mathbf{A} = \begin{bmatrix} 6 & 4 & 18 \\ 6 & 4 & -2 \\ 4 & 6 & 2 \end{bmatrix}. \tag{B.28}$$

Upon incorporating Eqs. (B.22) and (B.23) and using the value of Δ obtained in Eq. (B.14), we have

$$\mathbf{I} = \begin{bmatrix} i_1 \\ i_2 \\ i_3 \end{bmatrix} = \frac{1}{20} \begin{bmatrix} 6 & 4 & 18 \\ 6 & 4 & -2 \\ 4 & 6 & 2 \end{bmatrix} \begin{bmatrix} 10 \\ 0 \\ 5 \end{bmatrix}. \tag{B.29}$$

Standard matrix multiplication leads to

$$i_1 = \frac{1}{20} \begin{bmatrix} 6 & 4 & 18 \end{bmatrix} \begin{bmatrix} 10 \\ 0 \\ 5 \end{bmatrix}$$

$$= \frac{1}{20}(6 \times 10 + 4 \times 0 + 18 \times 5) = 7.5. \tag{B.30}$$

Similarly, multiplication using the second and third rows of adj **A** leads to $i_2 = i_3 = 2.5$.

B-3 MATLAB® or MathScript Solution

In MATLAB® or MathScript software, matrices **A** and **B** of Eq. (B.25) are entered as:

$$\mathbf{A} = [1\ 5\ {-4};\ {-1}\ {-3}\ 6;\ 1\ {-1}\ 0];$$
$$\mathbf{B} = [10\ 0\ 5];$$

The solution of $\mathbf{AI} = \mathbf{B}$ is obtained by entering the statement

$$\mathbf{I} = \text{inv}(\mathbf{A}) * \mathbf{B};$$

The MATLAB® or MathScript response would be

I =

7.5000

2.5000

2.5000.

APPENDIX C

Overview of Multisim

Included on the CD that accompanies this textbook is a brief tutorial for getting started with Multisim. As you will learn, Multisim is an extremely useful software for simulating and analyzing circuits. While the textbook introduces many of Multisim's fundamental concepts, in the interest of space, many others are left out. The tutorial strives to review as well as continue the textbook's coverage of Multisim. Importantly, the demos help clear up common stumbling blocks and the sometimes strange idiosyncrasies of the Multisim software.

The tutorial consists of 43 basic "Demos," which are mixtures of problems, solutions, investigations, and experiments in Multisim. The Demos are divided up into ten chapters (2 through 11) to correspond to the appropriate chapters in the textbook. Each demo attempts to focus on introducing one main concept of Multisim, although it is unavoidable that other concepts are introduced throughout. An index is included to allow for the quick referencing of the tutorial. In addition, scattered throughout the tutorial are five "Emphasis Demos," which focus less on working with Multisim and more on emphasizing a specific fundamental circuit concept through the use of Multisim.

The demos are intended to help you become proficient in Multisim via simple but powerful examples. As you study each chapter in class, it is a good idea to at least skim over them and do some of the sample problems in the CD. Multisim can be an invaluable tool when trying to understand how any circuit works!

The demos, grouped by chapter, cover the following material:

Chapter 2

2.1 Introduction to Multisim/The Three-way Switch

Reviews *layout basics* with a simple, yet elegant, circuit involving three-way switches and light bulbs.

2.2 Resistor Network Analysis

Introduces the Multimeter tool and resistor circuits with many resistors.

2.3 Thermal Sensing Wheatstone Bridge

Demonstrates a variable resistor sensor in a ***Wheatstone bridge circuit***.

2.4 The Wattmeter in Multisim

Introduces power measurement in resistive circuits using the Wattmeter tool.

2.5 A Study of Dependent Sources

Discusses the various ***dependent sources*** in Multisim, their limitations and uses.

2.6 An Introduction to ABM Sources

Discusses the ***Analog Behavioral Modeling (ABM) sources*** in Multisim, which allow for the creation of formula-based dependent sources.

Chapter 3

3.1 DC Circuit Analysis I

Discusses how to use the Measurement Probes and the Interactive Simulation mode to solve for the voltages and currents in simple circuits.

3.2 DC Circuit Analysis II

Discusses how to use the DC Operating Point Analysis tool to solve for the voltages and currents in simple circuits.

3.3 Multisim and Thévenin and Norton Circuits

Discusses a general technique, using the Measurement Probe, for determining ***Thévenin/Norton equivalents*** of any circuit.

3.4 Maximum Power Transfer

Uses the Wattmeter and the Interactive Simulation mode to examine power transfer in resistive circuits.

3.5* Plotting Power Transfer in Multisim

Demonstrates how to make an ***ABM source*** behave like a time-varying resistor and uses this component in a Parameter Sweep analysis to plot power transfer as a function of a varying resistance. This demo presents a very useful technique for plotting how a DC output changes as a function of a changing device parameter.

Chapter 4

4.1 Operational Amplifiers in Multisim

Introduces the DC Operating Point Analysis tool and uses it to analyze an ***inverting op-amp circuit***. Also provides very nice tips on making attractive, easy-to-read ***plots*** in Multisim.

4.2 Introducing the Function Generator and the Oscilloscope

Introduces the Function Generator and Oscilloscope instruments and uses them to measure the voltage gain of an op-amp circuit. Also discusses the Interactive Simulation tool in more detail.

4.3 Introduction to Signal Sources and the Transient Analysis

Introduces ***time-varying sources*** and the Transient Analysis tool in the context of a simple op-amp circuit.

4.4 Using an Operational Amplifier in a Simple Audio Mixer

Combines the lessons of Demos 4.1, 4.2, and 4.3 to build a three-channel op-amp ***audio mixer*** and analyze its time-dependent behavior.

Chapter 5

5.1 Introduction to Transient Circuits

Discusses how to build ***interactive switch-based transients*** in circuits with the Interactive Simulation tool.

5.2 Transient Analysis and First-Order Circuits

Discusses how to build ***voltage-controlled switches*** to generate transients with the Transient Analysis tool.

5.3 Inductors in Multisim

Extends Demos 5.1 and 5.2 to ***switch-driven transients*** in circuits with ***inductors***.

5.4* Time Constants in RC Circuits

Uses the Parameter Sweep tool to plot the response of an ***RC circuit*** as you vary the value of R.

Chapter 6

6.1 Parallel RLC Circuit Analysis

Applies the Oscilloscope and both types of switches discussed in Demos 5.1 and 5.2 to analyze the ***transient behavior of RLC circuits***.

6.2* An Over-, Under-, and Critically Damped Circuit

Applies ***ABM sources*** to modeling the three fundamental types of transient responses in an RLC circuit.

Chapter 7

7.1 Measuring Impedance with the Network Analyzer

A good introduction into the Network Analyzer tool.

7.2 Introduction to AC Analysis

Discusses how to produce ***frequency response plots*** using the AC Analysis tool. The demo uses an RLC circuit as an example, but the technique is useful for any type of circuit.

*Emphasis Demo

7.3 AC Thévenin Circuit Determination

Uses the AC Analysis tool and the Network Analyzer instrument to determine the open circuit voltage and complex impedance of an RLC circuit at a specific frequency. Using this data, the demo shows how to calculate the ***Thévenin equivalent*** circuit and demonstrates that the ***transient response*** of the original circuit and its Thévenin equivalent are the same.

7.4 Making an Impedance Purely Real

Uses the AC Analysis tool and the Network Analyzer instrument to adjust a circuit's frequency response.

7.5 Modeling an AC-to-DC Power Supply

Builds, tests, and analyzes a ***rectifier circuit***, very similar to that in Section 7-10, using Multisim.

7.6 Phase Shift Circuits in Multisim

A good companion demo to Example 7-19 in Chapter 7.

7.7 The Logic Analyzer: An Introduction

Describes the Logic Analyzer used in Example 7-19 of Section 7-11 in more detail.

Chapter 8

8.1 Introduction to RMS Values in Multisim

Discusses how to use the Multimeter instrument to obtain ***root-mean-square (rms)*** values for voltage, current, and power in a circuit.

8.2 AC Power Using AC Analysis

8.3 Power Factor in Multisim

These two demos (8.2 and 8.3) discuss the somewhat tricky business of plotting ***complex power*** and ***power factors*** as a function of frequency using the AC Analysis tool. Because of Multisim's variable and equation nomenclature, a user can easily spend a long time trying to enter the right equations in the analysis tools. This demo clarifies the jargon!

8.4 Three-Phase in Multisim

Demonstrates the ***three-phase source*** component in Multisim using the Measurement Probe and the Transient Analysis tool.

8.5* Maximizing Power Delivered to a Load in a Complex Circuit

Uses the AC Analysis tool to calculate the power delivered to a load as a function of frequency. A good companion demo to Section 8-7.

Chapter 9

9.1 Introduction to Filters in Multisim

Demonstrates how to use the Bode Plotter instrument and the AC Analysis tool to generate ***Bode plots*** of any circuit.

9.2 Modeling a (Very)-Low-Pass Filter in Multisim

Models a real-life application of a ***low-pass filter***. Frequency response and transient response are shown with the various Multisim analysis tools.

9.3 Speaker Crossover Circuit (Plotting Multiple Filters at Once)

This is a great companion demo to Technology Brief 18: Electrical Engineering and the Audiophile. It demonstrates how to design and test a basic ***audio crossover*** circuit with the AC Analysis tool.

9.4 AC Parameter Sweep in a Radio Tuner Circuit

Uses the Parameter Sweep tool and the AC Analysis tool to demonstrate how varying a capacitor's value adjusts a filter's response, thereby acting as a ***tuner*** (or "station selector") for a radio receiver.

9.5 60-Hz Active Notch Filter

Offers an analysis of a ***multi-stage op-amp bandstop*** (or notch) filter. A good companion demo to Sections 9-7 and 9-9.

Chapter 10

10.1 Piecewise Linear Sources

A good companion to Example 10-19 in Chapter 10 that provides more detail on the ***piecewise linear source***.

10.2 Exponential Sources

A good companion to Example 10-18 in Chapter 10 that provides more detail on the ***exponential source***.

Chapter 11

11.1 Introduction to the Spectrum Analyzer

11.2 Fourier Analysis in Multisim

These two demos (11.1 and 11.2) describe how to measure a signal's various frequency components using either the Spectrum Analyzer instrument and/or the Fourier Analysis tool.

11.3* Analysis of a Square Wave

This demo uses the Spectrum Analyzer and Oscilloscope to demonstrate the construction of a square wave from superimposed sinusoidal components at different frequencies (i.e., the sum of the Fourier components).

*Emphasis Demo

APPENDIX D

Mathematical Formulas

D-1 Trigonometric Relations

$$\sin x = \pm\cos(x \mp 90^\circ)$$

$$\cos x = \pm\sin(x \pm 90^\circ)$$

$$\sin x = -\sin(x \pm 180^\circ)$$

$$\cos x = -\cos(x \pm 180^\circ)$$

$$\sin(-x) = -\sin x$$

$$\cos(-x) = \cos x$$

$$\sin^2 x = \frac{1}{2}(1 - \cos 2x)$$

$$\cos^2 x = \frac{1}{2}(1 + \cos 2x)$$

$$\sin(x \pm y) = \sin x \cos y \pm \cos x \sin y$$

$$\cos(x \pm y) = \cos x \cos y \mp \sin x \sin y$$

$$2\sin x \sin y = \cos(x - y) - \cos(x + y)$$

$$2\sin x \cos y = \sin(x + y) + \sin(x - y)$$

$$2\cos x \cos y = \cos(x + y) + \cos(x - y)$$

$$\sin 2x = 2\sin x \cos x$$

$$\cos 2x = 1 - 2\sin^2 x$$

$$\sin x + \sin y = 2\sin\left(\frac{x+y}{2}\right)\cos\left(\frac{x-y}{2}\right)$$

$$\sin x - \sin y = 2\cos\left(\frac{x+y}{2}\right)\sin\left(\frac{x-y}{2}\right)$$

$$\cos x + \cos y = 2\cos\left(\frac{x+y}{2}\right)\cos\left(\frac{x-y}{2}\right)$$

$$\cos x - \cos y = -2\sin\left(\frac{x+y}{2}\right)\sin\left(\frac{x-y}{2}\right)$$

$$e^{jx} = \cos x + j\sin x \qquad \text{(Euler's identity)}$$

$$\sin x = \frac{e^{jx} - e^{-jx}}{2j}$$

$$\cos x = \frac{e^{jx} + e^{-jx}}{2}$$

$$\cos^2 x + \sin^2 x = 1$$

$$2\pi \text{ rad} = 360^\circ$$

$$1 \text{ rad} = 57.30^\circ$$

D-2 Indefinite Integrals

(a and b are constants)

$$\int \sin ax \, dx = -\frac{1}{a} \cos ax$$

$$\int \cos ax \, dx = \frac{1}{a} \sin ax$$

$$\int e^{ax} \, dx = \frac{1}{a} e^{ax}$$

$$\int \ln x \, dx = x \ln x - x$$

$$\int x e^{ax} \, dx = \frac{e^{ax}}{a^2} (ax - 1)$$

$$\int x^2 e^{ax} \, dx = \frac{e^{ax}}{a^3} (a^2 x^2 - 2ax + 2)$$

$$\int x \sin ax \, dx = \frac{1}{a^2} \sin ax - \frac{x}{a} \cos ax$$

$$\int x \cos ax \, dx = \frac{1}{a^2} \cos ax + \frac{x}{a} \sin ax$$

$$\int x^2 \sin ax \, dx = \frac{2x}{a^2} \sin ax - \frac{a^2 x^2 - 2}{a^3} \cos ax$$

$$\int x^2 \cos ax \, dx = \frac{2x}{a^2} \cos ax + \frac{a^2 x^2 - 2}{a^3} \sin ax$$

$$\int e^{ax} \sin bx \, dx = \frac{e^{ax}}{a^2 + b^2} (a \sin bx - b \cos bx)$$

$$\int e^{ax} \cos bx \, dx = \frac{e^{ax}}{a^2 + b^2} (a \cos bx + b \sin bx)$$

$$\int e^{ax} \sin^2 bx \, dx = \frac{e^{ax}}{a^2 + 4b^2} \left[(a \sin bx - 2b \cos bx) \sin bx + \frac{2b^2}{a} \right]$$

$$\int e^{ax} \cos^2 bx \, dx = \frac{e^{ax}}{a^2 + 4b^2} \left[(a \cos bx + 2b \sin bx) \cos bx + \frac{2b^2}{a} \right]$$

$$\int \sin ax \sin bx \, dx = \frac{\sin(a-b)x}{2(a-b)} - \frac{\sin(a+b)x}{2(a+b)}, \quad a^2 \neq b^2$$

$$\int \cos ax \cos bx \, dx = \frac{\sin(a-b)x}{2(a-b)} + \frac{\sin(a+b)x}{2(a+b)}, \quad a^2 \neq b^2$$

$$\int \sin ax \cos bx \, dx = -\frac{\cos(a-b)x}{2(a-b)} - \frac{\cos(a+b)x}{2(a+b)}, \quad a^2 \neq b^2$$

$$\int \sin^2 ax \, dx = \frac{x}{2} - \frac{\sin 2ax}{4a}$$

$$\int \cos^2 ax \, dx = \frac{x}{2} + \frac{\sin 2ax}{4a}$$

$$\int \frac{dx}{x^2 + a^2} = \frac{1}{a} \tan^{-1} \frac{x}{a}$$

$$\int \frac{dx}{(x^2 + a^2)^2} = \frac{1}{2a^2} \left(\frac{x}{x^2 + a^2} + \frac{1}{a} \tan^{-1} \frac{x}{a} \right)$$

$$\int \frac{x^2 \, dx}{a^2 + x^2} = x - a \tan^{-1} \frac{x}{a}$$

D-3 Definite Integrals

(m and n are integers)

$$\int_0^{2\pi} \sin nx\, dx = \int_0^{2\pi} \cos nx\, dx = 0$$

$$\int_0^{\pi} \sin^2 nx\, dx = \int_0^{\pi} \cos^2 nx\, dx = \frac{\pi}{2}$$

$$\int_0^{\pi} \sin nx \sin mx\, dx = 0, \quad n \neq m$$

$$\int_0^{\pi} \cos nx \cos mx\, dx = 0, \quad n \neq m$$

$$\int_0^{\pi} \sin nx \cos nx\, dx = 0$$

$$\int_0^{\pi} \sin nx \cos mx\, dx = \begin{cases} 0, & \text{if } m+n = \text{even and } m \neq n \\ \dfrac{2n}{n^2 - m^2}, & \text{if } m+n = \text{odd and } m \neq n \end{cases}$$

$$\int_0^{2\pi} \sin nx \cos mx\, dx = 0$$

$$\int_0^{\infty} \frac{\sin ax}{ax}\, dx = \frac{\pi}{2a}$$

D-4 Approximations for Small Quantities

For $|x| \ll 1$,

$$(1 \pm x)^n \simeq 1 \pm nx$$

$$(1 \pm x)^2 \simeq 1 \pm 2x$$

$$\sqrt{1 \pm x} \simeq 1 \pm \frac{x}{2}$$

$$\frac{1}{\sqrt{1 \pm x}} \simeq 1 \mp \frac{x}{2}$$

$$e^x = 1 + x + \frac{x^2}{2!} + \cdots \simeq 1 + x$$

$$\ln(1 + x) \simeq x$$

$$\sin x = x - \frac{x^3}{3!} + \frac{x^5}{5!} + \cdots \simeq x$$

$$\cos x = 1 - \frac{x^2}{2!} + \frac{x^4}{4!} + \cdots \simeq 1 - \frac{x^2}{2}$$

$$\lim_{x \to 0} \frac{\sin x}{x} = 1$$

APPENDIX E

Answers to Selected Problems

Chapter 1

1.1 **(b)** 0.000004 amps (A)
(d) 3.9×10^{11} volts (V)

1.3 **(e)** 36 MV

1.7 **(c)** $i(t) = 0.12e^{-0.4t}$ (pA)

1.9 **(b)** $\Delta Q(1, 12) = 2.948$ (C)

1.15 **(b)** $i = 0$ @ $t = 3$ s.

1.21 **(a)** $p(0) = 0$; $p(0.25 \text{ s}) = 0.5$ W

Chapter 2

2.1 $\ell \simeq 2$ km

2.3 **(b)** $R \simeq 1,174\ \Omega$

2.7 $R = 6.41\ \Omega$; $i = 17.2$ A

2.15 $I_x = 2.43$ A

2.19 $I_x = 3.57$ A; $I_y = 2.86$ A

2.23 $P = 0.32$ W

2.27 $I_x = 5$ A for $t < 0$; $I_x = 7.5$ A for $t > 0$

2.33 $R_{\text{eq}} = 9\ \Omega$

2.36 **(a)** $R_{\text{eq}} = 5.5\ \Omega$

2.41 $P = 40$ W

2.45 $P = 4$ W

2.49 **(a)** $R_3 = 1.5\ \Omega$

Chapter 3

3.1 $I = 3$ A

3.7 $I_x = -0.1$ A

3.11 $P = -24$ W

3.17 $V_x = 1.41$ V

3.23 $V = 5$ V

3.27 $V_x = 4$ V

3.31 $I_x = 6.25$ A

3.37 $V_x = 1.67$ V

3.41 $V_1 = 25.5$ V; $V_2 = 4.5$ V

3.47 $I = 0.05$ A

3.51 $V_{Th} = 4$ V; $R_{Th} = 5.2\ \Omega$

3.55 $V_{Th} = -7.6$ V; $R_{Th} = 1.6\ \Omega$

3.61 $I_N = 0.217$ A; $R_{Th} = 9.2\ \Omega$

3.67 $P_{max} = 2.09$ mW

3.73 $I_0 \simeq I_{REF}$

3.77 $V_{out} \simeq (R_L/R_E)V_{in}$

Chapter 4

4.1 $v_o = -10$ V

4.3 $v_o = -10$ V

4.9 $R_f = 16$ kΩ

4.13 $G = R_L(R_1 + R_2)/[R_1(R_3 + R_L)]$

4.19 $G = 0.33$; $-21\text{ V} \le v_s \le 21$ V

4.27 $v_o = (38 - 4v_s)$ V; $5.5\text{ V} \le v_s \le 13.5$ V

4.41 $v_o = -[(R_3/R_2)(R_1 + R_2)/(R_1 + R_s)]v_s$

4.45 $v_o = 8.5v_s$

4.51 $v_o = 2.5 - 10^4 v_s$

Chapter 5

5.2 **(b)** $v_2(t) = -2u(t+2) + 2u(t-2)$

5.7 $v_o = 12$ V; $\tau = 2$ s.

5.15 $v_1 = -12$ V; $v_2 = 6$ V; $v_3 = 2$ V

5.19 $C_{eq} = 2.95$ F

5.25 $i(t) = (0.25 - e^{-0.2t})$ A

5.29 $v_{C_1} = 20$ V; $v_{C_2} = 12$ V; $i_{L_1} = 0$; $i_{L_2} = 2$ A

5.33 **(c)** $i_C(\infty) = 0$; $v_C(\infty) = 8$ V

5.39 $v(t) = [-18 + 24e^{-1.25t}]$ V

5.43 $v(t) = 5.35e^{-1000(t-0.1)/45}$ V

5.49 $i_1(t) = 2.88e^{-10t}$ mA; $i_2(t) = 0.72e^{-20t}$ mA

5.53 $i(t) = [20 - 30e^{-500t}]$ mA

5.67 $W = 0.2\ \mu$m

Chapter 6

6.7 $i_1(0) = 1$ A; $i_2(0) = 2$ A

6.11 $i_L(t) = -90e^{-6t}$ A

6.15 $i_C(t) = -[(40/3)e^{-0.5t}\sin 0.375t]$ A

6.19 $i_C(t) = (12\sin t - 6\cos t)e^{-2t}$ A

6.27 $v_C(t) = (12 + 3e^{-60t})$ V

6.33 $i_L(t) = 1.5$ mA

6.39 $i_{C_1}(t) = (-0.0045e^{-83t} + 0.1045e^{-1917t})$ A for $0 \le t \le 1$ ms; $i_{C_2}(t) = (2.43 \times 10^{-3}\cos 400t - 0.017\sin 400t)$ A for $t > 1$ ms

6.45 $v_C(t) = 2.4 - (0.4\cos 3.74t - 0.428\cos 3.74t)e^{-6t}$ V

6.49 $v_{out}(t) = -8e^{-20t}\sin 74.83t$ V

Chapter 7

7.3 $v(t) = 12\cos(8\pi \times 10^3 t + 60°)$ V

7.9 $v(t) = 12\cos(100\pi t - 45°)$ V

7.10 **(c)** $\mathbf{z}_3 = 7.21e^{-j146.3°}$
(e) $\mathbf{z}_5 = 5e^{j53.13°}$

7.11 **(c)** $\mathbf{z}_3 = -1.22(1+j)$
(f) $\mathbf{z}_6 = 3 + j4$

7.15 **(c)** $\mathbf{z}_1\mathbf{z}_2^* = 10e^{-j105°}$

7.19 **(b)** $\mathbf{V}_2 = 2e^{j108°}$ V

7.25 $i_C(t) = 9.4\cos(2\pi \times 10^4 t - 21.48°)$ mA

7.27 **(b)** $\mathbf{Z}_2 = (98.5 + j1524)\ \Omega$

7.33 $L = 0$ or 2.5 mH

7.37 $\mathbf{V}_{Th} = -12$ V; $\mathbf{Z}_{Th} = 0$

7.45 $f = 795.8$ Hz

7.51 $\mathbf{I}_C = 1.93e^{j4.9°}$ A

7.57 $i_x(t) = 24.72\cos(5 \times 10^5 t - 74.06°)$ A

7.61 $v_{out}(t) = 11.32\cos(377t + 152.05°)$ V

Chapter 8

8.1 (a) $V_{av} = 2$ V

(b) $V_{rms} = 2.31$ V

8.7 (a) $V_{av} = 2$ V

(b) $V_{rms} = 3.79$ V

8.13 $P_{av} = 496.4$ mW

8.19 $P_{av} = 0.2$ mW

8.25 (a) $\mathbf{Z}_1 = (10.5 - j5.2)\ \Omega$; $\mathbf{Z}_2 = (7.23 + j5.05)\ \Omega$

Chapter 9

9.1 $\omega_0 = 10^4$ rad/s

9.5 $K_m = 50$; $K_f = 2 \times 10^4$

9.10 (b) -23 dB

(e) 20.81 dB

9.12 (c) 0.25

9.17 $\mathbf{H}(\omega) = -(10 + j\omega)(100 + j\omega)/50\omega$

9.23 $\omega_0 = 5$ rad/s; $Q = 20$; $B = 25$ rad/s; $\omega_{c_1} = 487.5$ rad/s; $\omega_{c_2} = 512.5$ rad/s

9.29 (a) $\mathbf{H}(\omega) = \dfrac{1}{2}\left(\dfrac{j\omega/\omega_c}{1 + j\omega/\omega_c}\right)$, with $\omega_c = 1.25 \times 10^4$ rad/s

9.35 (a) $\mathbf{H}(\omega) = \left[\dfrac{-j(\omega/\omega_{c_1})}{(1 + j\omega/\omega_{c_2})(1 + j\omega/\omega_{c_3})}\right]$, with

$\omega_{c_1} = 10^4$ rad/s, $\omega_{c_2} = 200$ rad/s, $\omega_{c_3} = 2 \times 10^6$ rad/s

9.40 77 to 92 MHz, or 97 to 112 MHz

Chapter 10

10.1 (d) $\mathbf{F}_4(\mathbf{s}) = 2(1 - e^{-2\mathbf{s}} - e^{-6\mathbf{s}} + e^{-7\mathbf{s}})/\mathbf{s}^2$

10.5 (a) $\mathbf{H}_1(\mathbf{s}) = \left(\dfrac{48}{\mathbf{s}+3} + \dfrac{12}{(\mathbf{s}+3)^2}\right) e^{-4\mathbf{s}}$

(c) $\mathbf{H}_3(\mathbf{s}) = \dfrac{60}{(\mathbf{s}+2)^4}$

10.9 $f(0^+) = 0$; $f(\infty) = 2$

10.15 $v(t) = [1.5 - 1.56e^{-4t} + 0.072e^{-12t}]\, u(t)$ V

10.19 $i_L(t) = [0.012e^{-0.13t} - 0.81e^{-3.3t}]\, u(t)$ mA

10.25 $i_L(t) = [2 + 7e^{-0.67t}\cos(1.43t - 106.5°)]\, u(t)$ A

10.31 $v_C(t) = [10(1 - e^{-t})\, u(t) - 10[1 - e^{-(t-2)}]\, u(t-2)]$

10.35 $v_{out}(t) = \left[\frac{1}{3} + \frac{8}{75} e^{-3t} + \frac{1}{10}\cos(4t - 53°)\right]$ V

10.41 (b) $y_2(t) = [2.4 + 4t - 4e^{-t} + 1.6e^{-5t}]\, u(t)$

10.47 (a) $\mathbf{H}(\mathbf{s}) = 10^8/(\mathbf{s} + 10^4)^2$

(b) $h(t) = te^{-at}\, u(t)$, with $a = 10^4\ \text{s}^{-1}$.

Chapter 11

11.5 (a) even and dc symmetry

(b) $f(t) = \displaystyle\sum_{n=1}^{\infty} \frac{160}{(n\pi)^2}\left[\cos\left(\frac{n\pi}{4}\right) - \cos\left(\frac{3n\pi}{4}\right)\right] \cdot \cos\left(\frac{n\pi t}{4}\right)$

11.9 (a) Odd and dc symmetry

(b) $f(t) = \displaystyle\sum_{n=1}^{\infty} \frac{10}{n\pi}\, [2 - \cos(n\pi)]\sin(n\pi t)$

11.16 $f_2(t) = -\dfrac{40}{\pi}\displaystyle\sum_{n=1}^{\infty} \frac{1}{n}\sin\left(\frac{n\pi t}{4}\right)$

11.26 (a) $v_{out}(t) = \displaystyle\sum_{n=1}^{\infty} 100 \times \left(\frac{4}{n\pi}\right)^2 \sin\left(\frac{n\pi}{2}\right)\cos(n\pi t + 90°)$ V

11.29 $P_{av} = 54.5$ mW; ac power fraction = 8.26%

11.33 $\mathbf{F}(\omega) = \dfrac{5}{3\omega^2}\,[(1 + j3\omega)e^{-j3\omega} - 1]$

11.39 $\mathbf{F}(\omega) = \dfrac{3}{\omega}\left[-2j - \dfrac{2j}{\omega}(\sin\omega - \sin 2\omega)\right]$

11.46 (b) $t\, f(t) \leftrightarrow \dfrac{5}{(2 + j\omega)^2}$

Index

A

D

E

F

G

H

I

J

K

L

N

O

P

Q

R

S